WELDING
Principles and Applications
SEVENTH EDITION

Larry Jeffus

DELMAR
CENGAGE Learning

Australia • Brazil • Japan • Korea • Mexico • Singapore • Spain • United Kingdom • United States

Welding: Principles and Applications
Seventh Edition

Larry Jeffus

Vice President, Editorial: **Dave Garza**

Director of Learning Solutions: **Sandy Clark**

Executive Editor: **Dave Boelio**

Managing Editor: **Larry Main**

Senior Product Manager: **Sharon Chambliss**

Editorial Assistant: **Jillian Borden**

Vice President, Marketing: **Jennifer Baker**

Executive Marketing Manager:
Deborah S. Yarnell

Marketing Specialist: **Mark Pierro**

Production Director: **Wendy Troeger**

Production Manager: **Mark Bernard**

Senior Content Project Manager: **Cheri Plasse**

Senior Art Director: **Joy Kocsis**

Technology Project Manager:
Christopher Catalina

For product information and technology assistance, contact us at
Professional & Career Group Customer Support, 1-800-648-7450

For permission to use material from this text or product,
submit all requests online at **cengage.com/permissions**
Further permissions questions can be e-mailed to
permissionrequest@cengage.com

Library of Congress Control Number: 2010941439

ISBN-13: 978-1-1110-3917-2

ISBN-10: 1-1110-3917-8

Delmar
5 Maxwell Drive
Clifton Park, NY 12065-2919
USA

Cengage Learning products are represented in Canada by Nelson Education, Ltd.

For your lifelong learning solutions, visit **delmar.cengage.com**

Visit our corporate website at **cengage.com**

Notice to the Reader

Publisher does not warrant or guarantee any of the products described herein or perform any independent analysis in connection with any of the product information contained herein. Publisher does not assume, and expressly disclaims, any obligation to obtain and include information other than that provided to it by the manufacturer. The reader is expressly warned to consider and adopt all safety precautions that might be indicated by the activities described herein and to avoid all potential hazards. By following the instructions contained herein, the reader willingly assumes all risks in connection with such instructions. The publisher makes no representations or warranties of any kind, including but not limited to, the warranties of fitness for particular purpose or merchantability, nor are any such representations implied with respect to the material set forth herein, and the publisher takes no responsibility with respect to such material. The publisher shall not be liable for any special, consequential, or exemplary damages resulting, in whole or part, from the readers' use of, or reliance upon, this material.

Printed in the U.S.A.
4 5 6 7 17 16 15 14

This book is dedicated to two very special people—my daughters Wendy and Amy.

Contents

SECTION 3
CUTTING AND GOUGING

SECTION 4
GAS SHIELDED WELDING

SECTION 5
RELATED PROCESSES

SECTION 6
RELATED PROCESSES AND TECHNOLOGY

Chapter 25   Welding Metallurgy

Chapter 26   Weldability of Metals

Chapter 27   Filler Metal Selection

Preface

Introduction

The welding industry presents a continuously growing and changing series of opportunities for skilled welders. Even with economic fluctuations, the job outlook for skilled welders is positive. Due to a steady growth in the demand for goods fabricated by welding, new welders are needed in every area of welding such as small shops, specialty fabrication shops, large industries, and construction. The student who is preparing for a career in welding will need to

- have excellent eye–hand coordination.
- work well with tools and equipment.
- know the theory and application of the various welding and cutting processes.
- be able to follow written and verbal instructions.
- work with or without close supervision.
- have effective written and verbal communications skills.
- be able to resolve basic mathematical problems.
- work well individually and in groups.
- read and interpret welding drawings and sketches.
- be computer literate.
- be alert and work safely.

A thorough study of *Welding: Principles and Applications* in a classroom/shop setting will help students prepare for the opportunities in welding technology. The comprehensive technical content provides the basis for the welding processes. The extensive descriptions of equipment and supplies, with in-depth explanations of their operation and function, are designed to familiarize students with the tools of the trade. The process descriptions, practices, and experiments coupled with actual performance teach the critical fabrication and welding skills required on the job. The text also discusses occupational opportunities in welding and explains the training required for certain welding occupations. The skills and personal traits recommended by the American Welding Society (AWS) for its Certified Welder program are included within the text. Students wishing to become certified under the AWS program must contact the American Welding Society for specific details.

The National Center for Welding Education and Training, known as Weld-Ed, is a partnership between business and industry, community and technical colleges, universities, the American Welding Society, and government to promote welding education.

Organization

The text is organized to guide the student's learning from an introduction to welding, through critical safety information, to details of specific welding and cutting processes, and on to the related areas of shop math, welding metallurgy, weldability of metals, reading technical drawings, fabrication, certification, testing and inspection of welds, and welding joint design, costs, and welding symbols.

Each section of the text introducing a welding process or processes begins with an introduction to the equipment and materials to be used in the process(es), including setup in preparation for welding. The remaining chapters for the specific process concentrate on the actual welding techniques in various applications and positions. The content progresses from basic concepts to the more complex welding technology. Once this technology is understood, the student is able to quickly master new welding tasks or processes.

The sections on welding processes are laid out so that they can be studied individually and in any order. This was done so students can study the process or processes that might relate to their job requirements. However, students are encouraged to study and learn all of the processes so they have the broadest possible future job opportunities.

Objectives listed at the beginning of each chapter tell the student and instructor what is to be learned while studying the chapter. A survey of the objectives will show that the student will have the opportunity to develop a full range of welding skills. Each major process is presented independently so that the instructor can include or exclude them to better meet the needs of the local area served by the program. However, the student can still learn all essential information needed for a thorough understanding of all processes studied.

Key Terms are listed at the beginning of the chapter. These key terms are **boldfaced** and defined throughout the chapters so students will recognize them as they appear. Terms and definitions used throughout the text are based on the American Welding Society's standards. Industry jargon has also been included when appropriate.

Cautions for the student are given throughout the text and point out potential safety concerns or give additional specific information that will make working safer.

Think Green text boxes contain information on both conserving materials, energy, and other natural resources and ways to avoid potential environmental contamination.

Metric equivalents are listed in parentheses for dimensions. When the standard unit is an approximation, the metric equivalent has been rounded to the nearest whole number; however, when the standard unit is an exact value, the metric conversions are more precise.

Illustrations consist of figures, tables, and charts. Figures include both photographs and line art. Numerous figures contain close-up full-color photos of actual welding, and others show welding products and equipment. The colorful detailed figure line art is used extensively throughout the text to help illustrate concepts and clarify the material. Tables and charts contain valuable technical information on materials, equipment setup, and welding process parameters. They are designed to help the student in class and serve later as an on-the-job reference.

Experiments and Practices are learning activities that are presented in most of the chapters. The end of each experiment is identified by the (♦) symbol and the end of each practice is identified by the (♦) symbol.

Experiments help the student learn the parameters of each welding process. Often, because it is hard both to perform the experiment and to observe the results closely, students may do most of the experiments in a small group. In the experiments, students change the parameters to observe the effect on the process. In this way, students learn to manipulate the variables to obtain the desired welding outcome for given conditions. The experiments provided in the chapters do not have right or wrong answers. They are designed to allow the student to learn the operating limitations or the effects of changes that may occur during the welding process.

Practices are included to enable the student to develop the required manipulative skills, using different materials and material thicknesses in different positions for each process. A sufficient number of practices is provided so that, after the basics are learned, the student may choose an area of specialization. Materials specified in the practices may be varied in both thickness and length to accommodate those supplies that students have in their lab. Changes within a limited range of both thickness and length will not affect the learning process designed for the practice.

Mechanical drawings are included with many of the welding practices. These drawings are included to help students better understand mechanical drawings and to show them how the metal is assembled. Most of the drawings are laid out in third-angle projection format, some are in the first-angle projection format, and a few are laid out with the side view shown in an alternate position. The third-angle projection format has been the standard used in the United States for years. However, because of the increasing interaction with the world economy and the fact that many other countries use the first-angle projection format, it has been included. All three drawing formats are commonly used and are included. Items not normally included on true mechanical drawings such as the weld, torch or electrode, and filler metal have been included to aid in students' understanding of the drawings.

Summaries at the end of each chapter recap the significant material covered in the chapter. This summary will help the student more completely understand the chapter material and will serve as a handy study tool.

Review questions at the end of each chapter can be used as indicators of how well the student has learned the material in each chapter.

Glossary definitions include the key terms listed at the beginning of each chapter and also other relevant welding terms. Included in the Glossary are bilingual terms in Spanish. Many definitions feature additional drawings to assist students in gaining a complete understanding of the terms.

Computers in Welding

As in every skilled trade in today's ever-changing world, computers are commonly used in welding. Some of the basic programs provide a cross-reference to welding filler metals, whereas others aid in weld symbol selection.

More complex programs allow welding engineers to design structures and test them for strength without ever building them. These programs aid in proper design and make more effective use of materials, resulting in better, more cost-effective construction. Some programs are helpful in writing Welding Procedure Specifications (WPSs), Procedure Qualification Records (PQRs), and Welder Qualification Test Records (WQTRs). These three types of documents are extensively used throughout the welding industry.

Most of the welding programs operate on a variety of platforms, but the most popular ones use a version of Microsoft Windows. Having a good basic understanding of the Windows operating platform will give you a great start with these programs. In addition, you should become familiar with one of the commonly used word processing programs, such as Microsoft Word. This will aid you in producing high-quality reports both in school and later on the job.

This seventh edition of *Welding: Principles and Applications* has been thoroughly revised and reorganized to reflect the latest welding technologies. Changes include the following:

- New chapters include "Shop Math and Weld Cost," "Reading Technical Drawings," and "Fabricating Techniques and Practices"
- Think Green boxed articles on environmental and conservation welding topics
- New welding processes and technologies such as friction stir welding
- Expanded material on processes such as plasma cutting, FCAW, GMAW, and others
- New Success Story vignettes from students, instructors, and welders telling their stories about learning to weld, jobs they have done, and other real-world experiences
- New feature stories at the end in many of the chapters
- New and updated illustrations and photographs in every chapter

The use of new, full-color, detailed close-up photographs and detailed colored line art make it much easier for the student to see what is expected to produce a quality weld.

Supplements

Accompanying the text is a carefully prepared supplements package, which includes an *Instructor's Guide,* an *Instructor's e-resource,* a *Study Guide/Lab Manual,* CourseMate, WebTutor Advantage, and the *Welding Principles and Practices on DVD* series.

The seventh edition Instructor Resources CD includes an instructor's guide in Microsoft Word, a computerized test bank in ExamView with hundreds of modifiable questions, chapter presentations in PowerPoint with full-color images and video clips, and a searchable Image Library of hundreds of full-color photos and line art from the core text.

The *Study Guide/Lab Manual* (ISBN: 1-1110-3918-6) has been updated to reflect changes made to the seventh edition. The *Study Guide/Lab Manual* is designed to reinforce student understanding of the concepts presented in the text. Each chapter starts with a review of the important topics discussed in the chapter. Students can then test their knowledge by answering additional questions. Lab exercises are included in those chapters (as appropriate) to reinforce the primary objectives of the lesson. Artwork and safety precautions are included throughout the manual.

The *Welding Principles and Practices on DVD* series explains the concepts and shows the procedures students need to understand to become proficient and professional welders. In-text references are provided for selected videos from the DVD series. Delmar Cengage Learning's *Welding Principles and Practices on DVD* combines previously released welding videos with new footage and revisions to bring the material completely up to date. Four DVDs cover shielded metal arc welding, gas metal arc welding, flux cored arc welding, and oxyacetylene welding in detail. The main subject areas are further broken down into subsections on each DVD for easy comprehension. This DVD set offers instructors and students the best welding multimedia learning tool at their fingertips.

The CourseMate for the seventh edition of *Welding: Principles and Applications* offers students and instructors access to important tools and resources, all in one online environment. The CourseMate includes an interactive e-Book for *Welding: Principles and Applications,* Seventh Edition; video clips; interactive quizzes; flashcards; an interactive glossary; and an Engagement Tracker tool for monitoring student progress in the CourseMate product.

Newly available for *Welding: Principles and Applications* is WebTutor Advantage for the Blackboard online course management system. The WebTutor includes chapter presentations in PowerPoint, end-of-chapter review questions, tests, discussion springboard topics, and more, all designed to enhance the classroom and shop experience.

FEATURES OF THE TEXT

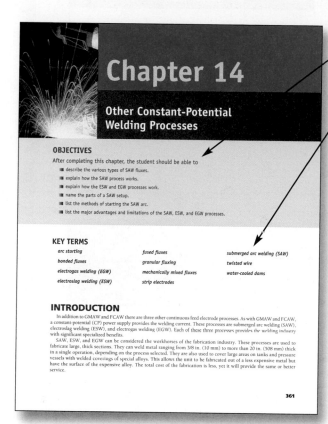

Objectives, found at the beginning of each chapter, are a brief list of the most important topics to study in the chapter.

Key Terms are the most important technical words you will learn in the chapter. These are listed at the beginning of each chapter following the Objectives and appear in color print where they are first defined. These terms are also defined in the Glossary at the end of the book.

The Welding: Principles and Practices on DVD series, explains the concepts and shows the procedures students need to understand to become proficient and professional welders. Four DVDs cover shielded metal arc welding, gas metal arc welding, flux cored arc welding, and oxyacetylene welding in detail.

Cautions summarize critical safety rules. They alert you to operations that could hurt you or someone else. Not only are they covered in the safety chapter, but you will find them throughout the text when they apply to the discussion, practice, or experiment.

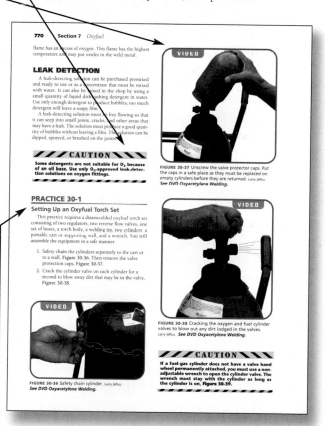

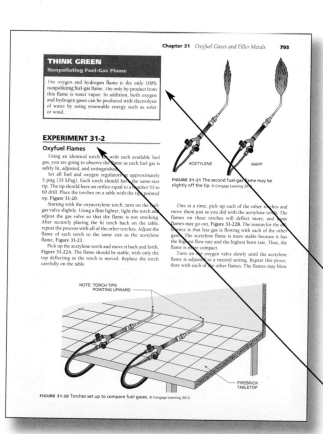

Think Green boxes contain information on both conserving materials, energy, and other natural resources and ways to avoid potential environmental contamination.

Practices are hands-on exercises designed to build your welding skills. Each practice describes in detail what skill you will learn and what equipment, supplies, and tools you will need to complete the exercise.

Experiments are designed to allow you to see what effect changes in the process settings, operation, or techniques have on the type of weld produced. Many are group activities and will help you learn as a team.

Summaries review the important points in the chapter and serve as a useful study tool.

Real-World Features at the end of all chapters present a story that describes a real-world application of the theory learned in the chapter. You will see how particular knowledge and skills are important to the world.

---

370 Section 4 Gas Shielded Welding

Summary

Submerged arc welding, electroslag welding, and electrogas welding processes are most often used for automated high-volume manufacturing facilities. These processes are used in the production of large welds in massive weldments. They are an ideal selection for rapidly producing such large, high-quality welds.

Manual SMAW is covered in this chapter so students who do not have access to the automated equipment can experiment with the process.

Because of the high amperages and large molten weld pools with these processes, it is not practical in most cases for welders to attempt these as manual processes. Therefore, most of the time these are produced using an automated machine, or robotic welding station.

High-Performance Steel Increasingly Used for Bridge Building

HPS 70W, a relatively new material for bridge construction, has many advantages, including corrosion and cracking resistance, high strength, and improved toughness characteristics.

The number of states using high-performance steel (HPS) 70W for bridge construction has steadily increased over the past several years. This is largely due to a law signed in 1988, the Transportation Equity Act for the 21st Century (TEA-21), which authorizes highway and other surface transportation programs for the next six years. Under TEA-21, states receive federal funds to improve highways, opening the door to many new construction projects across the United States. To encourage use of this new metal, the act provides additional funds for bridge projects constructed with HPS 70W.

The Evolution of HPS 70W for Bridge Construction

The military originally developed high-performance steel for use in submarine construction. In 1994, Congress established a steering committee composed of the U.S. Navy, the Federal Highway Administration (FHWA), and the American Iron and Steel Institute (AISI) to research ways HPS could be transferred from military technology to civilian applications. In particular, the committee was charged with finding new ways to use the steel for bridge construction.

HPS 70W (70,000 lb/in.^{2} yield strength steel) was considered for bridge work because it offers the industry many advantages. First and foremost, it has outstanding corrosion-resistance properties. The W in HPS 70W stands for "weathering" and means this material has controlled rusting characteristics that allow just enough corrosion to occur so a rust barrier is formed. Because of this barrier, painting is not required, meaning less maintenance for state highway crews. In contrast, a nonweathering steel often used in bridge construction requires constant painting and maintenance.

The 995-ft-long Clear Fork River Bridge in Tennessee was built with HPS 70W steel. Tennessee Department of Transportation

---

372 Section 4 Gas Shielded Welding

flux should be put in use or immediately transferred to the holding oven.

■ Flux handling. Eventually, flux particles will break down into fines (or dust). Flux recovery equipment cannot always separate out the fines, so operators should regularly purge the flux recovery equipment of all flux in the system and replace it with new flux. Operators should also take every precaution to ensure moisture does not contaminate flux during the welding process.

Future of HPS

Now that HPS 70W use is growing in the industry, what is next for the new steel?

Currently, both gas metal arc welding and flux cored arc welding consumables are being developed and tested to determine if these processes are suitable for use on HPS 70W. Second, fabricators of bridge girders are evolving and becoming more sophisticated in methods for preheating and maintaining interpass temperature, improved flux handling procedures, and trying alternate welding processes. Third, use of high-technology fabrication equipment, such as robotics, is being examined for the fabrication of innovative designs.

Since fabricating HPS 70W is relatively easy when the right consumables, process, and procedures are combined, more HPS 70W bridge projects are expected to be started.

Article courtesy of the Tennessee Department of Transportation.

Review

1. What protects the molten SAW pool from the atmosphere?
2. How can manual SA welding gun movement be performed?
3. What are the two methods of mechanical travel for SA welding?
4. How is the weld metal deposited in the molten weld pool of the SA welding process?
5. In what forms can SA welding filler metal be purchased?
6. How is the manganese range of the SA electrode noted in the AWS classification?
7. Why could a single SA welding flux have more than one AWS classification?
8. List the three groupings of SA welding fluxes according to their method of manufacturing.
9. Why are alloys not added to fused SA fluxes?
10. What is in bonded SA fluxes?
11. What must be done with SA fluxes to prevent contamination of the weld by hydrogen?
12. Why does the welder not have to wear a welding helmet?
13. What happens to the unfused SA welding flux?
14. Why is some form of mechanical guidance required with SA welding?
15. List the common methods used to start the SA arc.
16. Why would handheld SA welding be used?
17. What special characteristics must ES welding slags have?
18. What heats the ES welding flux?
19. How is an ES weld started?
20. Why do ES welds have large grain sizes?
21. List the advantages of ES welding.
22. What is the major difference between ESW and EGW?

Review questions help measure the skills and knowledge you learned in the chapter. Each question is designed to help you apply and understand the information in the chapter.

Success Stories are found at the beginning of each of the seven sections in the text. These stories are about real people who have become successful by using their welding skills. Each story is different, but one message is repeated by all story contributors: welding can be a rich and rewarding career.

Bilingual Glossary definitions provide a Spanish equivalent for each new term. Additional line art in the Glossary will also help you gain a greater understanding of challenging terms.

Success Story

Kassandra Rieke, who now lives in La Junta, Colorado, started her welding career at an early age. She was just 12 years old when her uncle first taught her to weld on a ranch in Meeker, Colorado. Never afraid of hard work, she got a job when she was just 18 years old running heavy equipment at a big gravel pit in Meeker for about a year.

Just after her 19th birthday, a friend took her to work as a welder's helper on oil and gas rigs. "Welding on oil rigs can be a lot of hard work—we usually worked six twelves (12 hours a day 6 days a week), but when you love what you are doing, it does not feel like work." She fondly recalls working 38 straight hours to get one rig back in service. "When one of the rigs is down, you just have to keep on welding as long as it takes to get it back working." Even with the long hours, she was back welding, a job she loved, and this was the beginning of her career.

By the time she moved with her family to Pueblo, she had graduated from high school and was looking for a welding school to hone her skills. She wanted to learn more about natural gas pipe welding. After visiting several schools and having heard some great things about it from welder friends, she decided to attend Trinidad State Junior College's Energy Production Industrial Construction Program.

As an outdoor enthusiast who loves hunting, fishing, and working outdoors and with classes over, she was anxious to get back out in the field and start welding again. She had her own welding rig consisting of a Dodge 1-ton dually with the custom fabricated bed she and a friend built, a Lincoln SA-200 Pipeliner welder generator, and torch set.

Even though she really enjoyed the opportunity to travel throughout many of the western states working as a welder-fabricator, she has moved back to southern Colorado. She currently works as an agricultural repair and fabrication welder in the La Junta area.

Recently, she added a plasma cutter to her current welding rig. She said it has really made cutting out parts so much easier than it was when she only had an oxyacetylene cutting torch. Running the plasma cutting torch off of the welder generator's AC power outlet has made it a lot easier to cut everything, especially on thin sheet metal. The plasma torch doesn't distort sheet metal like the acetylene torch did, so it is easier and faster to make a fabrication.

She loves to ride horses and rodeo a little and wants someday to save the money she earns from welding to buy a ranch and maybe help underprivileged kids by giving them the opportunity to experience the outdoor life that she has. ■

Acknowledgments

To bring a book of this size to publication requires the assistance of many individuals, and the author and publisher would like to thank the following for their unique contributions to this and/or prior editions:

- Marilyn K. Burris, for the years of work on this text and graphics.

- The American Welding Society, Inc., whose *Welding Journal* was an invaluable source for many of the special-interest articles.

- John L. Chastain, who worked with the author for many long hours to perfect the photographic techniques required to achieve the action photos.

- Dewayne Roy, Welding Department chairman at Mountain View College, Dallas, Texas, for his many contributions to this text.

- The author would like to express his deepest appreciation to Thermal Dynamics and Jay Jones for all the props and help they provided for the preparation of this text. In addition he would like to thank Garland Welding Supply Co. Inc. for the loan of material and supplies for photo shoots.

- Ernest Levert, welding engineer at Lockheed Martin for all of his great technical advice and for sharing his welding experiences.

- Special thanks are due to the following companies for their contributions to the text: Skills USA-VICA; Praxair; NASA Media Research Center; Miller Electric Co.; Caterpillar, Inc.; ESAB Welding & Cutting Products; Frommelt Safety Products; Hornell Speedglas, Inc.; Mine Safety Appliances, Co.; Lincoln Electric; Jackson Products/Thermadyne; Thermadyne Holdings; Hobart Brothers Co.; Concoa Controls Corp.; Stanley Works; Rexarc; Magnaflux Corp.; Buehler Ltd.; T.J. Snow Co., Inc.; Victor Equipment; E.O. Paton Electric Welding Institute; CRC-Evans Automatic Welding; Cherry Point Refinery; The Aluminum Assoc./Automotive & Light Truck Group; E.I. DuPont de Nemours & Co.; Philips Gmbh; Technical Systems; GWS Welding Supply Co.; Merrick Engineering, Inc.; Reynolds Metals Co.; Liquid Air Corp.; Alphagaz Div.; American Torch Tip; ARC Machines, Inc.; FANUX Robotics North America, Inc.; Alexander Binzel Corp.; Sciaky Brothers, Inc.; Aluminum Co. of America; National Machine Co.; Leybold Heraeus Vacuum Systems, Inc.; Sonobond Ultrasonics; Foster Instruments; Prince & Izant Company; United Association of the Journeymen and Apprentices of the Plumbing and Pipe Fitting Industry of the United States and Canada, Local No. 100; Atlas Copco Drilling Solutions Inc; Garland Welding Supply Co., Inc.; and the City of Garland Texas: Garland Power and Light.

- The following individuals, who reviewed the sixth edition in anticipation of the seventh. Their recommendations have been invaluable to the author: David Parker, Renton Technical College; Gonzalo Huerta, Imperial Valley College; and Mark Prosser, Blackhawk Technical College.

- The following individuals, who are featured in the Success Stories in the text and Online Companion; they are valuable contributors to the textbook and an inspiration for those entering the welding industry: Danielle DiBari, Jared Corley, Kassandra Rieke, Michael Crumpler, Mike Vaughan, Stephan Pawloski, and Jay Jones.

The author also would like to express his deepest appreciation to:

- The welding instructors at: Lexington Area Technical High School, SC; Worcester Technical High School, MA; Craven Community College, NC; Great Plains Technology Center, OK; Atlantic Technical Center, FL; El Camino Community College, CA; Wichita Area Technical College, KS; Antelope Valley College, CA; Blackhawk Technical College, WI; Wenatchee Valley College, WA; Tyler Junior College, TX; Midlands Technical College, SC; John A. Logan College, IL; Northwest Mississippi Community College, MS; Tarrant County College, TX; Greater Lowell Technical High School, MA; Long Beach City College, CA; Redding Area Community College, PA; College of the Ozarks, MO; Bessemer State Technical College, AL; York Technical College, SC; Lakeview High School, TX; Newberry County Career Center, SC; Palm Beach Community College, FL; Texas State Technical College, TX; Grand Rapids Community College, MI; Kilgore College, TX;

xviii **Acknowledgments**

Tulsa Technology Center, OK; Calcasieu Parish School, LA; Florence-Darlington Technical College, SC; Jefferson High School, TX; Coastal Carolina Community College, NC; Los Angeles Unified School District, CA; Vatterott College, MO; New River Community College, VA; New Hampshire Technical College at Manchester, NH; and Austin Community College, TX, for sharing with me their welding experiences, teaching experiences, and students' experiences, which

have helped form the basis for many of the updates in this edition.

- Tina Ivey, Sam Burris, David DuBois, and Traywick Duffie for all their help in the preparation of this edition

- To my wife, Carol, for all of her moral support, and to my daughters, Wendy and Amy, for all of the general office help they provided.

About the Author

In my sophomore year at New Bern High School in North Carolina, while taking drafting and wood shop, I'm proud to say I joined the Vocational Industrial Clubs of America (VICA), now SkillsUSA-VICA. SkillsUSA brings together educators, administrators, corporate America, labor organizations, trade associations, and government in a coordinated effort to address America's need for a globally competitive, skilled workforce. The mission of SkillsUSA is to help our students become world-class workers and responsible American citizens. Through my involvement in SkillsUSA, I learned a great deal about industry and business. I learned in SkillsUSA the value of integrity, responsibility, citizenship, service, and respect. In addition, I developed leadership skills, established goals, and learned the value of performing quality work. These are all things that I still use in my life today.

During my junior year of high school, I learned to weld in metal shop. I was taught basic welding principles and applications, and I was able to build a number of projects in shop using oxyacetylene welding, shielded metal arc welding, and torch brazing.

The practice welds helped me develop welding skills, and building the projects allowed me to start developing some fabrication skills. By the end of my junior year, I had become a fairly skilled welder.

In my senior year, I was given an opportunity to join Mr. Z.T. Koonce's first class in a new program called Industrial Cooperative Training (ICT). ICT is a cooperative work experience program that coordinates school experiences with real jobs. This allowed me to attend high school in the morning, where I completed my required English, math, and other academic courses for graduation. In my ICT class we were taught skills that would help us get a job—such as how to fill out a job application, how to interview, and so on. In the afternoons, I worked as a welder. After graduation, I started a full-time job as a welder at Barbour Boat Works, where I refined my welding skills and was allowed to work with the other welders in the shipyard. My first welding assignment was on a barge, making intermittent welds to attach the deck to the barge's ribs.

As my welding skills improved, my supervisor allowed me to apply my new welding skills to more difficult jobs. I welded on barges, military landing crafts, tugboats, PT boats, small tankers, and others. This is how I earned money toward my college education.

With my welding skills, I was able to get a job in a small welding shop in Madisonville, Tennessee, and attended Hiwassee College. After graduating from Hiwassee, I found other welding jobs that allowed me to continue my education at the University of Tennessee, where I earned a bachelor's degree. After four years, I had both a college degree and four years of welding experience, which together qualified me for my job as a vocational teacher.

During my career as a welder, I have welded on tanks, pressure vessels, oil well drilling equipment, farm equipment, buildings, race cars, and more. As a vocational teacher, I have taught in high schools, schools for special education, schools for the deaf, three colleges, and numerous industrial shops. I have also been a consultant to the welding industry.

Larry Jeffus has more than 40 years of experience as a dedicated classroom teacher and is the author of several Delmar Cengage Learning welding publications. Prior to retiring from teaching, Professor Jeffus taught at Eastfield College, part of the Dallas County Community College District. Since retiring from full-time teaching, he remains very active in the welding community, especially in the field of education. He serves on several welding program technical advisory committees and has visited high schools, colleges, universities, and technical campuses in more than 40 states and 4 foreign countries. Professor Jeffus was selected as Outstanding Post-Secondary Technical Educator in the State of Texas by the Texas Technical Society. He holds a bachelor of science degree and has completed postgraduate studies.

Professor Jeffus has served for seven years as a board member on the Texas Workforce Investment Council in the Texas Governor's office where he works to develop a skilled workforce and bring economic development to the state. He is a member of the Apprenticeship Project Leadership Team, where he helps establish apprenticeship training programs for the State of Texas, and he has made numerous trips to Washington, DC, lobbying for vocational and technical education.

He has been actively involved in the American Welding Society for more than 40 years, having served on the General Education Committee and as the chairman of the North Texas Section of the American Welding Society.

Index of Experiments and Practices

The following Experiments and Practices are listed in the order in which they appear in the chapter. It should be noted that not all chapters have Experiments and Practices.

Chapter 33

WELDING
Principles and Applications
SEVENTH EDITION

Larry Jeffus

SECTION 1

INTRODUCTION

Success Story

My name is Jared Corley and I have taken two years of welding technology in South Carolina at the Lexington Technology Center. I have many hobbies, such as hunting and fishing, but welding is my passion. I first became interested in welding by holding stuff for my grandpa when he was welding on miscellaneous projects such as trailers, deer stands, plant hangers, and so on. He worked as a welder at Reco USA for 40 years fabricating tanks and pressure vessels.

On my first day of class I did not know what to expect. I had never been in a real welding shop nor did I know anything about welding other than what I had seen my grandpa doing. After passing the welding safety test, we started welding. I was assigned a welding booth and was given instructions on how to strike and maintain an arc. Needless to say I had no idea what that meant or how to do that, but I was willing to give it a try. Soon my instructor, Mr. Ashley Black, noticed a big improvement in my welding skills. As he worked with me one-on-one, giving me advice and showing me techniques on how to do certain things, my welding got even better.

I'm in my second year of welding with Mr. Kevin Gratton. I decided that this is what I want to do for a living. I now pay even more attention everyday in class and work as hard as I can at welding. I know that what they are saying is the real deal. They have been there and done that.

I got a job welding and fabricating every day after school at D&T steel. My day consists of a math, English, and welding class, and then I go to work. I am learning so much more there just listening and watching older welders do the same thing I love. When I was hired, the owner said, "Boy, welding in a school welding booth and welding out here in the real world are two totally different things." He is right. After working there for about three months, I knew exactly what he was telling me. At school I weld on plates that are always placed just right to make it easier to weld; now I am having to stand, sit, squat, and lean over to make welds that are sometimes hard to reach or see.

Mr. Larry Jeffus came to my class one day, and I will always remember one of the things he said, "You cannot be average, you have to be good—cocky good. Anyone can pick up a helmet and burn some rods, but to be good you have to work at it and really concentrate." That's exactly what I do.

I am going to make something of myself. I know I can, and I have the right people there if I need help. I am cocky enough to say that I am the best in my class, but only because I work at it, and if you want to be better than me, I am going to make you work for it. ■

Chapter 1

Introduction to Welding

OBJECTIVES

After completing this chapter, the student should be able to

- explain how each one of the major welding processes works.
- list the factors that must be considered before a welding process is selected.
- discuss the history of welding.
- describe briefly the responsibilities and duties of the welder in various welding positions.
- define the terms *weld, forge welding, resistance welding, fusion welding, coalescence,* and *certification*.

KEY TERMS

American Welding Society (AWS)

automated operation

automatic operation

certification

coalescence ~~fuse together~~

flux cored arc welding (FCAW)

forge welding

fusion welding

TIG.
gas metal arc welding (GMAW)

gas tungsten arc welding (GTAW)

machine operation

manual operation

oxyfuel gas cutting (OFC)

oxyfuel gas welding (OFW)

qualification

resistance welding

semiautomatic operation

shielded metal arc welding (SMAW) STICK.

torch or oxyfuel brazing (TB)

weld

welding

INTRODUCTION

As methods of joining materials improved through the ages, so did the environment and mode of living for humans. Materials, tools, and machinery improved as civilization developed.

Fastening together the parts of work implements began when someone attached a stick to a stone to make a spear or axe. Egyptians used stone tools to create temples and pyramids that were fastened together with an adhesive of gypsum mortar. Some walls that still exist depict a space-oriented figure that was as appropriate then as now—an ibis-headed god named Thoth who protected the moon and was believed to cruise space in a vessel.

Other types of adhesives were used to join wood and stone in ancient times. However, it was a long time before the ancients discovered a method for joining metals. Workers in the Bronze and Iron Ages began to solve the problems of forming,

4

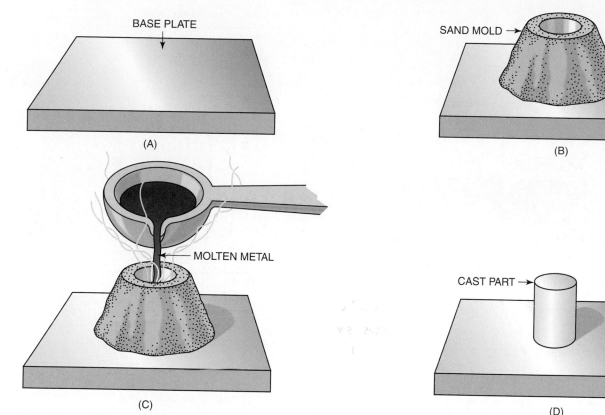

BASE PLATE

(A)

SAND MOLD →

(B)

MOLTEN METAL

(C)

CAST PART →

(D)

FIGURE 1-1 Direct casting: (A) base plate to have a part cast on it, (B) sand molded into shape desired, (C) pouring hot metal into mold, and (D) part cast is now part of the base plate. © Cengage Learning 2012

casting, and alloying metals. Welding metal surfaces was a problem that long puzzled metalworkers of that time period. Early metal-joining methods included such processes as forming a sand mold on top of a piece of metal and casting the desired shape directly on the base metal so that both parts fused together, forming a single piece of metal, **Figure 1-1**. Another metal-joining method used in early years was to place two pieces of metal close together and pour molten metal between them. When the edges of the base metal melted, the flow of metal was then dammed up and allowed to harden, **Figure 1-2**.

The Industrial Revolution, from 1750 to 1850, introduced a method of joining pieces of iron together known as **forge welding** or hammer welding. This process involved the use of a forge to heat the metal to a soft, plastic temperature. The ends of the iron were then placed together and hammered until fusion took place.

Forge welding remained as the primary welding method until Elihu Thomson, in the year 1886, developed the **resistance welding** technique. This technique provided a more reliable and faster way of joining metal than did previous methods.

As techniques were further developed, riveting was replaced in the United States and Europe by **fusion welding**. At that time the welding process was considered to be vital to military security. Welding repairs to the ships damaged during World War I were done in great secrecy. Even today some aspects of welding are closely guarded secrets.

Since the end of World War I, many welding methods have been developed for joining metals. These various welding methods are playing an important role in the expansion and production of the welding industry. Welding has become a dependable, efficient, and economical method for joining metal.

Welding Defined

A **weld** is defined by the American Welding Society (AWS) as "a localized **coalescence** (the fusion or growing together of the grain structure of the materials being welded) of metals or nonmetals produced either by heating the materials to the required welding temperatures, with or without the application of pressure, or by the application of pressure alone, and with or without the use of filler materials." **Welding** is defined as "a joining process that produces coalescence of materials by heating them to the welding temperature, with or without the application of pressure or by the application of pressure alone, and with or without the use of filler metal." In less technical language, a weld is made when separate pieces of material to be joined combine and form one piece when

- enough heat is applied to raise the temperature high enough to cause softening or melting and the pieces flow together,
- enough pressure is used to force the pieces together so that the surfaces coalesce, or

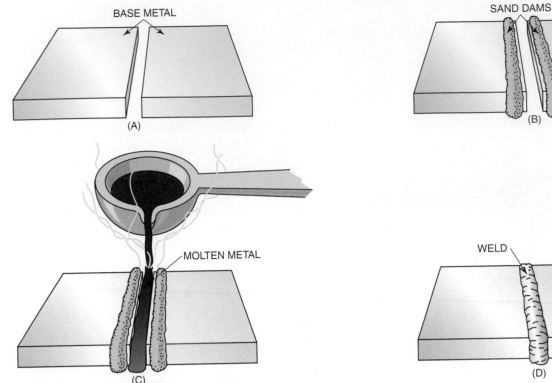

FIGURE 1-2 Flow welding: (A) two pieces of metal plate, (B) sand dams to hold molten metal in place, (C) molten metal poured between metal plates, and (D) finished welded plate. © Cengage Learning 2012

■ enough heat and pressure are used together to force the separate pieces of material to combine and form one piece.

A filler material may or may not be added to the joint to form a completed weld joint. It is also important to note that the word *material* is used because today welds can be made from a growing list of materials such as plastic, glass, and ceramics.

Uses of Welding

Modern welding techniques are employed in the construction of numerous products, **Figure 1-3** and **Figure 1-4**.

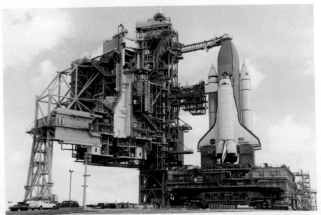

FIGURE 1-3 Space shuttle being made ready for its launch into space. Notice the large welded support structure used to prepare the shuttle for launch. NASA

Ships, buildings, bridges, and recreational rides are fabricated by welding processes. Welding is often used to produce the machines that are used to manufacture new products.

Welding has made it possible for airplane manufacturers to meet the design demands of strength-to-weight ratios for both commercial and military aircraft.

The exploration of space would not be possible without modern welding techniques. From the very beginning of early rockets to today's aerospace industry, welding has played an important role. The space shuttle's construction required the improvement of welding processes. Many of these improvements have helped improve our daily lives.

Many experiments aboard the space shuttle have involved welding and metal joining. We are currently building a permanent space station. Someday welders will be required to build such a large structure in the vacuum of space. **Figure 1-5** is a welding machine designed to be used in space. **Figure 1-6** shows a cosmonaut using the welder in open space. The specialized welder was developed at the E.O. Paton Electric Welding Institute in Kiev, Commonwealth of Independent States, the former Soviet Union. As the welding techniques are developed for this major project, we will see them being used here on Earth to improve our world.

Welding is used extensively in the manufacture of automobiles, farm equipment, home appliances, computer components, mining equipment, and construction equipment. Railway equipment, furnaces, boilers, air-conditioning units, and hundreds of other products we

Welded sculpture, Seattle, Washington.

Roller coaster at Silver Dollar City, Branson, Missouri.

Roller coaster at Silver Dollar City, Branson, Missouri.

Voyager of the Sea, Haiti.

Spiral staircase in Missouri City, Texas.

Voyager of the Sea dining room.

FIGURE 1-4 Welded joints are a critical component of structures. Larry Jeffus

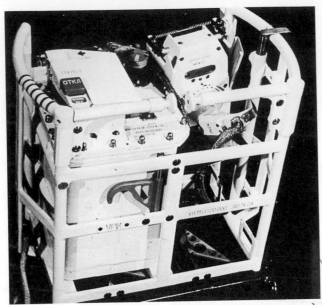

FIGURE 1-5 Machine designed to be used to weld in space. E.O. Paton Electric Welding Institute, Commonwealth of Independent States, the former Soviet Union

FIGURE 1-6 A cosmonaut makes a weld outside a spaceship. E.O. Paton Electric Welding Institute, Commonwealth of Independent States, the former Soviet Union

use in our daily lives are also joined together by some type of welding process.

Items ranging from dental braces to telecommunication satellites are assembled by welding. Very little in our modern world is not produced using some type of this versatile process.

Welding and Cutting Processes

Welding processes differ greatly in the manner in which heat, pressure, or both heat and pressure are applied and in the type of equipment used. **Table 1-1** lists

various welding and allied processes. Some 67 welding processes are listed, requiring hammering, pressing, or rolling to effect the coalescence in the weld joint. Other methods bring the metal to a fluid state, and the edges flow together.

The most popular welding processes are oxyacetylene welding (OAW); shielded metal arc welding (SMAW), often called stick welding; gas tungsten arc welding (GTAW); gas metal arc welding (GMAW); flux cored arc welding (FCAW); and **torch or oxyfuel brazing** (TB). The two most popular thermal cutting processes are oxyacetylene cutting (OAC) and plasma arc cutting (PAC).

The use of regional terms by skilled workers is a common practice in all trade areas, including welding. As an example, oxyacetylene welding is one part of the larger group of processes known as **oxyfuel gas welding (OFW)**. Some of the names used to refer to oxyacetylene welding (OAW) include *gas welding* and *torch welding*. Shielded metal arc welding (SMAW) is often called *stick welding, rod welding,* or just *welding*. As you begin your work career you will learn the various names used in your area, but you should always keep in mind and use the more formal terms whenever possible.

Shielded metal arc welding (SMAW) uses a 14-in. (355.6-mm)-long consumable stick electrode that conducts the welding current from the electrode holder to the work, and as the arc melts the end of the electrode away, it becomes part of the weld metal. The welding arc vaporizes the solid flux that covers the electrode so that it forms an expanding gaseous cloud to protect the molten weld metal. In addition to fluxes protecting molten weld metal, they also perform a number of beneficial functions for the weld, depending on the type of electrode being used.

SMA welding equipment can be very basic as compared to other welding processes. It can consist of a welding transformer and two welding cables with a work clamp and electrode holder, **Figure 1-7**. There are more types and sizes of SMA welding electrodes than there are filler metal types and sizes for any other welding process. This wide selection of filler metal allows welders to select the best electrode type and size to fit their specific welding job requirements. So a wide variety of metal types and metal thicknesses can be joined with one machine.

Gas tungsten arc welding (GTAW) uses a nonconsumable electrode made of tungsten. In GTA welding the arc between the electrode and the base metal melts the base metal and the end of the filler metal as it is manually dipped into the molten weld pool. A shielding gas flowing from the gun nozzle protects the molten weld metal from atmospheric contamination. A foot or thumb remote control switch may be added to the basic GTA welding setup to allow the welder better control, **Figure 1-8**. This remote control switch is often used to start and stop the welding current as well as make adjustments in the power level.

GTA welding is the cleanest of all the manual welding processes. But because there is no flux used to clean the weld in GTA welding, all surface contamination such as oxides, oil, dirt, and so on, must be cleaned from the part

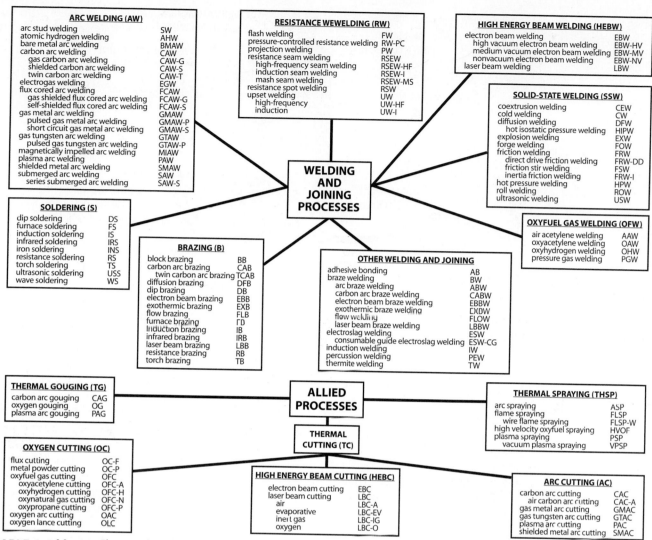

ARC WELDING (AW)

arc stud welding	SW
atomic hydrogen welding	AHW
bare metal arc welding	BMAW
carbon arc welding	CAW
gas carbon arc welding	CAW-G
shielded carbon arc welding	CAW-S
twin carbon arc welding	CAW-T
electrogas welding	EGW
flux cored arc welding	FCAW
gas shielded flux cored arc welding	FCAW-G
self-shielded flux cored arc welding	FCAW-S
gas metal arc welding	GMAW
pulsed gas metal arc welding	GMAW-P
short circuit gas metal arc welding	GMAW-S
gas tungsten arc welding	GTAW
pulsed gas tungsten arc welding	GTAW-P
magnetically impelled arc welding	MIAW
plasma arc welding	PAW
shielded metal arc welding	SMAW
submerged arc welding	SAW
series submerged arc welding	SAW-S

RESISTANCE WEWELDING (RW)

flash welding	FW
pressure-controlled resistance welding	RW-PC
projection welding	PW
resistance seam welding	RSEW
high-frequency seam welding	RSEW-HF
induction seam welding	RSEW-I
mash seam welding	RSEW-MS
resistance spot welding	RSW
upset welding	UW
high-frequency	UW-HF
induction	UW-I

HIGH ENERGY BEAM WELDING (HEBW)

electron beam welding	EBW
high vacuum electron beam welding	EBW-HV
medium vacuum electron beam welding	EBW-MV
nonvacuum electron beam welding	EBW-NV
laser beam welding	LBW

SOLID-STATE WELDING (SSW)

coextrusion welding	CEW
cold welding	CW
diffusion welding	DFW
hot isostatic pressure welding	HIPW
explosion welding	EXW
forge welding	FOW
friction welding	FRW
direct drive friction welding	FRW-DD
friction stir welding	FSW
inertia friction welding	FRW-I
hot pressure welding	HPW
roll welding	ROW
ultrasonic welding	USW

WELDING AND JOINING PROCESSES

SOLDERING (S)

dip soldering	DS
furnace soldering	FS
induction soldering	IS
infrared soldering	IRS
iron soldering	INS
resistance soldering	RS
torch soldering	TS
ultrasonic soldering	USS
wave soldering	WS

BRAZING (B)

block brazing	BB
carbon arc brazing	CAB
twin carbon arc brazing	TCAB
diffusion brazing	DFB
dip brazing	DB
electron beam brazing	EBB
exothermic brazing	EXB
flow brazing	FLB
furnace brazing	FB
induction brazing	IB
infrared brazing	IRB
laser beam brazing	LBB
resistance brazing	RB
torch brazing	TB

OTHER WELDING AND JOINING

adhesive bonding	AB
braze welding	BW
arc braze welding	ABW
carbon arc braze welding	CABW
electron beam braze welding	EBBW
exothermic braze welding	EXDW
flow welding	FLOW
laser beam braze welding	LBBW
electroslag welding	ESW
consumable guide electroslag welding	ESW-CG
induction welding	IW
percussion welding	PEW
thermite welding	TW

OXYFUEL GAS WELDING (OFW)

air acetylene welding	AAW
oxyacetylene welding	OAW
oxyhydrogen welding	OHW
pressure gas welding	PGW

THERMAL GOUGING (TG)

carbon arc gouging	CAG
oxygen gouging	OG
plasma arc gouging	PAG

ALLIED PROCESSES

THERMAL CUTTING (TC)

THERMAL SPRAYING (THSP)

arc spraying	ASP
flame spraying	FLSP
wire flame spraying	FLSP-W
high velocity oxyfuel spraying	HVOF
plasma spraying	PSP
vacuum plasma spraying	VPSP

OXYGEN CUTTING (OC)

flux cutting	OC-F
metal powder cutting	OC-P
oxyfuel gas cutting	OFC
oxyacetylene cutting	OFC-A
oxyhydrogen cutting	OFC-H
oxynatural gas cutting	OFC-N
oxypropane cutting	OFC-P
oxygen arc cutting	OAC
oxygen lance cutting	OLC

HIGH ENERGY BEAM CUTTING (HEBC)

electron beam cutting	EBC
laser beam cutting	LBC
air	LBC-A
evaporative	LBC-EV
inert gas	LBC-IG
oxygen	LBC-O

ARC CUTTING (AC)

carbon arc cutting	CAC
air carbon arc cutting	CAC-A
gas metal arc cutting	GMAC
gas tungsten arc cutting	GTAC
plasma arc cutting	PAC
shielded metal arc cutting	SMAC

TABLE 1-1 Master Charts of Welding and Joining Processes and of Allied Processes. *American Welding Society*

being welded and the filler metal so they do not contaminate the weld. Even though GTA welding is slower and requires a higher skill level as compared to other manual welding processes, it is still in demand because it can be used to make extremely high-quality welds in applications where weld integrity is critical. And there are metal alloys that can be joined only with the GTA welding process.

Gas metal arc welding (GMAW) uses a solid electrode wire that is continuously fed from a spool, through the welding cable assembly, and out through the gun.

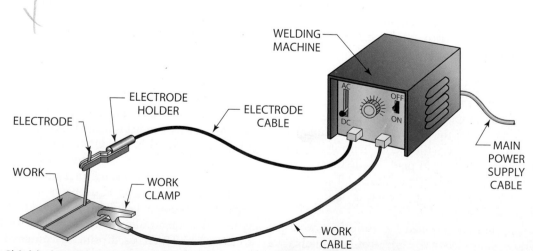

FIGURE 1-7 Shielded metal arc welding equipment. © Cengage Learning 2012

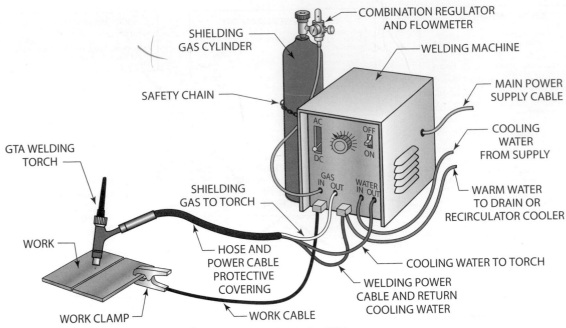

COMBINATION REGULATOR AND FLOWMETER

SHIELDING GAS CYLINDER

WELDING MACHINE

MAIN POWER SUPPLY CABLE

SAFETY CHAIN

COOLING WATER FROM SUPPLY

GTA WELDING TORCH

WARM WATER TO DRAIN OR RECIRCULATOR COOLER

SHIELDING GAS TO TORCH

WORK

HOSE AND POWER CABLE PROTECTIVE COVERING

COOLING WATER TO TORCH

WELDING POWER CABLE AND RETURN COOLING WATER

WORK CLAMP

WORK CABLE

FIGURE 1-8 Gas tungsten arc welding equipment. © Cengage Learning 2012

A shielding gas flows through a separate tube in the cable assembly, out of the welding gun nozzle, and around the electrode wire. The welding power flows through a cable in the cable assembly and is transferred to the electrode wire at the welding gun. The GMA weld is produced as the arc melts the end of the continuously fed filler electrode wire and the surface of the base metal. The molten electrode metal transfers across the arc and becomes part of the weld. The gas shield flows out of the welding gun nozzle to protect the molten weld from atmospheric contamination.

GMA welding is extremely fast and economical because it can produce long welds rapidly that require very little postweld cleanup. This process can be used to weld metal ranging in thickness from thin-gauge sheet metal to heavy plate by making only a few changes in the welding setup.

Flux cored arc welding (FCAW) uses a flux core electrode wire that is continuously fed from a spool, through the welding cable assembly, and out through the gun. The welding power also flows through the cable assembly. Some welding electrode wire types must be used with a shielding gas, as in GMA welding, but others have enough shielding produced as the flux core vaporizes. The welding current melts both the filler wire and the base metal. When some of the flux vaporizes, it forms a gaseous cloud that protects the surface of the weld. Some of the flux that melts travels across the arc with the molten filler metal where it enters the molten weld pool. Inside the molten weld metal, the flux gathers up impurities and floats them to the surface where it forms a slag covering on the weld as it cools.

Although slag must be cleaned from the FCA welds after completion, the process' advantage of high quality, versatility, and welding speed offset this requirement.

Gas metal arc welding and flux cored arc welding are very different welding processes, but they use very similar welding equipment, **Figure 1-9.** Both GMA and FCA welding are classified as semiautomatic processes because the filler metal is automatically fed into the welding arc, and the welder manually moves the welding gun along the joint being welded. GMA and FCA welding are the first choice for many welding fabricators because these processes are cost effective, produce high-quality welds, and are flexible and versatile. In addition to welding, supply stores, many others stores such as hardware stores, building supply stores, automotive supply stores, and others carry GMA/FCA welding equipment and filler metals.

Oxyacetylene Welding, Brazing, and Cutting

Oxyacetylene welding (OAW) and torch brazing (TB) can be done with the same equipment, and **oxyfuel gas cutting (OFC)** uses very similar equipment, **Figure 1-10.**

In OF welding and TB a high-temperature flame is produced at the torch tip by burning oxygen and a fuel gas. The most common fuel gas is acetylene; however, other combinations of oxygen and fuel gases (OF) can be used for welding, such as hydrogen, MAPP, or propane. In OF welding, the base metal is melted and a filler metal may be added to reinforce the weld. No flux is required to make an OF weld.

In TB, the metal is heated to a sufficient temperature but below its melting point so that a brazing alloy can be melted and bond to the hot base metal. A flux may be used to help the brazing alloy bond to the base metal. Both OF welding and TB are used primarily on smaller, thinner-gauge metals.

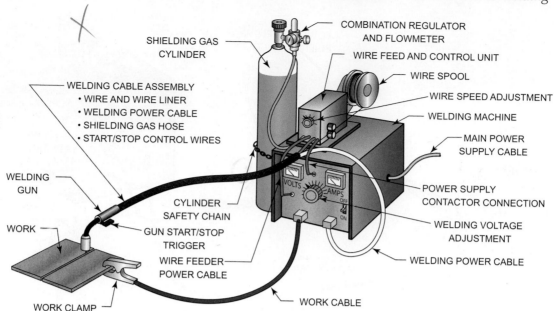

SHIELDING GAS CYLINDER

COMBINATION REGULATOR AND FLOWMETER

WIRE FEED AND CONTROL UNIT

WIRE SPOOL

WELDING CABLE ASSEMBLY
• WIRE AND WIRE LINER
• WELDING POWER CABLE
• SHIELDING GAS HOSE
• START/STOP CONTROL WIRES

WIRE SPEED ADJUSTMENT

WELDING MACHINE

MAIN POWER SUPPLY CABLE

WELDING GUN

POWER SUPPLY CONTACTOR CONNECTION

WORK

CYLINDER SAFETY CHAIN

GUN START/STOP TRIGGER

WIRE FEEDER POWER CABLE

WELDING VOLTAGE ADJUSTMENT

WELDING POWER CABLE

VOLTS AMPS OFF ON

WORK CLAMP

WORK CABLE

FIGURE 1-9 Gas metal arc welding equipment. © Cengage Learning 2012

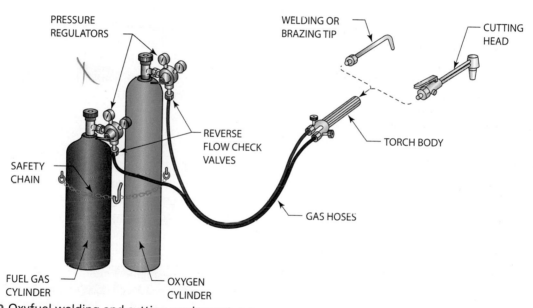

PRESSURE REGULATORS

WELDING OR BRAZING TIP

CUTTING HEAD

REVERSE FLOW CHECK VALVES

TORCH BODY

SAFETY CHAIN

GAS HOSES

FUEL GAS CYLINDER

OXYGEN CYLINDER

FIGURE 1-10 Oxyfuel welding and cutting equipment. © Cengage Learning 2012

Thermal Cutting Processes

There are a number of thermal cutting processes such as oxyfuel cutting (OFC) and plasma arc cutting (PAC). They are the most commonly used in most welding shops. Air carbon arc (AAC) cutting is also frequently used, and many larger fabrication shops have started using laser beam cutting (LBC).

Oxyfuel Gas Cutting Oxyfuel gas cutting uses the high-temperature flame to heat the surface of a piece of steel to a point where a forceful stream of oxygen flowing out a center hole in the tip causes the hot steel to burn away, leaving a gap or cut. Because OF cutting relies on the rapid oxidation of the base metal at elevated temperatures

to make a cut, the types of metals and alloys it can be used on are limited. OF cutting can be used on steel from a fraction of an inch thick to several feet depending on the capacity of the torch and tip being used.

Plasma Arc Cutting Plasma arc cutting (PAC) uses a stiff, highly ionized, extremely hot column of gas to almost instantly vaporize the metal being cut. Most ionized plasma is formed as high-pressure air is forced through a very small opening between a tungsten electrode and the torch tip, **Figure 1-11.** As the air is ionized, it heats up, expands, and exits the torch tip at supersonic speeds. PAC does not rely on rapid oxidation of the metal being cut, like OFC, so almost any metal or alloy can be cut.

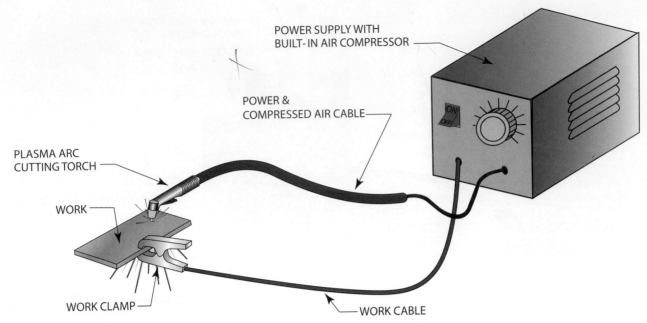

POWER SUPPLY WITH
BUILT- IN AIR COMPRESSOR

POWER &
COMPRESSED AIR CABLE

PLASMA ARC
CUTTING TORCH

WORK

WORK CLAMP

WORK CABLE

FIGURE 1-11 Plasm arc cutting equipment. © Cengage Learning 2012

PA cutting equipment consists of a transformer power supply, plasma torch and cable, work clamp and cable, and an air supply. Some PA cutting equipment has self-contained air compressors. Because the PA cutting process can be performed at some very high travel speeds, it is often used on automated cutting machines. The high travel speeds and very low heat input help to reduce or eliminate part distortion, a common problem with some OF cutting.

Selection of the Joining Process The selection of the joining process for a particular job depends upon many factors. No one specific rule controls the welding process to be selected for a certain job. Following are a few of the factors that must be considered when choosing a joining process.

- Availability of equipment—The types, capacity, and condition of equipment that can be used to make the welds.
- Repetitiveness of the operation—How many of the welds will be required to complete the job, and are they all the same?
- Quality requirements—Is this weld going to be used on a piece of furniture, to repair a piece of equipment, or to join a pipeline?
- Location of work—Will the weld be in a shop or on a remote job site?
- Materials to be joined—Are the parts made out of a standard metal or some exotic alloy?
- Appearance of the finished product—Will this be a weldment that is needed only to test an idea, or will it be a permanent structure?

- Size of the parts to be joined—Are the parts small, large, or different sizes, and can they be moved or must they be welded in place?
- Time available for work—Is this a rush job needing a fast repair, or is there time to allow for pre- and postweld cleanup?
- Skill or experience of workers—Do the welders have the ability to do the job?
- Cost of materials—Will the weldment be worth the expense of special equipment materials or finishing time?
- Code or specification requirements—Often the selection of the process is dictated by the governing agency, codes, or standards.

The welding engineer and/or the welder must not only decide on the welding process but must also select the method of applying it. The following methods are used to perform welding, cutting, or brazing operations.

- **Manual**—The welder is required to manipulate the entire process.
- **Semiautomatic**—The filler metal is added automatically, and all other manipulation is done manually by the welder.
- **Machine**—Operations are done mechanically under the observation and correction of a welding operator.
- **Automatic**—Operations are performed repeatedly by a machine that has been programmed to do an entire operation without interaction of the operator.
- **Automated**—Operations are performed repeatedly by a robot or other machine that is programmed flexibly to do a variety of processes.

Occupational Opportunities in Welding

The American welding industry has contributed to the widespread growth of the welding and allied processes. Without welding, much of what we use on a daily basis could not be manufactured. The list of these products grows every day, thus increasing the number of jobs for people with welding skills. The need to fill these well-paying jobs is not concentrated in major metropolitan areas but is found throughout the country and the world. Because of the diverse nature of the welding industry, the exact job duties of each skill area will vary. The following are general descriptions of the job classifications used in our profession; specific tasks may vary from one location to another.

Welders perform the actual welding. They are the skilled craftspeople who, through their own labor, produce the welds on a variety of complex products, **Figure 1-12**.

Tack welders, also skilled workers, often help the welder by making small welds to hold parts in place. The tack weld must be correctly applied so that it is strong enough to hold the assembly and still not interfere with the finished welding.

Welding operators, often skilled welders, operate machines or automatic equipment used to make welds.

Welders' helpers are employed in some welding shops to clean slag from the welds and help move and position weldments for the welder.

Welder assemblers, or **welder fitters**, position all the parts in their proper places and make them ready for the tack welders. These skilled workers must be able to interpret blueprints and welding procedures. They must also have knowledge of the effects of contraction and expansion of the various types of metals.

Welding inspectors are often required to hold a special **certification** such as the one supervised by the American Welding Society known as Certified Welding Inspector (CWI). To become a CWI, candidates must pass a test covering the welding process, blueprint reading, weld symbols, metallurgy, codes and standards, and inspection techniques. Vision screening is also required on a regular basis, once the technical skills have been demonstrated.

Welding shop supervisors may or may not weld on a regular basis, depending on the size of the shop. In addition to their welding skills, they must demonstrate good management skills by effectively planning jobs and assigning workers.

Welding salespersons may be employed by supply houses or equipment manufacturers. These jobs require a broad understanding of the welding process as well as good marketing skills. Good salespersons are able to provide technical information about their products to convince customers to make a purchase.

Welding shop owners are often welders who have a high degree of skill and knowledge of small-business management and prefer to operate their own businesses. These individuals may specialize in one field, such as hardfacing,

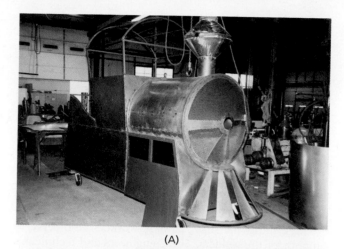
(A)

(B)

(C)

FIGURE 1-12 Amusement parks like Silver Dollar City in Branson, Missouri, require a lot of talented welders to produce such attractions as these. (a) Fabricating an antique train engine to be used in a parade. (b) Air-powered guns for launching toy balls. (c) The Branson Belle paddleboat. Larry Jeffus

repair and maintenance, or specialty fabrications, or they may operate as subcontractors of manufactured items. A welding business can be as small as one individual, one truck, and one portable welder or as large as a multimillion-dollar operation employing hundreds of workers.

Welding engineers design, specify, and oversee the construction of complex weldments. The welding engineer may work with other engineers in areas such as mechanics, electronics, chemicals, or civil engineering in the process of bringing a new building, ship, aircraft, or product into existence. The welding engineer is required to know all of the welding process and metallurgy as well as have good math, reading, communication, and design skills. This person usually has an advanced college degree and possesses a professional certification.

In many industries, the welder, welding operator, and tack welder must be able to pass a performance test to a specific code or standard.

The highest-paid welders are those who have the education and skills to read blueprints and do the required work to produce a weldment to strict specifications.

Large industrial firms employ workers who serve as support for the welders. These engineers and technicians must have knowledge of chemistry, physics, metallurgy, electricity, and mathematics. Engineers are responsible for research, design, development, and fabrication of a project. Technicians work as part of the engineering staff. These individuals may oversee the actual work for the engineer by providing the engineer with progress reports as well as chemical, physical, and mechanical test results. Technicians may also require engineers to build prototypes for testing and evaluation.

Another group of workers employed by industry does layouts or makes templates. These individuals have had drafting experience and have a knowledge of such operations as punching, cutting, shearing, twisting, and forming, among others. The layout is generally done directly on the material. A template is used for repetitive layouts and is made from sheet metal or other suitable materials.

The flame-cutting process is closely related to welding. Some operators use handheld torches, and others are skilled operators of oxyfuel cutting machines. These machines range from simple mechanical devices to highly sophisticated, tape, controlled, multiple-head machines that are operated by specialists, **Figure 1-13.**

Teamwork

In smaller welding shops welding is often performed by a single individual who may work without a direct supervisor. In larger welding shops a welder may work in a crew where most of the welding is performed in a work group or work team. Generally, a work group has an assigned leader who directs the member's activities, while a work team may not have a designated leader so they are more self-managed. However, in a work team a team member who has seniority or more expertise may serve as the understood team leader. Each member of the crew will have specific job responsibilities based on their job classification.

If there is only one welder, that welder may act as the leader directing the other crewmembers as well as performing the welding. The welder helper helps with the preparation and fitup of the weldment and other jobs as assigned. When welding is being done where other than a minor fire might be started, a crewmember must serve as a firewatch. The fire watch's primary responsibility is to make sure the site is safe so that no accidental fires get started. Other crewmembers may operate equipment to lift parts in place and other duties as required.

Leadership

A team leader must have knowledge of the Equal Employment Opportunity Laws regarding job discrimination and make sure these laws are followed. In addition a good leader should have a positive attitude, think and plan ahead, treat everyone with respect, show confidence in their crew members' abilities, and most importantly, communicate clearly to each member what is expected of them. It is important to recognize the various strengths and weaknesses of each crewmember so that they can be used in the most effective manner to get the job done. When necessary, coach crewmembers in how you want them to do their job. Often, demonstrating the proper way you want them to do something can be helpful.

A crew that works well together will be much more productive than one that is dysfunctional. As the crew leader, you need to be able to help resolve conflicts between team members. Identify the key issues that might be causing the conflict, and get the crew to agree on how the problem can be resolved. A work crew's performance is usually evaluated on the productiveness of the team; so hard work and cooperation are rewarded.

Training for Welding Occupations

Generally, several months of training are required to learn to weld. To become a skilled welder, both welding school and on-the-job experience are required. Because of the diverse nature of the welding industry, no single list of skills can be used to meet every job's requirements. However, there are specific skills that are required of most entry-level welders. This text covers those skill requirements.

In addition to welding skills, an entry-level welder must possess workplace skills. Workplace skills include a proficiency in reading, writing, math, communication, and science as well as good work habits and an acceptance of close supervision. Some welding jobs may also require a theoretical knowledge of welding, blueprint reading, welding symbols, metal properties, and electricity. A few of the jobs that require less skill can be learned after a few months of on-the-job training. However, the fabrication

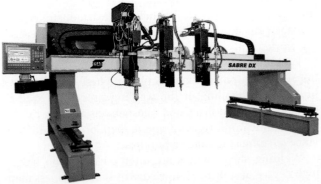

FIGURE 1-13 Numerical control oxygen cutting machine. ESAB Welding & Cutting Products

of certain alloys requires knowledge of metallurgical properties as well as the development of a greater skill in cutting and welding them.

Robotics and computer-aided manufacturing (CAM) both require more than a basic understanding of the welding process; they require that the student be computer literate.

A young person planning a career as a welder needs good eyesight, manual dexterity, hand-and-eye coordination, and understanding of welding technology. For entry into manual welding jobs, employers prefer to hire young people who have high school or vocational training in welding processes. Courses in drafting, blueprint reading, mathematics, and physics are also valuable.

Beginners in welding who have no training often start in manual welding production jobs that require minimum skill. Occasionally, they first work as helpers and are later moved into welding jobs. General helpers, if they show promise, may be given a chance to become welders by serving as helpers to experienced welders.

A formal apprenticeship is usually not required for general welders. A number of large companies have welding apprenticeship programs. The military, at several of its installations, has programs in welding.

Skill and technical knowledge requirements are higher in some industries. In the fields of atomic energy, aerospace, and pressure vessel construction, high standards for welders must be met to ensure that weldments will withstand the critical forces that they will be subjected to in use.

After two years of training at a vocational school or technical institute, the skilled welder may qualify as a technician. Technicians are generally involved in the interpretation of engineers' plans and instructions. Employment of welders is expected to increase rapidly as a result of the wider use of various welding processes, including the use of robots.

Many more skilled welders will be needed for maintenance and repair work in the expanding metalworking industries. The number of welders in production work is expected to increase in plants manufacturing sheet metal products, pressure vessels, boilers, railroads, storage tanks, air-conditioning equipment, and ships and in the field of energy exploration and energy resources. The construction industry will need an ever-increasing number of good welders as the use of welded steel buildings grows.

Employment prospects for welding operators are expected to continue to be favorable because of the increased use of machine, automatic, and automated welding in the manufacture of aircraft, missiles, railroad cars, automobiles, and numerous other products.

Before being assigned a job where service requirements of the weld are critical, welders usually must pass a certification test given by an employer.

In addition, some localities require welders to obtain a license for certain types of outside construction.

THINK GREEN

All welding and cutting processes consume large quantities of energy and materials, and some produce environmental pollution. It is important that you always look at ways to minimize the impact these processes have on our environment. For example, if a spill occurs, notify your supervisor and clean it up promptly and properly. Always look for ways to be better stewards of our environment.

After welders, welding operators, or tack welders have received a **certification** or **qualification** by passing a standardized test, they are only allowed to make welds covered by that specific test. The welding certification is very restrictive; it allows a welder to perform only code welds covered by that test. Certifications are usually good for a maximum of six months unless a welder is doing code-quality welds routinely. As a student, you should check into the acceptance of a welding qualification test before investing time and possibly money in the test.

The **American Welding Society (AWS)** has developed three levels of certification for welders. The first level, Entry Level Welder, is for the beginning welder, and Level II and Level III are for the more skilled welders. These certifications are gaining widespread acceptance by the industry. Certification allows welders to prove their skills on a standard test.

For additional information about the skills required for various welding job classifications, refer to **Table 1-2**.

Each year SkillsUSA sponsors a series of welding skill competitions for its student members. Students can begin by joining their local SkillsUSA chapter. They can then compete in local, regional, and state competitions. Each time, the students with the best welding skills and knowledge can advance to the next level of competition. Contestants are challenged with a written test and must show their proficiency in welding and fabrication. There is a national SkillsUSA Olympics competition held each year in Kansas City, Missouri. The winners at the national competition can then go on to the International Skill Olympics. The international competition is held in a different country each year. Like most professional organizations, SkillsUSA emphasizes community service and citizenship as key components to the philosophy of the organization.

Experiments and Practices

A number of the chapters in this book contain both experiments and practices. These are intended to help you develop your welding knowledge and skills.

The experiments are designed to allow you to see what effect changes in the process settings, operation, or techniques have on the type of weld produced. When you do

TABLE 1-2 Recommended Skills for Various Welding Occupations

Recommended Instructional Chapters for Welding Occupations*

Column group headers and chapter numbers:

General
- 1 Introduction
- 2 Safety

Oxyfuel Gas
- 31 Oxyfuel Gas
- 31 Gases
- 31 Filler Metals
- 30 Equipment and Setup
- 32 Plate Welding Flat Position
- 32 Plate Welding All Positions
- 32 Pipe Welding Flat Position
- 32 Pipe Welding All Positions
- 33 Soldering, Brazing, and Braze Welding
- 33 Filler Metals and Fluxes
- 33 Soldering
- 33 Brazing and Braze Welding

Shielded Metal Arc
- 7 Oxygen Cutting
- 7 Equipment and Setup
- 7 Plate and Pipe Oxygen Cutting
- 3 Shielded Metal Arc Welding
- 3 Equipment and Setup
- 27 Filler Metals
- 4 Plate Welding Flat Position
- 4 Plate Welding All Positions
- 5 Pipe Welding Flat Position
- 6 Pipe Welding All Positions

Gas Shielded Arc
- 15 Gas Tungsten Arc Welding
- 15 Equipment and Setup
- 27 Filler Metal
- 16 Plate Welding
- 17 Pipe Welding
- 10 Gas Metal Arc Welding
- 10 Equipment and Setup
- 27 Filler Metal
- 11 Plate Welding

Related Information
- 28 Automation and Robotics
- 29 Special Welding Processes
- 26 Weldability of Metals
- 23 Testing and Inspection
- 25 Welding Metallurgy
- 20 Welding Symbols
- 20 Joint Designs

Special Knowledge or Skill

Welding Occupations (rows):
Tack welder; Production SMA welder; Production GTA welder; Production GMA welder; Production machine operator (Numerical control); Construction welder; Welder assembler; Welder fitter; Maintenance welder (Surfacing); Owner/operator welder (Business); Welder craftsperson; Welding foreman (Leadership); Welding technician; Welding engineer (Business); Welding sales; Related occupations; Auto mechanic; Auto body (Surfacing); Machinist; Air conditioning/heating; Sheet metal; General maintenance; Farm/agricultural; Art/sculpture

Legend:

Principles
- ▢ (open box) Essential
- ╱ General knowledge of
- • Nice to know how to
- ■ Nonessential

Applications
- ▢ Pass appropriate weld test
- ╱ Make good welds
- • Nice to know how to
- ■ Nonessential

*Welding occupational job titles and specific responsibilities vary greatly. For a description of these job titles, refer to the glossary in this text.

an experiment, you should observe and possibly take notes on how the change affected the weld. Often as you make a weld it will be necessary for you to make changes in your equipment settings or your technique to ensure you are making an acceptable weld. By watching what happens when you make the changes in the welding shop, you will be better prepared to decide on changes required to make good welds on the job.

It is recommended that you work in a small group as you try the experiments. When doing the experiments in a small group, one person can be welding, one adjusting the equipment, and the others recording the machine settings and weld effects. It also allows you to watch the weld change more closely if someone is welding as you look on. Then, as a group member, changing places will reinforce your learning.

The practices are designed to build your welding skills. Each practice tells you in detail what equipment, supplies, and tools you will need as you develop the specific skill. In most chapters, the practices start off easy and become progressively harder. Welding is a skill that requires that you develop in stages from the basic to the more complex.

Each practice gives the evaluation or acceptable limits for the weld. All welds have some discontinuities, but if they are within the acceptable limits, they are not defects but are called flaws. As you practice your welding, keep in mind the acceptable limits so that you can progress to the next level when you have mastered the process and weld you are working on.

Welding Video Series

Delmar/Cengage Learning, in cooperation with the author, has produced a series of videotapes. Each of the four tape sets covers specific equipment setup and operation for welding, cutting, soldering, or brazing. When there are specific skills shown both in this textbook and on a videotape, you will see a framed shot from the video, as shown in **Figure 1-14**. Reading the material, watching the video, and practicing should help you to develop your welding skills more rapidly.

Metric Units

Both standard and metric (SI) units are given in this text. The SI units are in parentheses ( ) following the standard unit. When nonspecific values are used—for example, "set the gauge at 2 psig" where 2 is an approximate value—the SI units have been rounded off to the nearest whole number. Round off occurs in these cases to agree with the standard value and because whole numbers are easier to work with. The only time that SI units are not rounded off is when the standard unit is an exact measurement.

Often students have difficulty understanding metric units because exact conversions are used even when the standard measurement was an approximation. Rounding off the metric units makes understanding the metric system much easier, **Table 1-3**. Estimating the approximate conversion from one unit type to another makes it possible to quickly have an idea of how large or heavy an object is. When estimating a conversion, it is not necessary to be concise.

By using this approximation method, you can make most standard-to-metric conversions in your head without needing to use a calculator.

Once you have learned to use approximations for metric, you will find it easier to make exact conversions whenever necessary. Conversions must be exact in the shop when a part is dimensioned with one system's units and the other system must be used to fabricate the part. For that reason you must be able to make those conversions. **Table 1-4** and **Table 1-5** are set up to be used with or without the aid of a calculator. Many calculators today have built-in standard–metric conversions. Of course, it is a good idea to know how to make these conversions with and without these aids. Practice making such conversions whenever the opportunity arises.

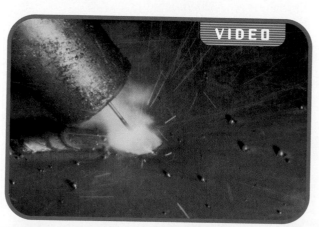

FIGURE 1-14 This GMA welding can be seen in the Gas Metal Arc Welding video series on tape 2. *Larry Jeffus*
See DVD 2 Gas Metal Arc Welding.

1/4 inch = 6 mm	
1/2 inch = 13 mm	
3/4 inch = 18 mm	
1 inch = 25 mm	
2 inches = 50 mm	
1/2 gal = 2 L	
1 gal = 4 L	
1 lb = 1/2 K	
2 lb = 1 K	
1 psig = 7 kPa	
1°F = 2°C	

TABLE 1-3 Conversion Approximations

TEMPERATURE

Units

°F (each 1° change)	=	0.555°C (change)
°C (each 1° change)	=	1.8°F (change)
32°F (ice freezing)	=	0°Celsius
212°F (boiling water)	=	100°Celsius
–460°F (absolute zero)	=	0°Rankine
–273°C (absolute zero)	=	0°Kelvin

Conversions

°F to °C _____ °F – 32 = _____ × .555 = _____ °C
°C to °F _____ °C × 1.8= _____ + 32 = _____ °F

LINEAR MEASUREMENT

Units

1 inch	=	25.4 millimeters
1 inch	=	2.54 centimeters
1 millimeter	=	0.0394 inch
1 centimeter	=	0.3937 inch
12 inches	=	1 foot
3 feet	=	1 yard
5280 feet	=	1 mile
10 millimeters	=	1 centimeter
10 centimeters	=	1 decimeter
10 decimeters	=	1 meter
1000 meters	=	1 kilometer

Conversions

in. to mm	_____ in.	×	25.4	= _____	mm
in. to cm	_____ in.	×	2.54	= _____	cm
ft to mm	_____ ft	×	304.8	= _____	mm
ft to m	_____ ft	×	0.3048	= _____	m
mm to in.	_____ mm	×	0.0394	= _____	in.
cm to in.	_____ cm	×	0.3937	= _____	in.
mm to ft	_____ mm	×	0.00328	= _____	ft
m to ft	_____ m	×	3.28	= _____	ft

AREA MEASUREMENT

Units

1 sq in.	=	0.0069 sq ft
1 sq ft	=	144 sq in.
1 sq ft	=	0.111 sq yd
1 sq yd	=	9 sq ft
1 sq in.	=	645.16 sq mm
1 sq mm	=	0.00155 sq in.
1 sq cm	=	100 sq mm
1 sq m	=	1000 sq cm

Conversions

sq in. to sq mm _____ sq in. × 645.16 = _____ sq mm
sq mm to sq in. _____ sq mm × 0.00155 = _____ sq in.

VOLUME MEASUREMENT

Units

1 cu in.	=	0.000578 cu ft
1 cu ft	=	1728 cu in.
1 cu ft	=	0.03704 cu yd
1 cu ft	=	28.32 L
1 cu ft	=	7.48 gal (U.S.)
1 gal (U.S.)	=	3.737 L
1 cu yd	=	27 cu ft
1 gal	=	0.1336 cu ft
1 cu in.	=	16.39 cu cm
1 L	=	1000 cu cm
1 L	=	61.02 cu in.
1 L	=	0.03531 cu ft
1 L	=	0.2642 gal (U.S.)
1 cu yd	=	0.769 cu m
1 cu m	=	1.3 cu yd

Conversions

cu in. to L	_____ cu in.	×	0.01638	= _____	L
L to cu in.	_____ L	×	61.02	= _____	cu in.
cu ft to L	_____ cu ft	×	28.32	= _____	L
L to cu ft	_____ L	×	0.03531	= _____	cu ft
L to gal	_____ L	×	0.2642	= _____	gal
gal to L	_____ gal	×	3.737	= _____	L

WEIGHT (MASS) MEASUREMENT

Units

1 oz	=	0.0625 lb
1 lb	=	16 oz
1 oz	=	28.35 g
1 g	=	0.03527 oz
1 lb	=	0.0005 ton
1 ton	=	2000 lb
1 oz	=	0.283 kg
1 lb	=	0.4535 kg
1 kg	=	35.27 oz
1 kg	=	2.205 lb
1 kg	=	1,000 g

Conversions

lb to kg	_____ lb	×	0.4535	= _____	kg
kg to lb	_____ kg	×	2.205	= _____	lb
oz to g	_____ oz	×	0.03527	= _____	g
g to oz	_____ g	×	28.35	= _____	oz

PRESSURE AND FORCE MEASUREMENTS

Units

1 psig	=	6.8948 kPa
1 kPa	=	0.145 psig
1 psig	=	0.000703 kg/sq mm
1 kg/sq mm	=	6894 psig
1 lb (force)	=	4.448 N
1 N (force)	=	0.2248 lb

Conversions

psig to kPa	_____ psig	×	6.8948	= _____	kPa
kPa to psig	_____ kPa	×	0.145	= _____	psig
lb to N	_____ lb	×	4.448	= _____	N
N to lb	_____ N	×	0.2248	= _____	psig

VELOCITY MEASUREMENTS

Units

1 in./sec	=	0.0833 ft/sec
1 ft/sec	=	12 in./sec
1 ft/min	=	720 in./sec
1 in./sec	=	0.4233 mm/sec
1 mm/sec	=	2.362 in./sec
1 cfm	=	0.4719 L/min
1 L/min	=	2.119 cfm

Conversions

ft/min to in./sec ____ ft/min × 720 = ____ in./sec
in./min to mm/sec ____ in./min × .4233 = ____ mm/sec
mm/sec. to in./min ____ mm/sec × 2.362 = ____ in./min
cfm to L/min ____ cfm × 0.4719 = ____ L/min
L/min to cfm ____ L/min × 2.119 = ____ cfm

TABLE 1-4 Table of Conversions: U.S. Customary (Standard) Units and Metric Units (SI)

U.S. Customer (Standard) Units

°F	=	degrees Fahrenheit
°R	=	degrees Rankine
	=	degrees absolute F
lb	=	pound
psi	=	pounds per square inch
	=	lb per sq in.
psia	=	pounds per square inch absolute
	=	psi + atmospheric pressure
in.	=	inches = in. = "
ft	=	foot or feet = ft = '
sq in.	=	square inch = in.
sq ft	=	square foot = ft
cu in.	=	cubic inch = in.
cu ft	=	cubic foot = ft
ft-lb	=	foot-pound
ton	=	ton of refrigeration effect
qt	=	quart

Metric Units (SI)

°C	=	degrees Celsius
°K	=	Kelvin
mm	=	millimeter
cm	=	centimeter
cm^2	=	centimeter squared

cm^3	=	centimeter cubed
dm	=	decimeter
dm^2	=	decimeter squared
dm^3	=	decimeter cubed
m	=	meter
m^2	=	meter squared
m^3	=	meter cubed
L	=	liter
g	=	gram
kg	=	kilogram
J	=	joule
kJ	=	kilojoule
N	=	newton
Pa	=	pascal
kPa	=	kilopascal
W	=	watt
kW	=	kilowatt
MW	=	megawatt

Miscellaneous Abbreviations

P	=	pressure	sec	=	seconds
h	=	hours	r	=	radius of circle
D	=	diameter	π	=	3.1416 (a constant
A	=	area			used in determining
V	=	volume			the area of a circle)
∞	=	infinity			

TABLE 1-5 Abbreviations and Symbols

Summary

Welding is a very diverse trade. Almost every manufactured product utilizes a welding or joining process in its production. Products that are produced by welding range from small objects, such as sunglasses and dental braces, to larger structures, such as buildings, ships, and space shuttles. Your knowledge and understanding of the various processes and their applications will provide you with employable skills that can result in a rich and rewarding career in the welding field. The art and science of joining metals has been around for centuries, and with changes and improvements in materials, equipment, and supplies, it will be with us through the remainder of the twenty-first century.

Welding at the Bottom of the World

Antarctica, the South Pole—the words summon images of a bleak, barren landscape; unimaginable cold; and incredible human drama.

Welder Walter Fischel lived and worked at the National Science Foundation Amundsen-Scott South Pole Station from November 3, 1998 until October 30, 1999, making him one of only about 1200 people who have wintered at the South Pole.

"Antarctica is a beautiful, pristine environment that should be preserved," Fischel said. "It is a harsh place—the highest, driest, windiest, darkest place on earth. It is like no other place on this planet. I did enjoy my time there. The work consumes much of your time and there was a good group of people there."

The now 32-year-old Fischel, from Columbia, Maryland, near Baltimore, never planned on becoming a welder or making it his career. Instead, he learned the trade in order to pay for his education, studying first at a Maryland community college and then at Colorado State University with the goal of becoming a geophysicist. He found he enjoyed welding, however, and eventually became qualified in the 3G position to do flat, horizontal, and vertical welding from 1/8 in. to unlimited thickness plates.

His journey to the South Pole began when he answered an advertisement in *the Washington Post* for construction workers in Antarctica. The United States maintains three stations on the continent—McMurdo, Palmer, and Amundsen-Scott South Pole. The South Pole station has been in continuous operation since 1956; it is supplied entirely by air from McMurdo. Housed in the dome since the early 1970s, South Pole station is in need of repair and modernization. A year-round construction project is underway, with an expected completion date of 2005. Fischel was hired by Antarctic Support Associates (ASA) to help build a new heavy equipment shop.

"I was selected and originally hired as an ironworker for a summer-only position at the South Pole," Fischel explained. "The project involved a lot more welding than anticipated, so I was given the winter-over contract late in the summer."

Approximately 200 people worked at the South Pole that season. During the perpetual light of summer, they worked in three overlapping nine-hour shifts, six days a week; in winter, the hours and days are the same, but there is only one shift. "You get one day off a week, which is enough because there's not a lot of activities going on," Fischel recalled.

At Work

Fischel explained there is no way to prepare yourself for working under the harsh conditions of the South Pole, but he found that, once welding was underway, the metal responded as expected. Because of the equipment available to him, he primarily used the shielded metal arc welding process with low-hydrogen electrodes. Given a choice, Fischel said he would have selected a wire feed process, such as flux cored arc welding or gas metal arc welding, to achieve higher deposition rates. He also did a large amount of oxyacetylene cutting, helped raise the structural steel, and shared the household duties involved in community living.

"Because of the lack of humidity, preheating the steel was not needed," he said. Postweld heat treatment was also unnecessary for the same reason. "The metal did react differently at first contact with the heat, which took a little getting used to. It seemed the metal was shocked at first

An aurora over the garage arch during winter 1999.
American Welding Society

A view of the heavy equipment garage Fischel helped build. American Welding Society

Fischel off duty during summer 1998. "Notice there's nothing there," he said. American Welding Society

Fischel changes the flag atop the dome in spring 1999 after a long winter. The temperature was −85°F with a windchill factor of −120°F. The South Pole's year includes only one day and one night, but with an extended dusk and dawn. American Welding Society

contact with the heat. It was hard to get a nice vertical flow, but once the metal warmed up it went smoothly. My first vertical pass there was a mess."

Those who work in Antarctica have found it takes up to 10 times longer to complete a task than it does at normal temperatures. First, workers suffer from chronic hypoxia (not enough oxygen in the blood) because of the high altitude. The South Pole station sits at an elevation of 9300 feet above sea level. Second, the multilayered protective clothing workers must wear restricts movement. Even with protective clothing, if the weather turns too severe no one can work outside for too long.

"My work for the summer was entirely outside the dome at temperatures from −20°F to −65°F. During the winter, it was primarily inside the structure we had erected. Some (winter) work was outside, but only for brief periods because of the extreme cold, −50°F to −105°F with no windchill included. At those temperatures, everything freezes, even the welding cable. There were also periods of no work outside because of whiteouts, where there was no distinction between land and sky at the horizon." (During a whiteout, people can easily become disoriented and be unable to find their way to shelter.)

Fischel's main projects during this time were to hermetically seal the new garage's floor and to work on the catchment pans that were part of the drainage system. He also made repairs to the station's equipment whenever necessary.

"In Antarctica, all the buildings have to be 100% sealed to prevent anything leaking into the ice," Fischel explained. As part of the Antarctic Treaty requirements, he had to weld all the floor joints to prevent oil or other fluids from the heavy equipment that will be stored in the garage from contaminating the environment. The garage's catchment pans were located between the floor beams. The beams had been cambered, so lighter weight beams could be used to help save shipping expenses. However, the catchment pans were overlooked and not cambered to match the beams. Fischel had to cut relief holes in the pans, bend them to match the beams, then weld them so they were completely sealed.

Back in the "World"

As for the future, Fischel said he wants to finish his degree eventually but does not know if he will switch careers. "I do enjoy welding," he said. "It's pretty much an international trade, and it's taken me a few places. I enjoy that. It's not like you need to speak the language, [because] welding's an international language. There's a certain way to do it no matter where you are."

Article courtesy of the American Welding Society.

Review

1. Explain how to forge weld.

2. What welding process is Elihu Thomson credited with developing in 1886?

3. During World War I, what process replaced riveting for ship repair?

4. Welding has become a _____, _____, and _____ method of joining metal.

5. What do we call the localized growing together of the grain structure during a weld?

6. Welding can be used to join both _____ and _____.

7. Some welding processes require both _____ and _____ to make a weld.

8. List six items that use welding in their construction.

9. Why have welding experiments been performed aboard the space shuttle?

10. What three things differ greatly from one welding process to another?

11. Which gases are most commonly used for the OFW process?

12. What is the technically correct name for *gas welding*?

13. Which welding process is the most commonly used to join metal?

14. What is the technically correct name for *stick welding*?

15. GTAW is the abbreviation for which process?

16. What is the ideal process for high welding rates on thin-gauge metal?

17. Flux inside the welding wire gives which process its name?

18. List six items that may be considered in selecting a welding process.

19. Which method of welding application requires a welder to manipulate the entire process?

20. Which method of welding requires the welder to control everything except the adding of filler metal?

21. Which welding process is repeatedly performed by a machine that has been programmed?

22. A _____ works with engineers to produce prototypes for testing.

23. A _____ places parts together in their proper position for the tack welder.

24. A _____ is a piece of sheet metal cut to the shape of a part so that it may be repetitively laid out.

25. What are the approximate standard units for the following SI values?
    a. 13 mm
    b. 4 L
    c. 100°C
    d. 4 K

26. What are the exact standard units for the following SI values?
    a. 13 mm
    b. 4 L
    c. 100°C
    d. 4 K

27. What are some advantages of a working environment that involves groups?

28. Give three examples of how you can benefit from following strong leadership in preparing for your career.

29. Are there positive attitudes you show in the workplace that could also be helpful in your role as a citizen in your community?

30. What technical skills would you as an employer look for in a potential new employee?

31. What personal qualities would you as an employer look for in a potential new employee?

32. How could good work habits create occupational opportunities for you?

33. How can employees' good work habits affect the profitability of the company they work for?

34. List three advantages of having a good understanding of the culture and common business practices of customers in foreign countries.

35. What are some professional benefits of working on an intercultural project?

Chapter 2

Safety in Welding

OBJECTIVES

After completing this chapter, the student should be able to

- describe the three classifications of burns and the emergency steps that should be taken to treat each of them.
- describe the dangers all three types of light pose to welding and how to protect yourself and others from these dangers.
- explain how to avoid eye and ear injuries.
- use a chart to select the correct eye and face protective devices for working and welding in a shop.
- list the safety points that should be covered in a training program for respiratory protection.
- describe the various types of respiratory protection equipment that are available.
- tell how to avoid dangerous fumes and gases by providing ventilation to the welding area.
- explain the purpose of material safety data sheets (MSDSs) and where they can be found.
- discuss the benefits of recycling waste material.
- describe what type of general work clothing should be worn in a welding shop.
- describe special protective clothing worn by welders to protect the hands, arms, body, waist, legs, and feet.
- describe the proper way to handle, secure, and store cylinders.
- discuss how to protect against the danger of fire when welding.
- explain why planned maintenance of tools and equipment is important.
- describe the most commonly used hand and power tools used by a welder.
- explain good electrical safety practices and list rules for extension cords and portable power tools.
- discuss the types of metal cutting machines.
- describe proper ways to safely lift heavy welded assemblies.
- list the rules for ladder safety.

KEY TERMS

acetone

acetylene

earmuffs

earplugs

electric shock

electrical ground

electrical resistance

exhaust pickups

flash burn ARC BURN

flash glasses

forced ventilation

full face shield

ground-fault circuit interpreter (GFCI)

goggles

infrared light

(Review Regular)

material safety data sheet (MSDS)

natural ventilation

safety glasses

type A fire extinguisher

type B fire extinguisher

type C fire extinguisher

type D fire extinguisher

ultraviolet light

valve protection cap

ventilation

visible light

warning label

welding helmet

INTRODUCTION

Accident prevention is the main intent of this chapter. The safety information included in this text is intended as a guide. There is no substitute for caution and common sense. A safe job is no accident; it takes work to make the job safe. Each person must take personal responsibility for their own safety and the safety of others on the job.

Welding is a very large and diverse industry. This chapter will concentrate on only that portion of welding safety related to the areas of light metal. You must read; learn; and follow all safety rules, regulations, and procedures for those areas.

Light welding fabrication, like all other areas of welding work, has a number of potential safety hazards. These hazards need not result in anyone being injured. Learning to work safely is as important as learning to be a skilled welding fabrication worker.

You must approach new jobs with your safety in mind. Your safety is your own responsibility, and you must take on that responsibility. It is not possible to anticipate all of the possible dangers in every job. There may be some dangers not covered in this text. You can get specific safety information from welding equipment manufacturers and their local suppliers, your local college or university, and the Internet.

If an accident does occur on a welding site, it can have consequences far beyond just the person injured. Serious accidents can result in local, state, or national investigations. For example, if the federal office of the Occupational Safety and Health Administration (OSHA) becomes involved, the job site may be closed for hours, days, weeks, months, or even permanently. While the job site is closed for the investigation, you may be off without pay. If it is determined that your intentional actions contributed to the accident, you may lose your job, be fined, or worse. Always follow the rules, and never engage in horseplay or play "practical jokes" while at work.

Burn Classification

Burns are one of the most common and painful injuries that occur in the welding shop. Burns can be caused by ultraviolet light rays as well as by contact with hot welding material. The chance of infection is high with burns because of the dead tissue. It is important that all burns receive proper medical treatment to reduce the chance of infection.

Burns are divided into three classifications, depending upon the degree of severity. The three classifications include first-degree, second-degree, and third-degree burns. Whether burns are caused by hot material or by light, they can be avoided if proper clothing and other protective gear are worn.

First-Degree Burns First-degree burns occur when the surface of the skin is reddish in color, tender, and painful and do not involve any broken skin. The first step in treating a first-degree burn is to immediately put the burned area under cold water (not iced) or apply cold water compresses (clean lint-free towel, washcloth, or handkerchief soaked in cold water) until the pain decreases. Then cover the area with sterile bandages or a clean cloth. Do not apply butter or grease. Do not apply

any other home remedies or medications without a doctor's recommendation. See **Figure 2-1**.

Second-Degree Burns Second-degree burns occur when the surface of the skin is severely damaged, resulting in the formation of blisters and possible breaks in the skin.

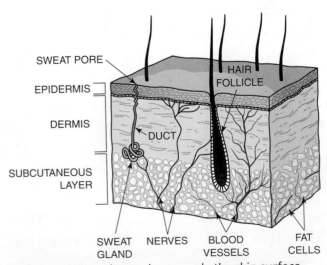

SWEAT PORE

EPIDERMIS

DERMIS

SUBCUTANEOUS LAYER

HAIR FOLLICLE

DUCT

SWEAT GLAND   NERVES   BLOOD VESSELS   FAT CELLS

FIGURE 2-1 First-degree burn—only the skin surface (epidermis) is affected. © Cengage Learning 2012

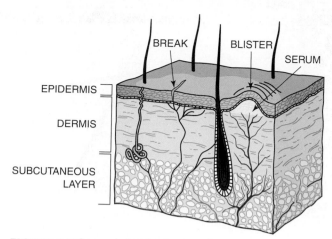

FIGURE 2-2 Second-degree burn—the epidermal layer is damaged, forming blisters or shallow breaks.
© Cengage Learning 2012

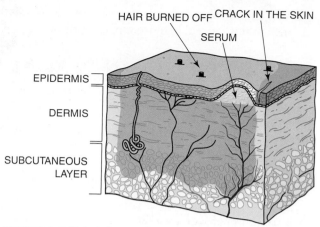

FIGURE 2-3 Third-degree burn—the epidermis, dermis, and the subcutaneous layers of tissue are destroyed.
© Cengage Learning 2012

Again, the most important first step in treating a second-degree burn is to put the area under cold water (not iced) or apply cold water compresses until the pain decreases. Gently pat the area dry with a clean lint-free towel, and cover the area with a sterile bandage or clean cloth to prevent infection. Seek medical attention. If the burns are around the mouth or nose, or involve singed nasal hair, breathing problems may develop. Do not apply ointments, sprays, antiseptics, or home remedies. Note: In an emergency any cold liquid you drink—for example, water, cold tea, soft drinks, or milk shake—can be poured on a burn. The purpose is to lower the skin temperature as quickly as possible to reduce tissue damage. See **Figure 2-2**.

Third-Degree Burns Third-degree burns occur when the surface of the skin and possibly the tissue below the skin appear white or charred. Initially, there may be little pain present because nerve endings have been destroyed. Do not remove any clothes that are stuck to the burn. Do not put ice water or ice on the burns; this could intensify the shock reaction. Do not apply ointments, sprays, antiseptics, or home remedies to burns. If the victim is on fire, smother the flames with a blanket, rug, or jacket. Breathing difficulties are common with burns around the face, neck, and mouth; be sure that the victim is breathing. Place a cold cloth or cool (not iced) water on burns of the face, hands, or feet to cool the burned areas. Cover the burned area with thick, sterile, nonfluffy dressings. Call for an ambulance immediately; people with even small third-degree burns need to consult a doctor. See **Figure 2-3**.

Burns Caused by Light Some types of light can cause burns. The three types of light include **ultraviolet, infrared,** and **visible.** Ultraviolet and infrared are not visible to the unaided human eye. They are the types of light that can cause burns. During welding, one or more of the three types of light may be present. Arc welding produces all three types of light, but gas welding produces visible and infrared light only.

The light from the welding process can be reflected from walls, ceilings, floors, or any other large surface. This reflected light is as dangerous as direct welding light. To reduce the danger from reflected light, the welding area, if possible, should be painted with a flat, dark-colored or black paint. Flat black will reduce the reflected light by absorbing more of it than any other color. When the welding is to be done on a job site, in a large shop, or in another area that cannot be painted, weld curtains can be placed to absorb the welding light, **Figure 2-4**. These special portable welding curtains may be either transparent or opaque. Transparent welding curtains are made of

FIGURE 2-4 Portable welding curtains.
Frommelt Safety Products

a special high-temperature, flame-resistant plastic that will prevent the harmful light from passing through.

> ## ⚠️ CAUTION
>
> **Welding curtains must always be used to protect other workers in the area who might be exposed to the welding light.**

Ultraviolet Light

Ultraviolet light waves are the most dangerous. They can cause first-degree and second-degree burns to a welder's eyes or to any exposed skin. Because a welder cannot see or feel ultraviolet light while being exposed to it, the welder must stay protected when in the area of any of the arc welding processes. The closer a welder is to the arc and the higher the current, the quicker a burn may occur. The ultraviolet light is so intense during some welding processes that a welder's eyes can receive a **flash burn** within seconds, and the skin can be burned within minutes. Ultraviolet light can pass through loosely woven clothing, thin clothing, light-colored clothing, and damaged or poorly maintained arc welding helmets.

Infrared Light

Infrared light is the light wave that is felt as heat. Although infrared light can cause burns, a person will immediately feel this type of light. Therefore, burns can easily be avoided. When you are welding you feel infrared light, and you are probably being exposed to ultraviolet light at the same time; therefore, protective action should be taken to cover yourself.

Visible Light

Visible light is the light that we see. It is produced in varying quantities and colors during welding. Too much visible light may cause temporary night blindness (poor eyesight under low light levels). Too little visible light may cause eyestrain, but visible light is not hazardous.

Face, Eye, and Ear Protection

Face and Eye Protection Eye protection must be worn in the shop at all times. Eye protection can be **safety glasses,** with side shields, (Figure 2-5), **goggles,** or a **full face shield.** To give better protection when working in brightly lit areas or outdoors, some welders wear **flash glasses,** which are special, lightly tinted safety glasses. These safety glasses provide protection from both flying debris and reflected light.

Suitable eye protection is important because eye damage caused by excessive exposure to arc light is not noticed. Welding light damage occurs often without warning, like a sunburn's effect that is felt the following day. Therefore, welders must take appropriate precautions in selecting filters or goggles that are suitable for the process being used, **Table 2-1.** Selecting the correct shade lens is also important

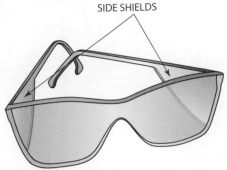

SIDE SHIELDS

FIGURE 2-5 Safety glasses with side shields. © Cengage Learning 2012

because both extremes of too light or too dark can cause eyestrain. New welders often select too dark a lens, assuming it will give them better protection, but this results in eyestrain in the same manner as if they were trying to read in a poorly lit room. In reality, any approved arc welding lenses will filter out the harmful ultraviolet light. Select a lens that lets you see comfortably. At the very least, the welder's eyes must not be strained by excessive glare from the arc.

Ultraviolet light can burn the eye in two ways. This light can injure either the white of the eye or the retina, which is the back of the eye. Burns on the retina are not painful but may cause some loss of eyesight. The whites of the eyes can also be burned by ultraviolet light, **Figure 2-6.** The whites of the eyes are very sensitive, and burns are very painful. The eyes are easily infected because, as with any burn, many cells are killed. These dead cells in the moist environment of the eyes will promote the growth of bacteria that cause infection. When the eye is burned, it feels as though there is something in the eye. Without a professional examination, however, it is impossible to tell if there is something in the eye. Because there may actually be something in the eye and because of the high risk of infection, home remedies or other medicines should never be used for eye burns. Any time you receive an eye injury, you should see a doctor.

Welding Helmets

Even with quality **welding helmets,** like that shown in **Figure 2-7,** the welder must check for potential problems that may occur from accidents or daily use. Small, undetectable leaks of ultraviolet light in an arc welding helmet can cause a welder's eyes to itch or feel sore after a day of welding. To prevent these leaks, make sure the lens gasket is installed correctly, **Figure 2-8.** The outer and inner clear lens must be plastic. As shown in **Figure 2-9,** the lens can be checked for cracks by twisting it between your fingers. Worn or cracked spots on a helmet must be repaired. Tape can be used as a temporary repair until the helmet can be replaced or permanently repaired.

Safety Glasses

Safety glasses with side shields are adequate for general use, but if heavy grinding, chipping, or overhead work is

1
Goggles, flexible fitting, regular ventilation

2
Goggles, flexible fitting, hooded ventilation

3
Goggles, cushioned fitting, rigid body

4
Spectacles

5
Spectacles, eyecup-type eye shields

6
Spectacles, semi-flat-fold side shields

7
Welding goggles, eyecup type, tinted lenses

7A
Chipping goggles, eyecup type, tinted lenses

8
Welding goggles, cover spec type, tinted lenses

8A
Chipping goggles, cover spec type, clear safety lenses

9
Welding goggles, cover spec type, tinted plate lens

10
Face shield, plastic or mesh window (see caution note)

11
Welding helmet

Non-side-shield spectacles are available for limited hazard use requiring only frontal protection.

Applications

Operation	Hazards	Protectors
Acetylene-Burning Acetylene-Cutting Acetylene-Welding	Sparks, Harmful Ray, Molten Metal, Flying Particles	7,8,9
Chemical Handling	Splash, Acid Burns, Fumes	2 (for severe exposure add 10)
Chipping	Flying Particles	1,2,4,5,6,7A,8A
Electric (Arc) Welding	Sparks, Intense Rays, Molten Metal	11 (in combination with 4,5,6 in tinted lenses advisable)
Furnace Operations	Glare, Heat, Molten Metal	7,8,9 (for severe exposure add 10)
Grinding Light	Flying Particles	1,3,5,6 (for severe exposure add 10)
Grinding-Heavy	Flying Particles	1,3,7A,8A (for severe exposure add 10)
Laboratory	Chemical Splash, Glass Breakage	2 (10 when in combination with 5,6)
Machining	Flying Particles	1,3,5,6 (for severe exposure add 10)
Molten Metals	Heat, Glare, Sparks, Splash	7,8 (10 in combination with 5,6 in tinted lenses)
Spot Welding	Flying Particles, Sparks	1,3,4,5,6 (tinted lenses advisable, for severe exposure add 10)

CAUTION:
Face shields alone do not provide adequate protection. Plastic lenses are advised for protection against molten metal splash.
Contact lenses, of themselves, do not provide eye protection in the industrial sense and shall not be worn in a hazardous environment without appropriate covering safety eyewear.

TABLE 2-1 Huntsman Selector Chart *Kedman Co., Huntsman Product Division*

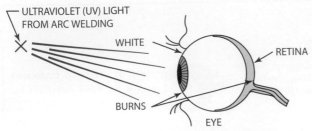

FIGURE 2-6 The eye can be burned on the white or on the retina by ultraviolet light. © Cengage Learning 2012

FIGURE 2-7 Typical arc welding helmets used to provide eye and face protection during welding. Larry Jeffus

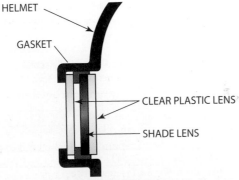

FIGURE 2-8 The correct placement of the gasket around the shade lens is important because it can stop ultraviolet light from bouncing around the lens assembly. © Cengage Learning 2012

being done, goggles or a full face shield should be worn in addition to safety glasses, **Figure 2-10**. Safety glasses are best for general protection. They must be worn under an arc welding helmet at all times.

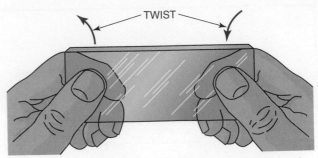

FIGURE 2-9 To check the shade lens for possible cracks, gently twist it. © Cengage Learning 2012

FIGURE 2-10 Full face shield. © Cengage Learning 2012

Ear Protection The welding environment can be very noisy. The sound level is at times high enough to cause pain and some loss of hearing if the welder's ears are unprotected. Hot sparks can also drop into an open ear, causing severe burns.

Ear protection is available in several forms. One form of protection is **earmuffs** that cover the outer ear completely, **Figure 2-11**. Another form of protection is **earplugs** that fit into the ear canal, **Figure 2-12**. Both of these protect a person's hearing, but only the earmuffs protect the outer ear from burns.

> **CAUTION**
>
> **Damage to your hearing caused by high sound levels may not be detected until later in life, and the resulting loss in hearing is permanent. Your hearing will not improve with time, and each exposure to high levels of sound will further damage your hearing.**

Respiratory Protection

All welding and cutting processes produce undesirable by-products, such as harmful dusts, fogs, fumes, mists, gases, smokes, sprays, or vapors. For your safety and the

FIGURE 2-11 Earmuffs provide complete ear protection and can be worn under a welding helmet. Mine Safety Appliances Company

FIGURE 2-12 Earplugs used as protection from noise only. Mine Safety Appliances Company

safety of others your primary objective will be to prevent these contaminants from forming and collecting in the shop atmosphere. This will be accomplished as much as possible by engineering and design control measures such as water tables for cutting, general and local **ventilation,** thorough cleaning of surface contaminants before starting work, and confinement of the operation to outdoor or open spaces.

Production of welding by-products cannot be avoided. They are created when the temperature of metals and fluxes is raised above the temperatures at which they vaporize or decompose. Most of the by-products are recondensed in the weld. However, some do escape into the atmosphere, producing the haze that occurs in improperly ventilated welding shops. Some fluxes used in welding electrodes produce fumes that may irritate the welder's nose, throat, and lungs.

When welders must work in an area where effective general controls to remove airborne welding by-products

are not feasible, respirators shall be provided by their employers when this equipment is necessary to protect their health. The respirators supplied by the welding shop must be applicable and suitable for the purpose intended. Where respirators are necessary to protect welders' health or whenever respirators are required by the welding shop, the shop will establish and implement a written respiratory protection program with worksite-specific procedures. Welders are responsible for following the welding shop's established written respiratory protection program. Guidelines for the respiratory protection program are available from the Occupational Safety and Health Administration (OSHA) office in Washington, DC.

Training Training must be a part of the welding shop's respiratory protection program. This training should include instruction on any and/or all of the following procedures for

- proper use of respirators, including techniques for putting them on and removing them,
- schedules for cleaning, disinfecting, storing, inspecting, repairing, discarding, and performing other aspects of maintenance of the respiratory protection equipment,
- selection of the proper respirators for use in the workplace, and any respiratory equipment limitations,
- procedures for testing the proper fitting of respirators,
- proper use of respirators in both routine and reasonably foreseeable emergency situations, and
- regular evaluation of the effectiveness of the program.

Equipment All respiratory protection equipment used in a welding shop should be certified by the National Institution for Occupational Safety and Health (NIOSH). Some of the types of respiratory protection equipment that may be used are the following:

- Air-purifying respirators have an air-purifying filter, cartridge, or canister that removes specific air contaminants by passing ambient air through the air-purifying element.
- Atmosphere-supplying respirators supply breathing air from a source independent of the ambient atmosphere; this includes both the supplied-air respirators (SARs) and self-contained breathing apparatus (SCBA)-type units.
- Demand respirators are atmosphere-supplying respirators that admit breathing air to the facepiece only when a negative pressure is created inside the facepiece by inhalation.
- Positive pressure respirators are respirators in which the pressure inside the respiratory inlet covering exceeds the ambient air pressure outside the respirator.

FRESH AIR

FIGURE 2-13 Filtered fresh air is forced into the welder's breathing area. © 2010 3M Company. All rights reserved.

- Powered air-purifying respirators (PAPRs) are air-purifying respirators that use a blower to force the ambient air through air-purifying elements to the inlet covering, **Figure 2-13**.

- Self-contained breathing apparatuses (SCBAs) are atmosphere-supplying respirators for which the breathing air source is designed to be carried by the user.

- Supplied-air respirators (SARs), or airline respirators, are atmosphere-supplying respirators for which the source of breathing air is not designed to be carried by the user.

Respiratory protection equipment used in many welding applications is of the filtering facepiece (dust mask) type, **Figure 2-14**. These mask types use the negative pressure as you inhale to draw air through a filter, which is an

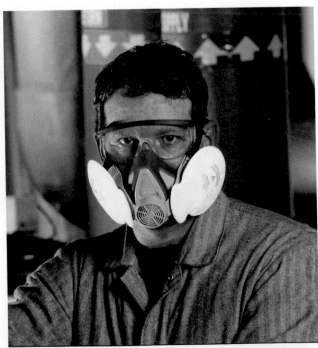

FIGURE 2-14 Typical respirator for contaminated environments. The filters can be selected for specific types of contaminants. Mine Safety Appliances Company

integral part of the facepiece. In areas of severe contamination you may use a hood-type respirator, which covers your head and neck and may even cover portions of your shoulders and torso.

▟▟▟ CAUTION ▙▙▙

Welding or cutting must never be performed on drums, barrels, tanks, vessels, or other containers until they have been emptied and cleaned thoroughly, eliminating all flammable materials and all substances (such as detergents, solvents, greases, tars, or acids) that might produce flammable, toxic, or explosive vapors when heated.

Fume Sources Some materials that can cause respiratory problems are used as paints, coating, or plating on metals to prevent rust or corrosion. Other potentially hazardous materials might be used as alloys in metals to give them special properties.

Before welding or cutting, any metal that has been painted or has any grease, oil, or chemicals on its surface must be thoroughly cleaned. This cleaning may be done by grinding, sandblasting, or applying an approved solvent. Metals that are plated or alloyed may not be able to be cleaned before welding or cutting begins.

Most paints containing lead have been removed from the market. But some industries, such as marine or ship applications, still use these lead-based paints. Often old machinery and farm equipment surfaces still have lead-based paint coatings. Solder often contains lead alloys. The welding and cutting of lead-bearing alloys or metals whose surfaces have been painted with lead-based paint can generate lead oxide fumes. Inhalation and ingestion of lead oxide fumes and other lead compounds will cause lead poisoning. Symptoms include a metallic taste in the mouth, loss of appetite, nausea, abdominal cramps, and insomnia. In time, anemia and a general weakness, chiefly in the muscles of the wrists, develop.

Both cadmium and zinc are plating materials used to prevent iron or steel from rusting. Cadmium is often used on bolts, nuts, hinges, and other hardware items, and it gives the surface a yellowish-gold appearance. Acute exposure to high concentrations of cadmium fumes can produce severe lung irritation. Long-term exposure to low levels of cadmium in air can result in emphysema (a disease affecting the lung's ability to absorb oxygen) and can damage the kidneys.

Zinc, often in the form of galvanizing, may be found on pipes, sheet metal, bolts, nuts, and many other types of hardware. Zinc plating that is thin may appear as a shiny, metallic patchwork or crystal pattern; thicker, hot-dipped zinc appears rough and may look dull. Zinc is used in large quantities in the manufacture of brass and is found in brazing rods. Inhalation of zinc oxide fumes can occur when welding or cutting on these materials. Exposure to these fumes is known to cause metal fume fever; its symptoms are very similar to those of common influenza.

Some concern has been expressed about the possibility of lung cancer being caused by some of the chromium compounds that are produced when welding stainless steels.

⚠ ///// CAUTION \\\\\

Extreme care must be taken to avoid the fumes produced when welding is done on dirty or used metal. Any chemicals that are on the metal will become mixed with the welding fumes, a combination that can be extremely hazardous. All metal must be cleaned before welding to avoid this potential problem.

Rather than take chances, welders should recognize that fumes of any type, regardless of their source, should not be inhaled. The best way to avoid problems is to provide adequate **ventilation.** If this is not possible, breathing protection should be used. Protective devices for use in poorly ventilated or confined areas are shown in Figure 2-13 and Figure 2-14.

Vapor Sources Potentially dangerous gases also can be present in a welding shop. Proper ventilation or respirators are necessary when welding in confined spaces, regardless of the welding process being used. Ozone is a gas that is produced by the ultraviolet radiation in the air in the vicinity of arc welding and cutting operations. Ozone is very irritating to all mucous membranes, with excessive exposure producing pulmonary edema, or fluid on the lung, making it difficult to breathe. Severe cases of pulmonary edema may require immediate care. Other effects of exposure to ozone include headache, chest pain, and dryness in the respiratory tract.

Phosgene is formed when ultraviolet radiation decomposes chlorinated hydrocarbon. Fumes from chlorinated hydrocarbons can come from solvents such as those used for degreasing metals and from refrigerants from air-conditioning systems. They decompose in the arc to produce a potentially dangerous chlorine acid compound. This compound reacts with the moisture in the lungs to produce hydrogen chloride, which in turn destroys lung tissue. For this reason, any use of chlorinated solvents should be well away from welding operations in which ultraviolet radiation or intense heat is generated. Any welding or cutting on refrigeration or air-conditioning piping must be done only after the refrigerant has been completely removed in accordance with Environmental Protection Agency (EPA) regulations.

Care also must be taken to avoid the infiltration of any fumes or gases, including argon or carbon dioxide, into a confined working space, such as when welding in tanks. The collection of some fumes and gases in a work area can go unnoticed by the welders. Concentrated fumes or gases can cause a fire or explosion if they are flammable, asphyxiation if they replace the oxygen in the air, or death if they are toxic.

Despite these fumes and other potential hazards in welding shops, welders have been found to be as healthy as workers employed in other industrial occupations.

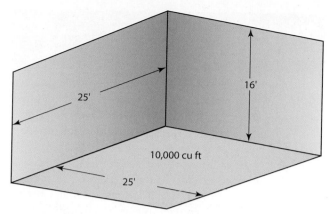

FIGURE 2-15 A room with a ceiling 16 ft (4.9 m) high may not require forced ventilation for one welder.
© Cengage Learning 2012

Ventilation

The actual welding area should be well ventilated. Excessive fumes, ozone, or smoke may collect in the welding area; ventilation should be provided for their removal. **Natural ventilation** is best, but forced ventilation may be required. Areas that have 10,000 cubic feet (283 cubic meters) or more per welder, or that have ceilings 16 feet (4.9 meters) high or higher, **Figure 2-15,** may not require forced ventilation unless fumes or smoke begin to collect.

Forced Ventilation Small shops or shops with large numbers of welders require forced ventilation. **Forced ventilation** can be general or localized using fixed or flexible **exhaust pickups, Figure 2-16.** General room ventilation must be at a rate of 2000 cu ft (56 m³) or more per person welding. Localized exhaust pickups must have a draft strong enough to provide 100 linear feet (30.5 m) per minute air velocity pulling welding fumes away from the welder. Local, state, or federal regulations may require that welding fumes be treated to remove hazardous components before they are released into the atmosphere.

Any system of ventilation should draw the fumes or smoke away before it rises past the level of the welder's face.

FIGURE 2-16 An exhaust pickup. Larry Jeffus

Welding shops generate a lot of waste material. Much of the waste is scrap metal. All scrap metal, including electrode stubs, can easily be recycled. Green practices like recycling metal are good for the environment and can generate revenue for your welding shop.

Some of the other waste, such as burned flux, cleaning solvents, and dust collected in shop air filtration systems, may be considered hazardous material. Check with the material manufacturer or an environmental consultant to determine if any waste material is considered hazardous. Throwing hazardous waste material into the trash, pouring it on the ground, or dumping it down the drain is illegal. Before you dispose of any welding shop waste that is considered hazardous, you must first consult local, state, and/or federal regulations. Protecting our environment from pollution is everyone's responsibility.

Forced ventilation is always required when welding on metals that contain zinc, lead, beryllium, cadmium, mercury, copper, austenetic manganese, or other materials that give off dangerous fumes.

Material Safety Data Sheets (MSDSs)

All manufacturers of potentially hazardous materials must provide to the users of their products detailed information regarding possible hazards resulting from the use of their products. These **material safety data sheets** are often called **MSDSs.** They must be provided to anyone using the product or anyone working in the area where the products are in use. Often companies will post these sheets on a bulletin board or put them in a convenient place near the work area. Some states have right-to-know laws that require specific training of all employees who handle or work in areas with hazardous materials.

////// **CAUTION** ◀◀◀◀

If you feel you have been injured while using a product, you should, if possible, take the material's MSDS with you when you are seeking medical treatment.

General Work Clothing

Special protective clothing cannot be worn at all times. It is, therefore, important to choose general work clothing that will minimize the possibility of getting burned because of the high temperature and amount of hot sparks, metal, and slag produced during welding, cutting, or brazing.

Work clothing must also stop ultraviolet light from passing through it. This is accomplished if the material chosen is a dark color, thick, and tightly woven. The best choice is 100% wool, but it is difficult to find. Another good choice is 100% cotton clothing, the most popular fabric used.

You must avoid wearing synthetic materials including nylon, rayon, and polyester. They can easily melt or catch fire. Some synthetics produce a hot, sticky residue that can make burns more severe. Others may produce poisonous gases.

The following are some guidelines for selecting work clothing:

- Shirts must be long-sleeved to protect the arms, have a high-buttoned collar to protect the neck, **Figure 2-17**, be long enough to tuck into the pants to protect the waist, and have flaps on the pockets to keep sparks out (or have no pockets).
- Pants must have legs long enough to cover the tops of the boots and must be without cuffs that would catch sparks.
- Boots must have high tops to keep out sparks, have steel toes to prevent crushed toes, **Figure 2-18**, and have smooth tops to prevent sparks from being trapped in seams.
- Caps should be thick enough to prevent sparks from burning the top of a welder's head.

All clothing must be free of frayed edges and holes. The clothing must be relatively tight-fitting to prevent excessive folds or wrinkles that might trap sparks.

Some welding clothes have pockets on the inside to prevent the pockets from collecting sparks. However, it is not safe to carry butane lighters or matches in these or any pockets while welding. Lighters and matches can easily catch fire or explode if they are subjected to the heat and sparks of welding.

FIGURE 2-17 The top button of the shirt worn by the welder should always be buttoned to avoid severe burns to that person's neck. Larry Jeffus

FIGURE 2-18 Safety boots with steel toes are required by many welding shops. © Cengage Learning 2012

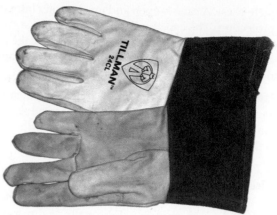

FIGURE 2-20 For welding that requires a great deal of manual dexterity, soft leather gloves can be worn. Larry Jeffus

CAUTION

There is no safe place to carry butane lighters or matches while welding or cutting. They can catch fire or explode if subjected to welding heat or sparks. Butane lighters may explode with the force of 1/4 stick of dynamite. Matches can erupt into a ball of fire. Both butane lighters and matches must always be removed from the welder's pockets and placed a safe distance away before any work is started.

Gauntlet gloves that have a cloth liner for insulation are best for hot work. Noninsulated gloves will give greater flexibility for fine work. Some leather gloves are available with a canvas gauntlet top, which should be used for light work only.

When a great deal of manual dexterity is required for gas tungsten arc welding, brazing, soldering, oxyfuel gas welding, and other delicate processes, soft leather gloves may be used, **Figure 2-20**. All-cotton gloves are sometimes used when doing very light welding.

Special Protective Clothing

General work clothing is worn by each person in the shop. In addition to this clothing, extra protection is needed for each person who is in direct contact with hot materials. Leather is often the best material to use, as it is lightweight, is flexible, resists burning, and is readily available. Synthetic insulating materials are also available. Ready-to-wear leather protection includes capes, jackets, aprons, sleeves, gloves, caps, pants, knee pads, and spats, among other items.

Hand Protection All-leather, gauntlet-type gloves should be worn when doing any welding, **Figure 2-19**.

Body Protection Full leather jackets and capes will protect a welder's shoulders, arms, and chest, **Figure 2-21**. A jacket, unlike the cape, protects a welder's back and

FIGURE 2-19 All leather, gauntlet-type welding gloves. Larry Jeffus

FIGURE 2-21 Full leather jacket. Larry Jeffus

complete chest. A cape is open and much cooler but offers less protection. The cape can be used with a bib apron to provide some additional protection while leaving the back cooler. Either the full jacket or the cape with a bib apron should be worn for any out-of-position work.

Waist and Lap Protection Bib aprons or full aprons will protect a welder's lap. Welders will especially need to protect their laps if they squat or sit while working and when they bend over or lean against a table.

Arm Protection For some vertical welding, a full or half sleeve can protect a person's arm, **Figure 2-22**. The sleeves work best if the work level is not above the welder's chest. Work levels higher than this usually require a jacket or cape to keep sparks off the welder's shoulders.

Leg and Foot Protection When heavy cutting or welding is being done and a large number of sparks are falling, leather pants and spats should be used to protect the welder's legs and feet. If the weather is hot and full leather pants are uncomfortable, leather aprons with leggings are available. Leggings can be strapped to the legs, leaving the back open. Spats will prevent sparks from burning through the front of lace-up boots.

Handling and Storing Cylinders

Oxygen and fuel gas cylinders or other flammable materials must be stored separately. The storage areas must be separated by 20 ft (6.1 m), or by a wall 5 ft high (1.5 m) with at least a 1/2-hour (hr) burn rating, **Figure 2-23**.

FIGURE 2-22 Full leather sleeve. Larry Jeffus

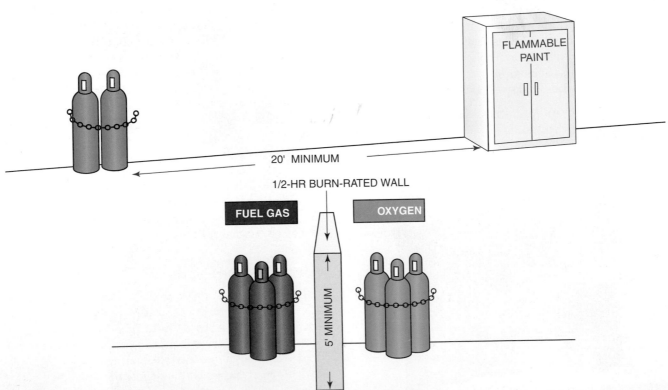

FIGURE 2-23 The minimum safe distance between stored fuel gas cylinders and any flammable material is 20 ft (6.1 m) or a wall 5 ft (1.5 m) high. © Cengage Learning 2012

The purpose of the distance or wall is to keep the heat of a small fire from causing the oxygen cylinder safety valve to release. If the safety valve were to release the oxygen, a small fire would become a raging inferno.

Inert gas cylinders may be stored separately or with oxygen cylinders.

Empty cylinders must be stored separately or with the same type of full cylinders in the same room or area. All cylinders must be stored vertically and have the protective caps screwed on firmly.

Securing Gas Cylinders Cylinders must be secured with a chain or other device so that they cannot be knocked over accidentally. Cylinders attached to a manifold or stored in a special room used only for cylinder storage should be chained.

Storage Areas Cylinder storage areas must be located away from halls, stairwells, and exits so that in case of an emergency they will not block an escape route. Storage areas should also be located away from heat, radiators, furnaces, and welding sparks. The location of storage areas should be such that unauthorized people cannot tamper with the cylinders. A **warning sign** that reads "Danger— No Smoking, Matches, or Open Lights," or similar wording, must be posted in the storage area, **Figure 2-24.**

Cylinders with Valve Protection Caps Cylinders equipped with a **valve protection cap** must have the cap in place unless the cylinder is in use. The protection cap prevents the valve from being broken off if the cylinder is knocked over. If the valve of a full high-pressure cylinder (argon, oxygen, CO_2, and mixed gases) is broken off, the cylinder can fly around the shop like a missile if it has not been secured properly.

FIGURE 2-25 Move a leaking fuel gas cylinder out of the building or any work area. The pressure should be slowly released after a warning is posted of the danger. © Cengage Learning 2012

Never lift a cylinder by the safety cap or the valve. The valve can easily break off or be damaged.

When moving cylinders, the valve protection cap must be on, especially if the cylinders are mounted on a truck or trailer for out-of-shop work. The cylinders must never be dropped or handled roughly.

General Precautions Use warm water (not boiling) to loosen cylinders that are frozen to the ground. Any cylinder that leaks, has a bad valve, or has damaged threads must be identified and reported to the supplier. A piece of soapstone can be used to write the problem on the cylinder. If the leak cannot be stopped by closing the cylinder valve, the cylinder should be moved to a vacant lot or an open area. The pressure should then be slowly released after posting a warning sign, **Figure 2-25.**

Acetylene cylinders that have been lying on their sides must stand upright for four hours or more before they are used. The acetylene is absorbed in **acetone,** and the acetone is absorbed in a filler. The filler does not allow the liquid to settle back away from the valve very quickly, **Figure 2-26.** If the cylinder has been in a horizontal position, using it too soon after it is placed in a vertical position may draw acetone out of the cylinder. Acetone lowers the flame temperature and can damage regulator or torch valve settings.

Fire Protection

Fire is a constant danger to the welder. Welding is considered to be "hot work" by the National Association of Fire Prevention. When performing welding outside of a shop, the welder may be required to obtain a hot work permit from the local fire marshal. During times when burn bans are in effect, performing hot work without a permit can be a violation of state or local laws. Even with a permit, welders can be held liable for any damage resulting from a fire caused by their welding. Highly combustible materials

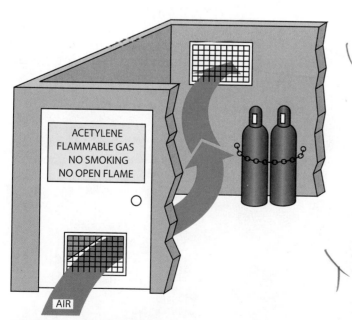

FIGURE 2-24 A separate room used to store acetylene must have good ventilation and should have a warning sign posted on the door. © Cengage Learning 2012

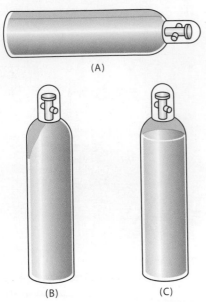

FIGURE 2-26 The acetone in an acetylene cylinder must have time to settle before the cylinder can be used safely. © Cengage Learning 2012

should be 35 ft (10.7 m) or more away from any welding. When it is necessary to weld within 35 ft (10.7 m) of combustible materials, when sparks can reach materials farther than 35 ft (10.7 m) away, or when anything more than a minor fire might start, a fire watch is needed.

Never weld outdoors when drought has resulted in a fire ban, **Figure 2-27**.

Fire Watch A fire watch can be provided by any person who knows how to sound the alarm and use a fire extinguisher. The fire extinguisher must be the type required to put out a fire for the type of combustible materials near the welding. Combustible materials that cannot be removed from the welding area should be soaked with water or covered with sand or noncombustible insulating blankets, whichever is available.

Fire Extinguishers The four types of fire extinguishers are type A, type B, type C, and type D. Each type is designed

FIGURE 2-27 You should not weld outside when the area is posted with a sign like this. Larry Jeffus

FIGURE 2-28 Type A fire extinguisher symbol. © Cengage Learning 2012

to put out fires on certain types of materials. Some fire extinguishers can be used on more than one type of fire. However, using the wrong type of fire extinguisher can be dangerous, either causing the fire to spread, causing electrical shock, or causing an explosion.

Type A Extinguishers

Type A fire extinguishers are used for combustible solids (articles that burn), such as paper, wood, and cloth. The symbol for a type A extinguisher is a green triangle with the letter *A* in the center, **Figure 2-28**.

Type B Extinguishers

Type B fire extinguishers are used for combustible liquids, such as oil, gas, and paint thinner. The symbol for a type B extinguisher is a red square with the letter *B* in the center, **Figure 2-29**.

Type C Extinguishers

Type C fire extinguishers are used for electrical fires. For example, they are used on fires involving motors, fuse boxes, and welding machines. The symbol for a type C extinguisher is a blue circle with the letter *C* in the center, **Figure 2-30**.

FIGURE 2-29 Type B fire extinguisher symbol. © Cengage Learning 2012

FIGURE 2-30 Type C fire extinguisher symbol. © Cengage Learning 2012

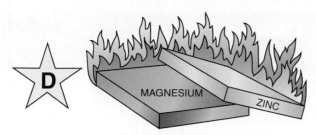

FIGURE 2-31 Type D fire extinguisher symbol.
© Cengage Learning 2012

Type D Extinguishers

Type D fire extinguishers are used on fires involving combustible metals, such as zinc, magnesium, and titanium. The symbol for a type D extinguisher is a yellow star with the letter *D* in the center, **Figure 2-31**.

Location of Fire Extinguishers Fire extinguishers should be of a type that can be used on the types of combustible materials located nearby, **Figure 2-32**. The extinguishers should be placed so that they can be easily removed without reaching over combustible material. They should also be placed at a level low enough to be easily lifted off the mounting, **Figure 2-33**. The location of fire extinguishers should be marked with red paint and signs, high enough so that their location can be seen from

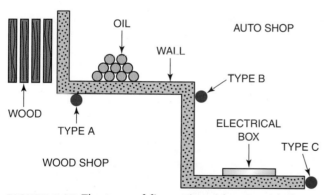

FIGURE 2-32 The type of fire extinguisher provided should be appropriate for the materials being used in the surrounding area. © Cengage Learning 2012

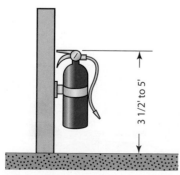

FIGURE 2-33 Mount the fire extinguisher so that it can be lifted easily in an emergency. © Cengage Learning 2012

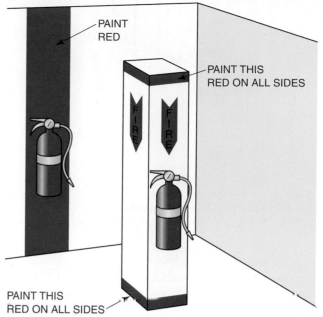

FIGURE 2-34 The location of fire extinguishers should be marked so they can be located easily in an emergency.
© Cengage Learning 2012

a distance over people and equipment. The extinguishers should also be marked near the floor so that they can be found even if a room is full of smoke, **Figure 2-34**.

Use A fire extinguisher works by breaking the fire triangle of heat, fuel, and oxygen. Most extinguishers both cool the fire and remove the oxygen. They use a variety of materials to extinguish the fire. The majority of fire extinguishers found in welding shops use foam, carbon dioxide, a pump tank, or dry chemicals.

When using a **foam** extinguisher, do not spray the stream directly into the burning liquid. Allow the foam to fall lightly on the base of the fire.

When using a **carbon dioxide** extinguisher, direct the discharge as close to the fire as possible, first at the edge of the flames and gradually to the center.

When using a **dry chemical** extinguisher, direct the extinguisher at the base of the flames. In the case of type A fires, follow up by directing the dry chemicals at the remaining material still burning. The extinguisher must be directed at the base of the fire where the fuel is located, **Figure 2-35**.

Equipment Maintenance

A routine schedule for planned maintenance (PM) of equipment will aid in detecting potential problems such as leaking coolant, loose wires, poor grounds, frayed insulation, or split hoses. Small problems, if fixed in time, can prevent the loss of valuable time due to equipment breakdown or injury.

Any maintenance beyond routine external maintenance should be referred to a trained service technician. In most areas, it is against the law for anyone but a

FIGURE 2-35 Point the extinguisher at the material burning, not at the flames. © Cengage Learning 2012

licensed electrician to work on arc welders and anyone but a factory-trained repair technician to work on regulators. Electrical shock and exploding regulators can cause serious injury or death.

Hoses Hoses must be used only for the gas or liquid for which they were designed. Green hoses are to be used only for oxygen, and red hoses are to be used only for acetylene or other fuel gases. Using unnecessarily long lengths of hoses should be avoided. Never use oil, grease, or other pipe-fitting compounds on any joints. Hoses should also be kept out of the direct line of sparks. Any leaking or bad joints in gas hoses must be repaired.

Work Area

The work area should be kept picked up and swept clean. Collections of steel, welding electrode stubs, wire, hoses, and cables are difficult to work around and easy to trip over. An electrode caddy can be used to hold the electrodes and stubs, **Figure 2-36**. Hooks can be made to hold hoses and cables, and scrap steel should be thrown into scrap bins.

Keeping outdoor work areas swept clean of welding or cutting debris will prevent the debris from being washed away by rainwater. Some debris that is washed away may contaminate our streams and lakes.

Arc welding areas should be painted with a flat dark-colored or black finish to absorb as much of the ultraviolet light as possible. Portable screens should be used whenever arc welding is to be done outside of a welding booth.

If a piece of hot metal is going to be left unattended, write the word *hot* on it before leaving. This procedure can also be used to warn people of hot tables, vises, firebricks, and tools.

Hand Tools

Hand tools are used by the welder to do necessary assembly and disassembly of parts for welding as well as to perform routine equipment maintenance.

The adjustable wrench is the most popular tool used by the welder. When using this wrench, it should be adjusted tightly on the nut and pushed so that most of the force is on the fixed jaw, **Figure 2-37**. When a wrench is being used on a tight bolt or nut, the wrench should be pushed with the palm of an open hand or pulled to prevent injuring the hand. If a nut or bolt is too tight to be loosened with a wrench, obtain a longer wrench. A cheater bar should not be used.

The fewer points a box end wrench or socket has, the stronger it is and the less likely it is to slip or damage the nut or bolt, **Figure 2-38**.

Striking a hammer directly against a hard surface such as another hammer face or anvil may result in chips flying off and causing injury.

The mushroomed heads of chisels, punches, and the faces of hammers should be ground off, **Figure 2-39**. Chisels and punches that are going to be hit harder than a

FIGURE 2-36 An easy-to-build electrode caddy can be used to hold both electrodes and stubs. Larry Jeffus

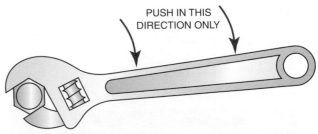

PUSH IN THIS DIRECTION ONLY

FIGURE 2-37 The adjustable wrench is stronger when used in the direction indicated. © Cengage Learning 2012

FIGURE 2-38 The fewer the points, the less likely the wrench is to slip. Larry Jeffus

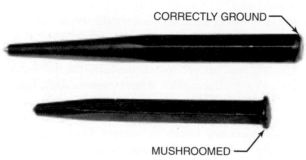

CORRECTLY GROUND

MUSHROOMED

FIGURE 2-39 Any mushroomed heads must be ground off. Larry Jeffus

slight tap should be held in a chisel holder or with pliers to eliminate the danger of injuring your hand.

A handle should be placed on the tang of a file to avoid injuring your hand, **Figure 2-40**. A file can be kept free of chips by rubbing a piece of soapstone on it before it is used.

It is important to remember to use the correct tool for the job. Do not try to force a tool to do a job it was not designed to do.

Hand Tool Safety Hand tools used in welding fabrication should be treated properly and not abused. Many accidents can be avoided by using the right tool for the job.

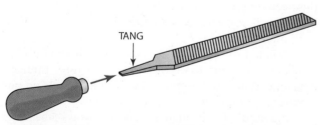

TANG

FIGURE 2-40 To protect yourself from the sharp tang of a file, always use a handle with a file. © Cengage Learning 2012

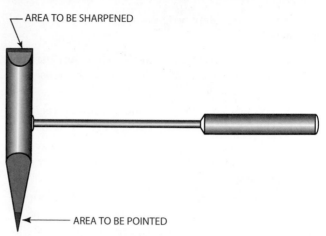

AREA TO BE SHARPENED

AREA TO BE POINTED

FIGURE 2-41 Welder's chipping hammers work best and are safer to use if the head is sharpened periodically. © Cengage Learning 2012

For instance, use a tool that is the correct size for the work rather than one that is too large or too small.

Keep hand tools clean to protect them against the damage caused by corrosion. Wipe off any accumulated dirt, mud, and grease. Occasionally dip the tools in cleaning fluids or solvents and wipe them clean. Lubricate adjustable and other moving parts to prevent wear and misalignment.

Make sure that hand tools are sharp, **Figure 2-41**. Sharp tools make work easier, improve the accuracy of the work, save time, and are safer than dull tools. When sharpening, redressing, or repairing tools, shape, grind, hone, file, fit, and set them properly using other tools suited to each purpose. For sharpening tools, either an oilstone or a grindstone is preferred. If grinding on an abrasive wheel is required, grind only a small amount at a time. Hold the tool lightly against the wheel to prevent overheating, and frequently dip the part being ground in water to keep it cool. This will protect the hardness of the metal and help to retain the sharpness of the cutting edge. Be sure to wear safety goggles when sharpening or redressing tools.

When carrying tools, protect the cutting edges and carry the tools in such a way that you will not endanger yourself or others. Carry pointed or sharp-edged tools in pouches or holsters.

Hammer Safety

Keep hammer handles secure and safe. Check wedges and handles frequently. Be sure heads are wedged tightly on handles. Keep handles smooth and free of rough or jagged edges. Do not rely on friction tape to secure split handles or to prevent handles from splitting. Replace handles that are split or chipped or that cannot be refitted securely.

When swinging a hammer, be absolutely certain that no one is within range or can come within range of the swing or be struck by flying material. Always allow plenty of room for arm and body movements when hammering on anything.

The following safety precautions generally apply to all hammers:

- Check to see that the handle is tight before using any hammer. Never use a hammer with a loose or damaged handle.
- Always use a hammer of suitable size and weight for the job.
- Discard or repair any tool if the face shows excessive wear, dents, chips, mushrooming, or improper redressing.
- Rest the face of the hammer on the work before striking to get the feel or aim; then, grasp the handle firmly with the hand near the end of the handle. Move the fingers out of the way before striking with force.
- A hammer blow should always be struck squarely, with the hammer face parallel to the surface being struck. Always avoid glancing blows and over-and-under strikes.
- For striking another tool (cold chisel, punch, wedge, etc.), the face of the hammer should be proportionately larger than the head of the tool. For example, a 1/2-in. (13-mm) cold chisel requires at least a 1-in. (25-mm) hammer face.
- Never use one hammer to strike another hammer.
- Do not use the end of the handle of any tool for tamping or prying; it might split.

Electrical Safety

Electric shock can cause injuries and even death unless proper precautions are taken. Most welding and cutting operations involve electrical equipment in addition to the arc welding power supplies. Grinders, electric motors on automatic cutting machines, and drills are examples. Most electrical equipment in a welding shop is powered by alternating-current (AC) sources having input voltages ranging from 115 volts to 460 volts. However, fatalities have occurred when working with equipment operating at less than 80 volts. Most electric shocks in the welding industry are a result of accidental contact with bare or poorly insulated conductors. **Electrical resistance** is lowered in the presence of water or moisture, so welders must take special precautions when working under damp or wet conditions, including perspiration. **Figure 2-42** shows a typical **warning label** shipped with the welding equipment.

The workpiece being welded and the frame or chassis of all electrically powered machines must be connected to a good **electrical ground.** The work lead from the welding power supply is not an electrical ground and is not sufficient. A separate lead is required to ground the workpiece and power source.

Electrical connections must be tight. Terminals for welding leads and power cables must be shielded from accidental contact by personnel or by metal objects. Cables must be used within their current carrying and duty cycle capacities; otherwise, they will overheat and break down the insulation rapidly. Cable connectors for lengthening leads must be insulated.

///// **CAUTION** \\\\\

Welding cables must never be spliced within 10 ft (3 m) of the electrode holder.

Cables must be checked periodically to be sure that they have not become frayed, and, if they have, they must be replaced immediately.

Welders should not allow the metal parts of electrodes or electrode holders to touch their skin or wet coverings on their bodies. Dry gloves in good condition must always be worn. Rubber-soled shoes are advisable. Precautions against accidental contact with bare conducting surfaces must be taken when the welder is required to work in cramped kneeling, sitting, or lying positions. Insulated mats or dry wooden boards are desirable protection from being grounded.

Welding circuits must be turned off when the workstation is left unattended. When working on the welder, welding leads, electrode holder, torches, wire feeder, guns, or other parts of the main power supply must be turned off and locked or tagged, to prevent electrocution. Since the electrode holder is energized when changing coated electrodes, the welder must wear dry gloves.

Electrical Safety Systems

For protection from electrical shock, the standard portable power tool is built with either of two equally safe systems: external grounding or double insulation.

A tool with external grounding has a wire that runs from the housing through the power cord to a third prong on the power plug. When this third prong is connected to a grounded, three-hole electrical outlet, the grounding wire will carry any current that leaks past the electrical insulation of the tool away from the user and into the ground. In most electrical systems, the three-prong plug fits into a three-prong, grounded receptacle. If the tool is operated at less than 150 volts, it has a plug like that shown in **Figure 2-43A.** If it is used at 150 to 250 volts, it has a plug like that shown in **Figure 2-43B.** In either type, the green (or green and yellow) conductor in the tool cord is the grounding wire. Never connect the grounding wire to a power terminal.

A double-insulated tool has an extra layer of electrical insulation that eliminates the need for a three-pronged plug and grounded outlet. Double-insulated tools do not require grounding and, therefore, have a two-prong plug. In addition, double-insulated tools are always labeled as such on their nameplate or case, **Figure 2-44.**

Welding Safety Checklist

Hazard	Factors to Consider	Precaution Summary
Electric shock can kill	• **Wetness** • **Welder in or on workpiece** • **Confined space** • **Electrode holder and cable insulation**	• Insulate welder from workpiece and ground using *dry* insulation. Rubber mat or dry wood. • Wear *dry, hole-free* gloves. (Change as necessary to keep dry.) • Do not touch electrically "hot" parts or electrode with bare skin or wet clothing. • If wet area and welder cannot be insulated from workpiece with dry insulation, use a semiautomatic, constant-voltage welder or stick welder with voltage reducing device. • Keep electrode holder and cable insulation in good condition. Do not use if insulation damaged or missing.
Fumes and gases can be dangerous	• **Confined area** • **Positioning of welder's head** • **Lack of general ventilation** • **Electrode types, i.e., manganese, chromium, etc. See MSDS** • **Base metal coatings, galvanize, paint**	• Use ventilation or exhaust to keep air breathing zone clear, comfortable. • Use helmet and positioning of head to minimize fume in breathing zone. • Read warnings on electrode container and material safety data sheet (MSDS) for electrode. • Provide additional ventilation/exhaust where special ventilation requirements exist. • Use special care when welding in a confined area. • Do not weld unless ventilation is adequate.
Welding sparks can cause fire or explosion	• **Containers which have held combustibles** • **Flammable materials**	• Do not weld on containers which have held combustible materials (unless strict AWS F4.1 procedures are followed). Check before welding. • Remove flammable materials from welding area or shield from sparks, heat. • Keep a fire watch in area during and after welding. • Keep a fire extinguisher in the welding area. • Wear fire retardant clothing and hat. Use earplugs when welding overhead.
Arc rays can burn eyes and skin	• **Process: gas-shielded arc most severe**	• Select a filter lens which is comfortable for you while welding. • Always use helmet when welding. • Provide non-flammable shielding to protect others. • Wear clothing which protects skin while welding.
Confined space	• **Metal enclosure** • **Wetness** • **Restricted entry** • **Heavier than air gas** • **Welder inside or on workpiece**	• Carefully evaluate adequacy of ventilation especially where electrode requires special ventilation or where gas may displace breathing air. • If basic electric shock precautions cannot be followed to insulate welder from work and electrode, use semiautomatic, constant-voltage equipment with cold electrode or stick welder with voltage reducing device. • Provide welder helper and method of welder retrieval from outside enclosure.
General work area hazards	• **Cluttered area**	• Keep cables, materials, tools neatly organized.
	• **Indirect work (welding ground) connection**	• Connect work cable as close as possible to area where welding is being performed. Do *not* allow alternate circuits through scaffold cables, hoist chains, ground leads.
	• **Electrical equipment**	• Use only double insulated or properly grounded equipment. • Always disconnect power to equipment before servicing.
	• **Engine-driven equipment**	• Use in only open, well ventilated areas. • Keep enclosure complete and guards in place. • See Lincoln service shop if guards are missing. • Refuel with engine off. • If using auxiliary power, OSHA may require GFI protection or assured grounding program (or isolated windings if less than 5KW).
	• **Gas cylinders**	• Never touch cylinder with the electrode. • Never lift a machine with cylinder attached. • Keep cylinder upright and chained to support.

FIGURE 2-42 Note the warning information contained on this typical label, which may be attached to welding equipment or in the equipment owner's manual. Lincoln Electric Company

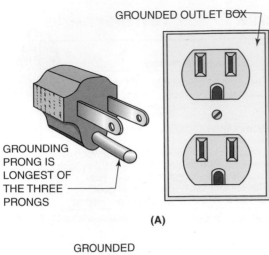

GROUNDED OUTLET BOX

GROUNDING PRONG IS LONGEST OF THE THREE PRONGS

(A)

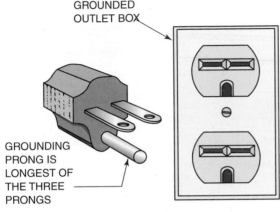

GROUNDED OUTLET BOX

GROUNDING PRONG IS LONGEST OF THE THREE PRONGS

(B)

FIGURE 2-43 (A) A three-prong grounding plug for use with up to 150-volt tools and (B) a grounding plug for use with 150- to 250-volt tools. © Cengage Learning 2012

1/2" Single-Speed Reversing Drill
120V AC - ONLY 5.0 A 600 RPM

(UL) DOUBLE INSULATED

CAUTION

FOR SAFETY OPERATION SEE INSTRUCTION MANUAL. WHEN SERVICING USE ONLY IDENTICAL REPLACMENT PARTS.

FIGURE 2-44 Typical portable power tool nameplate.
© Cengage Learning 2012

Voltage Warnings

Before connecting a tool to a power supply, be sure the voltage supplied is the same as that specified on the nameplate of the tool. A power source with a voltage greater than that specified for the tool can lead to serious injury to the user as well as damage to the tool. Using a power source with a voltage lower than the rating on the nameplate is harmful to the motor.

Tool nameplates also bear a figure with the abbreviation *amps* (for amperes, a measure of electric current). This refers to the current-drawing requirement of the tool. The higher the input current, the more powerful the motor.

Extension Cords

If there is some distance from the power source to the work area or if the portable tool is equipped with a stub power cord, an extension cord must be used. When using extension cords on portable power tools, the size of the conductors must be large enough to prevent an excessive drop in voltage. A voltage drop is the lowering of the voltage at the power tool from that of the voltage at the supply. This occurs because of resistance to electrical flow in the wire. A voltage drop causes loss of power, overheating, and possible motor damage. **Table 2-2** shows the correct size extension cord to use based on cord length and nameplate amperage rating. If in doubt, use the next larger size. The smaller the gauge number of an extension cord, the larger the cord.

Only three-wire, grounded extension cords connected to properly grounded, three-wire receptacles should be used. Two-wire extension cords with two-prong plugs should not be used. Current specifications require outdoor receptacles to be protected with **ground-fault circuit interpreter (GFCI)** devices. These safety devices are often referred to as a **GFI**.

When using extension cords, keep in mind the following safety tips:

- Always connect the cord of a portable electric power tool into the extension cord before the extension cord is connected to the outlet.

- Always unplug the extension cord from the receptacle before unplugging the cord of the portable power tool from the extension cord.

- Extension cords should be long enough to make connections without being pulled taut, creating unnecessary strain or wear, but should not be excessively long.

- Be sure that the extension cord does not come in contact with sharp objects or hot surfaces. The cords should not be allowed to kink, nor should they be dipped in or splattered with oil, grease, or chemicals.

Name-plate Amperes	Cord Length in Feet							
	25	50	75	100	125	150	175	200
1	16	16	16	16	16	16	16	16
2	16	16	16	16	16	16	16	16
3	16	16	16	16	16	16	14	14
4	16	16	16	16	16	14	14	12
5	16	16	16	16	14	14	12	12
6	16	16	16	14	14	12	12	12
7	16	16	14	14	12	12	12	10
8	14	14	14	14	12	12	10	10
9	14	14	14	12	12	10	10	10
10	14	14	14	12	12	10	10	10
11	12	12	12	12	10	10	10	8
12	12	12	12	12	10	10	8	8

TABLE 2-2 Recommended Extension Cord Sizes for Use with Portable Electric Tools

- Before using a cord, inspect it for loose or exposed wires and damaged insulation. If a cord is damaged, replace it. This also applies to the tool's power cord.

- Extension cords should be checked frequently while in use to detect unusual heating. Any cable that feels more than slightly warm to a bare hand placed outside the insulation should be checked immediately for overloading.

- See that the extension cord is positioned so that no one trips or stumbles over it.

- To prevent the accidental separation of a tool cord from an extension cord during operation, make a knot as shown in **Figure 2-45A** or use a cord connector as shown in **Figure 2-45B**.

- Extension cords that go through dirt and mud must be cleaned before storing.

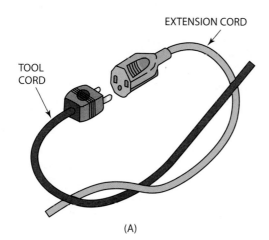

(A)

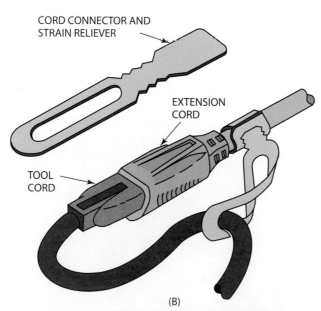

(B)

FIGURE 2-45 (A) A knot will prevent the extension cord from accidentally pulling apart from the tool cord during operation. (B) A cord connector will serve the same purpose. © Cengage Learning 2012

Safety Rules for Portable Electric Tools

In all tool operation, safety is simply the removal of any element of chance. Following are a few safety precautions that should be observed. These are general rules that apply to all power tools. They should be strictly obeyed to avoid injury to the operator and damage to the power tool.

- Know the tool. Learn the tool's applications and limitations as well as its specific potential hazards by reading the manufacturer's literature.

- Ground the portable power tool unless it is double insulated. If the tool is equipped with a three-prong plug, it must be plugged into a three-hole electrical receptacle. Never remove the third prong.

- Do not expose the power tool to water or rain. Do not use a power tool in wet locations.

- Keep the work area well lighted. Avoid chemical or corrosive environments.

- Because electric tools spark, portable electric tools should never be started or operated in the presence of propane, natural gas, gasoline, paint thinner, acetylene, or other flammable vapors that could cause a fire or explosion.

- Do not force a cutting tool to cut faster. It will do the job better and more safely if operated at the cutting rate for which it was designed.

- Use the right tool for the job. Never use a tool for any purpose other than that for which it was designed.

- Wear eye protectors. Safety glasses or goggles will protect the eyes while you operate power tools.

- Wear a face or dust mask if the operation creates dust.

- Take care of the power cord. Never carry a tool by its cord or yank it to disconnect it from the receptacle.

- Secure your work with clamps. It is safer than using your hands, and it frees both hands to operate the tool.

- Do not overreach when operating a power tool. Keep proper footing and balance at all times.

- Maintain power tools. Follow the manufacturer's instructions for lubricating and changing accessories. Replace all worn, broken, or lost parts immediately.

- Disconnect the tools from the power source when they are not in use.

- Form the habit of checking to see that any keys or wrenches are removed from the tool before turning it on.

- Avoid accidental starting. Do not carry a plugged-in tool with your finger on the switch. Be sure the switch is off when plugging in the tool.

- Be sure accessories and cutting bits are attached securely to the tool.

FIGURE 2-46 Always check to be sure that the grinding stone and the grinder are compatible before installing the stone. Larry Jeffus

- Do not use tools with cracked or damaged housings.
- When operating a portable power tool, give it your full and undivided attention; avoid dangerous distractions.
- Never use a power tool with its safeties or guards removed or are inoperable.

Grinders Grinding using a pedestal grinder or a portable grinder is required to do many welding jobs correctly. Often it is necessary to grind a groove, remove rust, or smooth a weld. Grinding stones have the maximum revolutions per minute (RPM) listed on the paper blotter, **Figure 2-46**. They must never be used on a machine with a higher-rated RPM. If grinding stones are turned too fast, they can explode.

Grinding Stone

Before a grinding stone is put on the machine, it should be tested for cracks. This is done by tapping the stone in four places and listening for a sharp ring, which indicates it is good, **Figure 2-47**. A dull sound indicates that the

FIGURE 2-48 Use a grinding stone redressing tool as needed to keep the stone in balance. Larry Jeffus

grinding stone is cracked and should not be used. Once a stone has been installed and has been used, it may need to be trued and balanced by using a special tool designed for that purpose, **Figure 2-48**. Truing keeps the stone face flat and sharp for better results.

Types of Grinding Stones

Each grinding stone is made for grinding specific types of metal. Most stones are for ferrous metals, meaning iron, cast iron, steel, and stainless steel, among others. Some stones are made for nonferrous metals such as aluminum, copper, and brass. If a ferrous stone is used to grind nonferrous metal, the stone will become glazed (the surface clogs with metal) and may explode due to frictional heat building up on the surface. If a nonferrous stone is used to grind ferrous metal, the stone will be quickly worn away.

When the stone wears down, keep the tool rest adjusted to within 1/16 in. (2 mm), **Figure 2-49**, so that the metal being ground cannot be pulled between the tool rest and the stone surface. Stones should not be used when they are worn down to the size of the paper blotter. If small parts become hot from grinding, pliers can be used to hold them. Gloves should never be worn when grinding. If a glove gets caught in a stone, the whole hand may be drawn in.

The sparks from grinding should be directed down and away from other people or equipment.

Drills Before starting to drill, secure the workpiece as necessary and fasten it in a vise or clamp. Holding a small

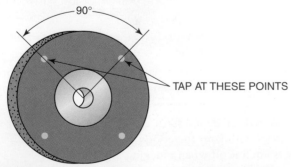

FIGURE 2-47 Grinding stones should be checked for cracks before they are installed. © Cengage Learning 2012

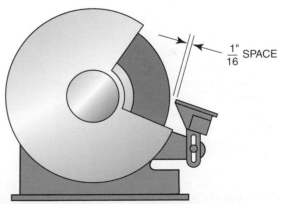

FIGURE 2-49 Keep the tool rest adjusted.
© Cengage Learning 2012

item in your hand can cause injury if it is suddenly seized by the bit and whirled from your grip. This is most likely to happen just as the bit breaks through the hole at the back side of the work. All sheet metal tends to cause the bit to grab as it goes through. This can be controlled by reducing the pressure on the drill just as the bit starts to go through the workpiece.

Carefully center the drill bit in the jaws of the chuck and securely tighten it. Do not insert the bit off center, because it will wobble and probably break when it is used. Drill bits that are 1/4 in. (6 mm) may be hand tightened in the drill chuck to prevent them from snapping if they are accidentally grabbed. Hand tightening the small bits allows them to spin in the chuck if necessary, thus reducing bit breakage. This technique does not always work because some chucks cannot hold the bits securely enough to prevent them from spinning during normal use. In these cases, the chuck must be tightened securely with a chuck key.

When possible, center-punch the workpiece before drilling to prevent the drill bit from moving across the surface as the drilling begins. After centering the drill bit tip on the exact point at which the hole is to be drilled, start the motor by pulling the trigger switch. Never apply a spinning drill bit to the work. With a variable-speed drill, run it at a very low speed until the cut has begun. Then gradually increase to the optimum drill speed.

Except when it is desirable to drill a hole at an angle, hold the drill perpendicular to the face of the work. Align the drill bit and the axis of the drill in the direction the hole is to go and apply pressure only along this line, with no sideways or bending pressure. Changing the direction of pressure will distort the dimensions of the hole and might snap a small drill bit.

Use just enough steady, even pressure to keep the drill cutting. Guide the drill by leading it slightly, if needed, but do not force it. Too much pressure can cause the bit to break or overheat. Too little pressure will keep the bit from cutting and dull its edges due to the friction created by sliding over the surface.

If the drill becomes jammed in the hole, release the trigger immediately, remove the drill bit from the work, and determine the cause of the stalling or jamming. Do not squeeze the trigger on or off in an attempt to free a stalled or jammed drill. When using a reversing-type model, the direction of the rotation may be reversed to help free a jammed bit. Be sure the direction of the rotation is reset before trying to continue the drilling.

Reduce the pressure on the drill just before the bit cuts through the work to avoid stalling in metal. When the bit has completely penetrated the work and is spinning freely, withdraw it from the work while the motor is still running, and then turn off the drill.

Metal Cutting Machines

Many types of mechanical metal cutting machines are used in the welding shop—for example, shears, punches, cut-off machines, and band saws. Their advantages over thermal cutting include little or no postcutting cleanup, the wide variety of metals that can be cut, and the fact that the metal is not heated.

////// **CAUTION** \\\\\\

Before operating any power equipment for the first time, you must read the manufacturer's safety and operating instructions and should be assisted by someone with experience with the equipment. Be sure your hands are clear of the machine before the equipment is started. Always turn off the power and lock it off before working on any part of the equipment.

Shears and Punches Welders frequently use shears and punches in the fabrication of metal for welding. These machines can be operated either by hand or by powerful motors. Hand-operated equipment is usually limited to thin sheet stock or small bar stock. Powered equipment can be used on material an inch or more in thickness and several feet wide, depending on its rating. Their power can be a potential danger if these machines are not used correctly. Both shears and punches are rated by the thickness, width, and type of metal that they can be safely used to work. Failure to follow these limitations can result in damage to the equipment, damage to the metal being worked, and injury to the operator.

Shears work like powerful scissors. The correct placement of the metal being cut is as close to the pivot pin as possible, **Figure 2-50**. The metal being sheared must be securely held in place by the clamp on the shear before it is cut. If you are cutting a long piece of metal that is not being supported by the shear table, then portable supports must be used. As the metal is being cut, it may suddenly

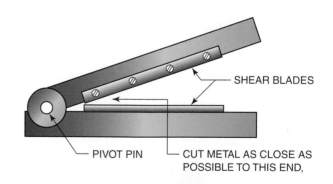
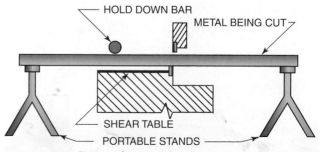

FIGURE 2-50 Power shear. © Cengage Learning 2012

move or bounce around; if you are holding on to it, this can cause a serious injury.

Power punches are usually either hydraulic or flywheel operated. Both types move quickly, but only the hydraulic can usually be stopped midstroke. Once the flywheel-type punch has been engaged, in contrast, it will make a complete cycle before it stops. Because punches move quickly or may not be stopped, it is very important that the operator's two hands be clear of the machine and that the metal be held firmly in place by the machine clamps before starting the punching operation.

Cut-Off Machines Cut-off machines may use abrasive wheels or special saw blades to make their cuts. Most abrasive cut-off wheels spin at high speeds (high RPMs) and are used dry (without coolant). Most saws operate much more slowly and with a liquid coolant. Both types of machines produce quality cuts in a variety of bar- or structural-shaped metals. The cuts require little or no postcut cleanup. Always wear eye protection when operating these machines. Before a cut is started, the metal must be clamped securely in the machine vise. Even the slightest movement of the metal can bind or break the wheel or blade. If the machine has a manual feed, the cutting force must be applied at a smooth and steady rate. Apply only enough force to make the cut without dogging down the motor. Use only reinforced abrasive cut-off wheels that have an RPM rating equal to or higher than the machine-rated speed.

Band Saws Band saws can be purchased as vertical or horizontal, and some can be used in either position. Some band saws can be operated with a cooling liquid and are called *wet saws*; most small saws operate dry. The blade guides must be adjusted as closely as possible to the metal being cut. The cutting speed and cutting pressure must be low enough to prevent the blade from overheating. When using a vertical band saw with a manual feed, you must keep your hands away from the front of the blade so that if your hand slips, it will not strike the moving blade. If the blade breaks, sticks, or comes off the track, turn off the power, lock it off, and wait for the band saw drive wheels

to come to a complete stop before touching the blade. Be careful of hot flying chips.

Material Handling

Proper lifting, moving, and handling of large, heavy welded assemblies are important to the safety of the workers and the weldment. Improper work habits can cause serious personal injury as well as cause damage to equipment and materials.

Lifting When you are lifting a heavy object, the weight of the object should be distributed evenly between both hands, and your legs should be used to lift, not your back, **Figure 2-51**. Do not try to lift a large or bulky object without help if the object is heavier than you can lift with one hand.

Hoists or Cranes The capacity of hoists or cranes should be checked before trying to lift a load. They can be accidentally overloaded with welded assemblies. Keep any load as close to the ground as possible while it is being moved. Pushing a load on a crane is better than pulling a load. It is advisable to stand to one side of ropes, chains, and cables that are being used to move or lift a load, **Figure 2-52**.

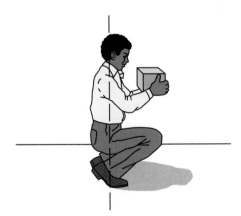

FIGURE 2-51 Lift with your legs, not your back.
© Cengage Learning 2012

FIGURE 2-52 Never stand in line with a rope, chain, or cable that is being used to move or lift a load. © Cengage Learning 2012

FIGURE 2-53 When moving a load overhead, stay out of the way of the load in case it falls. © Cengage Learning 2012

If they break and snap back, they will miss you. If it is necessary to pull a load, use a rope, **Figure 2-53**.

Ladder Safety

Improper use of ladders is often a factor in falls. Always keep this in mind when erecting a ladder; even short step stools can pose a potential fall hazard. Never approach a climb assuming that because it is not high, it cannot be that dangerous. All ladder usage poses a danger to your safety. Some welders think that if a ladder starts to fall they will just "jump clear." You cannot jump clear if the ladder under you has given way because there is nothing solid under your feet for you to jump from. When a ladder falls, you fall. Keep the area around the base of the ladder clear so if you do fall, it will not be into debris or equipment.

Types of Ladders Both stepladders and straight ladders are used extensively in welding fabrication. Straight ladders may be single section or extension-type ladders. Most ladders are made from wood, aluminum, or fiberglass, each type of which has its advantages and disadvantages, **Table 2-3**. All ladders used in welding should be listed with the American National Standards Institute

(ANSI) and Underwriters Laboratories (UL) to ensure that they are constructed to a standard of safety.

Ladder Inspection Over time, ladders can become worn or damaged and should be inspected each time they are used. Look for loose or damaged steps, rungs, rails, braces, and safety feet. Check to see that all hardware is tight, including hinges, locks, nuts, bolts, screws, and rivets. Wooden ladders must be checked for cracks, rot, or wood decay. Never use a defective ladder. Make any necessary repairs before it is used or, if it cannot be repaired, replace it.

Rules for Ladder Use Read the entire ladder manufacturer's list of safety rules before using the ladder for the first time. Stepladders must be locked in the full opened position with the spreaders. Straight or extension ladders must be used at the proper angle; either too steep or too flat is dangerous, **Figure 2-54**.

The following are general safety and usage rules for ladders:

- Follow all recommended practices for safe use and storage.
- Do not exceed the manufacturer's recommended maximum weight limit for the ladder.

Material	Advantages	Disadvantages
Wood	Electrically non-conductive	Long-term exposure to weather will cause rotting
Aluminum	Lightweight Weather resistant	Electrically conductive Shakier than wood or fiberglass
Fiberglass	Electrically nonconductive Weather resistant	Heavier than aluminum and wood Fiberglass splinters

TABLE 2-3 Major Advantages and Disadvantages of Typical Ladder Materials

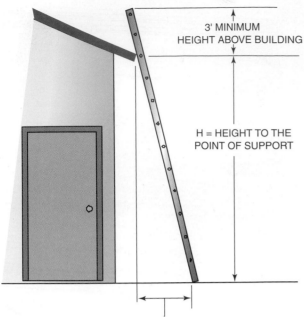

3' MINIMUM HEIGHT ABOVE BUILDING

H = HEIGHT TO THE POINT OF SUPPORT

THE BASE OF A LADDER SHOULD BE SET OUT A DISTANCE EQUAL TO 1/4 OF THE HEIGHT TO THE POINT OF SUPPORT (H/4)

FIGURE 2-54 Make sure the ladder is leaning at the proper angle. © Cengage Learning 2012

- Before setting up a ladder, make certain it will be erected on a level, solid surface.
- Never use a ladder in a wet or muddy area where water or mud will be tracked up the ladder's steps or rungs. Only climb or descend ladders with clean, dry shoes.
- Tie the ladder securely in place.
- Climb and descend the ladder cautiously.
- Do not carry tools and supplies in your hand as you climb or descend a ladder. Use a rope to raise or lower the items once you are safely in place.
- Never use ladders around live electrical wires.
- Never use a ladder that is too short for the job so you have to reach or stand on the top step.
- Wear well-fitted shoes or boots.

Summary

The safety of the welder working in industry is of utmost importance to the industry. A sizable amount of money is spent for the protection of welders. Usually, manufacturers have a safety department with one individual in charge of plant safety. The safety officer's job is to make sure that all welders comply with safety rules during production. The proper clothing, shoes, and eye protection to be worn are emphasized in these plants. Any worker who does not follow established safety rules is subject to dismissal.

If an accident does occur, it is important that appropriate and immediate first aid steps be taken. All welding shops should have established plans for actions to take in case of accidents. You should take time to learn the proper procedure for accident response and reporting before you need to respond in an emergency. After the situation has been properly taken care of, you should fill out an accident report.

Equipment is periodically checked to be sure that it is safe and in proper working condition. Maintenance workers are employed to see that the equipment is in proper working condition at all times.

Further safety information is available in *Safety for Welders,* by Larry F. Jeffus, published by Delmar/Cengage Learning, and from the American Welding Society or the U.S. Department of Labor (OSHA) regulations.

Heads Up on Safety: Use Proper Head and Eye Protection

American Welding Society

Virtually every worker required to wear a hard hat—the most familiar symbol of personal protective equipment (PPE)—now wears one, yet studies indicate about 30% of those users are not in compliance with Occupational Safety and Health Administration (OSHA) standards.

And the news is not as good for eye protection. According to the Industrial Safety Equipment Association (ISEA), 68% of all employees who should use protective eyewear do not. That is why an estimated 400,000 eye injuries occur on the job every year, according to the National Society to Prevent Blindness. We all know that OSHA has very strict rules regarding head and eye protection. For instance, one OSHA regulation mandates that welders use eye protection whenever exposed to flying objects, infrared (IR) and ultraviolet (UV) radiation, glare, liquids, molten metals, and other dangers. It is hard to think of any job that would not have some of these hazards.

Of course, we cannot simply rely on OSHA regulations to solve all our safety concerns; we have to assume responsibility for protecting ourselves. That means wearing hard hats, welding helmets, and safety glasses on the job.

Hard hats and eye protection are readily available. Some safety glasses can cost as little as $5, and basic hard hats are not much more. Following are some important basics for proper head and eye protection.

Head Protection

Always wear appropriate head protection. ANSI Z89 regulation groups hard hats into two categories, one to protect against falling objects and another to protect against side impact and contact exposure to low- or high-voltage conductors.

Make sure the hard hat fits well and has increased side impact protection. Hard hats should not fall off when the wearer looks down or tilts the head. If falling is a hazard in your workplace, the biggest danger is that your hard hat will not stay on when you fall. It is best to use a chinstrap to secure the hard hat.

Obey "Hard Hat Area" signs. These can alert you to dangers from falling and flying objects and possible side impacts.

Never wear a hard hat backward. Many welders often turn the safety cap in their welding helmets around and wear the shell backward. If a hard hat was not tested in this position, it will not meet ANSI Z89 standards.

Do not drill holes in the shell of a hard hat or use adhesives or paint on the shell. This may reduce impact resistance.

Hard hats generally do not have a service life. Obviously, a hard hat used every day will need to be replaced more quickly than one used for a once-a-week trip to a hard hat area.

Take advantage of accessories. Attachments such as earmuffs, visors, liners, and faceshields can make your job more pleasant, and proper headgear suspension will not only make your hard hat more comfortable, but it will also ensure the headgear stays where it belongs.

Maintain head protection. Clean hard hats with mild detergent and water, never with solvents. Inspect the suspension often and discard the hard hat if there is any deterioration.

Eye Protection

Never use regular prescription glasses as safety glasses. If you wear glasses, use protective eyewear that fits over your glasses, or have your prescription incorporated into your safety glasses. Always wear safety glasses under welding helmets, visors, and faceshields, which are considered secondary, not primary, protection.

Go for the best. Low-quality optics may not only distort your vision and cause nausea, headaches, and fatigue, but they may not be in compliance with safety standards. Safety glasses should comply with ANSI Z89 standards.

Always use a shaded lens when welding, brazing, cutting with a torch, and the like. Welding lenses and visors are manufactured in shades to protect against radiation hazards and are rated from 1 to 14—the higher the number, the darker the lens.

Do not use nonabsorptive, tinted lenses for welding. Many lenses are not absorptive lenses intended for welding. For example, Willson® TCG™ dark, gray lenses absorb 86% of infrared (IR) radiation but are not absorptive welding lenses. They are mainly used outdoors where the colors of wiring need to be recognized.

Safety spectacles, the primary protection device for the eyes, should be worn beneath welding helmets, face shields, and visors. A welding helmet, for example, can flip up, leaving the eyes unprotected if spectacles are not worn. American Welding Society

Lasers require special lenses. Shaded lenses used for protection during welding cannot be used with lasers, which require specialized protective eyewear, since they produce a different kind of light.

Take advantage of special lenses to make your job easier and safer. For example, if you move frequently from an indoor to an outdoor workplace, consider using a fog-resistant lens.

Make sure tinted lenses do not affect the way you see color. This is especially important if you are working with colored electrical wiring or are encountering traffic signals.

Use secondary protection if it is needed. If a job exposes you to the risk of impact or other hazards, use a faceshield or visor as a secondary safety device. OSHA requires safety spectacles to be worn even when a welding helmet is also being worn. This is because a welding helmet could accidentally flip up and expose your eyes to high glare or high UV or IR radiation.

Keep eyewear protection clean and in good shape. Use only mild soap to clean lenses, and avoid drying them with paper towels or anything else that could scratch. If the lenses become scratched or cracked, they should be replaced.

When in doubt, consult the standards. Both ANSI and OSHA have specific standards to help you choose the right protection for the task at hand. There is no reason that 68% of workers who need eye protection should not wear it when proper protection can be within reach. You and your coworkers will not become a statistic if you consult with manufacturers or suppliers on the right personal protection for the job.

Article courtesy of the American Welding Society.

Review

1. What is the key to preventing accidents in a welding shop?

2. Who is ultimately responsible for the welder's safety?

3. Describe the three classifications of burns.

4. What emergency steps should be taken to treat burns?

5. List the three types of light that may be present during welding.

6. Which type of light is the most likely to cause burns? Why?

7. What can be done on the job site to reduce the danger of reflected light?

8. In what two ways can ultraviolet light burn the eyes?

9. What is the name of the eye burn that can occur in a fraction of a second?

10. Why is it important to seek medical treatment for eye burns?

11. Why must eye protection be worn at all times in the welding shop?

12. According to Table 2-1 what eye and/or face protection should be used for each of the following:
    a. Acetylene welding
    b. Chipping
    c. Electric arc welding
    d. Spot welding

13. What types of injuries can occur to the ears during welding?

14. What types of protection are available to protect the ears during welding?

15. What types of information should be covered in a respirator training program?

16. Name two types of respirators and describe how they work.

17. List the materials that can give off dangerous fumes during welding and require forced ventilation.

Review Questions (continued)

18. Why must metal that has been used before be cleaned prior to welding?

19. Under what conditions can natural ventilation be used?

20. When must forced ventilation be used?

21. Who must be provided with material safety data sheets (MSDSs)?

22. Name two advantages of recycling scrap metal.

23. What fabric(s) are the best choice to wear as general work clothing in a welding shop?

24. Describe the ideal work shirt, pants, boots, and caps that should be worn in a welding shop.

25. Why is it unsafe to carry butane lighters or matches in your pockets while welding?

26. What special protective items can be worn to provide extra protection for a welder's hands, arms, body, waist, legs, and feet?

27. Describe an acceptable storage area for a cylinder of fuel gas.

28. How must high-pressure gas cylinders be stored so they cannot accidentally be knocked over?

29. What should be done with a leaking cylinder if the leak cannot be stopped?

30. Why is it important that acetylene cylinders not be stored horizontally?

31. How far away should highly combustible materials be from any welding or cutting?

32. List the four types of fire extinguishers and what type of material that is on fire they are used to extinguish.

33. What is hot work?

34. When is a fire watch needed?

35. Why is it important to have a planned maintenance program for tools and equipment?

36. Why is it important to keep a welding area clean?

37. What should you do if you have to leave a piece of hot metal unattended?

38. Why must a mushroomed chisel or hammer be reground?

39. What causes most electric shock in the welding industry?

40. What can happen if too much power is being carried by a cable?

41. Why must equipment be turned off and unplugged before working on the electrical terminals?

42. According to the Welding Safety Checklist in Figure 2-42, what are the factors necessary for a confined space hazard?

43. According to Table 2-2, what gauge wire size would be needed for a power tool that has a nameplate amperage of 9 and a cord length of 100 ft?

44. What is a GFCI?

45. Why is it important to not weld when everything is wet?

46. List five safety tips for safe extension cord use.

47. List 10 safety rules for the safe use of portable electric tools.

48. List two types of grinders used by welders.

49. How close to the grinding stone face should the tool rest be adjusted?

50. Name metal cutting machines used in the welding shop and what their advantages are.

51. Describe how a person should safely lift a heavy object.

52. List the things that should be inspected on a ladder.

53. List and explain five ladder use safety rules.

SECTION 2

SHIELDED METAL ARC WELDING

Success Story

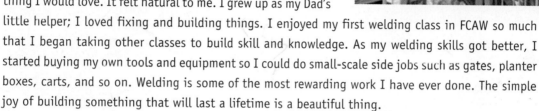

I'm Danelle DiBari, and I took my first welding class just for fun. When I struck my first arc, I knew this was something I would love. It felt natural to me. I grew up as my Dad's little helper; I loved fixing and building things. I enjoyed my first welding class in FCAW so much that I began taking other classes to build skill and knowledge. As my welding skills got better, I started buying my own tools and equipment so I could do small-scale side jobs such as gates, planter boxes, carts, and so on. Welding is some of the most rewarding work I have ever done. The simple joy of building something that will last a lifetime is a beautiful thing.

As a female working in a male-dominated skill area, you have to work hard and go above and beyond to earn respect and prove that you can haul your own steel and do your own work without burdening your partners or co-workers. No one wants to do double the work just because they are working alongside a 5'2", 105 lb female who just doesn't fit the typical stereotype for a welder. One upside to being a female welder is the "shock" factor. I'm always asked, "So, you're really a welder?" Sometimes I say, "Why yes I am." or "No, I'm just faking it, so please ignore the burn holes in my jeans." I've really had fun with it.

I was asked to teach a welding class at College of Southern Nevada, and I gladly accepted. My first semester of teaching FCAW is behind me now, and it was a blast. I must have done well because I was asked to come back and teach another semester —too cool! Teaching welding classes and sharing my experiences with students has been extremely rewarding. I believe that I bring a different and unique element to my students, and I find I'm learning a lot from them as well.

Though there have been many struggles along the way, the tough times have made me what I am. Working hard, staying motivated, and keeping a positive attitude have contributed to my success. I am extremely grateful to have acquired this skill. Because of technology and material changes, the welding world grows and changes constantly and presents new and exciting learning opportunities. My central goal is to learn all that I can and motivate others to do the same. I want to create works of art that no one has seen before that I can inspire others with. This is a legacy that I can pass down to my children and grandchildren (yes, I'm a grandma and loving it.) I am living proof that you can do whatever you set your mind to no matter what obstacles get in your way. I am constantly being asked, "Why welding?" My response: "I just love it." ■

Chapter 3

Shielded Metal Arc Equipment, Setup, and Operation

OBJECTIVES

After completing this chapter, the student should be able to

- describe the process of shielded metal arc welding (SMAW).
- list and define the three units used to measure a welding current.
- tell how adding chemicals to the coverings of the electrodes affects the arc.
- discuss the three different types of current used for welding.
- explain the types of welding power supplies and which type the shielded metal arc welding process requires.
- define *open circuit voltage* and *operating voltage*.
- explain arc blow, what causes it, and how to control it.
- tell what the purpose of a welding transformer is and what kind of change occurs to the voltage and amperage with a step-down transformer.
- compare generators and alternators.
- tell the purpose of a rectifier.
- read a welding machine duty cycle chart and explain its significance.
- demonstrate how to determine the proper welding cable size.
- demonstrate how to service and repair electrode holders.
- discuss the problems that can occur as a result of poor work lead clamping.
- describe the factors that should be considered when placing an arc welding machine in a welding area.

KEY TERMS

amperage

anode

cathode

duty cycle

electrons

inverter

magnetic flux lines

open circuit voltage

operating voltage

output

rectifier

step-down transformer

voltage

wattage

welding cables

welding leads

INTRODUCTION

Shielded metal arc welding (SMAW) is a welding process that uses a flux-covered metal electrode to carry an electrical current, **Figure 3-1**. The current forms an arc across the gap between the end of the electrode and the work. The electric arc creates sufficient heat to melt both the electrode and the work. Molten metal from the electrode travels across the arc to the molten pool on the base metal, where they mix together. The end of the electrode and molten pool of metal is surrounded, purified, and protected by a gaseous cloud and a covering of molten flux produced as the flux coating of the electrode burns or vaporizes. As the arc moves away, the mixture of molten electrode and base metal solidifies and becomes one piece. At the same time, the molten flux solidifies, forming a solid slag. Some electrode types produce heavier slag coverings than others.

SMAW is a widely used welding process because of its low cost, flexibility, portability, and versatility. The machine and the electrodes are low in cost. The machine itself can be as simple as a 110-volt, step-down transformer. The electrodes are available from a large number of manufacturers in packages from 1 lb (0.5 kg) to 50 lbs (22 kg).

The SMAW process is very versatile because the same SMA welding machine can be used to make a wide variety of weld joint designs in a wide variety of metal types and thicknesses, and in all positions:

- Joint designs—In addition to the standard butt, lap, tee, and outside corner joints, at some time SMAW has been certified to be used to weld every possible joint design.

- Metal types—Although mild steel is the most commonly SMA welded metal, stainless, aluminum and cast iron are easily SMA welded.

- Metal thickness—Metal as thin as 16 gauge, approximately 1/16 in. (2 mm) thick up to several feet thick, can be SMA welded.

- All position—The flat welding position is the easiest and most productive because large welds can be made fast using SMA welding, but the process can be used to make welds in any position.

SMAW is a very portable process because it is easy to move the equipment, and engine-driven generator-type welders are available. Also the limited amount of equipment required for the process makes moving easy.

The process is versatile, and it is used to weld almost any metal or alloy, including cast iron, aluminum, stainless steel, and nickel.

Welding Current

The source of heat for arc welding is an electric current. An electric current is the flow of **electrons**. Electrons flow through a conductor from negative (−) to positive (+), **Figure 3-2**. Resistance to the flow of electrons (electricity)

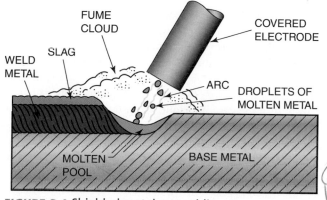

FUME CLOUD

COVERED ELECTRODE

SLAG

WELD METAL

ARC

DROPLETS OF MOLTEN METAL

MOLTEN POOL

BASE METAL

FIGURE 3-1 Shielded metal arc welding. © Cengage Learning 2012

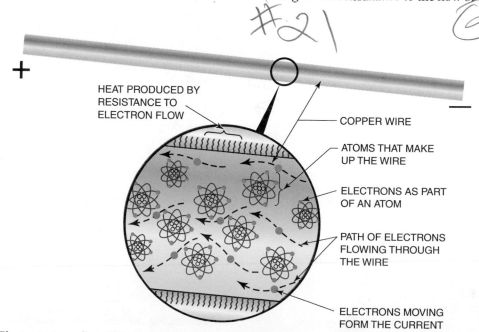

+

−

HEAT PRODUCED BY RESISTANCE TO ELECTRON FLOW

COPPER WIRE

ATOMS THAT MAKE UP THE WIRE

ELECTRONS AS PART OF AN ATOM

PATH OF ELECTRONS FLOWING THROUGH THE WIRE

ELECTRONS MOVING FORM THE CURRENT

FIGURE 3-2 Electrons traveling along a conductor. © Cengage Learning 2012

produces heat. The greater the resistance, the greater the heat. Air has a high resistance to current flow. As the electrons jump the air gap between the end of the electrode and the work, a great deal of heat is produced. Electrons flowing across an air gap produce an arc.

Electrical Measurement

Three units are used to describe any electrical current. The three units are voltage (V), amperage (A), and wattage (W).

- **Voltage**, or volts (V), is the measurement of electrical pressure in the same way that pounds per square inch is a measurement of water pressure. Voltage controls the maximum gap the electrons can jump to form the arc. A higher voltage can jump a larger gap. Welding voltage is associated with the welding temperature.

- **Amperage**, or amps (A), is the measurement of the total number of electrons flowing, in the same way that gallons is a measurement of the amount of water flowing. Amperage controls the size of the arc. Amperage is associated with the welding heat.

- **Wattage**, or watts (W), is a measurement of the amount of electrical energy or power in the arc. Watts are calculated by multiplying voltage (V) times amperes (A), **Figure 3-3**. Watts are associated with welding power or how much heat and temperature an arc produces, **Figure 3-4**.

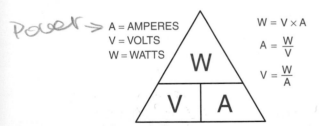

A = AMPERES
V = VOLTS
W = WATTS

$W = V \times A$

$A = \dfrac{W}{V}$

$V = \dfrac{W}{A}$

FIGURE 3-3 Ohm's law. © Cengage Learning 2012

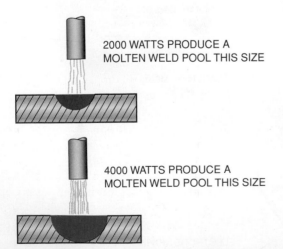

2000 WATTS PRODUCE A MOLTEN WELD POOL THIS SIZE

4000 WATTS PRODUCE A MOLTEN WELD POOL THIS SIZE

FIGURE 3-4 The molten weld pool size depends upon the energy (watts), the metal mass, and thermal conductivity. © Cengage Learning 2012

SMA Welding Arc Temperature and Heat

The term *temperature* refers to the degree or level of energy in a material and can be measured in degrees with a thermometer. The term *heat* refers to the quantity of energy in a material and cannot be easily measured. The quantity of heat in a material can be determined by knowing both the temperature and mass (weight) of an object. For example, the small, red-hot spark from a grinder and a red-hot weld may both be at the same temperature but the weld has more heat. Likewise, the temperature of an arc from a small diameter electrode is the same as the arc from a large diameter electrode, but the larger arc has more heat.

The temperature of a welding arc is dependent on the voltage, arc length, and atmosphere. The arc temperature can range from around 5500°F (3000°C) to above 36,000°F (20,000°C), but most SMA welding arcs have effective temperatures around 11,000°F (6100°C). The voltage and arc length are closely related. The shorter the arc, the lower the arc voltage and the lower the temperature produced, and as the arc lengthens, the resistance increases, thus causing a rise in the arc voltage and temperature.

Most shielded metal arc welding electrodes have chemicals added to their coverings to stabilize the arc. These arc stabilizers form conductive ions that make the arc more stable and reduce the arc resistance. This makes it easier to hold an arc. By lowering the resistance, the arc stabilizers also lower the arc temperature. Other chemicals within the gaseous cloud around the arc may raise or lower the resistance.

The amount of heat produced by the arc is determined by the amperage. The higher the amperage setting the higher the heat produced by the welding arc, and the lower the amperage setting the lower the heat produced. Each diameter of electrode has a recommended minimum and maximum amperage range and therefore a recommended heat range. If you were to try to put too many amps through a small diameter electrode it will overheat and could even melt. If the amperage setting is too low for an electrode diameter, the end of the electrode may not melt evenly, if at all.

Not all of the heat produced by an arc reaches the weld. Some of the heat is radiated away in the form of light and heat waves, **Figure 3-5**. Some additional heat is carried away with the hot gases formed by the electrode covering.

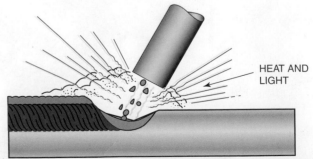

HEAT AND LIGHT

FIGURE 3-5 Energy is lost from the weld in the form of radiation and convection. © Cengage Learning 2012

Heat is also lost through conduction in the work. In total, about 50% of all heat produced by an arc is missing from the weld.

The 50% of the remaining heat the arc produces is not distributed evenly between both ends of the arc. This distribution depends on the composition of the electrode's coating and type of welding current.

Types of Welding Currents The three different types of current used for welding are alternating current (AC), direct-current electrode negative (DCEN), and direct-current electrode positive (DCEP). The terms *DCEN* and *DCEP* have replaced the former terms *direct-current straight polarity (DCSP)* and *direct-current reverse polarity (DCRP)*. DCEN and DCSP are the same currents, and DCEP and DCRP are the same currents. Some electrodes can be used with only one type of current. Others can be used with two or more types of current. Each welding current has a different effect on the weld.

DCEN

In direct-current electrode negative, the electrode is negative, and the work is positive, **Figure 3-6**. The electrons are leaving the electrode and traveling across the arc to the surface of the metal being welded. This results in approximately one-third of the welding heat on the electrode and two-thirds on the metal being welded. DCEN welding current produces a high electrode melting rate.

DCEP

In direct-current electrode positive, the electrode is positive, and the work is negative, **Figure 3-7**. The electrons are leaving the surface of the metal being welded and traveling across the arc to the electrode. This results in approximately two-thirds of the welding heat on the electrode and one-third on the metal being welded.

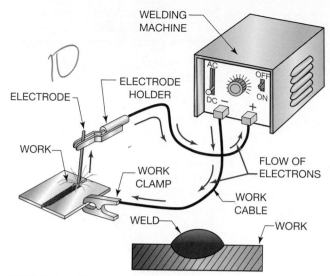

FIGURE 3-7 Direct-current electrode positive (DCEP), direct-current reverse polarity (DCRP) © Cengage Learning 2012

AC

In alternating current, the electrons change direction every 1/120 of a second so that the electrode and work alternate from **anode** to **cathode**, **Figure 3-8**. The positive side of an electrode arc is called the anode, and the negative side is called the cathode. The rapid reversal of the current flow causes the welding heat to be evenly distributed on both the work and the electrode—that is, half on the work and half on the electrode. The even heating gives the weld bead a balance between penetration and buildup, **Figure 3-9**.

Types of Welding Power

Welding power can be supplied as

- Constant voltage (CV)—The arc voltage remains constant at the selected setting even if the arc length and amperage increase or decrease.

- Rising arc voltage (RAV)—The arc voltage increases as the amperage increases.

- Constant current (CC)—The total welding current (watts) remains the same. This type of power is also called drooping arc voltage (DAV) because the arc voltage decreases as the amperage increases.

The shielded metal arc welding (SMAW) process requires a constant current arc voltage characteristic, illustrated by the constant current line in **Figure 3-10**. The shielded metal arc welding machine's voltage **output** decreases as current increases. This output power source provides a reasonably high open circuit voltage before the arc is struck. The high open circuit voltage quickly stabilizes the arc. The arc voltage rapidly drops to the lower closed circuit level after the arc is struck. Following this short starting surge, the power (watts) remains almost constant despite the changes in arc length. With a constant voltage output, small changes in arc length would cause the power (watts) to make large swings. The welder would lose control of the weld.

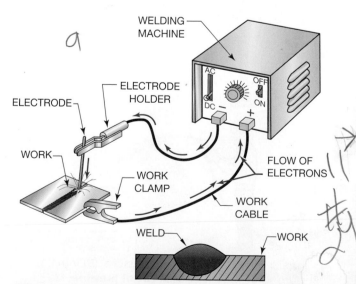

FIGURE 3-6 Direct-current electrode negative (DCEN), direct-current straight polarity (DCSP). © Cengage Learning 2012

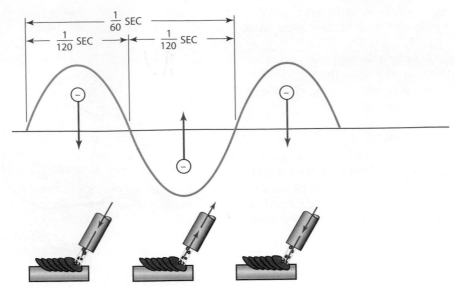

FIGURE 3-8 AC sine wave. © Cengage Learning 2012

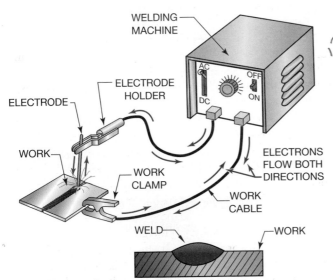

FIGURE 3-9 Alternating current (AC).
© Cengage Learning 2012

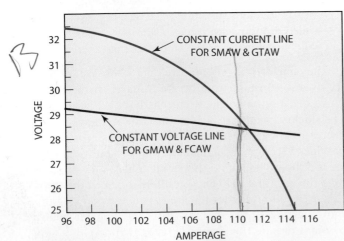

FIGURE 3-10 Constant voltage (CV) and constant current (CC). © Cengage Learning 2012

Open Circuit Voltage

Open circuit voltage is the voltage at the electrode before striking an arc (with no current being drawn). The open circuit voltage is much like the higher surge of pressure you might observe when a water hose nozzle is first opened, **Figure 3-11A and B**. It is easy to see that the

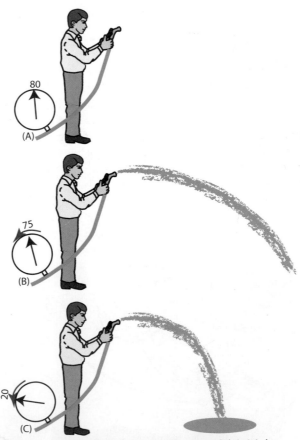

FIGURE 3-11 (A) Closed circuit pressure, (B) initial pressure surge, and (C) closed circuit pressure.
© Cengage Learning 2012

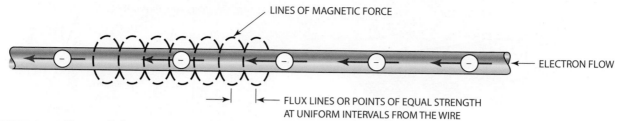

FIGURE 3-12 Magnetic force around a wire. © Cengage Learning 2012

initial pressure from the garden hose was higher than the pressure of the continuous flow of water. The open circuit voltage is usually between 50 V and 80 V. The higher the open circuit voltage, the easier it is to strike an arc because of the initial higher voltage pressure.

CAUTION

The maximum safe open circuit voltage for welders is 80 V. The higher the voltage, the greater the chance of an electrical shock.

Operating Voltage

Operating, welding, or closed circuit voltage is the voltage at the arc during welding. **Operating voltage** is much like the water pressure observed as the water hose is being used, **Figure 3-11C**. The operating voltage will vary with arc length, type of electrode being used, and type of current, and polarity. The welding voltage will be between 17 V and 40 V.

Arc Blow

When electrons flow, they create lines of magnetic force that circle around the path of flow, **Figure 3-12**. These lines of magnetic force are referred to as **magnetic flux lines**. They space themselves evenly along a current-carrying wire. If the wire is bent, the flux lines on one side are compressed together, and those on the other side are stretched out, **Figure 3-13**. The unevenly spaced flux lines try to straighten the wire so that the lines can be evenly spaced once again. The force that they place on the wire is usually small, so the wire does not move. However, when welding with very high amperages, 600 amperes or more, the force may actually cause the wire to move.

The welding current flowing through a plate or any residual magnetic fields in the plate will result in unevenly spaced flux lines. These uneven flux lines can, in turn, cause the arc between the electrode and the work to move during welding. The term *arc blow* refers to this movement of the arc. Arc blow makes the arc drift like a string would drift in the wind. Arc blow can be more of a problem when the magnetic fields are the most uneven such as when they are concentrated in corners, at the ends of plates, and when the work lead is connected to only one side of a plate, **Figure 3-14**.

The more complex a weldment becomes, the more likely arc blow will become a problem. Complex weldments can distort the magnetic lines of flux in unexpected ways. If you encounter severe arc blow during a weld, stop welding and take corrective measures to control or reduce the arc blow.

Arc blow can be controlled or reduced by connecting the work lead to the end of the weld joint, and then welding away from the work lead, **Figure 3-15**. Another way of controlling arc blow is to use two work leads, one on each side of the weld. The best way to eliminate arc blow is to use alternating current. Because alternating current changes directions, the flux lines do not become strong enough to bend the arc before the current changes direction. If it is impossible to move the work connection or to change to AC, a very short arc length can help control arc blow. A large tack weld or a change in the electrode angle can also help control arc blow.

Arc blow may not be a problem as you are learning to weld in the shop because most welding tables are all steel. However, if you are using a pipe stand to hold your welding practice plates, arc blow can become a problem. Try reclamping your practice plates.

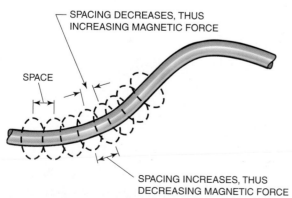

FIGURE 3-13 Magnetic forces concentrate around bends in wires. © Cengage Learning 2012

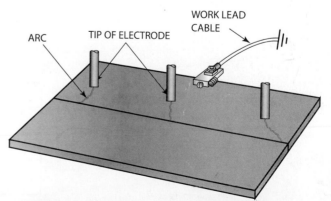

FIGURE 3-14 Arc blow. © Cengage Learning 2012

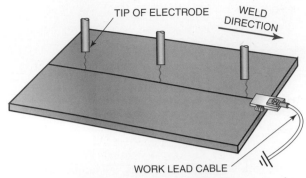

FIGURE 3-15 Correct current connections to control arc blow. © Cengage Learning 2012

Types of Power Sources

Two types of electrical devices can be used to produce the low-voltage, high-amperage current combination that arc welding requires. One type uses electric motors or internal combustion engines to drive alternators or generators. The other type uses step-down transformers. Because transformer-type welding machines are quieter, are more energy efficient, require less maintenance, and are less expensive, they are now the industry standards. However, engine-powered generators are still widely used for portable welding.

Transformer-Type Welding Machines A welding transformer uses the alternating current (AC) supplied to the welding shop at a high voltage to produce the low-voltage welding power. The heart of these welders is the step-down transformer. All transformers have the following three major components:

- Primary coil—the winding attached to the incoming electrical power
- Secondary coil—the winding that has the electrical current induced and is connected to the welding lead and work leads
- Core—made of laminated sheets of steel and used to concentrate the magnetic field produced in the primary winding into the secondary winding, **Figure 3-16**

As electrons flow through a wire, they produce a magnetic field around the wire. If the wire is wound into a coil, the weak magnetic field of each wire is concentrated

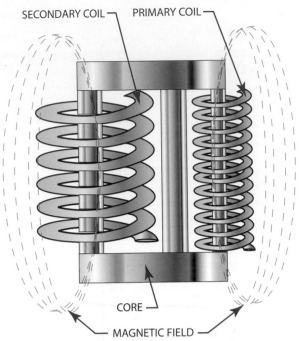

FIGURE 3-16 Parts of a step-down transformer. © Cengage Learning 2012

to produce a much stronger central magnetic force. Because the current being used is alternating or reversing each 1/120 of a second, the magnetic field is constantly being built and allowed to collapse. By placing a second or secondary winding of wire in the magnetic field produced by the first or primary winding, a current will be induced in the secondary winding. The placing of an iron core in the center of these coils will increase the concentration of the magnetic field, **Figure 3-17**.

A transformer with more turns of wire in the primary winding than in the secondary winding is known as a **step-down transformer.** A step-down transformer takes a high-voltage, low-amperage current and changes it into a low-voltage, high-amperage current. Except for some power lost by heat within a transformer, the power (watts) into a transformer equals the power (watts) out because the volts and amperes are mutually increased and decreased.

A transformer welder is a step-down transformer. It takes the high line voltage (110 V, 220 V, 440 V, etc.) and low-amperage current (30 A, 50 A, 60 A, etc.) and changes it into 17 V to 45 V at 190 A to 590 A.

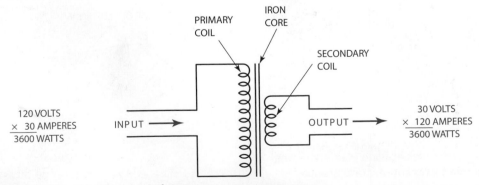

FIGURE 3-17 Diagram of a step-down transformer. © Cengage Learning 2012

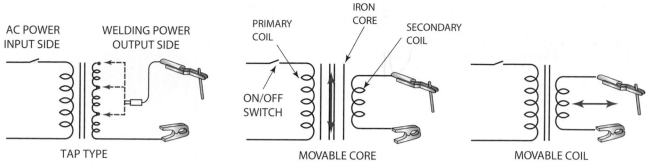

FIGURE 3-18 Major types of adjustable welding transformers. © Cengage Learning 2012

THINK GREEN
Conserve Electricity

Turn the power switch off any time you have stopped welding for a few minutes because electricity flows through the primary coil of a transformer-type welder any time the power switch is on, Figure 3-18. The amount of current flowing is not as much as it is when you are welding, but this practice will save electricity and can lengthen the life of your welding machine.

Welding machines can be classified by the method by which they control or adjust the welding current. The major classifications are multiple coil (called tap type), movable coil, movable core, (see **Figure 3-18**), and inverter type.

Multiple-Coil The multiple-coil machine, or tap-type machine, allows the selection of different current settings by tapping into the secondary coil at a different turn value. The greater the number of turns, the higher the amperage is induced in the turns. These machines may have a large number of fixed amperes, **Figure 3-19**, or

they may have two or more amperages that can be adjusted further with a fine adjusting knob. The fine adjusting knob may be marked in amperes, or it may be marked in tenths, hundredths, or in any other unit.

EXPERIMENT 3-1

Estimating Amperages

Using a pencil and paper, you will prepare a rough estimate of the amperage setting of a welding machine. **Figure 3-20** shows a welding machine with low, medium, and high tap amperage ranges. A fine adjusting knob is marked with 10 equal divisions, and each division is again divided by 10 smaller lines.

The machine is set on the medium range, 50 to 250 amperes, and the fine adjusting knob is turned until it points to the line marked 5 (halfway between 0 and 10). This means that the amperage is halfway from 50 to 250, or 150 amperes. If the fine adjusting knob points between 2 and 3, the resulting amperage is one-quarter of the way from 50 to 250, or about 100 amperes. If the knob points between 7 and 8, the amperage is three-quarters of the way from 50 to 250, or about 200 amperes. If the knob points at 4, the amperage is more than 100 but a little less than 150, or about 130 to 140 amperes. What is the amperage if the knob points at 6?

Since this is a method of estimating only, the amperage value obtained is close enough to allow an arc to be

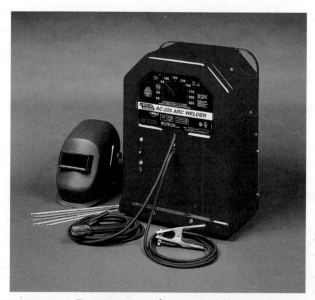

FIGURE 3-19 Tap-type transformer welding machine.
Lincoln Electric Company

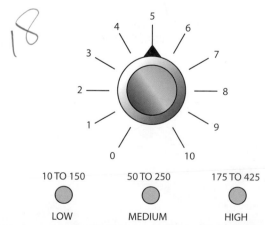

FIGURE 3-20 Welder power range taps and adjustment knob. © Cengage Learning 2012

struck. The welder can then finish the fine adjusting knob to obtain a good weld.

Complete a copy of the "Student Welding Report" listed in Appendix I or provided by your instructor. ◆

EXPERIMENT 3-2

Calculating the Amperage Setting

Using a pencil and paper or calculator, you will calculate the exact value for each space on the fine adjusting knob of a welding machine.

With the machine set on the medium range, from 50 to 250 amperes, first subtract the low amperage from the high amperage to get the amperage spread ($250 - 50 = 200$). Now divide the amperage spread by the number of units shown on the fine adjusting knob ($200 \div 10 = 20$). Each unit is equal to a 20-ampere increase, **Table 3-1**. When the knob points to 0, the amperage is 50; when the knob points to 1, the amperage is 70; and at 2, the amperage is 90, **Figure 3-21**. There are 100 small units on the fine adjusting knob. Dividing the amperage spread by the number of small units gives the amperage value for each unit ($200 \div 100 = 2$). Therefore, if the knob points to 6.1,

Setting	Value in Amperes
0 = 50 + 0,	or 50 A
1 = 50 + 20,	or 70 A
2 = 50 + 40,	or 90 A
3 = 50 + 60,	or 110 A
4 = 50 + 80,	or 130 A
5 = 50 + 100,	or 150 A
6 = 50 + 120,	or 170 A
7 = 50 + 140,	or 190 A
8 = 50 + 160,	or 210 A
9 = 50 + 180,	or 230 A
10 = 50 + 200,	or 250 A

TABLE 3-1 Example of a Table Used to Calculate the Amperage Setting

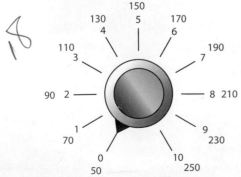

FIGURE 3-21 Fine adjusting knob. © Cengage Learning 2012

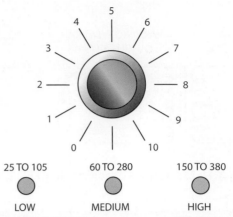

FIGURE 3-22 Practice 3-1. © Cengage Learning 2012

the amperage is set at a value of $50 + 120 + 2 = 172$ amperes. This method provides a good starting place for the current setting, but if the welding is to be made in accordance with a welding procedure's specific amperage setting, it will be necessary to use a calibrated meter to make the correct setting.

Complete a copy of the "Student Welding Report" listed in Appendix I or provided by your instructor. ◆

PRACTICE 3-1

Estimating Amperages

Using a pencil and paper and the amperage ranges given in this practice (or from machines in the shop), you will estimate the amperage when the knob is at the one-quarter, one-half, and three-quarter settings, **Figure 3-22**.

Complete a copy of the "Student Welding Report" listed in Appendix I or provided by your instructor. ◆

PRACTICE 3-2

Calculating Amperages

Using a pencil and paper or a calculator, and the amperage ranges given in this practice (or from machines in the shop), you will calculate the amperages for each of the following knob settings: 1, 4, 7, 9, 2.3, 5.7, and 8.5.

Complete a copy of the "Student Welding Report" listed in Appendix I or provided by your instructor. ◆

Movable Coil or Core Movable coil or movable core machines are adjusted by turning a handwheel that moves the internal parts closer together or farther apart. The adjustment may also be made by moving a lever or turning a knob, **Figure 3-23**. These machines may have a high and low range, but they do not have a fine adjusting knob. The closer the primary and secondary coils are, the greater is the induced current; the greater the distance between the coils, the smaller the induced current, **Figure 3-24**. Moving the core in concentrates more of the magnetic force on the secondary coil, thus increasing the current. Moving the core out allows the field to disperse, and the current is reduced, **Figure 3-25**.

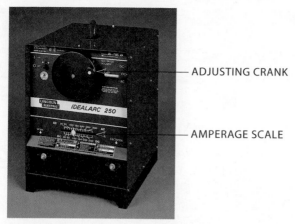

FIGURE 3-23 A movable, core-type welding machine.
Lincoln Electric Company

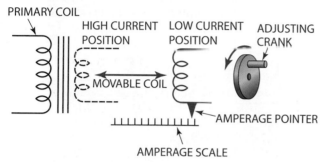

FIGURE 3-24 Movable coil. © Cengage Learning 2012

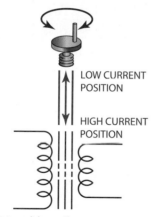

FIGURE 3-25 Movable coil. © Cengage Learning 2012

Inverter Inverter welding machines will be much smaller than other types of machines of the same amperage range. This smaller size makes the welder much more portable as well as increases the energy efficiency, **Figure 3-26**. In a standard welding transformer, the iron core used to concentrate the magnetic field in the coils must be a specific size. The size of the iron core is determined by the length of time it takes for the magnetic field to build and collapse. By using solid-state electronic parts, the incoming power in an inverter welder is changed from 60 cycles a second to several thousand cycles a second. This higher frequency allows the use of a transformer that may be as light as 7 lb (3 kg) and still do the work of a standard transformer weighing 100 lb

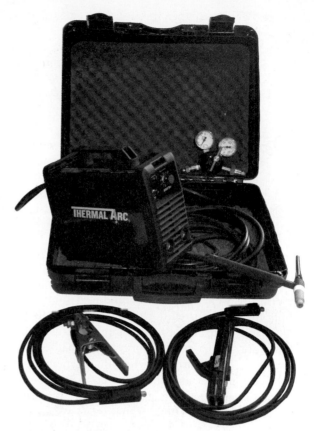

FIGURE 3-26 This 95-amp inverter SMA-GTA welder weighs less than 9 lbs and fits into its own briefcase-size carrier. Larry Jeffus

(45 kg). Additional electronic parts remove the high frequency for the output welding power.

The use of electronics in the inverter-type welder allows it to produce any desired type of welding power. Before the invention of this machine, each type of welding required a separate machine. Now a single welding machine can produce the specific type of current needed for shielded metal arc welding, gas tungsten arc welding, gas metal arc welding, and plasma arc cutting. Because the machine can be light enough to be carried closer to work, shorter welding cables can be used. The welder does not have to walk as far to adjust the machine. Welding machine power wire is cheaper than welding cables. Some manufacturers produce machines that can be stacked so that when you need a larger machine all you have to do is add another unit to your existing welder.

Generator- and Alternator-Type Welders

Generators and alternators both produce welding electricity from a mechanical power source. Both devices have an armature that rotates and a stator that is stationary. As a wire moves through a magnetic force field, electrons in the wire are made to move, producing electricity.

In an alternator, magnetic lines of force rotate inside a coil of wire, **Figure 3-27**. An alternator can produce AC only. In a generator, a coil of wire rotates inside a

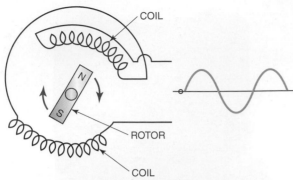

FIGURE 3-27 Schematic diagram of an alternator.
© Cengage Learning 2012

(A)

(B)

FIGURE 3-29 (A) Portable engine generator welder. (B) Lightweight portable SMA and GTA welder.
Lincoln Electric Company

magnetic field. A generator produces DC. It is possible for alternators to use diodes to change the AC to DC for welding. In generators, the welding current is produced on the armature and is picked up with brushes, **Figure 3-28**. In alternators, the welding current is produced on the stator, and only the small current for the electromagnetic force field goes across the brushes. Therefore, the brushes in an alternator are smaller and last longer. Alternators can be smaller in size and lighter in weight than generators and still produce the same amount of power.

Engine-driven generators and alternators may run at the welding speed all the time, or they may have an option that reduces their speed to an idle when welding stops. This option saves fuel and reduces wear on the welding machine. To strike an arc when using this type of welder, stick the electrode to the work for a second. When you hear the welding machine (welder) pick up speed, remove the electrode from the work and strike an arc. In general, the voltage and amperage are too low to start a weld, so shorting the electrode to the work should not cause the electrode to stick. A timer can be set to control the length of time that the welder maintains speed after the arc is broken. The time should be set long enough to change electrodes without losing speed.

Portable welders often have 110-volt or 220-volt plug outlets, which can be used to run grinders, drills, lights, and other equipment. The power provided may be AC or DC. If DC is provided, only equipment with brush-type motors or tungsten lightbulbs can be used. If the plug is

not specifically labeled 110 volts AC, check the owner's manual before using it for such devices as radios or other electronic equipment. Typical portable welders are shown in **Figure 3-29A** and **B**.

Routine Maintenance One of the major drawbacks to portable engine-driven welders is that they require more maintenance than do the other types of welding machines. Poor maintenance practices can lead to a variety of problems including starting and running difficulties, failure to maintain consistent welding power, higher operating cost, and significantly reduced engine life.

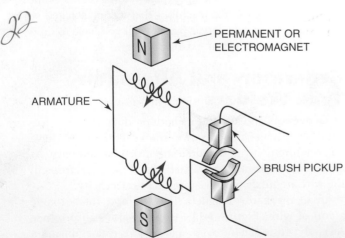

FIGURE 3-28 Diagram of a generator. © Cengage Learning 2012

THINK GREEN
Reduce Air Pollution

It is important to keep portable engine welders properly tuned up so they do not produce excessive air pollution. A properly maintained engine will produce less air pollution and burn less fuel.

Check Each Day before Starting
Oil level
Water level
Fuel level
Check Each Monday
Battery level
Cables
Fuel line filter
Check at Beginning of Month
Air filter
Belts and hoses
Change oil and filter
Check Each Fall
Antifreeze
Test battery
Pack wheel bearings
Change gas filter

TABLE 3-2 Portable Welder Checklist. The owner's manual should be checked for any additional items that might need attention

THINK GREEN
Recycle Used Oil and Batteries

Both used engine oil and batteries can be recycled. Used oil can be re-refined so it can be used again, and the lead and other parts of used batteries can be recycled to make new ones.

It is recommended that a routine maintenance schedule for portable welders be set up and followed. By checking the oil, coolant, battery, filters, fuel, and other parts, the life of the equipment can be extended. A checklist can be posted on the welder, **Table 3-2**.

Converting AC to DC

Alternating welding current can be converted to direct current by using a series of rectifiers. A **rectifier** allows current to flow in one direction only, **Figure 3-30**.

If one rectifier is added, the welding power appears as shown in **Figure 3-31**. It would be difficult to weld with pulsating power such as this. A series of rectifiers, known as a bridge rectifier, can modify the alternating current so that it appears as shown in **Figure 3-32**.

Rectifiers become hot as they change AC to DC. They must be attached to a heat sink and cooled by having air blown over them. The heat produced by a rectifier reduces the power efficiency of the welding machine. **Figure 3-33** shows the amperage dial of a typical machine. Notice that at the same dial settings for AC and

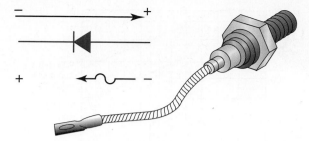

FIGURE 3-30 Rectifier. © Cengage Learning 2012

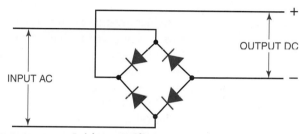

FIGURE 3-31 One rectifier in a welding power supply results in pulsating power. © Cengage Learning 2012

FIGURE 3-32 Bridge rectifier. © Cengage Learning 2012

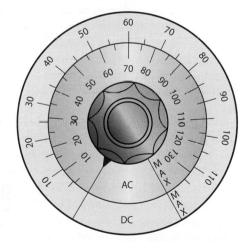

FIGURE 3-33 Typical dial on an AC-DC transformer rectifier welder. © Cengage Learning 2012

DC, the DC is at a lower amperage. The difference in amperage (power) is due to heat lost in the rectifiers. The loss in power makes operation with AC more efficient and less expensive compared to DC.

A DC adapter for small AC machines is available from manufacturers. For some types of welding, AC does not work properly.

Duty Cycle

Welding machines produce internal heat at the same time they produce the welding current. Except for automatic welding machines, welders are rarely used every minute for

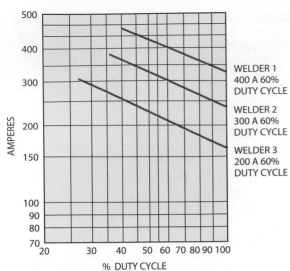

FIGURE 3-34 Duty cycle of a typical shielded metal arc welding machine. © Cengage Learning 2012

long periods of time. The welder must take time to change electrodes, change positions, or change parts. Shielded metal arc welding never continues for long periods of time.

The **duty cycle** is the percentage of time a welding machine can be used continuously. A 60% duty cycle means that out of any 10 minutes, the machine can be used for a total of 6 minutes at the maximum rated current. When providing power at this level, it must be cooled off for 4 minutes out of every 10 minutes. The duty cycle increases as the amperage is lowered and decreases for higher amperages, **Figure 3-34**. Most welding machines weld at a 60% rate or less. Therefore, most manufacturers list the amperage rating for a 60% duty cycle on the nameplate that is attached to the machine. Other duty cycles are given on a graph in the owner's manual.

The manufacturing cost of power supplies increases in proportion to their rated output and duty cycle. To reduce their price, it is necessary to reduce either their rating or their duty cycle. For this reason, some home-hobby welding machines may have duty cycles as low as 20% even at a low welding setting of 90 to 100 amperes. The duty cycle on these machines should never be exceeded because a buildup of the internal temperature can cause the transformer insulation to break down, damaging the power source.

PRACTICE 3-3

Reading the Duty Cycle Chart

Using a pencil and paper and the duty cycle chart in Figure 3-34 (or one from machines in the shop), you will determine the following:

Welder 1: Maximum welding amperage percent duty cycle at maximum amperage

Welder 2: Maximum welding amperage percent duty cycle at maximum amperage

Welder 3: Maximum welding amperage percent duty cycle at maximum amperage

Welder 1: Maximum welding amperage at 100% duty cycle

Welder 2: Maximum welding amperage at 100% duty cycle

Welder 3: Maximum welding amperage at 100% duty cycle

Complete a copy of the "Student Welding Report" listed in Appendix I or provided by your instructor. ◆

Welder Accessories

A number of items must be used with a welding machine to complete the setup. The major items are the welding cables, the electrode holders, and the work clamps.

Welding Cables The terms **welding cables** and **welding leads** mean the same thing. Cables to be used for welding must be flexible, well insulated, and the correct size for the job. Most welding cables are made from stranded copper wire. Some manufacturers sell a newer type of cable made from aluminum wires. The aluminum wires are lighter and less expensive than copper. Because aluminum as a conductor is not as good as copper for a given wire size, the aluminum wire should be one size larger than would be required for copper.

The insulation on welding cables will be exposed to hot sparks, flames, grease, oils, sharp edges, impact, and other types of wear. To withstand such wear, only specially manufactured insulation should be used for welding cable. Several new types of insulation are available that will give longer service against these adverse conditions.

As electricity flows through a cable, the resistance to the flow causes the cable to heat up and increase the voltage drop. To minimize the loss of power and prevent overheating, the electrode cable and work cable must be the correct size. **Table 3-3** lists the minimum size cable that is required for each amperage and length. Large welding lead sizes make electrode manipulation difficult. Smaller cable can be spliced to the electrode end of a large cable to make it more flexible. This whip-end cable must not be over 10 ft (3 m) long.

> ///// **CAUTION** /////
>
> **A splice in a cable should not be within 10 ft (3 m) of the electrode because of the possibility of electrical shock.**

PRACTICE 3-4

Determining Welding Lead Sizes

Using a pencil and paper and Table 3-3, "Copper and Aluminum Welding Lead Sizes," you will determine the following:

1. The minimum copper welding lead size for a 200-amp welder with 100-ft (30-m) leads

		Copper Welding Lead Sizes								
Amperes		**100**	**150**	**200**	**250**	**300**	**350**	**400**	**450**	**500**
ft	**m**									
50	15	2	2	2	2	1	1/0	1/0	2/0	2/0
75	23	2	2	1	1/0	2/0	2/0	3/0	3/0	4/0
100	30	2	1	1/0	2/0	3/0	4/0	4/0		
125	38	2	1/0	2/0	3/0	4/0				
150	46	1	2/0	3/0	4/0					
175	53	1/0	3/0	4/0						
200	61	1/0	3/0	4/0						
250	76	2/0	4/0							
300	91	3/0								
350	107	3/0								
400	122	4/0								

		Aluminum Welding Lead Sizes								
Amperes		**100**	**150**	**200**	**250**	**300**	**350**	**400**	**450**	**500**
ft	**m**									
50	15	2	2	1/0	2/0	2/0	3/0	4/0		
75	23	2	1/0	2/0	3/0	4/0				
100	30	1/0	2/0	4/0						
125	38	2/0	3/0							
150	46	2/0	3/0							
175	53	3/0								
200	61	4/0								
225	69	4/0								

TABLE 3-3 Copper and Aluminum Welding Lead Sizes

2. The minimum copper welding lead size for a 125-amp welder with 225-ft (69-m) leads

3. The maximum length aluminum welding lead that can carry 300 amps

Splices and end lugs are available from suppliers. Be sure that a good electrical connection is made whenever splices or lugs are used. A poor electrical connection will result in heat buildup, voltage drop, and poor service from the cable. Splices and end lugs must be well insulated against possible electrical shorting, **Figure 3-35**.

Complete a copy of the "Student Welding Report" listed in Appendix I or provided by your instructor. ◆

Electrode Holders The electrode holder should be of the proper amperage rating and in good repair for safe welding. Electrode holders are designed to be used at their maximum amperage rating or less. Higher amperage values will cause the holder to overheat and burn up. If the holder is too large for the amperage range being used, manipulation is hard and operator fatigue increases. Make sure that the correct amperage holder is chosen, **Figure 3-36**.

/ / / / **CAUTION** \ \ \ \

Never dip a hot electrode holder in water to cool it off. The problem causing the holder to overheat should be repaired.

A properly sized electrode holder can overheat if the jaws are dirty or loose, or if the cable is loose. If the holder heats up, welding power is being lost. In addition, a hot electrode holder is uncomfortable to work with.

Replacement springs, jaws, insulators, handles, screws, and other parts are available to keep the holder in good working order, **Figure 3-37**. To prevent excessive damage to the holder, welding electrodes should not be burned too short. A 2-in. (51-mm) electrode stub is short enough to minimize electrode waste and save the holder.

PRACTICE 3-5

Repairing Electrode Holders

Using the manufacturer's instructions for your type of electrode holder, required hand tools, and replacement parts, you will do the following:

/ / / / **CAUTION** \ \ \ \

Before starting any work, make sure that the power to the welder is off and locked off or the welding lead has been removed from the machine.

1. Remove the electrode holder from the welding cable.
2. Remove the jaw insulating covers.
3. Replace the jaw insulating covers.
4. Reconnect the electrode holder to the welding cable.
5. Turn on the welding power or reconnect the welding cable to the welder.
6. Make a weld to ensure that the repair was made correctly.

POWER LUG AND
CABLE ENDS ARE
INSULATED

FIGURE 3-35 Power lug protection is provided by insulators. ESAB Welding & Cutting Products

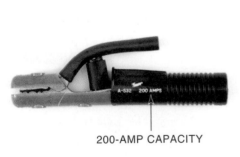

200-AMP CAPACITY

FIGURE 3-36 The amperage capacity of an electrode holder is often marked on its side. Larry Jeffus

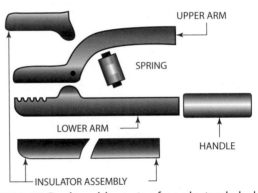

UPPER ARM

SPRING

LOWER ARM

HANDLE

INSULATOR ASSEMBLY

FIGURE 3-37 Replaceable parts of an electrode holder.
© Cengage Learning 2012

Complete a copy of the "Student Welding Report" listed in Appendix I or provided by your instructor. ◆

Work Clamps The work clamp must be the correct size for the current being used, and it must clamp tightly to the material. Heat can build up in the work clamp, reducing welding efficiency, just as was previously described for the electrode holder. Power losses in the work clamp are often overlooked. The clamp should be carefully touched occasionally to find out if it is getting hot.

In addition to power losses due to poor work lead clamping, a loose clamp may cause arcing that can damage a part. If the part is to be moved during welding, a swivel-type work clamp may be needed, **Figure 3-38**. It may be necessary to weld a tab to thick parts so that the work lead can be clamped to the tab, **Figure 3-39**.

Equipment Setup

Arc welding machines should be located near the welding site, but far enough away so that they are not covered with spark showers. The machines may be stacked to save space, but there must be enough room between the machines to ensure the air can circulate so as to keep the machines from overheating. The air that is circulated through the machine should be as free as possible of dust, oil, and metal filings. Even in a good location, the power should be turned off periodically and the machine blown out with compressed air, **Figure 3-40**.

The welding machine should be located away from cleaning tanks and any other sources of corrosive fumes that could be blown through it. Water leaks must be fixed and puddles cleaned up before a machine is used.

FIGURE 3-38 A work clamp may be attached to the workpiece. © Cengage Learning 2012

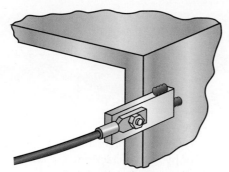

FIGURE 3-39 Tack welded ground to part. © Cengage Learning 2012

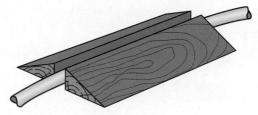

FIGURE 3-41 To prevent people from tripping, when cables must be placed in walkways, lay two blocks of wood beside the cables. © Cengage Learning 2012

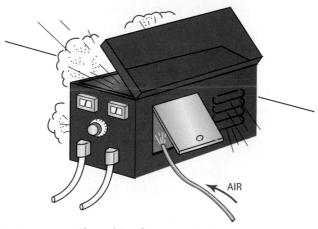

FIGURE 3-40 Slag, chips from grinding, and dust must be blown out occasionally so that they will not start a fire or cause a short-out or other types of machine failure. © Cengage Learning 2012

coiled. Cables should not be placed on the floor in aisles or walkways. If cables must cross a walkway, the cable must be installed overhead, or it must be protected by a ramp, **Figure 3-41**. The welding machine and its main power switch should be off while a person is installing or working on the cables.

The workstation must be free of combustible materials. Screens or curtains should be provided to protect other workers from the arc light.

The welding cable should never be wrapped around arms, shoulders, waist, or any other part of the body. If the cable was caught by any moving equipment, such as a forklift, crane, or dolly, a welder could be pulled off balance or more seriously injured. If it is necessary to hold the weight off the cable so that the welding can be more easily done, a free hand can be used. The cable should be held so that if it is pulled it can be easily released.

Power to the machine must be fused, and a power shut-off switch provided. The switch must be located so that it can be reached in an emergency without touching either the machine or the welding station. The machine case or frame must be grounded.

The welding cables should be sufficiently long to reach the workstation but not so long that they must always be

▨▨ /// CAUTION ▧▧

The cable should never be tied to scaffolding or ladders. If the cable is caught by moving equipment, the scaffolding or ladder may be upset, causing serious personal injury.

Check the surroundings before starting to weld. If heavy materials are being moved in the area around you, there should be a safety watch. A safety watch can warn a person of danger while that person is welding.

Summary

Understanding the scientific theory of electricity and magnetism aids you in understanding how the welding currents are produced and their reactions to changes in their physical surroundings. Understanding electromagnetic phenomena aids you in controlling arc blow. Failure to control arc blow can result in weld failures. In addition, understanding electricity helps you interpret information given on manufacturers' tables, charts, and equipment specifications.

Before starting any new job or welding operation, be sure to check the equipment manufacturer's safety guidelines for proper operation and maintenance. Follow all recommended guidelines.

Keeping your work area clean and orderly helps prevent accidents.

Experienced Welders Make a Difficult Offshore Weld Run Smoothly

Offshore structures are often located in rough seas and may often be damaged by an impact with supply or standby vessels. In most cases, the damage is not severe, but in one particular instance it was substantial. The damage was caused to a 25-year-old gas production platform in the North Sea when a supply boat collided with the platform with enough impact to buckle, then shear off, one end of a horizontal bracing from the jacket leg and then cause further damage at a node point.

After inspection and structural analysis, the jacket's structural integrity was found not to be impaired; however, analysis also confirmed that repair of the node weld cracks and of the sheared area on the leg needed to be performed as soon as practical. The damage occurred barely 6 feet from the level of lowest annual tide. A repair to the jacket structure so close to the waterline needed the weather to cooperate. Welding close to sea level presented

problems associated with preheating the weld area and the possibility of hydrogen cracking. Repairers decided to attach a patch plate over the damaged area using shielded metal arc welding (SMAW) with welding electrodes that would diffuse hydrogen more slowly. To make sure hydrogen levels were maintained at the lowest possible level, all electrodes were baked at 572°F to 662°F (300°C to 350°C), then transferred to a heated quiver at 302°F (150°C), in which they were delivered to the weld site.

All repairs were carried out by personnel experienced in rope-access techniques, since they needed to access the damaged area from the platform above. A predetermined welding sequence ensured the deposited weld metal was evenly distributed around the patched area, limiting the development of residual stresses. The locations of the repairs made it difficult to maintain the required preheat temperature of 300°F (150°C). Because conventional preheating using electric induction heating was impossible, preheating was performed using oxypropane heating equipment.

Damage sustained during impact by a supply vessel.
American Welding Society

Close-up of damage to node section and bracing.
American Welding Society

Completed repair to jacket leg. American Welding Society

To remove cracks on the node section of one of the legs, the repairers dressed (ground) each one to a depth of 0.08 in. (2 mm). If the crack was still visible, 0.08 in. (2 mm) more was removed by dressing. This process was repeated until the crack was removed. Afterward, these dressed areas were examined by dye penetrant inspection, a nondestructive examination, to determine that the defects were completely removed. Most cracks were superficial, but in cases where the depth exceeded 0.16 in. (4 mm) the cracks were repaired using the weld procedure qualified for the patch repair. After completion, all welds were ground to a smooth profile.

When the work was completed, the weld areas were inspected, again with dye penetrant inspection. Weather conditions were ideal during the repair work, and since ideal conditions continued, a second inspection was made after 48 hours to relieve any fears of delayed hydrogen-induced cracking. In addition, ultrasonic inspection using probes was used to make sure that no defects, such as lamellar tearing (a type of cracking in welds often caused by shrinkage), had occurred.

In the end, this difficult repair went smoothly. The repair area was, fortunately, above the waterline, and the weather cooperated perfectly. The repairs were done using rope access techniques, and using austenitic electrodes reduced the possibility of heat affected zone (HAZ) hydrogen cracking. Nondestructive examination confirmed that cracking and other defects did not exist. Indeed, this methodology is considered extremely practical, cost effective, and ideal for use on offshore structures with damage close to the waterline and with the need to execute immediate repairs.

Article courtesy of the American Welding Society.

Review

1. Describe the welding current.

2. What produces the heat during a shielded metal arc weld?

3. Voltage can be described as _____.

4. Amperage can be described as _____.

5. Wattage can be described as _____.

6. What determines the exact temperature of the shielded metal welding arc?

7. Does all of the heat produced by an SMA weld stay in the weld? Why or why not?

8. What do the following abbreviations mean: AC, DCEN, DCEP, DCSP, and DCRP?

9. Sketch a welding machine, an electrode lead, an electrode holder, an electrode, a work lead, and work connected for DCEN welding.

10. Sketch a welding machine, an electrode lead, an electrode holder, an electrode, a work lead, and work connected for DCEP welding.

11. Why is SMA welding current referred to as *constant current?*

12. What is the higher voltage at the electrode before the arc is struck called? What is its advantage to welding?

13. Referring to the graph in Figure 3-9, what would the voltage be for the CC power supply at 110 amps? What would the watts be?

14. How does arc blow affect welding?

15. How can arc blow be controlled?

16. How does a welding transformer work?

17. What are *taps* on a welding transformer?

18. What would the approximate amperage setting be if a welder were set to the high range (150 to 350 amps) and the fine adjustment knob were pointing at 5 on a 10-point scale?

19. What are the advantages of the inverter-type welding power supply?

20. What is the difference between the welding current produced by alternators and by generators?

21. What are the advantages of alternators over generators?

22. What must be checked before using the 110-volt power plug on a portable welder?

23. What is meant by a welder's *duty cycle?*

24. Why must a welding machine's duty cycle never be exceeded?

25. Why is copper better than aluminum for welding cables?

26. A splice in a welding cable should never be any closer than _____ to the electrode holder.

27. Why must the electrode holder be correctly sized?

28. What can cause a properly sized electrode holder to overheat?

29. What problem can occur if welding machines are stacked or placed too closely together?

30. Why must welding cables never be tied to scaffolding or ladders?

Chapter 4

Shielded Metal Arc Welding of Plate

OBJECTIVES

After completing this chapter, the student should be able to

- demonstrate safe work practices.
- demonstrate the ability to strike an arc at a specific point.
- list the problems that can result if the welding current is set too low or too high.
- discuss how to select the correct diameter of a welding electrode for a weld.
- describe the effects of overheating a weld by comparing the bead's shape for width, reinforcement, and appearance.
- define *arc length*, and describe the effects of using too short or too long an arc length.
- compare a leading electrode angle to a trailing electrode angle.
- tell what characteristics of the weld bead can be controlled by the movement or weaving of the welding electrode.
- demonstrate 10 weave patterns for weld beads.
- discuss the importance of positioning the welder and the plate properly before starting to weld.
- give the characteristics of the three filler metal groups E6010 and E6011, E6012 and E6013, and E7016 and E7018.
- define *stringer beads* and tell how they are used.
- demonstrate a vertical up stringer bead and a horizontal stringer bead.
- demonstrate how to make a welded square butt joint in the flat, vertical up, and horizontal positions.
- on an edge joint, demonstrate how to make a flat weld, a vertical down weld, a vertical up weld, a horizontal weld, and an overhead weld.
- on an outside corner joint, demonstrate how to make a flat weld, a vertical down weld, a vertical up weld, a horizontal weld, and an overhead weld.
- demonstrate how to make a welded lap joint in the flat position, a welded horizontal lap joint, a vertical up-welded lap joint, and an overhead-welded lap joint.
- demonstrate how to make a welded tee joint in the flat, horizontal, vertical, and overhead positions.

KEY TERMS

amperage range	*electrode angle*	*square butt joint*
arc length	*lap joint*	*stringer bead*
cellulose-based fluxes	*mineral-based fluxes*	*tee joint*
chill plate	*rutile-based fluxes*	*weave pattern*

INTRODUCTION

Shielded metal arc welding (SMAW), or stick welding, is a common method used to join plate. This method provides a high temperature and concentration of heat, which allows a small molten weld pool to be built up quickly. The addition of filler metal from the electrode adds reinforcement and increases the strength of the weld. SMAW can be performed on almost any type of metal 1/8 in. (3 mm) thick or thicker. A minimum amount of equipment is required, and it can be portable.

High-quality welds can be consistently produced on almost any type of metal and in any position. The quality of the welds produced depends largely upon the skill of the welder. Developing the necessary skill level requires practice. However, practicing the welds repeatedly without changing techniques will not aid in developing the required skills. Each time a weld is completed it should be evaluated, and then a change should be made in the technique to improve the next weld.

PRACTICE 4-1

Shielded Metal Arc Welding Safety

Using a welding workstation, welding machine, welding electrodes, welding helmet, eye and ear protection, welding gloves, proper work clothing, and any special protective clothing that may be required; demonstrate to your instructor and other students the safe way to prepare yourself and the welding workstation for welding. Include in your demonstration appropriate references to burn protection, eye and ear protection, material specification data sheets, ventilation, electrical safety, general work clothing, special protective clothing, and area cleanup.

Complete a copy of the "Student Welding Report" listed in Appendix I or provided by your instructor. ◆

EXPERIMENT 4-1

Striking the Arc

Using a properly set up and adjusted arc welding machine; the proper safety protection, as demonstrated in Practice 4-1; E6011 welding electrodes having a 1/8 in. (3 mm) diameter; and one piece of mild steel plate, 1/4 in. (6 mm) thick; you will practice striking an arc, **Figure 4-1**.

With the electrode held over the plate, lower your helmet. Scratch the electrode across the plate (like striking a large match), **Figure 4-2**. As the arc is established, slightly raise the electrode to the desired arc length. Hold the arc in one place until the molten weld pool builds to the desired size. Slowly lower the electrode as it burns off and move it forward to start the bead.

If the electrode sticks to the plate, quickly squeeze the electrode holder lever to release the electrode. Allow the electrode to briefly cool then break it free by bending it back and

---

THINK GREEN
Use Scrap Metal

Most welding shops have a lot of scrap metal that can be used for experiments in this chapter. Using scrap metal will not affect the experiment but will save material.

---

forth a few times. Do not touch the electrode without gloves because it will still be hot. If the flux breaks away very far from the end of the electrode, throw out the electrode because restarting the arc will be very difficult, **Figure 4-3**.

Once you are able to easily strike an arc and make a weld, try to strike the arc where it will be remelted by the weld you are making. Arc strikes on the metal's surface that are not covered up by the weld are considered to be weld defects by most codes.

Break the arc by rapidly raising the electrode after completing a 1-in. (25-mm) weld bead. Restart the arc as you did before, and make another short weld. Repeat this process until you can easily start the arc each time. Turn off the welding machine and clean up your work area when you are finished welding.

Complete a copy of the "Student Welding Report" listed in Appendix I or provided by your instructor. ◆

EXPERIMENT 4-2

Striking the Arc Accurately

Using the same materials and setup as described in Experiment 4-1, you will start the arc at a specific spot in order to prevent damage to the surrounding plate.

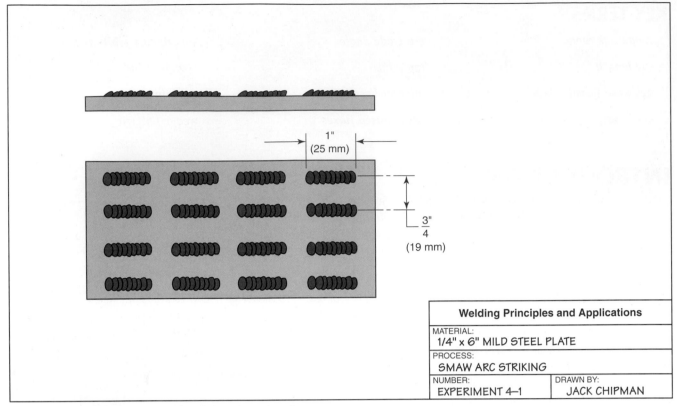

Welding Principles and Applications	
MATERIAL: 1/4" x 6" MILD STEEL PLATE	
PROCESS: SMAW ARC STRIKING	
NUMBER: EXPERIMENT 4–1	DRAWN BY: JACK CHIPMAN

FIGURE 4-1 Striking an arc and running short beads. © Cengage Learning 2012

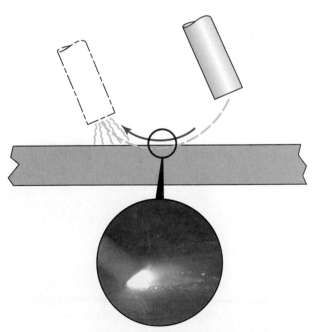

FIGURE 4-2 Striking the arc. Larry Jeffus

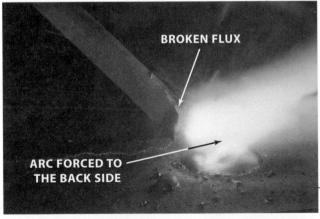

FIGURE 4-3 If the flux is broken off the end completely or on one side, the arc can be erratic or forced to the side. Larry Jeffus

Hold the electrode over the desired starting point. After lowering your helmet, swiftly bounce the electrode against the plate, **Figure 4-4.** A lot of practice is required to develop the speed and skill needed to prevent the electrode from sticking to the plate.

A more accurate method of starting the arc involves holding the electrode steady by resting it on your free hand like a pool cue. The electrode is rapidly pushed forward so that it strikes the metal exactly where it should. This is an excellent method of striking an arc. Striking an arc in an incorrect spot may cause damage to the base metal.

Practice starting the arc until you can start it within 1/4 in. (6 mm) of the desired location. Turn off the welding machine and clean up your work area when you are finished welding.

Complete a copy of the "Student Welding Report" listed in Appendix I or provided by your instructor. ◆

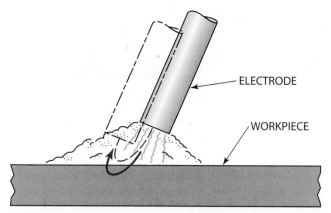

FIGURE 4-4 Striking the arc on a spot. © Cengage Learning 2012

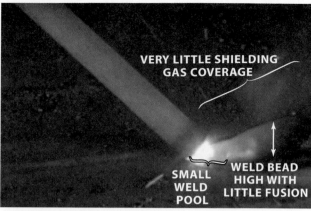

FIGURE 4-5 Welding with the amperage set too low. Larry Jeffus

Effect of Too High or Too Low Current Settings

Each welding electrode must be operated in a specified current (amperage) range, **Table 4-1**. Welding with the current set too low results in poor fusion and poor arc stability, **Figure 4-5**. The weld may have slag or gas inclusions because the molten weld pool was not fluid long enough for the flux to react. Little or no penetration of the weld into the base plate may also be evident. With the current set too low, the arc length is very short. A very short arc length results in frequent shorting and sticking of the electrode.

The core wire of the welding electrode is limited in the amount of current it can carry. As the current is increased, the wire heats up because of electrical resistance. This preheating of the wire causes some of the chemicals in the covering to be burned out too early, **Figure 4-6**. The loss of the proper balance of elements causes poor arc stability. This condition leads to spatter, porosity, and slag inclusions.

An increase in the amount of spatter is also caused by longer arc lengths. The weld bead made at a high amperage setting is wide and flat with deep penetration. The spatter is excessive and is mostly hard. The spatter is called hard because it fuses to the base plate and is difficult to remove, **Figure 4-7**. The electrode covering is discolored more than 1/8 in. (3 mm) to 1/4 in. (6 mm) from the end of the electrode. Extremely high settings may also cause the electrode to discolor, crack, glow red, or burn.

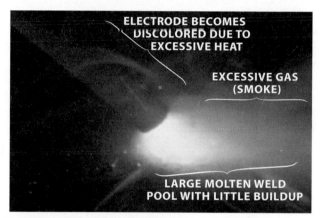

FIGURE 4-6 Welding with too high an amperage. Larry Jeffus

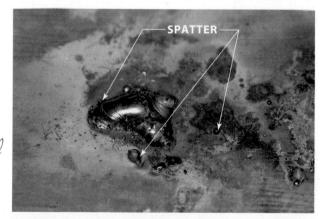

FIGURE 4-7 Hard weld spatter fused to base metal. Larry Jeffus

Electrode	Classification					
Size	E6010	E6011	E6012	E6013	E7016	E7018
3/32 in. (2.4 mm)	40–80	50–70	40–90	40–85	75–105	70–110
1/8 in. (3.2 mm)	70–130	85–125	75–130	70–120	100–150	90–165
5/32 in. (4 mm)	110–165	130–160	120–200	130–160	140–190	125–220

TABLE 4-1 Welding Amperage Range

EXPERIMENT 4-3

Effect of Amperage Changes on a Weld Bead

For this experiment, you will need an arc welding machine; welding gloves; safety glasses; a welding helmet; appropriate clothing; E6011 welding electrodes having a 1/8 in. (3 mm) diameter; and one piece of mild steel plate, 1/4 in. (6 mm) to 1/2 in. (13 mm) thick. You will observe what happens to the weld bead when the amperage settings are raised and lowered.

Starting with the machine set at approximately 90 A AC or DCRP, strike an arc and make a weld 1 in. (25 mm) long. Break the arc. Raise the current setting by 10 A, strike an arc, and make another weld 1 in. (25 mm) long. Repeat this procedure until the machine amperage is set at the maximum value.

Replace the electrode and reset the machine to 90 A. Make a weld 1 in. (25 mm) long. Stop and lower the current setting by 10 A. Repeat this procedure until the machine amperage is set at a minimum value.

Cool and chip the plate, comparing the different welds for width, buildup, molten weld pool size, spatter, slag removal, and penetration, **Figure 4-8A** and **B**. In addition, compare the electrode stubs. Turn off the welding machine and clean up your work area when you are finished welding.

Complete a copy of the "Student Welding Report" listed in Appendix I or provided by your instructor. ◆

////// **CAUTION** \\\\\\

Do not change the current settings during welding. A change in the setting may cause arcing inside the machine, resulting in damage to the machine.

(A)

(B)

FIGURE 4-8 (A) Weld before cleaning. (B) Weld after cleaning. Larry Jeffus

AMOUNT OF HEAT DIRECTED AT WELD	WELD POOL
TOO LOW	
CORRECT	
TOO HOT	

FIGURE 4-9 The effect on the shape of the molten weld pool caused by the heat input. © Cengage Learning 2012

Electrode Size and Heat

The selection of the correct size of welding electrode for a weld is determined by some or all of the following: the skill of the welder, the thickness of the metal to be welded, the size of the metal, and welding codes or standards. Using small diameter electrodes requires less skill than using large diameter electrodes. The deposition rate, or the rate that weld metal is added to the weld, is slower when small diameter electrodes are used. Small diameter electrodes will make acceptable welds on thick plate, but more time is required to make the weld.

Large diameter electrodes may overheat the metal if they are used with thin or small pieces of metal. To determine if a weld is too hot, watch the shape of the trailing edge of the molten weld pool, **Figure 4-9**. Rounded ripples indicate the weld is cooling uniformly and that the heat is not excessive. If the ripples are pointed, the weld is cooling too slowly because of excessive heat. Extreme overheating can cause a burnthrough. Once a burnthrough occurs, it is hard to repair.

To correct an overheating problem, a welder can turn down the amperage, use a shorter arc, travel at a faster rate, use a **chill plate** (a large piece of metal used to absorb excessive heat), or use a smaller electrode at a lower current setting.

EXPERIMENT 4-4

Excessive Heat

Using a properly set up and adjusted arc welding machine; the proper safety protection; E6011 welding electrodes having a 1/8 in. (3 mm) diameter; and three pieces of mild steel plate, 1/8 in. (3 mm), 3/16 in. (4.8 mm), and 1/4 in. (6 mm) thick; you will observe the effects of overheating on the weld. Make a stringer weld on each of the three plates using the same amperage setting, travel rate, and arc length for each weld. Cool and chip the welds. Then compare the weld beads for width, reinforcement, and appearance.

Using the same amperage settings, make additional welds on the 1/8-in. (3-mm) and 3/16-in. (4.8-mm) plates. Vary the arc lengths and travel speeds for these welds. Cool and chip each weld and compare the beads for width, reinforcement, and appearance. Make additional welds on the 1/8-in. (3-mm) and 3/16-in. (4.8-mm) plates, using the same arc length and travel speed as in the earlier part of this experiment but at a lower amperage setting. Cool and chip the welds and compare the beads for width, reinforcement, and appearance.

Welders often use the terms *heat* and *amperage* interchangeabley when they are speaking about making changes to the welding current. For example, welders may say "Turn up the heat a little," or "Turn up the amperage a little." In both cases, they are asking that the welding amperage be increased a little.

The plates should be cooled between each weld so that the heat from the previous weld does not affect the test results. Turn off the welding machine and clean up your work area when you are finished welding.

Complete a copy of the "Student Welding Report" listed in Appendix I or provided by your instructor. ◆

Arc Length

The **arc length** is the distance the arc must jump from the end of the electrode to the plate or weld pool surface. As the weld progresses, the electrode becomes shorter as it is consumed. To maintain a constant arc length, the electrode must be lowered continuously. Maintaining a constant arc length is important, as too great a change in the arc length will adversely affect the weld.

As the arc length is shortened, metal transferring across the gap may short out the electrode, causing it to stick to the plate. The weld that results from a short arc is narrow and has a high buildup, **Figure 4-10**.

Long arc lengths produce more spatter because the metal being transferred may drop outside of the molten weld pool. The weld is wider and has little buildup, **Figure 4-11**.

There is a narrow range for the arc length in which it is stable, metal transfer is smooth, spatter is minimized, and the bead shape is controlled. Factors affecting the length

FIGURE 4-11 Welding with too long an arc length.
Larry Jeffus

are the type of electrode, joint design, metal thickness, and current setting.

Some welding electrodes, such as E7024, have a thick flux covering. The rate at which the covering melts is slow enough to permit the electrode coating to be rested against the plate. The arc burns back inside the covering as the electrode is dragged along touching the joint, **Figure 4-12**. For that reason electrodes like E7024 are sometimes referred to as drag electrodes or drag rods. For this type of welding electrode, the arc length is maintained by the electrode covering.

An arc will jump to the closest metal conductor. On joints that are deep or narrow, the arc is pulled to one side and not to the root, **Figure 4-13**. As a result, the root fusion is reduced or may be nonexistent, thus causing a poor weld. If a very short arc is used, the arc is forced into the root for better fusion.

Because shorter arcs produce less heat and penetration, they are best suited for use on thin metal or thin-to-thick metal joints. Using this technique, metal as thin as 16 gauge can be arc welded easily. Higher amperage settings are required to maintain a short arc that gives good fusion with a minimum of slag inclusions. The higher settings, however, must be within the **amperage range** for the specific electrode.

FIGURE 4-10 Welding with too short an arc length.
Larry Jeffus

HIGH NARROW BEAD WITH A HEAVY SLAG COVER

FIGURE 4-12 Welding with a drag technique. Larry Jeffus

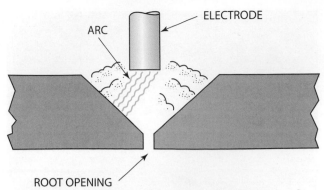

FIGURE 4-13 The arc may jump to the closest metal, reducing root penetration. © Cengage Learning 2012

Finding the correct arc length often requires some trial and adjustment. Most welding jobs require an arc length of 1/8 in. (3 mm) to 3/8 in. (10 mm) when using an 1/8 in. (3 mm) electrode, but this distance varies. It may be necessary to change the arc length when welding to adjust for varying welding conditions.

EXPERIMENT 4-5

Effect of Changing the Arc Length on a Weld

Using an arc welding machine; welding gloves; safety glasses; welding helmet; appropriate clothing; E6011 welding electrodes having a 1/8 in. (3 mm) diameter; and one piece of mild steel plate, 1/4 in. (6 mm) to 1/2 in. (13 mm) thick; you will observe the effect of changing the arc length on a weld.

Starting with the welding machine set at approximately 90 A AC or DCRP, strike an arc and make a weld 1 in. (25 mm) long. Continue welding while slowly increasing the arc length until the arc is broken. Restart the arc and make another weld 1 in. (25 mm) long. Welding should again be continued while slowly shortening the arc length

until the arc stops. Quickly break the electrode free from the plate, or release the electrode by squeezing the lever on the electrode holder.

Cool and chip both welds. Compare both welding beads for width, reinforcement, uniformity, spatter, and appearance. Turn off the welding machine and clean up your work area when you are finished welding.

Complete a copy of the "Student Welding Report" listed in Appendix I or provided by your instructor. ◆

Electrode Angle

The **electrode angle** is measured from the electrode to the surface of the metal. The term used to identify the electrode angle is affected by the direction of travel, generally leading or trailing, **Figure 4-14**. The relative angle is important because there is a jetting force blowing the metal and flux from the end of the electrode to the plate.

Leading Angle A leading electrode angle pushes molten metal and slag ahead of the weld, **Figure 4-15**. When welding in the flat position, caution must be taken to prevent cold lap and slag inclusions. The solid metal

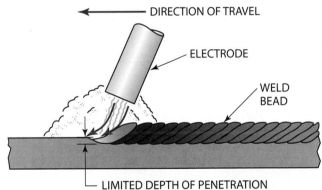

FIGURE 4-15 Leading, lag, or pushing electrode angle. © Cengage Learning 2012

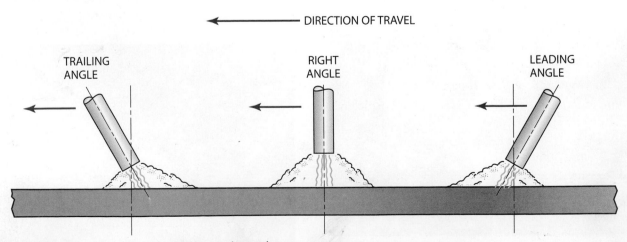

FIGURE 4-14 Direction of travel and electrode angle. © Cengage Learning 2012

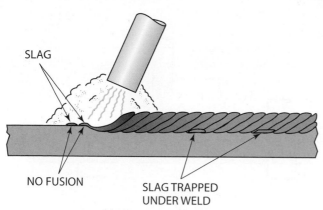

FIGURE 4-16 Some electrodes, such as E7018, may not remove the deposits ahead of the molten weld pool, resulting in discontinuities within the weld.
© Cengage Learning 2012

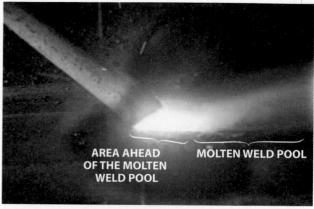

FIGURE 4-17 Metal being melted ahead of the molten weld pool helps to ensure good weld fusion. Larry Jeffus

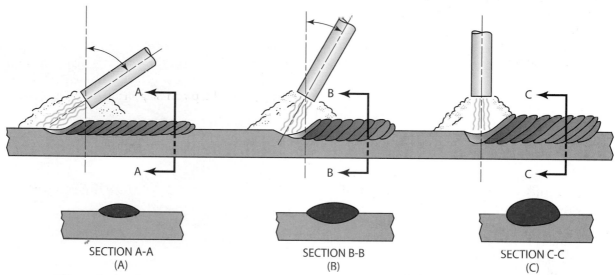

FIGURE 4-18 Effect of a leading angle on weld bead buildup, width, and penetration. As the angle increases toward the vertical position (C), penetration increases. © Cengage Learning 2012

ahead of the weld cools and solidifies the molten filler metal and slag before they can melt the solid metal. This rapid cooling prevents the metals from fusing together, **Figure 4-16.** As the weld passes over this area, heat from the arc may not melt it. As a result, some cold lap and slag inclusions are left.

The following are suggestions for preventing cold lap and slag inclusions:

- Use as little leading angle as possible.
- Ensure that the arc melts the base metal completely, **Figure 4-17.**
- Use a penetrating-type electrode that causes little buildup.
- Move the arc back and forth across the molten weld pool to fuse both edges.

A leading angle can be used to minimize penetration or to help hold molten·metal in place for vertical welds, **Figure 4-18.**

Trailing Angle A trailing electrode angle pushes the molten metal away from the leading edge of the molten weld pool toward the back where it solidifies, **Figure 4-19.** As the molten metal is forced away from the bottom of the weld, the arc melts more of the base metal, which results

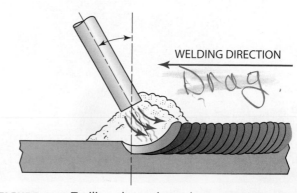

FIGURE 4-19 Trailing electrode angle. © Cengage Learning 2012

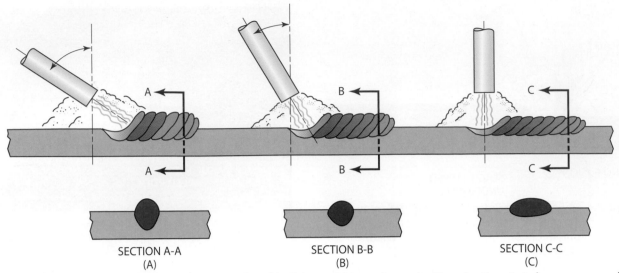

FIGURE 4-20 Effect of a trailing angle on weld bead buildup, width, and penetration. Section A-A shows more weld buildup due to a greater angle of the electrode. © Cengage Learning 2012

in deeper penetration. The molten metal pushed to the back of the weld solidifies and forms reinforcement for the weld, **Figure 4-20**.

EXPERIMENT 4-6

Effect of Changing the Electrode Angle on a Weld

Using a properly set up and adjusted arc welding machine; the proper safety protection; E6011 welding electrodes having a 1/8 in. (3 mm) diameter; and one piece of mild steel plate, 1/4 in. (6 mm) to 1/2 in. (13 mm) thick; you will observe the effect of changes in the electrode angle on a weld.

Start welding with a sharp trailing angle. Make a weld about 1 in. (25 mm) long. Closely observe the molten weld pool at the points shown in **Figure 4-21**. Slowly increase the electrode angle and continue to observe the weld.

When you reach a 90° electrode angle, make a weld about 1 in. (25 mm) long. Observe the parts of the molten weld pool as shown in Figure 4-21.

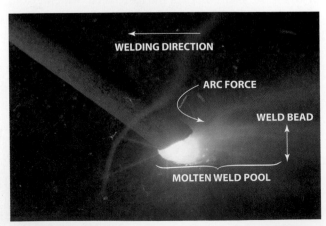

FIGURE 4-21 Welding with a trailing angle. Larry Jeffus

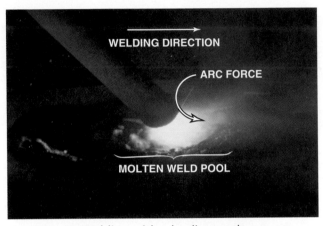

FIGURE 4-22 Welding with a leading angle. Larry Jeffus

Continue welding and change the electrode angle to a sharp leading angle. Observe the molten weld pool at the points shown in **Figure 4-22**.

During this experiment, you must maintain a constant arc length, travel speed, and weave pattern if the observations and results are to be accurate.

Cool and chip the weld. Compare the weld bead for uniformity in width, reinforcement, and appearance. Turn off the welding machine and clean up your work area when you are finished welding.

Complete a copy of the "Student Welding Report" listed in Appendix I or provided by your instructor. ◆

Electrode Manipulation

The movement or weaving of the welding electrode can control the following characteristics of the weld bead: penetration, buildup, width, porosity, undercut, overlap, and slag inclusions. The exact **weave pattern** for each weld is often the personal choice of the welder. However, some patterns are especially helpful for specific welding situations. The pattern selected for a flat (1G) butt joint is

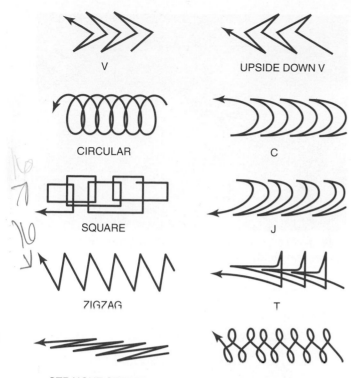

FIGURE 4-23 Weave patterns. © Cengage Learning 2012

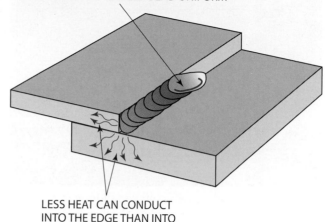

FIGURE 4-25 The "J" pattern allows the heat to be concentrated on the thicker plate. © Cengage Learning 2012

not as critical as is the pattern selection for other joints and other positions.

Many weave patterns are available for the welder to use. **Figure 4-23** shows 10 different patterns that can be used for most welding conditions.

The circular pattern is often used for flat position welds on butt, tee, and outside corner joints, and for buildup or surfacing applications. The circle can be made wider or longer to change the bead width or penetration, **Figure 4-24**.

The "C" and square patterns are both good for most 1G (flat) welds but can also be used for vertical (3G) positions. These patterns can also be used if there is a large gap to be filled when both pieces of metal are nearly the same size and thickness.

The "J" pattern works well on flat (1F) lap joints, all vertical (3G) joints, and horizontal (2G) butt and lap (2F) welds. This pattern allows the heat to be concentrated on the thicker plate, **Figure 4-25**. It also allows the reinforcement to be built up on the metal deposited during the first part of the pattern. As a result, a uniform bead contour is maintained during out-of-position welds.

The "T" pattern works well with fillet welds in the vertical (3F) and overhead (4F) positions, **Figure 4-26**. It also can be used for deep groove welds for the hot pass. The top of the "T" can be used to fill in the toe of the weld to prevent undercutting.

The straight step pattern can be used for stringer beads, root pass welds, and multiple pass welds in all positions. For this pattern, the smallest quantity of metal is molten at one time as compared to other patterns. Therefore, the weld is more easily controlled. At the same time that the electrode is stepped forward, the arc length is increased so that no metal is deposited ahead of the molten weld pool,

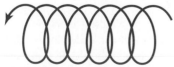

THIS WEAVE PATTERN RESULTS IN A NARROW BEAD WITH DEEP PENETRATION

THIS WEAVE PATTERN RESULTS IN A WIDE BEAD WITH SHALLOW PENETRATION

FIGURE 4-24 Changing the weave pattern width to change the weld bead characteristics. © Cengage Learning 2012

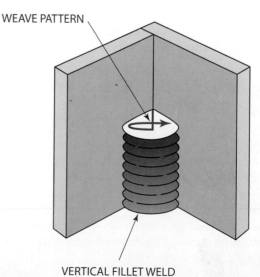

FIGURE 4-26 "T" pattern. © Cengage Learning 2012

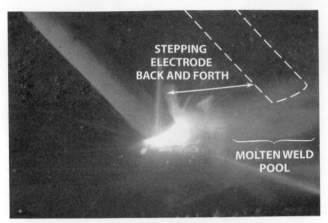

FIGURE 4-27 The electrode is moved slightly forward and then returned to the weld pool. Larry Jeffus

FIGURE 4-28 The electrode does not deposit metal or melt the base metal. Larry Jeffus

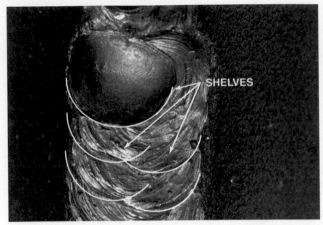

FIGURE 4-29 Using the shelf to support the molten pool for vertical welds. Larry Jeffus

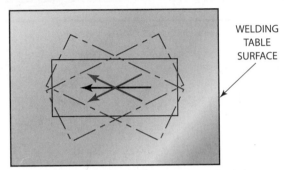

FIGURE 4-30 Change the plate angle to find the most comfortable welding position. © Cengage Learning 2012

Figure 4-27 and **Figure 4-28**. This action allows the molten weld pool to cool to a controllable size. In addition, the arc burns off any paint, oil, or dirt from the metal before it can contaminate the weld.

The figure-8 pattern and the zigzag pattern are used as cover passes in the flat and vertical positions. Do not weave more than 2 1/2 times the width of the electrode. These patterns deposit a large quantity of metal at one time. A shelf can be used to support the molten weld pool when making vertical welds using either of these patterns, **Figure 4-29**.

Positioning of the Welder and the Plate

The welder should be in a relaxed, comfortable position before starting to weld. A good position is important for both the comfort of the welder and the quality of the welds. Welding in an awkward position can cause welder fatigue, which leads to poor welder coordination and poor-quality welds. Welders must have enough freedom of movement so that they do not need to change position during a weld. Body position changes should be made only during electrode changes.

When the welding helmet is down, the welder is blind to the surroundings. Due to the arc, the field of vision of the welder is also very limited. These factors often cause the welder to sway. To stop this swaying, the welder should lean against or hold onto a stable object. When welding, even if a welder is seated, touching a stable object will make that welder more stable and will make welding more relaxing.

Welding is easier if the welder can find the most comfortable angle. The welder should be in either a seated or a standing position in front of the welding table. The welding machine should be turned off. With an electrode in place in the electrode holder, the welder can draw a straight line along the plate to be welded. By turning the plate to several different angles, the welder should be able to determine which angle is most comfortable for welding, **Figure 4-30**.

Practice Welds

Practice welds are grouped according to the type of joint and the type of welding electrode. The welder or instructor should select the order in which the welds are made. The stringer beads should be practiced first in each position before the welder tries the different joints in each position. Some time can be saved by starting with the stringer beads. If this is done, it is not necessary to cut or tack the plate together, and a number of beads can be made on the same plate.

Students will find it easier to start with butt joints. The lap, tee, and outside corner joints are all about the same level of difficulty.

Starting with the flat position allows the welder to build skills slowly so that out-of-position welds become easier to do. The horizontal tee and lap welds are almost as easy to make as the flat welds. Overhead welds are as simple to make as vertical welds, but they are harder to position. Horizontal butt welds are more difficult to perform than most other welds.

Electrodes Arc welding electrodes used for practice welds are grouped into three filler metal (F number) classes according to their major welding characteristics. The groups are E6010 and E6011, E6012 and E6013, and E7016 and E7018.

F3 E6010 and E6011 Electrodes

Both of these electrodes have **cellulose-based fluxes.** As a result, these electrodes have a forceful arc with little slag left on the weld bead.

F2 E6012 and E6013 Electrodes

These electrodes have rutile-based fluxes, giving a smooth, easy arc with a thick slag left on the weld bead.

F4 E7016 and E7018 Electrodes

Both of these electrodes have a mineral-based flux. The resulting arc is smooth and easy, with a very heavy slag left on the weld bead.

The cellulose- and rutile-based groups of electrodes have characteristics that make them the best electrodes for starting specific welds. The electrodes with the cellulose-based fluxes do not have heavy slags that may interfere with the welder's view of the weld. This feature is an advantage for flat tee and lap joints. Electrodes with the **rutile-based fluxes** (giving an easy arc with low spatter) are easier to control and are used for flat stringer beads and butt joints.

Unless a specific electrode has been required by a Welding Procedure Specification (WPS), welders can select what they consider to be the best electrode for a specific weld. Without a WPS, the welder has the final choice. An accomplished welder can make defect-free welds on all types of joints using all types of electrodes in any weld position.

Electrodes with **mineral-based fluxes** should be the last choice. Welds with a good appearance are more easily made with these electrodes, but strong welds are hard

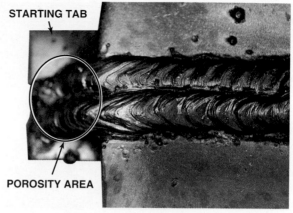
FIGURE 4-31 Porosity is found on the starting tab where it will not affect the weld. Larry Jeffus

to obtain. Without special care being taken during the start of the weld, porosity will be formed in the weld. **Figure 4-31** shows a starting tab used to prevent this porosity from becoming part of the finished weld. More information on electrode selection can be found in Chapter 27.

Stringer Beads

A straight weld bead on the surface of a plate, with little or no side-to-side electrode movement, is known as a **stringer bead.** Stringer beads are used by students to practice maintaining arc length and electrode angle so that their welds will be straight, uniform, and free from defects. Stringer beads, **Figure 4-32**, are also used to set the machine amperage and for buildup or surfacing applications. Stringer beads are the most commonly used type of bead for vertical, horizontal, and overhead welds.

The stringer bead should be straight. A beginning welder needs time to develop the skill of viewing the entire welding area. At first, the welder sees only the arc, **Figure 4-33.** With practice, the welder begins to see parts of the molten weld pool. After much practice, the welder will see the molten weld pool (front, back, and both sides), slag, buildup, and the surrounding plate, **Figure 4-34.** Often, at this skill level, the welder may not even notice the arc.

A straight weld is easily made once the welder develops the ability to view the entire welding zone. The welder will occasionally glance around to ensure that the weld is straight. In addition, it can be noted if the weld is uniform and free from defects. The ability of the welder to view the entire weld area is demonstrated by making consistently straight and uniform stringer beads.

FIGURE 4-32 Stringer bead. Larry Jeffus

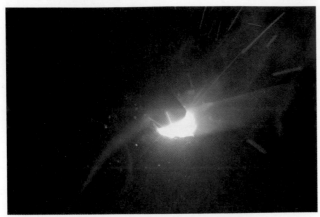

FIGURE 4-33 New welders frequently see only the arc and sparks from the electrode. Larry Jeffus

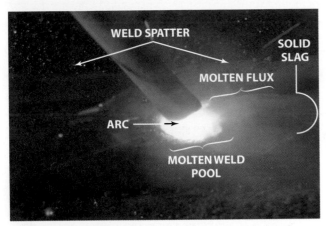

FIGURE 4-34 More experienced welders can see the molten pool, metal being transferred across the arc, and penetration into the base metal. Larry Jeffus

After making practice stringer beads, a variety of weave bead patterns should be practiced to gain the ability to control the molten weld pool when welding out of position.

PRACTICE 4-2

Straight Stringer Beads in the Flat Position Using E6010 or E6011 Electrodes, E6012 or E6013 Electrodes, and E7016 or E7018 Electrodes

Using a properly set up and adjusted arc welding machine; proper safety protection, as demonstrated in Practice 4-1; arc welding electrodes with a 1/8 in. (3 mm) diameter; and one piece of mild steel plate, 6 in. (152 mm) long and 1/4 in. (6 mm) thick; you will make straight stringer beads.

- Starting at one end of the plate, make a straight weld the full length of the plate.

- Watch the molten weld pool at this point, not the end of the electrode. As you become more skillful, it is easier to watch the molten weld pool.

- Repeat the beads with all three (F) groups of electrodes until you have consistently good beads.

- Cool, chip, and inspect the bead for defects after completing it. Turn off the welding machine and clean up your work area when you are finished welding.

Complete a copy of the "Student Welding Report" listed in Appendix I or provided by your instructor. ◆

PRACTICE 4-3

Stringer Beads in the Vertical Up Position Using E6010 or E6011 Electrodes, E6012 or E6013 Electrodes, and E7016 or E7018 Electrodes

Using the same setup, materials, and electrodes as listed in Practice 4-2, you will make vertical up stringer beads. Start with the plate at a 45° angle.

This technique is the same as that used to make a vertical weld. However, a lower level of skill is required at 45°, and it is easier to develop your skill. After the welder masters the 45° angle, the angle is increased successively until a vertical position is reached, **Figure 4-35**.

Before the molten metal drips down the bead, the back of the molten weld pool will start to bulge, **Figure 4-36**. When this happens, increase the speed of travel and the weave pattern.

Cool, chip, and inspect each completed weld for defects. Repeat the beads as necessary with all three (F) groups of electrodes until consistently good beads are obtained in this position. Turn off the welding machine and clean up your work area when you are finished welding.

Complete a copy of the "Student Welding Report" listed in Appendix I or provided by your instructor. ◆

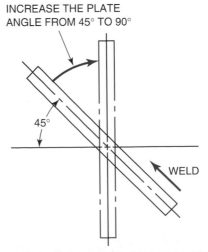

FIGURE 4-35 Once the 45° angle is mastered, the plate angle is increased successively until a vertical position (90°) is reached. © Cengage Learning 2012

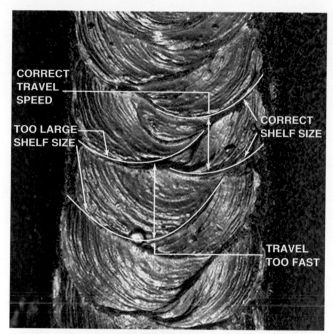

FIGURE 4-36 E7018 vertical up weld. Larry Jeffus

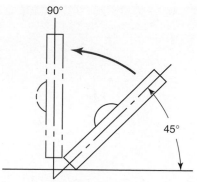

FIGURE 4-37 Change the plate angle as welding skill improves. © Cengage Learning 2012

PRACTICE 4-4

Horizontal Stringer Beads Using E6010 or E6011 Electrodes, E6012 or E6013 Electrodes, and E7016 or E7018 Electrodes

Using the same setup, materials, and electrodes as listed in Practice 4-2, you will make horizontal stringer beads on a plate.

When the welder begins to practice the horizontal stringer bead, the plate may be reclined slightly, **Figure 4-37**.

This placement allows the welder to build the required skill by practicing the correct techniques successfully. The "J" weave pattern is suggested for this practice. As the electrode is drawn along the straight back of the "J," metal is deposited. This metal supports the molten weld pool, resulting in a bead with a uniform contour, **Figure 4-38**.

Angling the electrode up and back toward the weld causes more metal to be deposited along the top edge of the weld. Keeping the bead small allows the surface tension to hold the molten weld pool in place.

Gradually increase the angle of the plate until it is vertical and the stringer bead is horizontal. Repeat the beads as needed with all three (F) groups of electrodes until consistently good beads are obtained in this position. Turn off the welding machine and clean up your work area when you are finished welding.

Complete a copy of the "Student Welding Report" listed in Appendix I or provided by your instructor. ◆

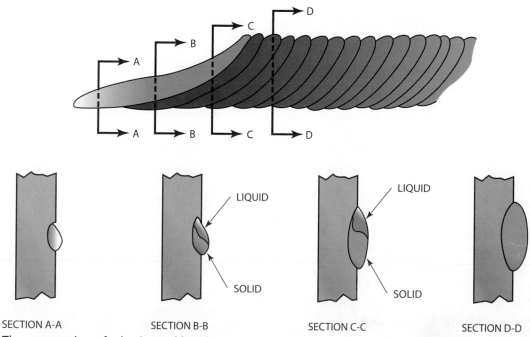

FIGURE 4-38 The progression of a horizontal bead. © Cengage Learning 2012

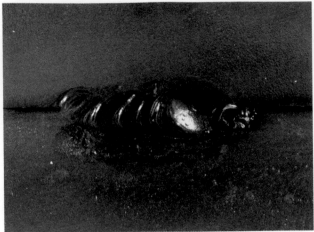

FIGURE 4-39 The tack weld should be small and uniform to minimize its effect on the final weld. © Cengage Learning 2012

FIGURE 4-41 After the plates are tack welded together, they can be forced into alignment by striking them with a hammer. © Cengage Learning 2012

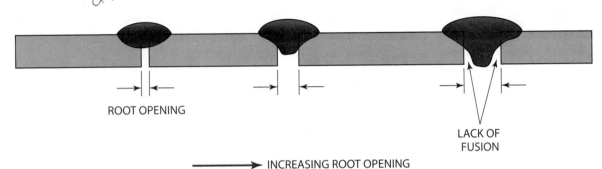

ROOT OPENING

LACK OF FUSION

INCREASING ROOT OPENING

FIGURE 4-40 Effect of root opening on weld penetration. © Cengage Learning 2012

Square Butt Joint

The **square butt joint** is made by tack welding two flat pieces of plate together, **Figure 4-39**. The space between the plates is called the root opening or root gap. Changes in the root opening will affect penetration. As the space increases, the weld penetration also increases. The root opening for most butt welds will vary from 0 in. (0 mm) to 1/8 in. (3 mm). Excessively large openings can cause burnthrough or a cold lap at the weld root, **Figure 4-40**.

After a butt weld is completed, the plate can be cut apart so it can be used for rewelding. The strips for butt welding should be no smaller than 1 in. (25 mm) wide. If they are too narrow, there will be a problem with heat buildup.

If the plate strips are no longer flat after the weld has been cut out, they can be tack welded together and flattened with a hammer, **Figure 4-41**.

PRACTICE 4-5

Welded Square Butt Joint in the Flat Position (1G) Using E6010 or E6011 Electrodes, E6012 or E6013 Electrodes, and E7016 or E7018 Electrodes

Using a properly set up and adjusted arc welding machine; proper safety protection; arc welding electrodes

having a 1/8 in. (3 mm) diameter; and two or more pieces of mild steel plate, 6 in. (152 mm) long and 1/4 in. (6 mm) thick; you will make a welded square butt joint in the flat position, **Figure 4-43**.

Tack weld the plates together and place them flat on the welding table. Starting at one end, establish a molten weld pool on both plates. Hold the electrode in the molten weld pool until it flows together, **Figure 4-44**. After the gap is bridged by the molten weld pool, start weaving the electrode slowly back and forth across the joint. Moving the electrode too quickly from side to side may result in slag being trapped in the joint, **Figure 4-45**.

Continue the weld along the 6-in. (152-mm) length of the joint. Normally, deep penetration is not required

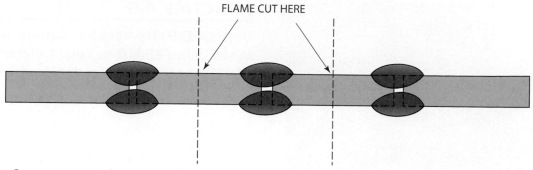

FIGURE 4-42 Conserve material; reuse practice plates when possible. © Cengage Learning 2012

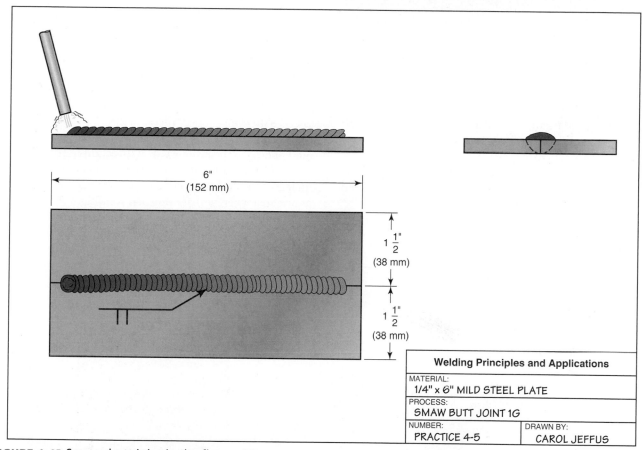

FIGURE 4-43 Square butt joint in the flat position. © Cengage Learning 2012

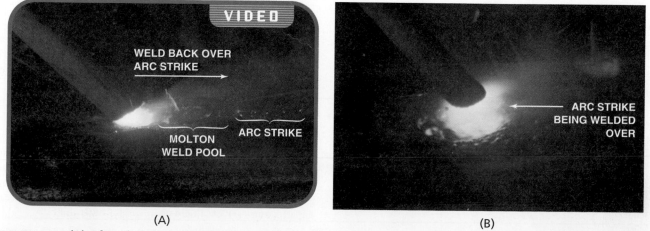

(A)   (B)

FIGURE 4-44 (A) After the arc is established, hold it in one area long enough to establish the size of molten weld pool desired. (B) Weld back over the arc strike to melt it into the weld. A & B Larry Jeffus *See DVD 1 Shielded Metal Arc Welding.*

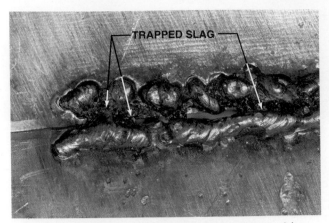

FIGURE 4-45 Moving the electrode from side to side too quickly can result in slag being trapped between the plates. Larry Jeffus

for this type of weld. If full plate penetration is required, the edges of the butt joint should be beveled or a larger-than-normal root gap should be used. Cool, chip, and inspect the weld for uniformity and soundness. Repeat the welds as needed to master all three (F) groups of electrodes in this position. Turn off the welding machine and clean up your work area when you are finished welding.

Complete a copy of the "Student Welding Report" listed in Appendix I or provided by your instructor. ◆

PRACTICE 4-6

Vertical (3G) Up-Welded Square Butt Weld Using E6010 or E6011 Electrodes, E6012 or E6013 Electrodes, and E7016 or E7018 Electrodes

Using the same setup, materials, and electrodes as listed in Practice 4-5, you will make vertical up-welded square butt joints.

With the plates at a 45° angle, start at the bottom and make the molten weld pool bridge the gap between the plates, **Figure 4-46**. Build the bead size slowly so that the molten weld pool has a shelf for support. The "C," "J," or square weave patterns work well for this joint.

As the electrode is moved up the weld, the arc is lengthened slightly so that little or no metal is deposited ahead of the molten weld pool. When the electrode is brought back into the molten weld pool, it should be lowered to deposit metal, **Figure 4-47**.

As skill is developed, increase the plate angle until it is vertical. Cool, chip, and inspect the weld for uniformity and defects. Repeat the welds with all three (F) groups of electrodes until you can consistently make welds free of defects. Turn off the welding machine and clean up your work area when you are finished welding.

Complete a copy of the "Student Welding Report" listed in Appendix I or provided by your instructor. ◆

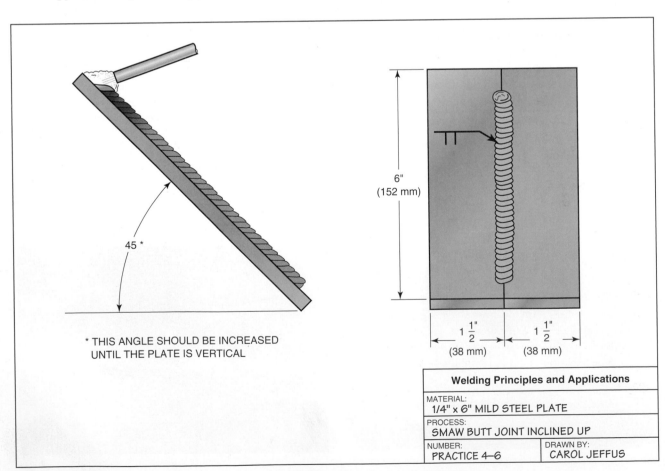

45 *

* THIS ANGLE SHOULD BE INCREASED UNTIL THE PLATE IS VERTICAL

6"
(152 mm)

1 1/2"
(38 mm)

1 1/2"
(38 mm)

Welding Principles and Applications

MATERIAL:	
1/4" x 6" MILD STEEL PLATE	
PROCESS:	
SMAW BUTT JOINT INCLINED UP	
NUMBER:	DRAWN BY:
PRACTICE 4–6	CAROL JEFFUS

FIGURE 4-46 Square butt joint in the vertical up position. © Cengage Learning 2012

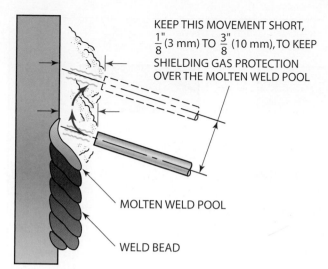

KEEP THIS MOVEMENT SHORT,
$\frac{1}{8}$" (3 mm) TO $\frac{3}{8}$" (10 mm), TO KEEP
SHIELDING GAS PROTECTION
OVER THE MOLTEN WELD POOL

MOLTEN WELD POOL

WELD BEAD

FIGURE 4-47 Electrode movement for vertical up welds.
© Cengage Learning 2012

PRACTICE 4-7

Welded Horizontal (2G) Square Butt Weld Using E6010 or E6011 Electrodes, E6012 or E6013 Electrodes, and E7016 or E7018 Electrodes

Using the same setup, materials, and electrodes as described in Practice 4-5, you will make a welded horizontal square butt joint.

- Start practicing these welds with the plate at a slight angle.
- Strike the arc on the bottom plate and build the molten weld pool until it bridges the gap.

If the weld is started on the top plate, slag will be trapped in the root at the beginning of the weld because of poor initial penetration. The slag may cause the weld to crack when it is placed in service.

The "J" weave pattern is recommended in order to deposit metal on the lower plate so that it can support the bead. By pushing the electrode inward as you cross the gap between the plates, deeper penetration is achieved.

As you acquire more skill, gradually increase the plate angle until it is vertical and the weld is horizontal.

- Cool, chip, and inspect the weld for uniformity and defects.
- Repeat the welds with all three (F) groups of electrodes until you can consistently make welds free of defects. Turn off the welding machine and clean up your work area when you are finished welding.

Complete a copy of the "Student Welding Report" listed in Appendix I or provided by your instructor. ◆

Edge Weld

An edge weld joint is made by placing the edges of the plate evenly, **Figure 4-48**. When assembling the edge

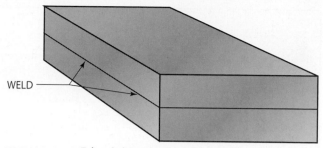

WELD

FIGURE 4-48 Edge joint. © Cengage Learning 2012

joint, the plates should be clamped tightly together; there should not be any gap between the plates. Both edges of the plate assembly can be welded. Make the tack welds to hold the plates together along the ends of the joint, **Figure 4-49**.

The size of the weld should equal the thickness of the plate being joined. A good indication that the weld is being made large enough is when the weld bead width is equal to the width of the joint, **Figure 4-50**. The weld bead should also have a slight buildup.

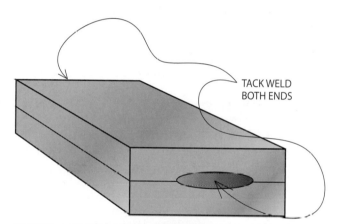

TACK WELD
BOTH ENDS

FIGURE 4-49 Make tack welds at the ends of the joint.
© Cengage Learning 2012

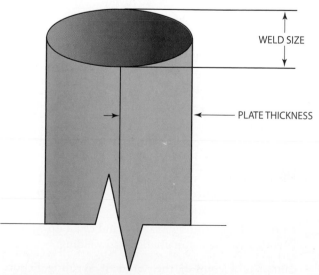

WELD SIZE

PLATE THICKNESS

FIGURE 4-50 Edge weld size. © Cengage Learning 2012

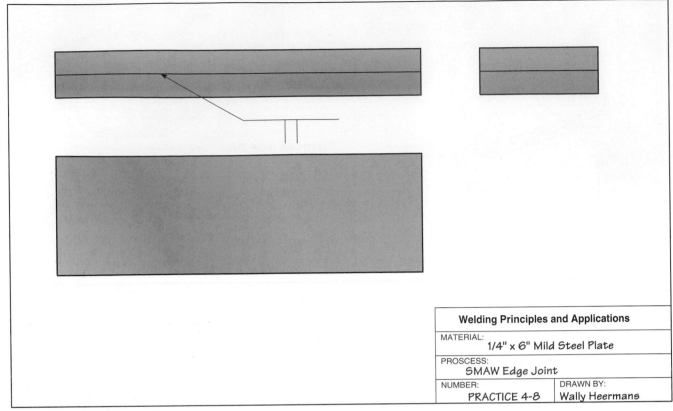

Welding Principles and Applications	
MATERIAL: 1/4" x 6" Mild Steel Plate	
PROSCESS: SMAW Edge Joint	
NUMBER: PRACTICE 4-8	DRAWN BY: Wally Heermans

FIGURE 4-51 Practice 4-8 edge joint. © Cengage Learning 2012

PRACTICE 4-8

Edge Weld in the Flat Position Using E6010 or E6011 Electrodes, E6012 or E6013 Electrodes, and E7016 or E7018 Electrodes

Using a properly set up and adjusted arc welding machine; proper safety protection, as demonstrated in Practice 4-1; arc welding electrodes with a 1/8 in. (3 mm) diameter; and two pieces of mild steel plate, 6 in. (152 mm) long and 1/4 in. (6 mm) thick; you will make a weld on an edge joint, **Figure 4-51**.

- Clamp the plates flat together and make a tack weld along each end of the plates.
- Starting at one end of the plate, make a straight weld the full length of the plate. Make the weld bead as wide as the width of the edge joint.
- Watch the molten weld pool, not the end of the electrode.
- Cool, chip, and inspect the weld for uniformity and defects.
- Repeat the welds as needed with all three (F) groups of electrodes until you can consistently make welds free of defects.
- Turn off the welding machine and clean up your work area when you are finished welding.

Complete a copy of the "Student Welding Report" listed in Appendix I or provided by your instructor. ◆

PRACTICE 4-9

Edge Joint in the Vertical Down Position Using E6010 or E6011 Electrodes, E6012 or E6013 Electrodes, and E7016 or E7018 Electrodes

Using the same setup, materials, and electrodes as listed in Practice 4-8, you will make a vertical down weld on an edge joint. Start with the plates at a 45° angle.

This technique is the same as that used to make vertical down welds. However, a lower level of skill is required at 45°, and it is easier to develop your skill. After you master the 45° angle, the angle is increased successively until a vertical position is reached, **Figure 4-52**.

- Make the weld bead as wide as the joint. Controlling a weld bead this size is more difficult, but you must develop the skill required to control this larger molten weld pool.
- Cool, chip, and inspect the weld for uniformity and defects.
- Repeat the welds as needed with all three (F) groups of electrodes until you can consistently make welds free of defects. Turn off the welding machine and clean up your work area when you are finished welding.

Complete a copy of the "Student Welding Report" listed in Appendix I or provided by your instructor. ◆

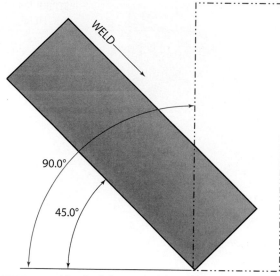

FIGURE 4-52 Vertical down weld. © Cengage Learning 2012

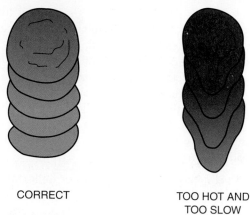

CORRECT

TOO HOT AND TOO SLOW

FIGURE 4-54 Watch the trailing edge of the weld pool to judge the correct travel speed. © Cengage Learning 2012

PRACTICE 4-10

Edge Joint in the Vertical Up Position Using E6010 or E6011 Electrodes, E6012 or E6013 Electrodes, and E7016 or E7018 Electrodes

Using the same setup, materials, and electrodes as listed in Practice 4-8, you will make a vertical up weld on an edge joint. Start with the plates at a 45° angle.

This technique is the same as that used to make vertical up welds. However, a lower level of skill is required at 45°, and it is easier to develop your skill. After you master the 45° angle, the angle is increased successively until a vertical position is reached, **Figure 4-53**.

Before the molten metal drips down the bead, the back of the molten weld pool will start to bulge, **Figure 4-54**. When this happens, increase the speed of travel and the weave pattern.

- Cool, chip, and inspect the weld for uniformity and defects.

- Repeat the welds as needed with all three (F) groups of electrodes until you can consistently make welds free of defects. Turn off the welding machine and clean up your work area when you are finished welding.

Complete a copy of the "Student Welding Report" listed in Appendix I or provided by your instructor. ◆

PRACTICE 4-11

Edge Joint in the Horizontal Position Using E6010 or E6011 Electrodes, E6012 or E6013 Electrodes, and E7016 or E7018 Electrodes

Using the same setup, materials, and electrodes as listed in Practice 4-8, you will make a horizontal weld on an edge joint. When you begin to practice the horizontal weld, the plate may be reclined slightly, **Figure 4-55**. This placement allows the welder to build the required skill by practicing the correct techniques successfully.

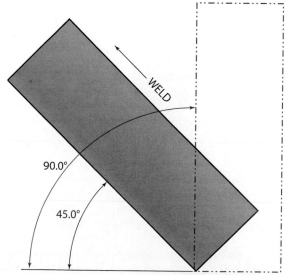

FIGURE 4-53 Vertical up weld. © Cengage Learning 2012

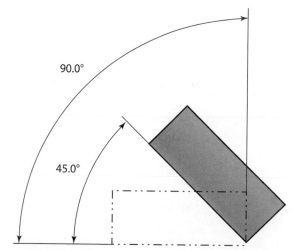

FIGURE 4-55 Incline angle. © Cengage Learning 2012

The "J" weave or stepped pattern is suggested for this practice. As the electrode is drawn back to the back edge of the weld pool, metal is deposited. Use the metal being deposited to support the molten weld pool.

Angling the electrode up and back toward the weld causes more metal to be deposited along the top edge of the weld. Keeping the bead small allows the surface tension to hold the molten weld pool in place.

Gradually increase the angle of the plate until it and the weld bead are horizontal.

- Cool, chip, and inspect the weld for uniformity and defects.

- Repeat the welds as needed with all three (F) groups of electrodes until you can consistently make welds free of defects. Turn off the welding machine and clean up your work area when you are finished welding.

Complete a copy of the "Student Welding Report" listed in Appendix I or provided by your instructor. ◆

PRACTICE 4-12

Edge Joint in the Overhead Position Using E6010 or E6011 Electrodes, E6012 or E6013 Electrodes, and E7016 or E7018 Electrodes

Using the same setup, materials, and electrodes as listed in Practice 4-8, you will make an overhead weld on an edge joint.

- With the electrode pointed in a slightly trailing angle, **Figure 4-56**, strike the arc in the joint.

- Keep a very short arc length.

- Use the stepped pattern and move the electrode forward slightly when the molten weld pool grows to the correct size, **Figure 4-57**.

As the molten weld pool gets larger it has a tendency to quickly become convex. If you keep the arc in the molten weld pool once the joint is filled and the weld face is flat, it will quickly overfill and become convex. This can result in the weld face forming drips of metal hanging from the weld like icicles, **Figure 4-58**.

- When the molten weld pool cools and begins to shrink, move the arc back near the center of the weld.

- Hold the arc in this new location until the molten weld pool again grows to the correct size.

- Step the electrode forward again and keep repeating this pattern until the weld progresses along the entire weld joint length.

- Cool, chip, and inspect the weld for uniformity and defects.

- Repeat the welds as needed with all three (F) groups of electrodes until you can consistently

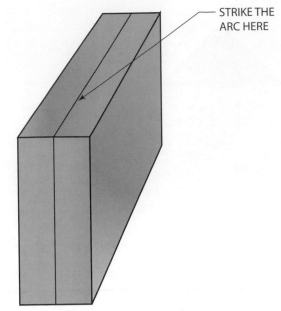

FIGURE 4-56 Strike the arc in the joint. © Cengage Learning 2012

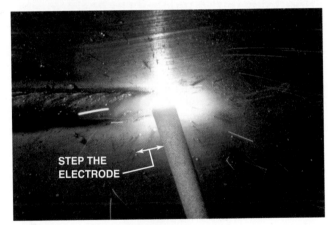

FIGURE 4-57 **Step the electrode.** Larry Jeffus

FIGURE 4-58 Welding too slow or with too high of an amperage setting will result in the weld metal dripping down like icicles. © Cengage Learning 2012

make welds free of defects. Turn off the welding machine and clean up your work area when you are finished welding.

Complete a copy of the "Student Welding Report" listed in Appendix I or provided by your instructor. ◆

Outside Corner Joint

An outside corner joint is made by placing the plates at a 90° angle to each other, with the edges forming a V-groove, **Figure 4-59**. There may or may not be a slight root opening left between the plate edges. Small tack welds should be made approximately 1/2 in. (13 mm) from both ends of the joint.

The weld bead should completely fill the V-groove formed by the plates and may have a slightly convex surface buildup. The back side of an outside corner joint can be used to practice fillet welds, or four plates can be made into a box tube shape, **Figure 4-60**.

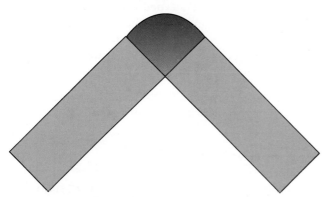

FIGURE 4-59 V formed by an outside corner joint.
© Cengage Learning 2012

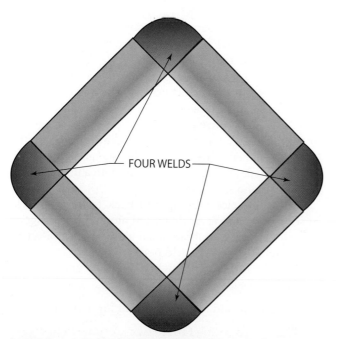

FOUR WELDS

FIGURE 4-60 Box tube made from four outside corner joint welds. © Cengage Learning 2012

PRACTICE 4-13

Outside Corner Joint in the Flat Position Using E6010 or E6011 Electrodes, E6012 or E6013 Electrodes, and E7016 or E7018 Electrodes

Using a properly set up and adjusted arc welding machine; proper safety protection, as demonstrated in Practice 4-1; arc welding electrodes with a 1/8 in. (3 mm) diameter; and two pieces of mild steel plate 6 in. (152 mm) long and 1/4 in. (6 mm) thick; you will make a weld on an outside corner joint.

- Starting at one end of the plate, make a straight weld the full length of the plate.
- Watch the molten weld pool at this point, not the end of the electrode. As you become more skillful, it is easier to watch the molten weld pool
- Cool, chip, and inspect the weld for uniformity and defects.
- Repeat the welds as needed with all three (F) groups of electrodes until you can consistently make welds free of defects. Turn off the welding machine and clean up your work area when you are finished welding.

Complete a copy of the "Student Welding Report" listed in Appendix I or provided by your instructor. ◆

PRACTICE 4-14

Outside Corner Joint in the Vertical Down Position Using E6010 or E6011 Electrodes, E6012 or E6013 Electrodes, and E7016 or E7018 Electrodes

Using the same setup, materials, and electrodes as listed in Practice 4-13, you will make a vertical down weld on an outside corner joint. Start with the plate at a 45° angle.

This technique is the same as that used to make vertical down welds. However, a lower level of skill is required at 45°, and it is easier to develop your skill. After you master the 45° angle, the angle is increased successively until a vertical position is reached, **Figure 4-61**.

- Cool, chip, and inspect the weld for uniformity and defects.
- Repeat the welds as needed with all three (F) groups of electrodes until you can consistently make welds free of defects. Turn off the welding machine and clean up your work area when you are finished welding.

Complete a copy of the "Student Welding Report" listed in Appendix I or provided by your instructor. ◆

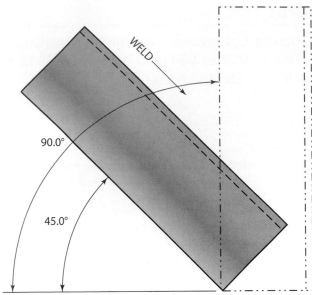

FIGURE 4-61 Start with a 45° angle and increase it to 90°. © Cengage Learning 2012

PRACTICE 4-15

Outside Corner Joint in the Vertical Up Position Using E6010 or E6011 Electrodes, E6012 or E6013 Electrodes, and E7016 or E7018 Electrodes

Using the same setup, materials, and electrodes as listed in Practice 4-13, you will make a vertical up weld on an outside corner joint. Start with the plate at a 45° angle.

This technique is the same as that used to make vertical up welds. However, a lower level of skill is required at 45°, and it is easier to develop your skill. After the welder masters the 45° angle, the angle is increased successively until a vertical position is reached, **Figure 4-62**.

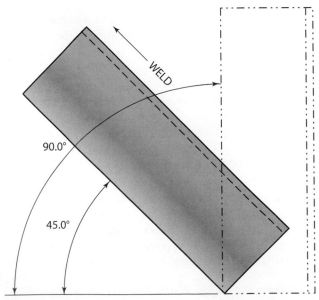

FIGURE 4-62 Vertical up weld. © Cengage Learning 2012

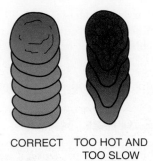

CORRECT   TOO HOT AND
TOO SLOW

FIGURE 4-63 Watch the trailing edge of the weld pool to judge the correct travel speed. © Cengage Learning 2012

Before the molten metal drips down the bead, the back of the molten weld pool will start to bulge, **Figure 4-63**. When this happens, increase the speed of travel and the weave pattern.

- Cool, chip, and inspect the weld for uniformity and defects.
- Repeat the welds as needed with all three (F) groups of electrodes until you can consistently make welds free of defects. Turn off the welding machine and clean up your work area when you are finished welding.

Complete a copy of the "Student Welding Report" listed in Appendix I or provided by your instructor. ◆

PRACTICE 4-16

Outside Corner Joint in the Horizontal Position Using E6010 or E6011 Electrodes, E6012 or E6013 Electrodes, and E7016 or E7018 Electrodes

Using the same setup, materials, and electrodes as listed in Practice 4-13, you will make a horizontal weld on an outside corner joint. When the welder begins to practice the horizontal weld, the joint may be reclined slightly, **Figure 4-64**. This placement allows the welder to build the required skill by practicing the correct techniques successfully. The "J" weave or stepped pattern is suggested for this practice. As the electrode is drawn back into the weld pool, metal is deposited. This metal supports the molten weld pool, resulting in a bead with a uniform contour, **Figure 4-65**.

Angling the electrode up and back toward the weld causes more metal to be deposited along the top edge of the weld. Keeping the bead small allows the surface tension to hold the molten weld pool in place.

Gradually increase the angle of the plate until it is vertical and the weld bead is horizontal.

- Cool, chip, and inspect the weld for uniformity and defects.
- Repeat the welds as needed with all three (F) groups of electrodes until you can consistently

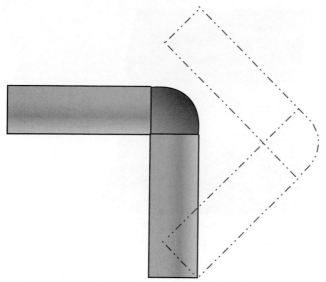

FIGURE 4-64 Incline angle. © Cengage Learning 2012

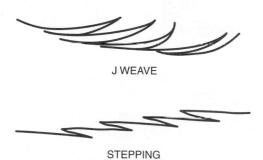

J WEAVE

STEPPING

FIGURE 4-65 "J" weave or stepping. © Cengage Learning 2012

make welds free of defects. Turn off the welding machine and clean up your work area when you are finished welding.

Complete a copy of the "Student Welding Report" listed in Appendix I or provided by your instructor. ◆

PRACTICE 4-17

Outside Corner Joint in the Overhead Position Using E6010 or E6011 Electrodes, E6012 or E6013 Electrodes, and E7016 or E7018 Electrodes

Using the same setup, materials, and electrodes as listed in Practice 4-13, you will make an overhead-welded outside corner joint.

- With the electrode pointed slightly into the joint, **Figure 4-66**, strike the arc in the joint.
- Keep a very short arc length.
- Use the stepped pattern and move the electrode forward slightly when the molten weld pool grows to the correct size, **Figure 4-67**.

As the molten weld pool gets larger, it has a tendency to quickly become convex. If you keep the arc in the

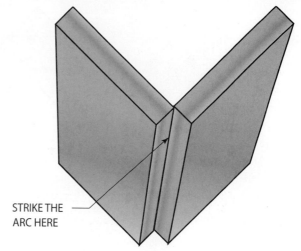

STRIKE THE ARC HERE

FIGURE 4-66 Strike arc in the joint. © Cengage Learning 2012

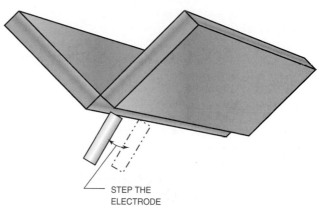

STEP THE ELECTRODE

FIGURE 4-67 Stepping the electrode to control weld size. © Cengage Learning 2012

molten weld pool once the joint is filled and the weld face is flat, it will quickly overfill and become convex. This can result in the weld face forming drips of metal hanging from the weld like icicles, **Figure 4-68**.

- When the molten weld pool cools and begins to shrink, move the arc back near the center of the weld.
- Hold the arc in this new location until the molten weld pool again grows to the correct size.
- Step the electrode forward again and keep repeating this pattern until the weld progresses along the entire weld joint length.
- Cool, chip, and inspect the weld for uniformity and defects.
- Repeat the welds as needed with all three (F) groups of electrodes until you can consistently make welds free of defects. Turn off the welding machine and clean up your work area when you are finished welding.

Complete a copy of the "Student Welding Report" listed in Appendix I or provided by your instructor. ◆

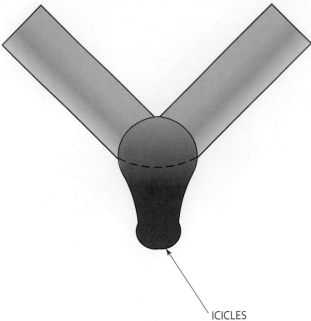

FIGURE 4-68 Welding too slowly or with too high of an amperage setting will result in the weld metal dripping down like icicles. © Cengage Learning 2012

Lap Joint

A **lap joint** is made by overlapping the edges of two plates, **Figure 4-69**. The joint can be welded on one side or both sides with a fillet weld. In Practice 4-7, both sides should be welded unless otherwise noted.

As the fillet weld is made on the lap joint, the buildup should equal the thickness of the plate, **Figure 4-70**.

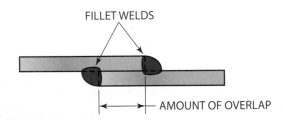

FIGURE 4-69 Lap joint. © Cengage Learning 2012

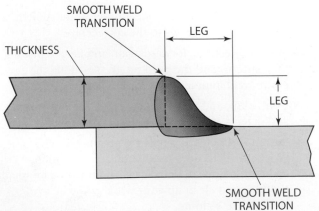

FIGURE 4-70 The legs of a fillet weld generally should be equal to the thickness of the base metal. © Cengage Learning 2012

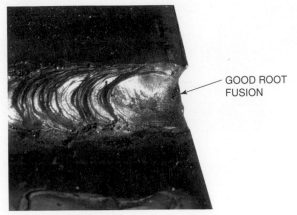

FIGURE 4-71 Watch the root of the weld bead to be sure there is complete fusion. Larry Jeffus

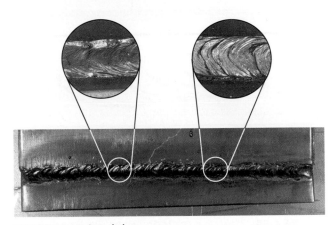

FIGURE 4-72 Lap joint. Larry Jeffus

A good weld will have a smooth transition from the plate surface to the weld. If this transition is abrupt, it can cause stresses that will weaken the joint.

Penetration for lap joints does not improve their strength; complete fusion is required. The root of fillet welds must be melted to ensure a completely fused joint. If the molten weld pool shows a notch during the weld, **Figure 4-71**, this is an indication that the root is not being fused together. The weave pattern will help prevent this problem, **Figure 4-72**.

PRACTICE 4-18

Welded Lap Joint in the Flat Position (1F) Using E6010 or E6011 Electrodes, E6012 or E6013 Electrodes, and E7016 or E7018 Electrodes

Using a properly set up and adjusted arc welding machine; proper safety protection; arc welding electrodes having a 1/8 in. (3 mm) diameter; and two or more pieces of mild steel plate, 6 in. (152 mm) long and 1/4 in. (6 mm) thick; you will make a welded lap joint in the flat position, **Figure 4-73**.

Hold the plates together tightly with an overlap of no more than 1/4 in. (6 mm). Tack weld the plates together.

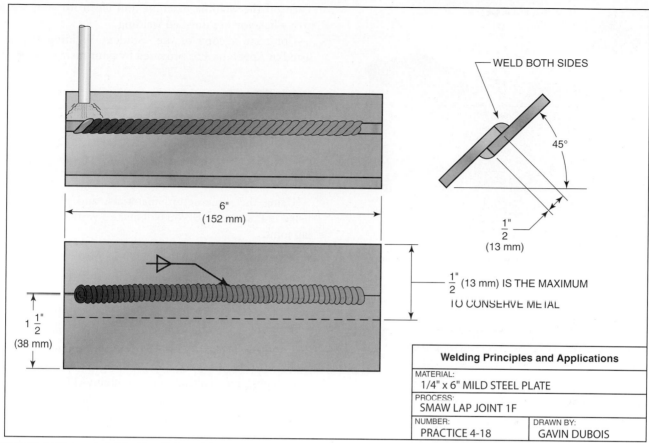

WELD BOTH SIDES

45°

$\frac{1}{2}$"
(13 mm)

$\frac{1}{2}$" (13 mm) IS THE MAXIMUM
TO CONSERVE METAL

6"
(152 mm)

$1\frac{1}{2}$"
(38 mm)

Welding Principles and Applications	
MATERIAL: 1/4" x 6" MILD STEEL PLATE	
PROCESS: SMAW LAP JOINT 1F	
NUMBER: PRACTICE 4-18	DRAWN BY: GAVIN DUBOIS

FIGURE 4-73 Lap joint in the flat position. © Cengage Learning 2012

A small tack weld may be added in the center to prevent distortion during welding, **Figure 4-74**. Chip the tacks before you start to weld.

The "J," "C," or zigzag weave patterns work well on this joint. Strike the arc and establish a molten pool directly in the joint. Move the electrode out on the bottom plate and then onto the weld to the top edge of the top plate, **Figure 4-75**. Follow the surface of the plates with the arc. Do not follow the trailing edge of the weld bead. Following the molten weld pool will not allow for good

root fusion and will cause slag to collect in the root. If slag does collect, a good weld is not possible. Stop the weld and chip the slag to remove it before the weld is completed. Cool, chip, and inspect the weld for uniformity and defects. Repeat the welds with all three (F) groups of electrodes until you can consistently make welds free of defects. Turn off the welding machine and clean up your work area when you are finished welding.

Complete a copy of the "Student Welding Report" listed in Appendix I or provided by your instructor. ◆

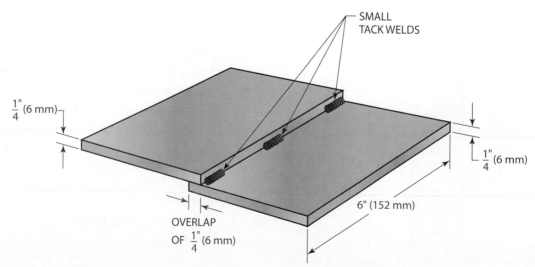

SMALL
TACK WELDS

$\frac{1}{4}$" (6 mm)

$\frac{1}{4}$" (6 mm)

6" (152 mm)

OVERLAP
OF $\frac{1}{4}$" (6 mm)

FIGURE 4-74 Tack welding the plates together. © Cengage Learning 2012

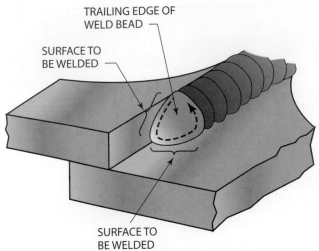

TRAILING EDGE OF
WELD BEAD

SURFACE TO
BE WELDED

SURFACE TO
BE WELDED

FIGURE 4-75 Follow the surface of the plate to ensure good fusion. © Cengage Learning 2012

PRACTICE 4-19

Welded Lap Joint in the Horizontal Position (2F) Using E6010 or E6011 Electrodes, E6012 or E6013 Electrodes, and E7016 or E7018 Electrodes

Using the same setup, materials, and electrodes as listed in Practice 4-18, you will make a welded horizontal lap joint.

The horizontal lap joint and the flat lap joint require nearly the same technique and skill to achieve a proper weld, **Figure 4-76.** Use the "J," "C," or zigzag weave patterns to make the weld. Do not allow slag to collect in the root. The fillet must be equally divided between both plates for good strength. After completing the weld, cool, chip, and inspect the weld for uniformity and defects. Repeat the welds using all three (F) groups of electrodes until you can consistently make welds free of defects.

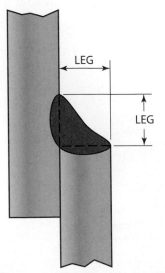

LEG

LEG

FIGURE 4-76 The horizontal lap joint should have a fillet weld that is equal on both plates. © Cengage Learning 2012

Turn off the welding machine and clean up your work area when you are finished welding.

Complete a copy of the "Student Welding Report" listed in Appendix I or provided by your instructor. ◆

PRACTICE 4-20

Lap Joint in the Vertical Position (3F) Using E6010 or E6011 Electrodes, E6012 or E6013 Electrodes, and E7016 or E7018 Electrodes

Using the same setup, materials, and electrodes as listed in Practice 4-18, you will make a vertical up-welded lap joint.

- Start practicing this weld with the plate at a 45° angle.
- Gradually increase the angle of the plate to vertical as skill is gained in welding this joint. The "J" or "T" weave patterns work well on this joint.
- Establish a molten weld pool in the root of the joint.
- Use the "T" pattern to step ahead of the molten weld pool, allowing it to cool slightly. Do not deposit metal ahead of the molten weld pool.
- As the molten weld pool size starts to decrease, move the electrode back down into the molten weld pool.
- Quickly move the electrode from side to side in the molten weld pool, filling up the joint.
- Cool, chip, and inspect the weld for uniformity and defects.
- Repeat the welds as necessary with all three (F) groups of electrodes until you can consistently make welds free of defects. Turn off the welding machine and clean up your work area when you are finished welding.

Complete a copy of the "Student Welding Report" listed in Appendix I or provided by your instructor. ◆

Practice 4-21

Lap Joint in the Overhead Position (4F) Using E6010 or E6011 Electrodes, E6012 or E6013 Electrodes, and E7016 or E7018 Electrodes

Using the same setup, materials, and electrodes as listed in Practice 4-18, you will make an overhead-welded lap joint.

- With the electrode pointed slightly into the joint, **Figure 4-77,** strike the arc in the inside corner of the lap joint.
- Keep a very short arc length.
- Use the stepped pattern and move the electrode forward slightly when the molten weld pool grows to the correct size, **Figure 4-78.**

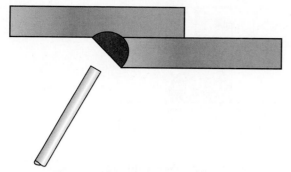

FIGURE 4-77 Point the electrode slightly toward the root of the joint. © Cengage Learning 2012

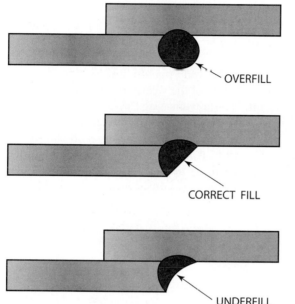

FIGURE 4-78 Correct fillet weld size for overhead welds. © Cengage Learning 2012

As the molten weld pool gets larger it has a tendency to quickly become convex. If you keep the arc in the molten weld pool once the joint is filled and the weld face is flat, it will quickly overfill and become convex. This can result in the weld face forming drips of metal hanging from the weld like icicles, **Figure 4-79**.

- When the molten weld pool cools and begins to shrink, move the arc back near the center of the weld.

- Hold the arc in this new location until the molten weld pool again grows to the correct size.

- Step the electrode forward again and keep repeating this pattern until the weld progresses along the entire weld joint length.

- Cool, chip, and inspect the weld for uniformity and defects.

- Repeat the welds as needed with all three (F) groups of electrodes until you can consistently make welds free of defects. Turn off the welding machine and clean up your work area when you are finished welding.

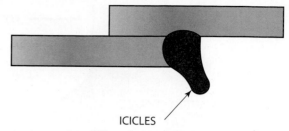

FIGURE 4-79 Overfilling the molten weld pool will result in drips of metal called icicles. © Cengage Learning 2012

Complete a copy of the "Student Welding Report" listed in Appendix I or provided by your instructor. ◆

Tee Joint

The **tee joint** is made by tack welding one piece of metal on another piece of metal at a right angle, **Figure 4-80**. After the joint is tack welded together, the slag is chipped from the tack welds. If the slag is not removed, it will cause a slag inclusion in the final weld.

The heat is not distributed uniformly between both plates during a tee weld. Because the plate that forms the stem of the tee can conduct heat away from the arc in only one direction, it will heat up faster than the base plate. Heat escapes into the base plate in two directions. When using a weave pattern, most of the heat should be directed to the base plate to keep the weld size more uniform and to help prevent an undercut.

A welded tee joint can be strong if it is welded on both sides, even without having deep penetration, **Figure 4-81**. The weld will be as strong as the base plate if the size of the two welds equals the total thickness of the base plate.

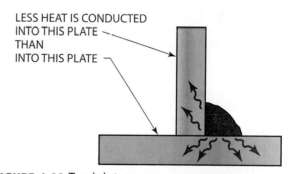

FIGURE 4-80 Tee joint. © Cengage Learning 2012

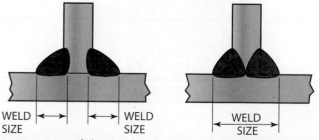

FIGURE 4-81 If the total weld sizes are equal, then both tee joints would have equal strength. © Cengage Learning 2012

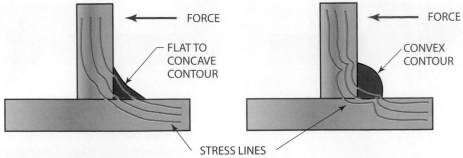

FIGURE 4-82 The stresses are distributed more uniformly through a flat or concave fillet weld. © Cengage Learning 2012

The weld bead should have a flat or slightly concave appearance to ensure the greatest strength and efficiency, **Figure 4-82**.

PRACTICE 4-22

Tee Joint in the Flat Position (1F) Using E6010 or E6011 Electrodes, E6012 or E6013 Electrodes, and E7016 or E7018 Electrodes

Using a properly set up and adjusted arc welding machine; proper safety protection; arc welding electrodes having a 1/8 in. (3 mm) diameter; and two or more pieces of mild steel plate, 6 in. (152 mm) long and 1/4 in. (6 mm) thick; you will make a welded tee joint in the flat position, **Figure 4-83**.

After the plates are tack welded together, place them on the welding table so the weld will be flat. Start at one end and establish a molten weld pool on both plates. Allow the molten weld pool to flow together before starting the bead. Any of the weave patterns will work well on this joint. To prevent slag inclusions, use a slightly higher-than-normal amperage setting.

When the 6-in. (152-mm)-long weld is completed, cool, chip, and inspect it for uniformity and soundness. Repeat the welds as needed for all these groups of electrodes until you can consistently make welds free of

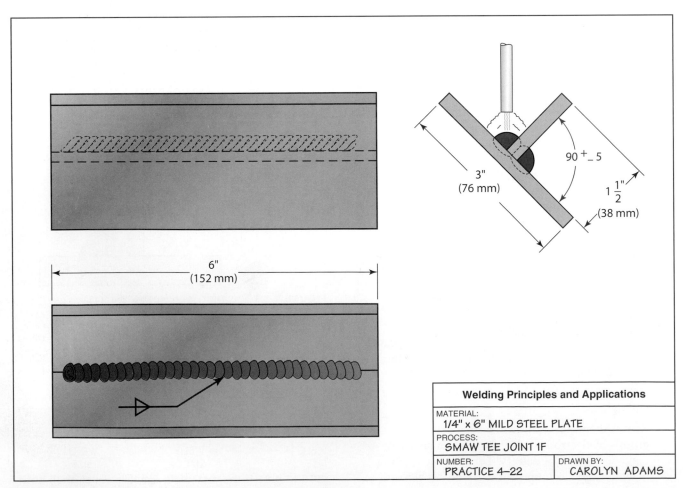

Welding Principles and Applications

MATERIAL:	
1/4" x 6" MILD STEEL PLATE	
PROCESS:	
SMAW TEE JOINT 1F	
NUMBER:	DRAWN BY:
PRACTICE 4–22	CAROLYN ADAMS

FIGURE 4-83 Tee joint in the flat position. © Cengage Learning 2012

defects. Turn off the welding machine and clean up your work area when you are finished welding.

Complete a copy of the "Student Welding Report" listed in Appendix I or provided by your instructor. ◆

PRACTICE 4-23

Tee Joint in the Horizontal Position (2F) Using E6010 or E6011 Electrodes, E6012 or E6013 Electrodes, and E7016 or E7018 Electrodes

Using the same setup, materials, and electrodes as listed in Practice 4-22, you will make a welded tee joint in the horizontal position.

Place the tack welded tee plates flat on the welding table so that the weld is horizontal and the plates are flat and vertical, **Figure 4-84**. Start the arc on the flat plate and establish a molten weld pool in the root on both plates. Using the "J" or "C" weave patterns, push the arc into the root and slightly up the vertical plate. You must keep the root of the joint fusing together with the weld metal. If the metal does not fuse, a notch will appear on the leading edge of the weld bead. Poor or incomplete root fusion will cause the weld to be weak and easily cracked under a load.

When the weld is completed, cool, chip, and inspect it for uniformity and defects. Undercut on the vertical plate is the most common defect. Repeat the welds with all three (F) groups of electrodes until you can consistently make welds free of defects. Turn off the welding machine and clean up your work area when you are finished welding.

Complete a copy of the "Student Welding Report" listed in Appendix I or provided by your instructor. ◆

PRACTICE 4-24

Tee Joint in the Vertical Position (3F) Using E6010 or E6011 Electrodes, E6012 or E6013 Electrodes, and E7016 or E7018 Electrodes

Using the same setup, materials, and electrodes as listed in Practice 4-22, you will make a welded tee joint in the vertical position.

Practice this weld with the plate at a 45° angle. This position will allow you to develop your skill for the vertical position. Start the arc and molten weld pool deep in

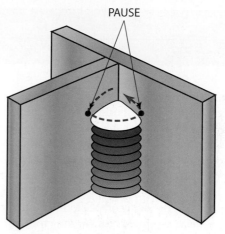

FIGURE 4-85 Pausing just above the undercut will fill it. This action also causes undercut, but that will be filled on the next cycle. © Cengage Learning 2012

the root of the joint. Build a shelf large enough to support the bead as it progresses up the joint. The square, "J," or "C" patterns can be used, but the "T" or stepped patterns will allow deeper root penetration.

For this weld, undercut is a problem on both sides of the weld. It can be controlled by holding the arc on the side long enough for filler metal to flow down and fill it, **Figure 4-85**. Cool, chip, and inspect the weld for uniformity and defects. Repeat the welds as necessary with all three (F) groups of electrodes until you can consistently make welds free of defects. Turn off the welding machine and clean up your work area when you are finished welding.

Complete a copy of the "Student Welding Report" listed in Appendix I or provided by your instructor. ◆

PRACTICE 4-25

Tee Joint in the Overhead Position (4F) Using E6010 or E6011 Electrodes, E6012 or E6013 Electrodes, and E7016 or E7018 Electrodes

Using the same setup, materials, and electrodes as listed in Practice 4-12, you will make a welded tee joint in the overhead position.

Start the arc and molten weld pool deep in the root of the joint. Keep a very short arc length. The stepped pattern will allow deeper root penetration.

For this weld, undercut is a problem on both sides of the weld with a high buildup in the center. It can be controlled by holding the arc on the side long enough for filler metal to flow in and fill it. Cool, chip, and inspect the weld for uniformity and defects. Repeat the welds as necessary with all three (F) groups of electrodes until you can consistently make welds free of defects. Turn off the welding machine and clean up your work area when you are finished welding.

Complete a copy of the "Student Welding Report" listed in Appendix I or provided by your instructor. ◆

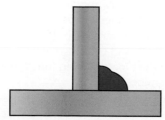

FIGURE 4-84 Horizontal tee. © Cengage Learning 2012

Summary

The shielded metal arc welding process is most often referred to in welding shops as stick welding. Some people say that it gets this name for one of two reasons. The first is most obviously as a result of the stick shape of the electrode. The second reason is experienced by all new welders; it is the tendency for the electrode to stick to the workpiece. All new welders experience this, and your ability to control the sticking of the electrode can be improved as you develop the proper arc-striking techniques.

For a new welder, it is often difficult to concentrate on anything other than the bright sparks and glow at the end of the electrode. But, with time, as you develop your skills, your visual field will increase, allowing you to see a much larger welding zone. This skill comes with time and practice. Developing this skill is essential for you to become a highly proficient welder. Nothing enhances your welding skills more than time under the hood, actually welding, cleaning off the weld, inspecting it, determining the necessary corrections to be made, and immediately trying to produce the next weld with a higher level of quality.

Keeping Shipshape through Underwater Welding

A Miami-based diving contractor helps customers avoid unscheduled dry docks by making top-quality underwater welding repairs.

Just like a bus, train, or airplane, a ship follows a time schedule. Any disruption to that schedule translates into thousands of dollars or more in lost revenue because cargo is not arriving on time or passengers cannot take their cruises. When underwater welding repairs are needed to get a vessel back on schedule, one of the companies that ship owners turn to is Miami Diver, Inc.

Based in Miami, Florida, the company specializes in underwater ship maintenance, including repairs, husbandry, and surveys. Six of their twelve Miami employees are diver/welders.

Being able to work "within the ship's schedule is actually the primary reason we're called in," said Kevin Peters, president of Miami Diver. That means the company's employees travel all over the world to repair ships, and, at times, if a ship can continue underway, stay aboard and work whenever it reaches a port of call.

Miami Diver performs dry repairs using hyperbaric welding chambers and closed cofferdams, as well as wet welding repairs. The welders must work not only to the requirements of ANSI/AWS D3.6M:1999 Specification for Underwater Welding, but also to those of the various ship classification societies.

Achieving "Surface-Quality" Wet Welds

While underwater wet welding offers advantages, such as speed, versatility, and cost effectiveness, the ship classification societies consider wet welds temporary repairs, that must be redone in dry dock.

AWS D3.6 divides underwater wet welds into four classes—A, B, C, and O—based on varying sets of required properties defined by mechanical tests, surface appearance, and nondestructive examination requirements. The specification defines the four classes as follows:

- Class A welds "are intended to be suitable for applications and design stresses comparable to their conventional surface welding counterparts by virtue of specifying comparable properties and testing requirements."
- Class B welds "are intended for less critical applications where lower ductility, moderate porosity, and other limited discontinuities can be tolerated."
- Class C welds "need only satisfy lesser requirements than class A, B, and O and are intended for applications where the load-bearing function is not a primary consideration."
- Class O underwater welds must also meet the requirements of another code or specification.

Two years ago, in anticipation of possible changes in the American Bureau of Shipbuilding's regulations and believing the company could provide added value to its customers by qualifying its welders to the requirements of the D3.6 standard, Peters began researching wet welding electrodes. While certain that he and the other welders could meet the B and C class requirements, Peters was not as optimistic about meeting the requirements of the all-weld-metal tensile test for A-class welds. The specification requires a minimum of 14% elongation. Less elongation can result in a lack of weld ductility.

Peters purchased quantities of each brand of underwater welding electrode. Welds were made in the training tank at the company's Miami office, then subjected to visual and destructive testing. "In the end, we found one electrode we had never used before called Hydro-Weld FS," Peters recalled. "We found that electrode exceeded any of our expectations as far as elongation."

A transition to the new electrode proved somewhat difficult, however. The electrode the company had been using was equivalent to an E7014 shielded metal arc welding electrode, while the new electrode was an E6013 equivalent. "Even topside, those have different arcs," Peters said. "When you're accustomed to a certain rod for so long, to get [the welders] to switch was somewhat of an obstacle."

Intensive Training

As they practiced with the new electrode, the welders found they were not getting enough penetration. Although adjustments were made following a series of phone calls between Miami and the United Kingdom, where Hydro-Weld is headquartered, the problem was not solved. Peters decided to bring in experts from the electrode manufacturer and conduct a training class.

Joining forces with Trident BV and Cores Diving, Peters shipped over a pallet of electrodes for the class. The company's goal was not only to become proficient in using the new electrode but also to become the first diving contractors in the underwater ship repair industry to offer "surface-quality, structural wet welds" that could be accepted as permanent wet welding repairs. Hydro-Weld wrote the wet welding procedure specifications the welders would follow and conducted the 10-day long training program.

Involving the Classification Societies

Miami Diver brought in representatives from the six largest ship classification societies—American Bureau of Shipping, Lloyds Register of Shipping, Det Norske Veritas, Bureau Veritas, Rina, and Germanischer Lloyd—to witness the qualification of the welding procedures and welder qualifications. An independent laboratory was hired to perform tests on the weld coupons.

Three positions were used to produce the coupons: 2F horizontal, 3F vertical, and 4F overhead. Groove weld

The pad eyes that will hold this cofferdam to the side of the ship were welded using A-class wet welding procedures. As a result, the welds were classified as a permanent repair, which will not have to be reworked when the vessel goes into dry dock. Miami Diver, Inc.

specimens for the Charpy impact and all-weld-metal tensile tests as well as a longitudinal fillet weld shear strength test specimen were produced.

In the end, the classification societies confirmed that the specimens complied with, or exceeded, the requirements of AWS D3.6M for class-A welds.

"When the whole thing was over, it was probably in the neighborhood of $75,000, $80,000, to run that course [because] you have to take that many men off hire and pay the manufacturer," Peters said. "The consumables alone were in excess of $10,000. Then [you have to pay] the class societies."

As a result, some wet welds previously considered temporary are now deemed permanent. The customer can also avoid unscheduled dry docks, even for repairs that continue to be classified as temporary. "You can get a temporary repair that will allow the vessel to trade on its normal charter until the next scheduled dry dock," Peters said.

Underwater Welding Jobs

The A-class weld procedures proved useful during a recent repair job in the Indian Ocean. The vessel, a semi-submersible drill rig modified into a platform from which rockets launch satellites, is paired with a command ship. The platform's dry-dock schedule is once every 10 years. The owner tries to avoid unscheduled dry docks because the vessel is so large the only available dry dock is in Asia. Moving the vessel there and back is extremely expensive.

A problem developed in the bow thrusters, an integral part of the platform's dynamic positioning system that limited its ability to pitch. The only way to fix the problem was in a dry environment, so workers from Miami Diver West (the California subsidiary of Miami Diver) welded 13-ft × 13-ft cofferdams, which could be attached to the side of the ship.

Jim Allen, president of Miami Diver West, explained they built the cofferdam large enough so that, if necessary, the entire bow thruster could be disassembled, removed, and replaced with a new one. Using an enclosed cofferdam to

The bolts are underwater wet welds that held the gagging plates to the housing of the propulsion pod.
Miami Diver, Inc.

A diver/welder welding the gagging plates to the Azipod.
Miami Diver, Inc.

create a dry welding environment has been done many times before, Allen said, but this one was different because they built interchangeable feet, which would allow it to meet the curvature of either the platform or the command ship. "That way we can utilize the same box for different repairs," Allen said.

The A-class wet welding procedures were key to the repair, Allen said, because Det Norske Veritas classified the wet welds used to join the 12 pad eyes that would hold the cofferdam to the hull as permanent. Classified as permanent, the welds did not need to be reworked in dry dock. Once the cofferdam was pumped dry, the pad eyes needed to hold 50 tons to the hull in a watertight seal.

"Obviously, the welds have been proven now that they've actually held the box under," Peters said.

Securing a Propulsion System

The company also followed the A-class wet welding procedures when it performed temporary repairs on a cruise ship stuck at the Port of Miami dock.

A leak had developed in the starboard Azipod, the part of a pod propulsion system that eliminates the need for a rudder or stern thrusters, which hangs below the vessel and can rotate 360°. Because of the leak, the ship needed to go to Newport News, Virginia, to be placed into dry dock for repair. While the ship could run on its own power using the port pod, any movement of the propeller blades inside the leaking pod could damage its armature and the ship's electric motor. Time was critical because the cruise line estimated each day the ship remained off hire cost the company $1 million in lost revenue.

It was first suggested that the propeller blades be removed, but no one was sure how long that would take. Miami Diver came up with what it thought was a better solution. "We told them, 'We have wet weld procedures with Lloyds, under A class,'" Peters recalled. "'We'll make gagging plates and wet weld them to the Azipod—to the outer housing of the pod—and we'll prevent the prop from rotating.'" The A-class wet welding procedures helped sell the concept, Peters said, because the classification society had already approved those procedures. The cruise line agreed to the proposal, as did the pod manufacturer, provided the welds could pass a 50-ton load test. The manufacturer was concerned because the pod was under warranty.

"We removed two of the main fastenings to the hub and we fabricated new bolts that were studs," Peters explained. "This enabled us to lay a 1 1/2-in. plate over the hub of the propeller and bolt it to the propeller. Then the remaining part of the plate, approximately 1 ft, crossed over the pod housing, and it was welded with a full 1-in. by 2-ft long fillet weld."

The company brought in four welders from Los Angeles, welders from Holland, and part of its Miami-based crew for the project. They worked around the clock—four hours at a time in the water—to complete the job. Much of that time was spent in preparation, such as cutting plates, machining the bolts, and removing an epoxy coating from the pods.

In all, they fit and welded four gagging plates onto the housing of the Azipod. "The load test was not even an issue with us," Peters said. "We put four plates on, but we could have held it with one plate." Once the propeller blades could no longer move, the cruise ship successfully made its journey to Newport News.

Article courtesy of the American Welding Society.

Review

1. Describe two methods of striking an arc with an electrode.

2. Why is it important to strike the arc only in the weld joint?

3. What problems may result by using an electrode at too low a current setting?

4. What problems may result by using an electrode at too high a current setting?

5. According to Table 4-1, what would the amperage range be for the following electrodes?

   a. 1/8 in. (3.2 mm),

   b. 5/32 in. (4 mm),

   E6010   E7016

   c. 3/32 in. (2.4 mm),

   d. 1/8 in. (3.2 mm),

   E7018   E6011

6. What makes some spatter "hard?"

7. Why should you never change the current setting during a weld?

8. What factors should be considered when selecting an electrode size?

9. What can a welder do to control overheating of the metal pieces being welded?

10. What effect does changing the arc length have on the weld?

11. What arc problems can occur in deep or narrow weld joints?

12. Describe the difference between using a leading and a training electrode angle.

13. Can all electrodes be used with a leading angle? Why or why not?

14. What characteristics of the weld bead does the weaving of the electrode cause?

15. What are some of the applications for the circular pattern in the flat position?

16. Using a pencil and paper, draw two complete lines of the weave patterns you are most comfortable making.

17. Why is it important to find a good welding position?

18. Which electrodes would be grouped in the following F numbers: F3, F2, F4?

19. Give one advantage of using electrodes with cellulose-based fluxes.

20. What are stringer beads?

21. Describe an ideal tack weld.

22. What effect does the root opening or root cap have on a butt joint?

23. What can happen if the fillet weld on a lap joint does not have a smooth transition?

24. Which plate heats up faster on a tee joint? Why?

25. Can a tee weld be strong if the welds on both sides do not have deep penetration? Why or why not?

Chapter 5

Shielded Metal Arc Welding of Pipe

OBJECTIVES

After completing this chapter, the student should be able to

- discuss three general categories of pipe welds, including how they are used and what type of weld root penetration and strength they require.
- compare pipe to tubing.
- discuss the advantages of welded pipe.
- discuss the preparation needed before welding pipe.
- explain the importance of not having arc strikes outside of the weld groove on pipe welds.
- explain the purpose of a hot pass.
- describe the purpose of the root, filler, and cover passes for a pipe weld.
- name advantages of the horizontal rolled pipe position.
- describe the vertical fixed position and give advantages and disadvantages.
- discuss how to make a weld in the horizontal fixed position.
- describe the 45° fixed inclined position.

KEY TERMS

concave root surface	land	root suck back
fixed inclined position	pipe	tubing
horizontal fixed pipe position	pressure range	vertical fixed pipe position
horizontal rolled pipe position	root face	welding uphill or downhill
icicles	root gap	

INTRODUCTION

The pipe welder is considered by some other welders to be one of the most skilled welders in the industry. Often pipe welders share a great deal of pride. Some finished welded piping systems are considered works of art. Mastering the skills required to be a master pipe welder often takes a large commitment on the part of the welding student, but this commitment is well rewarded by the industry. The rewards of being a quality pipe welder include earning better pay, working with the best equipment, and often having a helper to do the less glamorous jobs of cleanup and setup.

FIGURE 5-1 Refineries use miles of welded pipe.
Larry Jeffus

The shielded metal arc welding process is very well suited to the fabrication and repair of piping systems. Welded steel pipe is used in factories, power plants, refineries, and buildings to carry liquids and gases for a variety of purposes, **Figure 5-1**. Pipe is used to carry such materials as water, steam, chemicals, gases, petroleum, radioactive materials, and many others. Oil and gas are distributed through cross-country piping systems all over the United States, from Texas, California, Maine, and the northern slopes of Alaska. Welded pipe is used throughout ships, planes, and spacecraft to carry such liquids as fuels and hydraulic fluids. Pipe and tubing are also used for structures such as handrails, columns in buildings, light posts, and bicycle and motorcycle frames, as well as many other items we come in contact with each day.

The purpose for which pipe will be used largely determines how it is welded. This text groups pipe welds into the following three general categories:

- Low-pressure or light structural service
- Medium-pressure or medium structural service
- High-pressure or heavy structural service

Low-pressure, or noncritical, piping systems may be used to carry water, noncorrosive or noncombustible chemicals, and other nonhazardous materials used in industry. These types of noncritical piping assemblies are also found on such structural items as handrails, truck racks, columns for buildings, and other light-duty products. These pipe joints must be free from such defects as pinholes, undercut, slag inclusions, or any other defect that may cause the joints to leak or break prematurely. The weld does not require 100% penetration, although penetration should be uniform. Much of the strength of these pipe joints comes from the reinforcement. These welds should always be located so that they are easily repairable if necessary.

Medium-pressure piping is used for low-pressure steam heat, corrosive or flammable chemicals, waste disposal, ship plumbing, and medium- to heavy-service structural items such as highway signs, railings or light posts, trailer axles, and equipment frames or stands. These pipe joints must withstand heavy loads, but their failure will not be disastrous. Much of their strength is due to weld reinforcement, but there should be 100% root penetration around most of the joint. Often the root pass has been welded with E6010 or E6011 electrodes, and the other passes were welded with E7018 electrodes. But changes in the E7018 fluxes have enabled them to be used on the open root pass, so it is now common to make all of the weld passes with E7018.

High-pressure, or critical, piping systems are used for high-pressure steam, radioactive materials, the Alaskan pipeline, fired or unfired boilers, refinery reactor lines, aircraft airframes, offshore oil-rig jackets (legs), motorcycle frames, race car roll cages, truck axles, and several other critical, heavy-duty applications. The welds on critical piping systems must be as strong or stronger than the pipe itself. Often, the pipe used for these applications is extra heavy-duty pipe, with heavier wall thicknesses. The weld must have 100% root penetration over 100% of the joint. Root, filler, and reinforcement weld passes are made with an E7018 or stronger electrode. The welds are usually tested, and defects are repaired.

Pipe and Tubing

Although **pipe** and **tubing** are similar in some aspects, they have different types of specifications and uses. They are only sometimes interchangeable.

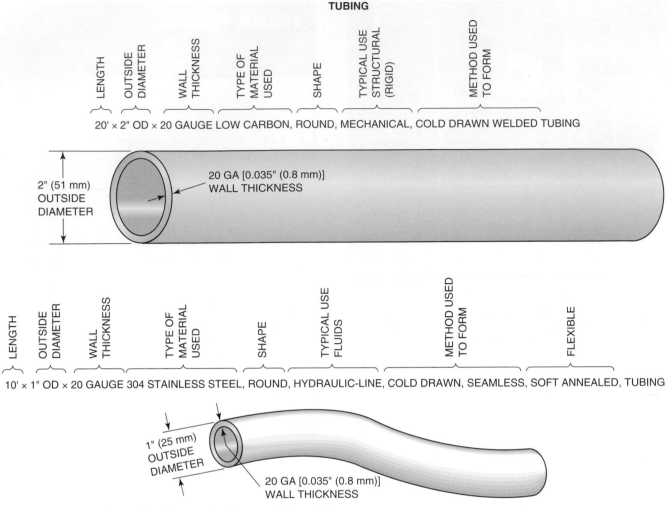

FIGURE 5-2 Typical specifications used when ordering tubing. © Cengage Learning 2012

The specifications for pipe sizes are given as the inside diameter for pipe that is 12 in. (305 mm) in diameter or smaller, and as the outside diameter for pipe larger than 12 in. (305 mm) in diameter. Tubing sizes are always given as the outside diameter. The desired shape of tubing, such as square, round, or rectangular, must also be listed with the ordering information, **Figure 5-2**.

The wall thickness of tubing is measured in inches (millimeters) or as U.S. standard sheet metal gauge thickness. The wall thickness for pipe is determined by its schedule, or **pressure range.** The larger the diameter of the pipe, the greater its area. As the area increases, so must the wall thickness in order for the wall to withstand the same pressure range, **Figure 5-3**.

The strength of pipe is given as a schedule. Schedules 10 through 180 are available; schedule 40 is often considered a standard strength. Tubing strength is the ability of tubing to withstand compression, bending, or twisting loads. Tubing should also be specified as rigid or flexible.

Pipe and tubing are both available as welded (seamed) or extruded (seamless).

Most pipe that will be shelded into a system is used to carry liquids or gases from one place to another. These systems may be designed to carry large or small quantities

or materials having a wide range of pressures. Small diameter pipe may be used for some structural applications, but usually only on a limited scale.

Small diameter flexible tubing is commonly used to carry pressurized liquids or gases. Ridged tubing is normally used for structural applications. Some tubing is designed for specific purposes, such as electrical mechanical tubing (EMT), which is used to protect electrical wiring. Tubing can be used to replace some standard structural shapes such as I beams, channels, and angles for buildings. Tubing is also available in sizes that will slide one inside the other to be used in places where telescoping tubing is required, **Figure 5-4**.

In this chapter, the term *pipe* will refer to pipe only. However, it should be understood that the welding sequence, procedures, and skills can also be used on thick-wall round tubing.

Advantages of Welded Pipe Most pipe 1 1/2 in. (38 mm) in diameter and all steel pipe 2 in. (51 mm) and larger are generally arc welded. Welded piping systems, compared to pipe joined by any other method, are stronger, require less maintenance, last for longer periods of time, allow smoother flow, and weigh less.

PIPE

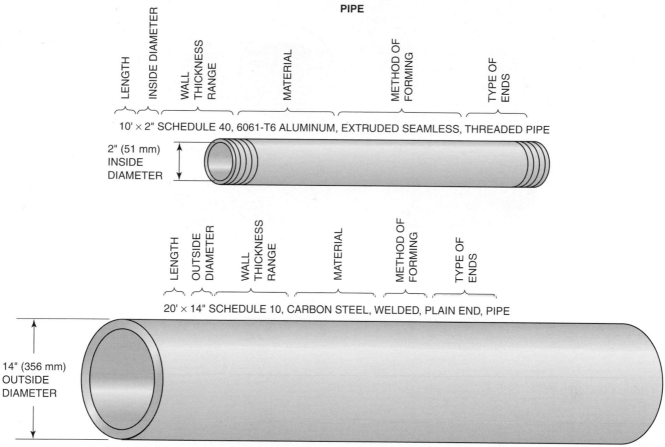

LENGTH INSIDE DIAMETER WALL THICKNESS RANGE MATERIAL METHOD OF FORMING TYPE OF ENDS

10' × 2" SCHEDULE 40, 6061-T6 ALUMINUM, EXTRUDED SEAMLESS, THREADED PIPE

2" (51 mm) INSIDE DIAMETER

LENGTH OUTSIDE DIAMETER WALL THICKNESS RANGE MATERIAL METHOD OF FORMING TYPE OF ENDS

20' × 14" SCHEDULE 10, CARBON STEEL, WELDED, PLAIN END, PIPE

14" (356 mm) OUTSIDE DIAMETER

FIGURE 5-3 Typical specifications used when ordering pipe. © Cengage Learning 2012

FIGURE 5-4 Space shuttle launch tower is constructed using round and rectangular tubing. NASA

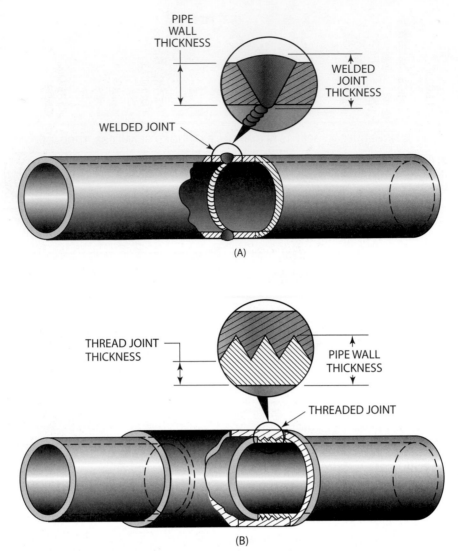

FIGURE 5-5 The welded joint (A) is thicker than the original pipe; the threaded joint (B) is thinner than the original pipe. © Cengage Learning 2012

Strength

The thickness of the pipe and fitting is the same when they are welded together. Threaded pipe is weakened because the threads reduce the wall thickness of the pipe, **Figure 5-5**.

Less Maintenance Required

Over much time and use, welded pipe joints are resistant to leaks.

Longer Lasting

Welded pipe joints resist corrosion caused by electrochemical reactions because all the parts are made of the same types of metal. Small cracks between the threads on threaded pipe are likely spots for corrosion to start.

Smoother Flow

The inside of a welded fitting is the same size as the pipe itself. As material flows through the pipe, less turbulence is caused by unequal diameters, **Figure 5-6**.

Large piping systems may be several miles in length. Lowered resistance to product flow can save on operating energy costs.

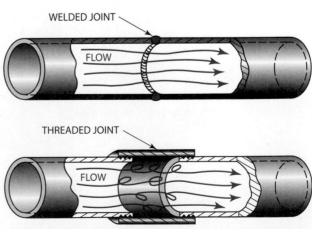

FIGURE 5-6 The flow along a welded pipe is less turbulent than that in a threaded pipe. © Cengage Learning 2012

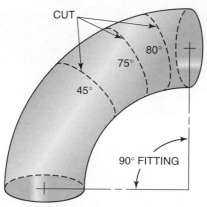

FIGURE 5-7 A standard 90° fitting can be cut to any special angle that is needed. © Cengage Learning 2012

Lighter Weight

Threaded fittings are larger and weigh more than welded fittings and joints. The weight savings when welded joints are used for an aircraft means that it can fly longer and faster and can carry a larger load for less money. Buildings and factories will also realize a savings because welded points are less expensive.

Other advantages of welded pipe include the following:

- Specially angled fittings can be made by cutting existing fittings, **Figure 5-7.**
- Odd-shaped parts can be fabricated.
- A lot of highly specialized equipment is not required for each different size of pipe.
- Alignment of parts is easier. It is not necessary to overtighten or undertighten fittings so that they will line up.
- Removing, replacing, or changing parts is easy because special connections are not needed to remove the parts.

Preparation and Fit-up

The ends of pipe must be beveled for maximum penetration and high joint strength. The end can be beveled by flame cutting, machining, grinding, or a combination process. It is important that the bevel be at the correct angle, about 37 1/2° depending on specifications, and that the end meet squarely with the mating pipe. The sharp or feathered inner edge of the bevel should be ground flat, forming a chamfer. This area is called a **root face** or **land.** Final shaping should be done with a grinder so that the **root gap** will be uniform.

The bevel on the end of the pipe can be machine cut using a portable pipe beveling machine, **Figure 5-8,** or a handheld torch. Chapter 7 describes how to set up and operate flame-cutting equipment. A turntable, similar to the one shown in **Figure 5-9,** can be made in the school shop and used for beveling short pieces of pipe. The turntable can be used vertically or horizontally. By turning the table slowly

FIGURE 5-8 Pipe beveling machine. Larry Jeffus

with the pipe held between the clamps, a hand torch can be used to produce smooth pipe bevels. Large-production welding shops may also use machines designed specifically for beveling pipe. These machines will accurately cut a 37 1/2° angle on the pipe.

The 37 1/2° angle allows easy access for the electrode with a minimum amount of filler metal required to fill the groove, **Figure 5-10.**

The root face will help a welder control both penetration and root suck back. Penetration control is improved because there is more metal near the edge to absorb excessive arc heat. This makes machine adjustments less

THINK GREEN
Conserve Material

Welded piping systems use less metal than threaded piping systems do. In a threaded piping system a metal coupling or fitting is required, but the fitting is not required for welded systems. The environmental savings come from the reduced use of raw materials and the elimination of the production and transportation fuel costs of the fittings.

PIPE HELD IN POSITION BY
CLAMPS FOR CUTTING

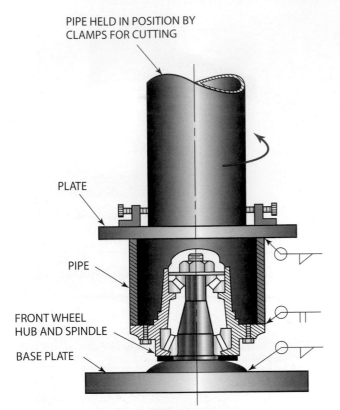

PLATE

PIPE

FRONT WHEEL
HUB AND SPINDLE

BASE PLATE

FIGURE 5-9 Turntable built from a front wheel assembly. © Cengage Learning 2012

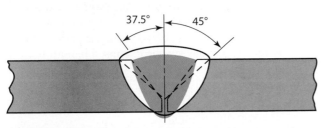

FIGURE 5-10 The 37 1/2° angled joint may use nearly 50% less filler metal, time, and heat as compared to the 45° angled joint. © Cengage Learning 2012

critical by allowing the molten weld pool to be quickly cooled between each electrode movement. **Root suck back** is caused by the surface tension of the molten metal trying to pull itself into a ball, forming a **concave root surface, Figure 5-11.** The root face allows a larger molten weld pool to be controlled, and because of the increased size of the molten weld pool, it is not so affected by surface tension, **Figure 5-12.**

Fitting pipe together and holding it in place for welding become more difficult as the diameter of the pipe gets larger. Devices for clamping and holding pipe in place are available, or a series of wedges and dogs can be used, **Figure 5-13.** In the Practices for this chapter, the pieces of pipe the welder will be using are about 1 1/2 in. (38 mm) in diameter. However, when welding on larger diameter

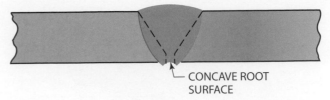

CONCAVE ROOT
SURFACE

FIGURE 5-11 Root surface concavity. © Cengage Learning 2012

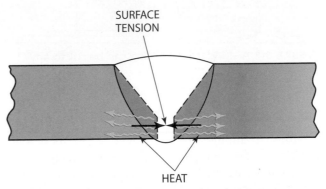

SURFACE
TENSION

HEAT

FIGURE 5-12 Heat is drawn out of the molten weld pool, and surface tension holds the pool in place. © Cengage Learning 2012

pipe sizes, the weld specimens must be larger than 1 1/2 in. (38 mm). Welds on larger pipe sizes need more metal to help absorb the higher heat required to make these welds.

A welder can use a vise to hold the pipes in place for tack welding. If the pipe is not round and does not align properly, first tack weld the pipe together and then quickly hit the tack while holding the pipe over the horn on an anvil, **Figure 5-14.** This action will force the pipe into alignment. For pipe that is too distorted to be forced into alignment in this manner, a welder must grind down the high points to ensure a good fit.

Practice Welds

One of the major problems to be overcome in pipe welding is learning how to make the transition from one position to another. The rate of change in welding position is slower with large diameter pipes, but the large diameter pipes require more time to weld. When a welder first starts welding, a large diameter pipe should be used in order to make learning this transition easier. As welders develop skill and the technique of pipe welding, they can change to the small diameter pipe sizes. Pipe as small as 3 in. (76 mm) can be welded quickly. It is large enough for the welder to be able to cut out test specimens.

Pipe used for these practice welds should be no shorter than 1 1/2 in. (38 mm). Pipe that is shorter than 1 1/2 in. (38 mm) rapidly becomes overheated, making welding more difficult.

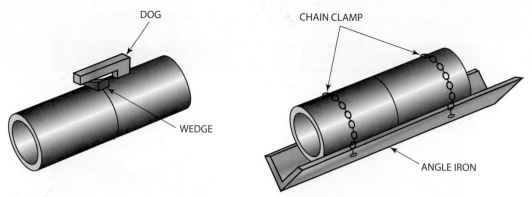

FIGURE 5-13 Shop fabrications used to align pipe joints. © Cengage Learning 2012

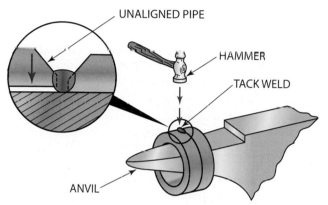

FIGURE 5-14 Hitting a hot tack weld can align a pipe joint. © Cengage Learning 2012

THINK GREEN
Accurate Fit-up Equals Savings

Properly fitted pipe requires less welding and grinding time and less filler metal and grinding disks. Even slight changes in the groove angle, root spacing, or alignment can make a difference; but very poor fit-up can result in the finished weld costing several times more than that of a properly fitted joint.

To progress more quickly with pipe welds, a welder should master grooved plate welds. Once plate welding is mastered in all positions, pipe skills are faster and easier to develop.

Pipe welding is either performed with E6010 or E6011 electrodes for the complete weld, or these electrodes are used for the root pass, and E7018 electrodes are used to complete the joint. Pipe welding can also be done using the E7018 electrode for the entire weld, **Figure 5-15**.

The practice pieces of pipe used in the school shop are much shorter than the pieces of pipe used in industry.

When learning to weld on short pipe, it is a good idea to avoid positioning oneself where longer pipe would eventually be located. Often it is easier to stand at the end of a pipe rather than to one side; however, this cannot be done if the weld is being made on a full length of pipe.

Weld Standards Weld quality is very important to the pipe welding industry. Like other welds, the major parts of the weld come under a higher level of inspection. However, the surface of the pipe on both sides of the weld is also important. No arc strikes should be made on this surface. Arc strikes outside of the weld groove are considered to be defects by much of the pipe welding industry. Arc strikes form small hardness spots, which, if not remelted by the weld, will crack as the pipe expands and contracts with heat

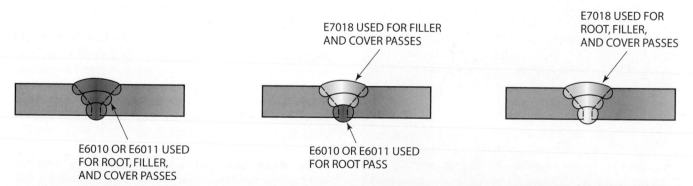

FIGURE 5-15 Single or multiple types of electrodes may be used when producing a pipe weld. The electrode selected is most often controlled by a code or specification. © Cengage Learning 2012

GOOD ROOT PASS

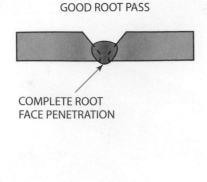

COMPLETE ROOT
FACE PENETRATION

POOR ROOT PASSES

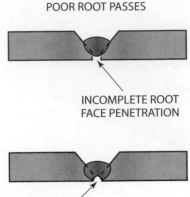

INCOMPLETE ROOT
FACE PENETRATION

CONCAVE ROOT FACE

FIGURE 5-16 Root pass. © Cengage Learning 2012

and pressure changes. Thus, they are a defect and must be removed and the area repaired. This removal under some standards may simply involve grinding them off, but under some codes they must be treated like any other defect, and a special weld repair procedure must be followed. Because of the importance of not having arc strikes outside the weld groove on pipe welds, you should try to avoid them from the beginning. In Chapter 4, Experiment 4-2, several techniques are described to avoid arc strikes outside of the welding zone. You may want to refer back to this section if you have difficulty in making arc starts accurately.

Root Weld A root weld is the first weld in a joint, **Figure 5-16.** It is part of a series of welds that make up a multiple pass weld. The root weld is used to establish the contour and depth of penetration. The most important part of a root weld is the internal root face, or, in the case of pipe, the inside surface. The face, or outside shape, or contour of the root weld is not so important.

For cleaners (pigs) to be used in a pipeline, the inside of each welded joint must be smooth, **Figure 5-17.**

FIGURE 5-17 The root face must be uniform. Larry Jeffus

Excessive penetration, known as **icicles,** gets in the way of cleaning, adds resistance to flow, and will cause weakened points on the weld. On some piping systems, consumable inserts or backing rings are used to control penetration and the inside contour. Most pipe welds are made without these devices to control penetration.

The face of a root weld is not important if the root surface is clean, smooth, and uniform. A grinder is used to remove excessive buildup and reshape the face of the root pass. This grinding removes slag along the sides of the weld bead and makes it easier to add the next pass. Not all root passes are ground. Pipe that is to be used in low- and medium-pressure systems is not usually ground. Grinding each root pass takes extra time and does not give the welder the experience of using a hot pass. Most slag must be completely removed by chipping and wire brushing before the hot pass is used.

If you need more experience or practice in making an open root weld, refer to Chapter 4.

Hot Pass The hot pass is used to quickly burn out small amounts of slag trapped along the edge of the root pass. This is slag that cannot be removed easily by chipping or wire brushing. The hot pass can also be used to reshape the root pass by using high current settings and a faster-than-normal travel speed.

Slag is mostly composed of silicon dioxide, which melts at about 3100°F (1705°C). Steel melts at approximately 2600°F (1440°C). A temperature of more than 500°F (270°C) hotter than the surrounding metal is required to melt slag. The slag can be floated to the surface by melting the surrounding metal. A high current will quickly melt enough surface to allow the slag to float free; a fast travel speed will prevent burnthrough. The fast travel speed forms a concave weld bead that is easy to clean for the welds that will follow.

Filler Pass After thoroughly removing slag from the weld groove by chipping, wire brushing, or grinding, it is ready to be filled. The filler pass(es) may be either a series of stringer beads, **Figure 5-18,** or a weave bead, **Figure 5-19.** Stringer beads require less welder skill because of the

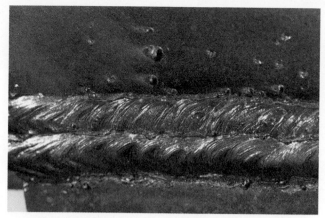

FIGURE 5-18 Filler pass using stringer beads. Larry Jeffus

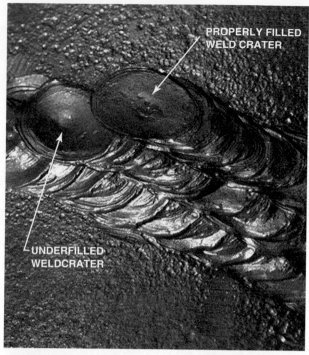

PROPERLY FILLED
WELD CRATER

UNDERFILLED
WELDCRATER

FIGURE 5-20 The weld crater should be filled to prevent cracking, and cleaned of slag before restarting the arc. Larry Jeffus

ELECTRODE DIAMETER	BEAD WIDTH
$\frac{1}{8}$" (3 mm)	$\frac{1}{4}$" (6 mm)
$\frac{5}{32}$" (4 mm)	$\frac{5}{16}$" (8 mm)
$\frac{3}{16}$" (4.8 mm)	$\frac{3}{8}$" (10 mm)

small amount of metal that is molten at one time. If done correctly, stringer beads are as strong as weave beads.

The weld bead crater must be cleaned before the next electrode is started. Failure to clean the crater will result in slag inclusions. On high-strength, high-pressure pipe welds, the crater should be slightly ground to ensure its cleanliness, **Figure 5-20**. When the bead has gone completely around the pipe, it should continue past the starting point so that good fusion is ensured, **Figure 5-21**. The

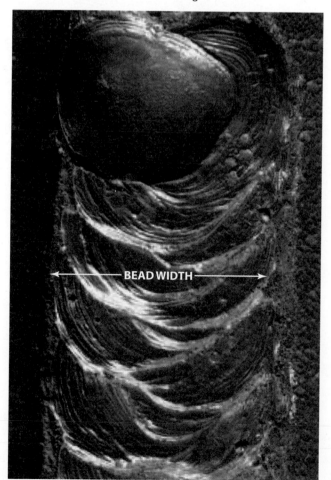

BEAD WIDTH

FIGURE 5-19 Filler pass using weave bead. The bead width should not be more than two times the rod diameter. Larry Jeffus

OTHER BEADS CONTINUE ON PAST

ONE BEAD STOPS HERE

FIGURE 5-21 Avoid starting and stopping all weld passes in the same area. Larry Jeffus

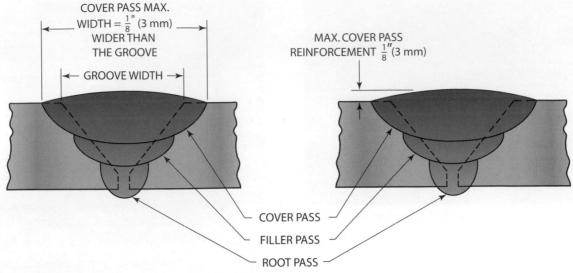

FIGURE 5-22 Excessively wide or builtup welds restrict pipe expansion at the joint, which may cause premature failure. Check the appropriate code or standard for exact specifications. © Cengage Learning 2012

FIGURE 5-23 Uniformity in each pass shows a high degree of welder skill and increases the probability that the weld will pass testing. Larry Jeffus

locations of starting and stopping spots for each weld pass must be staggered. The weld groove should be filled level with these beads so that it is ready for the cover pass.

Cover Pass The final covering on a weld is referred to as the cover pass or cap. It may be a weave or stringer bead. The cover pass should not be too wide or have too much reinforcement, **Figure 5-22**. Cover passes that are excessively large will reduce the pipe's strength, not increase it. A large cover pass will cause the stresses in the pipe to be concentrated at the sides of the weld. An oversized weld will not allow the pipe to expand and contract uniformly along its length. This concentration is similar to the restriction a rubber band would have on an inflated balloon if it were put around its center.

The cover pass should be kept as uniform and as neat looking as possible, **Figure 5-23**. A visual checking is often all that low- and medium-pressure welds receive, and a nice-looking cover will pass testing each time. A good cover pass, during a visual inspection, will indicate that the weld underneath is sound.

1G Horizontal Rolled Position

The **horizontal rolled pipe position** is commonly used in fabrication shops where structures or small systems can be

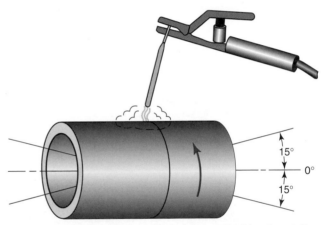

FIGURE 5-24 1G position. The pipe is rolled horizontally. © Cengage Learning 2012

positioned for the convenience of the welder, **Figure 5-24**. The consistent high quality and quantity of welds produced in this position make it very desirable for both the welder and the company.

The penetration and buildup of the weld are controlled more easily with the pipe in this position. Weld visibility and welder comfort are improved so that welder fatigue is less of a problem. The pipe can be rolled continuously with some types of positioners, and the weld can be made in one continuous bead.

Because of the ease in welding and the level of skill required, welders who are certified in this position may not be qualified to make welds in other positions.

PRACTICE 5-1

Beading, 1G Position, Using E6010 or E6011 Electrodes and E7018 Electrodes

Using a properly set up and adjusted arc welding machine, proper safety protection, E6010 or E6011 and

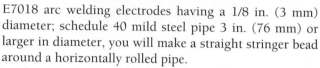

FIGURE 5-25 Angle iron pipe support. © Cengage Learning 2012

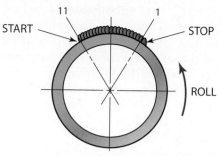

FIGURE 5-26 1G pipe welding area. © Cengage Learning 2012

E7018 arc welding electrodes having a 1/8 in. (3 mm) diameter; schedule 40 mild steel pipe 3 in. (76 mm) or larger in diameter, you will make a straight stringer bead around a horizontally rolled pipe.

Place the pipe horizontally on the welding table in a vee block made of angle iron, **Figure 5-25**. The vee block will hold the pipe steady and allow it to be moved easily between each bead. Strike an arc on the pipe at the 11 o'clock position. Make a stringer bead over the 12 o'clock position, stopping at the 1 o'clock position, **Figure 5-26**. Roll the pipe until the end of the weld is at the 11 o'clock position. Clean the weld crater by chipping and wire brushing.

Strike the arc again and establish a molten weld pool at the leading edge of the weld crater. With the molten weld pool reestablished, move the electrode back on the weld bead just short of the last full ripple, **Figure 5-27**. This action will both reestablish good fusion and keep the weld bead size uniform. Now that the new weld bead is tied into the old weld, continue welding to the 1 o'clock position again. Stop welding, roll the pipe, clean the crater, and resume welding. Keep repeating this procedure until the weld is completely around the pipe. Before the last weld is started, clean the end of the first weld so that the end and beginning beads can be tied together smoothly. When you reach the beginning bead, swing your electrode around on both sides of the weld bead. A poor beginning of a weld bead is always high and narrow and has little penetration, **Figure 5-28**. By swinging the weave pattern

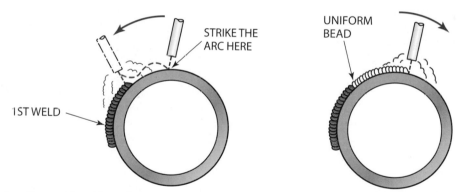

FIGURE 5-27 Keeping the weld uniform is important when restarting the arc. © Cengage Learning 2012

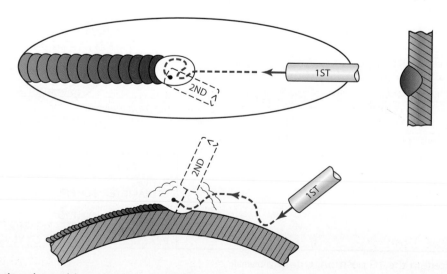

FIGURE 5-28 Restarting the weld. © Cengage Learning 2012

(the "C" pattern is best) on both sides of the bead, you can make the bead correctly so that the width is uniform. The added heat will give deeper penetration at the starting point. Hold the arc in the crater for a moment until it is built up but do not overfill the crater.

Cool, chip, and inspect the bead for defects. Repeat the beads as needed until they are mastered. Turn off the welding machine and clean up your work area when you are finished welding.

Complete a copy of the "Student Welding Report" listed in Appendix I or provided by your instructor. ◆

PRACTICE 5-2

Butt Joint, 1G Position, Using E6010 or E6011 Electrodes

Using a properly set up and adjusted arc welding machine, proper safety protection, E6010 or E6011 arc welding electrodes having a 1/8 in. (3 mm) diameter, and two or more pieces of schedule 40 mild steel pipe 3 in. (76 mm) or larger in diameter, you will make a pipe butt joint in the 1G horizontal rolled position, **Figure 5-29.**

Tack weld two pieces of pipe together as shown in **Figure 5-30.** Place the pipe horizontally in a vee block on the welding table. Start the root pass at the 11 o'clock position. Using a very short arc and high current setting, weld toward the 1 o'clock position. Stop and roll the pipe, chip the slag, and repeat the weld until you have completed the root pass.

Clean the root pass by chipping and wire brushing. The root pass should not be ground this time. Replace the pipe in the vee block on the table so that the hot pass can be done. Turn up the machine amperage, enough to remelt the root weld surface, for the hot pass. Use a stepped

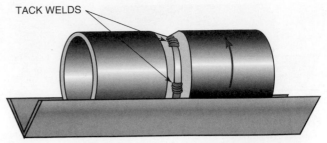

FIGURE 5-30 The tack welds are to be evenly spaced around the pipe. Use four tacks on small diameter pipe and six or more on large diameter pipe. © Cengage Learning 2012

electrode pattern, moving forward each time the molten weld pool washes out the slag, and returning each time the molten weld pool is nearly all solid, **Figure 5-31.** Weld from the 11 o'clock position to the 1 o'clock position before stopping, rolling, and chipping the weld. Repeat this procedure until the hot pass is complete.

The filler pass and cover pass may be the same pass on this joint. Turn down the machine amperage. Use a "T," "J," "C," or zigzag pattern for this weld. Start the weld at the 10 o'clock position and stop at the 12 o'clock position. Sweep the electrode so that the molten weld pool melts out any slag trapped by the hot pass. Watch the back edge of the bead to see that the molten weld pool is filling the groove completely. Turn, chip, and continue the bead until the weld is complete. Repeat this weld until you can consistently make welds free of defects. Turn off the welding machine and clean up your work area when you are finished welding.

Complete a copy of the "Student Welding Report" listed in Appendix I or provided by your instructor. ◆

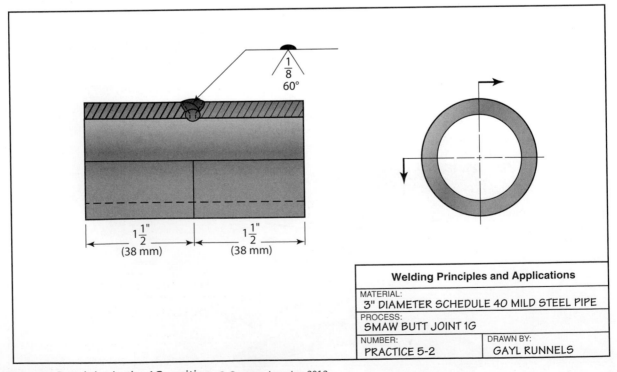

Welding Principles and Applications	
MATERIAL: 3" DIAMETER SCHEDULE 40 MILD STEEL PIPE	
PROCESS: SMAW BUTT JOINT 1G	
NUMBER: PRACTICE 5-2	DRAWN BY: GAYL RUNNELS

FIGURE 5-29 Butt joint in the 1G position. © Cengage Learning 2012

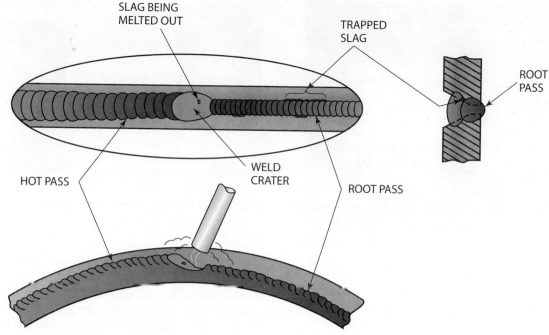

FIGURE 5-31 Hot pass. © Cengage Learning 2012

PRACTICE 5-3

Butt Joint, 1G Position, Using E6010 or E6011 Electrodes for the Root Pass with E7018 Electrodes for the Filler and Cover Passes

Using the same setup, materials, and procedures as described in Practice 5-2, you will make a horizontal rolled butt joint in pipe, **Figure 5-32**. The root pass is to have 100% penetration over 80% or more of the length of the weld.

Set the pipe in the vee block on the welding table and make the root pass as explained in Practice 5-2. Watch for 100% penetration with no icicles. A hot pass or grinder can be used to clean the face of the root pass. Use an E7018 electrode for the filler and cover passes. The E7018 electrode should not be weaved more than two-and-a-half times the diameter of the electrode. Excessively wide

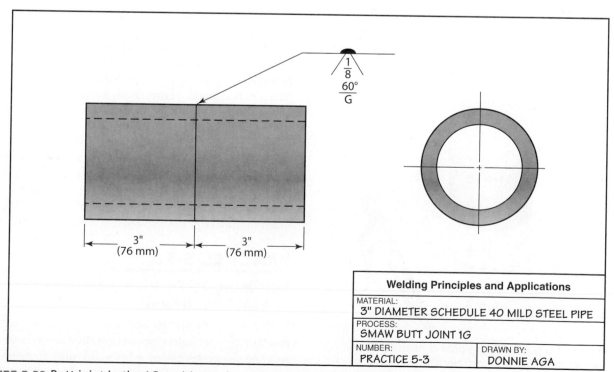

FIGURE 5-32 Butt joint in the 1G position to be tested. © Cengage Learning 2012

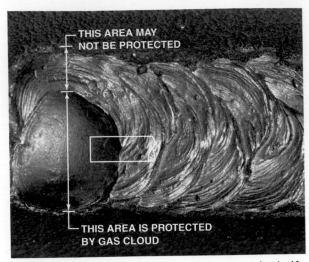

FIGURE 5-33 Weave beads more than two-and-a-half times the diameter of the electrode may be nice looking, but the atmosphere may contaminate the unprotected part of the molten weld pool. Larry Jeffus

weaving will allow the molten weld pool to become contaminated, **Figure 5-33.**

After the weld is completed, visually inspect it for 100% penetration around 80% of the root length. Check the weld for uniformity and visual defects on the cover pass. Repeat this weld until you can consistently make welds free of defects. Turn off the welding machine and clean up your work area when you are finished welding.

Complete a copy of the "Student Welding Report" listed in Appendix I or provided by your instructor. ◆

2G Vertical Fixed Position

In the 2G **vertical fixed pipe position,** the pipe is vertical and the weld is horizontal, **Figure 5-34.** With these welds, the welder does not need to change welding positions constantly. The major problem that faces welders

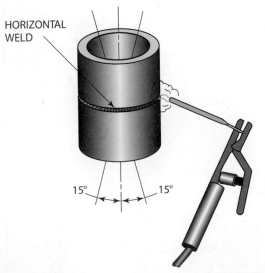

FIGURE 5-34 2G position. The pipe is fixed vertically, and the weld is made horizontally around it.

© Cengage Learning 2012

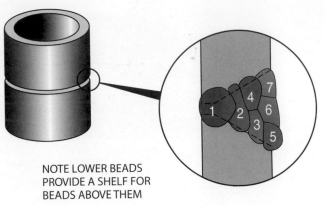

FIGURE 5-35 2G pipe welding position.

© Cengage Learning 2012

when welding pipe in this position is that the area to be welded is often located in corners. Because of this location, reaching the back side of the weld is difficult. In the practices that follow, you may turn the pipe between welds. As a welder gains more experience, welds in tight places will become easier.

The welds must be completed in the correct sequence, **Figure 5-35.** The root pass goes in as with other joints. To reduce the sagging of the bottom of the weld, increase the electrode-to-work angle. As long as the weld is burned in well and does not have cold lap on the bottom, the weld is correct. Each of the filler and cover welds that follow must be supported by the previous weld bead.

PRACTICE 5-4

Stringer Bead, 2G Position, Using E6010 or E6011 Electrodes and E7018 Electrodes

Using the same setup, materials, and electrodes as listed in Practice 5-1, you will make straight stringer beads on a pipe that is in the vertical position.

The "J" weave pattern should be used so that the molten weld pool will be supported by the lower edge of the solidified metal, **Figure 5-36.** Keep the electrode at an upward and trailing angle so the arc force will help to keep the weld in place.

Repeat these stringer beads as needed, with both groups of electrodes, until you can consistently make welds free of defects. Turn off the welding machine and clean up your work area when you are finished welding.

Complete a copy of the "Student Welding Report" listed in Appendix I or provided by your instructor. ◆

PRACTICE 5-5

Butt Joint, 2G Position, Using E6010 or E6011 Electrodes

Using a properly set up and adjusted arc welding machine, proper safety protection, E6010 or E6011 arc welding electrodes having a 1/8 in. (3 mm) diameter, and two or more pieces of schedule 40 mild steel pipe 3 in.

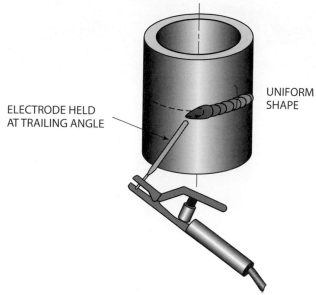

FIGURE 5-36 Electrode position and weave pattern for a weld on vertical pipe. © Cengage Learning 2012

(76 mm) or larger in diameter, you will make a butt joint on a pipe that is in the vertical position.

Place the pipe on the arc welding table. Strike an arc and make a root weld that is as long as possible. If the root gap is uniform, a step pattern must be used. After completing and cleaning the root pass, make a hot pass. The hot pass need only burn the root pass clean, **Figure 5-37**. Undercut on the top pipe is acceptable.

The filler and cover passes should be stringer beads. By keeping the molten weld pool size small, control is easier. Cool, chip, and inspect the completed weld for uniformity and defects. Repeat this weld until you can consistently make welds free of defects. Turn off the welding machine and clean up your work area when you are finished welding.

Complete a copy of the "Student Welding Report" listed in Appendix I or provided by your instructor. ◆

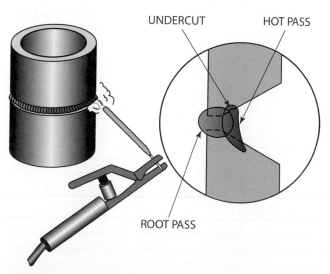

FIGURE 5-37 Hot pass. © Cengage Learning 2012

PRACTICE 5-6

Butt Joint, 2G Position, Using E6010 or E6011 Electrodes for the Root Pass and E7018 Electrodes for the Filler and Cover Passes

Using the same setup, materials, and procedures as described in Practice 5-4, you will make a vertical fixed pipe weld. The root pass is to have 100% penetration over 80% or more of the length of the weld.

Place the pipe vertically on the welding table. Hold the electrode at a 90° angle to the pipe axis and with a slight trailing angle, **Figure 5-38**. The electrode should be held tightly into the joint. If a burnthrough occurs, quickly push the electrode back over the burnthrough while increasing the trailing angle. This action forces the weld metal back into the opening. When the root pass is complete, chip the surface slag and then clean out the trapped slag by grinding or chipping, or use a hot pass.

Use E7018 electrodes for the filler and cover passes with a stringer pattern. The weave beads are not recommended with this electrode and position because they tend to undercut the top and overlap the bottom edge. After the weld is completed, visually inspect it for 100% penetration around 80% of the root length. Check the weld for uniformity and visual defects on the cover pass. Repeat the weld until you can consistently make welds free of defects. Turn off the welding machine and clean up your work area when you are finished welding.

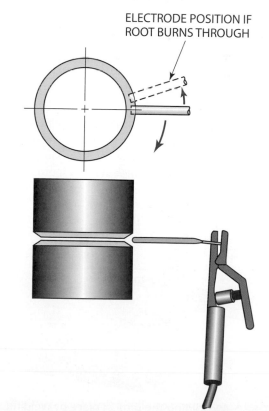

FIGURE 5-38 Electrode position and movement for the root pass. © Cengage Learning 2012

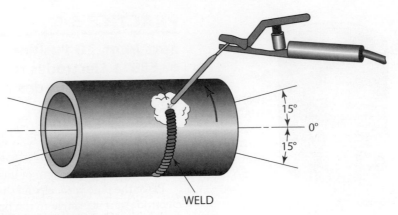

FIGURE 5-39 5G horizontal fixed position. © Cengage Learning 2012

Complete a copy of the "Student Welding Report" listed in Appendix I or provided by your instructor. ◆

5G Horizontal Fixed Position

The 5G **horizontal fixed pipe position** is the most often used pipe welding position. Welds produced in flat, vertical up or vertical down, and overhead positions must be uniform in appearance and of high quality.

When practicing these welds, mark the top of the pipe for future reference. Moving the pipe will make welding easier, but the same side must stay on the top at all times, **Figure 5-39.**

The root pass can be performed by **welding uphill or downhill.** In industry, the method used to weld the root pass is determined by established weld procedures. If there are no procedures requiring a specific direction, the choice is usually made based upon fit-up. A close parallel root opening can be welded uphill or downhill. A root opening that is wide or uneven must be welded uphill. In the following Practices, the welder can make the choice of direction, but both directions should be tried.

The pipe may be removed from the welding position for chipping, wire brushing, or grinding. The pipe can be held in place by welding a piece of flat stock to it and clamping the flat stock to a pipe stand, **Figure 5-40.**

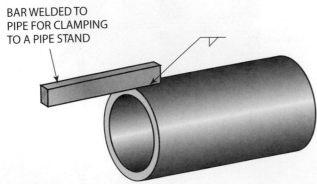

BAR WELDED TO PIPE FOR CLAMPING TO A PIPE STAND

FIGURE 5-40 Holding the pipe in place by welding a piece of flat stock to the pipe and then clamping the flat stock to a pipe stand. © Cengage Learning 2012

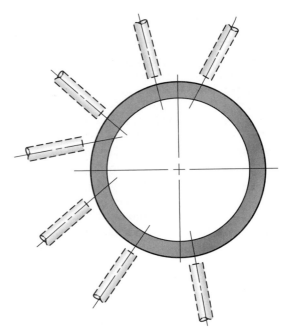

FIGURE 5-41 Electrode angle. © Cengage Learning 2012

The electrode angle should always be upward, **Figure 5-41.** Changing the angle toward the top and bottom will help control the bead shape. The bead, if welded downhill, should start before the 12 o'clock position and continue past the 6 o'clock position to ensure good fusion and tie-in of the welds. The arc must always be struck inside the joint preparation groove.

PRACTICE 5-7

Stringer Bead, 5G Position, Using E6010 or E6011 Electrodes and E7018 Electrodes

Using the same setup, materials, and electrodes as listed in Practice 5-1, you will make straight stringer beads in the horizontal fixed 5G position using both groups of electrodes.

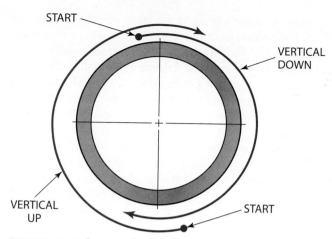

FIGURE 5-42 Stop at the 7 o'clock position.
© Cengage Learning 2012

Clamp the pipe horizontally between waist and chest level. Starting at the 11 o'clock position, make a downhill straight stringer bead through the 12 o'clock and 6 o'clock positions. Stop at the 7 o'clock position, **Figure 5-42.** Using a new electrode, start at the 5 o'clock position and make an uphill straight stringer bead through the 6 o'clock and 12 o'clock positions. Stop at the 1 o'clock position. Change the electrode angle to control the molten weld pool.

Repeat these stringer beads as needed, with each group of electrodes, until you can consistently make welds free of defects. Turn off the welding machine and clean up your work area when you are finished welding.

Complete a copy of the "Student Welding Report" listed in Appendix I or provided by your instructor. ◆

PRACTICE 5-8

Butt Joint, 5G Position, Using E6010 or E6011 Electrodes for the Root Pass and E7018 Electrodes for the Filler and Cover Passes

Using the same setup, materials, and procedures as listed in Practice 5-2, you will make a horizontal fixed 5G pipe weld. The root pass is to have 100% penetration over 80% or more of the length of the weld.

Mark the top of the pipe and mount it horizontally between waist and chest level. Weld the root pass uphill or downhill using E6010 or E6011 electrodes. Either grind the root pass or use a hot pass to clean out trapped slag.

Use E7018 electrodes for the filler and cover passes with stringer or weave patterns. When the weld is completed, visually inspect it for 100% penetration around 80% of the root length. Check the weld for uniformity and visual defects on the cover pass. Repeat the weld until you can consistently make welds free of defects. Turn off the welding machine and clean up your work area when you are finished welding.

Complete a copy of the "Student Welding Report" listed in Appendix I or provided by your instructor. ◆

PRACTICE 5-9

Butt Joint, 5G Position, Using E6010 or E6011 Electrodes

Using a properly set up and adjusted arc welding machine, proper safety protection, E6010 or E6011 arc welding electrodes having a 1/8 in. (3 mm) diameter, and two or more pieces of schedule 40 mild steel pipe 3 in. (76 mm) or larger in diameter, you will make a butt joint on a horizontally fixed 5G pipe.

Mark the top of the pipe and mount it between waist and chest level. Depending upon the root gap, make a root weld uphill or downhill using E6010 or E6011 electrodes. Check the root penetration to determine if it is better in one area than in another area. Chip and wire brush the weld and set the machine for a hot pass. Start the hot pass at the bottom and weld upward on both sides. The bead should be kept uniform with little buildup.

Using stringer or weave beads, make the filler and cover welds. If stringer beads are used, downhill welds can be made. Cool, chip, and inspect the weld for uniformity and defects. Repeat this weld until you can consistently make welds free of defects. Turn off the welding machine and clean up your work area when you are finished welding.

Complete a copy of the "Student Welding Report" listed in Appendix I or provided by your instructor. ◆

6G 45° Inclined Position

The 45° **fixed inclined position** is thought to be the most difficult pipe position. Qualifying in this position will certify the welder in the other positions using the same size electrodes and pipe sizes. The weld must be uniform and not have defects even though its position is changing in more than one direction at a time, **Figure 5-43.**

During this weld, it is necessary to continuously change the weld pattern, electrode angle, and weld speed. Small multipass stringer beads work best. With experience, however, weave beads are possible.

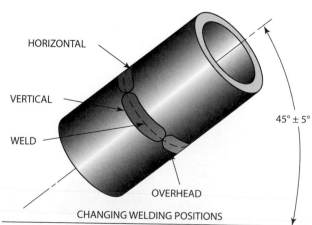

FIGURE 5-43 In the 6G position, the pipe is fixed at a 45° angle to the work surface. The effective welding angle changes as the weld progresses around the pipe.
© Cengage Learning 2012

PRACTICE 5-10

Stringer Bead, 6G Position, Using E6010 or E6011 Electrodes and E7018 Electrodes

Using the same setup, materials, and electrodes as listed in Practice 5-1, you will make straight stringer beads on a pipe in the 45° fixed inclined position.

Using the straight stepped "T" or whipping patterns, start at the bottom of the pipe with a keyhole technique. Keep the molten weld pool small and narrow for easier control. As the bead moves up to the side, the electrode angle should stay to the downhill side with a trailing (pulling) angle. When the weld passes beyond the side, the downhill and trailing angles are decreased. This is done so that, when the weld reaches the top, the electrode is perpendicular to the top of the pipe. Repeat this procedure on the opposite side.

Repeat these stringer beads as needed, with both groups of electrodes, until you can consistently make welds free of defects. Turn off the welding machine and clean up your work area when you are finished welding.

Complete a copy of the "Student Welding Report" listed in Appendix I or provided by your instructor. ◆

PRACTICE 5-11

Butt Joint, 6G Position, Using E6010 or E6011 Electrodes

Using a properly set up and adjusted arc welding machine, proper safety protection, E6010 or E6011 arc welding electrodes having a 1/8 in. (3 mm) diameter, and two or more pieces of schedule 40 mild steel pipe 3 in. (76 mm) or larger in diameter, you will make a butt joint on a pipe in the 45° fixed inclined position.

Starting at the top, make a vertical down root pass that ends just beyond the bottom. Repeat this weld on the other side. Chip and wire brush the slag so that an uphill hot pass can be made. The hot pass must be kept small and concave so that more slag will not be trapped along the downhill side. Clean the bead, turn down the machine amperage, and complete the joint with stringer beads, **Figure 5-44**.

Cool, chip, and inspect the completed weld for uniformity and defects. Repeat this weld until you can consistently make welds free of defects. Turn off the welding machine and clean up your work area when you are finished welding.

Complete a copy of the "Student Welding Report" listed in Appendix I or provided by your instructor. ◆

PRACTICE 5-12

Butt Joint, 6G Position, Using E6010 or E6011 Electrodes for the Root Pass and E7018 Electrodes for the Filler and Cover Passes

Using the same setup, materials, and procedures as listed in Practice 5-11, you will make a 45° fixed inclined pipe weld. The root pass is to have 100% penetration over 80% of the root length of the weld.

Make the root pass as either a vertical up or down weld, depending upon the root opening. Chip the slag and clean the weld by grinding or by using a hot pass. If a hot pass is used, chip and wire brush the joint. Using the E7018 electrode, start slightly before the center on the bottom and make a small stringer bead in an upward direction. Keep a trailing and a somewhat uphill electrode angle so that the weld is deposited on the bottom of the lower pipe. The next pass should use a downhill electrode angle so that the bead is on the uphill pipe. Alternate this process

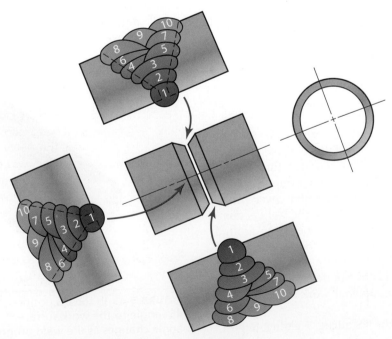

FIGURE 5-44 Weld bead positions for 6G pipe weld. © Cengage Learning 2012

until the bead is complete. Clean the weld and inspect it for 100% penetration over 80% of the root length. Check for uniformity and visual defects on the cover pass. Repeat the weld until you can consistently make welds free of defects. Turn off the welding machine and clean up your work area when you are finished welding.

Complete a copy of the "Student Welding Report" listed in Appendix I or provided by your instructor. ◆

Summary

Pipe welding is looked upon by many in the welding industry as the pinnacle of the welding trade. Pipe welding presents its own challenges to the welder because of the constantly changing weld position. With practice you learn to control the molten weld pool as it transitions through the various weld positions. As you will learn to recognize the subtle changes in the weld pool, you can make the welding technique changes as second nature.

As you observe a pipe welder in the field, you will notice that the welder spends a significant amount of time in the preparation of the equipment and pipe before any welding begins. Being properly set up and having the weld joint properly aligned before welding are essential in producing quality welded pipe joints.

Often there is a tendency to overweld pipe joints. You must also learn to control weld size. This is very important with pipe welds because too large a weld can put more stresses on the joint. Also, a good pipe welded joint may have many small weld passes.

Orbital Welding Helps NASA's X-34 Rocket Soar

Ever since the creation of NASA's X-1 research rocket plane, which first broke the sound barrier with Chuck Yeager at the controls, orbital tube welding has been the welding method of choice for aerospace. Now Orbital Sciences Corp., Dulles, Virginia, and NASA have teamed up to create the X-34 reusable launch vehicle (RLV), the newest generation of rocket planes. The X-34 is being designed to take the place of some of the more familiar—but also more expensive—spacecraft and boosters in use today.

The X-34 is air launched from an L-1011 Tri-Star jet to reduce launch costs. To reduce the cost of RLVs even further, the need for highly reliable, low-maintenance, leak-free tubing systems has driven the requirement for automated tube welding equipment.

Orbital Sciences chose the Liburdi Dimetrics Centaur III 150 PTW computer-controlled orbital tube welding machine for the job. The ability to butt-joint weld heavy-wall tubing in tight spaces was critical to the X-34's construction requirements. The X-34 quality team also needed to monitor repeatability and provide quality traceability for every weld. A computer-controlled orbital tube welding system could fulfill both requirements.

The X-34 spacecraft contains more than 2000 feet of tubing, ranging in diameter from 0.250 to 4.500 inches with wall thickness from 0.028 to 0.120 inches. More than 500 tube welds are needed to integrate the X-34's systems, and all welds must be square butt joints.

Challenges to Overcome

Material Chemistry. The X-34 uses only seamless 316L tubing. Fittings had to be made from billet or forging, and both tubing and fittings had to comply with ASTM specifications. One problem encountered was a welding phenomenon known as "pool shift," in which different surface-active elements in the parts being welded cause the weld pool to shift to one side of the joint being welded. A lower sulfur content on one side of a joint can have a stronger tension than the higher sulfur side, causing the pool to be drawn to one side of the joint.

If this happens, full penetration of the weld can still occur, but penetration of the weld bead can miss the joint. The weld could fully penetrate the work, but it would be shifted to one side of the joint, leaving the inner diameter unconsumed.

Material Weldability. Different heats of stainless steel tubing can make it difficult to weld by automatic fusion welding techniques. ASTM specifications for each type of stainless steel, such as 304, 316, 304L, and 316L, may vary in concentrations of alloying elements, such as chromium, nickel, molybdenum, copper, and sulfur, resulting in no two heats of 304 or 316 being exactly alike. Variations in alloying elements could dramatically affect the weld appearance and penetration of two samples of 316L stainless steel tubing, both with a 1-inch outer diameter and wall thickness of 0.083 inches, but neither having the same heats as a result of varied alloy concentrations.

Orbital Sciences had to exercise tight control over the purchase of stainless tubing intended for orbital welding to minimize any possible problems. The X-34 program has had to weld heats of tubing and fittings that had as much as a 0.021% difference in sulfur content, which can result in a significant pool shift. Despite the differences, developing techniques, procedures, and welding parameter controls kept the welds high quality and consistent.

Joint Type. The type of joint an application requires can be important when choosing weld equipment and approach. The X-34 required square butt-joint welds for many reasons. For example, no slag or flux is permitted to contaminate the tubing, and product scrap is reduced due to the typically lower porosity in these types of welds. Welds that require welding wire have a much greater chance of porosity due to all the possible sources of contamination. Since all X-34 welds are square butt-joint welds, no welding wire was used.

Shielding/Purge Gas. Shielding gas can be a critical ingredient in the success of a weld. Shielding gas minimizes porosity in a weld and can—in some cases, such as with mixed gases—almost eliminate porosity. Also, shielding gas is used to purge possible contaminants from a weld area, and manipulation of purge pressure can be used to support a weld while it is molten. Some of the more difficult applications in the X-34 demanded mixed welding gases to achieve the proper weld quality.

Electrodes. Electrode geometry has always been an important orbital welding parameter because it has such a strong effect on weld shape and penetration. The use of properly prepared tungsten helps ensure repeatable welds. The typical geometry is a 22° taper with a 0.010 to 0.020 flat tip. A tip without a flat point may create an unstable arc and produce welds that wander from side to side.

Tube Preparation. As in all welding, fit-up is critical to successfully producing repeatable welds. It is especially critical with orbital welding because specific parameters, such as travel speed, welding amperes, and arc volts, are preset. The tubing used for square butt joints must be cut square and the end face machined perpendicular to the tube centerline using a facing tool.

Why Welding Was Chosen for the X-34

NASA did not require welding for the X-34 RLV subsystems, but the RLV program required low maintenance and reliability. The company also wanted to achieve a low fluid leakage rate. The X-34 uses 5000-pound-per-square-inch pneumatic systems. If tubing were to fail, a considerable safety hazard would be presented. If a weld were to fail on a tube adjacent to the X-34's large pressure tanks, it would represent a major explosion hazard. The weld development work was done with this in mind.

Computer-controlled orbital tube welding was used on every weld to ensure repeatability and quality traceability during production. Orbital Sciences Corp.

The X-34 reusable launch vehicle attached to the underside of a Lockhead Tri-Star jet. Orbital Sciences Corp.

The engineers and technicians also relied on real-time data acquisition to ensure weld quality and repeatability. The team needed to have high confidence in the quality of the work done. Before long, the X-34 will take center stage among the lead stories in newspapers and on television news. It is expected to be an important element in NASA's more economical space launch program.

Article courtesy of the American Welding Society.

Review

1. List some applications that use low-pressure piping.

2. List some applications that use medium-pressure piping.

3. List some applications that use high-pressure piping.

4. Describe the differences between pipe and tubing.

5. What are the advantages of welded piping systems over other joining methods?

6. Why must the ends of pipe be beveled before being welded?

7. How can the ends of a pipe be beveled?

8. Why is the end of a pipe beveled at a 37 1/2° angle?

9. What causes root suck back or a concave root face?

10. Why are arc strikes outside the welding zone considered a problem on pipe?

11. What are the purposes of backing rings?

12. What is the purpose of a hot pass?

13. Why must the weld crater be cleaned before starting a new electrode?

14. What is the maximum width of the cover pass? Why?

15. When is the 1G welding position used?

16. How can the start of a weld be made less narrow and more uniform?

17. What supports the welds on a 2G pipe joint?

18. On 5G welds, what usually determines the direction of the root pass?

19. What three welding positions are incorporated in the 6G welding position?

20. Why are small multipass stringer beads best for 6G welds?

Chapter 6

Advanced Shielded Metal Arc Welding

OBJECTIVES

After completing this chapter, the student should be able to

- discuss how metal must be prepared before welding.
- describe the process, and demonstrate making the root pass, filler weld, and cover pass in all positions and techniques.
- explain the purpose of a hot pass.
- tell what should be checked with a visual inspection, and describe the appearance of an acceptable weld.
- demonstrate how to make
  - a root pass on plate in all positions.
  - a root pass on plate with an open root in all positions.
  - an open root weld on plate using the step technique in all positions.
  - a multiple pass filler weld on a V-joint in all positions using E7018 electrodes.
  - a cover bead in all positions.
  - a single V-groove open root butt joint with an increasing root opening.
  - a single V-groove open root butt joint with a decreasing root opening.
  - SMAW welds of plate to plate.
  - SMAW welds of pipe to pipe.

KEY TERMS

back gouging	*interpass temperature*	*root pass*
burnthrough	*key hole*	*wagon tracks*
cover pass	*molten weld pool*	*weld groove*
filler pass	*multiple pass weld*	*weld specimen*
guided bend	*postheating*	
hot pass	*preheating*	

INTRODUCTION

The SMAW process can be used to consistently produce high-quality welds. Sometimes it is necessary to make welds in less-than-ideal conditions. Knowing how to produce a weld of high strength in an out of position, difficult situation or in an unusual metal takes both practice and knowledge. A welder is frequently required to make these types of welds to a code or standard. This chapter covers the high-quality welding of plate, pipe, and plate to pipe. The practices are designed to give you the experience of taking code-type tests in a variety of materials and positions as well as to develop good workmanship.

Any time a code-quality weld requiring 100% joint penetration is to be made on metal thicker than 1/4 in. (6 mm), the metal edges must be prepared before welding. A joint is prepared for welding by cutting a groove in the metal along the edge. The preparation is done to allow deeper penetration into the joint of the weld for improved strength. Prepared joints often require more than one weld pass to complete them. By preparing the joint, metal several feet thick can be welded with great success, **Figure 6-1**. The same welding techniques are used when making prepared welds of any thickness.

The root pass is used to fuse the parts together and seal off possible atmospheric contamination from the filler weld. Once the root pass is completed and cleaned, a hot pass may be used to improve the weld contour and burn out small spots of trapped slag. For high-quality welds, a grinder should be used on the root pass to clean it.

Filler welds and cover welds are often made with low hydrogen electrodes such as E7018. These passes are used to fill and cap the weld groove.

Welders are often qualified by passing a required qualification test using a groove weld. The type of joint, thickness of metal, type and size of electrode, and position are all specified by agencies issuing codes and standards. Except for the American Welding Society's (AWS) Certified Welder program, taking a test according to one company or agency's specifications may not qualify a welder for another company or agency's testing procedures. But being able to pass one type of test will usually help the welder to pass other tests for the same type of joint, thickness of metal, type and size of electrode, and position. Information about the AWS Certified Welder program is available from the AWS's main office in Miami, Florida.

Root Pass

The **root pass** is the first weld bead of a **multiple pass weld.** The root pass fuses the two parts together and establishes the depth of weld metal penetration. A good root pass is needed to obtain a sound weld. The root may be either open or closed, using a backing strip or backing ring, **Figure 6-2**.

The backing strip used in a closed root may remain as part of the weld, or it may be removed. Because leaving the backing strip on a weld may cause it to fail due to concentrations of stresses along the backing strip, removable backup tapes have been developed. Backup tapes are made of high-temperature ceramics, **Figure 6-3**, that can be used to increase penetration and prevent burnthrough. The tape can be peeled off after the weld is completed. Most welds do not use backing strips.

On plates that have the joints prepared on both sides, the root face may be ground or gouged clean before another pass is applied to both sides, **Figure 6-4**. This practice has been applied to some large diameter pipes. However, welds that can be reached from only one side must be produced adequately, without the benefit of being able to clean and repair the back side.

The open root weld is widely used in plate and pipe designs. The face side of an open root weld is not so important as the root surface on the back or inside,

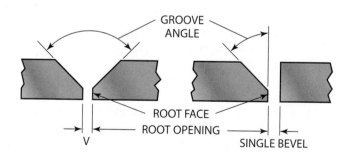

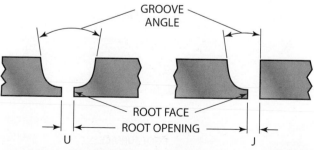

FIGURE 6-1 Standard grooves used to ensure satisfactory joint penetration. © Cengage Learning 2012

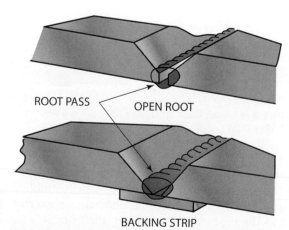

FIGURE 6-2 Root pass maximum deposit 1/4 in. (6 mm) thick. © Cengage Learning 2012

(A) FIBERGLASS

(B) WELD ROOT PASS MADE
USING CERAMIC BACKING TAPE

FIGURE 6-3 Welding backing tapes are available in different materials and shapes. (A) Fiberglass. (B) Weld root pass made using ceramic backing tape. © Cengage Learning 2012

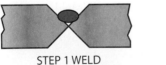

STEP 1 WELD

STEP 2 BACK GOUGE

STEP 3 BACK WELD

STEP 4 COMPLETE WELDING

FIGURE 6-4 Using back gouging to ensure a sound weld root. © Cengage Learning 2012

the weld will be evaluated from the root side only as long as there are not too many defects on the face. To practice the open root welds, the welder will be using mild steel plate that is 1/8 in. (3 mm) thick. The root face for most grooved joints will be about the same size. This thin plate will help the welder build skill without taking too much time beveling plate just to practice the root pass. Two different methods are used to make a root pass. One method

Figure 6-5. The face of a root weld may have some areas of poor uniformity in width, reinforcement, and buildup or have other defects such as undercut or overlap. As long as the root surface is correct, the front side can be ground, gouged, or burned out to produce a sound weld, **Figure 6-6.** For this reason, during the root pass practices,

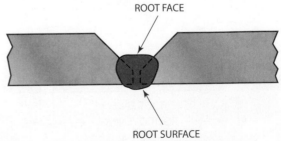

ROOT FACE

ROOT SURFACE

FIGURE 6-5 Ideal bead shape for the root pass.
© Cengage Learning 2012

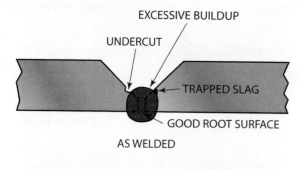

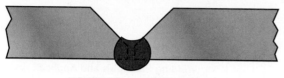

FIGURE 6-6 Grinding back the root pass to ensure a sound second pass. © Cengage Learning 2012

is used only on joints with little or no root gap. This method requires a high amperage and short arc length. The arc length is so short that the electrode flux may drag along on the edges of the joint. The setup for this method must be correct for it to work.

The other method can be used on joints with wide, narrow, or varying root gaps. A stepping electrode manipulation and key hole control the penetration. The electrode is moved in and out of the molten weld pool as the weld progresses along the joint. The edge of the metal is burned back slightly by the electrode just ahead of the molten weld pool, **Figure 6-7**. This is referred to as a **key hole,** and metal flows through the key hole to the root surface.

The key hole must be maintained to ensure 100% penetration. This method requires more welder skill and can be used on a wide variety of joint conditions. The face of the bead resulting from this technique often is defect free.

PRACTICE 6-1

Root Pass on Plate with a Backing Strip in All Positions

Using a properly set up and adjusted welding machine; proper safety protection; E6010 or E6011 arc welding electrodes having a 1/8 in. (3 mm) diameter; one or more pieces of mild steel plate, 1/8 in. (3 mm) thick × 6 in. (152 mm) long; and one strip of mild steel, 1/8 in. (3 mm) thick × 1 in. (25 mm) wide × 6 in. (152 mm) long, you will make a root weld in all positions, **Figure 6-8**. Tack weld the plates together with a 1/16-in. (2-mm) to 1/8-in. (3-mm) root opening. Be sure there are no gaps between the backing strip and plates when the pieces are tacked together, **Figure 6-9**. If there is a small gap between the backing strip and the plates, it can be removed by placing the assembled test plates on an anvil and striking the tack weld with a hammer. This will close up the gap by compressing the tack welds, **Figure 6-10**.

Use a straight step or "T" pattern for this root weld. Push the electrode into the root opening so that there is good fusion with the backing strip and bottom edge of the plates. Failure to push the penetration deep into the joint will result in a cold lap at the root, **Figure 6-11**.

Watch the **molten weld pool** and keep its size as uniform as possible. As the molten weld pool increases in size, move the electrode out of the weld pool. When the weld pool begins to cool, bring the electrode back into the molten weld pool. Use these weld pool indications to determine how far to move the electrode and when to return to the molten weld pool. After completing the weld, cut the plate, and inspect the cross section of the weld for good fusion at the edges. Repeat the welds as necessary until you can consistently make welds free of defects. Turn off the welding machine and clean up your work area when you are finished welding.

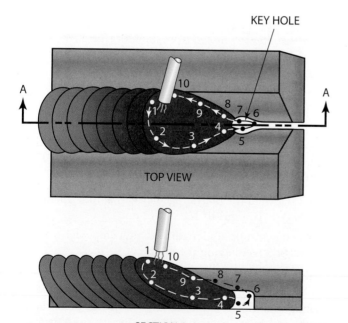

FIGURE 6-7 Electrode movement to open and use a key hole. © Cengage Learning 2012

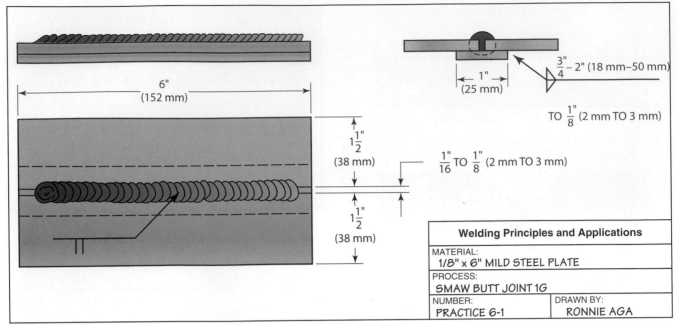

FIGURE 6-8 Square butt joint with a backing strip. © Cengage Learning 2012

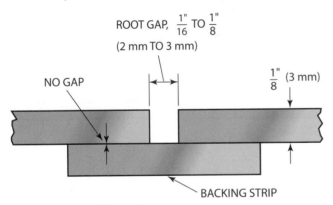

FIGURE 6-9 Backing strip. © Cengage Learning 2012

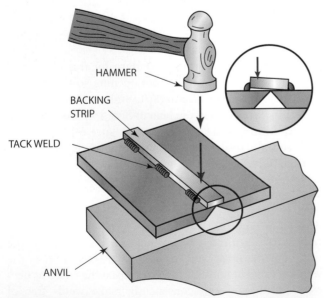

FIGURE 6-10 Using a hammer to align the backing strip and weld plates. © Cengage Learning 2012

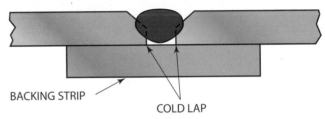

FIGURE 6-11 Incomplete root fusion. © Cengage Learning 2012

Complete a copy of the "Student Welding Report" listed in Appendix I or provided by your instructor. ◆

PRACTICE 6-2

Root Pass on Plate with an Open Root in All Positions

Using a properly set up and adjusted arc welding machine; proper safety protection; E6010 or E6011 arc welding electrodes with a 1/8 in. (3 mm) diameter; and two or more pieces of mild steel plate, 6 in. (152 mm) long × 1/8 in. (3 mm) thick, you will make a welded butt joint in all positions with 100% root penetration.

■ Tack weld the plates together with a root opening of 0 in. (0 mm) to 1/16 in. (2 mm).

■ Using a short arc length and high amperage setting, make a weld along the joint.

You can change the electrode angle to control penetration and burnthrough. As the trailing angle is decreased, making the electrode flatter to the plate, penetration, depth, and burnthrough decrease, **Figure 6-12**, because both the arc force and heat are directed away from the bottom of the joint back toward the weld. Surface tension

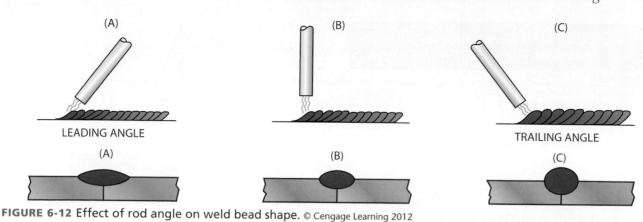

FIGURE 6-12 Effect of rod angle on weld bead shape. © Cengage Learning 2012

holds the metal in place, and the mass of the bead quickly cools the molten weld pool holding it in place. Increasing the electrode angle toward the perpendicular will increase penetration depth and possibly cause more burnthrough. The arc force and heat focused on the gap between the plates will push the molten metal through the joint.

The electrode holder can be slowly rocked from side to side while keeping the end of the electrode in the same spot on the joint, **Figure 6-13**. This will allow the arc force to better tie in the sides of the root to the base metal.

- When a burnthrough occurs, rapidly move the electrode back to a point on the weld surface just before the burnthrough.

- Lower the electrode angle and continue welding. If the burnthrough does not close, stop the weld, chip, and wire brush the weld.

- Check the size of the burnthrough. If it is larger than the diameter of the electrode, the root pass must be continued with the step method described in Practice 6-3. If the burnthrough is not too large, lower the amperage slightly and continue welding.

- Watch the color of the slag behind the weld. If the weld metal is not fusing to one side, the slag will be brighter in color on one side. The brighter color is caused by the slower cooling of the slag because there is less fused metal to conduct the heat away quickly.

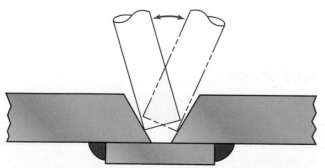

FIGURE 6-13 Rocking the top of the electrode while keeping the end in the same place helps control the bead shape. © Cengage Learning 2012

FIGURE 6-14 The root face (inside) appears uniform.
Larry Jeffus

- After the weld is completed, cooled, and chipped, check the back side of the plate for good root penetration. The root should have a small bead that will look as though it was welded from the back side, **Figure 6-14**. The penetration must be completely free of any drips of metal from the root face, called "icicles."

- Repeat the welds as necessary until you can consistently make welds free of defects. Turn off the welding machine and clean up your work area when you are finished welding.

Complete a copy of the "Student Welding Report" listed in Appendix I or provided by your instructor. ◆

PRACTICE 6-3

Open Root Weld on Plate Using the Step Technique in All Positions

Using the same setup, materials, and electrodes as described in Practice 6-2, you will make a welded butt joint in all positions with 100% root penetration.

	Amperage	Travel Speed	Electrode Size	Electrode Angle
To decrease puddle size	Decrease	Increase	Decrease	Leading
To increase puddle size	Increase	Decrease	Increase	Trailing

TABLE 6-1 Changes Affecting Molten Weld Bead Size

Tack weld the plates together with a root opening from 0 in. (0 mm) to 1/8 in. (3 mm). Using a medium amperage setting and a short stepping electrode motion, make a weld along the joint.

The electrode should be pushed deeply into the root to establish a key hole that will be used to ensure 100% root penetration. Once the key hole is established, the electrode is moved out and back in the molten weld pool at a steady, rhythmic rate. Watch the molten weld pool and key hole size to determine the rhythm and distance of electrode movement.

If the molten weld pool size decreases, the key hole will become smaller and may close completely. To increase the molten weld pool size and maintain the key hole, slow the rate of electrode movement and shorten the distance the electrode is moved away from the molten weld pool. This will increase the molten weld pool size and penetration because of increased localized heating.

If the molten weld pool becomes too large, metal may drip through the key hole, forming an icicle on the back side of the plate. Extremely large molten weld pool sizes can cause a large hole to be formed or cause **burnthrough.** Repairing large holes can require much time and skill. To keep the molten weld pool from becoming too large, increase the travel speed, decrease the angle, shorten the arc length, or lower the amperage, Table 6-1.

The distance the electrode is moved from the molten weld pool and the length of time in the molten weld pool are found by watching the molten weld pool. The molten weld pool size increases as you hold the arc in the molten weld pool until it reaches the desired size, about twice the electrode diameter, **Figure 6-15.** Move the electrode ahead of the molten weld pool, keeping the arc in the joint but being careful not to deposit any slag or metal ahead of the weld. To prevent metal and/or slag from transferring, raise the electrode to increase the arc length, **Figure 6-16.** Keep moving the electrode slowly forward as you watch the

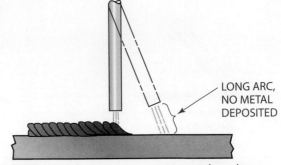

FIGURE 6-16 A long arc prevents metal or slag from being deposited ahead of the weld bead.
© Cengage Learning 2012

molten weld pool. The molten weld pool will suddenly start to solidify. At that time, move the electrode quickly back to the molten weld pool before it totally solidifies. Moving the electrode in a slight arc will raise the electrode ahead of the molten weld pool and automatically lower the electrode when it returns to the molten weld pool. Metal or slag deposited ahead of the molten weld pool may close the key hole, reduce penetration, and cause slag inclusions. Raising the end of the electrode too high or moving it too far ahead of the molten weld pool can cause all of the shielding gas to be blown away from the molten weld pool. If this happens, oxides can cause porosity. Keeping the electrode movement in balance takes concentration and practice.

Changing from one welding position to another requires an adjustment in timing, amperage, and electrode angle. The flat, horizontal, and overhead positions use about the same rhythm, but the vertical position may require a shorter time cycle for electrode movement. The amperage for the vertical position can be lower than that for the flat or horizontal, but the overhead position uses nearly the same amperage as flat and horizontal. The electrode angle for the flat and horizontal positions is about the same. For the vertical position, the electrode uses a sharper leading angle than does overhead, which is nearly perpendicular and may even be somewhat trailing.

Cool, chip, wire brush, and inspect both sides of the weld. The root surface should be slightly built up and look as though it was welded from that side (refer to Figure 6-14). Repeat the welds as necessary until you can consistently make welds free of defects. Turn off the welding machine and clean up your work area when you are finished welding.

Complete a copy of the "Student Welding Report" listed in Appendix I or provided by your instructor. ◆

Hot Pass

The surface of a root pass may be irregular, have undercut, overlap, slag inclusions, or other defects, depending upon the type of weld, the code or standards, and the condition of the root pass. The surface of a root pass can be cleaned by grinding or by using a hot pass.

On critical, high-strength code welds it is usually required that the root pass as well as each filler pass be ground (refer to Figure 6-6). This grinding eliminates weld

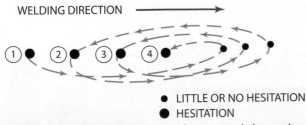

WELDING DIRECTION ⟶

● LITTLE OR NO HESITATION
● HESITATION

FIGURE 6-15 Weave pattern used to control the molten weld bead size. © Cengage Learning 2012

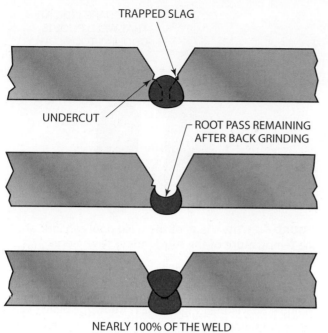

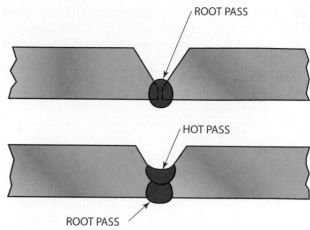

TRAPPED SLAG

UNDERCUT

ROOT PASS REMAINING AFTER BACK GRINDING

NEARLY 100% OF THE WELD DEPOSIT IS NOW THE HIGHER STRENGTH FILLER

FIGURE 6-17 Back grinding to remove both discontinuities; filler metal used for the root pass. © Cengage Learning 2012

ROOT PASS

HOT PASS

ROOT PASS

FIGURE 6-18 Using the hot pass to clean up the face of the root pass. © Cengage Learning 2012

discontinuities caused by slag entrapments. It also can be used to remove most of the E60 series weld metal so that the stronger weld metal can make up most of the weld. When high-strength, low-alloy welding electrodes are used, this grinding is important to remove most of the low-strength weld deposit. This will leave the weld made up of nearly 100% of the high-strength weld metal, **Figure 6-17**.

The fastest way to clean out trapped slag and make the root pass more uniform is to use a **hot pass**. The hot pass uses a higher-than-normal amperage setting and a fast

travel rate to reshape the bead and burn out the trapped slag. After chipping and wire brushing the root pass to remove all the slag possible, a welder is ready to make the hot pass. The ideal way to apply a hot pass is to rapidly melt a large surface area, **Figure 6-18**, so that the trapped slag can float to the surface. The slag, mostly silicon dioxide (SiO_2), may not melt itself so the surrounding steel must be melted to enable it to float free. The silicon dioxide may not melt because it melts at about 3100°F (1705°C), which is more than 500°F (270°C) hotter than the temperature at which the surrounding steel melts, around 2600°F (1440°C).

A very small amount of metal should be deposited during the hot pass so that the resulting weld is concave. A concave weld, compared to a convex weld, is more easily cleaned by chipping, wire brushing, or grinding. Failure to clean the convex root weld will result in a discontinuity showing up on an X-ray. Such discontinuities are called **wagon tracks**, Figure 6-19.

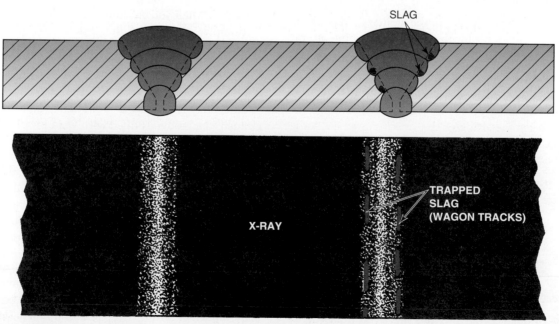

SLAG

TRAPPED SLAG (WAGON TRACKS)

X-RAY

FIGURE 6-19 Slag trapped between passes will show on an X-ray. © Cengage Learning 2012

The hot pass can also be used to repair or fill small spots of incomplete fusion or pinholes left in the root pass.

The normal weave pattern for a hot pass is the straight step or "T" pattern. The "T" can be used to wash out stubborn trapped slag better than the straight step pattern. The frequency of electrode movement is dependent upon the time required for the molten weld pool to start cooling. As with the root pass, metal or slag should not be deposited ahead of the bead. Do not allow the molten weld pool to cool completely or let the shielding gas covering be blown away from the molten weld pool.

The hot pass technique can also be used to clean some welds that may first require grinding or gouging for a repair. The penetration of the molten weld pool must be deep enough to free all trapped slag and burn out all porosity.

EXPERIMENT 6-1

Hot Pass to Repair a Poor Weld Bead

Using a properly set up and adjusted arc welding machine; proper safety protection; E6010 or E6011 arc welding electrodes having a 1/8 in. (3 mm) diameter; and two or more plates that have welds containing slag inclusions, lack of fusion, porosity, or other defects; you will make a hot pass to burn out the defects.

Chip and wire brush the weld bead. If necessary, use a punch to break apart large trapped slag deposits. The poorer the condition of the weld, the more vertical the joint should be for the hot pass. Large slag deposits tend to float around the molten weld pool and stay trapped in deep pockets in the flat position. With the weld in the vertical position, the slag can run out of the joint and down the face of the weld. Set the amperage as high as possible without overheating and burning up the electrode. Start at the bottom and weld upward using a combination of straight step and "T" patterns to keep the weld deposit uniform. Watch the back edge of the molten weld pool for size and the weld crater for the complete burning out of impurities, **Figure 6-20**.

The plate may start to become overheated because of the high heat input. If you notice that the weld bead is starting to cool too slowly and is growing in length, you

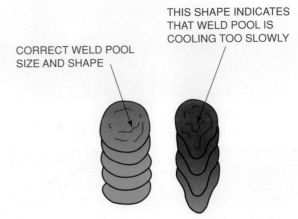

CORRECT WELD POOL
SIZE AND SHAPE

THIS SHAPE INDICATES
THAT WELD POOL IS
COOLING TOO SLOWLY

FIGURE 6-21 The shape of the weld pool can indicate the temperature of the surrounding base metal. © Cengage Learning 2012

should stop welding, **Figure 6-21**. Allow the plate to cool before continuing the weld.

/// /// CAUTION /// ///

This hot pass technique is designed to be used on noncritical, noncode welds only. It should not be used to cover bad welds, or as a means of repairing the work of a welder who is less skilled.

After the weld is completed, cool, chip, and inspect it for uniformity. The plate can be cut at places where you know large discontinuities existed before to see if they were repaired or only covered up. If you wish, this experiment can be repeated on other defects and joints.

Welds that have large defects in addition to excessive buildup may require some grinding to remove the buildup. Turn off the welding machine and clean up your work area when you are finished welding.

Complete a copy of the "Student Welding Report" listed in Appendix I or provided by your instructor. ◆

Filler Pass

After the root pass is completed and it has been cleaned, the groove is filled with weld metal. These weld beads make up the **filler passes.** More than one pass is often required.

Filler passes are made with stringer beads or weave beads. For multiple pass welds, the weld beads must overlap along the edges. They should overlap enough so that the finished bead is smooth, **Figure 6-22.** Stringer beads usually overlap about 50%, and weave beads overlap approximately 25%.

Each weld bead must be cleaned before the next bead is started. Slag left on the plate between welds cannot be completely burned out because filler welds should be made with a low amperage setting. Deep penetration will slow the rate of buildup in the joint. Deeply remelting the previous weld

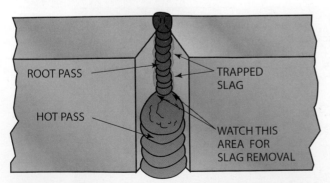

ROOT PASS

HOT PASS

TRAPPED SLAG

WATCH THIS AREA FOR SLAG REMOVAL

FIGURE 6-20 Burning out trapped slag by using a hot pass. © Cengage Learning 2012

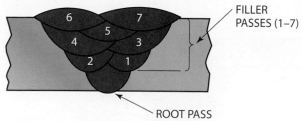

FIGURE 6-22 Filler passes—maximum thickness 1/8 in. (3 mm) each pass. © Cengage Learning 2012

metal may weaken the joint. All that is required of a filler weld is that it be completely fused to the base metal.

Chipping, wire brushing, and grinding are the best ways to remove slag between filler weld passes. After the weld is completed, it can be checked by ultrasonic or radiographic nondestructive testing. Most schools are not equipped to do this testing. Therefore, a quick check for soundness can be made by destructive testing. One method of testing the deposited weld metal is by cutting and cross sectioning the weld with an abrasive wheel and inspecting the weld. Another fast way to inspect filler passes is to cut a groove through the weld with a gouging tip. Watch the hot metal as it is washed away. The black spots that appear in the cut are slag inclusions. If only a few small spots appear, the weld probably will pass most tests. But if a long string or large pieces of inclusions appear, the weld will most likely fail.

PRACTICE 6-4

Multiple Pass Filler Weld on a V-Joint in All Positions

Using a properly set up and adjusted arc welding machine; proper safety protection; E6010 or E6011 arc welding electrodes having a 1/8 in. (3 mm) diameter; and two or more pieces of mild steel plate, 6 in. (152 mm) long × 3 in. (76 mm) wide × 3/8 in. (10 mm) thick, you will make a multiple pass filler weld on a V-joint.

Tack weld the plates together at the corner so that they form a V, **Figure 6-23**. Starting at one end, make a stringer bead along the entire length using the straight step or "T" weave pattern. Thoroughly clean off the slag from the weld before making the next bead. **Figure 6-24** shows the suggested sequence for locating the beads. Continue making welds and cleaning them until the weld is 1 in. (25 mm) or more thick. Both ends of the weld may taper down. If it is important that the ends be square, metal tabs are welded on the ends of the plate for starting and stopping, **Figure 6-25**. The tabs are removed after the weld is completed.

After the weld is completed, it can be visually inspected for uniformity. If nondestructive testing is available, it may be checked for discontinuities. The weld also may be inspected by sectioning it with an abrasive wheel or gouging out with a torch. Repeat these welds until they are mastered. Turn off the welding machine and clean up your work area when you are finished welding.

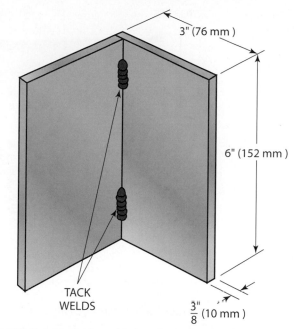

FIGURE 6-23 Tack weld the plates for the filler weld practice. © Cengage Learning 2012

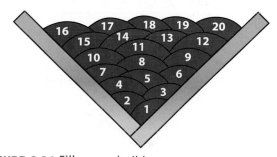

FIGURE 6-24 Filler pass buildup sequence. © Cengage Learning 2012

Complete a copy of the "Student Welding Report" listed in Appendix I or provided by your instructor. ◆

PRACTICE 6-5

Multiple Pass Filler Weld on a V-Joint in All Positions Using E7018 Electrodes

Using the same setup and materials as in Practice 6-4, and E7018 arc welding electrodes in place of E6010 or E6011 electrodes, you will make a multiple pass filler weld on a V-joint.

- Tack weld the two plates together at the corners so that they form a V.

- Using a slow, straightforward motion, with little or no stepping, "T" or inverted "V" motion, make a stringer bead along the root of the joint.

- Chip the slag and repeat the weld until there is a buildup of 1 in. (25 mm) or more.

STOP TAB   START TAB

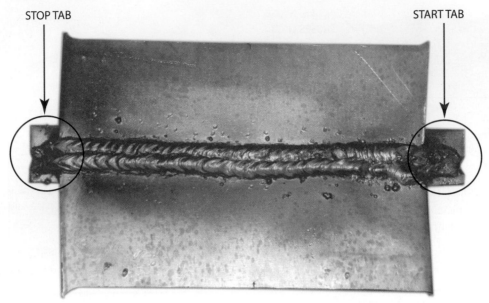

FIGURE 6-25 Run-off tabs help control possible underfill or burn back at the starting and stopping points of a groove weld. Larry Jeffus

If during the weld the buildup becomes uneven or large slag entrapments occur, they should be ground out. In industry, groove welds in plate 1 in. (25 mm) or more are normally repaired. Making these repairs now is good experience for the welder. All welders will at some time make a weld that may need repairing.

- After the weld is completed, visually test and non-destructive test the weld for external and internal discontinuities.
- Repeat the welds as necessary until you can consistently make welds free of defects. Turn off the welding machine and clean up your work area when you are finished welding.

Complete a copy of the "Student Welding Report" listed in Appendix I or provided by your instructor. ◆

Cover Pass

The last weld bead on a multipass weld is known as the **cover pass.** The cover pass may use a different electrode weave, or it may be the same as the filler beads. Keeping the cover pass uniform and neat looking is important. Most welds are not tested, and often the inspection program is only visual. Thus, the appearance might be the only factor used for accepting or rejecting welds.

The cover pass should be free of any visual defects such as undercut, overlap, porosity, or slag inclusions. It should be uniform in width and reinforcement. A cover pass should not be more than 1/8 in. (3 mm) wider than the groove opening, **Figure 6-26.** Cover passes that are too wide do not add to the weld strength.

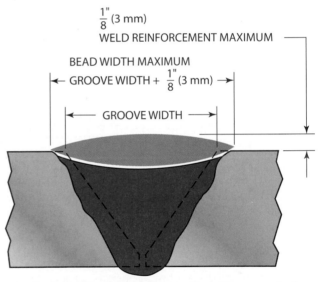

$\frac{1"}{8}$ (3 mm)

WELD REINFORCEMENT MAXIMUM

BEAD WIDTH MAXIMUM
← GROOVE WIDTH + $\frac{1"}{8}$ (3 mm) →

← GROOVE WIDTH →

FIGURE 6-26 The cover pass should not be excessively large. © Cengage Learning 2012

PRACTICE 6-6

Cover Bead in All Positions

Using a properly set up and adjusted arc welding machine; proper safety protection; E7018 arc welding electrodes having a 1/8 in. (3 mm) diameter; and one or more pieces of mild steel plate, 6 in. (152 mm) long × 1/4 in. (6 mm) thick, you will make a cover bead in each position.

THINK GREEN
Don't Overweld

Keeping the cover pass and reinforcement within the specifications is important for both weld strength and to save excessive expenses of both overwelding and later weld refinishing to remove excessive weld metal. Preventing overwelding will save a great deal of energy both during the welding process and the postweld cleanup.

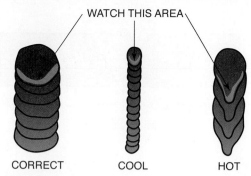

WATCH THIS AREA

CORRECT   COOL   HOT

FIGURE 6-28 Watch the back edge of the weld pool to determine the correct current. © Cengage Learning 2012

Remember, any time an E7018 low-hydrogen–type electrode is to be used, the weave pattern, if used, must not be any larger than two and a half times the diameter of the electrode. This weave cannot be any larger than 5/16 in. (7 mm) wide. Start welding at one end of the plate and weld to the other end. The weld bead should be about 5/16 in. (7 mm) wide having no more than 1/8 in. (3 mm) of uniform buildup, **Figure 6-27**. The weld buildup should have a smooth transition at the toe, with the plate and the face somewhat convex. Undercut at the toe and a concave or excessively built-up face are the most common problems. Watch the sides of the bead for undercut. When undercut occurs, keep the electrode just ahead of the spot until it is filled in. There should be a smooth transition between the weld and the plate (refer to Figure 6-26). The shape of the bead face can be controlled by watching the trailing edge of the molten weld pool. That trailing edge is the same as the finished bead, **Figure 6-28**.

Deep penetration is not required with this weld and may even result in some weakening. After the weld is completely cool, chip and inspect it for uniformity and defects. Repeat the welds as necessary until you can consistently make welds free of defects. Turn off the welding machine and clean up your work area when you are finished welding.

Complete a copy of the "Student Welding Report" listed in Appendix I or provided by your instructor. ◆

Plate Preparation

When welding on thick plate, it is impossible or impractical for the welder to try to get 100% penetration without preparing the plate for welding. The preparation of the plate is usually in the form of a **weld groove.** The groove can be cut into one side or both sides of the plate, and it may be cut into either just one plate or both plates of the joint, **Figure 6-29**. The type, depth, angle, and location of the groove are usually determined by a code standard that has been qualified for the specific job.

For SMA welds on plate 1/4 in. (6 mm) or thicker that need to have a weld with 100% joint penetration, the plate must be grooved. The groove may be ground, flame cut, plasma cut, gouged, or machined on the edge of the plate before or after the assembly. Bevels and V-grooves are best if they are cut before the parts are assembled. J-grooves and U-grooves can be cut either before or after assembly, **Figure 6-30**. The lap joint is seldom prepared with a groove because little or no strength can be gained by grooving this joint. The only advantage to grooving the lap joint design is to give additional clearance.

Plates that are thicker than 3/8 in. (10 mm) can be grooved on both sides but may be prepared on only one side. The choice to groove one or both sides is most times determined by joint design, position, and application. A tee joint in thick plate is easier to weld and will have less distortion if it is grooved on both sides. Plate in the flat position is usually grooved on only one side unless it can be repositioned. Welds that must have little distortion or that are going to be loaded equally from both sides are usually grooved on both sides. Sometimes plates are either grooved and welded or just welded on one side, and then back gouged and welded, **Figure 6-31. Back gouging** is a process of cutting a groove in the back side of a joint that has been welded. Back gouging can ensure 100% fusion at the root and remove discontinuities of the root pass. This process can also remove the root pass metal if the properties of the metal are not desirable to the finished

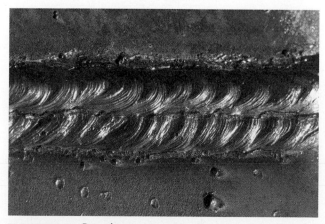

FIGURE 6-27 Practice cover pass. Larry Jeffus

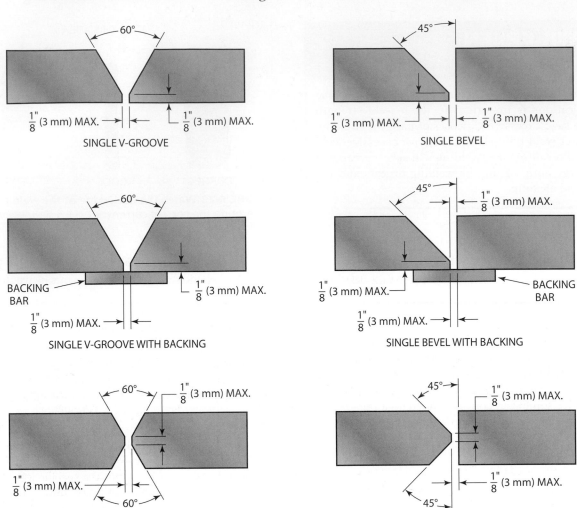

FIGURE 6-29 Typical butt joint preparations. © Cengage Learning 2012

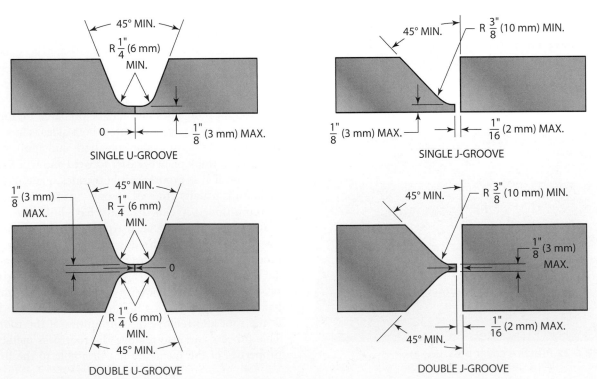

FIGURE 6-30 Typical butt joint preparations. © Cengage Learning 2012

STEP 1

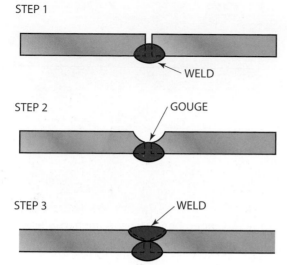

WELD

STEP 2

GOUGE

STEP 3

WELD

FIGURE 6-31 Back gouging sequence for a weld to ensure 100% joint penetration. © Cengage Learning 2012

weld, **Figure 6-32**. After back gouging, the groove is then welded. See Section 3 for more information on the various methods of gouging.

Heavy plate and pipe sections requiring preparations are often used in products manufactured under a code or standard. The American Welding Society (AWS), American Society of Mechanical Engineers (ASME), and American Bureau of Ships are a few of the agencies that issue codes and specifications. The AWS D1.1 and the ASME Boiler and Pressure Vessel (BPV) Section IX standards will be used in this chapter as the standards for multiple pass groove welds that will be tested. The groove depth and angle are determined by the plate or pipe thickness and process.

STEP 1. FINISH WELD ONE SIDE

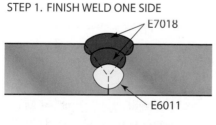

E7018

E6011

STEP 2. BACK GOUGE TO REMOVE ALL E6011 DEPOSITED IN FIRST WELD

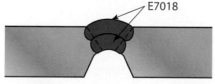

E7018

STEP 3. WELD U-GROOVE

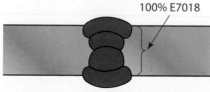

100% E7018

FIGURE 6-32 Back gouging to remove all weld metal used for the root pass or tacking. © Cengage Learning 2012

Chapter 22 covers these and other codes for welding plate and pipe.

Refer to Section 3 for information on beveling. After the plate is beveled, a grinder can be used to clean off oxides and improve the fit-up.

Preparing Specimens for Testing

The detailed preparation of specimens for testing in this chapter is based on the structural welding code AWS D1.1 and the ASME BPV Code, Section IX. The maximum allowable size of fissures (cracks or openings) in a **guided-bend** test specimen is given in codes for specific applications. Some of the standards are listed in ASTM E190 or AWS B4.0, AWS QC10, AWS QC11, and others. Copies of these publications are available from the appropriate organizations. More information on tests and testing can be found in Chapter 23.

Acceptance Criteria for Face Bends and Root Bends

The weld specimen must first pass visual inspection before it can be prepared for bend testing. Visual inspection looks to see that the weld is uniform in width and reinforcement. There should be no arc strikes on the plate other than those on the weld itself. The weld must be free of both incomplete fusion and cracks. The joint penetration must be either 100% or as required by the specifications. The weld must be free of overlap, and undercut must not exceed either 10% of the base metal or 1/32 in. (0.8 mm), whichever is less.

Correct **weld specimen** preparation is essential for reliable results. The weld must be uniform in width and reinforcement and have no undercut or overlap. The weld reinforcement and backing strip, if used, must be removed flush to the surface, **Figure 6-33**. They can be machined or

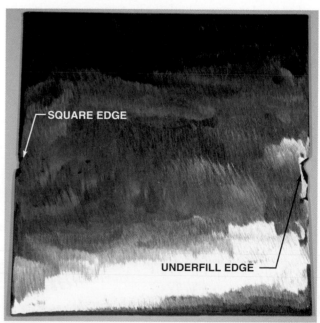

SQUARE EDGE

UNDERFILL EDGE

FIGURE 6-33 Plate ground in preparation for removing test specimens. Larry Jeffus

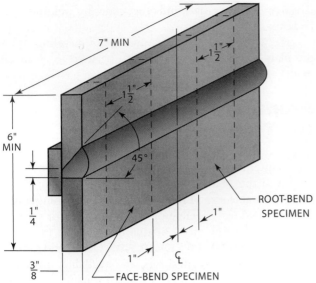

FIGURE 6-34 Sequence for removing guided bend specimens from the plate once welding is complete.
© Cengage Learning 2012

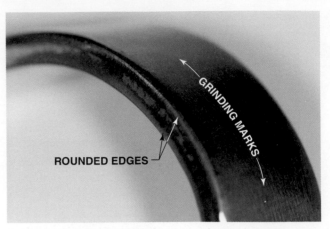

FIGURE 6-36 Guided-bend test specimen. Larry Jeffus

ground off. The plate thickness after removal must be a minimum of 3/8 in. (10 mm), and the pipe thickness must be equal to the pipe's original wall thickness. The specimens may be cut out of the test weldment using an abrasive disc, by sawing, or by cutting with a torch. Flame-cut specimens must have the edges ground or machined smooth after cutting. This procedure is done to remove the heat-affected zone caused by the cut, **Figure 6-34.**

All corners must be rounded to a radius of 1/8 in. (3 mm) maximum, and all grinding or machining marks must run lengthwise on the specimen, **Figure 6-35** and **Figure 6-36.** Rounding the corners and keeping all marks running lengthwise reduce the chance of good weld specimen failure due to poor surface preparation.

The weld must pass both the face and root bends to be acceptable. After bending, there can be no single defects larger than 1/8 in. (3.2 mm), and the sum of all defects larger than 1/32 in. (0.8 mm) but less than 1/8 in. (3.2 mm) must not exceed a total of 3/8 in. (9.6 mm) for each bend specimen. An exception is made for cracks that start at

the edge of the specimen and do not start at a defect in the specimen.

Restarting a Weld Bead

On all but short welds, the welding bead will need to be restarted after a welder stops to change electrodes. Because the metal cools as a welder changes electrodes and chips slag when restarting, the penetration and buildup may be adversely affected.

When a weld bead is nearing completion, it should be tapered so that when it is restarted the buildup will be more uniform. To taper a weld bead, the travel rate should be increased just before welding stops. A 1/4-in. (6-mm) taper is all that is required. The taper allows the new weld to be started and the depth of penetration reestablished without having excessive buildup, **Figure 6-37.**

The slag should always be chipped and the weld crater should be cleaned each time before restarting the weld. This is important to prevent slag inclusions at the start of the weld.

The arc should be restarted in the joint ahead of the weld. The electrodes must be allowed to heat up so that the arc is stabilized and a shielding gas cloud is reestablished to protect the weld. Hold a long arc as the electrode heats up so that metal is not deposited. Slowly bring the

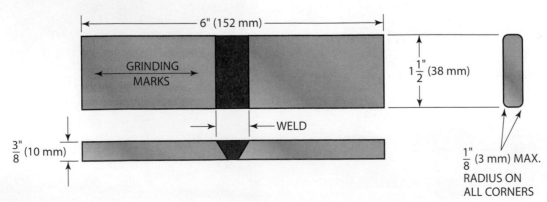

FIGURE 6-35 Guided-bend specimen. © Cengage Learning 2012

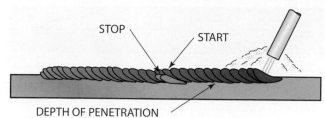

FIGURE 6-37 Tapering the size of the weld bead helps keep the depth of penetration uniform.
© Cengage Learning 2012

electrode downward and toward the weld bead until the arc is directly on the deepest part of the crater where the crater meets the plate in the joint, **Figure 6-38**. The electrode should be low enough to start transferring metal. Next, move the electrode in a semicircle near the back edge of the weld crater. Watch the buildup and match your speed in the semicircle to the deposit rate so that the weld is built up evenly, **Figure 6-39**. Move the electrode ahead and continue with the same weave pattern that was being used previously.

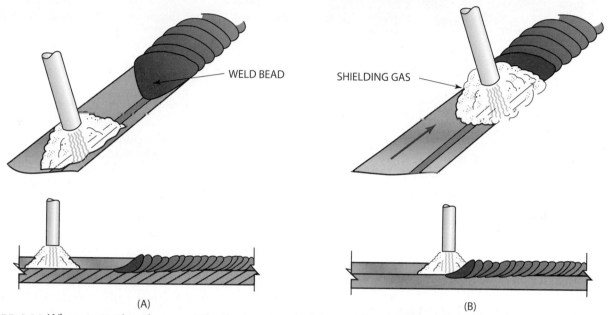

(A)

(B)

FIGURE 6-38 When restarting the arc, strike the arc ahead of the weld in the joint (A). Hold a long arc, and allow time for the electrode to heat up, forming the protective gas envelope. Move the electrode so that the arc is focused directly on the leading edge (root) of the previous weld crater (B). © Cengage Learning 2012

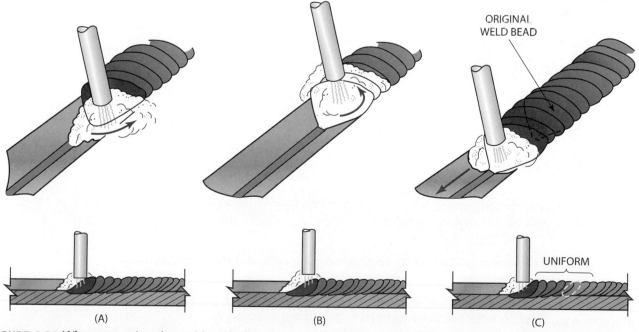

(A)

(B)

(C)

FIGURE 6-39 When restarting the weld pool after the root has been heated to the melting temperature, move the electrode upward along one side of the crater (A). Move the electrode along the top edge, depositing new weld metal (B). When the weld is built up uniformly with the previous weld, continue along the joint (C). © Cengage Learning 2012

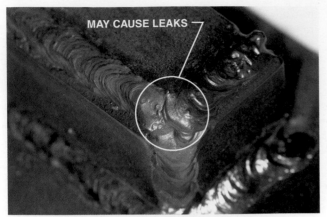

FIGURE 6-40 Correct method of welding through a corner. Stopping on a corner may cause leaks.

Larry Jeffus

Plate Thickness in. (mm)	Minimum Temperature	
	°F	°C
Up to 1/2 in. (13 mm)	70	21
1/2 in. (13 mm) to 1 in. (25 mm)	100	38
1 in. (25 mm) to 2 in. (51 mm)	200	95
Over 2 in. (51 mm)	300	150

*Metal should be above the dew point. Allow 1 hour for each inch in order to provide uniform heating or for localized preheating. Check to ensure that preheat temperatures are sufficient to melt thermal crayons a minimum of 3 in. (76 mm) in all directions from the area to be welded prior to welding arc application.

TABLE 6-2 Preheat Temperatures for Arc Welding on Low Carbon Steels

The movement to the root of the weld and back up on the bead serves both to build up the weld and reheat the metal so that the depth of penetration will remain the same. If the weld bead is started too quickly, penetration is reduced and buildup is high and narrow.

Starting and stopping weld beads in corners should be avoided. Tapering and restarting are especially difficult in corners, and this often results in defects, Figure 6-40.

Preheating and Postheating

Preheating is the application of heat to the metal before it is welded. This process helps to reduce cracking, hardness, distortion, and stresses by reducing the thermal shock from the weld and slowing the cooldown rate. Preheating is most often required on large, thick plates; when the plate is very cold; on days when the temperature is very cold; when small diameter electrodes are used; on high-carbon or manganese steels; on complex shapes; or with fast welding speeds.

With the practices that are to be tested in this chapter, preheating should be used if the base metal to be welded is very cold. It may also be used to reduce distortion on thick sections and to reduce hardness caused by the rapid cooling of the weld, which may occur in weld failure. Preheating the metal will slow the weld cooling rate, which results in a more ductile weld. **Table 6-2** lists the recommended preheat temperatures for plain carbon steels.

PRACTICE 6-7

Welding Procedure Specification (WPS)

Welding Procedures Specifications No.: PRACTICE 6-7

Date: _____

Title:
Welding <u>SMAW</u> of <u>plate</u> to <u>plate</u>.

Scope:
This procedure is applicable for <u>V-groove plate with a backing strip</u> within the range of <u>3/8 in. (10 mm)</u> through <u>3/4 in. (20 mm)</u>. Welding may be performed in the following positions: <u>1G, 2G, 3G, and 4G</u>.

Base Metal:
The base metal shall conform to <u>M1020 or A36</u>. Backing material specification <u>M1020 or A36</u>.

Filler Metal:
The filler metal shall conform to AWS specification No. <u>E6010 or E6011 root pass and E7018 for the cover pass</u> from AWS specification <u>A5.1</u>. This filler metal falls into F-numbers <u>F3 and F4</u> and A-number <u>A-1</u>.

Shielding Gas:
The shielding gas, or gases, shall conform to the following compositions and purity: <u>N/A</u>.

Joint Design and Tolerances:

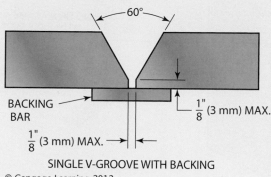

SINGLE V-GROOVE WITH BACKING
© Cengage Learning 2012

Preparation of Base Metal:

The bevel is to be flame or plasma cut on the edge of the plate before the parts are assembled. The beveled surface must be smooth and free of notches. Any roughness or notches that are deeper than 1/64 in. (0.4 mm) are unacceptable and must be ground smooth.

All hydrocarbons and other contaminants, such as cutting fluids, grease, oil, and primers, must be cleaned off all parts and filler metals before welding. This cleaning can be done with any suitable solvents or detergents. The backing strip, groove face, and inside and outside plate surfaces within 1 in. (25 mm) of the joint must be mechanically cleaned of slag, rust, and mill scale. Cleaning must be done with a wire brush or grinder down to bright metal.

Electrical Characteristics:

The current shall be <u>AC or DCRP</u>.

The base metal shall be on the <u>work lead or negative</u> side of the line.

Preheat:

The parts must be heated to a temperature higher than 70°F (21°C) before any welding is started.

Backing Gas:

N/A

Welding Technique:

Tack weld the plates together with the backing strip. There should be about a 1/8-in. (3-mm) root gap between the plates. Use the E6010 or E6011 arc welding electrodes to make a root pass to fuse the plates and backing strip together. Clean the slag from the root pass and use either a hot pass or grinder to remove any trapped slag.

Using the E7018 arc welding electrodes, make a series of filler welds in the groove until the joint is filled. Figure 6-44 shows the recommended location sequence for the weld beads.

Interpass Temperature:

The plate should not be heated to a temperature higher than 350°F (175°C) during the welding process. After each weld pass is completed, allow it to cool; the weldment must not be quenched in water.

Cleaning:

The slag can be chipped and/or ground off between passes but can only be chipped off of the cover pass.

Visual Inspection Criteria for Entry-Level Welders*:

There shall be no cracks, no incomplete fusion. There shall be no incomplete joint penetration in groove welds except as permitted for partial joint penetration welds.

The Test Supervisor shall examine the weld for acceptable appearance and shall be satisfied that the welder is skilled in using the process and procedure specified for the test.

Undercut shall not exceed the lesser of 10% of the base metal thickness or 1/32 in. (0.8 mm).

Where visual examination is the only criterion for acceptance, all weld passes are subject to visual examination at the discretion of the Test Supervisor.

The frequency of porosity shall not exceed one in each 4 in. (100 mm) of weld length, and the maximum diameter shall not exceed 3/32 in. (2.4 mm).

Welds shall be free from overlap.

Bend Test:

The weld is to be mechanically tested only after it has passed the visual inspection. Be sure that the test specimens are properly marked to identify the welder, the position, and the process.

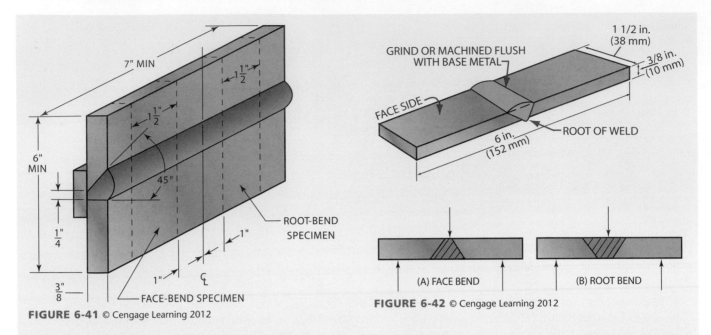

FIGURE 6-41 © Cengage Learning 2012

FIGURE 6-42 © Cengage Learning 2012

(A) FACE BEND

(B) ROOT BEND

Specimen Preparation:

For 3/8-in. (9.5-mm) test plates, two specimens are to be located in accordance with the requirements in **Figure 6-41**. One is to be prepared for a transverse face bend, and the other is to be prepared for a transverse root bend, **Figure 6-42**.

- *Transverse face bend.* The weld is perpendicular to the longitudinal axis of the specimen and is bent so that the weld face becomes the tension surface of the specimen.

- *Transverse root bend.* The weld is perpendicular to the longitudinal axis of the specimen and is bent so that the weld root becomes the tension surface of the specimen.

Acceptance Criteria for Bend Test*:

For acceptance, the convex surface of the face- and root-bend specimens shall meet both of the following requirements:

1. No single indication shall exceed 1/8 in. (3.2 mm) measured in any direction on the surface.

2. The sum of the greatest dimensions of all indications on the surface that exceed 1/32 in. (0.8 mm) but are less than or equal to 1/8 in. (3.2 mm) shall not exceed 3/8 in. (9.5 mm).

Cracks occurring at the corner of the specimens shall not be considered unless there is definite evidence that they result from slag or inclusions or other internal discontinuities.

Complete a copy of the "Student Welding Report" listed in Appendix I or provided by your instructor.

Repair:

No repairs of defects are allowed. ◆

Note the sequence in depositing the beads. © Cengage Learning 2012

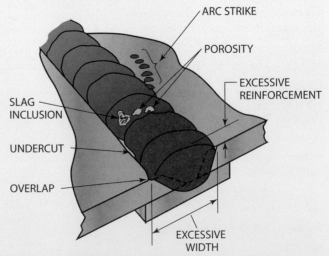

Common discontinuities found during a visual examination. © Cengage Learning 2012

*Courtesy of the American Welding Society.

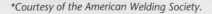

Sketches:

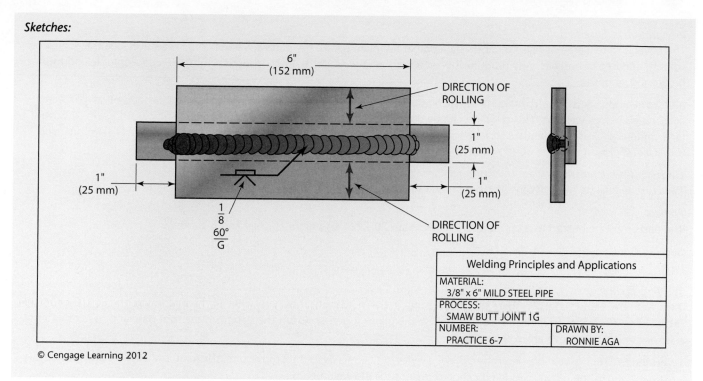

DIRECTION OF ROLLING

1" (25 mm)

1" (25 mm)

1" (25 mm)

6" (152 mm)

DIRECTION OF ROLLING

$\frac{1}{8}$
60°
G

Welding Principles and Applications	
MATERIAL: 3/8" x 6" MILD STEEL PIPE	
PROCESS: SMAW BUTT JOINT 1G	
NUMBER: PRACTICE 6-7	DRAWN BY: RONNIE AGA

© Cengage Learning 2012

PRACTICE 6-8

Welding Procedure Specification (WPS)

Welding Procedures Specifications No.: PRACTICE 6-8

Date: _____

Title:

Welding SMAW of plate to plate.

Scope:

This procedure is applicable for V-groove plate without a backing strip within the range of 3/8 in. (10 mm) through 3/4 in. (20 mm).

Welding may be performed in the following positions: 1G, 2G, 3G, and 4G.

Base Metal:

The base metal shall conform to M1020 or A36.

Backing material specification M1020 or A36.

Filler Metal:

The filler metal shall conform to AWS specification No. E6010 or E6011 root pass and E7018 for the cover pass from AWS specification A5.1. This filler metal falls into F-numbers F3 and F4 and A-number A-1.

Shielding Gas:

The shielding gas, or gases, shall conform to the following compositions and purity: N/A.

Joint Design and Tolerances:

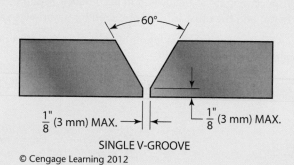

60°

$\frac{1}{8}$" (3 mm) MAX.

$\frac{1}{8}$" (3 mm) MAX.

SINGLE V-GROOVE

© Cengage Learning 2012

Preparation of Base Metal:
The bevel is to be flame or plasma cut on the edge of the plate before the parts are assembled. The beveled surface must be smooth and free of notches. Any roughness or notches that are deeper than 1/64 in. (0.4 mm) are unacceptable and must be ground smooth.

All hydrocarbons and other contaminants, such as cutting fluids, grease, oil, and primers, must be cleaned off all parts and filler metals before welding. This cleaning can be done with any suitable solvents or detergents. The backing strip, groove face, and inside and outside plate surface within 1 in. (25 mm) of the joint must be mechanically cleaned of slag, rust, and mill scale. Cleaning must be done with a wire brush or grinder down to bright metal.

Electrical Characteristics:
The current shall be <u>AC or DCRP</u>. The base metal shall be on the <u>work lead or negative</u> side of the line.

Preheat:
The parts must be heated to a temperature higher than 70°F (21°C) before any welding is started.

Backing Gas:
N/A

Welding Technique:
Tack weld the plates together; there should be about a 1/8-in. (3-mm) root gap between the plates. Use the E6010 or E6011 arc welding electrodes to make a root pass to fuse the plates together. Clean the slag from the root pass and use either a hot pass or grinder to remove any trapped slag.

Using the E7018 arc welding electrodes, make a series of filler welds in the groove until the joint is filled. The drawing on the next page shows the recommended location sequence for the weld beads.

Interpass Temperature:
The plate, outside of the heat-affected zone, should not be heated to a temperature higher than 350°F (175°C) during the welding process. After each weld pass is completed, allow it to cool; the weldment must not be quenched in water.

Cleaning:
The slag can be chipped and/or ground off between passes but can only be chipped off of the cover pass.

Visual Inspection Criteria for Entry-Level Welders*:
There shall be no cracks, no incomplete fusion. There shall be no incomplete joint penetration in groove welds except as permitted for partial joint penetration welds.

The Test Supervisor shall examine the weld for acceptable appearance and shall be satisfied that the welder is skilled in using the process and procedure specified for the test.

Undercut shall not exceed the lesser of 10% of the base metal thickness or 1/32 in. (0.8 mm).

Where visual examination is the only criterion for acceptance, all weld passes are subject to visual examination at the discretion of the Test Supervisor.

The frequency of porosity shall not exceed one in each 4 in. (100 mm) of weld length, and the maximum diameter shall not exceed 3/32 in. (2.4 mm).

Welds shall be free from overlap.

Bend Test:
The weld is to be mechanically tested only after it has passed the visual inspection. Be sure that the test specimens are properly marked to identify the welder, the position, and the process.

Specimen Preparation:
For 3/8-in. (9.5-mm) test plates, two specimens are to be located in accordance with the requirements in Figure 6-41. One is to be prepared for a transverse face bend, and the other is to be prepared for a transverse root bend, Figure 6-42.

- *Transverse face bend.* The weld is perpendicular to the longitudinal axis of the specimen and is bent so that the weld face becomes the tension surface of the specimen.

- *Transverse root bend.* The weld is perpendicular to the longitudinal axis of the specimen and is bent so that the weld root becomes the tension surface of the specimen.

*Courtesy of the American Welding Society.

Acceptance Criteria for Bend Test*:

For acceptance, the convex surface of the face- and root-bend specimens shall meet both of the following requirements:

1. No single indication shall exceed 1/8 in. (3.2 mm) measured in any direction on the surface.

2. The sum of the greatest dimensions of all indications on the surface that exceed 1/32 in. (0.8 mm) but are less than or equal to 1/8 in. (3.2 mm) shall not exceed 3/8 in. (9.5 mm).

Cracks occurring at the corner of the specimens shall not be considered unless there is definite evidence that they result from slag or inclusions or other internal discontinuities.

Repair:

No repairs of defects are allowed.

Sketches:

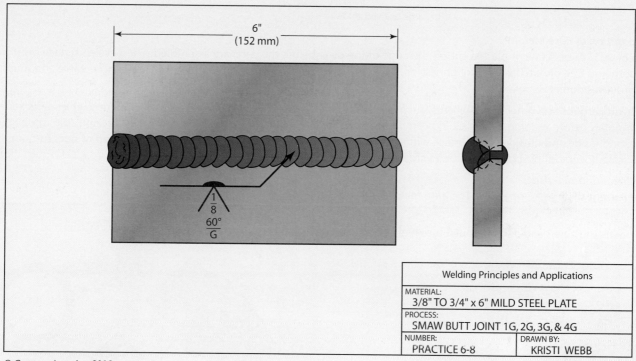

6"
(152 mm)

$\frac{1}{8}$
60°
G

Welding Principles and Applications	
MATERIAL: 3/8" TO 3/4" x 6" MILD STEEL PLATE	
PROCESS: SMAW BUTT JOINT 1G, 2G, 3G, & 4G	
NUMBER: PRACTICE 6-8	DRAWN BY: KRISTI WEBB

PRACTICE 6-9

Welding Procedure Specification (WPS)

Welding Procedures Specifications No.: PRACTICE 6-9

Date: _____

Title:

Welding <u>SMAW</u> of <u>pipe</u> to <u>pipe</u>.

Scope:

This procedure is applicable for <u>V-groove pipe</u> within the range of <u>3 in. (76 mm) schedule 40</u> through <u>4 in. (102 mm) schedule 40</u>.

Welding may be performed in the following positions: <u>1G, 2G, 5G, and 6G</u>.

Base Metal:

The base metal shall conform to <u>Carbon Steel A52</u>. Backing material specification: <u>N/A</u>.

Filler Metal:

The filler metal shall conform to AWS specification No. <u>E6010 or E6011 root pass and E7018 for the cover pass</u> from AWS specification <u>A5.1</u>. This filler metal falls into F-numbers <u>F3 and F4</u> and A-number <u>A-1</u>.

Shielding Gas:

The shielding gas, or gases, shall conform to the following compositions and purity: <u>N/A</u>.

Joint Design and Tolerances:

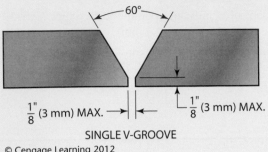

$\frac{1"}{8}$ (3 mm) MAX. →| |← $\frac{1"}{8}$ (3 mm) MAX.

SINGLE V-GROOVE
© Cengage Learning 2012

Preparation of Base Metal:

The bevel is to be flame or plasma cut on the edge of the plate before the parts are assembled. The beveled surface must be smooth and free of notches. Any roughness or notches that are deeper than 1/64 in. (0.4 mm) are unacceptable and must be ground smooth.

All hydrocarbons and other contaminants, such as cutting fluids, grease, oil, and primers, must be cleaned off all parts and filler metals before welding. This cleaning can be done with any suitable solvents or detergents. The backing strip, groove face, and inside and outside plate surface within 1 in. (25 mm) of the joint must be mechanically cleaned of slag, rust, and mill scale. Cleaning must be done with a wire brush or grinder down to bright metal.

Electrical Characteristics:

The current shall be <u>AC or DCRP</u>. The base metal shall be on the <u>work lead or negative</u> side of the line.

Preheat:

The parts must be heated to a temperature higher than 70°F (21°C) before any welding is started.

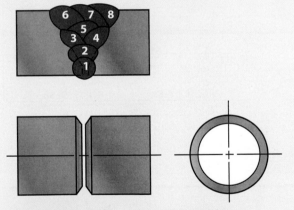

Weld bead sequence for 1G and 5G positions. © Cengage Learning 2012

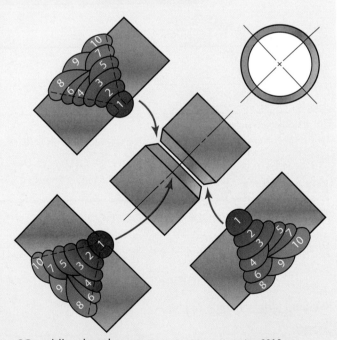

6G welding bead sequence. © Cengage Learning 2012

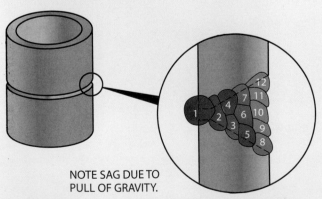

NOTE SAG DUE TO
PULL OF GRAVITY.

Correct sequence for applying welds to a pipe in the
2G position. © Cengage Learning 2012

Backing Gas:
N/A

Welding Technique:
Tack weld the pipes together; there should be about a 1/8-in. (3-mm) root gap between the pipe ends. Use the E6010 or E6011 arc welding electrodes to make a root pass to fuse the pipe ends together. Clean the slag from the root pass and use either a hot pass or grinder to remove any trapped slag.

Using the E7018 arc welding electrodes, make a series of filler welds in the groove until the joint is filled. Figure 6-50 shows the recommended location sequence for the weld beads for the 1G and 5G positions, Figure 6-51 for the 2G position, and Figure 6-52 for the 6G position.

Interpass Temperature:
The plate should not be heated to a temperature higher than 350°F (175°C) during the welding process. After each weld pass is completed, allow it to cool; the weldment must not be quenched in water.

Cleaning:
The slag can be chipped and/or ground off between passes but can only be chipped off of the cover pass.

Visual Inspection Criteria for Entry-Level Welders:*
There shall be no cracks, no incomplete fusion. There shall be no incomplete joint penetration in groove welds except as permitted for partial joint penetration welds.

The Test Supervisor shall examine the weld for acceptable appearance and shall be satisfied that the welder is skilled in using the process and procedure specified for the test.

Undercut shall not exceed the lesser of 10% of the base metal thickness or 1/32 in. (0.8 mm).

Where visual examination is the only criterion for acceptance, all weld passes are subject to visual examination at the discretion of the Test Supervisor.

The frequency of porosity shall not exceed one in each 4 in. (100 mm) of weld length, and the maximum diameter shall not exceed 3/32 in. (2.4 mm).

Welds shall be free from overlap.

Bend Test:
The weld is to be mechanically tested only after it has passed the visual inspection. Be sure that the test specimens are properly marked to identify the welder, the position, and the process.

Specimen Preparation:
For 3/8-in. (9.5-mm) test plates, two specimens are to be located in accordance with the requirements in Figure 6-41. One is to be prepared for a transverse face bend, and the other is to be prepared for a transverse root bend, Figure 6-42.

- *Transverse face bend.* The weld is perpendicular to the longitudinal axis of the specimen and is bent so that the weld face becomes the tension surface of the specimen.
- *Transverse root bend.* The weld is perpendicular to the longitudinal axis of the specimen and is bent so that the weld root becomes the tension surface of the specimen.

Acceptance Criteria for Bend Test:*
For acceptance, the convex surface of the face- and root-bend specimens shall meet both of the following requirements:

1. No single indication shall exceed 1/8 in. (3.2 mm) measured in any direction on the surface.
2. The sum of the greatest dimensions of all indications on the surface that exceed 1/32 in. (0.8 mm) but are less than or equal to 1/8 in. (3.2 mm) shall not exceed 3/8 in. (9.5 mm).

Cracks occurring at the corner of the specimens shall not be considered unless there is definite evidence that they result from slag or inclusions or other internal discontinuities.

Repair:
No repairs of defects are allowed. ◆

Courtesy of the American Welding Society.

Sketches:

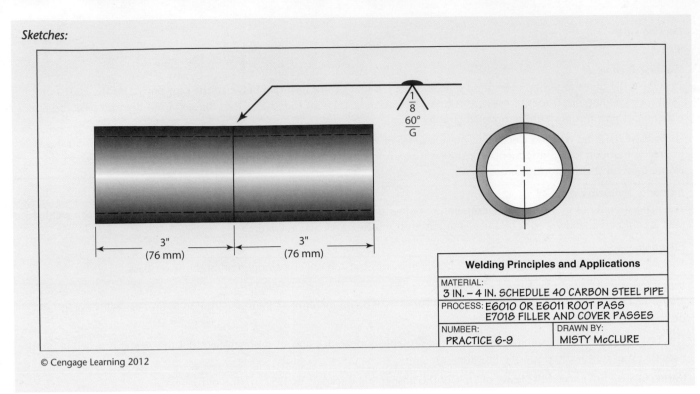

Welding Principles and Applications	
MATERIAL: 3 IN. – 4 IN. SCHEDULE 40 CARBON STEEL PIPE	
PROCESS: E6010 OR E6011 ROOT PASS E7018 FILLER AND COVER PASSES	
NUMBER: PRACTICE 6-9	DRAWN BY: MISTY McCLURE

© Cengage Learning 2012

Postheating is the application of heat to the metal after welding. Postheating is used to slow the cooling rate and reduce hardening.

Interpass temperature is the temperature of the metal during welding. The interpass temperature is given as a minimum and maximum. The minimum temperature is usually the same as the preheat temperature. If the plate cools below this temperature during welding, it should be reheated. The maximum temperature may be specified to keep the plate below a certain phase change temperature for the mild steel used in these practices. The maximum interpass temperature occurs when the weld bead cannot be controlled because of a slow cooling rate. When this happens, the plate should be allowed to cool down, but not below the minimum interpass temperature.

If, during the welding process, a welder must allow a practice weldment to cool so that the weld can be completed later, the weldment should be cooled slowly and then reheated before starting to weld again. A weld that is to be tested or that is done on any parts other than scrap should not be quenched in water.

Poor Fit-Up

Ideally, all welding will be performed on joints that are properly fitted. Most welds produced to a code or standard are properly fitted. Repair, prototype, and job shop welding, however, may not be cut and fitted properly. These welds must be performed under less-than-ideal conditions, but they still must be strong and have a good appearance.

To make a good weld on a poorly fitted joint requires some special skills. These welds also require a good welder, one whose skill is developed. A skilled welder can watch the molten weld pool and knows how to correct for problems before they develop into disasters. The welder must be able to read the molten weld pool correctly to make needed changes in amperage, current, electrode movement, electrode angle, and timing.

The amperage setting may have to be adjusted up or down by only a few amperes to make the necessary changes in molten weld pool size. Adjusting the machine is often preferable to lengthening the weave pattern excessively. The current may be changed from AC to DCSP or DCRP to vary the amount of heat input to the molten weld pool. Some electrodes can operate better than other electrodes with lower amperages on some currents. The current will also alter the forcefulness of some electrodes.

The "U," "J," and straight step patterns are usually the best to use, but they should not be moved more than required to close the gap or opening. On some poor-fitting joints, it is necessary to break and restart the arc to keep the molten weld pool under control. This will result in a weld with porosity, slag inclusions, and other defects. But

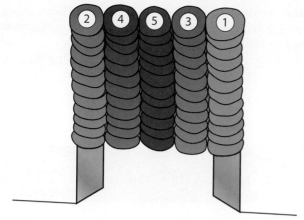

FIGURE 6-43 © Cengage Learning 2012

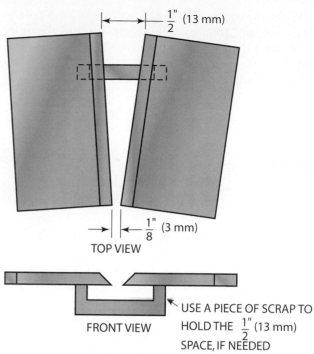

FIGURE 6-44 © Cengage Learning 2012

it is often better to have a poor weld than to have no weld. A poor root weld can be capped with a sound weld to improve joint strength.

Changing the electrode angle from leading to trailing improves poor fit. Sometimes a very flat angle will also help. The time interval that the electrode is moved into and out of the molten weld pool is critical in maintaining weld control. Returning to a molten weld pool too often or too soon can cause the molten weld pool to drop out of the joint. In most cases, a welder should return to the molten weld pool only after it has started to cool.

On some joints, it is possible to make stringer beads on both sides of the joint until the gap is closed, **Figure 6-43**. Note that the beads are made alternately from the edges of the joint to the center. Welds made in this manner can have good weld soundness and strength, but they require more time to complete.

PRACTICE 6-10

Single V-Groove Open Root Butt Joint with an Increasing Root Opening

For this practice, you will need a properly set up and adjusted arc welding machine proper safety protection; E6010 or E6011 and E7018 arc welding electrodes having a 1/8 in. (3 mm) diameter; and two or more pieces of mild steel plate, 3/8 in. (10 mm) thick × 4 in. (102 mm) wide × 12 in. (305 mm) long. You will weld a single V-groove open root butt joint that has a poor fit-up, starting from the close end.

Tack weld the plates together with a root opening of 1/8 in. (3 mm) at one end and 1/2 in. (13 mm) at the other end, **Figure 6-44**. Using the E6010 or the E6011 electrode, start the root pass at the narrow end and weld to the other end. As the root pass progresses along the widening root gap, care must be taken to maintain molten weld pool

control. The "J" or "U" weave pattern works best with low current settings. Long time intervals for electrode movements will give the best weld control.

When the root pass is completed, clean the weld and make a hot pass to burn out any trapped slag. Finish the weld with filler passes using the E7018 electrode. Cool, chip, and inspect the weld for uniformity and defects. Repeat these welds until you can consistently make welds free of defects. Turn off the welding machine and clean up your work area when you are finished welding.

Complete a copy of the "Student Welding Report" listed in Appendix I or provided by your instructor. ◆

PRACTICE 6-11

Single V-Groove Open Root Butt Joint with a Decreasing Root Opening

Using the same setup, equipment, and materials as described in Practice 6-10, you will weld a single V-groove open root butt joint that has a poor fit-up, starting from the wide end.

As in Practice 6-10, tack weld the plates together with a root opening of 1/8 in. (3 mm) at one end and 1/2 in. (13 mm) at the other end. Using the E6010 or E6011 electrode, start welding the root pass at the wide end. Both sides of the joint must be built up until it is possible to get the metal to flow together. The "J" or "U" weave pattern works best to control the bead.

When the root pass is completed, make a hot pass to clean out any trapped slag before making the filler passes with the E7018 electrode. After the weld is completed, cool, chip, and inspect it for uniformity and defects. Repeat these welds until you can consistently make welds free of defects. Turn off the welding machine and clean up your work area when you are finished welding.

Complete a copy of the "Student Welding Report" listed in Appendix I or provided by your instructor. ◆

Summary

Grooved welds on approximately 1/2-in. (13-mm) -thick plate are the most common test plates given to new welding applicants. Groove welds are used by many companies as the base welding skills performance test requirement for employment. The vertical and overhead positions are the most commonly used for the test. It is often assumed in the welding industry that a uniform weld free of visual defects will successfully pass destructive testing. This assumption has great basis in fact because, in most cases, such a weld reflects the welder's skills required to produce quality welds, so, in many cases, applicant test welds are evaluated only by the weld shop foreman or supervisor for visual defects. For this reason, you should always attempt to make your welds as uniform in appearance as possible. Learning how to make a "pretty" groove weld can often mean the difference between successfully earning the job and losing out to another applicant.

Artists Honored at International Institute of Welding (IIW) Assembly

Welding technology is a medium of expression for artists at a special exhibition organized by the Slovene Welding Society.

The Slovene Welding Society hosted the 54th IIW Assembly, in Ljubljana, Slovenia, in July 2001. The Slovene Welding Society also celebrated its fiftieth anniversary at this international gathering of welding experts.

International Institute of Welding

The International Institute of Welding (IIW) is a professional, independent, and nonprofit association founded in 1948 to promote the study of scientific phenomena related to welding and similar processes. The IIW has played an important role in the exchange of scientific and technical information on welding research and education, has assisted in the elaboration of international standards for welding, and has promoted the establishment of national welding associations.

The annual assembly attracted 500 to 700 delegates and experts from more than 30 countries to work on 16 technical commissions and several working units. Numerous papers on a variety of topics were also presented. Two days were available for the presentation of papers at the international conference. The special topic was the joining of dissimilar metals.

The Country of Slovenia

Slovenia is a Central European country that occupies 8000 square miles (20,551 sq km) on the sunny side of the Alps between the Adriatic Sea and the Pannonian Plain. Slovenia is a crossroads of important European routes heading from Central Europe to the Mediterranean Sea and from Italy to the Pannonian Plain.

The City of Ljubljana

Ljubljana, the capital of Slovenia, is a dynamic European city lying in a broad basin between the Alps and the Adriatic Sea. Ljubljana has been known by its present name for nine centuries; however, the ancient Romans established a city by the name of Emona on the present site of Ljubljana. Today, the city is a modern industrial, commercial, and university center with fine parks, art galleries, and theatres.

The IIW assembly in Ljubljana was held in Cankarjev Dom, a national cultural and congress center located in the heart of the city.

Welding: The Artist's Medium

On the occasion of the IIW annual assembly, welding experts had a unique opportunity to see an interesting exhibition of Slovene welding artists.

People have long used metal and have been aware of the material's unique properties. Metal has a special place among materials.

After World War II, when arc welding technology became broadly available, many artists started to use welding techniques for artistic and architectural purposes.

Lyre, by Aleksander Kovac.

Heaven below Triglav, by Marjan Cernigoj.

Our World, by Rudy Stopar.

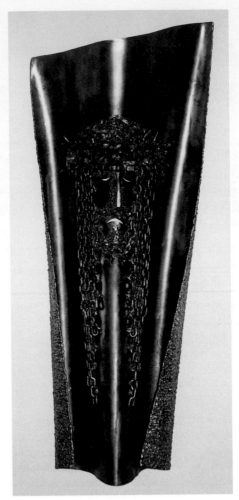

Suffering, by Lojze Pogorevc.

Artists have rapidly adopted welding technology, and Slovenia is no stranger to artistic expression through the use of this technology.

At this special exhibition, the participants of the annual assembly observed artistic works made with different welding techniques by many self-made artists. The artists had different professional backgrounds, and each one presented a unique artistic vision. The four artists profiled in the following section were representative of the exhibition.

The Artists

Aleksander Kovac was an opera singer in the Opera House of Maribor and retired in 1972. His work is based on intuition and a feeling for metal and flame. Kovac's most famous works are inspired by objects from the world of opera, such as the lyre, a trademark of actors.

Marjan Cernigoj was a professional welder and three times the GMA welding champion in the former Yugoslavia. After retirement, he devoted himself to experimental work in the field of welding. He has invented a special welding technique for making relief pictures. Cernigoj's most frequent theme depicts images from nature.

Rudi Stopar is a mechanic, textile designer, and poet, as well as a welding artist working with metals such as steel, aluminum, and copper. He works mainly with the oxyfuel process and with gas tungsten arc welding. In Slovenia, Stopar is best known for sculptures based on Slovene legends and myths. Abroad, however, he is best known for presenting images that allude to problems in civilization.

Lojze Pogorevc is a doctor and surgeon who has made welding sculptures his favorite hobby. In the hospital he uses a scalpel to save human lives, and in the afternoons and at night he uses a flame and his heart to bring humanity to steel. An examination of the mysteries of life and death is the artistic signature of Pogorevc's sculptures.

The works of these and many other Slovene artists exhibited during the annual assembly are an expression of artistic design with metals that is deeply rooted in the culture of Slovenia.

Article courtesy of the American Welding Society.

Review

1. Why are some weld joints grooved?

2. Sketch four of the standard grooves used for welded joints.

3. Why are some backing strips removed from the finished weld?

4. Why are backing tapes used on some joints?

5. Why is it very important to make a weld with a good root surface?

6. What are the two common methods of making a root pass on an open root joint?

7. How can small gaps between the weld plate and backing strip be closed?

8. What effect does changing from a trailing angle to a leading angle have on a weld?

9. What benefit would there be to the root pass if the electrode holder were rocked back and forth while keeping the electrode tip in the joint?

10. What might cause the bright color on the flux as a weld cools?

11. What can happen if the molten weld pool becomes too large on a root weld?

12. What can be done to increase the amount of high-strength welding electrode in the final weld if the root weld was made with a low alloy electrode?

13. What is the purpose of the hot pass?

14. Why should a filler weld pass not have deep penetration?

15. Why is it important to have a good cover pass?

16. What can watching the back edge of a weld pool help you determine?

17. Other than penetration, why would thick butt joints be grooved on both sides?

18. List the things that a weld must be inspected for before it is ground for bend testing.

19. What determines the acceptance or rejection of a bend specimen?

20. What technique can be used to make restarting weld beads easier and more uniform?

21. Why should some weldments be preheated before welding starts?

22. What is postheating used for?

23. How can a wide gap in a joint be closed by welding?

SECTION 3

CUTTING AND GOUGING

Success Story

My name is Mike Vaughan, and I live in Cartersville, VA. After high school, I was first introduced to welding while working in the maintenance department for Reynolds Metals Company. Welding sparked my interest. I felt I had a natural skill and enjoyed both repair welding and fabricating parts.

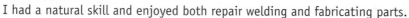

For the next 38 years, I worked for many industries including paper mills, heavy metal industry, construction companies, bridge work, and welded in everything, from fire departments to museums.

In 1991 I attended John Tyler Community College in Chester, Virginia in the evenings after work each day and earned a Welding Certificate in 1993. I then took several welding tests and earned AWS certifications.

As my welding skills and reputation grew, I was given the job to repair and remount a large ornate cast iron stairway that rose from ground level to the main second floor of the historic White House of the Confederacy in Richmond, VA.

Another job I got was to help save an old historic church that was scheduled for demolition by adding 20 feet to a flatbed trailer which was then used to move the church to its new permanent site.

As the welding shop supervisor at Powhatan Correctional Center in Virginia, I taught welding to inmates and supervised fabrication and maintenance throughout the center. Because the institution is self sufficient, my students fabricated such things as beds, tool cages, doors, gates, hand rails, locks, tables, trailers, stairs, and window bars. We repair welded everything from buses to high pressure steam lines. There never seemed to be a dull moment there.

I have enjoyed teaching since my first welding class that I taught as a part-time instructor at John Tyler Community College. I am now the Welding Program Head at J. Sargeant Reynolds Community College.

At Reynolds we have a dual enrollment program with the area high schools, so I get to teach both high school and college students. I also teach fly fishing classes here in the summer.

Being able to pass along my welding skills to the different student groups I have taught has been wonderfully rewarding to me. Knowing I have helped students obtain better jobs and pay so they and their families can live better lives, is even more rewarding.

I always teach my students to take pride in their work. I tell them that every weld they make has their signature on it.

Indeed, my career in welding has certainly made me feel like I have left my signature in this world. I am very proud to say that I am a welder. ▰

Chapter 7

Flame Cutting

OBJECTIVES

After completing this chapter, the student should be able to

- describe the oxyfuel gas cutting process and list the commonly used fuel gases.
- discuss which metals can be cut using the oxyfuel gas cutting process.
- describe the eye protection that must be used for flame cutting.
- tell how to determine the correct size and type of cutting tip for a specific job.
- demonstrate how to properly set up a cutting torch.
- demonstrate how to properly clean a cutting tip.
- demonstrate how to safely light the torch.
- tell what actions can be taken to ensure a smooth cut when using a hand torch.
- describe how to accurately lay out a line to be cut.
- demonstrate how to set the working gas pressure on the regulator.
- explain the chemical process that takes place during the burning away of the metal when using an oxyfuel gas cutting torch.
- explain what the kerf surface can reveal about what was correct or incorrect with the preheat flame, cutting speed, and oxygen pressure.
- demonstrate how to make a machine cut and then evaluate the results.
- describe soft slag and hard slag and what causes them.
- describe the various types of guides that can be used to guide the torch for a more accurate cut.
- demonstrate how to make a flat, straight cut in thin plate, thick plate, and sheet metal.
- demonstrate how to make a flame-cut hole.
- describe the two major methods of controlling distortion of the metal during the heating or cutting process.
- demonstrate how to make a straight line cut in the vertical and overhead positions.
- discuss the various factors that can become a problem when cutting.
- demonstrate how to cut out internal and external shapes.
- demonstrate how to make a square cut on pipe in the horizontal rolled position, the horizontal fixed position, and the vertical position.

KEY TERMS

cutting tip to work distance (handwritten annotation)

coupling distance	high-speed cutting tip	preheat flame
cutting lever	kindling point	preheat holes
cutting tips	machine cutting torch	slag
drag	MPS gases	soapstone
drag lines	orifice	soft slag
equal-pressure torches	oxyacetylene hand torch	tip cleaners
hard slag	oxyfuel gas cutting (OFC)	venturi

INTRODUCTION

Oxyfuel gas cutting (OFC) is a group of oxygen cutting processes that use heat from an oxyfuel gas flame to raise the temperature of the metal to its kindling temperature before a high-pressure stream of oxygen is directed onto the metal, causing it to be cut. The kindling temperature of a material is the temperature at which rapid oxidation (combustion) can begin. The kindling temperature of steel in pure oxygen is 1600° to 1800°F (870°C to 900°C). The processes in this group are identified by the type of fuel gas used with oxygen to produce the preheat flame. Oxyfuel gas cutting is most commonly performed with oxyacetylene cutting (OFC-A). **Table 7-1** lists a number of other fuel gases used for OFC. MPS (MAPP) gas is increasingly being used today for cutting and rivals acetylene's popularity in some areas of the country.

More people use the oxyfuel cutting torch than any other welding process. The cutting torch is used by workers in virtually all areas, including manufacturing, maintenance, automotive repair, railroad, farming, and more. It is unfortunately one of the most commonly misused processes. Most workers know how to light the torch and make a cut, but their cuts are very poor quality. Often, in addition to making bad cuts, they use unsafe torch techniques. A good oxyfuel cut should not only be straight and square but also should require little or no postcut cleanup.

Excessive postcut cleanup results in extra cost, which is an expense that cannot be justified.

Manual, mechanized, and automatic OFC processes are used in industry. Hand-controlled, manual cutting is done in short-run production and one-of-a-kind fabrication, as well as in demolition and scrapping operations. Manual cutting is also used in the field for steel construction. Mechanized or automatic cutting is widely used in production work where a large number of identical cuts are made over and over or where very precise cuts are required. In mechanized or automatic cutting, more than one cutting head may be mounted so several cuts can be made at the same time.

Various oxyfuel cutting specialties are found on the job, including flame cutting, gouging, beveling, scarfing, and the operation of an automated cutting machine. In addition to these cutting jobs, some welders work dismantling scrap metal, such as scrap autos or construction demolition.

Metals Cut by the Oxyfuel Process

Oxyfuel gas cutting is used to cut iron base alloys. Low carbon steels (up to 0.3% carbon) are easy to cut. Any metal that requires preheating for welding, such as high alloy and high alloy carbon steels should also be preheated before cutting. Failing to preheat some high-strength alloys before cutting can result in a very thin hardness zone on the cut surface. This hardness zone can cause cracks to start in the finished part. If high-strength steel flame-cut parts are to be bent or formed, the hardened edge may cause cracks to form. These surface cracks can actually cause the part to fracture and fail. High nickel steels, cast iron, and stainless steel are difficult to cut. Most nonferrous metals—such as brass, copper, and aluminum—cannot be cut by oxyfuel cutting. A few reactive nonferrous metals, such as titanium and magnesium, can be cut. These metals seldom are cut with the OFC process because of the extensive postcut cleanup required.

Fuel Gas	Flame (Fahrenheit)	Temperature* (Celsius)
Acetylene	5589°	3087°
MAPP®	5301°	2927°
Natural gas	4600°	2538°
Propane	4579°	2526°
Propylene	5193°	2867°
Hydrogen	4820°	2660°

*Approximate neutral oxyfuel flame temperature.

TABLE 7-1 Fuel Gases Used for Flame Cutting

Type of Cutting Operation	Hazard	Suggested Shade Number
Light cutting, up to 1 in.	Sparks, harmful rays, molten metal, flying particles	3 or 4
Medium cutting, 1–6 in.		4 or 5
Heavy cutting, over 6 in.		5 or 6

TABLE 7-2 A General Guide for the Selection of Eye and Face Protection Equipment

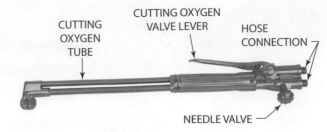

FIGURE 7-1 Dedicated oxyfuel cutting torch.
Victor Equipment Company

Eye Protection for Flame Cutting

The National Bureau of Standards has identified proper filter plates and uses. The recommended filter plates are identified by shade number and are related to the type of cutting operation being performed.

Goggles or other suitable eye protection must be used for flame cutting. Goggles should have vents near the lenses to prevent fogging. Cover lenses or plates should be provided to protect the filter lens. All lens glass should be ground properly so that the front and rear surfaces are smooth. Filter lenses must be marked so that the shade number can be readily identified, **Table 7-2**.

Cutting Torches

The **oxyacetylene hand torch** is the most common type of oxyfuel gas cutting torch used in industry. The hand torch, as it is often called, may be either a part of a combination welding and cutting torch set or a cutting torch only, **Figure 7-1**. The combination welding-cutting torch offers more flexibility because a cutting head, welding tip, or heating tip can be attached quickly to the same torch body, **Figure 7-2**. Combination torch sets are often used in schools, automotive repair shops, auto body shops, and small welding shops or with any job where flexibility in equipment is needed. A cut made with either type of torch has the same quality; however, the dedicated cutting torches are usually longer and have larger gas flow passages than the combination torches. The added length of the dedicated cutting torch helps keep the operator farther away from the heat and sparks and allows thicker material to be cut.

Oxygen is mixed with the fuel gas to form a high-temperature preheating flame. The two gases must be completely mixed before they leave the tip and create the flame. Two methods are used to mix the gases. One method uses a mixing chamber, and the other method uses an injector chamber.

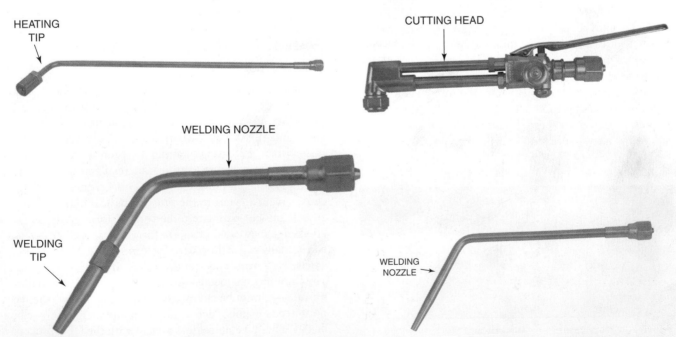

FIGURE 7-2 The attachments that are used for heating, cutting, welding, or brazing make the combination torch set flexible. Victor Equipment Company

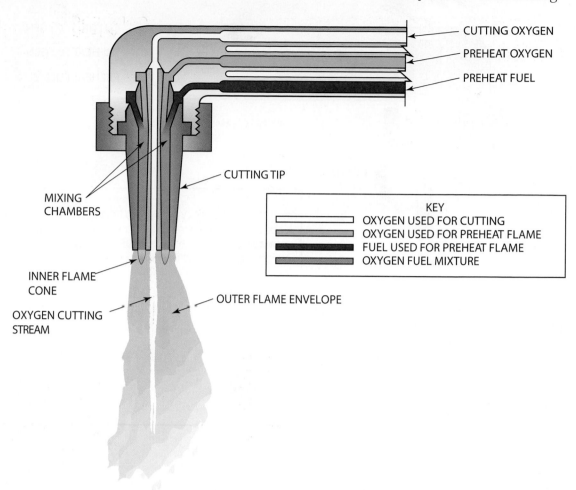

CUTTING OXYGEN

PREHEAT OXYGEN

PREHEAT FUEL

CUTTING TIP

MIXING CHAMBERS

KEY
OXYGEN USED FOR CUTTING
OXYGEN USED FOR PREHEAT FLAME
FUEL USED FOR PREHEAT FLAME
OXYGEN FUEL MIXTURE

INNER FLAME CONE

OUTER FLAME ENVELOPE

OXYGEN CUTTING STREAM

FIGURE 7-3 A mixing chamber located in the tip. © Cengage Learning 2012

The mixing chamber may be located in the torch body or in the tip, **Figure 7-3.** Torches that use a mixing chamber are known as **equal-pressure torches** because the gases must enter the mixing chamber under the same pressure. The mixing chamber is larger than both the gas inlet and the gas outlet. This larger size causes turbulence in the gases, resulting in the gases mixing thoroughly.

Injector torches will work with both equal gas pressures and low fuel-gas pressures, **Figure 7-4.** The injector allows the oxygen to draw the fuel gas into the chamber even if the fuel gas pressure is as low as 6 oz/in.^{2} (26 g/cm^2). The injector works by passing the oxygen through a **venturi,** which creates a low-pressure area that pulls the fuel gases in and mixes them together. An injector-type torch must be used if a low-pressure acetylene generator or low-pressure residential natural gas is used as the fuel gas supply.

The cutting head may hold the cutting tip at a right angle to the torch body or it may be held at a slight angle. Torches with the tip slightly angled are easier for the welder to use when cutting flat plate. Torches with a right-angle tip are easier for the welder to use when cutting

pipe, angle iron, I-beams, or other uneven material shapes. Both types of torches can be used for any type of material being cut, but practice is needed to keep the cut square and accurate.

The location of the **cutting lever** may vary from one torch to another, **Figure 7-5.** Most cutting levers pivot from the front or back end of the torch body. Personal preference will determine which one the welder uses.

A **machine cutting torch,** sometimes referred to as a line burner, operates in a similar manner to a hand-cutting torch. The machine cutting torch may require two oxygen regulators, one for the preheat oxygen and the other for the cutting oxygen stream. The addition of a separate cutting oxygen supply allows the flame to be more accurately adjusted. It also allows the pressures to be adjusted during a cut without disturbing the other parts of the flame. Various machine cutting torches are shown in **Figure 7-6, Figure 7-7,** and **Figure 7-8.**

Cutting Tips

Most **cutting tips** are made of copper alloy, but some tips are chrome. Chrome plating prevents spatter from

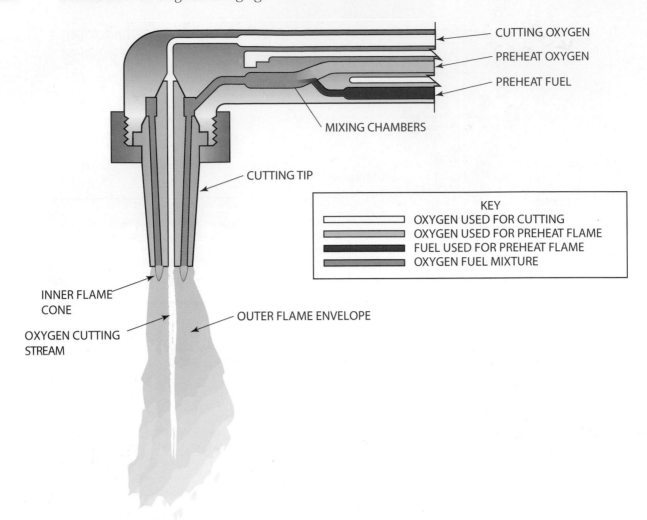

CUTTING OXYGEN

PREHEAT OXYGEN

PREHEAT FUEL

MIXING CHAMBERS

CUTTING TIP

KEY
OXYGEN USED FOR CUTTING
OXYGEN USED FOR PREHEAT FLAME
FUEL USED FOR PREHEAT FLAME
OXYGEN FUEL MIXTURE

INNER FLAME CONE

OXYGEN CUTTING STREAM

OUTER FLAME ENVELOPE

FIGURE 7-4 Injector mixing torch. © Cengage Learning 2012

sticking to the tip, thus prolonging its usefulness. Tip designs change for the different types of uses and gases, and from one torch manufacturer to another, **Figure 7-9**.

Tips for straight cutting are either standard or high-speed, **Figure 7-10**. The **high-speed cutting tip** is designed to allow a higher cutting oxygen pressure, which allows the torch to travel faster. High-speed tips are also available for different types of fuel gases.

The diameter, or size of the center cutting orifice, determines the thickness of the metal that can be cut. A larger

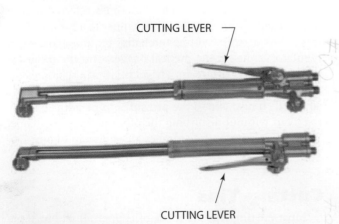

CUTTING LEVER

CUTTING LEVER

FIGURE 7-5 The cutting lever may be located on the front or back of the torch body. Victor Equipment Company

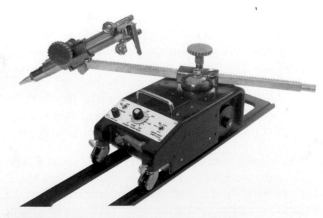

FIGURE 7-6 Portable flame-cutting machine. Victor Equipment Company

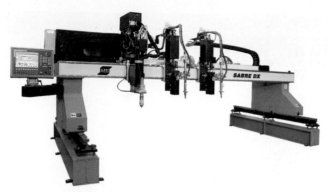

FIGURE 7-7 Multiple head cutting machine.
ESAB Welding & Cutting Products

FIGURE 7-9 Different cutting torch seals for different manufacturers' torches. Larry Jeffus

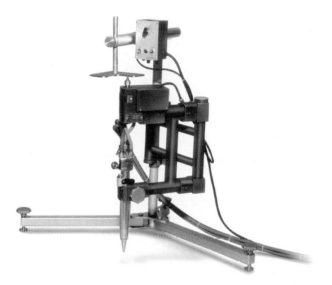

FIGURE 7-8 Portable cutting machine for highly complex shapes. ESAB Welding & Cutting Products

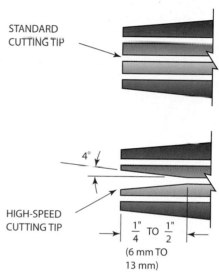

STANDARD CUTTING TIP

4°

HIGH-SPEED CUTTING TIP

$\frac{1}{4}$" TO $\frac{1}{2}$"
(6 mm TO 13 mm)

FIGURE 7-10 Comparison of standard and high-speed cutting tips. © Cengage Learning 2012

diameter oxygen orifice is required for cutting thick metal. There is no standard numbering system for sizing cutting tips. Each manufacturer uses its own system. Some systems are similar; some are not. **Table 7-3** lists several manufacturers' tip numbering systems. As a way of comparing the size of one manufacturer's tip size to another, the center hole diameter in inches is given below the tip number. For example, in Table 7-3 you can see that Airco's tip number 00 has a center orifice size equal to a number 70 drill size. In Table 7-3 you can see that this cutting tip is designed for cutting metal approximately 1/8 in. (3 mm) thick. Other manufacturer tip numbers

Manufacturer	Metal Thickness, Inches (mm)										
	1/8 (3)	1/4 (6)	1/2 (13)	3/4 (19)	1 (24)	1 1/2 (37)	2 (49)	2 1/2 (61)	3 (74)	4 (98)	5 (123)
Cutting orifice drill number	70	68	60	56	54	53	50	47	45	39	31
Airco	00	0	1	1	2	2	3	4	4	5	6
ESAB	1/4	1/4	1/2	1 1/2	1 1/2	1 1/2	4	4	4	4	8
Harris	000	00	0	1	1	2	2	3	3	3	4
Oxweld	2	3	4	6	6	6	8	8	8	8	8
Purox	3	3	4	4	5	5	7	7	7	7	9
Smith	00	0	1	2	2	3	3	4	4	4	5
Victor	000	00	0	1	2	2	3	4	4	5	6

TABLE 7-3 Comparison of Some Manufacturers' Oxyacetylene Cutting Tip Identifications

Tip Size	Metal Thickness, Inches (mm)										
	1/8 (3)	1/4 (6)	1/2 (13)	3/4 (19)	1 (24)	1 1/2 (37)	2 (49)	2 1/2 (61)	3 (74)	4 (98)	5 (123)
Cutting orifice drill number	70	68	60	56	54	53	50	4/	45	39	31
WYPO tip cleaner number*	10	10	15	18	22	24	26				
Campbell Hausfeld tip cleaner number*	3	3	6	9	10	11	12				
Oxygen pressure, psi**	20	20	25	30	35	35	40	40	40	45	45
	25	25	30	35	40	40	45	45	45	55	55
Oxygen pressure, kPa**	140	140	170	200	240	240	275	275	275	310	310
	170	170	200	240	275	275	310	310	310	380	380
Acetylene pressure, psi**	3	3	3	3	3	3	4	4	5	6	8
	5	5	5	5	5	5	8	8	11	13	14
Acetylene pressure, kPa**	20	20	20	20	20	20	30	30	35	40	55
	35	35	35	35	35	35	55	55	75	90	95

*There is no standard numbering system for tip cleaners, so numbers can differ from one manufacturer to another.
**Tip size and pressures are approximate. Use the manufacturer's specification for equipment being used when available.

TABLE 7-4 Center Cutting Orifice Size, Metal Thickness, and Gas Pressures for Oxyacetylene Cutting

designed for this thickness have the following numbers: 000, 00, 1/4, 2, and 3.

Finding the correctly sized tip for a job can be confusing, especially if you are using the cutting unit for the first time. To make it easier to select a tip, you can use a standard set of **tip cleaners** to find the size of the center cutting orifice. **Table 7-4** lists the material thickness being cut with the tip cleaner size.

If the manufacturers' recommendations for gas pressure are not available, you can use Table 7-4 to find the approximate pressures to be used with the tip. Actual gas pressures vary, depending on a number of factors, such as the equipment manufacturer, the condition of the equipment, hose length, hose diameter, regulator size, and operator skill. In all cases, start out with the pressure recommended by the particular manufacturer of the equipment being used. Adjust the pressure to fit the job being cut.

A wide variety of tip shapes are also available for specialized cutting jobs. Each tip, of course, also comes in several sizes. Some tips are specialized for the kind of fuel gas being used. Different means are used to attach the cutting tip to the torch head. Some tips screw in; others have a push fitting.

In addition to those tips for torch hand cutting, different designs are used for mechanized and automated cutting tips. Mechanized and automated cutting tips are designed for high-speed cutting with high-speed oxygen flow.

Always choose the correct type and size of tip for the specific cutting job. Check the manufacturer's literature for tip size and type recommendations. Make sure the tip is designed for the type of fuel gas being used. Inspect the tip before using it. If the tip is clogged or dirty, clean the tip and clean out the orifices with the proper size drill. Check to make sure there is no damage to the threads. If the threads or the tapered seat is damaged, do not use the tip.

The amount of **preheat flame** required to make a perfect cut is determined by the type of fuel gas used and by the material thickness, shape, and surface condition. Materials that are thick; are round; or have surfaces covered with rust, paint, oil, and so on require more preheat flame, **Figure 7-11**.

Different cutting tips are available for each of the major types of fuel gases. The differences in the type or number of **preheat holes** determine the type of fuel gas to be used in the tip. **Table 7-5** lists the fuel gas and range of preheat holes or tip designs used with each gas. Acetylene is used in tips having from one to six preheat holes. Some large acetylene cutting tips may have eight or more preheat holes.

//// **CAUTION** \\\\

Acetylene must be used in tips that are designed to be used with acetylene. If acetylene is used in tips designed for other fuel gases, the tip may overheat, causing a backfire or even causing the tip to explode.

MPS gases are used in tips having eight preheat holes or in a two-piece tip that is not recessed, **Figure 7-12**. These gases have a slower flame combustion rate (see Chapter 26) than acetylene. For tips with fewer than eight preheat holes, there may not be enough heat to start a cut, or the flame may pop out when the cutting lever is pressed.

Propane and natural gas should be used in two-piece tips that are typically deeply recessed, Figure 7-12. The flame burns at such a slow rate that it may not stay lit on any other tip.

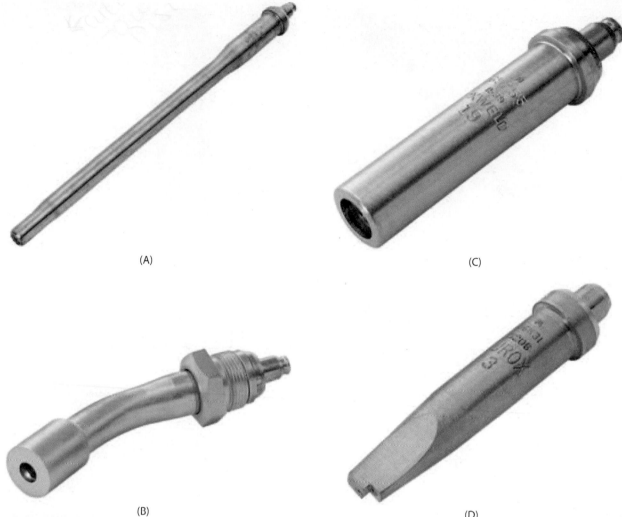

(A)

(C)

(B)

(D)

FIGURE 7-11 Special cutting tips come in a variety of shapes, for many purposes. They can have different sizes and numbers of preheat holes. (A) 10-in. long cutting tip, (B) water cooled cutting tip, (C) replaceable end cutting tips, and (D) sheet metal cutting tip. ESAB Welding & Cutting Products

Fuel Gas	Number of Preheat Holes
Acetylene	One to six
MPS (MAPP)	Eight- to two-piece tip
Propane and natural gas	Two-piece tip

TABLE 7-5 Fuel Gas and Number of Preheat Holes Needed in the Cutting Tip

Some cutting tips have metal-to-metal seals. When they are installed in the torch head, a wrench must be used to tighten the nut. Other cutting tips have fiber packing seats to seal the tip to the torch. If a wrench is used to tighten the nut for this type of tip, the tip seat may be damaged, **Figure 7-13**. A torch owner's manual should be checked or a welding supplier should be asked about the best way to tighten various torch tips.

When removing a cutting tip, if the tip is stuck in the torch head, tap the back of the head with a plastic hammer,

Figure 7-14. Any tapping on the side of the tip may damage the seat.

To check the assembled torch tip for a good seal, turn on the oxygen valve and spray the tip with a leak-detecting solution, **Figure 7-15**.

///// **CAUTION** \\\\\

Carefully handle and store the tips to prevent damage to the tip seats and to keep dirt from becoming stuck in the small holes.

If the cutting tip seat or the torch head seat is damaged, it can be repaired by using a reamer designed for the specific torch tip and head, **Figure 7-16**, or it can be sent out for repair. New fiber packings are available for tips with packings. The original leak-checking test should be repeated to be sure the new seal is good.

(A)

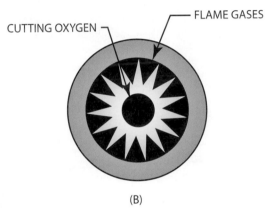

CUTTING OXYGEN   FLAME GASES

(B)

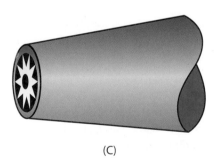

(C)

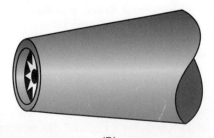

(D)

FIGURE 7-12 Parts of a two-piece cutting tip.
ESAB Welding & Cutting Products

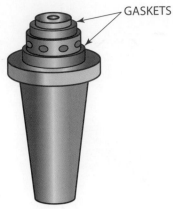

GASKETS

FIGURE 7-13 Some cutting tips use gaskets to make a tight seal. © Cengage Learning 2012

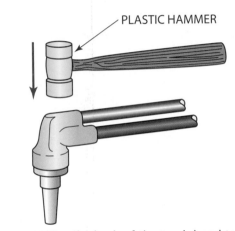

PLASTIC HAMMER

FIGURE 7-14 Tap the back of the torch head to remove a tip that is stuck. The tip itself should never be tapped. © Cengage Learning 2012

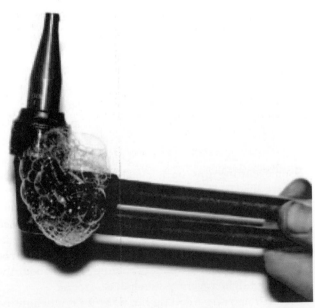

FIGURE 7-15 Checking a cutting tip for leaks. Larry Jeffus

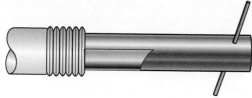

FIGURE 7-16 Damaged torch seats can be repaired by using a reamer. © Cengage Learning 2012

Oxyfuel Cutting, Setup, and Operation

The setting up of a cutting torch system is exactly like setting up oxyfuel welding equipment except for the adjustment of gas pressures. This chapter covers gas pressure adjustments and cutting equipment operations. Chapter 30, "Oxyfuel Welding and Cutting, Equipment, Setup and Operation," gives detailed technical information and instructions for oxyfuel systems. The chapter covers the following topics:

- Safety
- Pressure regulator setup and operation
- Welding and cutting torch design and service
- Reverse flow and flashback valves
- Hoses and fittings
- Types of flames
- Leak detection

PRACTICE 7-1

Setting Up a Cutting Torch

Demonstrate to other students and your instructor the proper method of setting up cylinders, regulators, hoses, and the cutting torch.

1. The oxygen and acetylene cylinders must be securely chained to a cart or wall before the safety caps are removed.

2. After removing the safety caps, stand to one side and crack (open and quickly close) the cylinder valves, being sure there are no sources of possible ignition that may start a fire. Cracking the cylinder valves is done to blow out any dirt that may be in the valves.

3. Visually inspect all of the parts for any damage, needed repair, or cleaning.

4. Attach the regulators to the cylinder valves and tighten them securely with a wrench.

5. Attach a reverse flow valve or flashback arrestor, if the torch does not have them built in, to the hose connection on the regulator or to the hose connection on the torch body, depending on the type of reverse flow valve in the set. Occasionally, test each reverse flow valve by blowing through it to make sure it works properly.

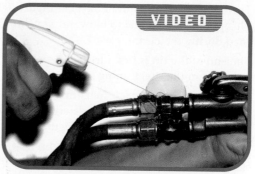

FIGURE 7-17 Leak-check all gas fittings. Larry Jeffus *See Welding Principles and Practices on DVD. DVD 4—Oxyacetylene Welding.*

6. If the torch you will be using is a combination-type torch, attach the cutting head at this time.

7. Last, install a cutting tip on the torch.

8. Before the cylinder valves are opened, back out the pressure regulating screws so that when the valves are opened the gauges will show zero pounds working pressure.

9. Stand to one side of the regulator's face as the cylinder valves are opened slowly.

10. The oxygen valve is opened all the way until it becomes tight, but do not overtighten, and the acetylene valve is opened no more than one-half turn.

11. Open one torch valve and then turn the regulating screw in slowly until 2 psig to 4 psig (14 kPag to 30 kPag) shows on the working pressure gauge. Allow the gas to escape so that the line is completely purged.

12. If you are using a combination welding and cutting torch, the oxygen valve nearest the hose connection must be opened before the flame adjusting valve or cutting lever will work.

13. Close the torch valve and repeat the purging process with the other gas.

14. Be sure there are no sources of possible ignition that may result in a fire.

15. With both torch valves closed, spray a leak-detecting solution on all connections, including the cylinder valves. Tighten any connection that shows bubbles, **Figure 7-17**.

Complete a copy of the "Student Welding Report" listed in Appendix I or provided by your instructor. ◆

PRACTICE 7-2

Cleaning a Cutting Tip

Using a cutting torch set that is assembled and adjusted as described in Practice 7-1, and a set of tip cleaners, you will clean the cutting tip.

1. Turn on a small amount of oxygen, **Figure 7-18**. This procedure is done to blow out any dirt loosened during the cleaning.

2. The end of the tip is first filed flat, using the file provided in the tip cleaning set, **Figure 7-19**.

3. Try several sizes of tip cleaners in a preheat hole until the correct size cleaner is determined. It should easily go all the way into the tip, **Figure 7-20**.

4. Push the cleaner in and out of each preheat hole several times. Tip cleaners are small, round files. Excessive use of them will greatly increase the **orifice** (hole) size.

FIGURE 7-18 Turn on the oxygen valve. Larry Jeffus

FIGURE 7-19 File the end of the tip flat. Larry Jeffus

FIGURE 7-20 A tip cleaner should be used to clean the flame and center cutting holes. Larry Jeffus

5. Next, depress the cutting lever and, by trial and error, select the correct size tip cleaner for the center cutting orifice.

A tip cleaner should never be forced. If the tip needs additional care, refer to the section on tip care in Chapter 30.

Complete a copy of the "Student Welding Report" listed in Appendix I or provided by your instructor. ◆

PRACTICE 7-3

Lighting the Torch

Wearing welding goggles, gloves, and any other required personal protective clothing, and with a cutting torch set that is safely assembled, you will light the torch.

1. Set the regulator working pressure for the tip size. If you do not know the correct pressure for the tip, start with the fuel set at 5 psig (35 kPag) and the oxygen set at 25 psig (170 kPag).

2. Point the torch tip upward and away from any equipment or other students.

3. Turn on just the acetylene valve and use only a spark lighter to ignite the acetylene. The torch may not stay lit. If this happens, close the valve slightly and try to relight the torch.

4. If the flame is small, it will produce heavy black soot and smoke. In this case, turn the flame up to stop the soot and smoke. The welder need not be concerned if the flame jumps slightly away from the torch tip.

5. With the acetylene flame burning smoke free, slowly open the oxygen valve and by using only the oxygen valve adjust the flame to a neutral setting, **Figure 7-21**.

6. When the cutting oxygen lever is depressed, the flame may become slightly carbonizing. This may occur because of a drop in line pressure due to the high flow of oxygen through the cutting orifice.

7. With the cutting lever depressed, readjust the preheat flame to a neutral setting.

The flame will become slightly oxidizing when the cutting lever is released. Since an oxidizing flame is hotter than a neutral flame, the metal being cut will be preheated faster. When the cut is started by depressing the lever, the flame automatically returns to the neutral setting and does not oxidize the top of the plate.

8. Extinguish the flame by first turning off the oxygen and then the acetylene.

Complete a copy of the "Student Welding Report" listed in Appendix I or provided by your instructor. ◆

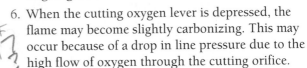

///// **CAUTION** \\\\\

Turning off the acetylene first will often cause a loud POP. This can often cause soot to clog the tip and torch.

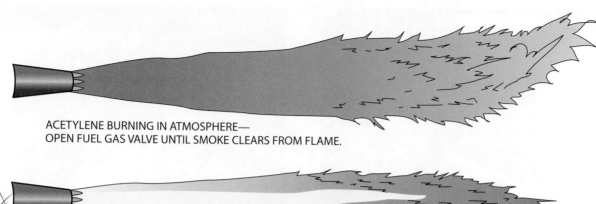

ACETYLENE BURNING IN ATMOSPHERE—
OPEN FUEL GAS VALVE UNTIL SMOKE CLEARS FROM FLAME.

CARBURIZING FLAME—
(EXCESS ACETYLENE WITH OXYGEN). PREHEAT FLAMES REQUIRE MORE OXYGEN

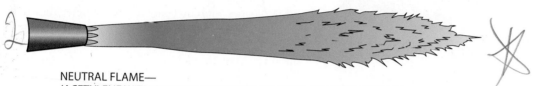

NEUTRAL FLAME—
(ACETYLENE WITH OXYGEN). TEMPERATURE 6300° F. PROPER PREHEAT ADJUSTMENT FOR ALL CUTTING.

NEUTRAL FLAME WITH CUTTING JET OPEN—
CUTTING JET MUST BE STRAIGHT AND CLEAR.

OXIDIZING FLAME—
(ACETYLENE WITH EXCESS OXYGEN). NOT RECOMMENDED FOR AVERAGE CUTTING.

FIGURE 7-21 Oxyacetylene flame adjustments for the cutting torch. © Cengage Learning 2012

The 3 Kind of Flames.

Hand Cutting

When making a cut with a hand torch, it is important for the welder to be steady in order to make the cut as smooth as possible. A welder must also be comfortable and free to move the torch along the line to be cut. It is a good idea for a welder to get into position and practice the cutting movement a few times before lighting the torch. Even when the welder and the torch are braced properly, a tiny movement such as a heartbeat will cause a slight ripple in the cut. Attempting a cut without leaning on the work, to brace oneself, is tiring and causes inaccuracies.

The torch should be braced with the left hand if the welder is right-handed or with the right hand if the welder is left-handed. The torch may be moved by sliding it toward you over your supporting hand, **Figure 7-22** and

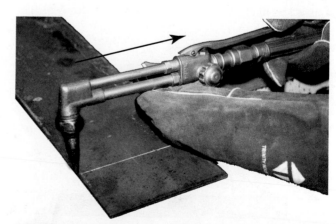

FIGURE 7-22 For short cuts, the torch can be drawn over the gloved hand. Larry Jeffus

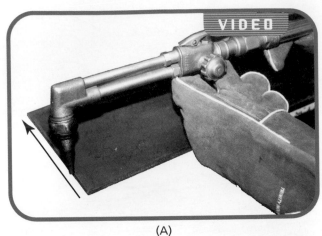

(A)

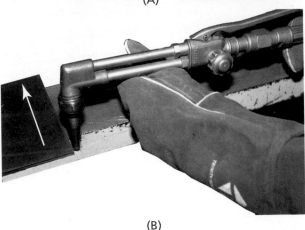

(B)

FIGURE 7-23 For longer cuts, the torch can be moved by sliding your gloved hand along the plate parallel to the cut: (A) start and (B) finish. Always check for free and easy movement before lighting the torch. Larry Jeffus
See Welding Principles and Practices on DVD.
DVD 4—Oxyacetylene Welding.

Figure 7-23. The torch can also be pivoted on the supporting hand. If the pivoting method is used, care must be taken to prevent the cut from becoming a series of arcs.

A slight forward torch angle helps the flame preheat the metal, keeps some of the reflected flame heat off the tip, aids in blowing dirt and oxides away from the cut, and keeps the tip clean for a longer period of time because slag is less likely to be blown back on it, **Figure 7-24.** The

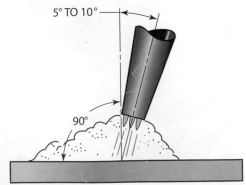

5° TO 10°

90°

FIGURE 7-24 A slight forward angle helps when cutting thin material. © Cengage Learning 2012

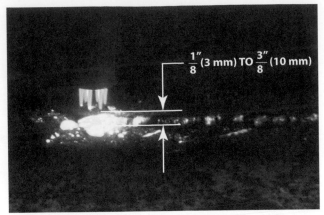

$\frac{1}{8}$" (3 mm) TO $\frac{3}{8}$" (10 mm)

FIGURE 7-25 Inner cone to work distance.
© Cengage Learning 2012

forward angle can be used only for a straight line square cut. If shapes are cut using a slight angle, the part will have beveled sides.

When making a cut, the inner cones of the flame should be kept 1/8 in. (3 mm) to 3/8 in. (10 mm) from the surface of the plate, **Figure 7-25.** This distance is known as the **coupling distance.**

To start a cut on the edge of a plate, hold the torch at a right angle to the surface or pointed slightly away from the edge, **Figure 7-26.** The torch must also be pointed so that the cut is started at the very edge. The edge of the plate heats up more quickly and allows the cut to be started sooner. Also, fewer sparks will be blown around the shop. Once the cut is started, the torch should be rotated back to a right angle to the surface or to a slight leading angle.

THE TORCH IS AT A SLIGHT ANGLE AWAY FROM THE EDGE OF THE PLATE

FIGURE 7-26 Starting a cut on the edge of a plate. Notice how the torch is pointed at a slight angle away from the edge. © Cengage Learning 2012

If a cut is to be started in a place other than the edge of the plate, the inner cones should be held as close as possible to the metal. Having the inner cones touch the metal will speed up the preheat time. When the metal is hot enough to allow the cut to start, the torch should be raised as the cutting lever is slowly depressed. When the metal is pierced, the torch should be lowered again, **Figure 7-27**. By raising the torch tip away from the metal, the amount of sparks blown into the air is reduced, and the tip is kept cleaner. If the metal being cut is thick, it may be necessary to move the torch tip in a small circle

THINK GREEN
Avoid Contaminating the Environment

Any residue of the chemicals that remain in a container must be cleaned out properly before cutting can begin. Refer to the material's manufacturer and material safety data sheet (MSDS) to determine how to safely clean out the container. Any spilled material must be cleaned promptly so it will not be washed away by rainwater where it could contaminate the environment.

In addition to the possible direct soil and water contamination that could be caused by the chemicals, the heat of the cutting torch can release air pollutants.

as the hole pierces the metal. If the metal is to be cut in both directions from the spot where it was pierced, the torch should be moved backward a short distance and then forward, **Figure 7-28**. This prevents slag from refilling

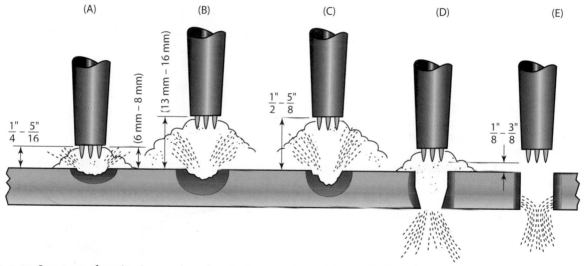

(A)   (B)   (C)   (D)   (E)

FIGURE 7-27 Sequence for piercing a plate for starting a cut or making a hole. © Cengage Learning 2012

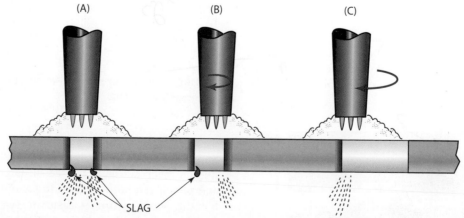

(A)   (B)   (C)

SLAG

FIGURE 7-28 (A) A short, backward movement, (B) before the cut is carried forward (C) clears the slag from the kerf. Slag left in the kerf may cause the cutting stream to gouge into the base metal, resulting in a poor cut.
© Cengage Learning 2012

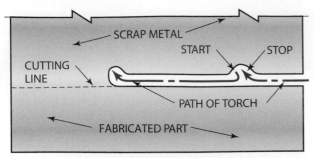

FIGURE 7-29 Turning out into scrap will make stopping and starting points smoother. © Cengage Learning 2012

the kerf at the starting point, thus making it difficult to cut in the other direction. The kerf is the space produced during any cutting process.

Starts and stops can be made more easily and better if one side of the metal being cut is scrap. When it is necessary to stop and reposition oneself before continuing the cut, the cut should be turned out, a short distance, into the scrap side of the metal, **Figure 7-29**. The extra space that this procedure provides will allow a smoother and more even start with less chance that slag will block the cut. If neither side of the cut is to be scrap, the forward movement should be stopped for a moment before releasing the cutting lever. This action will allow the **drag,** or the distance that the bottom of the cut is behind the top, to catch up before stopping, **Figure 7-30**. To restart, use the same procedure that was given for starting a cut at the edge of the plate.

The proper alignment of the preheat holes will speed up and improve the cut. The holes should be aligned so that one is directly on the line ahead of the cut and another is aimed down into the cut when making a straight line square cut, **Figure 7-31**. The flame is directed toward the smaller piece and the sharpest edge when cutting a bevel. For this reason, the tip should be changed so that at least two of the flames are on the larger plate and none of the flames is directed on the sharp edge, **Figure 7-32**. If the preheat flame is directed at the edge, it will be rounded off as it is melted off.

Layout

Laying out a line to be cut can be done with a piece of **soapstone** or a chalk line. To obtain an accurate line, a scribe or a punch can be used. If a piece of soapstone is

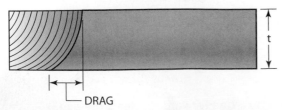

FIGURE 7-30 Drag is the distance by which the bottom of a cut lags behind the top. © Cengage Learning 2012

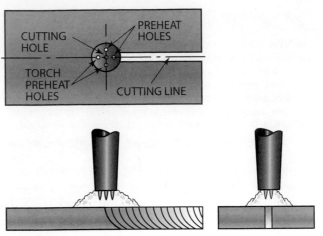

FIGURE 7-31 Tip alignment for a square cut. © Cengage Learning 2012

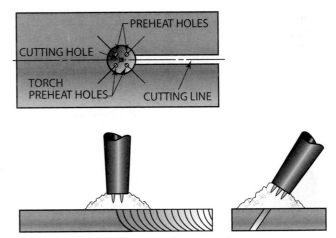

FIGURE 7-32 Tip alignment for a bevel cut. © Cengage Learning 2012

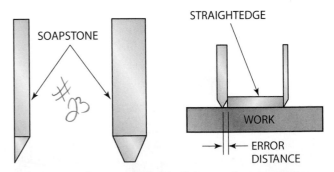

FIGURE 7-33 Proper method of sharpening a soapstone. © Cengage Learning 2012

used, it should be sharpened properly to increase accuracy, **Figure 7-33**. A chalk line will make a long, straight line on metal and is best used on large jobs. The scribe and punch can both be used to lay out an accurate line, but the punched line is easier to see when cutting. A punch can be held as shown in **Figure 7-34**, with the tip just above the surface of the metal. When the punch is struck with a lightweight hammer, it will make a mark.

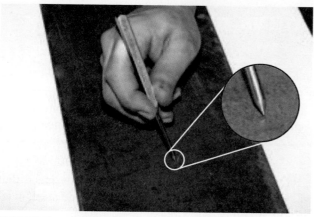

FIGURE 7-34 Holding the punch slightly above the surface allows the punch to be struck rapidly and moved along a line to mark it for cutting. Larry Jeffus

If you move your hand along the line and rapidly strike the punch, it will leave a series of punch marks for the cut to follow.

Selecting the Correct Tip and Setting the Pressure

Each welding equipment manufacturer uses its own numbering system to designate the tip size. It would be impossible to remember each of the systems. Each manufacturer, however, does relate the tip number to the numbered drill size used to make the holes. On the back of most tip cleaning sets, the manufacturer lists the equivalent drill size of each tip cleaner. By remembering approximately which tip cleaner was used on a particular tip for a metal thickness range, a welder can easily select the correct tip when using a new torch set. Using the tip cleaner that you are familiar with, try it in the various torch tips until you find the correct tip that the tip cleaner fits. **Table 7-6** lists the tip drill size, pressure range, and metal thickness range for which the tip can be used.

#24

Metal Thickness in. (mm)	Center Orifice Size No. Drill Size	Tip Cleaner No.*	Oxygen Pressure lb/in.² (kPa)	Acetylene lb/in.² (kPa)
1/8 (3)	60	7	10 (70)	3 (20)
1/4 (6)	60	7	15 (100)	3 (20)
3/8 (10)	55	11	20 (140)	3 (20)
1/2 (13)	55	11	25 (170)	4 (30)
3/4 (19)	55	11	30 (200)	4 (30)
1 (25)	53	12	35 (240)	4 (30)
2 (51)	49	13	45 (310)	5 (35)
3 (76)	49	13	50 (340)	5 (35)
4 (102)	49	13	55 (380)	5 (35)
5 (127)	45	**	60 (410)	5 (35)

*The tip cleaner number when counted from the small end toward the large end in a standard tip cleaner set.
**Larger than normally included in a standard tip cleaner set.

TABLE 7-6 Cutting Pressure and Tip Size

PRACTICE 7-4
Setting the Gas Pressures

Setting the working pressure of the regulators can be done by following a table, or it can be set by watching the flame. Following all the equipment manufacturer's operation and safety instructions, and using all required personal protective equipment, you are going to light and adjust the oxyacetylene flame.

1. To set the regulator by watching the flame, first set the acetylene pressure at 2 psig to 4 psig (14 kPag to 30 kPag) and then light the acetylene flame.

2. Open the acetylene torch valve one to two turns and reduce the regulator pressure by backing out the setscrew until the flame starts to smoke.

3. Increase the pressure until the smoke stops and then increase it just a little more.

This is the maximum fuel gas pressure the tip needs. With a larger tip and a longer hose, the pressure must be set higher. This is the best setting, and it is the safest one to use. With this lowest possible setting, there is less chance of a leak. If the hoses are damaged, the resulting fire will be much smaller than a fire burning from a hose with a higher pressure. There is also less chance of a leak with the lower pressure.

4. With the acetylene adjusted so that the flame just stops smoking, slowly open the torch oxygen valve.

5. Adjust the torch to a neutral flame. When the cutting lever is depressed, the flame will become carbonizing, not having enough oxygen pressure.

6. While holding the cutting lever down, increase the oxygen regulator pressure slightly. Readjust the flame, as needed, to a neutral setting by using the oxygen valve on the torch.

7. Increase the pressure slowly and readjust the flame as you watch the length of the clear cutting stream in the center of the flame, **Figure 7-35A**. The center stream will stay fairly long until a pressure is reached that causes turbulence, disrupting the cutting stream. This turbulence will cause the flame to shorten in length considerably, **Figure 7-35B**.

8. With the cutting lever still depressed, reduce the oxygen pressure until the flame lengthens once again. This is the maximum oxygen pressure that this tip can use without disrupting turbulence in the cutting stream. This turbulence will cause a very poor cut. The lower pressure also will keep the sparks from being blown a longer distance from the work, **Figure 7-36**.

Complete a copy of the "Student Welding Report" listed in Appendix I or provided by your instructor. ◆

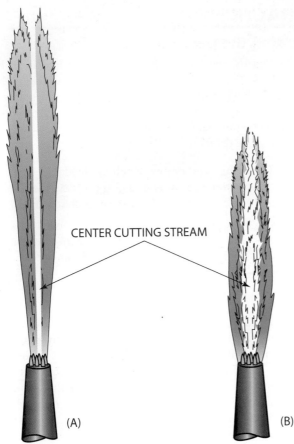

CENTER CUTTING STREAM

(A)　　　　　　　　(B)

FIGURE 7-35 A clean cutting tip will have a long, well-defined oxygen stream. © Cengage Learning 2012

The Chemistry of a Cut

The oxyfuel gas cutting torch works when the metal being cut rapidly oxidizes or burns. This rapid oxidization or burning occurs when a high-pressure stream of pure oxygen is directed on the metal after it has been preheated to a temperature above its kindling point. **Kindling point** is the lowest temperature at which a material will burn. The kindling temperature of iron is 1600°F (870°C), which is a dull red color. Note that iron is the pure element and cast iron is an alloy made primarily of iron and

carbon. The process will work easily on any metal that will rapidly oxidize, such as iron, low carbon steel, magnesium, titanium, and zinc.

///// CAUTION /////

Some metals release harmful oxides when they are cut. Extreme caution must be taken when cutting used, oily, dirty, or painted metals. They often produce very dangerous fumes when they are cut. You may need extra ventilation and a respirator to be safe. Check with the welding shop foreman or shop safety officer before cutting any metal you are unfamiliar with.

The process is most often used to cut iron and low carbon steels because unlike most of the metals, little or no oxides are left on the metal, and it can easily be welded.

The burning away of the metal is a chemical reaction with iron (Fe) and oxygen (O). The oxygen forms an iron oxide, primarily Fe_3O_4, that is light gray in color. Heat is produced by the metal as it burns. This heat helps carry the cut along. On thick pieces of metal, once a small spot starts burning (being cut), the heat generated helps the cut continue quickly through the metal. With some cuts the heat produced may overheat small strips of metal being cut from a larger piece. As an example, the center piece of a hole being cut will quickly become red hot and will start to oxidize with the surrounding air, **Figure 7-37**. This heat produced by the cut makes it difficult to cut out small or internal parts.

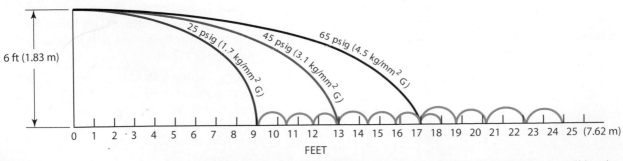

6 ft (1.83 m)

25 psig (1.7 kg/mm² G)　45 psig (3.1 kg/mm² G)　65 psig (4.5 kg/mm² G)

0　1　2　3　4　5　6　7　8　9　10　11　12　13　14　15　16　17　18　19　20　21　22　23　24　25 (7.62 m)

FEET

FIGURE 7-36 The sparks from cutting a mild steel plate, 3/8 in. (10 mm) thick, 6 ft (1.8 m) from the floor, will be thrown much farther if the cutting pressure is too high for the plate thickness. These cuts were made with a Victor cutting tip no. 0-1-101 using 25 psig (1.7 kg/mm²) as recommended by the manufacturer and by excessive pressures of 45 psig (3.1 kg/mm²) and 65 psig (4.5 kg/mm²). Larry Jeffus

FIGURE 7-37 As a hole is cut, the center may be overheated. ***See Welding Principles and Practices on DVD.*** **DVD 4—Oxyacetylene Welding.**

EXPERIMENT 7-1

Observing Heat Produced during a Cut

This experiment may require more skill than you have developed by this time. You may wish to observe your instructor performing the experiment or try it at a later time.

Using a properly lit and adjusted cutting torch; welding gloves; appropriate eye protection clothing, and all other required personal protective equipment; and one piece of clean mild steel plate 6 in. (152 mm) long × 1/4 in. (6 mm)

to 1/2 in. (13 mm) thick, you will make an oxyfuel gas cut without the preheat flame.

Place the piece of metal so that the cutting sparks fall safely away from you. With the torch lit, pass the flame over the length of the plate until it is warm, but not hot. Brace yourself and start a cut near the edge of the plate. When the cut has been established, have another student turn off the acetylene regulator. The cut should continue if you remain steady and the plate is warm enough. *Hint: Using a slightly larger tip size will make this easier.*

Complete a copy of the "Student Welding Report" listed in Appendix I or provided by your instructor. ◆

The Physics of a Cut

As a cut progresses along a plate, a record of what happened during the cut is preserved along both sides of the kerf. This record indicates to the welder what was correct or incorrect with the preheat flame, cutting speed, and oxygen pressure.

Preheat The size and number of preheat holes in a tip have an effect on both the top and bottom edges of the metal. An excessive amount of preheat flame results in the top edge of the plate being melted or rounded off. In addition, an excessive amount of hard-to-remove slag is deposited along the bottom edge. If the flame is too small, the travel speed must be slower. A reduction in speed may result in the cutting stream wandering from side to side. The torch tip can be raised slightly to eliminate some of the damage caused by too much preheat. However, raising the torch tip causes the cutting stream of oxygen to be less forceful and less accurate.

Speed The cutting speed should be fast enough so that the **drag lines** have a slight slant backward if the tip is held at a 90° angle to the plate, **Figure 7-38**. If the cutting

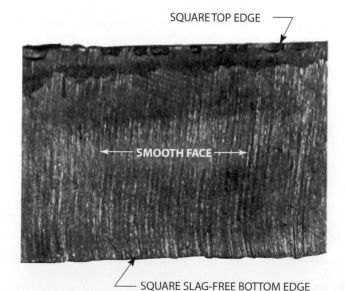

FIGURE 7-38 Correct cut. Larry Jeffus

FIGURE 7-39 Too fast a travel speed resulting in an incomplete cut; too much preheat and the tip is too close, causing the top edge to be melted and removed. Larry Jeffus

speed is too fast, the oxygen stream may not have time to go completely through the metal, resulting in an incomplete cut, **Figure 7-39**. Too slow a cutting speed results in the cutting stream wandering, thus causing gouges in the side of the cut, **Figure 7-40** and **Figure 7-41**.

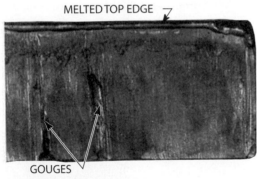

FIGURE 7-40 Too slow a travel speed results in the cutting stream wandering, thus causing gouges in the surface; preheat flame is too close, melting the top edge. Larry Jeffus

FIGURE 7-41 Too slow a travel speed at the start; too much preheat. Larry Jeffus

Pressure A correct pressure setting results in the sides of the cut being flat and smooth. A pressure setting that is too high causes the cutting stream to expand as it leaves the tip, resulting in the sides of the kerf being slightly dished, **Figure 7-42**. When the pressure setting is too low, the cut may not go completely through the metal.

EXPERIMENT 7-2

Effect of Flame, Speed, and Pressure on a Machine Cut

Using a properly lit and adjusted automatic cutting machine; welding gloves; appropriate eye protection; clothing, and all other required personal protective equipment; a variety of tip sizes, and one piece of mild steel plate 6 in. (152 mm) long × 1/2 in. (13 mm) to 1 in. (25 mm) thick, you will observe the effect of the preheat flame, travel speed, and pressure on the metal being cut.

Using the variety of tips, speeds, and oxygen pressures, make a series of cuts on the plate. As the cut is being made, listen to the sound it makes. Also look at the stream of sparks coming off the bottom. A good cut should have a smooth, even sound, and the sparks should come off the bottom of the metal more like a stream than a spray, **Figure 7-43**. When the cut is complete, look at the drag lines to determine what was correct or incorrect with the cut, **Figure 7-44**.

Repeat this experiment until you know a good cut by the sound it makes and the stream of sparks. A good cut has little or no slag left on the bottom of the plate.

Complete a copy of the "Student Welding Report" listed in Appendix I or provided by your instructor. ◆

EXPERIMENT 7-3

Effect of Flame, Speed, and Pressure on a Hand Cut

Using a properly lit and adjusted hand torch, welding gloves; appropriate eye protection; clothing, and all other required personal protective equipment; and the same tip sizes and mild steel plate, repeat Experiment 7-2 to note the effects of the preheat flame, travel speed, and pressure on hand cutting.

Complete a copy of the "Student Welding Report" listed in Appendix I or provided by your instructor. ◆

Slag The two types of **slag** produced during a cut are soft slag and hard slag. **Soft slag** is very porous, brittle, and easily removed from a cut. There is little or no unoxidized iron in it. It may be found on some good cuts. Hard slag may be mixed with soft slag. Hard slag is attached solidly to the bottom edge of a cut, and it requires a lot of chipping and grinding to be removed. There is 30% to 40% or more unoxidized iron in hard slag. The higher the unoxidized iron content, the more difficult the slag is to remove. Slag is found on bad cuts, due to dirty tips, too much preheat, too slow a travel speed, too short a coupling distance, or incorrect oxygen pressure.

CORRECT CUT

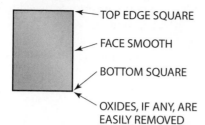

- TOP EDGE SQUARE
- FACE SMOOTH
- BOTTOM SQUARE
- OXIDES, IF ANY, ARE EASILY REMOVED

TRAVEL SPEED TOO SLOW

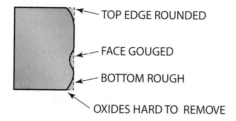

- TOP EDGE ROUNDED
- FACE GOUGED
- BOTTOM ROUGH
- OXIDES HARD TO REMOVE

TRAVEL SPEED TOO FAST

- TOP EDGE SHARP
- DRAG LINES PRONOUNCED
- BOTTOM ROUNDED

PREHEAT FLAMES TOO HIGH ABOVE THE SURFACE

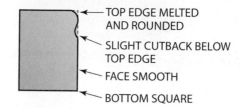

- TOP EDGE MELTED AND ROUNDED
- SLIGHT CUTBACK BELOW TOP EDGE
- FACE SMOOTH
- BOTTOM SQUARE

PREHEAT FLAMES TOO CLOSE TO THE SURFACE

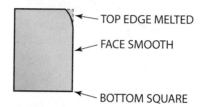

- TOP EDGE MELTED
- FACE SMOOTH
- BOTTOM SQUARE

CUTTING OXYGEN PRESSURE TOO HIGH

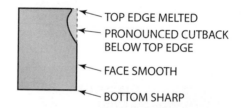

- TOP EDGE MELTED
- PRONOUNCED CUTBACK BELOW TOP EDGE
- FACE SMOOTH
- BOTTOM SHARP

FIGURE 7-42 Profile of flame-cut plates. © Cengage Learning 2012

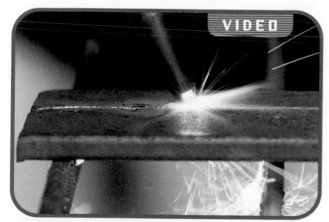

FIGURE 7-43 A good cut showing a steady stream of sparks flying out from the bottom of the cut. Larry Jeffus *See Welding Principles and Practices on DVD. DVD 4—Oxyacetylene Welding.*

FIGURE 7-44 Poor cut. The slag is backing up because the cut is not going through the plate. Larry Jeffus

The slag from a cut may be kept off one side of the plate being cut by slightly angling the cut toward the scrap side of the cut, **Figure 7-45.** The angle needed to force the slag away from the good side of the plate may be as small as 2° or 3°. This technique works best on thin sections; on thicker sections the bevel may show.

Plate Cutting

Low carbon steel plate can be cut quickly and accurately, whether thin-gauge sheet metal or sections more than 4 ft (1.2 m) thick are used. It is possible to achieve cutting speeds as fast as 32 in. per minute (13.5 mm/s), in 1/8-in. (3-mm) plate, and accuracy on machine cuts of ±3/64 in. Some very large hand-cutting torches with an

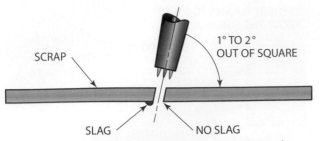

FIGURE 7-45 A slight angle on the torch will put the slag on the scrap side of the cut. © Cengage Learning 2012

oxygen cutting volume of 600 cfh (2830 L/min) can cut metal that is 4 ft (1.2 m) thick, **Figure 7-46.** Most hand torches will not easily cut metal that is more than 7 in. (178 mm) to 10 in. (254 mm) thick.

The thicker the plate, the more difficult the cut is to make. Thin plate, 1/4 in. (6 mm) or less, can be cut and the pieces separated even if poor techniques and incorrect pressure settings are used. Thick plate, 1/2 in. (13 mm) or thicker, often cannot be separated if the cut is not correct. For very heavy cuts, on plate 12 in. (305 mm) or thicker, the equipment and operator technique must be near perfection or the cut will be faulty.

Plate that is properly cut can be assembled and welded with little or no postcut cleanup. Poor-quality cuts require more time to clean up than is needed to make the required adjustments to make a good weld.

Cutting Table

Because of the nature of the torch cutting process, special consideration is given to the flame cutting support. Any piece being cut should be supported so the torch flame will not cut through the piece and into the table. Special cutting tables are used that expose only a small metal area to the torch flame. Some tables use parallel steel

bars of metal and others use cast-iron pyramids. All cutting should be set up so the flame and oxygen stream runs between the support bars or over the edge of the table.

If an ordinary welding table or another steel table is used, special care must be taken to avoid cutting through the tabletop. The piece being cut may be supported above the support table by firebrick. Another method is to cut the metal over the edge of the table.

Torch Guides

In manual torch cutting, a guide or support is frequently used to allow for better control and more even cutting. It takes a very skilled welder to make a straight, clean cut even when following a marked line. It is even more difficult to make a radius cut to any accuracy. Guides and supports allow the height and angle of the torch head to remain constant. The speed of the cut, which is very important to making a clean, even kerf, must be controlled by the welder.

Since the torch must be held in an exact position while making any accurate cut, the welder normally supports the torch weight with the hand. Supporting the torch weight this way not only allows for more accurate work but also cuts down on fatigue. A rest, such as a firebrick, is also used to support the torch.

Various types of guides can be used to guide the torch in a straight line. **Figure 7-47** shows one type of guide using angle iron. The edge of the angle is followed to make the straight cut. Bevel cuts can be made freehand with the torch, but it is very difficult to keep them uniform. More accurate bevel cuts are made by resting the torch against the angle side of an angle iron.

Special roller guides, **Figure 7-48A**, can also be attached to the torch head. The attachment holds the torch cutting tip at an exact height.

When cutting circles, a circle cutting attachment is used. **Figure 7-48B** shows how the attachment fits on the

FIGURE 7-46 Hand torches for thick sections. Victor Equipment Company

FIGURE 7-47 Using angle irons to aid in making cuts. © Cengage Learning 2012

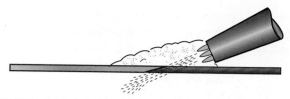

FIGURE 7-48 Devices that are used to improve hand cutting. Victor Equipment Company

torch head. The radius can be preset to any required distance. The cutter revolves around the center point when making the cut. The roller controls the torch tip height above the plate surface.

PRACTICE 7-5

Flat, Straight Cut in Thin Plate

Using a properly lit and adjusted cutting torch and one piece of mild steel plate 6 in. (152 mm) long × 1/4 in. (6 mm) thick, you will cut off 1/2-in. (13-mm) strips.

Using a straightedge and soapstone, make several straight lines 1/2 in. (13 mm) apart. Starting at one end, make a cut along the entire length of plate. The strip must fall free, be slag free, and be within ±3/32 in. (2 mm) of a straight line and ±5° of being square. Repeat this procedure until the cut can be made straight and slag-free. Turn off the cylinder valves, bleed the hoses, back out the pressure regulators, and clean up your work area when you are finished cutting.

Complete a copy of the "Student Welding Report" listed in Appendix I or provided by your instructor. ◆

PRACTICE 7-6

Flat, Straight Cut in Thick Plate

Using a properly lit and adjusted cutting torch and one piece of mild steel plate 6 in. (152 mm) long × 1/2 in. (13 mm) thick or thicker, you will cut off 1/2-in. (13-mm) strips. *Note: Remember that starting a cut in thick plate will take longer, and the cutting speed will be slower.* Lay out, cut, and evaluate the cut as was done in Practice 7-5. Repeat this procedure until the cut can be made straight

and slag-free. Turn off the cylinder valves, bleed the hoses, back out the pressure regulators, and clean up your work area when you are finished cutting.

Complete a copy of the "Student Welding Report" listed in Appendix I or provided by your instructor. ◆

PRACTICE 7-7

Flat, Straight Cut in Sheet Metal

Use a properly lit and adjusted cutting torch and a piece of mild steel sheet that is 10 in. (254 mm) long and 18 gauge to 11 gauge thick. Holding the torch at a very sharp leading angle, **Figure 7-49**, cut the sheet along the line. The cut must be smooth and straight with as little slag as possible. Repeat this procedure until the cut can be made flat, straight, and slag-free. Turn off the cylinder valves, bleed the hoses, back out the pressure regulators, and clean up your work area when you are finished cutting.

Complete a copy of the "Student Welding Report" listed in Appendix I or provided by your instructor. ◆

PRACTICE 7-8

Flame Cutting Holes

Using a properly lit and adjusted cutting torch, welding gloves, appropriate eye protection and clothing, and one piece of mild steel plate 1/4 in. (6 mm) thick, you will cut holes with diameters of 1/2 in. (13 mm) and 1 in. (25 mm). Using the technique described for piercing a hole, start in the center, and make an outward spiral until the hole is the desired size, **Figure 7-50**. The hole must be

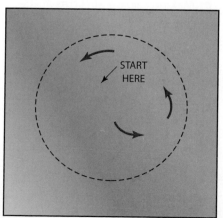

FIGURE 7-50 Start a cut for a hole near the middle.

within ±3/32 in. (2 mm) of being round and ±5° of being square. The hole may have slag on the bottom. Repeat this procedure until both small and large sizes of holes can be made within tolerance. Turn off the cylinder valves, bleed the hoses, back out the pressure regulators, and clean up your work area when you are finished cutting.

Complete a copy of the "Student Welding Report" listed in Appendix I or provided by your instructor. ◆

Distortion

Distortion is when the metal bends or twists out of shape as a result of being heated during the cutting process. This is a major problem when cutting a plate. If the distortion is not controlled, the end product might be worthless. There are two major methods of controlling distortion. One method involves making two parallel cuts on the same plate at the same speed and time, **Figure 7-51**. Because the plate is heated evenly, distortion is kept to a minimum, **Figure 7-52**.

The second method involves starting the cut a short distance from the edge of the plate, skipping other short tabs every 2 ft (0.6 m) to 3 ft (0.9 m) to keep the cut from separating. Once the plate cools, the remaining tabs are cut, **Figure 7-53**.

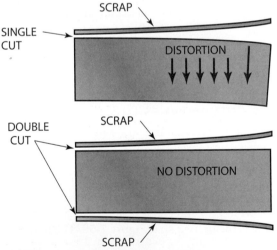

FIGURE 7-51 Making two parallel cuts at the same time will control distortion. © Cengage Learning 2012

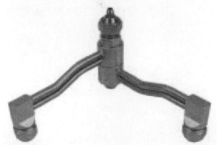

FIGURE 7-52 Slitting adaptor for cutting machine. It can be used for parallel cuts from 1 in. (38 mm) to 12 in. (500 mm). Ideal for cutting test coupons.

Victor Equipment Company

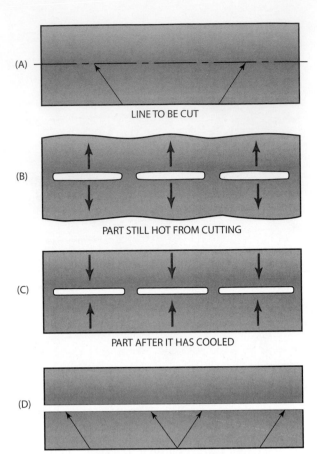

FIGURE 7-53 Steps used during cutting to minimize distortion. © Cengage Learning 2012

EXPERIMENT 7-4

Minimizing Distortion

Using a properly lit and adjusted cutting torch, welding gloves, appropriate eye protection and clothing, and two pieces of mild steel 10 in. (254 mm) long × 1/4 in. (6 mm) thick, you will make two cuts and then compare the distortion. Lay out and cut out both pieces of metal as shown in **Figure 7-54**. Allow the metal to cool, and then cut the remaining tabs. Compare the four pieces of metal for distortion.

Complete a copy of the "Student Welding Report" listed in Appendix I or provided by your instructor. ◆

PRACTICE 7-9

Beveling a Plate

Use a properly lit and adjusted cutting torch, welding gloves, appropriate eye protection and clothing, and one piece of mild steel plate 6 in. (152 mm) long × 3/8 in. (10 mm) thick. You will make a 45° bevel down the length of the plate.

Mark the plate in strips 1/2 in. (13 mm) wide. Set the tip for beveling and cut a bevel. The bevel should be within ±3/32 in. (2 mm) of a straight line and ±5° of a 45° angle.

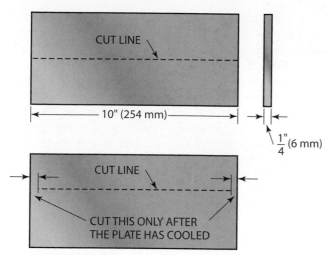

FIGURE 7-54 Making two cuts with minimum distortion. *Note: Sizes of these and other cutting projects can be changed to fit available stock.* © Cengage Learning 2012

There may be some soft slag, but no hard slag, on the beveled plate. Repeat this Practice until the cut can be made within tolerance. Turn off the cylinder valves, bleed the hoses, back out the pressure regulators, and clean up your work area when you are finished cutting.

Complete a copy of the "Student Welding Report" listed in Appendix I or provided by your instructor. ◆

PRACTICE 7-10

Vertical Straight Cut

For this Practice, you will need a properly lit and adjusted cutting torch, welding gloves, appropriate eye protection and clothing, and one piece of mild steel plate 6 in. (152 mm) long × 1/4 in. (6 mm) to 3/8 in. (10 mm) thick, marked in strips 1/2 in. (13 mm) wide and held in the vertical position. You will make a straight line cut. Make sure that the sparks do not cause a safety hazard and that the metal being cut off will not fall on any person or object.

Starting at the top, make one cut downward. Then, starting at the bottom, make the next cut upward. The cut must be free of hard slag and within ±3/32 in. (2 mm) of a straight line and ±5° of being square. Repeat these cuts until they can be made within tolerance. Turn off the cylinder valves, bleed the hoses, back out the pressure regulators, and clean up your work area when you are finished cutting.

Complete a copy of the "Student Welding Report" listed in Appendix I or provided by your instructor. ◆

PRACTICE 7-11

Overhead Straight Cut

Using a properly lit and adjusted cutting torch; welding gloves; appropriate eye protection and clothing; and one piece of mild steel plate 6 in. (152 mm) long × 1/4 in.

(6 mm) to 3/8 in. (10 mm) thick, marked in strips 1/2 in. (13 mm) wide, you will make a cut in the overhead position. When making overhead cuts, it is important to be completely protected from the hot sparks. In addition to the standard safety clothing, you should wear a leather jacket, leather apron, cap, ear protection, and a full face shield.

The torch can be angled so that most of the sparks will be blown away. The metal should fall free when the cut is completed. The cut must be within 1/8 in. (3 mm) of a straight line and ±5° of being square. Repeat this practice until the cut can be made within tolerance. Turn off the cylinder valves, bleed the hoses, back out the pressure regulators, and clean up your work area when you are finished cutting.

Complete a copy of the "Student Welding Report" listed in Appendix I or provided by your instructor. ◆

Cutting Applications

Making practice cuts on a piece of metal that will only become scrap is a good way to learn the proper torch techniques. If a bad cut is made, there is no loss. In a production shop, where each piece of metal is important, however, scrapped metal due to bad cuts decreases the shop's profits.

A number of factors can affect your ability to make a quality cut on a part that do not exist during practice cuts. The following are some of the things that can become problems when cutting:

- *Changing positions:* Often, parts are larger than can be cut from one position, so you may have to move to complete the cut. Stopping and restarting a cut can result in a small flaw in the cut surface. If this flaw exceeds the acceptable limits, the cut surface must be repaired before the part can be used. To avoid this problem, always try to stop at corners if the cut cannot be completed without moving.

- *Sparks:* You will often be making cuts in large plates. Even an ideal cut can create sparks that bounce around the plate surface. These sparks often find their way into your glove, under your arm, or to any other place that will become uncomfortable. Experienced welders will usually keep working if the sparks are not too large or too uncomfortable. With experience you will learn how to angle the torch, direct the cut, and position your body to minimize this problem.

- *Hot surfaces:* As you continue making cuts to complete the part, it will begin to heat up. Depending on the size of the part, the number of cuts per part, and the number of parts being cut, this heat can become uncomfortable. You may find it necessary to hold the torch farther back from the tip, but this will affect the quality of your cuts, **Figure 7-55**. Sometimes you might be able to rest your hand on a block to keep it off of the plate. Another problem

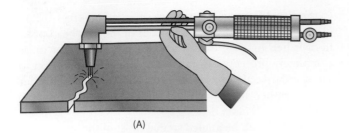

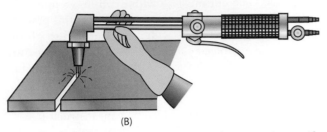

FIGURE 7-55 It is easier to make straight smooth cuts if you can brace the torch closer to the tip like cut (B).
© Cengage Learning 2012

with heat buildup is that it may become high enough to affect the cut quality. Heat becomes a problem when it causes the top edge of the plate to melt during a cut, as if the torch tip were too large. This is more of a problem when several cuts are being made in close proximity. Planning your cutting sequence and allowing cooling time will help control this potential problem.

- *Tip cleaning:* As with any cutting, the tip will catch small sparks and become dirty or clogged. You must decide how dirty or clogged you will let the tip get before you stop to clean it. Time spent cleaning the tip reduces productivity, unfortunately. On the other hand, if you do not stop occasionally to clean up, the quality of the cut will become so bad that postcut cleanup will become excessive. It is your responsibility to decide when and how often to clean the tip.

- *Blowback:* As a cut progresses across the surface of a large plate, it may cross supports underneath the plate. During practice cuts this seldom if ever happens, but, depending on the design of the cutting table, it will occur even under the best of conditions. If the support is small, the blowback may not cover you with sparks, plug the cutting tip, or cause a major flaw in the cut surface. If the support is large, then one or all of these events can occur. If you see that the blowback is not clearing quickly, it may be necessary to stop the cut. Stopping the cut halts the shower of sparks but leaves you with a restart problem.

PRACTICE 7-12

Cutting Out Internal and External Shapes

Using a properly lit and adjusted cutting torch, welding gloves, appropriate eye protection and clothing, and

one piece of plate 1/4 in. (6 mm) to 3/8 in. (10 mm) thick, you may lay out and cut out one of the sample patterns shown in **Figure 7-56**, one of the projects in Chapter 19, or any other design available.

Choose the pattern that best fits the piece of metal you have and mark it using a center punch. The exact size and shape of the layout are not as important as the accuracy of the cut. The cut must be made so that the center-punched line is left on the part and so that there is no more than 1/8 in. (3 mm) between the cut edge and the line, **Figure 7-57** and **Figure 7-58**. Repeat this practice until the cut can be made within tolerance. Turn off the cylinder valves, bleed the hoses, back out the pressure regulators, and clean up your work area when you are finished cutting.

Complete a copy of the "Student Welding Report" listed in Appendix I or provided by your instructor. ◆

Pipe Cutting

Freehand pipe cutting may be done in one of two ways. On small diameter pipe, usually under 3 in. (76 mm), the torch tip is held straight up and down and moved from the center to each side, **Figure 7-59**. This technique can also be used successfully on larger pipe.

For large diameter pipe, 3 in. (76 mm) and larger, the torch tip is always pointed toward the center of the pipe, **Figure 7-60**. This technique is also used on all sizes of heavy-walled pipe and can be used on some smaller pipe sizes.

The torch body should be held so that it is parallel to the centerline of the pipe. Holding the torch parallel helps to keep the cut square.

> **⚠ CAUTION**
>
> **When cutting pipe, hot sparks can come out of the end of the pipe nearest you, causing severe burns. For protection from hot sparks, plug up the open end of the pipe nearest you, put up a barrier to the sparks, or stand to one side of the material being cut.**

PRACTICE 7-13

Square Cut on Pipe, 1G (Horizontal Rolled) Position

Using a properly lit and adjusted cutting torch, welding gloves, appropriate eye protection and clothing, and one piece of schedule 40 steel pipe with a diameter of 3 in. (76 mm), you will cut off 1/2-in. (13-mm) long rings.

Using a template and a piece of soapstone, mark several rings, each 1/2 in. (13 mm) wide, around the pipe. Place the pipe horizontally on the cutting table. Start the cut at the top of the pipe using the proper piercing technique. Move the torch backward along the line and then forward; this will keep slag out of the cut. If the end of the cut closes in with slag, this will cause the oxygen to gouge the

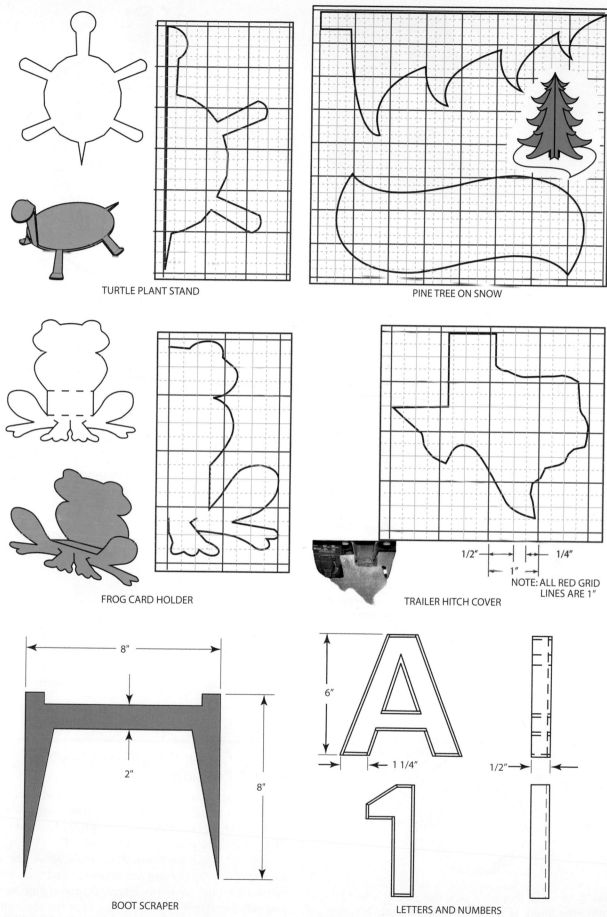

TURTLE PLANT STAND

PINE TREE ON SNOW

FROG CARD HOLDER

1/2"   1/4"
1"
NOTE: ALL RED GRID LINES ARE 1"

TRAILER HITCH COVER

8"

2"

8"

BOOT SCRAPER

6"

1 1/4"

1/2"

LETTERS AND NUMBERS

FIGURE 7-56 Suggested patterns for practice. © Cengage Learning 2012

FIGURE 7-57 Beginning a cut with the torch concentrating the flame on the top edge to speed starting.
Larry Jeffus

FIGURE 7-58 The torch is rotated to allow the preheating of the plate ahead of the cut. This speeds the cutting and also provides better visibility of the line being cut.
Larry Jeffus

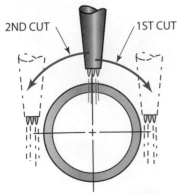

FIGURE 7-59 Small diameter pipe can be cut without changing the angle of the torch. After the top is cut, roll the pipe to cut the bottom. © Cengage Learning 2012

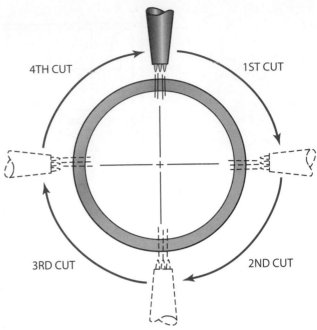

FIGURE 7-60 On large diameter pipe, the torch is turned to keep it at a right angle to the pipe. The pipe should be cut as far as possible before stopping and turning it. © Cengage Learning 2012

edge of the pipe when the cut is continued. Keep the tip pointed straight down. When you have gone as far with the cut as you can comfortably, quickly flip the flame away from the pipe. Restart the cut at the top of the pipe and cut as far as possible in the other direction. Stop and turn the pipe so that the end of the cut is on top and the cut can be continued around the pipe. When the cut is completed, the ring must fall free. When the pipe is placed upright on a flat plate, the pipe must stand within 5° of vertical and have no gaps higher then 1/8 in. (3 mm) under the cut. Repeat this procedure until the cut can be made within tolerance. Turn off the cylinder valves, bleed the hoses, back out the pressure regulators, and clean up your work area when you are finished cutting.

Complete a copy of the "Student Welding Report" listed in Appendix I or provided by your instructor. ◆

PRACTICE 7-14

Square Cut on Pipe, 1G (Horizontal Rolled) Position

Using the same equipment, materials, and markings as described in Practice 7-13, you will cut off the 1/2-in. (13-mm) long rings while keeping the tip pointed toward the center of the pipe.

Starting at the top, pierce the pipe. Move the torch backward to keep the slag out of the cut and then forward around the pipe, stopping when you have gone as far as you can comfortably. Restart the cut at the top and proceed with the cut in the other direction. Roll the pipe and continue the cut until the ring falls off freely. Stand the cut end of the pipe on a flat plate. The pipe must stand within 5° of vertical and have no gaps higher than 1/8 in. (3 mm). Repeat this practice until the cut can be made within tolerance. Turn off the cylinder valves, bleed the hoses, back out the pressure regulators, and clean up your work area when you are finished cutting.

Complete a copy of the "Student Welding Report" listed in Appendix I or provided by your instructor. ◆

PRACTICE 7-15

Square Cut on Pipe, 5G (Horizontal Fixed) Position

With the same equipment, materials, and markings as described in Practice 7-13, you will cut off 1/2-in. (13-mm) rings, using either technique, without rolling the pipe.

Start at the top and cut down both sides as far as you can comfortably. Reposition yourself and continue the cut under the pipe until the ring falls off freely. Stand the cut end of the pipe on a flat plate. The pipe must stand within 5° of vertical and have no gaps higher than 1/8 in. (3 mm). Repeat this practice until the cut can be made within tolerance. Turn off the cylinder valves, bleed the hoses, back out the pressure regulators, and clean up your work area when you are finished cutting.

Complete a copy of the "Student Welding Report" listed in Appendix I or provided by your instructor. ◆

PRACTICE 7-16

Square Cut on Pipe, 2G (Vertical) Position

With the same equipment, materials, and markings as listed in Practice 7-13, you will cut off 1/2-in. (13-mm)

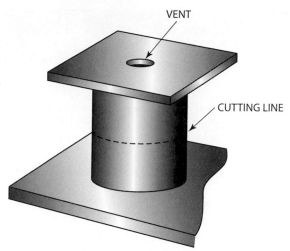

FIGURE 7-61 Place a plate on top of a short piece of pipe to keep the sparks from flying around the shop.
© Cengage Learning 2012

rings, using either technique, from a pipe in the vertical position.

Place a flat piece of plate over the open top end of the pipe to keep the sparks contained, **Figure 7-61.** Start on one side and proceed around the pipe until the cut is completed. Because of slag, the ring may have to be tapped free. Stand the cut end of the pipe on a flat plate. The pipe must stand within 5° of vertical and have no gaps higher than 1/8 in. (3 mm). Repeat this practice until the cut can be made within tolerance. Turn off the cylinder valves, bleed the hoses, back out the pressure regulators, and clean up your work area when you are finished cutting.

Complete a copy of the "Student Welding Report" listed in Appendix I or provided by your instructor. ◆

Summary

The oxyfuel cutting torch is one of the most commonly used (and misused) tools in the welding industry. When used properly it can produce almost machine-cut quality requiring no postcut cleanup. However, when misused the oxyfuel torch can produce some of the most difficult problems for the welding fabricator to overcome. As you learn to use the cutting torch efficiently, you can dramatically reduce postcut cleanup time and increase productivity.

Equipment setup and torch tip cleaning are essential elements required for the welder to produce quality oxyfuel cuts. Take your time each time you are setting up and preparing to make a cut; this is not wasted time. A good setup will ensure that the cut's quality will meet the fabricator's quality needs.

The travel speed for cutting will vary, depending on a number of factors, such as plate thickness or the surface condition of the metal being cut. A good, clean, quality cut will progress at its own rate. Do not try to rush through too quickly. Learn to develop a sense for the cutting rate that produces the best-quality cut.

Oxygen Cutting

The chemical reaction between oxygen and an oxidizing metal heated to a high temperature begins the cutting action of the oxygen cutting (OC) process. The temperature of the metal is maintained with a flame from the combustion of an oxygen-fuel gas mixture. A separate stream of oxygen does the cutting.

The cutting torch has internal chambers for mixing the fuel gas and oxygen, as well as a separate chamber for delivering high-purity oxygen to the cutting area. The torch nozzle or tip has numerous ports machined into it for regulating and concentrating the flame and the stream of cutting oxygen.

When iron is brought to a temperature above 1600°F (870°C) and combined with oxygen of 99.5% or greater purity, rapid oxidation occurs. This chemical reaction releases a tremendous amount of heat, which supports the cutting process by preheating the metal at the cutting point to its ignition temperature. Some of the iron adjacent to the cutting point also is melted and blown away by the oxygen stream. Oxygen with purity levels below 99.5% significantly decreases the cutting efficiency.

Since oxidation of iron is at the heart of the cutting process, metals with high levels of nickel or chromium and nonferrous metals cannot be cut effectively with this process.

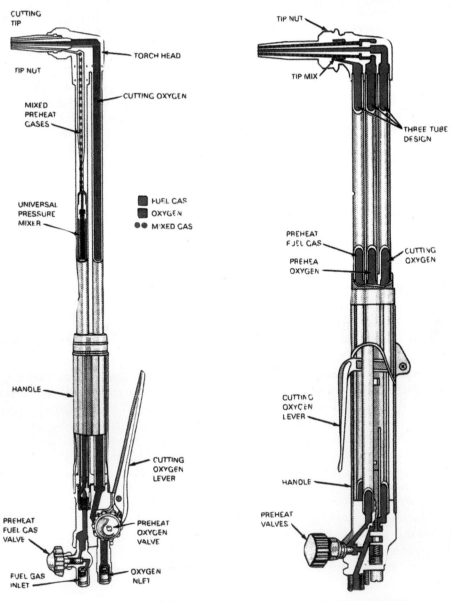

Schematics of two types of oxygen cutting torches. Art courtesy of the *Welding Handbook*, Vol. 2, 8th Edition.

A variety of fuel gases can be used in oxygen cutting. Acetylene, with its widespread availability and high flame temperature, is the common choice. Depending on the shape and color of the flame, the correct ratio of oxygen to acetylene for efficient cutting can be determined. Acetylene must not be used at pressures higher than 15 lb/in.² gauge pressure or 30 lb/in.² absolute pressure. When used at higher pressures, it can become explosive under heat or shock conditions.

Hoses with specially designed fittings are used in oxygen cutting. The hose for oxygen, which in the United States is green, has a right-hand threaded fitting with a smooth surface. The fuel-gas hose, which is normally red, has left-hand threads, and the nut is notched.

Article courtesy of the American Welding Society.

Review

1. Using Table 7-1, list the six different fuel gases in rank order according to their temperature.

2. What is a combination welding-cutting torch?

3. State one advantage of owning a combination welding-cutting torch as opposed to just having a cutting torch.

4. State one advantage of owning a dedicated cutting torch as compared to having a combination welding-cutting torch.

5. What is a mixing chamber? Where is it located?

6. Define the term *equal-pressure torch*. How does it work?

7. How does an injector-type mixing chamber work?

8. State the advantages of having two oxygen regulators on a machine cutting torch.

9. Why are some copper alloy cutting tips chrome plated?

10. What determines the amount of preheat flame requirements of a torch?

11. What can happen if acetylene is used on a tip designed to be used with propane or other such gas?

12. Why are some propane and natural gas tips made with a deep, recessed center?

13. What types of tip seals are used with cutting torch tips?

14. If a cutting tip should stick in the cutting head, how should it be removed?

15. How can cutting torch tip seals be repaired?

16. Why is the oxygen valve turned on before starting to clean a cutting tip?

17. Why does the preheat flame become slightly oxidizing when the cutting lever is released?

18. What causes the tiny ripples in a hand cut?

19. Why is a slight forward torch angle helpful for cutting?

20. Why should cans, drums, tanks, or other sealed containers not be opened with a cutting torch?

21. Why is the torch tip raised as the cutting lever is depressed when cutting a hole?

22. Why are the preheat holes not aligned in the kerf when making a bevel cut?

23. Sketch the proper end shape of a soapstone that is to be used for marking metal.

24. Using Table 7-4, answer the following:
    a. Oxygen pressure for cutting 1/4-in. (6-mm) thick metal
    b. Acetylene pressure for cutting 1-in. (25-mm) thick metal
    c. Tip cleaner size for a tip for 2-in. (51-mm) thick metal
    d. Drill size for a tip for 1/2-in. (13-mm) thick metal

25. What is the best way to set the oxygen pressure for cutting?

26. What metals can be cut with the oxyfuel gas process?

27. Why is it important to have extra ventilation and/or a respirator when cutting some used metal?

28. What factors regarding a cut can be read from the sides of the kerf after a cut?

29. What is hard slag?

30. Why is it important to make good-quality cuts?

31. Describe the methods of controlling distortion when making cuts.

32. How does cutting small diameter pipe differ from cutting large diameter pipe?

Chapter 8

Plasma Arc Cutting

OBJECTIVES

After completing this chapter, the student should be able to

- describe plasma and describe a plasma torch.
- explain how a plasma cutting torch works.
- list the advantages and disadvantages of using a plasma cutting torch.
- demonstrate an ability to set up and use a plasma cutting torch.

KEY TERMS

arc cutting

arc plasma

cup

dross

electrode setback

electrode tip

heat-affected zone

high-frequency alternating
current

ionized gas

joules

kerf

nozzle

nozzle insulator

nozzle tip

pilot arc

plasma

plasma arc

plasma arc gouging

stack cutting

standoff distance

water shroud

water table

INTRODUCTION

The plasma process was originally developed in the mid-1950s as an attempt to create an arc, using argon, that would be as hot as the arc created when using helium gas. The early gas tungsten arc welding process used helium gas and was called "heliarc." This early GTA welding process worked well with helium, but helium was expensive. The gas manufacturing companies had argon as a by-product from the production of oxygen. There was no good commercial market for this waste argon gas, but gas manufacturers believed there would be a good market if they could find a way to make argon weld similar to helium.

Early experiments found that by restricting the arc in a fast-flowing column of argon a plasma was formed. The plasma was hot enough to rapidly melt any metal. The problem was that the fast-moving gas blew the molten metal away. They could not find a way to control this scattering of the molten metal, so they decided to introduce this as a cutting process, not a welding process.

Plasma can cut sheet metal so easily that it has become popular to use it to cut out the smallest of decorations. Most often the smallest of animals, people, buildings, and scenery used on gates, fences, barns, and so forth, have been cut out using PAC, **Figure 8-2.**

Small plasma cutting machines can do many of the same cutting jobs that are done with an oxyacetylene torch, but without the expense of renting gas cylinders. Small plasma cutting machines can use 120 V electrical power from any standard wall plug or auxiliary power plug on a portable welder, **Figure 8-3.**

Plasma machines can cut any type of metal including aluminum, stainless steel, and cast iron. They can cut mild steel ranging from sheet metal up to about 3/8 in. (9 mm), which means that the plasma torch can do most of the cutting required in any welding shop. Larger, more powerful machines are available that can cut an inch (25 mm) or more, but their expense for most small welding shops would be hard to justify.

FIGURE 8-1 Portable plasma arc cutting machine.
ESAB Welding & Cutting Products

Several years later, with the invention of the gas lens, plasma was successfully used for welding. Today, the plasma arc can be used for plasma arc welding (PAW), plasma spraying (PSP), plasma arc cutting (PAC), and plasma arc gouging. Plasma arc cutting is the most often used plasma process.

PAC is very popular as the result of the introduction of smaller, less expensive plasma equipment to the welding field, **Figure 8-1.** A typical portable plasma cutting system can cut mild steel up to 1 1/2 in. (38 mm) thick. Plasma cutters have the unique ability to cut metals without making them very hot. This means that there is less distortion and heat damage than would be caused with an oxyacetylene cutting torch. Very intricate shapes can be cut out without warping.

FIGURE 8-3 Portable welder's auxiliary power plug can be used for plasma cutting. Larry Jeffus

FIGURE 8-2 Plasma-cut sheet metal panels decorate this gate. Larry Jeffus

PLASMA

The word **plasma** has two meanings: it is the fluid portion of blood and it is a state of matter that is found in the region of an electrical discharge (arc). All states of matter have their own characteristics. A solid has shape and form, a liquid seeks its own level and takes the shape of its container, and a gas has no distinct shape or volume and fills its container. Plasma is a matter that is highly conductive and easily controlled and shaped by a magnetic field. The plasma created by an arc is an **ionized gas** that has both electrons and positive ions whose charges are nearly equal to each other. For welding we use the electrical definition of *plasma*.

A plasma is present in any electrical discharge. A plasma consists of charged particles that conduct the electrons across the gap that is energized to a sufficiently high enough level.

A welding plasma results when a gas is heated to a high enough temperature to convert into positive and negative ions, neutral atoms, and negative electrons. The temperature of an unrestricted arc is about 11,000°F (6100°C), but the temperature created when the arc is concentrated to form a plasma is about 43,000°F (23,900°C), **Figure 8-4.** This is hot enough to rapidly melt or vaporize any metal it comes in contact with.

Arc Plasma The term **arc plasma** is defined as gas that has been heated to at least a partially ionized condition, enabling it to conduct an electric current.[1] The term **plasma arc** is the term most often used in the welding industry when referring to the arc plasma used in welding and cutting processes. The plasma arc produces both the

high temperature and intense light associated with all forms of arc welding and **arc cutting** processes.

Plasma Torch

The plasma torch is a device, depending on its design, that allows the creation and control of the plasma for welding or cutting processes. The plasma is created in both the cutting and welding torches in the same basic manner, and both torches have the same basic parts. A plasma torch supplies electrical energy to a gas to change it into the high-energy state of a plasma.

Torch Body The torch body, on a manual-type torch, is made of a special plastic that is resistant to high temperatures, ultraviolet light, and impact, **Figure 8-5.** The torch body is a place that provides a good grip area and protects the cable and hose connections to the head. The torch body is available in a variety of lengths and sizes. Generally, the longer, larger torches are used for the higher-capacity machines; however, sometimes you might want a longer or larger torch to give yourself better control or a longer reach. On machine torches the body is

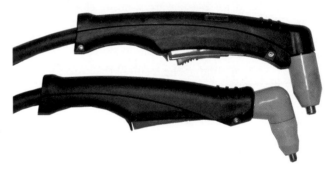

FIGURE 8-5 Examples of handheld plasma arc cutting torches. © Cengage Learning 2012

---

[1]ANSI/AWS A3.0-89—An American National Standard, Standard Welding Terms and Definitions

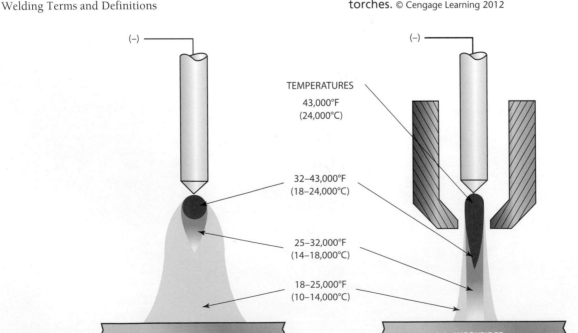

FIGURE 8-4 Approximate temperature differences between a standard arc and a plasma arc. American Welding Society

often called a barrel and may come equipped with a rack attached to its side. The rack is a flat gear that allows the torch to be raised and lowered manually to the correct height above the work.

Torch Head The torch head is attached to the torch body where the cables and hoses attach to the electrode tip, nozzle tip, and nozzle. The torch and head may be connected at any angle, such as 90°, 75°, or 180° (straight), or it can be flexible. The 75° and 90° angles are popular for manual operations, and the 180° straight torch heads are most often used for machine operations. Because of the heat in the head produced by the arc, some provisions for cooling the head and its internal parts must be made. This cooling for low-power torches may be either by air or water. Higher-power torches must be liquid cooled, **Figure 8-6**. It is possible to replace just the torch head on most torches if it becomes worn or damaged.

Power Switch Most handheld torches have a manual power switch, which is used to start and stop the power source, gas, and cooling water (if used). The switch most often used is a thumb switch located on the torch body, but it may be a foot control or located on the panel for machine-type equipment. The thumb switch may be molded into the torch body or it may be attached to the torch body with a strap clamp. The foot control must be rugged enough to withstand the welding shop environment. Some equipment has an automatic system that starts the plasma when the torch is brought close to the work.

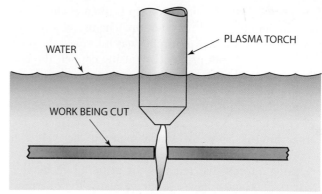

FIGURE 8-6 Water tables keep the torch cool and reduce the sound and fumes for high-powered torches.
© Cengage Learning 2012

Common Torch Parts The electrode tip, nozzle insulator, nozzle tip, nozzle guide, and nozzle are the parts of the torch that must be replaced periodically as they wear out or become damaged from use, **Figure 8-7**.

///// **CAUTION** \\\\\

Improper use of the torch or assembly of torch parts may result in damage to the torch body as well as the frequent replacement of these parts.

The metal parts are usually made out of copper, and they may be plated. The plating of copper parts will help them stay spatter-free longer.

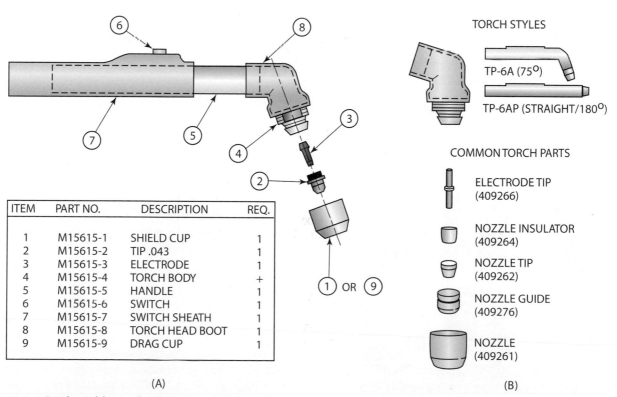

ITEM	PART NO.	DESCRIPTION	REQ.
1	M15615-1	SHIELD CUP	1
2	M15615-2	TIP .043	1
3	M15615-3	ELECTRODE	1
4	M15615-4	TORCH BODY	+
5	M15615-5	HANDLE	1
6	M15615-6	SWITCH	1
7	M15615-7	SWITCH SHEATH	1
8	M15615-8	TORCH HEAD BOOT	1
9	M15615-9	DRAG CUP	1

(A)

(B)

FIGURE 8-7 Replaceable torch parts. © Cengage Learning 2012

Electrode Tip The **electrode tip** is often made of copper with an imbedded tungsten tip. The use of a copper/tungsten tip in the newer torches has improved the quality of work they can produce. By using copper, the heat generated at the tip can be conducted away faster. Keeping the tip as cool as possible lengthens the life of the tip and allows for better-quality cuts for a longer time. The newer-designed torches are a major improvement over earlier torches, some of which required the welder to accurately grind the tungsten electrode into shape. If you are using a torch that requires the grinding of the electrode tip, you must have a guide to ensure that the tungsten is properly prepared.

Nozzle Insulator The **nozzle insulator** is between the electrode tip and the nozzle tip. The nozzle insulator provides the critical gap spacing and the electrical separation of the parts. The spacing between the electrode tip and the nozzle tip, called **electrode setback,** is critical to the proper operation of the system.

Nozzle Tip The **nozzle tip** has a small, cone-shaped, constricting orifice in the center. The electrode setback space, between the electrode tip and the nozzle tip, is where the electric current forms the plasma. The preset close-fitting parts provide the restriction of the gas in the presence of the electric current so the plasma can be generated, **Figure 8-8.** The diameter of the constricting orifice and the electrode setback are major factors in the operation of the torch. As the diameter of the orifice changes, the plasma jet action will be affected. When the setback distance is changed, the arc voltage and current flow will change.

Nozzle The **nozzle,** sometimes called the **cup,** is made of ceramic or any other high-temperature–resistant substance. This helps prevent the internal electrical parts from accidental shorting and provides control of the shielding gas or water injection if they are used, **Figure 8-9.**

Water Shroud A **water shroud** nozzle may be attached to some torches. The water surrounding the nozzle tip is

FIGURE 8-9 Different torches use different types of nozzle tips. Larry Jeffus

used to control the potential hazards of light, fumes, noise, or other pollutants produced by the process.

Replacement Parts The electrode nozzle and nozzle tip are designed to be replaceable because during normal use they will wear out. Manufacturers of PAC torches have replacement part kits that contain a variety of these consumable torch parts, **Figure 8-10.**

Power and Gas Cables

A number of power and control cables and gas and cooling water hoses may be used to connect the power supply with the torch, **Figure 8-11.** This multipart cable is usually covered to provide some protection to the cables and hoses inside and to make handling the cable easier. This covering is heat resistant but will not prevent damage to the cables and hoses inside if it comes in contact with hot metal or is exposed directly to the cutting sparks.

Power Cable The power cable must have a high-voltage–rated insulation, and it is made of finely stranded copper wire to allow for maximum flexibility of the torch, **Figure 8-12.** For all nontransfer-type torches and those that use a high-frequency pilot arc, there are two power conductors, one positive (+) and one negative (−). The size and current-carrying capacity of this cable are controlling factors to the power range of the torch. As the capacity of the equipment increases, the cable must be made large enough to carry the increased current. The larger cables are less flexible and more difficult to manipulate. To make the cable smaller on water-cooled torches, the cable is run inside the cooling water return line. Putting the power cable inside the return water line allows a smaller cable to carry more current. The water prevents the cable from overheating.

Gas Hoses There may be two gas hoses running to the torch. One hose carries the gas used to produce the plasma, and the other provides a shielding gas coverage. On some small-amperage cutting torches there is only one gas line. The gas line is made of a special heat-resistant, ultraviolet-light–resistant plastic. If it is necessary to replace the tubing because it is damaged, be sure to use the tubing provided by the manufacturer or a welding supplier. The tubing must be sized to carry the required

FIGURE 8-8 Nozzles are available in a variety of shapes for different types of cutting jobs. Larry Jeffus

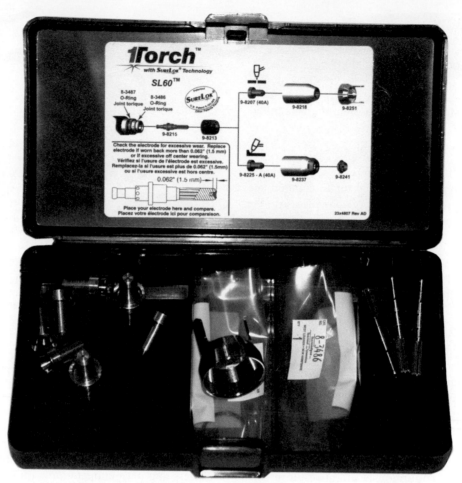

FIGURE 8-10 Torch part replacement kit. Larry Jeffus

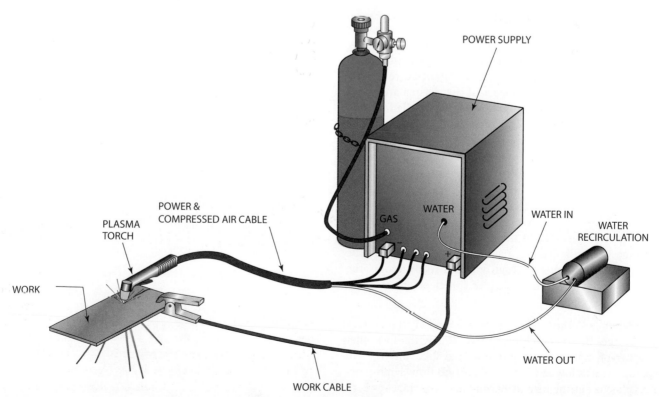

FIGURE 8-11 Typical manual plasma arc cutting setup. © Cengage Learning 2012

FIGURE 8-12 Thin strands of copper help to make the cables more flexible. Larry Jeffus

gas flow rate within the pressure range of the torch, and it must be free of solvents and oils that might contaminate the gas. If the pressure of the gas supplied is excessive, the tubing may leak at the fittings or rupture.

Control Wire The control wire is a two-conductor, low-voltage, stranded copper wire. This wire connects the power switch to the power supply. This allows the welder to start and stop the plasma power and gas as needed during the cut or weld.

Water Tubing Medium- and high-amperage torches may be water cooled. The water for cooling early model torches had to be deionized. Failure to use deionized water on these torches will result in the torch arcing out internally. This arcing may destroy or damage the torch's electrode tip and the nozzle tip. To see if your torch requires this special water, refer to the manufacturer's manual. If cooling water is required, it must be switched on and off at the same time as the plasma power. Allowing the water to circulate continuously might result in condensation in the torch. When the power is reapplied, the water will cause internal arcing damage.

Power Requirements

Voltage The production of the plasma requires a direct-current (DC), high-voltage, constant-current (drooping arc voltage) power supply. A constant-current–type machine allows for a rapid start of the plasma arc at the high open circuit voltage and a more controlled plasma arc as the voltage rapidly drops to the lower closed voltage level. The voltage required for most welding operations, such as shielded metal arc, gas metal arc, gas tungsten arc, and flux cored arc, ranges from 18 volts to 45 volts. The voltage for a plasma arc process ranges from 50 to 200 volts closed circuit and 150 to 400 volts open circuit. This higher electrical potential is required because the resistance of the gas increases as it is forced through a small orifice. The potential voltage of the power supplied must be high enough to overcome the resistance in the circuit in order for electrons to flow, **Figure 8-13.**

Amperage Although the voltage is higher, the current (amperage) flow is much lower than it is with most other welding processes. Some low-powered PAC torches will operate with as low as 10 amps of current flow. High-powered plasma cutting machines can have amperages as high as 200 amps, and some very large automated cutting

FIGURE 8-13 Inverter-type plasma arc cutting power supply. Larry Jeffus

machines may have 1000-ampere capacities. The higher the amperage capacity, the faster and thicker they will cut.

Watts The plasma process uses approximately the same amount of power, in watts, as a similar nonplasma process. Watts are the units of measure for electrical power. By determining the total watts used for both the nonplasma process and plasma operation, you can make a comparison. Watts used in a circuit are determined by multiplying the voltage times the amperage, **Figure 8-14.** For example, a 1/8-in. diameter E6011 electrode will operate at 18 volts and 90 amperes. The total watts used would be

$$W = V \times A$$
$$W = 18 \times 90$$
$$W = 1620 \text{ watts of power}$$

A low-power PAC torch operating with only 20 amperes and 85 volts would be using a total of

$$W = V \times A$$
$$W = 85 \times 20$$
$$W = 1700 \text{ watts of power}$$

Compressed Air

Most small shop plasma arc cutting torches use compressed air to form the plasma and to make the cut.

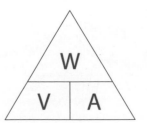

FIGURE 8-14 Ohm's law. © Cengage Learning 2012

Compressed air must be clean and dry, so a filter dryer must be used to prevent contaminants like oil, dirt, or moisture from entering the plasma torch. Any contamination entering the torch can cause internal arcing between the electrode and the nozzle. Compressed air can be supplied by either an external compressor or an internal compressor. Many of these PA cutting machines have air compressors built into the power supply. These internal compressors are designed to provide the correct airflow and pressure. Cutting machines with internal compressors are very convenient.

HEAT INPUT

Although the total power used by both plasma and nonplasma processes is similar, the actual energy input into the work per linear foot is less with plasma. The very high temperatures of the plasma process allow much higher traveling rates so that the same amount of heat input is spread over a much larger area. This has the effect of lowering the **joules** per inch of heat the weld or cut will receive. Table 8-1 shows the cutting performance of a typical plasma torch. Note the relationship among amperage, cutting speed, and metal thickness. The lower the amperage, the slower the cutting speed or the thinner the metal that can be cut.

A high travel speed with plasma cutting will result in a heat input that is much lower than that of the oxyfuel cutting process. A steel plate cut using the plasma process may have only a slight increase in temperature following the cut. It is often possible to pick up a part only moments after it is cut using plasma and find that it is cool to the touch. The same part cut with oxyfuel would be much hotter and require a longer time to cool off.

Distortion

Any time metal is heated in a localized zone or spot it expands in that area and, after the metal cools, it is no

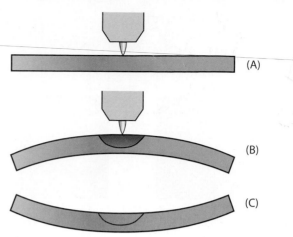

FIGURE 8-15 (A) When a flat piece of metal is heated, (B) it expands, bending the edges away from the heat. (C) When the spot cools, it shrinks, bending the edges toward the heat. © Cengage Learning 2012

longer straight or flat, **Figure 8-15**. If a piece of metal is cut, there will be localized heating along the edge of the cut, and, unless special care is taken, the part will not be usable as a result of its distortion, **Figure 8-16**. This distortion is a much greater problem with thin metals. By using a plasma cutter, an auto body worker can cut the thin, low-alloy sheet metal of a damaged car with little problem from distortion.

On thicker sections, the hardness zone along the edge of a cut will be reduced so much that it is not a problem. When using oxyfuel cutting of thick plate, especially higher-alloyed metals, this hardness zone can cause cracking and failure if the metal is shaped after cutting, **Figure 8-17**. Often the plates must be preheated before they are cut using oxyfuel to reduce the **heat-affected zone**. This preheating adds greatly to the cost of fabrication both in time and fuel costs. By being able to make most cuts without preheating, the plasma process will greatly reduce fabrication cost.

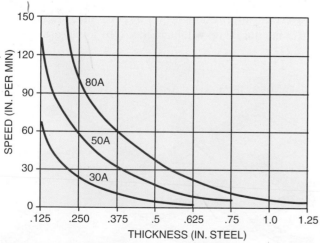

TABLE 8-1 Plasma Arc Cutting Perimeters
ESAB Welding & Cutting Products

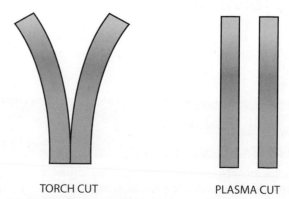

TORCH CUT   PLASMA CUT

FIGURE 8-16 The heat of a torch cut causes metal to bend, but the plasma cut is so fast that little or no bending occurs. © Cengage Learning 2012

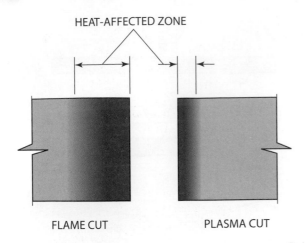

HEAT-AFFECTED ZONE

FLAME CUT   PLASMA CUT

FIGURE 8-17 A smaller heat-affected zone will result in less hardness or brittleness along the cut edge. © Cengage Learning 2012

FIGURE 8-18 Plasma arc machine cut in 2-in. thick mild steel. Larry Jeffus

Applications

Early plasma arc cutting systems required that either helium or argon gas be used for the plasma and shielding gases. As the development of the process improved, it was possible to start the PAC torch using argon or helium and switch to less expensive nitrogen. The use of nitrogen as the plasma cutting gas greatly reduced the cost of operating a plasma system. Because of its operating expense, plasma cutting was limited to metals not easily cut using oxyfuel. Aluminum, stainless steel, and copper were the metals most often cut using plasma.

As the development of the process improved, less expensive gases and even dry compressed air could be used, and the torches and power supplies improved. By the early 1980s, the PAC process had advanced to a point where it was used for cutting all but the thicker sections of mild steel.

Cutting Speed High cutting speeds are possible, up to 300 in./min (750 cm/min); that is, 25 ft a minute (7.5 meters/min), or about 1/4 mile an hour (3/8 km/h), **Figure 8-18**. The fastest oxyfuel cutting equipment could cut at only about one-fourth that speed. A problem with early high-speed machine cutting was that the cutting machines could not reliably make cuts as fast as the PAC torch. That problem has been resolved, and the new

machines and robots can operate at the upper limits of the plasma torch capacity. These machines and robots are capable of automatically maintaining the optimum torch standoff distance to the work. Some cutting systems will even follow the irregular surfaces of preformed part blanks.

Metals Any material that is electrically conductive can be cut using the PAC process. In a few applications nonconductive materials can be coated with conductive material so that they can be cut also. Although it is possible to make cuts in metal as thick as 7 in. (178 mm), it is not cost effective. The most popular materials cut are carbon steel up to 1 in. (25 mm), stainless steel up to 4 in. (102 mm), and aluminum up to 6 in. (152 mm). These are not the upper limits of the PAC process, but beyond these limits other cutting processes may be less expensive. A shop may PAC thicker material even if it is not cost effective because it does not have ready access to the alternative process.

Other metals commonly cut using PAC are copper; nickel alloys; high-strength, low-alloy steels; and clad materials. It is also used to cut expanded metals, screens, and other items that would require frequent starts and stops if the oxyfuel process were used, **Figure 8-19**.

Standoff Distance The **standoff distance** is the distance from the nozzle tip to the work, **Figure 8-20**. This distance is very critical to producing quality plasma arc cuts. As the distance increases, the arc force is diminished and tends to spread out. This causes the kerf to be wider, the top edge of the plate to become rounded, and the formation of more dross on the bottom edge of the plate. However, if this distance becomes too close, the working life of the nozzle tip will be reduced. In some cases an arc

EXPANDED METAL

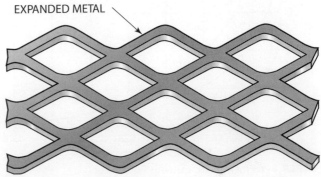

FIGURE 8-19 Expanded metal. © Cengage Learning 2012

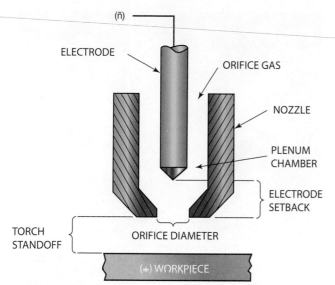

FIGURE 8-20 Conventional plasma torch terminology.
American Welding Society

FIGURE 8-21 A castle nozzle tip can be used to allow the torch to be dragged across the surface.
© Cengage Learning 2012

can form between the nozzle tip and the metal that instantly destroys the tip.

On some new torches, it is possible to drag the nozzle tip along the surface of the work without shorting it out. This is a large help when working on metal out of position or on thin sheet metal. Before you use your torch in this manner, you must check the owner's manual to see if it will operate in contact with the work, **Figure 8-21**. This technique will allow the nozzle tip orifice to become contaminated more quickly.

Starting Methods Because the electrode tip is located inside the nozzle tip, and a high initial resistance to current flow exists in the gas flow before the plasma is generated, it is necessary to have a specific starting method. Two methods are used to establish a current path through the gas.

The most common method uses a **high-frequency alternating current** carried through the conductor, the electrode, and back from the nozzle tip. This high-frequency current will ionize the gas and allow it to carry the initial current to establish a pilot arc, **Figure 8-22**. After the pilot arc has been started, the high-frequency starting circuit can be stopped. A **pilot arc** is an arc between the electrode tip and the nozzle tip within the torch head. This is a nontransfer arc, so the workpiece is not part of the current path. The low current of the pilot arc, although it is inside the torch, does not create enough heat to damage the torch parts. When the torch is brought close enough to the work, the primary arc will follow the pilot arc across the gap, to the work, and the main plasma is started. Once the main plasma is started, the pilot arc power can be shut off.

The second method of starting requires the electrode tip and nozzle tip to be momentarily shorted together. This is accomplished by automatically moving them together and immediately separating them again. The momentary shorting allows the arc to be created without damaging the torch parts.

Kerf The **kerf** is the space left in the workpiece as the metal is removed during a cut. The width of a PAC kerf is often wider than that of an oxyfuel cut. Several factors

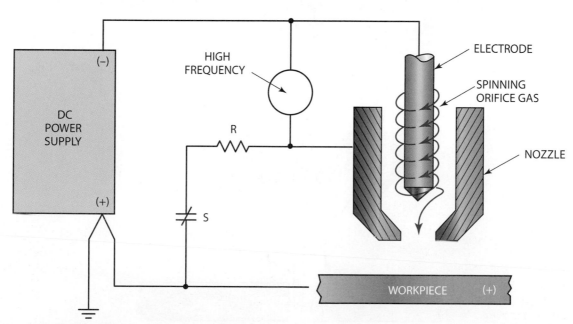

FIGURE 8-22 Plasma arc torch circuitry. American Welding Society

will affect the width of the kerf. A few of the factors are as follows:

- Standoff distance—The closer the torch nozzle tip is to the work, the narrower the kerf will be, **Figure 8-23**.

- Orifice diameter—Keeping the diameter of the nozzle orifice as small as possible will keep the kerf smaller.

- Power setting—Too high or too low a power setting will cause an increase in the kerf width.

- Travel speed—As the travel speed is increased, the kerf width will decrease; however, the bevel on the sides and the dross formation will increase if the speeds are excessive.

- Gas—The type of gas or gas mixture will affect the kerf width as the gas change affects travel speed, power, concentration of the plasma stream, and other factors.

- Electrode and nozzle tip—As these parts begin to wear out from use or are damaged, the PAC quality and kerf width will be adversely affected.

- Swirling of the plasma gas—On some torches, the gas is directed in a circular motion around the electrode before it enters the nozzle tip orifice. This swirling causes the plasma stream that is produced to be more dense with straighter sides. The result is an improved cut quality, including a narrow kerf, **Figure 8-24**.

- Water injection—The injection of water into the plasma stream as it leaves the nozzle tip is not the

FIGURE 8-23 When the standoff distance is correct, as with this machine cut, almost no sparks bounce back on the cutting tip. ESAB Welding & Cutting Products

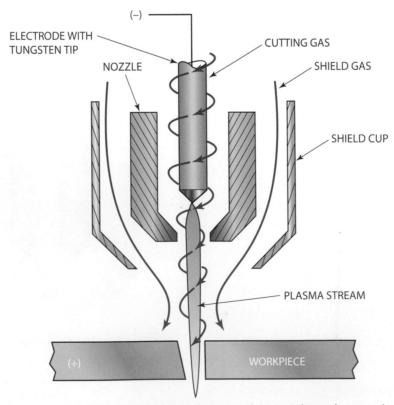

FIGURE 8-24 The cutting gas can swirl around the electrode to produce a tighter plasma column. American Welding Society

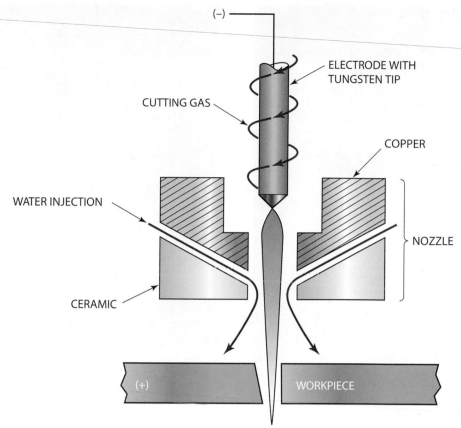

FIGURE 8-25 Water injection plasma arc cutting. Notice that the kerf is narrow, and one side is square.
American Welding Society

same as the use of a water shroud. Water injection into the plasma stream will increase the swirl and further concentrate the plasma. This improves the cutting quality; lengthens the life of the nozzle tip; and makes a squarer, narrower kerf, **Figure 8-25**.

Table 8-2 lists some standard kerf widths for several metal thicknesses. These are to be used as a guide for nesting of parts on a plate to maximize the material used and minimize scrap. The kerf size may vary from this depending on a number of variables with your PAC system. You should make test cuts to verify the size of the kerf before starting any large production cuts.

Because the sides of the plasma stream are not parallel as they leave the nozzle tip, there is a bevel left on the sides of all plasma cuts. This bevel angle is from 1/2° to 3° depending on metal thickness, torch speed, type of gas, standoff distance, nozzle tip condition, and other factors affecting a quality cut. On thin metals, this bevel is undetectable and offers no problem in part fabrication or finishing.

The use of a plasma swirling-type torch and the direction the cut is made can cause one side of the cut to be square and the scrap side to have all of the bevel, **Figure 8-26**. This technique is only effective provided that one side of the cut is to be scrap.

Gases Almost any gas or gas mixture can be used today for the PAC process. Changing the gas or gas mixture is one method of controlling the plasma cut. Although the type of gas or gases used will have a major effect on the cutting performance, it is only one of a number of changes that a technician can make to help produce a quality cut. Following are some of the effects on the cut that changing the PAC gas(es) will have:

- Force—The amount of mechanical impact on the material being cut; the density of the gas and its ability to disperse the molten metal.

- Central concentration—Some gases will have a more compact plasma stream. This factor will greatly affect the kerf width and cutting speed.

- Heat content—As the electrical resistance of a gas or gas mixture changes, it will affect the heat content of the plasma it produces. The higher the resistance, the higher the heat produced by the plasma.

- Kerf width—The ability of the plasma to remain in a tightly compact stream will produce a deeper cut with less of a bevel on the sides.

Plate Thickness		Kerf Allowance	
in.	**mm**	**in.**	**mm**
1/8 to 1	3.2 to 25.4	+3/32	+2.4
1 to 2	25.4 to 51.0	+3/16	+4.8
2 to 5	51.0 to 127.0	+5/16	+8.0

TABLE 8-2 Standard Kerf Widths for Several Metal Thicknesses

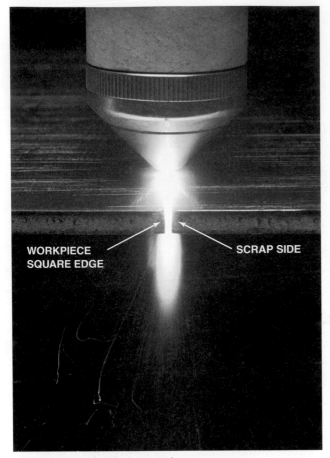

FIGURE 8-26 Plasma arc cutting. © Cengage Learning 2012

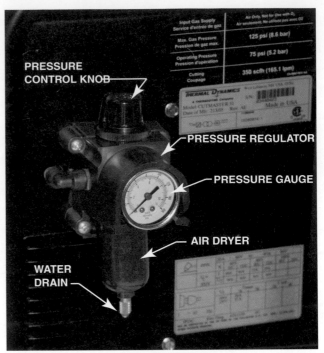

FIGURE 8-27 Controlling the pressure is one way of controlling gas flow. Some portable plasma arc cutting machines have their own air pressure regulator and dryer. Air must be dried to provide a stable plasma arc.
Larry Jeffus

- Dross formation—The dross that may be attached along the bottom edge of the cut can be controlled or eliminated.
- Top edge rounding—The rounding of the top edge of the plate can often be eliminated by correctly selecting the gas or gases that are to be used.
- Metal type—Because of the formation of undesirable compounds on the cut surface as the metal reacts to elements in the plasma, some metals may not be cut with specific gas(es).

Table 8-3 lists some of the popular gases and gas mixtures used for various PAC metals. The selection of a gas or gas mixture for a specific operation to maximize the

system performance must be tested with the equipment and setup being used. With constant developments and improvements in the PAC system, new gases and gas mixtures are continuously being added to the list. In addition to the type of gas, it is important to have the correct gas flow rate for the size tip, metal type, and thickness. Too low a gas flow will result in a cut having excessive dross and sharply beveled sides, **Figure 8-27**. Too high a gas flow will produce a poor cut because of turbulence in the plasma stream and waste gas. A flow measuring kit can be used to test the flow at the plasma torch for more accurate adjustments.

Stack Cutting Because the PAC process does not rely on the thermal conductivity between stacked parts, like the oxyfuel process, thin sheets can be stacked and cut efficiently. With the oxyfuel **stack cutting** of sheets, it is important that there not be any air gaps between layers. Also, it is often necessary to make a weld along the side of the stack in order for the cut to start consistently.

The PAC process does not have these limitations. It is recommended that the sheets be held together for cutting, but this can be accomplished by using standard C-clamps. The clamping needs to be tight because, if the space between layers is excessive, the sheets may stick together. The only problem that will be encountered is that, because of the kerf bevel, the parts near the bottom might be slightly larger if the stack is very thick. This problem can be controlled by using the same techniques as described for making the kerf square.

Metal	Gas
Carbon and low alloy steel	Nitrogen Argon with 0% to 35% hydrogen air
Stainless steel	Nitrogen Argon with 0% to 35% hydrogen
Aluminum and aluminum alloys	Nitrogen Argon with 0% to 35% hydrogen
All plasma arc gouging	Argon with 35% to 40% hydrogen

TABLE 8-3 Gases for Plasma Arc Cutting and Gouging

Dross Dross is the metal compound that resolidifies and attaches itself to the bottom of a cut. This metal compound is made up mostly of unoxidized metal, metal oxides, and nitrides. It is possible to make cuts dross-free if the PAC equipment is in good operating condition and the metal is not too thick for the size of the torch being used. Because dross contains more unoxidized metal than most OFC slag, often it is much harder to remove if it sticks to the cut. The thickness that a dross-free cut can be made is dependent on a number of factors, including the gas(es) used for the cut, travel speed, standoff distance, nozzle tip orifice diameter, wear condition of electrode tip and nozzle tip, gas velocity, and plasma stream swirl.

Stainless steel and aluminum are easily cut dross-free. Carbon steel, copper, and nickel-copper alloys are much more difficult to cut dross-free.

Machine Cutting

Almost any plasma torch can be attached to some type of semiautomatic or automatic device to allow it to make machine cuts. The simplest devices are oxyfuel portable flame cutting machines that run on tracks, **Figure 8-28.** These portable machines are good for mostly straight or circular cuts. Complex shapes can be cut with a pattern cutter that uses a magnetic tracing system to follow the template's shape, **Figure 8-29.**

High-powered PAC machines may have amperages up to 1000 amps. These machines must be used with semi-automatic or automatic cutting systems. The heat, light, and other potential hazards of these machines make them unsafe for manual operations.

Large, dedicated, computer-controlled cutting machines have been built specifically for PAC systems. These machines have the high travel speeds required to produce good-quality cuts and have a high volume of production. With these machines, the operator can input the specific cutting instructions such as speed, current, gas flow, location, and shape of the part to be cut, and the machine will make the cut with a high degree of accuracy once or any number of times.

FIGURE 8-29 Portable pattern cutter can cut shapes, circles, and straight lines. © Cengage Learning 2012

Robotic cutters are also available to perform high-quality, high-volume PAC, **Figure 8-30.** The advantage of using a robot is that, in most cases, the robot is capable of being set up for multitasking. When a robot is programmed, it can cut the part out; change and tool itself and weld the parts together; change the tool; and grind, drill, or paint the finished unit.

Water Tables Machine cutting lends itself to the use of water cutting tables, although they can be used with most hand torches. The **water table** is used to reduce the noise level, control the plasma light, trap the sparks, eliminate most of the fume hazard, and reduce distortion.

Water tables either support the metal just above the surface of the water or they submerge the metal about 3 in. (76 mm) below the water's surface. Both types of water tables must have some method of removing the cut parts, scrap, and slag that build up in the bottom. Often

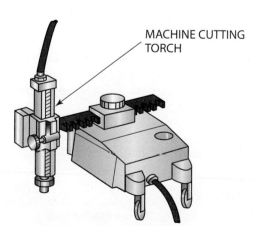

FIGURE 8-28 Machine cutting tool.
ESAB Welding & Cutting Products

MACHINE CUTTING
TORCH

THINK GREEN
Avoid Water Contamination

Heavy metal contamination of water systems caused by rainwater washing fine metal particles away that are left by PAC can be avoided by cleaning up after outdoor cutting.

FIGURE 8-30 Automated cutting. ESAB Welding & Cutting Products

the surface-type tables will have the PAC torch connected to a water shroud nozzle, **Figure 8-31.** By using a water shroud nozzle, the surface table will offer the same advantages to the PAC process as the submerged table offers. In most cases, the manufacturers of this type of equipment have made provisions for a special dye to be added to the water. This dye will help control the harmful light produced by the PAC. Check with the equipment's manufacturer for limitations and application of the use of dyes.

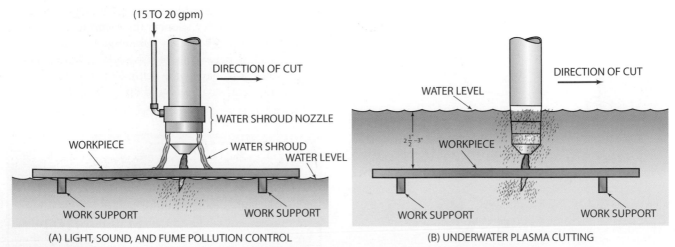

(A) LIGHT, SOUND, AND FUME POLLUTION CONTROL

(B) UNDERWATER PLASMA CUTTING

FIGURE 8-31 A water table can be used either with (A) a water shroud or (B) underwater torches.
ESAB Welding & Cutting Products

Manual Cutting

Manual plasma arc cutting is the most versatile of the PAC processes. It can be used in all positions, on almost any surface, and on most metals. This process is limited to low-power plasma machines; however, even these machines can cut up to 1 1/2-in. (38-mm) thick metals. The limitation to low power, 100 amperes or less, is primarily for safety reasons. The higher-powered machines have extremely dangerous open circuit voltages that can kill a person if accidentally touched.

Setup Wearing all of the required personal protection equipment and following all of the manufacturer's safety rules, most equipment can be set up using the following steps:

- Make any adjustments and changes to the electrode tip, nozzle tip, nozzle, or other torch component before the machine power is turned on because you can be shocked if the gun trigger is accidentally activated while you are servicing these parts.

//// CAUTION \\\\

The open circuit voltage on a plasma machine can be high enough to cause severe electrical shock or death.

- Make sure the work clamp is connected to a clean, unpainted spot on the metal you will be cutting.
- Check to see if there is anything behind the cut that will block the sparks from falling free of the cut.
- Check to see if there is anything that will be damaged or set on fire with the sparks.
- Set the cutting amperage to maximum.
- Make a practice cut to see that the material can be cut cleanly.
- Reduce the amperage, and make another practice cut. Repeat this process until you have the amperage set as low as possible while still making a clean cut.

Note: Setting the amperage to the lowest possible level will extend the life of the torch parts.

Safety

PAC has many of the same safety concerns as most other electric welding or cutting processes. Some special concerns are specific to this process.

- Electrical shock—Because the open circuit voltage is much higher for this process than for any other, extra caution must be taken. The chance that a fatal shock could be received from this equipment is much higher than from any other welding equipment.
- Moisture—Often water is used with PAC torches to cool the torch, improve the cutting characteristic, or as part of a water table. Any time water is used it

is very important that there be no leaks or splashes. The chance of electrical shock is greatly increased if there is moisture on the floor, cables, or equipment.

- Noise—Because the plasma stream is passing through the nozzle orifice at a high speed, a loud sound is produced. The sound level increases as the power level increases. Even with low-power equipment the decibel (dB) level is above safety ranges. Some type of ear protection is required to prevent damage to the operator and other people in the area of the PAC equipment when it is in operation. High levels of sound can have a cumulative effect on one's hearing. Over time, one's ability to hear will decrease unless proper precautions are taken. See the owner's manual for recommendations for the equipment in use.
- Light—The PAC process produces light radiation in all three spectrums. This large quantity of visible light, if the eyes are unprotected, will cause night blindness. The most dangerous of the lights is ultraviolet. Like other arc processes, this light can cause burns to the skin and eyes. The third light, infrared, can be felt as heat, and it is not as much of a hazard. Some type of eye protection must be worn when any PAC is in progress. **Table 8-4** lists the recommended lens shade numbers for various power-level machines.
- Fumes—This process produces a large quantity of fumes that are potentially hazardous. A specific means for removing them from the work space should be in place. A downdraft table is ideal for manual work, but some special pickups may be required for larger applications. The use of a water table and/or a water shroud nozzle will greatly help to control fumes. Often the fumes cannot be exhausted into the open air without first being filtered or treated to remove dangerous levels of contaminants. Before installing an exhaust system, you must first check with local, state, and federal officials to see if specific safeguards are required.
- Gases—Some of the plasma gas mixtures include hydrogen; because this is a flammable gas, extra care must be taken to ensure that the system is leakproof.
- Sparks—As with any process that produces sparks, the danger of an accidental fire is always present. This is a larger concern with PAC because the sparks are often thrown some distance from the work area and the operator's vision is restricted

Current Range A	Minimum Shade	Comfortable Shade
Less than 300	8	9
300 to 400	9	12
400 plus	10	14

TABLE 8-4 Recommended Shade Densities for Filter Lenses

by a welding helmet. If there is any possibility that sparks will be thrown out of the immediate work area, a fire watch must be present. A fire watch is a person whose sole job is to watch for the possible starting of a fire. This person must know how to sound the alarm and have appropriate firefighting equipment handy. **Never cut in the presence of combustible materials.**

■ Operator check out—Never operate any PAC equipment without first reading the manufacturer owner and operator's manual for the specific equipment to be used. It is a good idea to have someone who is familiar with the equipment go through the operation after you have read the manual.

Straight Cuts

Straight cuts are the most common type of cuts made with PAC torches. You can hold the torch close to the head because it does not get as hot as an oxyacetylene torch. This will help you keep the cut smoother. One common problem with making long, straight cuts is a tendency to make the cut with a slight arc when the torch is held at a 90° angle to the cut, **Figure 8-32**. If you slide your hand along the plate surface, you can eliminate some of this arcing. Another technique to help you keep your line straight is to use a guide such as an angle iron or straight edge, **Figure 8-33**. Because there are so few sparks when cutting thin metal, you may even be able to use a square as a guide without damaging it with sparks or heat. If you use a straightedge as a guide, make sure your cut is on the line as it is hard to see the line as it is being cut because of the size of the nozzle on many plasma torches.

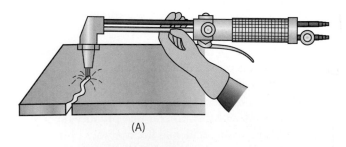

(A)

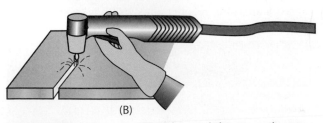

(B)

FIGURE 8-32 It is easier to make straight, smooth cuts if you can brace the torch closer to the tip, as in cut (B). American Welding Society

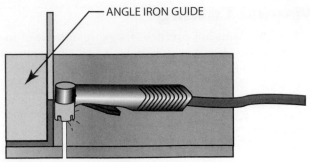

─ ANGLE IRON GUIDE

FIGURE 8-33 Use an angle iron as a guide to make a straight cut. © Cengage Learning 2012

PRACTICE 8-1

Flat, Straight Cuts in Thin Plate

Using a properly set-up and adjusted PAC machine; proper safety protection; and one or more pieces of mild steel, stainless steel, and aluminum 6 in. (152 mm) long and 16 gauge and 1/8 in. (3 mm) thick; you will cut off strips 1/2 in. wide (13 mm), **Figure 8-34**.

■ Starting at one end of the piece of metal that is 1/8 in. (3 mm) thick, hold the torch as close as possible to a 90° angle.

■ Lower your hood and establish a plasma cutting stream.

■ Move the torch in a straight line down the plate toward the other end, **Figure 8-35**.

■ If the width of the kerf changes, speed up or slow down the travel rate to keep the kerf the same size for the entire length of the plate.

Repeat the cut using both thicknesses of all three types of metals until you can make consistently smooth cuts that are within ±3/32 in. (2.3 mm) of a straight line and ±5° of being square. Turn off the PAC equipment and clean up your work area when you are finished cutting.

Complete a copy of the "Student Welding Report" listed in Appendix I or provided by your instructor. ◆

PRACTICE 8-2

Flat, Straight Cuts in Thick Plate

Using a properly set up and adjusted PAC machine; proper safety protection; and one or more pieces of mild steel, stainless steel, and aluminum 6 in. (152 mm) long and 1/4 in. (6 mm) and 1/2 in. (13 mm) thick; you will cut off strips 1/2 in. (13 mm) wide. Follow the same procedure as outlined in Practice 8-1.

Repeat the cut using both thicknesses of all three types of metals until you can make consistently smooth cuts that are within ±3/32 in. (2.3 mm) of a straight line and ±5° of being square. Turn off the PAC equipment and clean up your work area when you are finished cutting.

Complete a copy of the "Student Welding Report" listed in Appendix I or provided by your instructor. ◆

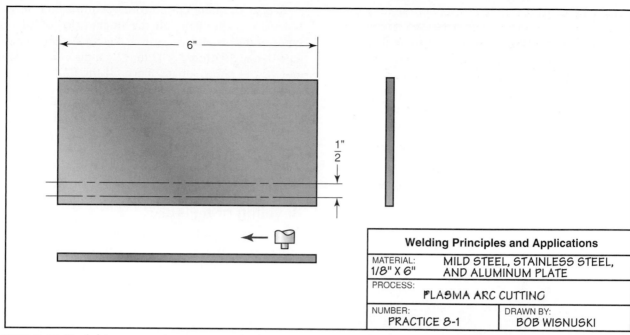

FIGURE 8-34 Straight, square plasma arc cutting. © Cengage Learning 2012

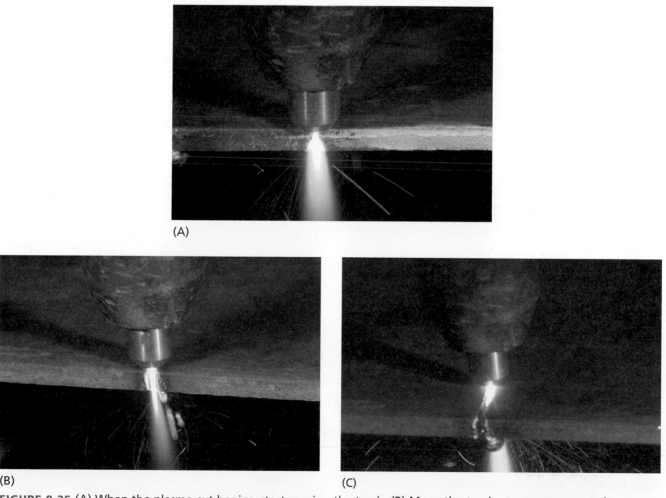

FIGURE 8-35 (A) When the plasma cut begins, start moving the torch. (B) Move the torch at a constant speed. (C) Continue moving the torch smoothly in a straight line toward the other end of the cut. Larry Jeffus

PRACTICE 8-3

Flat Cutting Holes

Using a properly set up and adjusted PAC machine; proper safety protection; and one or more pieces of mild steel, stainless steel, and aluminum 16 gauge, 1/8 in. (3 mm), 1/4 in. (6 mm), and 1/2 in. (13 mm) thick, you will cut 1/2-in. (13-mm) and 1-in. (25-mm) holes.

- Starting with the piece of metal that is 1/8 in. (3 mm) thick, hold the torch as close as possible to a 90° angle.

- Lower your hood and establish a plasma cutting stream.

- Move the torch in an outward spiral until the hole is the desired size, **Figure 8-36**.

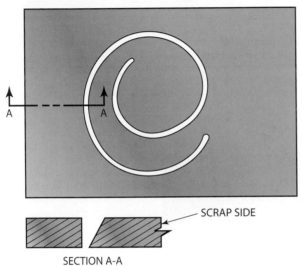

SECTION A-A

FIGURE 8-36 When cutting a hole, spiral the torch outward in a counterclockwise direction. © Cengage Learning 2012

Repeat the hole cutting process until both sizes of holes are made using all the thicknesses of all three types of metals and you can make consistently smooth cuts that are within ±3/32 in. (2.3 mm) of being round and ±5° of being square. Turn off the PAC equipment and clean up your work area when you are finished cutting.

Complete a copy of the "Student Welding Report" listed in Appendix I or provided by your instructor. ◆

PRACTICE 8-4

Beveling of a Plate

Using a properly set up and adjusted PAC machine; proper safety protection; and one or more pieces of mild steel, stainless steel, and aluminum 6 in. (152 mm) long and 1/4 in. (3 mm) and 1/2 in. (6 mm) thick, you will cut a 45° bevel down the length of the plate.

- Starting at one end of the piece of metal that is 1/4 in. (3 mm) thick, hold the torch as close as possible to a 45° angle.

- Lower your hood and establish a plasma cutting stream.

- Move the torch in a straight line down the plate toward the other end, **Figure 8-37**.

Repeat the cut using both thicknesses of all three types of metals until you can make consistently smooth cuts that are within ±3/32 in. (2.3 mm) of a straight line and ±5° of a 45° angle. Turn off the PAC equipment and clean up your work area when you are finished cutting.

Complete a copy of the "Student Welding Report" listed in Appendix I or provided by your instructor. ◆

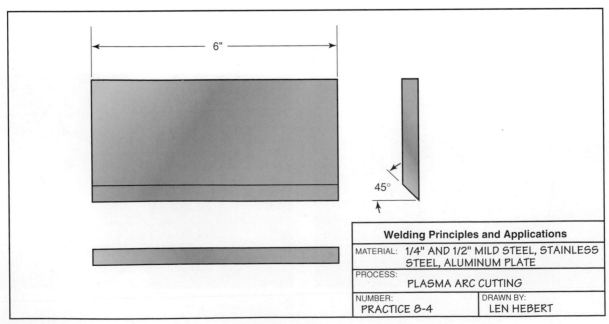

6"

45°

Welding Principles and Applications	
MATERIAL: 1/4" AND 1/2" MILD STEEL, STAINLESS STEEL, ALUMINUM PLATE	
PROCESS: PLASMA ARC CUTTING	
NUMBER: PRACTICE 8-4	DRAWN BY: LEN HEBERT

FIGURE 8-37 Beveled plasma arc cutting. © Cengage Learning 2012

Plasma Arc Gouging

Plasma arc gouging is a recent introduction to the PAC processes. The process is similar to that of air carbon arc gouging in that a U-groove can be cut into the metal's surface. The removal of metal along a joint before the metal is welded or the removal of a defect for repairing can easily be done using this variation of PAC, **Figure 8-38**. An advantage of using PA cutting to remove a weld is that any slag trapped in the weld will not affect the PAC gouging process.

The torch is set up with a less concentrated plasma stream. This will allow the washing away of the molten metal instead of thrusting it out to form a cut. The torch is held at approximately a 30° angle to the metal surface. Once the groove is started, it can be controlled by the rate of travel, torch angle, and side-to-side torch movement.

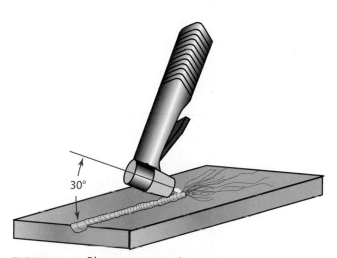

FIGURE 8-38 Plasma arc gouging a U-groove in a plate.
© Cengage Learning 2012

Plasma arc gouging is effective on most metals. Stainless steel and aluminum are especially good metals to gouge because there is almost no cleanup. The groove is clean, bright, and ready to be welded. Plasma arc gouging is especially beneficial with these metals because there is no reasonable alternative available. The only other process that can leave the metal ready to weld is to have the groove machined, and machining is slow and expensive compared to plasma arc gouging.

It is important to try to not remove too much metal in one pass. The process will work better if small amounts are removed at a time. If a deeper groove is required, multiple gouging passes can be used.

PRACTICE 8-5

U-Grooving of a Plate

Using a properly set up and adjusted PAC machine, proper safety protection, one or more pieces of mild steel, stainless steel, and aluminum 6 in. (152 mm) long and 1/4 in. (6 mm) or 1/2 in. (13 mm) thick, you will cut a U-groove down the length of the plate.

- Starting at one end of the piece of metal, hold the torch as close as possible to a 30° angle, **Figure 8-39**.
- Lower your hood and establish a plasma cutting stream.
- Move the torch in a straight line down the plate toward the other end.
- If the width of the U-groove changes, speed up or slow down the travel rate to keep the groove the same width and depth for the entire length of the plate.

Repeat the gouging of the U-groove using all three types of metals until you can make consistently smooth grooves that are within ±3/32 in. (2.3 mm) of a straight

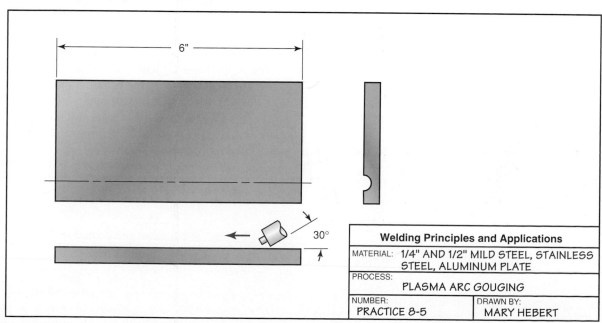

Welding Principles and Applications	
MATERIAL: 1/4" AND 1/2" MILD STEEL, STAINLESS STEEL, ALUMINUM PLATE	
PROCESS: PLASMA ARC GOUGING	
NUMBER: PRACTICE 8-5	DRAWN BY: MARY HEBERT

FIGURE 8-39 Plasma arc gouging. © Cengage Learning 2012

FIGURE 8-40 Cutting pipe with a plasma torch. © Cengage Learning 2012

line and uniform in width and depth. Turn off the PAC equipment and clean up your work area when you are finished cutting.

Complete a copy of the "Student Welding Report" listed in Appendix I or provided by your instructor. ◆

Cutting Round Stock

Often it is necessary to PA cut a round piece of metal such as a pipe, shaft, rod, or bolt. Round pieces of metal can be a challenge to cut because the cut starts out much like a gouged groove and transitions to something like piercing a hole. In addition, it is important to keep the plasma stream straight and in line with the line that is being cut. The plasma torch will cut in the direction it is pointed, so if it is not straight, the cut may have a beveled edge.

PRACTICE 8-6

Cutting Round Stock

Using a properly set up and adjusted PAC machine, proper safety protection, and one or more pieces of pipe and bar stock, you will cut off strips 1/2 in. (13 mm) wide.

- Hold the torch so it is pointed parallel to the round piece and in line with the line to be cut, **Figure 8-40**.
- Lower your hood, and pull the trigger to start the pilot arc and the cutting plasma stream.
- Move the torch onto the side of the round surface so that a groove is started.
- If the round stock is 1/2 in. (13 mm) or smaller, keep the torch pointed in the same direction, and move the torch across the piece.
- If the round stock is 1/2 in. (13 mm) or thicker than the PAC torch can cut through easily, then you must move the torch back and forth to make the kerf wider to allow the cut to go all the way through the round stock.

Repeat the cut using both thicknesses of all three types of metals until you can make consistently smooth cuts that are within ±3/32 in. (2.3 mm) of a straight line and ±5° of being square. Turn off the PAC equipment and clean up your work area when you are finished cutting.

Complete a copy of the "Student Welding Report" listed in Appendix I or provided by your instructor. ◆

Summary

Plasma arc cutting is quickly becoming one of the most popularly used cutting processes in the welding industry. Developments and changes in the equipment and torch design have extended the effective cutting life for the consumable torch parts, which has reduced the cutting cost and improved cut quality significantly. This process was once found only in large shops that cut stainless steel and aluminum; however, today it is being used by almost every segment of the industry, including small shops and home hobbyists.

One of the most difficult parts of a plasma cutting process to be mastered by the beginning student is the high rate of cutting speed. Developing an eye and ear for the sights and sounds the process exhibits as a high-quality cut is being produced will be a significant aid in your skill development.

Weld Shop Keeps U.S. Coast Guard Ready

With a history going back to the late 1700s, the United States Coast Guard stands always ready to respond to public needs in a wide variety of maritime activities. This includes maintaining safe and secure ports, protecting the marine environment, aiding people in distress, managing the waterways to ensure safe traffic, and enforcing the law.

The Integrated Support Command (ISC) industrial weld shop for the United States Coast Guard in Miami stands ready, too, with its reputation as a premier station for ship alterations and repair, using plasma arc cutting machines and inverter gas tungsten arc (GTA) and gas metal arc (GMA) welding machines, which are portable and have multiple-metal flexibility.

Ship Alterations

One of the primary vessels for law enforcement is the 110-ft (33-m) island-class patrol boat. These vessels are also involved in port security, search and rescue, and defense readiness operations. With excellent range and sea-keeping capabilities, an island-class patrol boat displaces 154 tons of water, has a maximum speed of more than 26 knots, and has an armament of one 25-mm and two .50-caliber machine guns.

When vessels need retrofitting, modification, and repair-ship alterations, or "ship alts" in shop language—they spend three weeks dockside at ISC. Ship alts may include having a total engine overhaul, reconfiguring the fuel tank, installing new piping for drains and sewer tanks, replacing fin stabilizers, doing mess hall work, reconfiguring deck lights, replacing the hydraulic lines for steering, installing new throttle cable assemblies, hull refitting (from filling in small indentations to replacing whole sections), and resurfacing the propeller, the propeller shaft, and the rudder.

The Miami ISC usually has ship alts scheduled seven months in advance. In addition to working on the 100-ft cutters, ISC has performed a complete overhaul on several 41-ft (12.3-m) utility boats (all vessels under 65 ft [19.5 m] in length are classified as boats; a "cutter" is basically any vessel 65 ft in length or greater).

On Location

Ship alts often involve cutting out sections of worn, outdated, or damaged metal parts. Metals commonly encountered include the marine grade (5086) aluminum of the hull, radar arch, railings, and cleats; British steel (HY 80), also for hulls; the steel and aluminum of the mess deck; and copper-nickel alloys for the plumbing. Because of the variety of metals encountered and the need to work quickly, ISC uses the plasma arc cutting process for most metal removal.

"If the metal can conduct electricity, we believe it's a job for a plasma cutting machine," Nathiel Roper, a welder at ISC's industrial weld shop, said.

A 110-ft U.S. Coast Guard patrol boat. American Welding Society

Plasma arc cutting is a process by which an arc is constricted by passing it through a small nozzle from the electrode to the workpiece. Plasma arc cutting provides numerous advantages over other cutting methods. It is faster than a saw. It cuts any electrically conductive metal, whereas oxyfuel cannot cut stainless steel or aluminum. Plasma cuts faster than oxyfuel. A preheat cycle is not required. The kerf width (the width of the cut) is small and more precise. Its smaller heat-affected zone (HAZ) prevents warping or damaging paint in the surrounding areas. In addition, it is cleaner, less expensive, safer, and more convenient because it uses ordinary compressed air for "fuel" instead of potentially harmful gases, a real concern when working in a ship's hull.

Conventional plasma arc cutting was introduced in 1957. For decades, the technology was restricted to the shop floor of industries because transformers within the machines made them heavy, bulky, and not very portable. In recent years, however, advanced inverter technology has reduced the size and weight of the machines and has freed them from the shop floor.

David Hall, who was ISC's industrial weld shop leader prior to a promotion, said his plasma arc cutting machine is one-eighth the size of the shop's older machines. The small machine can cut approximately 20 in./min on 3/8-in. (9.53-mm) aluminum.

Vic Miller, a welder at ISC, used the plasma arc machine to cut away a large section of a utility boat's hull where the 5/16-in. (7.94-mm) thick aluminum hull plating had eroded to less than 75% of its original thickness. Other than a brief pass with a disc sander, no additional finishing was required before fitting the new hull plating in place. The new sections of plating will also be cut with an inverter plasma arc machine, shaped on ISC's press brakes, and gas metal arc welded with the Miller XMT 304 inverter-based welding machine and XR-A push-pull wire feeder running 0.035-in. (0.889-mm) diameter 5356 wire. The whole setup is portable.

The shop uses the plasma arc to cut steel up to 1 1/8 in. (2.86 cm) thick. "I slow down my speed and listen for the arc coming out the bottom of the plate," Roper said, "and that ensures that I've got a clean cut."

For gas tungsten arc welding, Roper gets into tight, confined spaces, such as in the battery room, the aft steering space, or near the fuel cells. He especially uses the GTAW machine for pipe fitting, tacking the pipe for fit, and positioning while inside the cutter, then taking it back to the weld shop for finishing.

Repairing Hulls

The GTAW machine's AC frequency can be adjusted, allowing more heat into the work, higher penetration, and more cleaning. Miller said frequency adjustment helps when repairing the hulls and other aluminum structures on older vessels.

"See the pitting on the hull [refer to the photograph] where electrolysis has set in? Over the years," Miller said, "salt works its way into the metal, plus the aluminum oxides get in really deep. We need a more penetrating arc to penetrate down into the material. Then, when the aluminum melts, the impurities get raised to the surface of the molten metal. Then I can start adding filler rod."

Before welding, Miller uses a small die grinder to grind out the bad metal in small pits on the hull that sandblasting cannot remove. For especially deep pitting, he will cut out a small portion (a 3-in. or 5-in. [7.62-cm or 12.7-cm]-diameter circle) with the plasma arc machine and weld new plate in its place. Using a 2%-thoriated tungsten electrode, Miller boosts the frequency beyond 60 Hz and starts welding. This is possible on an inverter-based machine, which allows operators to adjust welding output frequency from 20 Hz to 250 Hz.

Increasing frequency produces a tight, focused arc cone, which narrows the weld bead and creates a penetrating arc. It also allows operators to direct the arc at the joint to avoid having the arc dance from plate to plate.

Roper uses increased frequency and an AC balance control when rebuilding a stuffing tube, which is Schedule 120 aluminum pipe that goes through the hull and holds the propeller shaft. Good welds are mandatory because the weld seals the stuffing tube against water entering the hull.

However, the welding is difficult. First, it involves joining new metal to old metal subjected to oil and salt. Second, since he cannot weld from the inside of the pipe, Roper relies on arc force and good joint preparation for complete penetration. He also allows about 2 in. (5.08 cm) of clearance between the bottom of the stuffing tube and the hull.

"You have to be a contortionist to weld a stuffing tube. I use a mirror to see where I'm welding," Roper said. For welding the top of the tube, he sets the frequency control to about 110 Hz and extends the balance control to nearly the maximum penetration setting.

A section of a utility boat's hull with pitting. A worn section (circular shape) with especially deep pitting has been removed with plasma arc cutting, and a new section has been GTA welded in its place. American Welding Society

"When working on top and on relatively clean metal, I can boost both the frequency and the penetration because the weld pool falls right into the joint," Roper said. "For dirtier metal, I set the balance control for more cleaning. The same thing holds true for the bottom of the weld. There, I must use a lower setting so that gravity doesn't suck the weld pool out of the joint. However, I keep the frequency high. This gives me a tight arc cone for working in that small space, and it gives me good penetration."

Article courtesy of the American Welding Society.

Review

1. What is an electrical plasma?

2. Approximately how many times hotter is the plasma arc than an unrestricted arc?

3. Why is the body on a manual plasma torch made from a special type of plastic?

4. How are the torch heads of a plasma torch cooled?

5. What three types of power switches are used on plasma torches?

6. Why are some copper parts plated?

7. Why have copper/tungsten tips helped the plasma torch?

8. What provides the gap between the electrode tip and the nozzle tip?

9. What are the advantages of a water shroud nozzle for plasma cutting?

10. How is water used to control power cable over-heating?

11. What factors must be considered when selecting a material for plasma arc gas hoses?

12. Why must the cooling water be turned off when the plasma cutting torch is not being used?

13. What type of voltage is required for a plasma cutting torch?

14. A plasma arc torch that is operating with 90 volts (closed circuit) and 25 amps is using how many watts of power?

15. Why do plasma cuts have little or no distortion?

16. What has limited the upper cutting speed limits of the plasma arc cutting process?

17. What metals can be commonly cut using the PAC process?

18. What happens to torches not using drag nozzle tips if the standoff distance is not maintained?

19. What are the two methods for establishing the plasma path to the metal being cut?

20. List eight items that will affect the quality of a PAC kerf.

21. List seven items that are affected by the choice of PAC gas or gases on the cut.

22. Describe *stack cutting*.

23. How does PAC dross compare with OFC slag?

24. Why must high-power PAC machines be used by a semiautomatic or automatic cutting system?

25. What is the advantage of using a water table for PAC?

26. How is plasma arc gouging performed?

Chapter 9

Related Cutting Processes

OBJECTIVES

After completing this chapter, the student should be able to

■ discuss the various cutting processes.

■ give examples of the types of applications that might be best suited to specific cutting processes.

■ list advantages and disadvantages of using the different cutting processes.

■ explain the safety considerations of each of the different cutting processes.

KEY TERMS

abrasives

air carbon arc cutting (CAC-A)

carbon electrode

cutting gas assist

delamination

exothermic gases

gas laser

gouging

graphite

laser beam cuts (LBC)

laser beam drilling (LBD)

laser beam welds (LBW)

monochromatic

oxygen lance cutting

solid state lasers

synchronized waveform

washing

water jet cutting

YAG laser

INTRODUCTION

The number of specialized cutting processes being developed and improved increases every year. To date there are a number of specialized cutting processes that have been perfected and are in use. The new cutting methods are being used by a much wider group than just the welding industry. These processes can be used to cut a wide variety of items, such as glass, plastic, printed circuit boards, cloth, and fiberglass insulation, and more items are being added almost daily. Material cutting and even hole drilling using a welding-related process have become common in the workplace.

Only a few of the more common cutting processes are covered in this chapter. These are the ones that a welder might be required to either be able to perform or have a working knowledge of.

FIGURE 9-1 Bar codes are read by lasers to input information into computers. © Cengage Learning 2012

WHITE LIGHT

LASER LIGHT

FIGURE 9-2 White light is made up of different frequencies (colors) of light waves. Laser light is made up of single-frequency light waves traveling parallel to each other. © Cengage Learning 2012

LASER BEAM CUTTING (LBC) AND LASER BEAM DRILLING (LBD)

Lasers have developed from the first bright red spot generated by the ruby rod to a multibillion-dollar industry. The use of lasers has become very common today. We see their use in our world every day. They help speed our checking out at stores when they are used to read the uniform product code (UPC) on our purchases, **Figure 9-1.** Lasers are used by surveyors to measure distances and to see that a building under construction is level. Doctors can use lasers to make very precise cuts during the most delicate operations. They are even used to entertain us at shows and concerts.

Manufacturers use the laser to do everything from marking products, joining parts, communicating between machines, guiding, cutting, or punching machines to cutting a variety of materials. The industrial laser can be used to guide a machine's cut accurately or to measure distances from a few hundred to thousands of miles.

This versatile tool has helped produce products that could not be produced without the laser's light. A laser can be used to drill holes (LBD) through the hardest materials, such as synthetic diamonds, tungsten carbide cutting tools, quartz, glass, or ceramics. Lasers are used for welding (LBW) materials that are too thin or too hard to be welded with other heat sources. They can be used to cut (LBC) materials that must be cut without overheating delicate parts that might be located just a few thousandths of an inch away.

The benefits and capacities of the laser have made it one of the most rapidly growing areas in manufacturing. This is a useful tool that will continue to be developed and improved for years to come.

Lasers

The term *laser* is an acronym for light amplification stimulation emission of radiation. A laser is a form of light that is **monochromatic,** a single-color wavelength that travels in parallel waves, **Figure 9-2.** This form of light is generated when certain types of materials are excited either by intense light or with an electric current. The atoms or molecules of the lasing material release their energy in the form of light. The laser light is produced as the atom or molecule slows down from a high-energy state to a lower-energy state, **Figure 9-3.** The laser light is then bounced back and forth between one fully reflective and one partially reflective surface at the ends of the lasing material. As the light is reflected, it begins to form a **synchronized waveform.** The light that passes through the partially reflective mirror is the laser light, and it is ready to do its job, **Figure 9-4.** The frequency of this vibration determines the color of the laser's light beam. Most lasers used for cutting, drilling, and welding produce a laser light beam in the infrared range.

Manufacturers use the laser to do everything from burning information on products as small and hard as diamonds to guiding machines to grind, cut, punch, drill, and cut to accuracies within a few thousandths of an inch. Fabricators are using lasers to cut, trim, weld, heat treat, clad, vapor deposit coatings, anneal, and shock harden metals. Lasers are often used to cut out thin sheet metal parts because lasers are more flexible and cost effective than the commonly used punch presses. When punch presses are used, expensive punch and die tools have to be made first, and over time, as these tools wear out, replacements must be made.

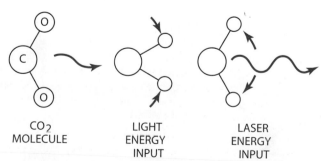

CO₂ MOLECULE   LIGHT ENERGY INPUT   LASER ENERGY INPUT

FIGURE 9-3 Energy is stored in a carbon dioxide (CO_2) molecule until the molecule can not hold any more. The energy is released suddenly like when a balloon pops. This quick release results in a burst of light energy.

© Cengage Learning 2012

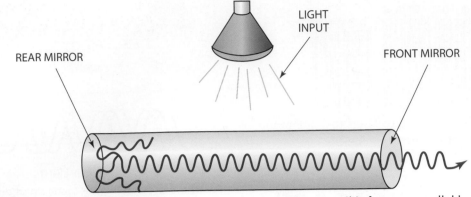

FIGURE 9-4 Light energy reflects back and forth between the end mirrors until it forms a parallel laser light beam.
© Cengage Learning 2012

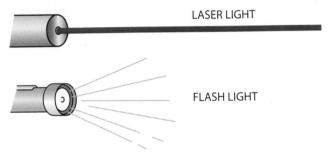

FIGURE 9-5 Laser light stays in a very tight column unlike most other lights. © Cengage Learning 2012

When the laser beam exits the laser generator, it can be treated as any other type of light. It is possible to focus, reflect, absorb, or defuse the light. By having the light waves traveling in such a uniform manner, they exhibit some specialized characteristics. The light beam tends to remain in a very tight column without spreading out, unlike ordinary light, **Figure 9-5**. This characteristic allows the laser beam to travel over some distance without significantly being affected by the distance it travels or the air it is traveling through.

Because the beam is not adversely affected by its travel, it can be used to carry information or transmit energy. When the laser wraps around a loaf of bread or a box of crackers, its image is so precise that the data in a uniform product code strip can accurately be received by the store's computer. The military uses lasers to aim weapons or identify targets so that they can be hit by missiles or bombs.

Laser Types The early lasers all used a synthetic ruby rod as the material that produced the laser light. Today, a large number of materials can be used to produce the laser. These materials include such common items as glass and such exotic items as neodymium-doped, yttrium aluminum garnet (Nd:YAG), often referred to as a **YAG laser, Figure 9-6**.

Lasers can be divided into two major types: lasers using a solid material for the laser and lasers using a gas. Each of these two types is divided into two groups based on their method of operation: lasers that operate continuously and lasers that are pulsed.

Solid State Lasers The first lasers were all **solid state lasers.** They used a ruby rod to produce the laser. Today, the most popular solid laser is the neodymium-doped,

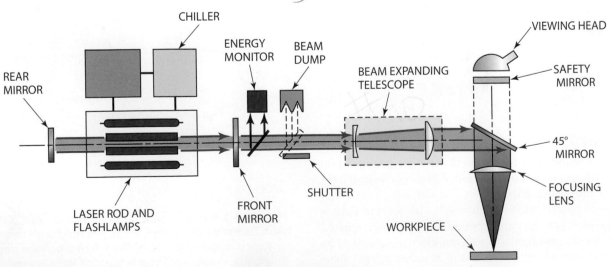

FIGURE 9-6 Schematic representation of the elements of an Nd:YAG laser. American Welding Society

yttrium aluminum garnet (YAG). This is a synthetic crystal that when exposed to the intense light from the flash tubes can produce high quantities of laser energy. The high-powered solid lasers have a problem because the internal temperatures of the laser rod increase with operation time. These lasers are most often used with a low-power continuous or high-powered pulse. The solid state laser is capable of generating the highest-powered laser pulses.

Gas Laser The **gas laser** uses one or more gases for the laser. Popular gas-type lasers use nitrogen, helium, carbon dioxide (CO_2), or mixtures of helium, nitrogen, and CO_2. Gas lasers either use a gas-charged cylinder where the gas is static or use a chamber that has a means of circulating the laser gas. The highest continuous power output-type lasers are the gas type with a blower to circulate the gas through a heat exchanger; they can be rapidly pulsed to improve their cutting capacity, **Figure 9-7**.

Applications

Laser light beams can be focused into a very compact area. This allows the photo energy that the light possesses to be converted to heat energy as it strikes the surface of a material. The highly concentrated energy can cause the instantaneous melting or vaporization of the material being struck by the laser beam. The equipment used to produce **laser beam welds (LBW), laser beam cuts (LBC),** or **laser beam drilling (LBD)** is similar in design and operation. Most laser welding and cutting operations use a gas laser, and most drilling operations use a solid state laser.

The ability of a material's surface to absorb or reflect the laser light affects the laser's operating efficiency. All materials will reflect some of the laser's light, some more than

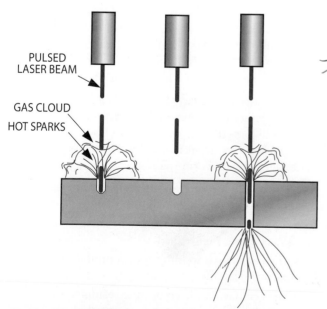

FIGURE 9-7 The gas cloud and hot sparks caused by the cutting laser beam dissipate between the pulses of the laser beam, which results in a better cut.

© Cengage Learning 2012

Assist Gases	Material
Air	Aluminum
	Plastic
	Wood
	Composites
	Alumina
	Glass
	Quartz
Oxygen	Carbon steel
	Stainless steel
	Copper
Nitrogen	Stainless steel
	Aluminum
	Nickel alloys
Argon	Titanium

TABLE 9-1 Various Cutting Assist Gases and the Materials They Cut

others. The absorption rate will increase for any material once the laser beam begins to heat the surface. Once the threshold of the surface temperature is reached, the process will continue at a much higher level of efficiency.

Laser Beam Cutting A laser weld and laser cut both bring the material to a molten state. In the welding process, the material is allowed to flow together and cool to form the weld metal. In the cutting process, a jet of gas is directed into the molten material to expel it through the bottom of the cut. Although the laser is used primarily to cut very thin materials, it can be used to cut up to 1 in. (2.5 cm) of carbon steel.

The **cutting gas assist** can be either a nonreactive gas or an exothermic. **Table 9-1** lists the various gases and the materials that they are used to cut. Nonreactive gases do not add any heat to the cutting process; they simply remove the molten material by blowing it out of the kerf. **Exothermic gases** react with the material being cut, like an oxyfuel cutting torch. The additional heat produced as the exothermic cutting gas reacts with the metal being cut helps blow the molten material out of the kerf.

Following are some advantages of laser cutting:

- Narrow heat-affected zone—Little or no heating of the surrounding material is observed. It is possible to make very close parallel cuts without damaging the strip that is cut out, **Figure 9-8**.

- No electrical conductivity required—The part being cut does not have to be electrically conductive, so materials like glass, quartz, and plastic can be cut. There is also no chance that a stray electrical charge might damage delicate computer chips while they are being cut using a laser.

- Noncontact—Nothing comes in contact with the part being cut except the laser beam. Small parts that may have finished surfaces or small surface details can be cut without the danger of disrupting or damaging the surface. It is also not necessary to hold the parts securely as it is when a cutting tool is used.

FIGURE 9-8 Detailed small part not much larger than a dime made with a laser. Larry Jeffus

- Narrow kerf—The width of the kerf is very small, which allows the nesting of parts in close proximity to each other, which will reduce waste of expensive materials, **Figure 9-9**.

- Automation and robotics—The laser beam can easily be directed through an articulated guide to the working end of an automated machine or a robot.

- Top edge—The top edge will be smooth and square without being rounded.

Following are some limitations of laser cutting:

- Stacked materials can have significantly different melting temperatures.

- Component placement on fabrications can limit the beam's access for some cuts.

- Materials thicker than 0.40 in. can often be cut more efficiently with another process.

Laser Beam Drilling The pulsed laser is the best choice for drilling operations. A very short burst of high laser energy is concentrated on a small spot. The intense heat vaporizes the small spot with enough force to thrust the material out, leaving a small crater. The repeated blasting to the pulsed laser results in the crater becoming increasingly

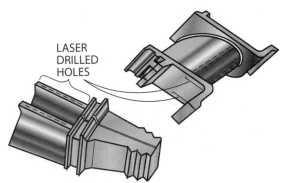

LASER DRILLED HOLES

FIGURE 9-10 Jet engine blades and a rotor component showing laser drilled holes. American Welding Society

deeper. This process continues until the hole is drilled through the part or to a desired depth, **Figure 9-10**.

A laser can be used to drill holes through the hardest materials, such as tungsten carbide cutting tools, quartz, glass, ceramics, or even synthetic diamonds. No other drilling process can match the precision, speed, or economy of the laser drill.

Lasers can drill up to 600 holes a minute through a 1/8-in. (3-mm) thick steel plate, and holes as small as 0.0001 in. (0.0025 mm) in diameter can be drilled. The limitation on the hole's depth is the laser's focal length. Most holes are less than 1 in. (2.5 cm) in depth.

Laser Beam Welding The laser beam can be used to melt the surface of the materials so they can flow together and cool to form the weld metal. Most laser beam welding is performed on thin materials. Following are some advantages of laser beam welding:

- The highly focused laser beam's heat allows it to make welds on very small thin parts.

- Welds can be made that are narrow and have deep penetration, **Figure 9-11**.

- The laser's high welding speeds produce very narrow heat-affected zones.

- The very fast localized heating results in little or no postweld distortion.

- There is no electric current that might damage sensitive electronic parts that are being welded on or nearby.

FIGURE 9-9 Because of the narrow kerf, small, detailed parts can be nested one inside the other. Larry Jeffus

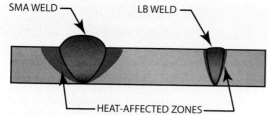

FIGURE 9-11 Smaller welds and heat-affected zones reduce weld distortion. © Cengage Learning 2012

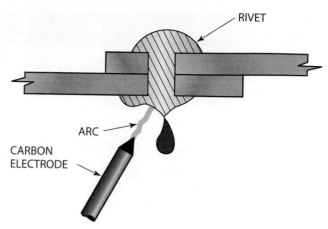

FIGURE 9-12 A carbon electrode was used to melt the head off rivets. © Cengage Learning 2012

Laser Equipment Most lasers range from 400 to 1500 watts in power. Some large machines have as much as 25 kW of power. Although the power of most lasers is relatively small when compared to other welding processes, it is the laser's ability to concentrate the power into a small area that makes it work so well. The power density of a cutting laser can be equal to 65 million watts per square inch.

Laser equipment is larger than most of the other welding or cutting power supplies. A typical unit will require about as much floor space as a large desk, about 10 sq ft.

Recent advancements, increased competition, and a rapidly increasing market have helped lower the high cost of the equipment. But even with the current cost of equipment, laser cutting and drilling have one of the lowest average hourly operating costs. The high reliability and long life of the equipment have also helped reduce operating costs.

AIR CARBON ARC CUTTING

The **air carbon arc cutting (CAC-A)** process was developed in the early 1940s, and was originally named air arc cutting (AAC). Air carbon arc cutting was an improvement of the carbon arc process. The carbon arc process was used in the vertical and overhead positions and removed metal by melting a large enough spot so gravity would cause it to drip off the base plate. This process was slow and could not be accurately controlled. A recurring problem when removing rivet heads, as shown in **Figure 9-12**, was that some of the molten rivet head could weld itself to the base metal in any position other than overhead. It was found that by using a stream of air the molten metal could be blown away. This greatly improved the speed, quality, and controllability of the process.

In the late 1940s, the first air carbon arc cutting torch was developed. Before this development the process required two welders, one to control the carbon arc and the other to guide the air stream. The new torch had both the carbon electrode holder and the air stream in the same unit. This basic design is still in use today, **Figure 9-13**.

Unlike the oxyfuel process, the air carbon arc cutting process does not require that the base metal be reactive with the cutting stream. Oxyfuel cutting can be performed only on metals that can be rapidly oxidized by the cutting stream of oxygen. The air stream in this process blows the molten metal away. This greatly increases the list of metals that can be cut, **Table 9-2**.

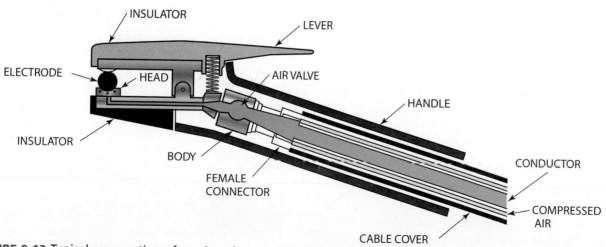

FIGURE 9-13 Typical cross section of an air carbon arc gouging torch. American Welding Society

Base Metals	Recommendations
Carbon steel and low alloy steel	Use DC electrodes with DCEP current. AC can be used but with a 50% loss in efficiency.
Stainless steel	Same as for carbon steel.
Cast iron, including malleable and ductile iron	Use of 13 mm or larger electrodes at the highest rated amperage is necessary. There are also special techniques that need to be used when gouging these metals. The push angle should be at least 70° and depth of cut should not exceed 13 mm per pass.
Copper alloys (copper content 60% and under)	Use DC electrodes with DCEN (electrode negative) at maximum amperage rating of the electrode.
Copper alloys (copper content over 60%, or size of workpiece is large)	Use DC electrodes with DCEN at maximum amperage rating of the electrode or use AC electrodes with AC.
Aluminum bronze and aluminum nickel bronze (special naval propeller alloy)	Use DC electrodes with DCEN.
Nickel alloys (nickel content is over 80%)	Use AC electrodes with AC.
Nickel alloys (nickel content less than 80%)	Use DC electrodes with DCEP.
Magnesium alloys	Use DC electrodes with DCEP. Before welding, surface of groove should be wire brushed.
Aluminum	Use DC electrodes with DCEP. Wire brushing with stainless wire brushes is mandatory prior to welding. Electrode extension (length of electrode between electrode torch and workpiece) should not exceed 76 mm for good-quality work. DC electrodes with DCEN can also be used.
Titanium, zirconium, hafnium, and their alloys	Should not be cut or gouged in preparation for welding or remelting without subsequent mechanical removal of surface layer from cut surface.

Note: Where preheat is required for welding, similar preheat should be used for gouging.

TABLE 9-2 Recommended Procedures for Air Carbon Arc Cutting of Different Metals

---

THINK GREEN

Clean Up Any Possible Contamination

Some types of metals removed by the CAC-A cutting process can become groundwater or storm water contaminants if they are not cleaned up following outdoor cutting. It is a good idea to keep outdoor areas swept clean so rainwater does not wash heavy metals into the environment.

Manual Torch Design

The air carbon arc cutting torch is designed differently than the shielded metal arc electrode holder. Following are the major differences between an electrode holder and an air carbon arc torch:

- The lower electrode jaw has a series of air holes.
- The jaw has only one electrode-locating groove.
- The electrode jaw can pivot.
- There is an air valve on the torch lead.

By having only one electrode-locating groove in the jaw and pivoting the jaw, the air stream will always be aimed correctly. The air must be aimed just under and behind the electrode and always in the same direction, **Figure 9-14**.

This ensures that the air stream will be directed at the spot where the electrode arcs to base the metal.

Torches are available in a number of amperage sizes. The larger torches have greater capacity but are less flexible to use on small parts.

The torch can be permanently attached to a welding cable and air hose or it can be attached to welding power by gripping a tab at the end of the cable with the shielded metal arc electrode holder, **Figure 9-15**. The temporary attachment can be made easier if the air hose is equipped with a quick disconnect. A quick disconnect on the air hose will allow it to be used for other air tools such as grinders or chippers. Greater flexibility for a workstation can be achieved with this arrangement.

Electrodes Air carbon arc cutting electrodes are available as copper-coated or plain (without a coating). The copper

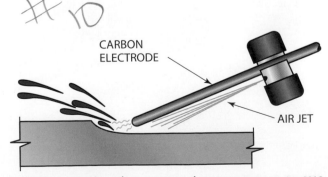

FIGURE 9-14 Air carbon arc gouging. © Cengage Learning 2012

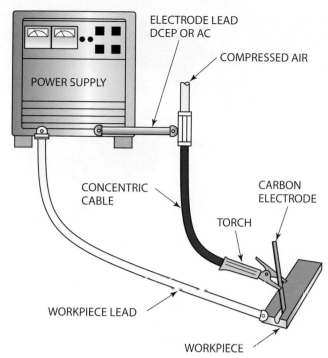

FIGURE 9-15 Air carbon arc gouging equipment setup.
American Welding Society

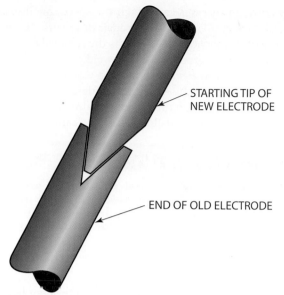

FIGURE 9-17 Air carbon arc electrode joint.
© Cengage Learning 2012

coating helps decrease the **carbon electrode** overheating by increasing its ability to carry higher currents and improving the heat dissipation. The copper coating provides increased strength, to reduce accidental breakage.

Electrodes can be round, flat, and semiround, **Figure 9-16**. The round electrodes are used for most gouging operations, and the flat electrodes are most often used to scarf off a surface. Round electrodes are also available in sizes ranging from 1/8 in. (3 mm) to 1 in. (25 mm) in diameter. Flat electrodes are available in 3/8 in. (10 mm) and 5/8 in. (16 mm) diameters.

Electrodes are available to be used on both direct-current electrode positive (DCEP) and alternating current. The DCEP electrodes are the most commonly used, and they are made of carbon in the form of **graphite.** The AC electrodes are less common; they have some elements added to the carbon to stabilize the arc, which is needed for the AC power.

To reduce waste, electrodes are made so that they can be joined together. The joint consists of a female tapered socket at the top end and a matching tang on the bottom

end, **Figure 9-17**. The connection of the new electrode to the remaining setup will allow the stub to be consumed with little loss of electrode stock. This connecting of electrodes is required for most track-type air carbon arc cutting operations to allow for longer cuts.

Power Sources Most shielded metal arc welding power supplies can be used for air carbon arc cutting. The operating voltage required for air carbon arc cutting needs to be 28 volts or higher. This voltage is slightly higher than that required for most SMA welding, but most welders will meet this requirement. Check the manufacturer owner's manual to see if your welder is approved for air carbon arc cutting. If the voltage is lower than the minimum, the arc will tend to sputter out, and it will be hard to make clean cuts.

Because most carbon arc cutting requires a high amperage setting, it may be necessary to stop some cuts so that the duty cycle of the welder is not exceeded. On large industrial welders this is not normally a problem.

Air Supply Air supplied to the torch should be between 80 and 100 psi (550 and 690 kPa). The minimum pressure is around 40 psi (276 kPa). The correct air pressure will result in cuts that are clean, smooth, and uniform. The airflow rate is also important. If the air line is too small or the compressor does not have the required capacity, there will be a loss in air pressure at the torch tip. This line loss will result in a lower-than-required flow at the tip. The resulting cut will be less desirable in quality.

Application

Air carbon arc cutting can cut a variety of materials. It is a relatively low-cost way of cutting most metals, especially stainless steel, aluminum, nickel alloys, and copper. Air carbon arc cutting is most often used for repair work.

ROUND   FLAT   SEMIROUND
FIGURE 9-16 Cross sections of carbon electrodes.
© Cengage Learning 2012

Few cutting processes can match the speed, quality, and cost savings of this process for repair or rework. In repair or rework, the most difficult part is the removal of the old weld or cutting a groove so a new weld can be made. The air carbon arc can easily remove the worst welds even if they contain slag inclusions or other defects. For repairs, the arc can cut through thin layers of paint, oil, or rust and make a groove that needs little, if any, cleanup.

///// **CAUTION** \\\\\

Never cut on any material that might produce fumes that would be hazardous to your health without taking proper safety precautions, including adequate ventilation and/or the wearing of a respirator.

The highly localized heat results in only slight heating of the surrounding metal. As a result, usually there is no need to preheat hardenable metals to prevent hardness zones. Cast iron is a metal that can be carbon arc gouged to prepare a crack for welding without causing further damage to the part by inputting excessive heat.

Air carbon arc cutting can be used to remove a weld from a part. The removal of welds can be accomplished with such success that often the part needs no postcut cleanup. The root of a weld can be back gouged so that a backing weld can be made ensuring 100% weld penetration, **Figure 9-18**.

The electrode should extend approximately 6 in. (152 mm) from the torch when starting a cut and, as the cut progresses, the electrode is consumed. Stop the cut and readjust the electrode when its end is approximately 3 in. (76 mm) from the electrode holder. This will reduce the damage to the torch caused by the intense heat of the operation.

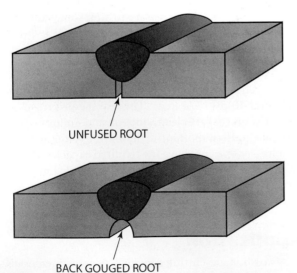

UNFUSED ROOT

BACK GOUGED ROOT

FIGURE 9-18 Back gouging the root of a weld made in thick metal can ensure that a weld with 100% joint fusion can be made. © Cengage Learning 2012

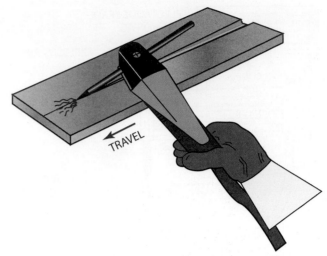

FIGURE 9-19 Manual air carbon arc gouging operation in the flat position. American Welding Society

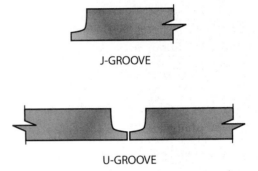

J-GROOVE

U-GROOVE

FIGURE 9-20 Air carbon arc gouging groove shapes.
© Cengage Learning 2012

Gouging **Gouging** is the most common application of the air carbon arc cutting processes. Arc gouging is the removal of a quantity of metal to form a groove or bevel, **Figure 9-19**. The groove produced along an edge of a plate is usually a J-groove. The groove produced along a joint between plates is usually a U-groove, **Figure 9-20**. Both grooves are used as a means to ensure that the weld applied to the joint will have the required penetration into the metal.

Washing **Washing** is the process sometimes used to remove large areas of metal so that hard surfacing can be applied, **Figure 9-21**. Washing can be used to remove large areas that contain defects, to reduce the transitional stresses of unequal-thickness plates, or to allow space for the capping of a surface with a wear-resistant material.

Safety

In addition to the safety requirements of shielded metal arc, air carbon arc requires several special precautions, such as the following:

- Sparks—The quantity and volume of sparks and molten metal spatter generated during this process are a major safety hazard. Extra precautions must

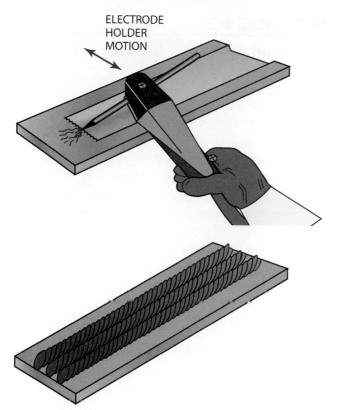

ELECTRODE
HOLDER
MOTION

FIGURE 9-21 Hard-surfacing weld applied in the carbon arc cut groove. American Welding Society

be taken to ensure that other workers, equipment, materials, or property in the area will not be affected by the spark stream.

■ Noise—This process produces a high level of sound. The sound level is high enough to cause hearing damage if proper ear protection is not used.

■ Light—The arc light produced is the same as that produced by the shielded metal arc welding process. But because the arc has no smoke to defuse the light and the amperages are usually much higher, the chances of receiving arc burns are much higher. Additional protection should be worn, such as thicker clothing, a leather jacket, and leather aprons.

■ Eyes—Because of the intense arc light, a darker welding filter lens for the helmet should be used.

■ Fumes—The combination of the air and the metal being removed results in a high volume of fumes. Special consideration must be made for the removal of these fumes from the work area. Before installing a ventilation system, check with local, state, and federal laws. Some of the fumes may have to be filtered before they can be released into the air.

■ Surface contamination—Often this process is used to prepare damaged parts so that they can be repaired. If the used parts have paint, oils, or other contaminants that might generate hazardous fumes, they must be removed in an acceptable manner before any cutting begins.

THINK GREEN
Control Sound Pollution

CAC-A can produce high sound levels. Many communities have noise ordinances that define noise as a sound that is unwanted or annoying, and they set limits on sound levels. Sound barriers can be used to block the noise from CAC-A adversely affecting neighboring businesses or homes.

■ Equipment—Check the manufacturer owner's manual for specific safety information concerning the power supply and the torch before you start any work with each piece of equipment for the first time.

U-Grooves

U-grooving is used to remove defective welds and to prepare a thick metal joint so that a full penetration weld can be made. Often the easiest way to prepare a joint for a full penetration weld is to use one of the thermal processes and cut a V-groove or bevel along the edge. However, sometimes parts must be fitted and assembled to ensure everything fits as designed before a groove can be cut. In these cases, a U-groove can easily be cut along the assembled edges of the joint.

PRACTICE 9-1

Air Carbon Arc Straight U-Groove in the Flat Position

Using an air carbon arc cutting torch and welding power supply that have been safely set up in accordance with the manufacturer's specific instructions in the owner's manual and wearing safety glasses, welding helmet, gloves, and any other required personal protection clothing, you will make a 6-in. (152 mm) long straight U-groove gouge in a carbon steel plate.

1. Adjust the air pressure to approximately 80 psi.
2. Set the amperage within the range for the diameter electrode you are using by referring to the box the electrodes came in.
3. Check to see that the stream of sparks will not start a fire or cause any damage to anyone or anything in the area.
4. Make sure the area is safe, and turn on the welder.
5. Using a good, dry, leather glove to avoid electrical shock, insert the electrode in the torch jaws so that about 6 in. is extending outward. Be sure not to touch the electrode to any metal parts, because it may short out.

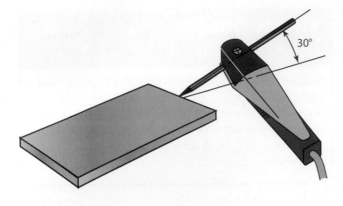

FIGURE 9-22 Air carbon arc U-groove gouging.
© Cengage Learning 2012

6. Turn on the air at the torch head.

7. Lower your arc welding helmet.

8. Slowly bring the electrode down at about a 30° angle so it will make contact with the plate near the starting edge, **Figure 9-22**. Be prepared for a loud, sharp sound when the arc starts.

9. Once the arc is struck, move the electrode in a straight line down the plate toward the other end. Keep the speed and angle of the torch constant.

10. When you reach the other end, lift the torch so the arc will stop.

11. Raise your helmet and stop the air.

12. Remove the remaining electrode from the torch so it will not accidentally touch anything.

When the metal is cool, chip or brush any slag or dross off of the plate. This material should remove easily. The groove must be within ±1/8 in. (3 mm) of being straight and within ±3/32 in. (2.4 mm) of uniformity in width and depth. Repeat this cut until it can be made within these tolerances. Turn off the CAC equipment and clean up your work area when you are finished cutting.

Complete a copy of the "Student Welding Report" listed in Appendix I or provided by your instructor. ◆

PRACTICE 9-2

Air Carbon Arc Edge J-Groove in the Flat Position

Using the same equipment, adjustments, setup, and materials as described in Practice 9-1, you are going to make a J-groove along the edge of the plate.

1. Adjust the air pressure to approximately 80 psi.

2. Set the amperage within the range for the diameter electrode you are using.

3. Check to see that the stream of sparks will not start a fire or cause any damage to anyone or anything in the area.

4. Make sure the area is safe, and turn on the welder.

5. Using a good, dry, leather glove to avoid electrical shock, insert the electrode in the torch jaws so that about 6 in. is extending outward. Be sure not to touch the electrode to any metal parts because it may short out.

6. Turn on the air at the torch head.

7. Lower your arc welding helmet.

8. Slowly bring the electrode down at about a 30° angle so it will make contact with the plate near the starting edge, **Figure 9-23**.

9. Once the arc is struck, move the electrode in a straight line down the edge of the plate toward the other end. Keep the speed and angle of the torch constant.

10. When you reach the other end, lift the torch so the arc will stop.

11. Raise your helmet and stop the air.

12. Remove the remaining electrode from the torch so it will not accidentally touch anything.

FIGURE 9-23 Air carbon arc J-groove gouging.
© Cengage Learning 2012

When the metal is cool, chip or brush any slag or dross off of the plate. This material should remove easily. The groove must be within ±1/8 in. (3 mm) of being straight and within ±3/32 in. (2.4 mm) of uniformity in width and depth. Repeat this cut until it can be made within these tolerances. Turn off the CAC equipment and clean up your work area when you are finished cutting.

Complete a copy of the "Student Welding Report" listed in Appendix I or provided by your instructor. ◆

PRACTICE 9-3

Air Carbon Arc Back Gouging in the Flat Position

Using the same equipment, adjustments, setup, and materials as described in Practice 9-1, you are going to make a U-groove along the root face of a weld joint on a plate, **Figure 9-24**.

Follow the same starting procedure as you did in Practice 9-1.

1. Start the arc at the joint between the two plates.
2. Once the arc is struck, move the electrode in a straight line down the edge of the plate toward the other end. Watch the bottom of the cut to see that it is deep enough. If there is a line along the bottom of the groove, it needs to be deeper. Once the groove depth is determined, keep the speed and angle of the torch constant.
3. When you reach the other end, break the arc off.
4. Raise your helmet and stop the air.
5. Remove the remaining electrode from the torch so it will not accidentally touch anything.

When the metal is cool, chip or brush any slag or dross off of the plate. This material should remove easily. The

groove must be within ±1/8 in. (3 mm) of being straight, but it may vary in depth so that all of the unfused root of the weld has been removed. Repeat this cut until it can be made within these tolerances. Turn off the CAC equipment and clean up your work area when you are finished cutting.

Complete a copy of the "Student Welding Report" listed in Appendix I or provided by your instructor. ◆

PRACTICE 9-4

Air Carbon Arc Weld Removal in the Flat Position

Using the same equipment, adjustments, setup, and materials as described in Practice 9-1, you are going to make a U-groove to remove a weld from a plate, **Figure 9-25**.

Follow the same starting procedure as you did in Practice 9-1.

1. Start the arc on the weld.
2. Once the arc is struck, move the electrode in a straight line down the weld toward the other end. Watch the bottom of the cut to see that it is deep enough. If there is not a line along the bottom of the groove, it needs to be deeper. Once the groove depth is determined, keep the speed and angle of the torch constant.
3. When you reach the other end, break the arc off.
4. Raise your helmet and stop the air.
5. Remove the remaining electrode from the torch so it will not accidentally touch anything.

When the metal is cool, chip or brush any slag or dross off of the plate. This material should remove easily. The groove must be within ±1/8 in. (3 mm) of being straight, but it may vary in depth so that all of the weld metal has

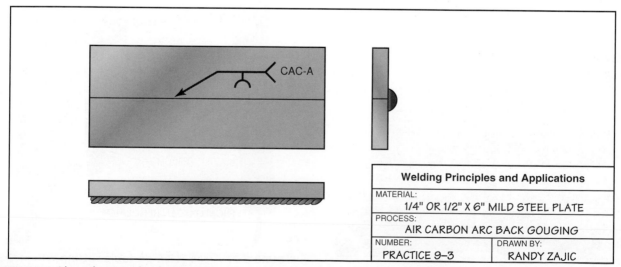

Welding Principles and Applications

MATERIAL:		
1/4" OR 1/2" X 6" MILD STEEL PLATE		
PROCESS:		
AIR CARBON ARC BACK GOUGING		
NUMBER:		DRAWN BY:
PRACTICE 9-3		RANDY ZAJIC

FIGURE 9-24 Air carbon arc back gouging. © Cengage Learning 2012

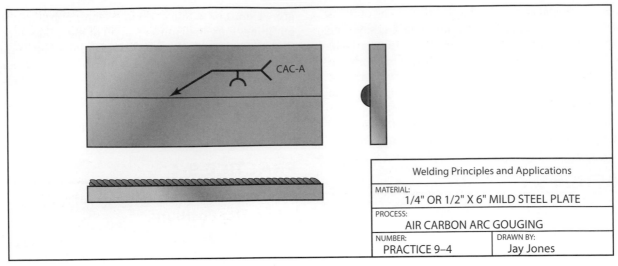

FIGURE 9-25 Air carbon arc weld gouging. © Cengage Learning 2012

been removed. Repeat this cut until it can be made within these tolerances. Turn off the CAC equipment and clean up your work area when you are finished cutting.

Complete a copy of the "Student Welding Report" listed in Appendix I or provided by your instructor. ◆

OXYGEN LANCE CUTTING

The **oxygen lance cutting** process uses a consumable alloy tube, **Figure 9-26**. The tip of the tube is heated to its kindling temperature. A high-pressure oxygen flow is started through the lance. The oxygen reacts with the hot lance tip, releasing sufficient heat to sustain the reaction, **Figure 9-27**.

The rod tip is heated up to a red-hot temperature using an oxyfuel torch or electric resistance. Once the oxygen stream is started, it reacts with the lance material, which results in the creation of both a high temperature and heat-releasing reaction.

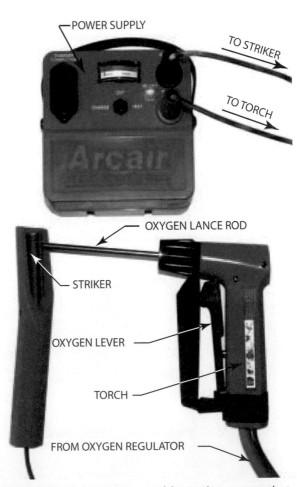

FIGURE 9-27 Arcair SLICE portable cutting system that uses sparks created between the striker and tube to ignite the oxygen lance rod for cutting. Larry Jeffus

FIGURE 9-26 Oxygen lance rods may be rolled out of flat strips of special alloys to form a tube. Larry Jeffus

The intense reaction of the lance allows it to be used to cut through a variety of materials. The hot metal leaving the lance tip has not completed its exothermic reaction. As this reactive mass impacts the surface of the material being cut, it releases a large quantity of energy into that surface. Thermal conductivity between the molten metal and the base material is a very efficient method of heat transfer. This, along with the continued burning of the lance material on the surface, causes the base material to become molten.

Once the base material is molten, it may react with the burning lance material, forming fumes or slag, which is then blown from the cut. Any molten material not becoming reactive is carried out of the cut with the slag or blown out with the oxygen stream.

The addition of steel rods or other metals to the center of the oxygen lance tube has increased their productivity. The improved lances last longer and cut faster.

Applications

The oxygen lance's unique method of cutting allows it to be used to cut material not normally cut using a thermal process. Films have portrayed the oxygen lance as a tool used by thieves to cut into safes. In reality, this would result in the valuables in the safe being destroyed. The oxygen lances can be used to cut reinforced concrete.

Cutting concrete has been used in the demolition of buildings. It allows the quick removal of thick sections of the building without the dangerous vibration caused by most conventional methods. This use of the oxygen lance has been a lifesaving factor for rescue work following earthquakes. The oxygen lance saved thousands of hours and countless lives in Mexico City following the devastation from the city's worst earthquake. Oxygen lances were used to cut large sections of concrete that fell from buildings into manageably sized pieces. Local and national news agencies showed building rubble being cut away by rescue workers using the oxygen lance.

The oxygen lance is also used to cut thick sections of cast iron, aluminum, and steel. Often in the production of these metals, thick sections must be cut. Occasionally, equipment failure will stop metal production. If the metal in production is allowed to cool, it may need to be cut in sections so it can be removed from the machine. The oxygen lance's process is very effective in this type of work.

Safety

It is important to follow all safety procedures when using this process. Manufacturers list specific safety precautions for the oxygen lances they produce. Read and follow those instructions carefully. The major safety concerns are as follows:

- Fumes—The large quantity of fumes generated is often a health hazard. An approved ventilation system must be provided if this work is to be done in a building or any other enclosed area.

- Heat—This operation produces both high levels of radiant heat and plumes of molten sparks and slag. The operators must wear special heat-resistant clothing.

- Noise—Sound is produced well above safety levels. Ear protection must be worn by anyone in the area.

WATER JET CUTTING

Water jet cutting is not a thermal cutting process. This method of cutting does not put any heat into the material being cut. The cut is accomplished by the rapid erosion of the material by a high-pressure jet of water, **Figure 9-28**. An abrasive powder may be added to the stream of water. **Abrasives** are added when hard materials such as metals are being cut.

The lack of heat input to the material being cut makes this process unique. Materials that heat might distort, make harder, or cause to delaminate are ideally suited to this process. The lack of heat distortion allows thin material to be cut with the edge quality of a laser cut and as distortion-free as a shear cut. **Delamination** is not a problem when cutting composite or laminated materials, such as carbon fibers, resins, or computer circuit boards, **Figure 9-29**.

Applications

The kerf width does not tend to change unless too high a travel speed is being used. This results in a square, smooth finish on the cut surface. Postcut cleanup of the parts is totally eliminated for most materials, and only slight work is needed on a few others, **Figure 9-30**. The quality of the cut surface can be controlled so that even parts for the aerospace industry can often be assembled as cut.

The addition of an abrasive powder can speed up the cutting, allow harder materials to be cut, and improve the surface finish of a cut. The powder most often used is garnet. It is also commonly used as an abrasive on sandpaper. If an abrasive is used, the small water jet orifice will wear out faster.

Materials that often gum up a cutting blade, such as plexiglass, ABS plastic, and rubber, can be cut easily. There is nothing for the material to adhere to that would disrupt the cut. The lack of heat also reduces the tendency of the material's cut surface to become galled.

Most water jet cutting is performed by some automated or robotic system. There are a few band-saw-type, hand-fed cutting machines that are used for single cuts or when limited production is required.

ARC CUTTING ELECTRODES

Arc cutting electrodes do not require any special equipment other than a standard SMA welding machine, and they fit into the standard electrode holder. Because the arc cutting electrodes work with any standard arc

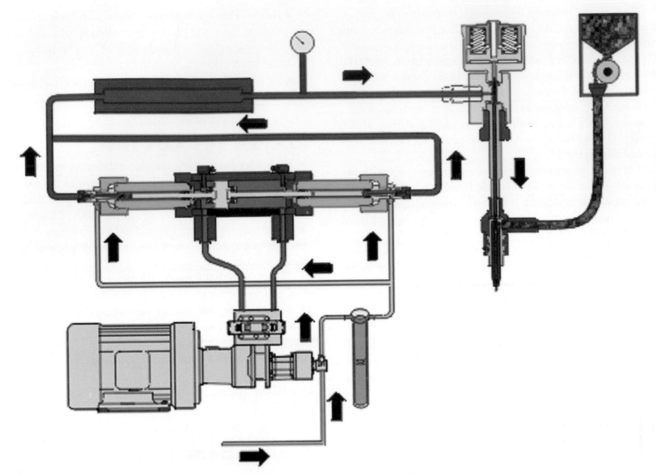

FIGURE 9-28 Basic diagram illustrating the elements of the water jet cutting system. © Cengage Learning 2012

FIGURE 9-29 Water jet cutting is very fast, clean, and free from heat distortion. Larry Jeffus

FIGURE 9-30 Notice how narrow and clean this cut is. Larry Jeffus

welder, this makes them the ideal choice for many cutting jobs. For example, if only an occasional cut is needed, the expense of having either an OFC or a PAC piece of equipment cannot be justified. Many types of arc cutting electrodes can make clean cuts, although they may not make as square, smooth, and clean a cut as many other cutting processes. But they are much easier and faster to use than a metal saw for cutting or to drill a hole in thick metal.

Arc cutting electrodes are available for cutting steel, stainless steel, cast iron, aluminum, copper alloys, and many other metals. Steel up to 1 1/2 in. (37 mm) thick can be cut using this process. Some arc cutting electrodes are available for underwater cutting where using an acetylene torch might be difficult or impossible.

Arc cutting electrodes differ from standard arc welding electrodes in that they burn back inside the outer flux covering, creating a small cavity, **Figure 9-31**. The heat of the arc causes the metal core and inner layer of flux to vaporize and rapidly expand. The small cavity acts like a small combustion chamber, and the hot vaporized material is blasted out like a small jet engine. Combined with the heat of the arc and the jetting action, the metal is cut.

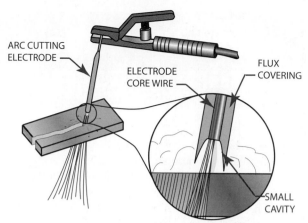

FIGURE 9-31 The small cavity at the end of the electrode creates enough jetting action to blow the molten metal formed by the arc away to form a cut. © Cengage Learning 2012

Applications

This cutting process is ideal any time that it is not practical or cost effective to bring along an oxygen acetylene cutting set and cylinders or a plasma cutting machine. Arc cutting electrodes can be used to cut metal, pierce holes, or to gouge a groove for welding or gouge out a defective weld.

Summary

The welding field's ability to provide industry with a wide variety of cutting processes has increased productivity for a variety of industries. An example of this diversity is the use of a laser beam to cut apart computer chips in the electronics field. Without these cutting processes, industry would have to rely on the much slower and more expensive mechanical or abrasive cutting process. There are no mechanical or abrasive cutting processes that can compete with the speed of these new cutting processes and none that can compete with their versatility.

Somewhat less attractive but nonetheless efficient processes such as carbon arc gouging and oxygen lance cutting fulfill specific industrial applications, such as salvage or scrap work. Few cutting processes can rival the metal-removing capacity of the air carbon arc or oxygen lance processes. In addition, the low cost of the equipment and the flexibilities and application of these processes have lent themselves very successfully to the salvage and scrap industry. Air carbon arc gouging is extensively used for weld removal and repair work. A skilled technician can produce a groove that requires little or no postcut cleanup prior to rewelding.

Lasers: The New Wave in Ship Construction

During the past 10 years, laser materials processing has been accepted into a number of industries. The automotive, medical device, electronics, and consumer products industries have developed laser materials processes and designs to improve product performance. With advances in hardware and processing, lasers are now making inroads into industries with new material and fabrication demands. These industries include heavy machinery, construction equipment, aerospace, oil/chemical, power generation, and shipbuilding.

Laser materials processing uses the energy of photon-matter interaction to heat a targeted material. The density and interaction time of the laser beam with the material determine what process is accomplished. At lower power densities and longer interaction times, the laser can be used to heat treat or thermally form materials. As the power density is increased, materials can be melted to form resolidified or clad surfaces. At higher power densities and shorter interaction times, the laser can produce a key hole in the molten material and achieve deep penetration welds with very high depth-to-width ratios. Further modification of the interaction can result in cutting and drilling with very little heat input into the part.

The advantages of using lasers are determined by what process is needed. At lower power densities, the laser has an advantage over induction and furnace heating because of its accuracy in treating specific areas, which can minimize thermal impact to surrounding materials. At higher power densities, heat input can be accurately controlled, permitting high-quality clads and direct material buildups. In welding, the laser permits deep welds to be made with very low heat input at very high speeds. The result is weld joints with low impact to the surrounding materials and low distortion. While other processes can cut faster (oxyfuel or plasma) or without heat (water jet), the laser is flexible and precise for a wide range of thicknesses and types of metals and nonmetals.

The possible use of lasers in shipbuilding has been of interest for a number of years. In the past, the U.S. Navy conducted studies on how lasers could be used for cutting and fabricating Navy combatants and submarines and what impact that would have on the materials used in naval structures. Most of these applications were based on the use of high-power CO_2 lasers and complex delivery systems. The Navy determined that equipment and operational costs, reliability, and return on investment for such applications did not justify their use.

Since these early studies, there have been improvements in laser systems that have had a positive impact on laser applications in thick-section materials. First, changes in laser designs have increased beam quality and have improved laser welding and cutting capabilities. Second, lasers have become more "industrialized" through improvements in

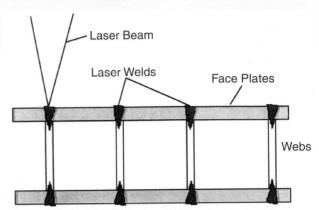

Diagram of "sandwich panel" being fabricated by Meyer Werft Shipyard for shipboard application.
American Welding Society

design and standardization of components. Third, the cost has decreased in real dollar terms. Other industries, such as automotive and sheet metal fabrication, where laser processing has been easier to justify, have driven these changes.

One factor that has limited the insertion of laser processing into ship applications has been the lack of standards and specifications. ASME and AWS have specifications for pressure structures and recommended practices, respectively, but there are few widely accepted standards for laser-fabricated structures. The International Organization for Standardization (ISO) has been developing specifications for laser cut quality and laser cutting systems. In the United States, the AWS C7C Committee is developing general laser processing specifications. The U.S. Navy, in previous efforts, has accepted laser processing only for specific applications. Efforts are underway to develop acceptance criteria for laser cut and welded structures for U.S. Navy applications.

Shipyard Laser Use around the World

A number of shipyards are starting to implement laser processing. As an example, Odense Lindo Steel, Meyer Werft, and Fincantieri are all involved in implementing laser beam welding in the fabrication of ship structures. Other European yards, such as Blohm & Voss and Vosper-Thornycroft, are involved in laser cutting. In Japan, a major producer of high-power laser cutting systems, a large number of shipyards have implemented laser cutting.

Work at the European shipyard Meyer Werft is of special interest. Meyer Werft is using the deep penetration characteristics of the laser to alter the design of the bulkhead and deck panels for improved productivity. Using faceplates on either side of a series of vertical webs, laser stake welds are made through each faceplate and into the vertical webs. The resulting "closed-cell" design is stiffer than a conventional plate/stiffener design and occupies less space. Also, the "flatness" of the finished panels makes outfitting (installation of flooring, insulation, wiring, and other piping) easier and faster. The reduced space also can mean a decrease in the overall height of the ship for the addition of an extra deck.

In the United States, there have been only a few cases in which lasers have actually been used in shipyard applications, although laser materials processing for shipboard applications has been under investigation for more than 20 years. This has been partially due to the low production rate of commercial ships in the U.S. and in part because of the qualification requirements for the U.S. Navy. Applications currently used in United States shipyards for laser-processed materials include cladding, welding, and cutting.

As an outcome of research accomplished by the U.S. Navy Manufacturing Technology (MANTECH) program, there have been a number of applications for laser cladding and welding. Laser cladding for repair has been accomplished with steel, nickel, aluminum, and cobalt-based alloys. Procedures have been qualified for the repair of such items as diesel cooling pump shafts, steam valve seats, aircraft carrier catapult components, and aluminum torpedo shell sections. In many cases, the laser process has been selected because of low dilution and heat input, which result in a "near-pure" deposition of material with little thermal impact on the part. The repairs are justified because of the high cost of replacing the parts and the long lead times required to qualify new suppliers and receive replacement parts.

Cutting is presently the most active U.S. shipyard application for lasers. Bender Shipbuilding and Repair of Mobile, Alabama, has purchased a 6-kW CO_2 laser cutting system with a cutting table for the processing of steel, stainless steel, and aluminum alloys. The system is part of a "first operation" that receives, cleans, primes, and cuts plates prior to fabrication in the shipyard. The laser system can cut steel up to 32 mm thick and aluminum up to 12 mm thick. The advantage of the laser has been the ability to "common line cut," which is not possible with plasma. The laser system can also be used to pierce and cut openings in the plate that would otherwise be drilled and manually cut as secondary processes. Bender has already used the laser system to cut steel for its ships and is looking ahead to other applications.

Laser Use Growing Worldwide

The use of lasers in shipyards is on the rise throughout the world. Laser cutting is quickly replacing more traditional oxy-fuel and plasma cutting techniques due to the cost savings and the laser's precision and flexibility. The advantage of laser welding for the fabrication of ship structures also appears promising. To take full advantage of these properties, new designs, lasers, and processing techniques are being developed. At the same time, other secondary processes, such as heat treating, forming, and cladding, are being developed for fabrication and repair. New specifications are being written that will cover the implementation of lasers into ship fabrication. All of these indicate lasers will be the new tool in the fabrication of ships and naval structures in the years to come.

Article courtesy of the American Welding Society.

Review

1. Describe the use of the following processes: LBD, LBW, and LBC.

2. How is laser light formed?

3. Other than the lasing material, what is the difference between solid state lasers and gas lasers?

4. What effect does a material's surface have on the laser beam?

5. Using Table 9-1, list the assist cutting gas and the material it can be used to cut.

6. What are reactive laser assist gases called? How do they work?

7. What type of laser beam is used for drilling? Why?

8. What is CAC-A?

9. Using Table 9-2, give the recommended procedure for air carbon arc gouging of carbon steel, magnesium alloys, and low alloy copper.

10. Why are some carbon arc electrodes copper-coated?

11. What may occur if an SMA welding machine has below minimum arc voltage for air carbon arc gouging?

12. What is washing, and how can it be used?

13. How are oxygen lance cuts usually started?

14. What unusual material can be cut with an oxygen lance?

15. What are the advantages of abrasive powder to the water jet cutting stream?

16. What applications are arc cutting electrodes used for?

SECTION 4

GAS SHIELDED WELDING

Success Story

Michael Crumpler is the welding instructor at Southeastern Technical College in Vidalia Georgia. His unplanned welding career began 25 years ago when his mother helped him buy his first car. Reality hit Michael when he learned he would have to get a job to pay for the gas and insurance. The job search began with his mother's prayers and her driving him to Imperial Fabricating in New Deal, TN. Because he had no skills or work experience, the application process took only a few minutes. The interviewer asked Michael if he could weld, read blueprints, or fabricate. He could do none of these things.

"It was hard for me to believe it when I heard that I was exactly what they were looking for," says Michael. "They said they needed someone they could train. I could not believe it. It seems as if I was destined for the welding industry. I have loved welding ever since."

He wanted to learn more about welding technology and to advance his career opportunities, so he earned a degree at Brewton-Parker College in Mount Vernon, Georgia and a diploma in Welding and Joining Technology from Heart of Georgia Technical College in Dublin, GA. He took additional courses for professional development at the Hobart Institute of Welding Technology in Troy, Ohio.

Michael has been a certified structural welder and has had a variety of welding job opportunities, including working as a pipe welder and boilermaker for B.E. & K. Industrial Contractors in Birmingham, Alabama. In addition to working in many states across the country, he has traveled to Managua, Nicaragua to supervise, fabricate, and weld during the construction of the country's very first peanut- shelling plant.

He has held certifications in numerous welding processes, including structural welding and pipe welding. In addition to being a welding instructor, he is a Certified Welding Educator, a Certified Welding Inspector, a Certified Welding Supervisor, and is a member of The American Welding Society.

His experience began as a full-time welding instructor in 2001 at Swainsboro Technical College. He also taught dual enrollment courses with David Emanuel Academy, Johnson County, Swainsboro and Jenkins County High Schools. In 2007, four of his students from Jenkins County won first place in the FFA welding competition; one student even made a perfect score on the welding test.

"I am teaching my students not only job skills, but life skills," says Michael. "Working with a young person and seeing everything come together and hearing them say, 'I can do this,' is my favorite part of the job." ■

Chapter 10

Gas Metal Arc Welding Equipment, Setup, and Operation

OBJECTIVES

After completing this chapter, the student should be able to

- list the various terms used to describe gas metal arc welding.
- discuss the various methods of metal transfer including the axial spray metal transfer process, globular transfer, pulsed-arc metal transfer, buried-arc transfer, and short-circuiting transfer GMAW-S.
- list shielding gases used for short-circuiting, spray, and pulsed-spray transfer.
- describe the more commonly used GMA welding filler metals.
- define *deposition efficiency*, and tell how a welder can control the deposition rate.
- define *voltage*, *electrical potential*, *amperage*, and *electrical current* as related to GMA welding.
- tell how wire-feed speed is determined and what it affects.
- discuss how the GMAW molten weld pool can be controlled by varying the shielding gas, power settings, weave pattern, travel speed, electrode extension, and gun angle.
- describe the backhand and forehand welding techniques.
- list and describe the basic GMAW equipment.
- explain how the arc spot weld produced by GMAW differs from electric resistance spot welding and the advantages of GMA spot welding.

KEY TERMS

axial spray metal transfer

buried-arc transfer

electrode extension (stickout)

globular transfer

pinch effect

pulsed-arc metal transfer

short-circuiting transfer

slope

synergic systems

transition current

INTRODUCTION

In the 1920s, a metal arc welding process using an unshielded wire was being used to assemble the rear axle housings for automobiles. The introduction of the shielded metal arc welding electrode rapidly replaced the bare wire. The shielded metal arc welding electrode made a much higher-quality weld. In 1948, the first inert gas metal arc welding (GMAW)

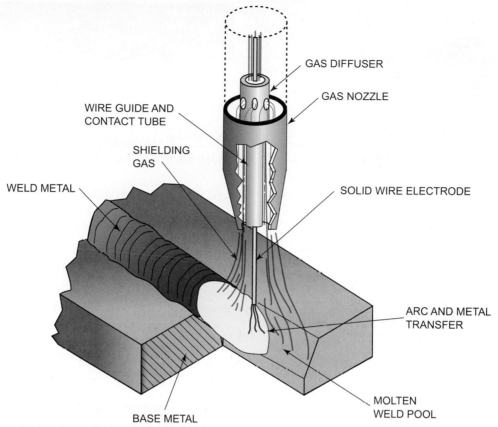

FIGURE 10-1 Gas shielded metal arc welding (GMAW). © Cengage Learning 2012

process, as it is known today, was developed and became commercially available, **Figure 10-1**. In the beginning, the GMAW process was used to weld aluminum using argon (Ar) gas for shielding. As a result, the process was known as MIG, which stands for metal inert gas welding. The later introduction of CO_2 and O_2 to the shielding gas has resulted in the American Welding Society's preferred term, *gas metal arc welding (GMAW)*. Although the American Welding Society uses the term *gas metal arc welding* to describe this process, it is known in the field by several other terms, such as

- *MIG,* which is short for *metal inert gas welding,*
- *MAG,* which is short for *metal active gas welding,* and
- *wire welding,* which describes the electrode used.

The GMAW process may be performed as semiautomatic (SA), machine (ME), or automatic (AU) welding, **Table 10-1**. The GMA welding process is commonly

performed as a semiautomatic process and is often mistakenly referred to as "semiautomatic welding." Equipment is available to perform most of the wire-feed processes semiautomatically, and the GMAW process can be fully automated. Robotic arc welding often uses GMAW because of the adaptability of the process in any position, **Figure 10-2**.

The rising use of all the various types of consumable wire welding processes has resulted in the increased sales of wire. At one time, wire made up less than 1% of the total market of filler metal. The total tonnage of filler metals used has grown and so has the percentage of wire. Today, wire exceeds 50% of the total tonnage of filler metals produced and used.

Much of the increase in the use of the wire welding processes is due to the increases in the quality of the welds produced. This improvement is due to an increased reliability of the wire-feed systems, improvements in the filler metal, smaller wire sizes, faster welding speed, higher

Function	Manual (MA) (Example: SMAW)	Semiautomatic (SA) (Example: GMAW)	Machine (ME) (Example: GMAW)	Automatic (AU) (Example: GMAW)
Maintain the arc	Welder	Machine	Machine	Machine
Feed the filler metal	Welder	Machine	Machine	Machine
Provide the joint travel	Welder	Welder	Machine	Machine
Provide the joint guidance	Welder	Welder	Welder	Machine

TABLE 10-1 Methods of Performing Welding Processes

(A)

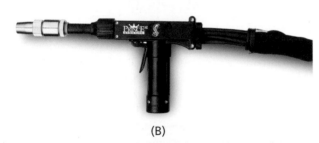

(B)

(C)

FIGURE 10-2 (A) Semiautomatic GMA welding setup. (B) Machine GMA welding gun with a friction drive, which provides both uniform nozzle-to-work distance and travel speed. (C) Automatic GMA welding.
Lincoln Electric Company

Electrode Diameter	Pounds per Hour		
Amperage	0.35	0.45	0.63
50	2.0	—	—
100	4.8	4.2	—
150	7.5	6.7	5.1
200	—	8.7	7.8
250	—	12.7	11.1
300	—	—	14.4

TABLE 10-2 GMA Weld Deposition Rates

weld deposition rates, less expensive shielding gases, and improved welding techniques. **Table 10-2** shows the typical weld deposition rates using the GMA welding process. The increased usage has led to a reduction in the cost of equipment. GMA welding equipment is now found even in small shops.

In this chapter, the semiautomatic GMA welding process will be covered. The skill required to set up and operate this process is basic to the understanding and operation of other wire-feed processes. The reaction of the weld to changes in voltage, amperage, feed speed, stickout, and gas is similar to that of most wire-feed processes.

WELD METAL TRANSFER METHODS

The GMAW process is unique in that there are several modes of transferring the filler metal from the wire to the weld. Each mode of metal transfer has its own unique characteristics. The mode of metal transfer is the mechanism by which the molten filler metal is transferred across the arc to the base metal. The modes of metal transfer are short-circuiting transfer (GMAW-S), axial spray transfer, globular transfer, and pulsed-arc transfer (GMAW-P). To change from one metal transfer mode to another, all you have to do is make the necessary changes in the voltage and amperage settings. In other words, it is the equipment setup and not the equipment or filler metal that determines the GMAW mode of metal transfer. Therefore, it is possible to make welds with each of the metal transfer modes by simply changing the voltage and amperage without even having to change the type of filler wire. In some cases it may be necessary to change the shielding gas; for example, CO_2 cannot be used to make welds in all three transfer modes, but $Ar+O_2$ can.

Selecting the mode of metal transfer used depends on the welding power source, the wire electrode size, type and thickness of material, type of shielding gas used, and the best welding position for the task.

Short-Circuiting Transfer GMAW-S

Low currents allow the liquid metal at the electrode tip to be transferred by direct contact with the molten weld pool. This process requires close interaction between the

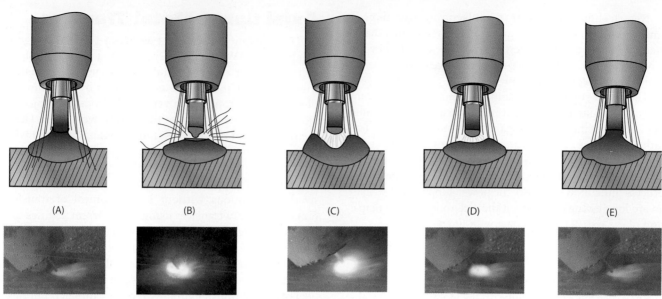

FIGURE 10-3 Schematic of short-circuiting transfer. Larry Jeffus

wire feeder and the power supply. This technique is called **short-circuiting transfer.**

The short-circuiting mode of transfer is the most common process used with GMA welding

- on thin or properly prepared thick sections of material,
- on a combination of thick to thin materials,
- with a wide range of electrode diameters, and
- with a wide range of shielding gases.

The 0.023, 0.030, 0.035, and 0.045 wire electrodes are the recommended diameters for the short-circuiting mode. Shielding gas used on carbon steel is carbon dioxide (CO_2) or a combination of 25% CO_2 and 75% argon (Ar). The amperage range may be as low as 35 for materials of 24 gauge or as high as 225 for materials up to 1/8 in. (3 mm) in thickness on square groove weld joints. Thicker base metals can be welded if the edges are beveled to accept a complete joint weld penetration.

The transfer mechanisms in this process are quite simple and straightforward, as shown schematically in **Figure 10-3**. To start, the wire is in direct contact with the molten weld pool, **Figure 10-3A**. Once the electrode touches the molten weld pool, the arc and its resistance are removed. Without the arc resistance, the welding amperage quickly rises as it begins to flow freely through the tip of the wire into the molten weld pool. The resistance to current flow is highest at the point where the electrode touches the molten weld pool. The resistance is high because both the electrode tip and weld pool are very hot. The higher the temperature, the higher the resistance to current flow. A combination of high current flow and high resistance causes a rapid rise in the temperature of the electrode tip.

As the current flow increases, the interface between the wire and molten weld pool is heated until it explodes into

a vapor (**Figure 10-3B**), establishing an arc. This small explosion produces sufficient force to depress the molten weld pool. A gap between the electrode tip and the molten weld pool (**Figure 10-3C**) immediately opens. With the resistance of the arc reestablished, the voltage increases as the current decreases.

The low current flow is insufficient to continue melting the electrode tip off as fast as it is being fed into the arc. As a result, the arc length rapidly decreases (**Figure 10-3D**) until the electrode tip contacts the molten weld pool (**Figure 10-3E**). The liquid formed at the wire tip during the arc-on interval is transferred by surface tension to the molten weld pool, and the cycle begins again with another short circuit.

If the system is properly tuned, the rate of short circuiting can be repeated from approximately 20 to 200 times per second, causing a characteristic buzzing sound. The spatter is low and the process easy to use. The low heat produced by GMAW-S makes the system easy to use in all positions on sheet metal, low carbon steel, low alloy steel, and stainless steel ranging in thickness from 25 gauge (0.02 in.; 0.5 mm) to 12 gauge (0.1 in.; 2.6 mm). The short-circuiting process does not produce enough heat to make quality welds in sections much thicker than 1/4 in. (6 mm) unless it is used for the root pass on a grooved weld or to fill gaps in joints. Although this technique is highly effective, lack-of-fusion defects can occur unless the process is perfectly tuned and the welder is highly skilled, especially on thicker metal.

Carbon dioxide works well with this short-circuiting process because it produces the forceful arc needed during the arc-on interval to displace the weld pool. Helium can be used as well. Pure argon is not as effective because its arc tends to be sluggish and not very fluid. However, a mixture of 25% carbon dioxide and 75% argon produces a less harsh arc and a flatter, more fluid, and desirable weld profile. Although more costly, this gas mixture is preferred.

New technology in wire manufacturing has allowed smaller wire diameters to be produced. These smaller diameters have become the preferred size even though they are more expensive. The short-circuiting process works better with a short electrode stickout.

The power supply is most critical. It must have a constant potential output and sufficient inductance to slow the time rate of current increase during the short-circuit interval. Too little inductance causes spatter due to high current surges. Too much inductance causes the system to become sluggish. The short-circuiting rate decreases enough to make the process difficult to use. Also, the power supply must sustain an arc long enough to premelt the electrode tip in anticipation of the transfer at recontact with the weld pool.

Globular Transfer

Globular transfer is generally used on thin materials and at a very low current range. In the globular transfer, the arc melts the end of the electrode, forming a molten ball of metal. When the ball of metal becomes so large, its surface tension cannot hold it onto the end of the wire. It falls across the arc, landing in the molten weld pool. Because there is little control over where the glob of metal lands and the weld pool tends to be a small landing target, this process is rarely used by itself. It is used more commonly in combination with pulsed-spray transfer. With this combination of processes, the molten weld pool is larger and the glob is more likely to land in the molten pool, **Figure 10-4.**

Axial Spray Metal Transfer

The freedom from spatter associated with the argon-shielded GMAW process results from a unique mode of metal transfer called **axial spray metal transfer, Figure 10-5.** This process is identified by the pointing of the wire tip from which very small drops are projected axially across the arc gap to the molten weld pool. There are hundreds of drops per second crossing from the wire to the base metal. These drops are propelled by arc forces at high velocity in the direction the wire is pointing. This projection of drops enables welding in the vertical and overhead positions without losing control of transfer. Some axial spray transfers using high current settings can produce a large molten weld pool that is very fluid, and the weld may be difficult to control in out-of-position welds. Because the drops are very small and directed at the molten weld pool, the process is spatter-free.

This spray transfer process requires three conditions: argon shielding (or argon-rich shielding gas mixtures), DCEP polarity, and a current level above a critical amount called the **transition current.** The shielding gas is usually a mixture of 95% to 98% argon and 2% to 5% oxygen. The added percentage of oxygen allows greater weld penetration. **Figure 10-6** illustrates how the rate of drops transferred changes in relationship to the welding current. At low currents, the drops are large and are transferred at rates below

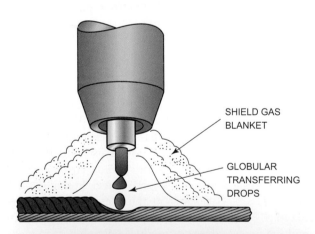

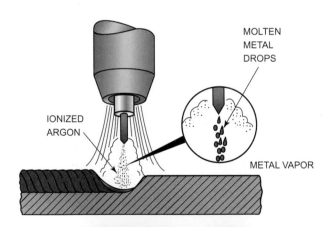

FIGURE 10-4 Globular metal transfer. Large drop is supported by arc forces. © Cengage Learning 2012

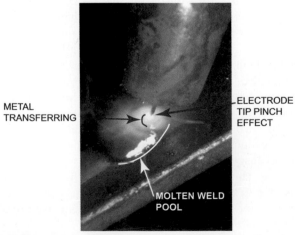

FIGURE 10-5 Axial spray metal transfer. Note the pinch effect of filler wire and the symmetrical metal transfer column. Larry Jeffus

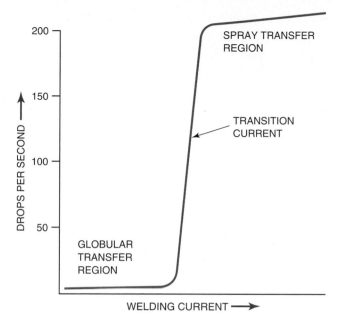

FIGURE 10-6 Desirable spray transfer shown schematically. © Cengage Learning 2012

10 per second. These drops move slowly, falling from the electrode tip as gravity pulls them down. They tend to bridge the gap between the electrode tip end and molten weld pool. This produces a momentary short circuit that throws off spatter. However, the mode of transfer changes very abruptly above the critical current, producing the desirable spray. This change in the rate of transfer as related to current is shown schematically in Figure 10-6.

The transition current depends on the alloy being welded. It also is proportional to the wire diameter, meaning that higher currents are needed with larger diameter wires. The need for high current density imposes some restrictions on the process. The high current hinders welding sheet metal because the high heat cuts through sheet metal. **Table 10-3** lists the welding parameters for a variety of gases, wire sizes, and metal thicknesses for GMA welding of mild steel.

The spray transfer process generally uses larger diameter wire electrodes with higher amperage ranges for greater production. The higher the amperage range, the

faster the weld bead progresses and the joint filled. The axial spray transfer process is very hot and virtually free of any spatter. The sound produced by axial spray transfer is a quiet, hissing sound, unlike the short-circuit process, which makes a raspy, frying sound. The high welding current produces a great deal of UV light, so you need extra burn protection for your eyes, hands, and arms.

//// **CAUTION** \\\\

The heat produced during axial spray welding using large diameter wire or high current may be intense enough to cause the filter lens in a welding helmet to shatter. Be sure the helmet is equipped with a clear plastic lens on the inside of the filter lens. Avoid getting your face too close to the intense heat.

Pulsed-Arc Metal Transfer

The current produced by the **pulsed-arc metal transfer** (GMAW-P) mode is a dual pulsed current. One pulse of high current is for the axial spray transfer mode, and the other lower pulse of current should not transfer any weld metal, **Figure 10-7**. This pulsing of the current levels permits the use of the high amperage transfer mode and the low amperage so that the total heat input to the weld is lower. The high current of the spray mode produces good penetration and fusion, and the low current allows the weld pool to cool and contract slightly, so it is easier to control. The ease of controlling the weld is a major advantage of GMAW-P. Experienced GMA welders can pick up this variation quickly, and new welders often can develop their skills faster on GMAW-P than on traditional GMA welding.

Pulsed-arc welding systems were developed in the mid-1960s, but this technology did not receive much attention until solid state electronics were developed to handle the high power required of welding power supplies. Solid state electronics provided a better, simpler, and more economical way to control the pulsing process. The newest generation of pulsed-arc systems interlocks the power supply and wire feeder so that the proper settings of the wire-feed end power supply are obtained for any given job by adjusting a single

Mild Steel	Wire-Feed Speed, in./min		Voltage, V				
Base-Material Thickness, in.	0.035-in.	0.045-in.	CO_2	75 Ar-25 CO_2	Ar	98 Ar-2 O_2	Current A
0.036	105–115	—	18	16	—	—	50–60
0.048	140–160	70	19	17	—	—	70–80
0.060	180–220	90–110	20	17.7	—	—	90–110
0.075	240–260	120–130	20.7	18	20	—	120–130
1/8	280–300	140–150	21.5	18.5	20.5	—	140–150
3/16	320–340	160–175	22	19	21.5	23.5	160–170
1/4	360–380	185–195	22.7	19.5	22.5	24.5	180–190
5/16	400–420	210–220	23.5	20.5	23.5	25	200–210
3/8	420–520	220–270	25	22	25	26.5	220–250
1/2 and up	—	375	28	26	29	31	300

TABLE 10-3 GMA Welding Parameters for Mild Steel

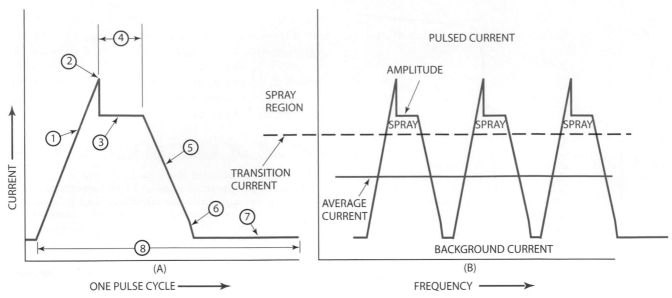

FIGURE 10-7 Mechanism of pulsed-arc spray transfer at a low average current. © Cengage Learning 2012

knob. Such systems have been termed **synergic systems.** Before pulsed-arc welding machines were programmable, welders had to adjust many of the pulse's components separately. Failure to get the correct setup resulted in a number of welding problems. The newer pulsed-arc welding systems are more complex, but because of the interconnectivity of the relationship between the wire-feeder and power supply settings, the computer programming makes the operation much easier. Some welders come preprogrammed with the most common settings for metals such as mild steel, stainless steel, and aluminum. Additional programs for welding on specialty metals such as titanium, Inconel, copper, and others can be added from the equipment manufacturer's library of programs or be created in-house to meet the shop's specific requirements.

GMAW-P can be used to make welds on a wide variety of metals. Most metals can be welded in thicknesses ranging from thin-gauge sheet metal to thick plate. Steel welds can be made using any of the common diameter filler wires with a 1% oxygen 99% argon shielding gas mixture. Most other metals are welded with 100% argon. The high penetration and great fusion characteristics of the pulsed-arc process work great on aluminum. The process can overcome the high conductivity of aluminum to make thick section welds without preheating.

When the pulsed-arc equipment is equipped with a hot start function, welds can be started without the need for a starting tab because the initial current can be high enough to ensure complete joint penetration and fusion, **Figure 10-8**.

Pulsed-Arc Metal Transfer Current Cycle

The graph in Figure 10-7 shows the pulse of welding power and identifies the eight components:

1 Ramp up
2 Overshoot
3 High pulse current
4 High pulse time
5 Ramp down
6 Step-off current
7 Background current
8 Pulse width

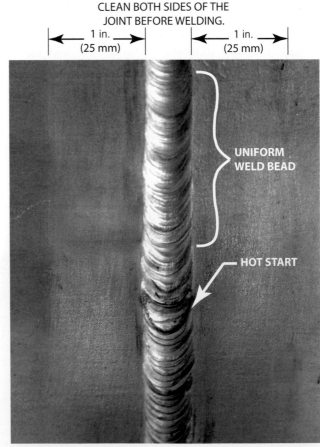

FIGURE 10-8 The hot start helps to make restarting the weld uniform and defect-free. © Cengage Learning 2012

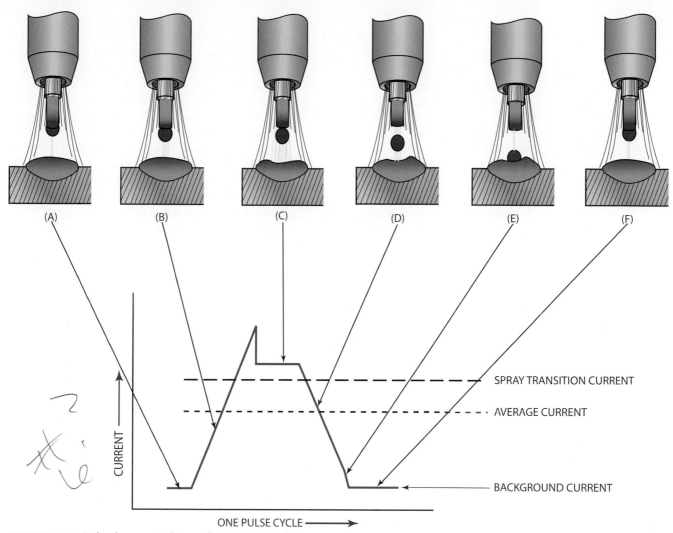

(A)   (B)   (C)   (D)   (E)   (F)

SPRAY TRANSITION CURRENT

AVERAGE CURRENT

BACKGROUND CURRENT

ONE PULSE CYCLE ➞

FIGURE 10-9 Pulsed-arc metal transfer current cycle. © Cengage Learning 2012

Figure 10-9 shows the relationship among welding current, metal transfer, and time for the pulses of a typical pulsed-arc transfer.

- **Ramp up** refers to how the electric current in a transformer takes a few milliseconds to build up the magnetic field to full strength once the coil is energized. The rate at which the current builds up the magnetic field and the resulting welding current is called slope and is measured in amps/milliseconds. The ramp up or slope can be adjusted on some welding machines. Faster rates are associated with a stiffer, louder, more forceful arc, and lower rates are associated with a softer-sounding arc. As the power ramps up, it starts forming the molten droplet of metal on the electrode tip.

- **Overshoot** refers to the surging electrical current passing through the transformer like a wave on water. The overshoot wave of current has a peak that is higher than the actual welding current. This small peak of power helps start the droplet formation and it adds to the pinch effect required for metal transfer. The higher the percentage the

power is overshot, the more rigid the arc and the less likely it is to be deflected by arc blow.

- **High pulse current** is the peak current that flows across the arc during the high current pulse. It has to be high enough to be above the spray transition current level so that the metal is not transferred in the globular mode. This is the part of the pulse that does the real work of welding. The higher the peak current is, the greater the weld penetration.

- **High current time** is the length of time the peak current is on. It ranges from less than a millisecond to several milliseconds. The longer the high current pulse time, the more likely that multiple droplets of metal will be transferred across the arc. Longer high pulse times will increase penetration and can allow the transfer to become a true spray transfer. If this transformation to spray occurs, it may make the molten weld pool too large to be controlled in out-of-position welds. The droplets are typically 1 to 1.2 times larger than the diameter of the electrode. As with conventional spray arc, the drops are propelled across the arc gap, allowing metal transfer in all positions.

■ **Ramp down** is a function of the time it takes the magnetic field in the transformer to collapse once the coil is de-energized. The magnetic field does not collapse at a uniform rate, but it tends to decay more rapidly at first and slows as it nears the step-off current point. Some manufacturers may depict this line as a concave curve and not linear as in Figure 10-7; however, for all practical purposes, whether it is drawn as a straight linear, logarithmic, or exponential representation, it's the actual de-energizing time in milliseconds that affects the weld. The longer the ramp down time, the more fluid the droplet and molten weld pool will be. Ramp down times range from less than a millisecond to several milliseconds.

> NOTE: The time it takes for the magnetic field to build and collapse in a transformer is determined by the size of the iron core and primary and secondary coils. Without inverter technology, which uses a much smaller transformer, the ramp up and ramp down times would be too long for pulsed-arc transfer to work efficiently. The transformer is around 7 lb (3 kg) in an inverter welder and around 100 lb (45 kg) in a conventional transformer welder.

■ **Step-off current** is the point where the final residual magnetic field collapses to the background current. Changes in the step-off current have been associated with arc stability for some higher electrical resistant filler metals such as stainless steel and nickel alloys.

■ **Background current** is the lower pulse current that keeps the arc alive between high pulse currents. The higher the background current is, the greater the weld penetration.

■ **Pulse width** is the time in milliseconds from the beginning of a single pulse cycle to the beginning of the next cycle. The number of pulses that occur in a second is known as the pulse frequency. Pulse frequency can be adjusted from a few pulses a second to over 1000 pulses a second.

Because of the complexity and interconnectivity of the pulsed arc's pulse configuration, most of the variables are preset by the manufacturer or are controlled by the welding machine's computer program.

Advantages of Pulsed-Arc Metal Transfer

■ **Lower average currents**—GMAW-P has a much lower average welding current as compared to other GMA welding processes for the same metal type, thickness, and joint design. Despite the lower average welding current, GMAW-P has excellent joint penetration. The average welding current is primarily affected by the weld cycle time, high pulse current, and background current settings. Even though there is less total current, the high pulse power level will let you make deep penetrating welds on thick sections of aluminum in many applications without preheating the part.

The lower total current and heat input provided by the pulsed-arc transfer permits its use on thinner-gauge base metals. Traditionally, thin-gauge metals were welded with smaller diameter electrodes using the short-circuiting metal transfer mode. Even with the welding machine set perfectly, welders had to use high travel speeds and/or very aggressive weave patterns to avoid burnthrough. Often the high travel rates resulted in areas along the joint where the joint was not fused, and slower welding speeds with aggressive weave patterns left the joint overwelded. However, the pulse metal transfer method can use larger diameter filler wires to make high-quality welds on steel as thin as 0.035 in. (0.9 mm).

> NOTE: A cost savings for using pulsed metal transfer on thin-gauge metals is that larger diameter filler wire can be used, and the larger diameter GMA welding wires are less expensive per pound than the thinner wires.

■ **All position**—Because of the lower average welding current, GMAW-P can be used in any position.

■ **Less distortion**—The lower the average current the weld has, the less heat input to the weld, which reduces distortion.

■ **Poor fit-up**—The molten weld pool is smaller and less fluid, so it is easier to control even when making welds with wider-than-average root gaps. Making high-quality welds without excessive burnthrough while bridging the gap of a less-than-ideal joint fit-up will save a tremendous amount of time to refit and possibly prevent scrapping parts.

■ **Reduced spatter**—There is little to no spatter with GMAW-P. The lack of any significant spatter will result in less postwelding cleanup of the parts, longer-lasting gun parts, and a reduction of spatter buildup on jigs and fixtures.

■ **High transfer efficiency**—The reduction in spatter means that more of the filler metal is transferred to the weld. In many applications, 98% transfer efficiencies can be obtained, which reduces wasted filler metal.

■ **High travel speeds**—On some thin sections, travel speeds greater than 50 in./min (1.27 m/min) can be obtained.

■ **Lower fume production**—There is little fume produced during welding, which means less shop air pollution.

■ **High-quality welds**—The weld deposits have little chance of hydrogen entrapment, better fusion, improved impact test values, and better weld appearance than any other GMA welding metal transfer method.

■ **Arc blow resistant**—The arc is very stable and resistant to arc blow. Part of the arc blow resistance comes from the pulsing current and some from the shorter arc length. The arc length must be just long enough so that the end of the electrode or metal being transferred across the arc does not short out. Once the arc length is set, slight variations in gun height will not affect the arc length. This is particularly beneficial when welding through an inside corner where gun access and welder visibility might be limited.

- **Welder appeal**—With only one or two knobs to adjust once the welding program has been selected, all the welder has to do is weld. On some pulsed-arc welding machines, once the program is selected for the filler metal size, metal type, and thickness, the only thing the welder needs to adjust is the arc length and possibly the welding arc cone width. The welding cone width is a function of the welding current and the duration of the high pulse power and has a substantial effect on the average welding current. Some pulse welding machines have a separate adjustment for arc cone width.

Welders find the welding environment with little spatter, fumes, and reduced heat more pleasant to work in than most other welding processes.

Disadvantages of Pulsed-Arc Metal Transfer A few issues should be considered before the purchase of a GMAW-P system, including the following:

- **Equipment cost**—The initial capital investment for pulsed-arc equipment is significantly higher than for traditional GMA welding equipment. Unless the shop has enough work to keep the pulsed-arc welding station busy, then buying it may not be a good investment.
- **System complexity**—There are a much larger number of system components, and they are much more complex as compared to traditional GMAW equipment. More complex systems can take more time, expertise, and money to keep them working properly.

Shielding Gases for Spray or Pulsed-Spray Transfer

One hundred percent argon shielding is used on all metals except ferrous metals such as mild steel and stainless steel. Argon shielding gas or mixtures of argon and other gases are required for the axial spray transfer process. Common argon shielding gas mixtures used for steel contain helium and/or oxygen.

Helium/argon mixtures may contain as much as 80% helium. The helium is added to the argon to increase the power in the arc without affecting the desirable qualities of the spray mode. With more helium, the transfer becomes progressively more globular, forcing the use of a different welding mode, to be described later. Since helium and argon gases are inert, they do not react chemically with any metals.

The cathodic cleaning action associated with argon at DCEP (DCRP) is also very important for fabricating metals such as aluminum. Aluminum forms a heavy undesirable surface oxide when heated and exposed to air.

This same cleaning action causes problems with steels. Iron oxide in and on the steel surface can be a good emitter of electrons that may attract the arc. But these oxides are not uniformly distributed, resulting in very irregular arc movement and, in turn, irregular weld deposits. This problem was solved by adding small amounts of oxygen to the argon. The reaction produces a uniform film of iron oxide on the weld pool and provides a stable site for the arc. This discovery enabled uniform welds in ferrous alloys and expanded the use of GMAW to welding those materials.

The amount of oxygen needed to stabilize arcs in steel varies with the alloy. Generally, 2% is sufficient for carbon and low alloy steels. In the case of stainless steels, about 0.5% should prevent a refractory scale of chromium oxide. Carbon dioxide can substitute for oxygen. More than 2% is needed, however, and 8% appears to be optimum for low alloy steels. In many applications, carbon dioxide is the preferred addition because the weld bead has a better contour and the arc appears to be more stable.

Buried-Arc Transfer

The arc in carbon dioxide is very forceful. Because of this, the wire tip can be driven below the surface of the molten weld pool. With the shorter arcs, the drop size is small, and any spatter produced as the result of short circuits is trapped in the cavity produced by the arc—hence the name **buried-arc transfer**, Figure 10-10. The resultant welds tend to be more highly crowned than those produced with open arcs, but they are relatively free of spatter and offer a decided advantage of welding speed. These characteristics make the buried-arc process useful for high-speed mechanized welding of thin sections, such as that found in compressor domes for hermetic air-conditioning and refrigeration equipment or for automotive components.

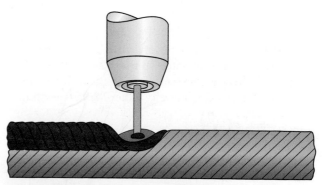

FIGURE 10-10 Buried-arc transfer. Wire tip is with the weld crater. Spatter is trapped. © Cengage Learning 2012

Base Metal Type	AWS Filler Metal Specification
Aluminum and aluminum alloys	A5.10
Copper and copper alloys	A5.6
Magnesium alloys	A5.19
Nickel and nickel alloys	A5.14
Stainless steel (austenitic)	A5.9
Steel (carbon)	A5.18
Titanium and titanium alloys	A5.16

TABLE 10-4 AWS Filler Metal Specifications for Different Base Metals

Base Metal	Electrode Diameter		Amperage
	Inch	Millimeter	Range
Carbon steel	0.023	0.6	35–190
	0.030	0.8	40–220
	0.035	0.9	60–280
	0.045	1.2	125–380
	1/16	1.6	275–450
Stainless steel	0.023	0.6	40–150
	0.030	0.8	60–160
	0.035	0.9	70–210
	0.045	1.2	140–310
	1/16	1.6	280–450

TABLE 10-5 Filler Metal Diameters and Amperage Ranges

GMAW Filler Metal Specifications

GMA welding filler metals are available for a variety of metals, **Table 10-4**. The most frequently used filler metals are AWS specification A5.18 for carbon steel and AWS specification A5.9 for stainless steel. Wire electrodes are produced in 0.023, 0.030, 0.035, 0.045, and 0.062 diameters. Other larger diameters are available for production work and can include wire diameter sizes such as 1/16, 5/64, and 7/64 in. **Table 10-5** lists the most common sizes and the amperage ranges for these electrodes. The amperage will vary depending on the method of metal transfer, type of shielding gas, and base metal thickness. Some steel wire electrodes have a thin copper coating. This coating provides some protection to the electrode from rusting and improves the electrical contact between the wire electrode and the contact tube. They may look like copper wire because of the very thin copper cladding. The amount of copper is so small that it either burns off or is diluted into the weld pool with no significant effect to the weld bead.

For more information on each of the GMA welding electrodes, refer to Chapter 27, "Filler Metal Selection."

Wire Melting and Deposition Rates

The wire melting rates, deposition rates, and wire-feed speeds of the consumable wire welding processes are affected by the same variables. Before discussing them,

however, these terms need to be defined. The wire melting rate, measured in inches per minute (in./min) or pounds per hour (lb/hr), is the rate at which the arc consumes the wire. The deposition rate, the measure of weld metal deposited, is nearly always less than the melting rate because not all of the wire is converted to weld metal. Some is lost as slag, spatter, or fume. The amount of weld metal deposited in ratio to the wire used is called the *deposition efficiency*.

Deposition efficiencies depend on the process, on the gas used, and even on how the welder sets welding conditions. With efficiencies of approximately 98%, solid wires with argon shields are best. Some of the self-shielded cored wires are poorest, with efficiencies as low as 80%.

Welders can control the deposition rate by changing the current, electrode extension, and diameter of the wire. To obtain higher melting rates, they can increase the current or wire extension or decrease the wire diameter. Knowing the precise constants is unimportant. However, it is important to know that current greatly affects melting rate and that the extension must be controlled if results are to be reproducible.

WELDING POWER SUPPLIES

To better understand terminology used to describe the different welding power supplies, you need to know the following electrical terms:

■ Voltage, or volts (V), is a measurement of electrical pressure, in the same way that pounds per square inch is a measurement of water pressure.

■ Electrical potential means the same thing as voltage and is usually expressed by using the term *potential* (P). The terms *voltage, volts,* and *potential* can all be interchanged when referring to electrical pressure.

■ Amperage, or amps (A), is the measurement of the total number of electrons flowing, in the same way that gallons are a measurement of the amount of water flowing.

■ Electrical current means the same thing as amperage and is usually expressed by using the term *current* (C). The terms *amperage, amps,* and *current* can all be interchanged when referring to electrical flow.

GMAW power supplies are the constant-voltage, constant-potential (CV, CP)-type machines, unlike SMAW power supplies, which are the constant-current (CC)-type machines and are sometimes called drooping arc voltage (DAV). It is impossible to make acceptable welds using the wrong type of power supply. Constant-voltage power supplies are available as transformer-rectifiers or as motor-generators, **Figure 10-11**. Some newer machines use electronics, enabling them to supply both types of power at the flip of a switch.

The relationships between current and voltage with different combinations of arc length or wire-feed speeds are called volt-ampere characteristics. The volt-ampere characteristics of arcs in argon with constant arc lengths or constant wire-feed speeds are shown in **Figure 10-12**. To maintain a constant arc length while increasing current, it is necessary to increase voltage. For example, with a 1/8-in. (3-mm) arc length, increasing current from 150 to 300 amperes requires a voltage increase from about 26 to 31 volts. The current increase illustrated here results from increasing the wire-feed speed from 200 in. to 500 in. per minute (5 m/min to 13 m/min).

Speed of the Wire Electrode

The wire-feed speed is generally recommended by the electrode manufacturer and is selected in inches per minute (ipm), or how fast the wire exits the contact tube. The welder uses a wire speed control dial on the wire-feed

FIGURE 10-11 Transformer-rectifier welding power supply. ESAB Welding & Cutting Products

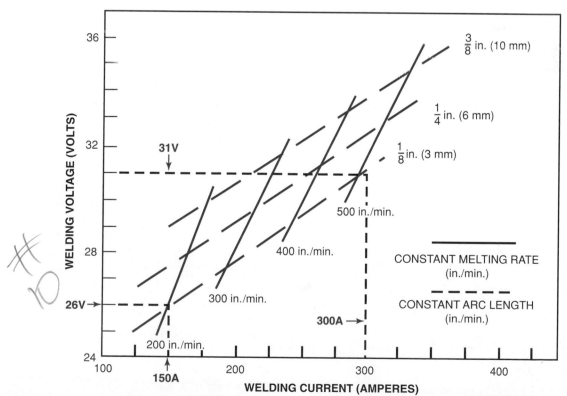

FIGURE 10-12 The arc length and arc voltage are affected by the welding current and wire-feed speed (0.045-in. [1.43-mm] wire; 1-in. [25-mm] electrode extension). © Cengage Learning 2012

Wire-Feed Speed* in./min (m/min)	Wire Diameter Amperages			
	0.030 in. (0.8 mm)	0.035 in. (0.9 mm)	0.045 in. (1.2 mm)	0.062 in. (1.6 mm)
100 (2.5)	40	65	120	190
200 (5.0)	80	120	200	330
300 (7.6)	130	170	260	425
400 (10.2)	160	210	320	490
500 (12.7)	180	245	365	—
600 (15.2)	200	265	400	—
700 (17.8)	215	280	430	—

*To check feed speed, run out wire for 1 minute and then measure its length.

TABLE 10-6 Typical Amperages for Carbon Steel

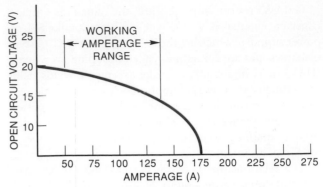

FIGURE 10-13 Constant potential welder slope.
© Cengage Learning 2012

unit to control ipm. It can be advanced or slowed to control the burn-off rate, or how fast the electrode transfers into the weld pool, to meet the welder's skill in controlling the weld pool, **Table 10-6.**

To accurately measure wire-feed ipm, snip off the wire at the contact tube. Squeeze the trigger for 6 seconds; release and snip off the wire electrode. Measure the number of inches of wire that was fed out in the 15 seconds. Now using basic shop math, multiply its total length in inches by 10. The result is how many inches of wire were fed per minute.

Power Supplies for Short-Circuiting Transfer

Although the GMA power source is said to have a constant potential (CP), it is not perfectly constant. The graph in **Figure 10-13** shows that there is a slight decrease in voltage as the amperage increases within the working range. The rate of decrease is known as **slope.** It is expressed as the voltage decrease per 100-ampere increase—for example, 10 V/100 A. For short-circuiting welding, some are equipped to allow changes in the slope by steps or continuous adjustment.

The slope, which is called the *volt-ampere curve,* is often drawn as a straight line because it is fairly straight within the working range of the machine. Whether it is drawn as a curve or a straight line, the slope can be found

by finding two points. The first point is the set voltage as read from the voltmeter when the gun switch is activated but no welding is being done. This is referred to as the *open circuit voltage.* The second point is the voltage and amperage as read during a weld. The voltage control is not adjusted during the test but the amperage can vary. The slope is the voltage difference between the first and second readings. The difference can be found by subtracting the second voltage from the first voltage. Therefore, for settings over 100 amperes, it is easier to calculate the slope by adjusting the wire feed so that you are welding with 100 amperes, 200 amperes, 300 amperes, and so on. In other words, the voltage difference can be simply divided by 1 for 100 amperes, 2 for 200 amperes, and so forth.

The machine slope is affected by circuit resistance. Circuit resistance may result from a number of factors, including poor connections, long leads, or a dirty contact tube. A higher resistance means a steeper slope. In short-circuiting machines, increasing the inductance increases the slope. This increase slows the current's rate of change during short circuiting and the arcing intervals, **Figure 10-14.** Therefore, slope and inductance become synonymous in this discussion. As the slope increases, both the short-circuit current and **pinch effect** are reduced. A flat slope has both an increased short-circuit current and a greater pinch effect.

The machine slope affects the short-circuiting metal transfer mode more than it does the other modes. Too much current and pinch effect from a flat slope cause a violent short and arc restart cycle, which results in

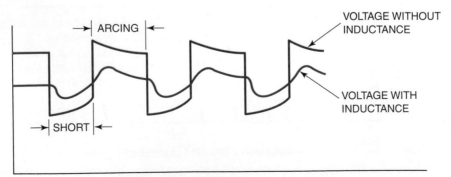

FIGURE 10-14 Voltage pattern with and without inductance. © Cengage Learning 2012

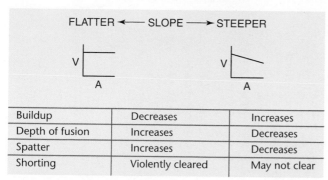

	FLATTER ← SLOPE → STEEPER	
Buildup	Decreases	Increases
Depth of fusion	Increases	Decreases
Spatter	Increases	Decreases
Shorting	Violently cleared	May not clear

TABLE 10-7 Effect of Slope

increased spatter. Too little current and pinch effect from a steep slope result in the short circuit not being cleared as the wire freezes in the molten pool and piles up on the work, **Table 10-7.**

The slope should be adjusted so that a proper spatter-free metal transfer occurs. On machines that have adjustable slopes, this is easily set. Experiment 11-5 describes a method of adjusting the circuit resistance to change the slope on machines that have a fixed slope. This is done by varying the contact tube-to-work distance. The GMA filler wire is much too small to carry the welding current and heats up due to its resistance to the current flow. The greater the tube-to-work distance, the greater the circuit resistance and the steeper the slope. By increasing or decreasing this distance, a proper slope can be obtained so that the short circuiting is smoother with less spatter.

MOLTEN WELD POOL CONTROL

The GMAW molten weld pool can be controlled by varying the following factors: shielding gas, power settings, weave pattern, travel speed, electrode extension, and gun angle.

Shielding Gas The shielding gas selected for a weld has a definite effect on the weld produced. The properties that can be affected include the method of metal transfer, welding speed, weld contour, arc cleaning effect, and fluidity of the molten weld pool.

In addition to the effects on the weld itself, the metal to be welded must be considered in selecting a shielding gas. Some metals must be welded with an inert gas such as argon or helium or mixtures of argon and helium. Other metals weld more favorably with reactive gases such as carbon dioxide or with mixtures of inert gases and reactive

gases such as argon and oxygen or argon and carbon dioxide, **Table 10-8.** The most commonly used shielding gases are 75% argon + 25% CO_2, argon + 1% to 5% oxygen, and carbon dioxide, **Figure 10-15.**

■ **Argon:** The atomic symbol for argon is *Ar*, and it is an inert gas. *Inert gases* do not react with any other substance and are insoluble in molten metal. One hundred percent argon is used on nonferrous metals such as aluminum, copper, magnesium, nickel, and their alloys; but 100% argon is not normally used for making welds on ferrous metals.

Because argon is denser than air, it effectively shields welds by pushing the lighter air away. Argon is relatively easy to ionize. Easily ionized gases can carry long arcs at lower voltages. This makes it less sensitive to changes in arc length.

Argon gas is naturally found in all air and is collected in air separation plants. There are two methods of separating air to extract nitrogen, oxygen, and argon. In the cryogenic process, air is super-cooled to temperatures that cause it to liquefy and the gases are separated. In the noncryogenic process, molecular sieves (strainers with very small holes) separate the various gases much like using a screen to separate sand from gravel.

■ **Argon gas blends:** Oxygen, carbon dioxide, helium, and nitrogen can be blended with argon to change argon's welding characteristics. Adding reactive gases (oxidizing), such as oxygen or carbon dioxide, to argon tends to stabilize the arc, promote favorable metal transfer, and minimize spatter. As a result, the penetration pattern is improved, and undercutting is reduced or eliminated. Adding helium or nitrogen gases (nonreactive or inert) increases the arc heat for deeper penetration.

The amount of the reactive gases, oxygen or carbon dioxide, required to produce the desired effects is quite small. As little as a half percent change in the amount of oxygen will produce a noticeable effect on the weld. Most of the time blends containing 1% to 5% of oxygen are used. Carbon dioxide may be added to argon in the range of 20% to 30%. Blends of argon with less than 10% carbon dioxide may not have enough arc voltage to give the desired results. The most commonly used argon CO_2 blend is 25% CO_2.

When using oxidizing shielding gases with oxygen or carbon dioxide added, a suitable filler wire containing deoxidizers should be used to prevent

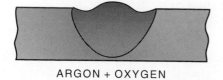

ARGON + OXYGEN

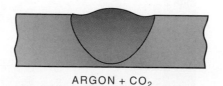

ARGON + CO_2

CARBON DIOXIDE

FIGURE 10-15 Effect of shielding gas on weld bead shape. © Cengage Learning 2012

GMAW Metals, Shielding Gases, and Gas Blends (A)

| Metals | Gases | | Blends of Two Gases | | | | | | | | | Blends of Three Gases | | |
	Argon (Ar)	CO₂	Ar + 1% O₂	Ar + 2% O₂	Ar + 5% O₂	Ar + 5% CO₂	Ar + 10% CO₂	Ar + 25% CO₂	Ar + 25% He	Ar + 50% He	Ar + 75% He	Ar + CO₂ + O₂	Ar + CO₂ + Nitrogen	Ar + CO₂ + Helium
Aluminum	X								X	X	X			X
Copper alloys	X	X							X	X	X			X
Stainless steel		X	X	X	X			X				X		
Steel			X	X	X	X	X	X				X	X	
Magnesium	X								X	X	X			
Nickel alloys	X								X	X	X			

GMAW Shielding Gases, Gas Blends, Metals, and Welding Process (B)

Gases/Blend	Gas Reaction	Application	Remarks
Argon (Ar)	Inert	Nonferrous metals	Provides spray transfer
Helium (He)	Inert	Aluminum and magnesium	Very hot arc for welds on thick sections, usually used in gas blends to increase the arc temperature and penetration
Ar + 1% O₂	Oxidizing	Stainless steel	Oxygen provides arc stability
Ar + 2% O₂	Oxidizing	Stainless steel	Oxygen provides arc stability
Ar + 5% O₂	Oxidizing	Mild and low alloy steel	Provides spray transfer
Ar + 5% CO₂	Oxidizing	Low alloy steel	Pulse spray and short-circuit transfer in out-of-position welds
Ar + 10% CO₂	Oxidizing	Low alloy steel	Same as above with a wider, more fluid weld pool
Ar + 25% CO₂	Oxidizing	Mild, low alloy steels and stainless steel	Smooth weld surface, reduces penetration with short-circuiting transfer
CO₂	Oxidizing	Mild, low alloy steels and stainless steel	Least expensive gas, deep penetration with short-circuiting or globular transfer
Nitrogen	Almost inert	Copper and copper alloys	Has high heat input with globular transfer
Ar + 25% He	Inert	Al, Mg, copper, nickel, and their alloys	Higher heat input than Ar, for thicker metal
Ar + 50% He	Inert	Al, Mg, copper, nickel, and their alloys	Higher heat in arc use on heavier thickness with spray transfer
Ar + 75% He	Inert	Copper, nickel, and their alloys	Highest heat input
Ar + CO₂ + O₂	Oxidizing	Low alloy steel and some stainless steels	All metal transfer for automatic and robotic applications
Ar + CO₂ + N	Almost inert	Stainless steel	All metal transfer, excellent for thin gauge material
He + 7.5% Ar + 2.5% CO₂	Almost inert	Stainless steel and some low alloy steels	Excellent toughness; excellent arc stability, wetting characteristics, and bead contour; little spatter with short-circuiting transfer

TABLE 10-8 (A) Metals Compared to GMAW Shielding Gases and Gas Blends; (B) GMAW Shielding Gases and Gas Blends Compared with Metals

porosity in the weld. The presence of oxygen in the shielding gas can also cause some loss of certain alloying elements, such as chromium, vanadium, aluminum, titanium, manganese, and silicon.

■ **Helium:** The atomic symbol for helium is *He,* and it is an inert gas that is a product of the natural gas industry. It is removed from natural gas as the gas undergoes separation (fractionation) for purification or refinement.

   Helium is lighter than air, thus its flow rates must be about twice as high as argon's for acceptable stiffness in the gas stream to be able to push air away from the weld. Proper protection is difficult in drafts unless high flow rates are used. It requires a higher voltage to ionize, which produces a much hotter arc. There is a noticeable increase in both the heat and temperature of a helium arc. This hotter arc makes it easier to make welds on thick sections of aluminum and magnesium.

   Small quantities of helium are blended with other heavier gases. These blends take advantage of the heat produced by the lightweight helium and weld coverage by the other heavier gas. Thus, each gas is contributing its primary advantage to the blended gas.

■ **Carbon dioxide:** Carbon dioxide is a compound made up of one carbon atom (C) and two oxygen atoms (O_2), and its molecular formula is CO_2. One hundred percent carbon dioxide is widely used as a shielding gas for GMA welding of steels. It allows higher welding speed, better penetration, good mechanical properties, and costs less than the inert gases. The chief drawback in the use of carbon dioxide is the less-steady arc characteristics and a considerable increase in weld spatter. The spatter can be kept to a minimum by maintaining a very short, uniform arc length. CO_2 can produce sound welds provided a filler wire having the proper deoxidizing additives is used.

■ **Nitrogen:** The atomic symbol for nitrogen is *N.* It is not an inert gas but is relatively nonreactive to the molten weld pool. It is often used in blended gases to increase the arc's heat and temperature. One hundred percent nitrogen can be used to weld copper and copper alloys.

THINK GREEN
Renewable Gases

Argon, nitrogen, and oxygen GMA shielding gases are refined from air in our atmosphere and are returned to the atmosphere in almost the exact condition as they leave the welding zone. Because argon is an inert gas, it is returned to the atmosphere in the exact same condition as it was. Nitrogen may pick up some oxygen atoms, and oxygen may pick up some metal oxides, but most of the shielding gas is unchanged.

Power Settings

As the power settings, voltage, and amperage are adjusted, the weld bead is affected. Making an acceptable weld requires a balancing of the voltage and amperage. If either or both are set too high or too low, the weld penetration can decrease. A GMA welding machine has no direct amperage settings. Instead, the amperage at the arc is adjusted by changing the wire-feed speed. As a result of the welding machine's maintaining a constant voltage when the wire-feed speed increases, more amperage flows across the arc. This higher amperage is required to melt the wire so that the same arc voltage can be maintained. The higher amperage is used to melt the filler wire and does not increase the penetration. In fact, the weld penetration may decrease significantly.

Increasing and decreasing the voltage changes the arc length but may not put more heat into the weld. Like changes in the amperage, these voltage changes may decrease weld penetration.

Weave Pattern

The GMA welding process is greatly affected by the location of the electrode tip and molten weld pool. During the short-circuiting process if the arc is directed to the base metal and outside the molten weld pool, the welding process may stop. Without the resistance of the hot molten metal, high amperage surges occur each time the electrode tip touches the base metal, resulting in a loud pop and a shower of sparks. It is something that occurs each time a new weld is started. So, when making the weave pattern, you must keep the arc and electrode tip directed into the molten weld pool. Other than the sensitivity to arc location, most of the SMAW weave pattern can be used for GMA welds.

Travel Speed

Because the location of the arc inside the molten weld pool is important, the welding travel speed cannot exceed the ability of the arc to melt the base metal. Too high a travel speed can result in overrunning of the weld pool and an uncontrollable arc. Fusion between the base metal and filler metal can completely stop if the travel rate is too fast. If the travel rate is too slow and the weld pool size increases excessively, it can also restrict fusion to the base plate.

Electrode Extension

The **electrode extension (stickout)** is the distance from the contact tube to the arc measured along the wire. Adjustments in this distance cause a change in the wire resistance and the resulting weld bead, **Figure 10-16**.

GMA welding currents are relatively high for the wire sizes, even for the low current values used in short-circuiting arc metal transfer, **Figure 10-17**. As the length of wire extending from the contact tube to the work increases, the voltage, too, should increase. Since this change is impossible with a constant-voltage power supply, the system compensates by reducing the current. In other

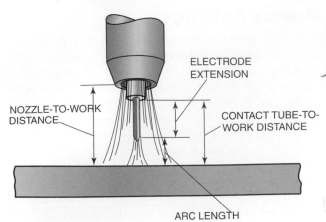

FIGURE 10-16 Electrode-to-work distances.

© Cengage Learning 2012

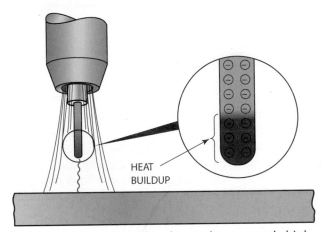

FIGURE 10-17 Heat buildup due to the extremely high current for the small conductor (electrode).

© Cengage Learning 2012

words, by increasing the electrode extension and maintaining the same wire-feed speed, the current has to change to provide the same resistance drop. This situation leads to a reduction in weld heat, penetration, and fusion, and an increase in buildup. On the other hand, as the electrode extension distance is shortened, the weld heats up, penetrates more, and builds up less, **Figure 10-18.**

Experiment 11-3 explains the technique of using varying extension lengths to change the weld characteristics. Using this technique, a welder can make acceptable welds on metal ranging in thickness from 16 gauge to 1/4 in. (6 mm) or more without changing the machine settings. When using this technique, the nozzle-to-work distance should be kept the same so that enough shielding gas coverage is provided. Some nozzles can be extended to provide coverage. Others must be exchanged with the correct-length nozzle, **Figure 10-19.**

Gun Angle

The *gun angle, work angle,* and *travel angle* are names used to refer to the relation of the gun to the work surface. The gun angle can be used to control the weld pool. The electric arc produces an electrical force known as the arc force. The arc force can be used to counteract the gravitational pull that tends to make the liquid weld pool sag or run ahead of the arc. By manipulating the electrode travel angle for the flat and horizontal position of welding to a 20° to 90° angle from the vertical, the weld pool can be controlled. A 40° to 50° angle from the vertical plate is recommended for fillet welds.

Changes in this angle will affect the weld bead shape and penetration. Shallower angles are needed when welding thinner materials to prevent burnthrough. Steeper, perpendicular angles are used for thicker materials.

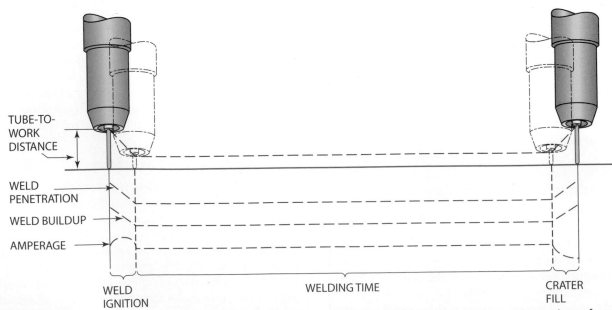

FIGURE 10-18 Using the changing tube-to-work distance to improve both the starting and stopping points of a weld.

© Cengage Learning 2012

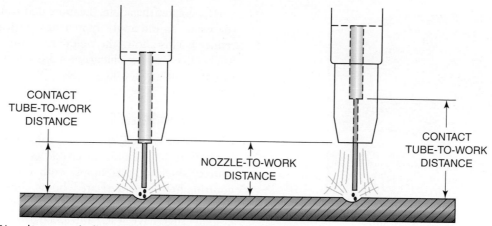

FIGURE 10-19 Nozzle-to-work distance can differ from the contact tube-to-work distance. © Cengage Learning 2012

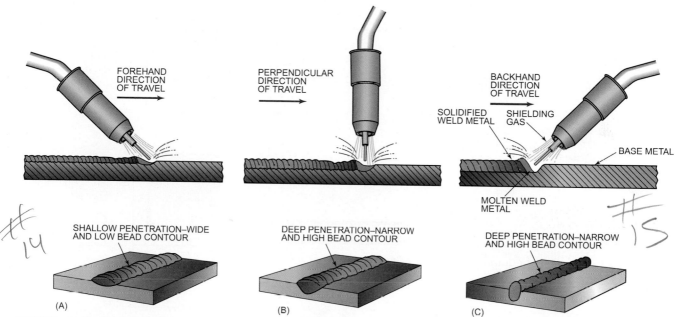

FIGURE 10-20 (A) Forehand welding or push angle, (B) perpendicular, and (C) backhand welding or drag angle.
© Cengage Learning 2012

Forehand/Perpendicular/Backhand Welding

Forehand, perpendicular, and *backhand* are the terms most often used to describe the gun angle as it relates to the work and the direction of travel. The forehand technique is sometimes referred to as pushing the weld bead, **Figure 10-20A**, and backhand may be referred to as pulling or dragging the weld bead, **Figure 10-20C**. The term *perpendicular* is used when the gun angle is at approximately 90° to the work surface, **Figure 10-20B**.

Forehand welding has good joint visibility and makes welds with less joint penetration; this technique works well on vertical up and overhead welds, **Figure 10-21**.

Perpendicular welding has a good balance between penetration and reinforcement and is used on automated welding.

Backhand welding has good bead visibility and makes welds with deeper joint penetration, **Figure 10-22**.

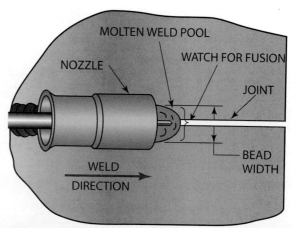

FIGURE 10-21 Forehand weld joint visibility.
© Cengage Learning 2012

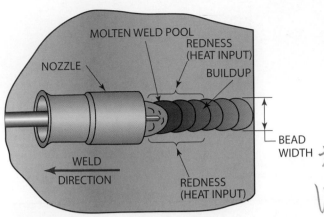

FIGURE 10-22 Backhand weld visibility.
© Cengage Learning 2012

The greater the angle, the more defined is the effect on the weld. As the angle approaches vertical, the effect is reduced. This allows the welder to change the weld bead as effectively as the changes resulting from adjusting the machine current settings.

EQUIPMENT

The basic GMAW equipment consists of the gun; electrode (wire) feed unit; electrode (wire) supply; power source; shielding gas supply with flowmeter/regulator; control circuit; and related hoses, liners, and cables, **Figure 10-23** and **Figure 10-24**. Larger, more complex systems may have water for cooling, solenoids for controlling gas flow, and carriages for moving the work or

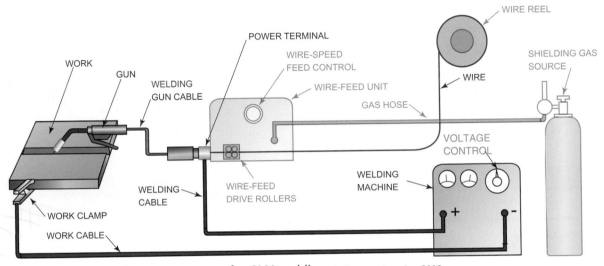

FIGURE 10-23 Schematic of equipment setup for GMA welding. © Cengage Learning 2012

FIGURE 10-24 Small 110-volt GMA welder. Thermal Arc, a Thermadyne Company

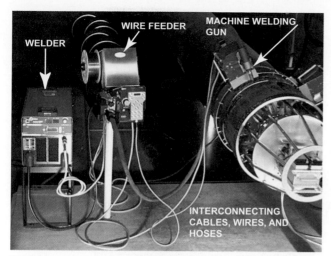

FIGURE 10-25 Typical interconnecting cables and wires for a automatic GMA welding station. Dyna Torque Technologies, Inc.

(A)

the gun or both, **Figure 10-25.** The system may be stationary or portable. In most cases, the system is meant to be used for only one process. Some manufacturers, however, do make power sources that can be switched over for other uses.

Power Source

The power source may be either a transformer-rectifier or generator type. The transformers are stationary and commonly require a three-phase power source. Engine generators are ideal for portable use or where sufficient power is not available.

The welding machine produces a DC welding current ranging from 40 amperes to 600 amperes with 10 volts to 40 volts, depending upon the machine. In the past, some GMA processes used AC welding current, but DCRP is used almost exclusively now. Typical power supplies are shown in **Figure 10-26.**

Because of the long periods of continuous use, GMA welding machines have a 100% duty cycle. This allows the machine to be run continuously without damage.

Electrode (Wire) Feed Unit

The purpose of the electrode feeder is to provide a steady and reliable supply of wire to the weld. Slight changes in the rate at which the wire is fed have distinct effects on the weld.

The motor used in a feed unit is usually a DC type that can be continuously adjusted over the desired range. **Figure 10-27** and **Figure 10-28** show typical wire-feed units and accessories.

Push-Type Feed System The wire rollers are clamped securely against the wire to provide the necessary friction to push the wire through the conduit to the gun. The pressure applied on the wire can be adjusted. A groove is provided in the roller to aid in alignment and to lessen the chance of slippage. Most manufacturers

(B)

FIGURE 10-26 An expensive 200-ampere constant-voltage power supply (A), and a 650-ampere constant-voltage and constant-current power supply (B) for multipurpose GMAW applications.

© Cengage Learning 2012

(A)

(B)

FIGURE 10-27 (A) A 90-ampere power supply and wire feeder for welding sheet steel with carbon dioxide shielding. (B) Modern wire feeder with digital preset and readout of wire feed speed and closed-loop control. © Cengage Learning 2012

provide rollers with smooth or knurled U-shaped or V-shaped grooves, **Figure 10-29.** Knurling (a series of ridges cut into the groove) helps grip larger diameter wires so that they can be pushed along more easily. Soft wires, such as aluminum, are easy to damage if knurled rollers are used. Soft wires are best used with U-grooved rollers. Even V-grooved rollers can distort the surface of the wire, causing problems. V-grooved rollers are best suited for hard wires, such as mild steel and stainless steel. It is also important to use the correct-size grooves in the rollers.

Variations of the push-type electrode wire feeder include the pull type and push-pull type. The difference is in the size and location of the drive rollers. In the push-type system, the electrode must have enough strength to be pushed through the conduit without kinking. Mild steel and stainless steel can be readily pushed 15 ft (4 m) to 20 ft (6 m), but aluminum is much harder to push over 10 ft (3 m).

Pull-Type Feed System In pull-type systems, a smaller but higher-speed motor is located in the gun to pull the wire through the conduit. Using this system, it is possible to move even soft wire over great distances. The disadvantages are that the gun is heavier and more difficult to use, rethreading the wire takes more time, and the operating life of the motor is shorter. Because of improvements in push-type wire feed systems, pull-type wire feed systems are not commonly used anymore.

Push-Pull–Type Feed System Push-pull–type feed systems use a synchronized system with feed motors located at both ends of the electrode conduit, **Figure 10-30.** This system

can be used to move any type of wire over long distances by periodically installing a feed roller into the electrode conduit. Compared to the pull-type system, the advantages of this system include the ability to move wire over longer distances, faster rethreading, and increased motor life due to the reduced load. A disadvantage is that the system is more expensive.

Linear Electrode Feed System Linear electrode feed systems use a different method to move the wire and change the feed speed. Standard systems use rollers that pinch the wire between the rollers. A system of gears is used between the motor and rollers to provide roller speed within the desired range. The linear feed system does not have gears or conventional-type rollers.

The linear feed system uses a small motor with a hollow armature shaft through which the wire is fed. The rollers are attached so that they move around the wire. Changing the roller pitch (angle) changes the speed at which the wire is moved without changing the motor speed. This system works in the same way that changing the pitch on a screw, either coarse threads or fine threads, affects the rate that the screw will move through a spinning nut.

The advantage of a linear system is that the bulky system of gears is eliminated, thus reducing weight, size, and wasted power. The motor operates at a constant high speed where it is more efficient. The reduced size allows the system to be housed in the gun or within an enclosure in the cable. Several linear wire feeders can be synchronized to provide an extended operating range. The disadvantage of a linear system is that the wire may become twisted as it is moved through the feeder.

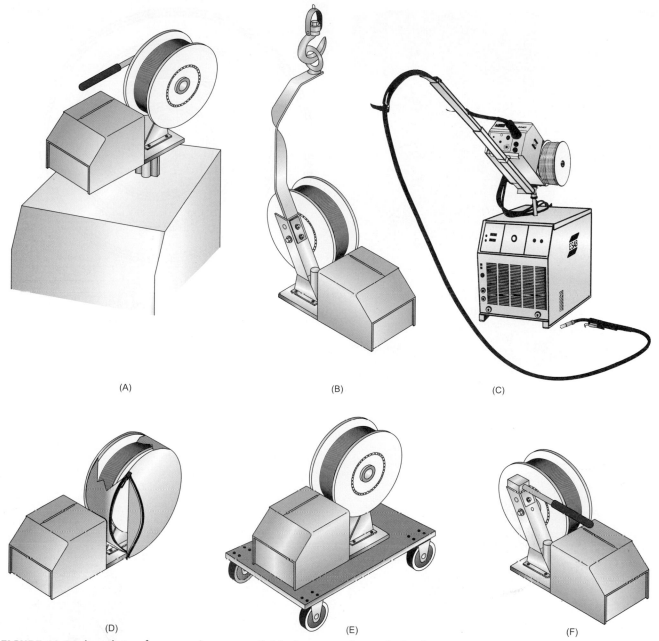

(A) (B) (C)

(D) (E) (F)

FIGURE 10-28 A variety of accessories are available for most electrode feed systems: (A) swivel post, (B) boom hanging bracket, (C) counterbalance mini-boom, (D) spool cover, (E) wire feeder wheel cart, and (F) carrying handle. © Cengage Learning 2012

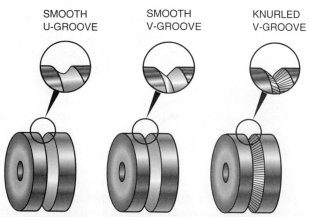

SMOOTH U-GROOVE   SMOOTH V-GROOVE   KNURLED V-GROOVE

FIGURE 10-29 Feed rollers. © Cengage Learning 2012

Spool Gun A spool gun is a compact, self-contained system consisting of a small drive system and a wire supply, **Figure 10-31.** This system allows the welder to move freely around a job with only a power lead and shielding gas hose to manage. The major control system is usually mounted on the welder. The feed rollers and motor are found in the gun just behind the nozzle and contact tube. Because of the short distance that the wire must be moved, very soft wires (aluminum) can be used. A small spool of welding wire is located just behind the feed rollers. The small spools of wire required in these guns are often very expensive. Although the guns are small, they feel heavy when being used.

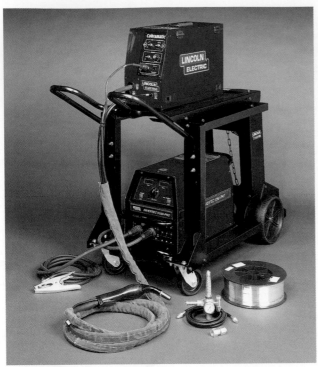

FIGURE 10-30 Wire-feed system that enables the wire to be moved through a longer cable. Lincoln Electric Company

FIGURE 10-31 Feeder/gun for GMA welding.
© Cengage Learning 2012

Electrode Conduit The electrode conduit or liner guides the welding wire from the feed rollers to the gun. It may be encased in a lead that contains the shielding gas.

Power cable and gun switch circuit wires are contained in a conduit that is made of a tightly wound coil having the needed flexibility and strength. The steel conduit may have a nylon or Teflon liner to protect soft, easily scratched metals, such as aluminum, as they are fed.

If the conduit is not an integral part of the lead, it must be firmly attached to both ends of the lead. Failure to attach the conduit can result in misalignment, which causes additional drag or makes the wire jam completely. If the conduit does not extend through the lead casing to make a connection, it can be drawn out by tightly coiling the lead, **Figure 10-32**. Coiling will force the conduit out

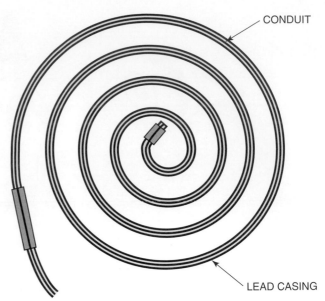

FIGURE 10-32 Tightly coiled lead casing will force the liner out of the gun. © Cengage Learning 2012

so that it can be connected. If the conduit is too long for the lead, it should be cut off and filed smooth. Too long a lead will bend and twist inside the conduit, which may cause feed problems.

Welding Gun The welding gun attaches to the end of the power cable, electrode conduit, and shielding gas hose, **Figure 10-33**. It is used by the welder to produce the weld. A trigger switch is used to start and stop the weld cycle. The gun also has a contact tube, which is used to transfer the welding current to the electrode moving through

FIGURE 10-33 A typical GMA welding gun used for most welding processes with a heat shield attached to protect the welder's gloved hand from intense heat generated when welding with high amperages.
© Cengage Learning 2012

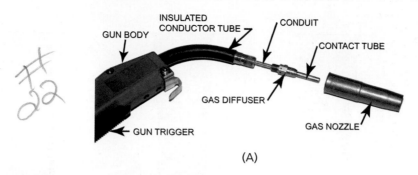

(A)

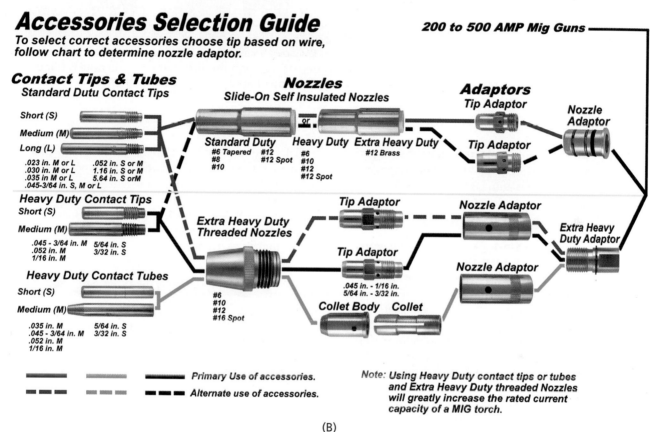

(B)

FIGURE 10-34 (A) Typical replaceable parts of a GMA welding gun. (B) Accessories and parts selection guide for a GMA welding gun. ESAB Welding & Cutting Products

the gun, and a gas nozzle, which directs the shielding gas onto the weld, **Figure 10-34.**

GMA SPOT WELDING

GMA can be used to make high-quality arc spot welds. Welds can be made using standard or specialized equipment. The arc spot weld produced by GMAW differs from electric resistance spot welding. The GMAW spot weld starts on one surface of one member and burns through to the other member, **Figure 10-35.** Fusion between the members occurs, and a small nugget is left on the metal surface.

GMA spot welding has some advantages such as the following: (1) welds can be made in thin-to-thick materials; (2) the weld can be made when only one side of the materials to be welded is accessible; and (3) the weld can

be made when there is paint on the interfacing surfaces. The arc spot weld can also be used to assemble parts for welding to be done at a later time.

Thin metal can be attached to thicker sections using an arc spot weld. If a thin-to-thick butt, lap, or tee joint is to be welded with complete joint penetration, often the thin material will burn back, leaving a hole, or there will not be enough heat to melt the thick section. With an arc spot weld, the burning back of the thin material allows the thicker metal to be melted. As more metal is added to the weld, the burnthrough is filled, (Figure 10-35).

The GMA spot weld is produced from only one side. Therefore, it can be used on awkward shapes and in cases where the other side of the surface being welded should not be damaged. This makes it an excellent process for auto body repair. In addition, because the metals are melted

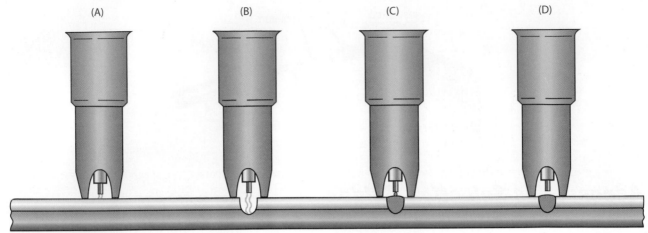

FIGURE 10-35 GMA spot weld: (A) The arc starts, (B) a hole is burned through the first plate, (C) the hole is filled with weld metal, and (D) the wire feed stops and the arc burns the electrode back. © Cengage Learning 2012

and the molten weld pool is agitated, thin films of paint between the members being joined need not be removed. This is an added benefit for auto body repair work.

⫽⫽⫽ CAUTION ⫻⫻⫻

Safety glasses and/or flash glasses must be worn to protect the eyes from flying sparks.

Specially designed nozzles provide flash protection, part alignment, and arc alignment, **Figure 10-36**. As a result, for some small jobs it may be possible to perform the weld with only safety glasses. Welders can shut their eyes and turn their head during the weld.

The optional control timer provides weld time and burn-back time. To make a weld, the amperage, voltage, and length of welding time must be set correctly. The burn-back time is a short period at the end of the weld when the wire feed stops but the current does not. This allows the wire to be burned back so it does not stick in the weld, Figure 10-35.

⫽⫽⫽ CAUTION ⫻⫻⫻

This is not advisable for any work requiring more than just a few spot welds. Prolonged exposure to the reflected ultraviolet light will cause skin burns.

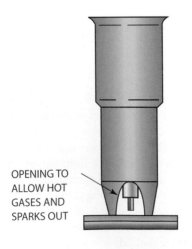

OPENING TO ALLOW HOT GASES AND SPARKS OUT

FLAT

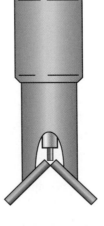

OUTSIDE CORNER

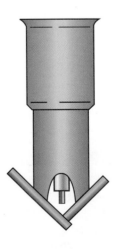

TEE OR FILLET

FIGURE 10-36 Specialized nozzles for GMA spot welding. © Cengage Learning 2012

Summary

The keys to producing quality GMA welds are equipment, setup, and adjustments. Once you have mastered these skills, the only remaining obstacle to your producing consistent, uniform, high-quality welds is your ability to follow, or track, the joint consistently. Some welders find that lightly dragging their glove along the metal surface or edge of the fabrication can aid them in controlling the weld consistency. One of the various advantages of the GMA welding process is its ability to produce long, uninterrupted welds. However, this often leads to welder fatigue. Finding a comfortable welding position that you can maintain for several minutes at a time will both improve your weld quality and reduce your fatigue.

Selecting the proper method of metal transfer—short arc, globular, or axial spray—is normally done by the welding shop foreman or supervisor. He or she makes these selections based on the material being welded, the welding position, and other such factors. A welder must be proficient with each of the various methods of metal transfer. It is important, therefore, that you spend time practicing and developing your skills with each of these processes.

Simple Steps to Achieving Better Gas Metal Arc Welding

Welders often ask the question "What is the difference between gas metal arc welding (GMAW) and metal inert gas (MIG) welding?" The term *metal inert gas* primarily means that the shielding gas will not combine with other elements in the weld pool. The American Welding Society (AWS) has adopted the term *gas metal arc welding* because the shielding gases utilized now are not necessarily inert. In many cases, the components of the shielding gas cause a chemical reaction, which promotes a combination in the weld pool that will improve strength, penetration, fusion, and weldability.

The reason *MIG* is still commonly used to describe the gas metal arc welding process is because most people find it much easier to say. The same holds true for gas tungsten arc welding. Even though the proper AWS term is *GTAW*, most people prefer to call the process *TIG*. This is not correct, just easier to say.

Advantages such as the capability to weld all commercial metals and alloys, high welding speeds, all-position welding, and ease in automation have combined to make GMAW a widely used welding process. However, the process also has limitations.

Improving Penetration

Some of the main problems with the GMAW process have been incomplete fusion and/or penetration. This can make manufacturers nervous when the process has been listed by one of their vendors. The perception that GMAW is difficult may have been added to by the American Welding Society's D1.1, Structural Welding Code—Steel. D1.1 allows the use of prequalified welding procedures for almost every process, except with the short-circuiting method of transfer with GMAW.

The capability of welding a wide variety of commercial metals and alloys, as well as other advantages, has made gas metal arc welding a widely used process.
American Welding Society

So, how can the negatives of GMAW be eliminated, or at least reduced to a minimum? Several changes can greatly improve penetration and fusion, and one does not need to be a rocket scientist to bring about these changes. Often only common sense is needed to improve welds. The first step to producing sound welds should be to take a look at what needs to be welded, the material type, and the end use of the welded product.

When welding heavier thicknesses of steel (3/8 in. and above), a good start is made with precleaning and preheating. Because GMAW is such a fast-fill, fast-freeze process, penetration can be a problem in the root pass when the steel is fairly cool. As the steel conducts heat away from the weld pool, the weld metal contracts, or shrinks, at a rapid rate, causing undue stress. The stressed steel is pulled in opposite directions, causing the weld to crack. A preheat temperature of about 150°F to 300°F should suffice on mild (low carbon) or medium carbon steel.

As with most processes, an increase in voltage can aid in fusion, but if the voltage is above the range for recommended amperage/voltage parameters, porosity and/or underbead cracking can occur.

Assume a 1/2-in. thick A-36 steel in a beveled groove joint is being welded. The joint is to be back ground and welded on the second side. The plates have been properly beveled, fitted, tacked, cleaned, and preheated. Assuming full penetration is desired, what type of welding wire and shielding gas should be used?

If this is to be a statically loaded connection that will undergo minimal stress, perhaps what is in order is an AWS ER70S-3 electrode with carbon dioxide (CO_2) as the shielding gas. The CO_2 shielding gas greatly aids penetration and is much less expensive than many other gases. This sounds great, but remember that for every advantage there is usually a disadvantage. While welders usually can count on CO_2 to produce very strong welds, it can also produce less attractive welds. Spatter and a coarse-looking weld can be a problem if good looks matter for the completed weld. If the weld must not only be strong but also look good, a wiser selection may be AWS ER70S-6 with a shielding gas mixture of 75% argon and 25% carbon dioxide.

A mixture of ER70S-6 with 75% Ar and 25% CO_2 not only is suitable for statically loaded structures but also serves well for dynamically loaded structures. This combination of electrode and shielding gas will provide excellent low-hydrogen characteristics.

The Effects of Hydrogen

Many welders are familiar with the term *low hydrogen* but really have no understanding as to why low hydrogen is desirable. During World War II, many of the United States Navy's rapid-fire gun mounts came apart due to underbead cracking. The underbead cracking was primarily caused by extremely high temperatures during the welding process. The high temperatures initiated the release of hydrogen into the molten weld pool, especially with the electrode used at the time. This hydrogen gas would then be trapped in the weld heat-affected zone (HAZ). Entrapment of gas in the grain boundaries would eventually spread into intergranular cracking under the weld. Ultimately, failure of the weld would be accelerated due to fatigue. The same problem holds true when welding martensitic steels. (Martensite is the hardest microstructure that can be formed in a carbon or alloy steel.)

Proper preheat, interpass temperature maintenance, and postweld heat treatment, plus proper electrode/shielding gas combination, can all but eliminate this problem. Heavy, argon-based shielding gases produce pleasing welds with low spatter. The only drawbacks are the higher cost and the fact that closer attention must be paid to the welder's technique to attain proper fusion and penetration.

Shielding Gas Flow Rate

No matter how well a joint is prepared or how well the filler metal and shielding gases are matched to the material, no matter what the preheat and whether interpass temperature monitoring is utilized, a weld is doomed to failure if improper shielding gas flow rates are used. Usually about 25–35 ft³/h is sufficient. If gas flow rate is too low, a lack of shielding will occur and the weld will be porous.

If windy conditions exist, shielding gas will be blown away. Increasing flow rate will not always solve the problem. If shielding gas flow rate is too high, the gas will become turbulent, drawing oxides and nitrides into the weld and causing porosity. The solution is to set up wind barricades.

Another problem with shielding gas coverage is spatter buildup, either in the nozzle or in the gas diffuser of the gun itself. Spatter buildup will prevent the gas from exiting the gun and providing shielding in the first place. If the shielding gas hose coming from the tank is cut or leaking from a loose connection, pressure is lost without necessarily registering as being low on the flowmeter.

Training

The most common problem with gas metal arc welding is caused by uninformed welders performing the welding operations. In most instances, the cure for GMAW-related welding problems is to train the welder.

While it is true that many variations exist for gas metal arc welding, keep in mind that the basic variables hold true no matter which shielding gases, techniques, and power supplies are used.

Article courtesy of the American Welding Society.

Review

1. Why is usage of the term *GMAW* preferable to *MIG* for gas metal arc welding?

2. Using Table 10-1, answer the following:
   a. What maintains the arc in machine welding?
   b. What feeds the filler metal in manual welding?
   c. What provides the joint travel in automatic welding?
   d. What provides the joint guidance in semiautomatic welding?

3. In what form is metal transferred across the arc in the axial spray metal transfer method of GMA welding?

4. What three conditions are required for the spray transfer process to occur?

5. Using Table 10-3, answer the following:
   a. What should the wire-feed speed and voltage ranges be to weld 1/8-in. (3 mm) metal with 0.035-in. (0.90-mm)wire using argon shielding gas?
   b. What should the amperage and voltage range be using 98% Ar + 2% O_2 to weld 1/4-in. (6-mm) metal with 0.045-in. (1.2-mm) wire?

6. What ranges does the pulsed-arc metal transfer shift between?

7. Why is helium added to argon when making some spray or pulsed-spray transfer welds?

8. Why does DCEP help with welds on metals such as aluminum?

9. Why is CO_2 added to argon when making GMA spray transfer welds?

10. Using Figure 10-12, what is the approximate voltage at 175 amps at 200 in./min?

11. Using Table 10-6, what would the amperage be for 0.035-in. (0.9-mm) wire at 200 in./min (5 m/min)?

12. What may happen if the GMA welding electrode is allowed to strike the base metal outside of the molten weld pool?

13. What effect does shortening the electrode extension have on weld penetration?

14. Describe the weld produced by a forehand welding angle.

15. Describe the weld produced by a backhand welding angle.

16. What components make up a GMA welding system?

17. Why must GMA welders have a 100% duty cycle?

18. What can happen if rollers of the wrong shape are used on aluminum wire?

19. Where is the drive motor located in a pull-type wire-feed system?

20. How is the wire-feed speed changed with a linear feed system?

21. What type of liner should be used for aluminum wire?

22. What parts of a typical GMA welding gun can be replaced?

23. Describe the spot welding process using a GMA welder.

Chapter 11

Gas Metal Arc Welding

OBJECTIVES

After completing this chapter, the student should be able to

- demonstrate how to properly set up a GMA welding installation.
- demonstrate how to thread the electrode wire on a GMAW machine.
- demonstrate how to set the shielding gas flow rate on a GMAW machine.
- use various settings on a GMA welding machine, and compare the effects on a weld.
- show the effect of changing the electrode extension on a weld.
- describe the effects of changing the welding gun angle on the weld bead.
- tell what must be considered when selecting the right shielding gas for a particular application.
- evaluate weld beads made with various shielding gas mixtures.
- tell why hot-rolled steel should be cleaned to bright metal before welding.
- demonstrate how to properly make GMA welds in butt joints, lap joints, and tee joints in all positions that can pass the specified standard.

KEY TERMS

bird-nesting

cast

conduit liner

contact tube

feed rollers

flow rate

spool drag

wire-feed speed

INTRODUCTION

Performing a satisfactory GMA weld requires more than just manipulative skill. The setup, voltage, amperage, electrode extension, and welding angle, as well as other factors, can dramatically affect the weld produced. The very best welding conditions are those that will allow a welder to produce the largest quantity of successful welds in the shortest period of time with the highest productivity. Because these are semiautomatic or automatic processes, increased productivity may require only that the welder increase the travel speed and current. This does not mean that the welder will work harder but, rather, that the welder will work more productively, resulting in a greater cost efficiency.

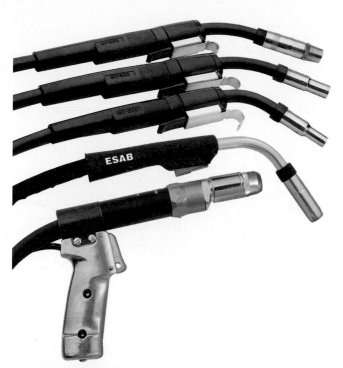

FIGURE 11-2 Make sure the gas cylinder is chained securely in place before removing the safety cap. Larry Jeffus

FIGURE 11-1 GMA welding guns are available in a variety of sizes and shapes. Note that the gun necks range from straight up to nearly a 90-degree angle. ESAB Welding & Cutting Products

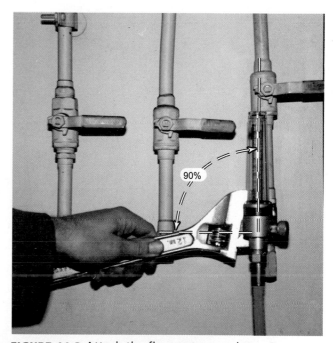

FIGURE 11-3 Attach the flowmeter regulator. Be sure the tube is vertical. Larry Jeffus

The more cost efficient welders can be, the more competitive they and their companies become. This can make the difference between being awarded a bid or a job and having or losing work.

Setup

The same equipment may be used for semiautomatic GMAW, FCAW, and SAW. Often, FCAW and SAW equipment will have a higher amperage range. In addition, equipment for FCAW and SAW is more likely to be automated than that for GMAW. However, GMA welding equipment can be automated easily.

The basic GMAW installation consists of the following: welding gun, gun switch circuit, electrode conduit-welding contractor control, electrode feed unit, electrode supply, power source, shielding gas supply, shielding gas flowmeter regulator, shielding gas hoses, and both power and work cables. Typical water-cooled and air-cooled guns are shown in **Figure 11-1**. The equipment setup in this chapter is similar to equipment built by other manufacturers, which means that any skills developed can be transferred easily to other equipment.

PRACTICE 11-1

GMAW Equipment Setup

For this practice, you will need a GMAW power source, a welding gun, an electrode feed unit, an electrode supply, a shielding gas supply, a shielding gas flowmeter regulator, electrode conduit, power and work leads, shielding

gas hoses, assorted hand tools, spare parts, and any other required materials. In this practice, you will properly set up a GMA welding installation.

If the shielding gas supply is a cylinder, it must be chained securely in place before the valve protection cap is removed, **Figure 11-2**. Standing to one side of the cylinder, quickly crack the valve to blow out any dirt in the valve before the flowmeter regulator is attached, **Figure 11-3**. With the flowmeter regulator attached securely to the cylinder valve, attach the correct hose from the flowmeter to the "gas-in" connection on the electrode feed unit or machine.

Install the reel of electrode (welding wire) on the holder and secure it, **Figure 11-4**. Check the feed roller size to ensure that it matches the wire size, **Figure 11-5**. The conduit liner size should be checked to be sure that it is compatible with the wire size. Connect the conduit to

FIGURE 11-4 When installing the spool of wire, check the label to be sure that the wire is the correct type and size. Larry Jeffus

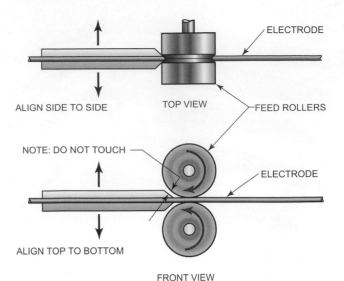

FIGURE 11-6 Feed roller and conduit alignment.
© Cengage Learning 2012

the feed unit. The conduit or an extension should be aligned with the groove in the roller and set as close to the roller as possible without touching, **Figure 11-6**. Misalignment at this point can contribute to a bird's nest, **Figure 11-7**. **Bird-nesting** of the electrode wire results

FIGURE 11-7 "Bird's nest" in the filler wire at the feed rollers. Larry Jeffus

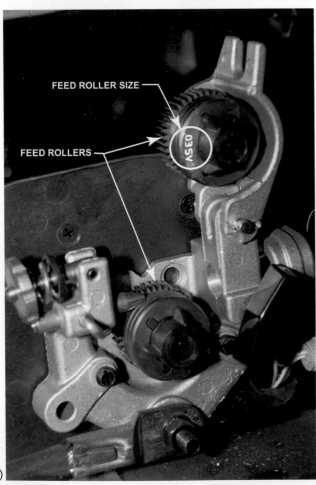

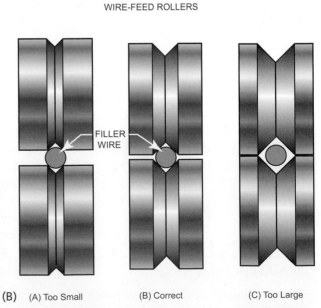

FIGURE 11-5 (A) Check to be certain that the feed rollers are the correct size for the wire being used. (B) If the wire-feed rollers are too small, the welding wire could be damaged. If the wire-feed rollers are too large, the rollers will not grip the wire. Larry Jeffus

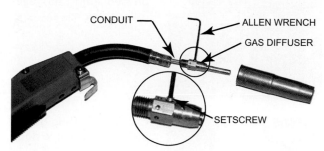

FIGURE 11-8 GMA welding gun assembly. Larry Jeffus

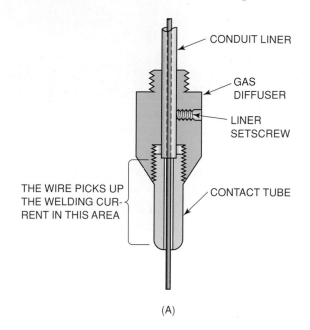

(A)

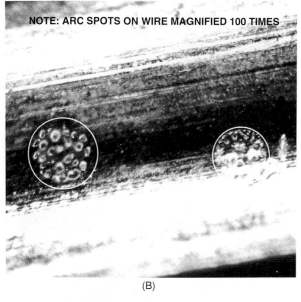

NOTE: ARC SPOTS ON WIRE MAGNIFIED 100 TIMES

(B)

when the **feed roller** pushes the wire into a tangled ball because the wire would not go through the outfeed side conduit, and appears to look like a bird's nest.

Be sure the power is off before attaching the welding cables. The electrode and work leads should be attached to the proper terminals. The electrode lead should be attached to the terminal marked electrode or positive (+). If necessary, it is also attached to the power cable part of the gun lead. The work lead should be attached to work or negative (−).

The shielding "gas-out" side of the solenoid is then also attached to the gun lead. If a separate splice is required from the gun switch circuit to the feed unit, it should be connected at this time. Check to see that the welding contactor circuit is connected from the feed unit to the power source.

The welding gun should be securely attached to the main lead cable and conduit, **Figure 11-8**. There should be a gas diffuser attached to the end of the conduit liner to ensure proper alignment. A **contact tube** (tip) of the correct size to match the electrode wire size being used should be installed, **Figure 11-9**. A shielding gas nozzle is attached to complete the assembly.

Recheck all fittings and connections for tightness. Loose fittings can leak; loose connections can cause added resistance, reducing the welding efficiency. Some manufacturers include detailed setup instructions with their equipment, **Figure 11-10**.

Complete a copy of the "Student Welding Report" listed in Appendix I or provided by your instructor. ◆

PRACTICE 11-2

Threading GMAW Wire

Using the GMAW machine that was properly assembled in Practice 11-1, you will turn on the machine and thread the electrode wire through the system.

Check to see that the unit is assembled correctly according to the manufacturer's specifications. Switch on the power and check the gun switch circuit by depressing the switch. The power source relays, feed relays, gas solenoid, and feed motor should all activate.

Cut the end of the electrode wire free. Hold it tightly so that it does not unwind. The wire has a natural curve that is known as its **cast**. The cast is measured by the diameter of the circle that the wire would make if it were loosely

(C)

FIGURE 11-9 The contact tube must be the correct size. (A) Too small a contact tube will cause the wire to stick. (B) Too large a contact tube can cause arcing to occur between the wire and tube. (C) Heat from the arcing can damage the tube. Larry Jeffus

Open the side cover.

With the gun trigger pressed, adjust the feed roller tension.

Remove the empty wire spool.

Check the setting guide inside the machine door.

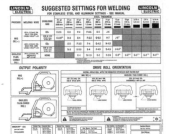

Release upper feed roller.

Set the voltage and wire feed for the metal you are going to be welding.

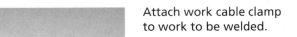

Reload the wire spool with the free end unreeling from the bottom.

Attach work cable clamp to work to be welded.

Thread wire through guide between rollers and into wire cable.

Connect gas to coupling at rear of case and turn on shielding gas.

Set the polarity as DCEP from GMA welding.

ALWAYS WEAR PROPER SAFETY EQUIPMENT. Pull trigger and weld.

Turn the input switch on.

FIGURE 11-10 Example of manufacturer's setup instructions. Lincoln Electric Company

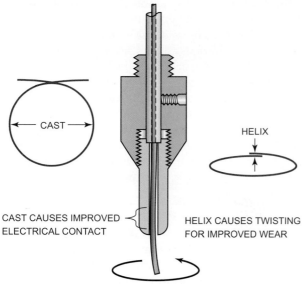

FIGURE 11-13 Adjust the wire-feed tensioner. Larry Jeffus

FIGURE 11-11 The cast of the welding wire causes it to rub firmly inside of the contact tube for good electrical contact. The helix causes the electrode to twist inside of the contact tube so that the tube is worn uniformly.
© Cengage Learning 2012

laid on a flat surface. The cast helps the wire make a good electrical contact as it passes through the contact tube, **Figure 11-11.** However, the cast can be a problem when threading the system. To make threading easier, straighten about 12 in. (305 mm) of the end of the wire and cut any kinks off.

Separate the wire-feed rollers and push the wire first through the guides, then between the rollers, and finally into the **conduit liner, Figure 11-12.** Reset the rollers so there is a slight amount of compression on the wire, **Figure 11-13.** Set the **wire-feed speed** control to a slow speed. Hold the welding gun so that the electrode conduit and cable are as straight as possible.

Press the gun switch. The wire should start feeding into the liner. Watch to make certain that the wire feeds

smoothly and release the gun switch as soon as the end comes through the contact tube.

///// ///// CAUTION \\\\\ \\\\\

If the wire stops feeding before it reaches the end of the contact tube, stop and check the system. If no obvious problem can be found, mark the wire with tape and remove it from the gun. It then can be held next to the system to determine the location of the problem.

With the wire feed running, adjust the feed roller compression so that the wire reel can be stopped easily by a slight pressure. Too light a roller pressure will cause the wire to feed erratically. Too high a pressure can turn a minor problem into a major disaster. If the wire jams at a high roller pressure, the feed rollers keep feeding the wire, causing it to bird-nest and possibly short out. With a light pressure, the wire can stop, preventing bird-nesting. This is very important with soft wires. The other advantage of a light pressure is that the feed will stop if something like clothing or gas hoses are caught in the reel.

With the feed running, adjust the **spool drag** so that the reel stops when the feed stops. The reel should not coast to a stop because the wire can be snagged easily. Also, when the feed restarts, a jolt occurs when the slack in the wire is taken up. This jolt can be enough to momentarily stop the wire, possibly causing a discontinuity in the weld.

When the test runs are completed, the wire can either be rewound or cut off. Some wire-feed units have a retract button. This allows the feed driver to reverse and retract the wire automatically. To rewind the wire on units without this retract feature, release the rollers and turn them backward by hand. If the machine will not allow the feed rollers to be released without upsetting the tension, you must cut the wire.

Complete a copy of the "Student Welding Report" listed in Appendix I or provided by your instructor. ◆

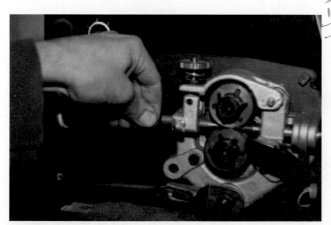

FIGURE 11-12 Push the wire through the guides by hand. Larry Jeffus

CAUTION

Do not discard pieces of wire on the floor. They present a hazard to safe movement around the machine. In addition, a small piece of wire can work its way into a filter screen on the welding power source. If the piece of wire shorts out inside the machine, it could become charged with high voltage, which could cause injury or death. Always wind the wire tightly into a ball or cut it into short lengths before discarding it in the proper waste container.

Gas Density and Flow Rates

Density is the chief determinant of how effective a gas is for arc shielding. The lower the density of a gas, the higher will be the **flow rate** required for equal arc protection. Flow rates, however, are not in proportion to the densities. Helium, with about one-tenth the density of argon, requires only twice the flow for equal protection.

THINK GREEN
Save Shielding Gas

Setting the shielding gas flow rate as low as possible within the welding procedure range for your welding conditions will reduce the wasting of this consumable material. Although most shielding gases are renewable, their purchase price is not, so using them sparingly will cut costs.

The correct flow rate can be set by checking welding guides that are available from the welding equipment and filler metal manufacturers. These welding guides list the gas flow required for various nozzle sizes and welding amperage settings. Some welders feel that a higher gas flow will provide better weld coverage, but that is not always the case. High gas flow rates waste shielding gases and may lead to contamination. The contamination comes from turbulence in the gas at high flow rates. Air is drawn into the gas envelope by the venturi effect around the edge of the nozzle. Also, the air can be drawn in under the nozzle if the torch is held at too sharp an angle to the metal, **Figure 11-14.**

NOTE: If you need more shielding gas coverage in a windy or drafty area, use both a larger diameter gas nozzle and a higher gas flow rate. The larger the nozzle size, the higher the permissible flow rate without causing turbulence. Larger nozzle sizes may restrict your visibility of the weld. You might also consider setting up a wind barrier to protect your welding from the wind, **Figure 11-15.**

Wire-Feed Speed

Because changes in the wire-feed speed automatically change the amperage, it is possible to set the amperage by using a chart and measuring the length of wire fed per minute, **Table 11-1.** The wire-feed speed is generally recommended by the electrode manufacturer and is selected in inches per minute (ipm), which can be measured by how fast the wire exits the contact tube. The welder uses a wire-feed speed control dial on the wire-feed unit to control the ipm.

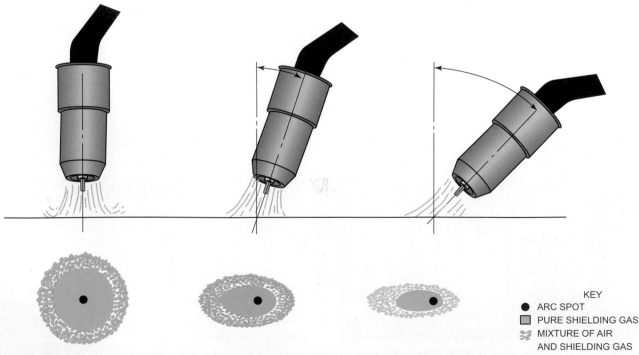

KEY
● ARC SPOT
▢ PURE SHIELDING GAS
▨ MIXTURE OF AIR AND SHIELDING GAS

FIGURE 11-14 The welding gun angle affects the shielding gas coverage for the molten weld pool. © Cengage Learning 2012

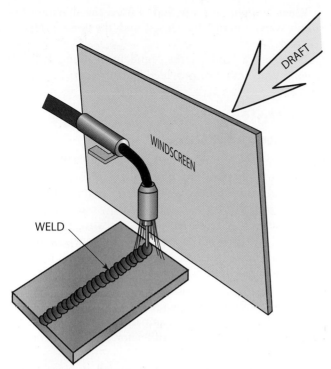

FIGURE 11-15 A windscreen can help prevent the shielding gas from being blown away. © Cengage Learning 2012

THINK GREEN
Rewind the Wire to Reduce Waste

Rewinding the electrode as opposed to cutting it off reduces both electrode waste and eliminates the potential hazards that long lengths of loose electrode wire can cause. Loose electrode wire can be a trip hazard as well as an electrical hazard if it comes in contact with the welder electrical terminals.

The wire-feed speed is given in a range. The range allows you to adjust the feed speed according to the welding conditions. The wire speed control dial can be advanced or slowed to control the burn weld size and deposition rate.

EXPERIMENT 11-1

Setting Wire-Feed Speed

Using the equipment setup as described in Practice 11-1 and the threaded machine as described in Practice 11-2, you will set the wire-feed speed. At high wire-feed speeds many feet of wire can be fed out during a full minute's wire-feed speed test, so a shorter time test is desirable. For example, in Table 11-1 the slowest wire-feed speed for a 0.030 spray arc would be 500 ipm which is 41 feet of wire per minute, and at the highest speed of 650 ipm, 54 feet of wire would be fed per minute. There are two commonly used shorter timed tests to reduce the wasting of electrode. One uses 15 seconds, and the resulting length is multiplied by 4 to determine the inches per minute. The second test uses 6 seconds, and the resulting length is multiplied by 10 to determine the inches per minute. The 15-second test is more accurate, but for most applications the 6-second test is adequate.

- Snip the end of the wire off at the contact tube.
- Turn on the welding machine.
- Point the welding gun away from metal that might accidentally complete the welding circuit, and squeeze the trigger for 6 seconds.
- Using a tape measure, check the length of wire.
- Multiply the length of the wire by 10.
- The result is how many inches of wire would be fed in a minute.
- Release the drive roller spring tensioner so that the electrode spool can be hand turned backward to draw the electrode back onto the spool.

Complete a copy of the "Student Welding Report" listed in Appendix I or provided by your instructor. ◆

EXPERIMENT 11-2

Setting Gas Flow Rate

Using the equipment setup as described in Practice 11-1 and the threaded machine as described in Practice 11-2, you will set the shielding gas flow rate.

RECOMMENDED GMA WELDING SETUP FOR CARBON STEEL						
	SHORT-CIRCUIT ARC			SPRAY METAL ARC		
ELECTRODE DIAMETER IN.	AMPS	VOLTS	WIRE-FEED SPEED* (IPM)	AMPS	VOLTS	WIRE-FEED SPEED (IPM)
0.023	45-**70**-90	14-**15**-16	150-**300**-380	100-**110**-125	23-**23**-25	400-**450**-620
0.030	60-**100**-140	14-**15**-16	150-**220**-350	160-**180**-200	24-**25**-26	500-**520**-650
0.035	90-**130**-160	15-**17**-19	180-**250**-300	180-**200**-230	25-**26**-27	400-**480**-550
0.045	130-**160**-200	17-**18**-19	125-**150**-200	260-**300**-340	25-**27**-30	300-**350**-500
0.052	150-**160**-200	17-**18**-20	135-**140**-190	275-**325**-400	26-**28**-33	265-**310**-390

Bold values represents the optimum setting using DCEP and a shielding gas flow rate from 35 to 45 CFM.
*To check the wire-feed speed, run out wire for 6 seconds, measure its length, and multiply the measurement by 10.

TABLE 11-1 Typical Amperages for Carbon Steel

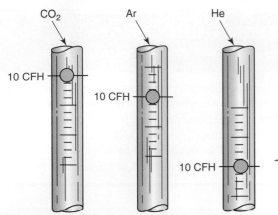

FIGURE 11-16 Each of these gases is flowing at the same cfh (L/min) rate. Because helium (He) is less dense, its indicator ball is the lowest. Be sure that you are reading the correct scale for the gas being used.

© Cengage Learning 2012

The exact flow rate required for a certain job will vary depending upon welding conditions. This experiment will help you determine how those conditions affect the flow rate. You will start by setting the shielding gas flow rate at 35 cfh (16 L/min).

Turn on the shielding gas supply valve. If the supply is a cylinder, the valve is opened all the way. With the machine power on and the welding gun switch depressed, you are ready to set the flow rate. Slowly turn in the adjusting screw and watch the float ball as it rises in a tube on a column of gas. The faster the gas flows, the higher the ball will float. A scale on the tube allows you to read the flow rate. Different scales are used with each type of gas being used. Since various gases have different densities (weights), the ball will float at varying levels even though the flow rates are the same, **Figure 11-16**. The line corresponding to the flow rate may be read as it compares to the top, center, or bottom of the ball, depending upon the manufacturer's instructions. There should be some marking or instruction on the tube or regulator to tell a person how it should be read, **Figure 11-17**.

Release the welding gun switch, and the gas flow should stop. Turn off the power and spray the hose fittings with a leak-detecting solution.

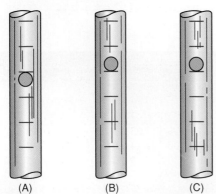

FIGURE 11-17 Three methods of reading a flowmeter: (A) top of ball, (B) center of ball, and (C) bottom of ball.

© Cengage Learning 2012

When stopping for a period of time, the shielding gas supply valve should be closed and the hose pressure released.

Complete a copy of the "Student Welding Report" listed in Appendix I or provided by your instructor. ◆

Arc-Voltage and Amperage Characteristics

The arc-voltage and amperage characteristics of GMA welding are different from most other welding processes. The voltage is set on the welder, and the amperage is set by changing the wire-feed speed. At any one voltage setting the amperage required to melt the wire must change as it is fed into the weld. It requires more amperage to melt the wire the faster it is fed, and less the slower it is fed.

Because changes in the wire-feed speed directly change the amperage, it is possible to set the amperage by using a chart and measuring the length of wire fed per minute, Table 11-1. The voltage and amperage required for a specific metal transfer method differ for various wire sizes, shielding gases, and metals.

The voltage and amperage setting will be specified for all welding done according to a welding procedure specification (WPS) or other codes and standards. However, most welding—like that done in small production shops, as maintenance welding, for repair work, in farm shops, and the like—is not done to specific code or standard and therefore no specific setting exists. For that reason, it is important to learn to make the adjustments necessary to allow you to produce quality welds.

EXPERIMENT 11-3

Setting the Current

Using a properly assembled GMA welding machine, proper safety protection, and one piece of mild steel plate approximately 12 in. (305 mm) long and 1/4 in. (6 mm) thick, you will change the current settings and observe the effect on GMAW.

On a scale of 0 to 10, set the wire-feed speed control dial at 5, or halfway between the low and high settings of the unit. The voltage is also set at a point halfway between the low and high settings. The shielding gas can be CO_2, argon, or a mixture. The gas flow should be adjusted to a rate of 35 cfh (16 L/min).

Hold the welding gun at a comfortable angle, lower your welding hood, and pull the trigger. As the wire feeds and contacts the plate, the weld will begin. Move the gun slowly along the plate. Note the following welding conditions as the weld progresses: voltage, amperage, weld direction, metal transfer, spatter, molten weld pool size, and penetration. Stop and record your observations in **Table 11-2**. Evaluate the quality of the weld as acceptable or unacceptable.

Reduce the voltage somewhat and make another weld, keeping all other weld variables (travel speed, stickout,

Weld Acceptability	Voltage	Amperage	Spatter	Molten Pool Size	Penetration
Good	20	75	Light	Small	Little

Electrode diameter   .035 in. (0.9 mm)
Shielding gas CO_2
Welding direction   Backhand

TABLE 11-2 Setting the Current

direction, amperage) the same. Observe the weld and upon stopping record the results. Repeat this procedure until the voltage has been lowered to the minimum value indicated on the machine. Near the lower end, the wire may stick, jump, or simply no longer weld.

Return the voltage indicator to the original starting position and make a short test weld. Stop and compare the results to those first observed. Then slightly increase the voltage setting and make another weld. Repeat the procedure of observing and recording the results as the voltage is increased in steps until the maximum machine capability is obtained. Near the maximum setting the spatter may become excessive if CO_2 shielding gas is used. Care must be taken to prevent the wire from fusing to the contact tube.

Return the voltage indicator to the original starting position and make a short test weld. Compare the results observed with those previously obtained.

Lower the wire-feed speed setting slightly and use the same procedure as before. First lower and then raise the voltage through a complete range and record your observations. After a complete set of test results is obtained from this amperage setting, again lower the wire-feed speed for a new series of tests. Repeat this procedure until the amperage is at the minimum setting shown on the machine. At low amperages and high voltage settings, the wire may tend to pop violently as a result of the uncontrolled arc.

Return the wire-feed speed and voltages to the original settings. Make a test weld and compare the results with the original tests. Slightly raise the wire speed and again run a set of tests as the voltage is changed in small steps. After each series, return the voltage setting to the starting point and increase the wire-feed speed. Make a new set of tests.

All of the test data can be gathered into an operational graph for the machine, wire type, size, and shielding gas. Use **Table 11-3** to plot the graph. The acceptable welds should be marked on the lines that extend from the appropriate voltages and amperages. Upon completion, the graph will give you the optimum settings for the operation of this particular GMAW setup. The optimum settings are along a line in the center of the acceptable welds.

Experienced welders will follow a much shorter version of this type of procedure anytime they are starting to work on a new machine or testing for a new job. This

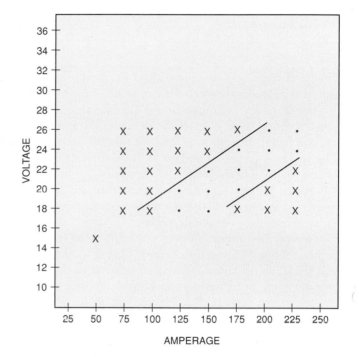

WIRE SIZE   0.045 IN. (1.2 MM)
SHIELDING GAS   Ar + 2% O_2
                FOREHAND

X UNACCEPTABLE WELD

• ACCEPTABLE WELD

TABLE 11-3 Graph for GMAW Machine Settings

experiment can be repeated using different types of wire, wire sizes, shielding gases, and weld directions. Turn off the welding machine and shielding gas and clean up your work area when you are finished welding.

Complete a copy of the "Student Welding Report" listed in Appendix I or provided by your instructor. ◆

Electrode Extension

Because of the constant-potential (CP) power supply, the welding current will change as the distance between the contact tube and the work changes. Although this

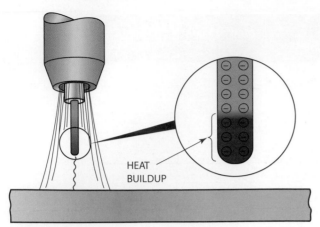

FIGURE 11-18 Heat buildup due to the extremely high current for the small conductor (electrode).

© Cengage Learning 2012

change is slight, it is enough to affect the weld being produced. The longer the electrode extension, the greater the resistance to the welding current flowing through the small welding wire. This results in some of the welding current being changed to heat at the tip of the electrode, **Figure 11-18.** With a standard SMA welding CC power supply this would also reduce the arc voltage, but with a CP power supply the voltage remains constant and the amperage increases. If the electrode extension is shortened, the welding current decreases.

The increase in current does not result in an increase in penetration because the current is being used to heat the electrode tip and not being transferred to the weld metal. Penetration is reduced and buildup is increased as the electrode extension is lengthened. Penetration is increased and buildup decreased as the electrode extension is shortened. Controlling the weld penetration and buildup by changing the electrode will help maintain weld bead shape during welding. It will also help you better understand what may be happening if a weld starts out correctly but begins to change as it progresses along the joint. You may be changing the electrode extension without noticing the change. Short electrode

stickout gives a hotter weld and long stickout results in a cooler weld.

EXPERIMENT 11-4

Electrode Extension

Using a properly assembled GMA welding machine; proper safety protection; and a few pieces of mild steel, each about 12 in. (305 mm) long and ranging in thickness from 16 gauge to 1/2 in. (13 mm), you will observe the effect of changing electrode extension on the weld.

Start at a low current setting. Using the graph developed in Experiment 11-2, set both the voltage and amperage. The settings should be equal to those on the optimum line established for the wire type and size being used with the same shielding gas.

Holding the welding gun at a comfortable angle and height, lower your helmet and start to weld. Make a weld approximately 2 in. (51 mm) long. Then reduce the distance from the gun to the work while continuing to weld. After a few inches, again shorten the electrode extension even more. Keep doing this in steps until the nozzle is as close as possible to the work. Stop and return the gun to the original starting distance.

Repeat the process just described but now increase the electrode extension, making welds of a few inches each. Keep increasing the electrode extension until the weld will no longer fuse or the wire becomes impossible to control.

Change the plate thickness and repeat the procedure. When the series has been completed with each plate thickness, raise the voltage and amperage to a medium setting and repeat the process. Upon completing this series of tests, adjust the voltage and amperage upward to a high setting. Make a full series of tests using the same procedures as before.

Record the results in **Table 11-4** after each series of tests. The final results can be plotted on a graph, as was done in **Table 11-5**, to establish the optimum electrode extension for each thickness, voltage, and amperage. Turn off the welding machine and shielding gas and clean up your work area when you are finished welding.

Weld Acceptability	Voltage	Amperage	Electrode Extension	Contact Tube-to-Work Distance	Bead Shape
Poor	20	100	1 in. (25 mm)	1 1/4 in. (31 mm)	Narrow, high, with little penetration

Electrode diameter   .035 in. (0.9 mm)
Shielding gas CO_2
Welding direction   Forehand

TABLE 11-4 Electrode Extension

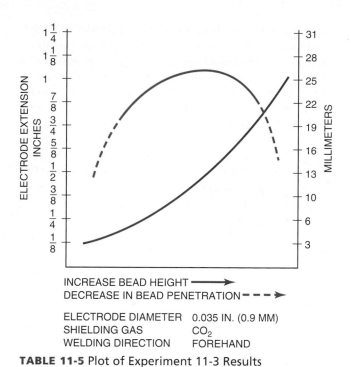

INCREASE BEAD HEIGHT ──────▶
DECREASE IN BEAD PENETRATION ─ ─ ─ ▶

ELECTRODE DIAMETER	0.035 IN. (0.9 MM)
SHIELDING GAS	CO_2
WELDING DIRECTION	FOREHAND

TABLE 11-5 Plot of Experiment 11-3 Results

Complete a copy of the "Student Welding Report" listed in Appendix I or provided by your instructor. ◆

Welding Gun Angle

The term *welding gun angle* refers to the angle between the GMA welding gun and the work as it relates to the direction of travel. Backhand welding, or dragging angle, **Figure 11-19**, produces a weld with deep penetration

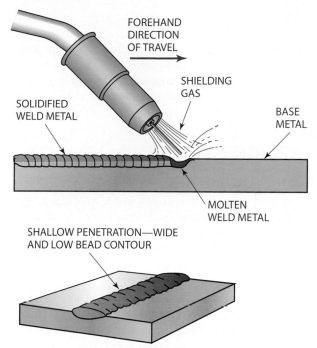

FIGURE 11-20 Forehand welding or pushing angle.
© Cengage Learning 2012

and higher buildup. Forehand welding, or pushing angle, **Figure 11-20**, produces a weld with shallow penetration and little buildup. Perpendicular welding has a good balance between penetration and reinforcement, **Figure 11-21**.

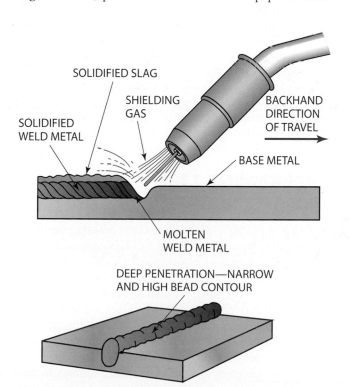

FIGURE 11-19 Backhand welding or dragging angle.
© Cengage Learning 2012

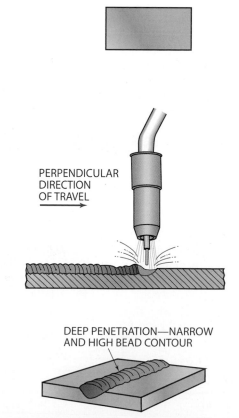

FIGURE 11-21 Perpendicular gun angle.
© Cengage Learning 2012

Slight changes in the welding gun angle can be used to control the weld as the groove spacing changes. A narrow gap may require more penetration, but as the gap spacing increases, a weld with less penetration may be required. Changing the electrode extension and welding gun angle at the same time can result in a quality weld being made with less-than-ideal conditions.

EXPERIMENT 11-5

Welding Gun Angle

Using a properly assembled GMA welding machine; proper safety protection; and some pieces of mild steel, each approximately 12 in. (305 mm) long and ranging in thickness from 16 gauge to 1/2 in. (13 mm), you will observe the effect of changing the welding gun angle on the weld bead.

Starting with a medium current setting and a plate that is 1/4 in. (6 mm) thick, hold the welding gun at a 30° angle to the plate in the direction of the weld, **Figure 11-22**. Lower your welding hood and depress the trigger. When the weld starts, move in a straight line and slowly pivot the gun angle as the weld progresses. Keep the travel speed, electrode extension, and weave pattern (if used) constant so that any change in the weld bead is caused by the angle change.

The pivot should be completed in the 12 in. (305 mm) of the weld. You will proceed from a 30° pushing angle to a 30° dragging angle. Repeat this procedure using different welding currents and plate thicknesses.

After the welds are complete, note the differences in width and reinforcement along the welds. Turn off the welding machine and shielding gas and clean up your work area when you are finished welding.

Complete a copy of the "Student Welding Report" listed in Appendix I or provided by your instructor. ◆

Effect of Shielding Gas on Welding

Shielding gases in the gas metal arc process are used primarily to protect the molten metal from oxidation and contamination. Other factors must be considered, however, in selecting the right gas for a particular application. Shielding gas can influence arc and metal transfer characteristics, weld penetration, width of fusion zone, surface shape patterns, welding speed, and undercut tendency. Inert gases such as argon and helium provide the necessary shielding because they do not form compounds with any other substance and are insoluble in molten metal. When used as pure gases for welding ferrous metals, argon and helium may produce an erratic arc action, promote undercutting, and result in other flaws.

It is therefore usually necessary to add controlled quantities of reactive gases to achieve good arc action and metal transfer with these materials. Adding oxygen or carbon dioxide to the inert gas tends to stabilize the arc, promote favorable metal transfer, and minimize spatter. As a result, the penetration pattern is improved and undercutting is reduced or eliminated.

Oxygen or carbon dioxide is often added to argon. The amount of reactive gas required to produce the desired effects is quite small. As little as 0.5% of oxygen will produce a noticeable change; 1% to 5% of oxygen is more common. Carbon dioxide may be added to argon in the 20% to 30% range. Mixtures of argon with less than 10% carbon dioxide may not have enough arc voltage to give the desired results.

Adding oxygen or carbon dioxide to an inert gas causes the shielding gas to become oxidizing. This in turn may cause porosity in some ferrous metals. In this case, a filler wire containing suitable deoxidizers should be used. The presence of oxygen in the shielding gas can also cause some loss of certain alloying elements, such as chromium, vanadium, aluminum, titanium, manganese, and silicon. Again, the addition of a deoxidizer to the filler wire is necessary.

Pure carbon dioxide has become widely used as a shielding gas for GMA welding of steels. It allows higher welding speed, better penetration, and good mechanical properties, and it costs less than the inert gases. The chief drawback in the use of carbon dioxide is the less-steady-arc characteristics and considerable weld-metal–spatter losses. The spatter can be kept to a minimum by maintaining a very short, uniform arc length. Consistently sound welds can be produced using carbon dioxide shielding, provided that a filler wire having the proper deoxidizing additives is used.

EXPERIMENT 11-6

Effect of Shielding Gas Changes

Using a properly assembled GMA welding machine; proper safety protection; a source of CO_2, argon, and oxygen gases or a variety of premixed shielding gases; two flowmeters (or one two-gas mixing regulator); and some pieces of mild steel plate, each about 12 in. (305 mm) long and ranging in thickness from 16 gauge to 1/2 in. (13 mm); you will observe the effect of various shielding gas mixtures on the weld.

Using a mixing flowmeter regulator will allow the gases to be mixed in any desired combination. A mixing ratio chart for use in arriving at the approximate gas percentages appears in **Table 11-6**. The exact ratios are not so important to you, as a student, as they are on code work.

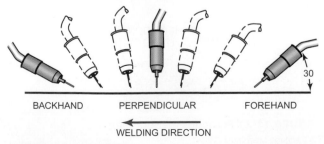

BACKHAND   PERPENDICULAR   FOREHAND

← WELDING DIRECTION

FIGURE 11-22 Welding gun angle. © Cengage Learning 2012

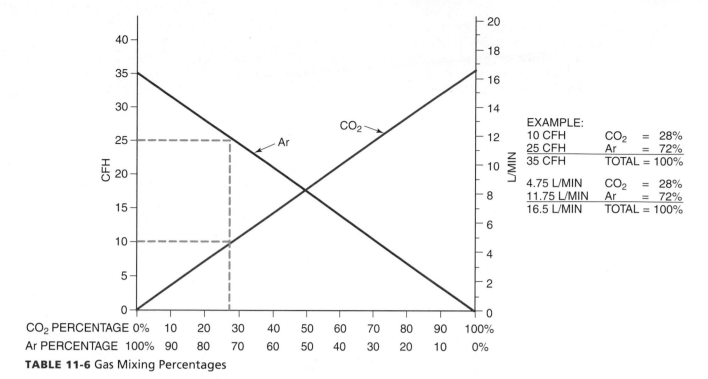

EXAMPLE:
10 CFH	CO₂	= 28%
25 CFH	Ar	= 72%
35 CFH	TOTAL	= 100%

4.75 L/MIN	CO₂	= 28%
11.75 L/MIN	Ar	= 72%
16.5 L/MIN	TOTAL	= 100%

TABLE 11-6 Gas Mixing Percentages

With a medium voltage and amperage setting and using a 100% carbon dioxide (CO_2) shielding gas, start making a weld. Either change the mixture after each weld or have another person changing the shielding gas during the weld. Keep the total flow rate the same by adding argon (Ar) while reducing the CO_2 to preserve the same flow rate. During the experiment, change over the shielding gas to 100% argon (Ar).

After the weld is complete, evaluate it for spatter, penetration, undercut, buildup, width, or other noticeable changes along its length. Using **Table 11-7**, record the results of your evaluation.

Repeat the procedure just explained two times more with both low and high power settings. Again, record your observations.

Starting with 100% argon (Ar), add oxygen (O_2) to the shielding gas. The oxygen percentage will range from 0% to 10%, **Table 11-8**. Very slight changes in the percentage will have dramatic effects on the weld. You will make three welds using low, medium, and high power settings. For each weld, you will record your observations.

During some of the welding tests, you will notice a change in the method of metal transfer, weld heat, and general weld performance without a change in the current settings. The shielding gas mixture can have major effects on the rate of metal transfer and the welding speed, as well as other welding variables. Higher speeds and greater production can be obtained by using some gas mixtures. However, the savings can be completely offset by the higher gas cost. Before making a final decision about the gas to be used, all the variables must be compared. **Table 11-9** lists premixed shielding gases and their uses. Turn off the welding machine and shielding gas and clean up your work area when you are finished welding.

Complete a copy of the "Student Welding Report" listed in Appendix I or provided by your instructor. ◆

Weld Acceptability	Voltage	Spatter	Penetration	Puddle Size	Bead Appearance
Good	75 Ar 25 CO₂	Very little	Deep	Large	Wide with little buildup

Electrode diameter .035 in. (0.9 mm)
Welding direction Forehand
Voltage 25
Amperage 150

TABLE 11-7 Shielding Gas Mixtures

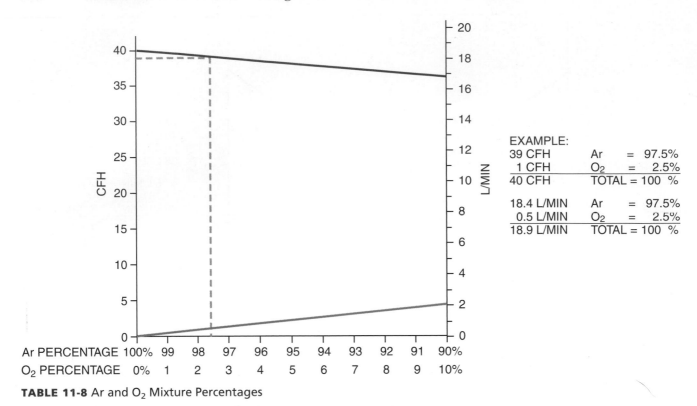

EXAMPLE:

39 CFH	Ar	=	97.5%
1 CFH	O$_2$	=	2.5%
40 CFH	TOTAL	=	100 %

18.4 L/MIN	Ar	=	97.5%
0.5 L/MIN	O$_2$	=	2.5%
18.9 L/MIN	TOTAL	=	100 %

TABLE 11-8 Ar and O$_2$ Mixture Percentages

Shielding Gas	Chemical Behavior	Uses and Usage Notes
1. Argon	Inert	Welding virtually all metals except steel
2. Helium	Inert	Al and Cu alloys for greater heat and to minimize porosity
3. Ar and He (20% to 80% to 50% to 50%)	Inert	Al and Cu alloys for greater heat and to minimize porosity but with quieter, more readily controlled arc action
4. N$_2$	Reducing	On Cu, very powerful arc
5. Ar + 25% to 30% N$_2$	Reducing	On Cu, powerful but smoother operating, more readily controlled arc than with N$_2$
6. Ar + 1% to 2% O$_2$	Oxidizing	Stainless and alloy steels, also for some deoxidized copper alloys
7. Ar + 3% to 5% O$_2$	Oxidizing	Plain carbon, alloy, and stainless steels (generally requires highly deoxidized wire)
8. Ar + 3% to 5% O$_2$	Oxidizing	Various steels using deoxidized wire
9. Ar + 20% to 30% O$_2$	Oxidizing	Various steels, chiefly with short-circuiting arc
10. Ar + 5% O$_2$ + 15% CO$_2$	Oxidizing	Various steels using deoxidized wire
11. CO$_2$	Oxidizing	Plain carbon and low alloy steels, deoxidized wire essential
12. CO$_2$ + 3% to 10% O$_2$	Oxidizing	Various steels using deoxidized wire
13. CO$_2$ + 20% O$_2$	Oxidizing	Steels

TABLE 11-9 Shielding Gases and Gas Mixtures Used for Gas Metal Arc Welding

Practices

The practices in this chapter are grouped according to those requiring similar techniques and setups. To make acceptable GMA welds consistently, the major skill required is the ability to set up the equipment and weldment. Changes such as variations in material thickness, position, and type of joint require changes both in technique and setup. A correctly set-up GMA welding station can, in many cases, be operated with minimum skill. Often, the only difference between a welder earning a minimum wage and one earning the maximum wage is the ability to make correct machine setups.

Ideally, only a few tests would be needed for the welder to make the necessary adjustments in setup and manipulation techniques to achieve a good weld. The previous welding experiments should have given the welder a graphic set of comparisons to help that welder make the correct changes. In addition to keeping the test data, you may want to keep the test plates for a more accurate comparison.

The grouping of practices in this chapter will keep the number of variables in the setup to a minimum. Often, the only change required before going on to the next weld is to adjust the power settings.

Deoxidizing Element	Strength
Aluminum (Al)	Very strong
Manganese (Mn)	Weak
Silicon (Si)	Weak
Titanium (Ti)	Very strong
Zirconium (Zr)	Very strong

TABLE 11-10 Sufficient Deoxidizing Elements Must Be Added to the Filler Wire to Minimize Porosity in the Molten Weld Pool

Figures that are given in some of the practices will give the welder general operating conditions, such as voltage, amperage, and shielding gas and/or gas mixture. These are general values, so the welder will have to make some fine adjustments. Differences in the type of machine being used and the material surface condition will affect the settings. For this reason, it is preferable to use the settings developed during the experiments.

Metal Preparation

All hot-rolled steel has an oxide layer, which is formed during the rolling process, called mill scale. *Mill scale* is a thin layer of dark gray or black iron oxide. Some hot-rolled steels that have had this layer removed either mechanically or chemically can be purchased. However, almost all of the hot-rolled steel used today still has this layer because it offers some protection from rusting.

Mill scale is not removed for noncode welding because it does not prevent most welds from being suitable for service. For practice welds that will be visually inspected, mill scale can usually be left on the plate. Filler metals and fluxes usually have deoxidizers added to them so that the adverse effects of the mill scale are reduced or eliminated, **Table 11-10**. But with GMA welding wire it is difficult to add enough deoxidizers to remove all effects of mill scale. The porosity that mill scale causes is most often confined to the interior of the weld and is not visible on the surface, **Figure 11-23**. Because it is not visible on the surface, it usually goes unnoticed and the weld passes visual inspection.

If the practices are going to be destructively tested or if the work is of a critical nature, then all welding surfaces within the weld groove and the surrounding surfaces within 1 in. (25 mm) must be cleaned to bright metal, **Figure 11-24**. Cleaning may be either grinding, filing, sanding, or blasting.

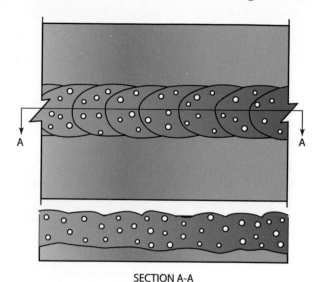

SECTION A-A

FIGURE 11-23 Uniformly scattered porosities. © Cengage Learning 2012

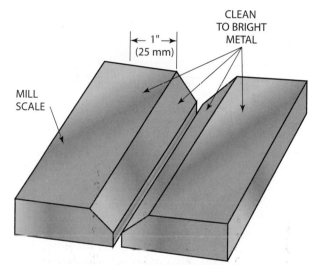

FIGURE 11-24 Clean all surfaces to bright metal before welding. © Cengage Learning 2012

Flat Position, 1G and 1F Positions

PRACTICE 11-3

Stringer Beads Using the Short-Circuiting Metal Transfer Method in the Flat Position

Using a properly set-up and adjusted GMA welding machine, **Table 11-11**, proper safety protection, 0.035-in. and/or 0.045-in. (0.9-mm and/or 1.2-mm) diameter wire,

Process	Wire Diameter	Amperage Range (Optimum)	Voltage Range (Optimum)	Shielding Gas
Short circuiting	0.030	60 (100) 140	14 (15) 16	100% CO_2
	0.035	90 (130) 150	16 (17) 20	75% Ar + 25% CO_2 98% Ar + 2% O

TABLE 11-11 Typical Welding Current Settings for Short-Circuiting Metal Transfer for Mild Steel

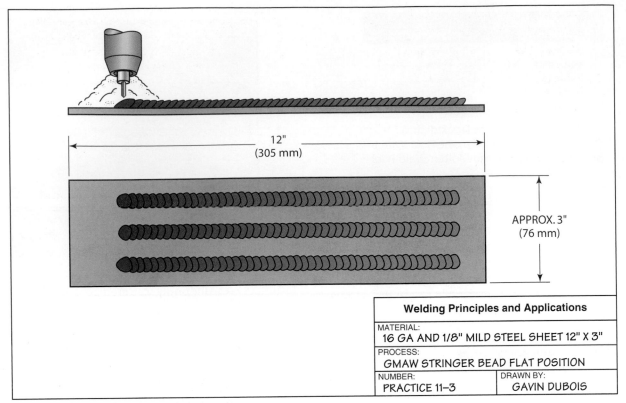

FIGURE 11-25 Stringer beads in the flat position. © Cengage Learning 2012

and two or more pieces of mild steel sheet 12 in. (305 mm) long and 16 gauge and 1/8 in. (3 mm) thick, you will make a stringer bead weld in the flat position, **Figure 11-25**.

Starting at one end of the plate and using either a pushing or dragging technique, make a weld bead along the entire 12-in. (305-mm) length of the metal. After the weld is complete, check its appearance. Make any needed changes to correct the weld (refer to Table 11-3 and Table 11-5). Repeat the weld and make additional adjustments. After the machine is set, start to work on improving the straightness and uniformity of the weld.

Keeping the bead straight and uniform can be hard because of the limited visibility due to the small amount of light and the size of the molten weld pool. The welder's view is further restricted by the shielding gas nozzle, **Figure 11-26**.

Even with limited visibility, it is possible to make a satisfactory weld by watching the edge of the molten weld pool, the sparks, and the weld bead produced. Watching the leading edge of the molten weld pool (forehand welding, pushing technique) will show you the molten weld pool fusion and width. Watching the trailing edge of the molten weld pool (backhand welding, dragging technique) will show you the amount of buildup and the relative heat input, **Figure 11-27**. The quantity and size of sparks produced can indicate the relative location of the filler wire in the molten weld pool. The number of sparks will increase as the wire strikes the solid metal ahead of the molten weld pool. The gun itself will begin to vibrate or bump as the wire momentarily pushes against the cooler, unmelted base metal before it melts. Changes in weld width, buildup, and proper joint

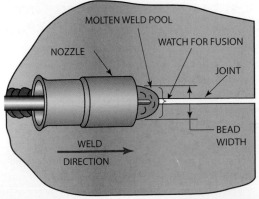

FIGURE 11-26 The shielding gas nozzle restricts the welder's view. © Cengage Learning 2012

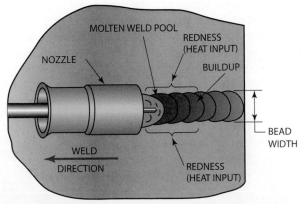

FIGURE 11-27 Watch the trailing edge of the molten weld pool. © Cengage Learning 2012

tracking can be seen by watching the bead as it appears from behind the shielding gas nozzle.

Repeat each type of bead as needed until consistently good beads are obtained. Turn off the welding machine and shielding gas and clean up your work area when you are finished welding.

Complete a copy of the "Student Welding Report" listed in Appendix I or provided by your instructor. ◆

PRACTICE 11-4

Flat Position Butt Joint, Lap Joint, and Tee Joint

Using the same equipment, materials, and procedures as listed in Practice 11-3, make welded butt joints, lap joints, and tee joints in the flat position, **Figure 11-28A, B, and C.**

- Tack weld the sheets together and place them flat on the welding table, **Figure 11-29.**

- Starting at one end, run a bead along the joint. Watch the molten weld pool and bead for signs that a change in technique may be required.

- Make any needed changes as the weld progresses. By the time the weld is complete, you should be making the weld nearly perfectly.

- Using the same technique that was established in the last weld, make another weld. This time, the entire 12 in. (305 mm) of weld should be flawless.

Repeat each type of joint with both thicknesses of metal as needed until consistently good beads are obtained. Turn

off the welding machine and shielding gas and clean up your work area when you are finished welding.

Complete a copy of the "Student Welding Report" listed in Appendix I or provided by your instructor. ◆

PRACTICE 11-5

Flat Position Butt Joint, Lap Joint, and Tee Joint, All with 100% Penetration

Using the same equipment, materials, and setup as listed in Practice 11-3, make a welded joint in the flat position with 100% penetration, along the entire 12-in. (305-mm) length of the welded joint. Repeat each type of joint as needed until consistently good beads are obtained. Turn off the welding machine and shielding gas and clean up your work area when you are finished welding.

Complete a copy of the "Student Welding Report" listed in Appendix I or provided by your instructor. ◆

PRACTICE 11-6

Flat Position Butt Joint, Lap Joint, and Tee Joint, All Welds to Be Tested

Using the same equipment, materials, and setup as listed in Practice 11-3, make each of the welded joints in the flat position, **Figure 11-30.** Each weld joint must pass the bend test. Repeat each type of weld joint until all pass the guided-bend test. Turn off the welding machine and shielding gas and clean up your work area when you are finished welding.

Complete a copy of the "Student Welding Report" listed in Appendix I or provided by your instructor. ◆

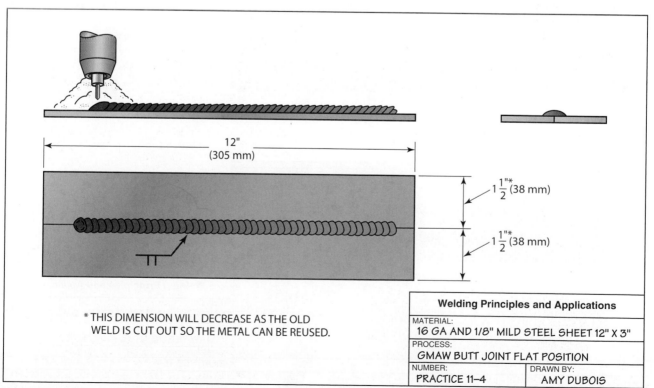

FIGURE 11-28 (A) Butt joint in the flat position. © Cengage Learning 2012

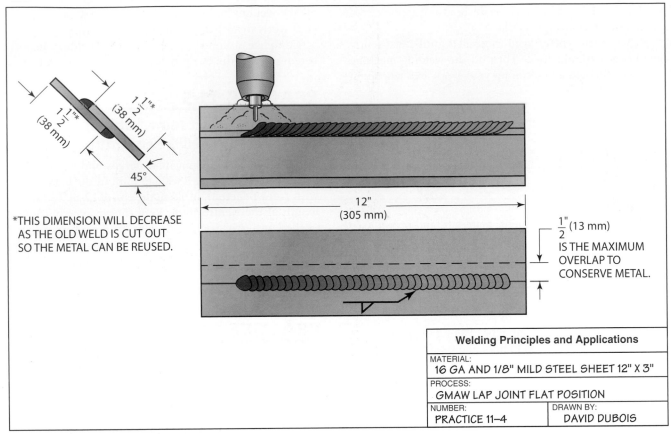

FIGURE 11-28 (B) Lap joint in the flat position. © Cengage Learning 2012

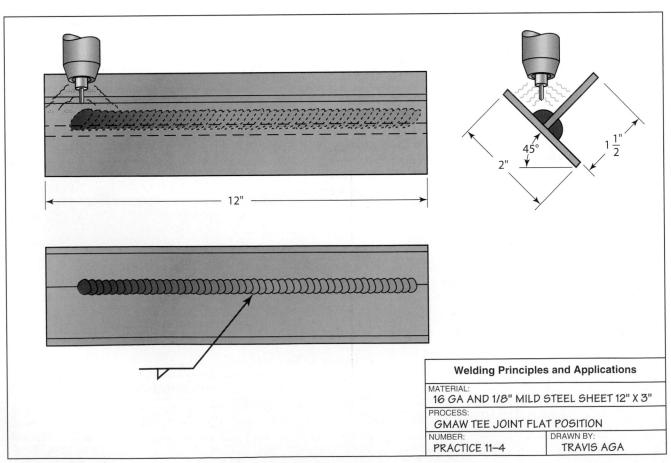

FIGURE 11-28 (C) Tee joint in the flat position. © Cengage Learning 2012

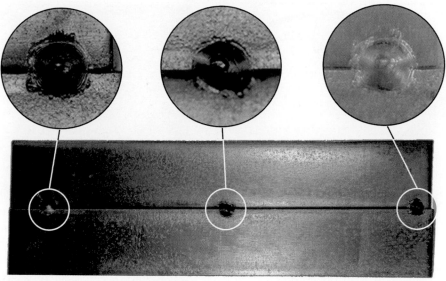

FIGURE 11-29 Use enough tack welds to keep the joint in alignment during welding. Small tack welds are easier to weld over without adversely affecting the weld. Larry Jeffus

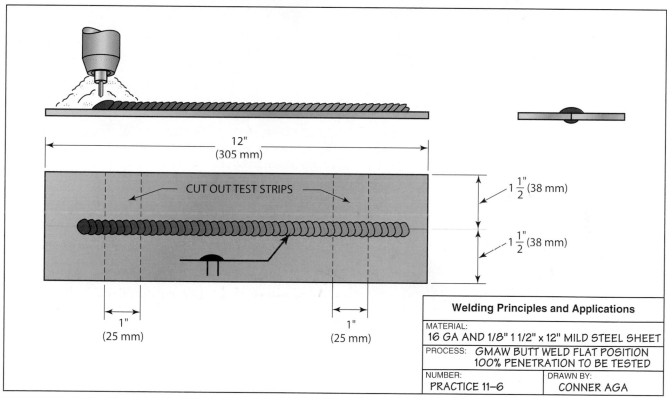

FIGURE 11-30 Butt joint in the flat position to be tested. © Cengage Learning 2012

Vertical Up 3G and 3F Positions

PRACTICE 11-7

Stringer Bead at a 45° Vertical Up Angle

Using the same equipment, materials, and setup as listed in Practice 11-3, you will make a vertical up stringer bead on a plate at a 45° inclined angle.

Start at the bottom of the plate and hold the welding gun at a slight pushing or upward angle to the plate, **Figure 11-31.** Brace yourself, lower your hood, and begin to weld. Depending upon the machine settings and type of shielding gas used, you will make a weave pattern.

If the molten weld pool is large and fluid (hot), use a "C" or "J" weave pattern to allow a longer time for the molten weld pool to cool, **Figure 11-32.** Do not make the weave so long or fast that the wire is allowed to strike the metal ahead of the molten weld pool. If this happens,

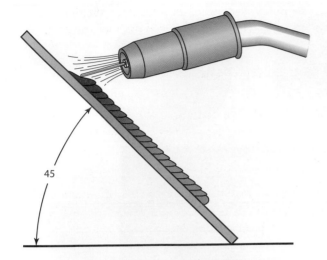

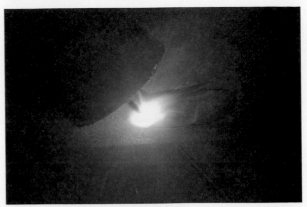

FIGURE 11-33 Burst of spatter caused by incorrect electrode contact with base metal. *Larry Jeffus*

FIGURE 11-31 Vertical up position. *Larry Jeffus*

FIGURE 11-34 Weld separated from the plate; there is no fusion between the weld and plate. *Larry Jeffus*

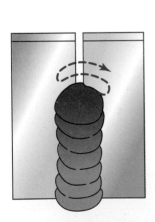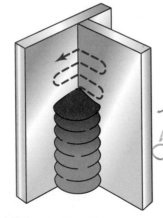

FIGURE 11-32 Vertical up welding weave patterns.
© Cengage Learning 2012

spatter increases and a spot or zone of incomplete fusion may occur, **Figure 11-33**.

If the molten weld pool is small and controllable, use a small "C," zigzag, or "J" weave pattern to control the width and buildup of the weld. A slower speed can also be used. Watch for complete fusion along the leading edge of the molten weld pool. **Figure 11-34** shows a weld that did not fuse with the plate.

A weld that is high and has little or no fusion is too "cold." Changing the welding technique will not correct

this problem. The welder must stop welding and make the needed adjustments.

As the weld progresses up the plate, the back or trailing edge of the molten weld pool will cool, forming a shelf to support the molten metal. Watch the shelf to be sure that molten metal does not run over, forming a drip. When it appears that the metal may flow over the shelf, either increase the weave lengths, lengthen the electrode extension, or stop and start the current for brief moments to allow the weld to cool. Stopping for brief moments will not allow the shielding gas to be lost.

Continue to weld along the entire 12-in. (305-mm) length of plate. Repeat this weld as needed until a straight and uniform weld bead is produced. Turn off the welding machine and shielding gas and clean up your work area when you are finished welding.

Complete a copy of the "Student Welding Report" listed in Appendix I or provided by your instructor. ◆

PRACTICE 11-8

Stringer Bead in the Vertical Up Position

Repeat Practice 11-7 and increase the angle of the plate until you have mastered a straight and uniform weld bead in the vertical up position. Turn off the welding machine and shielding gas and clean up your work area when you are finished welding.

Complete a copy of the "Student Welding Report" listed in Appendix I or provided by your instructor. ◆

PRACTICE 11-9

Butt Joint, Lap Joint, and Tee Joint in the Vertical Up Position at a 45° Angle

Using the same equipment, materials, and setup as listed in Practice 11-3, you will make vertical up welded joints on a plate at a 45° inclined angle.

Tack weld the metal pieces together and brace them in position. Check to see that you have free movement along the entire joint to prevent stopping and restarting during the weld. Avoiding stops and starts both speeds up the welding time and eliminates discontinuities.

The weave pattern should allow for adequate fusion on both edges of the joint. Watch the edges to be sure that they are being melted so that adequate fusion and penetration occur.

Repeat each type of joint as needed until consistently good beads are obtained. Turn off the welding machine and shielding gas and clean up your work area when you are finished welding.

Complete a copy of the "Student Welding Report" listed in Appendix I or provided by your instructor. ◆

PRACTICE 11-10

Butt Joint, Lap Joint, and Tee Joint in the Vertical Up Position with 100% Penetration

Using the same equipment, materials, and setup as listed in Practice 11-3, you will increase the plate angle gradually as you develop skill until you are making satisfactory welds in the vertical up position.

Repeat each type of joint as needed until consistently good beads are obtained. Turn off the welding machine and shielding gas and clean up your work area when you are finished welding.

Complete a copy of the "Student Welding Report" listed in Appendix I or provided by your instructor. ◆

PRACTICE 11-11

Butt Joint, Lap Joint, and Tee Joint in the Vertical Up Position, All Welds to Be Tested

Using the same equipment, materials, and setup as listed in Practice 11-3, you will make the welded joints in the vertical up position. Each weld must pass the bend test. Repeat each type of weld joint until all pass the bend test. Turn off the welding machine and shielding gas and clean up your work area when you are finished welding.

Complete a copy of the "Student Welding Report" listed in Appendix I or provided by your instructor. ◆

Vertical Down 3G and 3F Positions

The vertical down welding technique can be useful when making some types of welds. The major advantages of this technique are the following:

- Speed—Very high rates of travel are possible.
- Shallow penetration—Thin sections or root openings can be welded with little burnthrough.
- Good bead appearance—The weld has a nice width-to-height ratio and is uniform.

Vertical down welds are often used on thin sheet metals or in the root pass in grooved joints. The combination of controlled penetration and higher welding speeds makes vertical down the best choice for such welds. The ease with which welds having a good appearance can be made is deceiving. Generally, more skill is required to make sound welds with this technique than in the vertical up position. The most common problem with these welds is lack of fusion or overlap. To prevent these problems, the arc must be kept at or near the leading edge of the molten weld pool.

PRACTICE 11-12

Stringer Bead at a 45° Vertical Down Angle

Using the same equipment, materials, and setup as listed in Practice 11-3, you will make a vertical down stringer bead on a plate at a 45° inclined angle.

Holding the welding gun at the top of the plate with a slight dragging angle, **Figure 11-35**, will help to increase penetration, hold back the molten weld pool, and improve visibility of the weld. Be sure that your movements along the 12-in. (305-mm) length of plate are unrestricted.

Lower your hood and start the weld. Watch both the leading edge and sides of the molten weld pool for fusion. The leading edge should flow into the base metal, not curl over it. The sides of the molten weld pool should also show fusion into the base metal and not be flashed (ragged) along the edges.

The weld may be made with or without a weave pattern. If a weave pattern is used, it should be a "C" pattern. The "C" should follow the leading edge of the weld. Some changes on the gun angle may help to increase penetration. Experiment with the gun angle as the weld progresses.

Repeat these welds until you have established a rhythm and technique that work well for you. The welds must be straight and uniform and have complete fusion. Turn off the welding machine and shielding gas and clean up your work area when you are finished welding.

FIGURE 11-35 Vertical down position. Larry Jeffus

Complete a copy of the "Student Welding Report" listed in Appendix I or provided by your instructor. ◆

PRACTICE 11-13

Stringer Bead in the Vertical Down Position

Repeat Practice 11-12 and increase the angle of the plate until you have developed the skill to repeatedly make good welds in the vertical down position. The weld bead must be straight and uniform and have complete fusion. Turn off the welding machine and shielding gas and clean up your work area when you are finished welding.

Complete a copy of the "Student Welding Report" listed in Appendix I or provided by your instructor. ◆

PRACTICE 11-14

Butt Joint, Lap Joint, and Tee Joint in the Vertical Down Position

Using the same equipment, materials, and setup as listed in Practice 11-3, you will make vertical down welded joints.

Tack weld the pieces of metal together and brace them in position. Using the same technique developed in Practice 11-11, start at the top of the joint and weld down the length of the joint. When the weld is complete, inspect it for discontinuities and make any necessary changes in your technique. Repeat each type of joint as needed until consistently good welds are obtained. Turn off the welding machine and shielding gas and clean up your work area when you are finished welding.

Complete a copy of the "Student Welding Report" listed in Appendix I or provided by your instructor. ◆

PRACTICE 11-15

Butt Joint and Tee Joint in the Vertical Down Position with 100% Penetration

Using the same equipment, materials, and setup as listed in Practice 11-3, you will make welded joints with 100% weld penetration.

It may be necessary to adjust the root opening to meet the penetration requirements. The lap joint was omitted from this practice because little additional skill can be developed with it that is not already acquired with the tee joint. Repeat each type of joint as needed until consistently good welds are obtained. Turn off the welding machine and shielding gas and clean up your work area when you are finished welding.

Complete a copy of the "Student Welding Report" listed in Appendix I or provided by your instructor. ◆

PRACTICE 11-16

Butt Joint and Tee Joint in the Vertical Down Position, Welds to Be Tested

Using the same equipment, materials, and setup as listed in Practice 11-3, you will make the welded joints in the vertical down position. Each weld must pass the bend test. Repeat each type of weld joint until both pass the bend test. Turn off the welding machine and shielding gas and clean up your work area when you are finished welding.

Complete a copy of the "Student Welding Report" listed in Appendix I or provided by your instructor. ◆

Horizontal 2G and 2F Positions

PRACTICE 11-17

Horizontal Stringer Bead at a 45° Angle

Using the same equipment, materials, and setup as listed in Practice 11-3, you will make a horizontal stringer bead on a plate at a 45° reclined angle.

Start at one end with the gun pointed in a slightly upward direction, **Figure 11-36.** You may use a pushing or a dragging gun angle, depending upon the current setting and penetration desired. Undercutting along the top

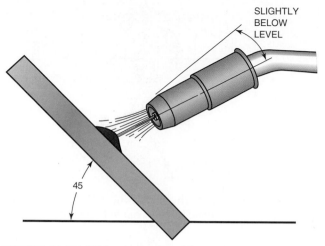

SLIGHTLY
BELOW
LEVEL

45

FIGURE 11-36 45° horizontal position. © Cengage Learning 2012

edge and overlap along the bottom edge are problems with both gun angles. Careful attention must be paid to the manipulation "weave" technique used to overcome these problems.

The most successful weave patterns are the "C" and "J" patterns. The "J" pattern is the most frequently used. The "J" pattern allows weld metal to be deposited along a shelf created by the previous weave, **Figure 11-37**. The length of the "J" can be changed to control the weld bead size. Smaller weld beads are easier to control than large ones.

Repeat these welds until you have established the rhythm and technique that work well for you. The weld must be straight and uniform and have complete fusion. Turn off the welding machine and shielding gas and clean up your work area when you are finished welding.

Complete a copy of the "Student Welding Report" listed in Appendix I or provided by your instructor. ◆

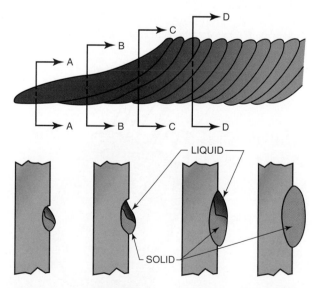

LIQUID

SOLID

SECTION A-A   SECTION B-B   SECTION C-C   SECTION D-D
FIGURE 11-37 The actual size of the molten weld pool remains small along the weld. © Cengage Learning 2012

PRACTICE 11-18

Stringer Bead in the Horizontal Position

Repeat Practice 11-17 and increase the angle of the plate until you have developed the skill to repeatedly make good horizontal welds on a vertical surface. The weld bead must be straight and uniform and have complete fusion. Turn off the welding machine and shielding gas and clean up your work area when you are finished welding.

Complete a copy of the "Student Welding Report" listed in Appendix I or provided by your instructor. ◆

PRACTICE 11-19

Butt Joint, Lap Joint, and Tee Joint in the Horizontal Position

Using the same equipment, materials, and setup as listed in Practice 11-3, you will make horizontal welded joints.

Tack weld the pieces of metal together and brace them in position using the same skills developed in Practice 11-17. Starting at one end, make a weld along the entire length of the joint. When making the butt or lap joints, it may help to recline the plates at a 45° angle until you have developed the technique required. Repeat each type of joint as needed until consistently good welds are obtained. Turn off the welding machine and shielding gas and clean up your work area when you are finished welding.

Complete a copy of the "Student Welding Report" listed in Appendix I or provided by your instructor. ◆

PRACTICE 11-20

Butt Joint and Tee Joint in the Horizontal Position with 100% Penetration

Using the same equipment, materials, and setup as listed in Practice 11-3, you will make overhead joints having 100% penetration in the horizontal position.

It may be necessary to adjust the root opening to meet the penetration requirements. Repeat each type of joint as needed until consistently good welds are obtained. Turn off the welding machine and shielding gas and clean up your work area when you are finished welding.

Complete a copy of the "Student Welding Report" listed in Appendix I or provided by your instructor. ◆

PRACTICE 11-21

Butt Joint and Tee Joint in the Horizontal Position, Welds to Be Tested

Using the same equipment, materials, and setup as listed in Practice 11-3, you will make the welded joints in the horizontal position. Each weld must pass the bend test. Repeat each type of weld joint until both pass the bend test. Turn off the welding machine and shielding

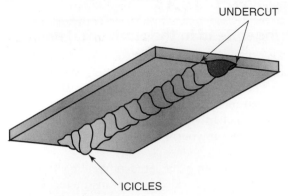

FIGURE 11-38 Overhead weld. © Cengage Learning 2012

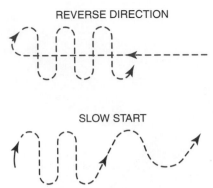

FIGURE 11-39 Two methods of concentrating heat at the beginning of a weld bead to aid in penetration depth. © Cengage Learning 2012

gas and clean up your work area when you are finished welding.

Complete a copy of the "Student Welding Report" listed in Appendix I or provided by your instructor. ◆

Overhead 4G and 4F Positions

Following are several advantages for the use of short-circuiting arc metal transfer in the overhead position:

■ Small molten weld pool size—The smaller size of the molten weld pool allows surface tension to hold it in place. Less molten weld pool sag results in improved bead contour with less undercut and fewer icicles, **Figure 11-38**.

■ Direct metal transfer—The direct metal transfer method does not rely on other forces to get the filler metal into the molten weld pool. This results in efficient metal transfer and less spatter and loss of filler metal.

PRACTICE 11-22

Stringer Bead Overhead Position

Using the same equipment, materials, and setup as listed in Practice 11-3, you will make a welded stringer bead in the overhead position.

The molten weld pool should be kept as small as possible for easier control. A small molten weld pool can be achieved by using lower current settings, by using a longer wire stickout, by traveling faster, or by pushing the molten weld pool. The technique used is the welder's choice. Often a combination of techniques can be used with excellent results.

Lower current settings require closer control of gun manipulation to ensure that the wire is fed into the molten weld pool just behind the leading edge. The low power will cause overlap and more spatter if this wire-to-molten weld pool contact position is not closely maintained.

Faster travel speeds allow the welder to maintain a high production rate even if multiple passes are required to complete the weld. Weld penetration into the base

metal at the start of the bead can be obtained by using a slow start or quickly reversing the weld direction. Both the slow start and reversal of weld direction put more heat into the start to increase penetration, **Figure 11-39**. The higher speed also reduces the amount of weld distortion by reducing the amount of time that heat is applied to a joint.

The pushing or trailing gun angle forces the bead to be flatter by spreading it out over a wider area as compared to the bead resulting from a dragging or backhand gun angle. The wider, shallow molten weld pool cools faster, resulting in less time for sagging and the formation of icicles.

When welding overhead, extra personal protection is required to reduce the danger of burns. Leather sleeves or leather jackets and caps should be worn.

Much of the spatter created during overhead welding falls into the shielding gas nozzle. The effectiveness of the shielding gas is reduced, **Figure 11-40**, and the contact tube may short out to the gas nozzle, **Figure 11-41**. Turbulence caused by the spatter obstructing the gas may

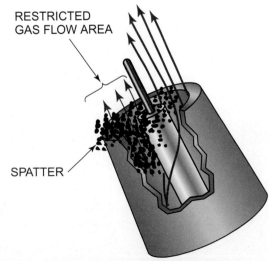

FIGURE 11-40 Shielding gas flow affected by excessive weld spatter in nozzle. © Cengage Learning 2012

FIGURE 11-41 Gas nozzle damaged after shorting out against the work. Larry Jeffus

lead to weld contamination. The shorted gas nozzle may arc to the work, causing damage both to the nozzle and to the plate. To control the amount of spatter, a longer stickout and/or a sharper gun-to-plate angle is required to allow most of the spatter to fall clear of the gas nozzle. The nozzle can be dipped, sprayed, or injected automatically, **Figure 11-42**, with anti-spatter to help prevent the spatter from sticking. Applying anti-spatter will not stop the spatter from building up, but it does make its removal much easier.

Make several short weld beads using various techniques to establish the method that is most successful and most comfortable for you. After each weld, stop and evaluate it before making a change. When you have decided on the technique to be used, make a welded stringer bead that is 12 in. (305 mm) long.

Repeat the weld until it can be made straight, uniform, and free from any visual defects. Turn off the welding machine and shielding gas and clean up your work area when you are finished welding.

Complete a copy of the "Student Welding Report" listed in Appendix I or provided by your instructor. ◆

PRACTICE 11-23

Butt Joint, Lap Joint, and Tee Joint in the Overhead Position

Using the same equipment, materials, and setup as listed in Practice 11-3, you will make an overhead welded joint.

Tack weld the pieces of metal together and secure them in the overhead position. Be sure you have an unrestricted view and freedom of movement along the joint. Start at one end and make a weld along the joint. Use the same technique developed in Practice 11-22.

Repeat the weld until it can be made straight, uniform, and free from any visual defects. Turn off the welding machine and shielding gas and clean up your work area when you are finished welding.

Complete a copy of the "Student Welding Report" listed in Appendix I or provided by your instructor. ◆

PRACTICE 11-24

Butt Joint and Tee Joint in the Overhead Position with 100% Penetration

Using the same equipment, materials, and setup as listed in Practice 11-3, you will make overhead welded joints having 100% penetration.

Tack weld the metal together. It may be necessary to adjust the root opening to allow 100% weld metal penetration. During these welds, it may be necessary to use a dragging or backhand torch angle. When used with a "C" or "J" weave pattern, this torch angle helps to achieve the

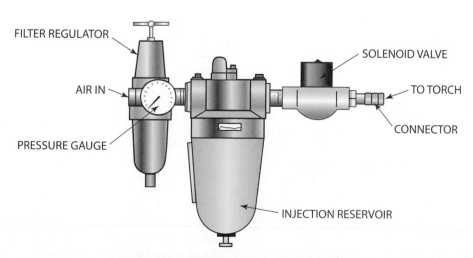

OPERATING INSTRUCTIONS AND PARTS MANUAL

FIGURE 11-42 Automatic anti-spatter system that can be added to a GMA welding gun. © Cengage Learning 2012

ROOT OPENING

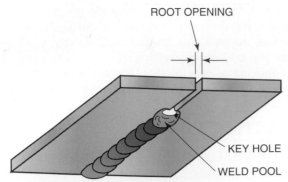

KEY HOLE

WELD POOL

FIGURE 11-43 Overhead welding. © Cengage Learning 2012

FIGURE 11-44 Weld bead made with GMAW globular metal transfer mode. Larry Jeffus

desired depth of penetration. A key hole just ahead of the molten weld pool is a good sign that the metal is being penetrated, **Figure 11-43**.

Repeat the weld until it can be made straight, uniform, and free from any visual defects. Turn off the welding machine and shielding gas and clean up your work area when you are finished welding.

Complete a copy of the "Student Welding Report" listed in Appendix I or provided by your instructor. ◆

PRACTICE 11-25

Butt Joint and Tee Joint in the Overhead Position, Welds to Be Tested

Using the same equipment, materials, and setup as listed in Practice 11-3, you will make a welded joint in the overhead position. Each weld must pass the bend test. Repeat each type of weld joint until both pass the bend test. Turn off the welding machine and shielding gas and clean up your work area when you are finished welding.

Complete a copy of the "Student Welding Report" listed in Appendix I or provided by your instructor. ◆

Globular Metal Transfer, 1G Position

PRACTICE 11-26

Stringer Bead

Using a properly set-up and adjusted GMA welding machine (see **Table 11-11**), proper safety protection, 0.035-in. and/or 0.045-in. (0.9-mm and/or 1.2-mm) diameter wire, and two or more pieces of mild steel plate 12 in. (305 mm) long and 1/4 in. (6 mm) thick, make a stringer bead weld in the flat position using the globular metal transfer method.

- Start at one end of the plate and use either a push or drag technique to make a weld bead along the entire 12-in. (305-mm) length of the metal.

- After completing the weld, check its appearance and make any changes needed to correct the weld, Figure 11-44.

- Repeat the weld and make additional adjustments as required in the frequency, amplitude, or pulse width.

- After the machine is set, start working on improving the straightness and uniformity of the weld.

The location of the arc in the molten weld pool is not as critical in this method of metal transfer as it is in the short-circuiting arc method. If the arc is too far back on the molten weld pool, however, fusion along the leading edge may be reduced, **Figure 11-45**. Moving the arc to the cool metal ahead of the molten weld pool often causes an increase in spatter.

The weave pattern selected should allow the arc to follow the leading edge of the molten weld pool if deep penetration is needed. A pattern that moves the arc back from the leading edge completely or in part results in reduced penetration.

Repeat the weld until it can be made straight, uniform, and free from any visual defects. Turn off the welding machine and shielding gas and clean up your work area when you are finished welding.

Complete a copy of the "Student Welding Report" listed in Appendix I or provided by your instructor. ◆

PRACTICE 11-27

Butt Joint

Using the same equipment, materials, and setup as listed in Practice 11-26, you will make a welded joint in the flat position, **Figure 11-46**.

The heat produced during GMA welding is not enough to force 100% penetration consistently through metal thicker than 3/16 in. (4.8 mm). On occasion, or with extra effort, it is possible to obtain 100% penetration. However, the method generally is not reliable enough for most industrial applications. To ensure that the joint is completely fused, it is prepared for welding with a groove. Any one of several groove designs can be used, **Figure 11-47**. The V-groove is most frequently used because of the ease with which it can be produced.

FIGURE 11-45 Molten globule of metal being transferred across the arc. Larry Jeffus

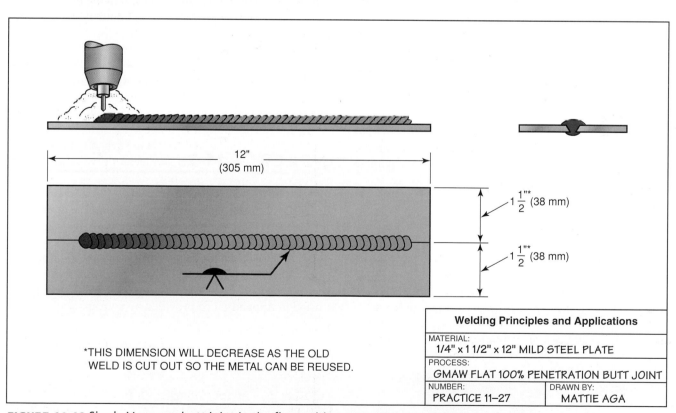

FIGURE 11-46 Single V-groove butt joint in the flat position. © Cengage Learning 2012

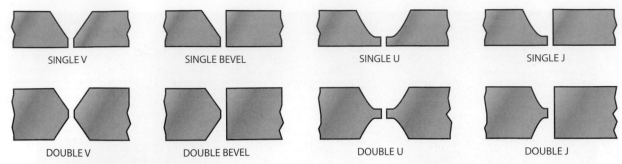

FIGURE 11-47 Typical groove designs. © Cengage Learning 2012

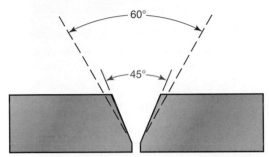

FIGURE 11-48 A smaller groove angle reduces both weld time and filler metal required to make the weld. © Cengage Learning 2012

The V-groove should be made with a 45° inclined angle, **Figure 11-48.** The small diameter filler metal allows the usual 60% inclined angle for SMAW to be reduced. This reduced groove size requires less filler metal and can be welded in less time, resulting in lower cost.

After the metal has been grooved, tack weld it together. The root opening between the plates should be 1/8 in. (3 mm) or less. Starting at one end, make a single pass weld along the entire joint. A weave pattern that follows the contour of the groove should be used to ensure complete fusion, **Figure 11-49.**

The completed weld should be uniform in width and reinforcement and have no visual defects. Repeat the weld until it can be made straight, uniform, and free from any visual defects. Turn off the welding machine and shielding gas and clean up your work area when you are finished welding.

Complete a copy of the "Student Welding Report" listed in Appendix I or provided by your instructor. ◆

PRACTICE 11-28

Butt Joint with 100% Penetration

Using the same equipment, materials, and setup as listed in Practice 11-16, you will make a groove weld having 100% penetration.

Using the technique developed in Practice 11-27 and making the necessary adjustments in the root gap, the weld metal must fuse 100% of the plate thickness. Watch the molten weld pool. If it appears to sink or does not increase in size, you are probably burning through. This action will cause excessive root penetration. To correct this problem, speed up the travel rate and/or increase the contact tube-to-work distance.

After the weld is complete, inspect the root surface for the proper appearance. Repeat the weld until it can be made straight, uniform, and free from any visual defects. Turn off the welding machine and shielding gas and clean up your work area when you are finished welding.

Complete a copy of the "Student Welding Report" listed in Appendix I or provided by your instructor. ◆

PRACTICE 11-29

Butt Joint to Be Tested

Using the same equipment, materials, and setup as listed in Practice 11-26, you will make a flat groove weld. The weld must pass the bend test. Repeat the weld until the test can be passed. Turn off the welding machine and shielding gas and clean up your work area when you are finished welding.

Complete a copy of the "Student Welding Report" listed in Appendix I or provided by your instructor. ◆

FIGURE 11-49 Uniform weld produced with a weave pattern. Larry Jeffus

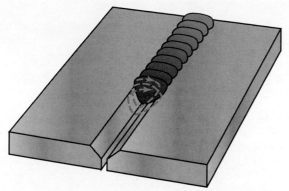

FIGURE 11-50 The electrode should be moved along the groove contour. © Cengage Learning 2012

PRACTICE 11-30

Tee Joint and Lap Joint in the IF Position

Using the same equipment, materials, and setup as listed in Practice 11-26, you will make a fillet weld in the flat position.

It is not necessary to groove the plates used for a tee or lap joint to obtain a sound weld. Fillet weld strength can be obtained by making the weld the proper size for the plate thickness.

The face or surface of a fillet weld should be as flat as possible. Welds with excessive buildup will waste metal. Welds with too little buildup will be weak, flat, or concave. Fillet weld beads have fewer stress points to cause weld failure during cyclical loading.

A weave pattern that follows the contour of the joint should be used to ensure adequate fusion, **Figure 11-50**. Repeat the weld until it can be made straight, uniform, and free from any visual defects. Turn off the welding machine and shielding gas and clean up your work area when you are finished welding.

Complete a copy of the "Student Welding Report" listed in Appendix I or provided by your instructor. ◆

PRACTICE 11-31

Tee Joint and Lap Joint in the 2F Position

Using the same equipment, materials, and setup as listed in Practice 11-26, you will make a fillet weld in the horizontal position.

Tack weld the metal together and place the assembly in the horizontal position. The globular pulsed-arc metal transfer method can be used for a horizontal weld. However, care must be taken to ensure that the legs of the

fillet are equal. Because of the size and fluidity of the molten weld pool, undercutting along the top edge and overlap along the bottom edge can also be problems.

To control or eliminate these defects, the beads must be small and quickly made. In addition, a proper weave pattern must be established. The pattern must follow the plate surfaces and establish a shelf to support the weld. After the weld is complete, inspect it for defects and measure it for uniformity. Repeat the weld until it can be made straight, uniform, and free from any visual defects. Turn off the welding machine and shielding gas and clean up your work area when you are finished welding.

Complete a copy of the "Student Welding Report" listed in Appendix I or provided by your instructor. ◆

Axial Spray

PRACTICE 11-32

Stringer Bead, 1G Position

Using a properly set-up and adjusted GMA welding machine (see **Table 11-12**), proper safety protection, 0.035-in. and/or 0.045-in. (0.9-mm and/or 1.2-mm) diameter wire, and two or more pieces of mild steel plate 12 in. (305 mm) long and 1/4 in. (6 mm) thick, you will make a welded stringer bead in the flat position.

Start at one end of the plate and use either a push or drag technique to make a weld bead along the entire 12-in. (305-mm) length of the metal using spray or pulsed-arc metal transfer. After the weld is complete, check its appearance and make any changes needed to correct the weld, **Figure 11-51**. Repeat the weld and make any additional

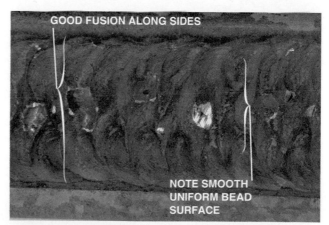

GOOD FUSION ALONG SIDES

NOTE SMOOTH UNIFORM BEAD SURFACE

FIGURE 11-51 Weld bead made with GMAW axial spray metal transfer. Larry Jeffus

Process	Wire Diameter	Amperage Range	Voltage Range	Shielding Gas
Axial spray	0.030	115–200	15–27	
	0.035	165–300	18–32	98% Ar + 2% O_2
	0.045	200–450	20–34	

TABLE 11-12 Typical Welding Current Settings for Axial Spray Metal Transfer for Mild Steel

adjustments required. After the machine is set, start working on improving the straightness and uniformity of the weld. Turn off the welding machine and shielding gas and clean up your work area when you are finished welding.

Complete a copy of the "Student Welding Report" listed in Appendix I or provided by your instructor. ◆

PRACTICE 11-33

Butt Joint, Lap Joint, and Tee Joint Using the Axial Spray Method

Using the same equipment, materials, and setup as listed in Practice 11-32, you will make a flat and horizontal weld using axial spray or pulsed axial spray metal transfer, **Figure 11-52**.

Tack weld the metal together and place the assembly in the flat position on the welding table. Start at one end and make a uniform weld along the entire 12-in. (305-mm) length of the joint. Watch the sides of the fillet weld for signs of undercutting.

Repeat the weld until it can be made straight, uniform, and free from any visual defects. Turn off the welding machine and shielding gas and clean up your work area when you are finished welding.

Complete a copy of the "Student Welding Report" listed in Appendix I or provided by your instructor. ◆

FIGURE 11-52 GMAW axial spray metal transfer. Larry Jeffus

PRACTICE 11-34

Butt Joint and Tee Joint

Using the same equipment, materials, and setup as listed in Practice 11-32, you will make a flat weld using axial spray metal transfer. Each weld must pass the guided-bend test. Repeat each type of weld joint as needed until the bend test can be passed. Turn off the welding machine and shielding gas and clean up your work area when you are finished welding.

Complete a copy of the "Student Welding Report" listed in Appendix I or provided by your instructor. ◆

Summary

Slight changes in welding gun angle and electrode extension can make significant differences in the quality of the weld produced. As a new welder you might find it difficult to tell the effect of these changes if they are slight. Therefore, as you start to learn this process it is a good idea to make more radical changes so it is easier for you to see their effects on the weld. Later as you develop your skills you can use these slight changes to aid in controlling the weld's quality and appearance as it progresses along the joint. Small adjustments in your welding technique are required to compensate for slight changes that occur along a welding joint, such as joint gap and the increasing temperature of the base metal.

Variations in conditions can significantly affect welding setup for the GMA process. Before starting an actual weld in the field you should practice to test your setup. Practice on scrap metal of a similar thickness and type of metal to be welded. A practice weld before you begin welding can significantly increase the chances that your weld will meet standards or specifications.

Making these sample or test welds is more important when you are welding in the field, since welds outside of the shop are more difficult to control and anticipate. Think of this much as an athlete warms up before competing.

You will find it beneficial when you are initially setting up your welder to have someone assist you so that he or she can make changes in the welding machine's settings as you are welding. This teamwork can significantly increase your setup accuracy and reduce setup time. Later on in the field, having developed a keen eye for watching the weld, you can then make these adjustments for yourself more rapidly and accurately. Working with another student in a group effort like this will also give you a better understanding of how other individuals' setup preferences affect their welds. Welding is an art, and therefore each welder may have slight differences in preference for voltage, amperage, gas flow, and other setup variables. This gives you an opportunity to learn more from others.

Aluminum Ferries Rely on Inverter Technology

If Rick Kirschman had believed the experts, he would be selling insurance or teaching welding at a community college.

"I still keep a sticky note from a welding consultant in my handbook next to the American Bureau of Shipping (ABS) specifications for an aluminum vessel," said Kirschman, project manager for Dakota Creek Industries, Inc., a shipyard in Anacortes, Washington. "The consultant's note said, 'These values cannot be attained. There are very few people in the world who can attain these values.'"

Fortunately for Dakota Creek, who had just won the contract to build two aluminum fast ferries, Kirschman did not believe either the consultant or in accepting failure. Instead, he believed in pulsed-gas metal arc welding (GMAW-P), inverter power sources, and programmable wire feeders.

After hundreds of test coupons, technical support, and six weeks of intense work, Kirschman developed procedures for pulsed GMA welding 5083 aluminum from 0.10 in. (2.5 mm) to more than 3 in. (75 mm) thick.

The welds produced have yield strengths of 27,250 lb/in^2 and tensile strengths of 45,270 lb/in^2 and have passed all bend tests. This exceeds ABS and Coast Guard standards by more than 21% on the yield and 12% on the tensile.

Not Too Hot, Not Too Cold

The 143-ft fast ferries can carry 300 passengers at speeds of 42 knots because their hulls, superstructures, and most other components are made from lightweight aluminum. The aluminum in the hull ranges from 0.19 in. to 0.024 in. (4.75 mm to 6 mm) near the bow and midsection and from 0.31 in. to 0.40 in. (8 mm to 10 mm) near the engines, as thicker sections better withstand vibration.

In other industries, most fabricators would use the GMAW process in the short-circuiting mode to weld thin aluminum and the spray transfer process to weld thicker sections. Dakota Creek, however, could not use either of these processes. ABS does not permit short-circuit transfer because of possible incomplete fusion. Spray transfer is accepted by ABS, but it would not work for the critical butt joint welds.

"Spray transfer hits the metal so hard, and aluminum absorbs heat so quickly, that this process destroys the temper of the metal," said Kirschman. As a result, welds fail yield and tensile strength tests. Only the GMAW-P process could provide a solution. ABS accepts the GMAW-P process because it is technically a modified spray transfer process.

The welding equipment chosen allows the pulse waveform components (peak amperage, background amperage, pulse width, and pulses per second) to be adjusted with the push of a few buttons.

Fast ferries fabricated by Dakota Creek can carry 300 passengers and attain speeds of 42 knots. American Welding Society

"Fine tuning pulsing values can be a deceptive process, since a beautiful looking weld does not necessarily indicate weld quality," noted Kirschman.

"A flat and shiny bead with good wet out probably won't pass ABS tests because it went in too hot," explained Kirschman. "Such a weld passes the bend test, but it probably won't have high enough tensile and yield values. Too much heat also degrades metal composition, so the weld won't pass an X-ray or ultrasonic test, either." But if the weld goes in too cool, the metal loses ductility and subsequently fails the bend test. Too little heat also produces incomplete fusion. "Aluminum dissipates heat so quickly that the base metal can rob heat from the weld pool, especially in the first few inches of the weld," said Kirschman.

Kirschman explained the latter concept from an experience he had when he visited another shipyard. The welding operator there did not believe him. The operator indicated some fillet welds and boasted of their excellent fusion. Kirschman proved his point by slamming a hammer against the weldment. The fillet welds popped off; they had failed to fuse to both surfaces.

Documentation and Patience Required

Kirschman stresses that creating a successful GMAW-P program requires a methodical approach. He changed one variable at a time, documented every change, and waited patiently for results after an ABS-certified lab calculated tensile, yield, and bend values.

The bow, hull, and superstructure of an aluminum ferry in various stages of fabrication. American Welding Society.

Operator Joe Gilden welds stainless steel pipe with the pulsed gas metal arc process. American Welding Society.

"Only after you get the lab results can you see how changing one factor affects the quality of the metal," he said. "That's really important when dealing with ABS specifications."

New Rules

Although the welding consultant the company hired underestimated Kirschman's patience and skill with GMAW-P programs, he did provide other good aluminum welding advice.

"Coming from a steel background, I didn't believe most of the things he told me at first," said Kirschman. "But after a month of testing, I understood." He says he learned that good mechanical strength in a butt joint weld demands using an included angle of more than 60°.

After completing the first side, aluminum also demands backgouging (from the base metal through to the weld bead) and creating another 60° included angle. Dakota Creek operators use a grinder with a heavy 4 1/2-in. (114.3-mm) blade designed to cut at 60°. "With anything less than a 60-degree angle, aluminum pulls apart prematurely in a bend test, even if the weld passes X-ray," stated Kirschman.

The company's welding procedures ensure that if the fast ferry's hull ever takes a hit, such as from a buoy or large log, the welds will remain as mechanically sound as possible. "Backgouging is critical with aluminum because, unlike malleable steel, aluminum does not retain its mechanical properties once overstressed," said Kirschman.

When welding aluminum, operators cannot strike the arc at the spot where they want the weld to begin. Instead, they must draw the gun slightly back, perhaps 1/2 in. to 1 in. (12.7 mm to 25.4 mm), as they pull the trigger and establish the arc. They then move forward and weld over the point of arc initiation, always using a forehand, or pushing, technique to ensure good gas coverage. Because a short circuit can occur during arc start, this method washes out any cold spot and ensures complete fusion.

Multiple Functions

For welding in a shipyard, a multiple process, inverter power source provides the most value and best return on investment. "We need to be able to stick weld, MIG, and pulsed MIG weld aluminum and stainless, TIG weld ferrous metals, and arc gouge with the same power supply," said Kirschman.

Since the initial two fast ferries, the company has completed a contract for two fast ferries for the State of Washington, a multimillion-dollar luxury yacht hull, and the stainless steel pipe systems on five tugboats, all manufactured with the GMAW-P process.

Article courtesy of the American Welding Society.

Review

1. What items make up a basic semiautomatic welding system?

2. What must be done to the shielding gas cylinder before the valve protection cap is removed?

3. Why is the shielding gas valve "cracked" before the flowmeter regulator is attached?

4. What causes the electrode to bird-nest?

5. Why must all fittings and connections be tight?

6. What parts should be activated by depressing the gun switch?

7. What benefit does a welding wire's cast provide?

8. What can be done to determine the location of a problem that stops the wire from being successfully fed through the conduit?

9. What are the advantages of using a feed roller pressure that is as light as possible?

10. Why should the feed roller drag prevent the spool from coasting to a stop when the feed stops?

11. Why must you always wind the wire tightly into a ball or cut it into short lengths before discarding it in the proper waste container?

12. What reactive gases are used with argon for GMA welding on steels?

13. What effect on the welding current does increasing the wire feed speed have?

14. Why would the flowmeter ball float at different heights with different shielding gases if the shielding gases are flowing at the same rate?

15. Using Table 11-1, determine the amperage if 400 ipm (10.2 m/min) of 0.45-in. (1.2-mm) steel wire is fed in 1 minute.

16. How is the amperage adjusted on a GMA welder?

17. What happens to the weld as the electrode extension is lengthened?

18. What is the effect on the weld of changing the welding angle from a dragging to a pushing angle?

19. What are the advantages of adding oxygen or CO_2 to argon for welds on steel?

20. What are the advantages of using CO_2 for making GMA welds on steel?

21. What is mill scale?

22. What type of porosity is most often caused by mill scale?

23. What should the welder watch if the view of the weld is obstructed by the shielding gas nozzle?

24. When making a vertical weld and it appears that the weld metal is going to drip over the shelf, what should you do?

25. What are the advantages of making vertical down welds?

26. How can small weld beads be maintained during overhead welds?

27. How can spatter be controlled on the nozzle when making overhead welds?

Chapter 12

Flux Cored Arc Welding Equipment, Setup, and Operation

OBJECTIVES

After completing this chapter, the student should be able to

- explain the FCA welding process.
- describe what equipment is needed for FCA welding.
- list the advantages of FCA welding, and explain its limitations.
- tell how electrodes are manufactured and explain the purpose of the electrode cast and helix.
- discuss what flux can provide to the weld and how fluxes are classified.
- explain what each of the digits in a standard FCAW electrode identification number mean.
- describe the proper care and handling of FCAW electrodes.
- list the common shielding gases used, and explain their benefits.
- explain how changing the welding gun angle affects the weld produced.
- identify the methods of metal transfer and describe each.
- explain the effect electrode extension has on FCA welding.
- tell what can cause weld porosity and how it can be prevented.

KEY TERMS

air-cooled gun ~shielding *[handwritten]*
grooving *[handwritten]*

carbon

coils

deoxidizers absorbs *[handwritten]*

⊗ dual shield flux, Inner shield *[handwritten]*

flux cored arc welding (FCAW)

forehand

lime-based flux

rutile-based flux

self-shielding

slag

smoke extraction nozzles

spools

water-cooled gun

INTRODUCTION

Flux cored arc welding (FCAW) is a fusion welding process in which weld heating is produced from an arc between the work and a continuously fed filler metal electrode. Atmospheric shielding is provided completely or in part by the flux sealed within the tubular electrode, **Figure 12-1**. Extra shielding may or may not be supplied through a nozzle in the same way as in GMAW.

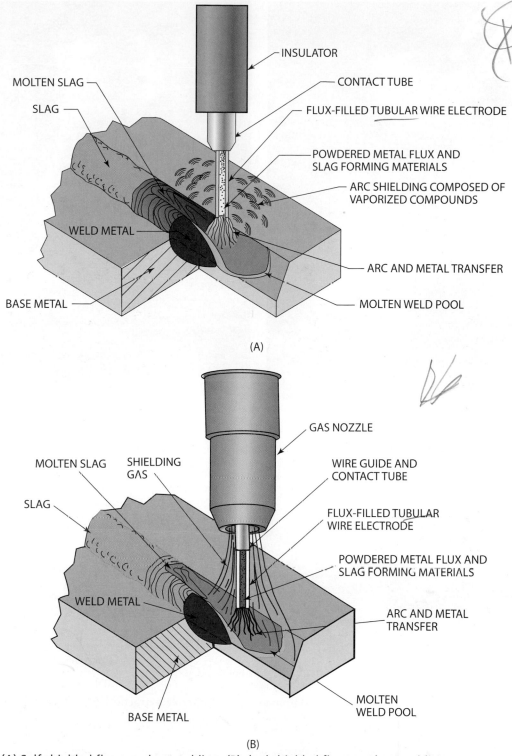

INSULATOR

CONTACT TUBE

FLUX-FILLED TUBULAR WIRE ELECTRODE

MOLTEN SLAG

SLAG

POWDERED METAL FLUX AND SLAG FORMING MATERIALS

ARC SHIELDING COMPOSED OF VAPORIZED COMPOUNDS

WELD METAL

ARC AND METAL TRANSFER

MOLTEN WELD POOL

BASE METAL

(A)

GAS NOZZLE

WIRE GUIDE AND CONTACT TUBE

MOLTEN SLAG

SHIELDING GAS

SLAG

FLUX-FILLED TUBULAR WIRE ELECTRODE

POWDERED METAL FLUX AND SLAG FORMING MATERIALS

WELD METAL

ARC AND METAL TRANSFER

MOLTEN WELD POOL

BASE METAL

(B)

FIGURE 12-1 (A) Self-shielded flux cored arc welding; (B) dual-shielded flux cored arc welding. American Welding Society

Although the process was introduced in the early 1950s, it represented less than 5% of the total amount of welding done in 1965. In 2005, it passed the 50% mark and is still rising. The rapid rise in the use of FCAW has been due to a number of factors. The improvements in the fluxes, smaller electrode diameters, reliability of the equipment, better electrode feed systems, and improved guns have all led to the increased usage. Guns equipped with **smoke extraction**

nozzles and electronic controls are the latest in a long line of improvements to this process, **Figure 12-2.**

Principles of Operation

FCA welding is similar in a number of ways to the operation of GMA welding, **Figure 12-3.** Both processes use a constant-potential (CP) or constant-voltage (CV)

(A)

(B)

(C)

(D)

FIGURE 12-2 (A) FCA welding without smoke extraction and (B) with smoke extraction. (C) Typical FCAW smoke extraction gun. (D) Typical smoke exhaust system. Lincoln Electric Company

power supply. *Constant-potential* and *voltage* are terms that have the same meaning. CP power supplies provide a controlled voltage (potential) to the welding electrode. The amperage (current) varies with the speed that the electrode is being fed into the molten weld pool. Just as in GMA welding, higher electrode feed speeds produce higher currents and slower feed speeds result in lower currents, assuming all other conditions remain constant.

The effects on the weld of electrode extension, gun angle, welding direction, travel speed, and other welder manipulations are similar to those experienced in GMA

welding. As in GMA welding, having a correctly set welder does not ensure a good weld. The skill of the welder is an important factor in producing high-quality welds.

The flux inside the electrode provides the molten weld pool with protection from the atmosphere, improves strength through chemical reactions and alloys, and improves the weld shape.

Atmospheric contamination of molten weld metal occurs as it travels across the arc gap and within the pool before it solidifies. The major atmospheric contaminations come from oxygen and nitrogen, the major elements

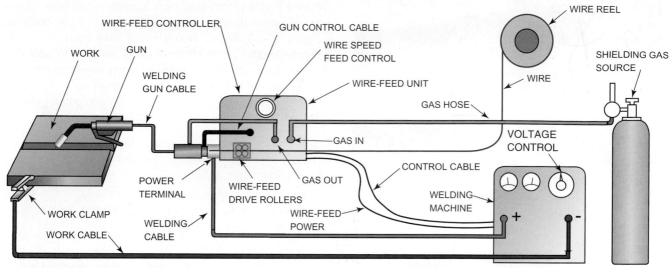

FIGURE 12-3 Large capacity wire-feed unit can be used with FCAW or GMAW. © Cengage Learning 2012

in air. The addition of fluxing and gas-forming elements to the core electrode reduces or eliminates their effects.

Improved strength and other physical or corrosive properties of the finished weld are improved by the flux. Small additions of alloying elements, deoxidizers, and gas-forming and slag agents can all improve the desired weld properties. **Carbon,** chromium, and vanadium can be added to improve hardness, strength, creep resistance, and corrosion resistance. Aluminum, silicon, and titanium all help remove oxides and/or nitrides in the weld. Potassium, sodium, and zirconium are added to the flux and form a slag.

A complete listing of weld metal additives and flux elements and their effects on the weld can be found in Chapter 25, "Welding Metallurgy," and Chapter 27, "Filler Metal Selection."

The flux core additives that serve as deoxidizers, gas formers, and slag formers either protect the molten weld pool or help to remove impurities from the base metal. Deoxidizers may convert small amounts of surface oxides like mill scale back into pure metal. They work much like the elements used to refine iron ore into steel.

Gas formers rapidly expand and push the surrounding air away from the molten weld pool. If oxygen in the air were to come in contact with the molten weld metal, the weld metal would quickly oxidize. Sometimes this can be seen at the end of a weld when the molten weld metal erupts in a shower of tiny sparks.

The slag covering of the weld is useful for several reasons. **Slag** helps the weld by protecting the hot metal from the effects of the atmosphere, controlling the bead shape by serving as a dam or mold, and serving as a blanket to slow the weld's cooling rate, which improves its physical properties, **Figure 12-4.**

Equipment

Power Supply The FCA welding power supply is the same type that is required for GMAW, called constant-potential, constant-voltage (CP, CV). The words *potential*

and *voltage* have the same electrical meaning and are used interchangeably. FCAW machines can be much more powerful than GMAW machines and are available with up to 1500 amperes of welding power.

Guns FCA welding guns are available as **water-cooled** or **air-cooled, Figure 12-5.** Although most of the FCA welding guns that you will find in schools are air-cooled, our industry often needs water-cooled guns because of the higher heat caused by longer welds made at higher currents. The water-cooled FCA welding gun is more efficient than an air-cooled gun at removing waste heat. The air-cooled gun is more portable because it has fewer hoses, and it may be made lighter so it is easier to manipulate than the water-cooled gun.

Also, the water-cooled gun requires a water reservoir or another system to give the needed cooling. There are two major ways that water can be supplied to the gun for cooling. Cooling water can be supplied directly from the building's water system, or it can be supplied from a recirculation system.

Cooling water supplied directly from the building's water system is usually dumped into a wastewater drain

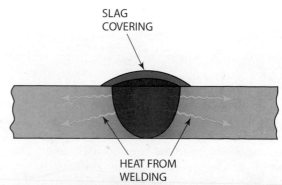

FIGURE 12-4 The slag covering keeps the welding heat from escaping quickly, thus slowing the cooling rate. © Cengage Learning 2012

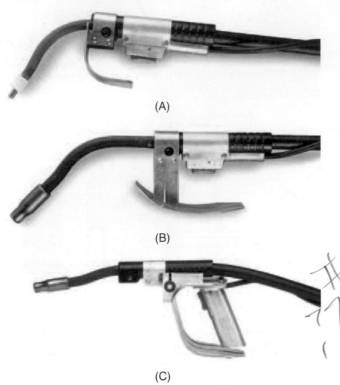

(A)

(B)

(C)

FIGURE 12-5 Typical FCA welding guns: (A) 350-ampere rating self-shielding, (B) 450-ampere rating gas shielding, and (C) 600-ampere rating gas shielding. Lincoln Electric Company

once it has passed through the gun. When this type of a system is used, a pressure regulator must be installed to prevent pressures that are too high from damaging the hoses. Water pressures higher than 35 psi (241 kg/mm^2) may cause the water hoses to burst. Check valves must also be installed in the supply line to prevent contaminated water from being drawn back into the water supply.

Recirculating cooling water systems eliminate any of the problems associated with open systems. Chemicals may be added to the water in recirculating systems to prevent freezing, to aid in pump lubrication, and to prevent algae growth. Only manufacturer-approved additives should be used in a recirculation system. Read all of the manufacturer's safety and data sheets before using these chemicals.

THINK GREEN
Conserve Water

Recirculating cooling water for FCA welding torches is important for water conservation. Some cities and states have laws that restrict the use of open systems because of the need for water conservation. Check with your city or state for any restrictions before installing an open water-cooling system.

Fume Extraction Nozzles Because of the large quantity of fumes that can be generated during FCA welding, Figure 12-2A, fume removal guns have been designed, Figure 12-2B. These systems use a vacuum to pull the smoke back into a specially designed smoke extraction nozzle on the welding gun. The disadvantage of having a slightly heavier gun is offset by the system's advantages. The advantages of the system are as follows:

- Cleaner air for the welder to breathe because the smoke is removed before it rises to the welder's face.
- Reduced heating and cooling cost because the smoke is concentrated, so less shop air must be removed with the smoke.

Electrode Feed Electrode feed systems are similar to those used for GMAW. The major difference is that larger FCAW machines that can use large diameter wire most often have two sets of feed rollers. The two sets of rollers help reduce the drive pressure on the electrode. Excessive pressure can distort the electrode wire diameter, which can allow some flux to be dropped inside the electrode guide tube.

Advantages

FCA welding offers the welding industry a number of important advantages.

High Deposition Rate High rates of depositing weld metal are possible. FCA welding deposition rates of more than 25 lb/hr (12 kg/hr) of weld metal are possible. This compares to about 10 lb/hr (6 kg/hr) for SMA welding using a very large diameter electrode of 1/4 in. (6 mm).

Minimum Electrode Waste The FCA method makes efficient use of filler metal; from 75% to 90% of the weight of the FCAW electrode is metal, the remainder being flux. SMAW electrodes have a maximum of 75% filler metal; some SMAW electrodes have much less. Also, a stub must be left at the end of each SMA welding electrode. The stub will average 2 in. (51 mm) in length, resulting in a loss of 11% or more of the SMAW filler electrode purchased. FCA welding has no stub loss, so nearly 100% of the FCAW electrode purchased is used.

Narrow Groove Angle Because of the deep penetration characteristic, no edge-beveling preparation is required on some joints in metal up to 1/2 in. (13 mm) in thickness. When bevels are cut, the joint-included angle can be reduced to as small as 35°, **Figure 12-6**. The reduced groove angle results in a smaller-sized weld. This can save 50% of filler metal with about the same savings in time and weld power used.

Minimum Precleaning The addition of **deoxidizers** and other fluxing agents permits high-quality welds to be made on plates with light surface oxides and mill scale.

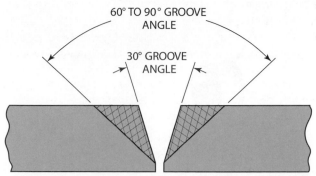

FIGURE 12-6 The narrower groove angle for FCAW saves on filler metal, welding time, and heat input into the part. © Cengage Learning 2012

This eliminates most of the precleaning required before GMA welding can be performed. Often it is possible to make excellent welds on plates in the "as cut" condition; no cleanup is needed.

All-Position Welding Small diameter electrode sizes in combination with special fluxes allow excellent welds in all positions. The slags produced assist in supporting the weld metal. This process is easy to use, and, when properly adjusted, it is much easier to use than other all-position arc welding processes.

Flexibility Changes in power settings can permit welding to be made on thin-gauge sheet metals or thicker plates using the same electrode size. Multipass welds allow joining unlimited thickness metals. This, too, is attainable with one size of electrode.

High Quality Many codes permit welds to be made using FCAW. The addition of the flux gives the process the high level of reliability needed for welding on boilers, pressure vessels, and structural steel.

Excellent Control The molten weld pool is more easily controlled with FCAW than with GMAW. The surface appearance is smooth and uniform even with less operator skill. Visibility is improved by removing the nozzle when using self-shielded electrodes.

THINK GREEN
FCA Welding Can Save Materials, Energy, and Time

The reduced groove angle will save weld metal, electrical energy, and time as compared to most other welding processes. In addition, the higher deposition rates and minimum electrode waste save manufacturing time and energy. FCA welding will help you work faster and more efficiently.

Limitations

The main limitation of flux cored arc welding is that it is confined to ferrous metals and nickle-based alloys. Generally, all low and medium carbon steels and some low alloy steels, cast irons, and a limited number of stainless steels are presently weldable using FCAW.

The equipment and electrodes used for the FCAW process are more expensive. However, the cost is quickly recoverable through higher productivity.

The removal of postweld slag requires another production step. The flux must be removed before the weldment is finished (painted) to prevent crevice corrosion.

With the increased welding output comes an increase in smoke and fume generation. The existing ventilation system in a shop might need to be increased to handle the added volume.

FCAW Electrodes

FCA welding electrodes are available as seamed- and seamless-type electrodes. Both types have tightly packed flux inside a metal outer covering. They differ in the way they are manufactured but have little if any difference in the way they weld. Seamless electrode flux cores are less likely to absorb moisture than the seamed electrode.

Methods of Manufacturing The electrodes have flux tightly packed inside. Seamed electrodes are made by first forming a thin sheet of metal into a U-shape, **Figure 12-7**. Then a measured quantity of flux is poured into the U-shape before it is squeezed shut. It is then passed through a series of dies to size it and further compact the flux.

Seamless electrodes start with a seamless tube. The tube is usually about 1 in. (25 mm) in diameter. One end of the tube is sealed, and the flux powder is poured into the open end. The tube is vibrated during the filling process to ensure that it fills completely. Once the tube is full, the open end is sealed. The tube is now sized using a series of dies, **Figure 12-8**.

In both these methods of manufacturing the electrode, the sheet and tube are made up of the desired alloy. Also in both cases the flux is compacted inside the metal skin. This compacting helps make the electrode operate smoother and more consistently.

Electrodes are available in sizes from 0.030 in. to 5/32 in. (0.8 mm to 3.9 mm) in diameter. Smaller diameter electrodes are much more expensive per pound than the same type in a larger diameter. Larger diameter electrodes produce such large welds they cannot be controlled in all positions. The most popular diameters range from 0.035 in. to 3/32 in. (0.9 mm to 2.3 mm).

The finished FCA filler metal is packaged in a number of forms for purchase by the end user. The electrode wire is wound (rolled) on spools, reels, coils made from metal,

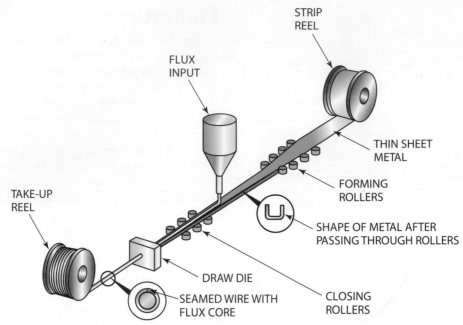

FIGURE 12-7 Putting the flux in the flux cored wire. © Cengage Learning 2012

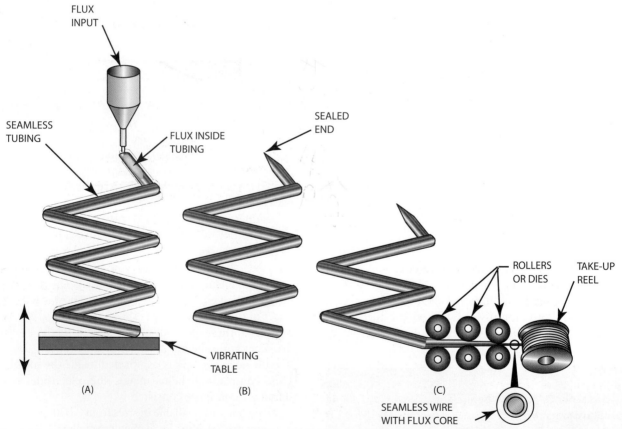

FIGURE 12-8 This shows one method of filling seamless FCA welding filler metal with flux. The vibration helps compact the granular flux inside the tube. © Cengage Learning 2012

wood, pressed fiber, or plastic, **Figure 12-9**. All of these are collectively called the electrode package. Sometimes drums are used with the wire coiled inside of the drum, **Table 12-1**. The American Welding Society (AWS) has a standard for the size of each of the package units. Although the dimensions of the packages are standard, the weight of

filler wire is not standard. More of the smaller diameter wire can fit into the same space than a larger diameter wire, so a package of 0.030 in. (0.8 mm) wire weighs more than the same-sized package of 3/32 in. (2.3 mm) wire.

Spools are made of plastic or fiberboard and are disposable. They are completely self-contained and are available

FIGURE 12-9 FCAW filler metal weights are approximate. They will vary by alloy and manufacturer. Lincoln Electric Company

Packaging	Outside Diameter	Width	Arbor (Hole) Diameter
Spools	4 in. (102 mm)	1 3/4 in. (44.5 mm)	5/8 in. (16 mm)
	8 in. (203 mm)	2 1/4 in. (57 mm)	2 1/16 in. (52.3 mm)
	12 in. (305 mm)	4 in. (102 mm)	2 1/16 in. (52.3 mm)
	14 in. (356 mm)	4 in. (102 mm)	2 1/16 in. (52.3 mm)
Reels	22 in. (559 mm)	12 1/2 in. (318 mm)	1 5/16 in. (33.3 mm)
	30 in. (762 mm)	16 in. (406 mm)	1 5/16 in. (33.3 mm)
Coils	16 1/4 in. (413 mm)	4 in. (102 mm)	12 in. (305 mm)
	Outside Diameter	**Inside Diameter**	**Height**
Drums	23 in. (584 mm)	16 in. (406 mm)	34 in. (864 mm)

TABLE 12-1 Packaging Size Specification for Commonly Used FCA Filler Wire

in approximate weights from 1 lb up to around 50 lb (0.5 kg to 25 kg). The smaller spools, 4 in. and 8 in. (102 mm and 203 mm), weighing from 1 lb to 7 lb, are most often used for smaller production runs or for home/hobby use; 12-in. and 14-in. (305-mm and 356-mm) spools are often used in schools and welding fabrication shops.

Coils come wrapped and/or wire tied together. They are unmounted, so they must be supported on a frame on the wire feeder in order to be used. Coils are available in weights of around 60 lb (27 kg). Because FCAW wires on coils do not have the expense of a disposable core, these wires cost a little less per pound, so they are more desirable for higher-production shops.

Reels are large wooden spools, and drums are shaped like barrels. Both reels and drums are used for high-production jobs. Both can contain approximately 300 lb to 1000 lb (136 kg to 454 kg) of FCAW wire. Because of their size, they are used primarily at fixed welding stations. Such stations are often associated with some form of automation, such as turntables or robotics.

Electrode Cast and Helix To see the cast and helix of a wire, feed out 10 ft (3 m) of wire electrode and cut it off. Lay it on the floor and observe that it forms a circle. The diameter of the circle is known as the cast of the wire, **Figure 12-10**.

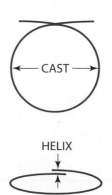

FIGURE 12-10 Method of measuring cast and helix of FCAW filler wire. © Cengage Learning 2012

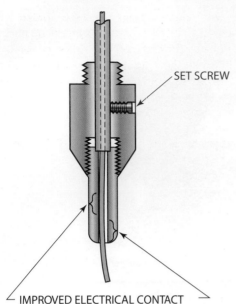

SET SCREW

IMPROVED ELECTRICAL CONTACT

FIGURE 12-11 Cast forces the wire to make better electrical contact with the tube. © Cengage Learning 2012

Note that the wire electrode does not lay flat. One end is slightly higher than the other. This height is the helix of the wire.

The AWS has specifications for both cast and helix for all FCA welding wires.

The cast and helix cause the wire to rub on the inside of the contact tube, **Figure 12-11.** The slight bend in the electrode wire ensures a positive electrical contact between the contact tube and filler wire.

FCA Welding Electrode Flux

The fluxes used are mainly rutile or lime based. The purpose of the fluxes is the same as in the SMAW process. That is, they can provide all or part of the following to the weld:

- *Deoxidizers:* Oxygen that is present in the welding zone has two forms. It can exist as free oxygen from the atmosphere surrounding the weld. Oxygen also can exist as part of a compound such as an iron oxide or carbon dioxide (CO_2). In either case, it can cause porosity in the weld if it is not removed or controlled. Chemicals are added that react to the presence of oxygen in either form and combine to form a harmless compound, **Table 12-2.** The new

Deoxidizing Element	Strength
Aluminum (Al)	Very strong
Manganese (Mn)	Weak
Silicon (Si)	Weak
Titanium (Ti)	Very strong
Zirconium (Zr)	Very strong

TABLE 12-2 Deoxidizing Elements Added to Filler Wire (to Minimize Porosity in the Molten Weld Pool)

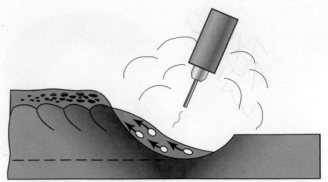

FIGURE 12-12 Impurities being floated to the surface by slag. © Cengage Learning 2012

compound can become part of the slag that solidifies on top of the weld, or some of it may stay in the weld as very small inclusions. Both methods result in a weld with better mechanical properties with less porosity.

- *Slag formers:* Slag serves several vital functions for the weld. It can react with the molten weld metal chemically, and it can affect the weld bead physically. In the molten state it moves through the molten weld pool and acts as a magnet or sponge to chemically combine with impurities in the metal and remove them, **Figure 12-12.** Slags can be refractory, become solid at a high temperature, and solidify over the weld, helping it hold its shape and slowing its cooling rate.

- *Fluxing agents:* Molten weld metal tends to have a high surface tension, which prevents it from flowing outward toward the edges of the weld. This causes undercutting along the junction of the weld and the base metal. Fluxing agents make the weld more fluid and allow it to flow outward, filling the undercut.

- *Arc stabilizers:* Chemicals in the flux affect the arc resistance. As the resistance is lowered, the arc voltage drops and penetration is reduced. When the arc resistance is increased, the arc voltage increases and weld penetration is increased. Although the resistance within the ionized arc stream may change, the arc is more stable and easier to control. It also improves the metal transfer by reducing spatter caused by an erratic arc.

- *Alloying elements:* Because of the difference in the mechanical properties of metal that is formed by rolling or forging and metal that is melted to form a weld bead, the metallurgical requirements of the two also differ. Some elements change the weld's strength, ductility, hardness, brittleness, toughness, and corrosion resistance. Powdered metal can be added to the flux and used as alloying elements or they may simply increase the deposition.

- *Shielding gas:* As elements in the flux are heated by the arc, some of them vaporize and form voluminous

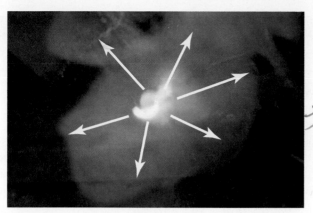

FIGURE 12-13 Rapidly expanding gas cloud. Larry Jeffus

gaseous clouds hundreds of times larger than their original volume. This rapidly expanding cloud forces the air around the weld zone away from the molten weld metal, **Figure 12-13.** Without the protection this process affords the molten metal, it would rapidly oxidize. Such oxidization would severely affect the weld's mechanical properties, rendering it unfit for service.

Types of FCAW Fluxes All FCAW fluxes are divided into two groups based on the acid or basic chemical reactivity of the slag. The AWS classifies T-1 as acid and T-5 as basic.

T-1 fluxes are **rutile-based fluxes** and are acidic. They produce a smooth, stable arc and a refractory high-temperature slag for out-of-position welding. These electrodes produce a fine drop transfer, a relatively low fume,

and an easily removed slag. The main limitation of the rutile fluxes is that their fluxing elements do not produce as high a quality deposit as do the T-5 systems.

T-5 fluxes are **lime-based fluxes** and are basic. They are very good at removing certain impurities from the weld metal, but their low melting temperature slag is fluid, which makes them generally unsuitable for out-of-position welding. These electrodes produce a more globular transfer, more spatter, more fume, and a more adherent slag than do the T-1 systems. These characteristics are tolerated when it is necessary to deposit very tough weld metal and for welding materials having a low tolerance for hydrogen.

Some rutile-based electrodes allow the addition of a shielding gas. With the weld being protected partially by the shielding gas, more elements can be added to the flux, which produces welds with the best of both flux systems, high-quality welds in all positions.

Some fluxes can be used on both single and multiple pass welds, and others are limited to single pass welds only. Using a single pass welding electrode for multipass welds may result in an excessive amount of manganese. The manganese is necessary to retain strength when making large, single pass welds. However, with the lower dilution associated with multipass techniques, it can strengthen the weld metal too much and reduce its ductility. In some cases, small welds that deeply penetrate the base metal can help control this problem.

Table 12-3 lists the shielding and polarity for the flux classifications of mild steel FCAW electrodes. The letter *G* is used to indicate an unspecified classification. The *G*

Classifications		Comments		Shielding Gas
T-1		Requires clean surfaces and produces little spatter. It can be used for single and multiple pass welds in the flat (1G and 1F) and horizontal (2F) positions.		Carbon dioxide (CO$_2$)
T-2		Requires clean surfaces and produces little spatter. It can be used for single pass welds in the flat (1G and 1F) and horizontal (2F) positions only.		Carbon dioxide (CO$_2$)
T-3		Used on thin-gauge steel for single pass welds in the flat (1G and 1F) and horizontal (2F) positions only.		None
T-4		Low penetration and moderate tendency to crack for single and multiple pass welds in the flat (1G and 1F) and horizontal (2F) positions.		None
T-5		Low penetration and a thin, easily removed slag, used for single and multiple pass welds in the flat (1G and 1F) positions only.		With or without carbon dioxide (CO$_2$)
T-6		Similar to T-5 without externally applied shielding gas.		None
T-G		The composition and classification of this electrode are not given in the preceding classes. It may be used for single or multiple pass welds.		With or without shielding

TABLE 12-3 Welding Characteristics of Seven Flux Classifications

Element	Reaction in Weld
Silicon (Si)	Ferrite former and deoxidizer
Chromium (Cr)	Ferrite and carbide former
Molybdenum (Mo)	Ferrite and carbide former
Columbium (Cb)	Strong ferrite former
Aluminum (Al)	Ferrite former and deoxidizer

TABLE 12-4 Ferrite-forming Elements Used in FCA Welding Fluxes

Metal	AWS Filler Metal Classification
Mild steel	A5.20
Stainless steel	A5.22
Chromium–molybdenum	A5.29

TABLE 12-5 Filler Metal Classification Numbers

means that the electrode has not been classified by the American Welding Society. Often the exact composition of fluxes is kept as a manufacturer's trade secret. Therefore, only limited information about the electrode's composition will be given. The only information often supplied is current, type of shielding required, and some strength characteristics.

As a result of the relatively rapid cooling of the weld metal, the weld may tend to become hard and brittle. This factor can be controlled by adding elements to the flux and the slag formed by the flux, **Table 12-4**. Ferrite is the softer, more ductile form of iron. The addition of ferrite-forming elements can control the hardness and brittleness of a weld. Refractory fluxes are sometimes called "fast-freeze" because they solidify at a higher temperature than the weld metal. By becoming solid first, this slag can cradle the molten weld pool and control its shape. This property is very important for out-of-position welds.

The impurities in the weld pool can be metallic or nonmetallic compounds. Metallic elements that are added to the metal during the manufacturing process in small quantities may be concentrated in the weld. These elements improve the grain structure, strength, hardness, resistance to corrosion, or other mechanical properties in the metal's as-rolled or formed state. But weld nugget is a small casting, and some alloys adversely affect the properties of this casting (weld metal). Nonmetallic compounds are primarily slag inclusions left in the metal from the fluxes used during manufacturing. The welding fluxes form slags that are less dense than the weld metal so that they will float to the surface before the weld solidifies.

Flux Cored Steel Electrode Identification The American Welding Society revised its A5.20 *Specification for Carbon Steel Electrodes for Flux Cored Arc Welding* in 1995 to reflect changes in the composition of the FCA filler metals. **Table 12-5** lists the AWS specifications for flux cored filler metals.

Mild Steel

The following electrode number is used as an example to explain the meaning for the specification of the electrode classification system *E70T-10* as follows (**Figure 12-14**):

E—Electrode.

7—Tensile strength in units of 10,000 psi for a good weld. This value is usually either *6* for 60,000 psi or *7* for 70,000 psi minimum weld strength. An exception is for the number *12*, which is used to denote filler metals having a range from 70,000 psi to 90,000 psi.

0—0 is used for flat and horizontal fillets only, and *1* is used for all-position electrodes.

T—Tubular (flux cored) electrode.

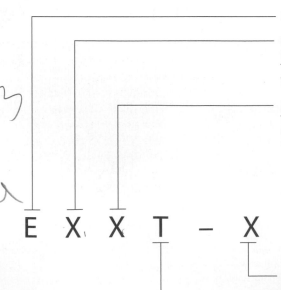

DESIGNATES AN ELECTRODE

INDICATES THE MINIMUM TENSILE STRENGTH OF THE DEPOSITED WELD METAL IN A TEST WELD MADE WITH THE ELECTRODE AND IN ACCORDANCE WITH SPECIFIED WELDING CONDITIONS

INDICATES THE PRIMARY WELDING POSITION FOR WHICH THE ELECTRODE IS DESIGNED:
0 – FLAT AND HORIZONTAL POSITIONS
1 – ALL POSITIONS

E X X T – X

INDICATES USABILITY AND PERFORMANCE CAPABILITIES

INDICATES A FLUX CORED ELECTRODE

FIGURE 12-14 Identification system for mild steel FCAW electrodes. American Welding Society

10—The number in this position can range from *1* to *14* and is used to indicate the electrode's shielding gas if any, number of passes, and other welding characteristics of the electrode. The letter *G* is used to indicate that the shielding gas, polarity, and impact properties are not specified. The letter *G* may or may not be followed with the letter *S*. *S* indicates an electrode suitable only for single pass welding.

The electrode classification *E70T-10* can have some optional identifiers added to the end of the number such as *E70T-10MJH8*. These additions are used to add qualifiers to the general classification so that specific codes or standards can be met. These additions have the following meanings:

M—Mixed gas, 75% to 80% Ar, balance CO_2. If there is no *M*, the shielding gas must be either CO_2 or the electrode is self-shielded.

J—Describes the Charpy V-notch impact test value of 20 ft-lb at –40°F.

H8—Describes the residual hydrogen levels in the weld: *H4* equals less than 4 ml/l00 g; *H8*, less than 8 ml/l00 g; *H16*, less than 16 ml/l00 g.

Stainless Steel Electrodes

The AWS classification for stainless steel for FCAW electrodes starts with the letter *E* as its prefix. Following the *E* prefix, the American Iron and Steel Institute's (AISI) three-digit stainless steel number is used. This number indicates the type of stainless steel in the filler metal.

To the right of the AISI number, the AWS adds a dash followed by a suffix number. The number 1 is used to indicate an all-position filler metal, and the number 3 is used to indicate an electrode to be used in the flat and horizontal positions only.

Metal Cored Steel Electrode Identification

The addition of metal powders to the flux core of FCA welding electrodes has produced a new classification of filler metals. The new filler metals evolved over time, and a new identification system was established by the AWS to identify these filler metals. Some of the earlier flux cored filler metals that already had powder metals in their core had their numbers changed to reflect the new designation. The designation was changed from the letter *T* for *tubular* to the letter *C* for *core*. For example, E70T-1 became E70C-3C. The complete explanation of the cored electrode *E70C-3C* follows:

E—Electrode.

7—Tensile strength in units of 10,000 psi for a good weld. This value is usually either *6* for 60,000 psi or *7* for 70,000 psi minimum weld strength. An exception is for the number *12*, which is used to denote filler metals having a range from 70,000 psi to 90,000 psi.

0—0 is used for flat and horizontal fillets only, and *1* is used for all-position electrodes.

C—Metal-cored (tubular) electrode.

3—3 is used for a Charpy impact of 20 ft-lb at 0°F, and *6* represents a Charpy impact of 20 ft-lb at –20°F.

C—The second letter *C* indicates CO_2, and the letter *M* indicates a mixed gas, 75% to 80% Ar, with the balance being CO_2. If there is no *M* or *C*, then the shielding gas is CO_2. The letter *G* is used to indicate that the shielding gas, polarity, and impact properties are not specified. The letter *G* may or may not be followed with the letter *S*. *S* indicates an electrode suitable only for single pass welding.

NOTE: The powdered metal added to the core flux can provide additional filler metal and/or alloys. This is one way the microalloys can be added in very small and controlled amounts, as low as 0.0005% to 0.005%. These are very powerful alloys that dramatically improve the metal's mechanical properties.

Care of Flux Core Electrodes Wire electrodes may be wrapped in sealed plastic bags for protection from the elements. Others may be wrapped in a special paper, and some are shipped in cans or cardboard boxes.

A small paper bag of a moisture-absorbing material, crystal desiccant, is sometimes placed in the shipping containers. It is enclosed to protect wire electrodes from moisture. Some wire electrodes require storage in an electric rod oven to prevent contamination from excessive moisture. Read the manufacturer's recommendations located in or on the electrode shipping container for information on use and storage.

Weather conditions affect your ability to make high-quality welds. Humidity increases the chance of moisture entering the weld zone. Water (H_2O), which consists of two parts hydrogen and one part oxygen, separates in the weld pool. When only one part of hydrogen is expelled, hydrogen entrapment occurs. Hydrogen entrapment can cause weld beads to crack or become brittle. The evaporating moisture will also cause porosity.

To prevent hydrogen entrapment, porosity, and atmospheric contamination, it may be necessary to preheat the base metal to drive out moisture. Storing the wire electrode in a dry location is recommended. The electrode may develop restrictions due to the tangling of the wire or become oxidized with excessive rusting if the wire electrode package is mishandled, thrown, dropped, or stored in a damp location.

/////CAUTION

Always keep the wire electrode dry and handle it as you would any important tool or piece of equipment.

Shielding Gas

FCA welding wire can be manufactured so that all of the required shielding of the molten weld pool is provided by the vaporization of some of the flux within the tubular electrode. When the electrode provides all of the shielding, it is called **self-shielding.** Other FCA welding wire must use an externally supplied shielding gas to provide the needed protection of the molten weld pool. When a shielding gas is added, it is called **dual shield.**

> NOTE: Sometimes the shielding gas is referred to as the shielding medium. For example, the shielding gas for E71T-5 is either argon 75% with 25% CO_2 or 100% CO_2.

Care must be taken to use the cored electrodes with the recommended gases or not to use gas at all with the self-shielded electrodes. Using a self-shielding flux cored electrode with a shielding gas may produce a defective weld. The shielding gas will prevent the proper disintegration of much of the deoxidizers. This results in the transfer of these materials across the arc to the weld. In high concentrations, the deoxidizers can produce slags that become trapped in the welds, causing undesirable defects. Lower concentrations may cause brittleness only. In either case, the chance of weld failure is increased. If these electrodes are used correctly, there is no problem.

The selection of a shielding gas will affect the arc and weld properties. The weld bead width, buildup, penetration, spatter, chemical composition, and mechanical properties are all affected as a result of the shielding gas selection.

Shielding gas comes in high-pressure cylinders. These cylinders are supplied with 2000 psi of pressure. Because of this high pressure, it is important that the cylinders be handled and stored safely. See Chapter 2 for specific cylinder safety instructions.

Gases used for FCA welding include CO_2 and mixtures of argon and CO_2. Argon gas is easily ionized by the arc. Ionization results in a highly concentrated path from the electrode to the weld. This concentration results in a smaller droplet size that is associated with the axial spray mode of metal transfer, **Figure 12-15.** A smooth stable arc

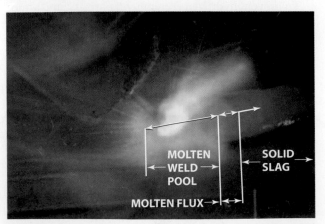

FIGURE 12-15 Axial spray transfer mode. Larry Jeffus

results and there is a minimum of spatter. This transfer mode continues as CO_2 is added to the argon until the mixture contains more than 25% of CO_2.

As the percentage of CO_2 increases in the argon mixture, weld penetration increases. This increase in penetration continues until a 100% CO_2 shielding gas is reached. But as the percentage of CO_2 is increased, the arc stability decreases. The less stable arc causes an increase in spatter. A mixture of 75% argon and 25% CO_2 works best for jobs requiring a mixed gas. This mixture is sometimes called C-25.

Straight CO_2 is used for some welding. But the CO_2 gas molecule is easily broken down in the welding arc. It forms carbon monoxide (CO) and free oxygen (O). Both gases are reactive to some alloys in the electrode. As these alloys travel from the electrode to the molten weld pool, some of them form oxides. Silicon and manganese are the primary alloys that become oxidized and lost from the weld metal.

Most FCA welding electrodes are specifically designed to be used with or without shielding gas and for a specific shielding gas or percentage mixture. For example, an electrode designed specifically for use with 100% CO_2 will have higher levels of silicon and manganese to compensate for the losses to oxidization. But if 100% argon or a mixture of argon and CO_2 is used, the weld will have an excessive amount of silicon and manganese. The weld will not have the desired mechanical or metallurgical properties. Although the weld may look satisfactory, it will probably fail prematurely.

> ///// **CAUTION** \\\\\
>
> **Never use an FCA welding electrode with a shielding gas it is not designated to be used with. The weld it produces may be unsafe.**

Welding Techniques

A welder can control weld beads made by FCA welding by making changes in the techniques used. The following explains how changing specific welding techniques will affect the weld produced.

Gun Angle The *gun angle, work angle,* and *travel angle* are terms used to refer to the relation of the gun to the work surface, **Figure 12-16.** The gun angle can be used to control the weld pool. The electric arc produces an electrical force known as the arc force. The arc force can be used to counteract the gravitational pull that tends to make the liquid weld pool sag or run ahead of the arc. By manipulating the electrode travel angle for the flat and horizontal position of welding to a 20° to 45° angle from the vertical, the weld pool can be controlled. A 40° to 50° angle from the vertical plate is recommended for fillet welds.

Changes in this angle will affect the weld bead shape and penetration. Shallower angles are needed when welding thinner materials to prevent burnthrough. Steeper, perpendicular angles are used for thicker materials.

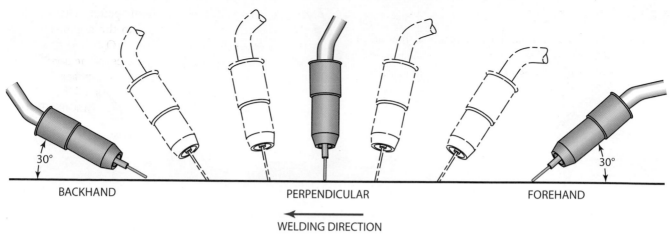

FIGURE 12-16 Welding gun angles. © Cengage Learning 2012

FCAW electrodes have a flux that is mineral based, often called low hydrogen. These fluxes are refractory and become solid at a high temperature. If too steep a forehand or pushing angle is used, slag from the electrode can be pushed ahead of the weld bead and solidify quickly on the cooler plate, **Figure 12-17**. Because the slag remains solid at higher temperatures than the temperature of the molten weld pool, it can be trapped under the edges of the weld by the molten weld metal. To avoid this problem, most flat and horizontal welds should be performed with a backhand angle.

Vertical up welds require a **forehand** gun angle. The forehand angle is needed to direct the arc deep into the groove or joint for better control of the weld pool and deeper penetration, **Figure 12-18**. Slag entrapment associated with most forehand welding is not a problem for vertical welds.

A gun angle around 90° to the metal surface either slightly forehand or backhand works best for overhead welds, **Figure 12-19**. The slight angle aids with visibility of the weld, and it helps control spatter buildup in the gas nozzle.

Forehand/Perpendicular/Backhand Techniques
Forehand, *perpendicular*, and *backhand* are the terms most

often used to describe the gun angle as it relates to the work and the direction of travel. The forehand technique is sometimes referred to as *pushing* the weld bead, and backhand may be referred to as *pulling* or *dragging* the weld bead. The term *perpendicular* is used when the gun angle is at approximately 90° to the work surface, **Figure 12-20**.

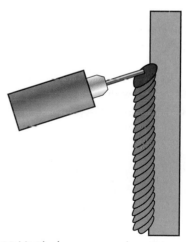

FIGURE 12-18 Vertical up gun angle. © Cengage Learning 2012

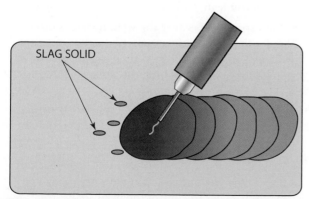

FIGURE 12-17 Large quantities of solid slag in front of a weld can cause slag to be trapped under the weld bead. © Cengage Learning 2012

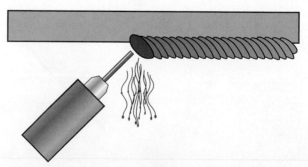

FIGURE 12-19 Weld gun position to control spatter buildup on an overhead weld. © Cengage Learning 2012

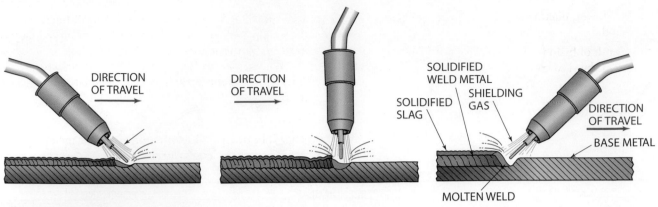

FIGURE 12-20 Changing the welding gun angle between forehand, perpendicular, or backhand angles will change the shape of the weld bead produced. © Cengage Learning 2012

Advantages of the Forehand Technique

The advantages of using the forehand welding technique are as follows:

- Joint visibility—You can easily see the joint where the bead will be deposited, **Figure 12-21.**

- Electrode extension—The contact tube tip is easier to see, making it easier to maintain a constant extension length.

- Less weld penetration—It is easier to weld on thin sheet metal without melting through.

- Out-of-position welds—This technique works well on vertical up and overhead joints for better control of the weld pool.

Disadvantages of the Forehand Technique

The disadvantages of using the forehand welding technique are as follows:

- Weld thickness—Thinner welds result because less weld reinforcement is applied to the weld joint.

- Welding speed—Because less weld metal is being applied, the rate of travel along the joint can be faster, which may make it harder to create a uniform weld.

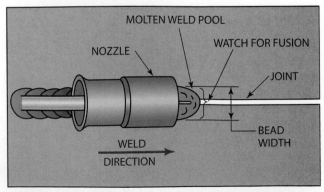

FIGURE 12-21 The shielding gas nozzle restricts the welder's view. © Cengage Learning 2012

- Slag inclusions—Some spattered slag can be thrown in front of the weld bead and be trapped or included in the weld, resulting in a weld defect.

- Spatter—Depending on the electrode, the amount of spatter may be slightly increased with the forehand technique.

Advantages of the Perpendicular Technique

The advantages of using the perpendicular welding technique are as follows:

- Machine and robotic welding—The perpendicular gun angle is used on automated welding because there is no need to change the gun angle when the weld changes direction.

- Uniform bead shape—The weld's penetration and reinforcement are balanced between those of forehand and backhand techniques.

Disadvantages of the Perpendicular Technique

The disadvantages of using the perpendicular welding technique are as follows:

- Limited visibility—Because the welding gun is directly over the weld, there is limited visibility of the weld unless you lean your head way over to the side.

- Weld spatter—Because the weld nozzle is directly under the weld in the overhead position, more weld spatter can collect in the nozzle causing gas flow problems or even shorting the tip to the nozzle.

Advantages of the Backhand Technique

The advantages of the backhand welding technique are as follows:

- Weld bead visibility—It is easy to see the back of the molten weld pool as you are welding, which makes it easier to control the bead shape.

- Travel speed—Because of the larger amount of weld metal being applied, the rate of travel may be slower, making it easier to create a uniform weld.

- Depth of fusion—The arc force and the greater heat from the slower travel rate both increase the depth of weld joint penetration.

Disadvantages of the Backhand Technique

The disadvantages of the backhand welding technique are as follows:

- Weld buildup—The weld bead may have a convex (raised or rounded) shaped weld face when you use the backhand technique.
- Postweld finishing—Because of the weld bead shape, more work may be required if the product has to be finished by grinding smooth.
- Joint following—It is harder to follow the joint because of your hand position with the FCAW gun being positioned over the joint, and you may wander from the seam, **Figure 12-22.**
- Loss of penetration—An inexperienced welder sometimes directs the wire too far back into the weld pool, causing the wire to build up in the face of the weld pool, reducing joint penetration.

Travel Speed The American Welding Society defines *travel speed* as the linear rate at which the arc is moved along the weld joint. Fast travel speeds deposit less filler metal. If the rate of travel increases, the filler metal cannot be deposited fast enough to adequately fill the path melted by the arc. This causes the weld bead to have a groove melted into the base metal next to the weld and left unfilled by the weld. This condition is known as undercut.

Undercut occurs along the edges or toes of the weld bead. Slower travel speeds will, at first, increase penetration and increase the filler weld metal deposited. As the filler metal increases, the weld bead will build up in the weld pool. Because of the deep penetration of flux cored wire, the angle at which you hold the gun is very important for a successful weld.

If all welding conditions are correct and remain constant, the preferred rate of travel for maximum weld penetration is a travel speed that allows you to stay within the selected welding variables and still control the fluidity of the weld pool. This is an intermediate travel speed, or progression, that is not too fast or too slow.

Another way to figure out correct travel speed is to consult the manufacturer's recommendations chart for the ipm burn-off rate for the selected electrode.

Mode of Metal Transfer The mode of metal transfer is used to describe how the molten weld metal is transferred across the arc to the base metal. The mode of metal transfer that is selected, the shape of the completed weld bead, and the depth of weld penetration depend on the welding power source, wire electrode size, type and thickness of material, type of shielding gas used, and best welding position for the task.

Spray Transfer—FCAW-G

The spray transfer mode is the most common process used with gas shielded FCAW, **Figure 12-23.**

As the gun trigger is depressed, the shielding gas automatically flows and the electrode bridges the distance from the contact tube to the base metal, making contact with the base metal to complete a circuit. The electrode shorts and becomes so hot that the base metal melts and forms a weld pool. The electrode melts into the weld pool and burns back toward the contact tube. A combination of high amperage and the shielding gas along with the electrode size produces a pinching effect on the molten electrode wire, causing the end of the electrode wire to spray across the arc.

The characteristic of spray-type transfer is a smooth arc, through which hundreds of small droplets per second are transferred through the arc from the electrode to the weld pool. At that moment, a transfer of metal is taking place. Spray transfer can produce a high quantity of metal droplets, up to approximately 250 per second above the transition current, or critical current. This means the current is dependent on the electrode size, composition of the electrode, and shielding gas so a spray transfer can take place. Below the transition current (critical current) globular transfer takes place.

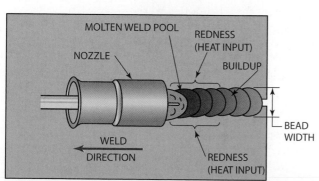

FIGURE 12-22 Watch the trailing edge of the molten weld pool. © Cengage Learning 2012

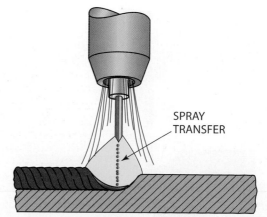

FIGURE 12-23 Spray transfer method. © Cengage Learning 2012

To achieve a spray transfer, high current and larger diameter electrode wire are needed. A shielding gas of carbon dioxide (CO_2), a mixture of carbon dioxide (CO_2) and argon (Ar), or an argon (Ar) oxygen (O_2) mixture is needed. FCAW-G is a welding process that, with the correct variables, can be used

- on thin or properly prepared thick sections of material,
- on a combination of thick to thin materials,
- with small or large electrode diameters, and
- with a combination of shielding gases.

Globular Transfer—FCAW-G

Globular transfer occurs when the welding current is below the transition current, **Figure 12-24**. The electrode forms a molten ball at its end that grows in size to approximately two to three times the original electrode diameter. These large molten balls are then transferred across the arc at the rate of several drops per second.

The arc becomes unstable because of the gravitational pull from the weight of these large drops. A spinning effect caused by a natural phenomenon takes place when argon gas is introduced to a large ball of molten metal on the electrode. This causes a spinning motion as the molten ball transfers across the arc to the base metal. This unstable globular transfer can produce excessive spatter.

Both FCAW-S and FCAW-G use DCEN when welding on thin-gauge materials to keep the heat in the base metal and the small diameter electrode at a controllable burn-off rate. The electrode can then be stabilized, and it is easier to manipulate and control the weld pool in all weld positions. Larger diameter electrodes are welded with DCEP because the larger diameters can keep up with the burn-off rates.

The recommended weld position means the position in which the workpiece is placed for welding. All welding positions use either spray or globular transfer, but for now we will concentrate on the flat and horizontal welding positions.

In the flat welding position the workpiece is placed flat on the work surface. In the horizontal welding position, the workpiece is positioned perpendicular to the workbench surface.

The amperage range may be from 30 amperes to 400 amperes or more for welding materials from gauge thickness up to 1 1/2 in (37 mm). On square-groove weld joints, thicker base metals can be welded with little or no edge preparation. This is one of the great advantages of FCAW. If edges are prepared and cut at an angle (beveled) to accept a complete joint weld penetration, the depth of penetration will be greatly increased. FCAW is commonly used for general repairs to mild steel in the horizontal, vertical, and overhead welding positions, sometimes referred to as out-of-position welding.

Electrode Extension The electrode extension is measured from the end of the electrode contact tube to the point the arc begins at the end of the electrode, **Figure 12-25**. Compared to GMA welding, the electrode extension required for FCAW is much greater. The longer extension is required for several reasons. The electrical resistance of the wire causes the wire to heat up, which can drive out moisture from the flux. This preheating of the wire also results in a smoother arc with less spatter.

Porosity FCA welding can produce high-quality welds in all positions, although porosity in the weld can be a persistent problem. Porosity can be caused by moisture in the flux, improper gun manipulation, or surface contamination.

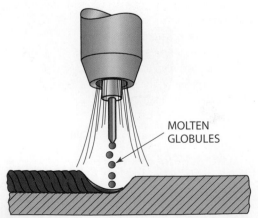

FIGURE 12-24 Globular transfer method.

© Cengage Learning 2012

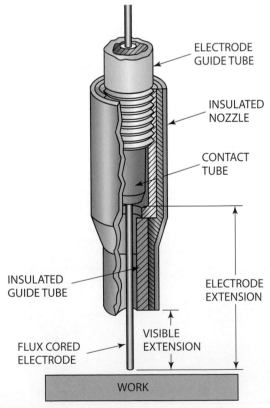

FIGURE 12-25 Self-shielded electrode nozzle.

American Welding Society

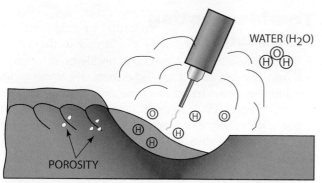

FIGURE 12-26 Water (H_2O) breaks down in the presence of the arc, and the hydrogen (H) is dissolved in the molten weld metal. © Cengage Learning 2012

The flux used in the FCA welding electrode is subject to picking up moisture from the surrounding atmosphere, so the electrodes must be stored in a dry area. Once the flux becomes contaminated with moisture, it is very difficult to remove. Water (H_2O) breaks down into free hydrogen and oxygen in the presence of an arc, **Figure 12-26**. The hydrogen can be absorbed into the molten weld metal, where it can cause postweld cracking. The oxygen is absorbed into the weld metal also, but it forms oxides in the metal.

If a shielding gas is being used, the FCA welding gun gas nozzle must be close enough to the weld to provide adequate shielding gas coverage. If there is a wind or if the nozzle-to-work distance is excessive, the shielding will be inadequate and cause weld porosity. If welding is to be done outside or in an area subject to drafts, the gas flow rate must be increased or a windshield must be placed to protect the weld, **Figure 12-27**.

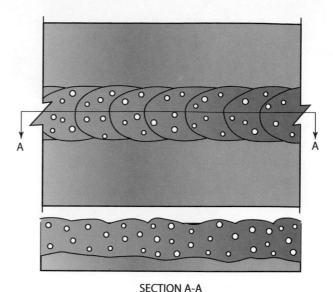

SECTION A-A

FIGURE 12-28 Uniformly scattered porosity.
© Cengage Learning 2012

A common misconception is that the flux within the electrode will either remove or control weld quality problems caused by surface contaminations. That is not true. The addition of flux makes FCA welding more tolerant to surface conditions than GMA welding, although it still is adversely affected by such contaminations.

New hot-rolled steel has a layer of dark gray or black iron oxide called *mill scale*. Although this layer is very thin, it may provide a source of enough oxygen to cause porosity in the weld. If mill scale causes porosity, it is usually uniformly scattered through the weld, **Figure 12-28**. Unless severe, uniformly scattered porosity is usually not visible in the finished weld. It is trapped under the surface as the weld cools.

Because porosity is under the weld surface, nondestructive testing methods, including X-ray, magnetic particle, and ultrasound, must be used to locate it in a weld. It can be detected by mechanical testing such as guided-bend, free-bend, and nick-break testing for establishing weld parameters. Often it is better to remove the mill scale before welding rather than risking the production of porosity.

All welding surfaces within the weld groove and the surrounding surfaces within 1 in. (25 mm) must be cleaned to bright metal, **Figure 12-29**. Cleaning may be either grinding, filing, sanding, or blasting.

Any time FCA welds are to be made on metals that are dirty, oily, rusty, or wet or that have been painted, the surface must be precleaned. Cleaning can be done chemically or mechanically.

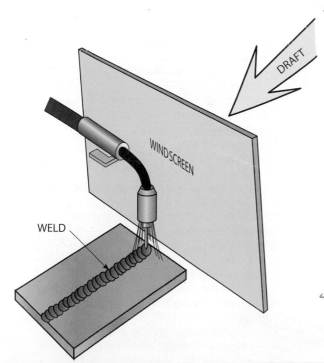

FIGURE 12-27 A windscreen can keep the welding shielding from being blown away. © Cengage Learning 2012

////// **CAUTION** //////

Chemically cleaning oil and paint off metal must be done according to the cleaner manufacturer's directions. The work must be done in an appropriate, approved area. The metal must be dry, and all residues of the cleaner must be removed before welding begins.

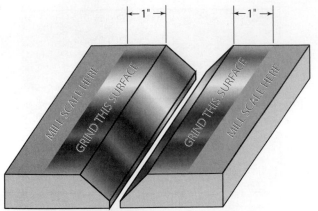

FIGURE 12-29 Grind mill scales off plates within 1 in. (25 mm) of the groove. © Cengage Learning 2012

One advantage of chemically cleaning oil and paint is that it is easier to clean larger areas. Both oil and paint smoke easily when heated, and such smoke can cause weld defects. They must be removed far enough from the weld so that weld heat does not cause them to smoke. In the case of small parts, the entire part may need to be cleaned.

Troubleshooting FCA Welding

Troubleshooting FCA welding problems is often a trial-and-error process. Trial and error is where you make one adjustment or change at a time and make a trial weld to see if the problem improved. If what you tried did not improve the problem or made it worse, reset the machine and try another adjustment. Keep doing this until the problem is resolved. Make only one adjustment or change at a time. Making two or more adjustments or changes at the same time can result in one improving the weld and the others causing new problems.

The most common causes of FCA welding problems are equipment setup. However, in the field, worn and dirty parts will from time to time develop problems. These worn or dirty parts can cause FCA welding problems similar to those caused by improper setup. Misdiagnosing the cause can result in the possible replacement of good parts. To reduce this time and expense, use the list of common FCA welding problems in **Table 12-6** to try to solve the weld problem before replacing parts. Often the equipment manufacturer will include a list of troubleshooting tips in the welder manufacturer's instruction booklet.

Welding Problem	Cause
Gun nozzle arcs to work	1. Weld spatter buildup in nozzle 2. Contact tube bent and touching nozzle
Wire feeds but no arc	1. Bad or missing work clamp (ground) 2. Loose jumper lead in welder 3. Electrode not contacting bare metal
Arc burns off wire at contact tube end	1. Feed rollers tension too loose 2. Wrong-sized feed rollers 3. Wire welded to contact tube 4. Wire liner worn or damaged 5. Out of wire 6. Worn or damaged contact tube
Wire feeds erratically	1. Feed rollers tension loose 2. Dirty liner 3. Worn or dirty contact tube
Arc pops and gun jerks during welding	1. Too high a wire-feed speed 2. Too low a voltage
Wire burns back and large globules of metal cross the arc	1. Too high a voltage 2. Too low a wire-feed speed 3. Wire slipping in feed rollers, not feeding smoothly
Weld does not burn into base metal	1. Too long an electrode extinction 2. Too low voltage and amperage settings
Weld burns through the base metal	1. Too short an electrode extinction 2. Too high voltage and amperage settings
Porosity in weld	1. Poor shielding gas coverage on dual-shield wire 2. Wrong shielding gas type being used 3. Shielding gas used on self-shielding wire 4. Single pass electrode used for multipass weld
Poor shielding gas coverage	1. Plugged or dirty gas diffuser 2. Too high or too low shielding gas flow rate 3. Shielding gas cylinder near empty

TABLE 12-6 FCAW Troubleshooting Table

Summary

Flux cored arc welding is used to produce more tons of welded fabrications than any other process. The ability to produce high-quality welds on a wide variety of material thicknesses and joint configurations has led to its popularity. As you learn and develop these skills, you will therefore be significantly increasing your employability and productivity in the welding industry.

A wide variety of flux cored arc welding filler metals and shielding gas combinations are available to you in industry. These various materials aid in producing welds of high quality under various welding conditions. Although the selection of the proper filler metal and gas coverage, if used, will significantly affect the finished weld's quality in the field, there are very few differences in manipulation and setup among these filler metals. Therefore, as you practice welding in a school or training program and learn to use a specific wire and shielding gas mixture, these skills are easily transferable to the next group of materials you will encounter on the job.

Ultrasonic Plastic Welding Basics

This ultrasonic welding primer will help you understand how joint design requirements and the selection of a welding machine frequency affect the finished product.

Manufacturers constantly search for equipment that will increase production, reduce rejects, and otherwise improve efficiency. Ultrasonic welding achieves those objectives. It is often used in the medical, electrical, automotive, packaging, toy, housewares, cosmetics, and other industries.

Ultrasonics can be used to insert metal fasteners in thermoplastic materials, to form a plastic rivet, and to cut and seal films and fabrics, as well as to join plastic parts together.

How does it work? Ultrasonic vibrational energy at the boundary of the plastic parts being joined causes the plastic material to soften and flow in a fraction of a second. When the material is pressed together and resolidifies, the bond is made.

No glues or solvents are needed. Tooling can be designed to secure and align the parts. Heating is confined to the interface area so the assembled part is not too hot to handle. Equipment can be integrated into automated lines.

There are two basic ultrasonic welding techniques: plunge welding and continuous welding. In plunge welding, the parts are placed under a tool or horn; the horn descends to the part under moderate pressure and the weld cycle begins. In the continuous welding process, the horn may "scan" the part, or the material is passed over or under the horn on a continuous basis.

Principles of Operation

Every ultrasonic unit contains the following four elements:

- A power supply, which takes line power at 50/60 cycles and changes it to a high ultrasonic frequency of 20,000 cycles per second or higher

- A converter or transducer, which changes the incoming high-frequency electrical signal to mechanical vibration of the same frequency

- A booster, which transmits the vibration energy and increases its amplitude

- An anvil or a nest to support the workpiece; for bonding of textiles, the pattern wheel replaces the anvil

Equipment

Systems have evolved into a range of sizes and styles to suit a wide range of applications. The most common type is a press. The welding press is equipped with a pneumatic system to supply the necessary contact force, and the head is mounted on a slide, so it can be raised and lowered to contact the part to be welded. It is important that the press be rigid, so bending deflections do not affect the weld consistency.

The ultrasonic energy can be started just before the horn contacts the part, after contact but before full pressure is reached, or when full preset pressure is reached.

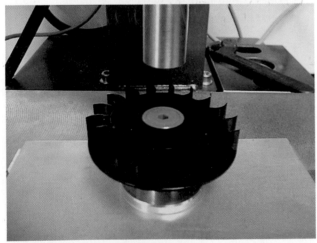

Real savings in the overall cost of manufacturing an automobile can be realized by extending the service life of resistance welding electrodes. American Welding Society

After the ultrasonic energy is stopped, there is usually a short delay before the pressure is released to permit the plastic to solidify.

For continuous welding of film and fabric, the ultrasonic system may be built into a table that resembles a sewing machine. The system is equipped with a rotating wheel, which can emboss or cut the fabric with a wide range of patterns.

Horns

The horn is the part of the ultrasonic system that contacts the parts to be joined. The horn is designed to resonate at the frequency of the ultrasonic system. When the horn vibrates, it stretches and shrinks in length by a small amount. This motion is referred to as the amplitude of the horn.

Weldable Materials

The term *plastic* may be used for thermoplastic or thermoset materials. Thermoset materials burn when heated and cannot be ultrasonically joined. Only thermoplastic materials are candidates for ultrasonic welding.

Joint Design

To ensure that plastic assemblies are properly joined, they should be designed from the beginning with a suitable joint design. Many factors are taken into consideration: the material to be bonded, the ultimate use of the product, the cost and ease of molding, and the location of the joint surface relative to the horn.

The joint geometry should be tailored for the use of the end product. The parts must not fit so tightly before joining that they inhibit the vibration needed to induce welding. Thin cross sections may crack under the action of vibration, and delicate parts, such as fine wires, may become damaged when their enclosures are welded. If the parts have thin walls that may bulge under pressure, it is advisable to support the part up to the joining zone. Obviously, the ideal conditions are not always attainable and compromises can be made.

Other Considerations

Part Size Part size has an influence on the power level required and the frequency of the welding machine selected. For large parts greater than 1 1/2 in. (3.811 cm) in diameter, or with any welding dimension longer than 2 in. (55 mm), select a 20-kHz welding machine. The power level—typically available at about 1000, 1500, 2000, or 3000 W—depends on the size of the weld area, the type of joint, and the material to be welded.

Part Features Sharp corners may fracture or melt when exposed to ultrasonic vibration. To reduce such stress fractures, corners and edges should be radiused.

Projections or tabs may fracture and even fall off. This tendency is reduced if the junction between the tab and the body of the part is radiused. Sometimes it is necessary to thicken the part, lightly clamp it, or, if possible, use a higher-frequency welding machine to reduce breakage.

Sizable holes, sharp angles, or bends within the part may also create problems because the ultrasonic energy may be deflected, leaving a section with little or no fusion.

Thin, unsupported sections on a part may vibrate or flex. If the flexing is severe, it may cause a hot spot in the material, even causing a hole in the part. Increasing the section thickness or switching to a higher-frequency welding machine may reduce this problem. Sometimes reducing the amplitude of vibration helps.

Applications

Ultrasonic welding is widely used for various component assembly applications in the medical, chemical, and electrical industries, as well as others. These include liquid-bearing vessels, IV components, hearing aids, filter assemblies, monitors, and diagnostic components.

Ultrasonic textile joining is used in medical gowns, booties, caps, face masks, hygiene products, bed protectors, surgical drapes, pillow covers, and filter media.

Article courtesy of the American Welding Society.

Review

1. List some factors that have led to the increased use of FCA welding.

2. How is FCAW similar to GMAW?

3. What does the FCA flux provide to the weld?

4. What are the major atmospheric contaminations of the molten weld metal?

5. How does slag help an FCA weld?

6. How can FCA welding guns be cooled?

7. What problems does excessive drive roller pressure cause?

8. List the advantages that FCA welding offers the welding industry.

9. Describe the two methods of manufacturing FCA electrode wire.

10. Why are the large diameter electrodes not used for all-position welding?

11. How do deoxidizers remove oxygen from the weld zone?

12. What do fluxing agents do for a weld?

13. Why are alloying elements added to the flux?

14. How does the flux form a shielding gas to protect the weld?

15. What are the main limitations of the rutile fluxes?

16. Why is it more difficult to use lime-based fluxed electrodes on out-of-position welds?

17. What benefit does adding an externally supplied shielding gas have on some rutile-based electrodes?

18. How do excessive amounts of manganese affect a weld?

19. Why are elements added that cause ferrite to form in the weld?

20. Why are some slags called refractory?

21. Why must a flux form a less dense slag?

22. Referring to Table 12-5, what is the AWS classification for FCA welding electrodes for stainless steel?

23. Describe the meaning of each part of the following FCA welding electrode identification: E81T-5.

24. What does the number 316 in E316T-1 mean?

25. What is the advantage of using an argon-CO_2 mixed shielding gas?

26. What can cause porosity in an FCA weld?

27. What happens to water in the welding arc?

28. What is the thin, dark gray or black layer on new hot-rolled steel? How can it affect the weld?

29. Why is uniformly scattered porosity hard to detect in a weld?

30. What cautions must be taken when chemically cleaning oil or paint from a piece of metal?

31. What can happen to slag that solidifies on the plate ahead of the weld?

32. How is the electrode extension measured?

Chapter 13

Flux Cored Arc Welding

OBJECTIVES

After completing this chapter, the student should be able to

- explain the purpose of setting up the FCA weld station properly.
- demonstrate how to properly set up an FCA welding station and how to thread the electrode wire through the system.
- discuss the disadvantages of having to bevel a plate before welding.
- describe how to make root, filler, and cover passes in FCA welding.
- demonstrate how to properly make FCA welds in butt joints, lap joints, and tee joints in all positions that can pass the specified standard.

KEY TERMS

Low—High Range.

$A = \dfrac{\omega}{V}$

amperage range	*feed rollers* U—V grove	*tee joint*
conduit liner	*lap joint*	*voltage range* pressure Range.
contact tube	*root face*	*weave bead*
critical weld	*stringer bead*	*wire-feed speed* = Amperage.

INTRODUCTION

Setup of the FCA weld station is the key to making quality welds. It may be possible, using a poorly set-up FCA welder, to make an acceptable weld in the flat position. The FCA welding process is often forgiving; thus, welds can often be made even when the welder is not set correctly. However, such welds will have major defects such as excessive spatter, undercut, overlap, porosity, slag inclusions, and poor weld bead contours. Setup becomes even more important for out-of-position welds. Making vertical and overhead welds can be difficult for a student welder with a properly set-up system, but it becomes impossible with a system that is out of adjustment.

Learning to set up and properly adjust the FCA welding system will allow you to produce high-quality welds at a high level of productivity.

FCAW is set up and manipulated in a manner similar to that of GMAW. The results of changes in electrode extension, voltage, amperage, and torch angle are essentially the same.

Although every manufacturer's FCA welding equipment is designed differently, all equipment is set up in a similar manner. It is always best to follow the specific manufacturer's recommendations regarding setup as provided in its

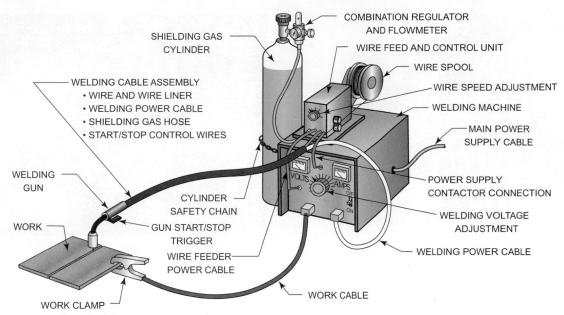

FIGURE 13-1 Basic FCA welding equipment identification. © Cengage Learning 2012

equipment literature. You will find, however, that, in the field, manufacturers' literature is not always available for the equipment you are asked to use. It is therefore important to have a good general knowledge and understanding of the setup procedure for FCA welding equipment. **Figure 13-1** shows all of the various components that make up an FCA welding station.

///// **CAUTION** \\\\\

FCA welding produces a lot of ultraviolet light, heat, sparks, slag, and welding fumes. Proper personal, protective clothing and special protective clothing must be worn to prevent burns from the ultraviolet light and hot weld metal. Eye protection must be worn to prevent injury from flying sparks and slag. Forced ventilation and possibly a respirator must be used to prevent fume-related injuries. Refer to the safety precautions provided by the equipment and electrode manufacturers and to Chapter 2, "Safety in Welding," for additional safety help.

Practices

The practices in this chapter are grouped according to those requiring similar techniques and setups. Plate welds are covered first, then sheet metal. The practices start with 1/4-in. (6-mm) mild steel plates; they are used because they require the least preparation times. The thicker 3/8-in. (9.5-mm) plates provide the basics of practicing groove welding. The 3/4-in. (19-mm) and thicker plates are used to develop the skills required to pass the unlimited thickness test often given to FCA welders. Sheet metal is grouped together because it presents a unique set of learning skills.

The major skill required for making consistently acceptable FCA welds is the ability to set up the welding system. Changes such as variations in material thickness, position, and type of joint require changes both in technique and setup. A correctly set-up FCA welding station can, in many cases, be operated by a less skilled welder. Often the only difference between a welder earning a minimum wage and one earning the maximum wage is the ability to correct machine setups.

For several reasons the FCA welding practice plates will be larger than most other practice plates. Welding heat and welding speed are the major factors that necessitate this increased size. FCA welding is both high energy and fast, and the welding energy (heat) input is so great that small practice plates may glow red by the end of a single weld pass. This would seriously affect the weld quality. To prevent this from happening, wider plates are used. Because of the higher welding speeds, longer plates are usually used.

Plates less than 1/2 in. (13 mm) thick will be 12 in. (305 mm) long for most practices. In addition to controlling the heat buildup, the longer plates are needed to give the welder enough time to practice welding. Learning to make longer welds is a skill that must also be practiced because the FCA welding process is used in industry to make long production welds.

Plates thicker than 1/2 in. (13 mm) can be shorter than 12 in. (305 mm). Most codes usually allow test plates of "unlimited thickness" to be as short as 7 in. (178 mm).

PRACTICE 13-1

FCAW Equipment Setup

For this practice, you will need a semiautomatic welding power source approved for FCA welding, welding gun, electrode feed unit, electrode supply, shielding gas supply,

FIGURE 13-2 Make sure the gas cylinder is chained securely in place before removing the safety cap. Larry Jeffus

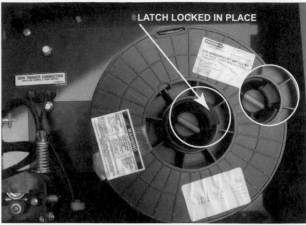

FIGURE 13-4 Wire reel may be secured by a center nut or locking lever. Larry Jeffus

shielding gas flowmeter regulator, electrode conduit, power and work leads, shielding gas hoses (if required), assorted hand tools, spare parts, and any other required materials. In this practice, you will demonstrate to a group of students and your instructor how to properly set up an FCA welding station. Some manufacturers include detailed setup instructions with their equipment. If such instructions are available for your equipment, follow them. Otherwise, use the following instructions.

If the shielding gas is to be used and it comes from a cylinder, the cylinder must be chained securely in place before the valve protection cap is removed, **Figure 13-2.** Stand to one side of the cylinder and quickly crack the valve to blow out any dirt in the valve before the flowmeter regulator is attached, **Figure 13-3.** Attach the correct hose from the regulator to the "gas-in" connection on the electrode feed unit or machine.

Install the reel of electrode (welding wire) on the holder and secure it, **Figure 13-4.** Check the feed roller size to ensure that it matches the wire size, **Figure 13-5.** The **conduit liner** size should be checked for compatibility with the wire size. Connect the conduit to the feed

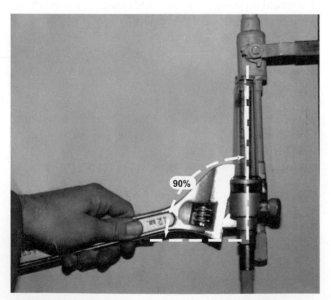

FIGURE 13-3 Attach the flowmeter regulator. Be sure the tube is vertical. Larry Jeffus

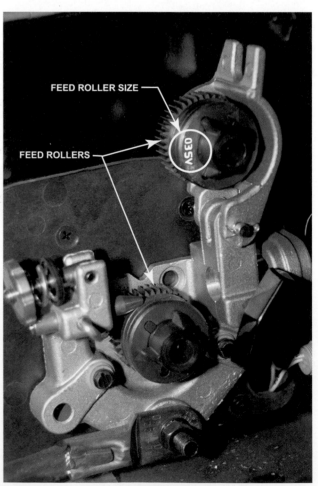

FIGURE 13-5 Check to be certain that the feed rollers are the correct size for the wire being used. Larry Jeffus

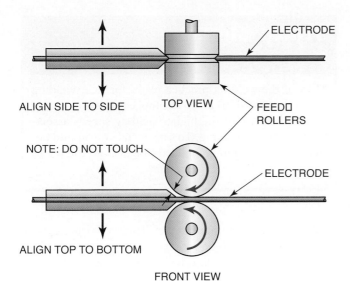

FRONT VIEW

FIGURE 13-6 Feed roller and conduit alignment.
© Cengage Learning 2012

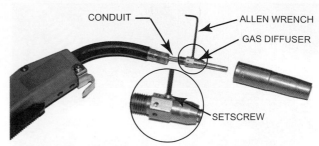

FIGURE 13-8 Securely attach conduit to gas diffuser and contact tube to prevent wire jams caused by misalignment. Larry Jeffus

unit. The conduit or an extension should be aligned with the groove in the roller and set as close to the roller as possible without touching, **Figure 13-6**. Misalignment at this point can contribute to a bird's nest, **Figure 13-7**. Bird-nesting of the electrode wire, so called because it looks like a bird's nest, results when the feed roller pushes the wire into a tangled ball because the wire would not go through the outfeed side conduit.

Be sure the power is off before attaching the welding cables. The electrode and work leads should be attached to the proper terminals. The electrode lead should be attached to the electrode or positive (+). If necessary, it is also attached to the power cable part of the gun lead. The work lead should be attached to the work or negative (−).

The shielding "gas-out" side of the solenoid is then also attached to the gun lead. If a separate splice is required from the gun switch circuit to the feed unit, it should be connected at this time. Check that the welding contacter circuit is connected from the feed unit to the power source.

The welding cable liner or wire conduit must be securely attached to the gas diffuser and **contact tube** (tip), **Figure 13-8**. The contact tube must be the correct size to match the electrode wire size being used. If a shielding gas is to be used, a gas nozzle would be attached to complete the assembly. If a gas nozzle is not needed for a shielding gas, it may still be installed. Because it is easy for a student to touch the work with the contact tube during welding, an electrical short may occur. This short-out of the contact tube will immediately destroy the tube. Although the gas nozzle may interfere with some visibility, it may be worth the trouble for a new welder. FCA welding is more sensitive to changes in arc voltage than is SMAW (stick) welding. Such variations in FCA welding voltage can dramatically and adversely affect your ability to maintain weld bead control. A loose or poor connection will result in increased circuit resistance and a loss of welding voltage. To be sure you have a good work connection, remove any dirt, rust, oil, or other surface contamination at the point the work clamp is connected to the weldment.

Complete a copy of the "Student Welding Report" listed in Appendix I or provided by your instructor. ◆

PRACTICE 13-2

Threading FCAW Wire

Using the FCAW machine that was properly assembled in Practice 11-1, you will turn on the machine and thread the electrode wire through the system.

Check that the unit is assembled correctly according to the manufacturer's specifications. Switch on the power and check the gun switch circuit by depressing the switch. The power source relays, feed relays, gas solenoid, and feed motor should all activate.

Cut off the end of the electrode wire if it is bent. When working with the wire, be sure to hold it tightly. The wire will become tangled if it is released. The wire has a natural curl known as *cast*. Straighten out about 12 in. (300 mm) of the curl to make threading easier.

Separate the wire **feed rollers** and push the wire first through the guides, then between the rollers, and finally into the conduit liner, **Figure 13-9**. Reset the rollers so there is a slight amount of compression on the wire,

FIGURE 13-7 "Bird's nest" in the filler wire at the feed rollers. Larry Jeffus

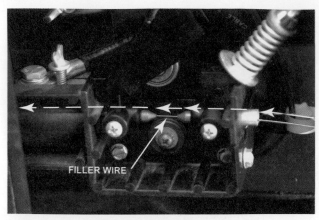

FIGURE 13-9 Push the wire through the guides by hand.
Larry Jeffus

Figure 13-10. Set the **wire-feed speed** control to a slow speed. Hold the welding gun so that the electrode conduit and cable are as straight as possible.

Press the gun switch. Pressing the gun switch to start the wire feeder is called triggering the gun. The wire should start feeding into the liner. Watch to make certain that the wire feeds smoothly and release the gun switch as soon as the end comes through the gun.

If the wire stops feeding before it reaches the end of the gun, stop and check the system. If no obvious problem can be found, mark the wire with tape and remove it from the gun. It then can be held next to the system to determine the location of the problem.

With the wire feed running, adjust the feed roller compression so that the wire reel can be stopped easily by a slight pressure. Too light a roller pressure will cause the wire to feed erratically. Too high a pressure can crush some wires, causing some flux to be dropped inside the wire liner. If this happens, you will have a continual problem with the wire not feeding smoothly or jamming.

With the feed running, adjust the spool drag so that the reel stops when the feed stops. The reel should not coast to a stop because the wire can be snagged easily. Also, when the feed restarts, a jolt occurs when the slack in the wire is taken up. This jolt can be enough to momentarily stop the wire, possibly causing a discontinuity in the weld.

FIGURE 13-10 Adjust the wire-feed tensioner. Larry Jeffus

When the test runs are completed, the wire can either be rewound or cut off. Some wire-feed units have a retract button. This allows the feed driver to reverse and retract the wire automatically. To rewind the wire on units without this retract feature, release the rollers and turn them backward by hand. If the machine will not allow the feed rollers to be released without upsetting the tension, you must cut the wire. Some wire reels have covers to prevent the collection of dust, dirt, and metal filings on the wire, **Figure 13-11**.

Complete a copy of the "Student Welding Report" listed in Appendix I or provided by your instructor. ◆

Flat-Position Welds
PRACTICE 13-3
Stringer Beads Flat Position

Using a properly set-up and adjusted FCA welding machine, **Table 13-1**; proper safety protection; 0.035-in. and/or 0.045-in. (0.9-mm and/or 1.2-mm) diameter E70T-1 and/or E70T-5 electrodes; and one or more pieces of mild steel plate, 12 in. (305 mm) long and 1/4 in. (6 mm) thick or thicker; you will make a stringer bead weld in the flat position, **Figure 13-12**.

Starting at one end of the plate and using a dragging technique, make a weld bead along the entire 12-in. (305-mm) length of the metal. After the weld is complete, check its appearance. Make any needed changes to correct the weld. Repeat the weld and make additional adjustments. After the machine is set, start to work on improving the straightness and uniformity of the weld. Use weave patterns of different widths and straight stringers without weaving.

Repeat with both classifications of electrodes as needed until beads can be made straight, uniform, and free from any visual defects. Turn off the welding machine and shielding gas and clean up your work area when you are finished welding.

Complete a copy of the "Student Welding Report" listed in Appendix I or provided by your instructor. ◆

Square-Groove Welds

One advantage of FCA welding is the ability to make 100% joint penetrating welds without beveling the edges of the plates. These full joint penetrating welds can be made in plates that are 1/4 in. (6 mm) or less in thickness.

(A)

(B)

FIGURE 13-11 (A) Covered wire reel. (B) Wire cover on a dual wire-feed system. Lincoln Electric Company

Electrode		Welding Power			Shielding Gas		Base Metal	
Type	Size	Amps	Wire-Feed Speed IPM (cm/min)	Volts	Type	Flow	Type	Thickness
E70T-1	0.035 in. (0.9 mm)	130 to 150	288 to 380 (732 to 975)	22 to 25	None	n/a	Low carbon steel	1/4 in. to 1/2 in. (6 mm to 13 mm)
E70T-1	0.045 in. (1.2 mm)	150 to 210	200 to 300 (508 to 762)	28 to 29	None	n/a	Low carbon steel	1/4 in. to 1/2 in. (6 mm to 13 mm)
E70-5	0.035 in. (0.9 mm)	130 to 200	288 to 576 (732 to 1463)	20 to 28	75% argon 25% CO_2	30 cfh	Low carbon steel	1/4 in. to 1/2 in. (6 mm to 13 mm)
E70T-5	0.045 in. (1.2 mm)	150 to 250	200 to 400 (508 to 1016)	23 to 29	75% argon 25% CO_2	35 cfh	Low carbon steel	1/4 in. to 1/2 in. (6 mm to 13 mm)

TABLE 13-1 FCA Welding Parameters for Use if Specific Settings Are Unavailable from Electrode Manufacturer

Welding on thicker plates risks the possibility of a lack of fusion on both sides of the **root face**, Figure 13-13.

There are several disadvantages of having to bevel a plate before welding:

- Beveling the edge of a plate adds an operation to the fabrication process.
- Both more filler metal and welding time are required to fill a beveled joint than are required to make a square-jointed weld.

- Beveled joints have more heat from the thermal beveling and additional welding required to fill the groove. The lower heat input to the square joint means less distortion.

The major disadvantage of making square-jointed welds is that as the plate thickness approaches 1/4 in. (6 mm) or the weld is out of position, a much higher level of skill is required. The skill required to make quality square welds can be acquired by practicing on thinner metal. It is much easier

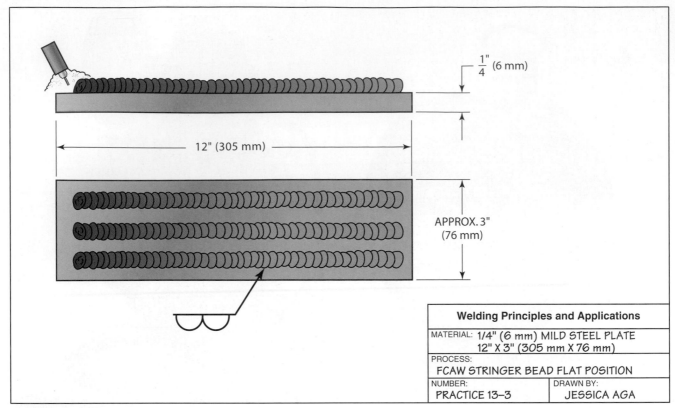

Welding Principles and Applications	
MATERIAL: 1/4" (6 mm) MILD STEEL PLATE 12" X 3" (305 mm X 76 mm)	
PROCESS: FCAW STRINGER BEAD FLAT POSITION	
NUMBER: PRACTICE 13–3	DRAWN BY: JESSICA AGA

FIGURE 13-12 FCAW stringer bead 1/4 in. (6 mm) mild steel flat position. © Cengage Learning 2012

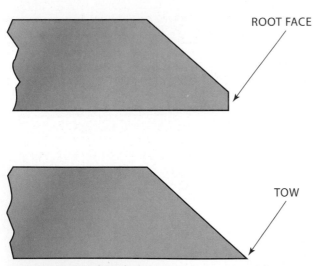

FIGURE 13-13 A beveled joint may or may not have a flat surface, called a root face. © Cengage Learning 2012

to make this type of weld in 1/8-in. (3-mm) thick metal and then move up in thickness as your skills improve.

PRACTICE 13-4

Butt Joint 1G

Using a properly set-up and adjusted FCA welding machine; proper safety protection; 0.035-in. and/or 0.045-in. (0.9-mm and/or 1.2-mm) diameter E70T-1 and/ or E70T-5 electrodes; and one or more pieces of mild steel plate,

12 in. (305 mm) long and 1/4 in. (6 mm) thick or less in thickness; you will make a groove weld in the flat position, **Figure 13-14**.

- Tack weld the plates together and place them in position to be welded.
- Starting at one end, run a bead along the joint. Watch the molten weld pool and bead for signs that a change in technique may be required.
- Make any needed changes as the weld progresses in order to produce a uniform weld.

Repeat with both classifications of electrodes as needed until defect-free welds can consistently be made in the 1/4-in. (6-mm) thick plate. Turn off the welding machine and shielding gas and clean up your work area when you are finished welding.

Complete a copy of the "Student Welding Report" listed in Appendix I or provided by your instructor. ◆

PRACTICE 13-5

Butt Joint 1G 100% to Be Tested

Using a properly set-up and adjusted FCA welding machine; proper safety protection; 0.035-in. and/or 0.045-in. (0.9-mm and/or 1.2-mm) diameter E70T-1 and/ or E70T-5 electrodes; and one or more pieces of mild steel plate, 12 in. (305 mm) long and 1/4 in. (6 mm) thick; you will make a groove weld in the flat position, **Figure 13-15**.

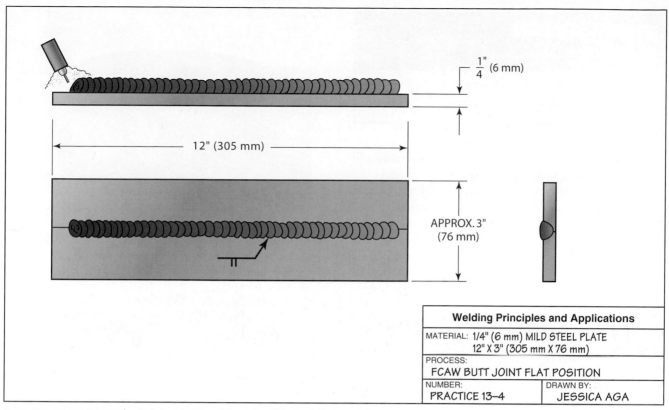

Welding Principles and Applications	
MATERIAL: 1/4" (6 mm) MILD STEEL PLATE 12" X 3" (305 mm X 76 mm)	
PROCESS: FCAW BUTT JOINT FLAT POSITION	
NUMBER: PRACTICE 13–4	DRAWN BY: JESSICA AGA

FIGURE 13-14 FCAW butt joint 1/4 in. (6 mm) mild steel flat position. © Cengage Learning 2012

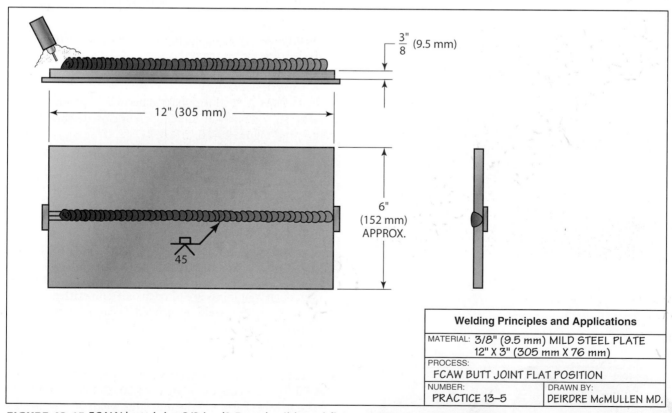

Welding Principles and Applications	
MATERIAL: 3/8" (9.5 mm) MILD STEEL PLATE 12" X 3" (305 mm X 76 mm)	
PROCESS: FCAW BUTT JOINT FLAT POSITION	
NUMBER: PRACTICE 13–5	DRAWN BY: DEIRDRE McMULLEN MD.

FIGURE 13-15 FCAW butt joint 3/8 in. (9.5 mm) mild steel flat position. © Cengage Learning 2012

Following the same instructions for the assembly and welding procedure outlined in Practice 13-4, repeat the weld using each electrode classification until welds using both electrodes can be made with 100% penetration that will pass a bend test. Turn off the welding machine and shielding gas and clean up your work area when you are finished welding.

Complete a copy of the "Student Welding Report" listed in Appendix I or provided by your instructor. ◆

V-Groove and Bevel-Groove Welds

Although for speed and economy engineers try to avoid specifying welds that require beveling the edges of plates, it is not always possible. Anytime the metal being welded is thicker than 1/4 in. (6 mm) and a 100% joint penetration weld is required, the edges of the plate must be prepared with a bevel. Fortunately, FCA welding allows a narrower groove to be made and still achieve a thorough thickness weld, **Figure 13-16**.

All FCA groove welds are made using three different types of weld passes, **Figure 13-17**.

- *Root pass:* The first weld bead of a multiple pass weld. The root pass fuses the two parts together and establishes the depth of weld metal penetration.

- *Filler pass:* Made after the root pass is completed and used to fill the groove with weld metal. More than one pass is often required.

- *Cover pass:* The last weld pass on a multipass weld. The cover pass may be made with one or more welds. It must be uniform in width, reinforcement, and appearance.

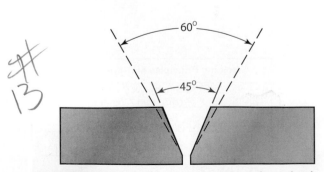

FIGURE 13-16 A smaller groove angle reduces both weld time and filler metal required to make the weld. © Cengage Learning 2012

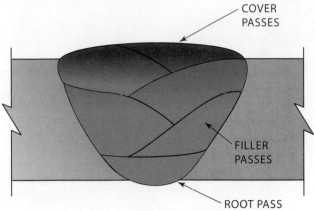

FIGURE 13-17 The three different types of weld passes that make up a weld. © Cengage Learning 2012

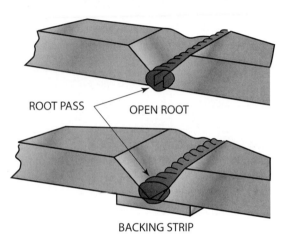

FIGURE 13-18 Root pass maximum deposit 1/4 in. (6 mm) thick. © Cengage Learning 2012

Root Pass A good root pass is needed to obtain a sound weld. The root may be either open or closed using a backing strip, **Figure 13-18**.

The backing strips are usually made from a piece of 1/4-in. (6-mm) thick, 1-in. (25-mm) wide metal that should be 2 in. (50 mm) longer than the base plates to serve as startup and runoff tabs for the weld. The strip is attached to the plate by tack welds made on the sides of the strip, **Figure 13-19**.

Most production welds do not use backing strips, so they are made as open root welds. Because of the difficulty in controlling FCA weld's root weld face contours, however, open root joints are often avoided on **critical welds.** If an open root weld is needed because of weldment design, the root pass may be put in with an SMAW electrode or the root face of the FCA weld can be retouched by grinding and/or back welding.

Care must be taken with any root pass not to have the weld face too convex, **Figure 13-20**. Convex weld faces tend to trap slag along the toe of the weld. FCA weld slag can be extremely difficult to remove in this area, especially if there is any undercutting. To avoid this, adjust the welding power settings, speed, and weave pattern so

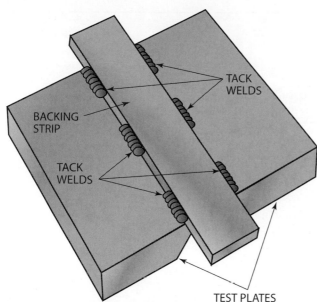

FIGURE 13-19 Securely tack weld the backing strip to the test plates. © Cengage Learning 2012

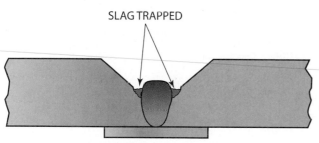

FIGURE 13-20 Slag trapped beside weld bead is hard to remove. © Cengage Learning 2012

that a flat or slightly concave weld face is produced, **Figure 13-21**.

Filler Pass Filler passes are made with either **stringer beads** or **weave beads** for flat or vertically positioned welds, but stringer beads work best for horizontal and overhead-positioned welds. When multiple pass filler welds are required, each weld bead must overlap the others along the edges. Edges should overlap smoothly enough so that the finished bead is uniform, **Figure 13-22**.

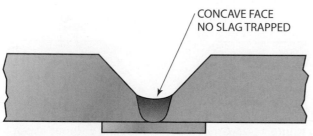

FIGURE 13-21 Flat or concave weld faces are easier to clean off. © Cengage Learning 2012

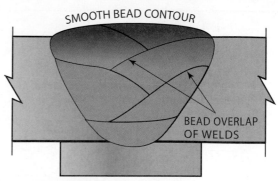

FIGURE 13-22 The surface of a multipass weld should be as smooth as if it were made by one weld. © Cengage Learning 2012

Stringer beads usually overlap about 25% to 50%, and weave beads overlap approximately 10% to 25%.

Each weld bead must be cleaned before the next bead is started. The filler pass ends when the groove has been filled to a level just below the plate surface.

Cover Pass The cover pass may or may not simply be a continuation of the weld beads used to make the filler pass(es). The major difference between the filler pass and the cover pass is the weld face importance. Keeping the face and toe of the cover pass uniform in width, reinforcement, and appearance and defect-free is essential. Most welds are not tested beyond a visual inspection. For that reason the appearance might be the only factor used for accepting or rejecting welds.

The cover pass must meet a strict visual inspection standard. The visual inspection looks to see that the weld is uniform in width and reinforcement. There should be no arc strikes on the plate other than those on the weld itself. The weld must be free of both incomplete fusion and cracks. The weld must be free of overlap, and undercut must not exceed either 10% of the base metal or 1/32 in. (0.8 mm), whichever is less. Reinforcement must have a smooth transition with the base plate and be no higher than 1/8 in. (3 mm), **Figure 13-23**.

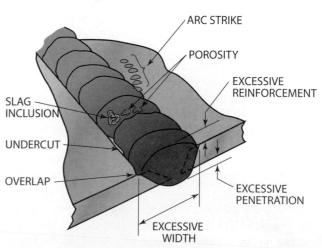

FIGURE 13-23 Common discontinuities found during a visual examination. © Cengage Learning 2012

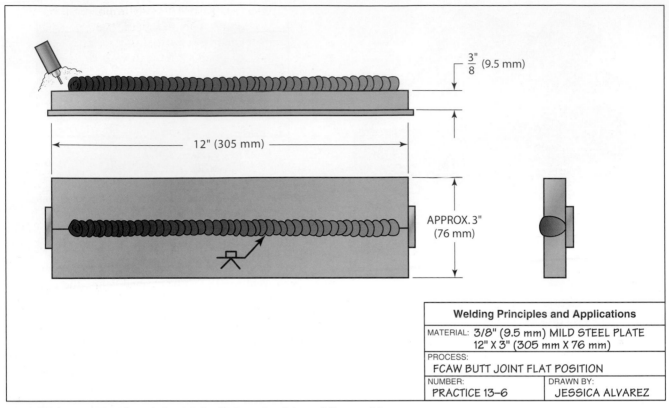

3/8" (9.5 mm)

12" (305 mm)

APPROX. 3" (76 mm)

Welding Principles and Applications

MATERIAL: 3/8" (9.5 mm) MILD STEEL PLATE
12" X 3" (305 mm X 76 mm)

PROCESS:
FCAW BUTT JOINT FLAT POSITION

NUMBER:
PRACTICE 13–6

DRAWN BY:
JESSICA ALVAREZ

FIGURE 13-24 FCAW butt joint 3/8 in. (9.5 mm) mild steel flat position. © Cengage Learning 2012

PRACTICE 13-6

Butt Joint 1G

Using a properly set-up and adjusted FCA welding machine, Table 13-1; proper safety protection; 0.035-in. and/or 0.045-in. (0.9-mm and/or 1.2-mm) diameter E70T-1 and/or E70T-5 electrodes; one or more pieces of mild steel plate, 12-in. (305-mm) long and 3/8-in. (9.5-mm) thick beveled plate; and a 14-in. (355-mm) long, 1-in. (25-mm) wide, and 1/4-in. (6-mm) thick backing strip; you will make a groove weld in the flat position, **Figure 13-24**.

Tack weld the backing strip to the plates. There should be a root gap between the plates that measures approximately 1/8 in. (3 mm). The beveled surface can be made with or without a root face, **Figure 13-25**.

Place the test plates in position at a comfortable height and location. Be sure you have complete and free movement along the full length of the weld joint. It is often a good idea to make a practice pass along the joint with the welding gun without power to make sure nothing will interfere with your making the weld. Be sure the welding cable is free and will not get caught on anything during the weld.

Start the weld outside the groove on the backing strip tab, **Figure 13-26**. This is done so that the arc is smooth and the molten weld pool size is established at the beginning of the groove. Continue the weld out onto the tab at

the outer end of the groove. This process ensures that the end of the groove is completely filled with weld.

Repeat with both classifications of electrodes as needed until consistently defect-free welds can be made. Turn off the welding machine and shielding gas and clean up your work area when you are finished welding.

Complete a copy of the "Student Welding Report" listed in Appendix I or provided by your instructor. ◆

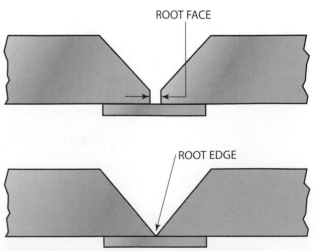

ROOT FACE

ROOT EDGE

FIGURE 13-25 Groove layout with and without a root face. © Cengage Learning 2012

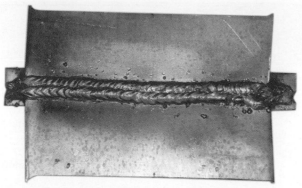

FIGURE 13-26 Run off tabs help control possible under-fill or burn back at the starting and stopping points of a groove weld. Larry Jeffus

PRACTICE 13-7

Butt Joint 1G 100% to Be Tested

Using a properly set-up and adjusted FCA welding machine; proper safety protection; 0.035-in. and/or 0.045-in. (0.9-mm and/or 1.2-mm) diameter E70T-1 and/or E70T-5 electrodes; one or more pieces of mild steel plate, 12-in. (305-mm) long and 3/8-in. (9.5-mm) thick beveled plate; and a 14-in. (355-mm) long, 1-in. (25-mm) wide, and 1/4-in. (6-mm) thick backing strip; you will make a groove weld in the flat position, **Figure 13-27.**

Following the same instructions for the assembly and welding procedure outlined in Practice 13-6, repeat the

weld using each electrode classification until welds using both electrodes can be made with 100% penetration that will pass a bend test. Turn off the welding machine and shielding gas and clean up your work area when you are finished welding.

Complete a copy of the "Student Welding Report" listed in Appendix I or provided by your instructor. ◆

PRACTICE 13-8

Butt Joint 1G

Using a properly set-up and adjusted FCA welding machine, **Table 13-2**; proper safety protection; 0.035-in. and/or through 1/16-in. (0.9-mm and/or through 1.6-mm) diameter E70T-1 and/or E70T-5 electrodes; one or more pieces of mild steel plate, 7-in. (178-mm) long and 3/4-in. (19-mm) thick or thicker beveled plate; and a 9-in. (230-mm) long, 1-in. (25-mm) wide, and 1/4-in. (6-mm) thick backing strip; you will make a groove weld in the flat position, **Figure 13-28.**

Following the same instructions for the assembly and welding procedure outlined in Practice 13-6, repeat the weld with both classifications of electrodes as needed until consistently defect-free welds can be made. Turn off the welding machine and shielding gas and clean up your work area when you are finished welding.

Complete a copy of the "Student Welding Report" listed in Appendix I or provided by your instructor. ◆

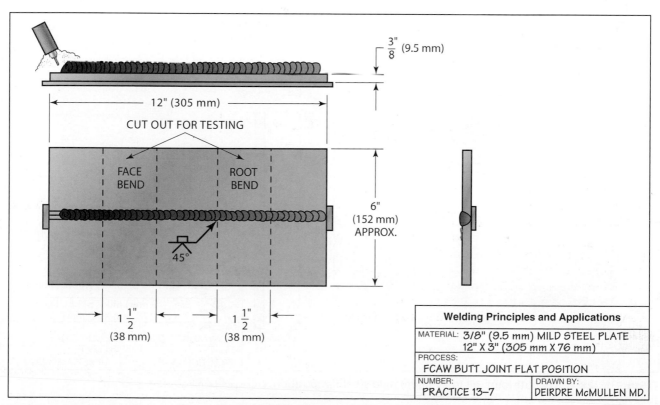

FIGURE 13-27 FCAW butt joint 3/8 in. (9.5 mm) mild steel flat position. © Cengage Learning 2012

Electrode		Welding Power			Shielding Gas		Base Metal	
Type	**Size**	**Amps**	**Wire-Feed Speed IPM (cm/min)**	**Volts**	**Type**	**Flow**	**Type**	**Thickness**
E70T-1	0.035 in. (0.9 mm)	130 to 150	288 to 380 (732 to 975)	22 to 25	None	n/a	Low carbon steel	1/2 in. to 3/4 in. (13 mm to 19 mm)
E70T-1	0.045 in. (1.2 mm)	150 to 210	200 to 300 (508 to 762)	28 to 29	None	n/a	Low carbon steel	1/2 in. to 3/4 in. (13 mm to 19 mm)
E70T-1	0.052 in. (1.4 mm)	150 to 300	150 to 350 (381 to 889)	25 to 33	None	n/a	Low carbon steel	1/2 in. to 3/4 in. (13 mm to 19 mm)
E70T-1	1/16 in. (1.6 mm)	200 to 400	150 to 300 (381 to 762)	27 to 33	None	n/a	Low carbon steel	1/2 in. to 3/4 in. (13 mm to 19 mm)
E70T-5	0.035 in. (0.9 mm)	130 to 200	288 to 576 (732 to 1463)	20 to 28	75% argon 25% CO_2	30 cfh	Low carbon steel	1/2 in. to 3/4 in. (13 mm to 19 mm)
E70T-5	0.045 in. (1.2 mm)	150 to 250	200 to 400 (508 to 1016)	23 to 29	75% argon 25% CO_2	35 cfh	Low carbon steel	1/2 in. to 3/4 in. (13 mm to 19 mm)
E70T-5	0.052 in. (1.4 mm)	150 to 300	150 to 350 (381 to 889)	21 to 32	75% argon 25% CO_2	35 cfh	Low carbon steel	1/2 in. to 3/4 in. (13 mm to 19 mm)
E70T-5	1/16 in. (1.6 mm)	180 to 400	145 to 350 (368 to 889)	21 to 34	75% argon 25% CO_2	40 cfh	Low carbon steel	1/2 in. to 3/4 in. (13 mm to 19 mm)

TABLE 13-2 FCA Welding Parameters for Use If Specific Settings Are Unavailable from Electrode Manufacturer

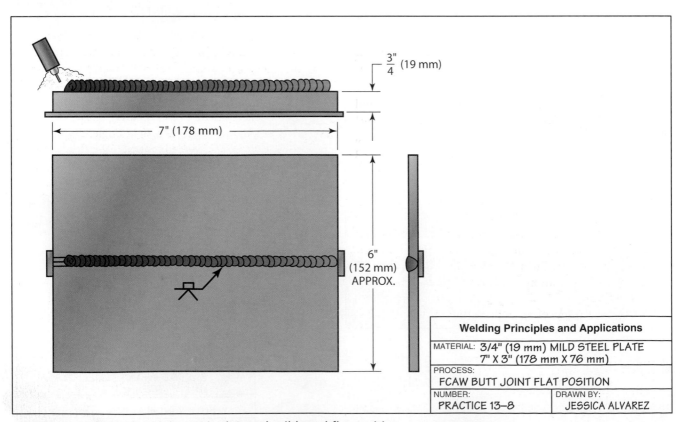

FIGURE 13-28 FCAW butt joint 3/4 in. (19 mm) mild steel flat position. © Cengage Learning 2012

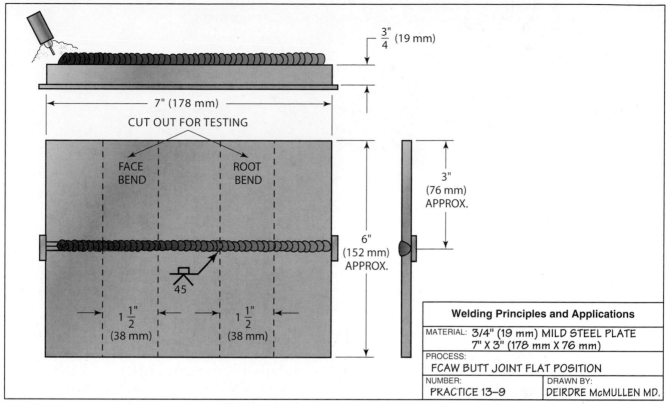

FIGURE 13-29 FCAW butt joint 3/4 in. (19 mm) mild steel flat position. © Cengage Learning 2012

PRACTICE 13-9

Butt Joint 1G 100% to Be Tested

Using a properly set-up and adjusted FCA welding machine; proper safety protection; 0.035-in. and/or through 1/16-in. (0.9-mm and/or through 1.6-mm) diameter E70T-1 and/or E70T-5 electrodes; one or more pieces of mild steel plate, 7-in. (178-mm) long and 3/4-in. (19-mm) thick or thicker beveled plate; and a 9-in. (230-mm) long, 1-in. (25-mm) wide, and 1/4-in. (6-mm) thick backing strip; you will make a groove weld in the flat position, **Figure 13-29.**

Following the same instructions for the assembly and welding procedure outlined in Practice 13-6, repeat the weld using each electrode classification until welds using both electrodes can be made with 100% penetration that will pass a bend test. Turn off the welding machine and shielding gas and clean up your work area when you are finished welding.

Complete a copy of the "Student Welding Report" listed in Appendix I or provided by your instructor. ◆

Fillet Welds

A fillet weld is the type of weld made on the **lap joint** and **tee joint.** It should be built up equal to the thickness of the plate, **Figure 13-30.** On thick plates the fillet must be made up of several passes as with a groove weld. The difference with a fillet weld is that a smooth transition from the plate surface to the weld is required. If this transition is abrupt, it can cause stresses that will weaken the joint.

The lap joint is made by overlapping the edges of the plates. They should be held together tightly before tack welding them together. A small tack weld may be added in the center to prevent distortion during welding, **Figure 13-31.** Chip and wire brush the tacks before you start to weld.

The tee joint is made by tack welding one piece of metal on another piece of metal at a right angle, **Figure 13-32.** After the joint is tack welded together, the slag is chipped from the tack welds. If the slag is not removed, it will cause a slag inclusion in the final weld.

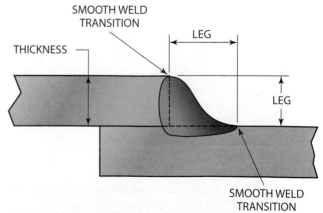

FIGURE 13-30 The legs of a fillet weld should generally be equal to the thickness of the base metal.
© Cengage Learning 2012

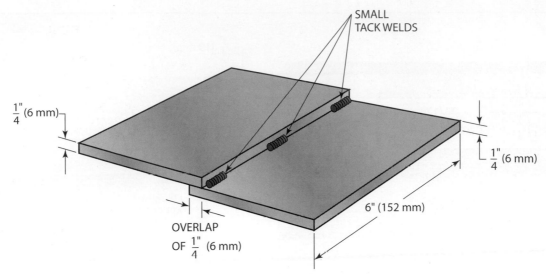

FIGURE 13-31 Tack welding the plates together. © Cengage Learning 2012

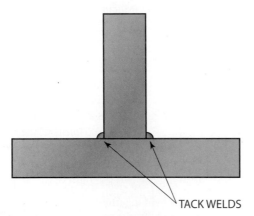

FIGURE 13-32 Tack welding both sides of a tee joint will help keep the tee square for welding. © Cengage Learning 2012

Holding thick plates tightly together on tee joints may cause underbead cracking or lamellar tearing, **Figure 13-33.** On thick plates the weld shrinkage can be great enough to pull the metal apart well below the bead or its heat-affected

zone. In production welds, cracking can be controlled by not assembling the plates tightly together. The space between the two plates can be set by placing a small wire spacer between them, **Figure 13-34.**

A fillet welded lap or tee joint can be strong if it is welded on both sides, even without having deep penetration, **Figure 13-35.** Some tee joints may be prepared for welding by cutting either a bevel or a J-groove in the vertical plate. This cut is not required for strength but may be necessary

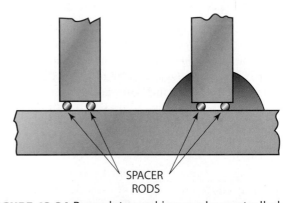

FIGURE 13-34 Base plate cracking can be controlled by placing spacers in the joint before welding.
© Cengage Learning 2012

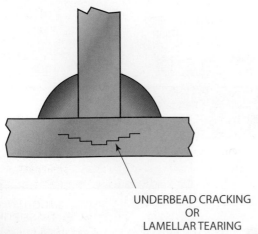

FIGURE 13-33 Underbead cracking or lamellar tearing of the base plate. © Cengage Learning 2012

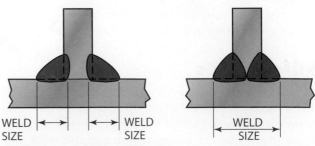

FIGURE 13-35 If the total weld sizes are equal, then both tee joints would have equal strength. © Cengage Learning 2012

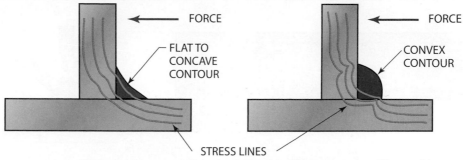

FIGURE 13-36 The stresses are distributed more uniformly through a flat or concave fillet weld. © Cengage Learning 2012

because of design limitations. Unless otherwise instructed, most fillet welds will be equal in size to the plates welded. A fillet weld will be as strong as the base plate if the size of the two welds equals the total thickness of the base plate. The weld bead should have a flat or slightly concave appearance to ensure the greatest strength and efficiency, **Figure 13-36**.

The root of fillet welds must be melted to ensure a completely fused joint. A notch along the root of the weld pool is an indication that the root is not being fused together, **Figure 13-37**. To achieve complete root fusion, move the arc to a point as close as possible to the leading edge of the weld pool, **Figure 13-38**. If the arc strikes the unmelted

FIGURE 13-37 Watch the root of the weld bead to be sure there is complete fusion. Larry Jeffus

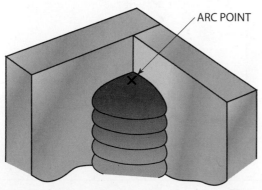

FIGURE 13-38 Moving the arc as close as possible to the leading edge of the weld will provide good root fusion.
© Cengage Learning 2012

plate ahead of the molten weld pool, it may become erratic, which will increase weld spatter.

PRACTICE 13-10

Lap Joint and Tee Joint 1F

Using a properly set-up and adjusted FCA welding machine; proper safety protection; 0.035-in. and/or 0.045-in. (0.9-mm and/or 1.2-mm) diameter E70T-1 and/or E70T-5 electrodes; and one or more pieces of mild steel plate, 12-in. (305-mm) long and 3/8-in. (9.5-mm) thick beveled plate; you will make a fillet weld in the flat position.

Tack weld the pieces of metal together and brace them in position. When making the lap or tee joints in the flat position, the plates must be at a 45° angle so that the surface of the weld will be flat, **Figure 13-39A** and **B**. Starting at one end, make a weld along the entire length of the joint.

Repeat each type of joint with both classifications of electrodes as needed until consistently defect-free welds can be made. Turn off the welding machine and shielding gas and clean up your work area when you are finished welding.

Complete a copy of the "Student Welding Report" listed in Appendix I or provided by your instructor. ◆

PRACTICE 13-11

Lap Joint and Tee Joint 1F 100% to Be Tested

Using a properly set-up and adjusted FCA welding machine; proper safety protection; 0.035-in. and/or 0.045-in. (0.9-mm and/or 1.2-mm) diameter E70T-1 and/or E70T-5 electrodes; and one or more pieces of mild steel plate, 12-in. (305-mm) long and 3/8-in. (9.5-mm) thick beveled plate; you will make a fillet weld in the flat position, on a lap joint, **Figure 13-40A**, and a tee joint, **Figure 13-40B**.

Following the same instructions for the assembly and welding procedure outlined in Practice 13-10, repeat the weld using each electrode classification until welds using both electrodes can be made with 100% penetration that will pass a bend test. Turn off the welding machine and shielding gas and clean up your work area when you are finished welding.

Complete a copy of the "Student Welding Report" listed in Appendix I or provided by your instructor. ◆

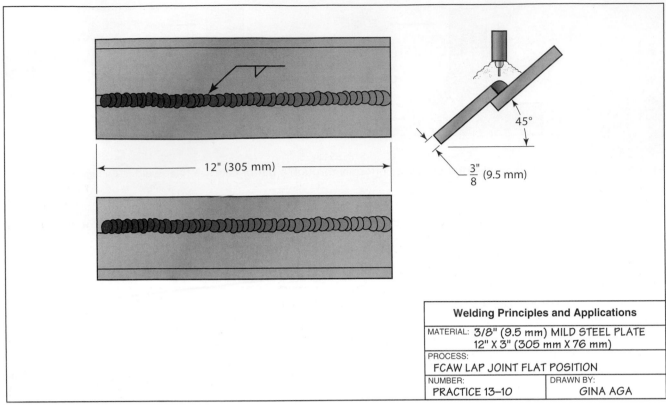

FIGURE 13-39 (A) FCAW lap joint 3/8 in. (9.5 mm) mild steel flat position. © Cengage Learning 2012

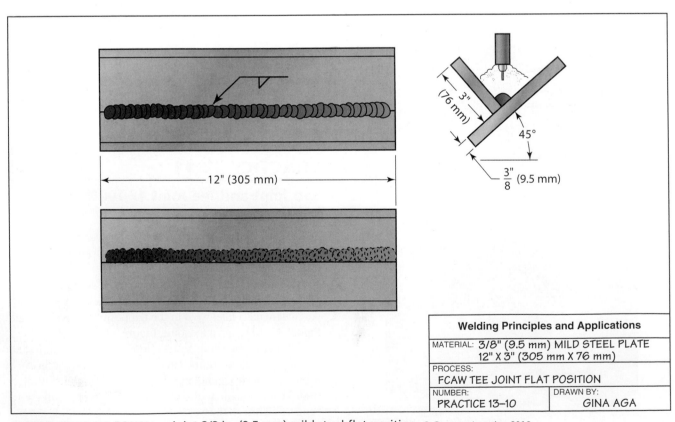

FIGURE 13-39 (B) FCAW tee joint 3/8 in. (9.5 mm) mild steel flat position. © Cengage Learning 2012

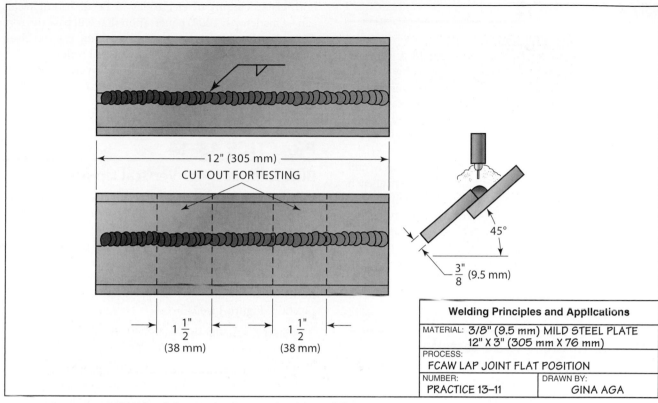

FIGURE 13-40 (A) FCAW lap joint 3/8 in. (9.5 mm) mild steel flat position. © Cengage Learning 2012

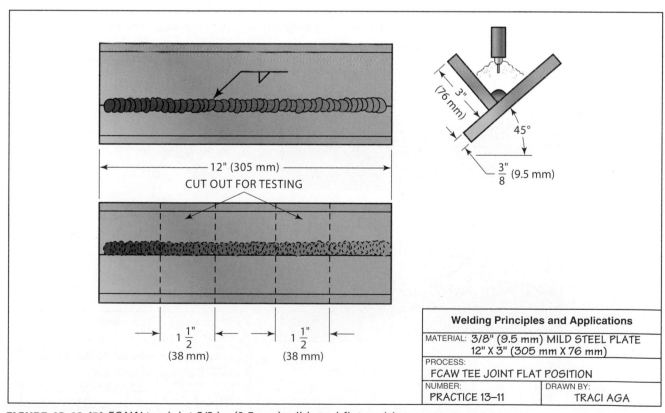

FIGURE 13-40 (B) FCAW tee joint 3/8 in. (9.5 mm) mild steel flat position. © Cengage Learning 2012

PRACTICE 13-12

Tee Joint 1F

Using a properly set-up and adjusted FCA welding machine; proper safety protection; 0.035-in. and/or through 1/16-in. (0.9-mm and/or through 1.6-mm) diameter E70T-1 and/or E70T-5 electrodes; and one or more pieces of mild steel plate, 7-in. (178-mm) long and 3/4-in. (19-mm) thick or thicker beveled plate; you will make a fillet weld in the flat position.

Following the same instructions for the assembly and welding procedure outlined in Practice 13-10, repeat each type of joint with both classifications of electrodes as needed until consistently defect-free welds can be made. Turn off the welding machine and shielding gas and clean up your work area when you are finished welding.

Complete a copy of the "Student Welding Report" listed in Appendix I or provided by your instructor. ◆

PRACTICE 13-13

Tee Joint 1F 100% to Be Tested

Using a properly set-up and adjusted FCA welding machine; proper safety protection; 0.035-in. and/or through 1/16-in. (0.9-mm and/or through 1.6-mm) diameter E70T-1 and/or E70T-5 electrodes; and one or more pieces of mild steel plate, 7-in. (178-mm) long and 3/4-in. (19-mm) thick or thicker beveled plate; you will make a fillet weld in the flat position, **Figure 13-41**.

Following the same instructions for the assembly and welding procedure outlined in Practice 13-10, repeat with

both classifications of electrodes as needed until welds can be made with 100% penetration that will pass the test. Turn off the welding machine and shielding gas and clean up your work area when you are finished welding.

Complete a copy of the "Student Welding Report" listed in Appendix I or provided by your instructor. ◆

Vertical Welds

PRACTICE 13-14

Butt Joint at a 45° Vertical Up Angle

Using a properly set-up and adjusted FCA welding machine; proper safety protection; 0.035-in. and/or 0.045-in. (0.9-mm and/or 1.2-mm) diameter E70T-1 and/or E70T-5 electrodes; and one or more pieces of mild steel plate, 12 in. (305 mm) long and 1/4 in. (6 mm) thick or thinner; you will increase the plate angle gradually as you develop skill until you are making satisfactory welds in the vertical up position, **Figure 13-42**.

- Start practicing this weld with the plate at a 45° angle.

- Gradually increase the angle of the plate to vertical as skill is gained in welding this joint. A straight stringer bead or slight zigzag will work well on this joint.

- Establish a molten weld pool in the root of the joint.

- Cool, chip, and inspect the weld for uniformity and defects.

It is easier to make a quality weld in the vertical up position if both the **amperage** and **voltage** are set at the

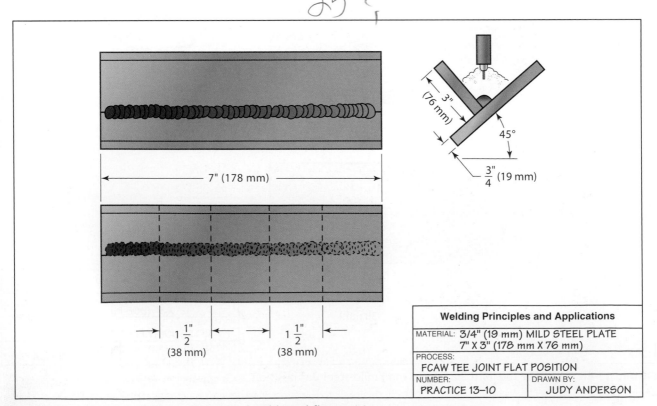

FIGURE 13-41 FCAW tee joint 3/4 in. (19 mm) mild steel flat position. © Cengage Learning 2012

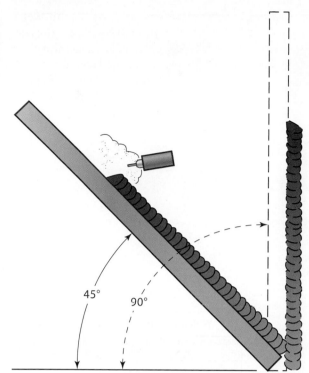

FIGURE 13-42 Start making welds with the plate at a 45° angle. As your skill develops, increase the angle until the plate is vertical. © Cengage Learning 2012

lower end of their **ranges.** This will make the molten weld pool smaller, less fluid, and easier to control. A problem with lower power settings is that the weld bead can often be very convex, **Figure 13-43.** Faster travel speed and/or slightly wider weave patterns can be used to control the bead shape.

Start at the bottom of the plate and hold the welding gun at a slight upward angle to the plate, **Figure 13-44.**

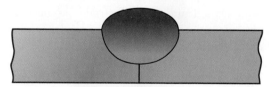

FIGURE 13-43 Low amperage causes too much buildup and not enough penetration. © Cengage Learning 2012

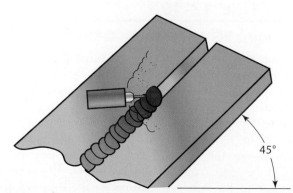

FIGURE 13-44 45° vertical up. © Cengage Learning 2012

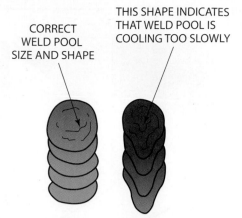

FIGURE 13-45 The shape of the weld pool can indicate the temperature of the surrounding base metal.
© Cengage Learning 2012

Brace yourself, lower your hood, and begin to weld. Depending on the machine settings and type of electrode used, you will make a weave pattern.

If the molten weld pool is large and fluid (hot), use a "C" or "J" weave pattern to allow a longer time for the molten weld pool to cool, **Figure 13-45.** Do not make the weave so long or fast that the electrode is allowed to strike the metal ahead of the molten weld pool. If this happens, spatter increases and a spot or zone of incomplete fusion may occur.

A weld that is high and has little or no fusion is too "cold." Changing the welding technique will not correct this problem. The welder must stop welding and make the needed adjustments to the power supply or electrode feeder. Continue to weld along the entire 12-in. (305-mm) length of plate.

Repeat welds with both electrodes as needed until defect-free welds can be consistently made vertically in the 1/4-in. (6-mm) thick plate. Turn off the welding machine and shielding gas and clean up your work area when you are finished welding.

Complete a copy of the "Student Welding Report" listed in Appendix I or provided by your instructor. ◆

PRACTICE 13-15

Butt Joint 3G

Using a properly set-up and adjusted FCA welding machine; proper safety protection; 0.035-in. and/or 0.045-in. (0.9-mm and/or 1.2-mm) diameter E70T-1 and/or E70T-5 electrodes; and one or more pieces of mild steel plate, 12 in. (305 mm) long and 1/4 in. (6 mm) thick or thinner; you will make a groove weld in the vertical position, **Figure 13-46.**

Following the same instructions for the assembly and welding procedure outlined in Practice 13-14, repeat with both classifications of electrodes as needed until defect-free welds can be consistently made in the 1/4-in. (6-mm) thick plate. Turn off the welding machine and

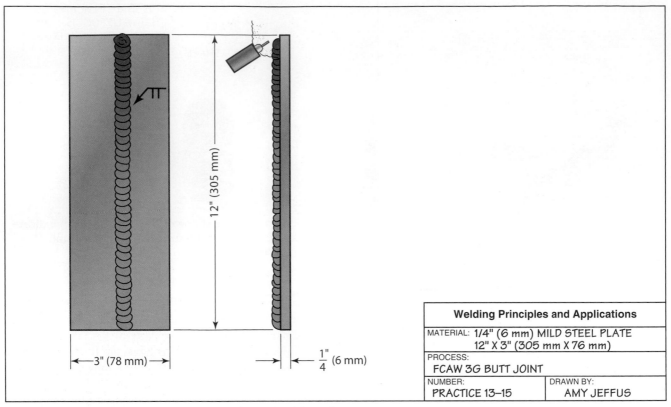

Welding Principles and Applications	
MATERIAL: 1/4" (6 mm) MILD STEEL PLATE 12" X 3" (305 mm X 76 mm)	
PROCESS: FCAW 3G BUTT JOINT	
NUMBER: PRACTICE 13–15	DRAWN BY: AMY JEFFUS

FIGURE 13-46 FCAW 3G butt joint 1/4 in. (6 mm) mild steel. © Cengage Learning 2012

shielding gas and clean up your work area when you are finished welding.

Complete a copy of the "Student Welding Report" listed in Appendix I or provided by your instructor. ◆

PRACTICE 13-16

Butt Joint 3G 100% to Be Tested

Using a properly set-up and adjusted FCA welding machine; proper safety protection; 0.035-in. and/or 0.045-in. (0.9-mm and/or 1.2-mm) diameter E70T-1 and/or E70T-5 electrodes; and one or more pieces of mild steel plate, 12 in. (305 mm) long and 1/4 in. (6 mm) thick; you will make a groove weld in the vertical position.

Following the same instructions for the assembly and welding procedure as outlined in Practice 13-14, repeat the weld using each electrode classification until welds using both electrodes can be made with 100% penetration that will pass a bend test. Turn off the welding machine and shielding gas and clean up your work area when you are finished welding.

Complete a copy of the "Student Welding Report" listed in Appendix I or provided by your instructor. ◆

PRACTICE 13-17

Butt Joint 3G

Using a properly set-up and adjusted FCA welding machine; proper safety protection; 0.035-in. and/or 0.045-in.

(0.9-mm and/or 1.2-mm) diameter E70T-1 and/or E70T-5 electrodes; one or more pieces of mild steel plate, 12-in. (305-mm) long and 3/8-in. (9.5-mm) thick beveled plate; and a 14-in. (355-mm) long, 1-in. (25-mm) wide, and 1/4-in. (6-mm) thick backing strip; you will make a groove weld in the vertical position.

Following the same instructions for the assembly and welding procedure as outlined in Practice 13-14, repeat with both classifications of electrodes as needed until defect-free welds can consistently be made. Turn off the welding machine and shielding gas and clean up your work area when you are finished welding.

Complete a copy of the "Student Welding Report" listed in Appendix I or provided by your instructor. ◆

PRACTICE 13-18

Butt Joint 3G 100% to Be Tested

Using a properly set-up and adjusted FCA welding machine; proper safety protection; 0.035-in. and/or 0.045-in. (0.9-mm and/or 1.2-mm) diameter E70T-1 and/or E70T-5 electrodes; one or more pieces of mild steel plate, 12-in. (305-mm) long and 3/8-in. (9.5-mm) thick beveled plate; and a 14-in. (355-mm) long, 1-in. (25-mm) wide, and 1/4-in. (6-mm) thick backing strip; you will make a groove weld in the vertical position, **Figure 13-47**.

Following the same instructions for the assembly and welding procedure outlined in Practice 13-14, repeat with

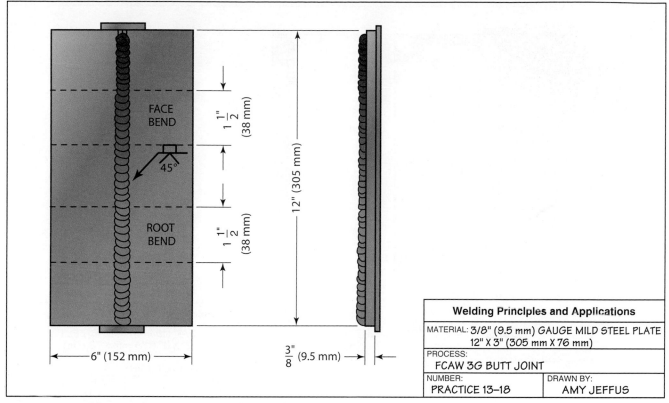

FIGURE 13-47 FCAW 3G butt joint 3/8 in. (9.5 mm) mild steel. © Cengage Learning 2012

electrodes as needed until welds can be made with 100% penetration that will pass a bend test. Turn off the welding machine and shielding gas and clean up your work area when you are finished welding.

Complete a copy of the "Student Welding Report" listed in Appendix I or provided by your instructor. ◆

PRACTICE 13-19

Butt Joint at a 45° Vertical Up Angle

Using a properly set-up and adjusted FCA welding machine; proper safety protection; 0.035-in. and/or through 1/16-in. (0.9-mm and/or through 1.6-mm) diameter E70T-1 and/or E70T-5 electrodes; one or more pieces of mild steel plate, 7-in. (178-mm) long and 3/4-in. (19-mm) thick or thicker beveled plate; and a 9-in. (230-mm) long, 1-in. (25-mm) wide, and 1/4-in. (6-mm) thick backing strip; you will increase the plate angle gradually as you develop skill until you are making satisfactory welds in the vertical up position.

Following the same instructions for the assembly and welding procedure outlined in Practice 13-14, repeat with both classifications of electrodes as needed until defect-free welds can consistently be made. Turn off the welding machine and shielding gas and clean up your work area when you are finished welding.

Complete a copy of the "Student Welding Report" listed in Appendix I or provided by your instructor. ◆

PRACTICE 13-20

Butt Joint 3G

Using a properly set-up and adjusted FCA welding machine; proper safety protection; 0.035-in. and/or through 1/16-in. (0.9-mm and/or through 1.6-mm) diameter E70T-1 and/or E70T-5 electrodes; one or more pieces of mild steel plate, 7-in. (178-mm) long and 3/4-in. (19-mm) thick or thicker beveled plate; and a 9-in. (230-mm) long, 1-in. (25-mm) wide, and 1/4-in. (6-mm) thick backing strip; you will make a groove weld in the vertical position.

Following the same instructions for the assembly and welding procedure outlined in Practice 13-14, repeat with both classifications of electrodes as needed until defect-free welds can consistently be made. Turn off the welding machine and shielding gas and clean up your work area when you are finished welding.

Complete a copy of the "Student Welding Report" listed in Appendix I or provided by your instructor. ◆

PRACTICE 13-21

Butt Joint 3G 100% to Be Tested

Using a properly set-up and adjusted FCA welding machine; proper safety protection; 0.035-in. and/or through 1/16-in. (0.9-mm and/or through 1.6-mm) diameter E70T-1 and/or E70T-5 electrodes; one or more pieces of mild steel plate, 7-in. (178-mm) long and 3/4-in. (19-mm) thick or thicker beveled plate; and a 9-in. (230-mm) long,

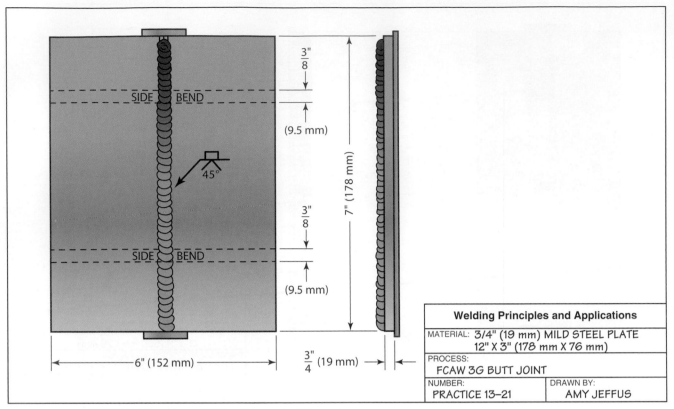

SIDE BEND

45°

SIDE BEND

3"
8

(9.5 mm)

7" (178 mm)

3"
8

(9.5 mm)

6" (152 mm)

3"
4 (19 mm)

Welding Principles and Applications

MATERIAL: 3/4" (19 mm) MILD STEEL PLATE
12" X 3" (178 mm X 76 mm)

PROCESS:
FCAW 3G BUTT JOINT

NUMBER:
PRACTICE 13–21

DRAWN BY:
AMY JEFFUS

FIGURE 13-48 FCAW 3G butt joint 3/4 in. (19 mm) mild steel. © Cengage Learning 2012

1-in. (25-mm) wide, and 1/4-in. (6-mm) thick backing strip; you will make a groove weld in the vertical position, **Figure 13-48**.

Following the same instructions for the assembly and welding procedure outlined in Practice 13-14, repeat the weld using each electrode classification until welds using both electrodes can be made with 100% penetration that will pass a bend test. Turn off the welding machine and shielding gas and clean up your work area when you are finished welding.

Complete a copy of the "Student Welding Report" listed in Appendix I or provided by your instructor. ◆

PRACTICE 13-22

Fillet Weld Joint at a 45° Vertical Up Angle

Using a properly set-up and adjusted FCA welding machine; proper safety protection; 0.035-in. and/or 0.045-in. (0.9-mm and/or 1.2-mm) diameter E70T-1 and/or E70T-5 electrodes; and one or more pieces of mild steel plate, 12 in. (305 mm) long and 3/8 in. (9.5 mm) thick; you will increase the plate angle gradually as you develop skill until you are making satisfactory welds in the vertical up position, **Figure 13-49**.

Tack weld the metal pieces together and brace them in position. Check to see that you have free movement along

the entire joint to prevent stopping and restarting during the weld. Avoiding stops and starts both speeds up the welding time and eliminates discontinuities.

It is easier to make a quality weld in the vertical up position if both the amperage and voltage are set at the

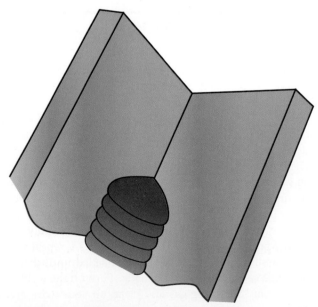

FIGURE 13-49 45° vertical up fillet weld.
© Cengage Learning 2012

lower end of their ranges. This will make the molten weld pool smaller, less fluid, and easier to control. A problem with the lower power settings is that the weld bead often is very convex. A convex face on a weld bead often makes it more difficult to remove the slag along the toe of the weld.

The weave pattern should allow for adequate fusion on both edges of the joint. Watch the edges to be sure that they are being melted so that adequate fusion and penetration occur.

Repeat with electrodes as needed until defect-free welds can consistently be made vertically. Turn off the welding machine and shielding gas and clean up your work area when you are finished welding.

Complete a copy of the "Student Welding Report" listed in Appendix I or provided by your instructor. ◆

PRACTICE 13-23

Lap Joint and Tee Joint 3F

Using a properly set-up and adjusted FCA welding machine; proper safety protection; 0.035-in. and/or 0.045-in. (0.9-mm and/or 1.2-mm) diameter E70T-1 and/or E70T-5 electrodes; and one or more pieces of mild steel plate, 12-in. (305-mm) long and 3/8-in. (9.5-mm) thick beveled plate; you will make a fillet weld in the vertical position.

Following the same instructions for the assembly and welding procedure outlined in Practice 13-22, repeat each type of joint with both classifications of electrodes as needed until defect-free welds can consistently be made. Turn off the welding machine and shielding gas and clean up your work area when you are finished welding.

Complete a copy of the "Student Welding Report" listed in Appendix I or provided by your instructor. ◆

PRACTICE 13-24

Lap Joint and Tee Joint 3F 100% to Be Tested

Using a properly set-up and adjusted FCA welding machine; proper safety protection; 0.035-in. and/or 0.045-in. (0.9-mm and/or 1.2-mm) diameter E70T-1 and/or E70T-5 electrodes; and one or more pieces of mild steel plate, 12-in. (305-mm) long and 3/8-in. (9.5-mm) thick beveled plate; you will make a fillet weld in the vertical position.

Following the same instructions for the assembly and welding procedure as outlined in Practice 13-22, repeat each type of joint with both classifications of electrodes as needed until welds can be made with 100% penetration that will pass the test. Turn off the welding machine and shielding gas and clean up your work area when you are finished welding.

Complete a copy of the "Student Welding Report" listed in Appendix I or provided by your instructor. ◆

PRACTICE 13-25

Tee Joint 3F

Using a properly set-up and adjusted FCA welding machine; proper safety protection; 0.035-in. and/or through 1/16-in. (0.9-mm and/or through 1.6-mm) diameter E70T-1 and/or E70T-5 electrodes; and one or more pieces of mild steel plate, 7-in. (178-mm) long and 3/4-in. (19-mm) thick or thicker beveled plate, you will make a fillet weld in the vertical position.

Following the same instructions for the assembly and welding procedure outlined in Practice 13-22, repeat each type of joint with both classifications of electrodes as needed until defect-free welds can consistently be made. Turn off the welding machine and shielding gas and clean up your work area when you are finished welding.

Complete a copy of the "Student Welding Report" listed in Appendix I or provided by your instructor. ◆

PRACTICE 13-26

Tee Joint 3F 100% to Be Tested

Using a properly set-up and adjusted FCA welding machine; proper safety protection; 0.035-in. and/or through 1/16-in. (0.9-mm and/or through 1.6-mm) diameter E70T-1 and/or E70T-5 electrodes; and one or more pieces of mild steel plate, 7-in. (178-mm) long and 3/4-in. (19-mm) thick or thicker beveled plate; you will make a fillet weld in the vertical position, **Figure 13-50**.

Following the same instructions for the assembly and welding procedure outlined in Practice 13-22, repeat with both classifications of electrodes as needed until welds can be made with 100% penetration that will pass the test. Turn off the welding machine and shielding gas and clean up your work area when you are finished welding.

Complete a copy of the "Student Welding Report" listed in Appendix I or provided by your instructor. ◆

Horizontal Welds

PRACTICE 13-27

Lap Joint and Tee Joint 2F

Using a properly set-up and adjusted FCA welding machine; proper safety protection; 0.035-in. and/or 0.045-in. (0.9-mm and/or 1.2-mm) diameter E70T-1 and/or E70T-5 electrodes; and one or more pieces of mild steel plate, 12-in. (305-mm) long and 3/8-in. (9.5-mm) thick beveled plate, you will make a fillet weld in the horizontal position, **Figure 13-51A** and **B**.

The root weld must be kept small so that its contour can be controlled. Too large a root pass can trap slag under overlap along the lower edge of the weld, **Figure 13-52**. Clean each pass thoroughly before the weld bead is started.

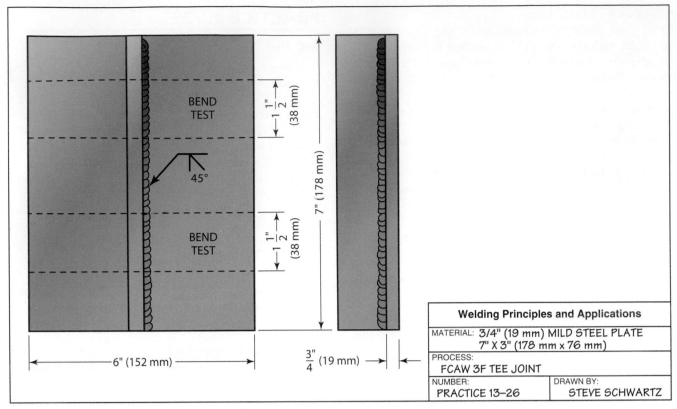

FIGURE 13-50 FCAW 3F tee joint 3/4 in. (19 mm) mild steel. © Cengage Learning 2012

Labels within image 1:

BEND TEST

BEND TEST

45°

$1\frac{1}{2}$" (38 mm)

$1\frac{1}{2}$" (38 mm)

7" (178 mm)

6" (152 mm)

$\frac{3}{4}$" (19 mm)

Welding Principles and Applications

MATERIAL: 3/4" (19 mm) MILD STEEL PLATE
7" X 3" (178 mm x 76 mm)

PROCESS:
FCAW 3F TEE JOINT

NUMBER: PRACTICE 13–26

DRAWN BY: STEVE SCHWARTZ

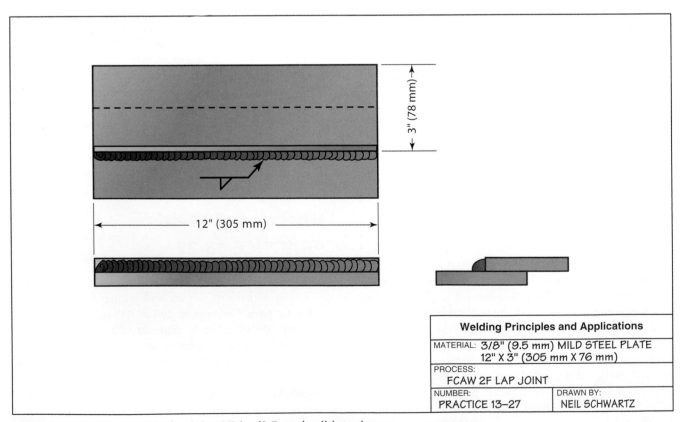

FIGURE 13-51 (A) FCAW 2F lap joint 3/8 in. (9.5 mm) mild steel. © Cengage Learning 2012

Labels within image 2:

3" (78 mm)

12" (305 mm)

Welding Principles and Applications

MATERIAL: 3/8" (9.5 mm) MILD STEEL PLATE
12" X 3" (305 mm X 76 mm)

PROCESS:
FCAW 2F LAP JOINT

NUMBER: PRACTICE 13–27

DRAWN BY: NEIL SCHWARTZ

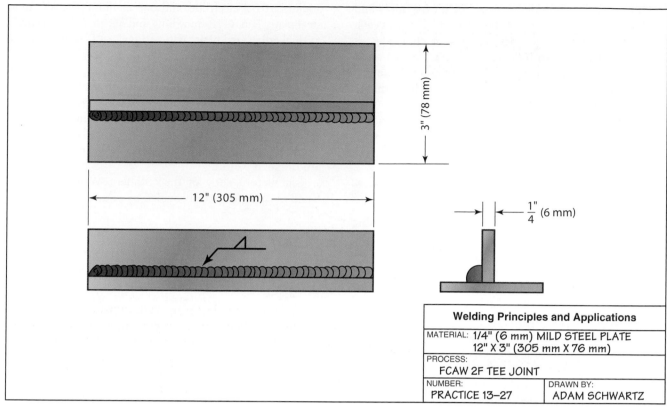

Welding Principles and Applications	
MATERIAL: 1/4" (6 mm) MILD STEEL PLATE 12" X 3" (305 mm X 76 mm)	
PROCESS: FCAW 2F TEE JOINT	
NUMBER: PRACTICE 13–27	DRAWN BY: ADAM SCHWARTZ

FIGURE 13-51 (B) FCAW 2F tee joint 1/4 in. mild steel. © Cengage Learning 2012

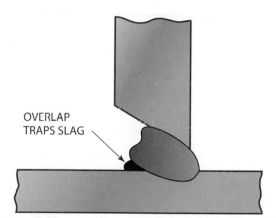

FIGURE 13-52 Slag can be trapped along the side of the root pass. © Cengage Learning 2012

FIGURE 13-53 FCAW weld bead positions for a 100% penetration grooved tee joint. © Cengage Learning 2012

Follow the weld bead sequence shown in **Figure 13-53**. Keeping all of the weld beads small will help control their contour.

Repeat each type of joint with both classifications of electrodes as needed until defect-free welds can consistently be made. Turn off the welding machine and shielding gas and clean up your work area when you are finished welding.

Complete a copy of the "Student Welding Report" listed in Appendix I or provided by your instructor. ◆

PRACTICE 13-28

Lap Joint and Tee Joint 2F 100% to Be Tested

Using a properly set-up and adjusted FCA welding machine; proper safety protection; 0.035-in. and/or 0.045-in. (0.9-mm and/or 1.2-mm) diameter E70T-1 and/or E70T-5 electrodes; and one or more pieces of mild steel plate, 12-in. (305-mm) long and 3/8-in. (9.5-mm) thick beveled plate; you will make a fillet weld in the horizontal position.

Following the same instructions for the assembly and welding procedure outlined in Practice 13-27, repeat each type of joint with both classifications of electrodes as needed until welds can be made with 100% penetration that will pass the test. Turn off the welding machine and shielding gas and clean up your work area when you are finished welding.

Complete a copy of the "Student Welding Report" listed in Appendix I or provided by your instructor. ◆

PRACTICE 13-29

Tee Joint 2F

Using a properly set-up and adjusted FCA welding machine; proper safety protection; 0.035-in. and/or through 1/16-in. (0.9-mm and/or through 1.6-mm) diameter E70T-1 and/or E70T-5 electrodes; and one or more pieces of mild steel plate, 7-in. (178-mm) long and 3/4-in. (19-mm) thick or thicker beveled plate; you will make a fillet weld in the horizontal position.

Following the same instructions for the assembly and welding procedure outlined in Practice 13-27, repeat each type of joint with both classifications of electrodes as needed until defect-free welds can consistently be made. Turn off the welding machine and shielding gas and clean up your work area when you are finished welding.

Complete a copy of the "Student Welding Report" listed in Appendix I or provided by your instructor. ◆

PRACTICE 13-30

Tee Joint 2F 100% to Be Tested

Using a properly set-up and adjusted FCA welding machine; proper safety protection; 0.035-in. and/or through 1/16-in. (0.9-mm and/or through 1.6-mm) diameter E70T-1

and/or E70T-5 electrode; and one or more pieces of mild steel plate, 7-in. (178-mm) long and 3/4-in. (19-mm) thick or thicker beveled plate; you will make a fillet weld in the horizontal position.

Following the same instructions for the assembly and welding procedure outlined in Practice 13-27, repeat with both classifications of electrodes as needed until welds can be made with 100% penetration that will pass the test. Turn off the welding machine and shielding gas and clean up your work area when you are finished welding.

Complete a copy of the "Student Welding Report" listed in Appendix I or provided by your instructor. ◆

PRACTICE 13-31

Stringer Bead at a 45° Horizontal Angle

Using a properly set-up and adjusted FCA welding machine; proper safety protection; 0.035-in. and/or 0.045-in. (0.9-mm and/or 1.2-mm) diameter E70T-1 and/or E70T-5 electrodes; and one or more pieces of mild steel plate, 12 in. (305 mm) long and 1/4 in. (6 mm) thick or thinner; you will increase the plate angle gradually as you develop skill until you are making satisfactory horizontal welds across the vertical face of the plate, **Figure 13-54.**

Repeat the weld using each electrode classification until welds using both electrodes can be made horizontally with uniform bead contours when the plates are vertical. Turn off the welding machine and shielding gas and clean up your work area when you are finished welding.

Complete a copy of the "Student Welding Report" listed in Appendix I or provided by your instructor. ◆

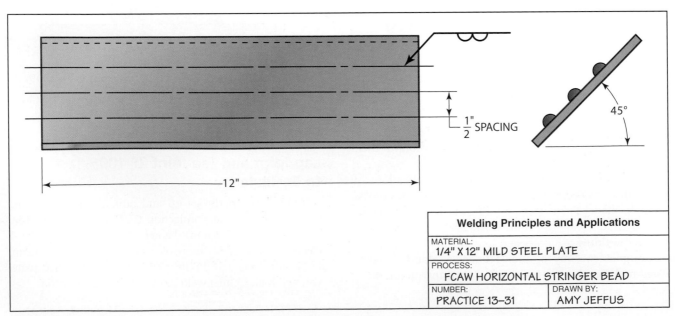

Welding Principles and Applications

MATERIAL:	
1/4" X 12" MILD STEEL PLATE	
PROCESS:	
FCAW HORIZONTAL STRINGER BEAD	
NUMBER:	DRAWN BY:
PRACTICE 13–31	AMY JEFFUS

FIGURE 13-54 FCAW horizontal stringer bead. © Cengage Learning 2012

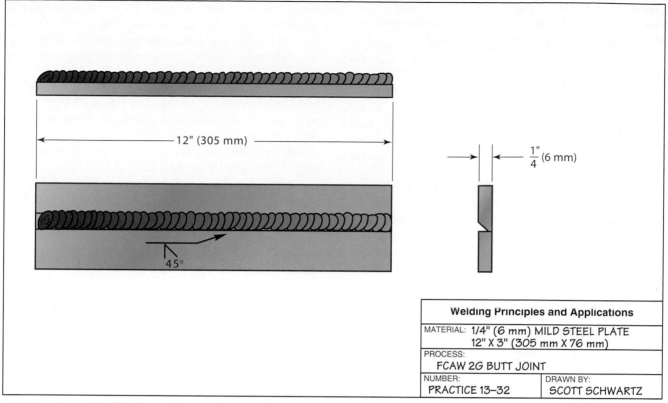

FIGURE 13-55 FCAW 2G butt joint 1/4 in. (6 mm) mild steel. © Cengage Learning 2012

PRACTICE 13-32

Butt Joint 2G

Using a properly set-up and adjusted FCA welding machine; proper safety protection; 0.035-in. and/or 0.045-in. (0.9-mm and/or 1.2-mm) diameter E70T-1 and/or E70T-5 electrodes; and one or more pieces of mild steel plate, 12 in. (305 mm) long and 1/4 in. (6 mm) thick or thinner; you will make a groove weld in the horizontal position, **Figure 13-55.**

Following the same instructions for the assembly and welding procedure outlined in Practice 13-31, repeat with both classifications of electrodes as needed until defect-free welds can consistently be made in the 1/4-in. (6-mm) thick plate. Turn off the welding machine and shielding gas and clean up your work area when you are finished welding.

Complete a copy of the "Student Welding Report" listed in Appendix I or provided by your instructor. ◆

PRACTICE 13-33

Butt Joint 2G 100% to Be Tested

Using a properly set-up and adjusted FCA welding machine; proper safety protection; 0.035-in. and/or 0.045-in. (0.9-mm and/or 1.2-mm) diameter E70T-1 and/or E70T-5 electrodes; and one or more pieces of mild steel plate, 12 in. (305 mm) long and 1/4 in. (6 mm) thick; you will make a groove weld in the horizontal position.

Following the same instructions for the assembly and welding procedure outlined in Practice 13-31, repeat the weld using each electrode classification until welds using both electrodes can be made with 100% penetration that will pass a bend test. Turn off the welding machine and shielding gas and clean up your work area when you are finished welding.

Complete a copy of the "Student Welding Report" listed in Appendix I or provided by your instructor. ◆

PRACTICE 13-34

Butt Joint 2G

Using a properly set-up and adjusted FCA welding machine; proper safety protection; 0.035-in. and/or 0.045-in. (0.9-mm and/or 1.2-mm) diameter E70T-1 and/or E70T-5 electrodes; one or more pieces of mild steel plate, 12-in. (305-mm) long and 3/8-in. (9.5-mm) thick beveled plate; and a 14-in. (355-mm) long, 1-in. (25-mm) wide, and 1/4-in. (6-mm) thick backing strip; you will make a groove weld in the horizontal position.

Following the same instructions for the assembly and welding procedure outlined in Practice 13-31, repeat with both classifications of electrodes as needed until defect-free welds can consistently be made. Turn off the welding machine and shielding gas and clean up your work area when you are finished welding.

Complete a copy of the "Student Welding Report" listed in Appendix I or provided by your instructor. ◆

PRACTICE 13-35

Butt Joint 2G 100% to Be Tested

Using a properly set-up and adjusted FCA welding machine; proper safety protection; 0.035-in. and/or 0.045-in. (0.9-mm and/or 1.2-mm) diameter E70T-1 and/or E70T-5 electrodes; one or more pieces of mild steel plate, 12-in. (305-mm) long and 3/8-in. (9.5-mm) thick beveled plate; and a 14-in. (355-mm) long, 1-in. (25-mm) wide and 1/4-in. (6-mm) thick backing strip; you will make a groove weld in the horizontal position.

Following the same instructions for the assembly and welding procedure outlined in Practice 13-31, repeat the weld using each electrode classification until welds using both electrodes can be made with 100% penetration that will pass a bend test. Turn off the welding machine and shielding gas and clean up your work area when you are finished welding.

Complete a copy of the "Student Welding Report" listed in Appendix I or provided by your instructor. ◆

PRACTICE 13-36

Butt Joint 2G

Using a properly set-up and adjusted FCA welding machine; proper safety protection; 0.035-in. and/or through 1/16-in. (0.9-mm and/or through 1.6-mm) diameter E70T-1 and/or E70T-5 electrodes; one or more pieces of mild steel plate, 7-in. (178-mm) long and 3/4-in. (19-mm) thick or thicker beveled plate; and a 9-in. (230-mm) long, 1-in. (25-mm) wide, and 1/4-in. (6-mm) thick backing strip; you will make a groove weld in the horizontal position, **Figure 13-56**.

Following the same instructions for the assembly and welding procedure as outlined in Practice 13-31, repeat with both classifications of electrodes as needed until defect-free welds can consistently be made. Turn off the welding machine and shielding gas and clean up your work area when you are finished welding.

Complete a copy of the "Student Welding Report" listed in Appendix I or provided by your instructor. ◆

PRACTICE 13-37

Butt Joint 2G 100% to Be Tested

Using a properly set-up and adjusted FCA welding machine; proper safety protection; 0.035-in. and/or through 1/16-in. (0.9-mm and/or through 1.6-mm) diameter E70T-1 and/or E70T-5 electrodes; one or more pieces of mild steel plate, 7-in. (178-mm) long and 3/4-in. (19-mm) thick or thicker beveled plate; and a 9-in. (230-mm) long, 1-in. (25-mm) wide, and 1/4-in. (6-mm) thick backing strip; you will make a groove weld in the horizontal position.

Following the same instructions for the assembly and welding procedure outlined in Practice 13-31, repeat the weld using each electrode classification until welds using

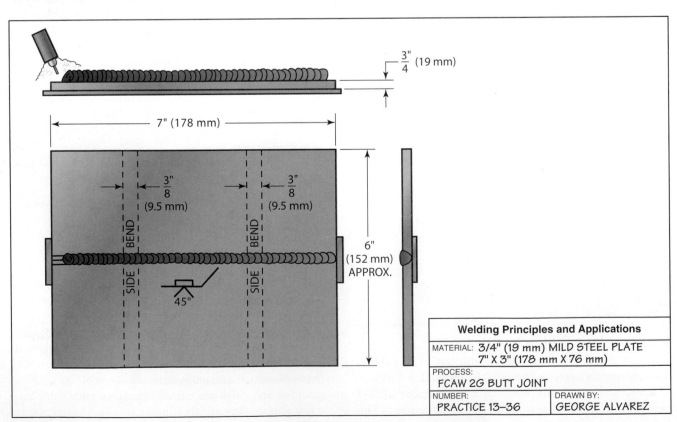

FIGURE 13-56 FCAW 2G butt joint 3/4 in. (19 mm) mild steel. © Cengage Learning 2012

both electrodes can be made with 100% penetration that will pass a bend test. Turn off the welding machine and shielding gas and clean up your work area when you are finished welding.

Complete a copy of the "Student Welding Report" listed in Appendix I or provided by your instructor. ◆

Overhead-Position Welds
PRACTICE 13-38

Butt Joint 4G

Using a properly set-up and adjusted FCA welding machine; proper safety protection; 0.035-in. and/or 0.045-in. (0.9-mm and/or 1.2-mm) diameter E70T-1 and/or E70T-5 electrodes; and one or more pieces of mild steel plate, 12 in. (305 mm) long and 1/4 in. (6 mm) thick or thinner; you will make a groove weld in the overhead position.

The molten weld pool should be kept as small as possible for easier control. A small molten weld pool can be achieved by using lower current settings and faster traveling speeds.

Lower current settings require closer control of gun manipulation to ensure that the electrode is fed into the molten weld pool just behind the leading edge. The low power will cause cold lap and more spatter if this electrode-to-molten weld pool contact position is not closely maintained.

Faster travel speeds allow the welder to maintain a high production rate even if multiple passes are required to complete the weld. Weld penetration into the base metal at the start of the bead can be obtained by using a slow start or quickly reversing the weld direction. Both the slow start and reversal of weld direction put more heat into the weld start to increase penetration. The higher speed also reduces the amount of weld distortion by reducing the amount of time that heat is applied to a joint.

When welding overhead, extra personal protection is required to reduce the danger of burns. Leather sleeves or leather jackets should be worn.

Much of the spatter created during overhead welding falls into or on the nozzle and contact tube. The contact tube may short-out to the gas nozzle. The shorted gas nozzle may arc to the work, causing damage both to the nozzle and to the plate. To control the amount of spatter, a longer stickout and/or a sharper gun-to-plate angle is required to allow most of the spatter to fall clear of the gun or nozzle, **Figure 13-57.**

Make several short weld beads using various techniques to establish the method that is most successful and most comfortable for you. After each weld, stop and evaluate it before making a change. When you have decided on the technique to be used, make a welded stringer bead that is 12 in. (305 mm) long.

Repeat with both classifications of electrodes as needed until defect-free welds can consistently be made in the 1/4-in. (6-mm) thick plate. Turn off the welding machine

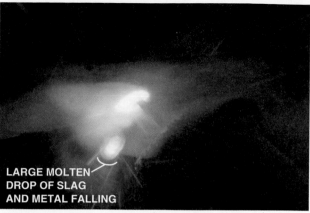

LARGE MOLTEN DROP OF SLAG AND METAL FALLING

FIGURE 13-57 Hold the gun so that weld spatter will not fall onto the gun. Larry Jeffus

and shielding gas and clean up your work area when you are finished welding.

Complete a copy of the "Student Welding Report" listed in Appendix I or provided by your instructor. ◆

PRACTICE 13-39

Butt Joint 4G 100% to Be Tested

Using a properly set-up and adjusted FCA welding machine; proper safety protection; 0.035-in. and/or 0.045-in. (0.9-mm and/or 1.2-mm) diameter E70T-1 and/or E70T-5 electrodes; and one or more pieces of mild steel plate, 12 in. (305 mm) long and 1/4 in. (6 mm) thick; you will make a groove weld in the overhead position.

Following the same instructions for the assembly and welding procedure outlined in Practice 13-38, repeat the weld using each electrode classification until welds using both electrodes can be made with 100% penetration that will pass a bend test. Turn off the welding machine and shielding gas and clean up your work area when you are finished welding.

Complete a copy of the "Student Welding Report" listed in Appendix I or provided by your instructor. ◆

PRACTICE 13-40

Butt Joint 4G

Using a properly set-up and adjusted FCA welding machine; proper safety protection; 0.035-in. and/or 0.045-in. (0.9-mm and/or 1.2-mm) diameter E70T-1 and/or E70T-5 electrodes; one or more pieces of mild steel plate, 12-in. (305-mm) long and 3/8-in. (9.5-mm) thick beveled plate; and a 14-in. (355-mm) long, 1-in. (25-mm) wide, and 1/4-in. (6-mm) thick backing strip; you will make a groove weld in the overhead position, **Figure 13-58.**

Following the same instructions for the assembly and welding procedure outlined in Practice 13-38, repeat with both classifications of electrodes as needed until defect-free

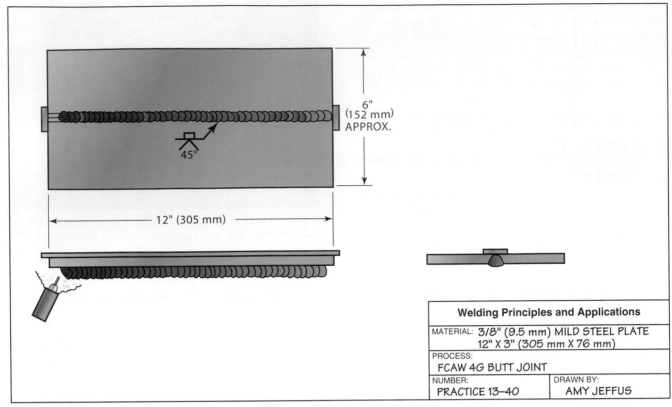

FIGURE 13-58 FCAW 4G butt joint 3/8 in. (9.5 mm) mild steel. © Cengage Learning 2012

welds can consistently be made. Turn off the welding machine and shielding gas and clean up your work area when you are finished welding.

Complete a copy of the "Student Welding Report" listed in Appendix I or provided by your instructor. ◆

PRACTICE 13-41

Butt Joint 4G 100% to Be Tested

Using a properly set-up and adjusted FCA welding machine; proper safety protection; 0.035-in. and/or 0.045-in. (0.9-mm and/or 1.2-mm) diameter E70T-1 and/or E70T-5 electrodes; one or more pieces of mild steel plate, 12-in. (305-mm) long and 3/8-in. (9.5-mm) thick beveled plate; and a 14-in. (355-mm) long, 1-in. (25-mm) wide, and 1/4-in. (6-mm) thick backing strip; you will make a groove weld in the overhead position.

Following the same instructions for the assembly and welding procedure outlined in Practice 13-38, repeat the weld using each electrode classification until welds using both electrodes can be made with 100% penetration that will pass a bend test. Turn off the welding machine and shielding gas and clean up your work area when you are finished welding.

Complete a copy of the "Student Welding Report" listed in Appendix I or provided by your instructor. ◆

PRACTICE 13-42

Butt Joint 4G

Using a properly set-up and adjusted FCA welding machine; proper safety protection; 0.035-in. and/or through 1/16-in. (0.9-mm and/or through 1.6-mm) diameter E70T-1 and/or E70T-5 electrodes; one or more pieces of mild steel plate, 7-in. (178-mm) long and 3/4-in. (19-mm) thick or thicker beveled plate; and a 9-in. (230-mm) long, 1-in. (25-mm) wide, and 1/4-in. (6-mm) thick backing strip; you will make a groove weld in the overhead position.

Following the same instructions for the assembly and welding procedure outlined in Practice 13-38, repeat with both classifications of electrodes as needed until defect-free welds can consistently be made. Turn off the welding machine and shielding gas and clean up your work area when you are finished welding.

Complete a copy of the "Student Welding Report" listed in Appendix I or provided by your instructor. ◆

PRACTICE 13-43

Butt Joint 4G 100% to Be Tested

Using a properly set-up and adjusted FCA welding machine; proper safety protection; 0.035-in. and/or through 1/16-in. (0.9-mm and/or through 1.6-mm) diameter E70T-1 and/or E70T-5 electrodes; one or more pieces of mild steel plate, 7-in. (178-mm) long and 3/4-in. (19-mm) thick or thicker beveled plate; and a 9-in. (230-mm) long, 1-in. (25-mm) wide, and 1/4-in. (6-mm) thick backing strip; you will make a groove weld in the overhead position.

Following the same instructions for the assembly and welding procedure outlined in Practice 13-38, repeat the weld using each electrode classification until welds using both electrodes can be made with 100% penetration that will pass a bend test. Turn off the welding machine and shielding gas and clean up your work area when you are finished welding.

Complete a copy of the "Student Welding Report" listed in Appendix I or provided by your instructor. ◆

PRACTICE 13-44

Lap Joint and Tee Joint 4F

Using a properly set-up and adjusted FCA welding machine; proper safety protection; 0.035-in. and/or 0.045-in. (0.9-mm and/or 1.2-mm) diameter E70T-1 and/or E70T-5 electrodes; and one or more pieces of mild steel plate, 12-in. (305-mm) long and 3/8-in. (9.5-mm) thick beveled plate; you will make a fillet weld in the overhead position, **Figure 13-59A** and **B.**

Following the same instructions for the assembly and welding procedure outlined in Practice 13-38, repeat

each type of joint with both classifications of electrodes as needed until defect-free welds can consistently be made. Turn off the welding machine and shielding gas and clean up your work area when you are finished welding.

Complete a copy of the "Student Welding Report" listed in Appendix I or provided by your instructor. ◆

PRACTICE 13-45

Lap Joint and Tee Joint 4F 100% to Be Tested

Using a properly set-up and adjusted FCA welding machine; proper safety protection; 0.035-in. and/or 0.045-in. (0.9-mm and/or 1.2-mm) diameter E70T-1 and/or E70T-5 electrodes; and one or more pieces of mild steel plate, 12-in. (305-mm) long and 3/8-in. (9.5-mm) thick beveled plate; you will make a fillet weld in the overhead position.

Following the same instructions for the assembly and welding procedure outlined in Practice 13-38, repeat each type of joint with both classifications of electrodes as needed until welds can be made with 100% penetration that will pass the test. Turn off the welding machine and shielding gas and clean up your work area when you are finished welding.

Complete a copy of the "Student Welding Report" listed in Appendix I or provided by your instructor. ◆

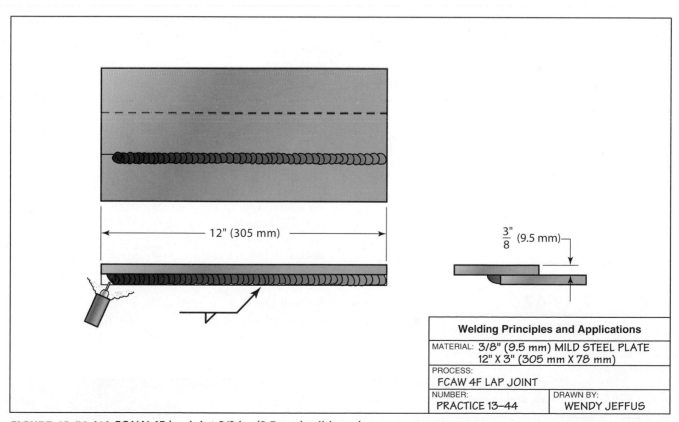

FIGURE 13-59 (A) FCAW 4F lap joint 3/8 in. (9.5 mm) mild steel. © Cengage Learning 2012

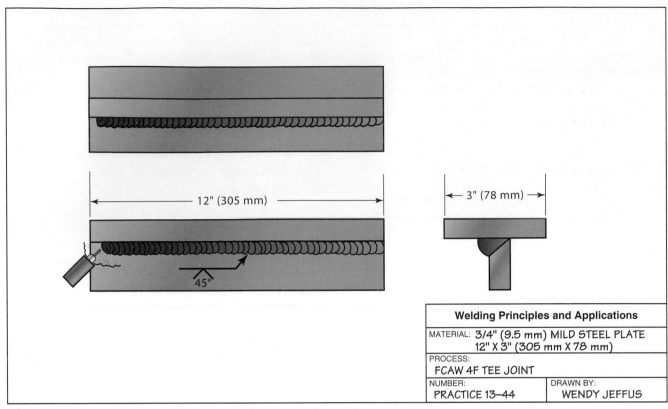

FIGURE 13-59 (B) FCAW 4F tee joint 3/4 in. (9.5 mm) mild steel. © Cengage Learning 2012

PRACTICE 13-46

Tee Joint 4F

Using a properly set-up and adjusted FCA welding machine; proper safety protection; 0.035-in. and/or through 1/16-in. (0.9-mm and/or through 1.6-mm) diameter E70T-1 and/or E70T-5 electrodes; and one or more pieces of mild steel plate, 7-in. (178-mm) long and 3/4-in. (19-mm) thick or thicker beveled plate; you will make a fillet weld in the overhead position.

Following the same instructions for the assembly and welding procedure outlined in Practice 13-38, repeat each type of joint with both classifications of electrodes as needed until defect-free welds can consistently be made. Turn off the welding machine and shielding gas and clean up your work area when you are finished welding.

Complete a copy of the "Student Welding Report" listed in Appendix I or provided by your instructor. ◆

PRACTICE 13-47

Tee Joint 4F 100% to Be Tested

Using a properly set-up and adjusted FCA welding machine; proper safety protection; 0.035-in. and/or through 1/16-in. (0.9-mm and/or through 1.6-mm) diameter E70T-1 and/or E70T-5 electrodes; and one or more pieces of mild steel plate, 7-in. (178-mm) long and 3/4-in. (19-mm) thick or thicker beveled plate; you will make a fillet weld in the overhead position.

Following the same instructions for the assembly and welding procedure as outlined in Practice 13-38, repeat with both classifications of electrodes as needed until welds can be made with 100% penetration that will pass the bend test. Turn off the welding machine and shielding gas and clean up your work area when you are finished welding.

Complete a copy of the "Student Welding Report" listed in Appendix I or provided by your instructor. ◆

Thin-Gauge Welding

The introduction of small electrode diameters has allowed FCA welding to be used on thin sheet metal. Usually these welds will be a fillet type. Fillet welds are the easiest weld to make on thin stock. An effort should be made when possible to design the weld so it is not a butt-type joint. A common use for FCA welding on thin stock is to join it to a thicker member, **Figure 13-60**. This type of weld is used to put panels in frames.

The following practices include some butt-type joints. You will find that the vertical down welds are the easiest ones to make. If it is possible to position the weldment, production speeds can be increased if butt joints are required.

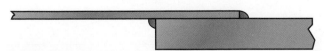

FIGURE 13-60 FCA welding thin to thick metal.
© Cengage Learning 2012

Electrode		Welding Power			Shielding Gas		Base Metal	
Type	**Size**	**Amps**	**Wire-Feed Speed IPM (cm/min)**	**Volts**	**Type**	**Flow**	**Type**	**Thick**
E70T-1	0.030 in. (0.8 mm)	40 to 145	90 to 340 (228 to 864)	20 to 27	None	n/a	Low carbon steel	16 gauge to 18 gauge
E70T-1	0.035 in. (0.9 mm)	130 to 200	288 to 576 (732 to 1463)	20 to 28	None	n/a	Low carbon steel	16 gauge to 18 gauge
E70T-5	0.035 in. (0.9 mm)	90 to 200	190 to 576 (483 to 1463)	16 to 29	57% argon 25% CO_2	35 cfh	Low carbon steel	16 gauge to 18 gauge

TABLE 13-3 FCA Welding Parameters for Use If Specific Settings Are Unavailable from Electrode Manufacturer

PRACTICE 13-48

Butt Joint 1G

Using a properly set-up and adjusted FCA welding machine, **Table 13-3**; proper safety protection; 0.030-in. and/or 0.035-in. (0.8-mm and/or 0.9-mm) diameter E70T-1 and/or E70T-5 electrodes; and one or more pieces of mild steel sheet, 12 in. (305 mm) long and 16 gauge to 18 gauge thick; you will make a butt weld in the flat position, **Figure 13-61**.

Do not leave a root opening for these welds. Even the slightest opening will result in a burnthrough. If a burnthrough occurs, the welder can be pulsed off and on so that the hole can be filled. This process will leave a larger-than-usual buildup. Excessive buildup can be ground off if necessary as part of the postweld cleanup.

Repeat with both classifications of electrodes as needed until defect-free welds can consistently be made. Turn off the welding machine and shielding gas and clean up your work area when you are finished welding.

Complete a copy of the "Student Welding Report" listed in Appendix I or provided by your instructor. ◆

PRACTICE 13-49

Butt Joint 1G 100% to Be Tested

Using a properly set-up and adjusted FCA welding machine; proper safety protection; 0.030-in. and/or 0.035-in.

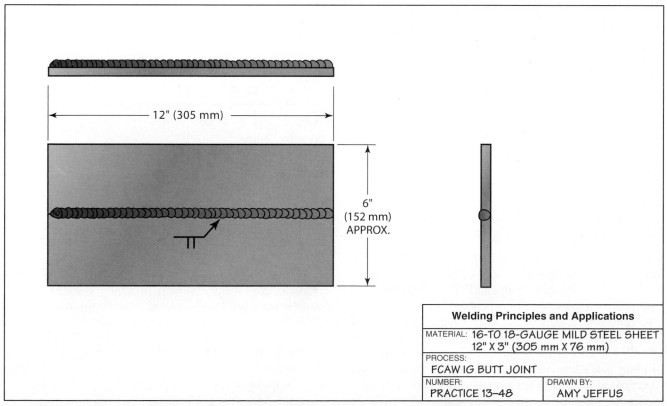

Welding Principles and Applications

MATERIAL: 16-TO 18-GAUGE MILD STEEL SHEET 12" X 3" (305 mm X 76 mm)

PROCESS: FCAW IG BUTT JOINT

NUMBER: PRACTICE 13–48

DRAWN BY: AMY JEFFUS

FIGURE 13-61 FCAW 1G butt joint 16- to 18-gauge mild steel. © Cengage Learning 2012

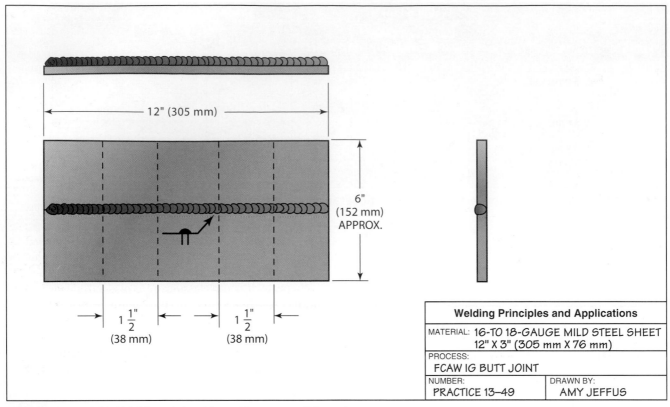

FIGURE 13-62 FCAW 1G butt joint 16- to 18-gauge mild steel. © Cengage Learning 2012

(0.8-mm and/or 0.9-mm) diameter E70T-1 and/or E70T-5 electrodes; and one or more pieces of mild steel sheet, 12 in. (305 mm) long and 16 gauge to 18 gauge thick; you will make a butt weld in the flat position, **Figure 13-62.**

Following the same instructions for the assembly and welding procedure outlined in Practice 13-47, repeat the weld using each electrode classification until welds using both electrodes can be made with 100% penetration that will pass a bend test. Turn off the welding machine and shielding gas and clean up your work area when you are finished welding.

Complete a copy of the "Student Welding Report" listed in Appendix I or provided by your instructor. ◆

PRACTICE 13-50

Lap Joint and Tee Joint 1F

Using a properly set-up and adjusted FCA welding machine; proper safety protection; 0.030-in. and/or 0.035-in. (0.8-mm and/or 0.9-mm) diameter E70T-1 and/or E70T-5 electrodes; and one or more pieces of mild steel sheet, 12 in. (305 mm) long and 16 gauge to 18 gauge thick; you will make a fillet weld in the flat position.

Following the same instructions for the assembly and welding procedure outlined in Practice 13-47, repeat each type of joint with both classifications of electrodes as needed until defect-free welds can consistently be

made. Turn off the welding machine and shielding gas and clean up your work area when you are finished welding.

Complete a copy of the "Student Welding Report" listed in Appendix I or provided by your instructor. ◆

PRACTICE 13-51

Lap Joint and Tee Joint 1F 100% to Be Tested

Using a properly set-up and adjusted FCA welding machine; proper safety protection; 0.030-in. and/or 0.035-in. (0.8-mm and/or 0.9-mm) diameter E70T-1 and/or E70T-5 electrodes; and one or more pieces of mild steel sheet, 12 in. (305 mm) long and 16 gauge to 18 gauge thick; you will make a fillet weld in the flat position, **Figure 13-63A** and **B.**

Following the same instructions for the assembly and welding procedure outlined in Practice 13-47, repeat each type of joint with both classifications of electrodes as needed until welds can be made with 100% penetration that will pass the bend test, **Figure 13-64A** and **B.** Turn off the welding machine and shielding gas and clean up your work area when you are finished welding.

Complete a copy of the "Student Welding Report" listed in Appendix I or provided by your instructor. ◆

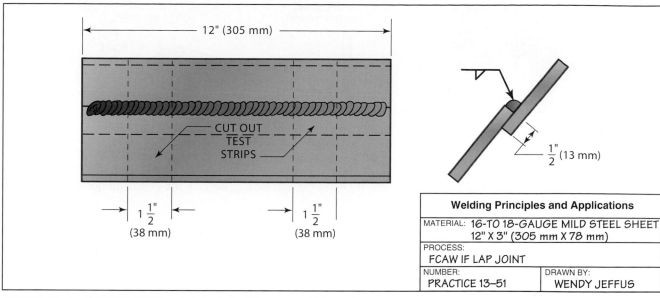

FIGURE 13-63 (A) FCAW 1F lap joint 16- to 18-gauge mild steel. © Cengage Learning 2012

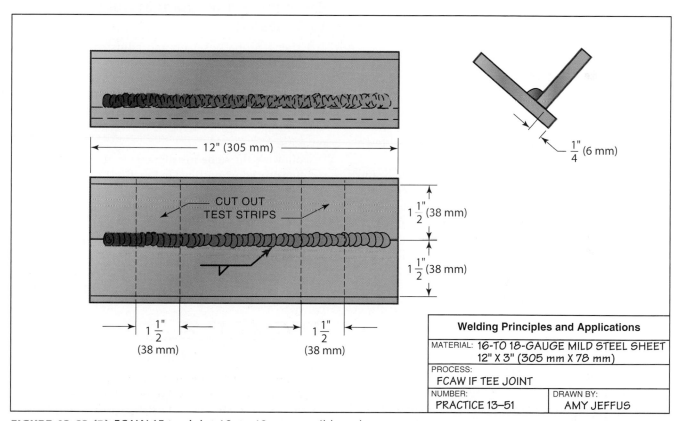

FIGURE 13-63 (B) FCAW 1F tee joint 16- to 18-gauge mild steel. © Cengage Learning 2012

PRACTICE 13-52

Butt Joint 3G

Using a properly set-up and adjusted FCA welding machine; proper safety protection; 0.030-in. and/or 0.035-in. (0.8-mm and/or 0.9-mm) diameter E70T-1 and/or E70T-5 electrodes; and one or more pieces of mild steel sheet, 12 in. (305 mm) long and 16 gauge to 18 gauge thick; you will make a butt weld in the vertical up or down position.

Following the same instructions for the assembly and welding procedure outlined in Practice 13-47, repeat with both classifications of electrodes as needed until defect-free welds can consistently be made. Turn off the welding

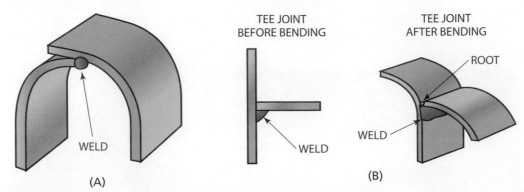

FIGURE 13-64 (A) 180° bend to test lap weld quality. (B) Bend the test strip to be sure the weld had good root fusion.
© Cengage Learning 2012

machine and shielding gas and clean up your work area when you are finished welding.

Complete a copy of the "Student Welding Report" listed in Appendix I or provided by your instructor. ◆

PRACTICE 13-53

Butt Joint 3G 100% to Be Tested

Using a properly set-up and adjusted FCA welding machine; proper safety protection; 0.030-in. and/or 0.035-in. (0.8-mm and/or 0.9-mm) diameter E70T-1 and/or E70T-5 electrodes; and one or more pieces of mild steel sheet, 12 in. (305 mm) long and 16 gauge to 18 gauge thick; you will make a butt weld in the vertical up or down position.

Following the same instructions for the assembly and welding procedure outlined in Practice 13-47, repeat the weld using each electrode classification until welds using both electrodes can be made with 100% penetration that will pass a bend test. Turn off the welding machine and shielding gas and clean up your work area when you are finished welding.

Complete a copy of the "Student Welding Report" listed in Appendix I or provided by your instructor. ◆

PRACTICE 13-54

Lap Joint and Tee Joint 3F

Using a properly set-up and adjusted FCA welding machine; proper safety protection; 0.030-in. and/or 0.035-in. (0.8-mm and/or 0.9-mm) diameter E70T-1 and/or E70T-5 electrodes; and one or more pieces of mild steel sheet, 12 in. (305 mm) long and 16 gauge to 18 gauge thick; you will make a fillet weld in the vertical up or down position.

Following the same instructions for the assembly and welding procedure outlined in Practice 13-47, repeat each type of joint with both classifications of electrodes as needed until defect-free welds can consistently be made. Turn off the welding machine and shielding gas and clean up your work area when you are finished welding.

Complete a copy of the "Student Welding Report" listed in Appendix I or provided by your instructor. ◆

PRACTICE 13-55

Lap Joint and Tee Joint 3F 100% to Be Tested

Using a properly set-up and adjusted FCA welding machine; proper safety protection; 0.030-in. and/or 0.035-in. (0.8-mm and/or 0.9-mm) diameter E70T-1 and/or E70T-5 electrodes; and one or more pieces of mild steel sheet, 12-in. (305-mm) long and 16 gauge to 18 gauge thick; you will make a fillet weld in the vertical up or down position.

Following the same instructions for the assembly and welding procedure outlined in Practice 13-47, repeat each type of joint with both classifications of electrodes as needed until welds can be made with 100% penetration that will pass the test. Turn off the welding machine and shielding gas and clean up your work area when you are finished welding.

Complete a copy of the "Student Welding Report" listed in Appendix I or provided by your instructor. ◆

PRACTICE 13-56

Lap Joint and Tee Joint 2F

Using a properly set-up and adjusted FCA welding machine; proper safety protection; 0.030-in. and/or 0.035-in. (0.8-mm and/or 0.9-mm) diameter E70T-1 and/or E70T-5 electrodes; and one or more pieces of mild steel sheet, 12 in. (305 mm) long and 16 gauge to 18 gauge thick; you will make a fillet weld in the horizontal position.

Following the same instructions for the assembly and welding procedure outlined in Practice 13-38, repeat each type of joint with both classifications of electrodes as needed until defect-free welds can consistently be made. Turn off the welding machine and shielding gas and clean up your work area when you are finished welding.

Complete a copy of the "Student Welding Report" listed in Appendix I or provided by your instructor. ◆

PRACTICE 13-57

Lap Joint and Tee Joint 2F 100% to Be Tested

Using a properly set-up and adjusted FCA welding machine; proper safety protection; 0.030-in. and/or 0.035-in. (0.8-mm and/or 0.9-mm) diameter E70T-1 and/or E70T-5 electrodes; and one or more pieces of mild steel sheet, 12 in. (305 mm) long and 16 gauge to 18 gauge thick; you will make a fillet weld in the horizontal position.

Following the same instructions for the assembly and welding procedure outlined in Practice 13-38, repeat each type of joint with both classifications of electrodes as needed until welds can be made with 100% penetration that will pass the bend test. Turn off the welding machine and shielding gas and clean up your work area when you are finished welding.

Complete a copy of the "Student Welding Report" listed in Appendix I or provided by your instructor. ◆

PRACTICE 13-58

Butt Joint 2G

Using a properly set-up and adjusted FCA welding machine; proper safety protection; 0.030-in. and/or 0.035-in. (0.8-mm and/or 0.9-mm) diameter E70T-1 and/or E70T-5 electrodes; and one or more pieces of mild steel sheet, 12 in. (305 mm) long and 16 gauge to 18 gauge thick; you will make a butt weld in the horizontal position.

Following the same instructions for the assembly and welding procedure outlined in Practice 13-38, repeat with both classifications of electrodes as needed until defect-free welds can consistently be made. Turn off the welding machine and shielding gas and clean up your work area when you are finished welding.

Complete a copy of the "Student Welding Report" listed in Appendix I or provided by your instructor. ◆

PRACTICE 13-59

Butt Joint 2G 100% to Be Tested

Using a properly set-up and adjusted FCA welding machine; proper safety protection; 0.030-in. and/or 0.035-in. (0.8-mm and/or 0.9-mm) diameter E70T-1 and/or E70T-5 electrodes; and one or more pieces of mild steel sheet, 12 in. (305 mm) long and 16 gauge to 18 gauge thick; you will make a butt weld in the horizontal position.

Following the same instructions for the assembly and welding procedure outlined in Practice 13-38, repeat the weld using each electrode classification until welds using both electrodes can be made with 100% penetration that will pass a bend test. Turn off the welding machine and shielding gas and clean up your work area when you are finished welding.

Complete a copy of the "Student Welding Report" listed in Appendix I or provided by your instructor. ◆

PRACTICE 13-60

Butt Joint 4G

Using a properly set-up and adjusted FCA welding machine; proper safety protection; 0.030-in. and/or 0.035-in. (0.8-mm and/or 0.9-mm) diameter E70T-1 and/or E70T-5 electrodes; and one or more pieces of mild steel sheet, 12 in. (305 mm) long and 16 gauge to 18 gauge thick; you will make a butt weld in the overhead position.

Following the same instructions for the assembly and welding procedure outlined in Practice 13-38, repeat with both classifications of electrodes as needed until defect-free welds can consistently be made. Turn off the welding machine and shielding gas and clean up your work area when you are finished welding.

Complete a copy of the "Student Welding Report" listed in Appendix I or provided by your instructor. ◆

PRACTICE 13-61

Butt Joint 4G 100% to Be Tested

Using a properly set-up and adjusted FCA welding machine; proper safety protection; 0.030-in. and/or 0.035-in. (0.8-mm and/or 0.9-mm) diameter E70T-1 and/or E70T-5 electrodes; and one or more pieces of mild steel sheet, 12 in. (305 mm) long and 16 gauge to 18 gauge thick; you will make a butt weld in the overhead position.

Following the same instructions for the assembly and welding procedure outlined in Practice 13-38, repeat the weld using each electrode classification until welds using both electrodes can be made with 100% penetration that will pass a bend test. Turn off the welding machine and shielding gas and clean up your work area when you are finished welding.

Complete a copy of the "Student Welding Report" listed in Appendix I or provided by your instructor. ◆

PRACTICE 13-62

Lap Joint and Tee Joint 4F

Using a properly set-up and adjusted FCA welding machine; proper safety protection; 0.030-in. and/or 0.035-in. (0.8-mm and/or 0.9-mm) diameter E70T-1 and/or E70T-5 electrodes; and one or more pieces of mild steel sheet, 12 in. (305 mm) long and 16 gauge to 18 gauge thick; you will make a fillet weld in the overhead position.

Following the same instructions for the assembly and welding procedure outlined in Practice 13-38, repeat each type of joint with both classifications of electrodes as needed until defect-free welds can consistently be made. Turn off the welding machine and shielding gas and clean up your work area when you are finished welding.

Complete a copy of the "Student Welding Report" listed in Appendix I or provided by your instructor. ◆

PRACTICE 13-63

Lap Joint and Tee Joint 4F 100% to Be Tested

Using a properly set-up and adjusted FCA welding machine; proper safety protection; 0.030-in. and/or 0.035-in. (0.8-mm and/or 0.9-mm) diameter E70T-1 and/or E70T-5 electrodes; and one or more pieces of mild steel sheet, 12 in. (305 mm) long and 16 gauge to 18 gauge thick; you will make a fillet weld in the overhead position.

Following the same instructions for the assembly and welding procedure outlined in Practice 13-38, repeat each type of joint with both classifications of electrodes as needed until welds can be made with 100% penetration that will pass the test. Turn off the welding machine and shielding gas and clean up your work area when you are finished welding.

Complete a copy of the "Student Welding Report" listed in Appendix I or provided by your instructor. ◆

Plug Welds

Plug welds are made by cutting or drilling a hole through the top plate and making a weld through that hole onto the plate that is directly behind the top plate, **Figure 13-65**. They are also used to join a thin section to

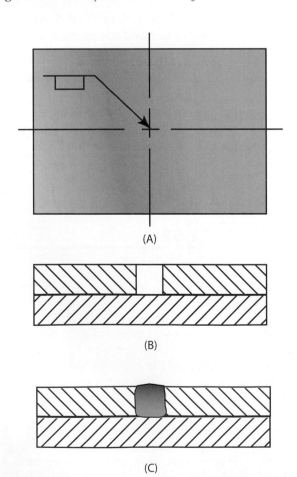

(A)

(B)

(C)

FIGURE 13-65 Plug weld. © Cengage Learning 2012

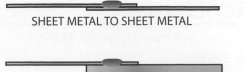

SHEET METAL TO SHEET METAL

SHEET METAL TO PLATE

FIGURE 13-66 Plug welds can be used to join sheet metal to sheet metal or sheet metal to plate.
© Cengage Learning 2012

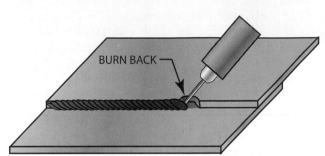

BURN BACK

FIGURE 13-67 Welding too slowly or not having the correct gun angle can result in the thin edge burning back on lap joints. © Cengage Learning 2012

another thin section or a thin section to a thicker section, **Figure 13-66**. Welding on a thin section's edge often causes that edge to melt away due to overheating of the thin edge, **Figure 13-67**. Sometimes plug welds are used to make blind welds that would not show after the weldment is finished.

PRACTICE 13-64

Plug Weld

Using a properly set-up and adjusted FCA welding machine, proper safety protection, 0.030-in. and/or 0.035-in. (0.8-mm and/or 0.9-mm) diameter E70T-1 and/or E70T-5 electrodes, and two pieces of mild steel plate that are 1/4 in. (6 mm) thick; you will make a plug weld in the flat position.

Using an OFC or PAC torch and proper safety protection, lay out and cut a 1-in. (25-mm) diameter hole through the top plate. Clean any slag off the back side of the cut, grinding it smooth if necessary so that the top plate will sit flat on the bottom plate. Clamp the plates together using a C-clamp, **Figure 13-68**.

NOTE: On thicker metal the plug weld sides are typically beveled to make it easier to join the circumference of the plug plate to the base plate.

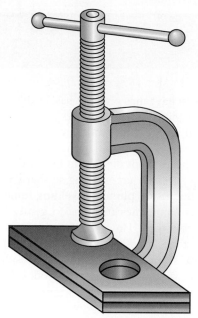

FIGURE 13-68 Clamping the plates together before making the plug weld will prevent gapping between the plates. © Cengage Learning 2012

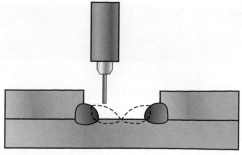

FIGURE 13-69 Start the plug weld around the outside edge of the hole. © Cengage Learning 2012

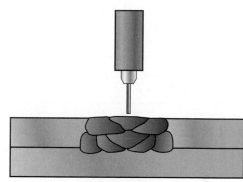

FIGURE 13-70 Complete the plug weld by making the necessary passes. It may be necessary to stop, chip, and clean the slag when making plug welds on thick metal sections. © Cengage Learning 2012

Be sure you have full range of movement so that you can keep the welding gun pointed at the root of the weld at approximately a 45° angle. Start the weld at the bottom inside edge of the hole, **Figure 13-69**, and make a full circumference of the plug weld joint. Pull the trigger and make the weld. Try to make the weld all the way around the bottom edge of the hole, but if you cannot, stop and chip the weld before starting the next weld. Once the weld has been completed all the way around the base, stop and chip the slag. If you do not chip the slag out of a deep plug weld, then the slag can build up excessively, making it impossible for you to maintain visibility of the weld and prevent slag inclusions.

The next weld pass will be mainly on the base plate, lapping halfway up on the previous weld, **Figure 13-70**. Hold the welding gun perpendicular to the plate, and make the weld all the way around. Continue the circular pattern, expanding it all the way out to the outside, and make two complete passes. Stop and chip the slag. Finish the weld by starting in the center, and build it up until the weld bead is flush with the surface.

Turn off the welding machine and shielding gas, and clean up your work area when you are finished welding.

Complete a copy of the "Student Welding Report" listed in Appendix I or provided by your instructor. ◆

Summary

In semiautomatic welding processes, the weld travel rate along the joint is controlled more by the process than by your welding technique. You must therefore learn how to travel at the proper rate to maintain the weld size. Flux cored arc welding is a relatively fast process. Therefore, your travel rates are much higher than for most other welding processes. This often causes new welders problems in that they have difficulty maintaining joint tracking as they are rapidly traveling along the groove. Practicing movement along the joint before you start is a good way of aiding in your development of these skills. The practice coupons in this section are 12 in. (305 mm) in length. However, if given the opportunity, you may want to weld longer joints after you have mastered the basic skills to further increase your joint tracking abilities.

Flux cored arc welding produces a large quantity of welding fumes. It is important that you position yourself so that your head is not directly in line with the rising fumes. Make sure you are welding so that your face is well out of this rising plume of welding fumes. In the field, welders sometimes use fans to gently blow the fumes away from them. However, if the fan is too close to the welding zone, excessive air velocity will blow the shielding away from the weld, which may result in weld porosity. Take precautions to protect yourself from any potential health hazards.

Welding Lends Architectural Flair to Airport Expansion

A steel fabricator takes on the challenge of constructing an intricate pipe canopy at Portland's International Airport.

More than 40 years ago, Joseph Fought had the foresight to build his steel fabricating facility on the outskirts of Portland, Oregon, long before the area was served by major highways. Today, with the outward expansion of Portland and Interstate 5 serving the area, his business is considered to be in a prime location. With three fabricating bays, Fought & Co. has 200,000 sq ft of building space spread over 12 acres.

Growth Spurs Airport Expansion

Sustained growth over four decades has brought Portland's population close to 1.5 million. With that growth comes the need to expand Portland's International Airport. Undergoing a major multiphase renovation to accommodate increased traffic, the airport's parking garage was expanded an additional 940,000 sq ft. Access to the garage was increased and seven additional elevators were put in to serve the garage. Fought & Co. provided steel fabrication work in all phases, but the company's most visible contribution to the expansion project is a 2 1/2-acre canopy, which welcomes visitors to the Portland International Airport. Although the company has worked on many memorable projects, the complexities associated with the fabrication and erection of this canopy place it high on the list of major accomplishments.

A Fabricating Challenge

The canopy project began in February 1999, two months later than planned, because of design challenges. The company quickly discovered it had to purchase special equipment to work with the different steel pipes. "Some of the pipes had to be machined on the inside so their heavy wall thickness would match up with the lighter wall thickness of other pipes," said Robert Waldhauser, production manager, Fought & Co.

Every complete penetration weld had to have a backup ring inside, and special jig fixtures had to be designed and built to accommodate the pipe. Waldhauser turned to a California-based company to build the machine necessary to oxygen cut the complete joint configuration. The fabrication of the machine and the design of the cutting program took nearly six months to complete.

An example of the complex joints required throughout the canopy. American Welding Society

Once the machine was in place, work began on the estimated 3,000 ft of pipe ends that had to be profile cut in sizes ranging from 5 in. to 8 in. wide. The customized machine produced the quality of cut required and put the company ahead in the fabrication schedule.

Welding Requirements

The welding was done to code, and, before moving ahead with the fabrication, numerous test pieces were welded for various joints. The welding process used was semiautomatic flux cored arc welding, a process the company uses often. The majority of joints required fillet and groove welds.

After trying various flux cored electrodes, Hobart's Excel-Arc 71 (E71T-1) flux cored wire was selected for the job, and it has performed well.

As one fabricating hurdle cleared, another popped up. Many of the joints in the canopy overlap, which means the joints must be welded halfway, then have the next section positioned over them.

"The fit-up is not always perfect because of the complexity of the joints. We've got pipes that are different widths with different wall thicknesses," said Waldhauser, "so there's a little bit of help required along the way with carbon arc gouging. But the project moved along well in spite of all of the complexities." The company put its best welders on the project, and the rejection rate was less than 3% on the weld joints in the canopy.

"Very little of the welding is done on the ground," said Waldhauser. "The welders have to get up on ladders and weld out of position. It's difficult work. Every joint is full penetration and all of the critical joints have to be 100% ultrasonic tested."

The canopy was classified as an architecturally exposed structure, so the welds have to be good from an appearance perspective. All of the welded joints were inspected and approved.

More Hurdles to Clear

Some of the sections of the canopy, which are the biggest structural members Fought & Co. has ever worked with, were shipped to the construction site in straight sections. Others had to be curved with the help of Albina Pipe Bending, a local company.

The project was as much of a challenge to Albina as it was to Fought & Co. Albina's difficulties arose as it was trying to handle a 60-ft long, 16-in. pipe in its fabricating bay.

To get it in the bay from the yard, a hole had to be cut in the wall and the pipe gradually taken in. "What happens then," said Waldhauser, "it sticks out the other side of the wall. The foreman at Albina says he can't bend it, and I tell him 'Yes, you can.' In order to get it done, they had to cut a hole in the wall of the foreman's office to bend the last 10 ft of the pipe."

Getting these sections of the canopy to the airport for assembly was yet another challenge. Because of the length (up to 60 ft) and width (up to 24 ft) of the load, transporting sections of the canopy was restricted to between Friday night and 5 A.M. Saturday morning and Saturday night to 5 A.M. Sunday morning. Several sections of the canopy were transported before it became routine to travel the 25 miles between the shop and the airport.

Once at the airport, the sections of each truss were assembled. When completed, each truss measured 240 ft in length and weighed 85 to 90 tons. The curvature offset on the straight line is 9 ft, but the truss itself is 15 ft deep and 12.6 ft wide.

"The canopy is like the icing on the cake after all of the expansion work that has been done on the airport," concluded Waldhauser. "Obviously, we're very proud of what we've accomplished on this project. People can design things like this on a computer and everything looks nice but it doesn't always work out as well when you go to build the thing. This one did, and a lot of people are going to enjoy our work."

Article courtesy of the American Welding Society.

In spite of the complexities of the project, weld rejection rates were less than 3%. American Welding Society

Review

1. Why is it important to make sure that an FCA welding system is set up properly if out-of-position welds are going to be made?

2. What major safety concerns should an FCA welder be cautious of?

3. Why should the FCA welding practice plates be large?

4. Why should the FCA welds be of substantial length?

5. What must be done to a shielding gas cylinder before its cap is removed?

6. What can happen to the wire if the conduit is misaligned at the feed rollers?

7. Why is it a good idea for a new student welder to use the gas nozzle even if a shielding gas is not used?

8. Why is the curl in the wire end straightened out?

9. What problems can a high-feed roller pressure cause?

10. Referring to Table 13-1, answer the following:
    a. What would the range of the feed speed be for an amperage of 150 at 25 volts for an E70T-1 0.035-in. (0.9-mm) electrode?
    b. What would the approximate amperage be for an E70T-5 0.045-in. (1.2-mm) electrode if it is being fed at a rate of 200 in. per minute (508 cm per minute)?

11. What are the disadvantages of beveling plates for welding?

12. FCA welds with 100% joint penetration can be made in plates up to what thickness?

13. What is the smallest V-groove angle that can be welded using the FCA welding process?

14. What is the purpose of the root pass?

15. Why is the FCA welding process not used for open-root critical welds?

16. Why should convex weld faces be avoided?

17. What bead pattern is best for overhead welds?

18. Why is the appearance of the cover pass so important?

19. What is the visual inspection standard's limitations of acceptance?

20. Why is the backing strip 2 in. (50 mm) longer than the test plate?

21. Why should there be a space between the plates when making a fillet weld on thick plates of a tee joint?

22. What would a small notch at the root weld's leading edge in a fillet weld mean?

23. What changes should be made in the setup for making a vertical up weld?

24. What problem must be overcome if the amperage and voltage are lowered to make a vertical weld?

25. How can higher welding speeds help control distortion?

26. How can spatter buildup on the welding gun be controlled in the overhead position?

Chapter 14

Other Constant-Potential Welding Processes

OBJECTIVES

After completing this chapter, the student should be able to

- describe the various types of SAW fluxes.
- explain how the SAW process works.
- explain how the ESW and EGW processes work.
- name the parts of a SAW setup.
- list the methods of starting the SAW arc.
- list the major advantages and limitations of the SAW, ESW, and EGW processes.

KEY TERMS

arc starting

bonded fluxes

electrogas welding (EGW)

electroslag welding (ESW)

fused fluxes

granular fluxing

mechanically mixed fluxes

strip electrodes

submerged arc welding (SAW)

twisted wire

water-cooled dams

INTRODUCTION

In addition to GMAW and FCAW there are three other continuous feed electrode processes. As with GMAW and FCAW, a constant-potential (CP) power supply provides the welding current. These processes are submerged arc welding (SAW), electroslag welding (ESW), and electrogas welding (EGW). Each of these three processes provides the welding industry with significant specialized benefits.

SAW, ESW, and EGW can be considered the workhorses of the fabrication industry. These processes are used to fabricate large, thick sections. They can weld metal ranging from 3/8 in. (10 mm) to more than 20 in. (508 mm) thick in a single operation, depending on the process selected. They are also used to cover large areas on tanks and pressure vessels with welded coverings of special alloys. This allows the unit to be fabricated out of a less expensive metal but have the surface of the expensive alloy. The total cost of the fabrication is less, yet it will provide the same or better service.

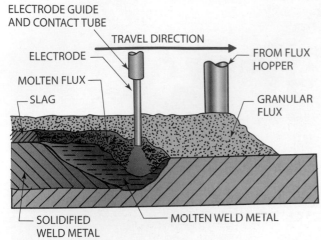

FIGURE 14-1 Submerged arc welding (SAW).

© Cengage Learning 2012

FIGURE 14-2 Automated submerged arc welding provides high-quality weld depositions. Lincoln Electric Company

Submerged Arc Welding (SAW)

The early 1930s saw the introduction of the submerged arc welding (SAW) process. The first of many processes to use a continuous wire as the electrode, its acceptance accelerated during World War II. Although the yearly tonnage of SAW electrode wire used has increased, its percentage of the total amount of filler metals used has remained below 10%.

Submerged arc welding (SAW) is a fusion welding process in which the heat is produced from an arc between the work and a continuously fed filler metal electrode. The molten weld pool is protected from the surrounding atmosphere by a thick blanket of molten flux and slag formed from the **granular fluxing** material preplaced on the work, **Figure 14-1. Figure 14-2** shows an application of the SAW process.

Weld Travel

Movement along a joint can be provided manually or mechanically. Handheld travel can be a welder manually moving the gun at a constant speed along the joint, or there might be a small gun-mounted motor and friction drive wheel to provide a more consistent travel rate.

Mechanical travel can be provided by either moving the gun along the joint or moving the work past a fixed gun.

THINK GREEN

Reduce Air Pollution

The SA welding process releases very little welding fumes into the atmosphere because the arc is completely covered with a heavy layer of granular flux.

Gun travel can be as simple as using a portable tractor-type carriage, **Figure 14-3**, or as complex as a robotic arm. Work travel can be provided using rollers or other types of positioners. In some cases, a positioner may be used in unison with some type of gun travel. These systems lend themselves to computer-controlled automation.

Electrode Feed

SAW electrode feed systems are similar to all of the other CP processes. The filler metal is provided on spools, on coils, or in bulk drums. The feed system consists of a variable speed drive motor, motor controller, and drive rollers. The drive rollers are usually the knurled V-groove type, although sometimes the smooth V-groove type rollers are used.

The filler metal is fed at a constant rate to the arc. The CP power provides the arc current to melt the wire at a rate matching the feed rate. The electrode melts and is transferred across the arc to the work through the molten pool of flux. The flux reacts with the metal both during the arc and within the molten weld pool. Weld cooling is somewhat slowed by the heat-retaining blanket provided by the slag. If the cooling rate is too fast in large welds, some impurities can be trapped in the center of the weld nugget. The rate of cooling can be slowed if preheat and postheat are used.

Contact Tip

A contact tip provides the source of welding current to the wire. This tip also directs the wire into the joint. The tip is usually made of beryllium-copper so it can withstand the heat of welding.

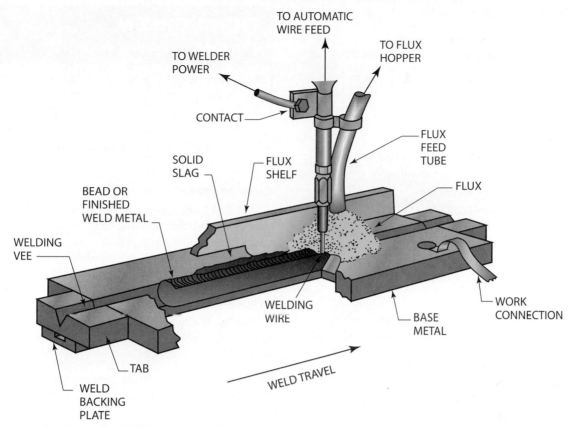

FIGURE 14-3 Schematic of a SAW setup in which the work is moved past the welding gun. © Cengage Learning 2012

Electrode

SAW filler metals are available as standard wire and in several special forms. The wire comes in sizes from 1/16 in. to 1/4 in. (1.6 mm to 6 mm). **Twisted wire** is used to give the arc some oscillating movement as the wire enters the weld, **Figure 14-4**. This oscillation helps fuse the toe of

the weld to the base metal. **Strip electrodes** are used for surfacing applications. The strips are available up to 3 in. (76 mm) wide and in several thicknesses.

Electrode Classification Electrode classifications are prefixed with the letter *E*, designating an electrode, **Figure 14-5**. The next letter, *L* (low), *M* (medium), or *H* (high), refers to the range of manganese. The next one or two digits indicate the normal carbon point of the wire. One carbon point equals 0.01% carbon. The last letter is *K* and may or may not be used. When it is used, it means the electrode was drawn from a silicon-killed, deoxidized steel.

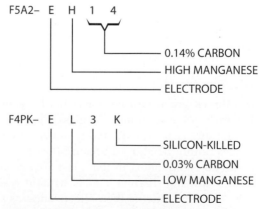

FIGURE 14-5 AWS filler wire identification system.
© Cengage Learning 2012

FIGURE 14-4 Twisted SAW filler wire. © Cengage Learning 2012

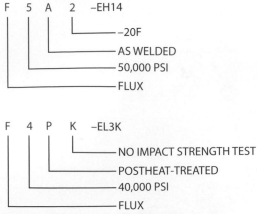

FIGURE 14-6 AWS flux identification system.
© Cengage Learning 2012

Flux

Fluxes are classified according to the mechanical properties of the weld metal deposited. The same chemically composed flux can have many different classifications, depending upon the classifications of the electrode it is used with and the condition of heat treatment given the weld for testing.

Flux Classification The basic flux classification is prefixed with the letter *F,* which designates it as a flux, **Figure 14-6.** This is followed by one digit, which represents 10,000 psi (69 MPa) minimum tensile strength of the weld. The digit is followed by the letter *A* or *P.* The *A* means the weld was tested and classified in the "as-welded condition." The *P* means the weld was tested and classified after the prescribed amount of postweld heat treatment. The next item in the classification is a single digit or the letter *Z.* The digit indicates the lowest temperature in divisions of 10°F, starting at 0°F units, that the weld metal will meet or exceed the required 20 foot-pound (27 J) impact strength test. For example, in the flux identified as **F 5 A "1" – xxxx,** the "1" is −10°F (−23°C) and in the flux identified as **F 5 A "3" – xxxx,** the "3" is −30°F (−34°C). The letter *Z* indicates that no impact strength test is required.

Flux Types Fluxes are grouped into three types according to their method of manufacture. These types are fused, bonded, and mechanically mixed. Granulated fluxes are available in bags or bulk containers.

Fused Fluxes

Fused fluxes are mixtures that have been heated until they melt into a solid metallic glass. They are then cooled and ground into the desired granular size range. Fused fluxes cannot be alloyed because they are a form of glass and all components in that glass are essentially oxides. They will not dissolve metals without reacting with them, thereby reducing their effectiveness as alloying materials.

Bonded Fluxes

Bonded fluxes are a mixture of fine particles of fluxing agents, deoxidizers, alloying elements, metal compounds, and a suitable binder that holds the mixture together in small, hard granules. Each granule is composed of all the ingredients in the correct proportions.

Mechanically Mixed

Mechanically mixed fluxes are mixtures of fused and bonded fluxes or a fine mixture of agents in a desired proportion for a certain job.

Flux Storage To prevent contamination of the weld by hydrogen, the flux must be kept dry and free from oils or other hydrocarbons. If flux becomes damp, it must be redried. Excessive levels of hydrogen in some steels can cause porosity. In hardenable steels, even small amounts of hydrogen can cause underbead cracking. Commercially available dryers are the best method of drying flux. Do not dry flux by using a direct flame. This may fuse the flux together; at the same time, the flame produces water vapor that might condense on the flux.

Advantages of SAW

The growth in the use of SAW has been due to its major advantages, as follows:

- Minimum operator protection required—The heavy blanket of granular flux covers all of the arc light except for an occasional flash. The absence of the arc light means that many welders can work close to each other without the problem of arc flash. The welder does not have to wear a welding helmet, so visibility and safety are improved. The granular flux also prevents most of the welding smoke from escaping. Even with a large number of welders operating in confined spaces, forced ventilation is virtually eliminated. The heating and cooling costs of the shop are lessened because the amount of ventilation required is greatly reduced.

- Highest deposition rate—Using large diameter wires, more than 40 lb/hr (18 kg/hr) can be deposited. This rate is nearly two times the rate of FCAW and four times that of SMAW. No process other than electroslag welding can come close to this deposition rate.

- Efficient use of materials—With SAW, there is no spatter to waste metal and cause cleanup problems. All of the electrode is transferred and becomes weld deposit. Only the melted flux needed for the weld is lost. Unfused granular flux can be retrieved and reused. The amount of flux consumed can be controlled by varying the arc length, which is done by changing the arc voltage.

- Weld size—Flat groove or fillet welds as large as 1 in. (25 mm) can be made in one pass using a single electrode. Larger sizes are possible with multiple electrodes.

- Easily adapted—With this process, the flux and wire are purchased separately. The flux can be used to change the alloys in the weld metal deposited from the electrode. By changing the flux, the properties of the weld are altered. The composition of the flux is easily changed to meet specific metallurgical properties. Two or more fluxes can be mixed, or granulated metal can be added to a flux or mixture to meet individual needs.

- High-quality welds—Many codes permit SAW to be used on structural iron, pressure vessels, cryogenic cylinders, and in many other critical applications.

Disadvantages of SAW

As with other processes, the submerged arc processes have several disadvantages:

- Restricted to flat position and horizontal fillets—Because the fluxes needed for submerged arc welding flow easily, welding is restricted to those positions in which the flux can produce a self-supporting blanket.

- Welding parameters need careful control—Because the flux hides the weld pool, welding conditions must be preset on the basis of experiments or with proven tabular information, including the contact tip-to-work distance, the current, the travel speed, and the voltage. Arc voltage must be carefully controlled to ensure the proper weld profile. Equally important, deviations in arc voltage can cause significant changes in the weld composition when using the fluxes as the source of alloys.

- Mechanical guidance is necessary—Without some sort of guidance, the arc could easily move away from the joint being welded. To take advantage of the deep penetration possible with the process, accurate positioning is very important. This need can become an expensive complication with other than perfectly straight joints. Even with good guidance systems, positioning problems can develop if the wire does not have a large and uniform cast (or little bend) and a negligible helix (twist). Significant variations in cast or a large helix will

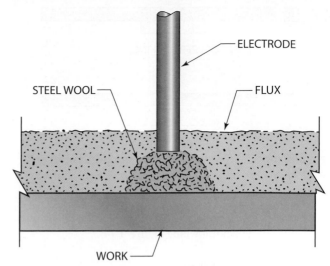

FIGURE 14-7 Steel wool placed between the electrode and work to aid in starting some SA welding.
© Cengage Learning 2012

cause the arc to wander. This is particularly important when a long wire electrode extension is used to increase deposition rates.

Arc Starting

There are six commonly used methods of **arc starting**.

- Steel wool ball starting—A small ball of steel wool is placed between the electrode and the work before the flux is added. The welding current causes the steel wool ball to quickly heat up, and the arc is started, **Figure 14-7**.

- High-frequency starting—A high-frequency current is sent through the electrode, and it establishes the arc when the welding power is turned on.

- Scratch starting—The electrode is dragged along the joint before the welding current is started. When the welding current makes contact with the moving base metal, it begins to spark, which starts the arc.

- Wire retract start—The wire is advanced until it touches the base metal and the welding current is applied. At that moment the electrode is withdrawn slightly to start the arc.

- Sharp wire start—If the end of the wire is cut to a point, that point will quickly arc when it contacts the base metal.

- Molten flux start—The arc will restart on its own if the electrode is lowered into a pool of molten slag and the welding power is restarted.

Weld Backing

Because the welds made with SA welding are usually very large, it is often necessary to support the root face of the weld. The support can be provided by placing something

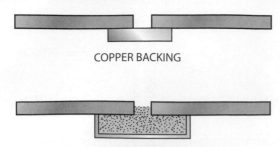

COPPER BACKING

FLUX BACKING

FIGURE 14-8 Two ways to provide backing for an open root SAW. © Cengage Learning 2012

under the joint, such as a strip of copper, a trough filled with flux, or a backing weld, **Figure 14-8**.

Handheld SAW

Handheld SA welding is increasing in usage for a variety of reasons. One of the most significant is that there are little if any fumes or smoke to exhaust. New local, state, and federal laws have put restrictions on the material that can be free-vented into the atmosphere. There are requirements in some locations that require the welding shop ventilation systems to have collectors and filtration equipment in operation. This process does not eliminate all fumes and smoke, but the reduction can significantly reduce the shop's ventilation costs. A shop using SA welding is a much cleaner place to work.

The arc light and spatter are blanketed by the flux covering so the welding technicians can wear lighter protective clothing. This reduces welder fatigue and increases productivity, **Figure 14-9**.

Experiments

Most welder training shops are not equipped to make SA welds. The gun and flux handling systems are the parts of the system that are not available. The welding power and wire feeder from a GMA welding station are the same as required for SAW.

With a little adaptation you can make SA welds using your GMAW equipment. This will allow you to experience on a limited basis the handheld method of SA welding. The welds you produce will not meet any code or standard, but the welding technique you will learn can be applied to SAW equipment in a fabrication shop.

FIGURE 14-9 Handheld SA welding gun.
Lincoln Electric Company

Because most SA welding wires are relatively large, they require high-amperage welders. The higher amperages only produce large welds; they do not help you learn. By using a smaller GMA welding wire, you will be making smaller, more controllable welds. The smaller welds enable you to see how your movements affect the weld.

Setup and Use For this series of experiments, set up the equipment in the same manner as for GMAW. If you have not completed Practice 11-1, "GMAW Equipment Setup," and Practice 11-2, "Threading GMAW Wire," you should do them now before you continue. Be sure to check the manufacturer's manual of the equipment to see that it is safely set up.

EXPERIMENT 14-1

SA Welding

Use a properly set-up and adjusted GMA welding machine; proper safety protection; 0.035-in. and/or 0.045-in. (0.9-mm and/or 1.2-mm) diameter electrodes; one or more weld experiment pieces fabricated out of mild steel plate, 3/8 in. (10 mm) or thicker, **Figure 14-10**. You are going to make a fillet weld in the flat position.

- Tack weld the plates together and place them flat on the welding table.
- You will not need a welding helmet, but you must wear flash glasses, gloves, and other safety equipment.
- Pour flux in the test specimen to a depth of 1 in. (25 mm).
- Leave the gas nozzle on the gun to prevent accidentally shorting the contact tube to the base metal. Leave the shielding gas off.
- Starting at one end, place the nozzle down to the top of the flux layer.
- Pull the gun trigger to start the welding current. You will feel the wire strike the joint, but the arc may not start. To start the arc if needed, use the scratch start method by moving the gun in a small circle. If the arc fails to start within a few moments, stop and cut the wire off and retry.
- When the arc starts, you will hear a smooth frying sound punctuated by an occasional pop.
- Move the nozzle at a steady speed along the joint until you are at the other end.
- Once the weld is cooled and flux and slag removed, inspect the weld for uniformity.

///// CAUTION \\\\\

The unfused flux on the weld is very hot. If not handled with care, it will cause severe burns. Using pliers after the weld has cooled significantly, pour the flux into a metal container so it can continue to cool. Moving the weld before it cools may result in a large quantity of molten slag to be dumped out.

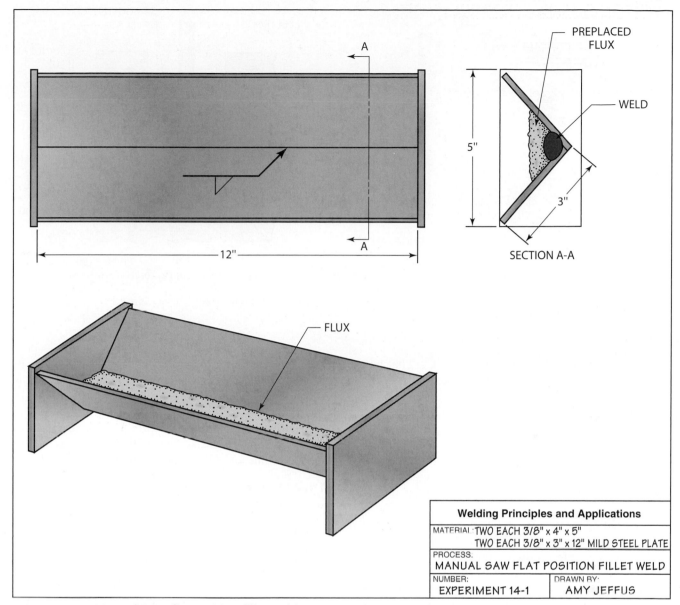

FIGURE 14-10 Manual SAW flat position fillet weld. © Cengage Learning 2012

Make changes in the current and voltage and try the weld again. Note the effect that these changes have on the weld. You may also draw a weave pattern in the surface of the flux with the gun nozzle as the weld progresses along the joint. By the time the weld is finished, you should be making the weld nearly perfectly.

Repeat each type of joint as needed until consistently good beads are obtained. Turn off the power to the welder and wire feeder; then clean up your work area.

Complete a copy of the "Student Welding Report" listed in Appendix I or provided by your instructor. ◆

Electroslag Welding (ESW)

Electroslag welding (ESW) is not an arc welding process. No arc is produced below the molten flux surface. The heat for welding is produced as a result of the electrical resistance of the molten flux. Fluxes used for

electroslag welding produce slags during the weld that have special characteristics. The slags are designed to be so highly conductive in the molten state that they can replace the arc without interrupting the electrical circuit. These slags are very fluid, and they must be relatively deep to maintain control of the circuit. Because the slags are fluid, some type of dam must be used to contain them.

The electroslag process is designed to contain the pool of fused slag within a cavity whose wells are the edges of the two plates being welded and the two **water-cooled dams** that bridge those plates, **Figure 14-11**. The wire or wires are fed into the cavity and guided by contact tips using equipment similar to that used with the submerged arc process. The need for a controlled pool size and the use of contact tip or tips within the cavity mean that only relatively heavy plate can be welded with this process.

Heated by its own resistance to the flow of current, the flux temperature is raised well above that of the plate to

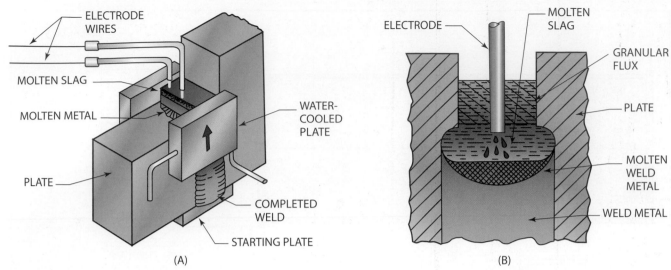

FIGURE 14-11 Diagram of electroslag welding process (A) section through workpieces and (B) weld during the making of the weld. © Cengage Learning 2012

be welded, generally above 3200°F (1760°C). That heat is used to prefuse the plate, which melts at temperatures below 2790°F (1532°C) so that metal being cast into it can bond with the plate edges. The flux also serves as a lubricant between the solidified metal and the dams. Additional flux must be added to the cavity to compensate for those losses. The wire is heated to its melting temperature by its internal resistance and remains fluid as it drops through the superheated flux to the molten weld pool.

The welding process is initiated by striking an arc between the wire and a bottom plate. After the arc energy melts enough flux to produce a pool of conductive slag, the arc is extinguished by increasing the distance between the wire tip and plate or by reducing the voltage from the power supply. One drawback of the process is that the metal produced during the starting process is defective and must be removed and repaired manually. This procedure must be followed at any position where welding has been interrupted and restarted.

Water-cooled copper dams confine the molten slag. The copper dams may extend the full length of the joint, or smaller dams may be moved up the joint as the weld metal solidifies. The rate of movement must match the speed at which the electrodes and the base metal are melted. The vertical progression is controlled automatically by sensing voltage changes with constant current power supplies or by sensing current with constant-voltage power supplies. The dams and the base metal plates serve as a cooling medium. The lower part of the metal bath solidifies progressively in an upward direction. Slag is added by the operator as needed.

Electroslag welding is suitable for welding joints in thick metal plates because the large molten weld pool is confined and molded by the dams. The joint is welded in one pass.

Electroslag welding tends to produce a large grain size for two reasons: (1) the large mass of metal that is molten at a given time and (2) the slow cooling of the metal.

Advantages The electroslag welding process has the following advantages:

- Minimum joint preparation.
- Welds can be made having a thickness of several feet (meters) in plain and alloy steels using multiple electrodes.
- High welding speed.
- Minimum distortion.
- The weld metal is highly purified by the metallurgical slag.

Disadvantages The electroslag welding process has the following disadvantages:

- Massive, expensive welding equipment and guidance systems are required.
- Lengthy setup times are needed.
- It cannot be used for welding the most commonly used thicknesses of plate.
- It can be used for only vertically positioned joints.
- It produces welds with very pronounced columnar microstructures and poor toughness.

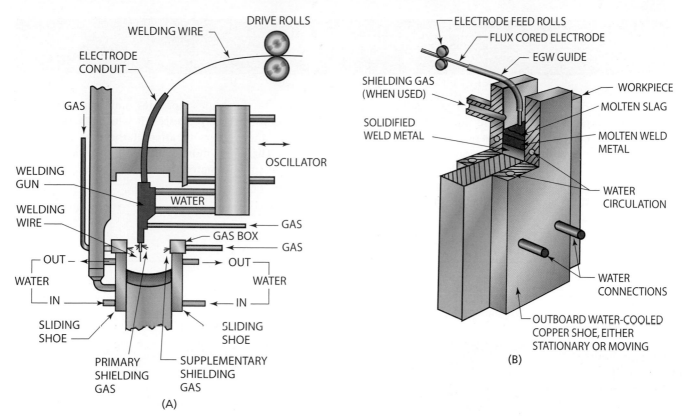

FIGURE 14-12 (A) Electrogas welding with a solid wire electrode. (B) Electrogas welding with a self-shielded flux cored electrode. © Cengage Learning 2012

Electrogas Welding (EGW)

Electrogas welding (EGW) is similar to electroslag welding in a number of ways. In addition to the power supply and wire feed, both processes are welded in the vertical up position using dams to contain the molten weld pool. Both processes can be used to weld on thick sections; however, EGW can weld plates as thin as 3/8 in. (10 mm).

The major difference in these processes is that there is an arc between the electrode and the molten weld pool with EGW. The molten weld pool is protected from the atmosphere by a shielding gas or by a flux core in the filler electrode, **Figure 14-12A** and **B**.

When a shielding gas is used, a mixture of carbon dioxide (CO_2) and argon (Ar) works well. The shielding gas flows through holes in the copper dam to cover the molten weld pool.

Figure 14-13 shows the AWS numbering system for EGW electrodes. There are two systems—one for the solid wire and the other for the tubular, flux-filled electrodes.

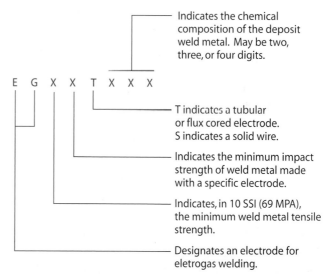

E G X X T X X X

- Indicates the chemical composition of the deposit weld metal. May be two, three, or four digits.
- T indicates a tubular or flux cored electrode. S indicates a solid wire.
- Indicates the minimum impact strength of weld metal made with a specific electrode.
- Indicates, in 10 SSI (69 MPA), the minimum weld metal tensile strength.
- Designates an electrode for eletrogas welding.

FIGURE 14-13 Electrogas welding electrode classification system. American Welding Society

Summary

Submerged arc welding, electroslag welding, and electrogas welding processes are most often used for automated high-volume manufacturing facilities. These processes are used in the production of large welds in massive weldments. They are an ideal selection for rapidly producing such large, high-quality welds.

Manual SMAW is covered in this chapter so students who do not have access to the automated equipment can experiment with the process.

Because of the high amperages and large molten weld pools with these processes, it is not practical in most cases for welders to attempt these as manual processes. Therefore, most of the time these are produced using an automated machine, or robotic welding station.

High-Performance Steel Increasingly Used for Bridge Building

HPS 70W, a relatively new material for bridge construction, has many advantages, including corrosion and cracking resistance, high strength, and improved toughness characteristics.

The number of states using high-performance steel (HPS) 70W for bridge construction has steadily increased over the past several years. This is largely due to a law signed in 1988, the Transportation Equity Act for the 21st Century (TEA-21), which authorizes highway and other surface transportation programs for the next six years. Under TEA-21, states receive federal funds to improve highways, opening the door to many new construction projects across the United States. To encourage use of this new metal, the act provides additional funds for bridge projects constructed with HPS 70W.

The Evolution of HPS 70W for Bridge Construction

The military originally developed high-performance steel for use in submarine construction. In 1994, Congress established a steering committee composed of the U.S. Navy, the Federal Highway Administration (FHWA), and the American Iron and Steel Institute (AISI) to research ways HPS could be transferred from military technology to civilian applications. In particular, the committee was charged with finding new ways to use the steel for bridge construction.

HPS 70W (70,000 lb/in.^{2} yield strength steel) was considered for bridge work because it offers the industry many

advantages. First and foremost, it has outstanding corrosion-resistance properties. The *W* in HPS 70W stands for "weathering" and means this material has controlled rusting characteristics that allow just enough corrosion to occur so a rust barrier is formed. Because of this barrier, painting is not required, meaning less maintenance for state highway crews. In contrast, a nonweathering steel often used in bridge construction requires constant painting and maintenance.

The 995-ft-long Clear Fork River Bridge in Tennesse was built with HPS 70W steel. Tennessee Department of Transportation

High-performance steel was used to build the 473-ft-long Martin Creek Bridge on State Route 53 in Tennessee. Tennessee Department of Transportation

The second advantage of HPS 70W is in the construction process itself. Because of its much greater yield strength, bridge decks that would require, for example, five girders with a 50,000-ksi steel may need only three or four with HPS. Fewer girders mean quicker erection and cost savings. The same is also true with bridge piers, which can be spaced farther apart because of the new metal's strength.

Although the cost of HPS 70W and accompanying construction materials is usually more than that of metals of lower yield strength, the fact that less material has to be erected can offset and even improve the overall cost factor.

The third advantage offered for bridge building is that HPS 70W has a very low carbon equivalent (CE) number, meaning it is resistant to cracking and hardening in the heat-affected zone (HAZ) after welding. For safety reasons, this is important because any cracking can affect the integrity of the bridge. HPS 70W also requires a lower preheat temperature, a plus because of the sheer size of most bridge girders. Reduced preheat saves money and time; it also means less energy is consumed in the preheating process.

Last, HPS 70W steel has improved toughness to retard the growth of discontinuities in the metal. Discontinuities can include weld cracking, porosity, undercutting, or poor bead shape. For bridge inspection, the objective is to have discontinuities grow slowly so that they can be caught without the bridge's soundness being compromised.

Welding HPS 70W

Submerged arc welding (SAW) has long been the most popular welding process in bridge girder fabrication because of its high quality, its relatively slow cooling rate, and the ability to weld without requiring a shielding gas. It is also an ideal process for welding the long, straight joints of bridge girders. Although some girder welding is done with shielded metal arc welding (SMAW), it is limited to difficult-to-automate welds because SMAW is a much slower process.

Welding Guidelines

Although in general HPS 70W is easy to weld, a few guidelines must be followed:

- *Turn voltage down.* With typical fluxes, operators turn up the voltage to enhance wetting, but for welding HPS 70W with the approved flux (see AWS D1.5, *Bridge Welding Code,* for further guidance on welding HPS 70W), voltage is reduced to improve wetting action.

- *Do not use electrode negative.* When welding with low-hydrogen fluxes, such as the one for HPS 70W, use DC electrode positive or AC polarity only. Do not use DC electrode negative.

- *Use larger diameter wire.* To produce a good, flat-faced weld profile in a horizontal fillet weld, use a larger diameter wire, such as 1/8 in. or 5/32 in.

- *Avoid welding in still air.* A chemical-type odor comes off the flux when it is consumed in the arc. A fan or an open door or window is recommended to improve air circulation.

- *Undermatched applications.* If there are instances when a 50W steel web, for example, will be welded to an HPS 70W flange, fabricators can use the requirements of the 50W set forth by AWS D1.5, *Bridge Welding Code,* rather than the requirements for the stronger HPS 70W material.

- *Proper preheat/interpass control.* Many girders used for bridge building can be 50 ft or more in length, which presents a preheating challenge. This is especially troublesome when using 50W and when multiple passes are required. The minimum interpass temperature must be maintained during all passes. The lower preheat and interpass temperatures required for HPS 70W are easier to maintain.

Tips for Flux Handling

Proper flux handling is especially critical with HPS 70W material. Follow these tips:

- *Flux conditioning.* There are two types of ovens needed to care for flux properly. The first, a holding oven, can keep the flux dry but cannot dry out flux that already contains moisture. The second type is a drying oven to drive moisture out of the flux. For HPS 70W, it is critical that a fabrication shop have both types of ovens because, when the flux is contaminated with moisture, it will need to be dried at temperatures between 250°F and 400°F (not to exceed 450°F). Once opened, hermetically sealed pails of flux should be put in use or immediately transferred to the holding oven.

- *Flux handling.* Eventually, flux particles will break down into fines (or dust). Flux recovery equipment cannot always separate out the fines, so operators should regularly purge the flux recovery equipment of all flux in the system and replace it with new flux. Operators should also take every precaution to ensure moisture does not contaminate flux during the welding process.

Future of HPS

Now that HPS 70W use is growing in the industry, what is next for the new steel?

Currently, both gas metal arc welding and flux cored arc welding consumables are being developed and tested to determine if these processes are suitable for use on HPS 70W. Second, fabricators of bridge girders are evolving and becoming more sophisticated in methods for preheating and maintaining interpass temperature, improved flux handling procedures, and trying alternate welding processes. Third, use of high-technology fabrication equipment, such as robotics, is being examined for the fabrication of innovative designs.

Since fabricating HPS 70W is relatively easy when the right consumables, process, and procedures are combined, more HPS 70W bridge projects are expected to be started.

Article courtesy of the Tennessee Department of Transportation.

Review

1. What protects the molten SAW pool from the atmosphere?

2. How can manual SA welding gun movement be performed?

3. What are the two methods of mechanical travel for SA welding?

4. How is the weld metal deposited in the molten weld pool of the SA welding process?

5. In what forms can SA welding filler metal be purchased?

6. How is the manganese range of the SA electrode noted in the AWS classification?

7. Why could a single SA welding flux have more than one AWS classification?

8. List the three groupings of SA welding fluxes according to their method of manufacturing.

9. Why are alloys not added to fused SA fluxes?

10. What is in bonded SA fluxes?

11. What must be done with SA fluxes to prevent contamination of the weld by hydrogen?

12. Why does the welder not have to wear a welding helmet?

13. What happens to the unfused SA welding flux?

14. Why is some form of mechanical guidance required with SA welding?

15. List the common methods used to start the SA arc.

16. Why would handheld SA welding be used?

17. What special characteristics must ES welding slags have?

18. What heats the ES welding flux?

19. How is an ES weld started?

20. Why do ES welds have large grain sizes?

21. List the advantages of ES welding.

22. What is the major difference between ESW and EGW?

Chapter 15

Gas Tungsten Arc Welding Equipment, Setup, Operation, and Filler Metals

OBJECTIVES

After completing this chapter, the student should be able to

- describe the gas tungsten arc welding process and list the other terms used to describe it.
- explain what makes tungsten a good electrode.
- tell how tungsten erosion can be limited.
- discuss how the various types of tungsten electrodes are used.
- tell how to shape the end of the tungsten electrode and how to clean it.
- demonstrate how to grind a point on a tungsten electrode using an electric grinder.
- demonstrate how to remove a contaminated tungsten end.
- demonstrate how to melt the end of the tungsten electrode into the desired shape.
- compare water-cooled GTA welding torches to air-cooled torches.
- tell the purposes of the three hoses connecting a water-cooled torch to the welding machine.
- discuss how to choose an appropriate nozzle for the job.
- tell what procedures must be followed to get an accurate reading on a flowmeter.
- compare the three types of welding current used for GTA welding.
- discuss the shielding gases used in the GTA welding process.
- define *preflow* and *postflow*.
- explain the problems that can occur as a result of an incorrect gas flow rate.
- demonstrate how to properly set up a GTA welder.
- demonstrate how to establish a GTA welding arc.

KEY TERMS

cleaning action	*inert gas*	*spark gap oscillator*
collet	*noble inert gases*	*tungsten*
flowmeter	*postflow*	
frequency	*preflow*	

INTRODUCTION

The gas tungsten arc welding (GTAW) process is sometimes referred to as TIG, or heliarc. The term *TIG* is short for tungsten inert gas welding, and the much older term *heliarc* was used because helium was the first gas used for the process. The aircraft industry developed the GTAW process for welding magnesium during the late 1930s and the early 1940s. During that time, helium was the primary shielding gas used, along with DCEP welding current. These caused many of the problems that limited application of the GTA welding process. But improvements in gas composition and a better understanding of the importance of polarity improved the effectiveness of the process and reduced its cost.

To use this process, an arc is established between a nonconsumable tungsten electrode and the base metal, which is called the work. Under the correct welding conditions, the electrode does not melt, although the work does at the spot where the arc impacts its surface and produces a molten weld pool. The filler metal is thin wire that is fed directly into the molten weld pool where it melts. Since hot tungsten is sensitive to oxygen contamination, a good inert shielding gas is required to keep the air away from the hot tungsten and molten weld metal. The **inert gas** provides the needed arc characteristics and protects the molten weld pool. Because fluxes are not used, the welds produced are sound, free of slags, and as corrosion resistant as the parent metal.

Before development of the GTAW process, welding aluminum and magnesium was difficult. The welds produced were porous and prone to corrosion.

When argon became plentiful and DCEN was recognized as more suitable than DCEP, the GTA process became more common. Until the development of GMAW in the late 1940s, GTAW was the only acceptable process for welding such reactive materials as aluminum, magnesium, titanium, and some grades of stainless steel, regardless of thickness. Reactive metals are ones that are easily contaminated when heated to their molten welding temperatures. Contamination can come from the air or be picked up from the surfaces such as oxides, oils, paints, or similar materials. Although economical for welding sheet metal, the process proved tedious and expensive for joining sections much thicker than 1/4 in. (6 mm). The eye–hand coordination required to make GTA welds is very similar to the coordination required for oxyfuel gas welding. GTA welding is often easier to learn when a person can oxyfuel gas weld.

Tungsten

Tungsten, atomic symbol W, has the following properties:

- High tensile strength: 500,000 lb/in.^{2} (3447 kg/mm^2)
- Hardness, Rockwell C45
- High melting temperature: 6170°F (3410°C)
- High boiling temperature: 10,700°F (5630°C)
- Good electrical conductivity

Tungsten is produced mainly by reduction of its trioxide with hydrogen. Powdered tungsten is then purified to 99.95+%, compressed, and sintered (heated to a temperature below melting where grain growth can occur) to make an ingot. The ingot is heated to increase ductility and then is swaged and drawn through dies to produce electrodes. These electrodes are available in sizes varying from 0.01 in. to 0.25 in. (0.25 mm to 6 mm) in diameter. The tungsten electrode, after drawing, has a heavy black oxide that is later chemically cleaned or ground off.

The high melting temperature and good electrical conductivity make tungsten the best choice for a nonconsumable electrode. The arc temperature, around 11,000°F (6000°C), is much higher than the melting temperature of tungsten but not much higher than its boiling temperature of 10,600°F (5900°C).

As the tungsten electrode becomes hot, the arc between the electrode and the work will stabilize. Because electrons are more freely emitted from a hot tungsten, the very highest temperature possible at the tungsten electrode tip is desired. Maintaining a balance between the heat required to have a stable arc and that high enough to melt the tungsten requires an understanding of the GTA torch and electrode.

The thermal conductivity of tungsten and the heat input are prime factors in the use of tungsten as an electrode. In general, tungsten is a good conductor of heat. This conductive property is what allows the tungsten electrode to withstand the arc temperature well above its melting temperature. The heat of the arc is conducted away from the electrode's end so fast that it does not reach its melting temperature. For example, a wooden match burns at approximately 3000°F (1647°C). Because aluminum melts at 1220°F (971°C), a match should easily melt an aluminum wire. However, a match will not even melt a 1/16-in. (2-mm) aluminum wire. The aluminum, like a tungsten electrode, conducts the heat away so quickly that it will not melt.

Because of the intense heat of the arc some erosion of the electrode will occur. This eroded metal is transferred across the arc, **Figure 15-1**. Slow erosion of the electrode results in limited tungsten inclusions in the weld, which are acceptable. Standard codes give the size and amount of tungsten inclusions that are allowable in various types of welds. The tungsten inclusions are hard spots that cause stresses to concentrate, possibly resulting in weld failure. Although tungsten erosion cannot be completely eliminated, it can be controlled. A few ways of limiting erosion include

- having good mechanical and electrical contact between the electrode and the collet,
- using as low a current as possible,
- using a water-cooled torch,

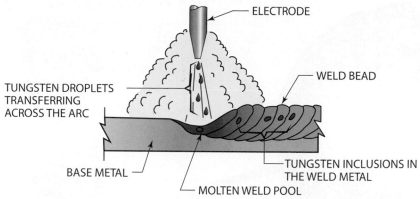

FIGURE 15-1 Some tungsten will erode from the electrode, be transferred across the arc, and become trapped in the weld deposit. © Cengage Learning 2012

- using as large a size of tungsten electrode as possible,
- using DCEN current,
- using as short an electrode extension from the collet as possible,
- using the proper electrode end shape, and
- using an alloyed tungsten electrode.

The torch end of the electrode is tightly clamped in a collet. The collet inside the torch is cooled by air or water. The **collet** is the cone-shaped sleeve that holds the electrode in the torch. Heat from both the arc and the tungsten electrode's resistance to the flow of current must be absorbed by the collet and torch. To ensure that the electrode is being cooled properly, be sure the collet connection is clean and tight. And for water-cooled torches, make sure water flow is adequate.

The difference in heat transfer efficiency between cleaned and ground surfaces of tungsten electrodes is shown in **Figure 15-2** and **Figure 15-3**.

Large diameter electrodes conduct more current because the resistance heating effects are reduced. However, excessively large sizes may result in too low a temperature for a stable arc.

In general, the current-carrying capacity at direct-current electrode negative (DCEN) is about 10 times greater than that at direct-current electrode positive (DCEP).

The preferred electrode tip shape impacts the temperature and erosion of the tungsten. With DCEN, a pointed tip concentrates the arc as much as possible and improves arc starting with either a high-voltage electrical discharge or a touch start. Because DCEN does not put much heat on the tip, it is relatively cool, the point is stable, and it can survive extensive use without damage, **Figure 15-4A**.

With alternating current, the tip is subjected to more heat than with DCEN. To allow a larger mass at the tip to withstand the higher heat, the tip is rounded. The melted end must be small to ensure the best arc stability, **Figure 15-4B**.

DCEP has the highest heat concentration on the electrode tip. For this reason a slight ball of molten tungsten

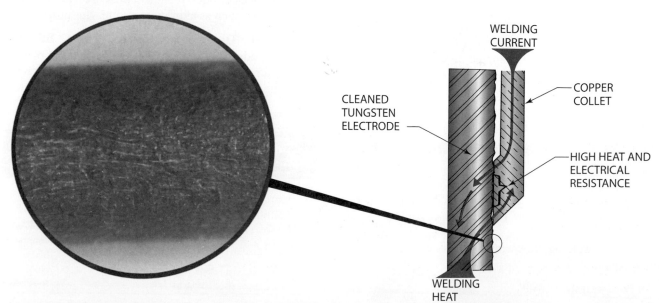

FIGURE 15-2 Irregular surface of a cleaned tungsten electrode (poor heat transfer to collet). © Cengage Learning 2012

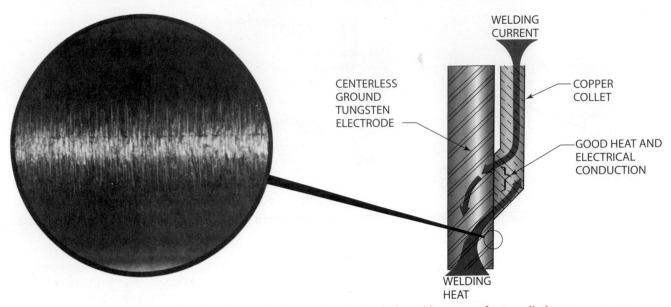

FIGURE 15-3 Smooth surface of a centerless ground tungsten electrode (good heat transfer to collet). © Cengage Learning 2012

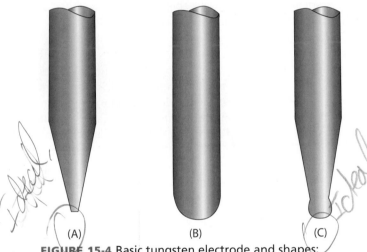

FIGURE 15-4 Basic tungsten electrode and shapes: pointed (A), rounded (B), and tapered with a balled end (C). © Cengage Learning 2012

is suspended at the end of a tapered electrode tip. The liquid tungsten surface tension with the larger mass above the molten ball holds it in place like a drop of water on your fingertip, **Figure 15-4C**.

Types of Tungsten Electrodes

Pure tungsten has a number of properties that make it an excellent nonconsumable electrode for the GTA welding process. These properties can be improved by adding cerium, lanthanum, thorium, or zirconium to the tungsten.

For GTA welding, tungsten electrodes are classified as the following:

- Pure tungsten, EWP
- 1% thorium tungsten, EWTh-1

- 2% thorium tungsten, EWTh-2
- 1/4% to 1/2% zirconium tungsten, EWZr
- 2% cerium tungsten, EWCe-2
- 1% lanthanum tungsten, EWLa-1
- Alloy not specified, EWG

See **Table 15-1**.

The type of finish on the tungsten must be specified as cleaned or ground. More information on composition and other requirements for tungsten welding electrodes is available in the AWS publication A5.12, *Specifications for Tungsten and Tungsten Alloy Electrodes for Arc Welding and Cutting*.

Pure Tungsten, EWP Pure tungsten has the poorest heat resistance and electron emission characteristics of all the tungsten electrodes. It has a limited use with AC welding of metals, such as aluminum and magnesium.

Thoriated Tungsten, EWTh-1 and EWTh-2 Thorium oxide (ThO_2), when added in percentages of up to 0.6% to tungsten, improves its current-carrying capacity. The addition of 1% to 2% of thorium oxide does not further

AWS Classification	Tungsten Composition	Tip Color
EWP	Pure tungsten	Green
EWTh-1	1% thorium added	Yellow
EWTh-2	2% thorium added	Red
EWZr	1/4% to 1/2% zirconium added	Brown
EWCe-2	2% cerium added	Orange
EWLa-1	1% lanthanum added	Black
EWG	Alloy not specified	Not specified

TABLE 15-1 Tungsten Electrode Types and Identification

—THORIUM SPIKE

FIGURE 15-5 Thorium spike on a balled end tungsten electrode. © Cengage Learning 2012

improve current-carrying capacities. It does, however, help with electron emission. This can be observed by a reduction in the electron force (voltage) required to maintain an arc of a specific length. Thorium also increases the serviceable life of the tungsten. The improved electron emission of the thoriated tungsten allows it to carry approximately 20% more current. This also results in a corresponding reduction in electrode tip temperature, resulting in less tungsten erosion and subsequent weld contamination.

Thoriated tungstens also provide a much easier arc starting characteristic than pure or zirconiated tungsten. Thoriated tungstens work well with DCEN. They can maintain a sharpened point well. They are very well suited for making welds on steel, steel alloys (including stainless), nickel alloys, and most other metals other than aluminum or magnesium.

Thoriated tungsten does not work well with alternating current (AC). It is difficult to maintain a balled end, which is required for AC welding. A thorium spike, **Figure 15-5**, may also develop on the balled end, disrupting a smooth arc.

CAUTION

Thorium is a very low-level radioactive oxide, but the level of radioactive contamination from a thorium electrode has not been found to be a health hazard during welding. It is, however, recommended that grinding dust be contained. Because of concern in other countries regarding radioactive contamination to the welder and welding environment, thoriated tungstens have been replaced with other alloys.

Zirconium Tungsten, EWZr Zirconium oxide (ZrO_2) also helps tungsten emit electrons freely. The addition of zirconium to the tungsten has the same effect on the electrode characteristic as thorium, but to a lesser degree. Because zirconium tungstens are more easily melted than thorium tungsten, ZrO_2 electrodes can be used with both AC and DC currents. Because of the difficulty in forming the desired balled end on thorium versus zirconium tungstens, they are normally the electrode chosen for AC welding of aluminum and magnesium alloys. Zirconiated

tungstens are more resistant to weld pool contamination than pure tungsten, thus providing excellent weld qualities with minimal contamination.

Zirconiated tungstens also have the advantage over thoriated tungsten in that they are not radioactive.

Cerium Tungsten, EWCe-2 Cerium oxide (CeO_2) is added to tungsten to improve the current-carrying capacity in the same manner as does thorium. These electrodes were developed as replacements for thoriated tungstens because they are not made of a radioactive material. Cerium oxide electrodes have a current-carrying capacity similar to that of pure tungsten; however, they have an improved arc starting and arc stability characteristic, similar to that of thoriated tungstens. They can also provide a longer life than most other electrodes, including thorium.

Cerium tungsten electrodes have a slightly higher arc voltage for a given length than does thoriated tungsten. This very slight increase in voltage does not cause problems for manual welding. The higher voltage, however, may require that a new weld test be performed to requalify welding procedures. Cerium tungsten may be used for both AC and DC welding currents. Cerium electrodes contain approximately 2% of cerium oxide.

Lanthanum Tungsten, EWLa-1 Lanthanum oxide (La_2O_3) in about 1% concentration is added to tungsten. Lanthanum oxide tungstens are not radioactive. They have current-carrying characteristics similar to those of the thorium tungstens, except that they have a slightly higher arc voltage than thorium and cerium tungstens. This does not normally pose a problem for manual arc welding; however, it will usually require that new test plates be produced to recertify weld procedures.

Alloy Not Specified, EWG The EWG classification is for tungstens whose alloys have been modified by manufacturers. Such alloys have been developed and tested by manufacturers to meet specific welding criteria. Specific alloy compositions are not normally available from manufacturers; however, they do provide welding characteristics for these electrodes.

Shaping the Tungsten

The desired end shape of a tungsten can be obtained by grinding, breaking, remelting the end, or using chemical compounds. Tungsten is brittle and easily broken. Welders must be sure to make a smooth, square break where they want it to be located.

Grinding A grinder is often used to clean a contaminated tungsten or to point the end of a tungsten. The grinder used to sharpen tungsten should have a fine, hard stone. It should be used for grinding tungsten only. Because of the hardness of the tungsten and its brittleness, the grinding stone chips off small particles of the electrode. A coarse grinding stone will result in more tungsten breakage and a poorer finish. If the grinder is used for metals other than tungsten, particles of these metals may become trapped on the tungsten as it is ground. The metal particles will quickly break free when the arc is started, resulting in contamination.

EXPERIMENT 15-1

Grinding the Tungsten to the Desired Shape

Using an electric grinder with a fine grinding stone, one piece of tungsten 2 in. (51 mm) long or longer, and safety glasses, you will grind a point on the tungsten electrode.

Because of the hardness of the tungsten, it will become hot. Its high thermal conductivity means that the heat will be transmitted quickly to your fingers. To prevent overheating, only light pressure should be applied against the grinding wheel. This will also reduce the possibility of accidentally breaking the tungsten.

Grind the tungsten so that the grinding marks run lengthwise, **Figure 15-6** and **Figure 15-7**. Lengthwise

FIGURE 15-6 Correct method of grinding a tungsten electrode. Larry Jeffus

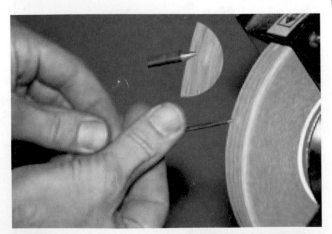

FIGURE 15-7 Incorrect method of grinding a tungsten electrode. Larry Jeffus

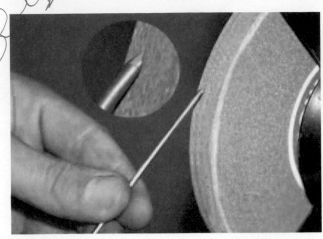

FIGURE 15-8 Correct way of holding a tungsten when grinding. Larry Jeffus

grinding reduces the amount of small particles of tungsten contaminating the weld. Move the tungsten up and down as it is twisted during grinding. This will prevent the tungsten from becoming hollow-ground.

Complete a copy of the "Student Welding Report" listed in Appendix I or provided by your instructor. ◆

///// CAUTION /////

When holding one end of the tungsten against the grinding wheel, the other end of the tungsten must not be directed toward the palm of your hand, Figure 15-8. This will prevent the tungsten from being stuck into your hand if the grinding wheel catches it and suddenly pushes it downward.

Breaking and Remelting Tungsten is hard but brittle, resulting in a low impact strength. If tungsten is struck sharply, it will break without bending. When it is held against a sharp corner and hit, a fairly square break will result. **Figure 15-9**, **Figure 15-10**, and **Figure 15-11** show ways to break the tungsten correctly on a sharp corner using a hammer, with two pliers, and wire cutters.

THINK GREEN
Recycle Tungsten

Tungsten is considered an exotic metal that can be recycled, so it is important to save the broken-off ends and short pieces. The scrap tungsten electrodes can be added to other tungsten scrap, such as worn-out tungsten carbide cutting tools, and recycled as mixed tungsten.

FIGURE 15-9 Breaking the contaminated end from a tungsten by striking it with a hammer. Larry Jeffus

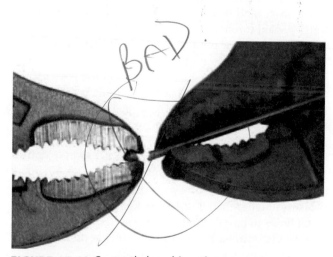

FIGURE 15-10 Correctly breaking the tungsten using two pairs of pliers. Larry Jeffus

Once the tungsten has been broken squarely, the end must be melted back so that it becomes somewhat rounded. This is accomplished by switching the welding current to DCEP and striking an arc under argon shielding on a piece of copper. If copper is not available, another piece of clean metal can be used. Do not use carbon, as it will contaminate the tungsten.

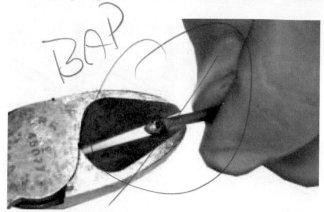

FIGURE 15-11 Using wire cutters to correctly break the tungsten. Larry Jeffus

EXPERIMENT 15-2

Removing a Contaminated Tungsten End by Breaking

Using short scrap pieces of tungsten, pliers or wire cutters, and a light machinist's hammer, you will break the end from the tungsten.

Break about 1/4 in. (6 mm) from the end of the tungsten using the appropriate method, depending upon the diameter of the tungsten. Observe the break; it should be square and relatively smooth, **Figure 15-12**. The end of the tungsten should be broken only if the tungsten is badly contaminated.

Complete a copy of the "Student Welding Report" listed in Appendix I or provided by your instructor. ◆

Chemical Cleaning and Pointing The tungsten can be cleaned and pointed using one of several compounds. The tungsten is heated by shorting it against the work. The tungsten is then dipped in the compound, a strong alkaline, which rapidly dissolves the hot tungsten. The chemical reaction is so fast that enough additional heat is produced to keep the tungsten hot,

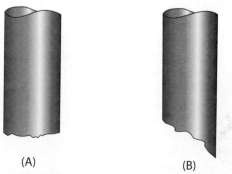

(A)   (B)

FIGURE 15-12 (A) Correctly broken tungsten electrode; (B) incorrectly broken tungsten electrode.
© Cengage Learning 2012

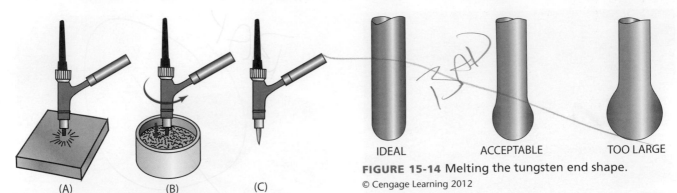

FIGURE 15-13 Chemically cleaning and pointing tungsten: (A) shorting the tungsten against the work to heat it to red hot, (B) inserting the tungsten into the compound and moving it around, and (C) cleaned and pointed tungsten ready for use.
© Cengage Learning 2012

IDEAL   ACCEPTABLE   TOO LARGE

FIGURE 15-14 Melting the tungsten end shape.
© Cengage Learning 2012

Figure 15-13. When the tungsten is removed, cooled, and cleaned, the end will be tapered to a fine point. If the electrode is contaminated, the chemical compound will dissolve the tungsten, allowing the contamination to fall free.

Pointing and Remelting The tapered tungsten with a balled end, a shape sometimes used for DCEP welding, is made by first grinding or chemically pointing the electrode. Using DCEP, as in the procedure for the remelted broken end, strike an arc on some copper under argon shielding and slowly increase the current until a ball starts to form on the tungsten. The ball should be made large enough so that the color of the end stays between dull red and bright red. If the color turns white, the ball is too small and should be made larger. To increase the size of the ball, simply apply more current until the end begins to melt. Surface tension will pull the molten tungsten up onto the tapered end. Lower the current and continue welding. DCEP is seldom used for welding. If the tip is still too hot, it may be necessary to increase the size of the tungsten.

EXPERIMENT 15-3

Melting the Tungsten End Shape

Using a properly set-up GTA welding machine, proper safety protection, one piece of copper or other clean piece of metal, and the tungsten that was sharpened and broken in Experiments 15-1 and 15-2, you will melt the end of the tungsten into the desired shape.

Properly install the tungsten, set the argon gas flow, switch the current to DCEP, and turn on the machine. Strike an arc on the copper and slowly increase the amperage. Watch the tungsten as it begins to melt, and stop the current when the desired shape has been obtained, **Figure 15-14**.

Complete a copy of the "Student Welding Report" listed in Appendix I or provided by your instructor. ◆

GTA Welding Equipment

Figure 15-15 and **Figure 15-16** show two industrial applications of gas tungsten arc welding.

Torches GTA welding torches are available water-cooled or air-cooled. The heat transfer efficiency for GTA welding may be as low as 20%. This means that 80% of the heat generated does not enter the weld. Much of this heat stays in the torch and must be removed by some type of cooling method.

The water-cooled GTA welding torch is more efficient than an air-cooled torch at removing waste heat. The water-cooled torch, as compared to the air-cooled torch, operates at a lower temperature, resulting in a lower tungsten temperature and less erosion.

The air-cooled torch is more portable because it has fewer hoses, and it may be easier to manipulate than the water-cooled torch. Also, the water-cooled torch requires a water reservoir or other system to give the needed cooling. The cooling water system should contain some type of safety device, **Figure 15-17**, to shut off the power if the water flow is interrupted. The power cable is surrounded by the return water to keep it cool, so a smaller-size cable can be used. Without the cooling water, the cable quickly overheats and melts through the hose.

The water can become stopped or restricted for a number of reasons, such as a kink in the hose, a heavy object set on the hose, or failure to turn on the system. Water pressures higher than 35 psi (241 kg/mm^2) may cause the water hoses to burst. When an open system is used, a pressure regulator must be installed to prevent pressures that are too high from damaging the hoses.

GTA welding torch heads are available in a variety of amperage ranges and designs, **Figure 15-18**. The amperage listed on a torch is the maximum rating and cannot be exceeded without possible damage to the torch. The various head angles allow better access in tight places. Some of the heads can be swiveled easily to new angles. The back cap that both protects and tightens the tungsten can be short or long, **Figure 15-19** and **Figure 15-20**.

Hoses A water-cooled torch has three hoses connecting it to the welding machine. The hoses are for shielding

FIGURE 15-15 Semiautomatic operation allows a stainless steel part to be GTA welded as it is turned past the torch.
Lincoln Electric Company

FIGURE 15-16 An operator GTA welds a cap ring on a pneumatic tank. Lincoln Electric Company

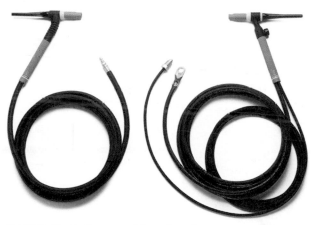

FIGURE 15-17 Power cable safety fuse.
ESAB Welding & Cutting Products

gas to the torch, cooling water to the torch and cooling the water return, and housing the power cables to the torch, **Figure 15-21.** Air-cooled torches may have one hose for shielding gas attached to the power cable, **Figure 15-22.**

The shielding gas hose must be made from a material such as plastic that will not contaminate the shielding gas. Rubber hoses contain oils that can be picked up by the gas, resulting in weld contamination.

The water-in hose may be made of any sturdy material. Water hose fittings have left-hand threads, and gas hose

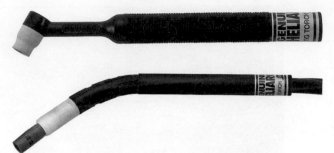

FIGURE 15-19 Short back caps are available for torches when space is a problem. ESAB Welding & Cutting Products

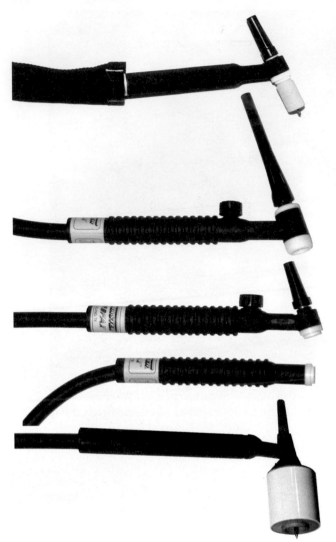

FIGURE 15-18 GTA welding torches. Larry Jeffus

FIGURE 15-20 Long back caps allow tungstens that are a full 7 in. (177 mm) long to be used.
ESAB Welding & Cutting Products

fittings have right-hand threads. This prevents the water and gas hoses from accidentally being reversed when attaching them to the welder. The return water hose also contains the welding power cable. This permits a much smaller-size cable to be used because the water keeps it cool.

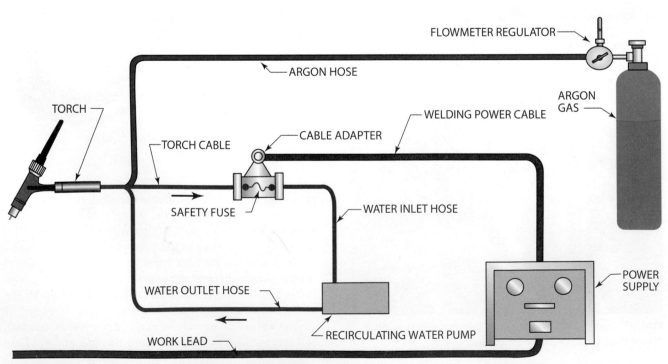

FIGURE 15-21 Schematic of a GTA welding setup with a water-cooled torch. © Cengage Learning 2012

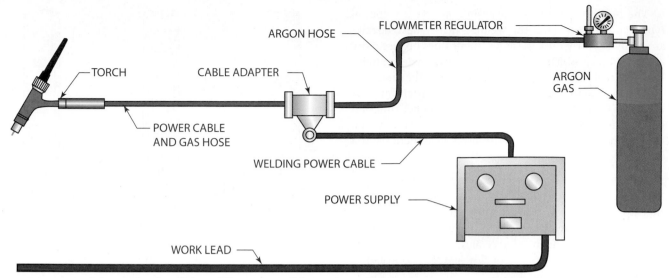

FIGURE 15-22 Schematic of a GTA welding setup with an air-cooled torch. © Cengage Learning 2012

The water must be supplied to the torch head and return around the cable. This allows the head to receive the maximum cooling from the water before the power cable warms it. Running the water through the torch first has another advantage. That is, when the water solenoid is closed, there is no water pressure in the hoses, which is particularly important. This feature also prevents condensation in the torch. If a water leak should occur during welding, the welding power is stopped, closing the water solenoid and thus stopping the leak.

A protective covering can be used to prevent the hoses from becoming damaged by hot metal, **Figure 15-23**. Even with this protection, the hoses should be supported, **Figure 15-24**, so that they are not underfoot on the floor. By supporting the hoses, the chance of their being damaged by hot sparks is reduced.

Nozzles The nozzle or cup is used to direct the shielding gas directly on the welding zone. The nozzle size is determined by the diameter of the opening and its length, **Table 15-2**. Nozzles may be made from a ceramic such as alumina or silicon nitride (opaque) or from fused quartz (clear). The nozzle may also have a gas lens to improve the gas flow pattern.

FIGURE 15-24 A bracket holds the leads off the floor.
Larry Jeffus

The nozzle size, both length and diameter, is often the welder's personal preference. Occasionally, a specific choice must be made based upon joint design or location. Small nozzle diameters allow the welder to better see the molten weld pool and can be operated with lower gas flow rates. Larger nozzle diameters can give better gas coverage, even in drafty places.

Ceramic nozzles are heat resistant and offer a relatively long life. The useful life of a ceramic nozzle is affected by the current level and proximity to the work. Silicon nitride nozzles will withstand much more heat, resulting in a longer useful life.

FIGURE 15-23 Zip-on protective covering also helps keep the hoses neat. ESAB Welding & Cutting Products

Tungsten Electrode Diameter		Nozzle Orifice Diameter	
in.	mm	in.	mm
1/16	2	1/4 to 3/8	6 to 10
3/32	2.4	3/8 to 7/16	10 to 11
1/8	3	7/16 to 1/2	11 to 13
3/16	4.8	1/2 to 3/4	13 to 19

TABLE 15-2 Recommended Cup Sizes

The fused quartz (glass) used in a nozzle is a special type that can withstand the welding heat. These nozzles are no more easily broken than ceramic ones but are more expensive. The added visibility with glass nozzles in tight, hard-to-reach places is often worth the added expense.

The longer a nozzle, the longer the tungsten must be extended from the collet. This can cause higher tungsten temperatures, resulting in greater tungsten erosion. When using long nozzles, it is better to use low amperages or a larger-size tungsten.

Flowmeter The **flowmeter** may be merely a flow regulator used on a manifold system, or it may be a combination flow and pressure regulator used on an individual cylinder. Some flowmeters use a ball that floats on top of the flowing gas stream in a clear tube to indicate the shielding gas flow rate; other flowmeters use a gauge, **Figure 15-25** and **Figure 15-26**.

In the tube- and ball-type flowmeter, the gas flow is controlled by opening a small valve at the base of the flowmeter.

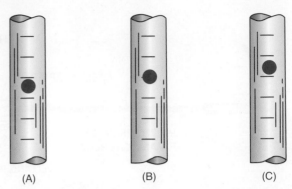

FIGURE 15-27 Three methods of reading a flowmeter: (A) top of ball, (B) center of ball, and (C) bottom of ball. © Cengage Learning 2012

The rate of flow is then read in units of CHF (cubic feet per hour), or L/min (liters per minute). The reading is taken from a fixed scale that is compared to a small ball floating on the stream of gas. Meters from various manufacturers may be read differently. For example, they may read from the top, center, or bottom of the ball, **Figure 15-27**. The ball floats on top of the stream of gas inside a tube that gradually increases in diameter in the upward direction. The increased size allows more room for the gas flow to pass by the ball. If the tube is not vertical, the reading is not accurate, but the flow is unchanged. Also, when using a line flowmeter, it is important to have the correct pressure. Changes in pressure will affect the accuracy of the flowmeter reading. To get accurate readings, be sure the gas being used is read on the proper flow scale. Less dense gases, such as helium and hydrogen, will not support the ball on as high a column with the same flow rate as a denser gas, such as argon.

The tube- and ball-type flowmeters must be mounted so that the tube is perfectly vertical so that the ball will float freely inside the tube. Because the gauge-type flowmeters will read correctly at any angle, they are often preferred over the tube and ball type.

Types of Welding Current

All three types of welding current, or polarities, can be used for GTA welding. Each current has individual features that make it more desirable for specific conditions or with certain types of metals.

The major differences among the currents are in their heat distributions and the presence or degree of arc cleaning. **Figure 15-28** shows the heat distribution for each of the three types of currents.

Direct-current electrode negative (DCEN), which used to be called direct-current straight polarity (DCSP), concentrates about two-thirds of its welding heat on the work and the remaining one-third on the tungsten. The higher heat input to the weld results in deep penetration. The low heat input into the tungsten means that a smaller-size tungsten can be used without erosion problems. The smaller-size electrode may not require pointing, resulting in a savings of time, money, and tungsten.

Direct-current electrode positive (DCEP), which used to be called direct-current reverse polarity (DCRP),

FIGURE 15-25 Flowmeter. Controls Corporation of America

FIGURE 15-26 Flowmeter regulator. Larry Jeffus

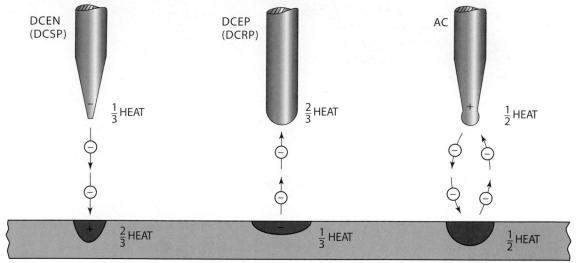

FIGURE 15-28 Heat distribution between the tungsten electrode and the work with each type of welding current.
© Cengage Learning 2012

concentrates only one-third of the arc heat on the plate and two-thirds of the heat on the electrode. This type of current produces wide welds with shallow penetration, but it has a strong cleaning action upon the base metal. The high heat input to the tungsten indicates that a large-size tungsten is required, and the end shape with a ball must be used. The low heat input to the metal and the strong cleaning action on the metal make this a good current for thin, heavily oxidized metals. The metal being welded will not emit electrons as freely as tungsten, so the arc may wander or be more erratic than DCEN.

There are many theories as to why DCEP has a **cleaning action.** The most probable explanation is that the electrons accelerated from the cathode surface lift the oxides that interfere with their movement. The positive ions accelerated to the metal's surface provide additional energy. In combination, the electrons and ions cause the surface erosion needed to produce the cleaning. Although this theory is disputed, it is important to note that cleaning does occur, that it requires argon-rich shield gases and DCEP polarity, and that it can be used to advantage, **Figure 15-29.**

Alternating current (AC) concentrates about half of its heat on the work and the other half on the tungsten.

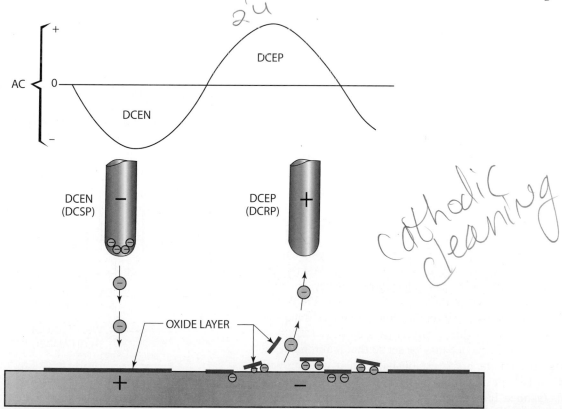

FIGURE 15-29 Electrons collect under the oxide layer during the DCEP portion of the cycle and lift the oxides from the surface. © Cengage Learning 2012

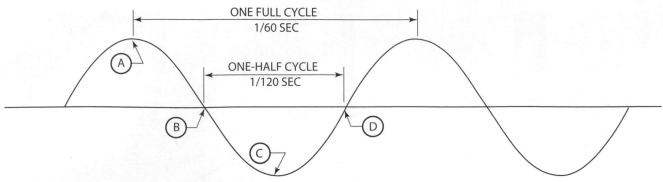

FIGURE 15-30 Sine wave of alternating current. © Cengage Learning 2012

Alternating current is DCEN half of the time and DCEP the other half of the time. The **frequency** at which the current cycles is the rate at which it makes a full change in direction, **Figure 15-30.** In the United States, the electric current cycles at the rate of 60 times per second, or 60 hertz (60 Hz). Referring again to Figure 15-30, the current is at its maximum peak at points A and B. The rate gradually decreases until it stops at points C and D. The arc at these points is extinguished and, as the current reversal begins, must be reestablished. This event requires the emission of electrons from the cathode to ionize the shielding gas. When the hot, emissive electrode becomes the cathode, reestablishing the arc is easy. However, it is often quite difficult to reestablish the arc when the colder and less emissive workpiece becomes the cathode. Because voltage from the power supply is designed to support a relatively low voltage arc, it may be insufficient to initiate electron flow. Thus, a voltage assist from another source is needed. A high-voltage but low-current **spark gap oscillator** commonly provides the assist at a relatively low cost. The high frequency ensures that a voltage peak will occur reasonably close to the current reversal in the welding arc, creating a low-resistance ionized path for the welding current to follow, **Figure 15-31A and B.** This same device is often used to initiate direct-current arcs, a particularly useful technique for mechanized welding.

The high-frequency current is established by capacitors discharging across a gap set on points inside the machine. Changing the point gap setting will change the frequency of the current. The closer the points are, the higher the frequency; the wider the spacing between the points, the lower the frequency. The voltage is stepped up with a transformer from the primary voltage supplied to the machine. The available amperage to the high-frequency circuit is very low. Thus, when the circuit is complete, the voltage quickly drops to a safe level. The high frequency is induced on the primary welding current in a coil.

The high frequency may be set so that it automatically cuts off after the arc is established, when welding with DC. It is kept on continuously with AC. When used in this manner, it is referred to as alternating current, high-frequency stabilized, or ACHF.

Shielding Gases

The shielding gases used for the GTA welding process are argon (Ar), helium (He), hydrogen (H), nitrogen (N), or a mixture of two or more of these gases. The purpose of the shielding gas is to protect the molten weld pool and the tungsten electrode from the harmful effects of air. The

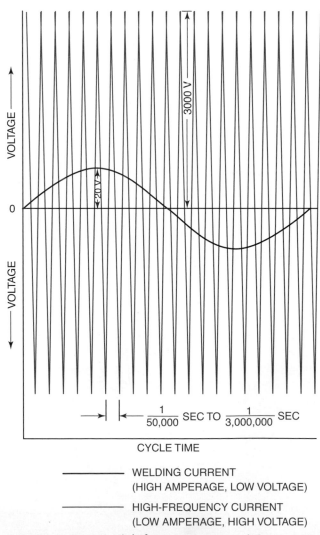

FIGURE 15-31 (A) High-frequency arc starting current shown over the low-frequency welding current. Larry Jeffus

FIGURE 15-31 (B) The high frequency first appears as a blue glow around the tungsten before the welding current starts its arc. Larry Jeffus

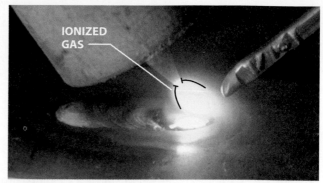

FIGURE 15-32 Highly concentrated ionized argon gas column. Larry Jeffus

shielding gas also affects the amount of heat produced by the arc and the resulting weld bead appearance.

Argon and helium are **noble inert gases.** This means that they will not combine chemically with any other material. Argon and helium may be found in mixtures but never as compounds. Because they are inert, they will not affect the molten weld pool in any way.

//// CAUTION \\\\

Never allow noninert gases such as O₂, CO₂, or N to come in contact with your inert gas system. Very small amounts can contaminate the inert gas, which may result in the weld failing.

Argon Argon makes up approximately 1% of air and is a by-product of the air reduction process that is used in the production of oxygen. Air can be separated cryogenically or noncryogenically. Cryogenic air separation plants cool air down to around $-300°F$ $(-185°C)$, a low enough temperature to cause it to liquefy. The various gases that make up air are separated using a distillation process. In the distillation process, the liquid is gradually allowed to warm so that each gas will boil off at its specific temperature.

There are several other types of noncryogenic processes used to separate oxygen and nitrogen from air. But only the molecular sieve process can produce the level of purity needed for some GTA welding. In the molecular sieve process, a material that has tiny pores, small enough to separate the different gasses based on their molecular size, is used. This process works much like sifting sand through a screen to separate rocks from the sand.

Because argon is denser than air, it effectively shields welds in deep grooves in the flat position. However, this higher density can be a hindrance when welding overhead because higher flow rates are necessary. The argon is relatively easy to ionize and thus suitable for alternating-current applications and easier starts. This property also permits fairly long arcs at lower voltages, making it virtually insensitive to changes in arc length. Argon is also the

only commercial gas that produces the cleaning discussed earlier. These characteristics are most useful for manual welding, especially with filler metals added, as shown in Figure 15-32.

Helium Helium is a by-product of the natural gas industry. It is removed from natural gas as the gas undergoes separation (fractionation) for purification or refinement.

Helium offers the advantage of deeper penetration. The arc force with helium is sufficient to displace the molten weld pool with very short arcs. In some mechanized applications, the tip of the tungsten electrode is positioned below the workpiece surface to obtain very deep and narrow penetration. This technique is especially effective for welding aged aluminum alloys prone to overaging. It also is very effective at high welding speeds, as for tube mills. However, helium is less forgiving for manual welding. With helium, penetration and bead profile are sensitive to the arc length, and the long arcs needed for feeding filler wires are more difficult to control.

Helium has been mixed with argon to gain the combined benefits of cathode cleaning and deeper penetration, particularly for manual welding. The most common of these mixtures is 75% helium and 25% argon.

Although the GTA process was developed with helium as the shielding gas, argon is now used whenever possible because it is much cheaper. Helium also has some disadvantages because it is lighter than air, thus preventing good shielding. Its flow rates must be about twice as high as argon's for acceptable stiffness in the gas stream, and proper protection is difficult in drafts unless high flow rates are used. It is difficult to ionize, necessitating higher voltages to support the arc and making the arc more difficult to ignite. Alternating-current arcs are very unstable. However, helium is not used with alternating current because the cleaning action does not occur.

Hydrogen Hydrogen is not an inert gas and is not used as a primary shielding gas. However, it can be added to argon when deep penetration and high welding speeds are needed. It also improves the weld surface cleanliness and bead profile on some grades of stainless steel that are very sensitive to oxygen. Hydrogen additions are restricted to stainless steels because hydrogen is the primary cause of

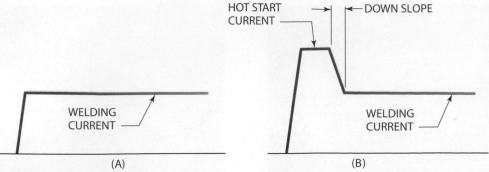

FIGURE 15-33 Standard method of starting welding current (A); hot start method of starting welding current (B). © Cengage Learning 2012

porosity in aluminum welds. It can cause porosity in carbon steels and, in highly restrained welds, underbead cracking in carbon and low alloy steels.

Nitrogen Nitrogen is not an inert gas. Like hydrogen, nitrogen has been used as an additive to argon. But it cannot be used with some materials, such as ferritic steels, because it produces porosity. In other cases, such as with austenitic stainless steels, nitrogen is useful as an austenite stabilizer in the alloy. It is used to increase penetration when welding copper. Unfortunately, because of the general success with inert gas mixtures and because of potential metallurgical problems, nitrogen has not received much attention as an additive for GTA welding.

Hot Start The hot start allows a controlled surge of welding current as the arc is started to establish a molten weld pool quickly. Establishing a molten weld pool rapidly on metals with a high thermal conductivity is often hard without this higher-than-normal current. Adjustments can be made in the length of time and the percentage above the normal current, **Figure 15-33**.

Preflow and Postflow

The **preflow** is the time during which gas flows to clear out any air in the nozzle or surrounding the weld zone. The welding torch must be held directly over the joint that is going to be welded so that the preflow shielding will force the air away from the welding zone before the arc starts.

The operator sets the length of time that the gas flows before the welding current is started, **Figure 15-34**. Because some machines do not have preflow, many welders find it hard to hold a position while waiting for the current to start. One solution to this problem is to use the postflow for preflow. Switch on the current to engage the postflow. Now, with the current off, the gas is flowing, and the GTA torch can be lowered to the welding position. The welder's helmet should be lowered and the current restarted before the postflow stops. This allows welders to have postflow and to start the arc when they are ready.

The **postflow** is the time during which the gas continues flowing after the welding current has stopped. This period serves to protect the molten weld pool, the filler rod, and the tungsten electrode as they cool to a temperature

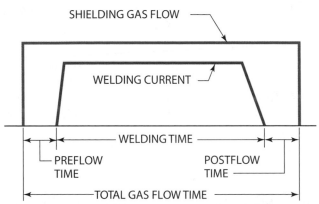

FIGURE 15-34 Welding time compared to shielding gas flow time. © Cengage Learning 2012

at which they will not oxidize rapidly. The time of the flow is determined by the welding current and the tungsten size, **Table 15-3**. During the postflow time, the torch must be held directly over the cooling weld pool so that it is protected from atmospheric contamination.

Shielding Gas Flow Rate

The shielding gas flow rate is measured in cubic feet per hour (CFH) or in metric measure as liters per minute (L/min). The rate of flow should be as low as possible and still give adequate coverage. High gas flow rates waste

Electrode Diameter		Postwelding Gas Flow Time*
in.	**mm**	
0.01	0.25	5 sec
0.02	0.5	5 sec
0.04	1.0	5 sec
1/16	2	8 sec
3/32	2.4	10 sec
1/8	3	15 sec
5/32	4	20 sec
3/16	4.8	25 sec
1/4	6	30 sec

*The time may be longer if either the base metal or the tungsten electrode does not cool below the rapid oxidation temperatures within the postpurge times shown.

TABLE 15-3 Postwelding Gas Flow Times

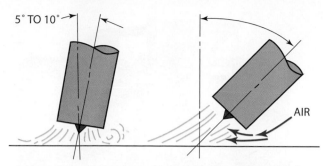

FIGURE 15-35 Too steep an angle between the torch and work may draw in air. © Cengage Learning 2012

5° TO 10°

AIR

shielding gases and may lead to contamination. The contamination comes from turbulence in the gas at high flow rates. Air is drawn into the gas envelope by a venturi effect around the edge of the nozzle. Also, the air can be drawn in under the nozzle if the torch is held at too sharp an angle to the metal, **Figure 15-35**.

The larger the nozzle size, the higher is the flow rate permissible without causing turbulence. **Table 15-4** shows the average and maximum flow rates for most nozzle sizes. A gas lens can be used in combination with the nozzle to stabilize the gas flow, thus eliminating some turbulence. A gas lens will add to the turbulence problem if there is any spatter or contamination on its surface, **Figure 15-36**.

Shielding gas dams may be placed on the sides of the joint to be welded so that the shielding gas will flood the weld zone better. The gas dams can be made by clamping metal pieces alongside the joint or by using a high temperature tape to hold thin foil or sheet metal in place. Shielding gas dams should be used anytime there is more than just a light draft or breeze and when reactive metals such as titanium are being welded.

Remote Controls

A remote control can be used to start the weld, increase or decrease the current, and stop the weld. The remote control can be either a foot-operated or hand-operated device. The foot control works adequately if the welder can be seated. Welds that must be performed away from a welding station may use a hand or thumb control or may not have any remote welding controls.

Most remote controls have an on-off switch that is activated at the first or last part of the control movement. A variable resistor increases the current as the control is pressed more. A variable resistor works in a manner similar to the accelerator pedal on a car to increase the power (current), **Figure 15-37**. The operating amperage range is

GAS LENS
GAS NOZZLE
TUNGSTEN ELECTRODE

FIGURE 15-36 Gas lens. Larry Jeffus

Nozzle Inside Diameter		Gas Flow*	
in.	mm	cfh	L/min
1/4	6	10–14	4.7–6.6
5/16	8	11–15	5.2–7.0
3/8	10	12–16	5.6–7.5
7/16	11	13–17	6.1–8.0
1/2	13	17–20	8.0–9.4
5/8	16	17–20	8.0–9.4

*The flow rates may need to be increased or decreased depending upon the conditions under which the weld is to be performed.

TABLE 15-4 Suggested Argon Gas Flow Rate for Given Cup Sizes

FIGURE 15-37 A foot-operated device can be used to increase the current. Larry Jeffus

FIGURE 15-38 GTA welding unit that can be added to a standard power supply so that it can be used for GTA welding. Lincoln Electric Company

FIGURE 15-39 Always be sure the power is off when making machine connections. Larry Jeffus

determined by the value that has been set on the main controls of the machine.

EXPERIMENT 15-4

Setting Up a GTA Welder

Using a GTA welding machine, remote control welding torch, gas flowmeter, gas source (cylinder or manifold), tungsten, nozzle, collet, collet body, cap, and any other hoses, special tools, and equipment required, you will set up the machine for GTA welding, **Figure 15-38**.

1. Start with the power switch off, **Figure 15-39**. Use a wrench to attach the torch hose to the machine. The water hoses should have left-hand threads to prevent incorrectly connecting them. Tighten the fittings only as tightly as needed to prevent leaks, **Figure 15-40**. Attach the cooling water "in" to the machine solenoid and the water "out" to the power block.

2. The flowmeter or flowmeter regulator should be attached next. If a gas cylinder is used, secure it in place with a safety chain. Then remove the valve protection cap and crack the valve to blow out any dirt, **Figure 15-41**. Attach the flowmeter so that the tube is vertical.

3. Connect the gas hose from the meter to the gas "in" connection on the machine.

4. With both the machine and main power switched off, turn on the water and gas so that the connection to the machine can be checked for leaks. Tighten any leaking fittings to stop the leak.

5. Turn on both the machine and main power switches and watch for leaks in the torch hoses and fittings.

6. With the power off, switch the machine to the GTA welding mode.

FIGURE 15-40 Tighten each fitting as it is connected to avoid missing a connection. Larry Jeffus

DIRT

FIGURE 15-41 During transportation or storage, dirt may collect in the valve. Cracking the valve is the best way to remove any dirt. © Cengage Learning 2012

FIGURE 15-42 Setting the current. Larry Jeffus

FIGURE 15-44 The high-frequency switch should be placed in the appropriate position. Larry Jeffus

FIGURE 15-43 Setting the amperage range. Larry Jeffus

FIGURE 15-45 Setting the remote control switch. Larry Jeffus

7. Select the desired type of current and amperage range, **Figure 15-42** and **Figure 15-43**.

8. Set the fine current adjustment to the proper range, depending upon the size of tungsten used, **Table 15-5**. Refer to Chapter 3 for more information on setting the fine current adjustment.

9. Place the high-frequency switch in the appropriate position, auto (HF start) for DC or continuous for AC, **Figure 15-44**.

10. The remote control can be plugged in and the selector switch set, **Figure 15-45**.

11. The collet and collet body should be installed on the torch first, **Figure 15-46**.

12. On the Linde or copies of Linde torches, installing the back cap first will stop the collet body from being screwed into the torch fully. A poor connection will result in excessive electrical and thermal resistance, causing a heat buildup in the head.

13. The tungsten can be installed and the end cap tightened to hold the tungsten in place.

14. Select and install the desired nozzle size. Adjust the tungsten length so that it does not stick out more than the diameter of the nozzle, **Figure 15-47**.

15. Check the manufacturer's operating manual for the machine to ensure that all connections and settings are correct.

Electrode Diameter		DCEN	DCEP	AC
in.	mm			
0.04	1	15–60	Not recommended	10–50
1/16	2	70–100	10–20	50–90
3/32	2.4	90–200	15–30	80–130
1/8	3	150–350	25–40	100–200
5/32	4	300–450	40–55	160–300

TABLE 15-5 Amperage Range of Tungsten Electrodes

FIGURE 15-46 Inserting collet and collet body. Larry Jeffus

FIGURE 15-47 Install the nozzle (cup) to the torch body. Larry Jeffus

16. Turn on the power, depress the remote control, and again check for leaks.

17. While the postflow is still engaged, set the gas flow by adjusting the valve on the flowmeter.

////// **CAUTION** \\\\\\

Turn off all power before attempting to stop any leaks in the water system.

The GTA welding system is now ready to be used. Complete a copy of the "Student Welding Report" listed in Appendix I or provided by your instructor. ◆

EXPERIMENT 15-5

Striking an Arc

Using a properly set-up GTA welding machine, proper safety gear, and clean scrap metal, you will strike a GTA welding arc.

1. Position yourself so that you are comfortable and can see the torch, tungsten, and plate while the tungsten tip is held about 1/4 in. (6 mm) above

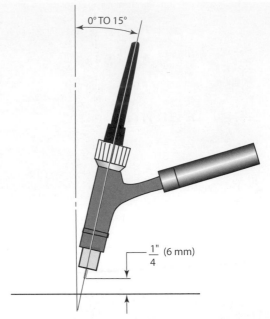

FIGURE 15-48 GTA torch position. © Cengage Learning 2012

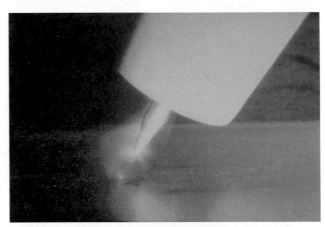

FIGURE 15-49 High-frequency starting before arc starts. Larry Jeffus

the metal. Try to hold the torch at a vertical angle ranging from 0° to 15°. Too steep an angle will not give adequate gas coverage, **Figure 15-48**.

2. Lower your arc welding helmet and depress the remote control. A high-pitched, erratic arc should be immediately jumping across the gap between the tungsten and the plate. If the high-frequency arc is not established, lower the torch until it appears, **Figure 15-49**.

3. Slowly increase the current until the main welding arc appears, **Figure 15-50**.

4. Observe the color change of the tungsten as the arc appears.

5. Move the tungsten around in a small circle until a molten weld pool appears on the metal.

FIGURE 15-50 Stable gas tungsten arc. Larry Jeffus

6. Slowly decrease the current and observe the change in the molten weld pool.
7. Reduce the current until the arc is extinguished.

8. Hold the torch in place over the weld until the postflow stops.
9. Raise your hood and inspect the weld.

//// **CAUTION** \\\\

Avoid touching the metal table with any unprotected skin or jewelry. The high frequency can cause an uncomfortable shock.

Repeat this procedure until you can easily start the arc and establish a molten weld pool using both AC and DCEN currents. Turn off the welding machine, water, and shielding gas when you are finished; then clean up your work area.

Complete a copy of the "Student Welding Report" listed in Appendix I or provided by your instructor. ◆

Summary

One of the prime considerations for gas tungsten arc welding equipment setup is the cleanliness of the equipment, supplies, base material or materials, and the welders themselves. When everything is clean, you will find that the welding process will proceed more easily and more successfully.

Another major factor affecting your ability to produce quality welds is the tungsten end or tip shape. As you practice making the various welds, you will find that keeping the tungsten electrode tip shaped appropriately will assist you in producing uniform welds.

Often, new welders feel that there is some sort of attraction between the tungsten electrode, filler metal, and base metal during the welding process because it seems to continually become contaminated. This almost continuous contamination can be very frustrating. At times it may seem overwhelming; however, with continued practice and diligence you will be able to control this problem. Even experienced welders in the field can be plagued from time to time with tungsten contamination. At other times, they can weld an entire day without contaminating the tungsten. It is often beneficial for students to realize that tungsten contamination is just part of the process, and they must therefore try to ignore the possibility of it happening and concentrate on producing the welds.

Welding a Pathway to the Stars

The striking Rose Center for Earth and Space, the American Museum of Natural History's "cosmic cathedral," uses its architecture to serve science.

It has captured the planets and stars in a glass cube, locking them up in a gleaming metal ball, where you and I can sit in the dark and peer into the vastness of the universe. The most sophisticated visualization tools available take us on this virtual tour of space at the Hayden Planetarium, centerpiece of the Frederick Phineas and Sandra Priest Rose Center for Earth and Space at the American Museum of Natural History in New York, New York.

Six years in the making and costing approximately $210 million, the Rose Center opened in February 2000. It replaced the original 1935 Hayden Planetarium and provides exhibition and research space.

"What I had originally conceived of as an iconic scientific tool evolved over the development of its design into a 'cosmic cathedral,' an intensely memorable spatial experience that is intended to awe and inspire visitors and to expand their understanding of the wonders of our universe and the power of scientific inquiry," explained James Stewart Polshek,

senior design partner of Polshek Partnership Architects, New York, New York, one of the building's designers.

The seven-story-tall Rose Center rests atop a one-story base of gray granite at Central Park West and West 81st Street. The cube measures approximately 120 ft per side. Visitors enter through a low, arched entranceway. The exterior walls facing south and east are solid; the north and west walls are made of an extremely clear glass called Pilkington water white. The 87-ft-diameter, 4-billion-lb Hayden Sphere appears to float inside. Visitors entering through the archway can see the bottom of the sphere but not the three pairs of steel columns that support it and inside only one set of columns is visible at any given point, resulting in the sphere appearing to be weightless. A 360-ft-long, 8-ft-wide ramp—the Heilbrunn Cosmic Pathway—encircles the lower half of the sphere one and a half times. The cantilevered spiral pathway serves as a time line, detailing the development of the universe.

Another pathway, the 400-ft-long Scales of the Universe, hugs the glass curtain wall along the second level of the Center. To illustrate the vast range of sizes in the universe, it uses the Hayden Sphere as a basis of comparison along with

The ring truss or compression ring located approximately midpoint is the main weight-bearing area for the sphere. It's where the legs attach. Netting was placed at the highest levels of the sphere to provide fall protection for workers during the field fabrication. American Welding Society

a variety of touchable models mounted along the walkway rail and suspended above.

The deceptively simple structure is, in reality, an architecturally and structurally complex building made primarily of steel, in which welding provides the support structure that makes the whole thing work.

Chicago Ornamental Iron Co. (COI), Melrose Park, Illinois, did most of the steel work for the Rose Center. Its work included the curtain wall and other elements of the glass support system, the ramp that emerges from the sphere, and the sphere itself. A. J. McNulty & Co. Inc., Queens, New York, was the on-site steel erector.

Horst Peppa, COI president, called the structure one of the most complex his company had ever worked on. In some ways, he said, even though everything worked in the end, "nothing made sense. If you look at the legs and the ramp, nothing squares up to anything. There's no rhyme or reason to half the things."

Chicago Ornamental Iron Co. began shop fabrication in December 1997, and the last truckload of sections headed for New York a year later. In the shop, COI exclusively used semiautomatic gas metal arc welding (GMAW) with straight CO_2 shielding gas. In the field, A. J. McNulty used shielded metal arc welding. According to Paul Miller, COI shop foreman, the company uses GMAW because it produces high-quality welds at high deposition rates, and it is versatile enough for use on a wide variety of metals.

The Rose Center for Earth and Space required an enormous amount of welding, Peppa recalled, so much welding that the 28 welders working on the job went through 3200 35-lb spools of 0.045-in. NS101 (AWS ER70S-3) welding wire from Lincoln Electric Co.—more than one semitractor trailer full.

COI's contract called for fabricating a mockup of part of the curtain wall and shipping it to Construction Research Laboratory in Miami, Florida, to undergo tests concerning water penetration, thermal expansion, and wind, among others. Chicago Ornamental Iron Co. built a 35-ft-tall corner unit section and a 21-ft-wide by 35-ft-tall wall section as one of the first parts of the job. The support structure for the curtain wall consists of a system of 6- and 8-in. tubular steel ladder trusses braced with high-strength stainless steel tension trusses. A series of steel "spiders" and bolts anchors the curtain wall.

Concurrently with the mockup, COI worked on finishing the drawings and beginning work on the ramp. One of the difficulties in building the ramp, Peppa said, was that it had to fit the requirements of the Americans with Disabilities Act. Therefore, it had to have flat areas every so many feet where a person using a wheelchair could stop and rest. "The ramp [floor] became a completely separate geometry from the rest," he said. "It's another spiral stair inside of a spiral stair."

The company also preassembled the compression ring—the main support structure for the sphere—located approximately at the sphere's equator. "The six legs have to hold onto something so the compression ring is the base and the rest of it is really somewhat falsework just to carry the shell and to give it shape," Peppa explained. "Structurally, everything is right here at the compression ring." With the weight of the ramp, the sphere, and the equipment inside the sphere as well as the weight of the occupants of the two theaters inside the sphere, the six legs support approximately 4 million pounds.

Shop foreman Miller considered the many T-Y-K connections the most difficult part of the job because there was so much welding to do in such a tight area. In fact, Peppa said, there were connections in which as many as 17 members joined at one location.

Inspection and Stress Relief

Lawrence C. Smetana is an AWS Certified Welding Inspector and the owner/operator of CWI Services, Bloomington, Illinois, and American Welding Services, Rockford, Illinois, two inspection and testing consulting businesses. Testwell Laboratories, Inc., Ossining, New York, hired Smetana to perform nondestructive examination of the welds at COI. He also worked on certain aspects of the drawings, such as the weld symbols and the prequalification of welding procedures. Smetana served as project manager for all the inspection operations.

Visitors walking along the Scales of the Universe walkway can compare the Hayden Sphere with touchable models mounted along the railing and suspended from the ceiling to reach an understanding of the size of things in space. *American Welding Society*

"We were responsible for all visual weld inspection," Smetana said. "We performed magnetic particle inspection on most all of the root passes on the full penetration welds, and on the cover passes we did ultrasonic examination on selected full-penetration welds both on the tubulars and on the structurals."

Random radiographic examination was also conducted on welds in some of the vertical truss columns for the curtain wall and some of the joints for the spiral ramp. Early on, radiography exposed a problem with some of the welds and prompted a change in joint design, he said.

"We discovered that the original joint design, which was a single-bevel groove with a backing, was giving us problems," Smetana explained. "In fact, 100% of the first group of X-rays were rejected for incomplete fusion. I worked with Horst Peppa and his people, and we decided to try a single V-groove." Follow-up radiographs on welds with the new joint design showed the problem had been solved. "We put an engineering change through with the design engineers, they approved it, and we went with that joint for the rest of the system. We did random radiography from that point on, and every joint we shot was acceptable, which bore out our mag particle results and our ultrasonic results, so we were very happy with that."

One step not initially anticipated was the need to stress relieve many of the full-penetration welds. "The sphere itself has a couple of ring girders that are the main support systems for the floors," Smetana explained, "and those members curved, twisted, and skewed all at the same time. There were some very complex, highly restrained joints in these members, and I was a little bit concerned with residual stress in these areas. They were too big to really postweld stress relieve in any reasonable fashion. We opted to use a vibratory stress relief system [during welding] and had some very, very good luck with that system."

Miller explained that postweld stress relief was performed on joints made prior to the acquisition of the vibratory equipment. Once the equipment was in place, stress relief was done as the welds were being made.

After the problems uncovered with the initial radiography were disposed of, any problems discovered were minor and sporadic, Smetana said. The nonconformance rate was very low. "I would say minor porosity, some incomplete fusion would be the predominant areas that we noted when we were testing."

Article courtesy of the American Welding Society.

Review

1. What early advancements made the GTA welding process more effective and reduced its cost?

2. What metals were weldable only by the GTAW process before GMAW was developed?

3. Which two of tungsten's properties make it the best choice for GTA welding?

4. Why must the tip of the tungsten be hot?

5. Why does some tungsten erosion occur?

6. What function regarding tungsten heat do the collet and torch play?

7. What problem can an excessively large tungsten cause?

8. What holds the molten ball of tungsten in place at the tip of the electrode during DCEP welding?

9. Using Table 15-1, answer the following:

   a. What color identifies EWTh-1?

   b. What is the composition of EWCe-2?

10. What does adding thorium oxide do for the tungsten electrode?

11. How can the end of a tungsten electrode be shaped?

12. Why should a grinding stone that is used for sharpening tungsten not be used for other metals?

13. Why should the grinding marks run lengthwise on the tungsten electrode end?

14. What are three ways of breaking off the contaminated end of a tungsten electrode?

15. What is the correct color to use on the balled end of a pointed and remelted tungsten tip on DCEP?

16. Why should the torch be as cool as possible?

17. What will happen to a water-cooled torch cable if the flow of cooling water stops?

18. Why must shielding gas hoses not be made from rubber?

19. What materials can be used to make nozzles?

20. What problem can a long nozzle cause to the tungsten?

21. Why must the tube of a flowmeter be vertical?

Review Questions (continued)

22. What is the heat distribution with DCEN welding current?

23. What is the heat distribution with DCEP welding current?

24. What is the heat distribution with AC welding current?

25. Why must AC welding power use high frequencies in order to work?

26. Why are argon and helium known as inert gases?

27. Why is argon's ease of ionization a benefit?

28. What makes helium difficult to use for manual welding?

29. What are the benefits of adding hydrogen to argon for welding?

30. What is the purpose of a hot start?

31. Using Table 15-3, determine the gas postflow time for a 3/32-in. (2.4-mm) tungsten.

32. What functions can a remote control provide the welder?

33. Using Table 15-4, determine the minimum gas flow rate for a 1/2-in. (13-mm) nozzle.

Chapter 16

Gas Tungsten Arc Welding of Plate

OBJECTIVES

After completing this chapter, the student should be able to

■ name the applications for which the gas tungsten arc welding (GTAW) process is more commonly used.

■ discuss the effects on the weld of varying torch angles.

■ explain why the filler rod end must be kept inside the protective zone of the shielding gas and how to accomplish this.

■ tell how tungsten contamination occurs and what should be done when it happens.

■ explain what can cause the actual welding amperage to change.

■ determine the correct machine settings for the minimum and maximum welding current for the machine used, the types and sizes of tungsten, and the metal types and thicknesses.

■ list factors that affect the gas preflow and postflow times required to protect the tungsten and the weld.

■ determine the minimum and maximum gas flow settings for each nozzle size, tungsten size, and amperage setting.

■ compare the characteristics of low carbon and mild steels, stainless steel, and aluminum with respect to GTA welding.

■ describe the metal preparation needed before GTA welding.

■ demonstrate how to properly make GTA welds in butt joints, lap joints, and tee joints in all positions that can pass the specified standard.

KEY TERMS

chill plate

contamination

gas coverage

oxide layer

protective zone

surface tension

INTRODUCTION

The gas tungsten arc welding process can be used to join nearly all types and thicknesses of metal. Welders can have a clear unobstructed view of the molten weld pool because GTA welding is fluxless, slagless, and smokeless. The clear view of the weld allows welders to make changes in their welding technique, current, travel speed, or rate of filler metal being added to the weld as the weld progresses to ensure that a quality weld is being made. This gives the welder a very fine control of the welding process.

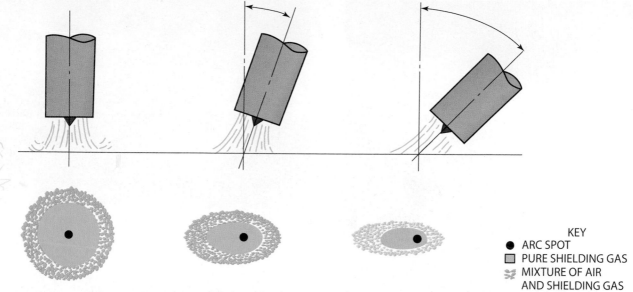

KEY
- ARC SPOT
- PURE SHIELDING GAS
- MIXTURE OF AIR AND SHIELDING GAS

FIGURE 16-1 Gas coverage patterns for different GTA torch angles. Note how the area covered by the shielding gas becomes narrower and elongates as the angle of the torch increases from the perpendicular. © Cengage Learning 2012

The fine control of the weld that is possible with GTA welding makes it an ideal process for very close-tolerance, high-quality welds. GTA welding is used to make critical welds such as those on aircraft structures, which if they fail, can cause serious injury, death, and/or significant loss of property. Sometimes GTA welding is used to make the critical root pass of a weld that will be completed using another faster process. It is also used when weld appearance is important to the look of the finished part such as some furniture, decorations, and/or sculptures.

The proper setup of GTA equipment can often affect the quality of the weld performed. Charts and graphs are available that give the correct amperage, gas flow rate, and time for various types of welds and metals. These charts are designed for optimum laboratory or classroom conditions. Actual conditions in the field will have an effect on these values. The experiments in this chapter are designed to help the welder understand the harmful effects on welding of less-than-ideal conditions. This will allow the welder to evaluate the appearance of a weld and make the necessary changes in technique or setup to improve the weld.

After a person has learned to weld in the lab, troubleshooting field welding problems will become much easier. The weld should be watched carefully to pick up changes that indicate a needed adjustment. When welders can do this, they have mastered the GTA process and have made themselves better potential employees. To make a weld is good; to solve a welding problem is better.

Torch Angle

The torch should be held as close to perpendicular as possible in relation to the plate surface. The torch may be angled from 0° to 15° from perpendicular for better visibility and still have the proper shielding **gas coverage.** As the gas flows out it must form a **protective zone** around the weld. Tilting the torch changes the shape of this protective zone, **Figure 16-1.** Too much tilting of the torch will cause the protective shielding gas zone to become so distorted that the weld may not be protected from contamination from the air. The closer the torch is held to perpendicular, the better the weld is shielded.

The velocity of the shielding gas also affects the protective zone as the torch angle changes. As the velocity increases, a low-pressure area develops behind the cup. When the low-pressure area becomes strong enough, air is pulled into the shielding gas. The sharper the angle and the higher the flow rate, the possibility of contamination is increased by the onset of turbulence in the gas stream. This causes air to become mixed with the shielding gas. Turbulence caused by the shielding gas striking the work will also cause air to mix with the shielding gas at high velocities.

Filler Rod Manipulation

The filler rod end must be kept inside the protective zone of the shielding gas, **Figure 16-2.** The end of the filler rod is hot, and if it is removed from the gas protection, it will oxidize rapidly. The oxide will then be added to the molten weld pool, **Figure 16-3.** When a weld is stopped so that the welder can change position, the shielding gas must be kept flowing around the rod end to protect it until it is cool. If the end of the rod becomes oxidized, it should be cut off before restarting. The following method can be used both to protect the rod end and reduce the possibility of crater cracking—that is, breaking the arc but keeping the torch over the crater while, at the same time, sticking the rod in the molten weld pool before it cools, **Figure 16-4.**

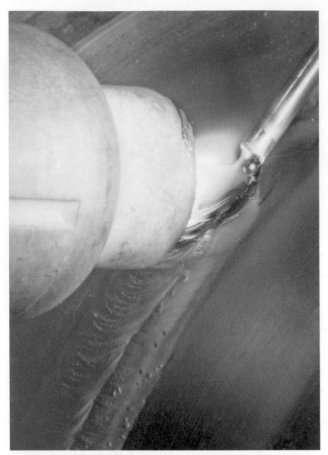

FIGURE 16-2 The hot filler rod end is well within the protective gas envelope. Larry Jeffus

FIGURE 16-3 (A) Filler properly protected, (B) some oxides on filler, (C) and excessive oxides caused by improper filler rod manipulation. Larry Jeffus

FIGURE 16-4 Filler being left in the molten weld pool as the arc is extinguished. Larry Jeffus

FIGURE 16-5 Filler being remelted as the weld is continued. Larry Jeffus

When the weld is restarted, the rod is simply melted loose again, **Figure 16-5.**

The rod should enter the shielding gas as close to the base metal as possible, **Figure 16-6.** A 15° angle or less to the plate surface prevents air from being pulled into the welding zone behind the rod, **Figure 16-7.** As an example, if a rod is held in a stream of running water, air can be pulled in. The faster the water flows or the steeper the angle at which the rod is held, the more air is pulled in.

FIGURE 16-6 Keep the filled metal at approximately a 15° angle. © Cengage Learning 2012

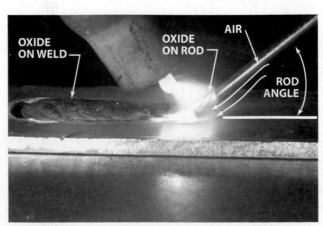

FIGURE 16-7 Too much filler rod angle has caused oxides to be formed on the filler rod end. Larry Jeffus

FIGURE 16-8 Contaminated tungsten. Larry Jeffus

The same action occurs with the shielding gas as its flow increases or as the rod angle increases.

Tungsten Contamination

For new welding students, the most frequently occurring and most time-consuming problem is tungsten **contamination.** The tungsten becomes contaminated when it touches the molten weld pool or when it is touched by the filler metal. When this happens, especially with aluminum, **surface tension** pulls the contamination up onto the hot tungsten, **Figure 16-8.** The extreme heat causes some of the metal to vaporize and form a large, widely scattered **oxide layer.** On aluminum, this layer is black. On iron (steel and stainless steel), this layer is a reddish color.

The contamination caused by the tungsten touching the molten weld pool or filler metal forms a weak weld. On a welding job, both the weld and the tungsten must be cleaned before any more welding can be done. The weld crater must be ground or chiseled to remove the tungsten contamination, and the tungsten end must be reshaped. Extremely tiny tungsten particles will show up if the weld is X-rayed. Failure to remove the contamination properly will result in the failure of the weld.

When starting to weld, the beginning student may save weld practice time by burning off the contamination. On a scrap, usually copper plate, strike an arc using a higher-than-normal amperage setting. The arc will be erratic and discolored at first, but, as the contamination vaporizes, the arc will stabilize. Contamination can also be knocked off by quickly flipping the torch head.

CAUTION

This procedure should never be used with heavy contaminations or when a welder is on the job in the field. It is designed only to help the new student in the first few days of training to save time and increase weld production.

Current Setting

The amperage set on a machine and the actual welding current are often not the same. The amperage indicated on the machine's control is the same as that at the arc only for the following conditions:

- The power to the machine is exactly correct.
- The lead length is very short.
- All cable connections are perfect with zero resistance.
- The arc length is exactly the right length.
- The remote current control is in the full on position.

If any one of these factors changes, the actual welding amperage will change.

In addition to the difference between indicated and actual welding amperage, there is a more significant difference between amperage and welding power. The welding power, in watts, is based on the formula $W = E \times I$, or volts (E) multiplied by amperes (I) equals watts (W). Thus, the indicated power to a weld from two different types of welding machines set at 100 amperes will vary depending upon the voltage of the machine.

The welding machine setting will vary within a range from low to high (cool to hot). The range for one machine may be different from that of another machine. The setting will also be different for various types and sizes of tungstens, polarities, types and thicknesses of metal, joint position or design, and shielding gas used.

A chart, such as the one in **Table 16-1,** and a series of tests can be used to set the lower and upper limits for the amperage settings. As students' welding skills improve with practice, they will become familiar with the machine settings so that a table for these settings is no longer needed. In the welding industry, some welders will mark a line on the dial of the machine to help in resetting the machine. If a welder is required to make a number of different machine setups, a list or chart can be made and taped to the machine. This practice is more professional than marking the machine dials.

Current and Tungsten Electrode Size	Amperage/Machine Setting				
	Too Low	Low	Good	High	Too High

TABLE 16-1 Sample Chart Used to Record GTA Welding Machine Settings

Experiments

Experiments are designed to help new welders learn some basic skills that will help them troubleshoot welding problems. If you do the experiments listed in this chapter, you will be better able to determine what is causing a problem with your weld. As you learn more about welding, subtle changes will become more noticeable. Even experienced welders make changes in the setup, current, or welding technique as they try to resolve a problem.

Experiment 16-1 will help the welder determine the correct machine settings for the minimum and maximum welding current for the machine used, the types and sizes of tungstens, and the metal types and thicknesses. Most welding will be performed with a medium-range or midrange machine setting. The exact setting is more important for machines without remote controls. The remote control allows changes in welding current to be made during the welding without having to stop.

EXPERIMENT 16-1

Setting the Welding Current

Using a properly set-up GTA welding machine and torch, proper safety protection, one of each available tungsten size and type, and 16-gauge mild steel 1/8 in. (3 mm) and 1/4 in. (6 mm) thick, you will work with a small group of students to develop a chart of the correct machine current setting for each type and size of tungsten.

Set the machine welding power switch for DCEN (DCSP) and the amperage control to its lowest setting, **Figure 16-9.** Sharpen a point on each tungsten and install one of the smaller diameter tungstens in the GTA torch. Select a nozzle with a 1/2-in. (13-mm) diameter hole and attach it to the torch head. Set the prepurge time to 0 and postpurge to 10 to 15 sec. Connect the remote control if it is available. Turn on the main power and hold the torch so that it cannot short out. Depress the remote controls to start the shielding gas so the flow rate can be set at 20 cfh (8 L/min). Switch the high frequency to start. All other

FIGURE 16-10 Melting first occurring. Larry Jeffus

functions, such as pulse, hot start, slope, and so on, should be in the off position.

Place the piece of 16-gauge sheet metal flat on the welding table. Hold the torch vertically with the tungsten about 1/4 in. (6 mm) above the metal. Lower your welding hood and fully depress the remote control. Watch the arc to see if it stabilizes and melts the metal. After a short period of time (15 to 30 sec), stop, raise your hood, and check the plate for a melted spot. If melting occurred, note the size of the spot and depth of penetration, **Figure 16-10.** Increase the amperage setting by 5 or 10 amperes, note the setting on the chart, and repeat the process.

After each test, observe and record the results. The important settings to note are

1. when the tungsten first heats up and the arc stabilizes,
2. when the metal first melts,
3. when 100% penetration of the metal first occurs,
4. when burnthrough first occurs, and
5. when the tungsten starts glowing white hot and/or melts.

The lowest (minimum) acceptable amperage setting is when the molten weld pool first appears on the base metal and the arc is stable. The highest (maximum) amperage setting is when the base metal burnthrough or melting of the tungsten occurs. Any current setting in between the high and low points is within the amperage range for that specific setup.

To establish the range for the next tungsten type or size, repeat the test. After each test, the metal should be cooled to prevent overheating. After each type and size of tungsten has been tested and an operating range established, repeat the procedure using the next thicker metal. Repeat this procedure until you have set up the operating ranges for all of the metals and tungstens you will be using. Turn off the welding machine, shielding gas, and cooling water and clean up your work area when you are finished welding.

Complete a copy of the "Student Welding Report" listed in Appendix I or provided by your instructor. ◆

FIGURE 16-9 Lower the welding current to zero or as low as possible. Larry Jeffus

Gas Flow

The gas preflow and postflow times required to protect both the tungsten and the weld depend upon the following factors:

- Wind or draft speed
- Nozzle size used
- Tungsten size used
- Amperage
- Joint design
- Welding position
- Type of metal welded

The weld quality can be adversely affected by improper gas flow settings. The lowest possible gas flow rates and the shortest preflow or postflow time can help reduce the cost of welding by saving the expensive shielding gas.

In Experiment 16-2, the minimum and maximum gas flow settings for each nozzle size, tungsten size, and amperage setting will be determined. The chart a welder prepares based on experiments is to improve that welder's skill and welding technique. Charts may differ slightly from one welder to another. As a welder's skill improves, the chart may change. As experience is gained, a welder will learn how to set the gas flow effectively without the need for this chart.

The minimum flow rates and times must be increased to weld in drafty areas or for out-of-position welds. The rates and times can be somewhat lower for tee joints or welds made in tight areas. The maximum flow rates must never be exceeded. Exceeding these flow rates causes weld contamination and increases the rejection rate.

EXPERIMENT 16-2

Setting Gas Flow

Using a properly set-up GTA welding machine and torch, proper safety protection, one of each available tungsten size, metal that is 16 gauge to 1/4 in. (6 mm) thick, and the welding current chart developed in Experiment 16-1, you will work with a small group of students to make a chart of the minimum and maximum flow rates and times for each nozzle size, tungsten size, and amperage setting. An assistant will also be needed to change and record the flow rate while you work.

Set the machine welding power switch for DCEN (DCSP). Set the amperage to the lowest setting for the size of tungsten used. Set the prepurge time to 0 and postpurge at 20 sec, **Figure 16-11**. Turn on the main power. With the torch held so that it cannot short out, depress the remote control to start the shielding gas flow and set the flow at 20 cfh (9 L/min). Switch the high frequency to start. All other functions, such as pulse, hot start, slope, and so on, should be in the off position.

Starting with the smallest nozzle and tungsten size, strike an arc and establish a molten pool on a piece of metal in the flat position. Watch the molten weld pool and

FIGURE 16-11 Setting the postpurge timer. Larry Jeffus

tungsten for signs of oxide formation as another person slowly lowers the gas flow rate. Have that person note this setting (where oxide formation begins), **Figure 16-12**, as the minimum flow rate on the chart next to the nozzle size and current setting. Now slowly increase the flow rate until the molten pool starts to be blown back or oxides start forming. This setting should be noted on the chart as the maximum flow rate for this current and nozzle size, **Table 16-2**. Lower the flow to a rate of 2 cfh or 3 cfh (1 L/min or 2 L/min) above the minimum value noted on the chart and then stop the arc. Record the length of time from the point when the arc stops and the tungsten stops glowing as the postflow time. Repeat this test at a medium and then high current setting for this nozzle and tungsten size. When using high current settings, it may be necessary to move the torch or use thicker plate to prevent burnthrough.

Repeat this test procedure with each available nozzle and tungsten size. Stainless steel or aluminum is preferred for this experiment because the oxides are more quickly noticeable than when mild steel is used. If aluminum is used, the welding current must be AC, and the high-frequency switch should be set on continuous.

To establish the minimum preflow time for each nozzle and tungsten size, set the amperage to a medium-high setting.

FIGURE 16-12 Oxides forming due to inadequate gas shielding. Larry Jeffus

Electrode and Nozzle Size	Flow Rate					Postflow Time		
	Too Low	**Low**	**Good**	**High**	**Too High**	**Too Short**	**OK**	**Too Long**

TABLE 16-2 Sample Chart for Setting Shielding Gas Flow Rate and Time

Hold the torch above the metal so that an arc will be instantly started. Set the preflow timer to 0 and the gas flow to just above the minimum value noted on the chart. Quickly strike an arc on metal thin enough to cause a weld pool to form instantly at that power setting. Stop the arc and examine the weld pool and tungsten for oxides. Repeat this procedure, increasing the preflow time until no oxides are formed on either the plate or tungsten. Record this time on the chart as the minimum preflow time. Repeat this test with each available nozzle and tungsten size. Turn off the welding machine, shielding gas, and cooling water and clean up your work area when you are finished welding.

Complete a copy of the "Student Welding Report" listed in Appendix I or provided by your instructor. ◆

Practice Welds

The practice welds are grouped according to the weld position and type of joint and not by the type of metal. The order in which a person decides to do the welds is that person's choice. It is suggested that the stringer beads be done in each metal and position before the different joints are tried. Each metal has its own characteristics that may make one metal easier for a person to work on than another metal.

Mild steel is inexpensive and requires the least amount of cleaning. Slight changes in the metal have little effect on the welding skill required. Stainless steel is somewhat affected by cleanliness, requiring little preweld cleaning. However, the weld pool shows overheating or poor gas coverage. With aluminum, cleanliness is a critical factor. Oxides on aluminum may prevent the molten weld pool from flowing together. The surface tension helps hold the metal in place, giving excellent bead contour and appearance.

The degree of difficulty a welder encounters with each of these metals depends upon the individual's experience. Try each weld with each metal to determine which metal will be easiest to master first. The type of welding machine and materials used will also affect a welder's progress. Practice will help welders overcome any obstacle to their progress.

Low Carbon and Mild Steels Low carbon and mild steel are two basic steel classifications. These steels are the most common type of steels a new GTA welding student will experience welding. Carbon is the primary alloy in these classifications of steel, and it ranges from 0.15% or less for low carbon and 0.15% to 0.30% for mild steel. The GTA welding techniques required for welding steels

in both classifications are the same. You start with an EWP, EWTh-1, EWTh-2, or EWCe-2 pointed tungsten, **Figure 16-13**, with the welding machine set for DCEN (DCSP) welding current. **Table 16-3** lists the types of filler metal used for both low carbon and mild steels.

During the manufacturing process, sometimes small pockets of primarily carbon dioxide gas become trapped inside low carbon and mild steels. There are only a few molecules of gas trapped inside the microscopic pockets within the steel, so they do not affect the steel strength. During most other types of welding, fluxes on the filler metal capture these gas pockets and they are removed. However, because GTA filler metals do not have fluxes, when these gas pockets become hot during welding, they

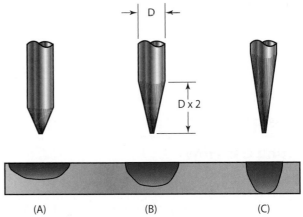

FIGURE 16-13 Tungsten tip shape for mild steel or stainless steel. © Cengage Learning 2012

SAE No.	Carbon %	AWS Filler Metal No.
Low Carbon		
1006	0.08 max	RG60 or ER70S-3
1008	0.10 max	RG60 or ER70S-3
1010	0.08 to 0.15	RG60 or ER70S-3
Mild Steel		
1015	0.11 to 0.16	RG60 or ER70S-3
1016	0.13 to 0.18	RG60 or ER70S-3
1018	0.15 to 0.20	RG60 or ER70S-3
1020	0.18 to 0.23	RG60 or ER70S-3
1025	0.22 to 0.29	RG60 or ER70S-3

TABLE 16-3 Filler Metals for Low Carbon and Mild Steels

THINK GREEN
Conserve Filler Metal

It is hard to use up the full length of filler rod. Depending on your skill, the joint type, or metal thickness, there may be from 3 in. to 6 in. (75 mm to 150 mm) of filler rod left as a stub. Welding mild steel stubs together is easier than welding stainless steel and aluminum stubs, but all three can be joined in the welding lab. However, it is not acceptable to join rod stubs in many welding shops that are working to a high standard or code. In this case, too many impurities and oxides might result in a weld failing inspection.

expand and can sometimes cause weld porosity. You are most likely to see porosity caused by these gases when you are not using a filler metal. Most GTA filler metals have some alloys, called deoxidizers, that can help prevent porosity caused by gases trapped in the base metal. RG45 gas welding rods do not have these deoxidizers and are not recommended for GTA welding.

Stainless Steel The setup and manipulation techniques required for stainless steel are nearly the same as those for low carbon and mild steels, and skills transfer is easy. The major difference is that most welds on steels do not show the effects of contamination as easily as do welds on stainless steels. To make a weld on stainless steel you must do a better job of precleaning the base metal and filler metal; make sure you have adequate shielding gas coverage and do not overheat the weld.

The most common sign that there is a problem with a stainless steel weld is the bead color after the weld. The greater the contamination, the darker the color. The exposure of the weld bead to the atmosphere before it has cooled will also change the bead color. It is impossible, however, to determine the extent of contamination of a weld with only visual inspection. Both light-colored and dark-colored welds may not be free from oxides. Thus, it is desirable to take the time and necessary precautions to make welds that are no darker than dark blue, **Table 16-4**. Welds with only slight oxide layers are better for multiple passes.

Using a low arc current setting with faster travel speeds is important when welding stainless steel because some stainless steels are subject to carbide precipitation. Carbide precipitation, the combining of carbon with chromium, occurs in some stainless steels when they are kept at a temperature between 800°F and 1500°F (625°C and 815°C) for a long time. There are a number of ways of controlling carbide precipitation by adding alloys to the stainless steel or by lowering the percentage of carbon. During welding, special alloy filler metals can be used to control the problem (see Chapter 27), but the most important thing a welder can do is travel fast and use as little welding heat as possible. For more information on controlling carbide precipitation in stainless steel, see Chapter 25.

Black crusty spots may appear on weld beads. These spots are often caused by improper cleaning of the filler rod or failure to keep the end of the rod inside the shielding gas.

Table 16-5 lists some common types of stainless steels and the recommended filler metals; see Chapter 27 for a more complete listing.

Aluminum Aluminum is GTA welded using an EWP, EWZr, EWCe-2, or EWLa-1 rounded tip tungsten, **Figure 16-14**, with the welding machine set for ACHF welding current. The alternating current provides good arc cleaning, and the continuous high frequency restarts the arc as the current changes direction.

The molten aluminum weld pool has high surface tension, which allows large weld beads to be controlled easily. **Table 16-6** lists some basic types of filler metal used for aluminum welding; see Chapter 27 for a more complete listing.

AISI No.	AWS Filler No.	AISI No.	AWS Filler No.
303	ER308	310	ER310
304	ER308	316	ER316L
304L	ER308L	316L	ER316L
309	ER309	410	ER410

TABLE 16-5 Filler Metals for Stainless Steels

FIGURE 16-14 Tungsten tip shape for aluminum.
© Cengage Learning 2012

Surface Color	Approximate Temperature at Which Color Is Formed	
	°F	(°C)
Light straw	400	(200)
Tan	450	(230)
Brown	525	(275)
Purple	575	(300)
Dark blue	600	(315)
Black	800	(425)

TABLE 16-4 Temperatures at Which Various Colored Oxide Layers Form on Steel

AISI No.	AWS Filler No.	AISI No.	AWS Filler No.
1100	ER1100	3004	ER4043
3003	ER4043	6061	ER4043

TABLE 16-6 Filler Metals for Aluminum Alloys

The high thermal conductivity of the metal may make starting a weld on thick sections difficult without first preheating the base metal. In most cases, the preheat temperature is around 300°F (150°C) but will vary depending on metal thickness and alloy type. Specific preheat temperatures are available from the metal supplier.

The processes of cleaning and keeping the metal clean take a lot of time. Removal of the oxide layer is easy using a chemical or mechanical method. Ten minutes after cleaning, however, the oxide layer may again be thick enough to require recleaning. The oxide that forms reduces the ability of the weld pool to flow together. Keep your hands and gloves clean and oil free so the base metal or filler rods do not become recontaminated.

Although aluminum resists oxidation at room temperature, it rapidly oxidizes at welding temperatures. If the filler rod is not kept inside the shielding gas, it will quickly oxidize, but, because of the low melting temperature of the filler rod, the end will melt before it is added to the weld pool if it is held too closely to the arc, **Figure 16-15** and **Figure 16-16**.

Metal Preparation Both the base metal and the filler metal used in the GTAW process must be thoroughly cleaned before welding. Contamination left on the metal will be deposited in the weld because there is no flux to remove it. Oxides, oils, and dirt are the most common types of contaminants. They can be removed mechanically or chemically. Mechanical metal cleaning may be done by grinding, wire brushing, scraping, machining, or filing. Chemical cleaning may be done by using acids, alkalies, solvents, or detergents.

> **CAUTION**
>
> The manufacturer's recommendations for using these products must be followed. Failure to do so may result in chemical burns, fires, fumes, or other safety hazards that could lead to serious injury. If anyone comes in contact with any chemicals, immediately refer to the material safety data sheet (MSDS) for the proper corrective action.

FIGURE 16-15 Aluminum filler being correctly added to the molten weld pool. Larry Jeffus

FIGURE 16-16 Filler rod being melted before it is added to the molten pool. Larry Jeffus

PRACTICE 16-1

Stringer Beads, Flat Position, on Mild Steel

Using a properly set-up and adjusted GTA welding machine on DCEN, proper safety protection, and one or more pieces of mild steel 6 in. (152 mm) long and 16 gauge and 1/8 in. (3 mm) thick, you will push a weld pool in a straight line down the plate, **Figure 16-17**. Maintain uniform weld pool size and penetration.

- Starting at one end of the piece of metal that is 1/8 in. (3 mm) thick, hold the torch as close as possible to a 90° angle.

- Lower your hood, strike an arc, and establish a weld pool.

- Move the torch in a stepping or circular oscillation pattern down the plate toward the other end, **Figure 16-18**.

- If the size of the weld pool changes, speed up or slow down the travel rate to keep the weld pool the same size for the entire length of the plate.

The ability to maintain uniformity in width and keep a straight line increases as you are able to see more than just the weld pool. As your skill improves, you will relax, and your field of vision will increase.

Repeat the process using both thicknesses of metal until you can consistently make the weld visually defect free. Turn off the welding machine, shielding gas, and cooling water and clean up your work area when you are finished welding.

Complete a copy of the "Student Welding Report" listed in Appendix I or provided by your instructor. ◆

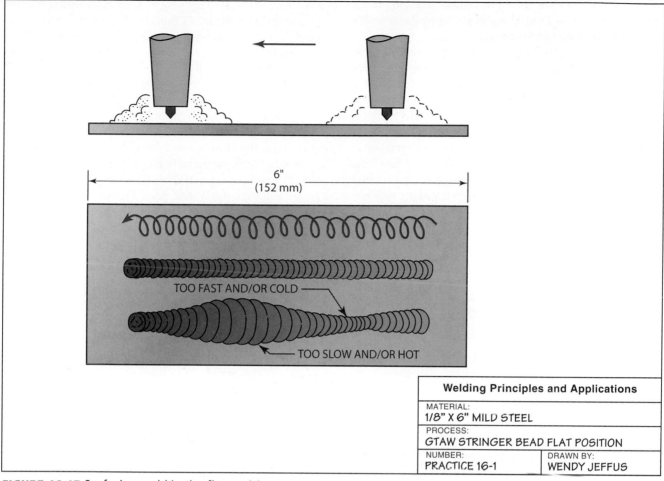

Welding Principles and Applications		
MATERIAL: 1/8" X 6" MILD STEEL		
PROCESS: GTAW STRINGER BEAD FLAT POSITION		
NUMBER: PRACTICE 16-1		DRAWN BY: WENDY JEFFUS

FIGURE 16-17 Surfacing weld in the flat position. © Cengage Learning 2012

FIGURE 16-18 Surfacing weld. Larry Jeffus

will push a molten weld pool in a straight line down the plate, keeping the width and penetration uniform.

To keep the formation of oxides on the bead to a minimum, a **chill plate** (a thick piece of metal used to absorb heat) may be required. Another method is to make the bead using as low a heat input as possible. When the weld is finished, the weld bead should be no darker than dark blue.

Repeat the process using both thicknesses of metal until you can consistently make the weld visually defect free. Turn off the welding machine, shielding gas, and cooling water and clean up your work area when you are finished welding.

Complete a copy of the "Student Welding Report" listed in Appendix I or provided by your instructor. ◆

PRACTICE 16-2

Stringer Beads, Flat Position, on Stainless Steel

Using the same equipment and material thicknesses as listed in Practice 16-1 and one or more pieces of stainless steel 6 in. (152 mm) long and 1/4 in. (6 mm) thick, you

PRACTICE 16-3

Stringer Beads, Flat Position, on Aluminum

Using the same equipment, setup for AC, and procedure as listed in Practice 16-1 and one or more pieces of aluminum 6 in. (152 mm) long and 1/16 in. (2 mm),

1/8 in. (3 mm), and 1/4 in. (6 mm) thick, you will push a weld pool in a straight line, maintaining uniform width and penetration for the length of the plate.

A high current setting will allow faster travel speeds. The faster speed helps control excessive penetration. Hot cracking may occur on some types of aluminum after a surfacing weld. This is not normally a problem when filler metal is added. If hot cracking occurs during this practice, do not be concerned.

Repeat the process using all thicknesses of metal until you can consistently make the weld visually defect free. Turn off the welding machine, shielding gas, and cooling water and clean up your work area when you are finished welding.

Complete a copy of the "Student Welding Report" listed in Appendix I or provided by your instructor. ◆

PRACTICE 16-4

Flat Position, Using Mild Steel, Stainless Steel, and Aluminum

For this practice, you will need a properly set-up and adjusted GTA welding machine; proper safety protection;

filler rods 36 in. (0.9 m) long × 1/16 in. (2 mm), 3/32 in. (2.4 mm), and 1/8 in. (3 mm) in diameter; one or more pieces of mild steel, stainless steel, and aluminum, 6 in. (152 mm) long × 1/16 in. (2 mm) and 1/8 in. (3 mm) thick; and aluminum plate 1/4 in. (6 mm) thick, **Table 16-7**, **Table 16-8**, and **Table 16-9**. In this practice, you will make a straight stringer bead, 6 in. (152 mm) long, that is uniform in width, reinforcement, and penetration, **Figure 16-19**. Use DCEN current on the steel and AC current on the aluminum.

Starting with the metal that is 1/8 in. (3 mm) thick and the filler rod having a 3/32-in. (2.4-mm) diameter, strike an arc and establish a weld pool, **Figure 16-20**. Move the torch in a circle as in the practice beading. When the torch is on one side, add filler rod to the other side of the molten weld pool, **Figure 16-21A and B**. The end of the rod can be held lightly in the leading edge of the molten weld pool, or it can be dipped into the molten weld pool. If you are using the dipping method, be sure not to allow the tip to melt and drip into the weld pool, **Figure 16-22**. Change to another size filler rod and determine its effect on the weld pool.

Maintain a smooth and uniform rhythm as filler metal is added. This will help to keep the bead uniform. Vary

Tungsten Electrode			Welding Power			Shielding Gas		Nozzle	Filler Metal	
Type	Size	Tip	Amp	Current	HF	Type	Flow	Size	Type	Size
EWTh-1 or EWTh-2	1/16" (2 mm)	Point	50 to 100	DCEN DCSP	Start or auto	Argon	16 cfh 7 L/min	3/8" (10 mm)	RG60 or ER70S-3	1/16"–3/32" (2–2.4 mm)
EWTh-1 or EWTh-2	3/32" (2.4 mm)	Point	70 to 150	DCEN DCSP	Start or auto	Argon	16 cfh 7 L/min	3/8" (10 mm)	RG60 or ER70S-3	1/16"–3/32" (2–2.4 mm)
EWTh-1 or EWTh-2	1/8" (3 mm)	Point	90 to 250	DCEN DCSP	Start or auto	Argon	20 cfh 9 L/min	1/2" (13 mm)	RG60 or ER70S-3	3/32"–1/8" (2.4–3 mm)

TABLE 16-7 Suggested Setting for GTA Welding of Mild Steel

Tungsten Electrode			Welding Power			Shielding Gas		Nozzle	Filler Metal	
Type	Size	Tip	Amp	Current	HF	Type	Flow	Size	Type	Size
EWTh-1 or EWTh-2	1/16" (2 mm)	Point	70 to 100	DCEN DCSP	Start or auto	Argon	16 cfh 7 L/min	3/8" (10 mm)	ER308 or ER316	1/16"–3/32" (2–2.4 mm)
EWTh-1 or EWTh-2	3/32" (2.4 mm)	Point	70 to 150	DCEN DCSP	Start or auto	Argon	16 cfh 7 L/min	3/8" (10 mm)	ER308 or ER316	1/16"–3/32" (2–2.4 mm)
EWTh-1 or EWTh-2	1/8" (3 mm)	Point	90 to 250	DCEN DCSP	Start or auto	Argon	20 cfh 9 L/min	1/2" (13 mm)	ER308 or ER316	3/32"–1/8" (2.4–3 mm)

TABLE 16-8 Suggested Setting for GTA Welding of Stainless Steel

Tungsten Electrode			Welding Power			Shielding Gas		Nozzle	Filler Metal	
Type	Size	Tip	Amp	Current	HF	Type	Flow	Size	Type	Size
EWP or EWZr	1/16"	Round 2 mm	50 to 90	AC	Continues or on	Argon	17 cfh 8 L/min	7/16" (11 mm)	ER1100 or ER4043	1/16"– 3/32" (2–2.4 mm)
EWP or EWZr	3/32"	Round 2.4 mm	80 to 130	AC	Continues or on	Argon	20 cfh 9 L/min	1/2" (13 mm)	ER1100 or ER4043	1/16"– 3/32" (2–2.4 mm)
EWP or EWZr	1/8"	Round 3 mm	100 to 200	AC	Continues or on	Argon	20 cfh 9 L/min	5/8" (16 mm)	ER1100 or ER4043	3/32"– 1/8" (2.4–3 mm)

TABLE 16-9 Suggested Setting for GTA Welding of Aluminum

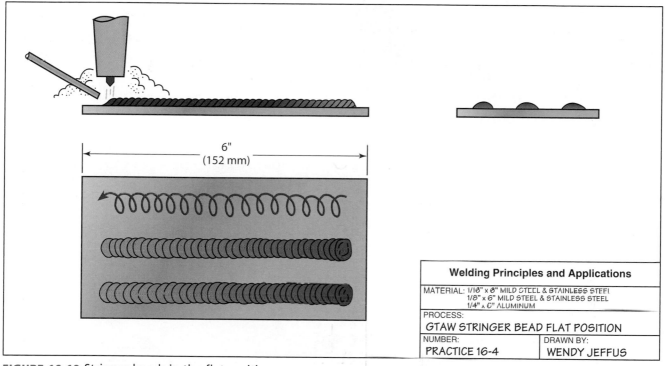

FIGURE 16-19 Stringer beads in the flat position. © Cengage Learning 2012

FIGURE 16-20 Establish a molten weld pool and dip the filler rod into it. Larry Jeffus

the rhythms to determine which one is easiest for you. If the rod sticks, move the torch toward the rod until it melts free.

When the full 6-in. (152-mm) long weld bead is completed, cool and inspect it for uniformity and defects. Repeat the process using all thicknesses of metal until you can consistently make the weld visually defect free. Turn off the welding machine, shielding gas, and cooling water and clean up your work area when you are finished welding.

Complete a copy of the "Student Welding Report" listed in Appendix I or provided by your instructor. ◆

FIGURE 16-21 Note the difference in the weld produced when different size filler rods are used. Larry Jeffus

MOVE THE FILLER METAL BACK AND FORTH

MOVE THE TORCH BACK AND FORTH

FIGURE 16-22 Move the electrode back as the filler rod is added. Larry Jeffus

PRACTICE 16-5

Outside Corner Joint, 1G Position, Using Mild Steel, Stainless Steel, and Aluminum

Using the same equipment and materials as listed in Practice 16-4, weld an outside corner joint in the flat position, **Figure 16-23**.

- Place one of the pieces of metal flat on the table and hold or brace the other piece of metal horizontally on it.
- Tack weld both ends of the plates together, **Figure 16-24**.
- Set the plates up and add two or three more tack welds on the joint as required, **Figure 16-25**.

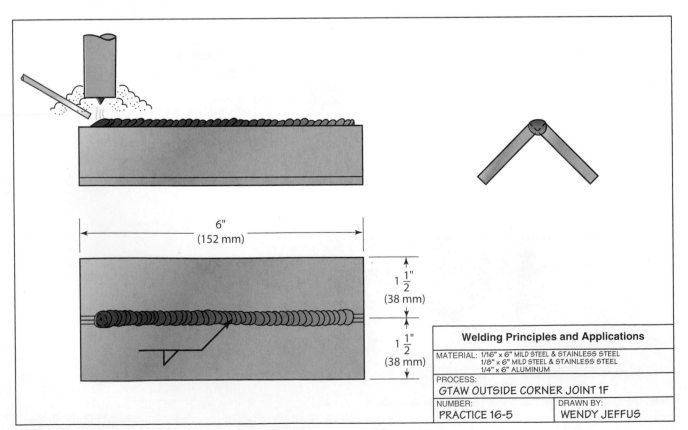

6"
(152 mm)

1 1/2"
(38 mm)

1 1/2"
(38 mm)

Welding Principles and Applications

MATERIAL: 1/16" x 6" MILD STEEL & STAINLESS STEEL 1/8" x 6" MILD STEEL & STAINLESS STEEL 1/4" x 6" ALUMINUM	
PROCESS: GTAW OUTSIDE CORNER JOINT 1F	
NUMBER: PRACTICE 16-5	DRAWN BY: WENDY JEFFUS

FIGURE 16-23 Outside corner joint in the flat position. © Cengage Learning 2012

FIGURE 16-24 Tack weld. Note the good fusion at the start and crater fill at the end. Larry Jeffus

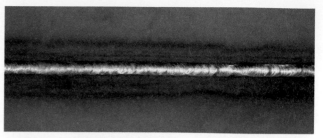

FIGURE 16-26 Outside corner joint. Note precleaning along weld. Larry Jeffus

FIGURE 16-25 Outside corner tack welded together. Larry Jeffus

PRACTICE 16-6

Butt Joint, 1G Position, Using Mild Steel, Stainless Steel, and Aluminum

Using the same equipment and materials as listed in Practice 16-4, you will weld a butt joint in the flat position, **Figure 16-27**.

Place the metal flat on the table and tack weld both ends together, **Figure 16-28**. Two or three additional tack welds can be made along the joint as needed. Starting at one end, make a uniform weld along the joint. Add filler metal as required to make a uniform weld.

■ Starting at one end, make a uniform weld, adding filler metal as needed. In **Figure 16-26**, note the metal areas that are precleaned before the weld is made.

Repeat each weld as needed until all are mastered. Turn off the welding machine, shielding gas, and cooling water and clean up your work area when you are finished welding.

Complete a copy of the "Student Welding Report" listed in Appendix I or provided by your instructor. ◆

Repeat the process using all thicknesses of metal until you can consistently make the weld visually defect free. Turn off the welding machine, shielding gas, and cooling water and clean up your work area when you are finished welding.

Complete a copy of the "Student Welding Report" listed in Appendix I or provided by your instructor. ◆

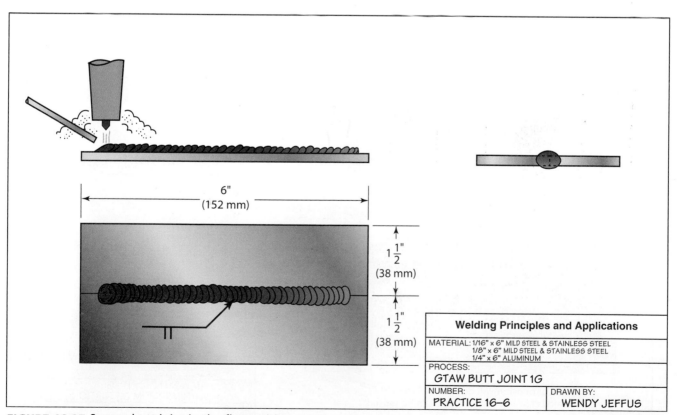

6"
(152 mm)

$1\frac{1}{2}$"
(38 mm)

$1\frac{1}{2}$"
(38 mm)

Welding Principles and Applications

MATERIAL: 1/16" x 6" MILD STEEL & STAINLESS STEEL
1/8" x 6" MILD STEEL & STAINLESS STEEL
1/4" x 6" ALUMINUM

PROCESS:
GTAW BUTT JOINT 1G

NUMBER:
PRACTICE 16-6

DRAWN BY:
WENDY JEFFUS

FIGURE 16-27 Square butt joint in the flat position. © Cengage Learning 2012

FIGURE 16-28 Tack weld on butt joint. Larry Jeffus

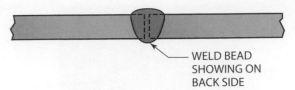

FIGURE 16-29 100% weld penetration. © Cengage Learning 2012

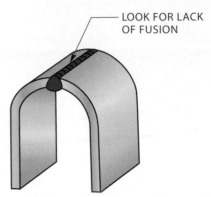

FIGURE 16-30 Bend the 1-in. (25-mm) strip of butt joint backward and look at the root for 100% penetration. © Cengage Learning 2012

PRACTICE 16-7

Butt Joint, 1G Position, with 100% Penetration, to Be Tested, Using Mild Steel, Stainless Steel, and Aluminum

Using the same equipment and materials as listed in Practice 16-4, you will weld a butt joint with 100% penetration, **Figure 16-29**, along the entire 6-in. (152-mm) length of the joint. After the weld is completed, shear out strips 1 in. (25 mm) wide and bend-test them as shown in **Figure 16-30**.

Repeat each weld until all have 100% root penetration. Turn off the welding machine, shielding gas, and cooling water and clean up your work area when you are finished welding.

Complete a copy of the "Student Welding Report" listed in Appendix I or provided by your instructor. ◆

PRACTICE 16-8

Butt Joint, 1G Position, with Minimum Distortion, Using Mild Steel, Stainless Steel, and Aluminum

Using the same equipment and materials as listed in Practice 16-4, you will weld a flat butt joint, while controlling both distortion and penetration, **Figure 16-31**.

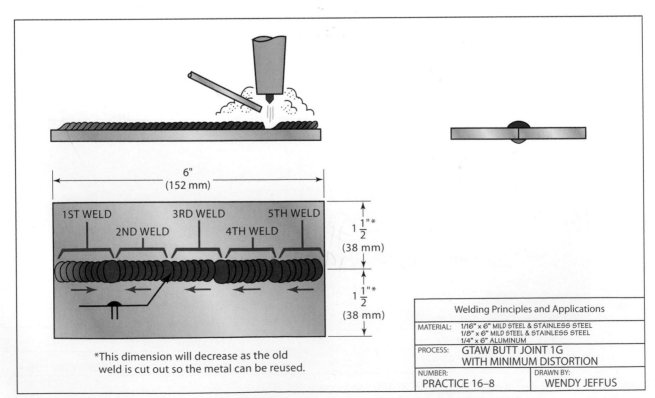

*This dimension will decrease as the old weld is cut out so the metal can be reused.

Welding Principles and Applications		
MATERIAL:	1/16" x 6" MILD STEEL & STAINLESS STEEL 1/8" x 6" MILD STEEL & STAINLESS STEEL 1/4" x 6" ALUMINUM	
PROCESS:	GTAW BUTT JOINT 1G WITH MINIMUM DISTORTION	
NUMBER: PRACTICE 16–8		DRAWN BY: WENDY JEFFUS

FIGURE 16-31 Square butt joint in the flat position with minimum distortion. © Cengage Learning 2012

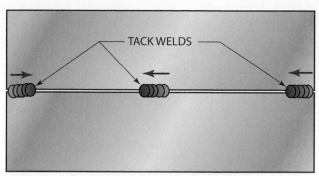

FIGURE 16-32 Tack welds on a butt joint. © Cengage Learning 2012

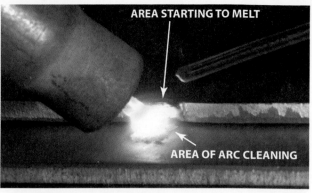

FIGURE 16-34 Be sure both the top and bottom pieces are melted before adding filler metal. Larry Jeffus

Tack weld the plates together as shown in **Figure 16-32**. Using a back-stepping weld sequence, make a series of welds approximately 1 in. (25 mm) long along the joint. Be sure to fill each weld crater adequately to reduce crater cracking.

Repeat the process using all thicknesses of metal until you can consistently make the weld visually defect free. Turn off the welding machine, shielding gas, and cooling water and clean up your work area when you are finished welding.

Complete a copy of the "Student Welding Report" listed in Appendix I or provided by your instructor. ◆

PRACTICE 16-9

Lap Joint, 1F Position, Using Mild Steel, Stainless Steel, and Aluminum

Using the same equipment and materials as listed in Practice 16-4, you will weld a lap joint in the flat position, **Figure 16-33**.

Place the two pieces of metal flat on the table with an overlap of 1/4 in. (6 mm) to 3/8 in. (10 mm). Hold the pieces

of metal tightly together and tack weld them as shown in **Figure 16-34** and **Figure 16-35**. Starting at one end, make a uniform fillet weld along the joint. Both sides of the joint can be welded.

Repeat the process using all thicknesses of metal until you can consistently make the weld visually defect free. Turn off the welding machine, shielding gas, and cooling water and clean up your work area when you are finished welding.

Complete a copy of the "Student Welding Report" listed in Appendix I or provided by your instructor. ◆

PRACTICE 16-10

Lap Joint, 1F Position, to Be Tested, Using Mild Steel, Stainless Steel, and Aluminum

Using the same equipment and materials as listed in Practice 16-4, you will make a fillet weld on one side of a lap

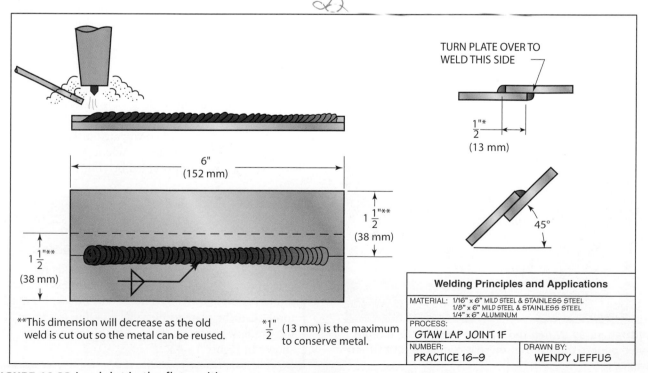

**This dimension will decrease as the old weld is cut out so the metal can be reused.

*$\frac{1}{2}$" (13 mm) is the maximum to conserve metal.

FIGURE 16-33 Lap joint in the flat position. © Cengage Learning 2012

FIGURE 16-35 Oxides form during tack welding. Do not complete the tack welds. These oxides will become part of the finished weld if the tack is completed. Larry Jeffus

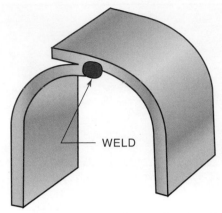

FIGURE 16-38 Bend the 1-in. (25-mm) strip of lap joint backward and look at the root for 100% penetration.
© Cengage Learning 2012

joint and test it for 100% root penetration, **Figure 16-36**, **Figure 16-37**, and **Figure 16-38**. After the weld is completed, shear out strips 1 in. (25 mm) wide on one side of the joint and bend-test them as shown in **Figure 16-39**.

Repeat each weld until all have 100% root penetration. Turn off the welding machine, shielding gas, and cooling water and clean up your work area when you are finished welding.

Complete a copy of the "Student Welding Report" listed in Appendix I or provided by your instructor. ◆

PRACTICE 16-11

Tee Joint, 1F Position, Using Mild Steel, Stainless Steel, and Aluminum

Using the same equipment and materials as listed in Practice 16-4, you will weld a tee joint in the flat position, **Figure 16-40**.

Place one of the pieces of metal flat on the table and hold or brace the other piece of metal horizontally on it. Tack weld both ends of the plates together, **Figure 16-41**. Set up the plates in the flat position and add two or three more tack welds to the joint as required, **Figure 16-42**.

On the metal that is 1/16 in. (1.5 mm) thick, it may not be possible to weld both sides, but on thicker material a fillet weld can usually be made on both sides. The exception to this is if carbide precipitation occurs on the stainless steel during welding.

Starting at one end, make a uniform weld, adding filler metal as needed.

Repeat the process using all thicknesses of metal until you can consistently make the weld visually defect free. Turn off the welding machine, shielding gas, and cooling water and clean up your work area when you are finished welding.

Complete a copy of the "Student Welding Report" listed in Appendix I or provided by your instructor. ◆

FIGURE 16-36 A notch indicates that the root was not properly melted and fused. Larry Jeffus

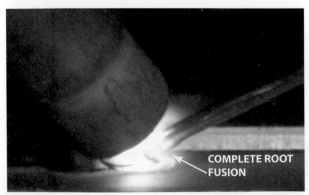

FIGURE 16-37 Watch the leading edge of the molten weld pool to ensure that there is complete root fusion.
Larry Jeffus

THINK GREEN
Recycle Weld Coupon

Mild steel weld coupons can have the welds removed using an OFC torch, but welds on stainless and aluminum coupons will have to be cut out with a PAC torch. Cut the plate as close as possible to the weld; for example, on a tee joint cut the vertical part as shown in **Figure 16-43**.

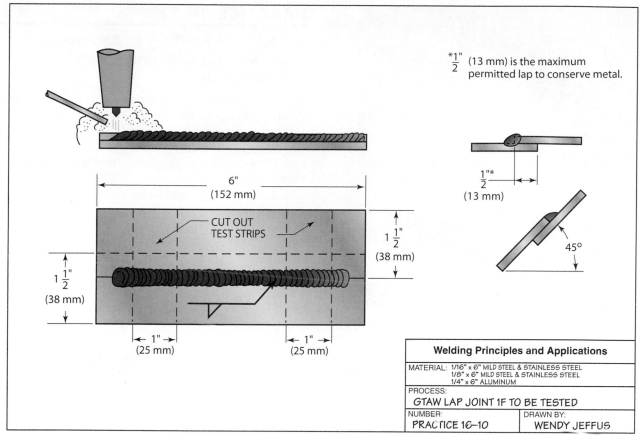

FIGURE 16-39 Remove strips and test for root fusion. © Cengage Learning 2012

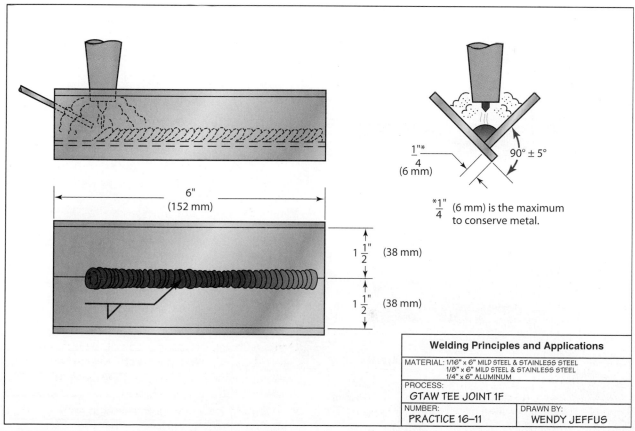

FIGURE 16-40 Tee joint in the flat position. © Cengage Learning 2012

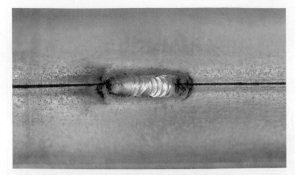

FIGURE 16-41 Tack weld on a tee joint. Larry Jeffus

FIGURE 16-42 Keep the tack welds small so that they will not affect the weld. Larry Jeffus

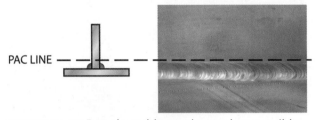

PAC LINE

FIGURE 16-43 Recycle weld test plates when possible. Larry Jeffus

PRACTICE 16-12

Tee Joint, 1F Position, to Be Tested, Using Mild Steel, Stainless Steel, and Aluminum

Using the same equipment and materials as listed in Practice 16-4, you will weld a tee joint and test it for 100% root penetration, **Figure 16-44**. After the weld is completed, cut or shear out strips 1 in. (25 mm) wide and bend-test them as shown in **Figure 16-45** and **Figure 16-46**.

Repeat each weld until all have 100% root penetration. Turn off the welding machine, shielding gas, and cooling water and clean up your work area when you are finished welding.

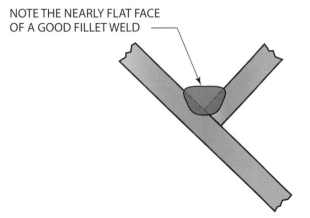

NOTE THE NEARLY FLAT FACE OF A GOOD FILLET WELD

FIGURE 16-44 100% weld penetration. © Cengage Learning 2012

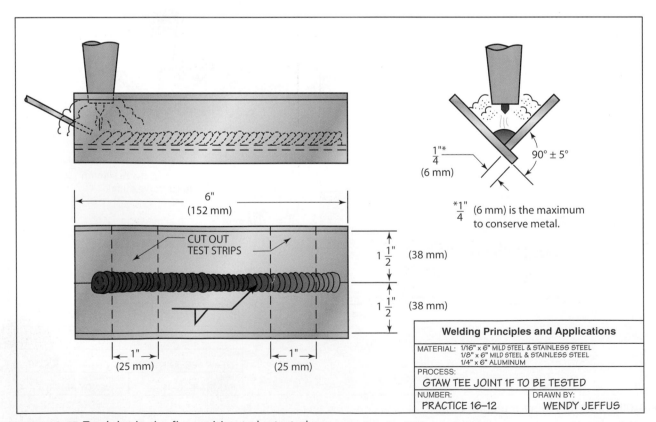

6" (152 mm)

CUT OUT TEST STRIPS

$1\frac{1}{2}$" (38 mm)

$1\frac{1}{2}$" (38 mm)

1" (25 mm)

1" (25 mm)

$\frac{1}{4}$"* (6 mm)

90° ± 5°

*$\frac{1}{4}$" (6 mm) is the maximum to conserve metal.

Welding Principles and Applications

MATERIAL:	1/16" x 6" MILD STEEL & STAINLESS STEEL
	1/8" x 6" MILD STEEL & STAINLESS STEEL
	1/4" x 6" ALUMINUM

PROCESS:
GTAW TEE JOINT 1F TO BE TESTED

| NUMBER: | DRAWN BY: |
| PRACTICE 16-12 | WENDY JEFFUS |

FIGURE 16-45 Tee joint in the flat position to be tested. © Cengage Learning 2012

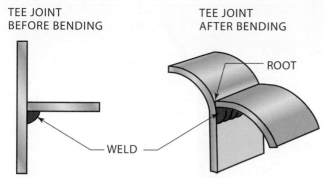

FIGURE 16-46 Bend the 1-in. (25-mm) strip of the tee joint backward and look at the root for 100% penetration. © Cengage Learning 2012

Complete a copy of the "Student Welding Report" listed in Appendix I or provided by your instructor. ◆

PRACTICE 16-13

Stringer Bead at a 45° Vertical Angle, Using Mild Steel, Stainless Steel, and Aluminum

Using the same equipment and materials as listed in Practice 16-4, you will make a stringer bead in the vertical up position.

- Starting at the bottom and welding in an upward direction, add the filler metal to the top edge of the weld pool and move the torch in a circle or "C" pattern, **Figure 16-47**. If the weld pool size starts to increase, the "C" pattern can be increased in length or the power can be decreased.

- Watch the weld pool and establish a rhythm of torch movement and addition of rod to keep the weld uniform.

Repeat the process using all thicknesses of metal until you can consistently make the weld visually defect free. Turn off the welding machine, shielding gas, and cooling water and clean up your work area when you are finished welding.

Complete a copy of the "Student Welding Report" listed in Appendix I or provided by your instructor. ◆

PRACTICE 16-14

Stringer Bead, 3G Position, Using Mild Steel, Stainless Steel, and Aluminum

Repeat Practice 16-13. Gradually increase the angle as you develop skill until the weld is being made in the vertical up position, **Figure 16-48**. Repeat the process using all thicknesses of metal until you can consistently make the weld visually defect free. Turn off the welding machine, shielding gas, and cooling water and clean up your work area when you are finished welding.

Complete a copy of the "Student Welding Report" listed in Appendix I or provided by your instructor. ◆

PRACTICE 16-15

Butt Joint at a 45° Vertical Angle, Using Mild Steel, Stainless Steel, and Aluminum

Using the same equipment and materials as listed in Practice 16-4, you will weld a butt joint in the vertical up position.

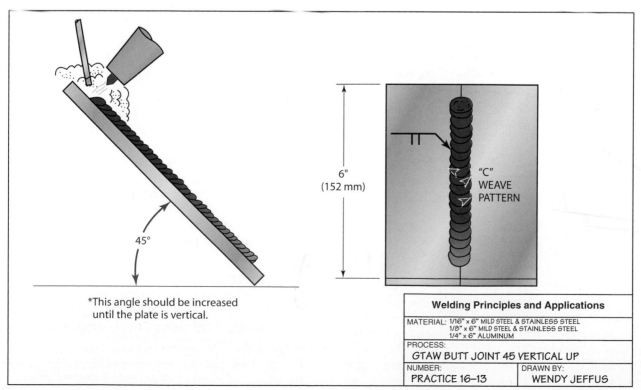

*This angle should be increased until the plate is vertical.

6"
(152 mm)

"C" WEAVE PATTERN

Welding Principles and Applications

MATERIAL:	1/16" x 6" MILD STEEL & STAINLESS STEEL
	1/8" x 6" MILD STEEL & STAINLESS STEEL
	1/4" x 6" ALUMINUM

| PROCESS: |
| GTAW BUTT JOINT 45 VERTICAL UP |

| NUMBER: | DRAWN BY: |
| PRACTICE 16–13 | WENDY JEFFUS |

FIGURE 16-47 45° vertical up. © Cengage Learning 2012

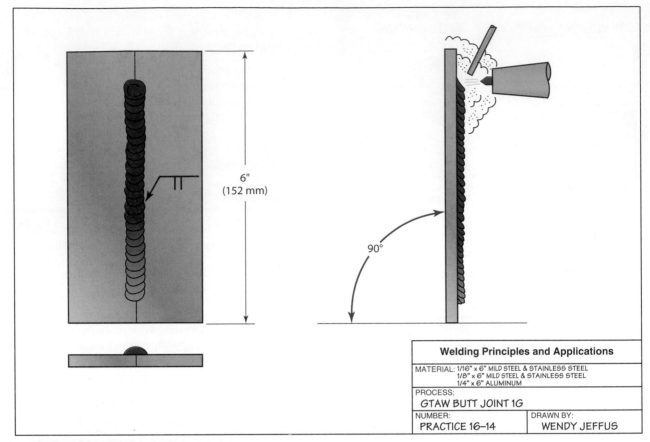

6"
(152 mm)

90°

Welding Principles and Applications

MATERIAL: 1/16" x 6" MILD STEEL & STAINLESS STEEL
1/8" x 6" MILD STEEL & STAINLESS STEEL
1/4" x 6" ALUMINUM

PROCESS:
GTAW BUTT JOINT 1G

NUMBER:	DRAWN BY:
PRACTICE 16–14	WENDY JEFFUS

FIGURE 16-48 Vertical up position. © Cengage Learning 2012

After tack welding the plates together, start the weld at the bottom and weld in an upward direction. The same rhythmic torch and rod movement practiced for the 45° stringer bead should be used to control the weld.

Repeat the process using all thicknesses of metal until you can consistently make the weld visually defect free. Turn off the welding machine, shielding gas, and cooling water and clean up your work area when you are finished welding.

Complete a copy of the "Student Welding Report" listed in Appendix I or provided by your instructor. ◆

PRACTICE 16-16

Butt Joint, 3G Position, Using Mild Steel, Stainless Steel, and Aluminum

Repeat Practice 16-15. Gradually increase the plate angle after each weld. As you develop skill, continue increasing the angle until the weld is being made in the vertical up position. Repeat the process using all thicknesses of metal until you can consistently make the weld visually defect free. Turn off the welding machine, shielding gas, and cooling water and clean up your work area when you are finished welding.

Complete a copy of the "Student Welding Report" listed in Appendix I or provided by your instructor. ◆

PRACTICE 16-17

Butt Joint, 3G Position, with 100% Penetration, to Be Tested, Using Mild Steel, Stainless Steel, and Aluminum

Repeat Practice 16-16. Make the needed changes in the root opening to allow 100% penetration, **Figure 16-49**. It may be necessary to provide a backing gas to protect the root from atmospheric contamination. After the weld is completed, visually inspect it for uniformity and defects. Then shear out strips 1 in. (25 mm) wide and bend-test them.

Repeat each weld until all have 100% root penetration. Turn off the welding machine, shielding gas, and cooling water and clean up your work area when you are finished welding.

Complete a copy of the "Student Welding Report" listed in Appendix I or provided by your instructor. ◆

PRACTICE 16-18

Lap Joint at a 45° Vertical Angle, Using Mild Steel, Stainless Steel, and Aluminum

Using the same equipment and materials as listed in Practice 16-4, you will make a vertical up fillet weld on a lap joint.

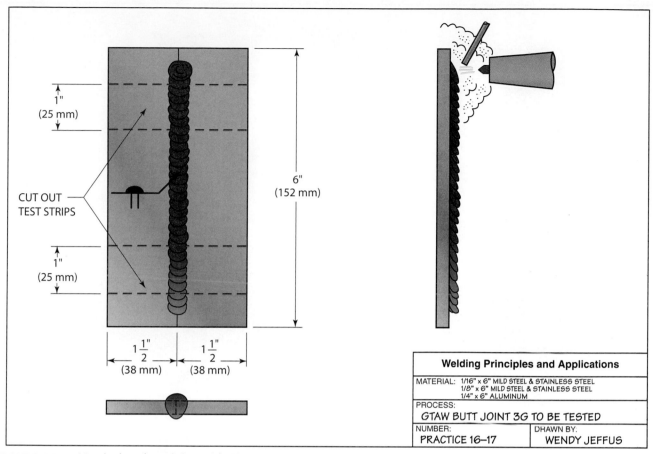

FIGURE 16-49 Vertical up butt joint with 100% penetration. Strips are to be cut for testing. © Cengage Learning 2012

After tack welding the plates together, start the weld at the bottom and weld in an upward direction. It is important to maintain a uniform weld rhythm so that a nice-looking weld bead is formed. It may be necessary to move the torch in and around the base of the weld pool to ensure adequate root fusion, **Figure 16-50**. The filler metal should be added along the top edge of the weld pool near the top plate.

Repeat the process using all thicknesses of metal until you can consistently make the weld visually defect free. Turn off the welding machine, shielding gas, and cooling water and clean up your work area when you are finished welding.

Complete a copy of the "Student Welding Report" listed in Appendix I or provided by your instructor. ◆

PRACTICE 16-19

Lap Joint, 3F Position, Using Mild Steel, Stainless Steel, and Aluminum

Repeat Practice 16-18. Gradually increase the plate angle after each weld as you develop your skill. Increase the angle until the weld is being made in the vertical up position, **Figure 16-51**.

Repeat the process using all thicknesses of metal until you can consistently make the weld visually defect free. Turn off the welding machine, shielding gas, and cooling water and clean up your work area when you are finished welding.

Complete a copy of the "Student Welding Report" listed in Appendix I or provided by your instructor. ◆

WATCH BUILDUP HERE.

FIGURE 16-50 Lap joint. Larry Jeffus

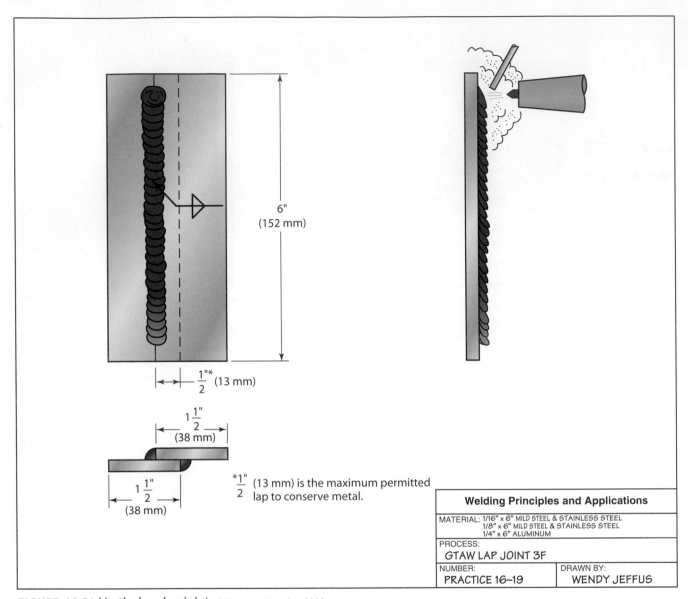

Welding Principles and Applications

MATERIAL: 1/16" x 6" MILD STEEL & STAINLESS STEEL
1/8" x 6" MILD STEEL & STAINLESS STEEL
1/4" x 6" ALUMINUM

PROCESS:
GTAW LAP JOINT 3F

NUMBER:	DRAWN BY:
PRACTICE 16–19	WENDY JEFFUS

FIGURE 16-51 Vertical up lap joint. © Cengage Learning 2012

PRACTICE 16-20

Lap Joint, 3F Position, with 100% Root Penetration, to Be Tested, Using Mild Steel, Stainless Steel, and Aluminum

Repeat Practice 16-19 and make the fillet weld in a lap joint with 100% root penetration. After the weld is completed, visually inspect it for uniformity and defects before shearing out strips 1 in. (25 mm) wide and bend-testing them.

Repeat each weld until all have 100% root penetration. Turn off the welding machine, shielding gas, and cooling water and clean up your work area when you are finished welding.

Complete a copy of the "Student Welding Report" listed in Appendix I or provided by your instructor. ◆

PRACTICE 16-21

Tee Joint at a 45° Vertical Angle, Using Mild Steel, Stainless Steel, and Aluminum

Using the same equipment and materials as listed in Practice 16-4, you will make a vertical up fillet weld on a tee joint.

■ After tack welding the plates together, start the weld at the bottom and weld in an upward direction. The edge of the side plate, **Figure 16-52**, will heat up more quickly than the back plate. This rapid heating often leads to undercutting along this edge of the weld.

■ To control undercutting, keep the arc on the back plate and add the filler metal to the weld pool near the side plate.

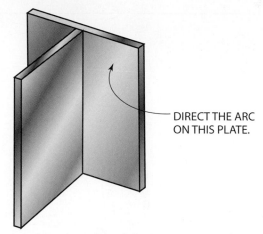

DIRECT THE ARC
ON THIS PLATE.

FIGURE 16-52 The edge of the intersecting plate will heat up faster than the base plate if the heat is not directed away from it. © Cengage Learning 2012

Repeat the process using all thicknesses of metal until you can consistently make the weld visually defect free. Turn off the welding machine, shielding gas, and cooling water and clean up your work area when you are finished welding.

Complete a copy of the "Student Welding Report" listed in Appendix I or provided by your instructor. ◆

PRACTICE 16-22

Tee Joint, 3F Position, Using Mild Steel, Stainless Steel, and Aluminum

Repeat Practice 16-21. Gradually increase the plate angle after each weld as you develop your skill. Increase the angle until the weld is being made in the vertical up position.

Repeat the process using all thicknesses of metal until you can consistently make the weld visually defect free. Turn off the welding machine, shielding gas, and cooling water and clean up your work area when you are finished welding.

Complete a copy of the "Student Welding Report" listed in Appendix I or provided by your instructor. ◆

PRACTICE 16-23

Tee Joint, 3F Position, with 100% Root Penetration, to Be Tested, Using Mild Steel, Stainless Steel, and Aluminum

Repeat Practice 16-21 and make the fillet weld in a lap joint with 100% root penetration. After the weld is completed, visually inspect it for uniformity and defects before shearing out strips 1 in. (25 mm) wide and bend-testing them.

Repeat each weld until all have 100% root penetration. Turn off the welding machine, shielding gas, and cooling water and clean up your work area when you are finished welding.

Complete a copy of the "Student Welding Report" listed in Appendix I or provided by your instructor. ◆

PRACTICE 16-24

Stringer Bead at a 45° Reclining Angle, Using Mild Steel, Stainless Steel, and Aluminum

Using the same equipment and materials as listed in Practice 16-4, you will make a weld bead on a plate at a 45° reclining angle, **Figure 16-53**. Add the filler metal along the top leading edge of the weld pool. Surface tension will help hold the weld pool on the top if the bead is not too large. The weld should be uniform in width and reinforcement.

Repeat the process using all thicknesses of metal until you can consistently make the weld visually defect free. Turn off the welding machine, shielding gas, and cooling water and clean up your work area when you are finished welding.

Complete a copy of the "Student Welding Report" listed in Appendix I or provided by your instructor. ◆

PRACTICE 16-25

Stringer Bead, 2G Position, Using Mild Steel, Stainless Steel, and Aluminum

Repeat Practice 16-24. Gradually increase the plate angle as you develop your skill until the weld is being made in the horizontal position on a vertical plate.

Repeat the process using all thicknesses of metal until you can consistently make the weld visually defect free. Turn off the welding machine, shielding gas, and cooling water and clean up your work area when you are finished welding.

Complete a copy of the "Student Welding Report" listed in Appendix I or provided by your instructor. ◆

PRACTICE 16-26

Butt Joint, 2G Position, Using Mild Steel, Stainless Steel, and Aluminum

Using the same equipment and materials as listed in Practice 16-4, you will weld a butt joint in the horizontal position.

The welding techniques are the same as those used in Practice 16-25. Add the filler metal to the top plate, and keep the bead size small so it will be uniform.

Repeat the process using all thicknesses of metal until you can consistently make the weld visually defect free. Turn off the welding machine, shielding gas, and cooling water and clean up your work area when you are finished welding.

Complete a copy of the "Student Welding Report" listed in Appendix I or provided by your instructor. ◆

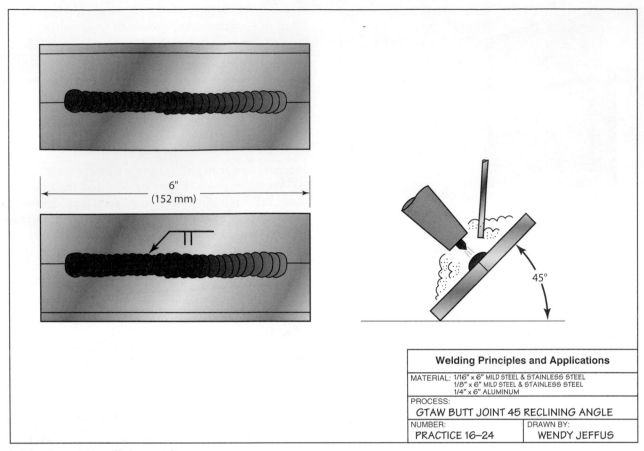

FIGURE 16-53 45° reclining angle. © Cengage Learning 2012

PRACTICE 16-27

Butt Joint, 2G Position, with 100% Penetration, to Be Tested, Using Mild Steel, Stainless Steel, and Aluminum

Repeat Practice 16-26. It may be necessary to increase the root opening to ensure 100% penetration. A backing gas may be required to prevent atmospheric contamination. Using a "J" weave pattern will help to maintain a uniform weld bead. After the weld is completed, visually inspect it for uniformity and defects. Then shear out strips 1 in. (25 mm) wide and bend-test them.

Repeat each weld until all have 100% root penetration. Turn off the welding machine, shielding gas, and cooling water and clean up your work area when you are finished welding.

Complete a copy of the "Student Welding Report" listed in Appendix I or provided by your instructor. ◆

PRACTICE 16-28

Lap Joint, 2F Position, Using Mild Steel, Stainless Steel, and Aluminum

Using the same equipment and materials as listed in Practice 16-4, make a horizontal fillet weld on a lap joint.

- After tack welding the plates together, start the weld at one end. The bottom plate will act as a shelf to support the molten weld pool, Figure 16-54.

- Add the filler metal along the top edge of the weld pool to help control undercutting.

Repeat the process using all thicknesses of metal until you can consistently make the weld visually defect free. Turn off the welding machine, shielding gas, and cooling water and clean up your work area when you are finished welding.

Complete a copy of the "Student Welding Report" listed in Appendix I or provided by your instructor. ◆

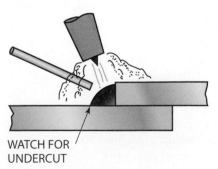

WATCH FOR UNDERCUT

FIGURE 16-54 Horizontal lap joint. © Cengage Learning 2012

PRACTICE 16-29

Lap Joint, 2F Position, with 100% Root Penetration, to Be Tested, Using Mild Steel, Stainless Steel, and Aluminum

Repeat Practice 16-28. Be sure the weld is penetrating the root 100%. After the weld is completed, visually inspect it for uniformity and defects before shearing out strips 1 in. (25 mm) wide and bend-testing them.

Repeat each weld until all have 100% root penetration. Turn off the welding machine, shielding gas, and cooling water and clean up your work area when you are finished welding.

Complete a copy of the "Student Welding Report" listed in Appendix I or provided by your instructor. ◆

PRACTICE 16-30

Tee Joint, 2F Position, Using Mild Steel, Stainless Steel, and Aluminum

Using the same equipment and materials as listed in Practice 16-4, you will make a horizontal fillet weld on a tee joint.

After tack welding the plates together, start the weld at one end. The bottom plate will act as a shelf to support the molten weld pool, **Figure 16-55**. As with the horizontal lap joint, add the filler metal along the top leading edge of the weld pool. This will help control undercutting.

Repeat the process using all thicknesses of metal until you can consistently make the weld visually defect free. Turn off the welding machine, shielding gas, and cooling

water and clean up your work area when you are finished welding.

Complete a copy of the "Student Welding Report" listed in Appendix I or provided by your instructor. ◆

PRACTICE 16-31

Tee Joint, 2F Position, with 100% Root Penetration, to Be Tested, Using Mild Steel, Stainless Steel, and Aluminum

Repeat Practice 16-30. Be sure the weld is penetrating the root 100%. After the weld is completed, visually inspect it for uniformity and defects before shearing out strips 1 in. (25 mm) wide and bend-testing them.

Repeat each weld until all have 100% root penetration. Turn off the welding machine, shielding gas, and cooling water and clean up your work area when you are finished welding.

Complete a copy of the "Student Welding Report" listed in Appendix I or provided by your instructor. ◆

PRACTICE 16-32

Stringer Bead, 4G Position, Using Mild Steel, Stainless Steel, and Aluminum

Using the same equipment and materials as listed in Practice 16-4, you will make a weld bead on a plate in the overhead position.

The surface tension on the molten metal will hold the welding bead on the plate providing that it is not too large. Add the filler metal along the leading edge of the

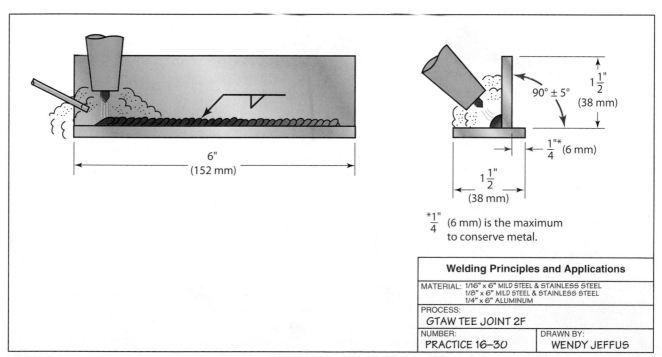

FIGURE 16-55 Tee joint in the horizontal position. © Cengage Learning 2012

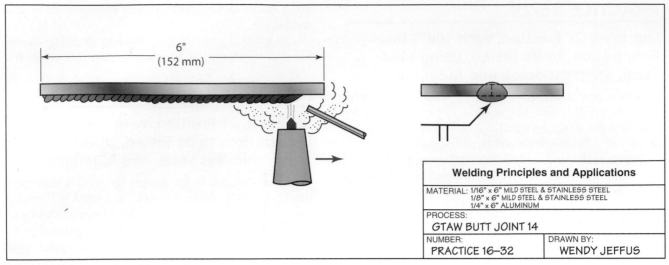

FIGURE 16-56 Overhead butt joint. © Cengage Learning 2012

weld pool, **Figure 16-56.** A wide weld with little buildup will be easier to control and less likely to undercut along the edge.

Repeat the process using all thicknesses of metal until you can consistently make the weld visually defect free. Turn off the welding machine, shielding gas, and cooling water and clean up your work area when you are finished welding.

Complete a copy of the "Student Welding Report" listed in Appendix I or provided by your instructor. ◆

PRACTICE 16-33

Butt Joint, 4G Position, Using Mild Steel, Stainless Steel, and Aluminum

Using the same equipment and materials as listed in Practice 16-4, you will weld a butt joint in the overhead position.

The same techniques used to make the stringer beads in Practice 16-32 are also used with the butt joint. The size of the bead should be kept small enough so that you can control the weld. The completed weld should be uniform and free from defects.

Repeat the process using all thicknesses of metal until you can consistently make the weld visually defect free. Turn off the welding machine, shielding gas, and cooling water and clean up your work area when you are finished welding.

Complete a copy of the "Student Welding Report" listed in Appendix I or provided by your instructor. ◆

PRACTICE 16-34

Lap Joint, 4F Position, Using Mild Steel, Stainless Steel, and Aluminum

Using the same equipment and materials as listed in Practice 16-4, you will make a fillet weld on a lap joint in the overhead position.

The major concentration of heat and filler metal should be on the top plate. Gravity and an occasional sweep of the torch along the bottom plate will pull the weld pool down. Undercutting along the top edge of the weld can be controlled by putting most of the filler metal along the top edge. The completed weld should be uniform and free from defects.

Repeat the process using all thicknesses of metal until you can consistently make the weld visually defect free. Turn off the welding machine, shielding gas, and cooling water and clean up your work area when you are finished welding.

Complete a copy of the "Student Welding Report" listed in Appendix I or provided by your instructor. ◆

PRACTICE 16-35

Tee Joint, 4F Position, Using Mild Steel, Stainless Steel, and Aluminum

Using the same equipment and materials as listed in Practice 16-4, you will make a fillet weld on a tee joint in the overhead position.

The same techniques used to make the overhead lap weld in Practice 16-34 are used with the tee joint. As with the lap joint, most of the heat and filler metal should be concentrated on the top plate. A "J" weave pattern will help pull down any needed metal to the side plate. The completed weld should be uniform and free from defects.

Repeat the process using all thicknesses of metal until you can consistently make the weld visually defect free. Turn off the welding machine, shielding gas, and cooling water and clean up your work area when you are finished welding.

Complete a copy of the "Student Welding Report" listed in Appendix I or provided by your instructor. ◆

Summary

One of the most difficult aspects of learning to produce gas tungsten arc welds is positioning yourself so that you can both control the electrode filler metal and see the joint at the same time. Beginning welders assume they must see the tungsten tip as they make the weld. Experienced welders, however, realize that they need to see only the leading edge of the molten weld pool to know how the weld is being produced. A good view of the leading edge will tell you how the base metal is being melted, or depth of penetration. You can even tell from this small portion whether the filler metal is being added at an appropriate rate. As you learn how to control the weld and develop your skills, it is a good idea to gradually reduce your need for seeing 100% of the molten weld pool. Increasing this part of your skill will be a significant advantage in the field because, unlike welding in the classroom, which is typically done on a comfortable-height table, welding in the field may have to be done out of position.

The Great Master's Horse Returns Home after 500 Years

Welding played an important role in producing a modern version of Leonardo da Vinci's unfinished masterpiece.

The United States recently presented Italy with *Il Cavallo,* the world's largest bronze sculpture of a horse. The gift honors the Italian people for 2000 years of cultural heritage and commemorates Leonardo da Vinci, the Renaissance's greatest universal man. To fully understand the significance of *Il Cavallo* (Italian for "the horse"), its history must be traced back more than 500 years. In 1482, the Duke of Milan commissioned Leonardo da Vinci to design the world's largest equestrian sculpture. Da Vinci used his keen grasp of animal anatomy to design and produce a 24-ft (7.2-m) clay model of his planned bronze sculpture. But on September 10, 1499, France invaded Milan and French soldiers destroyed da Vinci's clay model by using it for target practice. The bronze allotted for the sculpture was reassigned for making cannons, and the Duke's commission was dissolved. Da Vinci was reported to have wept on his deathbed for his unfinished masterpiece.

Charles Dent, a retired airline pilot, read the story about da Vinci's horse in an article in *National Geographic* in 1977 and conceived the idea to "give Leonardo his horse." Dent founded the nonprofit organization Leonardo da Vinci's Horse, Inc. (LDVHI) and worked with sculptors, art historians, da Vinci scholars, and equine experts to produce a modern version of da Vinci's dream. The sculpture was unveiled in Milan on September 10, 1999, exactly 500 years from the day da Vinci's original model was destroyed.

"The whole project is a beautiful concept," Peter Homestead said. Homestead is president of Tallix Foundry, Beacon, New York, the art foundry that produced *Il Cavallo.* "It is a beautiful idea to do a modern interpretation of something the great master, Leonardo da Vinci, never had a chance to see to completion."

To produce *Il Cavallo,* Tallix worked from a design by American artist Nina Akamu. Once an 8-ft (2.4-m) model was produced, it was put on an enlarging apparatus, which mechanically worked the 24-ft clay model to scale. Working the entire full-size model took one year.

Molding and Casting

A "mother mold" of the horse was made by spraying the model with liquid rubber mold material and then applying a half-inch layer of polyester resin and fiberglass to the rubber coating for a sturdy backing. When it hardened, the rubber sections were removed.

Most of the horse was sand-cast. Sand-casting involves making positive plaster castings from the mother mold and placing them, one at a time, in an 8-ft steel molding box. Next, a sand and binder mixture is packed around the section. Once the mixture has hardened, the sides of the molding box are separated, and the plaster is removed to create a negative image in the sand. The molding box is closed again, and molten bronze (at about 2000°F [1093°C]) is poured in through pour spouts.

When the bronze has cooled, the sand mold is broken and the result is the bronze casting. Lost wax casting was used on the more intricate areas, such as the mane, tail, and face, to pick up small details.

The materials used in casting and molding may have changed, but the methods have remained somewhat unchanged since da Vinci's time. Although Leonardo planned to cast the entire horse in one piece, experts today say it would have been impossible, because bronze is difficult to keep at constant temperature while it flows into the many crevices of a large mold. Tallix's experience in large metal sculpture dictated casting the sculpture in 60 pieces, all about 4 ft square (1.2 m) and comprised of 1/4- to 3/8-in. (6.3- to 9.5-mm) thick silicon bronze.

Assembly in the Foundry

The castings were matched, tack welded together, and then gas metal arc (GMA) welded on the inside of each joint using Deltaweld® power sources and XR™—a push-pull wire feeder from Miller Electric Mfg. Co., Appleton, Wisconsin. The wire feeder was chosen because it could reach the most difficult places within *Il Cavallo,* sometimes up to 30 ft (9 m) from the power source.

Tallix welders used 0.045-in. (1.14-mm) diameter silicon bronze wire (ER CuSi-A per AWS/SFA A5.7) at traveling speeds of 60 in./min (1.5 m/min).

"I'll bet we used 500 lb to 600 lb (225 kg to 270 kg) of wire, with a total of about 1 mile of welds," Homestead said.

The 24-ft tall bronze sculpture, *Il Cavallo,* is a modern-day interpretation of the sculpture Leonardo da Vinci designed and planned to build for the Duke of Milan. Here, workers position the head and neck section in preparation for bolting from the inside and then welding onto the body prior to its unveiling in September in Milan. American Welding Society

Gas tungsten arc welding (GTAW) was used on the external sides of the joints, where cosmetics were more important. Gas tungsten arc welding gives a more attractive weld because it produces less spatter and does not warp or distort the metal. Tallix used Miller Syncrowave® 250 GTAW machines and 1/16- to 3/32-in. (1.58- to 2.38-mm) diameter silicon bronze filler metal.

The welds were finished by a process called chasing, which blends the welds to the "rough rake" texture of the sculpture and prepares the surface for the patina (the chemical that colors the bronze) and wax.

A structural engineer analyzed the sculpture and set structural specifications based on loads from natural forces, such as wind and earthquakes, as well as the structure's own weight load. From those specifications, an armature of Type 304 stainless steel was designed to support the sculpture from the inside. The 3-in. (7.6 cm) diameter armature tubing was cut with a plasma arc cutting machine to create a "ribbing," which fit the exact curves and dimensions of the horse. To gas metal arc weld the armature to the bronze skin, Tallix used 0.045-in. diameter ERCuAl-A2 aluminum bronze wire.

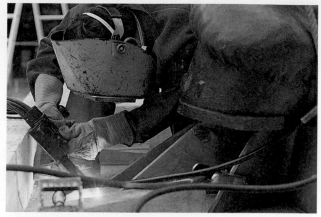

A welder uses the GMAW process to attach the stainless steel and titanium pole fitted inside two of the sculpture's legs to steel plates embedded in the concrete pedestal on which it now. American Welding Society

The two legs on which the sculpture rests were each fitted with an 8-in. (20.3-cm) diameter, 1-in. (2.54-cm) thick stainless steel/titanium pole, which extends 18 in. (45.7 cm) below the hooves and was eventually GMA welded to steel match plates embedded in the concrete pedestal in Milan. The poles reinforce the 15 tons of weight put on two legs—12 tons from the bronze skin, 3 tons from the armature.

Once the sculpture was completed, it was unbolted at the joints connecting the seven main parts (head/neck, body, four legs, and tail) and flown to Milan for final assembly at the Hippodrome's Cultural Park. Three welders, a finisher, and a patina worker from Tallix traveled to Milan to begin the one-week assembly on *Il Cavallo.* When they arrived, the same power sources and wire feeder they had used in the United States were waiting for them, provided by Miller Europe, which has facilities in Milan.

Assembly in Milan

To assemble the sculpture, the horse's body was set on its side, and the four legs were bolted and then welded on. Next, the body and legs were lifted by a crane, placed upright, and set on the concrete pedestal. The head and tail were lifted by the crane, positioned, and bolted on from the inside before being welded onto the body. Once all inside welding was completed, the trapdoor on the horse's belly was welded shut and finished.

Since its unveiling, *Il Cavallo* has received notice from people all over the world. To everyone involved in the making of *Il Cavallo,* it is much more than just a sculpture. "In the future, our culture will be studied through the sculptures and artifacts we produce today," Homestead said, "just as cultures have been studied by their art throughout history."

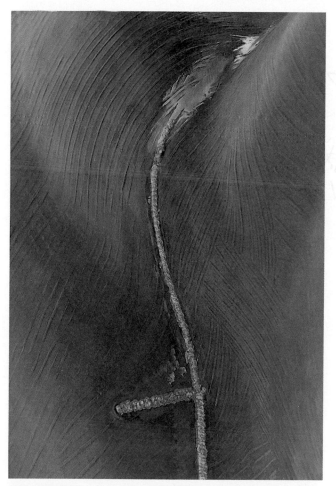

Gas tungsten arc welds were used on the outside of the sculpture for cosmetic reasons. American Welding Society

Article courtesy of the American Welding Society.

Review

1. What effect does torch angle have on the shielding gas protective zone?

2. Why must the end of the filler rod be kept in the shielding gas protective zone?

3. What can cause tungsten contamination?

4. What determines the correct current setting for a GTA weld?

5. What is the lowest acceptable amperage setting for GTA welding?

6. List the factors that affect the gas flow setting for GTA welding.

7. When should the minimum gas flow rates be increased?

8. What is the minimum gas flow rate for a nozzle size?

9. What is the maximum gas flow rate for a nozzle size?

10. Which incorrect welding parameters does stainless steel show clearly?

11. Using Table 16-4, determine the approximate temperature of metal that has formed a dark blue color.

12. Using Table 16-3, Table 16-5, and Table 16-6, list the filler metals for the following metals.
    a. 1020 low carbon steel
    b. 309 stainless steel

13. Why is it possible to control a large aluminum weld bead?

14. What may happen to the end of the aluminum welding rod if it is held too close to the arc?

15. What should be done if someone comes in contact with a cleaning chemical?

16. Using Table 16-7, determine the suggested setting for GTA welding of mild steel using a 3/32-in. (2.4-mm) tungsten.

17. What can be done to limit oxide formation on stainless steel?

18. How should the filler metal be added to the molten weld pool?

19. How can the rod be freed if it sticks to the plate?

20. How is an outside corner joint assembled?

21. What must be done with the weld craters when back stepping a weld? Why?

22. How is the lap joint tested for 100% penetration?

23. What can prevent both sides of a stainless steel tee joint from being welded?

24. How is the filler metal added for a 3F weld?

25. What can cause undercutting on a 3F tee joint?

26. What helps hold the weld in place on a 2F lap joint?

27. What helps hold the weld in place on a 4G weld?

Chapter 17

Gas Tungsten Arc Welding of Pipe

OBJECTIVES

After completing this chapter, the student should be able to

- describe how a pipe joint is prepared for welding.
- list the four most common root defects and the causes of each defect.
- discuss when and why a backing gas is used.
- explain the uses of a hot pass.
- sketch a single V-groove and indicate the location and sequence of welds for each position.
- make a single V-groove butt welded joint on a pipe in any position.

KEY TERMS

backing gas

concavity

consumable inserts

grapes

incomplete fusion

root

root penetration

root reinforcement

stress points

INTRODUCTION

Gas tungsten arc welding of pipe is used when the welded joint must have a high degree of integrity. A weld with integrity is one that is strong and free from defects that may cause the premature failure of the welded joint. Many industries, such as oil, gas, nuclear, chemical, and several others, require this kind of joint to be made, **Figure 17-1**. Failures in these types of pipe joints can be disastrous in addition to being extremely expensive to repair.

GTA welding gives industry the type of joint it needs. Welders who are skilled in the GTA welding process have the ability to make consistently high-quality welds with a low rejection rate.

GTA pipe welders are among the highest paid workers in the welding industry. The positions available are among the safest and most prestigious and have excellent working conditions. Preparing for such a job requires the development of high levels of skill and technical knowledge through much practice and study.

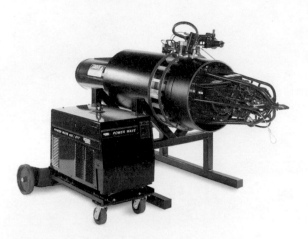

FIGURE 17-1 Automatic GTA outside pipe welding machine. Lincoln Electric Company

Practices

The practices in this chapter are all performed on mild steel pipe. Mild steel is the most readily available material on which to learn. The skill and techniques learned in these practices can be easily transferred to any other piping material. In most cases, the only change is in the composition of the filler material and possibly the current.

Joint Preparation

The ends of pipe must be grooved before welding. This pipe end preparation is to ensure a sound weld. The type

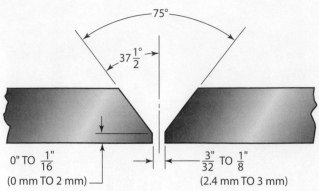

FIGURE 17-3 Typical V-groove preparation dimensions.
© Cengage Learning 2012

of groove used in preparing the ends of a pipe for welding will vary depending upon the pipe material, thickness, and application. The most often used grooves on mild steel are the single V-, single U-, single J-, and single bevel-grooves, **Figure 17-2**. Both the single bevel-groove and single V-groove can be easily flame cut or ground. This factor makes them the most frequently used grooves during training.

The end of the pipe is prepared with a 37 1/2° bevel, leaving a root face of 1/16 in. (2 mm) to 1/8 in. (3 mm), **Figure 17-3**. When both pipes are prepared in this manner, they form a 75° single V-groove. The groove angle allows good visibility of the weld with a minimum amount of filler metal required to fill the groove. The grinding or

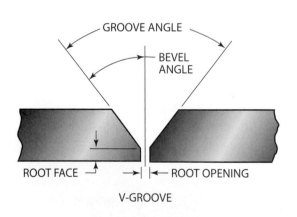

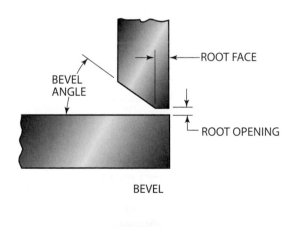

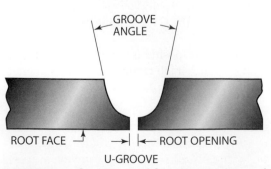

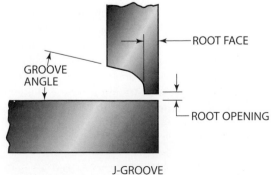

FIGURE 17-2 Grooves used for pipe-to-pipe and pipe-to-flange welds. © Cengage Learning 2012

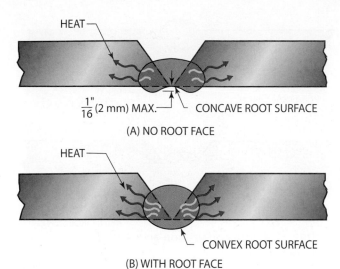

(A) NO ROOT FACE

(B) WITH ROOT FACE

FIGURE 17-4 (A) This joint cannot conduct enough heat away from the root pass to provide support for the molten root pass. Note the concave surface of the root face caused by suck back. (B) The larger mass of metal helps to cool the puddle quickly and provides for sufficient surface tension, which also helps control the root surface. © Cengage Learning 2012

machining of the sharp edge left on the pipe after beveling produces the root face. The root face is the small flat surface at the root of the groove. Failure to remove the root edge may result in the weld burning back the material, leaving too large a key hole to be filled by the weld. This may result in a concave root surface, sometimes referred to as *suck back*, **Figure 17-4**.

Before the joint is assembled, the welding surfaces must be cleaned and smoothed so that they are uniform and free of contaminants. To remove any possible sources of contamination to the weld, clean a 1-in. (25-mm) wide or wider band both inside and outside of the pipe, **Figure 17-5**. The band can be cleaned by grinding, brushing, or filing. The prepared edges must be clean and free from tears, cracks, slivers, or any other defects. Once the pipe surface has been cleaned, it is important that it not be touched with your hand because oil on your hand can contaminate the weld, **Figure 17-6**.

NOTE: When making welds on critical piping systems or when welding piping systems made with reactive metals such as titanium, you must clean both the filler metal and weld surface. Using a lint-free cloth and a solvent such as isopropyl alcohol, acetone, or methyl ethyl ketene (MEK), clean the filler metal and joint surface just before welding begins. And once the material is cleaned, always wear oil-free chemically resistant gloves, such as nitrile gloves, when handling the cleaned material.

The pipe should be tack welded together with a root opening of 3/32 in. (2.4 mm) to 1/8 in. (3 mm). Four or more tack welds evenly spaced around the pipe should be

(A)

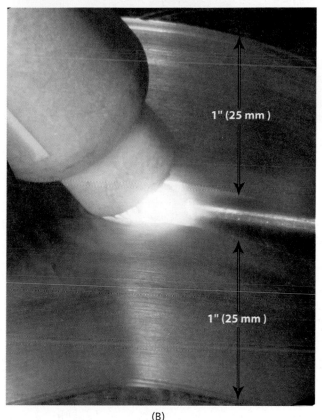

(B)

FIGURE 17-5 (A) The pipe is machined clean on both the inside and outside surfaces. (B) The groove must be cleaned, both outside and inside the pipe joint. Larry Jeffus

made. The tack welds should be located so that they do not interfere with the starting or stopping of the root pass. Starting or stopping the root pass on a tack weld can result in poor root penetration, **Figure 17-7**.

Root

The **root** of a weld is the deepest point into the joint where fusion between the base metal and filler metal occurs. The **root penetration** is the distance measured between the original surface of the joint and the deepest point of fusion. The **root reinforcement** is the amount

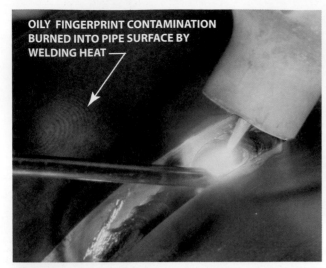

FIGURE 17-6 Oily fingerprint contamination. © Cengage Learning 2012

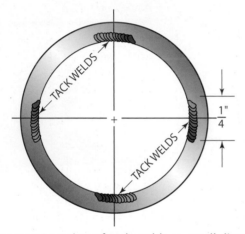

FIGURE 17-7 Location of tack welds on small diameter pipe. © Cengage Learning 2012

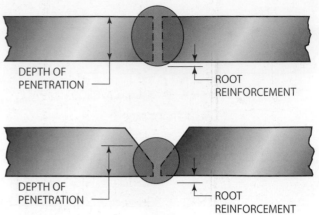

FIGURE 17-8 Root penetration and reinforcement. © Cengage Learning 2012

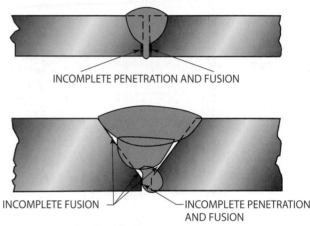

FIGURE 17-9 Lack of fusion. © Cengage Learning 2012

of metal deposited on the back side of a welded joint, **Figure 17-8.**

The depth to which a weld must penetrate in a pipe joint, or the amount of root reinforcement, is given in the code or standards being used. Most codes or standards for GTA pipe welding require 100% root penetration and no more than 1/16 in. (2 mm) of root reinforcement. Uniformity and the lack of defects in a weld are also important. The four most common root defects are incomplete fusion, concave root surface (suck back), excessive root reinforcement, and root contamination.

Incomplete Fusion Sometimes a weld does not completely penetrate the joint, or there may be a lack of fusion on one or both sides of the root. These conditions are caused by not enough heat penetrating the back side of the work, **Figure 17-9.** The problem with incomplete fusion is that it can form a **stress point.** Stress points can result in the premature cracking or failure of the weld at a load well under its expected strength.

Incomplete fusion can be observed if the weld is subjected to a root-bend test. A root bend subjects a welded joint to a bending force so that the root of the weld is in tension, **Figure 17-10.** One test that can be used is a standard test procedure (guided bend or free bend). Another test that can be used is to simply secure the welded joint in a vise and use a hammer to bend the metal, **Figure 17-11.**

FIGURE 17-10 Guided-bend test specimen. Larry Jeffus

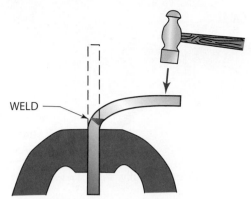

FIGURE 17-11 Bend test using a hammer and vise.
© Cengage Learning 2012

FIGURE 17-12 Root contamination. Larry Jeffus

Concave Root Surface A concave root surface (suck back) occurs when the back side of the root weld is concave in shape, Figure 17-4A. Common causes for this condition are when insufficient filler metal is added to the joint, or excessive heat is used in the overhead position. The **concavity** of the root surface results in reduced thickness (actual throat) of the weld, which in turn causes a lower joint strength.

Concave root surface can be visually inspected (VI) readily without destructive testing. The amount of concavity can be given

- as a percentage of the total weld length,
- as the average number and largest size of spots per inch (mm) of weld, or
- as a percentage of the effective throat. Normally, 1/16 in. (2 mm) is the maximum allowable root concavity.

Excessive Root Reinforcement Excessive root reinforcement or burnthrough is the excessive buildup of metal on the back side of a weld. Uneven buildup and burnthrough are sometimes referred to as **grapes.** This condition results from excessive heat, temperature, and filler metal during welding. Excessive root reinforcement can cause reduced material flow, result in clogged pipes, or form stress points that will result in premature weld failure.

The root of the weld should have some reinforcement (buildup), but this should not exceed 1/16 in. (1.5 mm). VI can be used to verify this condition. The amount of excessive reinforcement is given

- as a percentage of the total weld length,
- as the average number and largest size of spots per inch (mm) of weld, or
- as the maximum distance from the metal surface to the longest spot of excessive reinforcement.

Root Contamination The back side of the molten weld pool can be contaminated, causing porosity, embrittlement, oxide inclusions, and/or oxide layers. Contamination is caused by overheating, poor cleaning procedures, or improper protection from the surrounding atmosphere, **Figure 17-12.** Root contamination can lead to faster corrosion, oxide flaking, weld brittleness, leaks, stress points, or all of these.

Root surface contamination can be visually inspected. Heavy flakes and large deposits of oxides are signs of root surface contamination. Ideally, the color of the root should be close to that of the base metal. Internal porosity, oxides, or embrittlement can be detected readily by a root-bend test. The weld deposit should be as ductile (easily bent) as the surrounding base metal.

Backing Gas

Root contamination caused by the surrounding atmosphere is not a major problem when welding on mild steel pipe. Some type of protection, however, is needed when welding on alloy steel, stainless steel, aluminum, copper, and most other types of pipe. The easiest method of protecting the root from atmospheric contamination is to use a **backing gas.**

The type of gas used for backing will depend upon the type of pipe being welded. Nitrogen and CO_2 are often acceptable, depending upon the code or intended use of the pipe system. Argon, although expensive, can be used to back any type of pipe if a welder is unsure about a less expensive substitution.

There are several methods in common use for containing the backing gas in the pipe. On small diameters or short sections of pipe, the ends of the pipe are capped, **Figure 17-13.** The gas is allowed to purge the complete pipe section of air. This method requires too much purging time and gas to be practical on large diameter pipe. For larger diameters, the pipe is plugged on both sides of the joint to be welded so that a smaller area can be purged. If the piping system is complex, consisting of valves and numerous turns, water-soluble plugs or soft plastic gas bags are suggested. They can be blown out with air or water when the system is completed.

When a backing gas is used, the gas must have enough time to purge the pipe completely of air. When welding on large diameter pipes or long pipe sections, it is necessary

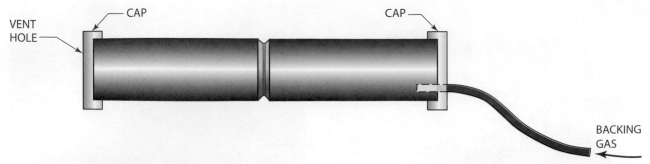

FIGURE 17-13 Backing gas can be fed into a pipe for welding by capping the ends externally. © Cengage Learning 2012

to estimate the length of time required to purge the air. First, estimate the interior volume of the pipe. Second, convert the flow rate into cubic inches per minute (cm^3/min). Third, to get the time required divide the estimated volume by the flow rate. Last, round any fractions of minutes off to the next highest minute.

Standard Units

1st Volume (in.^{3}) = length (in in.) $\times \pi \times$ radius^2 (in in.)
2nd Flow rate (in.^{3}/min) = cfh $\times$ 29*
3rd Flow time (min) = volume $\div$ flow rate

*29 (28.8) cubic inches per minute flow rate equals a 1 cubic foot per hour flow rate.

SI (Metric Units)

1st Volume (cubic cm) =
   length (in cm) $\times \pi \times$ radius^2 (in cm)
2nd Flow rate (cm^3/min) = L/min $\times$ 1000
3rd Flow time (min) = volume $\div$ flow rate

Standard Unit Example

How long would it take to purge the air out of a 10-ft (305-cm) long section of 4-in. (10-cm) diameter pipe if the flow rate is 20 cfh (9.4 L/min)?

Solution

Volume = 120 $\times$ 3.14 $\times$ 2^2
Volume = 1507 in.^{3}
Flow rate = 20 $\times$ 29
Flow rate = 580 in.^{3}/min
Flow time = 1507 $\div$ 580
Flow time = 2.59 min (approximately 3 min)

SI (Metric Unit) Example

How long would it take to purge the air out of a 3-m (10-ft) long section of 10-cm (4-in.) diameter pipe if the flow rate is 9 L/min (20 cfh)?

Volume = 300 $\times$ 3.14 $\times$ 5^2
Volume = 23,550 cm^3
Flow rate = 9 $\times$ 1000
Flow rate = 9000 cm^3/min
Flow time = 23,550 $\div$ 9000

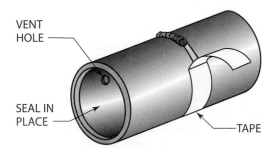

FIGURE 17-14 Using a special tape to cover the unwelded part of a pipe joint will reduce the need for high flow rates for the backing gas. © Cengage Learning 2012

Flow time = 2.61 min (approximately 3 min)
π = 3.14
Radius = 1/2 diameter*
Length in in. = length in ft $\times$ 12
Length in cm = length in m $\times$ 100

Taping over the joint prevents the gas from being blown out through the joint. This will allow a slower flow rate to be used on the purging gas once the pipe has been purged. The tape is removed just ahead of the weld, **Figure 17-14**.

Filler Metal

The addition of filler metal to the joint can be done by dipping or by preplacement. Dipping can be made easier if the rod is bent to the radius of the pipe, **Figure 17-15**. This will allow the welder to keep the rod angle close to the pipe to minimize contamination. This also makes it possible for welders to reach around the pipe and brace themselves better. The groove can also be used to guide the filler rod, reducing tungsten contamination.

---

*For the purposes of estimating, either the inside diameter (ID) or outside diameter (OD) of the pipe can be used.

Notice in the examples that the lengths, diameters, and flow rates are about the same and so are the minutes required for purging.

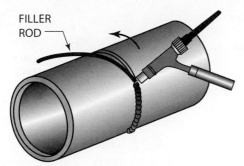

FIGURE 17-15 Filler rod curved to surface of pipe for ease in welding. © Cengage Learning 2012

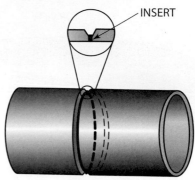

FIGURE 17-17 Specialized backing rings. © Cengage Learning 2012

Preplacing the filler rod allows the root pass and some filler passes to be made without rod manipulation. The filler rod, usually 1/8 in. (3 mm) in diameter, is bent in a radius slightly smaller than the pipe being welded. Tension from the wire will hold it in place during the weld. If it does not stay in place, a small tack weld can be made at one end. Both hands can be used to control the torch, making the weld easier and more accurate. This technique is accepted in most codes or standards for welding.

Consumable Inserts **Consumable inserts** are preplaced filler metal, which is used for the root pass when consistent, high-quality welds are required. These inserts help to reduce the number of repairs or rejections when welding under conditions that are less than ideal, such as in a limited space. Although most inserts are used on pipe, they are available as strips for flat plate.

Inserts are classified by their cross-sectional shape, as shown in **Figure 17-16**, and listed as follows:

- Class 1, A-shape
- Class 2, J-shape
- Class 3, rectangular shape
- Class 4, Y-shape
- Class 5, rectangular shape (contoured edges)

These are the most frequently used designs, **Figure 17-17**. Other shapes can be obtained from manufacturers upon request.

When ordering consumable inserts, the welder must specify the classification, size, style, and pipe schedule. Other information is available in the AWS Publication A5.30, *Specification for Consumable Inserts.*

PRACTICE 17-1

Tack Welding Pipe

Using a properly set-up and adjusted GTA welding machine; proper safety protection; two or more pieces of 3-in. (76-mm) to 10-in. (254-mm) diameter schedule 40 mild steel pipe, prepared with single V-grooved joints; and a few filler rods having 1/16-in. (2-mm), 3/32-in. (2.4-mm), and 1/8-in. (3-mm) diameters; you will tack weld a butt pipe joint.

Bend the rods with the 1/16-in. (2-mm) and 3/32-in. (2.4-mm) diameters into a U-shape. These rods will be used to set the desired root opening, **Figure 17-18**. Lay the pipe in an angle iron cradle. Slide the ends together so that the desired diameter wire is held between the ends. Hold the torch cup against the beveled sides of the groove. The torch should be nearly parallel with the pipe. Bring the filler rod end in close and rest it on the joint just ahead of the torch, **Figure 17-19**.

Lower your helmet and switch on the welding current by using the foot or hand control. Slowly straighten the torch so that the tungsten is brought closer to the groove. This pivoting of the torch around the nozzle will keep the

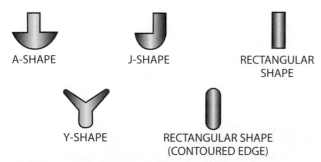

FIGURE 17-16 Consumable insert standard shapes. © Cengage Learning 2012

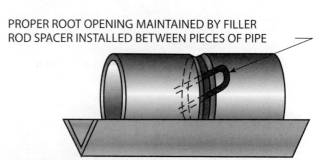

PROPER ROOT OPENING MAINTAINED BY FILLER ROD SPACER INSTALLED BETWEEN PIECES OF PIPE

FIGURE 17-18 Use a spacer like a piece of filler metal to set the root gap. © Cengage Learning 2012

TOP VIEW

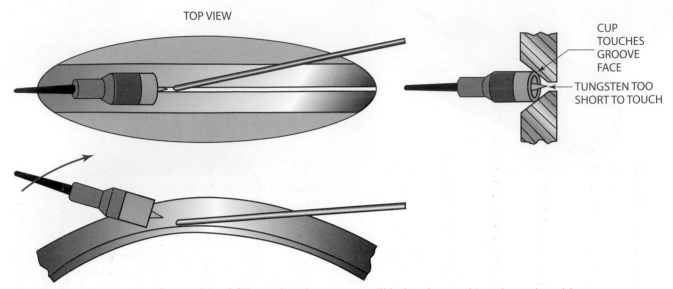

CUP
TOUCHES
GROOVE
FACE

TUNGSTEN TOO
SHORT TO TOUCH

FIGURE 17-19 Supporting the torch and filler rod in the groove will help when making the tack welds. © Cengage Learning 2012

torch aligned with the joint and help prevent arc starts outside of the joint. When the tungsten is close enough, an arc will be established. Increase the current by depressing the foot or hand control until a molten weld pool is established on both root faces.

Hold the filler rod tip in the molten weld pool as the torch is pivoted from side to side, **Figure 17-20**. Slowly move ahead and repeat this step until you have made a tack weld approximately 1/4 in. (6 mm) long. At the end of the tack weld, slowly reduce the current and fill the weld crater. If the tack weld crater is properly filled and it does not crack, grinding may not be required, **Figure 17-21**.

Roll the pipe 180° to the opposite side. Check the root opening and adjust the opening if needed. Make a tack weld 1/4 in. (6 mm) long. Roll the pipe 90° to a spot halfway between the first two tacks. Check and adjust the root opening if needed. Make a tack weld. Roll the pipe 180° to the opposite side and make a tack weld halfway between the first two tack welds.

The completed joint should have four tack welds evenly spaced around the pipe (at 90° intervals). The root

FIGURE 17-21 Good stopping point for a weld that will be restarted. Larry Jeffus

surfaces of the welds should be flat or slightly convex. The tacks should have good fusion into the base metal with no cold lap at the start, **Figure 17-22**.

Continue making tack welds until this procedure is mastered. Turn off the welding machine, shielding gas, and cooling water and clean up your work area when you are finished welding.

Complete a copy of the "Student Welding Report" listed in Appendix I or provided by your instructor. ◆

Cup Walking

Although the cup walking technique can be used on plate, it is most often used by pipe welders. Plate welders do not often use cup walking because it can be slower than

FIGURE 17-20 Larry Jeffus

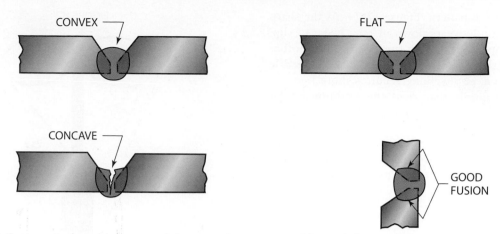

FIGURE 17-22 Concave tack welds may crack because they cannot withstand the stresses during cooling. © Cengage Learning 2012

traditional welding techniques; however, many pipe welders like to use it for several reasons:

- Reduced welder fatigue—Because the welding torch nozzle and filler metal are both resting on the pipe, welders can relax because they are less likely to touch the electrode to the weld or filler metal.
- Better welding torch control—The small uniform stepping movement of the torch nozzle along the weld results in a weld appearance that is more uniform and defect free than manual welding.
- Longer welds—Often welders can weld from the bottom to the top of the pipe without having to stop and reposition. Stops and starts are a primary cause of GTA weld defects.
- Higher weld quality—Less tungsten contamination, more uniform welds, and fewer stops and starts results in high-quality pipe welds.

Cup Walking Setup The tungsten length should be set so it extends out of the nozzle no more than the diameter of the nozzle opening, **Figure 17-23**. If the tungsten sticks out too much, it is more likely to become contaminated by touching the weld or filler metal. When making the root pass and the first one or two filler passes, the nozzle diameter should be large enough to touch both sides of the groove edges, **Figure 17-24**. A larger nozzle size may be needed for the last filler pass and cover pass.

Because the filler rod is rested in the welding joint, it is important that it should be slightly larger in diameter than the root opening so it does not fall through the root during welding. For example, when welding on a pipe that has a 3/32-in. (2.4-mm) root opening, a 1/8-in. (3-mm) or larger filler rod should be used.

The outside edge of the nozzle must be sharp so it does not slip as it is stepped along the joint when the torch is

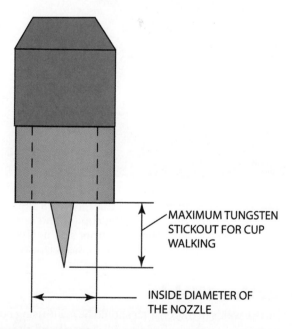

FIGURE 17-23 The maximum recommended tungsten extension for cup walking. © Cengage Learning 2012

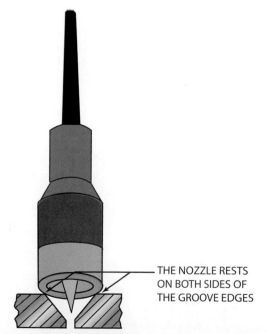

FIGURE 17-24 Rest the nozzle on the groove edges. © Cengage Learning 2012

moved in a figure-8 pattern, Figure 17-24. As the sharp corner of the nozzle wears down, it may begin to slip. When this happens, it can be turned so that another part of the nozzle edge can be used. Because the nozzle edge wears down, cup walking requires more frequent changes of the nozzle than does the traditional GTA welding technique.

> NOTE: For the best results with cup walking, you should use a high-temperature ceramic nozzle such as silicon nitride.

Cup Walking Technique During cup walking, both the nozzle and the filler metal are rested on the pipe joint, **Figure 17-25**. The nozzle is held firmly against the pipe while only a very light pressure is applied to the filler metal. The firm cup pressure helps the edge of the nozzle to grip the pipe so that it can be stepped forward as the figure-8 pattern is made. The firm pressure does not cause a problem with most piping materials; however, on softer metals such as aluminum, the nozzle can leave marks that can be deep enough to be considered defects. So, cup walking is not typically used on aluminum pipe.

Too much pressure on the filler rod can cause it to push through the molten weld metal, causing root surface problems such as excessive root penetration or buildup. The rod is held parallel to the joint so that the end is just at the leading edge of the molten weld pool. The end of the rod is melted by the figure-8 movement of the torch and flows into the leading edge of the molten weld pool.

The figure-8 movement, **Figure 17-26A**, is made so that one edge of the nozzle is lifted off of the pipe joint edge, **Figure 17-26B**, and moved slightly forward, **Figure 17-26C**. Each time the torch moves through the figure 8 it steps slightly forward. Moving your whole arm is less tiring than moving just your wrist, **Figure 17-27**.

After the weld has been built up with the root pass and several filler passes, the torch nozzle will begin touching the weld face as well as both sides of the groove. At this time you should change to a larger diameter nozzle to make it easier to walk the cup. The walking technique remains the same even though now three points are being contacted by the torch nozzle.

The nozzle will have only one contact point as the cover pass is welded. Some welders will use an even larger nozzle for the cover pass while others may stay with the same nozzle size they used for the last filler pass.

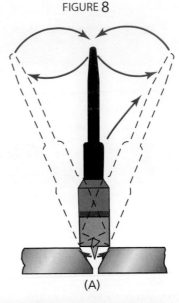

FIGURE **8**

(A)

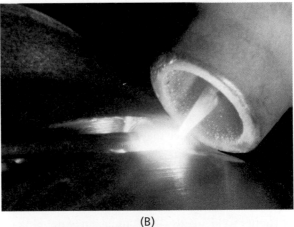

(B)

(C)

FIGURE 17-26 Moving the top of the torch in a figure-8 pattern (A) first lifts one side of the cup (B) and lets you move it slightly forward. Then the other side of the cup is lifted (C), and it is moved forward. © Cengage Learning 2012

FIGURE 17-25 Resting the cup and filler metal on the pipe groove. Larry Jeffus

Practice Welds

Practice welds in this chapter should be made using both the traditional method of adding filler metal and the cup walking technique. Both techniques are important, and as a skilled pipe welder you must be proficient with each.

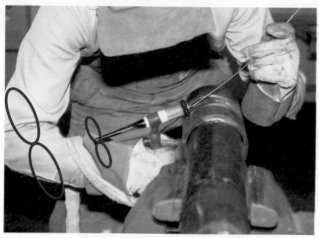

FIGURE 17-27 Moving your whole arm in the figure-8 pattern is less tiring than moving just your wrist. Larry Jeffus

PRACTICE 17-2

Root Pass, Horizontal Rolled Position (1G)

For this practice, you will need a properly set-up and adjusted GTA welding machine; proper safety protection; two or more pieces of 3-in. (76-mm) to 10-in. (254-mm) diameter schedule 40 mild steel pipe, prepared with single V-grooved joints and tack welded as described in Practice 17-1; and a few filler rods having 1/16-in. (2-mm), 3/32-in. (2.4-mm), and 1/8-in. (3-mm) diameters. You will make a root welding pass on a pipe in the horizontal rolled (1G) position, **Figure 17-28**.

Place the pipe securely in an angle iron vee block. Using the same procedure as practiced for the tack welds, place the cup against the beveled sides of the joint with the torch at a steep angle (refer to Figure 17-19). Start the weld near the 3 o'clock position and weld toward the

12 o'clock position. Stop welding just past the 12 o'clock position and roll the pipe. Starting the weld above the 3 o'clock position is easier, and starting the weld below this position is harder, **Figure 17-29**. Brace yourself against the pipe and try moving through the length of the weld to determine if you have full freedom of movement.

Hold the filler rod close to the pipe so it is protected by the shielding gas, thus preventing air from being drawn into the inert gas around the weld. The end of the rod must be far enough away from the starting point so it will not be melted immediately by the arc. However, it must be close enough so that it can be seen by the light of the arc.

When you are comfortable and ready to start, lower your helmet. Switch on the welding current by depressing the foot or hand control. Slowly pivot the torch until the arc starts. Increase the current to establish a new molten weld pool on both sides of the root face. Place the filler rod in the root of the weld as the torch is rocked from side to side and moved slowly ahead. The side-to-side motion can be used to walk the nozzle along the groove. The cup walking will ensure good fusion in both root faces and help make a very uniform weld, **Figure 17-30**. When walking the nozzle along the groove, the tip of the torch cap will make a figure-8 motion.

Nozzle walking the torch along a groove can be practiced without welding power so you can see how to move the torch. This technique can be used on most GTA groove welds in pipe or plate, **Figure 17-31**. A skilled welder can also use it on the filler and cover passes.

Watch the molten weld pool to make sure that it is balanced between both sides. If the molten weld pool is not balanced, then one side is probably not as hot, nor is it penetrating as well.

To ensure good fusion and penetration of the weld at the tack welds, or as the weld is completed, a special tie-in technique is required. Stop adding filler metal to the

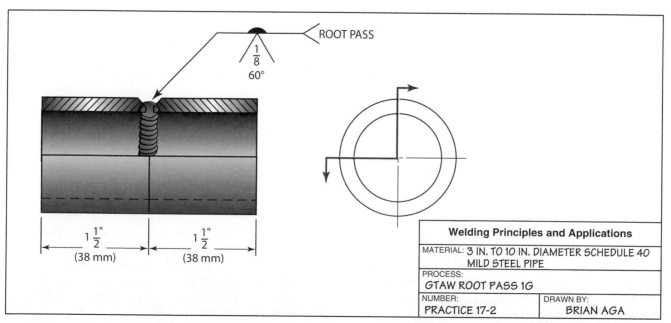

	Welding Principles and Applications	
	MATERIAL: 3 IN. TO 10 IN. DIAMETER SCHEDULE 40 MILD STEEL PIPE	
	PROCESS: GTAW ROOT PASS 1G	
	NUMBER: PRACTICE 17-2	DRAWN BY: BRIAN AGA

FIGURE 17-28 Root pass horizontal rolled position butt joint. © Cengage Learning 2012

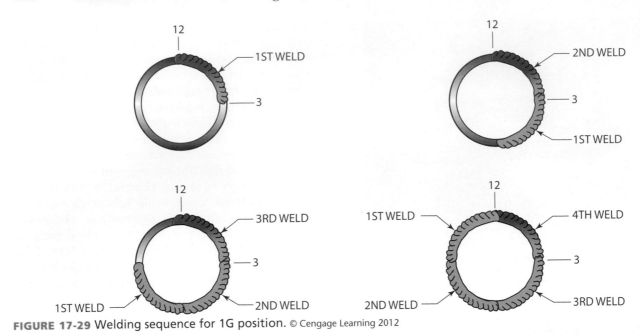

FIGURE 17-29 Welding sequence for 1G position. © Cengage Learning 2012

FIGURE 17-30 Leaving the filler metal fused to the weld pool makes it easier to continue the weld after you reposition yourself. Larry Jeffus

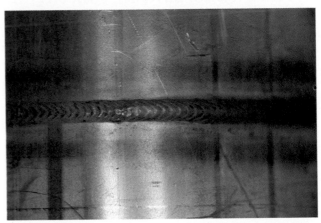

FIGURE 17-31 Taper down the weld when completing it to avoid a large crater. Larry Jeffus

molten weld pool when it is within 1/16 in. (2 mm) of the tack weld. At this time the key hole will be much smaller or completely closed.

Continue to rock the torch from side to side as you move ahead. Watch the molten weld pool. It should settle down or sink in when the weld has completely tied itself to the tack weld. Failure to do this will cause incomplete fusion or cold lap at the root. You should cross tack welds, carry the molten weld pool across the tack without adding filler, and start adding rod at the other end. If you are at

the closing end of the weld, slowly decrease the welding current as you add more welding rod to fill the weld crater or taper it down, **Figure 17-32.**

After the weld is complete around the pipe, visually check the root for any defects. Repeat this practice as needed until it is mastered. Turn off the welding machine, shielding gas, and cooling water and clean up your work area when you are finished welding.

Complete a copy of the "Student Welding Report" listed in Appendix I or provided by your instructor. ◆

Hot Pass

A hot pass can be used to correct some of the problems caused by a poor root pass. The hot pass may be a hotter-than-normal filler pass or it may be made without adding extra filler. It can be used to correct incomplete root fusion or excessive concavity of the root surface, **Figure 17-33.**

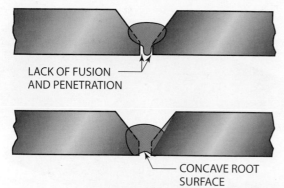

LACK OF FUSION
AND PENETRATION

CONCAVE ROOT
SURFACE

FIGURE 17-32 A hot pass can be used to correct incomplete root fusion or excessive concavity of the root surface. © Cengage Learning 2012

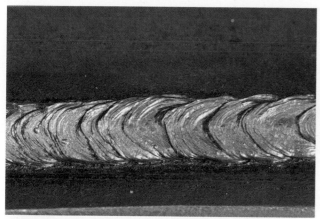

FIGURE 17-33 Inside root surface with proper penetration after the hot pass. Larry Jeffus

Concave root surfaces are generally caused by insufficient filler metal for the joint. To correct this condition, a hot filler pass is used to add both the needed metal and heat. During the weld, watch the molten weld pool surface to see when it sinks slightly (preferably equal to the needed reinforcement). Move the torch forward. Watch the molten weld pool sink and then add filler. Continue this process along the area of root concavity.

Incomplete fusion generally is caused by insufficient heat and temperature for the joint. To correct this condition, a hot pass is used without adding more filler metal. As before, watch the molten weld pool. When it sinks, move the torch ahead. Continue this process along the area of incomplete root fusion.

EXPERIMENT 17-1

Repairing a Root Pass Using a Hot Pass

Using a properly set-up and adjusted GTA welding machine; proper safety protection; two or more pieces of schedule 40 mild steel pipe 3 in. (76 mm) to 10 in. (254 mm) in diameter, prepared with single V-grooved joints with poor root weld passes; and some filler rods

having diameters of 1/16 in. (2 mm), 3/32 in. (2.4 mm), and 1/8 in. (3 mm); you will make a hot pass to correct defects in the root weld.

The first step is to determine the root weld defects and mark the location and type of defect on the outside of the pipe. Most defective root welds will have areas of one type of defect followed by another area of a different type of defect. This occurs most commonly as the student makes overadjustments to correct a problem or as the root opening and root face dimensions change.

Mark the groove with a punch or an alpha or numeric die stamp to indicate the type of defect below. Place the pipe with the defect in the 12 o'clock welding position and deposit the weld. The hot pass may vary in size as it progresses around the pipe. The weld can go faster across areas that do not require root repair. Turn off the welding machine, shielding gas, and cooling water and clean up your work area when you are finished welding.

Complete a copy of the "Student Welding Report" listed in Appendix I or provided by your instructor. ◆

PRACTICE 17-3

Stringer Bead, Horizontal Rolled Position (1G)

Using a properly set-up and adjusted GTA welding machine; proper safety protection; one or more pieces of mild steel pipe 3 in. (76 mm) to 10 in. (254 mm) in diameter; and some filler rods having diameters of 1/16 in. (2 mm), 3/32 in. (2.4 mm), and 1/8 in. (3 mm); you will make a straight stringer bead around a pipe in the horizontal rolled position.

Start by cleaning a strip 1 in. (25 mm) wide around the pipe. Brace yourself against the pipe so that you will not sway during the weld, **Figure 17-34.** The torch should have a slight upward slope, about 5° to 10° from perpendicular to the pipe surface, **Figure 17-35.** The filler wire should be held so that it enters the molten weld pool from the top center at a right angle to the torch.

FIGURE 17-34 Bracing the gloved hand against the pipe can help you control both the arc and filler metal. Note the precleaning of the pipe. Larry Jeffus

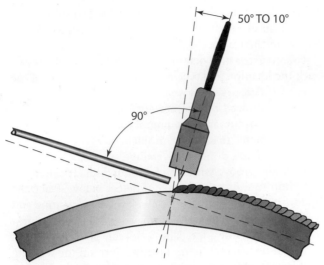

FIGURE 17-35 Alignment of torch and rod for pipe welding. © Cengage Learning 2012

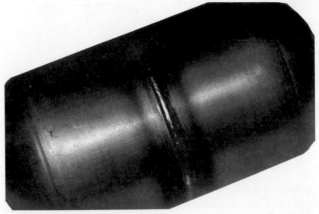

FIGURE 17-37 Beading on pipe. Larry Jeffus

FIGURE 17-38 The filler rod should be melted in the molten weld pool but not allowed to melt and drop into the weld pool. Larry Jeffus

Start the weld near the 3 o'clock position and weld upward to a point just past the 12 o'clock position. To minimize stops and starts, try to make the complete weld without stopping. With the power off and your helmet up, move the torch through the weld to be sure you have full freedom of movement.

With the torch held in position so that the tungsten is just above the desired starting spot, lower your helmet. Start the welding current by depressing the foot or hand control switch. Establish a molten weld pool and dip the filler rod in it. Use a straight forward and back movement with both the torch and rod, **Figure 17-36** and **Figure 17-37**. However, never move farther ahead than the leading edge of the molten weld pool. Too large a movement will lead to inadequate molten weld pool coverage by the shielding gas. The short motion is to allow the rod to be melted into the weld pool without the heat from the arc melting the rod back, **Figure 17-38**.

As the weld progresses from a vertical position to a flat position, the frequency of movement and pause times will change. Generally, the more vertical the weld, the faster the movement and the shorter the pause times. As the weld becomes flatter, the movement is slower and the pause times are longer, **Figure 17-39**. In the vertical position, gravity tends to pull the weld toward the back center of the bead. This makes a high crown on the weld with the possibility of undercutting along the edge. Flatter positions cause the weld to be pulled out, resulting in a thinner weld with less apparent reinforcement, **Figure 17-40**.

When the weld reaches the 12 o'clock position, stop and roll the pipe: To stop, slowly decrease the current and add welding rod to fill the weld crater. Keep the torch held over the weld until the postflow stops. This precaution will prevent the air from forming oxides on the hot weld bead.

After the weld is complete around the pipe, it should be visually inspected for straightness, uniformity, and defects. Repeat the process until you can consistently make the weld visually defect free. Turn off the welding machine, shielding gas, and cooling water and clean up your work area when you are finished welding.

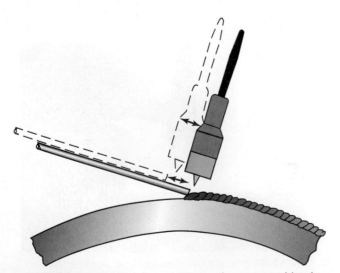

FIGURE 17-36 The torch and filler rod are moved back and forth together. © Cengage Learning 2012

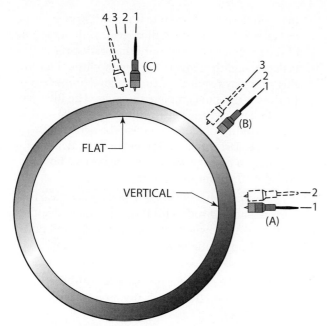

FIGURE 17-39 Counting to oneself is one way of setting the welding rhythm. As the weld progresses from vertical (A) to flat (C), the rhythm slows. © Cengage Learning 2012

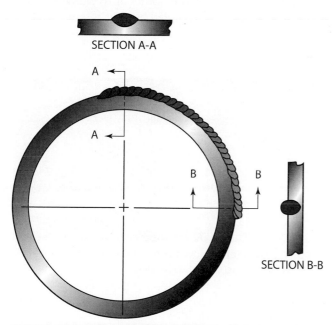

FIGURE 17-40 The weld bead contour, if uncontrolled, can change a great deal from the vertical to flat positions. © Cengage Learning 2012

Complete a copy of the "Student Welding Report" listed in Appendix I or provided by your instructor. ◆

PRACTICE 17-4

Weave and Lace Beads, Horizontal Rolled Position (1G)

Using a properly set-up and adjusted GTA welding machine; proper safety protection; one or more pieces of mild steel pipe 3 in. (76 mm) to 10 in. (254 mm) in diameter; and some filler rods having diameters of 1/16 in. (2 mm), 3/32 in. (2.4 mm), and 1/8 in. (3 mm); you will make a straight weave or lace bead around a pipe in the horizontal rolled position.

Start by cleaning a strip 1 in. (25 mm) wide around the pipe. Brace yourself and check to see that you have enough freedom of movement to make a weld from the 3 to 12 o'clock positions. Hold the torch with a 5° to 10° upward angle. The filler rod is to be held at a right angle to the torch.

Lower your helmet and establish a molten weld pool. Add filler metal as you move the torch to the side. Slowly build a shelf to support the molten weld pool or bead, **Figure 17-41.**

To make a weave bead, move the torch in a "C," "U," or zigzag pattern across the molten weld pool. The pattern should be no wider than one-half the diameter of the cup. Too large a weave means that part of the molten weld pool will not have a cover of the shielding gas. This results in atmospheric contamination of the molten weld pool.

To make a lace bead, slowly move the torch in a zigzag pattern across the weld bead. The molten weld pool should never be larger than that for a stringer bead. A lace bead can be made any desired width because the small molten weld pool is always protected from atmospheric contamination.

Continue the weave or lace bead up the pipe until you reach the top. To stop slowly, lower the current and add welding rod until the crater is filled. Turn the pipe and continue the bead around the pipe.

After the bead is complete around the pipe, visually inspect it for uniformity in width and reinforcement. The bead must also be free of visible defects. Repeat the process until you can consistently make the weld visually defect free. Turn off the welding machine, shielding gas, and cooling water and clean up your work area when you are finished welding.

Complete a copy of the "Student Welding Report" listed in Appendix I or provided by your instructor. ◆

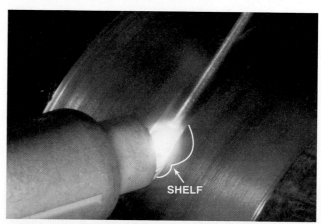

FIGURE 17-41 The shelf will support the molten pool during the vertical portion of the weld. Larry Jeffus

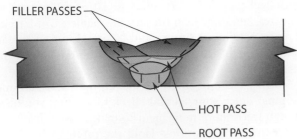

FIGURE 17-42 The filler pass(es) should fill the groove as much as possible but not more than to a 1/4 in. (6 mm) thickness at a time. © Cengage Learning 2012

Filler Pass

The filler pass is the next weld layer(s) to be made after the hot pass. The filler pass should be used to fill up the groove quickly. There should be complete fusion but little penetration. Deep penetration serves only to slow the completion of the joint and to subject the pipe to more heat (BTUs) and temperature (degrees) than needed, **Figure 17-42**.

Often only one GTA filler pass is used to protect the thin root pass from burnthrough. Because of the longer welding time required for the GTA process, pipe joints are often completed using another process. Most other processes, such as SMAW, GMAW, or FCAW, are much hotter and most likely will burn through the root without the added protection of a GTA hot pass and filler pass.

PRACTICE 17-5

Filler Pass (1G Position)

Using a properly set-up and adjusted GTA welding machine; proper safety protection; two or more pieces of 3-in. (76-mm) to 10-in. (254-mm) diameter schedule 40 mild steel pipe, prepared with single V-grooved joints and having a root pass (as described in Experiment 17-1); and some filler rods having 1/16-in. (2-mm), 3/32-in. (2.4-mm), and 1/8-in. (3-mm) diameters; make a filler welding pass in the groove in the horizontal rolled (1G) position.

- Place the pipe securely in an angle iron vee block.
- Using the same procedure practiced for the tack welds and root pass, place the cup against the beveled sides of the joint with the torch at a 5° to 10° upward angle.
- Brace yourself and check for full freedom of movement along the welded joint.
- Start the weld near the 1 o'clock position and end just beyond the 12 o'clock position.
- Lower your helmet and establish a molten weld pool. Using the same forward and backward motion

that you developed when making the stringer bead in Practice 17-3, add the filler metal at the top center of the molten weld pool until the bead surface is flat or slightly convex. Adding too little filler to this pass may result in burnthrough if another process is used to complete the joint. The filler pass should have as little penetration as possible so that the maximum reinforcement can be added with each pass. Deep penetration beyond the depth required to fuse the weld to the bevel and hot pass will slow the joint fill-up rate.

- To stop the molten weld pool, slowly decrease the current and add rod to fill the weld crater. Another method is to slowly decrease the current and pull the bead up on the beveled side of the joint, **Figure 17-43**.
- Turn the pipe and repeat this bead until the joint is filled up flush with the pipe surface, **Figure 17-44**.

Turn off the welding machine, shielding gas, and cooling water and clean up your work area when you are finished welding.

Complete a copy of the "Student Welding Report" listed in Appendix I or provided by your instructor. ◆

FIGURE 17-43 Taper down the weld when completing it to avoid a large crater. Larry Jeffus

FIGURE 17-44 The filler pass should fill the groove flush with the surface. Larry Jeffus

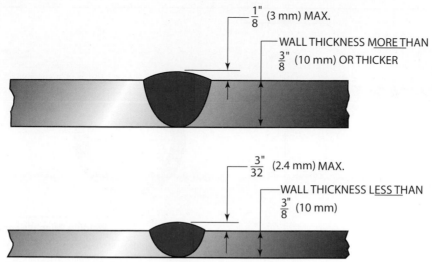

FIGURE 17-45 Excessively large weld beads prevent the pipe from uniformly expanding under pressure. This restriction from expanding can cause a failure to occur in or near the weld. © Cengage Learning 2012

Cover Pass

The stringer bead can be continued to cap the weld. If this technique is used, the bead should overlap the pipe surface no more than 3/32 in. (2.4 mm). Total reinforcement must not exceed 3/32 in. (2.4 mm) on pipe having walls less than 3/8 in. (10 mm) thick, **Figure 17-45**.

After the weld is complete, visually inspect it for uniformity in width and reinforcement and for any defects.

PRACTICE 17-6

Cover Pass (1G Position)

Using a properly set-up and adjusted GTA welding machine; proper safety protection; two or more pieces of 3-in. (76-mm) to 10-in. (254-mm) diameter schedule 40 mild steel pipe that are single V-grooved and are welded flush with filler passes (as described in Experiment 17-1); and some filler rods having 1/16-in. (2-mm), 3/32-in. (2.4-mm), and 1/8-in. (3-mm) diameters; you will cap the weld with a cover pass made in the horizontal rolled (1G) position, **Figure 17-46**.

Using the same techniques and skill developed in Practice 17-5, make a cover pass. The weld should not be more than 3/32 in. (2.4 mm) wider than the groove and should have a buildup of no more than 3/32 in. (2.4 mm).

After the weld is complete, visually inspect it for uniformity in width and reinforcement and for any defects. Repeat the process until you can consistently make the weld visually defect free. Turn off the welding machine, shielding gas, and cooling water and clean up your work area when you are finished welding.

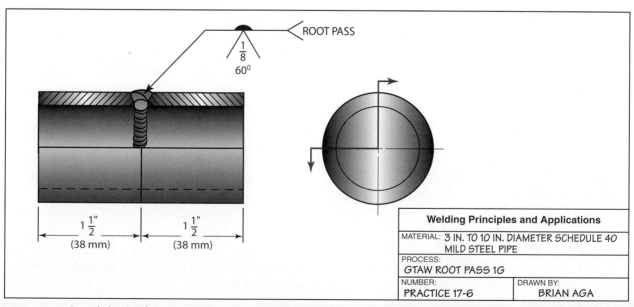

FIGURE 17-46 Butt joint in the 1G position. © Cengage Learning 2012

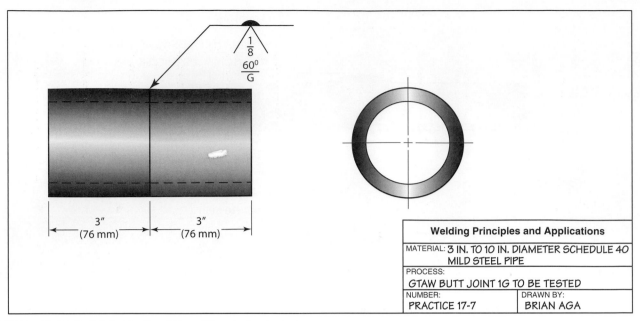

FIGURE 17-47 Butt joint in the 1G position to be tested. © Cengage Learning 2012

Complete a copy of the "Student Welding Report" listed in Appendix I or provided by your instructor. ◆

PRACTICE 17-7

Single V-Groove Pipe Weld, 1G Position, to Be Tested

Using the same equipment, setup, and materials as listed in Practice 17-1, you will make a welded butt joint in the horizontal rolled (1G) position. Then you will test the weld for defects, **Figure 17-47**.

Make the tack welds, root pass, filler pass, and cover pass as described in Practices 17-1, 17-2, 17-5, and 17-6.

When the weld is complete, inspect it for defects. If it passes visual inspection, cut out guided-bend test specimens as shown in **Figure 17-48**. Test the specimens for weld defects. Repeat this practice until the test is passed. Turn off the welding machine, shielding gas, and cooling water and clean up your work area when you are finished welding.

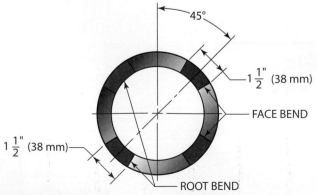

FIGURE 17-48 Guided-bend specimens. © Cengage Learning 2012

Complete a copy of the "Student Welding Report" listed in Appendix I or provided by your instructor. ◆

PRACTICE 17-8

Stringer Bead, Horizontal Fixed Position (5G)

Using the same equipment, setup, and materials as listed in Practice 17-1, make a straight stringer bead around an ungrooved pipe in the horizontal fixed position.

- Clean a strip 1 in. (25 mm) wide around the pipe.
- Mark the top for future reference and clamp the pipe at a comfortable work height.
- Brace yourself so you are steady and move through the weld to make sure that you have freedom of movement.
- Lower your helmet and establish a molten weld pool at the 6 o'clock position. Keep the molten weld pool size small so that it will stay uniform.
- Dip the filler metal into the front leading edge of the molten weld pool.
- Move the torch forward and backward as the rod is added. The frequency of movement will increase as the weld becomes more vertical.
- Continue the weld without stopping, if possible, to the 12 o'clock position.
- Repeat the weld up the other side in the same manner.

When the weld is complete, visually inspect it for straightness and uniformity, and for any defects. Repeat the process until you can consistently make the weld visually defect free. Turn off the welding machine, shielding gas, and cooling water and clean up your work area when you are finished welding.

Complete a copy of the "Student Welding Report" listed in Appendix I or provided by your instructor. ◆

PRACTICE 17-9A AND 17-9B

Single-V Butt Joint (5G Position)
A. Root Penetration May Vary
B. 100% Root Penetration to Be Tested

Using the same equipment, setup, and materials as listed in Practice 17-1, make a weld on a pipe in the horizontal fixed (5G) position.

The pipes should be tack welded, and the tacks should be made in the following clock positions: 12:00, 3:00, 6:00, and 9:00, **Figure 17-49** and **Figure 17-50**.

■ Start the root weld at the 6:30 o'clock position and weld uphill around one side of the pipe to the 12:30 o'clock position. Start with the cup against

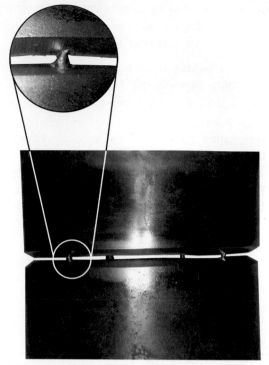

FIGURE 17-50 Tack welded pipe joint. Larry Jeffus

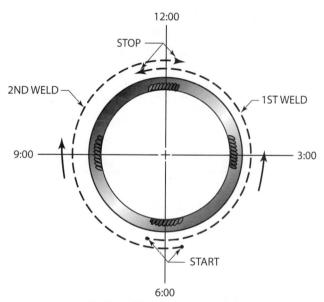

FIGURE 17-49 Tack welds on pipe in 5G position and the welding sequence. © Cengage Learning 2012

the pipe bevels, as was done before, and add the rod at the leading edge of the molten weld pool.

■ Repeat this process up the other side of the pipe. The starts and stops of these welds should overlap slightly to ensure a good tie-in.

The filler passes should also start near the 6:30 o'clock position and go up both sides of the pipe to a point near the 12:30 o'clock position, **Figure 17-51**. Staggering the location of these beads will help prevent defects arising from discontinuities caused at the starting and stopping points. Keeping these beads small will help you to control them. A large number of good weld beads is more desirable than a few poorly controlled beads.

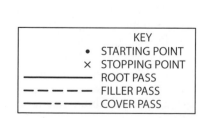

KEY	
•	STARTING POINT
✕	STOPPING POINT
———	ROOT PASS
– – –	FILLER PASS
–·–·–	COVER PASS

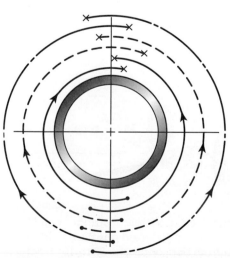

FIGURE 17-51 Stagger the starting and stopping points of each bead. © Cengage Learning 2012

The cover pass may be made as either a weave or lace bead. The lace bead is usually more easily controlled because the molten weld pool is smaller.

- After the weld is complete, visually inspect it for uniformity in width and reinforcement.

- Check also for visible defects and root penetration.

- If the joint passes all visual tests and has 100% root penetration, guided-bend test specimens can be cut out (refer to Figure 17-10).

- After bending, evaluate the specimens for defects. Repeat this practice until both parts A and B can be passed.

Turn off the welding machine, shielding gas, and cooling water and clean up your work area when you are finished welding.

Complete a copy of the "Student Welding Report" listed in Appendix I or provided by your instructor. ◆

PRACTICE 17-10

Stringer Bead, Vertical Fixed Position (2G)

Using the same equipment, setup, and materials as listed in Practice 17-1, make a straight stringer bead around a pipe in the vertical fixed position.

- Hold the torch at a 5° angle from horizontal and 5° to 10° from perpendicular to the pipe, **Figure 17-52**. This will allow good visibility and simplify the process of adding the rod at the top of the bead.

- Establish a small molten weld pool and filler along the top front edge, **Figure 17-53**. Gravity will tend to pull the metal down, especially if the molten weld pool becomes too large. When the molten

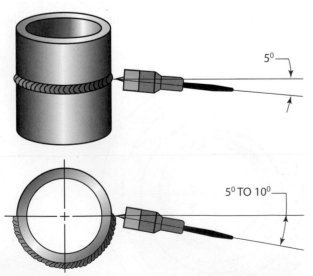

FIGURE 17-52 When making a straight stringer bead around a pipe in the horizontal fixed position, hold the torch at a 5° angle from horizontal and 5° to 10° from perpendicular to the pipe. © Cengage Learning 2012

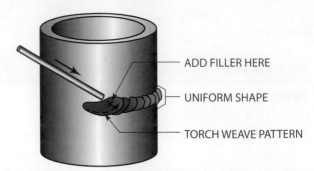

FIGURE 17-53 Rod and torch movement for a horizontal weld. © Cengage Learning 2012

weld pool sags, undercut appears along the top edge of the bead. A slight "J" pattern may help to control weld bead sag by making a small shelf to support the molten weld pool while it cools.

When the weld is completed, visually inspect it for uniformity, straightness, and any defects.

Repeat the process until you can consistently make the weld free of visual defects. Turn off the welding machine, shielding gas, and cooling water and clean up your work area when you are finished welding.

Complete a copy of the "Student Welding Report" listed in Appendix I or provided by your instructor. ◆

PRACTICE 17-11A AND 17-11B

Single-V Butt Joint (2G Position)
A. Root Penetration May Vary
B. 100% Root Penetration to Be Tested

Using the same equipment, setup, and materials as listed in Practice 17-1, you will weld a butt joint on a pipe in the vertical fixed (2G) position.

Tack weld the pipes and start the root pass at a point between the tack welds. The root weld should be small enough so that surface tension will hold it in place. Too large a root weld bead will cause the top root face to be slightly concave.

The filler passes can be larger if they are made along the lower beveled surface, **Figure 17-54**. This will support the filler passes and allow a faster joint buildup. The next pass is on the top side of the first. This process of resting the next pass on the last pass continues until the joint is filled.

The cover pass is started around the lower side of the joint, overlapping the pipe surface by no more than 1/8 in. (3 mm), **Figure 17-55**. The next pass covers about one-half of the first cover pass and should be slightly larger. This process of making each pass larger continues until the center weld is complete. Then, each weld is smaller than the last. This process builds a uniform reinforcement having a good profile.

After the weld is complete, visually inspect it for uniformity and defects. If it passes, inspect the root for 100% penetration. If the root passes, cut out guided-bend test

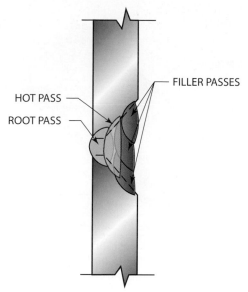

FIGURE 17-54 Filler pass(es) over root weld on pipe in 2G position. © Cengage Learning 2012

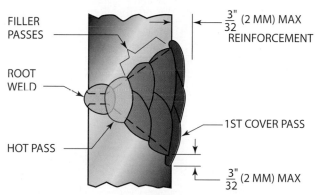

FIGURE 17-55 Cover passes on pipe in 2G position. © Cengage Learning 2012

specimens. Test the specimens and evaluate the results. Repeat this weld until both parts A and B can be passed. Turn off the welding machine, shielding gas, and cooling water and clean up your work area when you are finished welding.

Complete a copy of the "Student Welding Report" listed in Appendix I or provided by your instructor. ◆

PRACTICE 17-12

Stringer Bead on a Fixed Pipe at a 45° Inclined Angle (6G Position)

Using the same equipment, setup, and materials as listed in Practice 17-1, make a straight stringer bead around a fixed, ungrooved pipe at a 45° inclined angle.

- Starting at the 6:30 o'clock position with the torch at a slight downward angle, establish a small molten weld pool, **Figure 17-56.**
- Add the rod at the upper leading edge of the molten weld pool. A slight "J" pattern can help control the bead size. As the weld progresses around the pipe, the weld becomes more vertical. The rate of movement should increase.
- Keep the size of the weld small so it can be controlled.

After the weld is complete, visually inspect it for uniformity and for any defects. Repeat the process until you can consistently make the weld free of visual defects. Turn off the welding machine, shielding gas, and cooling water and clean up your work area when you are finished welding.

Complete a copy of the "Student Welding Report" listed in Appendix I or provided by your instructor. ◆

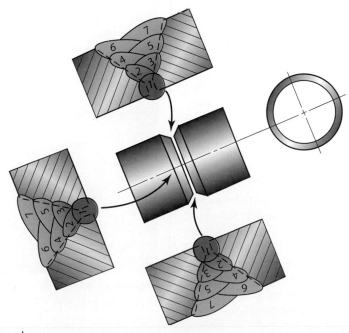

FIGURE 17-56 6G welding bead sequence. © Cengage Learning 2012

PRACTICE 17-13A AND 17-13B

Single-V Butt Joint (6G Position)
A. Root Penetration May Vary
B. 100% Root Penetration to Be Tested

Using the same equipment, setup, and materials as listed in Practice 17-1, you will make a welded butt joint on a pipe fixed at a 45° angle (6G) position.

Starting at the bottom, make the root pass upward to the 12:30 o'clock position. The size of the root pass should stay the same; however, it may be off to one side more than the other. As long as the root surface is uniform, this unbalanced appearance is acceptable. The filler passes should correct this problem.

The filler passes are applied to the downhill side first so that the upper side will be supported, **Figure 17-57**. The starts and stops should be staggered so that they do not increase the possibility of defects at any one point on the weld. A slight "J" weave pattern will help hold the shape of the bead as it changes from overhead to vertical to horizontal.

The cover pass is easy to control if it is a series of stringer beads. Start on the lower side and build up the cover pass as in the 2G position. Each weld should overlap the preceding weld so that the finished weld is uniform.

After the weld is complete, visually inspect it for uniformity and for any defects. If the weld passes, inspect

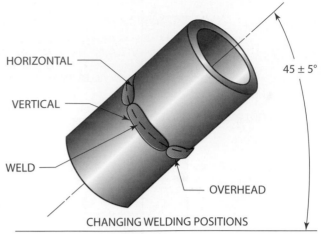

FIGURE 17-57 In the 6G position, the pipe is at a 45° angle. The effective welding angle changes as the weld progresses around the pipe. © Cengage Learning 2012

the root for 100% penetration. If the root passes this inspection, cut out guided-bend test specimens. Test the specimens and evaluate the results. Repeat this practice until both parts A and B can be passed. Turn off the welding machine, shielding gas, and cooling water and clean up your work area when you are finished welding.

Complete a copy of the "Student Welding Report" listed in Appendix I or provided by your instructor. ◆

Summary

Often gas tungsten arc welding of pipe is considered to be the most prestigious of the welding processes. It has obtained this status primarily because of where much of the GTA welding of pipe is done. High-quality gas tungsten arc pipe welding is performed on nuclear reactors, in petrochemical plants, and in the fabrication of the international space station. There are many other applications for this process, and they enjoy the same prestige, although they are often not nearly as critical.

The constantly changing weld position is the most challenging part of gas tungsten arc welding. Learning how to gently rest the nozzle on the surface of the pipe as you make your welds can be a tremendous benefit. The ceramic cup does not slide very easily on the metal surface, so it takes a very light touch to learn this technique. Once you have developed this light touch, you can begin working on the more complicated cup rocking technique. A skilled welder using the cup rocking technique can produce welds that appear machinelike in quality.

Access to the welding joint can be an additional challenge that the GTA welder must overcome, **Figure 17-58**.

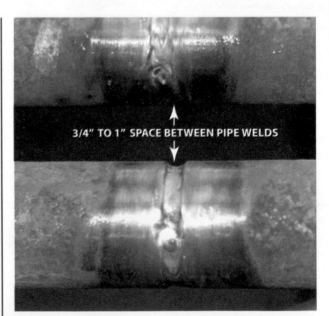

FIGURE 17-58 Restricted access during welding can challenge a welder's skills. Larry Jeffus

Summary (continued)

FIGURE 17-59 Typical restricted space on boiler tube welds. Larry Jeffus

The limited access on these boiler tubes is typical of the restricted access that can be found in both new and repaired GTA welding. Thousands of these restrictive access welds can be required to repair and/or update a power plant boiler, **Figure 17-59.** Each weld must pass inspection because any weld that fails can result in the plant having to be shut down for repair work.

Hot Tap Weld Prevents Offshore Piping System from Shutting Down

Seawater injection is commonly used in offshore facilities to maintain reservoir pressure and sustain oil and gas production. Before the seawater is injected into the reservoir, it is first processed through a deaerator to reduce the oxygen content of the water.

Seawater contains sulphates in the form of salts that, if not removed, react with barium in the reservoir to form barium sulphate. This compound is deposited within tubing as scale and eventually cuts off the flow of hydrocarbons to the wellhead. Removing barium sulphate scale from production tubing is difficult and, in extreme situations, may mean abandoning the well. Once a problem with scale has been identified, it is essential to add facilities within the water injection system to remove it.

For a recent North Sea project, the sulphate removal facilities were built onshore, then delivered offshore so that a platform crane could lift them from a supply ship. Once installed on the platform, the new equipment was hooked up by attaching a 12-in. (30.48-cm) diameter branch to the existing 16-in. (40.64-cm) water injection line by welding, using a procedure known as hot tapping.

Materials

The existing water injection pipe system was manufactured from super duplex stainless steel.

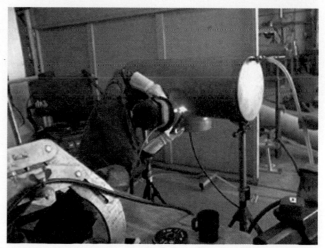

Welding the 12-in. branch to the 16-in. diameter water injection pipe mockup. American Welding Society

The engineers for the new facility selected a similar grade of material.

Welding Procedures

Welding procedures were carried out in accordance with Section IX of the American Society for Mechanical Engineering (ASME) Boiler and Pressure Vessel Code using gas tungsten arc welding (GTAW). The welding consumable selected was Metrode Zeron 100X, and all wires were stored in a dry condition before use.

The following procedures were qualified for the welding of the branch to the water injection pipe.

■ Butt joint weld procedure. This procedure was carried out to qualify the weld metal and heat-affected zone (HAZ) properties. Procedure tests were carried out on a 6-in. (152.4-mm) diameter 0.24-in. (6-mm) thick pipe, which featured a butt joint weld.

■ Hot tap simulation procedure. To establish a suitable weld procedure for hot tap welding of a 12-in. diameter branch to a 16-in. diameter water injection pipe, the production weld was simulated onshore. This would also allow the welder assigned to the offshore weld the opportunity to establish the best welding practice.

In order to establish the procedure for the offshore weld, it was decided to flow water through the 16-in. diameter pipe at 59°F (15°C) to simulate the actual water injection process.

Mechanical Properties

Mechanical tests selected from the weld procedure were cross-tensile, impact, and bend tests.

Bend tests from the surface and root areas of the weld procedure tests were carried out, and all test specimens were found to be satisfactory.

Hardness Tests

Hardness surveys using an (11-kg) load were carried out on the butt joint weld and the simulated hot tap weld. Hardness specimens were removed from the surface and the root area of the butt joint weld and the 3 o'clock and 12 o'clock positions of the simulated hot tap weld. The maximum hardness of the HAZ and weld metal of the procedure test plate recorded values of 283 for both positions. Hardness results from the simulated hot tap weld recorded values of 287 and 299 for the weld metal and HAZ, respectively. The hardness values recorded satisfied the contract requirements of a 300 Hardness Value.

Precautions were also taken to make sure the weld joint was free from moisture, which could result in absorption of hydrogen into the weld metal.

Corrosion Tests

Corrosion test specimens were selected from the root area of the butt joint weld and simulated hot tap. The examination

The completed weld at the offshore facility containing the saddle and reinforcing plates. American Welding Society

revealed evidence of pitting on the cut surface of each specimen, but the loss was considered acceptable within the code requirements.

Hot Tapping Offshore

The field weld was carried out to meet the procedures determined onshore. All tack welds were removed during welding of the root run. Precautions taken during welding included making sure the weld joint area was free from condensation, as any moisture present during welding could result in the absorption of hydrogen into the weld metal. Nondestructive examination (NDE) of the completed weld consisted of visual inspection and liquid penetrant. The welds for the saddle and reinforcing plates supporting the 12-in. diameter branch were inspected with nondestructive examination, similar to that of the other welds. To complete the branch connection, a 12-in. diameter flange section was welded to the 12-in. diameter branch. On completion, NDE was carried out to ensure the weld quality satisfied the requirements of ASME standard B31.3.

Summary

Hot tapping super duplex stainless steel pipe is not a common practice offshore. It is essential to weld a mockup of the pipe joints onshore to ensure the welding process is the right choice and the welder chosen to do the job is familiar with the weld position and welding technique needed for this kind of welding.

Article courtesy of the American Welding Society.

Review

1. What types of industries require GTAW-quality pipe welds?

2. What are the grooves most often used for welding on mild steel pipe?

3. What condition must the prepared edges of a pipe be in before welding?

4. How is the depth of root penetration measured?

5. What problem can incomplete fusion cause?

6. What are the common causes of a concave root surface?

7. What problems can excessive root reinforcement cause?

8. What can cause root contamination?

9. What gases are used as backing gases?

10. How long would it take to purge a 15-ft section of 6-in. diameter pipe using a flow rate of 35 cfh?

11. How long would it take to purge a 10-mm section of a 20-cm diameter pipe using a flow rate of 17 L/min?

12. How can filler metal be added to make a root pass?

13. What information must be supplied when ordering consumable inserts?

14. How is the tack weld ended so that it will not crack?

15. What is the 1G pipe position?

16. After the weld is set up and ready to begin, how might the welder want to check for freedom of movement?

17. What type of weld can be made using the nozzle walking technique?

18. What two problems can be corrected by a hot pass?

19. Why must you brace yourself when making a weld?

20. What can happen to the weld pool if the torch movement is too long?

21. What technique should be used when stopping a weld?

22. What is the maximum width that a weave pattern should be?

23. Why should the filler pass penetration be limited in depth?

24. How much filler metal should be added to a filler bead?

25. How large should a cover pass be on a pipe with a 1/4-in. (6-mm) wall thickness?

26. What is the 5G pipe position?

27. Why should the starting point of the beads be staggered?

28. How should the torch be held to allow good visibility on a 2G weld?

29. Why should the root bead be small on a 2G weld?

30. Why are stringer beads in the 6G position kept small?

31. Why are the downhill side filler welds put in first on 6G pipe?

SECTION 5

RELATED PROCESSES

Success Story

K assandra Rieke, who now lives in La Junta, Colorado, started her welding career at an early age. She was just 12 years old when her uncle first taught her to weld on a ranch in Meeker, Colorado. Never afraid of hard work, she got a job when she was just 18 years old running heavy equipment at a big gravel pit in Meeker for about a year.

Just after her 19th birthday, a friend took her to work as a welder's helper on oil and gas rigs. "Welding on oil rigs can be a lot of hard work—we usually worked six twelves (12 hours a day 6 days a week), but when you love what you are doing, it does not feel like work." She fondly recalls working 38 straight hours to get one rig back in service. "When one of the rigs is down, you just have to keep on welding as long as it takes to get it back working." Even with the long hours, she was back welding, a job she loved, and this was the beginning of her career.

By the time she moved with her family to Pueblo, she had graduated from high school and was looking for a welding school to hone her skills. She wanted to learn more about natural gas pipe welding. After visiting several schools and having heard some great things about it from welder friends, she decided to attend Trinidad State Junior College's Energy Production Industrial Construction Program.

As an outdoor enthusiast who loves hunting, fishing, and working outdoors and with classes over, she was anxious to get back out in the field and start welding again. She had her own welding rig consisting of a Dodge 1-ton dually with the custom fabricated bed she and a friend built, a Lincoln SA-200 Pipeliner welder generator, and torch set.

Even though she really enjoyed the opportunity to travel throughout many of the western states working as a welder-fabricator, she has moved back to southern Colorado. She currently works as an agricultural repair and fabrication welder in the La Junta area.

Recently, she added a plasma cutter to her current welding rig. She said it has really made cutting out parts so much easier than it was when she only had an oxyacetylene cutting torch. Running the plasma cutting torch off of the welder generator's AC power outlet has made it a lot easier to cut everything, especially on thin sheet metal. The plasma torch doesn't distort sheet metal like the acetylene torch did, so it is easier and faster to make a fabrication.

She loves to ride horses and rodeo a little and wants someday to save the money she earns from welding to buy a ranch and maybe help underprivileged kids by giving them the opportunity to experience the outdoor life that she has. ∎

Chapter 18

Shop Math and Weld Cost

OBJECTIVES

After completing this chapter, the student should be able to

- solve basic welding fabrication math problems.
- round numbers.
- convert mixed units, fractions, and decimal fractions.
- reduce fractions and decimal fractions.

KEY TERMS

common denominator

converting or conversion

decimal fractions

denominator

dimensioning

equation

formula

fractions

mixed units

numerator

rounding

sequence of mathematical operations

tolerance

whole numbers

INTRODUCTION

#1 The most common use of math in welding shops is for **dimensioning** and pricing. For dimensioning, most welding shops use the standard or English system with feet, inches, and fractions. A few shops use metric dimensioning (SI). The metric system, sometimes abbreviated as SI, comes from the French term *le Système International d'unités*. This system is made up of seven base units, which include units for length, temperature, weight, and so on, **Table 18-1**. For cost estimation, calculations may include labor costs hours and minutes; material costs in pounds, ounces, feet, and inches; and overhead costs as a percentage of the company's cost for insurance, electricity, rent, mortgage payments, and so on.

Although most shops use the standard system, the math functions of addition, subtraction, multiplication, and division of dimensions are easier in the metric system than they are in the standard system because the metric system is a decimal system. The advantage of a decimal system is that a calculator can more easily be used than with the mixed-numbered standard system.

Almost all welding shop dimensioning uses **mixed units.** An example of a mixed unit is feet and inches, with inches based on 12 in. to the foot. So, the largest number of inches is 11 because when you add one more inch it becomes 12 in., which is expressed as 1 ft. Other common mixed units are pounds and ounces, and hours and minutes. Mixed units present unique problems for addition, subtraction, multiplication, and division because each type of unit must be worked separately.

Abbreviations for Units				
Units		**Standard**		**Metric**
Temperature	F	Fahrenheit	C	Celsius
Length	yd	yard	m	meter
	ft	feet	cm	centimeter
	in.	inches	mm	millimeter
Weight	lb	pounds	kg	kilogram
	oz	ounces	g	grams
Liquid	gal	gallons	L	liters
	qt	quarts		
	pt	pints		
	oz	ounces		
Pressure	psi	pounds per square in.	kPa	kilopascal

TABLE 18-1 Common Abbreviations for Units of Measure

For example, you cannot add feet to inches without first converting the feet to inches.

Most welding shops encourage the use of calculators because they eliminate arithmetic errors. There are a few calculators that can add, subtract, multiply, and divide standard fractional dimensions, and some can even work mixed units of feet and inches. However, most math using fractions must be worked manually.

For pricing and cost estimating, welders must use math to enable their welding business to operate profitably. The owner or manager must be able to make cost-effective welding decisions. A number of factors affect the cost of producing weldments. Some of these factors include the following:

- Material
- Weld design
- Welding processes
- Finishing
- Labor
- Overhead

SHOP MATH

Mathematics has many branches—calculus, geometry, trigonometry, and so on—but arithmetic is the basic branch of mathematics that is primarily involved with combining numbers by addition, subtraction, multiplication, and division. Arithmetic is the math that most welders use on a daily basis. Welding layout and to some extent fitup may involve geometry and trigonometry. Both of these branches of mathematics deal with angles and points and are related to laying out and fitting up complexly shaped weldments.

Types of Numbers

Welders use only a few types of numbers on a regular basis. They most commonly use the following types of numbers:

- **Whole numbers—Whole numbers** are numbers used to express units in increments of 1, so they can be divided evenly by the number 1. Examples of whole numbers are 1, 2, 3, 4, 5, 6, 7, 8, 9, 10. They are the easiest type of numbers to use with all types of mathematical functions.

- **Decimal fractions—A decimal fraction** is a number that uses a decimal point to denote a unit that is smaller than 1. Examples of decimal fractions are 0.1, 0.35, 0.9518, 1.1, 5.4, 3.14, and 125.1234. Decimal fractions are expressed in units that are 10 times, 100 times, 1000 times, and so on, smaller than 1, **Figure 18-1.** They can be added and subtracted as easily as whole numbers. Decimal fractions can be used with whole numbers.

- **Mixed units**—Mixed units are measurements containing numbers that are expressed in two or more different units. An example of a mixed unit is a linear measurement such as 2 ft 6 in., with part of the measurement expressed in feet (2) and the other part in inches (6). Other examples of mixed units are angular measurements such as 45° 0', weight measurements such as 8 lb 4 oz, and time measurements such as 3 hrs 20 min 15 sec. The most common types of mixed units used in welding fabrication are linear dimensions, angular dimensions, weight, and time. Linear dimensions use units of feet and inches (SI: meters, centimeters, and millimeters); angular dimensions use degrees, minutes, and seconds; weight uses pounds and ounces (SI: kilograms and grams); and time uses hours, minutes, and seconds. The different units may be separated with a space ( ) or dash (–). It is important to keep the different number units straight when performing mathematical functions.

- **Fractions—A fraction** is two or more numbers used to express a unit smaller than one. Examples of fractions are 1/2, 3/4, 5/16, 2 3/8, and 9 1/2. A dash (–) or slash (/) is used to separate the top and bottom numbers of a fraction. The **denominator** is the bottom number of a fraction, and the **numerator** is the top number, **Figure 18-2.** Fractions can be the hardest type of number to work with because they often cannot be added, subtracted, multiplied, or divided without first having to make some type of **conversion.** Fractions can include whole numbers, such as the "9" in 9 1/2.

GENERAL MATH RULES

All of the math problems in this chapter will be set up and worked in the same manner.

1st Step   The equation or formula will be stated on the first line.

2nd Step   "Where": explains the meaning of any variables.

3rd Step   State the problem's known values and what answer is needed.

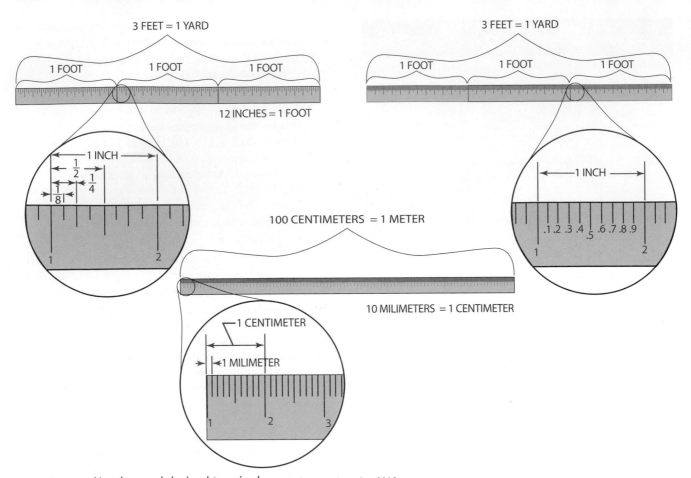

FIGURE 18-1 Number and decimal terminology. © Cengage Learning 2012

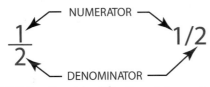

FIGURE 18-2 Fraction identification. © Cengage Learning 2012

4ᵗʰ Step Write the equation or formula.

5ᵗʰ Step Write the known values in place of the variables.

6ᵗʰ Step One mathematical calculation will be performed per line.

7ᵗʰ Step Add as many lines as necessary to complete the problem's calculations.

8ᵗʰ Step Give the answer, including units.

9ᵗʰ Step Explain the answer in a written statement.

When working a math problem by hand, write the down in a vertical series of lines with each individual step on a line. Keeping the equal symbol on each line lined up vertically will reduce confusion and make it easier to refer back to the problem later. Working math problems in this very structured way makes it easier for you to look back at the examples in this textbook and work new problems you might encounter in a welding shop. Years later, it can be very frustrating to know the equation or formula but not remember the sequence of mathematical steps.

Letters, numbers, and symbols are often used as expressions in equations and formulas to make it easier to write out the formula. For example, to find the area of a plate you would multiply length times width. However, if you assigned a letter to each of the variables, writing the formula can be made clearer. So, if we assign the letter A for area, the letter L for length, and the letter W for width, the formula can be written $A = L \times W$.

Sometimes formula expressions will have a superscript. Superscripts can be numbers or letters written to the upper right of a formula expression such as the 2 in the expression, 3^2. In this case, it means that the number 3 is squared, or multiplied by itself two times. Expressions may also have a subscript. Subscripts can be numbers or letters written to the lower right of an expression. They are often used to define the expression such as the term *weld* in the expression CS_{weld}. In this case, it is identifying the CS (cross section) as that of the weld. Some of the formula expressions used in this textbook are shown in **Table 18-2**.

EQUATIONS AND FORMULAS

An **equation** is a mathematical statement in which both sides are equal to each other; for example, $2X = 1Y$. In this equation the value of X is always going to be 1/2 of the value

Expression		Meaning
b	=	base dimension of weld
BD	=	bead depth
BOR	=	burn-off rate in pounds per hour
CS_{root}	=	cross-sectional area of the root opening
CS_{weld}	=	cross-sectional area of the weld
DE	=	deposit efficiency
DR	=	deposition rate in pounds per hour
EL	=	electrode length
GV	=	groove volume
h	=	height dimension of weld
LOCH	=	labor and overhead cost per hour
l	=	length dimension of weld
MD	=	metal density (weight of metal in pounds per cubic inch)
OF	=	operating factor in decimal percent
%DE	=	percentage of weld deposition efficiency
PT	=	plate thickness
RG	=	root gap
RO	=	root opening
SL	=	stub loss
TCS	=	total cross-sectional area
TLOC	=	total labor and overhead cost
$Wt_{electrode\ used}$	=	weight of electrode used
$Wt_{weld\ metal}$	=	weight of weld metal
LH	=	weld leg height
LW	=	weld leg width
WL	=	weld length

TABLE 18-2 Formula Expressions Used in This Textbook

of Y. An example of an equation used in metal fabrication would be: *the number of hours worked (hrs)* $\times$ *pay per hour ($)* = *total labor bill (T)*, or *hrs* $\times$ *$ = T*. If either the hours or the pay rate goes up, the total bill goes up too and vice versa.

To find the labor cost, use the following equation:

$$hrs \times \$ = T$$

Where:

hrs = hours worked
$ = hourly rate
T = total labor bill in dollars

Find the total labor bill for 4 hours of work at $15 per hour, **Figure 18-3**.

$$hrs \times \$ = T$$
$$4 \times 15 = T$$
$$60 = T$$

The total pay (T) for the 4 hours of work at $15 per hour would be $60.

A **formula** is a mathematical statement of the relationship of items. It also defines how one cell of data relates to another cell of data; for example, wt = [(l" $\times$ w" $\times$ t") ÷ 1728] $\times$ wt/ft. In this formula you must first find the volume of material by multiplying length (l) times width (w) times thickness (t) before dividing that number by the number of cubic inches in a cubic foot (1728), then multiply that number by the material's weight per cubic foot (wt/ft). The **sequence of mathematical operations** is important when working formulas and equations. For example, 6 $\times$ 4 ÷ 2 $\times$ 5 = 60, but (6 $\times$ 4) ÷ (2 $\times$ 5) = 2.4. When a formula has more than one mathematical operation, the operations must be performed in the following order:

1st perform all operations within parentheses.

2nd resolve any exponents.

3rd do all multiplication and division working from left to right.

4th do all addition and subtraction working from left to right.

JOB CARD		
Job *Weld out Go Cart Frame*	Date _____	Welder *lfj*
Starting Time	Ending Time	Total Time
7:00 am	*11:00 am*	*4 hrs*
	Total Hours	*4 hrs*
	Hourly Rate	x *$15/hr*
	Total Labor	*$60.00*

FIGURE 18-3 Bill of materials. © Cengage Learning 2012

To find the total weight of a piece of material, use the following formula:

$$wt = [(l'' \times w'' \times t'') \div 1728] \times wt/ft$$

Where:
wt = total weight
l" = length in inches
w" = width in inces
t" = thickness in inces
1728 = number of cubic inches in a cubic foot
wt/ft = weight of material in pounds per cubic foot

Find the total weight of a piece of steel that is 144 in. (365.76 cm) long, 12 in. (30.48 cm) wide, and 2 in. (5.08 cm) thick if steel weighs 490 lb per cubic foot (7847 kg per cubic meter):

Standard Units

$$wt = [(l'' \times w'' \times t'') \div 1728^*] \times wt/ft$$
$$wt = [(144 \times 12 \times 2) \div 1728] \times 490$$
$$wt = [(1728 \times 2) \div 1728] \times 490$$
$$wt = [(3456) \div 1728] \times 490$$
$$wt = 2 \times 490$$
$$wt = 980$$

SI Units

$$wt = [(l \text{ cm} \times w \text{ cm} \times t \text{ cm}) \div 10{,}000^*] \times wt/m$$

$$wt = [(365.76 \text{ cm} \times 30.48 \text{ cm} \times 5.08 \text{ cm})$$
$$\div 1{,}000{,}000] \times 7847$$
$$wt = [(11{,}148.4 \times 5.08) \div 1{,}000{,}000] \times 7847$$
$$wt = [(56{,}633.7) \div 100{,}000] \times 7847$$
$$wt = 0.5663 \times 7847$$
$$wt = 444.4$$

1,000,000 is the number of cubic centimeters in a cubic meter. The total weight of this piece of steel is 980 lb (444.4 kg).

MIXED UNITS

The problems caused while working with mixed numbers in the standard system do not exist in the metric system because the metric system is based on 10. In the metric system, to change the base unit all you have to do is move the decimal point. For example 25.4 mm is 2.54 cm and 100 mg is 10.0 g. However, in the standard system converting inches to feet or ounces to pounds is more complex because inches to feet is base 12 and ounces to pounds is base 16. So, when adding mixed units such as feet and inches, you have to add each type of number together first. For example, you would add the inches to inches and feet to feet.

Adding and Subtracting Mixed Units

An example of a mixed unit problem you might find in welding would be to see how many feet of metal stock you need if one piece is 10 ft 6 in. long and the other is 3 ft 5 in.

---

*1728 is the number of cubic inches in a cubic foot.

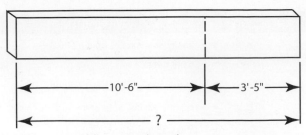

FIGURE 18-4 Adding mixed numbers. © Cengage Learning 2012

long, **Figure 18-4**. The first step would be to write the numbers in columns with feet over feet and inches over inches. The second step would be to add the inches to the inches and feet to the feet.

Add   10' 6" + 3' 5"

1st Step	Feet Column	Inch Column
	10'	6"
2nd Step +	3'	5"
	13'	11"

or

Add   10' 6" + 3' 5"

1st Step   10' + 3' = 13'
2nd Step 6" + 5" = 11"
3rd Step 13' + 11" = 13' 11"

When subtracting mixed units, use the same steps as were used for adding. For example, to see how many feet of scrap pipe you have left from a 7 ft 8 in. piece when 5 ft 3 in. will be cut off, **Figure 18-5**, you would do the following:

Subtract   7' 8" − 5' 3"

1st Step	Feet Column	Inch Column
	7'	8"
2nd Step −	5'	3"
	2'	5"

or

Subtract   7' 8" − 5' 3"

1st Step   7' − 5' = 2'
2nd Step 8" − 3" = 5"
3rd Step 2' + 5" = 2' 5"

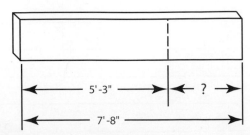

FIGURE 18-5 Subtracting mixed numbers. © Cengage Learning 2012

Feet and Inches		Pounds and Ounces	
Ft Fraction	**Inches**	**Lb Fraction**	**Ounces**
1/12	1	1/16	1
2/12	2	2/16	2
3/12	3	3/16	3
4/12	4	4/16	4
5/12	5	5/16	5
6/12	6	6/16	6
7/12	7	7/16	7
8/12	8	8/16	8
9/12	9	9/16	9
10/12	10	10/16	10
11/12	11	11/16	11
12/12	12	12/16	12
		13/16	13
		14/16	14
		15/16	15
		16/16	16

TABLE 18-3 Feet and Inches; Pounds and Ounces

When some mixed units are added, the sum can be reduced. For example, if you add 7 in. to 7 in., the total would be 14 in., which is the same as 1 ft 2 in. To reduce any inches equal to or larger than 12 in. to feet and inches, you divide the inches by 12. For example, how long will a piece of steel bar need to be if you are going to cut both a 7 ft 8 in. piece and a 6 ft 6 in. piece from it? Reduce the answer in inches to feet and inches using **Table 18-3**. Pounds and ounces are also mixed units, which can be reduced by dividing the ounces by 16.

Add   7' 8" + 6' 6"

	Feet Column	Inch Column
1st Step	7'	8"
2nd Step +	6'	6"
	13'	14"
Reduce and Add		
3rd Step	13'	14 ÷ 12
4th Step	13 + 1	2
5th Step	14'	2"

or

Add   7' 8" + 6' 6"

1st Step 7' + 6' = 13'

2nd Step 8" + 6" + 14"

3rd Step 13' + 14" = 13' 14"

Reduce 14" to feet and inches

4th Step 14 ÷ 12 = 1' and 2/12

5th Step 2/12 = 2"

Re-add   13' 0" + 1' 2"

6th Step 13' + 1' = 14'

7th Step 0" + 2" = 2"

8th Step 14' + 2" = 14' 2"

When subtracting one mixed unit from another and the small unit being subtracted is larger than the small unit it is being subtracted from—for example, 1 ft 4 in. from 2 ft 2 in.—you must make that unit larger. You can increase the small unit in a mixed number by subtracting one whole large unit by dividing it into the smaller units. For example, 2 ft 2 in. (**Figure 18-6A**) is the same dimension as 1 ft 14 in. (**Figure 18-6B**) and 4 lb 8 oz (**Figure 18-6C**) is the same weight as 3 lb 24 oz (**Figure 18-6D**).

Convert   2' 2"

1st Step 2' = 1' 12"

2nd Step 1' 12" + 2" = 1' 14"

Convert   4 lb 8 oz

1st Step 4 lb = 3 lb 16 oz

2nd Step 3 lb 16 oz + 8 oz = 3 lb 24 oz

With 2 ft 2 in. converted to 1 ft 14 in., you can now subtract 1 ft 4 in. the same way as you subtracted mixed numbers before.

Subtract   1' 14" − 1' 4"

1st Step 1' − 1' = 0'

2nd Step 14" − 4" = 10"

3rd Step 0' + 10" = 10"

To subtract 2 lb 10 oz from 4 lb 8 oz, you have to convert 1 lb to 16 oz and add that to the 8 oz as shown in Figure 18-6C and Figure 18-6D. Now 4 lb 8 oz has become 3 lb 24 oz.

Subtract   3 lb 24 oz − 2 lb 10 oz

1st Step 3 lb − 2 lb = 1 lb

2nd Step 24 oz − 10 oz = 14 oz

3rd Step 1 lb + 12 oz = 1 lb 14 oz

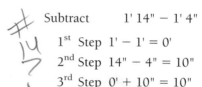

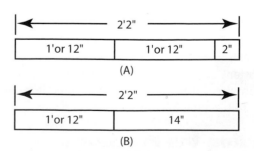

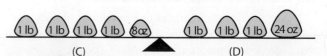

FIGURE 18-6 Mixed units. © Cengage Learning 2012

When adding multiple mixed numbers, add all of the inches first and then all of the feet. Then reduce the inches to feet and add the new mixed number to get the final answer. Follow the same process when adding pounds and ounces. Find the total length of angle iron needed if the following pieces are to be cut out.

Add   2' 6" + 5' 6" + 7' 3" + 8' 2" + 1' 1" + 3' 9"

1st Step	Feet Column	Inch Column
	2'	6"
	5'	6"
	7'	3"
	8'	2"
	1'	1"
2nd Step +	3'	9"
	26'	27"

Reduce and Add

3rd Step	26'	27 ÷ 12
4th Step	26 + 2	3
5th Step	28'	3"

or

Add   2' 6" + 5' 6" + 7' 3" + 8' 2" + 1' 1" + 3' 9"

1st Step   2' + 5' + 7' + 8' + 1' + 3' = 26'

2nd Step   6" + 6" + 3 + 2" + 1" + 9" = 27"

Reduce 27" to feet and inches

3rd Step   27 ÷ 12" = 2' 3/12"

4th Step   3/12" = 3"

Add   26' 0" + 2' 3"

5th Step   26' + 2' = 28'

6th Step   0" + 3" = 3"

7th Step   28' + 3" = 28' 3"

FRACTIONS

Fractions are commonly used in welding fabrication for dimensioning a distance that is less than an inch. **Figure 18-7** shows the common inch fractions most often used in welding fabrication. Because it is difficult for most manual welding to work with fractions smaller than 1/16 of an inch, such as 1/32 and 1/64 of an inch, these smaller dimensions are not commonly used in fabrication. When they are used, it is most often with some form of automated or machine welding or cutting process.

Finding the Fraction's Common Denominator

When the denominators of two fractions to be added or subtracted are different, one or both must be converted so that both denominators are the same. To convert a denominator, multiply both the numerator and the denominator

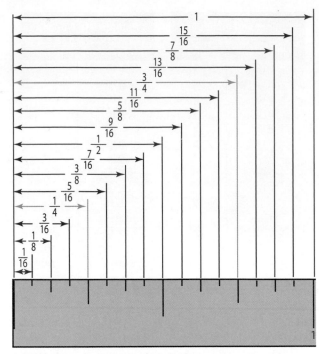

FIGURE 18-7 Fractions of an inch. © Cengage Learning 2012

of the fraction by the same number. For example, to convert 1/4 to 16ths, you would multiply both the numerator and denominator by 4: $4 \times 1 = 4$ and $4 \times 4 = 16$, which is 4/16. There are conversion tables available to convert the denominators and numerators, **Table 18-4**. To add 1/2 and 1/16 where both denominators are different, you must first find the **common denominator.** In this case it is 16.

Convert the Fraction $\dfrac{1}{2} = \dfrac{8}{16}$

Add $\dfrac{8}{16} + \dfrac{1}{16}$

$\dfrac{8 + 1}{16}$

$\dfrac{9}{16}$

Inch	Half	Fourth	Eighths	Sixteenths
				1/16
			1/8	2/16
				3/16
		1/4	2/8	4/16
				5/16
			3/8	6/16
				7/16
	1/2	2/4	4/8	8/16
				9/16
			5/8	10/16
				11/16
		3/4	6/8	12/16
				13/16
			7/8	14/16
				15/16
1	2/2	4/4	8/8	16/16

TABLE 18-4 Converting Fractions to a Common Denominator

Reducing Fractions

Some fractions can be reduced to a smaller denominator. For example, 4/8 is the same as 1/2, although when you are working with a rule or tape measure on the job, you can locate either measurement. Reductions may be necessary only when you are adding or subtracting several different dimensions or various fractional units.

> NOTE: It is important to be able to communicate clearly on the job. For example, you would want to reduce the fraction and ask for a 3/8-in. thick piece of steel, not a 6/16-in. thick piece of steel.

The normal way to reduce a fraction is to find the largest number that can be divided into both the denominator and numerator. If you are good at math, that can be easily done, but if you are not good at doing math in your head, there is an alternate way of reducing fractions. For reducing fractions in a welding fabrication shop, it is often easiest to divide both the numerator and denominator by 2. This method will simplify the reduction because all the fractional units found on shop rules and tapes are divisible by 2, for example, halves, fourths, eighths, sixteenths, and thirty-seconds. Using this method may require more than one reduction, but the simplicity of dividing by 2 offsets the time needed to repeat the reduction. For example, both the denominator and numerator of 4/8 can be divided by 2, so 4 ÷ 2 = 2 and 8 ÷ 2 = 4. That would make the new fraction 2/4, which can be reduced again by dividing the denominator and numerator one more time by 2. This last division results in 2/4 being reduced to 1/2. Reduction of fractions becomes easier with practice. Reduce 14/16 and 4/16 to their lowest common denominators.

Multiplying and Dividing Fractions

Although there is a way to multiply and divide fractions, the easiest way is to convert the fraction to a decimal fraction and, if needed, convert it back to a fraction once you have an answer. By converting the fraction to a decimal fraction, you can easily use a calculator to do the math. One slight problem with doing that is that some fractions when converted to decimal fractions and then divided can result in recurring decimals; for example, 1/3 becomes 0.333333333. If you add 0.333333333 + 0.333333333 + 0.333333333, you get 0.999999999, which is not equal to 1, but if you add 1/3 + 1/3 + 1/3, you get 3/3, which is 1. But realistically, 0.999999999 is only 0.000000001 less than 1, and most welding fabrication uses a dimensional **tolerance** of ±1/16 to ±1/8, which is a lot larger than that. **Table 18-5** lists the conversions for most fractions to decimal fractions. Also, converting fractions to decimal fractions and decimal fractions back to fractions is covered later in this chapter.

Inch Fraction	Inch Decimal	mm
1/16	1.5875	0.062500
1/8	3.1750	0.125000
3/16	4.7625	0.187500
1/4	6.3500	0.250000
5/16	7.9375	0.312500
3/8	9.5250	0.375000
7/16	11.1125	0.437500
1/2	12.7000	0.500000
9/16	14.2875	0.562500
5/8	15.8750	0.625000
11/16	17.4625	0.687500
3/4	19.0500	0.750000
13/16	20.6375	0.812500
7/8	22.2250	0.875000
15/16	23.8125	0.937500
1	25.4000	1.000000

TABLE 18-5 Conversion of Fractions, Decimals, and Millimeters

CONVERTING NUMBERS

Dimensions on a drawing are usually given in a consistent format and unit type. For example, everything would be standard units or metric units. Very seldom will you find that you must make conversions when reading a drawing. However, you may be asked to install or mount a unit such as a winch on a truck bumper or motor on a piece of equipment. If the part's drawing is dimensioned in metric and you are working in standard units, you may be able to simply measure the part using a tape or rule of the same measurement type as your drawing. Sometimes you will be working from the manufacturer's drawing, and then you may have to make conversions of measurements.

Conversion of measurements is also often used to make it easier to lay out the part. Conversions can also be used to reduce confusion with your coworkers on a job. It is much easier to understand 1 1/8 in. rather than trying to find 9/8 in. on a scale, even though they are the same.

Converting Fractions to Decimals

From time to time it may be necessary to convert fractional numbers to decimal numbers. A fraction-to-decimal conversion is needed before most calculators can be used to solve problems containing fractions. There are some calculators that will allow the inputting of fractions without converting them to decimals.

RULE: To convert a fraction to a decimal, divide the numerator (top number in the fraction) by the denominator (bottom number in the fraction) or use a conversion chart.

To convert 3/4 to a decimal:
$$3 \div 4 = 0.75$$

To convert 7/8 to a decimal:
$$7 \div 8 = 0.875$$

Tolerances All measuring, whether on a part or on the drawing, is in essence an estimate because no matter how accurately the measurement was made, there could always be a more accurate way of making it. Dimensioning tolerance is the difference between the exact dimension as shown on a drawing and the actual acceptable size of a part. The more accurate the measurement, the more time it takes. Drawings may state a dimensioning tolerance, the amount by which the part can be larger or smaller than the stated dimensions and still be acceptable. Tolerances are usually expressed as plus (+) and minus (−). If the tolerance is the same for both the plus and the minus, it can be written using the symbol ± (pronounced plus or minus), **Table 18-6.** In addition to the tolerance for a part, there may be an overall tolerance for the completed weldment. This dimension ensures that, if all the parts are either too large or too small, their cumulative effect will not make the completed weldment too large or too small. Most

Drawing Dimensions		Acceptable Dimensions	
Dimension	**Tolerance**	**Minimum**	**Maximum**
12"	±1/8"	11 7/8"	12 1/8"
2' 8"	±1/4"	2' 7 3/4"	2' 8 1/4"
10'	±1/8"	9' 11 7/8"	10' 1/8"
11"	±0.125	10.875"	11.125"
6'	±0.25	5' 11.75"	6' 0.25"
250 mm	±5 mm	245 mm	255 mm
300 mm	+ 5 mm–0 mm	300 mm	305 mm
175 cm	±10 mm	174 cm	176 cm

TABLE 18-6 Examples of Tolerances

weldments use a tolerance of ±1/16 in. or ±1/8 in., **Figure 18-8.**

Converting Decimals to Fractions

This process is less exact than the conversion of fractions to decimals. Except for specific decimals, the conversion will leave a remainder unless a small enough

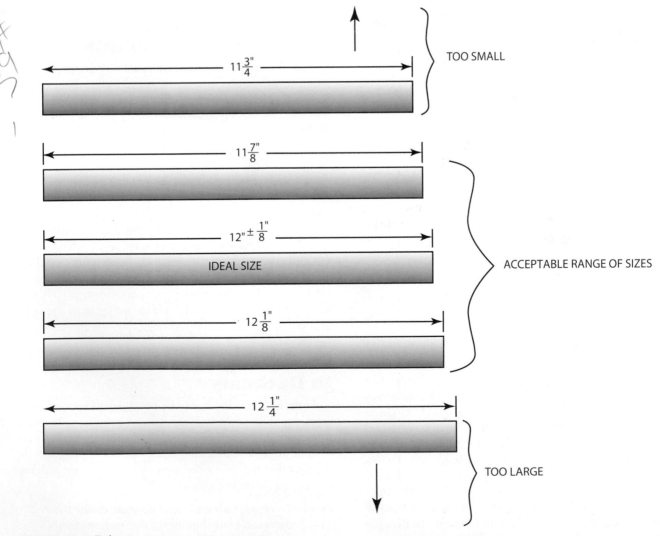

FIGURE 18-8 Tolerances. © Cengage Learning 2012

fraction is selected. For example, the decimal 0.765 is very close to the decimal 0.75, which easily converts to the fraction 3/4. The difference between 0.765 and 0.75 is 0.015 (0.765 − 0.75 = 0.015), which is within the acceptable tolerance for most welding applications. If you are working to a ±1/8-in. (3-mm) tolerance that has up to a 1/4 in. (6 mm) difference from the minimum to maximum dimensions, a measurement of 3/4 is acceptable. More accurately, 0.765 can be converted to 49/64 in., a dimension that would be hard to lay out and impossible to cut using a hand torch.

RULE: To convert a decimal to a fraction, multiply the decimal by the denominator of the fractional units desired; that is, for eighths (1/8) use 8, for fourths (1/4) use 4, and so on. After multiplying, place the product (dropping or **rounding** off the decimal remainder) over the fractional denominator used as the multiplier.

To convert 0.75 to fourths:
    0.75 × 4 = 3.0 or 3/4

To convert 0.75 to eighths:
    0.75 × 8 = 6.0 or 6/8, which will reduce to 3/4

To convert 0.51 to fourths:
    0.51 × 4 = 2.04 or 2/4, which will reduce to 1/2

Conversion Charts

Occasionally a welder must convert the units used on the drawing to the type of units used on the layout rule or tape. Fortunately, charts are available that can be used to easily convert between fractions, decimals, and metric units. To use these charts, Table 18-5, locate the original dimension and then look at the dimension in the adjacent column(s) of the new units needed. Both metric-to-standard conversions and standard-to-metric conversions result in answers that often contain long strings of decimal numbers. Often this new converted number, because of the decimals now attached to it, cannot be easily located on the rule or tape. In addition, most of the layout and fabrication work welders perform does not require such levels of accuracy. When a weldment's specifications do call for an accuracy that is more critical or accurate than can be expected to be laid out with a steel rule and marker, the parts are often machined to size after welding.

Measuring

Measuring for most welded fabrications does not require accuracies greater than what can be obtained with a steel rule or a steel tape, **Figure 18-9**. Tape measures and steel rules are available in standard units, decimal fractions, and metric units. Some may even have two or more different measuring units on the same tape. Using tapes and rules with multiple measuring units can make laying out weldments much easier when you are working with drawings and parts that have different unit measuring systems, **Figure 18-10**.

FIGURE 18-9 Measuring tape. © Cengage Learning 2012

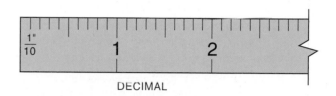

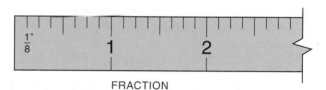

FIGURE 18-10 Decimal and fraction rules. © Cengage Learning 2012

WELDING COSTS

Estimating the costs of welding can be a difficult task because of the many variables involved. One approach is to have a welding engineer specify the type and size of weld to withstand the loads that the weldment must bear. Then the welding engineer must select the welding process and filler metal that will provide the required welds at the least possible cost. This method is used in large shops, where welding on a product can range from a few thousand dollars to well over a million dollars. With competition for work resulting in small profit margins, each job must be analyzed carefully.

The second approach, used by smaller shops, is to get a price for the materials and then estimate the production time. Process and filler metal costs are considered, but little thought is given to the hidden costs of equipment depreciation, joint efficiency, power, and so on. With the majority of the welding jobs in these shops taking from a few hours to a few days and with costs ranging from a few hundred dollars to a few thousand dollars, extensive cost analysis cannot be justified. Extensive estimating may take more time than the job itself. To remain profitable *and* competitive, some cost estimation is required. Only the cost considerations that a small shop should consider when estimating a job will be covered in this section.

Cost Estimation

A number of factors affect welding cost. These factors can be divided into two broad categories: fixed and variable. *Fixed costs* are those expenses that must be paid each and every day, week, month, or year, regardless of work or production. Examples of fixed costs include rent, taxes, insurance, and advertising. *Variable costs* are those expenses that change with the quantity of work being produced. Examples of variable costs include supplies, utilities, labor, and equipment leases. Expenses within both categories must be considered when making welding job estimates. These cost areas include the following:

- *Material cost:* The cost of new stock required to produce the weldment is fixed by the supplier. It is often possible to help control these costs by getting bids from several suppliers and combining as many jobs as possible in order to get any discount for larger quantities purchased.

- *Scrap cost:* Scrap is an inevitable part of any project. It costs a company in two ways: by wasting expensive resources and by requiring cleanup and removal. Reducing scrap production through proper planning will result in a direct saving for any project.

- *Process cost:* The major welding processes—SMAW, GMAW, and FCAW—differ widely in their cost of equipment, operating supplies, and production efficiency. SMAW has the lowest initial cost and has excellent flexibility but also a higher total cost for large jobs.

- *Filler metal:* The cost of filler metal per pound is only a small part of its actual cost. The major welding processes (SMAW, GMAW, FCAW, and GTAW) have widely varying deposition and efficiency rates.

- *Labor cost:* Total labor cost includes wages and benefits. Insurance, sick leave, vacation, social security, retirement, and other benefits can range from 25% to 75% of the total labor cost. Because labor costs are figured on an hourly basis, they can be controlled only by increasing productivity.

- *Overhead costs:* Overhead costs are often intangible costs related to doing business. These costs include building rent or mortgage, advertising, insurance, utilities, taxes, licenses, governmental fees, accounting, loan payments, and property upkeep.

- *Finishing cost:* Postwelding cleanup, grinding, painting, or other finishing adds to the weldment's final production cost. Many of these finishing processes can produce some level of health hazard, and a major concern to everyone is the environment. Complying with local, state, and federal environmental laws can add significantly to the cost of painting, dipping, and plating. New environmentally friendly paints, low-pressure spray guns, and water-based products are a few of the advancements in finishing that have helped reduce environmental compliance costs.

Joint Design

Joint design is an important consideration when estimating weld cost. The root opening, root face thickness, and bevel angles must be studied carefully when making design decisions. All these factors affect the weld dimensions, **Figure 18-11**, which in turn determine the amount of weld metal needed to fill the joint. Plate thickness is a major factor that, in most cases, cannot be changed. Increasing the root opening increases cost. But larger root openings often allow the bevel angle to be reduced while maintaining good access to the weld root.

To keep welding costs down, the joints should have the smallest possible root opening and the smallest reasonable bevel angle. These conditions are more easily achievable with

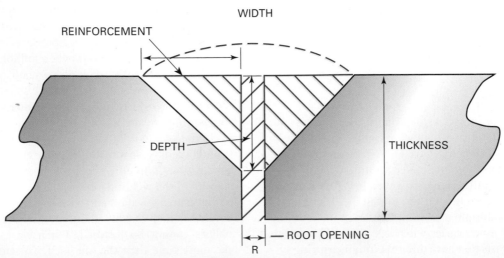

FIGURE 18-11 Calculation of weld requirements depending on joint design. © Cengage Learning 2012

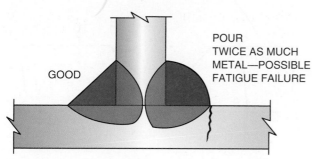

FIGURE 18-12 Overwelding can be harmful as well as costly. © Cengage Learning 2012

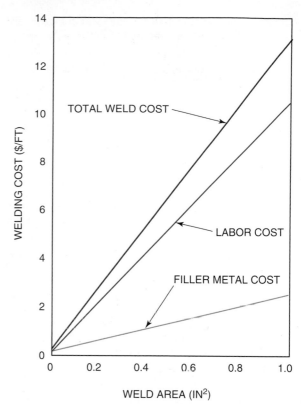

FIGURE 18-13 Increasing the weld size increases cost. Note that labor is 80% of the total cost (based on typical modern welding rates and efficiencies).
© Cengage Learning 2012

welding processes that provide deep penetration. The other benefit of deep penetration is the option of a deeper root face and its significant effect on the volume of filler metal needed. In addition, the amount of reinforcement affects welding costs. Some reinforcement is unavoidable in order to ensure a full thickness weld. However, too much reinforcement requires extra time and material, which can reduce the weld's strength and fatigue life.

The weld should be approximately the same size as the metal is thick. Welds that are undersized or oversized can cause joint failure. Welds that are undersized do not have enough area to hold the parts under load. Welds that are oversized can make the joint too stiff. The lack of flexibility of the weld causes the metal near the joint to be highly stressed, **Figure 18-12**. The same will happen to a piece of wire bent between two pliers: it will break. But a piece of wire bent between your fingers will withstand more bending before it breaks.

Welds made on parts of unequal size should allow enough joint flexibility to prevent a crack from forming along the edge of the weld. It is possible to taper the thicker metal to reduce the thickness or to build up the thinner metal.

Overwelding also contributes to welding costs. Welders often believe "if a little is good, a lot is better, and too much is just right." Too often it is assumed that a large reinforcement means greater strength, but that is never true.

Cutting weld volume also reduces the labor cost. It should be remembered that the labor content of welds almost always exceeds the cost of the expendable. Reducing the amount of filler metal needed also cuts the time needed to make the welds. This significant effect on labor costs, **Figure 18-13**, justifies the price of more expensive filler metals or welding processes. This assumes that the metals and processes provide enough increase either in penetration, to allow joint designs requiring less filler metal, or in deposition rates, to reduce welding times.

Groove Welds For groove welds the bevel angle greatly affects the filler metal volume. As the groove angle increases, a larger volume of filler metal is required to fill it during welding. Because the volume is in proportion to the angle, the wider the angle, the greater the volume. Knowing the bevel angle, the filler metal volume can be calculated.

The first thing in determining the filler metal volume required during welding is to find the cross-sectional area of the weld groove. The cross-sectional area is equal to one-half of the root opening times the bevel depth. Use the following formula to find the cross-sectional area of the weld:

Formula 18-1

$$CS_{weld} = \frac{RO \times BD}{2}$$

Where:

CS_{weld} = cross-sectional area of the weld
RO = root opening
BD = bead depth

Some large grooved joints have a root opening that will require a substantial volume of filler metal. To determine the volume of filler metal required for this part of the weld, you must first find the cross-sectional area of the root opening. The cross-sectional area of the root opening is equal to the plate thickness times the root gap. Use the following formula to find the cross-sectional area of the root:

Formula 18-2

$$CS_{root} = PT \times RG$$

Where:

CS_{root} = cross-sectional area of the root opening
PT = plate thickness
RG = root gap

The total cross-sectional area of the weld in Figure 18-11 is the sum of the cross-sectional area of the weld plus the cross-sectional area of the root opening. Use the following formula to find the total cross-sectional area of the weld:

Formula 18-3

$$TCS = CS_{weld} + CS_{root}$$

Where:

TCS = total cross-sectional area
CS_{weld} = cross-sectional area of the weld
CS_{root} = cross-sectional area of the root opening

The total groove volume is then determined by multiplying the total cross-sectional area of the groove by the weld length. Use the following formula to find the total groove volume of the weld:

Formula 18-4

$$GV = TCS \times WL$$

Where:

GV = groove volume
TCS = total cross-sectional area
WL = weld length

PRACTICE 18-1

Finding Weld Groove Volume

Using a pencil, paper, and calculator, determine the total volume of the following groove welds, **Figure 18-14**.

1. V-groove joint with the following dimensions:

   Width, 3/8 in.

   Depth, 3/8 in.

   Root opening, 1/8 in.

   Thickness, 1/2 in.

   Weld length, 144 in.

   Bevel joint with the following dimensions:

   Width, 0.25 in.

   Depth, 0.375 in.

   Root opening, 0.062 in.

   Thickness, 0.5 in.

   Weld length, 96 in.

2. V-groove joint with the following dimensions:

   Width, 12 mm

   Depth, 12 mm

   Root opening, 2 mm

   Thickness, 15 mm

   Weld length, 3600 mm

Complete a copy of the "Student Welding Report" listed in Appendix I or provided by your instructor. ◆

Fillet Welds The deep penetration processes of fillet welds offer lower costs and improved weld quality. Smaller fillet welds with deeper penetration also have the potential

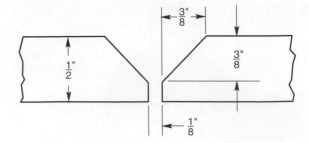

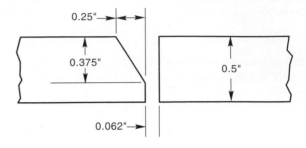

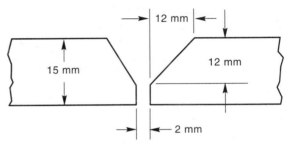

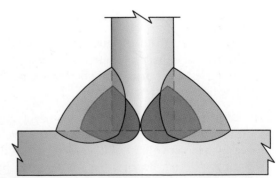

FIGURE 18-14 Find the weld groove volume. © Cengage Learning 2012

for yielding much stronger welds, **Figure 18-15**. The cross-sectional area of a fillet weld is equal to 1/2 of the weld leg height times the weld leg width:

Formula 18-5

#26?

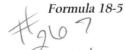

$$CS_{weld} = \frac{LH \times LW}{2}$$

Where:

CS_{weld} = cross-sectional area of the weld
LH = weld leg height
LW = weld leg width

FIGURE 18-15 The red 1/4-in. (6-mm) fillet weld is stronger than the other fillet weld and contains about one-half the amount of filler metal. © Cengage Learning 2012

Material	Weight lb/in.³	Weight g/cm³
Aluminum	0.096	2.73
Steel	0.287	7.945

TABLE 18-7 Density of Metals

The fillet weld volume is determined in the same manner as the groove weld, by multiplying the area times the length:

Formula 18-6

$$GV = CS_{weld} \times WL$$

Where:

GV = groove volume
CS_{weld} = cross-sectional area of the weld
WL = weld length

WELD METAL COST

In their technical data sheets, manufacturers of filler metal provide information regarding the welding metal. The number of electrodes per pound or the length of wire per pound can be used to determine the pounds of electrodes needed to produce a weld. To make this determination, the weight of weld metal required to fill the groove or make the fillet weld must be determined. The weight of weld metal is determined by multiplying the weld volume times the density of the metal, **Table 18-7**:

Formula 18-7

$$Wt_{weld\ metal} = GV \times MD$$

Where:

$Wt_{weld\ metal}$ = weight of weld metal
GV = groove volume
MD = metal density (weight of metal in pounds per cubic inch)

PRACTICE 18-2

Finding Weld Weight of Filler Metal

Using a pencil, paper, and a calculator, determine the weight of metal required for each of the welds described in Practice 18-1. Calculate the weight for both steel and aluminum base and filler metals. Using the weight of weld metal deposited allows for better comparisons when a number of different welds are being made. The weight of weld metal can either be determined for the welding prints or measured as the welder uses up supplies, **Figure 18-16**.

Complete a copy of the "Student Welding Report" listed in Appendix I or provided by your instructor. ◆

Cost of Electrodes, Wires, Gases, and Flux

To estimate the cost of electrodes, wire, gases, and flux, you must obtain the current cost per pound of electrode or welding wire plus the cost of shielding gas and flux

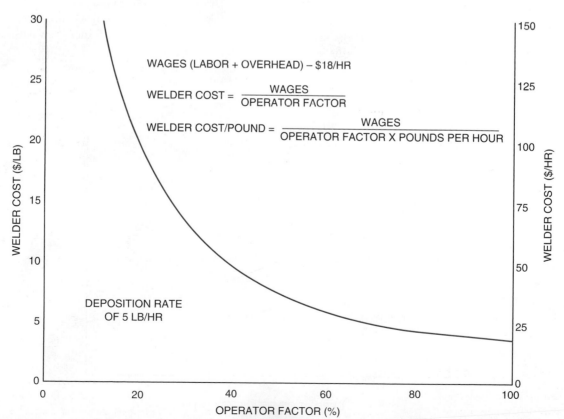

FIGURE 18-16 Low operator factors are costly. © Cengage Learning 2012

	GMAW		FCAW		
Wire diameter	0.035"	0.045"–1/16"	0.045"	1/16"	5/64"–1/8"
CFH	30	35	35	40	45

TABLE 18-8 Approximate Shielding Gas Flow Rate Cubic Feet per Hour

from a supplier or from the Internet. To determine how much shielding gas will be used, you must know the flow rate. Shielding gas flow rates vary slightly with the kind of gas used. The flow rates in **Table 18-8** are average values whether the shielding gas is an argon mixture or pure CO_2. Use these rates in your calculations if the actual flow rate is not available.

In the submerged arc process (SAW), the ratio of flux to wire consumed in the weld is approximately 1 to 1 by weight. When the loss, which is due to flux handling and flux recovery systems is considered, the average ratio of flux to wire is approximately 1.4 pounds of flux for each pound of wire consumed. If the actual flux-to-wire ratio is unknown, use 1.4 for cost estimating.

Deposition Efficiency

Not every pound of electrode filler metal purchased is converted into weld metal. Some portion of every electrode is lost as slag, spatter, and/or fume. Some, such as SMAW electrode stub ends, are unused. The amount of raw electrode deposited as weld metal is measured as deposition efficiency. If all is deposited, the deposition efficiency is 100%. If half is lost, the deposition efficiency is 50%. The argon-shielded GMA process is about 95% efficient. The SMAW process is between 40% and 60% efficient, depending on the electrode and the welder using it. Thus, a relatively costly filler metal with high efficiency can be as cost effective as one that appears to be cheaper,

Figure 18-17. The efficiency can then be calculated by the following formula:

Formula 18-8

$$DE = \frac{Wt_{weld\ metal}}{Wt_{electrode\ used}} \text{ or } DE = \frac{DR}{BOR}$$

Where:

$$DE = \text{deposition efficiency}$$
$$Wt_{weld\ metal} = \text{weight of weld metal}$$
$$Wt_{electrode\ used} = \text{weight of electrode used}$$
$$DR = \text{deposition rate in pounds per hour}$$
$$BOR = \text{burn-off rate in pounds per hour}$$

The deposition efficiency tells us how many pounds of weld metal can be produced from a given weight of the electrode of welding wire. As an example, 100 lb of a flux cored electrode with an efficiency of 85% will produce approximately 85 lb of weld metal. One hundred pounds of coated electrode with an efficiency of 65% will produce approximately 65 lb of weld metal less the weight of the stubs discarded. Note that electrodes priced at $0.65 actually cost about $1.30 per pound as weld metal because about 50% is wasted. The more expensive GMA wire at $0.85 costs about $0.90 per pound as weld metal when deposited because only 5% is lost as spatter and fume. Even with the added $0.15 for shield gas, the final cost of $1.05 is less than that of welds made with covered electrodes.

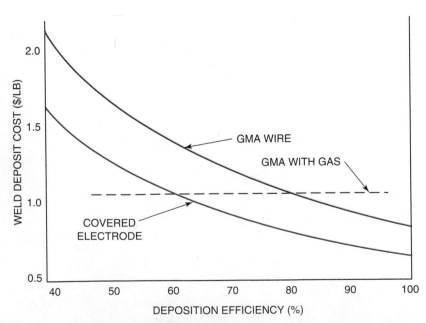

FIGURE 18-17 Weld metal cost is affected by the deposition efficiency of the process. © Cengage Learning 2012

Deposition Rate

The deposition rate is the rate at which weld metal can be deposited by a given electrode or welding wire, expressed in pounds per hour. It is based on continuous operation, with no time allowed for stops and starts for inserting a new electrode, cleaning slag, terminating the weld, or other reasons. The deposition rate will increase as the welding current is increased. When using solid or flux cored wires, the deposition rate will increase as the electrical stickout is increased and the same welding current is maintained. True deposition rates for each welding filler metal, whether it is a coated electrode or a solid or flux cored wire, can only be established by an actual test. The weldment is weighed before and after welding as the whole process is timed. **Table 18-9, Table 18-10, Table 18-11A, and B** contain average values for the deposition rate of the various welding filler metals based on welding laboratory tests and published data.

Deposition Data Tables

The deposition efficiency of a welding process refers to the percentage of the welding filler material that actually becomes part of the weld deposit. Some welding processes, such as SMAW, lose part of the filler material as the result of weld spatter while others, such as GTAW, do not. Even within a single process, the deposition efficiencies can change depending on the electrode classification used. Deposition data tables provide an average percentage of weld deposit efficiencies for each welding process.

Coated Electrodes The deposition efficiency of coated electrodes does not subtract the unused electrode stub that is discarded. This is understandable, since the stub length can vary with the operator and the application. Long, continuous welds are usually conducive to short stubs, while on short intermittent welds the stub length tends

E6010			
Electrode Diameter	Amperes	Deposition Rate, lb/hr	Efficiency %
1/8	100	2.1	76.3
	130	2.3	68.8
5/32	140	2.8	73.6
	170	2.9	64.1
3/16	160	3.3	74.9
	190	3.5	69.7
7/32	190	4.5	76.9
	230	5.1	73.1
1/4	220	5.9	77.9
	260	6.2	76.2

E6011			
Electrode Diameter	Amperes	Deposition Rate, lb/hr	Efficiency %
1/8	120	2.3	70.7
5/32	150	3.7	77.0
3/16	180	4.1	73.4
7/32	210	5.0	74.2
1/4	250	5.6	71.9

E6012			
Electrode Diameter	Amperes	Deposition Rate, lb/hr	Efficiency %
1/8	130	2.9	81.8
5/32	165	3.2	78.8
	200	3.4	69.0
3/16	220	4.0	77.0
	250	4.2	74.5
7/32	320	5.6	69.8
1/4	320	5.6	70.0
	360	6.6	67.7
	380	7.1	66.0
5/16	400	8.1	70.2

E6013			
Electrode Diameter	Amperes	Deposition Rate, lb/hr	Efficiency %
5/32	140	2.6	75.6
	160	3.0	74.1
	180	3.5	71.2
3/16	180	3.2	73.9
	200	3.8	71.1
	220	4.1	72.9
7/32	250	5.3	71.3
	270	5.7	73.0
	290	6.1	72.7
1/4	290	6.2	75.0
	310	6.5	73.5
	330	7.1	72.1
5/16	360	8.6	70.7
	390	9.4	71.8
	450	10.3	71.3

E7014			
Electrode Diameter	Amperes	Deposition Rate, lb/hr	Efficiency %
1/8	120	2.4	63.9
	150	3.1	61.1
5/32	160	3.0	71.9
	200	3.7	67.0
3/16	230	4.5	70.9
	270	5.5	73.2
7/32	290	5.8	67.2
	330	7.1	70.3
1/4	350	7.1	68.7
	400	8.7	69.9
5/16	440	8.9	62.2
	500	11.1	65.4

TABLE 18-9 Deposition Data

E7016			
Electrode Diameter	Amperes	Deposition Rate, lb/hr	Efficiency %
5/32	140	3.0	70.5
	160	3.2	69.1
	190	3.6	66.0
3/16	175	3.8	71.0
	200	4.2	71.0
	225	4.4	70.0
	250	4.8	65.8
1/4	250	5.9	74.5
	275	6.4	74.1
	300	6.8	73.2
	350	7.6	71.5
5/16	325	8.0	77.3
	375	9.0	76.3
	425	10.2	76.7

E7024			
Electrode Diameter	Amperes	Deposition Rate, lb/hr	Efficiency %
1/8	140	4.2	71.8
	180	5.1	70.7
5/32	180	5.3	71.3
	210	6.3	72.5
	240	7.2	69.4
3/16	245	7.5	69.2
	270	8.3	70.5
	290	9.1	68.0
7/32	320	9.4	72.4
	360	11.6	69.1
1/4	400	12.6	71.7

Low Alloy, Iron Powder Electrodes of the Types E7018, E8018, E9018, E10018, E11018, and E12018			
Electrode Diameter	Amperes	Deposition Rate, lb/hr	Efficiency %
3/32	70	1.37	70.5
	90	1.65	66.3
	110	1.73	64.4
1/8	120	2.58	71.6
	140	2.74	70.9
	160	2.99	68.1
5/32	140	3.11	75.0
	170	3.78	73.5
	200	4.31	73.0
3/16	200	4.85	76.4
	250	5.36	74.6
	300	5.61	70.3
7/32	250	6.50	75.0
	300	7.20	74.0
	350	7.40	73.0
1/4	300	7.72	78.0
	350	8.67	77.0
	400	9.04	74.0

TABLE 18-10 Deposition Data

Flux Cored Electrodes—Gas Shielded Types E70T-1, E71T-1, E70T-2, and All Low Alloy Types			
Electrode Diameter	Amperes	Deposition Rate, lb/hr	Efficiency %
0.45	180	5.3	85.0
	200	5.5	86.0
	240	6.9	84.0
	280	13.0	83.0
0.52	190	4.8	85.0
	210	5.3	83.5
	270	7.6	83.0
	300	9.8	85.0
1/16	200	5.2	85.0
	275	10.1	85.0
	300	11.5	85.0
	350	13.3	86.0
5/64	250	6.4	85.0
	350	10.5	85.0
	450	14.8	85.0
3/32	400	12.7	85.0
	450	15.0	86.0
	500	18.5	86.0
7/64	550	17.1	85.0
	625	19.6	86.0
	700	23.0	86.0
1/8	600	16.2	86.0
	725	22.5	86.0
	850	29.2	85.0

Flux Cored Electrodes— Self-Shielded			
Type and Diameter	Amperes	Deposition Rate, lb/hr	Efficiency %
E70T-3			
3/32	450	14.0	88
E70T-4			
3/32	400	18.0	85
.120	450	20.0	81
E70T-6			
5/64	350	11.9	86
3/32	480	14.7	81
E70T-7			
3/32	325	11.4	80
7/64	450	18.0	86
E71T-7			
.068	200	4.2	76
5/64	300	8.0	84
E71T-8			
5/64	220	4.4	77
3/32	300	6.7	77
EG1T8-K6			
5/64	235	4.3	76
E71T8-Ni1			
5/64	235	4.3	77
3/32	345	8.2	84
E70T-10			
3/32	400	13.0	69
E71T-11			
5/64	240	4.5	87
3/32	250	5.0	91
E70T4-K2			
3/32	300	14.0	83

NOTE: Values shown are optimum for each type and size.

TABLE 18-11A Deposition Data

Gas Metal Arc Welding Solid Wires			
Diameter, in.	Amperes	Melt-Off Rate, lb/h	Efficiency % See Note
.030	75	2.0	
	100	2.7	
	150	4.2	
	200	7.0	
.035	80	3.2	
	100	2.8	
	150	4.3	
	200	6.3	
	250	9.2	
.045	100	2.1	
	125	2.9	
	150	3.7	
	200	5.7	
	250	7.4	
	300	10.4	
	350	13.5	
1/16	250	6.7	
	275	8.6	
	300	9.2	
	350	11.5	
	400	14.3	
	450	17.8	

Submerged Arc Welding (Values for 1" Stickout)			
Diameter, in.	Amperes	Melt-Off Rate, lb/hr	Efficiency %
5/64	300	7.0	Assume
	400	10.2	99%
	500	15.0	Efficiency
3/32	400	9.4	Assume
	500	13.0	99%
	600	17.2	Efficiency
1/8	400	8.5	Assume
	500	11.5	99%
	600	15.0	Efficiency
	700	19.0	
5/32	500	11.3	Assume
	600	14.6	99%
	700	18.4	Efficiency
	800	22.0	
	900	26.1	
3/16	600	13.9	Assume
	700	17.5	99%
	800	21.0	Efficiency
	900	25.0	
	1000	29.2	
	1100	34.0	

NOTE: Efficiencies for GMAW with the following shielding gases:
  98% efficient with 98% Ar and 2% O_2
  98% efficient with 75% Ar and 25% CO_2
  93% efficient with straight CO_2

TABLE 18-11B Deposition Data

to be longer. **Figure 18-18** illustrates how the stub loss influences the electrode efficiency when using coated electrodes. In Figure 18-18, a 14-in. long, 5/32-in. diameter E7018 electrode at 140 amperes is considered. It is 75% efficient and a 2-in. stub loss is assumed. The 75% efficiency applies only to the 12 in. of the electrode consumed in making the weld, not to the 2-in. stub. When the 2-in. stub loss and the 25% lost to slag, spatter, and

fumes are considered, the efficiency minus stub loss is lowered to 64.3%. This means that, for each 100 lb of electrodes, you can expect an actual deposit of approximately 64.3 lb of weld metal if all electrodes are used to a 2-in. stub length.

The formula for efficiency including stub loss is important. It must always be used when estimating the cost of depositing weld metal by the SMAW method.

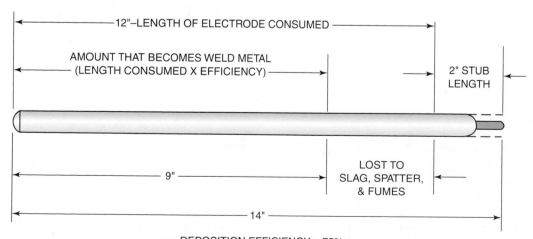

DEPOSITION EFFICIENCY = 75%
ACTUAL EFFICIENCY, INCLUDING STUB LOSS = 9 ÷ 14 = 64.3%

FIGURE 18-18 Deposition efficiency and stub loss. © Cengage Learning 2012

The formula used to establish the efficiency of coated electrodes including stub loss is based on the electrode length and is slightly inaccurate. That is, it does not consider that electrode weight is not evenly distributed because of flux removed from the electrode holder end (indicated by the dotted lines in Figure 18-18). Use of the formula will result in a 1.5% to 2.3% error that varies with electrode size, coating thickness, and stub length. However, the formula is acceptable for estimating purposes.

To find the percentage of the weld deposition efficiency for the values given in Figure 18-13, the formula is as follows:

Formula 18-9

$$\%DE = \left[\frac{(EL - SL) \times DE}{EL}\right] \times 100$$

Where:

$\%DE$ = percentage of weld deposition efficiency
EL = electrode length
SL = stub loss
DE = deposit efficiency

$$\%DE = \left[\frac{(EL - S) \times DE}{EL}\right] \times 100$$

$$\%DE = \left[\frac{(14 - 2) \times 0.75}{14}\right] \times 100$$

$$\%DE = \left[\frac{12 \times 0.75}{14}\right] \times 100$$

$$\%DE = \left[\frac{9}{14}\right] \times 100$$

$$\%DE = 0.6429 \times 100$$

$$\%DE = 64.29\%$$

In this example, the electrode length is known, the stub loss must be estimated, and the efficiency is taken from Table 18-9 and Table 18-10. Use an average stub loss and 3 in. for coated electrodes if the actual shop practices concerning stub loss are not known.

Efficiency of Flux Cored Wires Flux cored wires have a lower-power flux-to-metal ratio than coated electrodes and therefore a higher deposition efficiency. Stub loss need not be considered, since the wire is continuous. The gas shielded wires of the E70T-1 and E70T-2 types have efficiencies of 83% to 88%. The gas shielded basic slag wire (E70T-5) is 85% to 90% efficient with CO_2 as the shielded gas. The efficiency can reach 92% when a 75% argon and 25% CO_2 gas mixture is used. Use the efficiency figures in Table 18-11A for your calculations if the actual values are not known. The efficiency of self-shielded flux cored wires has more variation because of the large assortment of available types designed for specific applications.

Shielding Gas	Efficiency Rang	Average Efficiency
Pure CO_2	88% to 95%	93%
75% A–25% CO_2	94% to 98%	96%
98% A–2% O_2	97% to 98.5%	98%

TABLE 18-12 Deposition Efficiencies—Gas Metal Arc Welding Carbon and Low Alloy Steel Wires

The efficiency of the high-deposition, general-purpose type, such as E70T-4, is 81% to 86%, depending on wire size and electrical stickout. The chart in Table 18-11A shows the optimum conditions for each wire size and may be used in your calculations.

Efficiency of Solid Wire for GMAW The efficiency of solid wires in GMAW is very high and will vary with the shielding gas or gas mixture used, Table 18-11B. Using CO_2 will produce the most spatter, and the average efficiency will be about 93%. Using a 75% argon and 25% CO_2 gas mixture will result in somewhat less spatter and an efficiency of approximately 96%. A 98% argon and 2% oxygen mixture will produce even less spatter and the average efficiency will be about 98%. Stub loss need not be considered, since the wire is continuous. **Table 18-12** shows the average efficiencies to use in your calculations if the actual efficiency is not known.

Efficiency of Solid Wires for SAW In submerged arc welding there is no spatter loss, and an efficiency of 99% may be assumed. The only loss during welding is the short piece that the operator must clip off the end of the wire to remove the fused flux that forms at the termination of each weld. This is done to ensure a good start on the next weld.

Operating Factor

Operating factor is the percentage of a welder's working day actually spent on welding. It is the arc time in hours divided by the total hours worked. A 45% (0.45) operating factor means that only 45% of the welder's day is actually spent on welding. The rest of this time is spent installing a new electrode or wire, cleaning slag, positioning the weldment, cleaning spatter from the welding gun, and so on.

When using coated electrodes (SMAW), the operating factor can range from 15% to 40% depending on material handling, fixturing, and operator dexterity. If the actual operating factor is not known, an average of 30% may be used for cost estimates involving the shielded metal arc welding process.

When welding with solid wires (GMAW) using the semiautomatic method, operating factors ranging from 45% to 55% are easily attainable. For cost estimating purposes, use a 45% operating factor. The estimated

Welding Process			
SMAW	GMAW	FCAW	SAW
30%	50%	45%	40%

TABLE 18-13 Approximate Operating Factor

operating factor of FCAW is about 5% lower than that of GMAW to allow for slag removal time.

In semiautomatic submerged arc welding, slag removal and loose flux handling must be considered. A 40% operating factor is typical for this process.

Automatic welding using the GMAW, FCAW, and SAW processes requires that each application be studied individually. Operating factors ranging from 50% to 100% may be obtained depending on the degree of automation.

Table 18-13 shows average operating factor values for the various welding processes. These figures may be used for cost estimating when the actual operating factor is not known.

Knowing the productivity of welders is necessary when determining the cost of the finished part. In some shops, this cost is passed directly on to the customer in the form of cost plus. It may also be used to determine

at what level the shop can bid on new work and still make a profit. It is not often used to promote or penalize welders.

The number of parts produced is useful if there are a number of welders making the same or similar parts in a production shop. A comparison of the productivity can be made because there should be little difference in the average time compared with each welder's actual time.

The length of weld produced is a useful tool when a lot of the same welding is required. Welding on items like ships, tanks, or large vessels may take weeks, months, or years. The length that a welder produces in this type of production can be recorded on a regular basis. The deposition rates of the process also affect costs. Processes with high deposition rates can be very cost effective, **Figure 18-19**. A compelling reason for replacing covered electrodes with small diameter cored wires is that the deposition rates in the vertical position can be increased from about 2 lb per hour to over 5 lb per hour. Thus, the welder cost in this example drops from over $30 per pound to about $12 per pound of weld metal deposited.

Since operating factors and deposition rates interact strongly, their effects on weld costs are examined together. The time spent preparing and positioning weld joints for submerged arc welding is costly, but that process's high

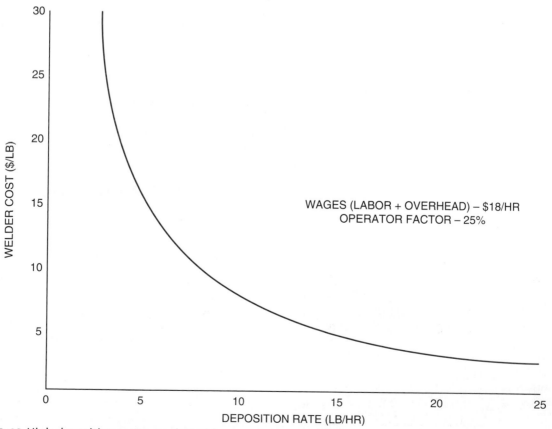

FIGURE 18-19 High deposition rates are desirable. © Cengage Learning 2012

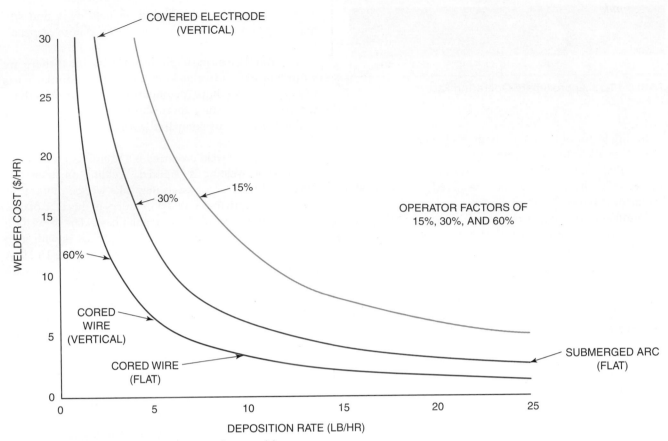

FIGURE 18-20 Newer processes can reduce welder costs. © Cengage Learning 2012

deposition rates justify the time, **Figure 18-20**. Covered electrodes, however, cannot compete with most other processes unless the setup time and other factors can be reduced. The speed with which alloys and electrode types can be changed explains why covered electrodes have remained competitive in small job shops, especially when typical welds are quite short. Changeover times with GMA processes can be lengthy, and if the welding jobs are small, the operator factor can drop below 15%. Excessively high deposition rates may be needed to compensate for that deficit.

Labor and Overhead Labor and overhead may be considered jointly in your calculations. Labor is the welder's hourly rate of pay including wages and benefits. Overhead includes allocated portions of plant operating and maintenance costs. Weld shops in manufacturing plants normally have established labor and overhead rates for each department. Labor and overhead rates can vary greatly from plant to plant and with location. **Table 18-14**

shows how labor and overhead can vary and suggests an average value to use in your calculations when the actual value is unknown.

Cost of Power Cost of electrical power is a very small part of the cost of depositing weld metal and in most cases is less than 1% of the total. It will be necessary for you to know the power cost expressed in dollars per kilowatt-hour ($/kWh) if required for a total cost estimate.

Other Useful Formulas

The following formulas will assist you in making other useful calculations:

For Formula 18-9 use the values from Example 1 to find the total weld metal weight. Formula 18-10 is used to find the number of hours required to complete the work.

Formula 18-10

$$TWt_{total\ weld\ metal} = \frac{Wt_{weld\ metal} \times WL}{DF}$$

$$TWt_{total\ weld\ metal} = \frac{0.814 \times 1280}{0.639}$$

$$TWt_{total\ weld\ metal} = 1631\ lbs$$

Small shops	$27.50 to $45.00
Large shops	$47.00 to $96.00
Average	$54.00

TABLE 18-14 Approximate Labor and Overhead Rates

Formula 18-11

$$W_{time} = \frac{Wt_{weld\ metal} \times WL}{DR \times OF}$$

$$W_{time} = \frac{0.814 \times 1280}{5.36 \times 0.30}$$

$$W_{time} = \frac{1042}{1608}$$

$$W_{time} = 648\ hrs$$

Use Formula 18-11 to find the number of hours required to make the same 1631 lb of weld if GMAW, FCAW, and SAW are used.

- GMAW equipment setup is 0.30-in. diameter wire at 100 amperes.
- FCAW equipment setup is 0.45-in. diameter E70T-1 at 180 amperes.
- SAW equipment setup is 1/8-in. diameter wire at 400 amperes.

Refer to Table 18-10 and Table 18-13 for the necessary data.

Summary

Math is a very important part of welding metal fabrication. You will often use your math skills as a welder fabricator to locate parts, lay out parts, calculate material needs, and determine cost. There is an old saying—measure twice, cut once. You never want to say, "I've cut it off twice, and it's still too short." Math is a skill, and like the skill of welding, it requires practice to become an expert. As you weld on projects, take every opportunity to practice your math. As you practice math, you will find shortcuts to some math problem solving.

One of the most important things when first laying out any weldment is to make sure that it is being laid out square. On smaller weldments you can use a squaring tool; however, on large weldments, such as an 18-foot by 7-foot (5.4 m by 2.1 m) trailer frame, even the largest squaring tools can be too small to accurately lay out the corners squarely. Using the Pythagorean Theorem for right triangles is often the best way to square large weldments. This theorem states, "In any right triangle, the area of the square whose side is the hypotenuse is equal to the sum of the areas of the square whose sides are the two legs," as shown by the following formula:

Formula 18-11
$$a^2 + b^2 = c^2$$

Where:

a^2 = side one of a right triangle
b^2 = side two of a right triangle
c^2 = hypotenuse of a right triangle

In practical terms if you make a mark three feet (3 m) along one side of the trailer frame and four feet (4 m) along the other side, then the distance between these two marks should measure 5 feet (5 m) if the corner is square, Figure 18-21. The theorem will work with any numbers; however, it is easier when the square root of the sum of the two sides is even, such as 3, 4, 5 and 6, 8, 10, Figure 18-22.

Once you have learned the mathematical processes, a calculator can be a big help. A word of caution about using a calculator: Do a quick estimate of the answer you expect from a problem. That way, when you get an answer from the calculator, you can compare the two. Sometimes we hit the wrong key on a calculator by mistake so the answer is not even close to being correct. If it differs greatly, recheck your work.

Controlling the cost of a weldment can be as important as producing a quality welded product, because if you spend too much time in preparing and producing the weld to a quality standard far above that required by the industry, the end product may be excessively expensive and unmarketable. An example of a product requiring a relatively low level of welding skills is yard art. For these decorative or ornamental pieces, the customer is most frequently looking at cost. Welding on these items must merely hold them together to meet market demands. An example of a product requiring a high level of welding skills is the engines on rockets. All of these welds must be precise at any expense. Most welding requirements obviously fall somewhere between these two extremes. For both you and your company's benefit, you must learn to meet their needs and standards in the most cost-effective manner.

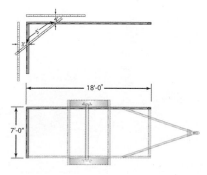

FIGURE 18-21 Squaring the trailer frame.

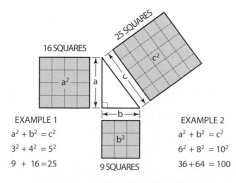

FIGURE 18-22 Pythagorean Theorem

Metal Cored Welding Wire Comes through on Heavy Weldments

Industries where high production rates make a difference in profitability are taking a hard look at replacing solid wire with metal cored wire. Welding applications of 1/16 in. or heavier mild steel with single or multiple passes are good candidates for switching from solid welding wire to metal cored welding wire if the following problems are occurring:

1. Poor fit up or uneven root opening

2. Melt through

3. Incomplete fusion (cold lap)

At a given current, metal cored wires can operate at a higher wire-feed speed than solid wire, which typically produces higher deposition rates. When using spray transfer, metal cored wires can be applied at a lower current than solid wire.

Metal cored wires also produce low fume and require little or no cleanup, qualities common to solid wire.

Recently, a metal cored welding wire was put to the test by a structural steel fabricator in a challenging application.

The Application

Welding a 5-in. thick steel plate extension, measuring 8 ft long by 7 in. wide, to a 5-in. thick column flange is the most challenging double-bevel groove weld Bill Sheffield has seen in his company's 20 years of fabrication experience. Sheffield is owner, president, and chief executive officer of Met-Con, Inc., a structural fabrication company in Cocoa, Beach Florida.

That weld, coupled with another double-bevel groove weld joining a column to an 11-in. thick rocket launch base, is a crucial joint in the construction of the rocket launch tower for Space Launch Complex 37, the Kennedy Space Center site for the U.S. Air Force's Evolved Expendable Launch Vehicle (EELV) program. The goal of the program, initiated in 1995, is to replace medium- and heavy-lift vehicles and supporting infrastructure with a family of expendable launch vehicles, while maintaining or improving the system's operability and reliability.

Restructuring for New Rockets

Space Launch Complex 37 was simply known as Pad 37 during the 1960s. It was used for the Saturn-1 and Apollo-5 and AS-203 missions. The site is being reconstructed to launch the Delta IV, Boeing's newest family of rockets. These vehicles will transport a wide range of payload classes for government and commercial customers. The largest Delta

IV will carry multiple satellites and lift approximately 29,100 lb into orbit.

Boeing awarded a turnkey contract to Raytheon Engineers and Constructors to design and build the new Space Launch Complex 37, which will include the launch pad, control center, fixed umbilical tower, mobile service tower, and propellant facility.

In turn, Raytheon awarded a subcontract to Met-Con, Inc., to construct the mobile service tower that will handle the boost in rocket launches. Met-Con's projects include numerous NASA space launch facility upgrades at Cape Canaveral Air Force Station, Kennedy Space Center, and Titusville, as well as extensive structural work for Walt Disney World, Sea World, the Dolphin Hotel, and recently the Orange County Courthouse.

The service tower, which moves up to 345 ft and is approximately 30 stories high, has been designed in four sequences. "The thickness of the components along with their sheer size made quality, maneuverability, and weld access major issues for us," said Sheffield.

The most difficult of the four sections of the tower is believed to be the first section. The heaviest welds are at this level. Lighter components are higher, which means more bolting than welding.

Constructing the Tower

The first section of tower construction is about 30 ft tall and consists of an 11-in. thick base and four quadrants,

Postweld cleanup is performed on welds for the column flange extension and on the rocket launch base.
American Welding Society

View down the fabricating bay showing work being performed on the launch tower columns. American Welding Society

each containing four columns. The steel being used in the structure is A572 Grade 50. Girders fit between the columns to form a frame, and 12- and 14-in.-diameter pipes form diagonal bracing. Horizontal extensions measuring 20 ft long are being welded at the bottom of the columns and will be attached to the launch base during the erection phase in the field. Additional pieces also will be fit into the structure on-site and then permanently welded. Miscellaneous iron hangs between sections to assemble as the next levels of 16 columns are added.

Sheffield's staff experts—Fred McDaniel and Rick Proulx from quality control and Brian Greer, the shop superintendent—were directed to examine available technologies and to select the most efficient, highest-quality welding process. They narrowed the options to submerged arc welding (SAW) and welding with flux cored wires and metal cored wires.

The joint geometry of the rocket tower presented demanding physical obstacles for the company's welders.

Estimates showed it would take about 100 passes to accomplish the full penetration welds required to join the two 5-in. thick components. McDaniel wanted the high deposition and performance offered by submerged arc welding and flux cored wires, but without the cleanup. Slag is one thing he specifically needed to avoid. "The possibility of any slag inclusion in the tower columns would have highly jeopardized the quality required for this project," said McDaniel.

Choosing and Using a Metal Cored Welding Wire

A metal cored welding wire was recommended: Tri Marks' Metalloy 70 in 1/16-in. diameter using a gas mixture of 75% argon and 25% CO_2. This particular metal cored wire is claimed to deposit approximately 18 lb/h with a 96% efficiency at 400 A and 36 V. Metal cored wires produce minimal spatter and are essentially slag free.

For welders, metal cored wires produce a brighter arc, so they switched from a No. 10 shade filter for their helmets to a No. 11 or 12. The welders adjusted to the darker lens and were producing straight welds in a few hours. The use of air-cooled helmets helped them adjust to the increased heat of the wire.

McDaniel is pleased with the quality and deposited weld metal efficiency of the metal cored welding wire.

During the first year of operation, the facility will have a capacity to launch up to four Delta IVs, with the full capability to handle 15 to 18 launches annually by the second year.

The Air Force's EELV program aims to reduce space launch costs up to 50% from today's rate of approximately $12,000 per pound of payload to orbit.

Article courtesy of the American Welding Society.

Review

1. What are two ways math is most commonly used in the welding shop?

2. What is the two-letter abbreviation for the metric system?

3. List factors that affect the cost of producing weldments.

4. List three examples of whole numbers.

5. List three examples of decimal fractions.

6. List three examples of a mixed unit.

7. List three examples of fractions.

8. What would the labor cost be if 20 hours were worked at an hourly rate of $25?

9. What is the first step in the sequence of mathematical operations?

10. If you need two pieces of pipe—one must be 15 ft and the other 10 ft—what is the total amount of pipe needed?

11. How many total feet of metal stock would you need if one piece is 12 ft 5 in. long and the other is 7 ft 3 in. long?

Review (continued)

12. How many total feet of metal stock would you need if one piece is 11 ft 9 in. long and the other is 6 ft 5 in. long?

13. How many feet of scrap pipe will you have left from a 9 ft 6 in. piece when 4 ft 2 in. is cut off?

14. How much scrap pipe will you have once you cut out 5.5 ft from a 20-ft length of pipe?

15. When the denominators of two fractions to be added or subtracted are different, what must be done before they can be added?

16. How thick will the finished part be if two pieces of metal are welded together if one is 3/4 in. thick and the other is 5/16 in. thick?

17. How much metal is left if 1/8 in. is ground off a 5/16-in. thick plate?

18. What is a dimensioning tolerance?

19. What is the minimum and maximum length a part can be if it is shown as needing to be 5 7/8 in. ± 1/8?

20. List examples of fixed and variable costs that must be considered when estimating a job.

21. List examples of overhead costs that a welding shop might have.

22. When estimating weld cost, what weld joint design factors should be considered?

23. When a weld is oversized, what joint failure problem can result?

24. How does the bevel angle in a groove weld affect the filler metal volume?

25. What is the cross-sectional area of a V-groove weld that is 6 mm wide and 8 mm deep on a 10-mm thick plate having a 2-mm root opening? What would the area be in square inches?

26. What is the cross-sectional area of a fillet weld that has an equal leg of 1/2 in.? What is the SI area?

27. What two amounts must be multiplied to determine the weight of weld metal required to fill a groove or make a fillet weld?

28. How many pounds of steel electrode are required to make a weld that has a volume of 18 in.^{3}?

29. Why is not every pound of electrode filler metal used converted into weld metal?

30. What does it mean if an electrode has a 50% deposition efficiency?

31. What is the meaning of the term *deposition rate*?

32. What factor is not included in the deposition efficiency of coated electrodes?

33. Why do flux cored wires have a higher deposition efficiency than coated electrodes?

34. If a welding project has a 45% operating factor, what does that mean?

Chapter 19

Reading Technical Drawings

OBJECTIVES

After completing this chapter, the student should be able to

- list the types of drawings that can be found in a set of drawings and what information is contained on each of them.
- sketch 10 types of lines, identify each, and explain how they are used on mechanical drawings.
- explain the difference between mechanical and pictorial drawings.
- name all of the various views that can be shown on drawings.
- read a set of drawings and explain each item shown and its dimensioning.
- discuss why a drawing may be scaled.
- compare the differences between sketches and mechanical drawings.
- demonstrate the ability to make a sketched drawing.
- illustrate how to use graph paper to make a scaled drawing.
- list the advantages of using computer-aided drafting software to make mechanical drawings.

KEY TERMS

alphabet of lines	*extension line*	*project routing information*
bill of materials	*hidden line*	*scale*
break line	*isometric drawings*	*section line*
cavalier drawings	*leaders and arrows*	*section view*
centerline	*mechanical drawings*	*set of drawings*
cut-a-ways	*object line*	*sketching*
cutting plane line	*orthographic projections*	*specifications*
detail views	*phantom lines*	*title box*
dimension line	*pictorial drawings*	*vector lines*

INTRODUCTION

Drawings are the tools that let us accurately communicate with each other in a very technical way. It is said that "a picture is worth a thousand words, but a mechanical drawing is worth millions." How many words would you have to use to tell someone how to build a jet airliner, a tanker ship, or even something as simple as a door hinge? Try it. How would you describe the thickness, diameter, and angle of the countersink of the screw holes and their locations? What would be the radius of the hinge pin? And the list of things you need to describe about the making of a hinge goes on and on. We can do all of that and more with easily understood mechanical drawings.

As you look at a basic mechanical drawing, you can see the object's shape, size, and location of its parts. However, as the drawing becomes more and more complex, it can be more difficult to see how everything relates. This chapter will help you see the basic layout of mechanical drawing and how the various parts of a drawing relate.

MECHANICAL DRAWINGS

Mechanical drawings have been around for centuries. Leonardo da Vinci (1452–1519) used mechanical drawings extensively in his inventive works. Many of his drawings still exist today and are as easily understood now as when they were drawn. For that reason mechanical drawings have been called the universal language; they are produced in a similar format worldwide. Despite the few differences in how the views are laid out, **Figure 19-1**, the drawings are understandable. Notwithstanding different languages and measuring systems, the basic shape of an object and location of components can be determined from any good drawing.

A group of drawings, known as a **set of drawings**, should contain enough information to enable a welder to produce the weldment. The set of drawings may contain various pages showing different aspects of the project to aid in its fabrication. The pages may include the following: title page, pictorial, assembly drawing, detailed drawing, and exploded view, **Figure 19-2**.

In addition to the actual shape as described by the various lines, a set of drawings may contain additional information such as the title box, specifications, project routing information, and bill of materials. The **title box**, which will appear in one corner of the drawing, may contain the name of the part, company name, scale of the drawing, date of the drawing, who made the drawing, drawing number, number of drawings in this set, and tolerances.

The **specifications** detail the type and grade of material to be used, including base metal, consumables such as filler metal, and hardware such as nuts and bolts.

The **project routing information** is used in large shops where an assembly line process is used. This information lets you know where the parts are to be sent once you have completed your part of the assembly process.

A **bill of materials** can also be included in the set of drawings. This is a list of the various items that will be needed to build the weldment, **Figure 19-3**.

Lines

To understand drawings, you need to know what the different types of lines represent. The language of drawing uses lines to represent its alphabet and the various parts of the object being illustrated. Different types of lines are used to represent various parts of the object being illustrated. The various line types are collectively known as the **alphabet of lines**, Table 19-1 and **Figure 19-4**.

(A) **Object lines**—Object lines show the edge of an object, the intersection of surfaces that form corners or edges, and the extent of a curved surface, such as the sides of a cylinder.

(B) **Hidden lines**—Hidden lines show the same features as object lines except that the corners, edges, and curved surfaces cannot be seen because they are hidden behind the surface of the object.

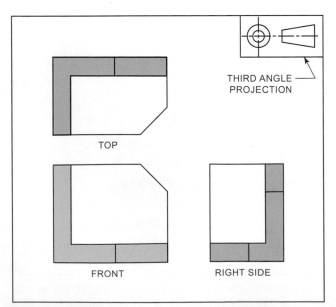

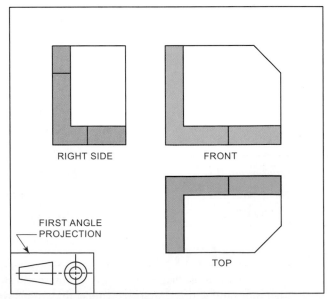

FIGURE 19-1 Of the two methods for locating the views, the third angle projection method is the most commonly used for welding drawings. © Cengage Learning 2012

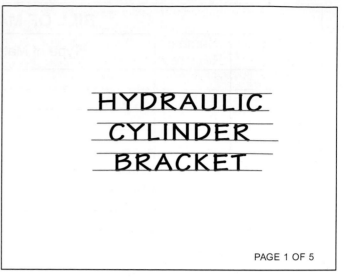

HYDRAULIC
CYLINDER
BRACKET

PAGE 1 OF 5

TITLE PAGE

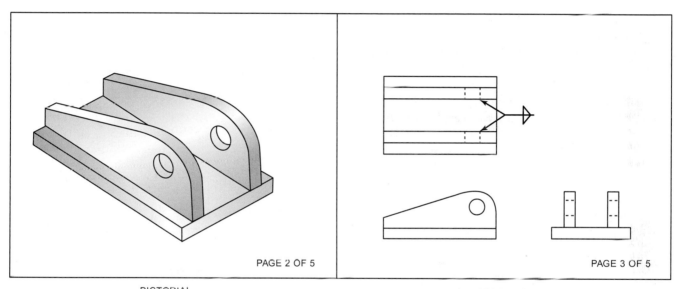

PAGE 2 OF 5

PICTORIAL

PAGE 3 OF 5

ASSEMBLY

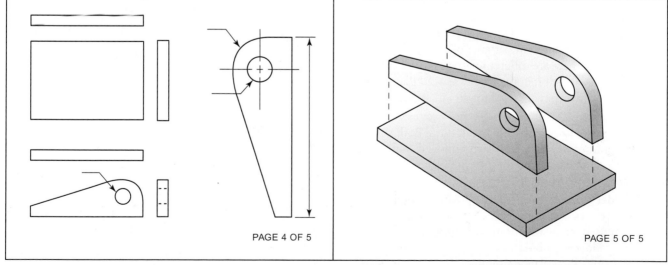

PAGE 4 OF 5

DETAIL

PAGE 5 OF 5

EXPLODED

FIGURE 19-2 Examples of some of the pages that can be found as part of a set of drawings. © Cengage Learning 2012

BILL OF MATERIALS

Part	Number Required	Type of Material	Size	
			Standard Units	SI Units

FIGURE 19-3 Example of a bill of materials. © Cengage Learning 2012

(C) **Centerlines**—Centerlines show the center point of circles, arcs, round, or symmetrical objects. They also locate the center point for holes, irregular curves, and bolts.

(D) **Dimension lines**—Dimension lines are drawn so that their ends touch the object being measured, or they may touch the extension line extending from the object being measured. Numbers in the dimension line or next to it give the size or length of an object.

(E) **Extension lines**—Extension lines are the lines extending from an object that locate the points being dimensioned.

(F) **Cutting plane lines**—Cutting plane lines represent an imaginary cut through the object. They are used to expose the details of internal parts that would not be shown clearly with hidden lines.

(G) **Section lines**—Section lines show the surface that has been imaginarily cut away with a cutting plane line to show internal details. Different types of materials, such as steel and cast iron, can be identified by using different patterns of section lines. Although there are different patterns for different materials, most often the evenly spaced diagonal lines for cast iron are used for all cut surfaces.

(H) **Break lines**—There are two types of break lines—long break lines and short break lines. Both show that part of an object has been removed. This is often done when a long uniform object needs to be shortened to fit the drawing page.

(I) **Leaders and arrows**—Leaders (the straight part) and arrows (the pointed end) point to a part to identify it, show the location, and/or are the basis of a welding symbol.

(J) **Phantom lines**—Phantom lines show an alternate position of a moving part or the extent of motion such as the on/off position of a light switch. They can also be used as a place holder for a part that will be added later.

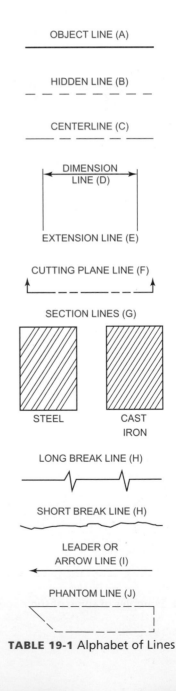

TABLE 19-1 Alphabet of Lines

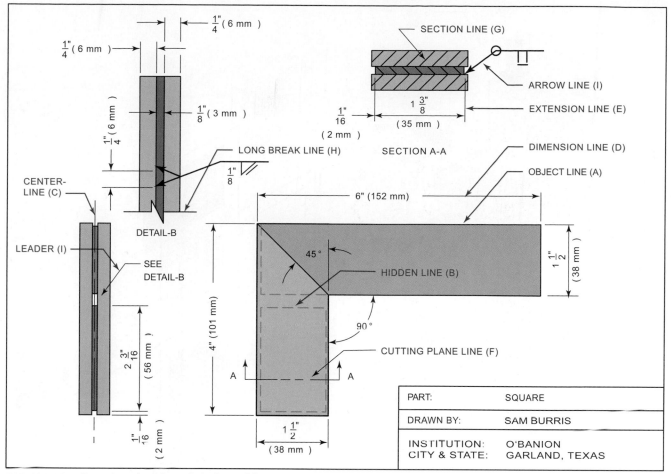

FIGURE 19-4 Drawing showing alphabet of lines. © Cengage Learning 2012

Types of Drawings

Drawings used for most welding projects can be divided into two categories—**orthographic projections** (mechanical drawings) and pictorial. The projection drawings are made as if you were looking through the sides of a glass box at the object and tracing its shape on the glass, **Figure 19-5A**. If all of the sides of the object were traced and the box unfolded and laid out flat, **Figure 19-5B**, there would be six basic views shown, **Figure 19-5C**.

Pictorial drawings present the object in a more realistic or understandable form. These drawings usually appear as one of two types, isometric or cavalier, **Figure 19-6**. The more realistic perspective drawing form is seldom used for welding projects.

Mechanical Drawings Usually, not all of the six views are required to build the weldment. Only those needed are normally provided, usually only the front, right side, and top views. Sometimes only one or two of these views are needed.

The front view is not necessarily the front of the object. The front view is selected because when the object is viewed from this direction, its overall shape is best described. As an example, the front view of a car or truck would probably be the side of the vehicle because viewing the vehicle from its front may not show enough detail to let you know whether it is a car, light truck, SUV, or van. From the front most vehicles may look very similar.

Pictorial Drawings Of the two types of pictorial drawings, the **isometric drawing** is more picture-like. Isometric drawings are drawn at a 30° angle so it appears you are looking at one corner. As with all drawings, all the lines that are parallel on the object appear parallel on the drawing.

On **cavalier drawings,** one surface, usually the front, is drawn flat to the page. It appears just like the front view of a mechanical drawing. The lines for the top and side surfaces are drawn back at an angle, usually 30°, 45°, or 60°.

Special Views

Special views may be included on a drawing to help describe the object so it can be made accurately. Special views on some drawings may include the following:

- *Section view:* The **section view** is drawn as if part of the object were sawn away to reveal internal details, **Figure 19-7**. This view is useful when the internal details would not be as clear if they were shown as

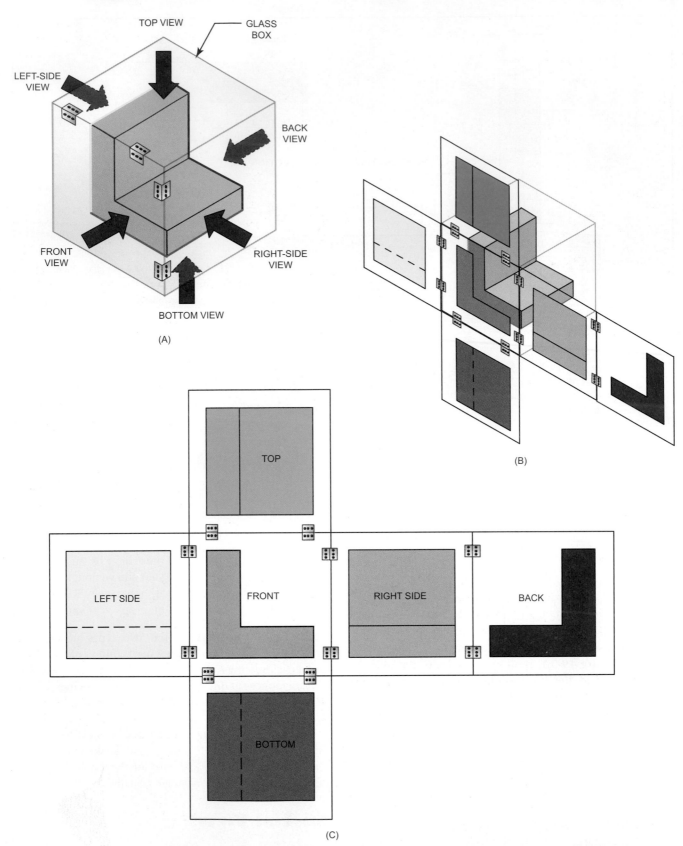

FIGURE 19-5 (A) Viewing an object as if it were inside a hinged glass box. (B) This is what you would see if you traced the views seen through the glass box and unfolded it. (C) Arrangement of the six views of the sides of the box as shown by the third angle projection method. © Cengage Learning 2012

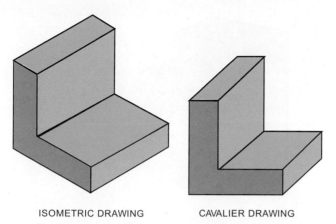

ISOMETRIC DRAWING   CAVALIER DRAWING

FIGURE 19-6 Two common ways of showing a pictorial drawing. © Cengage Learning 2012

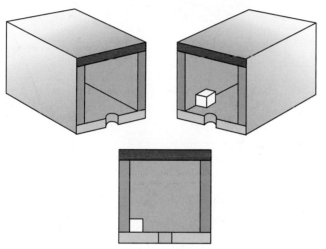

FIGURE 19-7 Section drawing. © Cengage Learning 2012

hidden lines. Sections can be either fully across the object or just partially across it. The imaginary cut surface is set off from other noncut surfaces by section lines drawn at an angle on the cut surfaces.

Some drawings use specific types of section lines to illustrate the type of material that the part was made with. The location of this imaginary cut is shown using a cutting plane line, **Figure 19-8**.

- *Cut-a-ways:* The **cut-a-way** view is used to show detail within a part that would be obscured by the part's surface. Often a freehand break line is used to outline the area that has been imaginarily removed to reveal the inner workings.

- *Detail views:* The **detail view** is usually an external view of a specific area of a part. Detail views show small details of a part's area and negate the need to draw an enlargement of the entire part. If only a small portion of a view has significance, this area can be shown in a detail view, either at the same scale or larger if needed. By showing only what is needed within the detail, the part drawn can be clearer and does not require such a large page.

- *Rotated views:* A rotated view can be used to show a surface of the part that would not normally be drawn square to any of the six normal view planes. If a surface is not square to the viewing angle, then lines may be distorted. For example, when viewed at an angle, a circle looks like an ellipse, **Figure 19-9**.

Dimensioning

Often it is necessary to look at other views to locate all of the dimensions required to build the object. By knowing how the views are arranged, it becomes easier to locate dimensions. Length dimensions can be found on the front and top views. Height dimensions can be found on the front and right side views. Width dimensions can be found on the top and right side views, **Figure 19-10**. The locating of dimensions on these views is consistent with both the first angle perspective or third angle perspective layouts.

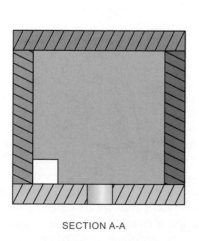

SECTION A-A

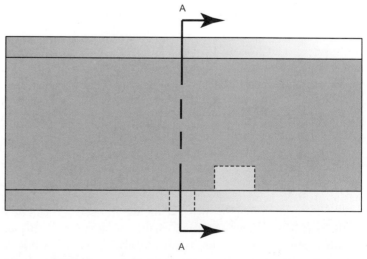

FIGURE 19-8 Cutting plane line and section. © Cengage Learning 2012

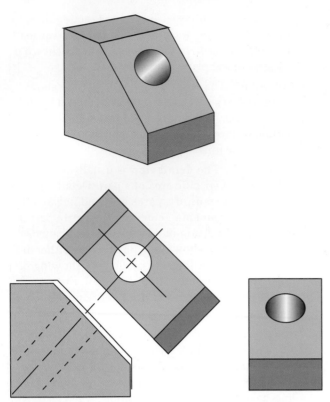

FIGURE 19-9 Notice that the round drilled hole looks misshapen or elliptical in the right-side view. © Cengage Learning 2012

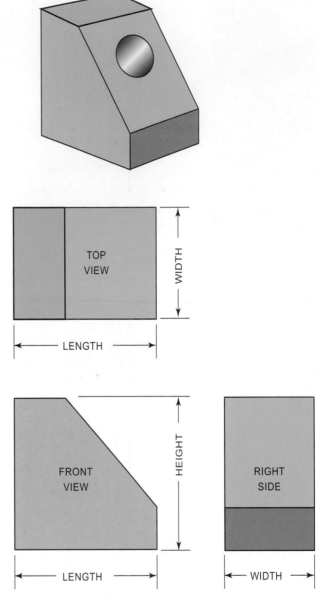

FIGURE 19-10 Locations where length, width, and height dimensions can be found on the three views of a drawing. © Cengage Learning 2012

If the needed dimensions cannot be found on the drawings, do not try to obtain them by measuring the drawing itself. Even if the original drawing was made very accurately, the paper it is on changes size with changes in humidity. Copies of the original drawing are never the exact same size. The most acceptable way of determining missing dimensions is to contact the person who made the drawing.

Keep the drawing clean and well away from any welding. Avoid writing or doing calculations on the drawing unless you are noting changes. Often there is a need to make a change in the part as it is being fabricated. When these changes are added to the "as drawn" drawing, the drawing is referred to as the "as built" drawing. It is important to keep these "as built" drawings so that they will be filed following the project for use at a later date. The better care you take with the drawings, the easier it will be for someone else to use them.

Drawing Scale

It would not be possible to make every drawing the same size as the parts being made, and for that reason you must change the scale. When we use a **scale,** we are saying that the part being drawn is drawn smaller or larger than it really is. An easy scale to use is 1 in. equals 1 ft, so that if we draw a line that is 10 in. long, it represents an actual distance of 10 ft. To aid in making and reading

scaled drawings, you can use a drafting tool called a *scale,* **Figure 19-11.** A scale is a special type of ruler that is marked with different units. There are two commonly used scales, the architectural scale and the engineering scale. The architectural scale is the one that is most often used for mechanical welding drawings.

Architectural scales are divided into fraction of inches. Some of the common units are: 1/8, 3/32, 1/4, 3/16, 3/8, and 3/4. These scales can be used to represent different lengths and units of measure. For example, the 1/4 scale can be used with inches so that 1/4 in. equals 1 in. or to represent feet as in 1/4 in. equals 1 ft. It could be used with yards, meters, miles, or any other standard unit.

Engineering scales are divided into decimals of inches. Some of the common units are 10th, 20th, 30th, 40th,

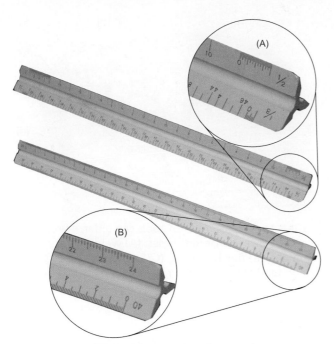

FIGURE 19-11 (A) Architectural scales may have 12 different measuring units on a single triangular-shaped scale. (B) Engineering scales usually have only six different measuring units on their scales.
Larry Jeffus

50th, and 60th. As with the architectural scale, these units can be used to represent any number of distances.

Not all drawings are scaled down; some are made larger, **Figure 19-12A**. Most often, details are drawn at a larger scale so that the important parts can be seen more clearly, **Figure 19-12B**.

READING MECHANICAL DRAWINGS

The pictorial drawing of the block in the glass cube shown in Figure 19-5A is color-coded. The same color-coding has been used for all of the multiview drawings in this chapter to help you identify the views. **Table 19-2** lists the colors and views used. The front surface in all the views is colored orange, the top view is green, the right side is blue, the left side is yellow, and the bottom is brown. Mechanical drawings you work with in the field are not color-coded.

In addition, the lines that represent the hidden surfaces of the block in the glass cube have been color coded. For example, you can see the red line for the back visible on the top and right-side views.

The color-coding is intended to help you learn how to read a mechanical drawing. The important element to

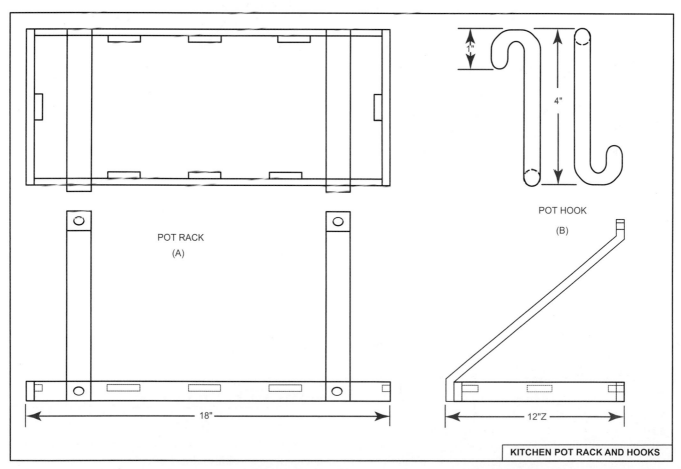

FIGURE 19-12 (A) The pot rack was scaled down to fit the drawing page, and (B) the detail of the pot hook was scaled up to show more detail. © Cengage Learning 2012

MECHANCIAL DRAWING VIEW COLOR CODES			
VIEW NAME	VIEW COLOR	VIEW NAME	VIEW COLOR
FRONT	ORANGE	LEFT SIDE	YELLOW
TOP	GREEN	BACK	RED
RIGHT SIDE	BLUE	BOTTOM	BROWN

TABLE 19-2 View Color Code

remember is that the front view is the main view. The standard views include the front, top, and right-side views. Not all drawings use all of the standard views and may or may not include additional views. For that reason, you need to know where each of the views are and how they relate to each other.

SKETCHING

Sketching is a quick and easy way of producing a drawing that can be used in the welding shop. Sketches and mechanical drawings have some similarities and some differences. They are similar in that they both contain the necessary information to produce a welded project. The main way they are different is that a sketch is a quick way of drawing an object and sketches may not be drawn to scale. Mechanical drawings take more time and are drawn to scale. They both should contain all the necessary information to build the desired weldment.

> NOTE: When graph paper is used, sketches can be easily drawn to scale, **Figure 19-13.**

A sketch can be any drawing that is made without the extensive use of drafting instruments or computer-aided design (CAD). Sketches are generally drawings that are made in the shop by the welder, shop supervisor, or by the customer. A sketched drawing can be made with or without the use of drawing tools such as scales, straightedges, curves, and circle templates, **Figure 19-14.** When drawing

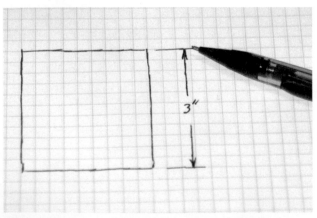

FIGURE 19-13 Graph paper makes it easier to sketch to scale. Larry Jeffus

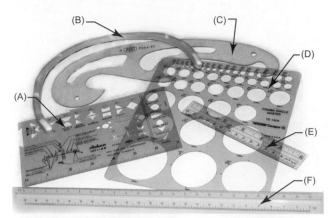

FIGURE 19-14 A number of tools can aid in drawing. These are a few different types: (A) welding symbol template, (B) flexible curve, (C) French curve, (D) circle template, (E) straightedge/rule, and (F) drafting scale. Larry Jeffus

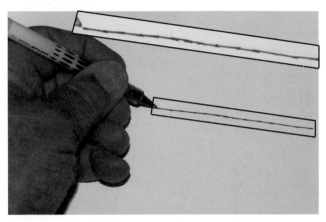

FIGURE 19-15 When sketching a line, rest your hand on the paper and make quick, short strokes as shown in the insert. Larry Jeffus

tools are used, the drawing may have straighter lines but take longer to produce. Straight lines are not always necessary to the actual production of the parts being welded as long as the drawing is clear.

Sketching is the process of making a line on a drawing by making a series of quick, short strokes with the pencil or pen, **Figure 19-15.** The technique of sketching a line may seem slow and awkward at first as compared to just holding the pencil on the paper and drawing a line. But a sketched line can be much straighter and faster once you have developed the skill. Each stroke may not be straight, but the line produced can be straight.

PRACTICE 19-1

Sketching Straight Lines

Using a pencil and unlined paper, you are going to sketch a series of 6-in. long straight lines.

Practice sketching from the right to the left and left to the right. You may find that it is easier for you to go one

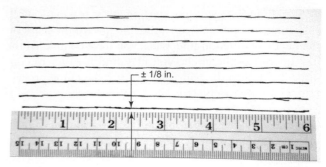

FIGURE 19-16 It is OK if the width of a sketched line varies slightly as a result of your making slight changes as you draw the line. © Cengage Learning 2012

direction than the other. The direction you sketch is not as important as your ability to make straight lines.

Start by making a small mark or dot approximately 6 in. from the point you plan on starting your sketched line. Don't measure; just estimate the distance. Part of being able to make quick sketches is the ability to judge lengths. The mark will give you an aiming point. Look at the point as you start the series of short sketched marks. Make each sketch mark about 1/2 in. to 3/4 in. long and make them in a quick, smooth series. Overlap each mark so they form a solid line, **Figure 19-16**. You may want to make the sketched line very light initially and go back over it to make it darker. It may be easier to make the sketch marks in the direction of travel or in the opposite direction—try both.

Once you have completed six or eight lines, lay a straightedge next to the lines and see how straight you were able to make them. Keep practicing sketching straight lines until you are able to make 6-in. long lines that are within ± 1/8 in. of being straight, **Figure 19-17.** ◆

PRACTICE 19-2

Sketching Circles and Arcs

Using a pencil and unlined paper, you are going to sketch a series of circles.

FIGURE 19-17 Practice 19-1: Sketch 6-in. long straight lines. Larry Jeffus

Start by sketching two light construction lines that cross at right angles. Construction lines are often used in drawings as guides to the finished line. Construction lines should be very light so they can be left on the drawing or easily erased.

Make two marks on each of the construction lines about 1/2 in. from the center point. These points will serve as your aiming points as you sketch the circle. The circle you sketch will be tangent to these points. A tangent straight line is one that meets a circle at a point where the circular line and straight line are going in the same direction. When a 12-in. ruler is placed on a round pipe, where the ruler and pipe meet is the tangent point. If you were to make a short straight line at the tangent point and keep doing this all the way around the pipe, you would wind up with a circle drawn from a series of short straight lines, **Figure 19-18**.

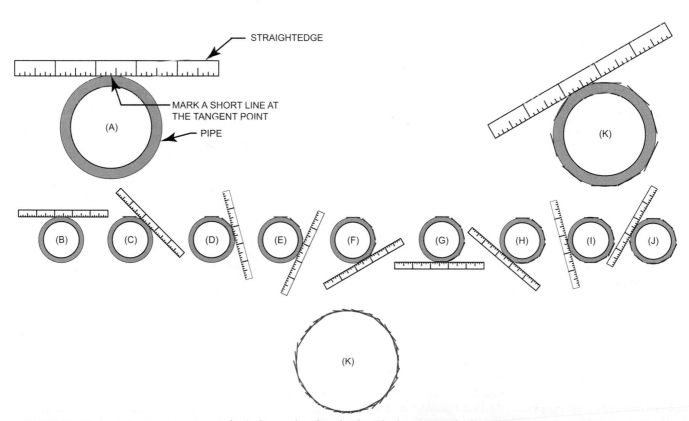

FIGURE 19-18 As you can see, a smooth circle can be sketched with short straight lines. © Cengage Learning 2012

Sketching a tangent line starting at the top mark, keep sketching and gradually turn the line toward the mark on the next construction line. Once you have completed the first quarter of the circle, you may find it easier to continue if you turn the paper. Repeat the sketching process until you have completed sketching the circle.

Repeat this process making several different size circles. On larger circles it may be helpful to make more construction lines. Using a circle template, check your circles for accuracy. Continue making sketched circles until you can draw them in several sizes within ±1/8 in. of round. ◆

NOTE: Sometimes it is easier to draw small circles in a square box. You can do this by first drawing a box using construction lines then drawing the circle inside the box.

PRACTICE 19-3

Sketching a Block

Using a pencil and unlined paper, you are going to sketch a mechanical drawing showing three views of a block as shown in **Figure 19-19**.

NOTE: In the mechanical drawing type called *orthographic projection,* the views must be arranged properly. The top view is drawn straight above the front view, and the right-side view is aligned to the right. This arrangement makes it possible for you to see distinguishing lines in one view and locate the corresponding lines in another view. This is how you find all of the location dimensions or even how you can identify material. For example, in one view an angle iron might be shown as parallel lines while in another view you would see the distinctive "L" shape. The same would apply to a cylinder that would appear as parallel lines in one view but as a circle in a different view.

Start by sketching construction lines as shown in **Figure 19-20A**. These lines will form the boxes for the front, top, and right-side views. Darken the lines that make up the object's lines so it is easier to see, **Figure 19-20B**.

Repeat this practice using the shaped objects shown in **Figure 19-21**. ◆

PRACTICE 19-4

Sketch a Candlestick Holder

Using a pencil and unlined paper, you are going to sketch a three-view mechanical drawing of the candlestick holder shown in **Figure 19-22**.

To lay out the angle for the candlestick holder, draw a vertical centerline that is 8 in. long. Draw the 2-in. long top line centered on the centerline. Measure down the correct distance from the top and draw the 4-in. long bottom line centered on the centerline. Connecting the endpoints of

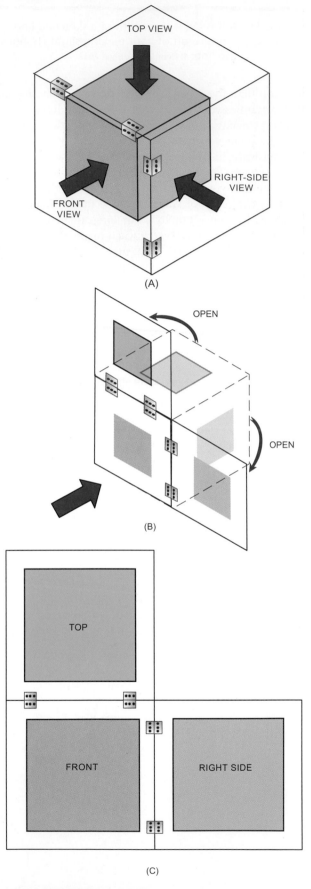

FIGURE 19-19 Practice 19-3: Sketch three views of the block; remember to keep the views in alignment as if they were traced on the sides of a hinged glass block.

© Cengage Learning 2012

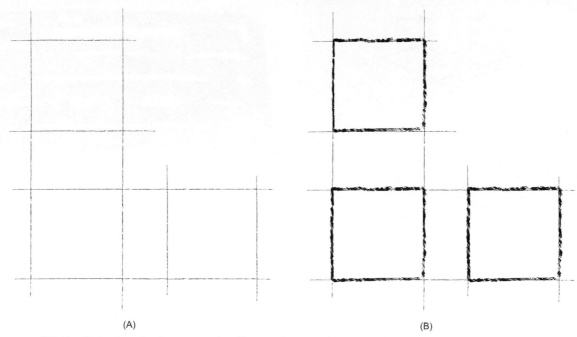

(A)   (B)

FIGURE 19-20 (A) Use light sketched construction lines to lay out the views. (B) If the construction lines are light enough, they may be left once the block is sketched darker. © Cengage Learning 2012

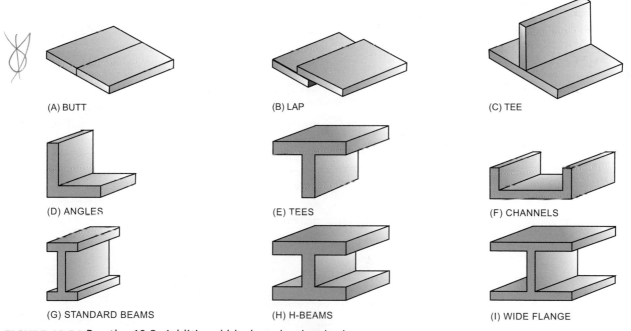

(A) BUTT   (B) LAP   (C) TEE

(D) ANGLES   (E) TEES   (F) CHANNELS

(G) STANDARD BEAMS   (H) H-BEAMS   (I) WIDE FLANGE

FIGURE 19-21 Practice 19-3: Additional blocks to be sketched. © Cengage Learning 2012

the top and bottom lines will automatically give the angle. But most important, remember that sketches do not have to be exactly to scale as long as all of the needed dimensions are shown.

Repeat the process using the candlestick holder shown in **Figure 19-23.** ◆

Erasers and Erasing

Most pencil erasers have an abrasive action on the paper as they are used to erase pencil marks. This can

sometimes cause the top surface of the paper to be roughed up or rubbed off. If you are not careful in erasing with a pencil eraser, you can damage the paper's surface, making it hard to redraw over that area.

Plastic erasers are usually white, and these erasers do not have an abrasive and will not damage the paper's surface like pencil erasers. Plastic erasers are very effective in removing unwanted pencil lines.

Sometimes you need to erase a small part of a line without removing or smudging a nearby line. There are thin metal tools called "eraser shields" that are used in

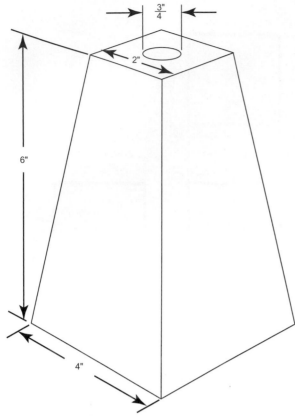

FIGURE 19-22 Practice 19-4: Sketch three views of this candlestick holder. © Cengage Learning 2012

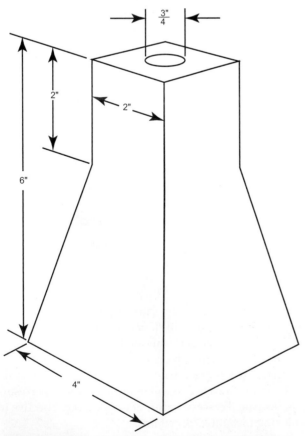

FIGURE 19-23 Practice 19-4: Additional candlestick holder to sketch. © Cengage Learning 2012

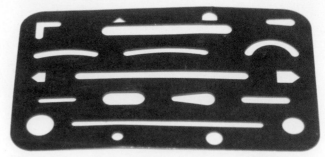

FIGURE 19-24 Thin metal eraser shield. Larry Jeffus

drafting to protect the neighboring line from erasure, **Figure 19-24.** An easy substitute for this tool is any scrap piece of paper. Simply cover the line you do not want to erase, and rub the eraser away from the edge so the edge is not wrinkled uncovering the line being protected.

NOTE: A Post-it note makes a great temporary eraser shield.

Both correction tape and correction fluid do not erase errors but cover them up so corrections can be made. They are easy to use and work well; however, if you are drawing with a pencil, they may be more difficult to draw over than an erased line.

Graph Paper

Making a sketch on graph paper is a way of both making your drawing more accurate and speeding up the sketching process. Graph paper is available with grid sizes ranging from 1/8 in. to 1/4 in. squares, **Figure 19-25.** Other sizes are available. Many copiers do not copy the light blue or light green lines on graph paper; so, if you need the lines later, you may need to make the copy machine copy darker or use another color of lined paper.

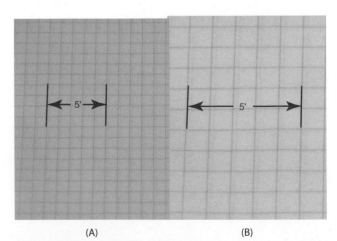

(A)   (B)

FIGURE 19-25 A sketched 5-ft long line would be much shorter on the (A) 1/8-in. graph paper than it is on the (B) 1/4-in. graph paper. © Cengage Learning 2012

Usually, the inch lines are a little darker on the graph paper, which makes it easier to count when measuring.

Even though you have lines to follow on graph paper, you may find that you can make a better-looking drawing by sketching over the grid lines rather than trying to just follow the grid line with your pencil. Graph paper does lend itself to the use of straightedges and other drafting tools, but with practice, sketching is faster and works well.

PRACTICE 19-5

Sketching Curves and Irregular Shapes

Using a pencil and graph paper, you are going to sketch a front view of the plant hanger shown in **Figure 19-26**.

Curves and irregular shapes can be easily drawn using a grid such as that found on graph paper. The first thing you need to do is locate a series of points on the graph paper that coincide with points on the curve you are copying. Start with the easy points where the lines on the paper cross at a point on the object. For example, one end of the curve starts at the intersection of lines E-3, so put a dot there. Next, the curve is tangent to lines F-2, so put a dot there. Follow the curve around, putting additional dots at the other intersecting points.

Once all of the easy dots are located, you are going to have to make some estimates for the next series of dots. For example, the curve almost touches the 1 line as it crosses the E line. Put a dot there and at similar points where the curve crosses other lines.

After you have located all of the points for the curve, sketch a line through all of the points, **Figure 19-27**. Refer to the figure to see how the line should pass through the point. If you are not sure, you can always add additional points that are not on lines to help guide your sketching.

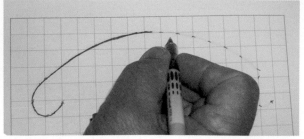

FIGURE 19-27 Connect the dots on the graph paper to copy the irregularly shaped curve. Larry Jeffus

NOTE: The curve you made is called free-form, and as the name implies, it is not an exacting process. If you are working on an exacting curve, there will often be center points for you to follow, **Figure 19-28**. You would use a compass to lay out this curve.

Repeat this practice and make a three-view drawing of the candlestick holder shown in **Figure 19-29**. ◆

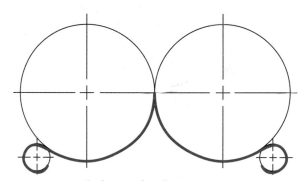

FIGURE 19-28 Circles can be drawn to create some curves. © Cengage Learning 2012

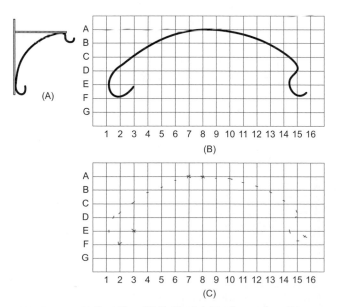

FIGURE 19-26 Practice 19-5: Sketching irregular shapes. © Cengage Learning 2012

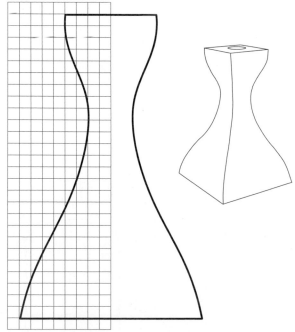

FIGURE 19-29 Practice 19-5: Additional curves to be sketched. © Cengage Learning 2012

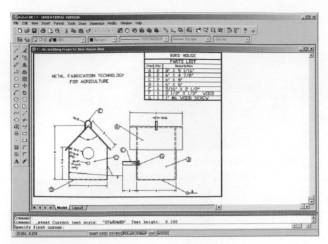

FIGURE 19-30 AutoCAD LT drafting program used to design a welded project. Larry Jeffus

COMPUTERS AND DRAWINGS

Computers have made it much easier to draw plans for projects. The welded birdhouse project shown in **Figure 19-30** was drawn on AutoCAD LT. There are a number of computer drawing programs like AutoCAD LT available. Drafting programs use **vector lines,** which is different from most drawing programs, which use raster art. Computers see vector lines as lines, and they see raster lines as if they were part of a picture. Because vector drawings are seen by the computer as lines, you can zoom in and out, measure, resize, reshape, or rotate the drawing and the lines stay crisp and sharp. For example, as the lines on the roof of the vector-drawn birdhouse in **Figure 19-31** are magnified 300 and 500 times, they stay sharp.

Computers see raster images as a series of small squares called pixels. Raster drawings are commonly known as bitmap drawings because the computer maps the location of every little bit (pixel) of the drawing. When these pixels are very small, your eye sees them as a line, but as you zoom in, they start looking like a bunch of colored squares. For example, as the lines on the roof of the bitmap-drawn birdhouse in **Figure 19-32** are magnified 300 and 500 times, they look like a group of colored squares not even recognizable as lines. Bitmap lines have a softer appearance, and they work best in art and photographic programs. The sharp, crisp lines of vector drawings work best for mechanical drawings.

Two-dimensional drafting programs, abbreviated 2D, like AutoCAD LT, allow you to make mechanical drawings accurately for projects. Vector drafting programs allow you to draw trailers, barns, or other large projects more accurately than they can be built. Using the pull-down and dimensioning options, you could set the precision for this birdhouse drawing at 1/256 in. (0.01 mm), **Figure 19-33.** Accuracy is important when parts must fit together and move without interfering with each other. It is also helpful when planning projects like the trailer in **Figure 19-34.** The back of the trailer has ramps built on that can be flipped down for easier loading of equipment. The ramps are hinged to the back of the trailer so that when they are in the up position, the dropped back end of the bed is level. AutoCAD LT makes designing the trailer easier, and leveling the bed makes it easier to haul hay, **Figure 19-35.**

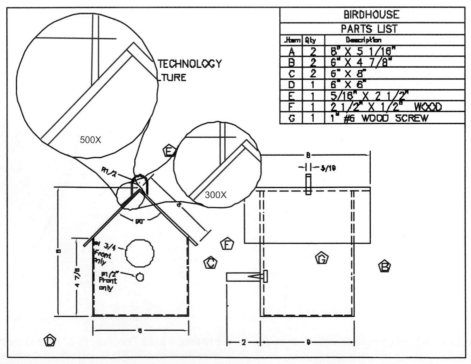

FIGURE 19-31 Vector line art drawing. © Cengage Learning 2012

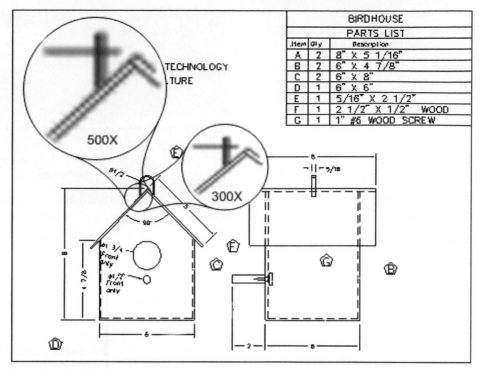

BIRDHOUSE
PARTS LIST

Item	Qty	Description
A	2	8" X 5 1/16"
B	2	6" X 4 7/8"
C	2	6" X 8"
D	1	6" X 6"
E	1	5/16" X 2 1/2"
F	1	2 1/2" X 1/2" WOOD
G	1	1" #6 WOOD SCREW

FIGURE 19-32 Bitmap line drawing. © Cengage Learning 2012

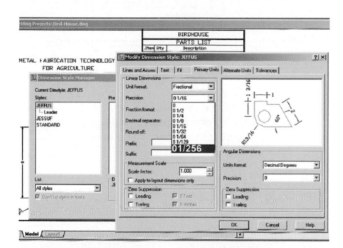

FIGURE 19-33 The drawing precision can be set on AutoCAD LT using the dimension style manager.
Larry Jeffus

FIGURE 19-34 Well-designed agricultural trailer for hauling both hay and equipment. Larry Jeffus

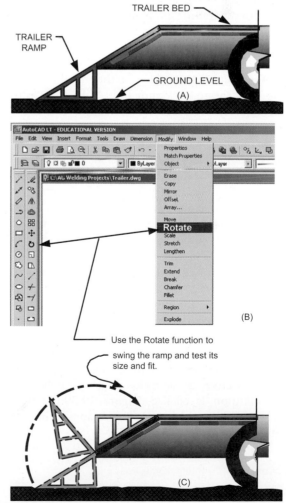

FIGURE 19-35 The AutoCAD LT Rotate feature can be used to check part alignment and fit, such as the trailer ramp on this drawing. Larry Jeffus

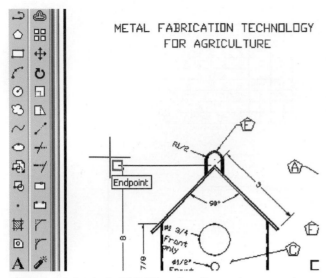

FIGURE 19-36 AutoCAD LT makes it easy to connect lines with its endpoints connecting feature. Larry Jeffus

FIGURE 19-37 Hovering above an icon will provide a one- or two-word hint. Larry Jeffus

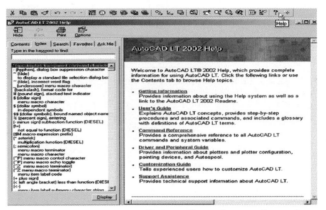

FIGURE 19-38 On-screen help will guide you through many of the computer program's features. Larry Jeffus

Lines on a mechanical drawing connect to other lines on the drawing. It is that connection between lines that allows the object being drawn to take shape. Very rarely would a mechanical drawing line just be drawn by itself. Computerized drafting programs make it easy to make lines connect. As the cursor nears the end of a line in AutoCAD LT, as with many vector drafting programs, the end of the line changes color, **Figure 19-36**. If you move closer, the line will "jump" to join the end of the first line drawn. This feature of "joining lines," like other features, can be adjusted or turned off as needed. This ability to customize the settings on a drafting program makes the program much easier and faster to use.

The readability and ease of understanding a drawing is dependent on the drawing's layout and location of dimensions and notes. Drafting programs allow you to move things around to make them clearer and easier to understand. This feature is important because on pencil drawings erasing and redrawing can result in a messy drawing, which can make it harder to understand.

On-screen help with computer drafting is available in several general ways. One method of getting on-screen help is by hovering the curser over a function button and a one- or two-word description to the button will appear, **Figure 19-37**. A more complete listing of help is available by clicking the "?" button on the tool bar at the top of the screen, **Figure 19-38**. Some programs like AutoCAD LT provide an Active Assistant that pops up the first time a function button is clicked, **Figure 19-39**. The Active Assistant provides step-by-step instructions on how to use that function.

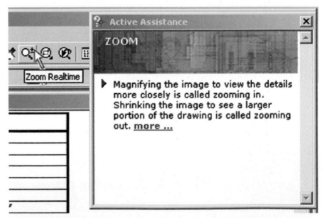

FIGURE 19-39 Active Assistant details the specific functions available. Larry Jeffus

Summary

Almost every welding fabrication will have a set of drawings detailing exactly how the assembly is to be made. Welders who can read and accurately follow these drawings have a skill that is valuable to any welding shop. Even the slightest error in the measuring, cutting, placing, or welding of the fabrication can cause it to be scrapped or necessitate expensive reworking.

All welders need to have a good understanding and working knowledge of mechanical drawings even if they are not directly involved with fabricating parts. Even though you may not have been the one to assemble the weldment, you should still be able to do a quick check to see that it was assembled according to the drawings.

Resistance Seam Welding Benefits Stainless Application

A redesigned production system increased productivity by 50% and reduced scrap rate to 1% on catalytic converter shells.

Full of expensive metal and formed to a precision shape, the stainless steel shell cannot leak. And even though it is a "no-show" part, it must be scratch free. These are some of the challenges that Walker Manufacturing had to solve to produce a quality catalytic converter shell.

Walker Manufacturing, Litchfield, Michigan, is a supplier to the automotive industry with expertise in manufacturing exhaust systems. An ongoing program to reduce waste and costs led to a reengineering of the process for manufacturing catalytic converter shells. Throughput was increased by 50%, the scrap rate was reduced by two-thirds, productivity was increased, and a coolant previously used was eliminated.

Doing It the Old Way

Originally, catalytic converter shells were produced on a gas tungsten arc (GTA) mill. The shells were formed from coil-fed 409 stainless steel into a tubular shape, notched for the cut-off process, and welded with the GTAW process. They were then cut to a specified length and sized to a final shape, and the shell edges were deburred. It was a labor-intensive process, according to Steve Sherwood, senior process engineer.

Two people were needed for the welding mill, two people to wash each part of the coolant used in the cutting process, and four people to check length and shape and deburr the shell. Deburring both ends of the component was crucial and often required 100% inspection. The shells were also checked for proper tolerances. Tooling changeover and replacement of deburring brushes required a lot of time.

There was also a recurring problem with the welding operation. At times a pinhole leak developed when the arc was initiated with the GTAW process. This was an important quality issue because hot exhaust gases could not leak from the shell. With leak problems and burrs, scrap rates approached 3%.

New Design Streamlines Operation

To solve these problems and improve the operation, a mash seam resistance welding system was developed by Newcor Bay City Division, Bay City, Michigan. The system automatically takes a flat stock stainless steel blank, forms it into a shell, seam welds it, and then sizes it to the final dimensions.

How It Works

Flat Type 409 stainless steel blanks ranging in thickness from 0.05 to 0.12 in. (1.4 to 3 mm) are loaded on a stacker at one end of the welding machine. The blank is picked up by suction cups and placed on a conveyor, which moves it through the shell forming area. The blank is then formed into a shell shape, overlapping the edges so it can next be resistance seam welded. It is then planished, a process that rolls the weld flat and reduces its height to within 15% of the metal's original thickness. It is placed into a station for forming its final size and shape to a tolerance of ±0.020 in. (±0.5 mm). The whole operation gives a stronger weld, the right shape, and cosmetically a more attractive part.

Big Improvements

Scrap rates were reduced to 1% and the 2300–2500 parts produced in an eight-hour shift represent a 50% increase in productivity. Mash seam welding eliminated the deburring process and the coolant that was previously needed. Tooling changeover now takes only 30 minutes.

Overall, quality is up, costs are down, and inventory and shipping schedules are more controllable.

Article courtesy of the American Welding Society.

Resistance welding unit showing wheel electrode used to join the overlapping edge of the shell. American Welding Society

After the shell is welded it is placed into a station where the component is finished to its final size and shape. American Welding Society

Review

1. Give examples of drawing pages that might be included in a set of drawings.

2. The title box on a drawing can contain what information?

3. Where would a welder look in the set of drawings to find a list of the various items that would be needed to build a weldment?

4. Sketch an object that contains all of the necessary elements so that you can use the following line types: object line, hidden line, centerline, extension line, and dimension line. Label each line type using leaders and arrows.

5. What type of lines can be used in a drawing to show that part of an object has been removed?

6. Describe what phantom lines are in a drawing.

7. Compare an orthographic projection (mechanical drawing) to a pictorial drawing.

8. A mechanical drawing usually includes which three views of an object?

9. List two types of pictorial drawings that can be included with a set of mechanical drawings.

10. What view might be used to show part of an object as if it were sawn away to reveal internal details?

11. What kind of information does a detail view on a drawing provide?

12. The height of an object would be found on which of the standard views in a set of drawings?

Review (continued)

13. What is the difference between "as drawn" and "as built" drawings?

14. What tool can be used to draw an object so that it is smaller or larger than it really is?

15. What is the difference between architectural scales and engineering scales?

16. Why is knowing how to sketch a drawing a valuable skill in the welding shop?

17. Which type of computer graphics program lines stay crisp and clean no matter how much you zoom in and out of the image?

Chapter 20

Welding Joint Design and Welding Symbols

OBJECTIVES

After completing this chapter, the student should be able to

- understand the basics of joint design.
- list the five major types of joints.
- list seven types of weld grooves.
- identify the major parts of a welding symbol.
- explain the parts of a groove preparation.
- describe how nondestructive test symbols are used.

KEY TERMS

combination symbol

edge preparation

groove (G) and fillet (F)

joint dimensions

joint type

weld joint

weld location

weld types

welding symbols

INTRODUCTION

The joint design affects the quality and cost of the completed weld. Selecting the most appropriate joint design for a welding job requires special attention and skill. The eventual design selection can be influenced by a number of factors.

Every weld joint selected for a job requires some compromises. For example, the compromises may be between strength and cost, equipment available, and welder skill, or any two, three, or more variables. Because there are so many factors, a good design requires experience. Even with experience, trial welds are necessary before selecting the final joint configuration and welding parameters.

This chapter familiarizes welders with the most important factors and gives them some appreciation of joint design and fabrication. Experience in the welding field will help a welder become a better joint designer and fabricator.

Welding symbols are the language used to let the welder know exactly what welding is needed. The welding symbol is used as shorthand and can provide the welder with all of the required information to make the correct weld. The emphasis in this chapter is on using and interpreting welding symbols so the welder will develop a welder's "vocabulary."

WELD JOINT DESIGN

The term *weld joint design* refers to the way pieces of metal are put together or aligned with each other. The five basic joint designs, or **joint types,** are butt joints, lap joints, tee joints, outside corner joints, and edge joints. **Figure 20-1** illustrates the way the joint members come together:

- Butt joint—In a butt joint the edges of the metal meet so that the thickness of the joint is approximately equal to the thickness of the metal. The metal surfaces are usually parallel with each other, although there can be some difference in thickness and/or misalignment of the plates. Butt joints can be welded from one side or both sides with some form of groove weld.

- Lap joint—In a lap joint the edges of the metal overlap so the thickness of the joint is approximately equal to the combined thickness of both pieces of metal. The distance the surfaces overlap each other may vary from a fraction of an inch to several inches or even feet. Lap welds are usually joined by making a fillet weld along the edge of one plate joining it to the surface of the other. There are several alternate ways of welding lap joints where the weld is made through one or both pieces of metal joining the lap in the center of the overlap. Some examples of this would be plug welds, seam welds, and stir welds. Welds can be made on one side or both sides of the joint.

- Tee joint—In a tee joint the edge of a piece of metal is placed on the surface of another piece of metal. Usually the parts are placed at a 90° angle with each other. Tee joints can be welded with a fillet weld applied to the surfaces or a weld can be made in a precut groove in the edge of the joining plate. In a few cases, a fillet weld can be made on top of a groove weld on a tee joint. Welds can be made on one side or both sides of the joint.

- Outside corner joint—In an outside corner joint the edges of the metal are brought together at an angle, usually around 90° to each other. The edges can meet at the corner evenly or they can overlap, **Figure 20-2.** The outside corner joint can be welded on both sides, with the outside being made as a groove weld and the inside being made as a fillet weld.

- Edge joint—In an edge joint the metal surfaces are placed together so that the edges are even. One or both plates may be formed by bending them at an angle, **Figure 20-3.** Edge joints are usually welded on only one side.

Welding drawings and specifications usually tell you exactly which joint design will be used for all of the welds to be made. Often a welding engineer or designer has determined the best type of joint to be used. However, on small projects or on some repair welding jobs you will be making the decision as to the joint design to be used. The way the pieces of metal fit together may determine the joint design that must be used. For example, the part shown in **Figure 20-4** can only be made using tee joints. Joints on every weldment are not as easily determined.

If you are the one choosing the weld joint design, you must consider a number of factors. Some of the factors

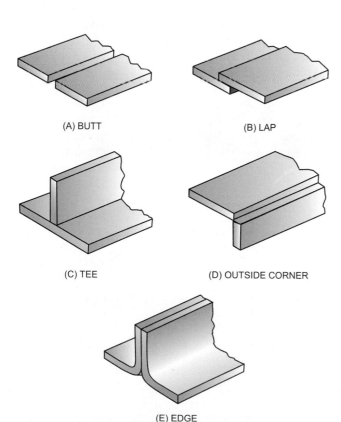

(A) BUTT   (B) LAP

(C) TEE   (D) OUTSIDE CORNER

(E) EDGE

FIGURE 20-1 Types of joints. © Cengage Learning 2012

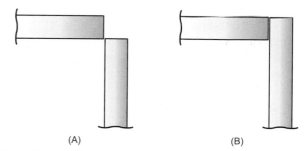

(A)   (B)

FIGURE 20-2 Two ways of fitting up an outside corner joint. © Cengage Learning 2012

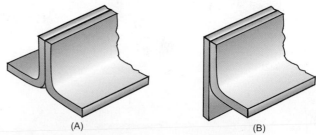

(A)   (B)

FIGURE 20-3 Edge joints. © Cengage Learning 2012

STEP 1
PREBEND BASE PLATE

STEP 2
CLAMP BASE PLATE
TO JIG

STEP 3
TACK WELD SUPPORTS
AT A PRESET ANGLE

STEP 4
WELD

STEP 5
REMOVE FROM JIG

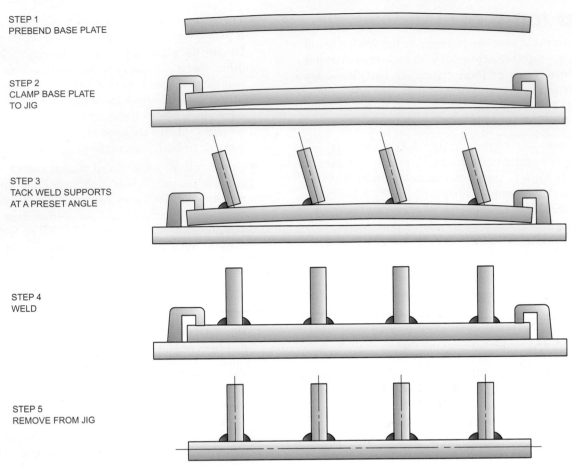

FIGURE 20-4 A method of controlling distortion. © Cengage Learning 2012

involve the type and thickness of metal being welded, the welding position, the welding process, finished weld properties, and any code requirements. The selection of the best joint design for a specific weldment requires that you carefully consider all of the various factors. Each factor, if considered alone, could result in a part that might not be able to be fabricated or might not meet the strength requirements. For example, a narrower joint angle requires less filler metal, and that results in lower welding cost. But if the angle is too small for the welding process being used,

the weld cannot be made strong enough. A large weld may be stronger, but it may result in the part being distorted so badly that it becomes useless.

The purpose of a **weld joint** is to join parts together so that the stresses are distributed. The forces causing stresses in welded joints are tensile, compression, bending, torsion, and shear, **Figure 20-5**. The ability of a welded joint to withstand these forces depends upon both the joint design and the weld integrity. Some joints can withstand some types of forces better than others.

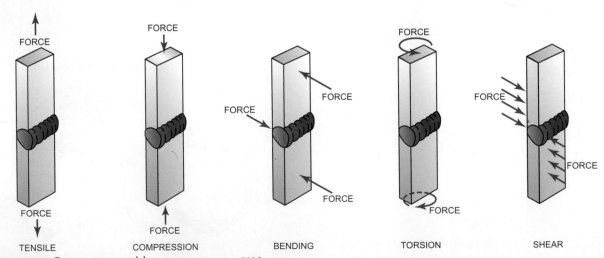

FORCE

FORCE

FORCE

FORCE

FORCE

FORCE

FORCE

FORCE

FORCE

FORCE

FORCE

FORCE

FORCE

TENSILE   COMPRESSION   BENDING   TORSION   SHEAR

FIGURE 20-5 Forces on a weld. © Cengage Learning 2012

Welding Process

The welding process to be used has a major effect on the selection of the joint design. Each welding process has characteristics that affect its performance. Some processes are easily used in any position; others may be restricted to one or more positions. The rate of travel, penetration, deposition rate, and heat input also affect the welds used on some joint designs. For example, a square butt joint can be made in very thick plates using either electroslag or electrogas welding, but not many other processes can be used on such a joint design.

Edge Preparation

The area of the metal's surface that is melted during the welding process is called the *faying surface.* The faying surface can be shaped before welding to increase the weld's strength; this is called **edge preparation.** The edge preparation may be the same on both members of the joint, or each side can be shaped differently, **Figure 20-6.** Reasons for preparing the faying surfaces for welding include the following:

- Codes and standards—Some codes and standards require specific edge preparations.

- Metals—To successfully weld some metals, they must be grooved, such as thick magnesium, which must be U-grooved; or cast-iron cracks, which must be drill stopped and grooved, **Figure 20-7.**

- Deeper weld penetration—With the metal removed by grooving or beveling the metal's edge, it is easier for the molten weld metal to completely fuse through the joint. In some cases, it is possible to make a through thickness weld from one side.

- Smooth appearance—The weld's surface can be ground smooth with the base metal so that the weld "disappears." This can be done for appearance or so that the weld does not interfere with the sliding or moving of parts along the surface.

- Increased strength—A weld should be as strong as or stronger than the base metal being joined. By having 100% joint fusion and an appropriate amount of weld reinforcement, the weld can meet its strength requirement.

Joint Dimensions

In some cases, the exact size, shape, and angle can be specified for a groove, **Figure 20-8.** If exact dimensions are not given, you may make the groove any size you feel necessary; but remember, the wider the groove, the more welding it will require to complete.

Metal Thickness

As the metal becomes thicker, you must change the joint design to ensure a sound weld. On thin sections it is

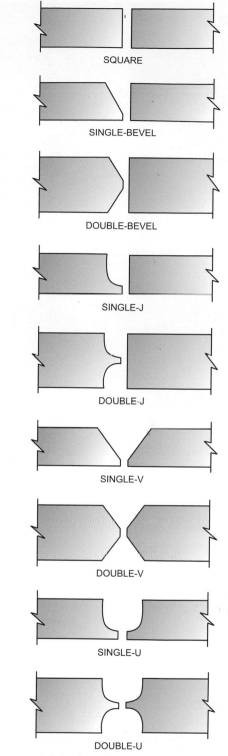

FIGURE 20-6 Joint edge preparation. © Cengage Learning 2012

often possible to make full penetration welds using a square butt joint. Square butt joints take less preparation time and less welding time. But with thicker plates or pipe, the edge must be prepared with a groove on one or both sides. The edge may be shaped with either a bevel, V-groove, J-groove, or U-groove.

When welding on thick plate or pipe, it is often impossible for the welder to get 100% penetration without some

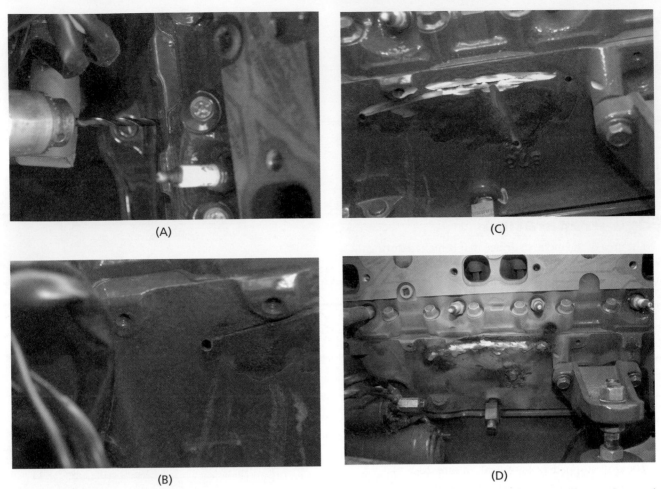

(A)

(C)

(B)

(D)

FIGURE 20-7 Steps used to repair a cast-iron crack. (A) Drill each end of the crack. (B) The drill hole will stop the crack from lengthening as it is being repaired. (C) Grind a U-groove all the way along the crack. (D) Grind the weld before finishing. Larry Jeffus

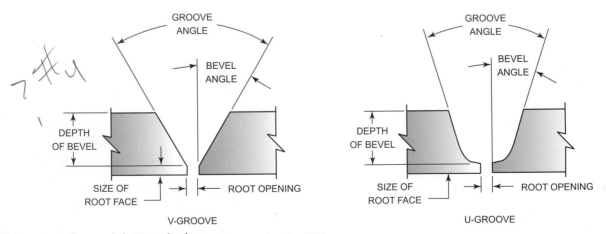

FIGURE 20-8 Groove joint terminology. © Cengage Learning 2012

type of groove being used. The groove may be cut into either just one of the plates or pipes or both. On some plates it can be cut both inside and outside of the joint. The groove may be ground, flame cut, gouged, sawed, or machined on the edge of the plate before or after the assembly. Bevels and V-grooves are best if they are cut before the parts are assembled, **Figure 20-9.** J-grooves and U-grooves

can be cut either before or after assembly, **Figure 20-10.** The lap joint is seldom prepared with a groove because little or no strength can be gained by grooving this joint.

For most welding processes, plates that are thicker than 3/8 in. (10 mm) may be grooved on both the inside and outside of the joint. Plate in the flat position is usually grooved on only one side unless it can be repositioned

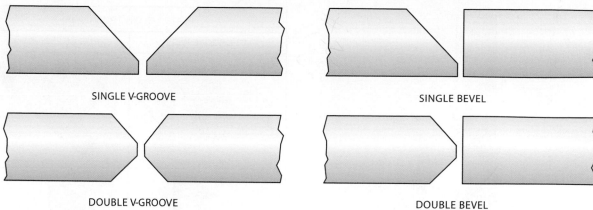

SINGLE V-GROOVE

SINGLE BEVEL

DOUBLE V-GROOVE

DOUBLE BEVEL

FIGURE 20-9 V-groove and bevel joint types. © Cengage Learning 2012

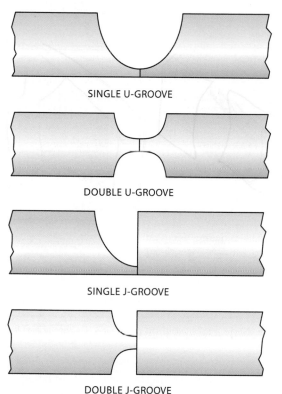

SINGLE U-GROOVE

DOUBLE U-GROOVE

SINGLE J-GROOVE

DOUBLE J-GROOVE

FIGURE 20-10 U-groove and J-groove joint types.
© Cengage Learning 2012

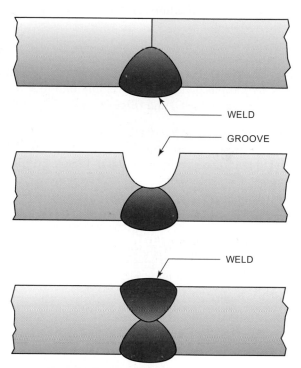

WELD

GROOVE

WELD

FIGURE 20-11 Back gouging a weld joint to ensure 100% joint penetration. © Cengage Learning 2012

or it is required to be welded on both sides. Tee joints in thick plate are easier to weld and will have less distortion if they are grooved on both sides.

Sometimes plates are either grooved and welded or just welded on one side and then back gouged and welded, **Figure 20-11**. Back gouging is a process of cutting a groove in the back side of a joint that has been welded. Back gouging can ensure 100% joint fusion at the root and can remove discontinuities of the root pass.

Metal Type

Because some metals have specific problems with thermal expansion, crack sensitivity, or distortion, the joint design selected must help control these problems. For example, magnesium is very susceptible to postweld stresses, and the U-groove works best for thick sections.

Welding Position

The most ideal welding position for most joints is the flat position because it allows for larger molten weld pools to be controlled. Usually the larger that a weld pool can be, the faster the joint can be completed. When welds are made in any position other than the flat position, they are referred to as being done *out of position*. Some types of grooves work better in out-of-position welding than others; for example, the bevel joint is often the best choice for horizontal butt welding, **Figure 20-12**.

FIGURE 20-12 Weld position for a horizontal joint.
© Cengage Learning 2012

Plate Welding Positions The American Welding Society (AWS) has divided plate welding into four basic positions for **grooves (G)** and **fillet (F)** welds as follows:

- Flat 1G or 1F—When welding is performed from the upper side of the joint, and the face of the weld is approximately horizontal, **Figure 20-13A** and **B**.

- Horizontal 2G or 2F—The axis of the weld is approximately horizontal, but the type of the weld dictates the complete definition. For a fillet weld, welding is performed on the upper side of an approximately vertical surface. For a groove weld, the face of the weld lies in an approximately vertical plane, **Figure 20-13C** and **D**.

- Vertical 3G or 3F—The axis of the weld is approximately vertical, **Figure 20-13E** and **F**.

- Overhead 4G or 4F—When welding is performed from the underside of the joint, **Figure 20-13G** and **H**.

Pipe Welding Positions The American Welding Society has divided pipe welding into five basic positions:

- Horizontal rolled 1G—The pipe is rolled either continuously or intermittently so that the weld is performed within 0° to 15° of the top of the pipe, **Figure 20-14A**.

- Horizontal fixed 5G—The pipe is parallel to the horizon, and the weld is made vertically around the pipe, **Figure 20-14B**.

- Vertical 2G—The pipe is vertical to the horizon, and the weld is made horizontally around the pipe, **Figure 20-14C**.

- Inclined 6G—The pipe is fixed in a 45° inclined angle, and the weld is made around the pipe, **Figure 20-14D**.

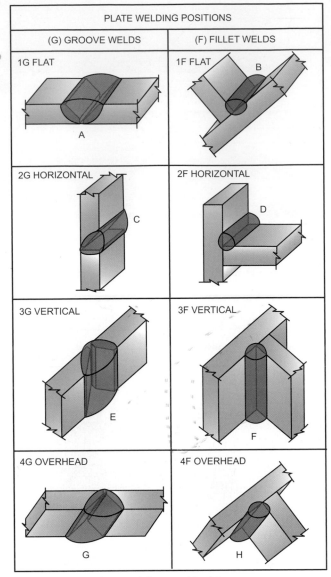

PLATE WELDING POSITIONS	
(G) GROOVE WELDS	(F) FILLET WELDS
1G FLAT	1F FLAT
2G HORIZONTAL	2F HORIZONTAL
3G VERTICAL	3F VERTICAL
4G OVERHEAD	4F OVERHEAD

FIGURE 20-13 Plate welding positions. © Cengage Learning 2012

- Inclined with a restriction ring 6GR—The pipe is fixed in a 45° inclined angle, and there is a restricting ring placed around the pipe below the weld groove, **Figure 20-14E**.

CODE OR STANDARDS REQUIREMENTS

The type, depth, angle, and location of the groove are usually determined by a code or standard that has been qualified for the specific job. Organizations such as the American Welding Society, the American Society of Mechanical Engineers (ASME), and the American Bureau of Ships (ABS) are among the agencies that issue such codes and specifications. The most common code or standards are the AWS D1.1 and the ASME Boiler and Pressure Vessel (BPV), Section IX.

The joint design for a specific set of specifications often must be what is known as prequalified. These joints have

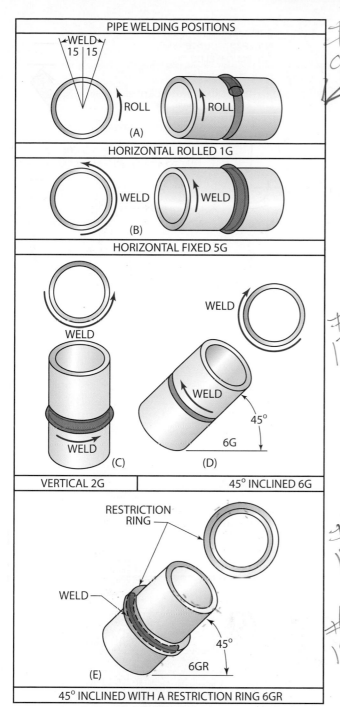

FIGURE 20-14 Pipe welding positions. © Cengage Learning 2012

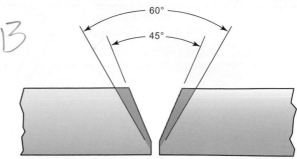

FIGURE 20-15 A smaller groove angle reduces both weld time and weld metal. © Cengage Learning 2012

welder to see the molten weld pool or room to get the electrode or torch into the joint.

Acceptable Cost

Almost any weld can be made in any material in any position. A number of factors can affect the cost of producing a weld. Joint design is one major way to control welding cost. Changes in the design can reduce cost yet still meet the weldment's strength requirements. Reducing the groove angle can also help, **Figure 20-15**. It will decrease the welding filler metal required to complete the weld as well as decrease the time required to fill the larger groove opening. Joint design must be a consideration for any project to be competitive and cost effective.

WELDING SYMBOLS

The use of welding symbols enables a designer to indicate clearly to the welder important detailed information regarding the weld. The information in the welding symbol can include the following details for the weld: length, depth of penetration, height of reinforcement, groove type, groove dimensions, location, process, filler metal, strength, number of welds, weld shape, and surface finishing. All this information would normally be included on the welding assembly drawings.

Welding symbols are a shorthand language for the welder. They save time and money and serve to ensure understanding and accuracy. Welding symbols have been standardized by the American Welding Society. Some of the more common symbols for welding are represented in this chapter. If more information is desired about symbols or how they apply to all forms of manual and automatic machine welding, these symbols can be found in the complete manual *Standard Symbols for Welding, Brazing and Nondestructive Examination,* ANSI/AWS A2.4, published as an American National Standard by the American Welding Society.

Figure 20-16 shows the basic components of welding symbols, consisting of a reference line with an arrow on one end. Other information relating to various features of the weld is shown by symbols, abbreviations, and figures located around the reference line. A tail is added to the basic symbol as necessary for the placement of specific information.

been tested and found to be reliable for the weldments for specific applications. The joint design can be modified, but the cost to have the new design accepted under the standard being used is often prohibitive.

Welder Skill

Often the skills or abilities of the welder are a limiting factor in joint design. A joint must be designed in a manner so that the welders can reliably reproduce it. Some joints have been designed without adequate room for the

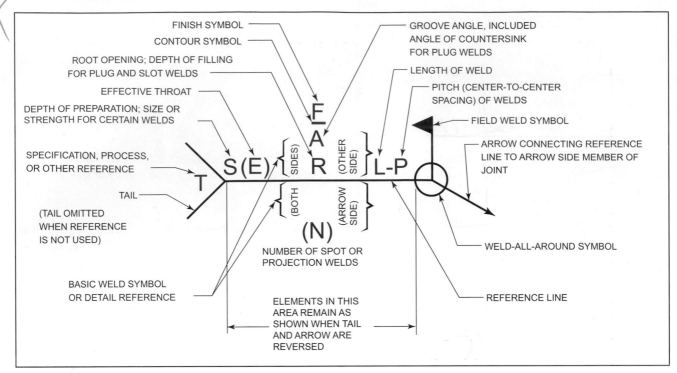

FIGURE 20-16 Standard location of elements of a welding symbol. American Welding Society

Indicating Types of Welds

Weld types are classified as follows: fillets, grooves, flange, plug or slot, spot or projection, seam, back or backing, and surfacing. Each type of weld has a specific symbol that is used on drawings to indicate the weld. A fillet weld, for example, is designated by a right triangle. A plug weld is indicated by a rectangle. All of the basic symbols are shown in **Figure 20-17**.

Weld Location

Welding symbols are applied to the joint as the basic reference. All joints have an arrow side (near side) and other

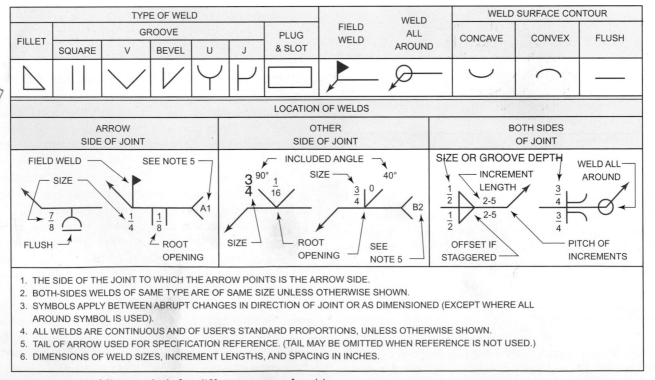

FIGURE 20-17 Welding symbols for different types of welds. American Welding Society

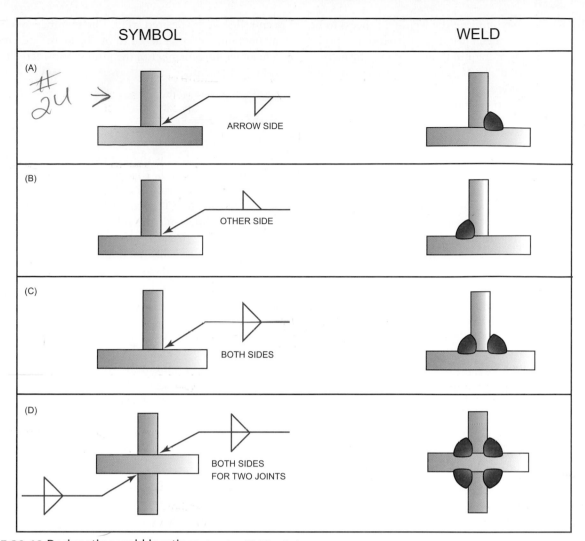

FIGURE 20-18 Designating weld locations. American Welding Society

side (far side). Accordingly, the terms *arrow side, other side,* and *both sides* are used to indicate the **weld location** with respect to the joint. The reference line is always drawn horizontally. An arrow line is drawn from one end or both ends of a reference line to the location of the weld. The arrow line can point to either side of the joint and extend either upward or downward.

If the weld is to be deposited on the arrow side of the joint (near side), the proper weld symbol is placed below the reference line, **Figure 20-18A.**

If the weld is to be deposited on the other side of the joint (far side), the weld symbol is placed above the reference line, Figure 20-18B. When welds are to be deposited on both sides of the same joint, the same weld symbol appears above and below the reference line, **Figure 20-18C** and **D**, along with all detailed information.

The tail is added to the basic welding symbol when it is necessary to designate the welding specifications, procedures, or other supplementary information needed to make the weld, **Figure 20-19**. The notation placed in the

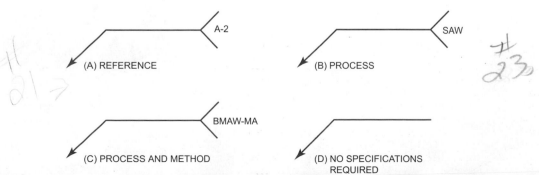

FIGURE 20-19 Locations of specifications, processes, and other references on weld symbols. American Welding Society

tail of the symbol may indicate the welding process to be used, the type of filler metal needed, whether or not peening or root chipping is required, and other information pertaining to the weld. If notations are not used, the tail of the symbol is omitted.

For joints that are to have more than one weld, a symbol is shown for each weld.

Location Significance of Arrow

In the case of fillet and groove welding symbols, the arrow connects the welding symbol reference line to one side of the joint. The surface of the joint the arrow point actually touches is considered to be the arrow side of the joint. The side opposite the arrow side of the joint is considered to be the other (far) side of the joint.

On a drawing, when a joint is illustrated by a single line and the arrow of a welding symbol is directed to the line, the arrow side of the joint is considered to be the near side of the joint.

For welds designated by the plug, slot, spot, seam, resistance, flash, upset, or projection welding symbols, the arrow connects the welding symbol reference line to the outer surface of one of the members of the joint at the centerline of the desired weld. The member to which the arrow points is considered to be the arrow side member. The remaining member of the joint is considered to be the other side member.

Fillet Welds

Dimensions of fillet welds are shown on the same side of the reference line as the weld symbol and are shown to the left of the symbol, **Figure 20-20A**. When both sides of a joint have the same size fillet welds, they are dimensioned as shown in **Figure 20-20B**. When both sides of a joint have different size fillet welds, both are dimensioned, **Figure 20-20C**. When the dimensions of one or both welds differ from the dimensions given in the general notes, both welds are dimensioned. The size of a fillet weld with unequal legs is shown in parentheses to the left of the weld symbol, **Figure 20-20D**. The length of a fillet weld, when indicated on the welding symbol, is shown to the right of the weld symbol, **Figure 20-20E**. In intermittent fillet welds, the length and pitch increments are placed to the right of the weld symbol, **Figure 20-21**. The first number represents the length of the weld, and the second number represents the pitch or the distance between the centers of two welds.

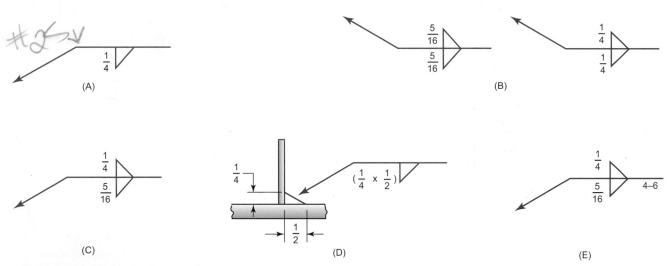

FIGURE 20-20 Dimensioning the fillet weld symbol. American Welding Society

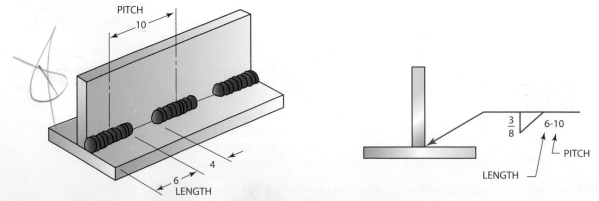

FIGURE 20-21 Dimensioning intermittent fillet welds. American Welding Society

Plug Welds

Holes in the arrow side member of a joint for plug welding are indicated by placing the weld symbol below the reference line. Holes in the other side member of a joint for plug welding are indicated by placing the weld symbol above the reference line, **Figure 20-23**. Refer to Figure 20-23 for the location of the dimensions used on plug welds. The diameter or size is located to the left of the symbol (A). The angle of the sides of the hole, if not square, is given above the symbol (B). The depth of buildup, if not completely flush with the surface, is given in the symbol (C). The center-to-center dimensioning or pitch is located on the right of the symbol (D).

Spot Welds

Dimensions of resistance spot welds are indicated on the same side of the reference line as the weld symbol, **Figure 20-24**. Such welds are dimensioned either by size or strength. The size is designated as the diameter of the weld expressed in fractions or in decimal hundredths of an inch. The size is shown with or without inch marks to the left of the weld symbol. The center-to-center spacing (pitch) is shown to the right of the symbol.

The strength of spot welds is shown as the minimum shear strength in pounds (newtons) per spot and is shown to the left of the symbol, **Figure 20-25A**. When a definite

FIGURE 20-22 Intermittent welds were used to help prevent cracks from spreading due to the severe vibration and stress during the launching of the Saturn V booster rocket, which was used to launch astronauts to the moon. Larry Jeffus

Intermittent welds can be used to reduce the amount of welding and possible weld distortion and to prevent a crack from spreading. It is easier for a crack to propagate through a continuous weld than it is on an intermittent weld where it has to restart at the beginning of each weld, **Figure 20-22**.

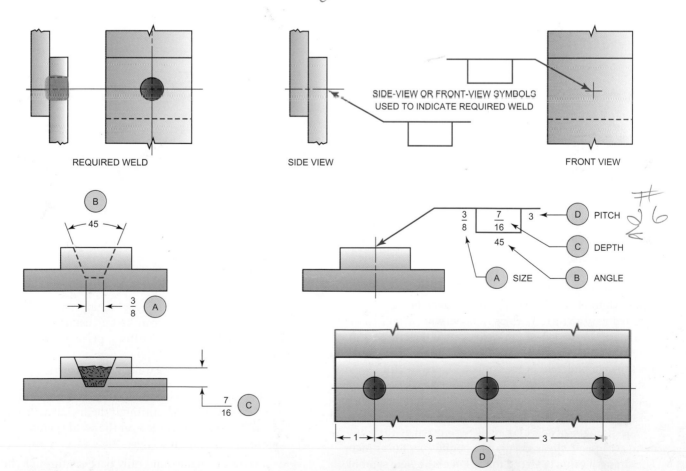

FIGURE 20-23 Applying dimensions to plug welds. American Welding Society

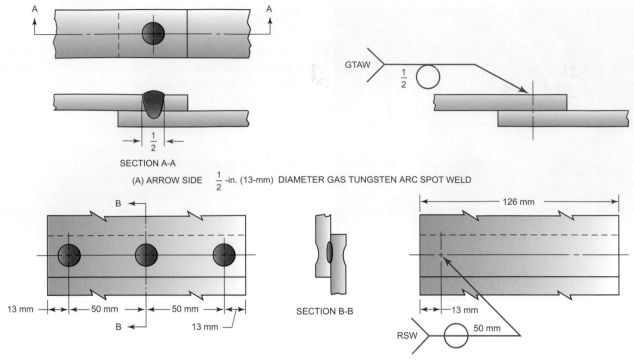

SECTION A-A

(A) ARROW SIDE $\frac{1}{2}$-in. (13-mm) DIAMETER GAS TUNGSTEN ARC SPOT WELD

SECTION B-B

(B) 50-mm PITCH ON A RESISTANCE SPOT WELD

FIGURE 20-24 Spot welding symbols. American Welding Society

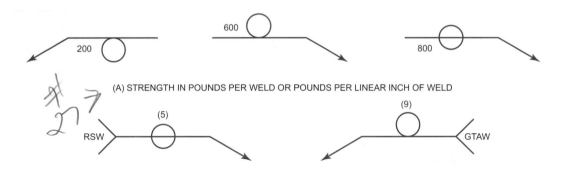

(A) STRENGTH IN POUNDS PER WELD OR POUNDS PER LINEAR INCH OF WELD

(B) QUANTITY OF SPOT WELDS

FIGURE 20-25 Designating strength and number of spot welds. American Welding Society

number of spot welds is desired in a certain joint, the quantity is placed above or below the weld symbol in parentheses, **Figure 20-25B**.

Seam Welds

Dimensions of seam welds are shown on the same side of the reference line as the weld symbol. Dimensions relate to either size or strength. The size of seam welds is designated as the width of the weld expressed in fractions or decimal hundredths of an inch. The size is shown with or without the inch marks to the left of the weld symbol, **Figure 20-26A**. When the length of a seam weld is indicated on the symbol, it is shown to the right of the symbol, **Figure 20-26B**. When seam welding extends for the full distance between abrupt changes in the direction of welding, a length dimension is not required on the welding symbol.

The strength of seam welds is designated as the minimum acceptable shear strength in pounds per linear inch. The strength value is placed to the left of the weld symbol, **Figure 20-27**.

Groove Welds

Joint strengths can be improved by making some type of groove preparation before the joint is welded. There are seven types of grooves. The groove can be made in one or both plates or on one or both sides. By cutting the groove in the plate, the weld can penetrate deeper into the joint. This helps to increase the joint strength without restricting weldment flexibility.

The grooves can be cut in base metal in a number of different ways. The groove can be cut using an oxyfuel cutting torch, air carbon arc cutting, plasma arc cutting, machining, or saws.

The various types of groove welds are classified as follows:

- Single-groove and symmetrical double-groove welds that extend completely through the members being joined. No size is included on the weld symbol, **Figure 20-28A** and **B**.

- Groove welds that extend only partway through the parts being joined. The size as measured from the

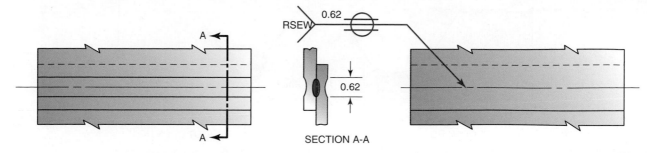

FIGURE 20-26 Designating the size of a seam weld. American Welding Society

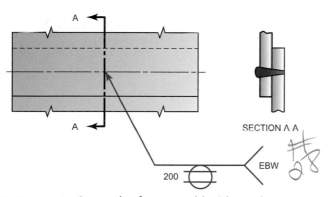

FIGURE 20-27 Strength of seam weld with an electron beam. American Welding Society

top of the surface to the bottom (not including reinforcement) is included to the left of the welding symbol, **Figure 20-28C.**

■ The size of groove welds with a specified *effective throat* is indicated by showing the depth of groove preparation with the effective throat appearing in parentheses and placed to the left of the weld symbol, **Figure 20-28D.** The size of square groove welds is indicated by showing the root penetration. The depth of chamfering and the root penetration are read in that order from left to right along the reference line.

■ The root opening of groove welds is the user's standard unless otherwise indicated. The root opening of groove welds, when not the user's standard, is shown inside the weld symbol, **Figure 20-28E and F.**

■ The root face's main purpose is to minimize the burn-through that can occur with a feather edge. The size of the root face is important to ensure good root fusion, **Figure 20-29.**

■ The size of flare groove welds is considered to extend only to the tangent points of the members, **Figure 20-30.**

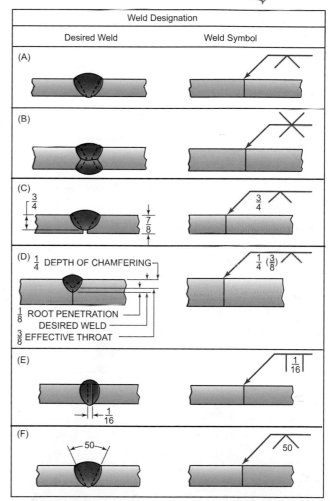

FIGURE 20-28 Designating groove weld location, size, and root penetration. American Welding Society

Backing

A backing (strip) is a piece of metal that is placed on the back side of a weld joint. The backing must be thick enough to withstand the heat of the root pass as it is burned in. A backing strip may be used on butt joints, tee joints, and outside corner joints, **Figure 20-31.**

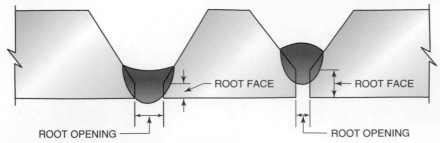

FIGURE 20-29 Effect root dimensioning can have on groove weld penetration. © Cengage Learning 2012

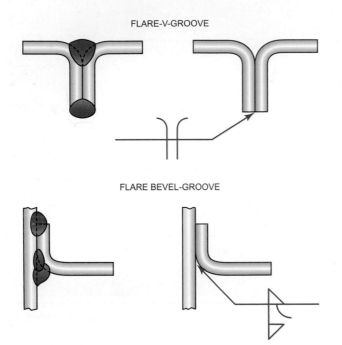

FLARE-V-GROOVE

FLARE BEVEL-GROOVE

FIGURE 20-30 Designating flare V- and flare bevel-groove welds. © Cengage Learning 2012

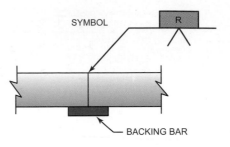

SYMBOL

BACKING BAR

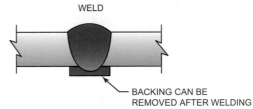

WELD

BACKING CAN BE REMOVED AFTER WELDING

FIGURE 20-32 Butt weld with backing plate.
© Cengage Learning 2012

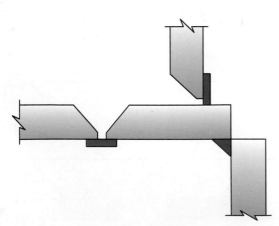

FIGURE 20-31 Backing strips. © Cengage Learning 2012

The backing may be either left on the finished weld or removed following welding. If the backing is to be removed, the letter R is placed in the backing symbol, **Figure 20-32.** The backing is often removed for a finished weld because it can be a source of stress concentration and a crevice to promote rusting.

Flanged Welds

The following welding symbols are used for light-gauge metal joints where the edges to be joined are bent to form flange or flare welds.

- Edge flange welds are shown by the edge flange weld symbol.
- Corner flange welds are indicated by the corner flange weld symbol.
- Dimensions of flange welds are shown on the same side of the reference line as the weld symbol and are placed to the left of the symbol, **Figure 20-33.** The radius and height above the point of tangency are indicated by showing both the radius and the height separated by a plus sign.
- The size of the flange weld is shown by a dimension placed outward from the flanged dimensions.

Nondestructive Testing Symbols

The increased use of nondestructive testing (NDT) as a means of quality assurance has resulted in the development of standardized symbols. They are used by the designer or

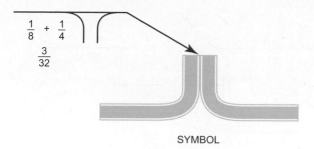

SYMBOL

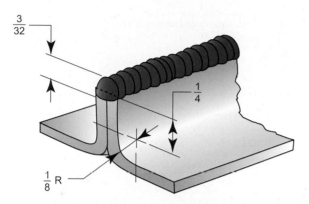

$\frac{3}{32}$

$\frac{1}{4}$

$\frac{1}{8}$ R

DESIRED WELD

FIGURE 20-33 Applying dimensions to flange welds.

American Welding Society

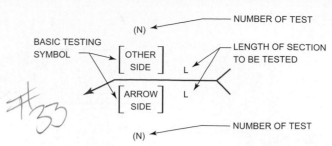

FIGURE 20-34 Basic nondestructive testing symbol.

© Cengage Learning 2012

Type of Nondestructive Test	Symbol
Visual	VT
Penetrant	PT
Dye penetrant	DPT
Fluorescent penetrant	FPT
Magnetic particle	MT
Eddy current	ET
Ultrasonic	UT
Acoustic emission	AET
Leak	LT
Proof	PRT
Radiographic	RT
Neutron radiographic	NRT

TABLE 20-1 Standard Nondestructive Testing Symbols

engineer to indicate the area to be tested and the type of test to be used. The inspection symbol uses the same basic reference line and arrow as the welding symbol, **Figure 20-34.**

The symbol for the type of nondestructive test to be used, **Table 20-1,** is shown with a reference line. The location above, below, or on the line has the same significance as it does with a welding symbol. Symbols above the line indicate other side, symbols below the line indicate arrow side, and symbols on the line indicate no preference for the side to be tested, **Figure 20-35.** Some tests may be performed on both sides; therefore, the symbol appears on both sides of the reference line.

Two or more tests may be required for the same section of weld. **Figure 20-36** shows methods of combining testing symbols to indicate more than one type of test to be performed.

The length of weld to be tested or the number of tests to be made can be noted on the symbol. The length either may be given to the right of the test symbol, usually in inches, or can be shown by the arrow line, **Figure 20-37.** The number of tests to be made is given in parentheses ( ) above or below the test symbol, **Figure 20-38.** The welding symbols and nondestructive testing symbols both can be combined into one symbol, **Figure 20-39.** The **combination symbol** may help both the welder and inspector to identify welds that need special attention. A special symbol can be used to show the direction of radiation used in a radiographic test, **Figure 20-40.**

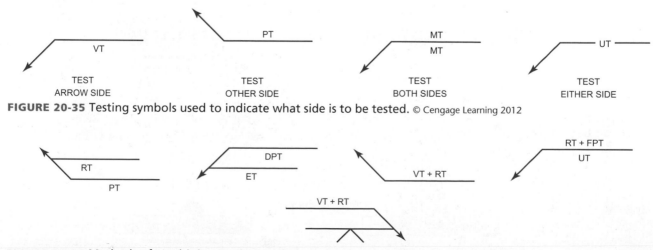

FIGURE 20-35 Testing symbols used to indicate what side is to be tested. © Cengage Learning 2012

FIGURE 20-36 Methods of combining testing symbols. © Cengage Learning 2012

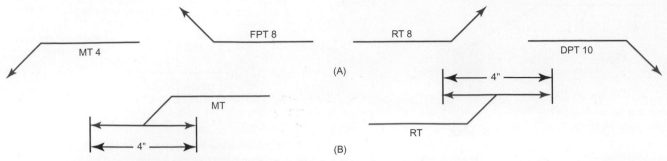

FIGURE 20-37 Two methods of designating the length of weld to be tested. © Cengage Learning 2012

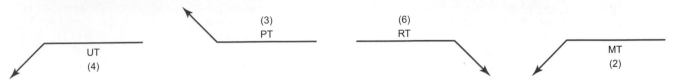

FIGURE 20-38 Method of specifying the number of tests to be made. © Cengage Learning 2012

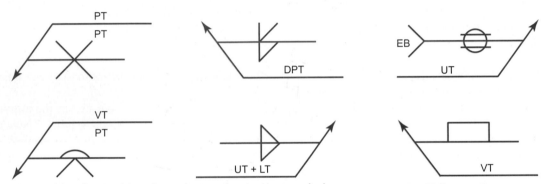

FIGURE 20-39 Combination weld and nondestructive testing symbols. © Cengage Learning 2012

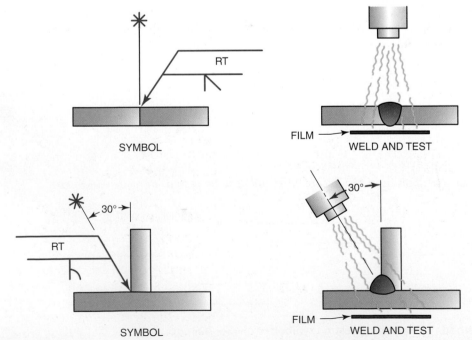

FIGURE 20-40 Combination symbol for weld and radiation source location for testing. © Cengage Learning 2012

Summary

Understanding the physics of joint design is essential for the welder so that you can recognize and anticipate the various forces that will be applied to a weldment in the field. In this way, you can select the proper weld joint configuration to withstand these forces. Engineers use static and dynamic loading computer programs to anticipate the weldment's strength requirements. However, from time to time a welder will be asked to make changes in structures as part of a modification or repair. In the field, the welder is expected to understand the types of forces being applied to the weldment and to determine the best joint design to prevent these forces from causing a structural failure.

You must also be able to interpret the meaning of welding symbols. Understanding the significance of a welding symbol will prevent one of the most common problems in the field—that of overwelding. A weld that is made excessively large can cause a structural failure as easily as one that is undersized. Welded structures must often flex under load. Weldments must be flexible within limits so they can give, so they are not brittle, and so they will not break. Overwelding can cause a structure to be too rigid and subject to a brittle fracture. *Do not overweld.*

Extending the Life of Resistance Welding Electrodes

Taking measures to extend spot electrode life can reduce unscheduled downtime and lower costs.

With the widespread use of coated steels, especially galvanized steels, electrode life for resistance welding has become more of a concern to automakers than ever before. The principal concern is cost. What does it cost to make a resistance spot weld? How much of that cost is traced to the electrode? What is the selling price of the electrode? How much time is spent dressing and changing the electrode, and how do those minutes translate into dollars and cents? What is the cost of the labor needed to make the change? Finding answers to such questions is a top priority for automakers.

It All Adds Up

Some estimate it costs automakers $.05 to make one spot weld. Since there are up to 3000 welds in each passenger car, as much as $150 is spent in the spot welding of one automobile. Between one-half and three-quarters of the cost of a spot weld is the electrode cost. With proper planning, an electrode or electrode cap actually might make it through the production run of one automobile, but greater savings will be realized if electrode life can be extended.

Electrode life is defined as the number of welds made with a pair of electrodes while maintaining weld button diameters above a specified minimum value. That button diameter is the main focus of attention in the automotive industry because it controls the cost of the resistance spot welding operation.

Galvanized Steel

When galvanized steel was first used on the automotive assembly lines, there were severe problems with electrode sticking. Engineers associated with resistance welding looked for ways to stop such sticking. From the standpoint of operating cost, electrode sticking was especially intolerable for robotic welding. Glid-Cop® dispersion-strengthened copper electrodes and titanium carbide coated electrodes offered some solutions.

Real savings in the overall cost of manufacturing an automobile can be realized by extending the service life of resistance welding electrodes. American Welding Society

Galvanized steels have become conventional materials on modern automotive assembly lines but are not as easy to spot weld as uncoated steels. Certain precautions have to be observed when using these zinc-coated materials. As a rule, higher currents are needed. The coatings tend to deteriorate the spot welding electrodes by rapid wear or by alloying the electrode itself. When copper fuses with zinc, the metallurgical end result is often brass. Electrode sticking will often occur as a result.

Weld schedules for galvanized steels include notable differences from those used for uncoated steels. Specifically, electrode forces are higher, weld times longer, and hold times shorter.

Aluminum's Story

Aluminum and electrode life are another world. According to David Coldwell—applications engineer, Alcan Automotive Products, Farmington Hills, Michigan—the most practical approaches to extending electrode life on aluminum can be found in attention to geometry and improved tip dressing. One solution, he said, is the use of a special burnishing tool as an alternative to a conventional dresser that maintains the surface of the electrode. The tool will remove any aluminum from the tip of the electrode, but it does not remove any of the copper alloy from the electrode itself.

One of the big three auto companies requires that the resistance spot welding electrode perform satisfactorily on aluminum for at least 2000 welds without increasing the welding current or dressing the electrodes.

Numerous aluminum alloys are being resistance spot welded in automotive production. Among the nonheat-treatable 5000 series alloys, such grades as 5052, 5182, and 5754 aluminum have become popular. In the heat-treatable aluminum alloys, favorites in the 2000 series include 2008, 2010, and 2036, while 6009, 6010, 6020, 6061, 6063, and 6111 are used from the 6000 series. The most popular grade of all, however, is 6111. On the lift gate of GM's Suburban SUV, the inner skin of nonheat-treatable 5182-0 aluminum is spot welded to the outer panel of heat-treatable 6111-T4P aluminum. Caps formed from copper-zirconium, and later copper-cadmium, have been used for this type of production welding.

Titanium Carbide-coated Electrodes

For several years, Huys Industries Limited, Weston, Ontario, Canada, has introduced its titanium carbide-coated electrodes (TiCaps™) to many plants in the automotive industry. According to Nigel Scotchmer, president of Huys Industries, these electrodes last longer, stick less, and offer a wider welding window. The company currently has one coating for steel and another for aluminum. Other coatings are under development.

Titanium carbide was first applied to copper about fifteen years ago, but not on resistance welding electrodes. As a new spot welding electrode, it was preceded, in fact, by GlidCop® technology.

According to Huys Industries, the Nissan Motor Manufacturing Corp. estimated a U.S. plant was saving more than $100,000 per year on a specific production line by using TiC-coated caps. Because of the higher conductivity of titanium, $55,260 per year was being saved in reduced utility costs. Also, engineers found tip dressing was not required when using coated electrodes, amounting to savings of $51,660 per year. A number of automotive plants report no electrode sticking when TiC-coated caps are used.

According to Scotchmer, the titanium carbide coating is actually welded onto the caps using a patented low-voltage welding process. "It creates a metallurgical bond between the sintered titanium carbide powder and the copper alloy substrate," he said.

The Heart of the Matter

According to Professor Norman Zhou at the University of Waterloo, Ontario, Canada, many aspects have to be analyzed in order to improve the life of the resistance welding electrode. He indicated metallurgy was at the heart of the overall problem. Industry has already developed a number of electrodes for the welding of steels, including precipitation-hardened copper electrodes such as copper-chromium and copper-zirconium alloys, as well as the dispersion-strengthened copper electrode.

Work is currently underway to study whether further improvements can be made when applying these electrodes to welding aluminum. New electrode materials (especially in the area of composites) may be used in the future, Zhou believes. Another area of improvement may come from applying coating materials on tip surfaces such as the titanium carbide coating developed for welding coated steels. The goal of one current research project involves finite element analysis of resistance spot welding of aluminum to reduce both stress and temperature distributions by determining the ideal design of electrode geometry.

The Importance of Weld Schedules

Shorter electrode life can be expected when welding galvanized, as opposed to uncoated, steels. In making the transition from uncoated steel to galvanized steel, weld schedules have to be different. Higher electrode forces, for example, need to be employed when welding galvanized steel. Also, weld time needs to be longer, and hold time is usually shorter.

Coated steels require higher currents than uncoated steels. Most coatings increase the area of contact with the electrode face. The increased contact area results in decreased current density as compared with uncoated steel. The higher currents for welding coated steels can lower electrode life due to increased electrode heating. In other words, the window is a lot smaller in the welding of galvanized steels.

Article courtesy of the American Welding Society.

Review

1. List the five joint types used in welding.
2. What stresses must a welded joint withstand?
3. Sketch and label five edge preparations used for welding joints.
4. Sketch a V-grooved butt joint, and label all of the joint's dimensions.
5. Sketch a weld on plates in the 1G and 1F positions.
6. Sketch a weld on plates in the 2G and 2F positions.
7. Sketch a weld on plates in the 3G and 3F positions.
8. Sketch a weld on plates in the 4G and 4F positions.
9. Sketch a weld on a pipe in the 1G position.
10. Sketch a weld on a pipe in the 5G position.
11. Sketch a weld on a pipe in the 2G position.
12. Sketch a weld on a pipe in the 6G position.
13. Sketch a weld on a pipe in the 6GR position.
14. Why are some joints back gouged?
15. Why is it usually better to make a weld in the flat position?
16. What is a prequalified joint?
17. Why is cost a consideration in joint design?
18. Why are welding symbols used?
19. What types of information can be included on a welding symbol?
20. Why is a tail added to the basic welding symbol?
21. What types of information may appear on the reference line of a welding symbol?
22. What are the different classifications of welds that a symbol can indicate?
23. How is the reference line always drawn?
24. What is meant if the weld symbol is placed below the reference line?
25. How are the dimensions for a fillet weld given?
26. What dimensions can be given for a plug weld?
27. What two units are used to show the minimum shear strength of a spot weld?
28. How is the strength of a seam weld specified?
29. How can the groove be cut on the edge of a plate?
30. Sketch and dimension a V-groove weld symbol for a weld on the arrow side, with 1/8-in. root opening, 3/4 in. in size, and having a groove angle of 45°.
31. How is the removal of the backing strip noted on a welding symbol?
32. How are flanged edges formed?
33. Sketch two NDT symbols illustrating different methods that can be used to indicate multiple test requirements for the same section of weld.

Chapter 21

Fabricating Techniques and Practices

OBJECTIVES

After completing this chapter, the student should be able to

- explain the various safety issues related to fabrication.
- list the advantages of using custom fabrication parts.
- demonstrate an understanding of the proper placement of tack welds.
- demonstrate the use of location and alignment points when assembling a project.
- explain how to adjust parts to meet the tolerance.
- describe how to control weld distortion.
- lay out and trace parts.
- identify common sizes and shapes of metals used in weldments.
- describe how to assemble and fit up parts for welding.

KEY TERMS

assembly	*fitting*	*tack welds*
contour marker	*fixtures*	*tolerance*
custom fabrication	*kerf*	*weldment*
fabrication	*nesting*	

INTRODUCTION

The first step in almost every welding operation is the **assembly** of the parts to be joined by welding. At the very basic level, assembly can be just placing two pieces of metal flat on a table and tack welding them together for practice welding. At a higher level is the assembly of complex equipment, buildings, ships, or other large welded structures. The important thing to remember, however, is that no matter how large or complicated the welded structure, it is assembled one piece at a time. That is true for a simple project you build as part of your welding shop learning in school or for the ship, rocket engine, or building you might construct one day.

FIGURE 21-1 Petrol chemical refinery near Point Comfort, Texas, along the South Texas coast. Larry Jeffus

FABRICATION

In addition to straight welding, welders are often required to fabricate a weldment. **Fabrication** is the process of assembling the parts to form a weldment. It can include layout, measuring, cutting, grinding, fitting, tack welding, and so on. The term **weldment** is a general term that refers to anything that was created primarily by welding. It may form a completed project or may only be part of a larger structure. Some weldments are composed of two or three parts; others may have hundreds or even thousands of individual parts, **Figure 21-1**. Even the largest weldments start by placing two parts together.

The number and type of steps required to take a plan and create a completed project vary depending on the complexity and size of the finished weldment. All welding projects start with a plan. This plan can range from a simple one that exists only in the mind of the welder to a very complex plan comprising a set of drawings. As a beginning welder, you must learn how to follow a set of drawings to produce a finished weldment.

Soon we will be fabricating large structures in space, **Figure 21-2** and **Figure 21-3**. The International Space Station is being assembled in space from large sections built here on Earth. Most of these assemblies required some type of welding. Someday we expect to be welding

FIGURE 21-3 The International Space Station was assembled in space. NASA

in space. Research for welding in space dates back to the 1960s with experiments done on board the U.S. Skylab. Today, that research continues with experiments on the International Space Station.

Safety

As with any welding, safety is of primary concern for fabrication of weldments. Fabrication may present some potential safety problems not normally encountered in straight shop welding. Unlike most practice welding, much of the larger fabrication work may need to be performed outside an enclosed welding booth. Additionally, several welders may be working on a structure at the same time. When you are welding around other workers who are not welders, you must let them know that you are going to be welding so they can protect themselves from the arc light, sparks, and other possible hazards. Tell them about the hazards because you cannot assume that they know about the hazards of welding. Extra care must be taken to ensure that burns do not occur on you or the other welders from the arc or hot sparks. When possible, you should erect portable welding curtains, **Figure 21-4**.

Ventilation is also important because the normal shop ventilation may not extend to the fabrication area.

FIGURE 21-2 The neutral buoyancy tank allows divers to work in spacesuits to simulate the microgravity of space. NASA

FIGURE 21-4 Portable welding curtains help protect other workers from arc flash. PIPCO

THINK GREEN

Anytime you are working outside of a welding shop, you must be careful not to contaminate the environment. Weld spatter and grinding dust from some metals like stainless steel, which contains chromium, can be hazardous to the environment, especially if they are carried away with storm water runoff. You must guard against contamination when you are doing other outdoor jobs such as cleaning parts with solvents, sandblasting, painting, and so on.

A portable fan may be needed to help blow the welding fumes away from the work area. Be sure the fan blows the fumes away from you and others.

Often you will be working in an area that has welding cables, torch hoses, extension cords, and other trip hazards lying around on the floor. These must be flat on the floor and should be covered if they are in a walkway to prevent accidental tripping. Keep all of the scrap metal and other debris picked up; a neat work area is a safe work area.

As the fabrication grows in size, it will become heavier. Make sure that it is stable and not likely to fall. A weldment that starts out stable and well supported can become unstable and likely to fall as it grows in size. Keeping it well supported is important especially if you have to crawl under it to weld on the bottom side. Check with your supervisor or shop safety officer before working under any weldment.

These and other safety concerns are covered in Chapter 2, "Welding Safety." You should also read any safety booklets supplied with the equipment before starting any project.

Parts and Pieces

Welded fabrications can be made from precut and preformed parts, or they can be made from hand-cut and hand-formed parts, **Figure 21-5A, B, and C.** In most

FIGURE 21-5 (B) Supporting yourself and your arms makes it easier to make smooth cuts. Larry Jeffus

FIGURE 21-5 (C) The guide helps you make the cut straight. Larry Jeffus

FIGURE 21-5 (A) Magnetic pipe wrap helps guide the plasma arc cutting torch. Larry Jeffus

cases, weldments are made using a variety of precut and preformed parts along with handmade pieces.

Preshaped pieces may be precut, bent, machined, or otherwise prepared before you receive them. This is a common practice in large shops and on large-run projects. Large shops may have an entire department dedicated to material and part preparation. When a large number of the same items are made in a large-run shop, it may outsource some of the parts to shops that specialize in mass-producing items. When making an assembly with precut and preformed parts, little or no on-the-job fitting may be required. That, of course, depends on how accurately the parts were prepared.

FIGURE 21-6 Custom tower base for a wind energy generator. Larry Jeffus

The opposite end of the spectrum from preformed parts is **custom fabrication** in which all or most of the assembly is handmade, **Figure 21-6.** This might include cutting, bending, grinding, drilling, or other similar processes, **Figure 21-7.** Almost all weldments were produced by hand until the introduction of automated equipment for cutting, bending, and machining. Today, many items, including some large machines and almost all repair work, are still custom fabricated.

Using preformed parts has a number of advantages:

- Cost—Shops that specialize in cutting out mass numbers of similar parts can do it less expensively than the same parts can be made one by one by hand.

- Speed—High-speed cutting and forming machines can produce a large number of items quickly.

- Accuracy—Automated equipment can make parts more accurately than they can be made by hand.

- Less waste—The wise use of materials is important to both control cost and to conserve natural resources.

Custom fabricating parts has a number of advantages:

- Originals—It is not practical to set up an automated process when there will be only one of a kind or a limited number of an item produced.

- Prototypes—Often, even if there are going to be thousands or even tens of thousands of a weldment produced, the first one, the prototype, must be made by hand to be sure that everything works as it was planned.

- Repairs—Seldom would it be necessary to make a large number of the same part or piece when making a repair on a damaged or worn item.

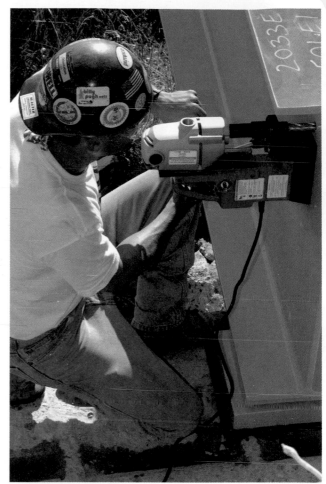

FIGURE 21-7 Magnetic drill makes it easier to drill holes accurately. Hougen

- Custom jobs—Sometimes people want to have something special or unique built or modified just for them.

Layout

Parts for fabrication may require that the welder lay out lines and locate points for cutting, bending, drilling, and assembling. Lines may be marked with a soapstone or a chalk line, scratched with a metal scribe, or punched with a center punch, **Figure 21-8.** If a piece of soapstone is used, it should be sharpened properly to increase accuracy, **Figure 21-9.** A chalk line will make a long, straight line on metal and is best used on large jobs, **Figure 21-10.** Both the scribe and punch can be used to lay out an accurate line, but the punched line is easier to see when cutting. A punch can be held as shown in **Figure 21-11,** with the tip just above the surface of the metal. When the punch is struck with a lightweight hammer, it will make a mark. If you move your hand along the line and rapidly strike the punch, it will leave a series of punch marks for the cut to follow.

Always start a layout as close to a corner of the material as possible. By starting in a corner or along the edge, you can take advantage of the preexisting cut as well as reduce wasted material.

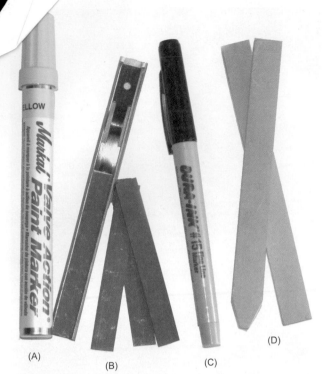

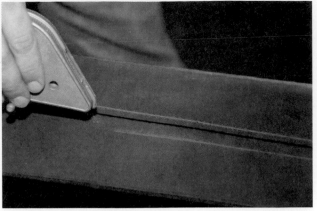

FIGURE 21-10 (B) Check to see that the line is dark enough to be easily seen. Larry Jeffus

FIGURE 21-8 (A) Paint marker, (B) grease marker, (C) felt-tip marker, and (D) soapstone. Larry Jeffus

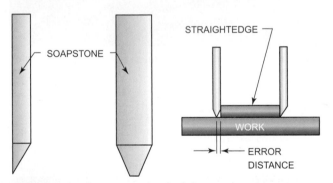

FIGURE 21-9 Proper method of sharpening a soapstone. © Cengage Learning 2012

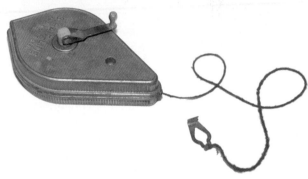

FIGURE 21-10 (C) Chalk line reel, powdered chalks are available in several different colors. Larry Jeffus

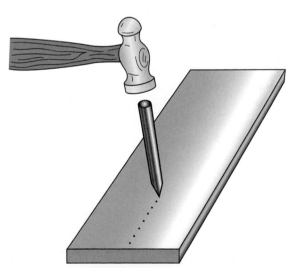

FIGURE 21-11 Holding the punch slightly above the surface allows it to be struck repeatedly, making punched marks to form a punched line that is easily seen for cutting. © Cengage Learning 2012

FIGURE 21-10 (A) Pull the chalk line tight and then snap the line. Larry Jeffus

It is easy to cut the wrong line. In welding shops, one person may lay out the parts and another may make the cuts. Even when one person does both jobs, it is easy to cut the wrong line, either because of the restricted view through cutting goggles or because of the large number

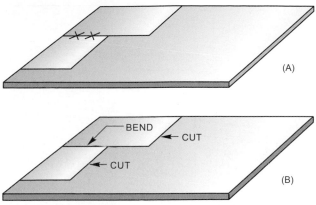

FIGURE 21-12 Identifying layout lines helps prevent mistakes during cutting. © Cengage Learning 2012

of lines on a part. To avoid making a cutting mistake, always identify whether lines are being used for cutting, for locating bends, as drill centers, or as assembly locations. The lines not to be cut may be marked with an *X*, or they may be identified by writing directly on the part. Mark the side of the line that is scrap so that when the kerf is removed from that side the part will be the proper size, **Figure 21-12**. Any lines that have been used for constructing the actual layout line or to locate points for drilling or are made in error must be erased completely or clearly marked to avoid confusion during cutting and assembly.

Some shops have their own shorthand methods for identifying layout lines, or you can develop your own system. Failure to develop and use a system for identifying lines will ultimately result in a mistake, **Figure 21-13**. In a welding shop you will find only those who have made the wrong cut and those who will make the wrong cut. When it does happen, check with the welding shop supervisor to see what corrective steps can be taken. One advantage for

most welding assemblies is that many cutting errors can be repaired by welding. Prequalified procedures are often established for just such an event, so check before deciding to scrap the part.

The process of laying out a part may be affected by the following factors:

- *Material shape:* **Figure 21-14** lists the most common metal shapes used for fabrication. Flat stock such as sheets and plates are easiest to lay out, and pipes and round tubing are the most difficult shapes to work with.

- *Part shape:* Parts with square and straight cuts are easier to lay out than parts with angles, circles, curves, and irregular shapes.

- *Tolerance:* The smaller or tighter the tolerance that must be maintained, the more difficult the layout.

- *Nesting:* The placement of parts together in a manner that will minimize the waste created is called **nesting**.

Parts with square or straight edges are the easiest to lay out. Simply measure the distance and use a square or straightedge to lay out the line to be cut, **Figure 21-15**. Straight cuts that are to be made parallel to an edge can be drawn by using a combination square and a piece of soapstone or scriber. Set the combination square to the correct dimension and drag it along the edge of the plate while holding the soapstone or scriber at the end of the combination square's blade, **Figure 21-16**.

PRACTICE 21-1

Laying Out Square, Rectangular, and Triangular Parts

Using a piece of metal or paper, soapstone or pencil, tape measure, and square, you will lay out the parts shown in **Figure 21-17**. The parts must be laid out within ± 1/16 in. of the dimensions. Convert the dimensions into SI metric units of measure.

Circles, arcs, and curves can be laid out by using either a compass or a circle template, **Figure 21-18**. The diameter is usually given for a hole or round part, and the radius is usually given for arcs and curves, **Figure 21-19**. The center of the circle, arc, or curve may be located using dimension lines and centerlines. Curves and arcs that are to be made tangent to another line may be dimensioned with only their radiuses, Figure 21-19.

Complete a copy of the "Student Welding Report" listed in Appendix I or provided by your instructor. ◆

PRACTICE 21-2

Laying Out Circles, Arcs, and Curves

Using a piece of metal or paper, soapstone or pencil, tape measure, compass, or circle template and square, you

FIGURE 21-13 Marking parts makes it less likely that the wrong one is welded in place. Larry Jeffus

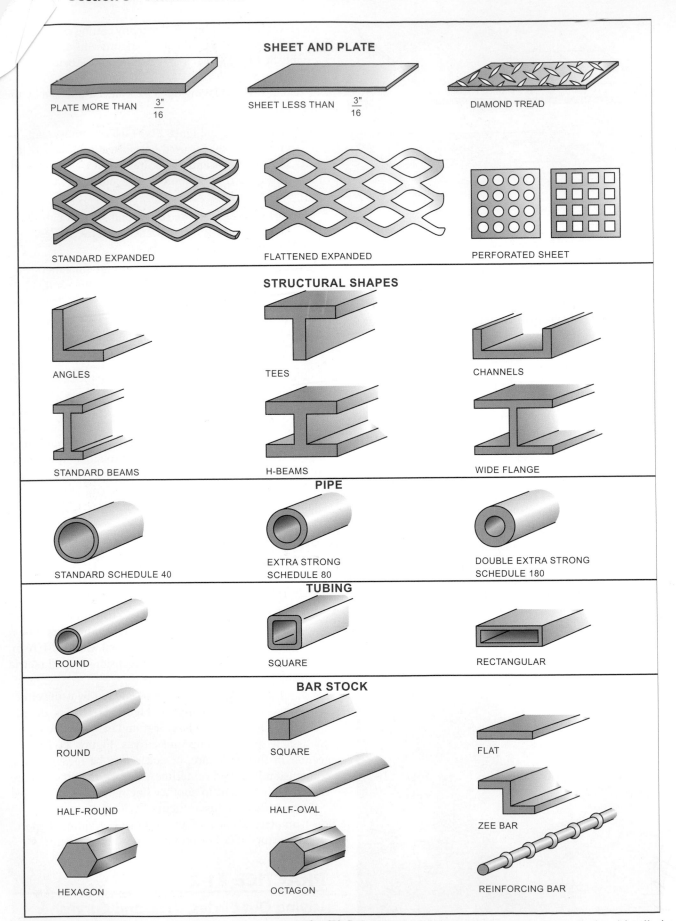

SHEET AND PLATE

PLATE MORE THAN $\frac{3"}{16}$

SHEET LESS THAN $\frac{3"}{16}$

DIAMOND TREAD

STANDARD EXPANDED

FLATTENED EXPANDED

PERFORATED SHEET

STRUCTURAL SHAPES

ANGLES

TEES

CHANNELS

STANDARD BEAMS

H-BEAMS

WIDE FLANGE

PIPE

STANDARD SCHEDULE 40

EXTRA STRONG
SCHEDULE 80

DOUBLE EXTRA STRONG
SCHEDULE 180

TUBING

ROUND

SQUARE

RECTANGULAR

BAR STOCK

ROUND

SQUARE

FLAT

HALF-ROUND

HALF-OVAL

ZEE BAR

HEXAGON

OCTAGON

REINFORCING BAR

FIGURE 21-14 Standard metal shapes; most are available with different surface finishes, such as hot-rolled, cold-rolled, or galvanized. © Cengage Learning 2012

FIGURE 21-15 Using a square to draw a straight line.

Larry Jeffus

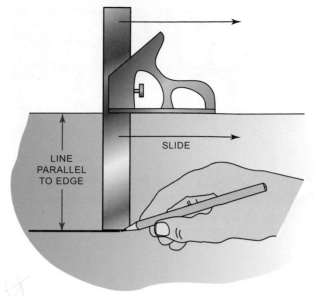

FIGURE 21-16 Using a combination square to lay out a strip of metal. © Cengage Learning 2012

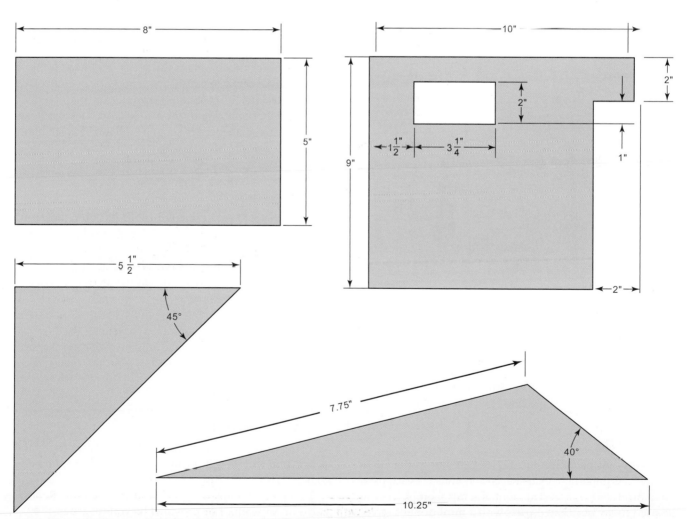

FIGURE 21-17 Lay out parts for Practice 21-1. © Cengage Learning 2012

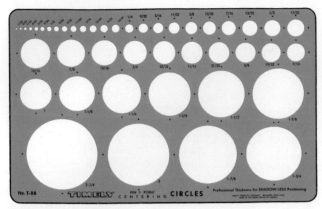

FIGURE 21-18 (A) Circle template. Larry Jeffus

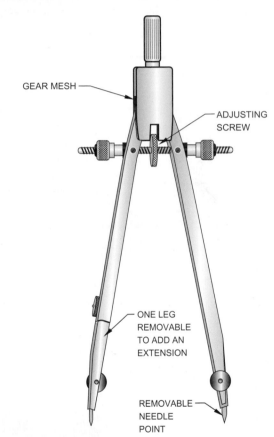

FIGURE 21-18 (B) Compass. © Cengage Learning 2012

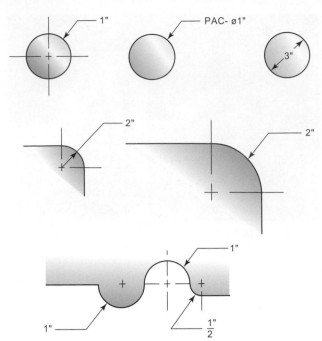

FIGURE 21-19 Dimensioning for arcs, curves, radii, and circles. © Cengage Learning 2012

Some computerized cutting machines can also be programmed to nest parts.

Manual nesting of parts may require several tries at laying out the parts to achieve the lowest possible scrap. On any cuts other than straight lines it is important to leave some material between parts. This material helps control the cut and prevents the cut from being interrupted or blowing out part of the metal, **Figure 21-21**. Even the finest laser cuts require that a small piece of scrap be laid out between parts.

PRACTICE 21-3

Nesting Layout

Using metal or paper that is 8 1/2 in. × 11 in., soapstone or pencil, tape measure, and square, you will lay out the parts shown in **Figure 21-22** in a manner that will result in the least scrap; assume a 0-in. kerf width. Use as many 8 1/2-in. × 11-in. pieces of stock as would be necessary to produce the parts using your layout. The parts must be laid out within ±1/16 in. of the dimensions. Convert the dimensions into SI metric units of measure.

Complete a copy of the "Student Welding Report" listed in Appendix I or provided by your instructor. ◆

PRACTICE 21-4

Bill of Materials

Using the parts laid out in Practice 21-3 and paper and pencil, you will fill out the bill of materials form shown in **Table 21-1**.

Complete a copy of the "Student Welding Report" listed in Appendix I or provided by your instructor. ◆

will lay out the parts shown in **Figure 21-20**. The parts must be laid out within ±1/16 in. of the dimensions. Convert the dimensions into SI metric units of measure.

Complete a copy of the "Student Welding Report" listed in Appendix I or provided by your instructor. ◆

Nesting

Laying out parts so that the least amount of scrap is produced is important. Odd-shaped and unusual-sized parts often produce the largest amount of scrap. Computers can be used to lay out nested parts with a minimum of scrap.

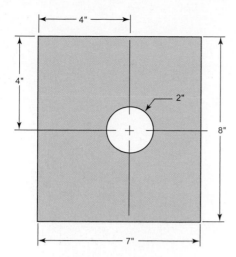

FIGURE 21-21 Parts nested for cutting; note the small blank space left between the parts. MultiCamPix

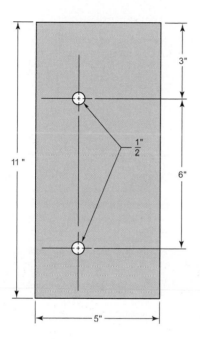

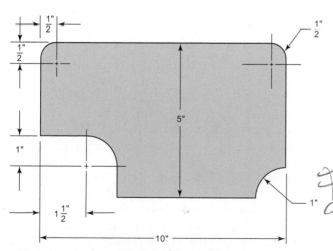

FIGURE 21-20 Lay out parts for Practice 21-2. © Cengage Learning 2012

Kerf Space

Because all cutting processes, except for some shearing, produce a kerf during the cut, this space must be included in the layout when parts are laid out side by side. Many angle iron cutting shears remove a thin strip of metal so that both sides of the cut are square, **Figure 21-23.** The **kerf** is the space created as material is removed during a cut. The width of a kerf varies depending on the cutting process used. Of the cutting processes used in most shops, the metal saw will produce one of the smallest kerfs and the handheld oxyfuel cutting torch can produce one of the widest.

When only one or two parts are being cut, the kerf width may not need to be added to the part dimension. This space may be taken up during assembly by the root gap required for a joint. If a large number of parts are being cut out of a single piece of stock, the kerf width can add up and increase the stock required for cutting out the parts, **Figure 21-24.**

PRACTICE 21-5

Allowing Space for the Kerf

Using a pencil, 8 1/2-in. × 11-in. paper, measuring tape or rule, and square, you will lay out four rectangles 2 1/2 in. × 5 1/4 in. down one side of the paper, leaving 3/32 in. for the kerf.

Two ways can be used to provide for the kerf spacing. One method is to draw a double line on the side of the part where the kerf is to be made, **Figure 21-25.** The other way is to lay out a single line and place an *X* on the side of the line on which the cut is to be made, **Figure 21-26.** Note that no kerf space need be left along the sides made next to the edge of the paper or next to the scrap. What is the total length and width of material needed to lay out these four parts?

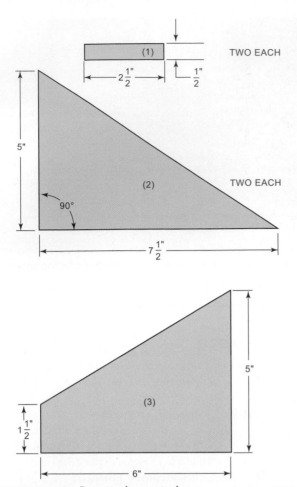

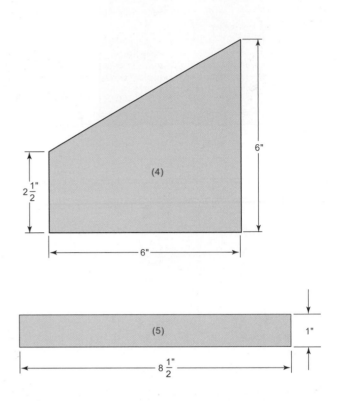

FIGURE 21-22 Parts to be nested. © Cengage Learning 2012

Part	Number Required	Type of Material	Size (Standard Units)	(SI Units)
Base	1	Hot-roll steel	1/2" × 5" × 8"	12.7 mm × 127 mm × 203.2 mm
Cleat	2	Hot-roll steel	1/2" × 4" × 8"	12.7 mm × 101.6 mm × 203.2 mm

TABLE 21-1 Bill of Materials Form

FIGURE 21-23 Angle iron shear. Scotchman

Parts can be laid out by tracing either an existing part or a template, **Figure 21-27**. When using either process, be sure the line you draw is made as tight as possible to the part's edge, **Figure 21-28**. The inside edge of the line is the exact size of the part. Make the cut on the line or to the outside so that the part will be the correct size once it is cleaned up, **Figure 21-29**. Sometimes a template is made of a part. Templates are useful if the part is complex and needs to fit into an existing weldment. They are also helpful when a large number of the same part are to be made or when the part is only occasionally used. The advantage of using templates is that once the detailed layout work is completed, exact replicas can be made anytime they are needed. Templates can be made out of heavy paper, cardboard, wood, sheet metal, or other appropriate material. The sturdier the material, the longer the template will last.

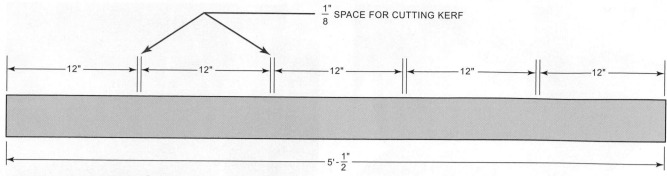

FIGURE 21-24 Because of the kerf, an additional 1/2 in. (13 mm) of stock would be required to make these five 1-ft (0.3048-mm) pieces. © Cengage Learning 2012

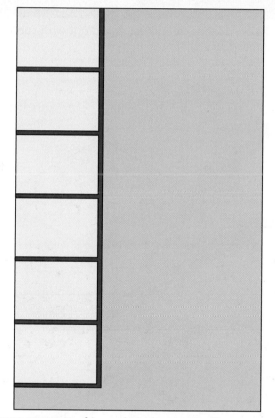

FIGURE 21-25 Kerf is made between the lines. © Cengage Learning 2012

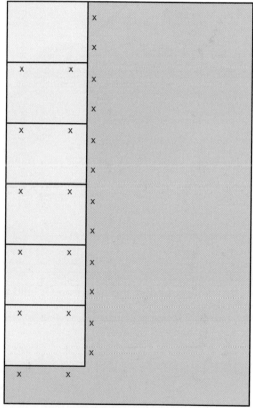

FIGURE 21-26 Xs mark the side of the line on which the kerf is to be made. © Cengage Learning 2012

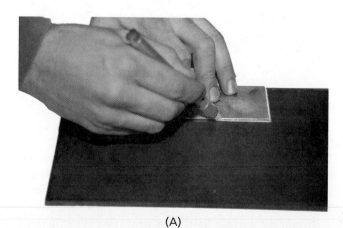

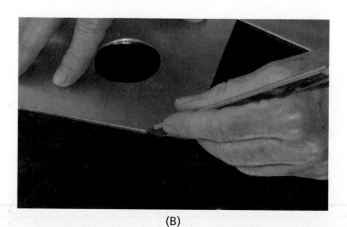

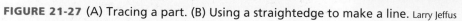

(A) (B)

FIGURE 21-27 (A) Tracing a part. (B) Using a straightedge to make a line. Larry Jeffus

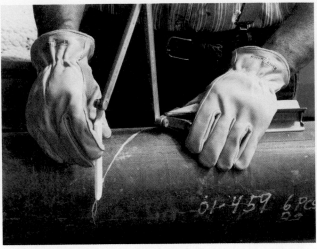

FIGURE 21-30 Pipe lateral being laid out with contour marker. Larry Jeffus

FIGURE 21-28 Be sure that the soapstone is held tightly into the part being traced. Larry Jeffus

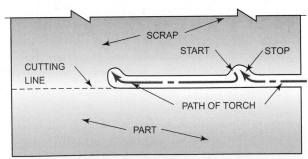

FIGURE 21-29 Turning out into scrap to make stopping and starting points smoother. © Cengage Learning 2012

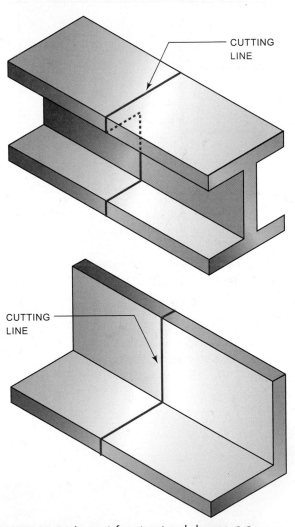

FIGURE 21-31 Layout for structural shapes. © Cengage Learning 2012

Special tools have been developed to aid in laying out parts; one such tool is the **contour marker**, Figure 21-30. These markers are highly accurate when properly used, but they do require a certain amount of practice. Once familiar with this tool, the user can lay out an almost infinite variety of joints within the limits of the tool being used. One advantage of tools like the contour marker is that all sides of a cut in structural shapes and pipe can

be laid out from one side without relocating the tool, **Figure 21-31**.

Complete a copy of the "Student Welding Report" listed in Appendix I or provided by your instructor. ◆

MATERIAL SHAPES

Metal stock can be purchased in a wide variety of shapes, sizes, and materials. Weldments may be constructed from combinations of sizes and/or shapes of metals. Only a single type of metal is usually used in most weldments unless a special property such as corrosion resistance is needed. In those cases, dissimilar metals may be joined into the fabrication at such locations as needed. The most common metal used is carbon steel, and the most common shapes used are plate, sheet, pipe, tubing, and angles. For that reason, most of the fabrication covered in this chapter will concentrate on carbon steel in those commonly used shapes. Transferring the fabrication skills learned in this chapter to the other metals and shapes should require only a little practice time.

Bill of Materials Form

Plate is usually 3/16 in. (4.8 mm) or thicker and measured in inches and fractions of inches. Plates are available in widths ranging from 12 in. (305 mm) up to 96 in. (2438 mm) and lengths from 8 ft (2.4 m) to 20 ft (6 m). Thickness ranges up to 12 in. (305 mm).

Sheets are usually 3/16 in. (4.7 mm) or less and measured in gauge or decimals of an inch. Several different gauge standards are used. The two most common are the Manufacturer's Standard Gauge for Sheet Steel, used for carbon steels, and the American Wire Gauge, used for most nonferrous metals such as aluminum and brass.

Pipe is dimensioned by its diameter and schedule or strength. Pipe that is smaller than 12 in. (305 mm) is dimensioned by its inside diameter, and the outside diameter is given for pipe that is 12 in. (305 mm) in diameter and larger, **Figure 21-32**. The strength of pipe is given as a schedule. Schedules 10 through 180 are available; schedule 40 is often considered a standard strength. The wall thickness for pipe is determined by its schedule (pressure range). The larger the diameter of the pipe, the greater its area. Pipe is available as welded (seamed) or extruded (seamless).

Tubing sizes are always given as the outside diameter. The desired shape of tubing, such as square, round, or rectangular, must also be listed with the ordering information.

The wall thickness of tubing is measured in inches (millimeters) or as Manufacturer's Standard Gauge for Sheet Metal. Tubing should also be specified as rigid or

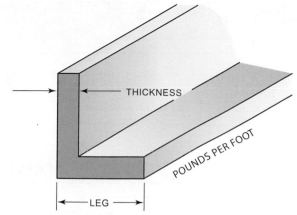

FIGURE 21-33 Specifications for sizing angles. © Cengage Learning 2012

flexible. The strength of tubing may also be specified as the ability of tubing to withstand compression, bending, or twisting loads.

Angles, sometimes referred to as angle iron, are dimensioned by giving the length of the legs of the angle and their thickness, **Figure 21-33**. Stock lengths of angles are 20 ft, 30 ft, 40 ft, and 60 ft (6 m, 9.1 m, 12.2 m, and 18.3 m).

OVERALL TOLERANCE

When fabricating a weldment that is made up of a number of parts all welded together, there is a potential problem that the overall size of the weldment can be wrong. Every part manufactured has a tolerance. **Tolerance** is the amount that a part can be bigger or smaller than it should be and still be acceptable. The more exact a part's tolerance, the more time it takes to make and, therefore, the more it costs. In most cases, welding engineers have calculated the effect of these slight variations in size when they designed a weldment. As the weldment fabricator, you must take these tolerances into consideration as you make the assembly to ensure that the overall size of the weldment is within its tolerance.

As the number of parts that make up a weldment increases, the problem of compounding the errors increases. For example, if there are 8 parts and each part is 1/8 in. larger than its ideal size but within its ±1/8 in. tolerance, the overall length of the finished weldment could be 8/8ths or 1 in. too long. Likewise, if each of the parts were 1/8 in. shorter, the overall length would be 1 in. too short, **Figure 21-34**. Therefore, an assembler must be mindful of both the size of each part and the overall size of the assembly. In addition to tolerances for size, parts also have angle tolerances. For example, each of the 10 pieces that make up the star in **Figure 21-35** are off by only 1°. But as you can see when they are assembled, the last corner does not fit, making the weldment unacceptable.

Ideally, all the parts for a weldment fit up perfectly; however, in reality that does not always happen. You cannot just throw out all of the parts that do not fit in order to find the ones that would make the perfect star. That is especially true if the parts are made within the correct tolerance.

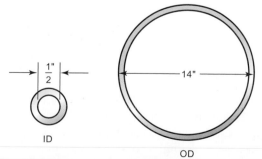

FIGURE 21-32 Inside diameter (ID); outside diameter (OD).
© Cengage Learning 2012

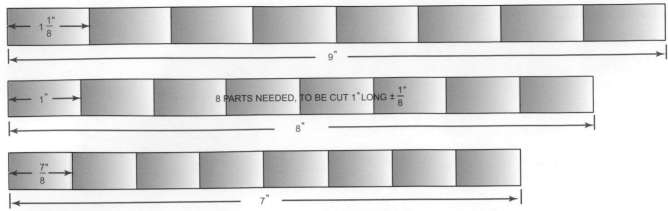

FIGURE 21-34 Lay out parts to avoid compounding dimensioning errors. © Cengage Learning 2012

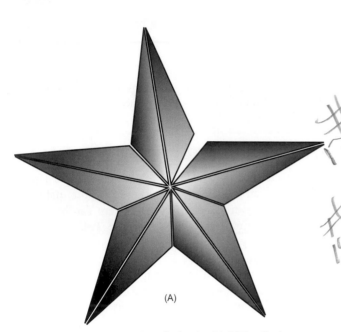

(A)

Rats! The puzzle pieces for the star didn't fit together!

(B)

That's much better.

FIGURE 21-35 Small errors on lots of parts can become a big error on the finished assembly. © Cengage Learning 2012

When parts like the ones in this star are made on the shorter side of the tolerance, you might be able to "loosen up" the joint tolerance to make the overall star work. Welded joints, like parts, have tolerances. By slightly adjusting the alignment of each of the 10 pieces, the star can be made within its overall acceptable tolerance. In this case, by making sure all the joints stay within tolerance, the complete star can be made without having to recut any of the parts. If you can make this assembly by adjusting the joint tolerance, you can assemble it faster; and since each of the parts is exactly the same, it will look perfect, Figure 21-35.

Whenever possible, try to get the parts to fit without having to recut or grind them; but remember, the finished weldment must be within tolerance. You want to avoid recutting and grinding because both will add time and cost to the finished project. However, you must remember that in some cases, the only way the weldment can be assembled within overall tolerance is to recut or grind some or all of the parts. If the finished weldment is not within tolerance, it may be unusable. If you must grind a part to fit, try to do as little grinding as possible to get the parts to fit up. Hand grinding is a time-consuming operation, and as Benjamin Franklin once said, "Remember that time is money."

Where you recut or grind a part can sometimes greatly affect the time required. For example, if the pieces used in the star need to be ground to fit, you might want to grind along the short side, which would be faster, **Figure 21-36A**. In addition it might be possible to grind only part of the edge to get the parts to fit up within tolerance. In the case of the parts shown in **Figure 21-36B** the required root opening tolerance of 1/4 in. ± 1/8 in. will allow the part's edge to be ground unevenly as long as the root opening tolerance is maintained, **Figure 21-36C**. Note how the root opening varies but stays within the acceptable tolerance. The root opening is 1/4 in. at one end, which is acceptable; but at the point where it becomes too close (less than 1/8 in.), begin grinding. The result as shown in **Figure 21-36D** meets the part's fit-up specifications and requires a minimum amount of grinding.

If the root opening is too wide, you may be able to make the weld, but it would be too large. Larger welds result in more filler metal being added and more heat input to the base metal. Larger welds may cause greater

root opening as specified in the welding procedure; to do so is wrong.

ASSEMBLY

The assembling process, bringing together all the parts of the weldment, requires a proficiency in several areas. You must be able to read the drawing and interpret the information provided there to properly locate each part. An assembly drawing has the necessary information, both graphically and dimensionally, to allow the various parts to be properly located as part of the weldment. If the assembly drawings include either pictorial or exploded views, this process is much easier for the beginning assembler; however, most assembly drawings are done as two, three, or more orthographic views, **Figure 21-37**. Orthographic

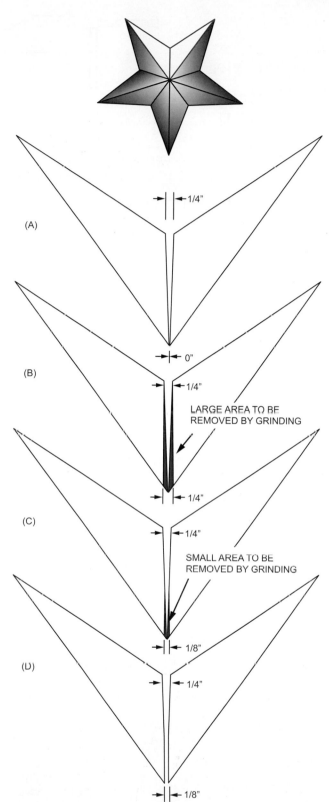

(A)
1/4"

(B)
0"

1/4"
LARGE AREA TO BE REMOVED BY GRINDING

(C)
1/4"

1/4"
SMALL AREA TO BE REMOVED BY GRINDING

(D)
1/8"

1/4"

1/8"

FIGURE 21-36 Trimming parts efficiently can save time.
© Cengage Learning 2012

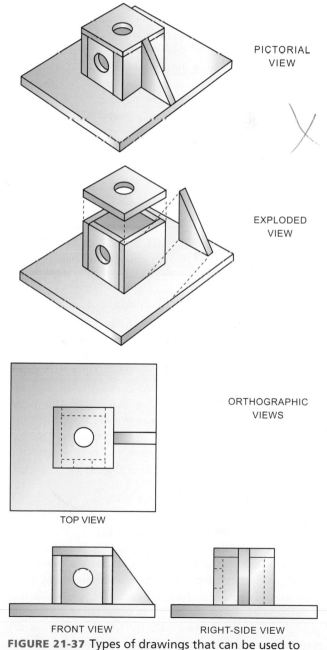

PICTORIAL VIEW

EXPLODED VIEW

ORTHOGRAPHIC VIEWS

TOP VIEW

FRONT VIEW   RIGHT-SIDE VIEW

FIGURE 21-37 Types of drawings that can be used to show a weldment assembly. © Cengage Learning 2012

weld distortion and have larger heat-affected zones. Both can result in a weld that will not withstand the part's designed strength specifications.

Even a quick visual inspection by a welding inspector would tell that the weld is unacceptable and must be repaired. You cannot deviate beyond the tolerance for the

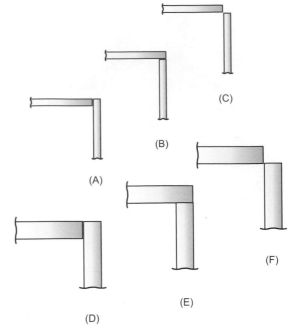

FIGURE 21-38 Laying out parts squarely is easier when they are placed along an edge or at the corner of the plate. © Cengage Learning 2012

views will be more difficult to interpret until you have developed an understanding of their various elements.

On very large projects such as buildings or ships, a corner or centerline is established as a baseline. This is the point where all measurements for all part location begins. When working with smaller weldments, a single part may be selected as such a starting point. Often, selecting the base part is automatic because all other parts are to be joined to this central part. On other weldments, however, the selection of a base part is strictly up to the assembler.

To start the assembly, select the largest or most central part to be the base for your assembly. All other parts will then be aligned to this one part. When possible, use the edge or corner as the base or starting point for locating parts because it makes alignment easier, **Figure 21-38**. A base also helps to prevent location and dimension errors. Otherwise, a slight misalignment of one part, even within tolerances, will be compounded by the misalignment of other parts, resulting in an unacceptable weldment. Using a baseline or base part will result in a more accurate assembly.

You must look at the drawing to see how the edges are fitted. This is much more important on thicker materials than it is with thin stock, **Figure 21-39A**. Of course, even on thin material it can be important if the part is to be built to a very tight tolerance. In that case, if the joint should be assembled as shown in Figure 21-39A but is assembled as shown in **Figure 21-39B**, the overall length in one direction decreases by the thickness of the material; and in the other direction the dimension increases by the thickness of the material.

On thicker materials it is easy to see how the overall dimensions of a weldment could change if the parts are not properly aligned during fit up. But in addition to not being the correct size, sometimes the weld itself will not be as strong if the joint is not aligned properly. The reason the weld might not be as strong as it is designed to be is because the tensile strength of thick metal plate differs depending on the direction the load is placed on the plate as compared to the rolling direction of the plate, **Figure 21-40A**. Metal plate, much like wood, will break in one direction easier than in another. In this aspect steel plate is similar to a wooden board in that the direction of the rolled grain of a plate and the direction of the wood grain in a board affect their strength. Many common materials have a grain; for example, when you tear a newspaper down the page, it tears fairly easily and straight. However, when you try to tear it across the page, the tear is much more jagged, **Figure 21-40B**.

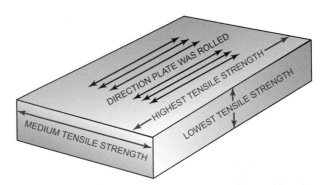

FIGURE 21-39 Part placement during assembly can affect the overall dimension, especially when the metal being assembled is thick. © Cengage Learning 2012

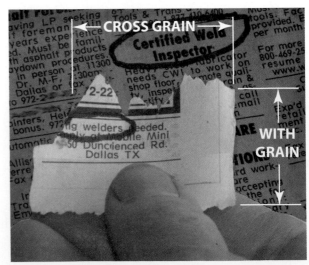

FIGURE 21-40 (A) A metal's strength is affected by the direction it is rolled because of its grain structure. © Cengage Learning 2012

FIGURE 21-40 (B) The effect of grain structure can be seen on a newspaper as it is torn. *Larry Jeffus*

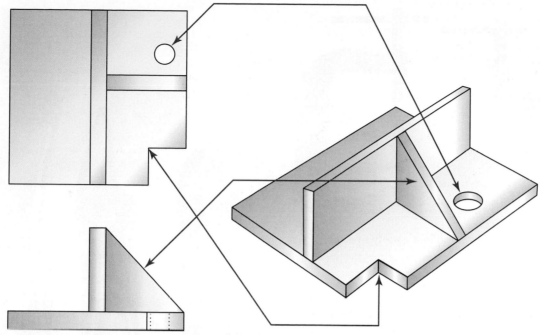

FIGURE 21-41 Identify unique points to aid in assembly. © Cengage Learning 2012

Identify each part of the assembly and mark each piece for future reference. If needed, you can hold the parts together and compare their orientation to the drawing. Locate points on the parts that can be easily identified on the drawing such as holes and notches, **Figure 21-41.** Now mark the location of these parts—top, front, or other such orientation—so you can locate them during the assembly.

Layout lines and other markings can be made on the base to locate other parts. Using a consistent method of marking will help prevent mistakes. One method is to draw parallel lines on both parts where they meet, **Figure 21-42.**

After the parts have been identified and marked, they can be either held or clamped into place. Holding the parts in alignment by hand for tack welding is fast but often leads to errors and thus is not recommended for beginning assemblers. Experienced assemblers recognize that clamping the parts in place before tack welding is a much more accurate method, **Figure 21-43.**

Assembly Tools

A variety of tools are used to make assembly easier. Both general tools and job-specific assembly tools are commonly used.

Clamps A variety of clamps can be used to temporarily hold parts in place so that they can be tack welded.

C-clamps, one of the most commonly used clamps, come in a variety of sizes, **Figure 21-44.** Some C-clamps have been specially designed for welding. Some of these clamps have a spatter cover over the screw, and others have their screws made of spatter-resistant materials such as copper alloys.

#18

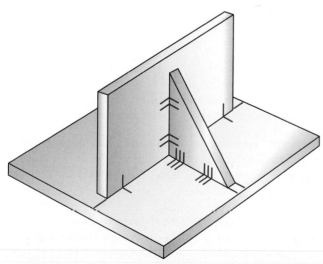

FIGURE 21-42 Lay out markings to help locate the parts for tack welding. © Cengage Learning 2012

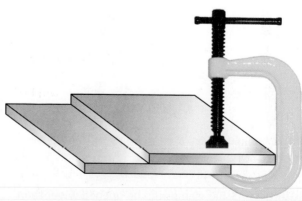

FIGURE 21-43 C-clamp being used to hold plates for tack welding. Larry Jeffus

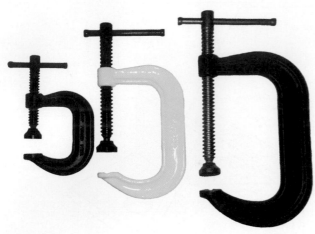

FIGURE 21-44 C-clamps come in a variety of sizes.
Larry Jeffus

- *Bar clamps* are useful for clamping larger parts. Bar clamps have a sliding lower jaw that can be positioned against the part before tightening the screw-clamping end, **Figure 21-45**. They are available in a variety of lengths.

- *Pipe clamps* are very similar to bar clamps. The advantage of pipe clamps is that the ends can be attached to a section of standard 1/2-in. (13 mm) pipe. This feature allows for greater flexibility in length, and the pipe can easily be changed if it becomes damaged.

- *Locking pliers* are available in a range of sizes with a number of various jaw designs, **Figure 21-46**. The versatility and gripping strength make locking pliers very useful. Some locking pliers have a self-adjusting feature that allows them to be moved between different thicknesses without the need to readjust them.

- *Cam-lock clamps* are specialty clamps that are often used in conjunction with a jig or a fixture. They can be preset, allowing for faster work, **Figure 21-47**.

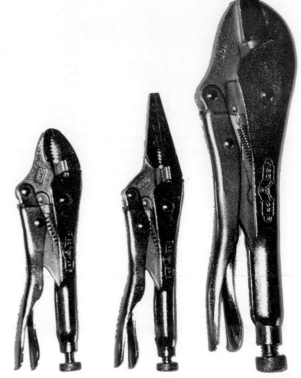

FIGURE 21-46 Three common jaw types on locking pliers.
Larry Jeffus

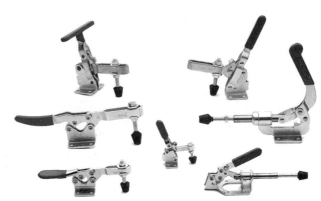

FIGURE 21-47 Toggle-type clamps. Strong Hand Tools & Good Hand, Inc.

- *Specialty clamps* such as those used for pipe welding, **Figure 21-48**, are available for many different types of jobs. Such specialty clamps make it possible to do faster and more accurate assembling.

Fixtures **Fixtures** are devices that are made to aid in assemblies and fabrication of weldments. When a number of similar parts are to be made, fixtures are helpful. They can increase speed and accuracy in the assembly of parts. Fixtures must be strong enough to support the weight of the parts, be able to withstand the rigors of repeated assemblies, and remain in tolerance. They may have clamping devices permanently attached to speed up their use, **Figure 21-49**. Often, locating pins or other devices are used to ensure proper part location. A well-designed fixture allows adequate room

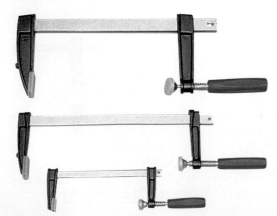

FIGURE 21-45 Bar clamps can be opened wider than most C-clamps to be used on larger weldments.
Strong Hand Tools & Good Hand, Inc.

(A)

(B)

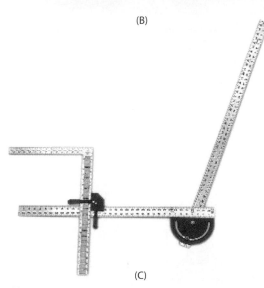

(C)

FIGURE 21-48 Specialty pipe clamps and alignment tools. Larry Jeffus

for the welder to make the necessary tack welds. Some parts are left in the fixture through the entire welding process to reduce distortion. Making fixtures for every job is cost prohibitive and not necessary for a skilled assembler.

FIGURE 21-49 Wooden jig holds the base plate in place while magnetic squares keep it aligned for welding. Larry Jeffus

FITTING

Fitting is the process of adjusting the parts of a weldment so that they meet the overall tolerance because not all parts fit exactly as they were designed. There may be slight imperfections in cutting or distortion of parts due to welding, heating, or mechanical damage. Some problems can be solved by grinding away the problem area. Hand grinders are most effective for this type of problem, **Figure 21-50.** Other situations may require that the parts be forced into alignment.

///// **CAUTION** >>>

Never operate a hand grinder without the safety guard and eye protection.

A simple way of correcting slight alignment problems is to make a small tack weld in the joint and then use a hammer and possibly an anvil to pound the part into place, **Figure 21-51.** Small tacks applied in this manner will become part of the finished weld. Be sure not to strike the part in a location that will damage the surface and render the finished part unsightly or unusable.

More aligning force can be applied using cleats or dogs with wedges or jacks. Cleats or dogs are pieces of metal

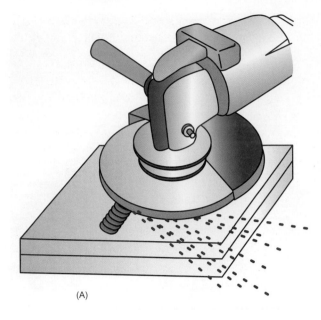

(A)

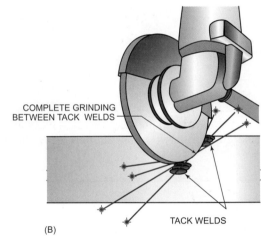

COMPLETE GRINDING
BETWEEN TACK WELDS

TACK WELDS

(B)

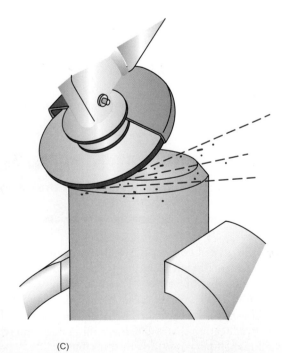

(C)

FIGURE 21-50 Abrasive grinding disk can be used to (A) remove excessive weld metal, (B) cut a groove, or (C) prepare a bevel for welding. © Cengage Learning 2012

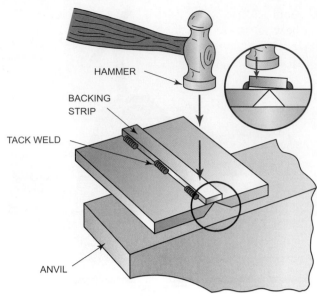

HAMMER

BACKING
STRIP

TACK WELD

ANVIL

FIGURE 21-51 A hammer can be used to bend metal into alignment after a tack weld has been made.
© Cengage Learning 2012

that are temporarily attached to the weldment's parts to enable them to be forced into place. Jacks will do a better job if the parts must be moved more than about 1/2 in. (13 mm), **Figure 21-52.** Anytime cleats or dogs are used, they must be removed and the area ground smooth.

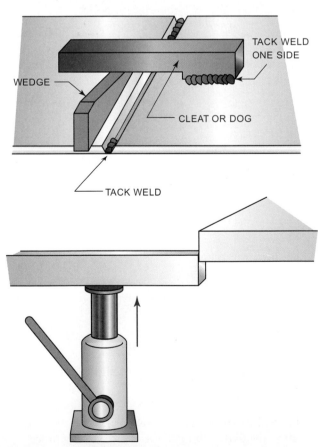

TACK WELD
ONE SIDE

WEDGE

CLEAT OR DOG

TACK WELD

FIGURE 21-52 (A) Cleat and wedge used to align parts. (B) Hydraulic jack used to realign a part. © Cengage Learning 2012

Some codes and standards will not allow cleats or dogs to be welded to the base metal. In these cases, more expensive and time-consuming fixtures must be constructed to help align the parts if needed.

TACK WELDS

Tack welds are the welds, usually small in size, that are made during the assembly to hold all of the parts of a weldment together so they can be finished welded. Making good tack welds is one of the keys to assembly work. Tack welds must also be small enough to be incorporated into the final weld without causing a discontinuity in its size or shape, **Figure 21-53.** They must be strong enough to hold the parts in place for welding but small enough so they become an unseen part of the finished weld. Deciding on the number, size, and location of tack

welds takes some planning. Following are some of the factors to consider regarding the number of tack welds:

- Thickness of the metal—A large number of very small tack welds should be used on thin metal sections, while a few large tack welds may be used for thicker metal parts.

- Length and shape of the joint—Obviously, short joints take fewer welds; but some long, straight joints may have very few tack welds compared to a shorter joint that is very curvy.

- Welding stresses—All welds create stress in the surrounding metal as they cool and shrink. Larger welds produce greater stresses that might pull tack welds loose from an adjoining part if the tack welds are not strong enough to withstand the welding stresses, **Figure 21-54.**

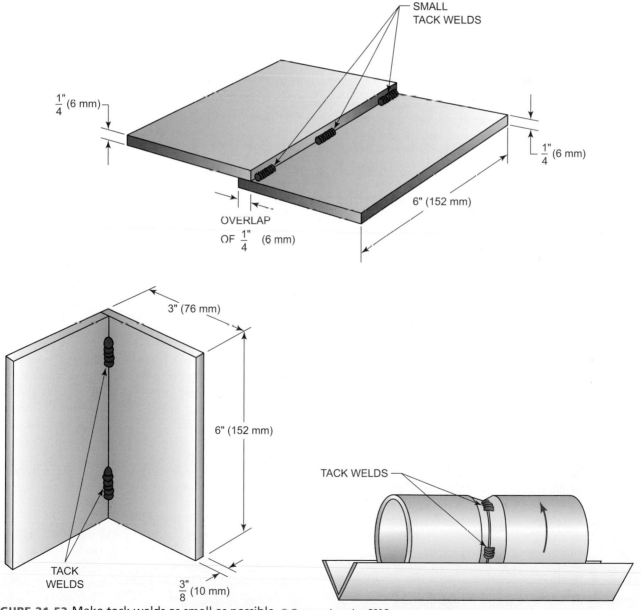

FIGURE 21-53 Make tack welds as small as possible. © Cengage Learning 2012

FIGURE 21-54 Tack weld cracked due to welding stress.
Larry Jeffus

- Tolerance—The more exacting the tolerance for the finished weldment, the more tack welds are required.
- Fit up—When custom bending parts during the fit-up process, it may be necessary to use a large number of small tack welds to keep the parts in alignment and make the bends more uniform.

NOTE: Often it is necessary to change between wearing a welding helmet, cutting goggles, and safety shield as a part is being assembled. Whenever you take off your helmet, remember to place it faceup and not facedown. Placing your helmet faceup will prevent weld spatter, grinding sparks, and other shop dirt from collecting inside. If your helmet gets filled with debris, it will fall all over your head and face the next time you put it on. Keep your cutting goggles and safety shield faceup too.

Tack welds must be made in accordance with any welding procedure with an appropriate filler metal. They must be located well within the joint so that they can be completely remelted into the finished weld, **Figure 21-55**. Post tack welding cleanup is required to remove any slag or impurities that may cause finished weld flaws. Sometimes the ends of a tack weld must be ground down to a taper to improve its tie-in to the finished weld metal.

Make sure that your tack welds are not going to be weld defects in the completed weldment. A good tack weld is one that does its job by holding parts in place yet is undetectable in the finished weld.

Do not assume that a broken tack weld has no effect and continue to weld. A broken tack can allow parts to shift well out of tolerance. On harder metals like steel you can often hear a tack weld break; however, for some soft metals like aluminum the tack weld may separate quietly. Depending on the type of metal and its size and thickness, a breaking tack weld can make a small sharp snap or deep resounding thump. You might hear one break while you

FIGURE 21-55 Tack weld made in line with the bolt hole so it will be out of the way when the flange is fitted. Larry Jeffus

are welding or sometime afterward. If you continue to weld, it may be impossible to pull the loose part back into position, which could result in the weldment not meeting its specifications. Sometimes this is referred to as "making scrap metal" and not a weldment.

WELDING

Good welding requires more than just filling up the joints with metal. The order and direction in which welds are made can significantly affect distortion in the weldment. Generally, welding on an assembly should be staggered from one part to another. This will allow both the welding heat and welding stresses to dissipate.

Keep the arc strikes in the welding joint so that they will be remelted as the weld is made. This will make the finished weldment look neater and reduce postweld cleanup. Some codes and standards do not allow arcs to be struck outside of the welding joint.

Striking the arc in the correct location on an assembly is more difficult than working on a welding table. When working on an assembly, you will often be in an awkward position, which makes it harder to strike the arc correctly. Several techniques will help you improve your arc-starting accuracy. You can use your free hand to guide the electrode or weld gun into the correct spot. Resting your arm, shoulder, or hip against the weldment can also help, **Figure 21-56**. Practicing starting the weld with the power off is sometimes helpful.

Be sure that you have enough freedom of movement to complete the weld joint. Check to see that your welding leads will not snag on anything that would prevent you from making a smooth weld. If you are welding out of position, be sure that welding sparks will not fall on you or other workers, **Figure 21-57**. If the weldment is too large to fit into a welding booth, portable welding screens should be used to protect other workers in the area from sparks and welding light.

FIGURE 21-56 Bracing yourself, using both hands, and wearing auto-darkening helmets make tack welding easier and more accurate. Larry Jeffus

FIGURE 21-57 When making out-of-position welds, make sure to position yourself so that sparks and slag will not fall directly on you. Larry Jeffus

Follow all safety and setup procedures for the welding process. Practice the weld to be sure that the machine is set up properly before starting on the weldment.

FINISHING

Depending on the size of the shop, the welder may be responsible for some or all of the finish work. Such work may vary from chipping, cleaning, or grinding the welds to applying paint or other protective surfaces.

Grinding of welds should be avoided if possible by properly sizing the weld as it is made. Grinding can be an expensive process, adding significant cost to the finished weldment. Sometimes it is necessary to grind for fitting purposes or for appearance, but even in these cases it should be minimized if possible.

**/// // CAUTION \\\ **

When using a portable grinder, be sure that it is properly grounded and that the sparks will not fall on others, cause damage, or start a fire. Always maintain control to prevent the stone from catching and gouging the part or yourself.

**/// // CAUTION \\\ **

Be sure that any stone or sandpaper used is rated for a speed in revolutions per minute (RPM) that is equal to or greater than the speed of the grinder itself. Using a stone with a lower-rated RPM can result in its flying apart with an explosive force.

Most grinding is done with a hand angle grinder, **Figure 21-58.** These grinders can be used with a flat or cupped grinding stone or sandpaper. As the grinder is used, the stone will wear down and must be discarded once it has worn down to the paper backing. It is a good practice to hold the grinder at an angle so that if anything is thrown off the stone or metal surface, it will not strike you or others in the area. Because of the speed of the grinding stone, any such impact can cause serious injury.

The grinder must be held securely so that there is a constant pressure on the work. If the pressure is too great, the grinder motor will overheat and may burn out. If the

FIGURE 21-58 Wire brushes and grinding stones used to clean up welds. Larry Jeffus

pressure is too light, the grinder may bounce, which could crack the grinding stone. Move the grinder in a smooth pattern along the weld. Watch the weld surface as it begins to take the desired shape and change your pattern as needed.

Painting and other finishes release fumes such as volatile organic compounds (VOCs), which are often regulated by local, state, and national governments. Special ventilation is required for most paints. Such a ventilation system will remove harmful fumes from the air before it is released back into the environment. Check with your local, state, or national regulating authority before using such products. Read and follow all manufacturer's instructions for the safe use of its product.

//// // CAUTION \\\\

Most paints are flammable and must be stored well away from any welding.

Summary

One of the greatest experiences as a welder/fabricator is completing work on a piece of equipment, building, trailer, or other structure. You can proudly point to it and say, "I helped to make that." Learning layout and fabrication techniques will let you someday be able to experience the sense of pride when you are then able to say, "I built that all by myself." Welded structures are an enduring monument to your skill as a craftsman, so it is important that every time you build a project, you do it as if it were going to be on display, because it is.

Welder Certification: Many Thrusts, Few Agree

U.S. welders must submit to constant testing and retesting to satisfy the numerous certification requirements in the industry. Are they victims of too many standards?

Times have changed. The old days when welders spent their entire careers with one employer are gone. Projects are now short-term. Welders are being forced to move from one company to another many times during their careers, but movement can be complicated. The various standards created by such organizations as ASME, AWS, and API—each with its own separate body of codes and standards—have created a situation in which companies must adhere to one or more sets of standards, depending upon the industry.

Welders are caught in the middle. Their performance is held accountable to a particular standard required for a particular job, so, if a welder certified by AWS standards for one job decides to move to another company or another industry, he or she will often be faced with the prospect of being retested according to ASME, or some other standard, in order to be properly certified to weld on the new job.

The reality of undergoing constant tests and retests has led many welders to the conclusion that their destinies are controlled by others, that they have no voice and no freedom of choice when it comes to employment. This disenchantment comes at a time when there is a growing shortage of welders in many parts of the country.

One Company, Multiple Standards

Trinity Industries, Inc., Dallas, Texas, is a case in point. Trinity builds bridges, barges, freight cars, propane tanks, and tank railcars. As a result of this varied product mix, the engineers are working with ASME, AWS, ABS, AAR, API, and other standards. The industry often has to subject its employees to redundant testing.

Trinity employs between 2000 and 4000 welders. The cost to the company of administering these tests can be extremely high, especially if such factors as material, lost productivity, administrative personnel to conduct and witness the test, and the cost of evaluating the welder's test sample are included.

Some Industries Test In-House

Welders who work for the Norfolk Southern Corporation, Roanoke, Virginia, are qualified to ASW D15.1, according to Ken R. Rollins, system welding coordinator for the major railroad. Welders are tested for competency in shielded metal arc welding, flux cored arc welding, and gas metal arc welding processes. They are expected to pass tests that show they can weld materials of unlimited thickness in all weld positions. For example, plate tests are conducted in the 3G and 4G positions, and pipe welds are made with the SMAW process.

Weld testing is administered by one of the company's certified welding instructors. Most instructors hold certifications as AWS Certified Welding Inspectors and Educators and have experience doing welding. Visual and bend-tests are used to qualify each welder's tests. Once a welder fulfills qualification requirements, he or she will be able to work within the company's various mechanical departments. The qualification is good for 180 days, and welders are expected to update that qualification prior to its expiration.

New welders joining Norfolk Southern's workforce must go through a four-week long training course in McDonough, Georgia. Each candidate welder is trained in shielded metal arc and flux cored arc welding of steel plate and in shielded metal arc welding of steel pipe.

Others Tap National Test Facilities

Welders who belong to the International Association of Bridge, Structural, Ornamental, and Reinforcing Iron Workers, AFL-CIO (the Ironworkers) union in Washington, DC, are qualified at 35 AWS-accredited test facilities throughout the country, said Ray Robertson, general vice president and executive director of apprenticeship and training. The program is known as the Ironworkers Welder Certification Program of North America. So far, more than 5000 welders have been qualified.

The International Training Institute (ITI) for the sheet metal and air-conditioning industry and AWS joined forces to sponsor the ITI/AWS Certified Welder Program. This program is for members of the Sheet Metal Workers International Association and Air Conditioning National Contractors Association. ITI uses more than 70 AWS-certified test facilities throughout the country, where welder qualification tests are performed. Welder certification training prior to testing is available at any of 162 training centers run by the local Joint Apprenticeship and Training Committees. There are now

more than 1000 qualified welders in the ITI/AWS database, according to S. L. Raymond, welding assessor for ITI.

Sheet metal welders move into a wide variety of industries, typically construction, pharmaceutical, chip-making and power plants. Processes covered include but are not limited to shielded metal arc, gas metal arc, and gas tungsten arc welding; soldering; and orbital welding. Depending on the situation, welds are either bend-tested, X-rayed, or visually inspected.

Stanching the Need to Retest

Efforts are being made to make it easier for welders to move from job to job. For example, more than 7000 welders have been qualified by Common Arc, a nonprofit corporation established by the National Association of Construction Boilermaker Employers with help from the International Brotherhood of Boilermakers, headquartered in Geneva, Illinois. Initiated in 1988, the Common Arc welder certification program provides precertification that will be recognized by multiple contractors for specific processes. This eliminates the need for retesting at each new job.

Common Arc is used mainly to qualify welders to ASME Section XI. In this setup, representatives from as many as 20 contractors will travel to a test facility to witness the testing of welders. Each weld test in this arrangement is given a bend-test. The advantage of this approach, according to supporters of Common Arc, is that a welder can become qualified with multiple Common Arc contractors simply by taking just one test. The contractor has an immediate source of skilled welders who meet ASME requirements, in this way eliminating the cost of job-site testing and the time constraints presented when a contractor needs qualified welders during emergency situations such as power outages.

In the United Association (UA) Pipefitters' Welder Certification Program (WCP), more than 300 training schools, owned and operated by the UA, prepare and test welders

A pipefitter prepares a pipe for welding. American Welding Society

An instructor at the United Association Pipefitters program observes an apprentice performing a shielded metal arc welding procedure. American Welding Society

Welders qualified by the UA Pipefitters become expert in various processes, including gas tungsten arc welding. American Welding Society

who are candidates for WCP qualification. Large facilities for training instructors are located in Ypsilanti, Michigan; Colton, California; and Charleston, South Carolina. Additional facilities are under construction at Peekskill, New York, and Jackson, Mississippi.

Much of the work in this program is directed at welders seeking qualification on jobs covered by the ASME Code for Pressure Piping, B31. The United Association's goal is to train welders with qualifications that will be interchangeable among manufacturers and contractors.

Those who successfully complete the program enter the job market as pretested, certified, and immediately available journeymen welders for the piping industry in the United States and Canada, at no cost to the employer. The United Association underwrites the cost of testing and qualifying its welders by independent testers, auditors, and other third parties involved in this sphere of activity.

Testing Is Part of the Job

Some organizations accept the reality that welder qualification standards are varied and welders will have to hold a variety of certifications to remain employable. Lyle E. Binns, an instructor at the Lincoln Electric Welding School, Cleveland, Ohio, also advises welding students to accept the fact they will be tested often. "We tell our students that certification is like a diploma, that it's a means of getting a job interview, and you have to prove to a prospective employer that you know how to weld. The employer will test you again. We tell them to expect to be tested throughout their careers. When they change from one process to another, they will probably be tested. We tell our students to become familiar with as many processes and with as many positions as practicable."

Article courtesy of the American Welding Society.

Review

1. List precautions you must take when welding around workers who are not welders.

2. What is a cost advantage of using preformed parts in a fabrication?

3. Why is a prototype custom fabricated before producing multiple weldments?

4. How would a small or tight tolerance affect the layout?

5. How can circles, arcs, and curves be laid out?

6. Which cutting process does not produce a kerf space?

7. What is kerf space?

8. Which of the cutting processes can leave the smallest kerf?

9. What are two common ways to provide for the kerf spacing?

10. What can a template be made of?

Review (continued)

11. What is the difference between plate and sheet material?

12. What is a part's tolerance?

13. What can you do to make parts fit without having to recut or regrind them?

14. Why should you avoid recutting or grinding parts whenever possible?

15. What problem can result if the root opening of a weld is too wide?

16. What is the benefit of identifying a single part as the base during assembly?

17. List three types of clamps used to temporarily hold parts in place so that they can be tack welded.

18. What makes some C clamps better for welding than others?

19. What is a simple way of correcting slight alignment problems?

20. What precautions should be taken if a hammer is used to shift a part into alignment?

21. What is the purpose of tack welds?

22. How many tack welds should be used on thin metal compared to thicker metal parts?

23. Does it matter if one of the tack welds breaks during welding and why?

24. Why is it important not to strike the arc outside of the weld joint?

25. What can happen if a grinding stone with a lower-rated RPM than the grinder is used?

Chapter 22

Welding Codes and Standards

OBJECTIVES

After completing this chapter, the student should be able to

- explain the difference between qualification and certification.
- list the major considerations for selecting a code or standard.
- write a welding procedure and specification.
- identify the three most common codes and describe their major uses.
- outline the steps required to certify a weld and welder.
- explain how a tentative WPS becomes a certified WPS.

KEY TERMS

API Standard 1104

ASME Section IX

AWS D1.1

code

Procedure Qualification
Record (PQR)

specification

standard

Welding Procedure
Specification (WPS)

Welding Schedule

INTRODUCTION

It is important to know that any weld produced is going to be the best one for the job. A method is also needed to ensure that each weld made in the same plant or on the same type of equipment in another plant will be of the same quality.

To meet these requirements various agencies have established codes and standards. These detailed written outlines explaining exactly how a weld is to be laid out, performed, and tested have made consistent quality welds possible. By having the required information, skilled welders in shops all around the city, state, country, or world can make the same weld to the same level of safety, strength, and reliability.

A testing procedure to certify the welder ensures that the welder has the skills to make the weld. Passing a weld test is much easier when all of the detailed information is provided.

Selecting the code or standard to be used to judge a weld is equally as important as having a skilled welder. Not every product welded needs to be manufactured to the same level. The decision on the appropriate code or standard can be one of the most important aspects of welding fabrication. If the wrong one is selected, the cost of fabrication can be too high, or the parts might not stand up to the service.

CODES, STANDARDS, PROCEDURES, AND SPECIFICATIONS

A number of organizations publish codes or specifications that cover a wide variety of welding conditions and applications. The selection of the specific code to be used is made by the engineers, designers, or governmental requirements. Codes and specifications are intended to be guidelines only and must be qualified for specific applications by testing.

A welding **code** or **standard** is a detailed listing of the rules or principles that are to be applied to a specific classification or type of product.

A welding **specification** is a detailed statement of the legal requirements for a specific classification or type of weld to be made on a specific product. Products manufactured to code or specification requirements commonly must be inspected and tested to ensure compliance.

A number of agencies and organizations publish welding codes and specifications. The selection of the particular code or specification to a weldment can be the result of one or more of the following requirements:

■ Local, state, or federal government regulations—Many governing agencies require that a specific code or standard be followed.

■ Bonding or insuring company—The weld must be shown to be fit for service requirements as established through testing. A bonding or insuring company must feel that the product is the safest that can be produced.

■ End user (customer) requirements—The manufacturer considers cost and reliability; that is, as stricter standards are applied to the welding, the cost of the weldments increases. The more lax the standard, the lower the cost, but the reliability and possibly the safety also decrease.

■ Standard industrial practices—The code or standard used is considered to be the standard one for the industry and has been in use for some time.

Following are the three most commonly used codes:

■ **API Standard 1104,** American Petroleum Institute—Used for pipelines
■ **ASME Section IX,** American Society of Mechanical Engineers—Used for pressure vessels and nuclear components
■ **AWS D1.1,** American Welding Society—Used for bridges, buildings, and other structural steel

The following organizations publish welding codes and/or specifications. Most can be contacted for additional information and current price list either directly or through the World Wide Web.

AAR

Association of American Railroads
425 Third Street, SW, Suite 1000
Washington, DC 20024-3228

AASHTO

American Association of State Highway
and Transportation Officials
444 North Capitol Street, NW, Suite 249
Washington, DC 20001-1539

AIA

Aerospace Industries Association
1000 Wilson Boulevard, Suite 1700
Arlington, VA 22209-3928

AISC

American Institute of Steel Construction
1 East Wacker Drive, Suite 700
Chicago, IL 60601-1802

ANSI

American National Standards Institute
1819 L Street, NW
Washington, DC 20036-3807

API

American Petroleum Institute
1220 L Street, NW
Washington, DC 20005-4070

AREMA

American Railway Engineering and Maintenance-
of-Way Association
10003 Derekwood Lane, Suite 210
Lanham, MD 20706-4875

ASME

American Society of Mechanical Engineers
3 Park Avenue
New York, NY 10016-5990

AWS

American Welding Society
550 NW LeJeune Road
Miami, FL 33126-5649

AWWA

American Water Works Association
6666 West Quincy Avenue
Denver, CO 80235-3098

MIL

Department of Defense Military Specification
Washington, DC 20301-0001

SAE

Society of Automotive Engineers

400 Commonwealth Drive

Warrendale, PA 15086-1600

WELDING PROCEDURE QUALIFICATION

Welding Procedure Specification (WPS)

A welding procedure specification is a set of written instructions by which a sound weld is made. Normally, the procedure is written in compliance with a specific code, specification, or definition.

Welding Procedure Specification (WPS) is the standard terminology used by the American Welding Society (AWS) and the American Society of Mechanical Engineers (ASME). **Welding Schedule** is the standard federal government, military, or aerospace terminology denoting a WPS. The shortened term *welding procedures* is the most common term used by the industry to denote a WPS.

The WPS lists all of the parameters required to produce a sound weld to the specific code, specifications, or definition. Specific parameters such as welding process, technique, electrode or filler, current, amperage, voltage, preheat, and postheat should also be included. The procedure should list a range or set of limitations on each, such as amps = 110–150, voltage = 17–22, and so on, with the more essential or critical parameters more closely defined or limited.

The WPS should give enough detail and specific information so that any qualified welder could follow it and produce the desired weld. The WPS should always be prepared as a tentative document until it is tested and qualified.

Qualifying the Welding Procedure Specification

The WPS must be qualified to prove or verify that the list of variables—amperage, voltage, filler, and so on—will provide a sound weld. Sample welds are prepared using the procedure and specifications listed in the tentative WPS. A record of all the parameters used to produce the test welds must be kept. Be sure to record the specifics for the parameters such as voltage, amperage, and so on. This information should be recorded on a form called the **Procedure Qualification Record (PQR).**

In most cases, the inspection agency, inspector, client, or customer will request a copy of both the WPS and the PQR before allowing production welding to begin.

Qualifying and Certifying

The process of qualifying and then certifying both the WPS and welders requires a number of specific requirements. The requirements may vary from one code or standard to

another, but the general process is the same for most. Before you invest in the testing required to qualify and certify processes and welders under a code, you must first obtain a copy of the code you are planning to use. The requirements of codes and standards change from time to time, and it is important that your copy is the most recent version.

The following is a generic schedule of required activities you might follow when qualifying and certifying the welding process, the welder(s), and/or welding operator.

1. A tentative welding procedure is prepared by a person knowledgeable of the process and technique to be used and the code or specification to be satisfied.

2. Test samples are welded in accordance with the tentative WPS, and the welding parameters are recorded on the PQR. The test must be witnessed by an authorized person from an independent testing lab, the customer, an insurance company, or other individual(s) as specified by the code or listing agency.

3. The test samples are tested under the supervision of the same individuals or group that witnessed the test by the applicable requirements, codes, or specifications.

4. If the test samples pass the applicable test, the procedure has completed qualification. It is then documented as qualified/finalized and is released for use in production.

5. If the test samples do not pass the applicable test, the tentative WPS value parameters are changed as deemed feasible. Test samples are then rewelded and retested to determine if they do or do not meet applicable requirements. This process is repeated until the test samples pass applicable requirements, and the procedure is finalized and released.

6. The welder making the test samples to be used in qualifying the procedure is normally considered qualified and is then certified in the specific procedure.

7. Other welders to be qualified weld test samples per the WPS, and the samples are tested per applicable requirements. If the samples pass, the welder is qualified to the specific procedure and certified accordingly.

8. A qualified WPS is usable for an indefinite length of time, usually until a process considered more efficient for a particular production weld is found.

9. The welder's qualification is normally considered effective for an indefinite period of time, unless the welder is not engaged in the specific process of welding for which he or she is qualified for a period exceeding six months; then the welder must requalify. Also a welder will need to requalify if for some reason the qualification is questioned.

Figure 22-1 and **Figure 22-2** are two examples of test records used to qualify a WPS and a welder for plate, PQR.

WELDING PROCEDURE SPECIFICATION (WPS)

Welding Procedures Specifications No: _____ Date: _____

TITLE:
Welding _____ of _____ to _____

SCOPE:
This procedure is applicable for _____
within the range of _____ through _____

Welding may be performed in the following positions _____

BASE METAL:
The base metal shall conform to _____

Backing material specification _____

FILLER METAL:
The filler metal shall conform to AWS classification No. _____ _____ from AWS
specification _____. This filler metal falls into F-number _____
and A-number _____.

SHIELDING GAS:
The shielding gas, or gases, shall conform to the following compositions and purity:
_____

JOINT DESIGN AND TOLERANCES:

PREPARATION OF BASE METAL:

ELECTRICAL CHARACTERISTICS:
The current shall be _____.

The base metal shall be on the _____ _____ side of the line.

PREHEAT:

BACKING GAS:

SAFETY:

WELDING TECHNIQUE:

INTERPASS TEMPERATURE:

CLEANING:

INSPECTION:

REPAIR:

SKETCHES:

BEND TEST: Specimen preparation:
Acceptance criteria for bend test:

FIGURE 22-1 Welding Procedure Specification (WPS). © Cengage Learning 2012

PROCEDURE QUALIFICATION RECORD (PQR)

Welding Qualification Record No: _____(1)_____ WPS No: _____(2)_____ Date: _____(3)_____

Material specification _____(4)_____ to _____

P-No. _____(5)_____ to P-No. _____ Thickness and O.D. _____(6)_____

Welding process: Manual _____(7)_____ Automatic _____(8)_____

Thickness Range _____(9)_____

Filler Metal

Specification No. _____(10)_____ Classification _____(11)_____ F-number _____(12)_____

A-number _____(13)_____ Filler Metal Size _____(14)_____ Trade Name _____(15)_____

Describe filler metal (if not covered by AWS specification) _____(16)_____

_____

Flux or Atmosphere

Shielding Gas _____(17)_____ Flow Rate _____(18)_____ Purge _____(19)_____

Flux Classification _____(20)_____ Trade Name _____(21)_____

Welding Variables

Joint Type _____(22)_____	Position _____(29)_____
Backing _____(23)_____	Preheat _____(30)_____
Passes and Size _____(24)_____	Bead Type _____(31)_____
No. of Arcs _____(25)_____	Current _____(32)_____
Ampere _____(26)_____	Volts _____(33)_____
Travel Speed _____(27)_____	Oscillation _____(34)_____

Interpass Temperature Range _____(28)_____

Weld Results

Appearance _____(35)_____ Weld Size _____(36)_____

Guided-Bend Test

Type	Result	Type	Result
(37)	(38)		

Tensile Test

Specimen No.	Dimensions Width|Thickness	Area	Ultimate Total Load, lb.	Ultimate Unit Stress, psi	Character of Failure and Location
(39)	(40)	(41)	(42)	(43)	(44)

Welder's Name _____(45)_____ Identification No. _____(46)_____ Laboratory Test No._____

By virtue of these test welder meets performance requirements.

Test Conducted by _____(47)_____ Address _____

per_____(48)_____ Date_____(49)_____

We certify that the statements in this record are correct and that the test welds performed and tested are in accordance with the WPS.

Manufacture _____(50)_____

Signed by _____

Date _____

FIGURE 22-2 Procedure Qualification Record (PQR). © Cengage Learning 2012

GENERAL INFORMATION

Normally, the format of the WPS is not dictated by the code or specification. Any format is acceptable as long as it lists the parameters or variables (essential or nonessential, amps, volts, filler identification, etc.) listed by the code or specification. Most codes or specifications appear in an acceptable or recommended format.

Ideally, the WPS should include all of the information required to make the weld. A welder should be able to be given the WPS without additional instructions and produce the weld. To help with this, it is often a good idea to include supplementary information with each WPS. The information might be basic instructions for the process. With some WPSs you might include several pages as attachments that can give the welder a little review of the setup, operation, testing, inspecting, and so on, which will help to ensure accuracy and uniformity in the welds.

Essential variables are those parameters in which a change is considered to affect the mechanical properties of the weldment to the point of requiring requalification of the procedure. Nonessential variables are those parameters in which a change may be made without requiring requalification of the procedure. However, a change in nonessential variables usually requires a revision to be made.

There are large differences among various codes. The AWS D1.1, *Structural Welding Code Steel,* allows some prequalified weld joints for specific processes (SMAW, SAW, FCAW, and GMAW). A written procedure is required for these joints, but since the procedure is tentative, it does not require support via a written PQR, **Figure 22-3.**

The Procedure Qualification Requirements regarding positions for groove welds in plate differ among codes. Some codes may require a written procedure for each position. The ASME Section IX, however, qualifies a welder for the 1G position when the welder qualifies for 2G, 3G, or 4G.

Ordinarily, the welder must be qualified/certified in accordance with a specific WPS. The welder's qualifying test plate may be examined radiographically or ultrasonically in place of bend tests. Specific codes or specifications must be referenced for details of the number of actual tensile, bend, or other type of test specimens and tests to be performed. For example, AWS D1.1 and ASME Section IX do not require a "Nick-Break Test Specimen," but API Standard 1104 does require it.

PRACTICE 22-1

Writing a Welding Procedure Specification (WPS)

Using the form provided and following the example, **Figure 22-3,** you will write a welding procedure specification. **Figure 22-4** is a form that is a composite of sample WPS forms provided by AWS, ASME, and API codes. You may want to obtain a copy of one of the codes or standards and compare a weld you made to the standard. Most of the unique information is provided in this short outline. Additional information that may be required for this form can be found in figures in this chapter. You may need to refer to some of the chapters on welding or to your notes to establish the actual limits of the welding variables (voltage, amperage, gas flow rates, nozzle size, etc.).

> NOTE: Not all of the blanks will be filled in on the forms. The forms are designed to be used with a large variety of weld procedures, so they have spaces that will not be used each time.

1. The WPS number is usually made up following a system established by the company. This number may or may not include coded information relating to the date it was written, who wrote it, material or process data, and so on.

2. Date that the WPS was written or effective.

3. The welding process(es) that will be used to perform the weld, such as SMAW, GMAW, GTAW, and so on.

4. The actual material type and thickness or pipe type and diameter and/or wall thickness. If all the material or pipe being joined is the same, then the same information will appear before and after "to."

5. Fillet or groove weld and the joint type, such as butt, lap, tee, and so on.

6. Thickness range qualified or diameter range qualified: For both plate and pipe, a weld performed successfully on one thickness qualifies a welder to weld on material within that range. See **Table 22-1** for a list of thickness ranges.

7. Material position: 1G, 2G, 3G, 4G, 1F, 2F, 3F, 4F, 5G, 6G, 6GR, Figure 22-3.

8. Base metal specification: This is the ASTM specification for the type and grade of material, including the P-number, **Table 22-2.**

9. If a backing material is used, then its ASTM or other specification information must be included here.

10. Classification number: This is the standard number found on the electrode or electrode box, such as E6010, E7018, E316-15, ER70S-3, or E70T-1.

11. Filler metal specification number: The AWS has specifications for chemical composition and physical properties for electrodes. Some of these specifications are listed in **Table 22-3.**

12. F-number: A specific grouping number for several classifications of electrodes having similar composition and welding characteristics. **Table 22-4** lists the F-number corresponding to the electrode used.

WELDING PROCEDURE SPECIFICATION (WPS)
FOR
ABC, INC.

Welding Procedures Specifications No: _____WPS-1A_____   Date: ____7-11-91____

TITLE:
Welding ____GTAW____ of _____3" SCHEDULE 80 316 SEAMLESS PIPE_____ to
___3" SCHEDULE 80 316 PIPE___.

SCOPE:
This procedure is applicable for _____V-GROOVED WELDS IN PIPING_____
within the range of _____3" SCHEDULE 80_____ through _____3" SCHEDULE 80_____.
Welding may be performed in the following positions ____1G____.

BASE METAL:
The base metal shall conform to ASTM A376 GRADE TP-316 SEAMLESS PIPE SCHEDULE 80 P8.
Backing material specification: _____N/A_____.

FILLER METAL:
The filler metal shall conform to AWS classification No. ____ER 316L____ from AWS
specification ____A5.9____. This filler metal falls into F-number _6_ and A-number _8_.

SHIELDING GAS:
The shielding gas, or gases, shall conform to the following compositions and purity:
_____WELDING GRADE ARGON_____.

JOINT DESIGN AND TOLERANCES:

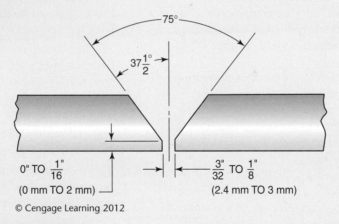

© Cengage Learning 2012

PREPARATION OF BASE METAL:
The edges of parts to be joined shall be prepared by machining. All parts to be joined must be cleaned prior to welding of all hydrocarbons and other contaminants, such as cutting fluids, grease, oil and primers, by suitable solvents. Both the inside and outside surfaces within 2" of the joint must be mechanically cleaned by using a stainless steel wire brush that has not been used for other purposes or pickled with 10% to 20% nitric acid solution. Joint alignment shall be maintained by four (4) tack welds equally spaced around the joint. The tack welds are to be made using the same GTA welding process used for the root pass. Both ends of each tack weld are to be ground to a taper prior to the beginning of the root pass.

ELECTRICAL CHARACTERISTICS:

The current shall be_____DIRECT CURRENT ELECTRODE NEGATIVE_____ .

The base metal shall be on the _____POSITIVE_____ side of the line.

PREHEAT:

The parts shall not be welded if they are below 70°F.

BACKING GAS:

To protect the inside of the root surface from the formation of oxides during welding, a continuous flow of argon into the part is required. The open end of the part must be capped (Figure 1) and the unwelded joint must be taped prior to the beginning of any welding. The backing gas must have a flow rate of 10 CFH to 15 CFH, and the flow must begin 2 minutes before welding starts and continue until the part has cooled to room temperature. The backing gas may be stopped between welds only if the part is allowed to cool to room temperature.

WELDING TECHNIQUE:

The GTA welding process is to be used for making the weld.

Electrode	1/8-inch diameter EWTh-2
Electrode tip geometry	Tapered 2 to 3 times length to diameter
Nozzle	1/2 inch diameter
Shielding gas	Argon
Shielding gas flow rate	20 CFH to 45 CFH
Current, A	90 to 150
Polarity	DCEN
Arc voltage, V	12 to 15
Filler metal type	ER 316L
Filler metal size	3/32 inch to 1/8 inch diameter
Backing gas	Argon
Backing gas flow rate	10 to 15 CFH
Preheat	70 degrees min.
Interpass temp.	500 degrees max.
Travel speed	As required

TACK WELDS:

With the pipe securely clamped into welding jig and the flange fitting properly located with the correct root gap, the four tack welds are to be performed (Figure 2). Holding the electrode so that it is very close to the root face but not touching (Figure 3), slowly increase the current until the arc starts and a molten weld pool is formed. Adding filler metal to maintain a slight convex weld face and a flat or slightly concave root face (Figure 4). When it is time to end the tack weld, lower the current slowly so that the molten weld pool can be tapered down in size (Figure 5). When all four tack welds are complete, allow the pipe to cool. Using a grinding wheel that has never been used on any metal other than 316 stainless steel, feather the ends of the tack welds.

FIGURE 22-3 Welding Procedure Specification (WPS) *(continued)*. © Cengage Learning 2012

ROOT WELD:

Holding the electrode so that it is very close to the root face but not touching, slowly increase the current until the arc starts and a molten weld pool is formed. As the weld progresses, add filler metal as required to maintain a slightly concave root face. When it is necessary to stop the weld, to reposition the part, or the weld is completed, the current must be lowered slowly so that the molten weld pool can be tapered down in size.

FILLER AND COVER WELDS:

Position the pipe so that the weld is in the 1G position. Connect the purge gas and start the flow 2 minutes before welding begins. A 5 CFH flow must be maintained during the remainder of the welding. The backing gas may be stopped if welding is to be discontinued for more than 15 minutes. A slightly higher welding current level will be required for the remaining welds. Start each of the filler welds so that the starting and ending points will be staggered (Figure 6). The size of each of the filler passes and the cover passes must not be larger than 1/4 inch wide. The weld bead size must be small so as to minimize the heat input in the weld joint and so as to allow the weld to stress relieve itself. The finished weld contour must be within tolerance (Figure 7).

INSPECTION:

The weld is to be inspected visually in accordance to AWS B1, Guide for Nondestructive Inspection of Welds. It must be found to be within tolerance as stated herein.

REPAIR:

Only slight surface discontinuities may be repaired with the approval of the QC department.

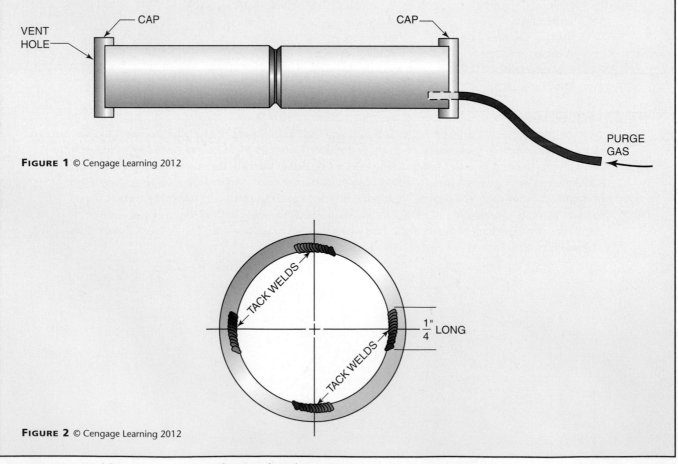

FIGURE 1 © Cengage Learning 2012

FIGURE 2 © Cengage Learning 2012

FIGURE 22-3 Welding Procedure Specification (WPS) *(continued)*. © Cengage Learning 2012

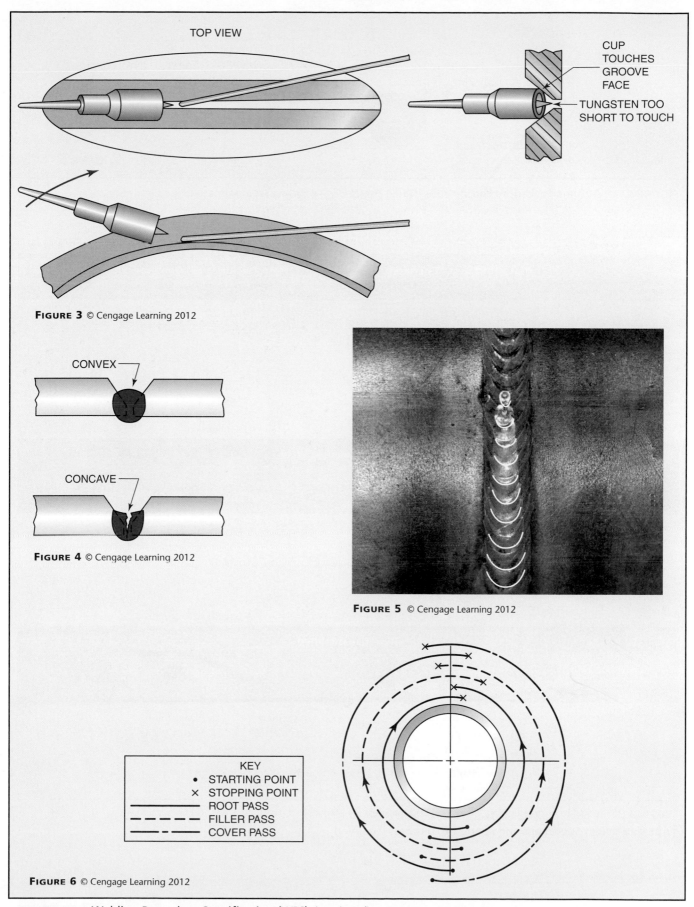

FIGURE 3 © Cengage Learning 2012

FIGURE 4 © Cengage Learning 2012

FIGURE 5 © Cengage Learning 2012

FIGURE 6 © Cengage Learning 2012

TOP VIEW

CUP TOUCHES GROOVE FACE

TUNGSTEN TOO SHORT TO TOUCH

CONVEX

CONCAVE

KEY
- STARTING POINT
× STOPPING POINT
— ROOT PASS
--- FILLER PASS
—·— COVER PASS

FIGURE 22-3 Welding Procedure Specification (WPS) *(continued).* © Cengage Learning 2012

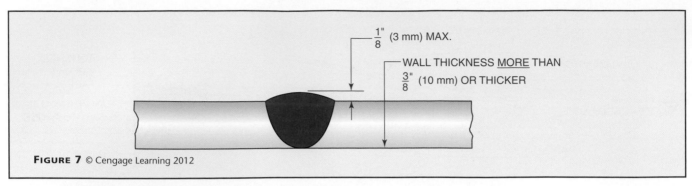

FIGURE 7 © Cengage Learning 2012

FIGURE 22-3 Welding Procedure Specification (WPS) *(concluded)*. © Cengage Learning 2012

Plate Thickness (T) Tested in. (mm)	Plate Thickness (T) Qualified in. (mm)
1/8 ≥ T ≤ 3/8* (3.1 ≥ T ≤ 9.5)	1/8 to 2T (3.1 to 2T)
3/8 (9.5)	3/4 (19.0)
3/8 ≥ T ≤ 1 (9.5 ≥ T ≤ 25.4)	2T 2T
1 and over (25.4 and over)	Unlimited Unlimited

Pipe Size of Sample Weld	
Diameter in. (mm)	**Wall Thickness, T**
2 (50.8) or	Sch. 80
3 (76.2)	Sch. 40
6 (152.4) or	Sch. 120
8 (203.2)	Sch. 80

Pipe Size Qualified		
Diameter in. (mm)	**Wall Thickness in. (mm)**	
	Minimum	Maximum
3/4 (19.0) through 4 (101.6)	0.063 (1.6)	0.674 (17.1)
4 (101.6) and over	0.187 (4.7)	Any

*Thickness (T) is equal to or greater than 1/8 in. (≥) and thickness (T) is equal to or less than 3/8 in. (≤).

TABLE 22-1 Test Specimens and Range of Thickness Qualified

WELDING PROCEDURE SPECIFICATION (WPS)

Welding Procedures Specifications No: _____(1)_____   Date: _____(2)_____

TITLE:
Welding _____(3)_____ of _____(4)_____ to _____(4)_____.

SCOPE:
This procedure is applicable for _____(5)_____
within the range of _____(6)_____ through _____(6)_____.

Welding may be performed in the following positions _____(7)_____.

BASE METAL:
The base metal shall conform to _____(8)_____.

Backing material specification _____(9)_____.

FILLER METAL:
The filler metal shall conform to AWS classification No. _____(10)_____ from
AWS specification _____(11)_____. This filler metal falls into F-number
_____(12)_____ and A-number _____(13)_____.

SHIELDING GAS:
The shielding gas, or gases, shall conform to the following compositions and purity:
_____(14)_____

JOINT DESIGN AND TOLERANCES:
                                            (15)
PREPARATION OF BASE METAL:
                                            (16)
ELECTRICAL CHARACTERISTICS:
The current shall be _____(17)_____.

The base metal shall be on the _____(18)_____ side of the line.

PREHEAT: (19)

BACKING GAS: (20)

WELDING TECHNIQUE: (21)

INTERPASS TEMPERATURE: (22)

CLEANING: (23)

INSPECTION: (24)

REPAIR: (25)

SKETCHES: (26)

FIGURE 22-4 Welding Procedure Specification (WPS). © Cengage Learning 2012

	Type of Material
P-1	Carbon steel
P-3	Low alloy steel
P-4	Low alloy steel
P-5	Alloy steel
P-6	High alloy steel—predominantly martensitic
P-7	High alloy steel—predominantly ferritic
P-8	High alloy steel—austenitic
P-9	Nickel alloy steel
P-10	Specialty high alloy steels
P-21	Aluminum and aluminum-base alloys
P-31	Copper and copper alloy
P-41	Nickel

TABLE 22-2 P-Numbers

A5.10	Aluminum—bare electrodes and rods
A5.3	Aluminum—covered electrodes
A5.8	Brazing filler metal
A5.1	Steel, carbon, covered electrodes
A5.20	Steel, carbon, flux cored electrodes
A5.17	Steel-carbon, submerged arc wires and fluxes
A5.18	Steel-carbon, gas metal arc electrodes
A5.2	Steel—oxyfuel gas welding
A5.5	Steel—low alloy covered electrodes
A5.23	Steel—low alloy electrodes and fluxes—submerged arc
A5.28	Steel—low alloy filler metals for gas shielded arc welding
A5.29	Steel—low alloy, flux cored electrodes

TABLE 22-3 Specification Numbers

Group Designation	Metal Types	AWS Electrode Classification
F1	Carbon steel	EXX20, EXX24, EXX27, EXX28
F2	Carbon steel	EXX12, EXX13, EXX14
F3	Carbon steel	EXX10, EXX11
F4	Carbon steel	EXX15, EXX16, EXX18
F5	Stainless steel	EXXX15, EXXX16
F6	Stainless steel	ERXXX
F22	Aluminum	ERXXXX

TABLE 22-4 F-Numbers

13. A-number: The classification of weld metal analysis. **Table 22-5** lists the A-numbers.

14. Shielding gas(es) and flow rate for GMAW, FCAW, or GTAW

Complete a copy of the "Student Welding Report" listed in Appendix I or provided by your instructor. ◆

PRACTICE 22-2

Procedure Qualification Record (PQR)

Following the procedure you wrote in Practice 22-1, you are going to make the weld to see if your tentative welding procedure and specification can be certified. Complete a copy of the form provided to record all of the appropriate information, Figure 22-2.

Complete a copy of the "Student Welding Report" listed in Appendix I or provided by your instructor. ◆

1. The PQR number is usually made up following a system established by the company. This number may or may not include coded information relating to the product being welded, material or process data, and the like.

2. The WPS number on which the PQR is based.

3. The date on which the welding took place.

4. Base metal specification: This is the ASTM specification for the type and grade of material.

5. Table 22-2 lists some commonly used metals and their P-numbers.

6. Test material thickness (or) test pipe outside diameter (OD) (and) wall thickness.

7. Manual welding processes are used to qualify a welder. Specify the process, such as GMAW, FCAW, SMAW, GTAW, and so on.

8. Automatic welding processes are used to qualify a welding operator. Specify the process (SAW, ESW, etc.).

				Analysis			
A No.	Types of Weld Deposit	C %	Mn %	Si %	Mo %	Cr %	Ni %
1	Mild steel	0.15	1.6	1.0	—	—	—
2	Carbon-moly	0.15	1.6	1.0	0.4—0.65	0.5	—
3	Chrome (0.4 to 2%)-moly	0.15	1.6	1.0	0.4—0.65	0.4—2.0	—
4	Chrome (2 to 6%)-moly	0.15	1.6	2.0	0.4—1.5	2.0—6.0	—
5	Chrome (6 to 10.5%)-moly	0.15	1.2	2.0	0.4—1.5	6.0—10.5	—
6	Chrome-martensitic	0.15	2.0	1.0	0.7	11.0—15.0	—
7	Chrome-ferritic	0.15	1.0	3.0	1.0	11.0—30.0	—
8	Chromium-nickel	0.15	2.5	1.0	4.0	14.5—30.0	7.5—15.0
9	Chromium-nickel	0.30	2.5	1.0	4.0	25.0—30.0	15.0—37.0
10	Nickel to 4%	0.15	1.7	1.0	0.55	—	0.8—4.0
11	Manganese-moly	0.17	1.25—2.25	1.0	0.25—0.75	—	0.85
12	Nickel-chrome-moly	0.15	0.75—2.25	1.0	0.25—0.8	1.5	1.25—2.25

TABLE 22-5 A-Number Classification of Ferrous Metals

9. Thickness range qualified (or) diameter range qualified: For both plate and pipe, a weld performed successfully on one thickness qualifies a welder to weld on material within that range. See Table 22-1 for a list of thickness ranges.

10. Filler metal specification number: The AWS has specifications for chemical composition and physical properties for electrodes. Some of these specifications are listed in Table 22-3.

11. Classification number: This is the standard number found on the electrode or electrode box, such as E6010, E7018, E316-15, or ER1100.

12. F-number: A specific grouping number for several classifications of electrodes having similar composition and welding characteristics. See Table 22-4 for the F-number corresponding to the electrode used.

13. A-number: The classification of weld metal analysis. See Table 22-5 for a list of A-numbers.

14. Give the diameter of electrode used.

15. Give the manufacturer's identification name or number.

16. List the manufacturer's chemical composition and physical properties as provided if the filler metal is not covered by an AWS specification.

17. Shielding gas or gas mixture for GMAW, FCAW, or GTAW.

18. Flow rate in cubic feet per hour (cfh).

19. The amount of time that the shielding gas is to flow to purge air from the welding zone.

20. SAW flux classification.

21. The manufacturer's identification name or number for the SAW flux.

22. Butt, lap, tee, or other joint type.

23. Backing strip material specification: This is the ASTM specification number.

24. The number of passes and the size.

25. Usually 1 except for some automatic SAW processes that may use multiple electrodes with multiple arcs.

26. The amount of current in amps used to make the weld. If the machine being used for the weld does not have an amp meter, a meter must be attached to the welding lead, within 2 ft (1.5 m) of the electrode holder, to get this reading.

27. The travel speed in inches per minute is usually given for machine or automatic welds.

28. This is the maximum temperature that the base metal is allowed to reach during the weld. Welding must stop and the part allowed to cool if this temperature is reached.

29. Test position: 1G, 2G, 3G, 4G, 1F, 2F, 3F, 4F, 5G, 6G, 6GR, **Figure 22-5.**

30. This is the minimum temperature that the base metal must be before welding can start.

31. Groove or fillet weld.

32. AC, DCEP, or DCEN.

33. The voltage is included for all welding processes.

34. The type of electrode movement used when making the weld.

35. Visually inspect the weld and record any flaws.

36. Record the legs and reinforcement dimensions. Measure and record the depth of the root penetration.

37. Four (4) test specimens are used for 3/8-in. (10-mm) or thinner metal. Two (2) will be root bent and two (2) face bent. For thicker metal all four (4) will be side bent.

38. Visually inspect the specimens after testing and record any discontinuities.

39. Identification number that was marked on the specimen.

40. Width and thickness of test section of the specimen.

41. Cross-sectional area of specimen in the test area.

42. The load at which the specimen failed.

43. The maximum load divided by the specimen's original area converts the ultimate total load for the specimen to the pounds per square inch (psi) that was required to break the material.

44. The type of failure, whether it was ductile or brittle, and where the failure occurred relative to the weld.

45. Welder's name: The person who performed the weld.

46. Identification number: On a welding job, every person has an identification number that is used on the time card and paycheck. In this space, you can write the class number or section number, since you do not have a clock number.

47. The name of the person who interpreted the results.

48. Qualifications of the test interpreter: This is usually a Certified Welding Inspector or other qualified person.

49. Date the results of the test were completed.

50. The name of the company that requested the test.

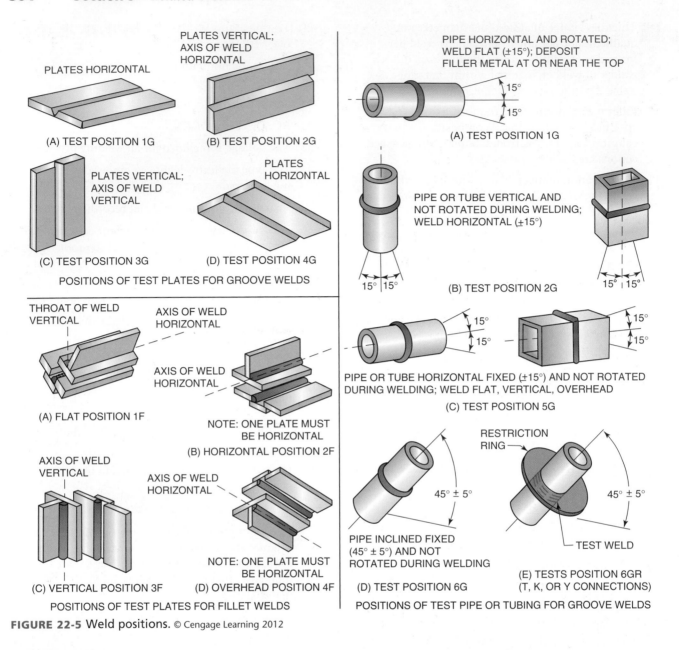

FIGURE 22-5 Weld positions. © Cengage Learning 2012

Summary

Over the years, often through trial and error, welders and welding engineers have developed standards, codes, and specifications that, when followed, will produce welds that will be sound and provide years of service. It is important to know that, under such codes and standards, not all welds must be perfect. Some levels of imperfection are acceptable and, through years of experience, such minor flaws have been determined not to be critical or to result in structural failure.

Being able to follow such codes and standards is important in that it helps control welding costs. The more precise the weld produced, the more expensive the weld is to produce. As a welder attempts perfection, welding preparation time, welding time, postweld cleanup time, and unnecessary rewelding time all increase, thus increasing the cost of the weld. Knowing what is "fit for service" as set by the code or standard is essential. Likewise, failing to produce a weld to the required code or standard can cause structural failure. Therefore, it is important that you familiarize yourself with the standards for your company's requirements.

Resistance Seam Welding Benefits Stainless Steel Application

A redesigned production system increased productivity by 50% and reduced scrap rate to 1% on catalytic converter shells.

Full of expensive metal and formed to a precision shape, the stainless steel shell cannot leak. And even though it is a "no-show" part, it must be scratch free. These are some of the challenges that Walker Manufacturing had to solve to produce a quality catalytic converter shell.

Walker Manufacturing, Litchfield, Michigan, is a supplier to the automotive industry with expertise in manufacturing exhaust systems. An ongoing program to reduce waste and costs led to a reengineering of the process for manufacturing catalytic converter shells. Throughput was increased by 50%, the scrap rate was reduced by two-thirds, productivity was increased, and a coolant previously used was eliminated.

Doing It the Old Way

Originally, catalytic converter shells were produced on a gas tungsten arc (GTA) mill. The shells were formed from coil-fed 409 stainless steel into a tubular shape, notched for the cutoff process, and welded with the GTAW process. They were then cut to a specified length and sized to a final shape, and the shell edges were deburred. It was a labor-intensive process, according to Steve Sherwood, senior process engineer.

Two people were needed for the welding mill, two people to wash each part of the coolant used in the cutting process, and four people to check length and shape and deburr the shell. Deburring both ends of the component was crucial and often required 100% inspection. The shells were also checked for proper tolerances. Tooling changeover and replacement of deburring brushes required a lot of time.

There was also a recurring problem with the welding operation. At times a pinhole leak developed when the arc was initiated with the GTAW process. This was an important quality issue because hot exhaust gases could not leak from the shell. With leak problems and burrs, scrap rates approached 3%.

New Design Streamlines Operation

To solve these problems and improve the operation, a mash seam resistance welding system was developed by Newcor Bay City Division, Bay City, Michigan. The system automatically takes a flat stock stainless steel blank, forms it into a shell, seam welds it, and then sizes it to the final dimensions.

Resistance welding unit showing wheel electrode used to join the overlapping edge of the shell. American Welding Society

After the shell is welded it is placed into a station where the component is finished to its final size and shape. American Welding Society

How It Works

Flat Type 409 stainless steel blanks ranging in thickness from 0.05 in. to 0.12 in. (1.4 mm to 3 mm) are loaded on a stacker at one end of the welding machine. The blank is picked up by suction cups and placed on a conveyor, which moves it through the shell forming area. The blank is then formed into a shell shape, overlapping the edges so it can next be resistance seam welded. It is then planished, a process that rolls the weld flat and reduces its height to within 15% of the metal's original thickness. It is placed into a station for forming its final size and shape to a tolerance of ±0.020 in. (0.5 mm). The whole operation gives a stronger weld, the right shape, and cosmetically a more attractive part.

Big Improvements

Scrap rates were reduced to 1% and the 2300–2500 parts produced in an 8-hour shift represent a 50% increase in productivity. Mash seam welding eliminated the deburring process and the coolant that was previously needed. Tooling changeover now takes only 30 minutes.

Overall, quality is up, costs are down, and inventory and shipping schedules are more controllable.

Article courtesy of the American Welding Society.

Review

1. Why is it important to select the correct welding code or standard?

2. What are codes and standards?

3. What is the difference between welding codes or standards and welding specifications?

4. What might influence the selection of a particular code or specification for welding?

5. As stricter standards are applied to the welding process, what happens to the cost of the weldment?

6. What can happen if welding standards are lax?

7. What are three commonly used codes?

8. What is a WPS?

9. What information should be included in a WPS?

10. What is the form called that records the parameters used to produce a test weld?

11. Who should witness the test welding being performed for a tentative WPS?

Chapter 23

Testing and Inspection of Welds

OBJECTIVES

After completing this chapter, the student should be able to

■ describe the difference between mechanical or destructive and nondestructive testing.

■ list the 12 most common discontinuities and the nondestructive methods of locating them.

■ discuss how both mechanical or destructive and nondestructive testing are performed.

■ explain why welds are tested.

■ evaluate a weld according to a given standard or code.

KEY TERMS

Brinell hardness tester

defect

discontinuities

eddy current inspection (ET)

etching

mechanical testing (DT)

nondestructive testing (NDT)

quality control

radiographic inspection (RT) *ex-Ray*

Rockwell hardness tester

shearing strength

tolerance

ultrasonic inspection (UT) *sound*

INTRODUCTION

It is important to know that a weld will meet the requirements of the company and/or codes or standards. It is also necessary to ensure the quality, reliability, and strength of a weldment. To meet these demands, an active inspection program is needed. The extent to which a welder and product are subjected to testing and inspection depends upon the intended service of the product. Items that are to be used in light, routine-type service, such as ornamental iron, fence posts, gates, and so forth, are not inspected as critically as products in critical use. Some of the items in critical use include a main nuclear reactor containment vessel, oil refinery high-pressure vessels, aircraft airframes, bridges, and so on. The type of inspection required is then very much dependent upon the type of service that the welded part will be required to withstand. The quality of the weld that will pass or be acceptable for one welding application may not meet the needs of another.

QUALITY CONTROL (QC)

Once a code or standard has been selected, a method is chosen for ensuring that the product meets the specifications. The two classifications of methods used in product **quality control** are destructive, or mechanical, testing and nondestructive testing. These methods can be used individually, or a combination of the two methods can be used. **Mechanical testing (DT)** methods, except for hydrostatic testing, result in the product being destroyed. **Nondestructive testing (NDT)** does not destroy the part being tested.

Mechanical testing is commonly used to qualify welders or welding procedures. It can be used in a random sample testing procedure in mass production. In many cases, a large number of identical parts are made, and a chosen number are destroyed by mechanical testing. The results of such tests are valid only for welds made under the same conditions because the only weld strengths known are the ones resulting from the tested pieces. It is then assumed that the strengths of the nontested pieces are the same.

Nondestructive testing is used for welder qualification, welding procedure qualification, and product quality control. Since the weldment is not damaged, all the welds can be tested and the part can actually be used for its intended purpose. Because the parts are not destroyed, more than one testing method can be used on the same part. Frequently, only part of the welds is tested to save time and money. The same comparison of random sampling applies to these tests as it does for mechanical testing. Critical parts or welds are usually 100% tested.

DISCONTINUITIES AND DEFECTS

Discontinuities and flaws are interruptions in the typical structure of a weld. They may be a lack of uniformity in the mechanical, metallurgical, or physical characteristics of the material or weld. All welds have discontinuities and flaws, but they are not necessarily defects.

A **defect,** according to AWS, is "a discontinuity or discontinuities that by nature or accumulated effect render a part or product unable to meet minimum applicable acceptance standards or specifications. The term designates rejectability."

In other words, many acceptable products may have welds that contain discontinuities. But no products may have welds that contain defects. The only difference between a discontinuity and a defect is when the discontinuity becomes so large or when there are so many small discontinuities that the weld is not acceptable under the standards for the code for that product. Some codes are more strict than others, so that the same weld might be acceptable under one code but not under another.

Ideally, a weld should not have any discontinuities, but that is practically impossible. The difference between

| American Bureau of Shipping |
| American Petroleum Institute |
| American Society of Mechanical Engineers |
| American Society for Testing and Materials |
| American Welding Society |
| British Welding Institute |
| United States government |

*A more complete listing of agencies with addresses is included in the appendix.
TABLE 23-1 Major Code Issuing Agencies

what is acceptable, fit for service, and perfection is known as **tolerance.** In many industries, the tolerances for welds have been established and are available as codes or standards. **Table 23-1** lists a few of the agencies that issue codes or standards. Each code or standard gives the tolerance that changes a discontinuity to a defect.

When evaluating a weld, it is important to note the type of discontinuity, the size of the discontinuity, and the location of the discontinuity. Any one of these factors or all three can be the deciding factors that, based on the applicable code or standard, change a discontinuity to a defect.

The 12 most common discontinuities are as follows:

- Porosity
- Inclusions
- Inadequate joint penetration
- Incomplete fusion
- Arc strikes
- Overlap (cold lap)
- Undercut
- Cracks
- Underfill
- Laminations
- Delaminations
- Lamellar tears

Some welding processes are more likely than others to cause some of the discontinuities. For example, laser welding would never produce an arc strike discontinuity and resistance spot welding could never produce a weld with underfill. **Table 23-2** lists the common discontinuities and the welding processes that might cause each.

Porosity

Porosity results when gas that was dissolved in the molten weld pool forms bubbles that are trapped as the metal cools to become solid. The bubbles that make up porosity form within the weld metal; for that reason, they cannot be seen as they form. These gas pockets form in the same way that bubbles form in a carbonated drink as it warms up or as air dissolved in water forms bubbles in the center of a cube of ice. Porosity appears in either spherical

	Shielded metal arc welding (SMAW)	Gas metal arc welding (GMAW)	Flux cored arc welding (FCAW)	Gas tungsten arc welding (GTAW)	Oxyacetylene welding (OAW)	Oxyhydrogen welding (OHW)	Submerged arc welding (SAW)	Laser beam welding (LBW)	Plasma arc welding (PAW)	Electron beam welding (EBW)	Carbon arc welding (CAW)	Pressure gas welding (PGW)	Electroslag welding (ESW)	Thermite welding (TW)
Porosity	X	X	X	X	X	X	X	X	X	X	X	X	X	X
Inclusions	X	X	X				X				X		X	X
Inadequate joint penetration	X	X	X		X		X		X	X			X	
Incomplete fusion	X	X	X	X	X	X	X	X	X	X	X	X	X	X
Arc strikes	X	X	X	X										
Overlap (cold lap)	X	X	X	X	X	X	X	X						X
Undercut	X	X	X	X	X	X	X	X			X			
Crater cracks	X	X	X	X	X	X	X	X			X			
Underfill	X	X	X	X	X	X	X	X			X			

The nine most common discontinuities.

TABLE 23-2 Common Discontinuities and the Joint Types They Might Be Found On

(ball-shaped) or cylindrical (tube- or tunnel-shaped) form. Cylindrical porosity is called wormhole. The rounded edges tend to reduce the stresses around them. Therefore, unless porosity is extensive, there is little or no loss in strength.

Porosity is most often caused by improper welding techniques, contamination, or an improper chemical balance between the filler and base metals.

Improper welding techniques may result in shielding gas not properly protecting the molten weld pool. For example, the E7018 electrode should not be weaved wider than two and a half times the electrode diameter because very little shielding gas is produced. As a result, parts of the weld are unprotected. Nitrogen from the air that dissolves in the weld pool and then becomes trapped during escape can produce porosity.

The intense heat of the weld can decompose paint, dirt, or oil from machining and rust or other oxides, producing hydrogen. This gas, like nitrogen, can also become trapped in the solidifying weld pool, producing porosity. When it causes porosity, hydrogen can also diffuse into the heat-affected zone, producing underbead cracking in some steels. The level needed to crack welds is below that necessary to produce porosity.

Porosity can be grouped into the following four major types:

- Uniformly scattered porosity is most frequently caused by poor welding techniques or faulty materials, **Figure 23-1**.

- Clustered porosity is most often caused by improper starting and stopping techniques, **Figure 23-2**.

- Linear porosity is most frequently caused by contamination within the joint, root, or interbead boundaries, **Figure 23-3**.

- Piping porosity, or wormhole, is most often caused by contamination at the root, **Figure 23-4**. This

porosity is unique because its formation depends on the gas escaping from the weld pool at the same rate as the pool is solidifying.

Refer to Table 23-2 for a listing of welds that may produce porosity.

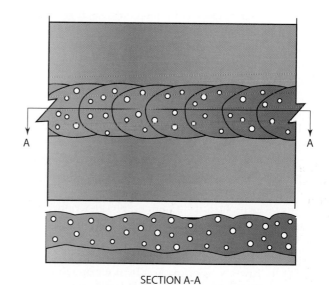

SECTION A-A

FIGURE 23-1 Uniformly scattered porosities. © Cengage Learning 2012

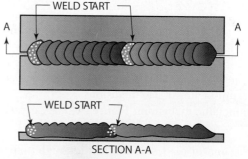

WELD START

WELD START

SECTION A-A

FIGURE 23-2 Clustered porosity. © Cengage Learning 2012

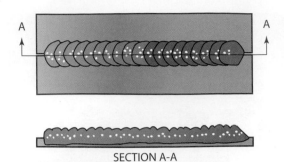

SECTION A-A

FIGURE 23-3 Linear porosity. © Cengage Learning 2012

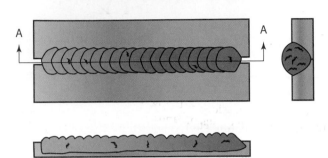

SECTION A-A

FIGURE 23-4 Piping or wormhole porosity.
© Cengage Learning 2012

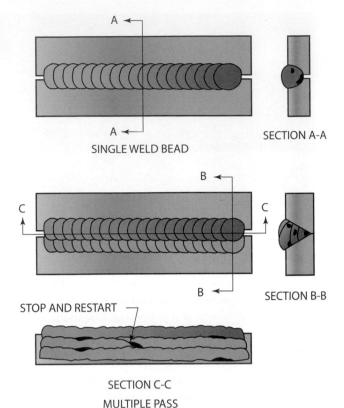

SINGLE WELD BEAD

SECTION A-A

SECTION B-B

STOP AND RESTART

SECTION C-C
MULTIPLE PASS

Inclusions

Inclusions are nonmetallic materials, such as slag and oxides, that are trapped in the weld metal, between weld beads, or between the weld and the base metal. Inclusions sometimes are jagged and irregularly shaped. Also, they can form in a continuous line. This causes stresses to concentrate and reduces the structural integrity (loss in strength) of the weld.

Although not visible, their development can be expected if prior welds were improperly cleaned or had a poor contour. Unless care is taken in reading radiographs, the presence of slag inclusions can be interpreted as other defects.

Linear slag inclusions in radiographs generally contain shadow details; otherwise, they could be interpreted as lack-of-fusion defects. These inclusions result from a lack of slag control caused by poor manipulation that allows the slag to flow ahead of the arc; by not removing all the slag from previous welds; or by welding highly crowned, incompletely fused welds.

Scattered inclusions can resemble porosity but, unlike porosity, they are generally not spherical. These inclusions can also result from inadequate removal of earlier slag deposits and poor manipulation of the arc. Additionally, heavy mill scale or rust serves as their source, or they can result from unfused pieces of damaged electrode coatings falling into the weld. In radiographs some detail will appear, unlike linear slag inclusions.

Nonmetallic inclusions, **Figure 23-5**, are caused under the following conditions:

- Slag and/or oxides do not have enough time to float to the surface of the molten weld pool.

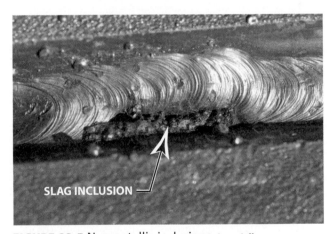

SLAG INCLUSION

FIGURE 23-5 Nonmetallic inclusions. Larry Jeffus

- There are sharp notches between weld beads or between the weld bead and the base metal that trap the material so that it cannot float out.

- The joint was designed with insufficient room for the correct manipulation of the molten weld pool.

Refer to Table 23-2 for a listing of welds that may produce nonmetallic inclusions.

Inadequate Joint Penetration

Inadequate joint penetration occurs when the depth that the weld penetrates the joint, **Figure 23-6**, is less than that needed to fuse through the plate or into the preceding weld. A defect usually results that could reduce the required cross-sectional area of the joint or become a source of stress concentration that leads to fatigue failure. The importance

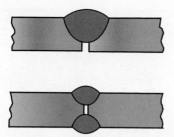

FIGURE 23-6 Inadequate joint penetration.
© Cengage Learning 2012

of such defects depends on the notch sensitivity of the metal and the factor of safety to which the weldment has been designed. Generally, if proper welding procedures are developed and followed, such defects do not occur.

Following are the major causes of inadequate joint penetration:

- Improper welding technique—The most common cause is a misdirected arc. Also the welding technique may require that both starting and run-out tabs be used so that the molten weld pool is well established before it reaches the joint. Sometimes, a failure to back gouge the root sufficiently provides a deeper root face than allowed for, **Figure 23-7**.

- Not enough welding current—Metals that are thick or have a high thermal conductivity are often preheated so that the weld heat is not drawn away so quickly by the metal that it cannot penetrate the joint.

- Improper joint fit up—This problem results when the weld joints are not prepared or fitted accurately. Too small a root gap or too large a root face will keep the weld from penetrating adequately.

- Improper joint design—When joints are accessible from both sides, back gouging is often used to ensure 100% root fusion.

Table 23-2 lists the welding processes that might cause inadequate joint penetration.

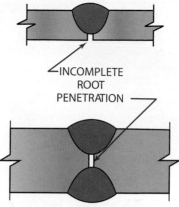

FIGURE 23-7 Incomplete root penetration.
© Cengage Learning 2012

Incomplete Fusion

Incomplete fusion is the lack of coalescence between the molten filler metal and previously deposited filler metal and/or the base metal, **Figure 23-8**. The lack of fusion between the filler metal and previously deposited weld metal is called *interpass cold lap*. The lack of fusion between the weld metal and the joint face is called *lack of sidewall fusion*. Both of these problems usually travel along all or most of the weld's length.

Following are some major causes of lack of fusion:

- Inadequate agitation—Lack of weld agitation to break up oxide layers. The base metal or weld filler metal may melt, but a thin layer of oxide may prevent coalescence from occurring.

- Improper welding techniques—Poor manipulation, such as moving too fast or using an improper electrode angle.

- Wrong welding process—For example, the use of short-circuiting transfer with GMAW to weld plate thicker than 1/4 in. (6 mm) can cause the problem because of the process's limited heat input to the weld.

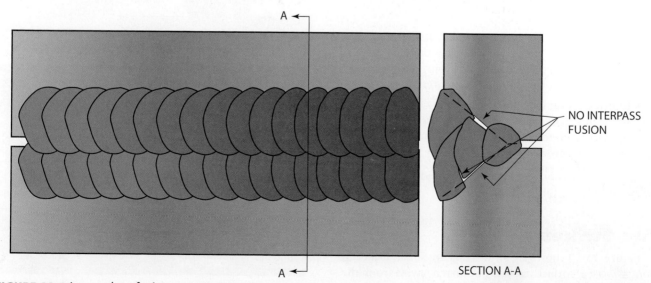

FIGURE 23-8 Incomplete fusion. © Cengage Learning 2012

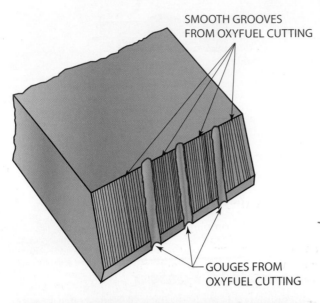

SMOOTH GROOVES
FROM OXYFUEL CUTTING

GOUGES FROM
OXYFUEL CUTTING

CUT GOUGES ON PLATE EDGE GROUND
SMOOTH

FIGURE 23-9 Remove gouges along the surface of the joint before welding. Larry Jeffus

- Improper edge preparation—Any notches or gouges in the edge of the weld joint must be removed. For example if a flame-cut plate has notches along the cut, they could result in a lack of fusion in each notch, **Figure 23-9.**

- Improper joint design—Incomplete fusion may also result from not enough heat to melt the base metal, or too little space allowed by the joint designer for correct molten weld pool manipulation.

- Improper joint cleaning—Failure to clean oxides from the joint surfaces resulting from the use of an oxyfuel torch to cut the plate, or failure to remove slag from a previous weld.

Incomplete fusion can be found in welds produced by all major welding processes.

Arc Strikes

Figure 23-10 shows arc strikes that are small, localized points where surface melting occurred away from the joint. These spots may be caused by accidentally striking

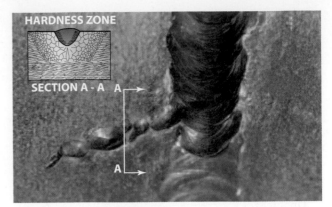

FIGURE 23-10 Arc strikes. Larry Jeffus

the arc in the wrong place and/or by faulty ground connections. Even though arc strikes can be ground smooth, they cannot be removed. These spots will always appear if an acid etch is used. They also can be localized hardness zones or the starting point for cracking. Arc strikes, even when ground flush for a guided bend, will open up to form small cracks or holes.

Overlap

Overlap, also called cold lap, occurs in fusion welds when weld deposits are larger than the joint is conditioned to accept. The weld metal then flows over the surface of the base metal without fusing to it, along the toe of the weld bead, **Figure 23-11.** It generally occurs on the horizontal

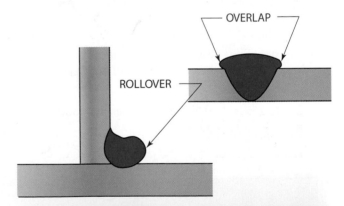

OVERLAP

ROLLOVER

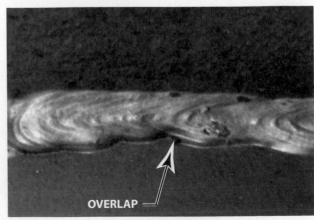

OVERLAP

FIGURE 23-11 Rollover or overlap. Larry Jeffus

leg of a horizontal fillet weld under extreme conditions. It can also occur on both sides of flat-positioned capping passes. With GMA welding, overlap occurs when using too much electrode extension to deposit metal at low power. Misdirecting the arc into the vertical leg and keeping the electrode nearly vertical will also cause overlap. To prevent overlap, the fillet weld must be correctly sized to less than 3/8 in. (9.5 mm), and the arc must be properly manipulated.

Undercut

Undercut is a groove melted into the base metal adjacent to the weld toe or weld root and left unfilled by weld metal, **Figure 23-12**. It can result from excessive current. It is a common problem with GMA welding when insufficient oxygen is used to stabilize the arc. Incorrect welding technique, such as incorrect electrode angle or excessive weave, can also cause undercut. To prevent undercutting, the welder can weld in the flat position by using multiple instead of single passes, changing the shield gas,

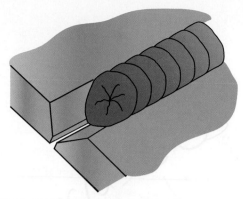

FIGURE 23-13 Crater or star cracks. © Cengage Learning 2012

and improving manipulative techniques to fill the removed base metal along the toe of the weld bead.

Crater Cracks

Crater cracks are the tiny cracks that develop in the weld craters as the weld pool shrinks and solidifies, **Figure 23-13**. Materials with a low melting temperature are rejected toward the crater center while freezing. Since these materials are the last to freeze, they are pulled apart or separated, as a result of the weld metal's shrinking as it cools. The high shrinkage stresses aggravate crack formation. Crater cracks can be minimized, if not prevented, by not interrupting the arc quickly at the end of a weld. This allows the arc to lengthen, the current to drop gradually, and the crater to fill and cool more slowly. Some GMAW equipment has a crater filling control that automatically and gradually reduces the wire-feed speed at the end of a weld. For all other welding processes, the most effective way of preventing crater cracking is to slightly pull the weld back, allowing it to pool up on the weld bead before breaking the arc, **Figure 23-14**.

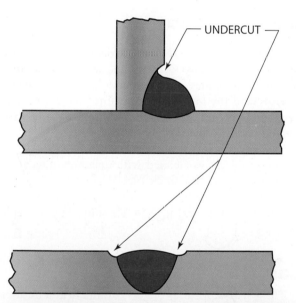

Underfill

Underfill on a groove weld is when the weld metal deposited is inadequate to bring the weld's face or root surfaces to a level equal to that of the original plane or plate surface. For a fillet weld it is when the weld deposit has an insufficient effective throat, **Figure 23-15**. This problem can usually be corrected by slowing down the travel rate or making more weld passes.

Plate-Generated Problems

Not all welding problems are caused by weld metal, the process, or the welder's lack of skill in depositing that metal. The material being fabricated can be at fault, too. Some problems result from internal plate defects that the welder cannot control. Others are the result of improper welding procedures that produce undesirable hard metallurgical structures in the heat-affected zone, as discussed

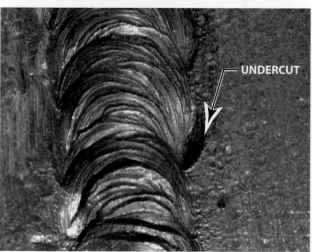

FIGURE 23-12 Undercut. Larry Jeffus

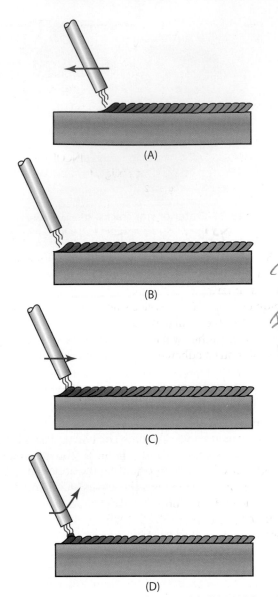

FIGURE 23-14 (A) Make a uniform weld. (B) Weld to the end of the plate. (C) Hold the arc for a second at the end of the weld. (D) Quickly, so you deposit as little weld metal as possible, move the arc back up onto the weld and break the arc. © Cengage Learning 2012

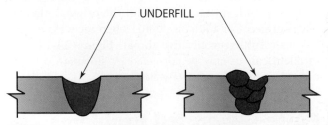

FIGURE 23-15 Underfill. © Cengage Learning 2012

in other chapters. The internal defects are the result of poor steelmaking practices. Steel producers try to keep their steels as sound as possible, but the mistakes that occur in steel production are blamed, too frequently, on the welding operation.

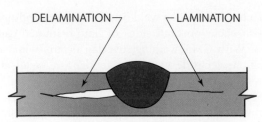

FIGURE 23-16 Lamination and delamination. © Cengage Learning 2012

Lamination

Laminations differ from lamellar tearing because they are more extensive and involve thicker layers of nonmetallic contaminants. Located toward the center of the plate, Figure 23-16, laminations are caused by insufficient cropping (removal of defects) of the pipe in ingots. The slag and oxides in the pipe are rolled out with the steel, producing the lamination. Laminations can also be caused when the ingot is rolled at too low a temperature or pressure.

Delamination

When laminations intersect a joint being welded, the heat and stresses of the weld may cause some laminations to become delaminated. Contamination of the weld metal may occur if the lamination contained large amounts of slag, mill scale, dirt, or other undesirable materials. Such contamination can cause wormhole porosity or lack-of-fusion defects.

The problems associated with delaminations are not easily corrected. If a thick plate is installed in a compression load, an effective solution can be to weld over the lamination to seal it. A better solution is to replace the steel.

Lamellar Tears

These tears appear as cracks parallel to and under the steel surface. In general, they are not in the heat-affected zone, and they have a steplike configuration. They result from the thin layers of nonmetallic inclusions that lie beneath the plate surface and have very poor ductility. Although barely noticeable, these inclusions separate when severely stressed, producing laminated cracks. These cracks are evident if the plate edges are exposed, Figure 23-17.

A solution to the problem is to redesign the joints in order to impose the lowest possible strain throughout the plate thickness. This can be accomplished by making smaller welds so that each subsequent weld pass heat-treats the previous pass to reduce the total stress in the finished weld, Figure 23-18. The joint design can be changed to reduce the stress on the through thickness section of the plate, Figure 23-19. Also refer to Chapter 20 "Welding Joint Design and Welding Symbols."

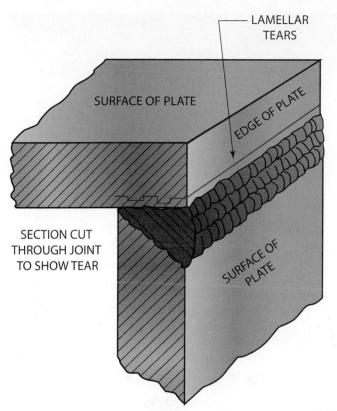

FIGURE 23-17 Example of lamellar tearing. © Cengage Learning 2012

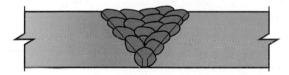

FIGURE 23-18 Using multiple welds to reduce weld stresses. © Cengage Learning 2012

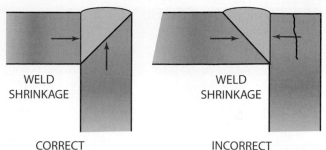

FIGURE 23-19 Correct joint design to reduce lamellar tears. © Cengage Learning 2012

DESTRUCTIVE TESTING (DT)

Because most destructive testing results in some degree of damage or total destruction of the part being tested, only an inference can be made about whether the other weldments are fit for service since they were not actually tested. However, a high reliability between the destructive test and the other weldments can be assumed provided there was strict adherence to the welding procedures.

Tensile Testing

Tensile tests are performed with specimens prepared as round bars or flat strips. The simple round bars are often used for testing only the weld metal, sometimes called "all weld metal testing." This test can be used on thick sections where base metal dilution into all of the weld metal is not possible. Round specimens are cut from the center of the weld metal. The flat bars are often used to test both the weld and the surrounding metal. Flat bars are usually cut at a 90° angle to the weld, **Figure 23-20**. **Table 23-3** shows how a number of standard smaller-size bars can be used, depending on the

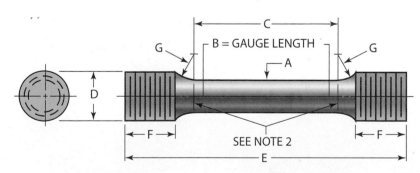

NOTE 1: DIMENSIONS A, B, AND C SHALL BE AS SHOWN, BUT ALTERNATE SHAPES OF ENDS MAY BE USED AS ALLOWED BY ASTM SPECIFICATION E-8.

NOTE 2: IT IS DESIRABLE TO HAVE THE DIAMETER OF THE SPECIMEN WITHIN THE GAUGE LENGTH SLIGHTLY SMALLER AT THE CENTER THAN AT THE ENDS. THE DIFFERENCE SHALL NOT EXCEED 1% OF THE DIAMETER.

FIGURE 23-20 Tensile testing specimen. © Cengage Learning 2012

Specimen	Dimensions of Specimen						
	in./mm	in./mm	in./mm	in./mm	in./mm	in./mm	in./mm
	A	B	C	D	E	F	G
C-1	0.500/12.7	2/50.8	2.25/57.1	0.750/19.05	4.25/107.9	0.750/19.05	0.375/9.52
C-2	0.437/11.09	1.750/44.4	2/50.8	0.625/15.8	4/101.6	0.750/19.05	0.375/9.52
C-3	0.357/9.06	1.4/35.5	1.750/44.4	0.500/12.7	3.500/88.9	0.625/15.8	0.375/9.52
C-4	0.252/6.40	1.0/25.4	1.250/31.7	0.375/9.52	2.50/63.5	0.500/12.7	0.125/3.17
C-5	0.126/3.2	0.500/12.7	0.750/19.05	0.250/6.35	1.750/44.4	0.375/9.52	0.125/3.17

TABLE 23-3 Dimensions of Tensile Testing Specimens

thickness of the metal to be tested. Bar size also depends on the size of the tensile testing equipment available for the testing.

Two flat specimens are used, commonly for testing thinner sections of metal. When testing welds, the specimen should include the heat-affected zone and the base plate. If the weld metal is stronger than the plate, failure occurs in the plate; if the weld is weaker, failure occurs in the weld. This test, then, is open to interpretation.

After the weld section is machined to the specified dimensions, it is placed in the tensile testing machine and pulled apart. A specimen used to determine the strength of a welded butt joint for plate is shown in **Figure 23-21**.

The tensile strength, in pounds per square inch, is obtained by dividing the maximum load, required to break the specimen, by the original cross-sectional area of the specimen at the middle. The cross-sectional area is obtained by using either formula $A = \pi r^2$ or $A = D^2 \times 0.785$ (r^2 is the same as multiplying the radius times itself; D^2 is the same as multiplying the diameter times itself). Using either formula will give you the same answer.

The elongation is found by fitting the fractured ends of the specimen together, measuring the distance between gauge marks, and subtracting the gauge length. The percent

of elongation is found by dividing the elongation by the gauge length and multiplying by 100.

$$E_1 = \frac{L_f - L_o}{L_o} \times 100$$

Where:
  E_1 = % of elongation
  L_f = final gauge length
  L_o = original gauge length

Fatigue Testing

Fatigue testing is used to determine how well a weld can resist repeated fluctuating stresses or cyclic loading. The maximum value of the stresses is less than the tensile strength of the material. Fatigue strength can be lowered by improperly made weld deposits, which may be caused by porosity, slag inclusions, lack of penetration, or cracks. Any one of these discontinuities can act as a point of stress, eventually resulting in the failure of the weld.

In the fatigue test, the part is subjected to repeated changes in applied stress. This test may be performed in one of several ways, depending upon the type of service the tested part must withstand. The results obtained are usually reported as the number of stress cycles that the part will resist without failure and the total stress used.

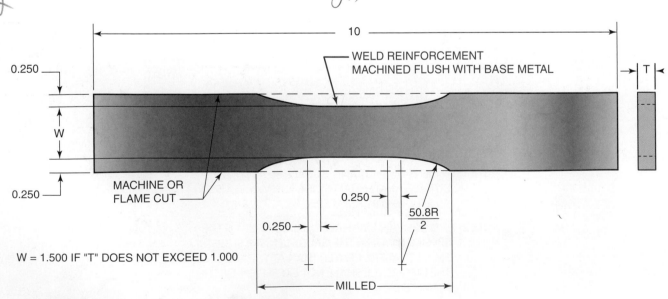

FIGURE 23-21 Tensile specimen for flat plate weld. © Cengage Learning 2012

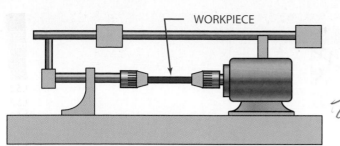

FIGURE 23-22 Fatigue testing. The specimen is placed in chucks of the machine. The machine is turned on, and as it rotates, the specimen is alternately bent twice for each revolution. © Cengage Learning 2012

In one type of test, the specimen is bent back and forth. This test subjects the part to alternating compression and tension. A fatigue testing machine is used for this test, **Figure 23-22**. The machine is turned on, and, as it rotates, the specimen is alternately bent twice for each revolution. In this case, failure is usually rapid.

Shearing Strength of Welds

The two forms of **shearing strength** of welds are transverse shearing strength and longitudinal shearing strength. To test transverse shearing strength, a specimen is prepared as shown in **Figure 23-23**. The width of the specimen is measured in inches or millimeters. A tensile load is applied, and the specimen is ruptured. The maximum load in pounds or kilograms is then determined.

The shearing strength of the weld, in pounds per linear inch, is obtained by dividing the maximum force by twice the width of the specimen.

$$\text{Shearing strength lb/in. (kg/mm)} = \frac{\text{maximum force}}{2(\text{width of specimen})}$$

To test longitudinal shearing strength, a specimen is prepared as shown in **Figure 23-24**. The length of each weld is measured in inches or millimeters. The specimen is then ruptured under a tensile load, and the maximum force in pounds or kilograms is determined.

The shearing strength of the weld in pounds per linear inch or kilograms/millimeter is obtained by dividing the maximum force by the sum of the length of welds that ruptured.

$$\text{Shearing strength lb/in. (kg/mm)} = \frac{\text{maximum force}}{\text{length of ruptured weld}}$$

Welded Butt Joints

The three methods of testing welded butt joints are (1) the nick-break test, (2) the guided-bend test, and (3) the free-bend test. It is possible to use variations of these tests.

Nick-Break Test A specimen for this test is prepared as shown in **Figure 23-25A**. The specimen is supported as shown in **Figure 23-25B**. A force is then applied, and the specimen is ruptured by one or more blows of a hammer.

CONVERSION TABLE – MILLIMETERS TO INCHES

DIM-mm	TOL	DIM-in.
9.52		0.375
9.52	±1.58	0.375
12.70		0.500
19.05		0.750
50.60		2.000
63.50		2.500
228.60		9.000
114.30		4.500

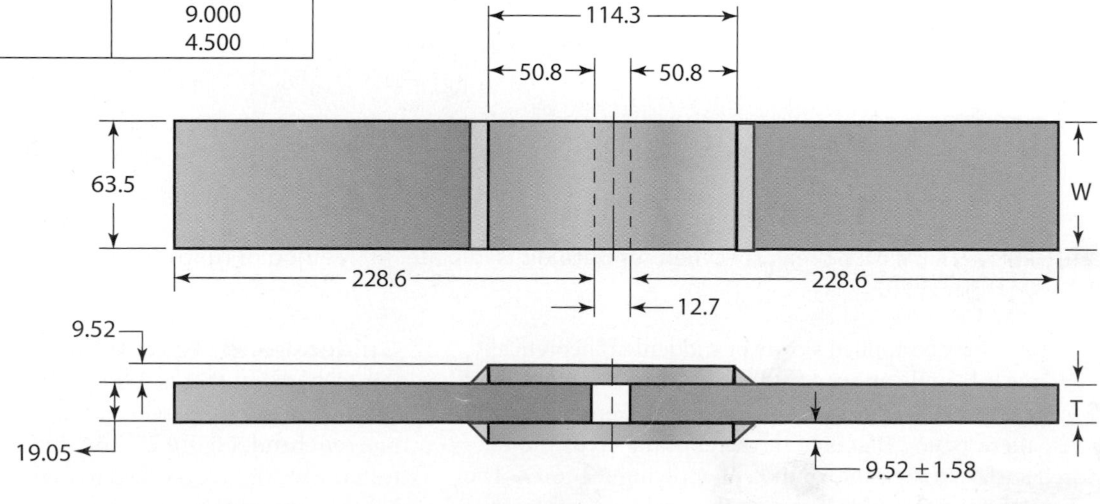

METRIC

FIGURE 23-23 Transverse fillet weld shearing specimen after welding. © Cengage Learning 2012

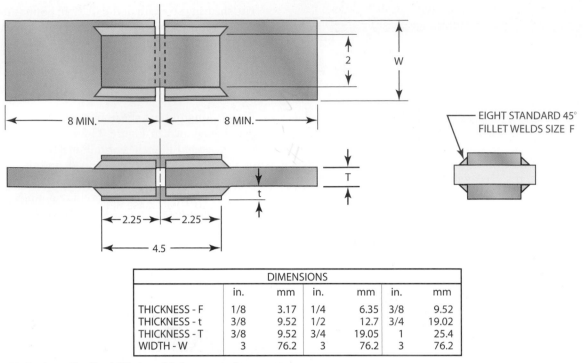

DIMENSIONS						
	in.	mm	in.	mm	in.	mm
THICKNESS - F	1/8	3.17	1/4	6.35	3/8	9.52
THICKNESS - t	3/8	9.52	1/2	12.7	3/4	19.02
THICKNESS - T	3/8	9.52	3/4	19.05	1	25.4
WIDTH - W	3	76.2	3	76.2	3	76.2

FIGURE 23-24 Longitudinal fillet weld shear specimen. © Cengage Learning 2012

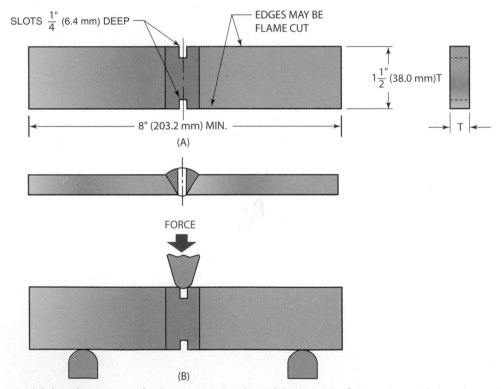

FIGURE 23-25 (A) Nick-break specimen for butt joints in plate. (B) Method of rupturing nick-break specimen.
© Cengage Learning 2012

The force may be applied slowly or suddenly. Theoretically, the rate of application could affect how the specimen breaks, especially at a critical temperature. Generally, however, there is no difference in the appearance of the fractured surface due to the method of applying the force. The surfaces of the fracture should be checked for soundness of the weld.

Guided-Bend Test To test welded, grooved butt joints on metal that is 3/8 in. (10 mm) thick or less, two specimens are prepared and tested—one face bend and one root bend, **Figure 23-26A** and **B**. If the welds pass this test, the welder is qualified to make groove welds on plate having a thickness range from 3/8 in. to 3/4 in. (10 mm to 19 mm). These welds need to be machined as shown in

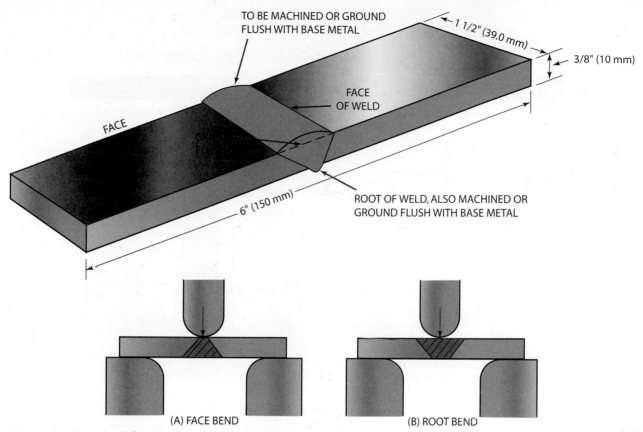

TO BE MACHINED OR GROUND
FLUSH WITH BASE METAL

1 1/2" (39.0 mm)

3/8" (10 mm)

FACE
OF WELD

FACE

6" (150 mm)

ROOT OF WELD, ALSO MACHINED OR
GROUND FLUSH WITH BASE METAL

(A) FACE BEND

(B) ROOT BEND

FIGURE 23-26 Root- and face-bend specimens for 3/8-in. (10-mm) plate. © Cengage Learning 2012

Figure 23-26A. If these specimens pass, the welder will also be qualified to make fillet welds on materials of any (unlimited) thicknesses. For welded, grooved butt joints on metal 1/2 in. (13 mm) thick, two side-bend specimens are prepared and tested, **Figure 23-27B**. If the welds pass

this test, the welder is qualified to weld on metals of unlimited thickness.

When the specimens are prepared, caution must be taken to ensure that all grinding marks run longitudinally to the specimen so that they do not cause stress cracking.

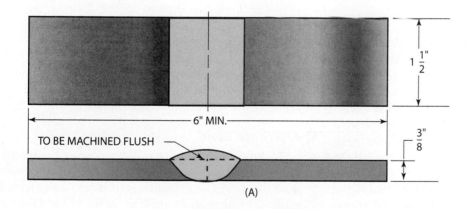

1 1/2"

6" MIN.

3"
8

TO BE MACHINED FLUSH

(A)

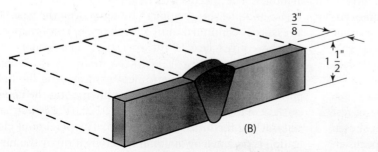

3"
8

1 1/2"

(B)

mm	CONVERSION in.
1.5	1/16
3.1	1/8
9.5	3/8
12.7	1/2
38.0	1 1/2
152.0	6

FIGURE 23-27 (A) Root- and face-bend specimens. (B) Side-bend specimen. © Cengage Learning 2012

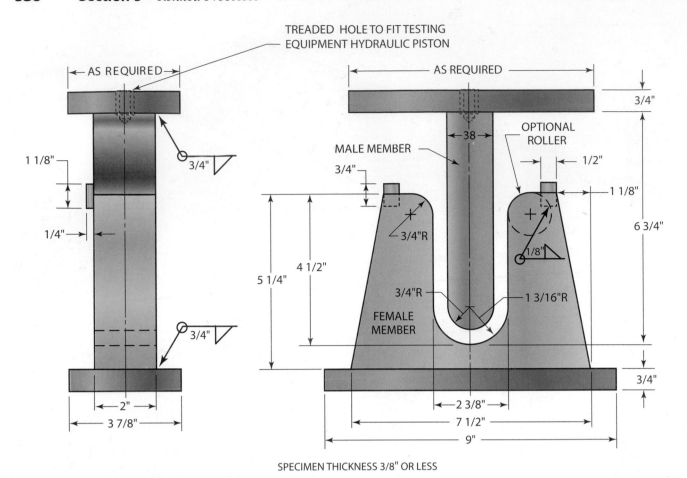

SPECIMEN THICKNESS 3/8" OR LESS

FIGURE 23-28 Fixture for guided-bend test. © Cengage Learning 2012

In addition, the edges must be rounded to reduce cracking that tends to radiate from sharp edges. The maximum ratio of this rounded edge is 1/8 in. (3 mm).

Procedure

The jig shown in **Figure 23-28** is commonly used to bend most specimens. Not all guided-bend testers have the same bending radius. Codes specify different bending radii depending on material type and thickness. Place the specimens in the jig with the weld in the middle. Face-bend specimens should be placed with the face of the weld toward the gap. Root-bend specimens should be positioned so that the root of the weld is directed toward the gap. Side-bend specimens are placed with either side facing up. The guided-bend specimen must be pushed all the way through open (roller-type) bend testers and within 1/8 in. (3 mm) of the bottom on fixture-type bend testers.

Once the test is completed, the specimen is removed. The convex surface is then examined for cracks or other discontinuities and judged acceptable or unacceptable according to specified criteria. Some surface cracks and openings are allowable under codes.

Free-Bend Test The free-bend test is used to test welded joints in plate. A specimen is prepared as shown in **Figure 23-29**. Note that the width of the specimen is 1.5 multiplied by the thickness of the specimen. Each

corner lengthwise should be rounded in a radius not exceeding one-tenth the thickness of the specimen. Tool marks should run the length of the specimen.

Gauge lines are drawn on the face of the weld, **Figure 23-30**. The distance between the gauge lines is 1/8 in. (3.17 mm) less than the face of the weld. The initial bend of the specimen is completed in the device illustrated in **Figure 23-31**. The gauge line surface should be directed toward the supports. The weld is located in the center of the supports and loading block.

Alternate Bend

The initial bend may be made by placing the specimen in the jaws of a vise with one-third the length projecting from the jaws. The specimen is then bent away from the gauge lines through an angle of 30° to 45° by blows of a hammer. The specimen is then inserted into the jaws of a vise and pressure is applied by tightening the vise. The pressure is continued until a crack or depression appears on the convex face of the specimen. The load is then removed.

The elongation is determined by measuring the minimum distance between the gauge lines along the convex surface of the weld to the nearest 0.01 in. (0.254 mm) and subtracting the initial gauge length. The percent of elongation is obtained by dividing the elongation by the initial gauge length and multiplying by 100.

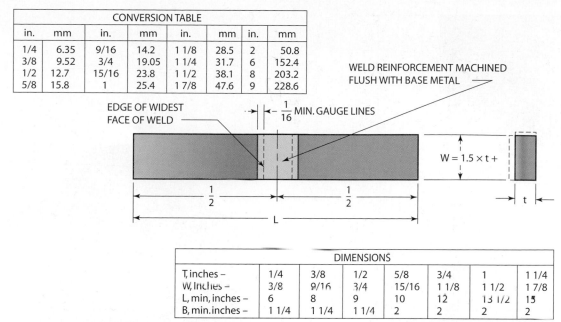

CONVERSION TABLE							
in.	mm	in.	mm	in.	mm	in.	mm
1/4	6.35	9/16	14.2	1 1/8	28.5	2	50.8
3/8	9.52	3/4	19.05	1 1/4	31.7	6	152.4
1/2	12.7	15/16	23.8	1 1/2	38.1	8	203.2
5/8	15.8	1	25.4	1 7/8	47.6	9	228.6

DIMENSIONS							
T, inches –	1/4	3/8	1/2	5/8	3/4	1	1 1/4
W, inches –	3/8	9/16	3/4	15/16	1 1/8	1 1/2	1 7/8
L, min, inches –	6	8	9	10	12	13 1/2	15
B, min. inches –	1 1/4	1 1/4	1 1/4	2	2	2	2

FIGURE 23-29 Free-bend test specimen. © Cengage Learning 2012

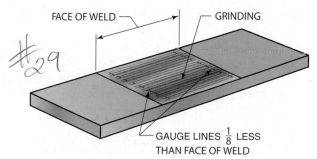

FIGURE 23-30 Gauge lines are drawn on the weld face of a free-bend specimen. © Cengage Learning 2012

Fillet Weld Break Test

The specimen for this test is made as shown in **Figure 23-32A**. In **Figure 23-32B**, a force is applied to the specimen until the specimen ruptures. Any convenient means of applying the force may be used, such as an arbor press, a testing machine, or hammer blows. The break surface should then be examined for soundness—that is, slag inclusions, overlap, porosity, lack of fusion, or other discontinuities.

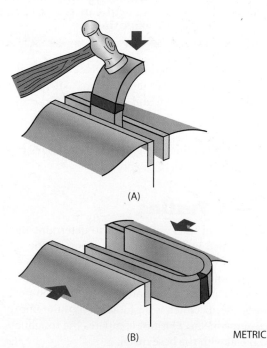

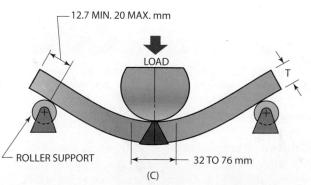

CONVERSION TABLE	
mm	in.
12.7	0.500
20.0	0.787
32.0	1.25
76.0	3.000

FIGURE 23-31 Free-bend test: (A) Initial bend can be made in this manner; (B) a vise can be used to make the final bend; and (C) another method used to make the bend. © Cengage Learning 2012

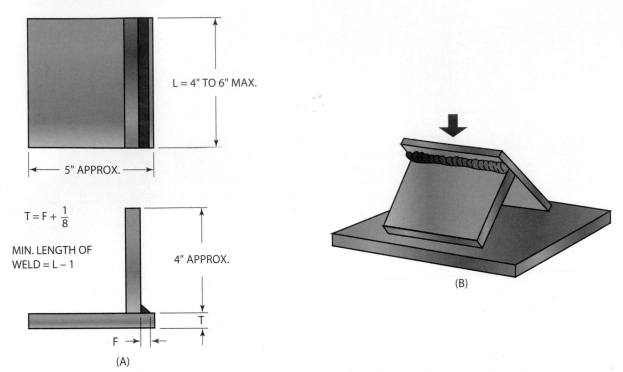

L = 4" TO 6" MAX.

5" APPROX.

$T = F + \dfrac{1}{8}$

MIN. LENGTH OF
WELD = L − 1

4" APPROX.

T

F

(A)

(B)

FIGURE 23-32 (A) Fillet weld break test, and (B) method of rupturing fillet weld break specimen. © Cengage Learning 2012

Testing by Etching

Specimens to be tested by **etching** are etched for two purposes: (1) to determine the soundness of a weld or (2) to determine the location of a weld.

A test specimen is produced by cutting a portion from the welded joint so that a complete cross section is obtained. The face of the cut is then filed and polished with fine abrasive cloth. The specimen can then be placed in the etching solution. The etching solution or reagent makes the boundary between the weld metal and base metal visible, if the boundary is not already distinctly visible.

The most commonly used etching solutions are hydrochloric acid, ammonium persulphate, and nitric acid.

Hydrochloric Acid Equal parts by volume of concentrated hydrochloric (muriatic) acid and water are mixed. The welds are immersed in the reagent at or near the boiling temperature. The acid usually enlarges gas pockets and dissolves slag inclusions, enlarging the resulting cavities.

///CAUTION

When mixing the muriatic acid into the water, be sure safety glasses and gloves are worn to prevent any injuries from occurring.

Ammonium Persulphate A solution is prepared consisting of one part of ammonium persulphate (solid) to nine parts of water by weight. The surface of the weld is rubbed with cotton saturated with this reagent at room temperature.

Nitric Acid A great deal of care should be exercised when using nitric acid because severe burns can result if it is used carelessly. One part of concentrated nitric acid is mixed with nine parts of water by volume.

///CAUTION

When diluting an acid, always pour the acid slowly into the water while continuously stirring the water. Carelessly handling this material or pouring water into the acid can result in burns, excessive fuming, or explosion.

The reagent is applied to the surface of the weld with a glass stirring rod at room temperature. Nitric acid has the capacity to etch rapidly and should be used on polished surfaces only.

After etching, the weld is rinsed in clear, hot water. Excess water is removed, and the etched surface is then immersed in ethyl alcohol and dried.

Impact Testing

A number of tests can be used to determine the impact capability of a weld. One common test is the Izod test, **Figure 23-33A**, in which a notched specimen is struck by an anvil mounted on a pendulum. The energy in foot-pounds read on the scale mounted on the machine required to break the specimen is an indication of the impact resistance of the metal. This test compares the toughness of the weld metal with the base metal.

Another type of impact test is the Charpy test. This test is similar to the Izod test. The major differences between the

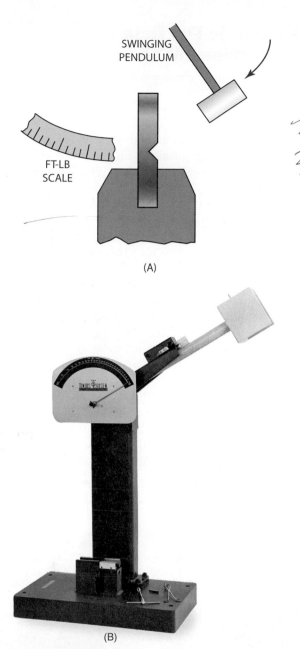

FIGURE 23-33 Impact testing: (A) specimen mounted for Izod impact toughness, and (B) a typical impact tester used for measuring the toughness of metals.
Tinius Olsen Testing Machine Co., Inc.

tests are that the Izod test specimen is gripped on one end and is held vertically and usually tested at room temperature and the Charpy test specimen is held horizontally, supported on both ends, and is usually tested at a specific temperature. All impact test specimens must be produced according to ASTM specifications. A typical impact tester is shown in **Figure 23-33B**.

NONDESTRUCTIVE TESTING (NDT)

Nondestructive testing of welds is a method used to test materials for surface defects such as cracks, arc strikes, undercuts, and lack of penetration. Internal or subsurface defects can include slag inclusions, porosity, and unfused metal in the interior of the weld.

Visual Inspection (VT)

Visual inspection is the most frequently used nondestructive testing method and is the first step in almost every other inspection process. The majority of welds receive only visual inspection. In this method, if the weld looks good it passes; if it looks bad it is rejected. This procedure is often overlooked when more sophisticated nondestructive testing methods are used. However, it should not be overlooked.

An active visual inspection schedule can reduce the finished weld rejection rate by more than 75%. Visual inspection can easily be used to check for fit up, interpass acceptance, welder technique, and other variables that will affect the weld quality. Minor problems can be identified and corrected before a weld is completed. This eliminates costly repairs or rejection.

Visual inspection should be used before any other nondestructive or mechanical tests are used to eliminate (reject) the obvious problem welds. Eliminating welds that have excessive surface discontinuities that will not pass the code or standards being used saves preparation time.

The AWS has set a number of visual inspection standards for entry-level welders to meet or exceed. To pass Level I certification standards, each weld and/or weld pass is to be inspected visually using the following criteria:

- There shall be no cracks or incomplete fusion.
- There shall be no incomplete joint penetration in groove welds except as permitted for partial joint penetration groove welds.
- The Test Supervisor shall examine the weld for acceptable appearance and shall be satisfied that the welder is skilled in the process and procedure specified for the test.
- Undercut shall not exceed the lesser of 10% of the base metal thickness or 1/32 in. (0.8 mm).
- Where visual examination is the only criterion for acceptance, all weld passes are subject to visual examination at the direction of the Test Supervisor.
- The frequency of porosity shall not exceed one in each 4 in. (100 mm) of weld length, and the maximum diameter shall not exceed 3/32 in. (2.4 mm).
- Welds shall be free from overlap.*

Penetrant Inspection (PT)

Penetrant inspection is used to locate minute surface cracks and porosity. Two types of penetrants are now in use, the color-contrast and the fluorescent versions.

---

*American Welding Society AWS QC10, *Specification for Qualification and Certification for Entry Level Welders.*

Color-contrast, often red, penetrants contain a colored dye that shows under ordinary white light. Fluorescent penetrants contain a more effective fluorescent dye that shows under black light.

The following steps outline the procedure to be followed when using a penetrant.

1. Precleaning. The test surface must be clean and dry. Suspected flaws must be cleaned and dried so that they are free of oil, water, or other contaminants.

2. The test surface must be covered with a film of penetrant by dipping, immersing, spraying, or brushing.

3. The test surface is then gently wiped, washed, or rinsed free of excess penetrant. It is dried with cloths or hot air.

4. A developing powder applied to the test surface acts as a blotter to speed the tendency of the penetrant to seep out of any flaws open to the test surface.

5. Depending upon the type of penetrant applied, visual inspection is made under ordinary white light or near-ultraviolet black light when viewed under this light, **Figure 23-34**. The penetrant fluoresces to a yellow-green color, which clearly defines the defect.

Magnetic Particle Inspection (MT)

Magnetic particle inspection uses finely divided ferromagnetic particles (powder) to indicate defects open to the surface or just below the surface on magnetic materials.

A magnetic field is induced in a part by passing an electric current through or around it. The magnetic field is always at right angles to the direction of current flow. Ferromagnetic powder registers an abrupt change in the resistance in the path of the magnetic field, such as would be caused by a crack lying at an angle to the direction of the magnetic poles at the crack. Finely divided ferromagnetic particles applied to the area will be attracted and outline the crack.

In **Figure 23-35**, the flow or discontinuity interrupting the magnetic field in a test part can be either longitudinal or circumferential. A different type of magnetization is used to detect defects that run down the axis, as opposed to those occurring around the girth of a part. For some applications you may need to test in both directions.

In **Figure 23-36A**, longitudinal magnetization allows detection of flaws running around the circumference of a part. The user places the test part inside an electrified coil. This induces a magnetic field down the length of

1. PRECLEAN INSPECTION AREA. SPRAY ON CLEANER/REMOVER –WIPE OFF WITH CLOTH.

2. APPLY PENETRANT, ALLOW SHORT PENETRATION PERIOD.

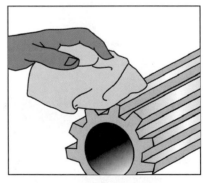

3. SPRAY CLEANER/REMOVER ON WIPING TOWEL AND WIPE SURFACE CLEAN.

4. SHAKE DEVELOPER CAN AND SPRAY ON A THICK, UNIFORM FILM OF DEVELOPER.

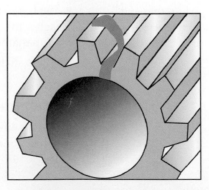

5. INSPECT. DEFECTS WILL SHOW AS BRIGHT RED LINES IN WHITE DEVELOPER BACKGROUND.

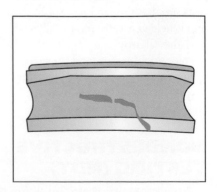

FIGURE 23-34 Penetrant testing. Adapted from Magnaflux Corporation

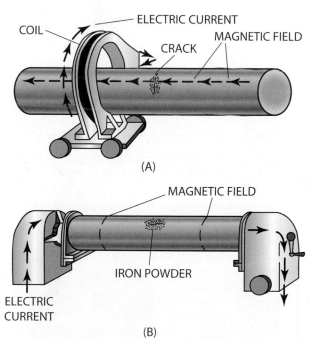

PARTICLES CLING TO THE DEFECT LIKE TACKS TO A SIMPLE MAGNET.

N SPECIMEN DEFECT S

MAGNETIC FIELD

BY INDUCING A MAGNETIC FIELD WITHIN THE PART TO BE TESTED, AND APPLYING A COATING OF MAGNETIC PARTICLES, SURFACE CRACKS ARE MADE VISIBLE, THE CRACKS IN EFFECT FORMING NEW MAGNETIC POLES.

FIGURE 23-35 Flaws and discontinuities interrupt magnetic fields. Adapted from Magnaflux Corporation

COIL
ELECTRIC CURRENT
MAGNETIC FIELD
CRACK

(A)

MAGNETIC FIELD
IRON POWDER
ELECTRIC CURRENT

(B)

FIGURE 23-36 (A) Longitudinal magnetic field. (B) Circumferential magnetic field.
Adapted from Magnaflux Corporation

the test part. In **Figure 23-36B**, circumferential magnetization allows detection of flaws occurring down the length of a test part. An electric current is sent down the length of the part to be inspected. The magnetic field thus induced allows defects along the length of the part to be detected.

Radiographic Inspection (RT)

Radiographic inspection (RT) is a method for detecting flaws inside weldments. Radiography gives a picture of all discontinuities that are parallel (vertical) or nearly parallel to the source. Discontinuities that are perpendicular (flat) or nearly perpendicular to the source may not be seen on the X-ray film. Instead of using visible light rays, the operator uses invisible, short-wavelength rays developed by X-ray machines, radioactive isotopes (gamma rays), and variations of these methods. These rays are capable of penetrating solid materials and reveal most flaws in a weldment on an X-ray film or a fluorescent screen. Flaws are revealed on films as dark or light areas against a contrasting background after exposure and processing, **Figure 23-37**.

The defect images in radiographs measure differences in how the X-rays are absorbed as they penetrate the weld. The weld itself absorbs most X-rays. If something less dense than the weld is present, such as a pore or lack of fusion defect, fewer X-rays are absorbed, darkening the film. If something more dense is present, such as heavy ripples on the weld surface, more X-rays will be absorbed, lightening the film.

Therefore, the foreign material's relative thickness (or lack of it) and differences in X-ray absorption determine the radiograph image's final shape and shading. Skilled readers of radiographs can interpret the significance of the light and dark regions by their shape and shading. The X-ray image is a shadow of the flaw. The farther the flaw is from the X-ray film, the fuzzier the image appears. Those skilled at interpreting weld defects in radiographs must also be very knowledgeable about welding.

Figure 23-38 shows samples of common weld defects and a representative radiograph for each.

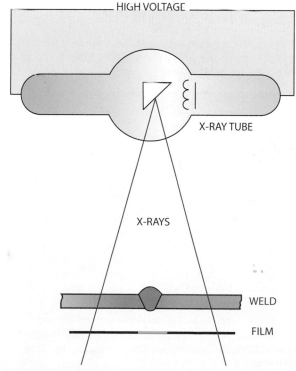

HIGH VOLTAGE

X-RAY TUBE

X-RAYS

WELD

FILM

FIGURE 23-37 Schematic of an X-ray system.
© Cengage Learning 2012

WELDING DEFECT: Lack of sidewall fusion (LOF). Elongated voids between the weld beads and the joint surfaces to be welded.

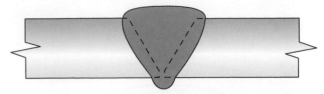

WELDING DEFECT: Excessive penetration (icicles, drop-thru). Extra metal at the bottom (root) of the weld.

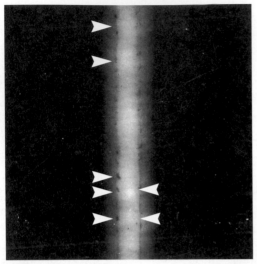

RADIOGRAPHIC IMAGE: Elongated parallel, or single, darker density lines sometimes with darker density spots dispersed along the LOF lines which are very straight in the lengthwise direction and not winding like elongated slag lines. Although one edge of the LOF lines may be very straight like LOP, lack of sidewall fusion images will not be in the center of the width of the weld image.

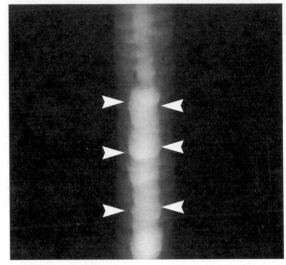

RADIOGRAPHIC IMAGE: A lighter density in the center of the width of the weld image either extended along the weld or in isolated circular "drops."

WELDING DEFECT: External concavity or insufficient fill. A depression in the top of the weld, or cover pass, indicating a thinner more normal section thickness.

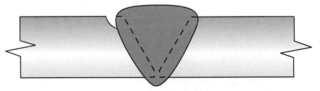

WELDING DEFECT: External undercut. A gouging out of the piece to be welded, alongside the edge of the top or "external" surface of the weld.

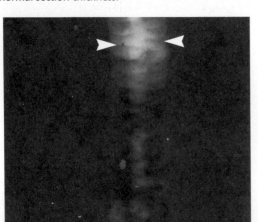

RADIOGRAPHIC IMAGE: A weld density darker than the density of the pieces being welded and extending across the full width of the weld image.

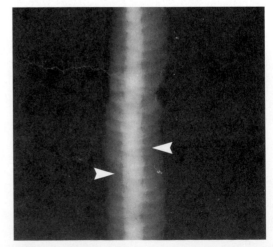

RADIOGRAPHIC IMAGE: An irregular darker density along the edge of the weld image. The density will always be darker than the density of the pieces being welded.

FIGURE 23-38 Welding defects with radiographic images. Reprinted with permission of E. I. DuPont de Nemours & Co., Inc. *(continued)*

WELDING DEFECT: Internal (root) undercut. A gouging out of the parent metal, alongside the edge of the bottom or "internal" surface of the weld.

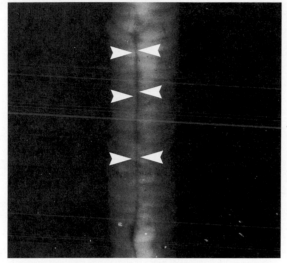

RADIOGRAPHIC IMAGE: An irregular darker density near the center of the width of the weld image and along the edge of the root pass image.

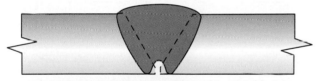

WELDING DEFECT: Internal concavity (suck back). A depression in the center of the surface of the root pass.

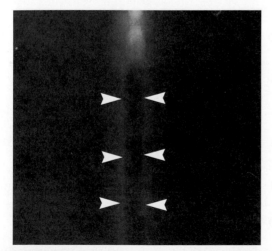

RADIOGRAPHIC IMAGE: An elongated, irregular darker density with fuzzy edges in the center of the width of the weld image.

WELDING DEFECT: Incomplete or lack of penetration (LOP). The edges of the pieces have not been welded together, usually at the bottom of single V-groove welds.

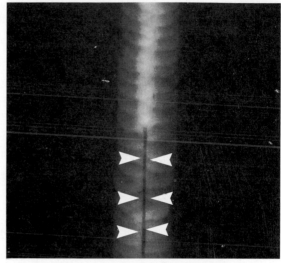

RADIOGRAPHIC IMAGE: A darker density band, with very straight parallel edges, in the center of the width of the weld image.

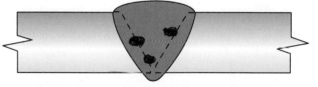

WELDING DEFECT: Interpass slag inclusions. Usually nonmetallic impurities that solidified on the weld surface and were not removed between weld passes.

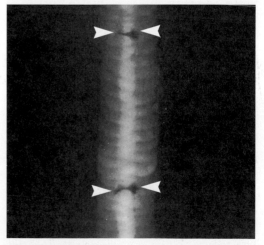

RADIOGRAPHIC IMAGE: An irregularly shaped darker density spot, usually slightly elongated and randomly spaced.

FIGURE 23-38 Welding defects with radiographic images. Reprinted with permission of E. I. DuPont de Nemours & Co., Inc. *(continued)*

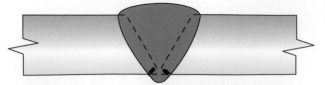

WELDING DEFECT: Elongated slag lines (wagon tracks). Impurities that solidified on the surface after welding and were not removed between passes.

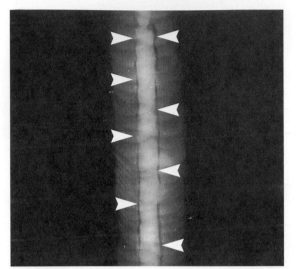

RADIOGRAPHIC IMAGE: Elongated, parallel, or single darker density lines, irregular in width and slightly winding in the length-wise direction.

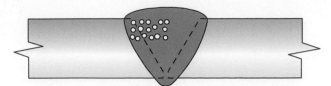

WELDING DEFECT: Cluster porosity. Rounded or slightly elongated voids grouped together.

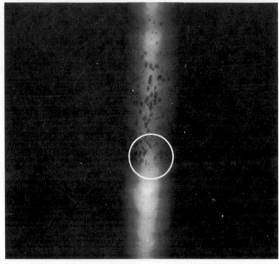

RADIOGRAPHIC IMAGE: Rounded or slightly elongated darker density spots in clusters with the clusters randomly spaced.

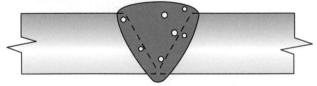

WELDING DEFECT: Scattered porosity. Rounded voids random in size and location.

RADIOGRAPHIC IMAGE: Rounded spots of darker densities random in size and location.

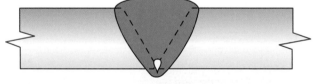

WELDING DEFECT: Root pass aligned porosity. Rounded and elongated voids in the bottom of the weld aligned along the weld centerline.

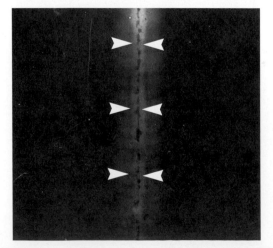

RADIOGRAPHIC IMAGE: Rounded and elongated darker density spots, which may be connected, in a straight line in the center of the width of the weld image.

FIGURE 23-38 Welding defects with radiographic images. Reprinted with permission of E. I. DuPont de Nemours & Co., Inc. *(continued)*

WELDING DEFECT: Transverse crack. A fracture in the weld metal running across the weld.

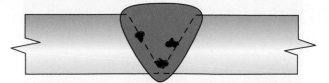

WELDING DEFECT: Tungsten inclusions. Random bits of tungsten fused, but not melted, into the weld metal.

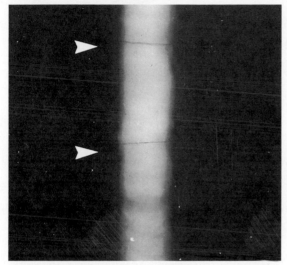

RADIOGRAPHIC IMAGE: Feathery, twisting line of darker density running across the width of the weld image.

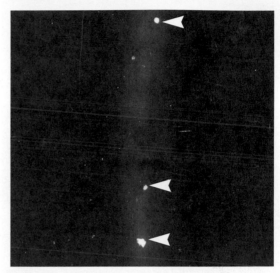

RADIOGRAPHIC IMAGE: Irregularly shaped lower density spots randomly located in the weld image.

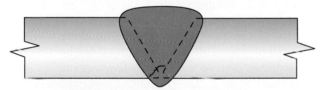

WELDING DEFECT: Longitudinal root crack. A fracture in the weld metal at the edge of the root pass.

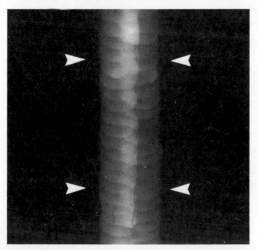

RADIOGRAPHIC IMAGE: Feathery, twisting lines of darker density along the edges of the image of the root pass. The "twisting" feature helps to distinguish the root crack from incomplete root penetration.

FIGURE 23-38 Welding defects with radiographic images. Reprinted with permission of E. I. DuPont de Nemours & Co., Inc.

FIGURE 23-39 New mobile X-Ray equipment. CMOS X-ray

FIGURE 23-40 Preparing to test the quality of a weld on a pipe using X-ray equipment. CMOS X-Ray

Four factors affect the selection of the radiation source:

- Thickness and density of the material
- Absorption characteristics
- Time available for inspection
- Location of the weld

Portable equipment is available for examining fixed or hard-to-move objects. The selection of the correct equipment for a particular application is determined by specific voltage required, degree of utility, economics of inspection, and production rates expected, **Figure 23-39** and **Figure 23-40**.

Ultrasonic Inspection (UT)

Ultrasonic inspection (UT) is fast and uses few consumable supplies, which makes it inexpensive for schools to use. However, because of the time required for most UT testing, it is not as economical in the field as the nondestructive testing method. This inspection method employs electronically produced high-frequency sound

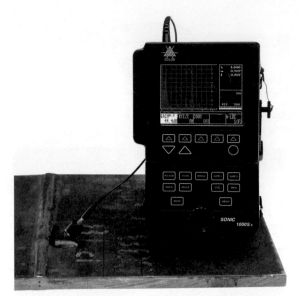

FIGURE 23-41 Portable ultrasonic inspection unit. Olympus NDT

waves, which penetrate metals and many other materials at speeds of several thousand feet (meters) per second.

The two types of ultrasonic equipment are pulse and resonance. The pulse-echo system, most often employed in the welding field, uses sound generated in short bursts or pulses. Since the high-frequency sound used is at a relatively low power, it has little ability to travel through air, so it must be conducted from the probe into the part through a medium such as oil or water. A portable ultrasonic inspection unit is shown in **Figure 23-41**.

Sound is directed into the part with a probe held in a preselected angle or direction so that flaws will reflect some energy back to the probe. These ultrasonic devices operate very much like depth sounders, or "fish finders." The speed of sound through a material is a known quantity. These devices measure the time required for a pulse to return from a reflective surface. Internal computers calculate the distance and present the information on a display screen where an operator can interpret the results. The signals can be "monitored" electronically to operate alarms, print systems, or recording equipment. Sound not reflected by flaws continues into the part. If the angle is correct, the sound energy will be reflected back to the probe from the opposite side. Flaw size is determined by plotting the length, height, width, and shape using trigonometric rules.

Figure 23-42 shows the path of the sound beam in butt welding testing. The operator must know the exit point of the sound beam, the exact angle of the refracted beam,

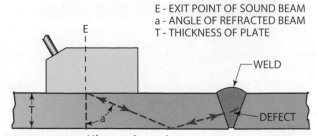

FIGURE 23-42 Ultrasonic testing. © Cengage Learning 2012

and the thickness of the plate when using shear wave-forms and compression waveforms.

Leak Checking

Leak checking can be performed by filling the welded container with either a gas or liquid. There may or may not be additional pressure applied to the material in the weldment. Water is the most frequently used liquid, although sometimes a liquid with a lower viscosity is used. If gas is used, it may be either a gas that can be detected with an instrument when it escapes through a flaw in the weld or an air leak that is checked with bubbles.

Eddy Current Inspection (ET)

Eddy current inspection (ET) is another nondestructive test. This method is based upon magnetic induction in which a magnetic field induces eddy currents within the material being tested. An eddy current is an induced electric current circulating wholly within a mass of metal. This method is effective in testing nonferrous and ferrous materials for internal and external cracks, slag inclusions, porosity, and lack of fusion that are on or very near the surface. Eddy current cannot locate flaws that are not near the surface.

A coil-carrying high-frequency alternating current is brought close to the metal to be tested. A current is produced in the metal by induction. The part is placed in or near high-frequency alternating-current coils. The magnitude and phase difference of these currents is, in turn, indicated in the actual impedance value of the pickup coil. Careful measurement of this impedance is the revealing factor in detecting defects in the weld.

Hardness Testing

Hardness is the resistance of metal to penetration and is an index of the wear resistance and strength of the metal. Hardness tests can be used to determine the relative hardness of the weld with the base metal. The two types of hardness testing machines in common use are the Rockwell and the Brinell testers.

The **Rockwell hardness tester,** Figure 23-43, uses a 120° diamond cone for hard metals and a 1/16-in. (1.58-mm) and 1/8-in. (3.175-mm) diameter hardened steel ball for softer metals. The method is based upon resistance-to-penetration measurement. The hardness is read directly from a dial on the tester. The depth of the impression is measured instead of the diameter. The tester has two scales for reading hardness, known as the B-scale and the C-scale. The C-scale is used for harder metals, the B-scale for softer metals.

The **Brinell hardness tester** measures the resistance of material to the penetration of a steel ball under constant pressure (about 3000 kg) for a minimum of approximately 30 sec, **Figure 23-44**. The diameter is measured microscopically, and the Brinell number is checked on a standard chart. Brinell hardness numbers are obtained by dividing the applied load by the area of the surface indentation.

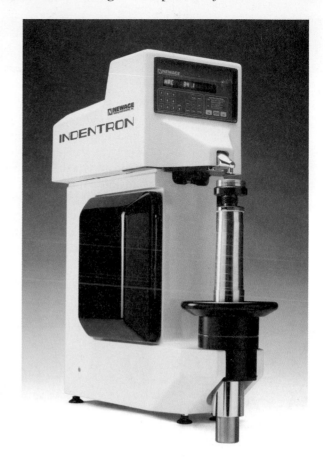

FIGURE 23-43 Rockwell hardness tester. Newage Testing Instruments, Inc.

FIGURE 23-44 Brinell hardness tester. Newage Testing Instruments, Inc.

Summary

It is impossible to "inspect in" quality; no amount of inspection will produce a quality product. Quality is something that must be built into a product. The purpose of testing and inspection is to verify that the required level of quality is being maintained. The most important standard that a weld must meet is that it must be fit for service. A weld must be able to meet the demands placed on the weldment without failure. No weld is perfect. It is important for both the welder and welding inspector to know the appropriate level of weld discontinuities. Producing or inspecting welds to an excessively high standard will result in a product that is excessively expensive.

Knowing the cause of weld defects and discontinuities can aid you in producing higher-quality welds by recognizing the cause and effect of such problems. Some welders believe that high-quality welds are a matter of luck, which is not true. Producing high-quality welds is a matter of skill and knowledge, which you must develop. Good welders often know whether welds they produce will or will not pass inspection. Inspecting your welds as part of your training program will help you develop this skill.

Development of Titanium Inspection Tools Based on Weld Color

A workmanship kit consisting of weld samples, color photographs, and an inspection guide was developed based on surface colors produced by different weld shielding conditions.

In recent years, the decreasing cost of titanium has led to a major increase in the use of this material. Its high strength-to-weight ratio and superior corrosion resistance make it attractive for many chemical, marine, and military applications. At the same time, the demanding nature of titanium fabrication has brought about the need for new and improved welding and inspection technologies to ensure the performance of the welded structure. These new technologies, combined with proper training, are needed to minimize the occurrence of weld contamination and to ensure proper assessment when contamination does occur.

Contamination of titanium weld metal by elements (oxygen, nitrogen, carbon, and hydrogen) reduces ductility and toughness while increasing strength and hardness. In some cases, weld cracking may result. Other types of contamination can also result in cracking from various mechanisms. For example, salts from fingerprints, chlorides from degreasers, or other compounds on heated surfaces in the vicinity of welds can increase stress corrosion cracking. Poor preparation and cleaning of the joint and filler materials can cause contamination before and during welding, poor shielding of the weld zone, or impurities in the shielding gas.

The weld pool is the most vulnerable to contamination, since diffusion of elements in molten titanium is very rapid. Weld color is an indicator of the degree of atmospheric contamination of the solidified weld and, sometimes, the weld pool. Other forms of contamination, such as moisture in shielding gases and contaminants on weld joint surfaces and filler metal, may not necessarily be indicated by color.

Importance of Titanium Weld Color

Weld color is the result of oxidation that can occur while welding titanium and titanium alloys. Titanium is a reactive material and will easily combine with oxygen. This is part of the reason that titanium oxidation or scaling is common at high temperatures. Rapid oxidation in titanium begins at temperatures above 840°F (450°C). As the temperature is increased, the oxidation rate becomes more rapid because oxygen from the titanium oxide that is formed on the surface of the metal is quickly dissolved

Color Inspection
Standard for Titanium

GLOSSAY SILVER

ACCEPTABLE

Weld Color Inspection
Standard for Titanium

LIGHT STRAW

ACCEPTABLE

Weld Color Inspection
Standard for Titanium

DARK STRAW

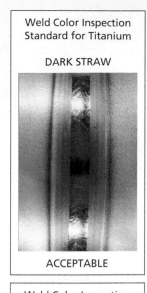

ACCEPTABLE

Weld Color Inspection
Standard for Titanium

PURPLE

REJECTABLE

Weld Color Inspection
Standard for Titanium

BLUE

REJECTABLE

Weld Color Inspection
Standard for Titanium

YELLOW

REJECTABLE

Weld Color Inspection
Standard for Titanium

GRAY

REJECTABLE

Weld Color Inspection
Standard for Titanium

WHITE

REJECTABLE

Weld Color Inspection
Standard for Titanium

BRUSHED

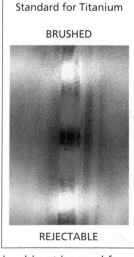

REJECTABLE

Front sides of weld inspection cards. Note: Images are for representational purposes only and should not be used for weld color inspection because of possible color distortion in the print reproduction processes. American Welding Society

into the bulk solid. This action results in a multilayer surface consisting of titanium oxide as the outermost layer. (This layer forms colors.)

Need for Weld Color Inspection Tools

There is a need for improved weld color inspecting tools because visual inspection of color can often be somewhat prejudiced. In welding, the color patterns formed from weld contamination are complex, since the thermal cycle and oxidation conditions vary across a weld from the far heat-affected zone into the fusion zone. Often a gradient of colors exists when some contamination has occurred during welding because the oxide formed under a variety of time, temperature, and environmental conditions. As a result, an inspector can be confused when interpreting the various colors, the patterns formed, and the proximity of certain colors to the weld. To clear up much of this confusion, a standard set of welds, each having a unique color, was sent out to all United States Navy shipyard personnel. These welds were made under precisely controlled laboratory conditions to ensure uniformity among all welds of a given color.

The weld samples came with a training guide, which describes the procedures of weld color inspection. One important drawback is that weld color does not tell whether the weld pool was exposed to air while in the molten state, affecting the through-thickness properties or after solidification, resulting only in surface contamination. For this reason, welding inspectors must take a conservative approach and assume the through-thickness properties have been damaged. The only reliable repair of welds with unacceptable color is complete weld replacement.

Weld Color Inspection Tools

Welding procedures were developed under carefully controlled laboratory conditions to produce specific colors that were uniform along the length of the welds and repeatable from one welded plate to another. All welds were photographed and reproduced on photographic paper to avoid color distortion. The photos were made into the laminated inspection cards shown.

Weld colors represented in the weld color samples are listed as follows in order of increasing contamination:

- Glossy silver—acceptable
- Light straw—acceptable (requires removal of color and correction of cause of contamination)
- Dark straw—acceptable (requires removal of color and correction of cause of contamination)
- Purple—rejectable
- Blue—rejectable
- Yellow—rejectable
- Gray—rejectable
- White—rejectable
- Brushed—rejectable (welds that have been wire brushed before inspection are rejectable regardless of color before brushing)

Use of the Inspection Tools

This workmanship kit was developed to ensure all welding and inspection personnel receive adequate training using identical weld samples. Each United States Navy shipyard was supplied with sets of weld samples, so that inspectors and welders could be trained using actual welds.

All welds in the workmanship kit illustrate steady-state weld color conditions—that is, the colors do not change along the length of the welds. This condition is not normally encountered in most actual welding operations, since disruptions in shielding gas flow are typically irregular. Since there could be an endless possibility of examples showing different types of contamination scenarios, the workmanship kit was designed to illustrate the colors themselves, rather than all of the possible patterns or color combinations. The main goals of the kit are to establish a basis for agreement among personnel related to titanium weld color and to remove much of the subjectivity related to color identification.

Article courtesy of the American Welding Society.

Review

1. Why are all welds not inspected to the same level or standard?

2. Why is the strength of all production parts not known if a sample number of parts is mechanically tested?

3. Why is it possible to do more than one nondestructive test on a weldment?

4. What is a discontinuity?

5. What is a defect?

6. What is tolerance?

7. What are the 12 most common discontinuities?

8. How can porosity form in a weld and not be seen by the welder?

9. What welding process can cause porosity to form?

10. How is piping porosity formed?

11. What are inclusions, and how are they caused?

12. When does inadequate joint penetration usually become defective?

13. How can a notch cause incomplete fusion?

14. How can an arc strike appear on a guided-bend test?

15. What is overlap?

16. What is undercut?

17. What causes crater cracks?

18. What is underfill?

19. What is the difference between a lamination and a delamination?

20. How can stress be reduced through a plate's thickness to reduce lamellar tearing?

21. What would be the tensile strength in pounds per square inch of a specimen measuring 0.375 in. thick and 1.0 in. wide if it failed at 27,000 lb?

22. What would be the elongation for a specimen for which the original gauge length was 2 in. and final gauge length was 2.5 in.?

23. How are the results of a stress test reported?

24. What would be the transverse shear strength per inch of weld if a specimen that was 2.5 in. wide withstood 25,000 lb?

25. What would be the longitudinal shearing strength per millimeter of a specimen that was 50.8 mm wide and 116 mm long and withstood 23,000 kg/mm?

26. What are the three methods of destructive testing of a welded butt joint?

27. How are the specimens bent for a guided-, root-, face-, and side-bend test?

28. How wide should a specimen be if the material thickness is
    a. 0.375 in.?
    b. 6.35 mm?

29. Why are guide lines drawn on the surface of a free-bend specimen?

30. What part of a fillet weld break test is examined?

31. What can happen if acids are handled carelessly?

32. What information about the weld does an impact test provide?

33. Which nondestructive test is most commonly used?

34. List the five steps to be followed when using a penetrant test.

35. What properties must metal have before it can be tested with the MT process?

36. Why will some flaws appear larger on an X-ray than they are in the weld?

37. How is the size of a flaw determined using ultrasonic inspection?

38. What is the major limitation of eddy current inspection?

39. What information does a hardness test reveal?

Chapter 24

Welder Certification

OBJECTIVES

After completing this chapter, the student should be able to

- explain welder qualification and certification.
- outline the steps required to certify a welder.
- make welds that meet a standard.
- explain the information found on a typical Welding Procedure Specification.

KEY TERMS

certified welders

entry-level welder

transverse face bend

transverse root bend

weld test

welder certification

welder performance
qualification

*Performance
of skills.*

INTRODUCTION

Welding, in most cases, is one of the few professions that requires job applicants to demonstrate their skills even if they are already certified. Some other professions—doctor, lawyer, and pilot, for example—do take a written test or require a license initially. But welders are often required to demonstrate their knowledge and their skills before being hired since welding, unlike most other occupations, requires a high degree of eye-hand coordination.

A method commonly used to test a welder's ability is the qualification or certification test. Welders who have passed such a test are referred to as qualified welders; if proper written records are kept of the test results, they are referred to as **certified welders.** Not all welding jobs require that the welder be certified. Some merely require that a basic **weld test** be passed before applicants are hired.

Welder certification can be divided into two major areas. The first area covers the traditional welder certification that has been used for years. This certification is used to demonstrate welding skills for a specific process on a specific weld, to qualify for a welding assignment, or as a requirement for employment.

The American Welding Society has developed the second, newer area of certification. This certification has three levels. The first level is primarily designed for the new welder needing to demonstrate **entry-level welder** skills. The other levels cover advanced welders and expert welders. This chapter covers the traditional certification and the AWS QC10, *Specification for Qualification and Certification for Entry-Level Welders.*

Qualified and Certified Welders

Welder qualification and welder certification are often misunderstood. Sometimes it is assumed that a qualified or certified welder can weld anything. Being certified does not mean that a welder can weld everything, nor does it mean that every weld that is made is acceptable. It means that the welder has demonstrated the skills and knowledge necessary to make good welds. To ensure that a welder is consistently making welds that meet the standard, welds are inspected and tested. The more critical the welding, the more critical the inspection and the more extensive the testing of the welds.

All welding processes can be tested for qualification and certification. The testing can range from making spot welds with an electric resistant spot welder to making electron beam welds on aircraft. Being qualified or certified in one area of welding does not mean that a welder can make quality welds in other areas. Most qualifications and certifications are restricted to a single welding process, position, metal, and thickness range.

Individual codes control test requirements. Within these codes, changes in any one of a number of essential variables can result in the need to recertify—for example,

- **Process:** Welders can be certified in each welding process such as SMAW, GMAW, FCAW, GTAW, EBW, and RSW. Therefore, a new test is required for each process.
- **Material:** The type of metal—such as steel, aluminum, stainless steel, and titanium—being welded will require a change in the certification. Even a change in the alloy within a base metal type can require a change in certification.
- **Thickness:** Each certification is valid on a specific range of thickness of base metal. This range is dependent on the thickness of the metal used in the test. For example, if a 3/8-in. (9.5-mm) plain carbon steel plate is used, then under some codes the welder would be qualified to make welds in plate thickness ranges from 3/16 in. to 3/4 in. (4.8 mm to 19 mm).
- **Filler metal:** Changes in the classification and size of the filler metal can require recertification.
- **Shielding gas:** If the process requires a shielding gas, then changes in gas type or mixture can affect the certification.
- **Position:** In most cases, a weld test taken in the flat position would limit certification to flat and possibly horizontal welding. A test taken in the vertical position, however, would usually allow the welder to work in the flat, horizontal, and vertical positions, depending on the code requirements.

- **Joint design:** Changes in weld type such as groove or fillet welds require a new certification. Additionally, variations in joint geometry, such as groove type, groove angle, and number of passes, can also require retesting.
- **Welding current:** In some cases, changing from AC to DC or changes such as to pulsed power and high frequency can affect the certification.

Welder Performance Qualification

Welder performance qualification is the demonstration of a welder's ability to produce welds meeting very specific prescribed standards. The form used to document this test is called the Welding Qualification Test Record. The detailed written instructions to be followed by the welder are called the Welder Qualification Procedure. Welders passing this certification are often referred to as being qualified welders or as qualified.

Welder Certification

A Welder Certification document is the written verification that a welder has produced welds meeting a prescribed standard of welder performance. A welder holding such a written verification is often referred to as being certified or as a certified welder.

AWS Entry-Level Welder Qualification and Welder Certification

The AWS entry-level welder qualification and certification program specifies a number of requirements not normally found in the traditional welder qualification and certification process. The additions to the AWS program have broadened the scope of the test. Areas such as practical knowledge have long been an assumed part of most certification programs but have not been a formal part of the process. Most companies have assumed that welders who could produce code-quality welds could understand enough of the technical aspects of welding.

Practical Knowledge

A written test must be passed with a minimum grade of 75% on all areas except safety. The safety questions must be answered with a minimum accuracy of 90%. The following subject areas, covered in the given chapters of this text, are included in the test:

Welding and cutting theory (Chapters 3, 7, 8, 9, 10, 12, 14, 15, 30, and 33)

Welding and cutting inspection and testing (Chapters 22 and 23)

Welding and cutting terms and definitions (Glossary)

Base and filler metal identification (Chapters 25, 26, and 27)

Common welding process variables (Chapters 4, 5, 6, 11, 13, 16, 17, and 32)

Electrical fundamentals (Chapter 3)

Drawing and welding symbol interpretation (Chapters 19 and 20)

Fabrication principles and practices (Chapter 21)

Safe practices (Chapter 2)

Refer to the specific chapters that relate to each of the required knowledge areas for the information required to pass that area of the test.

Welder Qualification and Certification Test Instructions for Practices

After you have mastered the welding and cutting knowledge and skills, you should be ready to start the required assemblies and welding and become certified. You can now take a qualification welding and cutting skills test for the AWS entry-level certification. If you have passed the knowledge and safety test and successfully pass the two bend tests and/or any one of the several workmanship assembly weldments, you can be certified. **Figure 24-1** and **Figure 24-2** are pictorial representations of the workmanship assembly weldments used in the following certifying practices.

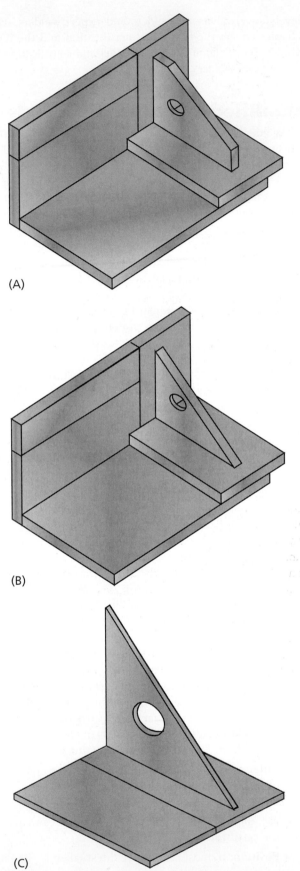

(A)

(B)

(C)

FIGURE 24-2 (A) FCAW carbon steel workmanship sample. (B) GMAW-S and GTAW carbon steel workmanship sample. (C) GMAW spray metal transfer on carbon steel and GTAW on stainless steel and aluminum workmanship samples. American Welding Society

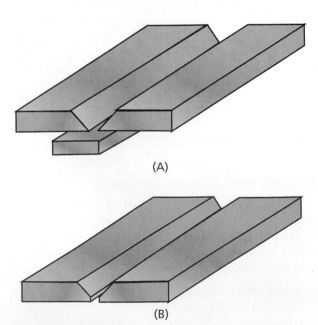

(A)

(B)

FIGURE 24-1 Carbon steel test plate (A) with backing and (B) open root (without backing). American Welding Society

Preparing Specimens for Testing

The detailed preparation of specimens for testing in this chapter is based on the structural welding code AWS D1.1 and the ASME BPV Code, Section IX. The maximum allowable size of fissures (cracks or openings) in a guided-bend test specimen is given in codes for specific applications. Some of the standards are listed in ASTM E190 or AWS B4.0, AWS QC10, AWS QC11, and others. Copies of these publications are available from the appropriate organizations. More information on tests and testing can be found in Chapter 23.

Acceptance Criteria for Face Bends and Root Bends The weld specimen must first pass visual inspection before it can be prepared for bend testing. Visual inspection looks to see that the weld is uniform in width and reinforcement. There should be no arc strikes on the plate other than those on the weld itself. The weld must be free of both incomplete fusion and cracks. The joint penetration must be either 100% or as specified by the specifications. The weld must be free of overlap, and undercut depth must not exceed either 10% of the base metal or 1/32 in. (0.8 mm), whichever is less.

Correct weld specimen preparation is essential for reliable results. The weld specimen must be uniform in width and reinforcement and have no undercut or overlap. The weld reinforcement and backing strip, if used, must be removed flush to the surface, **Figure 24-3**. They can be machined or ground off. The plate thickness after removal must be a minimum of 3/8 in. (9.5 mm), and the pipe thickness must be equal to the pipe's original wall thickness. The specimens may be cut out of the test weldment using an abrasive disc, by sawing, or by cutting with a torch. Flame-cut specimens must have the edges ground or machined smooth after cutting. This procedure is done to remove the heat-affected zone caused by the cut, **Figure 24-4**.

All corners must be rounded to a radius of 1/8 in. (3.2 mm) maximum, and all grinding or machining marks must run lengthwise on the specimen, **Figure 24-5** and **Figure 24-6**. Rounding the corners and keeping all marks running lengthwise reduce the chance of good weld specimen failure due to poor surface preparation.

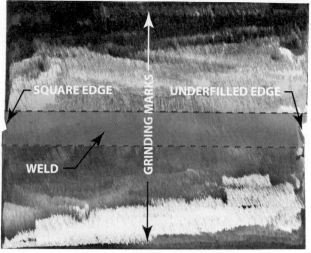

FIGURE 24-3 Plate ground in preparation for removing test specimens. Larry Jeffus

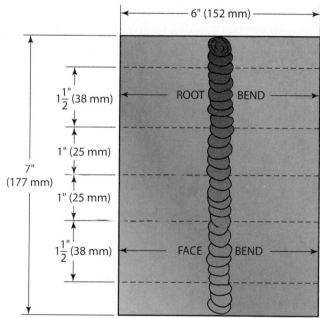

FIGURE 24-4 Sequence for removing guided-bend specimens from the plate once welding is complete. American Welding Society

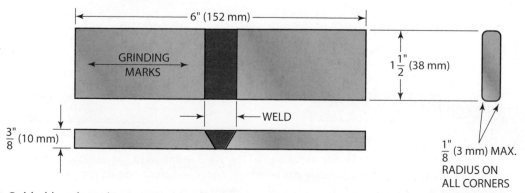

FIGURE 24-5 Guided-bend specimen. © Cengage Learning 2012

FIGURE 24-6 Guided-bend test specimen. Larry Jeffus

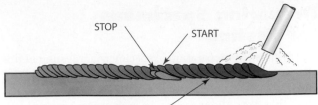

FIGURE 24-7 Tapering the size of the weld bead helps keep the depth of penetration uniform. © Cengage Learning 2012

The weld must pass both the face and root bends to be acceptable. After bending there can be no single defects larger than 1/8 in. (3.2 mm), and the sum of all defects larger than 1/32 in. (0.8 mm) but less than 1/8 in. (3.2 mm) must not exceed a total of 3/8 in. (9.5 mm) for each bend specimen. An exception is made for cracks that start at the edge of the specimen and do not start at a defect in the specimen.

Restarting a Weld Bead

On all but short welds, the welding bead will need to be restarted after a welder stops to change electrodes. Because the metal cools as a welder changes electrodes and chips slag when restarting, the penetration and buildup may be adversely affected.

When a weld bead is nearing completion, it should be tapered so that when it is restarted the buildup will be more uniform. To taper a weld bead, the travel rate should be increased just before welding stops. A 1/4-in. (6.4-mm) taper is all that is required. The taper allows the new weld to be started and the depth of penetration reestablished without having excessive buildup, **Figure 24-7**.

The slag should always be chipped and the weld crater should be cleaned each time before restarting the weld. This is important to prevent slag inclusions at the start of the weld.

The arc should be restarted in the joint ahead of the weld. The electrodes must be allowed to heat up so that the arc is stabilized and a shielding gas cloud is reestablished to protect the weld. To prevent weld metal from being deposited as the electrode is heated up and the arc stabilizes, hold a longer arc for a second or two. Slowly bring the electrode downward and toward the weld bead until the arc is directly on the deepest part of the crater where the crater meets the plate in the joint, **Figure 24-8**.

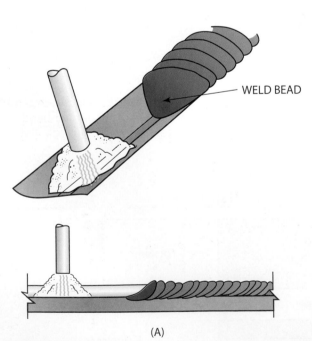

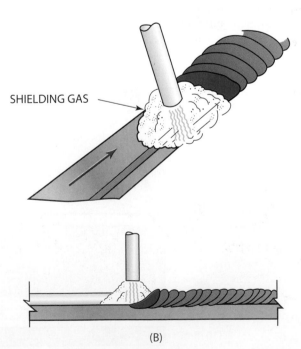

(A)   (B)

FIGURE 24-8 When restarting the arc, strike the arc ahead of the weld in the joint (A). Hold a long arc and allow time for the electrode to heat up, forming the protective gas envelope. Move the electrode so that the arc is focused directly on the leading edge (root) of the previous weld crater (B). © Cengage Learning 2012

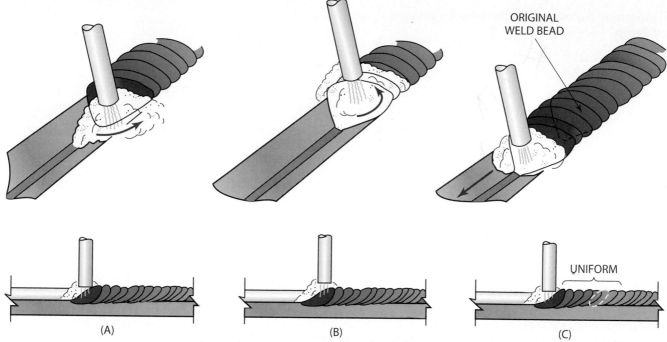

ORIGINAL
WELD BEAD

UNIFORM

(A)   (B)   (C)

FIGURE 24-9 When restarting the weld pool after the root has been heated to the melting temperature, move the electrode upward along one side of the crater (A). Move the electrode along the top edge, depositing new weld metal (B). When the weld is built up uniformly with the previous weld, continue along the joint (C). © Cengage Learning 2012

The electrode should be low enough to start transferring metal. Next, move the electrode in a semicircle near the back edge of the weld crater. Watch the buildup and match your speed in the semicircle to the deposit rate so that the weld is built up evenly, **Figure 24-9**. Move the electrode ahead and continue with the same weave pattern that was being used previously.

The movement to the root of the weld and back up on the bead serves both to build up the weld and reheat the metal so that the depth of penetration will remain the same. If the weld bead is started too quickly, penetration is reduced and buildup is high and narrow. Starting and stopping weld beads in corners should be avoided because this often results in defects, **Figure 24-10**.

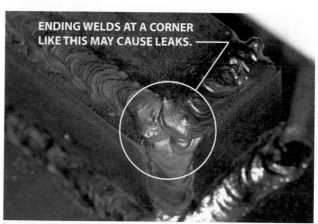

ENDING WELDS AT A CORNER
LIKE THIS MAY CAUSE LEAKS.

FIGURE 24-10 Incorrect method of welding a corner. Stopping on a corner like this may cause leaks. Larry Jeffus

PRACTICE 24-1

SMAW Workmanship Sample and Welder Qualification Test Plate for Limited Thickness Horizontal and Vertical Positions with E7018 Electrodes

Welding Procedure Specification (WPS) No.: Practice 24-1.

Title:
Welding SMAW of plate to plate.

Scope:
This procedure is applicable for single-bevel or V-groove plate with a backing strip within the range of 3/16 in. (4.8 mm) through 3/4 in. (20 mm).

Welding may be performed in the following positions:
2G and 3G (uphill).

Base Metal:
The base metal shall conform to carbon steel M-1 or P-1, Group 1 or 2.

- 2G test plates: two (2); each 3/8 in. (9.5 mm) thick, 3 in. (75 mm) wide, and 7 in. (175 mm) long, one having a 45° bevel along one edge.

- 3G test plates: two (2); each 3/8 in. (9.5 mm) thick, 3 in. (75 mm) wide, and 7 in. (175 mm) long, having a 45° included bevel.

Backing Strip Specification:
Carbon steel M-1 or P-1, Group 1, 2, or 3, each either 1/4 in. (6.4 mm) or 3/8 in. (9.5 mm) thick, 1 in. (25 mm) wide, and 9 in. (228 mm) long.

Filler Metal:
The filler metal shall conform to AWS specification no. E7018 from AWS specification A5.1. This filler metal falls into F-number F-4 and A-number A-1.

Shielding Gas:
N/A.

Joint Design and Tolerances:
Refer to the drawing in **Figure 24-11** for the joint layout specifications.

Preparation of Base Metal:
The bevel is to be flame or plasma cut on the edge of the plate before the parts are assembled. The beveled surface must be smooth and free of notches. Any roughness or notches that are deeper than 1/64 in. (0.4 mm) are unacceptable and must be ground smooth.

All hydrocarbons and other contaminants, such as cutting fluids, grease, oil, and primers, must be cleaned off all parts and filler metals before welding. This cleaning can be done with any suitable solvents or detergents. The backing strip, groove face, and inside and outside plate surface within 1 in. (25 mm) of the joint must be mechanically cleaned of slag, rust, and mill scale. Cleaning must be done with a wire brush or grinder down to bright metal.

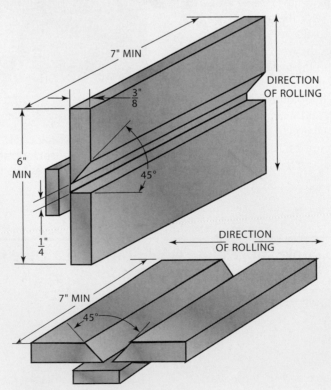

FIGURE 24-11 Practice 24-1 joint design. American Welding Society

Weld	Filler Metal Dia.	Current	Amperage Range
Tack	3/32 in. (2.4 mm)	DCEP	70 to 110
Root	1/8 in. (3.2 mm)	DCEP	90 to 165
Filler	5/32 in. (4 mm)	DCEP	125 to 220

TABLE 24-1 E7018 Current Settings

Electrical Characteristics:

The current shall be direct-current electrode positive (DCEP), **Table 24-1.** The base metal shall be on the negative side of the line.

Preheat:

The parts must be heated to a temperature higher than 50°F (10°C) before any welding is started.

Backing Gas:

N/A.

Safety:

Proper protective clothing and equipment must be used. The area must be free of all hazards that may affect the welder or others in the area. The welding machine, welding leads, work clamp, electrode holder, and other equipment must be in safe working order.

Welding Technique:

Tack weld the plates together with the backing strip. There should be about a 1/4-in. (6.4-mm) root gap between the plates. Use the E7018 arc welding electrodes to make a root pass to fuse the plates and backing strip together. Clean the slag from the root pass, being sure to remove any trapped slag along the sides of the weld.

Using the E7018 arc welding electrodes, make a series of stringer or weave filler welds, no thicker than 1/4 in. (6.4 mm), in the groove until the joint is filled.

Interpass Temperature:

The plate should not be heated to a temperature higher than 350°F (175°C) during the welding process. After each weld pass is completed, allow it to cool but never to a temperature below 50°F (10°C). The weldment must not be quenched in water.

Cleaning:

 The slag must be cleaned off between passes. The weld beads may be cleaned by a hand wire brush, a chipping hammer, a punch and hammer, or a needle scaler. All weld cleaning must be performed with the test plate in the welding position.

Visual Inspection Criteria for Entry Welders*:

There shall be no cracks, no incomplete fusion. There shall be no incomplete joint penetration in groove welds except as permitted for partial joint penetration welds.

The Test Supervisor shall examine the weld for acceptable appearance and shall be satisfied that the welder is skilled in using the process and procedure specified for the test.

Undercut shall not exceed the lesser of 10% of the base metal thickness or 1/32 in. (0.8 mm).

Where visual examination is the only criterion for acceptance, all weld passes are subject to visual examination at the discretion of the Test Supervisor.

The frequency of porosity shall not exceed one in each 4 in. (100 mm) of weld length, and the maximum diameter shall not exceed 3/32 in. (2.4 mm).

Welds shall be free from overlap.

Bend Test:

The weld is to be mechanically tested only after it has passed the visual inspection. Be sure that the test specimens are properly marked to identify the welder, the position, and the process.

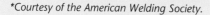

*Courtesy of the American Welding Society.

Specimen Preparation:

For 3/8-in. (9.5-mm) test plates, two specimens are to be located in accordance with the requirements in **Figure 24-12.** One is to be prepared for a **transverse face bend,** and the other is to be prepared for a **transverse root bend, Figure 24-13.**

- *Transverse face bend.* The weld is perpendicular to the longitudinal axis of the specimen and is bent so that the weld face becomes the tension surface of the specimen.

- *Transverse root bend.* The weld is perpendicular to the longitudinal axis of the specimen and is bent so that the weld root becomes the tension surface of the specimen.

Acceptance Criteria for Bend Test*:

For acceptance, the convex surface of the face and root bend specimens shall meet both of the following requirements:

1. No single indication shall exceed 1/8 in. (3.2 mm) measured in any direction on the surface.

2. The sum of the greatest dimensions of all indications on the surface that exceed 1/32 in. (0.8 mm) but are less than or equal to 1/8 in. (3.2 mm) shall not exceed 3/8 in. (9.5 mm).

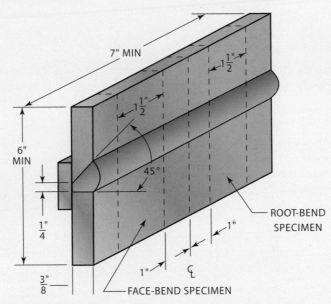

FIGURE 24-12 Test plate specimen location. American Welding Society

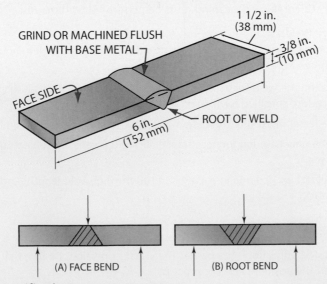

FIGURE 24-13 Test specimen specifications. American Welding Society

*Courtesy of the American Welding Society.

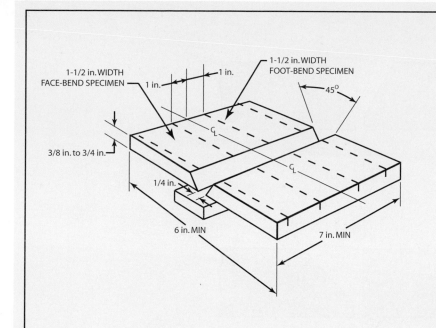

NOTES:
1. 3/8 in. thickness carbon steel material.
2. Performance Qualification #1 = 2G. Performance Qualification #2 = 3G, Uphill.
3. All welding done in position, according to applicable performance qualification requirements.
4. The backing thickness shall be 1/4 in. min. to 3/8 in. max.; backing width 1 in. min.
5. All parts may be mechanically cut or machine OFC.
6. Use WPS AWS EDU SMAW-01 for PQ#1-2G, and AWS EDU SMAW-02 for PQ#2-3G uphill. (See AWS QC10, Table 2.)
7. Visual examination in accordance with requirements of AWS QC10, Table 3.
8. Bend test in accordance with the requirements of QC10, Table 4.

ID	QTY	SIZE	METRIC CONVERSION	American Welding Society		
				Entry Welder Performance Qualification		
				SMAW Carbon Steel Test Plates		
				DATE:	SCALE:	DWG #: AWS EDU-6
				DR BY:	Tolerances: (Unless otherwise specified) DRAWING NOT TO SCALE	
				APP BY:	Fractions: ±1/16" Angles: + 10°, 5°	

FIGURE 24-14 AWS EDU-6 SMAW carbon steel plate workmanship sample. American Welding Society

Cracks occurring at the corner of the specimens shall not be considered unless there is definite evidence that they result from slag or inclusions or other internal discontinuities.

Sketches:

2G and 3G test plate drawings, **Figure 24-14.**

Paperwork:

Compete a "Bill of Materials," "Time Sheet," and "Performance Qualification Test Record" in Appendix I or as provided by your instructor. ◆

PRACTICE 24-2

Limited Thickness Welder Performance Qualification Test Plate without Backing

Welding Procedure Specification (WPS) No.: Practice 24-2.

Title:

Welding SMAW of plate to plate.

Scope:

This procedure is applicable for bevel and V-groove plate with a backing strip within the range of 3/8 in. (9.5 mm) through 3/4 in. (19 mm).

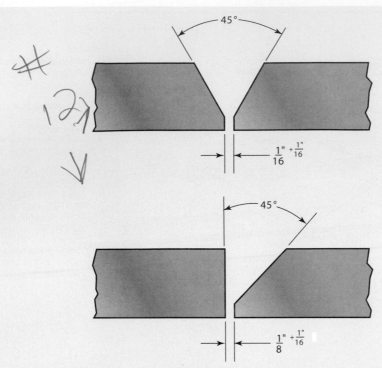

FIGURE 24-15 Practice 24-2 joint design. © Cengage Learning 2012

Welding may be performed in the following positions:
1G, 2G, 3G, and 4G.

Base Metal:
The base metal shall conform to M1020 or A36.

Backing Material Specification:
M1020 or A36.

Filler Metal:
The filler metal shall conform to AWS specification no. E6010 or E6011 root pass and E7018 for the cover pass from AWS specification A5.1. This filler metal falls into F-number F-3 and F-4 and A-number A-1.

Shielding Gas:
N/A.

Joint Design and Tolerances:
Refer to the drawing in **Figure 24-15** for the joint layout specifications.

Preparation of Base Metal:
The V-groove is to be flame or plasma cut on the edge of the plate before the parts are assembled. Prior to welding, all parts must be cleaned of all contaminants, such as paints, oils, grease, or primers. Both inside and outside surfaces within 1 in. (25 mm) of the joint must be mechanically cleaned using a wire brush or grinder.

Electrical Characteristics:
The current shall be AC or DC with the electrode positive (DCEP), **Table 24-2.**

AWS CLASSIFICATION AND POLARITY	ELECTRODE DIAMETER AND AMPERAGE RANGE		
	$\frac{3"}{32}$	$\frac{1"}{8}$	$\frac{5"}{32}$
E6010 - DCEP	40-80	70-130	110-165
E6011 - AC, DCEP	50-70	85-125	130-160
E7018 - AC, DCEP	70-110	90-165	125-220

TABLE 24-2 E6010, E6011, and E7018 Current Settings

Preheat:
The parts must be heated to a temperature higher than 70°F (21°C) before any welding is started.

Backing Gas:
N/A.

Welding Technique:
You can use a small piece of metal across the ends of the plate as a spacer to make it easier to tack weld the plates together; there should be about a 1/8-in. (3.2-mm) root gap between the plates. Use the E6010 or E6011 arc welding electrodes to make a root pass to fuse the plates together. Clean the slag from the root pass, and use either a hot pass or grinder to remove any trapped slag.

Using the E7018 arc welding electrodes, make a series of filler welds in the groove until the joint is filled.

Interpass Temperature:
The plate, outside of the heat-affected zone, should not be heated to a temperature higher than 400°F (205°C) during the welding process. After each weld pass is completed, allow it to cool; the weldment must not be quenched in water.

Cleaning:
The slag can be chipped and/or ground off between passes but can only be chipped off of the cover pass.

Inspection:
Visually inspect the weld for uniformity and discontinuities by using the criteria found in Practice 24-2. If the weld passes the visual inspection, then it is to be prepared and guided-bend tested according to the "guided-bend test" criteria found in Practice 24-2. Repeat each of the welds as needed until you can pass this test.

Postheating is the application of heat to the metal after welding. Postheating is used to slow the cooling rate and reduce hardening.

Interpass temperature is the temperature of the metal during welding. The interpass temperature is given as a minimum and maximum. The minimum temperature is usually the same as the preheat temperature. If the plate cools below this temperature during welding, it should be reheated. The maximum temperature may be specified to keep the plate below a certain phase change temperature for the mild steel used in these practices. The maximum interpass temperature occurs when the weld bead cannot be controlled because of a slow cooling rate. When this happens, the plate should be allowed to cool down but not below the minimum interpass temperature.

If, during the welding process, a welder must allow a practice weldment to cool so that the weld can be completed later, the weldment should be cooled slowly and then reheated before starting to weld again. A weld that is to be tested or that is done on any parts other than scrap should not be quenched in water.

Sketches:
1G, 2G, 3G, and 4G test plate drawing, **Figure 24-16.**

Paperwork:
Compete a "Bill of Materials," "Time Sheet," and "Performance Qualification Test Record" in Appendix I or as provided by your instructor. ◆

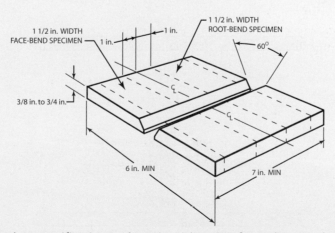

FIGURE 24-16 Open root test plate specification and specimen location for carbon steel sample. American Welding Society

PRACTICE 24-3

Gas Metal Arc Welding—Short-Circuit Metal Transfer (GMAW-S) Workmanship Sample

Welding Procedure Specification (WPS) No.: Practice 24-3.

Title:
Welding GMAW-S of plate to plate.

Scope:
This procedure is applicable for square groove and fillet welds within the range of 10 gauge (3.4 mm) through 14 gauge (1.9 mm).

Welding may be performed in the following positions:
All.

Base Metal:
The base metal shall conform to carbon steel M-1, P-1, and S-1 Group 1 or 2.

Backing Material Specification:
None.

Filler Metal:
The filler metal shall conform to AWS specification no. E70S-X from AWS specification A5.18. This filler metal falls into F-number F-6 and A-number A-1.

Shielding Gas:
The shielding gas, or gases, shall conform to the following compositions and purity: CO_2 at 30 to 50 cfh or 75% Ar/25% CO_2 at 30 to 50 cfh.

Joint Design and Tolerances:
Refer to the drawing in **Figure 24-17** for the joint layout specifications.

Preparation of Base Metal:
All parts may be mechanically cut or machine PAC unless specified as manual PAC.

All hydrocarbons and other contaminants, such as cutting fluids, grease, oil, and primers, must be cleaned off all parts and filler metals before welding. This cleaning can be done with any suitable solvents or detergents. The groove face and inside and outside plate surface within 1 in. (25 mm) of the joint must be mechanically cleaned of slag, rust, and mill scale. Cleaning must be done with a wire brush or grinder down to bright metal.

Electrical Characteristics:
Set the voltage, amperage, wire-feed speed, and shielding gas flow according to **Table 24-3**.

Preheat:
The parts must be heated to a temperature higher than 50°F (10°C) before any welding is started.

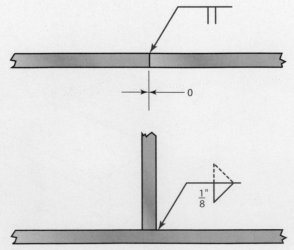

FIGURE 24-17 Practice 24-3 joint design. © Cengage Learning 2012

Electrode		Welding Power			Shielding Gas		Base Metal	
Type	Size	Amps	Wire-Feed Speed IPM (cm/min)	Volts	Type	Flow	Type	Thickness
E70S-X	0.035 in. (0.9 mm)	90 to 120	180 to 300 (457 to 762)	15 to 19	CO_2 or 75%Ar/25%CO_2	30 to 50	Low carbon steel	1/4 in. to 1/2 in. (6 mm to 13 mm)
E70S-X	0.045 in. (1.2 mm)	130 to 200	125 to 200 (318 to 508)	17 to 20	CO_2 or 75%Ar/25%CO_2	30 to 50	Low carbon steel	1/4 in. to 1/2 in. (6 mm to 13 mm)

TABLE 24-3 GMAW-S Machine Settings

Backing Gas:
N/A.

Safety:
Proper protective clothing and equipment must be used. The area must be free of all hazards that may affect the welder or others in the area. The welding machine, welding leads, work clamp, electrode holder, and other equipment must be in safe working order.

Welding Technique:
Using a 1/2-in. (13-mm) or larger gas nozzle for all welding, first tack weld the plates together according to the drawing. Use the E70S-X filler metal to fuse the plates together. Clean any silicon slag, being sure to remove any trapped silicon slag along the sides of the weld.

Using the E70S-X arc welding electrodes, make a series of stringer beads, no thicker than 3/16 in. (4.8 mm). The 1/8-in. (3.2-mm) fillet welds are to be made with one pass. All welds must be placed in the orientation shown in the drawing.

Interpass Temperature:
The plate should not be heated to a temperature higher than 350°F (175°C) during the welding process. After each weld pass is completed, allow it to cool but never to a temperature below 50°F (10°C). The weldment must not be quenched in water.

Cleaning:
Any slag must be cleaned off between passes. The weld beads may be cleaned by a hand wire brush, a chipping hammer, a punch and hammer, or a needle scaler. All weld cleaning must be performed with the test plate in the welding position.

Visual Inspection:
Visually inspect the weld for uniformity and discontinuities.

1. There shall be no cracks, no incomplete fusion.
2. There shall be no incomplete joint penetration in groove welds except as permitted for partial joint penetration welds.
3. The Test Supervisor shall examine the weld for acceptable appearance and shall be satisfied that the welder is skilled in using the process and procedure specified for the test.
4. Undercut shall not exceed the lesser of 10% of the base metal thickness or 1/32 in. (0.8 mm).
5. Where visual examination is the only criterion for acceptance, all weld passes are subject to visual examination at the discretion of the Test Supervisor.
6. The frequency of porosity shall not exceed one in each 4 in. (100 mm) of weld length, and the maximum diameter shall not exceed 3/32 in. (2.4 mm).
7. Welds shall be free from overlap.

Sketches:
GMAW Short-Circuit Metal Transfer Workmanship Sample drawing, **Figure 24-18.**

Paperwork:
Compete a "Bill of Materials," "Time Sheet," and "Performance Qualification Test Record" in Appendix I or as provided by your instructor. ◆

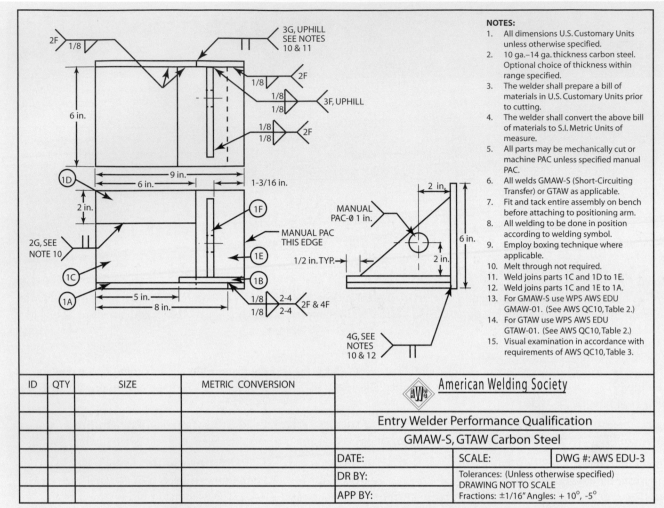

NOTES:
1. All dimensions U.S. Customary Units unless otherwise specified.
2. 10 ga.–14 ga. thickness carbon steel. Optional choice of thickness within range specified.
3. The welder shall prepare a bill of materials in U.S. Customary Units prior to cutting.
4. The welder shall convert the above bill of materials to S.I. Metric Units of measure.
5. All parts may be mechanically cut or machine PAC unless specified manual PAC.
6. All welds GMAW-S (Short-Circuiting Transfer) or GTAW as applicable.
7. Fit and tack entire assembly on bench before attaching to positioning arm.
8. All welding to be done in position according to welding symbol.
9. Employ boxing technique where applicable.
10. Melt through not required.
11. Weld joins parts 1C and 1D to 1E.
12. Weld joins parts 1C and 1E to 1A.
13. For GMAW-S use WPS AWS EDU GMAW-01. (See AWS QC10, Table 2.)
14. For GTAW use WPS AWS EDU GTAW-01. (See AWS QC10, Table 2.)
15. Visual examination in accordance with requirements of AWS QC10, Table 3.

ID	QTY	SIZE	METRIC CONVERSION	American Welding Society
				Entry Welder Performance Qualification
				GMAW-S, GTAW Carbon Steel
				DATE: / SCALE: / DWG #: AWS EDU-3
				DR BY: / Tolerances: (Unless otherwise specified) DRAWING NOT TO SCALE
				APP BY: / Fractions: ±1/16" Angles: + 10°, -5°

FIGURE 24-18 AWS EDU-3 GMAW Short-Circuit Metal Transfer Workmanship for carbon steel. American Welding Society

PRACTICE 24-4

Gas Metal Arc Welding Spray Transfer (GMAW) Workmanship Sample

Welding Procedure Specification (WPS) No.: Practice 24-4.

Title:
Welding GMAW of plate to plate.

Scope:
This procedure is applicable for fillet welds within the range of 1/8 in. (3.2 mm) through 1 1/2 in. (38 mm).

Welding may be performed in the following positions:
1F and 2F.

Base Metal:
The base metal shall conform to carbon steel M-1, P-1, and S-1, Group 1 or 2.

Backing Material Specification:
None.

Filler Metal:
The 0.035-in. (0.90-mm) to 0.045-in. (1.2-mm) diameter filler metal shall conform to AWS specification no. E70S-X from AWS specification A5.18. This filler metal falls into F-number F-6 and A-number A-1.

Shielding Gas:
The shielding gas, or gases, shall conform to the following compositions: 98% Ar/2% O_2 or 90% Ar/10% CO_2.

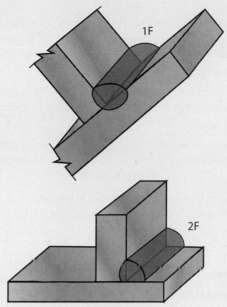

FIGURE 24-19 Practice 24-4 joint design. American Welding Society

Joint Design and Tolerances:
Refer to the drawing in **Figure 24-19** for the joint layout specifications.

Preparation of Base Metal:
The bevels are to be flame or plasma cut on the edges of the plate before the parts are assembled. The beveled surface must be smooth and free of notches. Any roughness or notches deeper than 1/64 in. (0.4 mm) must be ground smooth.

All hydrocarbons and other contaminants, such as cutting fluids, grease, oil, and primers, must be cleaned off all parts and filler metals before welding. This cleaning can be done with any suitable solvents or detergents. The groove face and inside and outside plate surface within 1 in. (25 mm) of the joint must be mechanically cleaned of slag, rust, and mill scale. Cleaning must be done with a wire brush or grinder down to bright metal.

Electrical Characteristics:
Set the voltage, amperage, wire-feed speed, and shielding gas flow according to **Table 24-4.**

Preheat:
The parts must be heated to a temperature higher than 50°F (10°C) before any welding is started.

Backing Gas:
N/A.

Safety:
Proper protective clothing and equipment must be used. The area must be free of all hazards that may affect the welder or others in the area. The welding machine, welding leads, work clamp, electrode holder, and other equipment must be in safe working order.

Electrode		Welding Power			Shielding Gas		Base Metal	
Type	Size	Amps	Wire-Feed Speed IPM (cm/min)	Volts	Type	Flow	Type	Thickness
E70S-X	0.035 in. (0.9 mm)	180 to 230	400 to 550 (1016 to 1397)	25 to 27	Ar plus 2% O_2 or 90% Ar/10% CO_2	30 to 50	Low carbon steel	1/4 in. to 1/2 in. (6 mm to 13 mm)
E70S-X	0.045 in. (1.2 mm)	260 to 340	300 to 500 (762 to 1270)	25 to 30	Ar plus 2% O_2 or 90% Ar/10% CO_2	30 to 50	Low carbon steel	1/4 in. to 1/2 in. (6 mm to 13 mm)

TABLE 24-4 GMAW-S Machine Settings

Welding Technique:

Using a 3/4-in. (19-mm) or larger gas nozzle for all welding, first tack weld the plates together according to the drawing. There should be about a 1/16-in. (1.6-mm) root gap between the plates with V-grooved or beveled edges. Use the E70S-X arc welding electrodes to make a root pass to fuse the plates together. Clean any silicon slag from the root pass, being sure to remove any trapped silicon slag along the sides of the weld.

Using the E70S-X arc welding electrodes, make a series of stringer or weave filler welds, no thicker than 1/4 in. (6.4 mm) in the groove until the joint is filled. The 1/4-in. (6.4-mm) fillet welds are to be made with one pass.

Interpass Temperature:

The plate should not be heated to a temperature higher than 350°F (175°C) during the welding process. After each weld pass is completed, allow it to cool but never to a temperature below 50°F (10°C). The weldment must not be quenched in water.

Cleaning:

Any slag must be cleaned off between passes. The weld beads may be cleaned by a hand wire brush, a chipping hammer, a punch and hammer, or a needle scaler. All weld cleaning must be performed with the test plate in the welding position.

Visual Inspection*:

Visually inspect the weld for uniformity and discontinuities.

1. There shall be no cracks, no incomplete fusion.

2. There shall be no incomplete joint penetration in groove welds except as permitted for partial joint penetration welds.

3. The Test Supervisor shall examine the weld for acceptable appearance and shall be satisfied that the welder is skilled in using the process and procedure specified for the test.

4. Undercut shall not exceed the lesser of 10% of the base metal thickness or 1/32 in. (0.8 mm).

5. Where visual examination is the only criterion for acceptance, all weld passes are subject to visual examination at the discretion of the Test Supervisor.

6. The frequency of porosity shall not exceed one in each 4 in. (100 mm) of weld length, and the maximum diameter shall not exceed 3/32 in. (2.4 mm).

7. Welds shall be free from overlap.

Sketches:

Gas Metal Arc Welding Spray Transfer (GMAW) Workmanship Sample drawing, **Figure 24-20**.

Paperwork:

Compete a "Bill of Materials," "Time Sheet," and "Performance Qualification Test Record" in Appendix I or as provided by your instructor. ◆

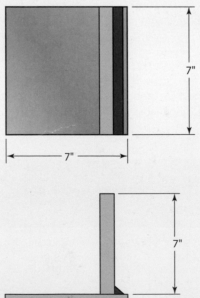

FIGURE 24-20 GMAW spray transfer fillet weld carbon steel workmanship sample. American Welding Society

Courtesy of the American Welding Society.

PRACTICE 24-5

Gas Metal Arc Welding—Short-Circuit Metal Transfer (GMAW-S) Limited Thickness Welder Performance Qualification Test Plate All Positions without Backing

Welding Procedure Specification (WPS) No.: Practice 24-5.

Title:
Welding GMAW-S of plate to plate.

Scope:
This procedure is applicable for V-groove or single-bevel welds within the range of 1/8 in. (3.2 mm) through 3/4 in. (19 mm).

Welding may be performed in the following positions:
All.

Base Metal:
The base metal shall conform to carbon steel M-1, P-1, and S-1, Group 1 or 2.

Backing Material Specification:
None.

Filler Metal:
The filler metal shall conform to AWS specification no. E70S-X from AWS specification A5.18. This filler metal falls into F-number F-6 and A-number A-1.

Shielding Gas:
The shielding gas, or gases, shall conform to the following compositions and purity: CO_2 at 30 to 50 cfh or 75% Ar/25% CO_2 at 30 to 50 cfh.

Joint Design and Tolerances:
Refer to the drawing in **Figure 24-21** for the joint layout specifications.

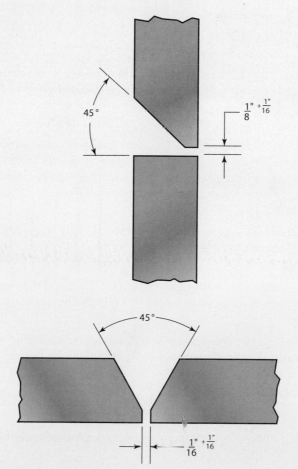

FIGURE 24-21 Practice 24-5 joint design. American Welding Society

Electrode		Welding Power			Shielding Gas		Base Metal	
Type	Size	Amps	Wire-Feed Speed IPM (cm/min)	Volts	Type	Flow	Type	Thickness
E70S-X	0.035 in. (0.9 mm)	90 to 160	180 to 300 (457 to 762)	15 to 19	Ar plus 2% O_2 or 90% Ar/10% CO_2	30 to 50	Low carbon steel	1/4 in. to 1/2 in. (6 mm to 13 mm)
E70S-X	0.045 in. (1.2 mm)	130 to 200	125 to 200 (317 to 508)	17 to 19	Ar plus 2% O_2 or 90% Ar/10% CO_2	30 to 50	Low carbon steel	1/4 in. to 1/2 in. (6 mm to 13 mm)

TABLE 24-5 GMAW Spray Metal Transfer Machine Settings

Preparation of Base Metal:

The bevels are to be flame or plasma cut on the edges of the plate before the parts are assembled. The beveled surface must be smooth and free of notches. Any roughness or notches deeper than 1/64 in. (0.4 mm) must be ground smooth.

All hydrocarbons and other contaminants, such as cutting fluids, grease, oil, and primers, must be cleaned off all parts and filler metals before welding. This cleaning can be done with any suitable solvents or detergents. The groove face and inside and outside plate surface within 1 in. (25 mm) of the joint must be mechanically cleaned of slag, rust, and mill scale. Cleaning must be done with a wire brush or grinder down to bright metal.

Electrical Characteristics:

Set the voltage, amperage, wire-feed speed, and shielding gas flow according to **Table 24-5**.

Preheat:

The parts must be heated to a temperature higher than 50°F (10°C) before any welding is started.

Backing Gas:

N/A.

Safety:

Proper protective clothing and equipment must be used. The area must be free of all hazards that may affect the welder or others in the area. The welding machine, welding leads, work clamp, electrode holder, and other equipment must be in safe working order.

Welding Technique:

Use a 1/2-in. (13-mm) or larger gas nozzle for all welding, and a small piece of metal across the ends of the plate as a spacer to make it easier to tack weld the plates together according to the drawing. There should be about a 1/8-in. (3.2-mm) root gap between the plates with V-grooved or beveled edges and 1/16-in. (1.6-mm) root faces. Use the E70S-X filler wire to make a root pass to fuse the plates together. Clean any silicon slag from the root pass, being sure to remove any trapped silicon slag along the sides of the weld.

Using the E70S-X filler wire, make a series of stringer or weave filler welds, no thicker than 1/4 in. (6.4 mm) in the groove until the joint is filled. Note: The horizontal (2G) weldment should be made with stringer beads only.

Interpass Temperature:

The plate should not be heated to a temperature higher than 350°F (175°C) during the welding process. After each weld pass is completed, allow it to cool but never to a temperature below 50°F (10°C). The weldment must not be quenched in water.

Cleaning:

Any slag must be cleaned off between passes. The weld beads may be cleaned by a hand wire brush, a chipping hammer, a punch and hammer, or a needle scaler. All weld cleaning must be performed with the test plate in the welding position.

Visual Inspection*:

Visually inspect the weld for uniformity and discontinuities.

1. There shall be no cracks, no incomplete fusion.

2. There shall be no incomplete joint penetration in groove welds except as permitted for partial joint penetration welds.

*Courtesy of the American Welding Society.

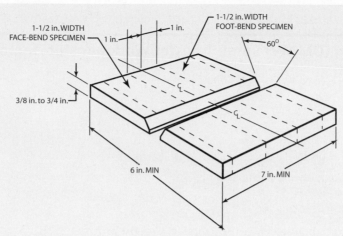

FIGURE 24-22 GMAW-S V-groove open root carbon steel workmanship sample. American Welding Society

3. The Test Supervisor shall examine the weld for acceptable appearance and shall be satisfied that the welder is skilled in using the process and procedure specified for the test.

4. Undercut shall not exceed the lesser of 10% of the base metal thickness or 1/32 in. (0.8 mm).

5. Where visual examination is the only criterion for acceptance, all weld passes are subject to visual examination at the discretion of the Test Supervisor.

6. The frequency of porosity shall not exceed one in each 4 in. (100 mm) of weld length and the maximum diameter shall not exceed 3/32 in. (2.4 mm).

7. Welds shall be free from overlap.

Bend Test:
The weld is to be mechanically tested only after it has passed the visual inspection. Be sure that the test specimens are properly marked to identify the welder, the position, and the process.

Specimen Preparation:
For 3/8 in. (9.5-mm) test plates, two specimens are to be located in accordance with the requirements of **Figure 24-22.** One is to be prepared for a transverse face bend, and the other is to be prepared for a transverse root bend.

- *Transverse face bend.* The weld is perpendicular to the longitudinal axis of the specimen and is bent so that the weld face becomes the tension surface of the specimen.

- *Transverse root bend.* The weld is perpendicular to the longitudinal axis of the specimen and is bent so that the weld root becomes the tension surface of the specimen.

Acceptance Criteria for Face and Root Bends:*
For acceptance, the convex surface of the face- and root-bend specimens shall meet both of the following requirements:

1. No single indication shall exceed 1/8 in. (3.2 mm) measured in any direction on the surface.

2. The sum of the greatest dimensions of all indications on the surface that exceed 1/32 in. (0.8 mm) but are less than or equal to 1/8 in. (3.2 mm) shall not exceed 3/8 in. (9.6 mm).

Cracks occurring at the corner of the specimens shall not be considered unless there is definite evidence that they result from slag inclusion or other internal discontinuities.

Sketches:
Gas Metal Arc Welding Short-Circuit Metal Transfer (GMAW-S) Workmanship Sample drawing, Figure 24-22.

Paperwork:
Compete a "Bill of Materials," "Time Sheet," and "Performance Qualification Test Record" in Appendix I or as provided by your instructor. ◆

*Courtesy of the American Welding Society.

PRACTICE 24-6

Gas Metal Arc Welding (GMAW) Spray Transfer Workmanship Sample

Welding Procedure Specification (WPS) No.: Practice (24-6).

Title:
Welding GMAW of plate to plate.

Scope:
This procedure is applicable for V-groove and fillet welds.

Welding may be performed in the following positions:
1G and 2F.

Base Metal:
The base metal shall conform to carbon steel M-1, P-1, and S-1, Group 1 or 2.

Backing Material Specification:
None.

Filler Metal:
The filler metal shall conform to AWS specification no. 0.035-in. (0.90-mm) to 0.045-in. (1.2-mm) diameter E70S-X from AWS specification A5.18. This filler metal falls into F-number F-6 and A-number A-1.

Shielding Gas:
The shielding gas, or gases, shall conform to the following compositions and purity: 98% Ar/2% O_2 or 90% Ar/10%CO_2.

Joint Design and Tolerances:
Refer to the drawing in **Figure 24-23** for the joint layout specifications.

Preparation of Base Metal:
The bevels are to be flame or plasma cut on the edges of the plate before the parts are assembled. The beveled surface must be smooth and free of notches. Any roughness or notches deeper than 1/64 in. (0.4 mm) must be ground smooth.

All hydrocarbons and other contaminants, such as cutting fluids, grease, oil, and primers, must be cleaned off all parts and filler metals before welding. This cleaning can be done with any suitable solvents or detergents. The groove face and inside and outside plate surface within 1 in. (25 mm) of the joint must be mechanically cleaned of slag, rust, and mill scale. Cleaning must be done with a wire brush or grinder down to bright metal.

Electrical Characteristics:
Set the voltage, amperage, wire-feed speed, and shielding gas flow according to **Table 24-6.**

Preheat:
The parts must be heated to a temperature higher than 50°F (10°C) before any welding is started.

Backing Gas:
N/A.

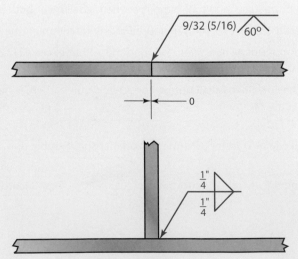

FIGURE 24-23 Practice 24-6 joint design. American Welding Society

Electrode		Welding Power			Shielding Gas		Base Metal	
Type	Size	Amps	Wire-Feed Speed IPM (cm/min)	Volts	Type	Flow	Type	Thickness
E70S-X	0.035 in. (0.9 mm)	180 to 230	400 to 550 (1016 to 1397)	25 to 27	Ar plus 2% O_2 or 90% Ar/10% CO_2	30 to 50	Low carbon steel	1/4 in. to 1/2 in. (6 mm to 13 mm)
E70S-X	0.045 in. (1.2 mm)	260 to 340	300 to 500 (762 to 1270)	25 to 30	Ar plus 2% O_2 or 90% Ar/10% CO_2	30 to 50	Low carbon steel	1/4 in. to 1/2 in. (6 mm to 13 mm)

TABLE 24-6 GMAW Spray Metal Transfer Machine Settings

Safety:

Proper protective clothing and equipment must be used. The area must be free of all hazards that may affect the welder or others in the area. The welding machine, welding leads, work clamp, electrode holder, and other equipment must be in safe working order.

Welding Technique:

Using a 3/4-in. (19-mm) or larger gas nozzle for all welding, first tack weld the plates together, **Figure 24-24.** Use the E70S-X arc welding electrodes to make the welds.

Using the E70S-X arc welding electrodes, make a series of stringer filler welds, no thicker than 1/4 in. (6.4 mm) in the groove until the joint is filled.

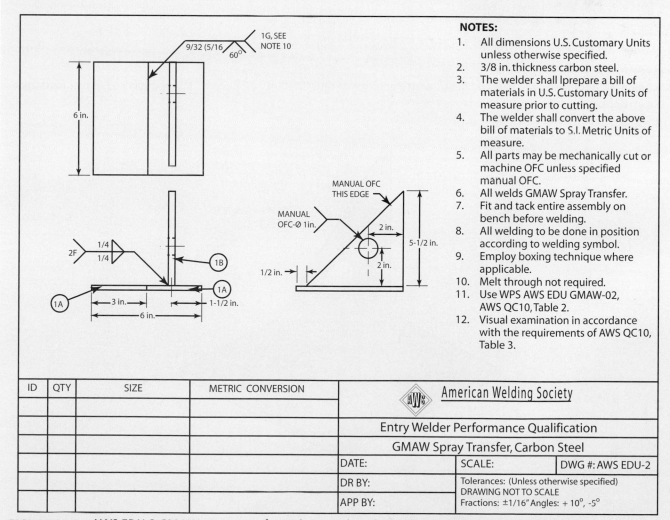

NOTES:
1. All dimensions U.S. Customary Units unless otherwise specified.
2. 3/8 in. thickness carbon steel.
3. The welder shall lprepare a bill of materials in U.S. Customary Units of measure prior to cutting.
4. The welder shall convert the above bill of materials to S.I. Metric Units of measure.
5. All parts may be mechanically cut or machine OFC unless specified manual OFC.
6. All welds GMAW Spray Transfer.
7. Fit and tack entire assembly on bench before welding.
8. All welding to be done in position according to welding symbol.
9. Employ boxing technique where applicable.
10. Melt through not required.
11. Use WPS AWS EDU GMAW-02, AWS QC10, Table 2.
12. Visual examination in accordance with the requirements of AWS QC10, Table 3.

ID	QTY	SIZE	METRIC CONVERSION	American Welding Society
				Entry Welder Performance Qualification
				GMAW Spray Transfer, Carbon Steel
				DATE: / SCALE: / DWG #: AWS EDU-2
				DR BY: / Tolerances: (Unless otherwise specified) DRAWING NOT TO SCALE
				APP BY: / Fractions: ±1/16" Angles: + 10°, -5°

FIGURE 24-24 AWS EDU-2 GMAW spray transfer carbon steel workmanship sample. American Welding Society

Interpass Temperature:

The plate should not be heated to a temperature higher than 350°F (175°C) during the welding process. After each weld pass is completed, allow it to cool but never to a temperature below 50°F (10°C). The weldment must not be quenched in water.

Cleaning:

Any slag must be cleaned off between passes. The weld beads may be cleaned by a hand wire brush, a chipping hammer, a punch and hammer, or a needle scaler. All weld cleaning must be performed with the test plate in the welding position.

Visual Inspection*:

Visually inspect the weld for uniformity and discontinuities.

1. There shall be no cracks, no incomplete fusion.

2. There shall be no incomplete joint penetration in groove welds except as permitted for partial joint penetration welds.

3. The Test Supervisor shall examine the weld for acceptable appearance and shall be satisfied that the welder is skilled in using the process and procedure specified for the test.

4. Undercut shall not exceed the lesser of 10% of the base metal thickness or 1/32 in. (0.8 mm).

5. Where visual examination is the only criterion for acceptance, all weld passes are subject to visual examination at the discretion of the Test Supervisor.

6. The frequency of porosity shall not exceed one in each 4 in. (100 mm) of weld length and the maximum diameter shall not exceed 3/32 in. (2.4 mm).

7. Welds shall be free from overlap.

Sketches:

Gas Metal Arc Welding Spray Transfer (GMAW) Workmanship Sample drawing, **Figure 24-25**.

Paperwork:

Compete a "Bill of Materials," "Time Sheet," and "Performance Qualification Test Record" in Appendix I or as provided by your instructor. ◆

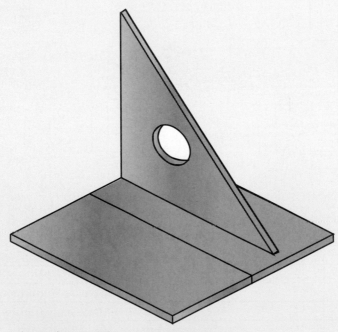

FIGURE 24-25 Practice 24-6 workmanship sample. American Welding Society

*Courtesy of the American Welding Society.

PRACTICE 24-7

AWS SENSE Entry-Level Welder Workmanship Sample for Flux Cored Arc Welding, Gas-Shielded (FCAW)

Welding Procedure Specification (WPS) No.: Practice 24-7.

Title:

Welding FCAW of plate to plate.

Scope:

This procedure is applicable for V-groove, bevel, and fillet welds within the range of 1/8 in. (3.2 mm) through 1 1/2 in. (38 mm).

Welding may be performed in the following positions:

All.

Base Metal:

The base metal shall conform to carbon steel M-1, P-1, and S-1, Group 1 or 2.

Backing Material Specification:

None.

Filler Metal:

The filler metal shall conform to AWS specification no. E71T-1 from AWS specification A5.20. This filler metal falls into F-number F-6 and A-number A-1.

Shielding Gas:

The shielding gas, or gases, shall conform to the following compositions and purity: CO_2 at 30 to 50 cfh or 75% Ar/25% CO_2 at 30 to 50 cfh.

Joint Design and Tolerances:

Refer to the drawing and specifications in **Figure 24-26** for the workmanship sample layout.

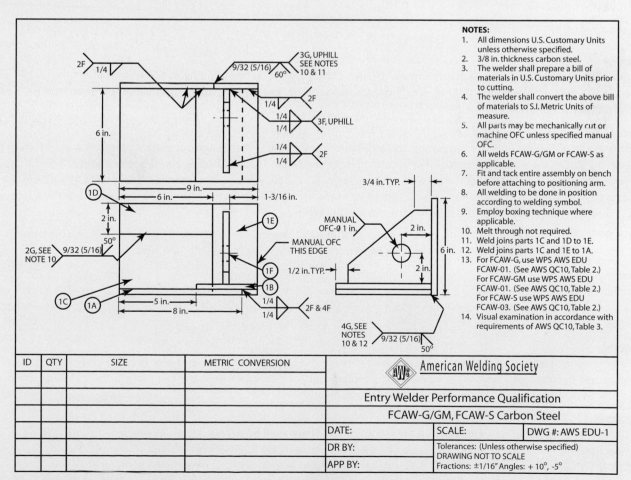

FIGURE 24-26 AWS EDU-1 FCAW carbon steel workmanship sample. American Welding Society

Electrode		Welding Power			Shielding Gas		Base Metal	
Type	Size	Amps	Wire-Feed Speed IPM (cm/min)	Volts	Type	Flow	Type	Thickness
E71T-1	0.035 in. (0.9 mm)	130 to 150	288 to 380 (732 to 975)	22 to 25	CO_2 or 75% Ar/25% CO_2	30 to 50	Low carbon steel	1/4 in. to 1/2 in. (6 mm to 13 mm)
E71T-1	0.045 in. (1.2 mm)	150 to 210	200 to 300 (508 to 762)	28 to 29	CO_2 or 75% Ar/25% CO_2	30 to 50	Low carbon steel	1/4 in. to 1/2 in. (6 mm to 13 mm)

TABLE 24-7 FCAW Gas Shielded Machine Settings

Preparation of Base Metal:

The bevels are to be flame or plasma cut on the edges of the plate before the parts are assembled. The beveled surface must be smooth and free of notches. Any roughness or notches deeper than 1/64 in. (0.4 mm) must be ground smooth.

All hydrocarbons and other contaminants, such as cutting fluids, grease, oil, and primers, must be cleaned off all parts and filler metals before welding. This cleaning can be done with any suitable solvents or detergents. The groove face and inside and outside plate surface within 1 in. (25 mm) of the joint must be mechanically cleaned of slag, rust, and mill scale. Cleaning must be done with a wire brush or grinder down to bright metal.

Electrical Characteristics:

Set the voltage, amperage, wire-feed speed, and shielding gas flow according to **Table 24-7**.

Preheat:

The parts must be heated to a temperature higher than 50°F (10°C) before any welding is started.

Backing Gas:

N/A.

Safety:

Proper protective clothing and equipment must be used. The area must be free of all hazards that may affect the welder or others in the area. The welding machine, welding leads, work clamp, electrode holder, and other equipment must be in safe working order.

Welding Technique:

Using a 1/2-in. (13-mm) or larger gas nozzle and a distance from contact tube to work of approximately 3/4 in. (19 mm) for all welding, first tack weld the plates together according to Figure 24-26. There should be a root gap of about 1/8 in. (3.2 mm) between the plates with V-grooved or beveled edges. Use an E71T-1 arc welding electrode to make a weld. If multiple pass welds are going to be made, a root pass weld should be made to fuse the plates together. Clean the slag from the root pass, being sure to remove any trapped slag along the sides of the weld.

Using an E71T-1 arc welding electrode, make a series of stringer or weave filler welds, no thicker than 1/4 in. (6.4 mm) in the groove until the joint is filled. The 1/4-in. (6.4-mm) fillet welds are to be made with one pass.

Interpass Temperature:

The plate should not be heated to a temperature higher than 350°F (175°C) during the welding process. After each weld pass is completed, allow it to cool but never to a temperature below 50°F (10°C). The weldment must not be quenched in water.

Cleaning:

The slag must be cleaned off between passes. The weld beads may be cleaned by a hand wire brush, a chipping hammer, a punch and hammer, or a needle scaler. All weld cleaning must be performed with the test plate in the welding position. A grinder may not be used to remove weld control problems such as undercut, overlap, or trapped slag.

Inspection:

Visually inspect the weld for uniformity and discontinuities. There shall be no cracks, no incomplete fusion, and no overlap. Undercut shall not exceed the lesser of 10% of the base metal thickness or 1/32 in. (0.8 mm). The frequency of porosity shall not exceed one in each 4 in. (100 mm) of weld length, and the maximum diameter shall not exceed 3/32 in. (2.4 mm).

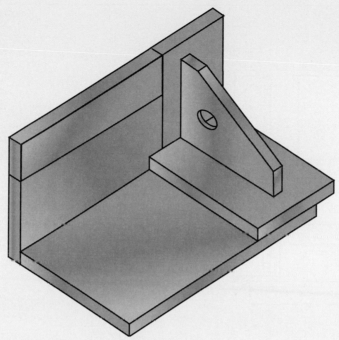

FIGURE 24-27 Practice 24-7 workmanship sample. American Welding Society

Sketches:
Flux Cored Arc Welding (FCAW) Gas Shielded Workmanship Sample drawing, **Figure 24-27.**

Paperwork:
Compete a "Bill of Materials," "Time Sheet," and "Performance Qualification Test Record" in Appendix I or as provided by your instructor. ◆

PRACTICE 24-8

AWS SENSE Entry-Level Welder Workmanship Sample for Flux Cored Arc Welding Self-Shielded (FCAW)

Welding Procedure Specification (WPS) No.: Practice 24-8.

Title:
Welding FCAW of plate to plate.

Scope:
This procedure is applicable for V-groove, bevel, and fillet welds within the range of 1/8 in. (3.2 mm) through 1 1/2 in. (38 mm).

Welding may be performed in the following positions:
All.

Base Metal:
The base metal shall conform to carbon steel M-1, P-1, and S-1, Group 1 or 2.

Backing Material Specification:
None.

Filler Metal:
The filler metal shall conform to AWS specification no. 0.035 in. (0.90 mm) to 0.0415 in. (1.2 mm) diameter E71T-11 from AWS specification A5.20. This filler metal falls into F-number F-6 and A-number A-1.

Shielding Gas:
None.

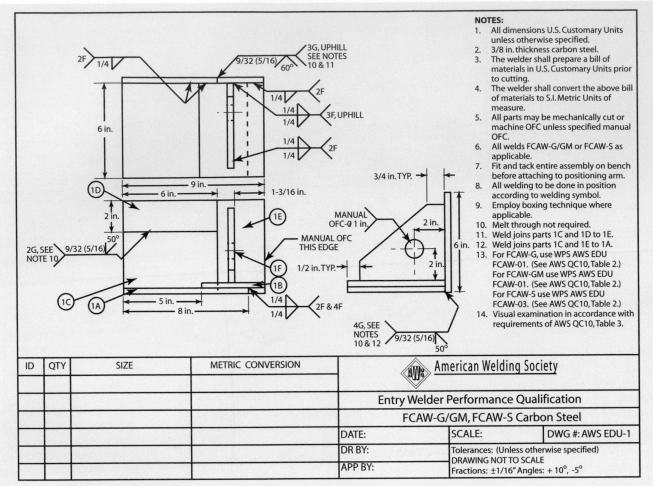

NOTES:
1. All dimensions U.S. Customary Units unless otherwise specified.
2. 3/8 in. thickness carbon steel.
3. The welder shall prepare a bill of materials in U.S. Customary Units prior to cutting.
4. The welder shall convert the above bill of materials to S.I. Metric Units of measure.
5. All parts may be mechanically cut or machine OFC unless specified manual OFC.
6. All welds FCAW-G/GM or FCAW-S as applicable.
7. Fit and tack entire assembly on bench before attaching to positioning arm.
8. All welding to be done in position according to welding symbol.
9. Employ boxing technique where applicable.
10. Melt through not required.
11. Weld joins parts 1C and 1D to 1E.
12. Weld joins parts 1C and 1E to 1A.
13. For FCAW-G, use WPS AWS EDU FCAW-01. (See AWS QC10, Table 2.)
For FCAW-GM use WPS AWS EDU FCAW-01. (See AWS QC10, Table 2.)
For FCAW-S use WPS AWS EDU FCAW-03. (See AWS QC10, Table 2.)
14. Visual examination in accordance with requirements of AWS QC10, Table 3.

ID	QTY	SIZE	METRIC CONVERSION	American Welding Society		
				Entry Welder Performance Qualification		
				FCAW-G/GM, FCAW-S Carbon Steel		
				DATE:	SCALE:	DWG #: AWS EDU-1
				DR BY:	Tolerances: (Unless otherwise specified) DRAWING NOT TO SCALE	
				APP BY:	Fractions: ±1/16" Angles: + 10°, -5°	

FIGURE 24-28 AWS EDU-3 FCAW carbon steel workmanship sample. American Welding Society

Joint Design and Tolerances:

Refer to the drawing and specifications in **Figure 24-28** for the workmanship sample layout.

Preparation of Base Metal:

The bevels are to be flame or plasma cut on the edges of the plate before the parts are assembled. The beveled surface must be smooth and free of notches. Any roughness or notches deeper than 1/64 in. (0.4 mm) must be ground smooth.

All hydrocarbons and other contaminants, such as cutting fluids, grease, oil, and primers, must be cleaned off all parts and filler metals before welding. This cleaning can be done with any suitable solvents or detergents. The groove face and inside and outside plate surface within 1 in. (25 mm) of the joint must be mechanically cleaned of slag, rust, and mill scale. Cleaning must be done with a wire brush or grinder down to bright metal.

Electrical Characteristics:

Set the voltage, amperage, and wire-feed speed flow according to **Table 24-8.**

ELECTRODE TYPE	DIAMETER	VOLTS	AMPS	WIRE-FEED SPEED IPM (cm/min)	ELECTRODE STICKOUT (INCH)
SELF-SHIELDED E70T-11 or E71T-11	0.030	15	40	69 (175)	3/8
		16	100	175 (445)	3/8
		16	160	440 (1118)	3/8
	0.035	15	80	81 (206)	3/8
		17	120	155 (394)	3/8
		17	200	392 (996)	3/8
	0.045	15	95	54 (137)	1/2
		17	150	118 (300)	1/2
		18	225	140 (356)	1/2

TABLE 24-8 FCAW Self-Shielded Machine Settings

Preheat:

The parts must be heated to a temperature higher than 50°F (10°C) before any welding is started.

Backing Gas:

N/A.

Safety:

Proper protective clothing and equipment must be used. The area must be free of all hazards that may affect the welder or others in the area. The welding machine, welding leads, work clamp, electrode holder, and other equipment must be in safe working order.

Welding Technique:

Using a 1/2-in. (13-mm) or larger gas nozzle and a distance from contact tube to work of approximately 3/4 in. (19 mm) for all welding, first tack weld the plates together according to Figure 24-28. There should be about a root gap of about 1/8 in. (3.2 mm) between the plates with V-grooved or beveled edges. Use an E71T-11 arc welding electrode to make a weld. If multiple pass welds are going to be made, a root pass weld should be made to fuse the plates together. Clean the slag from the root pass, being sure to remove any trapped slag along the sides of the weld.

Using an E71T-11 arc welding electrode, make a series of stringer or weave filler welds, no thicker than 1/4 in. (6.4 mm) in the groove until the joint is filled. The 1/4-in. (6.4-mm) fillet welds are to be made with one pass

Interpass Temperature:

The plate should not be heated to a temperature higher than 350°F (175°C) during the welding process. After each weld pass is completed, allow it to cool but never to a temperature below 50°F (10°C). The weldment must not be quenched in water.

Cleaning:

The slag must be cleaned off between passes. The weld beads may be cleaned by a hand wire brush, a chipping hammer, a punch and hammer, or a needle scaler. All weld cleaning must be performed with the test plate in the welding position. A grinder may not be used to remove weld control problems such as undercut, overlap, or trapped slag.

Inspection:

Visually inspect the weld for uniformity and discontinuities. There shall be no cracks, no incomplete fusion, and no overlap. Undercut shall not exceed the lesser of 10% of the base metal thickness or 1/32 in. (0.8 mm). The frequency of porosity shall not exceed one in each 4 in. (100 mm) of weld length, and the maximum diameter shall not exceed 3/32 in. (2.4 mm).

Sketches:

Flux Cored Arc Welding (FCAW) Self-Shielded Workmanship Sample drawing, **Figure 24-29.**

Paperwork:

Compete a "Bill of Materials," "Time Sheet," and "Performance Qualification Test Record" in Appendix I or as provided by your instructor. ◆

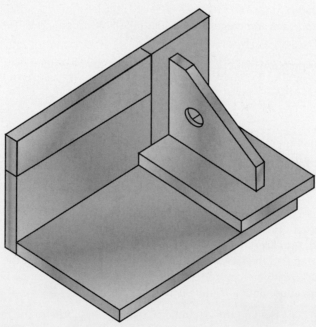

FIGURE 24-29 Practice 24-8 workmanship sample. American Welding Society

PRACTICE 24-9

Gas Tungsten Arc Welding (GTAW) on Plain Carbon Steel Workmanship Sample

Welding Procedure Specification (WPS) No.: Practice 24-9.

Title:
Welding GTAW of sheet to sheet.

Scope:
This procedure is applicable for square groove and fillet welds within the range of 18 gauge through 10 gauge.

Welding may be performed in the following positions:
1G and 2F.

Base Metal:
The base metal shall conform to carbon steel M-1, Group 1.

Backing Material Specification:
None.

Filler Metal:
The filler metal shall conform to AWS specification no. #70S-3 for 1/16 in. (1.6 mm) to 3/32 in. (2.4 mm) diameter as listed in AWS specification A5.18. This filler metal falls into F-number F-6 and A-number A-1.

Electrode:
The tungsten electrode shall conform to AWS specification no. EWTh-2, EWCe-2, or EWLa from AWS specification A5.12. The tungsten diameter shall be 1/8 in. (3.2 mm) maximum.

The tungsten end shape shall be tapered at two to three times its length to its diameter.

Shielding Gas:
The shielding gas, or gases, shall conform to the following compositions and purity: welding grade argon.

Joint Design and Tolerances:
Refer to the drawing and specifications in **Figure 24-30** for the workmanship sample layout.

Preparation of Base Metal:
All hydrocarbons and other contaminants, such as cutting fluids, grease, oil, and primers, must be cleaned off all parts and filler metals before welding. This cleaning can be done with any suitable solvents or detergents. The joint face and inside and outside plate surface within 1 in. (25 mm) of the joint must be mechanically cleaned of slag, rust, and mill scale. Cleaning must be done with a wire brush or grinder down to bright metal.

Electrical Characteristics:
Set the welding current to DCEN and the amperage and shielding gas flow according to **Table 24-9**.

Preheat:
The parts must be heated to a temperature higher than 50°F (10°C) before any welding is started.

Backing Gas:
None.

Safety:
Proper protective clothing and equipment must be used. The area must be free of all hazards that may affect the welder or others in the area. The welding machine, welding leads, work clamp, electrode holder, and other equipment must be in safe working order.

Welding Technique:
TACK WELDS: With the parts securely clamped in place with the correct root gap, the tack welds are to be performed. Holding the electrode so that it is very close to the root face but not touching, slowly increase the current until the arc starts and a molten weld pool is formed. Add filler metal as required to maintain a slightly convex weld face and a flat or slightly concave root face. When it is time to end the tack weld, lower the current slowly so that the molten weld pool can be tapered down in size. When all tack welds are complete, allow the parts to cool as needed before assembling the remaining parts. Repeat the tack welding procedure until the entire part is assembled.

SQUARE GROOVE AND FILLET WELDS: Holding the electrode so that it is very close to the metal surface but not touching, slowly increase the current until the arc starts and a molten weld pool is formed. As the weld progresses, add filler metal as

PRACTICE 24-10

Gas Tungsten Arc Welding (GTAW) on Stainless Steel Workmanship Sample

Welding Procedure Specification (WPS) No.: Practice 24-10.

Title:
Welding GTAW of sheet to sheet.

Scope:
This procedure is applicable for square groove and fillet welds within the range of 18 gauge through 10 gauge.

Welding may be performed in the following positions:
1G and 2F.

Base Metal:
The base metal shall conform to austenitic stainless steel M-8 or P-8.

Backing Material Specification:
None.

Filler Metal:
The filler metal shall conform to AWS specification no. ER3XX from AWS specification A5.9. This filler metal falls into F-number F-6 and A-number A-8.

Electrode:
The tungsten electrode shall conform to AWS specification no. EWTh-2, EWCe-2, or EWLa from AWS specification A5.12. The tungsten diameter shall be 1/8 in. (3.2 mm) maximum. The tungsten end shape shall be tapered at two to three times its length to its diameter.

Shielding Gas:
The shielding gas, or gases, shall conform to the following compositions and purity: welding grade argon.

Joint Design and Tolerances:
Refer to the drawing and specifications in **Figure 24-32** for the workmanship sample layout.

Preparation of Base Metal:
All hydrocarbons and other contaminants, such as cutting fluids, grease, oil, and primers, must be cleaned off all parts and filler metals before welding. This cleaning can be done with any suitable solvents or detergents. The joint face and inside and outside plate surface within 1 in. (25 mm) of the joint must be cleaned of slag, oxide, and scale. Cleaning can be mechanical or chemical. Mechanical metal cleaning can be done by grinding, stainless steel wire brushing, scraping, machining, or filing. Chemical cleaning can be done by using acids, alkalies, solvents, or detergents. Cleaning must be done down to bright metal.

Electrical Characteristics:
Set the welding current to DCEN and the amperage and shielding gas flow according to **Table 24-10.**

Preheat:
The parts must be heated to a temperature higher than 50°F (10°C) before any welding is started.

Backing Gas:
None.

Safety:
Proper protective clothing and equipment must be used. The area must be free of all hazards that may affect the welder or others in the area. The welding machine, welding leads, work clamp, electrode holder, and other equipment must be in safe working order.

Welding Technique:
TACK WELDS: With the parts securely clamped in place with the correct root gap, the tack welds are to be performed. Holding the electrode so that it is very close to the root face but not touching, slowly increase the current until the arc starts and a molten weld pool is formed. Add filler metal as required to maintain a slightly convex weld face and a flat or slightly concave root face. When it is time to end the tack weld, lower the current slowly so that the molten weld pool can be tapered down in size. When all tack welds are complete, allow the parts to cool as needed before assembling the remaining parts. Repeat the tack welding procedure until the entire part is assembled.

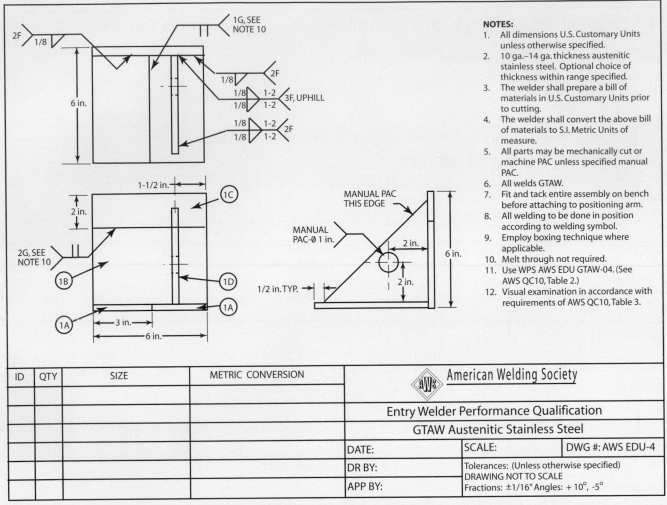

NOTES:
1. All dimensions U.S. Customary Units unless otherwise specified.
2. 10 ga.–14 ga. thickness austenitic stainless steel. Optional choice of thickness within range specified.
3. The welder shall prepare a bill of materials in U.S. Customary Units prior to cutting.
4. The welder shall convert the above bill of materials to S.I. Metric Units of measure.
5. All parts may be mechanically cut or machine PAC unless specified manual PAC.
6. All welds GTAW.
7. Fit and tack entire assembly on bench before attaching to positioning arm.
8. All welding to be done in position according to welding symbol.
9. Employ boxing technique where applicable.
10. Melt through not required.
11. Use WPS AWS EDU GTAW-04. (See AWS QC10, Table 2.)
12. Visual examination in accordance with requirements of AWS QC10, Table 3.

ID	QTY	SIZE	METRIC CONVERSION			
				American Welding Society		
				Entry Welder Performance Qualification		
				GTAW Austenitic Stainless Steel		
				DATE:	SCALE:	DWG #: AWS EDU-4
				DR BY:	Tolerances: (Unless otherwise specified) DRAWING NOT TO SCALE	
				APP BY:	Fractions: ±1/16" Angles: + 10°, -5°	

FIGURE 24-32 AWS EDU-4 GTAW austenitic stainless steel workmanship sample. American Welding Society

Metal Specifications		Gas Flow			Nozzle Size in. (mm)	Amperage Min. Max.
Thickness	**Diameter of ER3XX* (in.)**	**Rates cfm (L/min)**	**Preflow Times**	**Postflow Times**		
18 ga	1/16 in. (2 mm)	15 to 20 (7 to 9)	10 to 15 sec	10 to 25 sec	1/4 to 3/8 (6 to 10)	35 to 60
17 ga	1/16 in. (2 mm)	15 to 20 (7 to 9)	10 to 15 sec	10 to 25 sec	1/4 to 3/8 (6 to 10)	40 to 65
16 ga	1/16 in. (2 mm)	15 to 20 (7 to 9)	10 to 15 sec	10 to 25 sec	1/4 to 3/8 (6 to 10)	40 to 75
15 ga	1/16 in. (2 mm)	15 to 20 (7 to 9)	10 to 15 sec	10 to 25 sec	1/4 to 3/8 (6 to 10)	50 to 80
14 ga	3.32 in. (2.4 mm)	20 to 25 (9 to 12)	10 to 20 sec	10 to 30 sec	3/8 to 5/8 (10 to 16)	50 to 90
13 ga	3.32 in. (2.4 mm)	20 to 25 (9 to 12)	10 to 20 sec	10 to 30 sec	3/8 to 5/8 (10 to 16)	55 to 100
12 ga	3.32 in. (2.4 mm)	20 to 25 (9 to 12)	10 to 20 sec	10 to 30 sec	3/8 to 5/8 (10 to 16)	60 to 110
11 ga	3.32 in. (2.4 mm)	20 to 25 (9 to 12)	10 to 20 sec	10 to 30 sec	3/8 to 5/8 (10 to 16)	65 to 120
10 ga	3.32 in. (2.4 mm)	20 to 25 (9 to 12)	10 to 20 sec	10 to 30 sec	3/8 to 5/8 (10 to 16)	70 to 130

*Other ER3XX stainless steel A5.9 filler metal may be used.

TABLE 24-10 GTAW Stainless Steel Machine Settings

SQUARE GROOVE AND FILLET WELDS: Holding the electrode so that it is very close to the metal surface but not touching, slowly increase the current until the arc starts and a molten weld pool is formed. As the weld progresses, add filler metal as required to maintain a flat or slightly convex weld face. If it is necessary to stop the weld or to reposition yourself or if the weld is completed, the current must be lowered slowly so that the molten weld pool can be tapered down in size.

Interpass Temperature:

The plate should not be heated to a temperature higher than 350°F (180°C) during the welding process. After each weld pass is completed, allow it to cool but never to a temperature below 50°F (10°C). The weldment must not be quenched in water.

Cleaning:

Recleaning may be required if the parts or filler metal become contaminated or oxidized to a degree that the weld quality will be affected. Reclean using the same procedure used for the original metal preparation.

Visual Inspection:

Visual inspection criteria for entry welders*:

1. There shall be no cracks, no incomplete fusion.

2. There shall be no incomplete joint penetration in groove welds except as permitted for partial joint penetration groove welds.

3. The Test Supervisor shall examine the weld for acceptable appearance and shall be satisfied that the welder is skilled in using the process and procedure specified for the test.

4. Undercut shall not exceed the lesser of 10% of the base metal thickness or 1/32 in. (0.8 mm).

5. Where visual examination is the only criterion for acceptance, all weld passes are subject to visual examination at the discretion of the Test Supervisor.

6. The frequency of porosity shall not exceed one in each 4 in. (100 mm) of weld length and the maximum diameter shall not exceed 3/32 in. (2.4 mm).

7. Welds shall be free from overlap.

Sketches:

Gas Tungsten Arc Welding (GTAW) Workmanship Sample drawing for Stainless Steel, **Figure 24-33**.

Paperwork:

Compete a "Bill of Materials," "Time Sheet," and "Performance Qualification Test Record" in Appendix I or as provided by your instructor. ◆

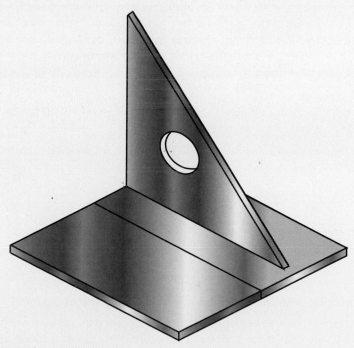

FIGURE 24-33 Practice 24-10 workmanship sample. American Welding Society

*Courtesy of the American Welding Society.

PRACTICE 24-11

Gas Tungsten Arc Welding (GTAW) on Aluminum Workmanship Sample

Welding Procedure Specification (WPS) No.: Practice 24-11.

Title:
Welding GTAW of sheet to sheet.

Scope:
This procedure is applicable for square groove and fillet welds within the range of 18 gauge through 10 gauge.

Welding may be performed in the following positions:
1G and 2F.

Base Metal:
The base metal shall conform to aluminum M-22 or P-22.

Backing Material Specification:
None.

Filler Metal:
The filler metal shall conform to AWS specification no. ER4043 from AWS specification A5.10. This filler metal falls into F-number F-22 and A-number A-5.10.

Electrode:
The tungsten electrode shall conform to AWS specification no. EWCe-2, EWZr, EWLa, or EWP from AWS specification A5.12. The tungsten diameter shall be 1/8 in. (3.2 mm) maximum. The tungsten end shape shall be rounded.

Shielding Gas:
The shielding gas, or gases, shall conform to the following compositions and purity: welding grade argon.

Joint Design and Tolerances:
Refer to the drawing and specifications in **Figure 24-34** for the workmanship sample layout.

Preparation of Base Metal:
All hydrocarbons and other contaminants, such as cutting fluids, grease, oil, and primers, must be cleaned off all parts and filler metals before welding. This cleaning can be done with any suitable solvents or detergents. The joint face and inside and outside plate surface within 1 in. (25 mm) of the joint must be mechanically or chemically cleaned of oxides. Mechanical cleaning may be done by stainless steel wire brushing, scraping, machining, or filing. Chemical cleaning may be done by using acids, alkalies, solvents, or detergents. Because the oxide layer may reform quickly and affect the weld, welding should be started within 10 minutes of cleaning.

Electrical Characteristics:
Set the welding current to AC high-frequency stabilized and the amperage and shielding gas flow according to **Table 24-11**.

Preheat:
The parts must be heated to a temperature higher than 50°F (10°C) before any welding is started.

Backing Gas:
N/A.

Safety:
Proper protective clothing and equipment must be used. The area must be free of all hazards that may affect the welder or others in the area. The welding machine, welding leads, work clamp, electrode holder, and other equipment must be in safe working order.

Welding Technique:
The welder's hands or gloves must be clean and oil free to prevent contamination of the metal or filler rods.

TACK WELDS: With the parts securely clamped in place with the correct root gap, the tack welds are to be performed. Holding the electrode so that it is very close to the root face but not touching, slowly increase the current until the arc starts and a molten weld pool is formed. Add filler metal as required to maintain a slightly convex weld face and a flat or slightly concave root face. When it is time to end the tack weld, lower the current slowly so that the molten weld pool can be tapered down in size. When all tack welds are complete, allow the parts to cool as needed before assembling the remaining parts. Repeat the tack welding procedure until the entire part is assembled.

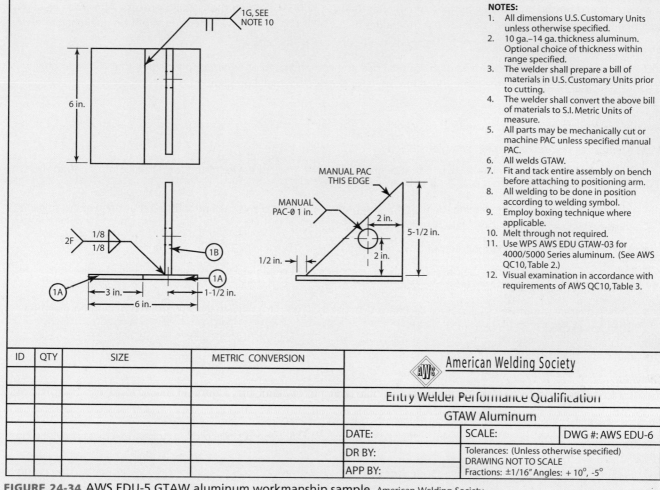

NOTES:
1. All dimensions U.S. Customary Units unless otherwise specified.
2. 10 ga.–14 ga. thickness aluminum. Optional choice of thickness within range specified.
3. The welder shall prepare a bill of materials in U.S. Customary Units prior to cutting.
4. The welder shall convert the above bill of materials to S.I. Metric Units of measure.
5. All parts may be mechanically cut or machine PAC unless specified manual PAC.
6. All welds GTAW.
7. Fit and tack entire assembly on bench before attaching to positioning arm.
8. All welding to be done in position according to welding symbol.
9. Employ boxing technique where applicable.
10. Melt through not required.
11. Use WPS AWS EDU GTAW-03 for 4000/5000 Series aluminum. (See AWS QC10, Table 2.)
12. Visual examination in accordance with requirements of AWS QC10, Table 3.

ID	QTY	SIZE	METRIC CONVERSION

American Welding Society

Entry Welder Performance Qualification

GTAW Aluminum

DATE:	SCALE:	DWG #: AWS EDU-6
DR BY:	Tolerances: (Unless otherwise specified) DRAWING NOT TO SCALE	
APP BY:	Fractions: ±1/16" Angles: + 10°, -5°	

FIGURE 24-34 AWS EDU-5 GTAW aluminum workmanship sample. American Welding Society

Metal Specifications		Gas Flow			Nozzle Size in. (mm)	Amperage Min. Max.
Thickness	Diameter of ER4043*	Rates cfm (L/min)	Preflow Times	Postflow Times		
18 ga	3/32 in. (2.4 mm)	20 to 30 (9 to 14)	10 to 15 sec	10 to 25 sec	1/4 to 3/8 (6 to 10)	40 to 60
17 ga	3/32 in. (2.4 mm)	20 to 30 (9 to 14)	10 to 15 sec	10 to 25 sec	1/4 to 3/8 (6 to 10)	50 to 70
16 ga	3/32 in. (2.4 mm)	20 to 30 (9 to 14)	10 to 15 sec	10 to 25 sec	1/4 to 3/8 (6 to 10)	60 to 75
15 ga	3/32 in. (2.4 mm)	20 to 30 (9 to 14)	10 to 15 sec	10 to 25 sec	1/4 to 3/8 (6 to 10)	65 to 85
14 ga	3/32 in. (2.4 mm)	20 to 30 (9 to 14)	10 to 15 sec	10 to 25 sec	1/4 to 3/8 (6 to 10)	75 to 90
13 ga	1/8 in. (3 mm)	25 to 40 (12 to 19)	10 to 20 sec	10 to 30 sec	3/8 to 5/8 (10 to 16)	85 to 100
12 ga	1/8 in. (3 mm)	25 to 40 (12 to 19)	10 to 20 sec	10 to 30 sec	3/8 to 5/8 (10 to 16)	90 to 110
11 ga	1/8 in. (3 mm)	25 to 40 (12 to 19)	10 to 20 sec	10 to 30 sec	3/8 to 5/8 (10 to 16)	100 to 115
10 ga	1/8 in. (3 mm)	25 to 40 (12 to 19)	10 to 20 sec	10 to 30 sec	3/8 to 5/8 (10 to 16)	100 to 125

*Other aluminum AWS A5.10 filler metal may be used if needed.

TABLE 24-11 GTAW Aluminum Machine Settings

SQUARE GROOVE AND FILLET WELDS: Holding the electrode so that it is very close to the metal surface but not touching, slowly increase the current until the arc starts and a molten weld pool is formed. As the weld progresses, add filler metal as required to maintain a flat or slightly convex weld face. If it is necessary to stop the weld or to reposition yourself or the weld is completed, the current must be lowered slowly so that the molten weld pool can be tapered down in size.

Interpass Temperature:

The plate should not be heated to a temperature higher than 120°F (49°C) during the welding process. After each weld pass is completed, allow it to cool but never to a temperature below 50°F (10°C). The weldment must not be quenched in water.

Cleaning:

Recleaning may be required if the parts or filler metal become contaminated or oxidized to a degree that the weld quality will be affected. Reclean using the same procedure used for the original metal preparation.

Visual Inspection:

Visual inspection criteria for entry welders*:

1. There shall be no cracks, no incomplete fusion.

2. There shall be no incomplete joint penetration in groove welds except as permitted for partial joint penetration groove welds.

3. The Test Supervisor shall examine the weld for acceptable appearance and shall be satisfied that the welder is skilled in using the process and procedure specified for the test.

4. Undercut shall not exceed the lesser of 10% of the base metal thickness or 1/32 in. (0.8 mm).

5. Where visual examination is the only criterion for acceptance, all weld passes are subject to visual examination at the discretion of the Test Supervisor.

6. The frequency of porosity shall not exceed one in each 4 in. (100 mm) of weld length, and the maximum diameter shall not exceed 3/32 in. (2.4 mm).

7. Welds shall be free from overlap.

Sketches:

Gas Tungsten Arc Welding (GTAW) Workmanship Sample drawing for Aluminum, **Figure 24-35.**

Paperwork:

Compete a "Bill of Materials," "Time Sheet," and "Performance Qualification Test Record" in Appendix I or as provided by your instructor. ◆

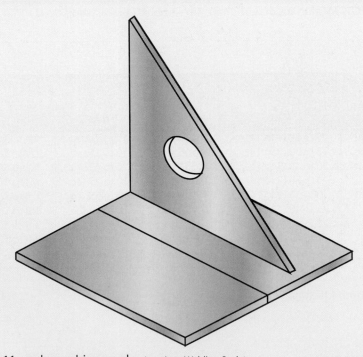

FIGURE 24-35 Practice 24-11 workmanship sample. American Welding Society

*Courtesy of the American Welding Society.

PRACTICE 24-12

Welder and Welder Operator Qualification Test Record (WQR)

Using a completed weld such as the ones from Practice 24-1 or 24-2 and the following list of steps, you will complete the test record shown in **Figure 24-36**. This form is a composite of sample test recording forms provided by AWS, ASME, and API codes. You may want to obtain a copy of one of the codes or standards and compare a weld you made to the standard. This form is useful when you are testing one of the practice welds in this text.

NOTE: Not all of the blanks will be filled in on the forms. The forms are designed to be used with a large variety of weld procedures, so they have spaces that will not be used each time.

---

1. Welder's name: The person who performed the weld.

2. Identification (WPS) No.: On a welding job, every person has an identification number that is used on the time card and paycheck. In this space, you can write the class number or section number since you do not have a clock number.

3. Welding process(es): Was the weld performed with SMAW, GMAW, or GTAW?

4. How was the weld accomplished: Manually, semi-automatically, or automatically?

5. Test position: 1G, 2G, 3G, 4G, 1F, 2F, 3F, 4F, 5G, 6G, 6GR, **Figure 24-37**.

6. What WPS was used for this test?

7. Base metal specification: This is the ASTM specification number, **Table 24-12**.

8. Test material thickness (or) test pipe diameter (and) wall thickness: The actual thickness of the welded material or pipe diameter and wall thickness.

9. Thickness range qualified (or) diameter range qualified: For both plate and pipe, a weld performed successfully on one thickness qualifies a welder to weld on material within that range. See **Table 24-13** for a list of thickness ranges.

10. Filler metal specification number: The AWS has specifications for chemical composition and physical properties for electrodes. Some of these specifications are listed in **Table 24-14**.

11. Classification number: This is the standard number found on the electrode or electrode box, such as E6010, E7018, E316-15, ER1100, and so forth.

12. F-number: A specific grouping number for several classifications of electrodes having similar composition and welding characteristics. See **Table 24-15** for the F-number corresponding to the electrode used.

13. Give the manufacturer's chemical composition and physical properties as provided.

14. Backing strip material specification: This is the ASTM specification number.

15. Give the diameter of electrode used and the manufacturer's identification name or number.

16. Flux for SAW or shielding gas(es) and flow rate for GMAW, FCAW, or GTAW.

Welder and Welding Operator Qualification Test Record (WQR)

Welder or welding operator's name: _____(1)_____   Identification No: _____(2)_____

Welding process _____(3)_____ Manual _____(4)_____ Semiautomatic _____(4)_____ Machine _____(4)_____

Position: _____(5)_____

(Flat, horizontal, overhead or vertical—if vertical, state whether up or down In accordance with welding procedure specification no:) _____(6)_____

Material specification: _____(7)_____

Diameter and wall thickness (if pipe)—otherwise, joint thickness: _____(8)_____

Thickness range this qualifies: _____(9)_____

Filler Metal

Specification No.: _____(10)_____   Classification _____(11)_____   F-number _____(12)_____

Describe filler metal (if not covered by AWS specification): _____(13)_____

Is backing strip used? _____(14)_____

Filler metal diameter and trade name: _____(15)_____ Flux for submerged arc or gas for gas metal arc or flux cored arc welding: _____(16)_____

FIGURE 24-36 Welder and Welding Operator Qualification Test Record (WQR). American Welding Society

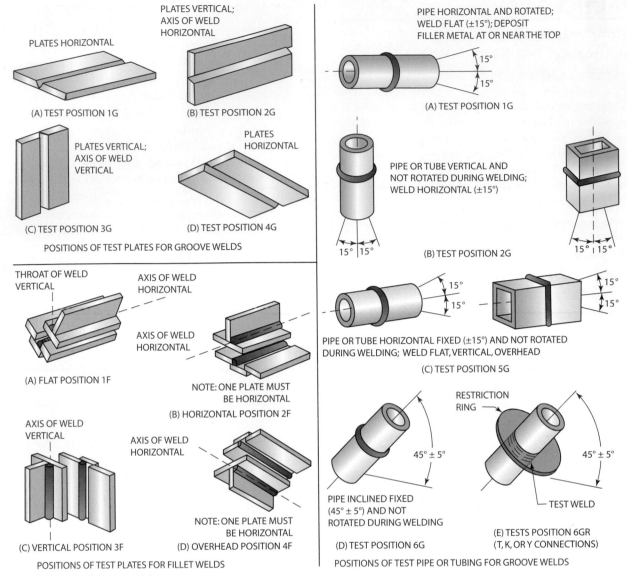

FIGURE 24-37 Welding positions. American Welding Society

Type of Material

P-1	Carbon steel
P-3	Low alloy steel
P-4	Low alloy steel
P-5	Alloy steel
P-6	High alloy steel–predominantley martensitic
P-7	High alloy steel–predominantley ferritic
P-8	High alloy steel–austenitic
P-9	Nickel alloy steel
P-10	Specialty high alloy steels
P-21	Aluminum and aluminum - base alloys
P-31	Copper and copper alloy
P-41	Nickel

TABLE 24-12 P-Numbers

If the weld is a groove weld, follow steps 17 through 22 and then skip to step 27. If the weld test is a fillet weld, skip to step 22.

17. Visually inspect the weld and record any flaws.

18. Record the weld face, root face, and reinforcement dimensions.

19. Four (4) test specimens are used for 3/8-in. (10-mm) or thinner metal. Two (2) will be root bent and two (2) face bent. For thicker metal, all four (4) will be side bent.

20. Visually inspect the specimens after testing and record any discontinuities.

21. Who witnessed the welding for verification that the WPS was followed?

22. The identification number assigned by the testing.

If the weld is a fillet weld, follow steps 23 through 28.

23. Visually inspect the weld and record any flaws.

24. Record the legs and reinforcement dimensions.

25. Measure and record the depth of the root penetration.

26. Polish the side of the specimens and apply an acid to show the complete outline of the weld.

27. Who witnessed the welding for verification that the WPS was followed?

28. The identification number assigned by the testing.

Plate Thickness (T) Tested in. (mm)	Plate Thickness (T) Qualified in. (mm)
1/8 ≤ T < 3/8* (3.1 ≤ T < 9.5)	1/8 to 2T (3.1 to 2T)
3/8 (9.5)	3/4 (19.0)
3/8 > T ≤ 1 (9.5 > T≤ 25.4)	2T 2T
1 and over (25.4 and over)	Unlimited Unlimited

Pipe Size of Sample Weld	
Diameter in. (mm)	Wall Thickness, T
2 (50.8) or 3 (76.2)	Sch. 80 Sch. 40
6 (152.4) or 8 (203.2)	Sch. 120 Sch. 80

Pipe Size Qualified		
Diameter in. (mm)	Wall Thickness, in. (mm)	
3/4 (19.0) through 4 (101.6)	Minimum 0.063 (1.6)	Maximum 0.674 (17.1)
4 (101.6) and over	0.187 (4.7)	Any

*Thickness (T) is equal to or greater than 1/8 in. (≥) and thickness (T) is equal to or less than 3/8 in. (≤).

TABLE 24-13 Test Specimen and Range of Thickness Qualified

If a radiographic test is used, follow steps 29 through 33. If this test is not used, go to step 34.

29. The number the lab placed on the X-ray film before it was exposed on the weld.
30. Record the results of the reading of the film.

Group Designation	Metal Types	AWS Electrode Classification
F1	Carbon steel	EXX20, EXX24, EXX27, EXX28
F2	Carbon steel	EXX12, EXX13, EXX14
F3	Carbon steel	EXX10, EXX11
F4	Carbon steel	EXX15, EXX16, EXX18
F5	Stainless steel	EXXX15, EXXX16
F6	Stainless steel	ERXXX
F22	Aluminum	ERXXXX

TABLE 24-15 F-Numbers

A-Number	Metal and Process(es)
A5.10	Aluminum—bare electrodes and rods
A5.3	Aluminum—covered electrodes
A5.8	Brazing filler metal
A5.1	Steel, carbon, covered electrodes
A5.20	Steel, carbon, flux cored electrodes
A5.17	Steel-carbon, submerged arc wires and fluxes
A5.18	Steel-carbon, gas metal arc electrodes
A5.2	Steel—oxyfuel gas welding
A5.5	Steel—low alloy covered electrodes
A5.23	Steel—low alloy electrodes and fluxes— submerged arc
A5.28	Steel—low alloy filler metals for gas shielded arc welding
A5.29	Steel—low alloy, flux cored electrodes

TABLE 24-14 A-Numbers

31. Whether the test passed or failed the specific code.
32. Who witnessed the welding for verification that the WPS was followed?
33. The number assigned by the testing.
34. The name of the company that requested the test.
35. The name of the person who interpreted the results. This is usually a Certified Welding Inspector (CWI) or other qualified person.
36. Date the results of the test were completed.

Summary

Becoming a certified welder establishes your credentials in the industry. Not every industry requires certification; however, all of the welding fields recognize the importance of being certified. Advertisements in the newspaper, the Internet, and outside of welding shops prominently display the words *Certified Welder*. Even people outside of the welding industry recognize the significance of someone having obtained the educational level and proficiency required to become a certified welder.

An important part of passing a certification test is your ability to follow all of the very specific details required by the certification process for which you are being tested. Read the qualification test procedures carefully. There may be specific requirements in this test you have not experienced before; do not assume you know what is required. As you have learned in this chapter, there are many specific things that must be done in preparation for the certification test to ensure that the test results are valid and to ensure its successful completion. By diligently following the procedures you will certainly enhance your chance of passing the certification. Often your first certification is the most difficult, but it is part of the learning process. Once you have learned the proper techniques and methods of performing certification welding, additional certification tests will become much easier. Experienced certified welders in the field have no difficulty routinely passing certification testing.

Liberty Ships of World War II

The 2710 all-welded Liberty cargo ships built between 1941 and 1945 played a significant role in winning World War II. Most ships that were built before 1941 were riveted and not welded. To change over from riveting ships to welding ships took tens of thousands of new welders. These men and women were expected to learn welding and ship fitting in a very short time so these ships could be built as fast as they were needed. One ship was actually built in 4 days and 15 ½ hours. It took a major effort in recruiting, training, and certifying of the welders to meet the needs of this tremendous undertaking.

As a young welder in the mid-1960s, I worked with some of the welders who had, 20 years earlier, helped build the Liberty Ships. At that time we were all working at Barbour Boat Works in New Bern, North Carolina. It was common on cold damp winter mornings for everyone to show up 20 to 30 minutes before work and stand around an open fire in a barrel, drinking coffee, and talking. I remember them talking about the long hours they spent stick welding on the Liberty Ships. To meet the production schedule large subassemblies were fabricated in the yard and then had to be fitted to the main ship's hull. Since this was the first time anything like this had been done, the welders said fitting was often challenging.

A lot was learned about building all-welded ships during the construction of the Liberty Ships. Many of the ships were sunk during the war, but others survived and later entered commercial shipping. The SS *Hellas Liberty* and SS *John W. Brown* are two Liberty Ships that have been refitted and are still in service today. They are a testament to the skill and craftsmanship of the many welders who built them.

I would like to thank the North Carolina Museum of History for allowing me to reprint these photos and this story about the Liberty Ships. (Larry Jeffus).

As the United States entered World War II, the military found itself ill prepared for large-scale naval operations. By 1941, German submarines were sinking American and British ships at a rate far exceeding their production. The government asked certain large shipbuilding companies, including Newport News Shipbuilding Company in Virginia, to produce on government contract much needed cargo ships for the war effort. In January of 1941 the Newport News Shipbuilding Company in Virginia announced the creation of a subsidiary company in Wilmington, North Carolina, for the purpose of building Liberty Ships. This emergency shipyard, along with 8 others, geared up to produce 260 ships in 1941—a tall order with a desperate purpose.

The new shipyards had templates from which to manufacture prefabricated ships. The Merchant Marine Act of 1936 established a national maritime commission, which produced designs for three standard types of cargo ships. These designs proved invaluable to the high-speed ship production required in 1941. Welding, a new and for the most

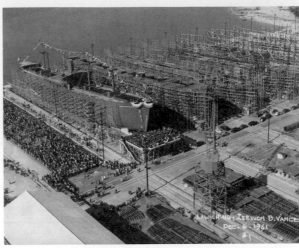

Subassemblies waiting to be fitted to the hull are laid out in the lower right of this photo. North Carolina Museum of History

Eight Liberty Ships under construction are lined up in an assembly line. North Carolina Museum of History

part untested technique in shipbuilding, replaced the laborious and time-consuming method of riveting steel plates to ship frames.

Each cargo ship measured 441 feet long and 56 feet wide. Two oil-burning boilers fed a three-cylinder, reciprocating steam engine, propelling the ship at a speed of up to 11 knots. The vessels could carry more than 9000 tons of cargo in addition to transport airplanes, tanks, and locomotives lashed to their decks. They could carry 2840 jeeps, 440 tanks, or 230 million rounds of rifle ammunition. Newspapers dubbed the ships "ugly ducklings," and President Franklin D. Roosevelt called them "dreadful-looking objects." Attempting to present a more positive image, the U.S. Maritime Commission referred to the ships as the Liberty Fleet and proclaimed

The completed Liberty Ship steams out of the Wilmington harbor. North Carolina Museum of History

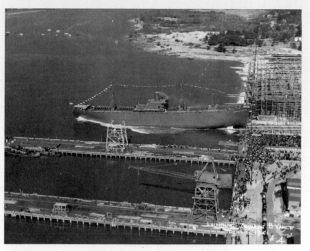

The Liberty Ship *Zebulon B. Vance* slips down the dry dock way to begin its Merchant Marine Service. North Carolina Museum of History

September 27, 1941, the day the first 15 ships were launched, Liberty Fleet Day.

Most Liberty ships took the names of eminent people or of cities that had bought a large number of war bonds. On December 6, 1941, the *Zebulon B. Vance*, named for North Carolina's governor from 1877 to 1879, became the first Liberty Ship launched from Wilmington. Wilmington's population swelled during the war as people arrived to work for

the North Carolina Shipbuilding Company. In 1940, about 35,000 people lived in Wilmington. In 1943, at its peak, the shipyard counted more than 20,000 employees. The workforce included many women, embodied by the fictional poster character Rosie the Riveter. The Wilmington shipyard built 40 Liberty ships in 1942. By the end of the war, 243 ships had come off the line.

Review

1. Does a welder have to retest if a different welding process is to be used?

2. Does every welder need to be certified to get a job?

3. What does having a certification document mean?

4. What are the three levels of AWS's certification?

5. List the steps that should be taken when an SMA weld is being stopped and restarted to avoid weld defects.

6. What can happen if the slag is not chipped off a weld crater before the weld is continued?

7. What problem can occur if weld beads are stopped in corners?

8. What is the acceptable limit for roughness or notches in the surface of a groove that is to be welded?

9. List the types of contaminants that must be cleaned off all parts and filler metals before welding.

10. List the ways that a plate's surface can be cleaned prior to welding.

11. How can SMAW slag be cleaned off between passes?

12. Sketch an open root V-groove joint and show the minimum and maximum root openings allowed as shown in Figure 24-15.

13. What is the maximum undercut allowed on a short-circuit metal transfer workmanship sample?

14. What is the maximum interpass temperature for a GMAW spray transfer workmanship sample?

15. What shielding gas and flow rates should be used with GMAW-S test plate welds without a backing strip?

16. What is the maximum interpass temperature for a plain carbon steel GTAW workmanship sample?

17. What is the AWS identification for the type of tungsten electrode that should be used for welding on the stainless steel workmanship sample?

18. What is the AWS identification for the type of tungsten electrode that should be used for welding on the aluminum workmanship sample?

SECTION 6

RELATED PROCESSES AND TECHNOLOGY

Success Story

My name is Stephan Pawloski and I live in Langley BC Canada. In high school I had the guidance of an exceptional metalwork teacher, Richard Kenny, who did wonders to spark our young minds to the possibilities of working with metal. But then things departed a bit from the conventional. I started university, majoring in the social sciences, but welding never left my mind. Partway through my first year I was going stir crazy not being able to get my hands dirty on a regular basis.

I couldn't find a local welding college where I could double time my courses—Social Sciences during the day and Welding courses at night. I decided to self-train and took a welding job at a local logging outfit. Welding at the logging company was a great experience because they never seemed to run out of broken equipment. Every spare moment I could find was spent reading about welding or practicing welding. Larry Jeffus's *Welding Principles and Applications* would later become a big help in my learning process.

Using this "take control" attitude allowed me to really hone in on the various areas needing more time and practice. At first I cut my welds apart watching the torch wash through the layers looking for any defects. Later, I built my own guided-bend tester and would spend hours welding samples in various positions and cutting them apart to see where I was missing. Those little tig start-and-stop tie-ins can really show up on something like that if you're not careful and let yourself get sloppy.

People would see my welds holding up in the field and it wasn't long before I had built a solid reputation for being a high-quality general welder in our area. I built my own mobile welding rig and started my own company. My main welding equipment consists of a Miller 300-amp engine drive welder, a Lincoln SA-200 Pipeliner, a Lincoln PowerMig, and a Miller Dynasty 300DX tig welder; so there wasn't much that came through the doors I couldn't weld.

Throughout the years, I have come to learn that it doesn't really matter how a person goes about learning. What matters most is that they find a subject that interests them enough that they take control of their education, whether it be through a formal apprenticeship, in the field training, or a self-guided approach such as what I started with. I have come to know a diverse range of professional welders from those who never saw a day of school past grade nine, to those that took applied science courses in the trade. When we stop comparing ourselves on the imaginary socio-status ledger boards and realize that we are all a small part of a much larger project, our welding trade as a whole benefits. ■

Chapter 25

Welding Metallurgy

OBJECTIVES

After completing this chapter, the student should be able to

- list the crystalline structures of metals and explain how grains form.
- work with phase diagrams.
- list the five mechanisms used to strengthen metals.
- explain why steels are such versatile materials.
- describe the types of weld heat-affected zones.
- discuss the problems hydrogen causes during steel welding.
- discuss the heat treatments used in welding.
- explain the cause of corrosion in stainless steel welds.

KEY TERMS

allotropic	face-centered cubic (FCC)	pearlite
allotropic transformation	ferrite	phase diagrams *change matter.*
austenite	grain refinement *make better.*	precipitation hardening *Releasing*
body-centered cubic (BCC)	heat-affected zone (HAZ)	recrystallization temperature
cementite	heat treatments	solid solutions
crystal lattices	martensite	spheroidized microstructure
crystalline structures	metallurgy	tempering
eutectic composition	(needle-like) acicular structure	unit cell

change in matter, Liquid to solid.

where it first starts.

INTRODUCTION

Skilled welders need to know more than just how to establish an arc and manipulate the electrode in order to consistently deposit uniform, high-quality welds. It is important for a competent welder to understand the materials being welded. With this knowledge, the welder can select the best processes and procedures to produce a weldment that is as strong and tough as possible. The welder needs to learn **metallurgy** to recognize that special attention might be needed when welding certain types of steel and to understand the kind of care required.

Metals gain their desirable mechanical and chemical properties as a result of alloying and heat-treating. Welding operations heat the metals, and that heating will certainly change not only the metal's initial structure but its properties as well. A skilled welder can minimize the effects these changes will have on the metal and its properties.

HEAT, TEMPERATURE, AND ENERGY

Heat and *temperature* are both terms used to describe the quantity and level of thermal energy present. To better comprehend what takes place during a weld you must understand the differences between heat and temperature. Heat is the quantity of thermal energy, and temperature is the level of thermal activity.

Although both *heat* and *temperature* are used to describe the thermal energy in a material, they are independent values. A material can have a large quantity of heat energy in it but the material can be at a low temperature. Conversely, a material can be at a high temperature but have very little heat.

Heat

Heat is the amount of thermal energy in matter. All matter contains heat down to absolute zero ($-460°F$ or $-273°C$). The basic U.S. unit of measure for heat is the British thermal unit (BTU). One BTU is defined as the amount of heat required to raise 1 pound of water 1 degree Fahrenheit, and the SI unit of heat is the joule (J).

There are two forms of heat. One is called *sensible* because as it changes, a change in temperature can be sensed or measured. The other form of heat is called *latent*. Latent heat is the heat required to change matter from one state to another, and it does not result in a temperature change. For example, if a pot containing water is heated on a stove, the water picks up heat from the burner and the water's temperature increases. The more heat put into the water, the higher its temperature until it begins to boil. Once the water reaches $212°F$ ($100°C$), its temperature stops rising. As long as the pot is on the burner, heat is being put into the water, but there is no increase in sensible heat. The heat is all going into the latent heat required to change the water from the liquid state to a gaseous state, **Figure 25-1**.

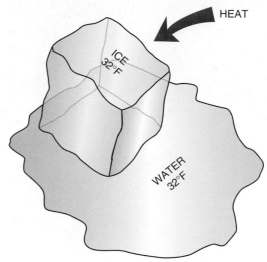

FIGURE 25-1 There is no change in temperature when there is a change in state. © Cengage Learning 2012

Latent heat is absorbed by a material as it changes from a solid to a liquid state and from a liquid to a gaseous state. When matter changes from a gaseous to a liquid state or from a liquid to a solid state, latent heat must be removed, **Figure 25-2**. A change in a material's latent heat also occurs when there is a change in the structure of

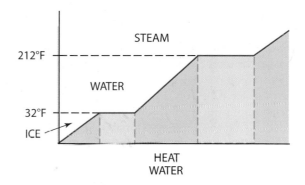

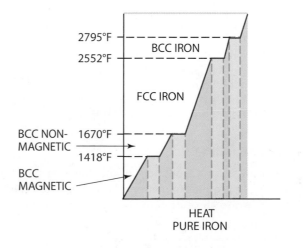

KEY
□ AREA OF SENSIBLE HEAT CHANGE
□ AREA OF LATENT HEAT CHANGE

FIGURE 25-2 Sensible and latent heat. © Cengage Learning 2012

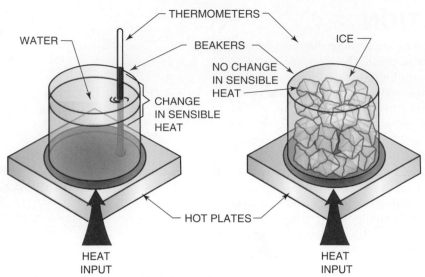

FIGURE 25-3 Both beakers have heat being added, but only one has a change in temperature. © Cengage Learning 2012

the material. For example, when the crystal lattice of a metal changes, a change in latent heat occurs.

EXPERIMENT 25-1

Latent and Sensible Heat

In this experiment you will be working in a small group to observe latent and sensible heat. Using two beakers, two hot plates, two thermometers, a cup of ice, a cup of ice water, gloves, safety glasses, and any other required safety protection, you are going to observe the effect of latent heat on the temperature rise of water, **Figure 25-3**.

Put 1 cup of ice water 32°F (0°C) in one beaker and 1 cup of ice 32°F (0°C) in the other beaker. Place both beakers on a hot plate. Slowly stir the contents of each beaker using the thermometers. Observe the change in temperatures each thermometer measures. Record the following information:

1. What was the temperature when you started?

2. What was the temperature after one minute?

3. What was the temperature of the water without the ice when the ice in the other cup was all gone?

4. How long did it take for all of the ice to melt?

Complete a copy of the "Student Welding Report" listed in Appendix I or provided by your instructor. ◆

Temperature

Temperature is a measurement of the vibrating speed or frequency of the atoms in matter. The atoms in all matter vibrate down to absolute zero. The basic unit of measure is the degree. The U.S. unit is degrees Fahrenheit, and the SI unit is degrees Celsius.

As matter becomes warmer, its atoms vibrate at a higher frequency. As matter cools, the vibrating frequency slows. This vibration of the atoms is what gives off the infrared light that comes from all objects that are above

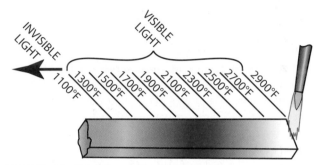

FIGURE 25-4 Visible and invisible light.
© Cengage Learning 2012

absolute zero. When the object becomes hot enough, the vibrating frequency of the atoms gives off visible light. We see that light as a dull red glow when the surface reaches a little above 1000°F. As the surface becomes even hotter, we can see the color light it gives off change until it glows "white hot," **Figure 25-4**.

We can tell the temperature of an object by the frequency of the light that its vibrating atoms produce. That is how scientists tell the temperature of distant stars.

EXPERIMENT 25-2

Temper Colors

In this experiment you will be working in a small group to observe the formation of temper colors as metal is heated. Using a piece of mild steel, a safely set-up oxy-fuel torch, gloves, safety glasses, and required personal protection, you are going to observe the changing temperature's effect on the color of the steel, **Figure 25-5**.

Light the torch and hold it near one side of the mild steel. Observe the other side of the metal to see when it begins to change color. Record the following information:

1. What was the first color that you could see?

2. What were the second, third, fourth, and so on, colors that you could see?

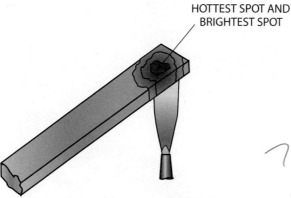

HOTTEST SPOT AND
BRIGHTEST SPOT

FIGURE 25-5 As the temperature of the metal increases, it begins to glow. © Cengage Learning 2012

3. What was the last color that appeared before the metal melted?

4. Compare the colors you saw with the colors and temperatures shown in Figure 25-4.

Repeat this experiment using different metals and see what colors they produce.

Complete a copy of the "Student Welding Report" listed in Appendix I or provided by your instructor. ◆

MECHANICAL PROPERTIES OF METAL

In learning welding skills, an understanding of the mechanical properties of metals is most important. The mechanical properties of a metal can be described as those quantifiable properties that enable the metal to resist externally imposed forces without failing. If the mechanical properties of a metal are known, a product can be constructed that will meet specific engineering specifications. As a result, a safe and sound structure can be constructed.

All of a metal's properties interact with one another. Some properties are similar and complement each other, but others tend to be opposites. For example, a metal cannot be both hard and ductile. Some metals are hard and brittle, and others are hard and tough. Probably the most outstanding property of metals is the ability to have their properties altered by some form of heat treatment. Metals can be soft and then made hard, brittle, or strong by the correct heat treatment, yet other heat treatments can return the metal back to its original, soft form.

It is the responsibility of the metallurgist or engineer to select a metal that has the best group of properties for any specific job. Except in very unusual cases, a metallurgist would not create a new alloy for a job but merely select one from the tens of thousands of metal alloys already available. Often metallurgists must make difficult choices when designs call for properties that are usually not found together. Additionally, the more unique the alloy, the greater its cost and often the more difficult it is to weld.

This section describes some of the significant mechanical properties of metals. The next section describes how various heat treatments can be used to change a metal's properties.

The sections that follow describe some of the significant mechanical properties that the welder should be familiar with to do a successful job of welding fabrication.

Hardness

Hardness may be defined as resistance to penetration. Files and drills are made of metals that rank high in hardness when properly heat-treated. The hardness property may in many metals be increased or decreased by heat-treating methods and increased in other metals by cold working. Since hardness is proportional to strength, it is a quick way to determine strength. It is also useful in determining whether the metal received the proper heat treatment, since heat treatment also affects strength. Hardness is measurable in a number of ways. Most methods quantify a metal's resistance to highly localized deformation.

Brittleness

Brittleness is the ease with which a metal will crack or break apart without noticeable deformation. Glass is brittle, and when broken, all of the pieces fit back together because it did not bend (deform) before breaking. Some types of cast iron are brittle and once broken will fit back together like a puzzle's parts. Brittleness is related to hardness in metals. Generally, as the hardness of a metal is increased, the brittleness is also increased. Brittleness is not measured by any testing method. It is the absence of ductility.

Ductility

Ductility is the ability of a metal to be permanently twisted, drawn out, bent, or changed in shape without cracking or breaking. Ductile metals include aluminum, copper, zinc, and soft steel. Ductility is measurable in a number of ways. Ductility in tensile tests is usually measured as a percentage of elongation and as a percentage of reduction in area. It also can be measured with bend tests.

Toughness

Toughness is the property that allows a metal to withstand forces, sudden shock, or bends without fracturing. Toughness may vary considerably with different methods of load application and is commonly recognized as resistance to shock or impact loading.

Toughness is measured most often with the Charpy test. This test yields information about the resistance of a metal to sudden loading in the presence of a severe notch. Because only a small specimen is required, the test is faulted for not providing a general picture of a component's toughness. Unfortunately, tests on a larger scale require very expensive equipment and are very time consuming.

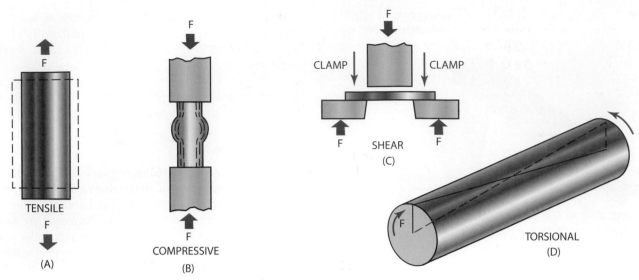

FIGURE 25-6 Types of forces (F) applied to metal. © Cengage Learning 2012

Strength

Strength is the property of a metal to resist deformation. Common types of strength measurements are tensile, compressive, shear, and torsional, **Figure 25-6.**

- **Tensile strength:** *Tensile strength* refers to the property of a material that resists forces applied to pull metal apart. Tension has two parts: yield strength and ultimate strength. *Yield strength* is the amount of strain needed to permanently deform a test specimen. The *yield point* is the point during tensile loading when the metal stops stretching and begins to be permanently made longer by deforming. Like a rubber band that stretches and returns to its original size, metal that stretches before the yield point is reached will return to its original shape. After the yield point is reached, the metal is usually longer and thinner. Some metals stretch a great deal before they yield, and others stretch a great deal before and after the yield point. These metals are considered to have high ductility. *Ultimate strength* is a measure of the load that breaks a specimen. Some metals may become work-hardened as they are stretched during a tensile test. These metals will actually become stronger and harder as a result of being tested. Other metals lose strength once they pass the yield point and fail at a much lower force. Metals that do not stretch much before they break are brittle. The tensile strength of a metal can be determined by a tensile testing machine.

- **Compressive strength:** Compressive strength is the property of a material to resist being crushed. The compressive strength of cast iron, rather brittle material, is three to four times its tensile strength.

- **Shear strength:** Shear strength of a material is a measure of how well a part can withstand forces acting to cut or slice it apart.

- **Torsional strength:** Torsional strength is the property of a material to withstand a twisting force.

Other Mechanical Concepts

Strain is deformation caused by stress. The part shown in **Figure 25-7** is under stress and was strained (deformed) by the external load. The deformation is in the form of a bend.

Elasticity is the ability of a material to return to its original form after removal of the load. The yield point of a material is the limit to which the material can be loaded and still return to its original form after the load has been removed, **Figure 25-8.**

BEAM RETURNS TO ORIGINAL FORM

FIGURE 25-7 Effect of excessive stress causing permanent strain in a beam. © Cengage Learning 2012

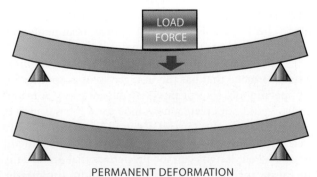

PERMANENT DEFORMATION

FIGURE 25-8 Reaction of an elastic beam to a force.
© Cengage Learning 2012

Elastic limit is defined as the maximum load, per unit of area, to which a material will respond with a deformation directly proportional to the load. When the force on the material exceeds the elastic limit, the material will be deformed permanently. The amount of permanent deformation is proportional to the stress level above the elastic limit. When stressed below its elastic limit, the metal returns to its original shape.

Impact strength is the ability of a metal to resist fracture under a sudden load. An example of a material that is ductile and yet has low impact strength is Silly Putty. If it is pulled slowly, it stretches easily, but if it is pulled quickly, it forms a brittle fracture.

STRUCTURE OF MATTER

All solid matter exists in one of two basic forms. Solid matter is either crystalline or amorphous in form. Solids that are crystalline in form have an orderly arrangement of their atoms. Each crystal making up the solid can be very small, too small to be seen without a microscope. Examples of materials that are crystalline in form include metals and most minerals, such as table salt, **Table 25-1**. Amorphous materials have no orderly arrangement of their atoms into crystals. Examples of amorphous materials include glass and silicon, **Table 25-2**. Both crystalline and amorphous materials look and feel like solids, so without sophisticated testing equipment, you cannot tell the difference between them

Metal	Crystal Type
Aluminum	fcc
Chromium	bcc
Copper	fcc
Gold	fcc
Iron (alpha)	fcc
Iron (delta)	bcc
Iron (gamma)	fcc
Lead	fcc
Nickel	fcc
Silver	fcc
Tungsten	bcc
Zinc	hcp

KEY
fcc = Face-centered cubic
bcc = Body-centered cubic
hcp = Hexagonal close-packed

TABLE 25-1 Crystal Structure of Common Metals

Materials		Crystal Type
Copper acetylide	Cu_2C_2	None
Gadolinium oxide	Gd_2O_3	None
Iron hydroxide	$Fe(OH)_2$	None
Lead oxide	Pb_2O	None
Nickel monosulfide	NiS	None
Silicone dioxide	SiO_2	None

TABLE 25-2 Amorphous Materials

Crystalline Structures of Metal

The fundamental building blocks of all metals are atoms arranged in very precise three-dimensional patterns called **crystal lattices.** Each metal has a characteristic pattern that forms these crystal lattices. The smallest identifiable group of atoms is the **unit cell.** The unit cells that characterize all commercial metals are illustrated in **Figure 25-9**, **Figure 25-10**, and **Figure 25-11**. It may take

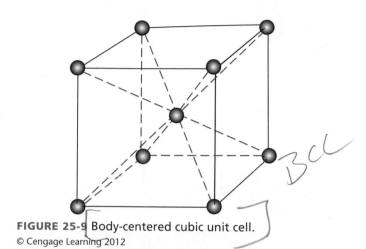

FIGURE 25-9 Body-centered cubic unit cell.
© Cengage Learning 2012

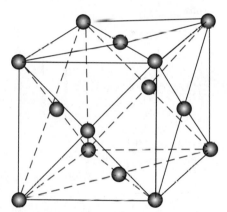

FIGURE 25-10 Face-centered cubic unit cell.
© Cengage Learning 2012

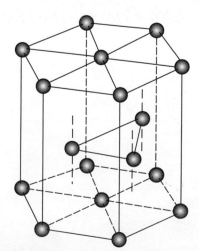

FIGURE 25-11 Hexagonal close-packed cubic unit cell.
© Cengage Learning 2012

millions of these individual unit cells to form one crystal. Although some metals have identical atomic arrangements, the dimensions between individual atoms vary from metal to metal. Some metals change their lattice structure when heated above a specific temperature.

Crystals develop and grow by the attachment of atoms to the submicroscopic unit cells forming the metal's characteristic crystal structure. The final individual crystal size within the metal depends on the length of time the material is at the crystal-forming temperature and any obstructions to its growth. In most cases, solid crystals continue to grow larger randomly from a liquid as it cools, until they encounter other crystals growing in a similar fashion. The result is solid metal composed of microscopic crystals with the same structure but different orientations. Their crystalline shapes depend on how they grew and how other crystals interrupted that growth. Crystals growing from liquid develop a dendritic or columnar structure, **Figure 25-12** and **Figure 25-13**.

The crystal structures are studied by polishing small pieces of the metals, etching them in dilute acids, and examining the etched structures with a microscope. Such examinations enable metallurgists to determine how the metal formed and to observe changes caused by heat treating and alloying. These microscopic examinations reveal various combinations of three different phases: pure metal, **solid solutions** of two or more metals dissolved in one another, and intermetallic compounds.

PHASE DIAGRAMS

If the metalworking industries used only pure metals, the only required information about their crystalline structure could be a list of their melting temperatures and **crystalline structures.** However, most engineering materials are alloys, not pure metals. An *alloy* is a metal with one or more elements added to it, resulting in a significant change in the metal's properties. It is inconvenient, if not impossible, to list all phases and temperatures at which alloys exist. That kind of information is summarized in graphs called **phase diagrams.** Phase diagrams are also known as *equilibrium* or *constitution diagrams,* and the terms are used interchangeably. These diagrams do not necessarily describe what happens with rapid changes in temperature since metals are sluggish in response to temperature fluctuations. They do describe the constituents present at temperature equilibrium.

Lead-Tin Phase Diagram

The chart in **Figure 25-14** is a phase diagram representing the changes brought about by alloying lead (Pb) with tin (Sn). Although this phase diagram is simpler than the iron-carbon phase diagram used for steel, it has many similarities. On the charts, temperature is vertical and alloy percent; in this case, lead and tin are horizontal.

Metallurgy uses Greek letters to identify different crystal structures. On this chart, the Greek letter "a" (alpha, or α) is used for one crystal form and the Greek letter "b" (beta, or β) is used to represent the other crystal form. The chart has four different areas identified:

1. *Liquid phase:* The area at the top with the highest temperatures is where all the metal is a molten liquid.

2. *Solid phase:* The area in the lower center with the lowest temperatures is a solid mixture of (α- and β-type crystals.

3. *Liquid-solid phase:* The two triangular-shaped areas that touch in the center contain a slurry or paste made up of liquid and a specific type of solid crystals.

4. *Solid-solution phase:* The two triangular-shaped areas that stand vertical, one on each side next to the temperature scales, are solid crystals in either α form on the left side or β form on the right side.

Notice on the chart that although 100% lead becomes a liquid at 620°F (327°C) and 100% tin becomes a liquid

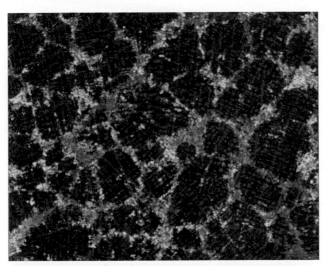

FIGURE 25-12 Cell boundaries in gray cast iron. Buehler Ltd.

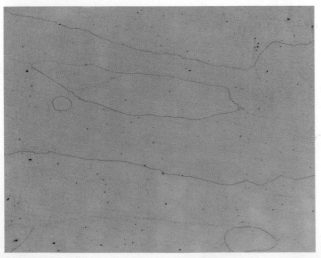

FIGURE 25-13 Columnar structure. Buehler Ltd.

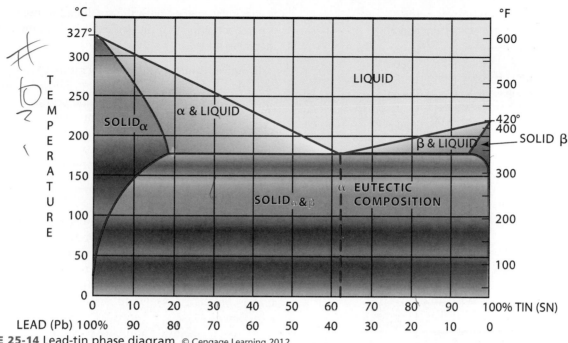

FIGURE 25-14 Lead-tin phase diagram. © Cengage Learning 2012

at 420°F (232°C), a mixture of 38.1% lead and 61.9% tin becomes a liquid at 362°F (183°C). The mixture has a lower melting temperature than either of the two metals in the mixture. The mixture melts at a temperature of 258°F (144°C) below 100% lead and 58°F (49°C) below 100% tin. This mixture is called the eutectic composition of lead and tin. A **eutectic composition** is the lowest possible melting temperature of an alloy.

On the chart, the broadest temperature for a slurry or paste is a mixture of approximately 80% (80.5%) lead and 20% (19.5%) tin. This mixture remains partially liquid and solid during a 173°F (78°C) temperature change. While a metal is in the liquid-solid phase, it is very weak and any movement will cause cracks to form. For this reason, some aluminum alloys crack when a GTA weld is made without adding filler metal and many other metals form crater cracks at the end of a weld. In both cases, as the metal cools it shrinks and pulls itself apart in the center of the weld. The addition of filler metal changes the alloy so it is not as subject to hot cracking. The only metals that do not go through the liquid-solid phase are pure metals and those eutectic composition alloys.

As a 90% lead and 10% tin mixture cools from the 100% liquid phase, it forms an α solid crystal in a liquid. As cooling continues all of the liquid forms into the α crystal. But as solid β crystals cool to around 300°F (150°C), some of them change into the β crystal form. This type of solid-solution phase change occurs in many metals. Steel goes through several such changes as it is heated and cooled even though it never melted. It is these changes in steel that allow it to be hardened and softened by heating, *quenching, tempering,* and *annealing.*

Phase diagrams for other metal alloys are more complicated, but they are used in exactly the same way. They describe the effects of changes in temperature or alloying on different phases.

> NOTE: *The following information relates to Figure 25-15.*

Iron-Carbon Phase Diagram

The iron-carbon phase diagram is illustrated in **Figure 25-15.** The iron-carbon phase diagram is more complex than that for the lead-tin—it has more lines with more solid-solution phase changes—but it is read in the same way. In the iron-carbon diagram used here, the percentage of iron starts at 100% and goes down to 99.1%, while the percentage of carbon goes from 0.0% to 0.9%. Unlike the lead-tin alloys that go from 0.0% to 100% mixtures, very small changes in the percentage of carbon produce major changes in the alloy's properties.

Iron is a pure metal element containing no measurable carbon and is relatively soft. This soft metal when alloyed with as little as 0.80% carbon can become tool steel, **Table 25-3.** Other alloying elements are added to iron to enhance its properties. No other alloying element has such a dramatic effect as carbon.

Iron is called an **allotropic** metal, because it exists in two different crystal forms in the solid state. It changes between the different crystal forms as its temperature changes. The changes in the crystal structure occur at very precise temperatures. Pure iron forms the **body-centered cubic (BCC)** crystal below a temperature of 1675°F (913°C). Iron changes to form the **face-centered cubic (FCC)** crystal above 1675°F (913°C). The body-centered cubic form of iron is called *alpha ferrite,* abbreviated α-Fe. The face-centered cubic form of iron is called austenite, abbreviated γ-Fe.

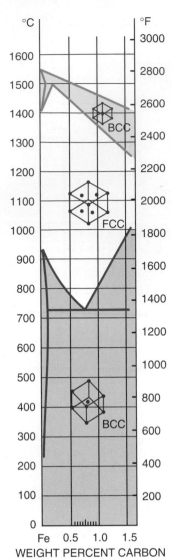

FIGURE 25-15 Iron-carbon phase diagram.
© Cengage Learning 2012

is triangular-shaped with the small end to the right. This is in the direction that the percentage of carbon in the iron increases.

2. **Lower transformation temperature (A_1)**—Termed Ac_1 on heating, Ar_1 on cooling. Below Ac_1 structure ordinarily consists of ferrite and pearlite (see 2a and 2b in this list). On heating through Ac_1 these constituents begin to dissolve in each other to form austenite (see 2d), which is nonmagnetic. This dissolving action continues on heating through the transformation range until the solid solution is complete at the upper transformation temperature.[1]

We do not think of solids as being able to dissolve into each other, but above the lower transformation temperature (A_1) the carbon-rich cementite (see 2c) and the carbon-lean ferrite, both solids, dissolve into each other. The dissolving of one material into another as the mixture is heated is similar to dissolving salt into water, Experiment 25-3.

Ferrite, cementite, and austenite are all crystalline forms of iron and iron-carbon alloys. Their formation is based on three factors: the carbon content, temperature, and time.

- *Carbon content:* Before any of the crystal forms can be created, the correct carbon percentage must exist in the alloy.

- *Temperature:* With the correct carbon content, an iron-carbon alloy will form a specific crystal at the approximate temperature for that specific alloy, given enough time.

- *Time:* Crystal formation requires time. The more time at the correct temperature, the larger the crystals can grow. Heating is a relatively slow process compared to quenching, which can occur in a fraction of a second. Therefore, crystals tend to grow larger or change forms during heating and stay frozen in the larger size or form when cooled.

2a. **Ferrite** is practically pure iron (in plain carbon steels) existing below the lower transformation temperature. It is magnetic and has very slight solid solubility, less than 0.02%, of carbon. Very low carbon

NOTE: *The following information relates to **Figure 25-16**.* The numbered paragraphs can be found printed in a column on the right side of the figure.

1. **Transformation range**—In this range steels undergo internal atomic changes that radically affect the properties of the material.[1]

You will notice that the transformation range (1) is between lines (2) and (3). This area (1) on **Figure 25-17**

[1]*Basic Guide to Ferrous Metallurgy*, Copyright 1954 by Tempil, Hamilton Boulevard, South Plainfield, NJ 07080.

Alloy Name	% Carbon*	Major Properties
Iron	0.0% to 0.03%	Soft, easily formed, not hardenable
Low carbon	0.03% to 0.30%	Strong, formable
Medium carbon	0.30% to 0.50%	High strength, tough
High carbon	0.50% to 0.90%	Hard, tough
Tool steel	0.80% to 1.50%	Hard, brittle
Cast iron	2% to 4%	Hard, brittle, most types resist oxidation

TABLE 25-3 Iron-Carbon Alloys *Carbon is not the only alloying element added to iron.

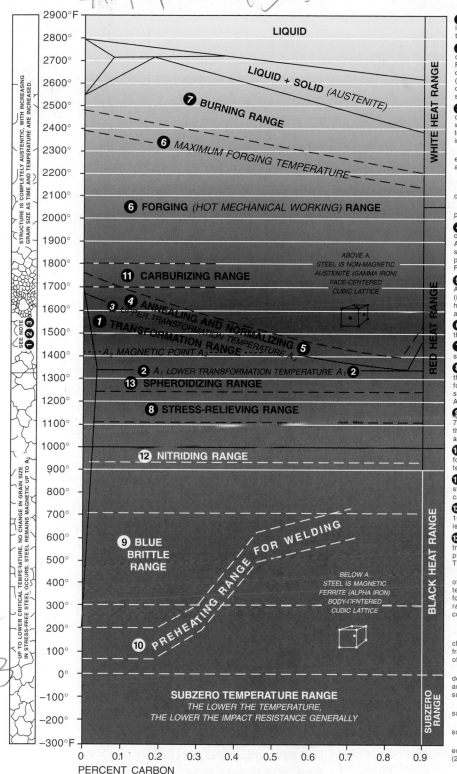

❶ TRANSFORMATION RANGE. In this range steels undergo internal atomic changes which radically affect the properties of the material.

❷ LOWER TRANSFORMATION TEMPERATURE (A₁). Termed Ac₁ on heating, Ar₁ on cooling. Below Ac₁ structure ordinarily consists of FERRITE and PEARLITE (see below). On heating through Ac₁ these constituents begin to dissolve in each other to form AUSTENITE (see below) which is nonmagnetic. This dissolving action continues on heating through the TRANSFORMATION RANGE until the solid solution is complete at the upper transformation temperature.

❸ UPPER TRANSFORMATION TEMPERATURE (A₃). Termed Ac₃ on heating, Ar₃ on cooling. Above this temperature the structure consists wholly of AUSTENITE which coarsens with increasing time and temperature. Upper transformation temperature is lowered as carbon increases to 0.85% (eutectoid point).

• FERRITE is practically pure iron (in plain carbon steels) existing below the lower transformation temperature. It is magnetic and has very slight solid solubility for carbon.

• PEARLITE is a mechanical mixture of FERRITE and CEMENTITE.

• CEMENTITE or IRON CARBIDE is a compound of iron and carbon, Fe₃C.

• AUSTENITE is the nonmagnetic form of iron and has the power to dissolve carbon and alloying elements.

❹ ANNEALING, frequently referred to as FULL ANNEALING, consists of heating steels to slightly above Ac₃, holding for AUSTENITE to form, then slowly cooling in order to produce small grain size, softness, good ductility and other desirable properties. On cooling slowly the AUSTENITE transforms to FERRITE and PEARLITE.

❺ NORMALIZING consists of heating steels to slightly above Ac₃, holding for AUSTENITE to form, then followed by cooling (in still air). On cooling, AUSTENITE transforms giving somewhat higher strength and hardness and slightly less ductility than in annealing.

❻ FORGING RANGE extends to several hundred degrees above the UPPER TRANSFORMATION TEMPERATURE.

❼ BURNING RANGE is above the FORGING RANGE. Burned steel is ruined and cannot be cured except by remelting.

❽ STRESS RELIEVING consists of heating to a point below the LOWER TRANSFORMATION TEMPERATURE, A₁, holding for a sufficiently long period to relieve locked-up stresses, then slowly cooling. This process is sometimes called PROCESS ANNEALING.

❾ BLUE BRITTLE RANGE occurs approximately from 300°F to 700°F. Peening or working of steels should not be done between these temperatures, since they are more brittle in this range than above or below it.

❿ PREHEATING FOR WELDING is carried out to prevent crack formation. See TEMPIL PREHEATING CHART for recommended temperatures for various steels and nonferrous metals.

⓫ CARBURIZING consists of dissolving carbon into surface of steel by heating to above transformation range in presence of carburizing compounds.

⓬ NITRIDING consists of heating certain special steels to about 1000°F for long periods in the presence of ammonia gas. Nitrogen is absorbed into the surface to produce extremely hard "skins."

⓭ SPHEROIDIZING consists of heating to just below the lower transformation temperature, A₁, for a sufficient length of time to put the CEMENTITE constituent of PEARLITE into globular form. This produces softness and in many cases good machinability.

• MARTENSITE is the hardest of the transformation products of AUSTENITE and is formed only on cooling below a certain temperature known as the M₃ temperature (about 400°F to 600°F for carbon steels). Cooling to this temperature must be sufficiently rapid to prevent AUSTENITE from transforming to softer constituents at higher temperatures.

• EUTECTOID STEEL contains approximately 0.85% carbon.

• FLAKING occurs in many alloy steels and is a defect characterized by localized microcracking and "flake-like" fracturing. It is usually attributed to hydrogen bursts. Cure consists of cycle cooling to at least 600°F before air-cooling.

• OPEN OR RIMMING STEEL has not been completely deoxidized and the ingot solidifies with a sound surface ("rim") and a core portion containing blowholes which are welded in subsequent hot rolling.

• KILLED STEEL has been deoxidized at least sufficiently to solidify without appreciable gas evolution.

• SEMIKILLED STEEL has been partially deoxidized to reduce solidification shrinkage in the ingot.

• A SIMPLE RULE: Brinell Hardness divided by 2, times 1000, equals approximate Tensile Strength in pounds per square inch. (200 Brinell ÷ 2 x 1000 = approx. 100,000 Tensile Strength, p.s.i.)

FIGURE 25-16 Basic guide to ferrous metallurgy. Tempil, An Illinois Tools Works Co.

content alloys produce more ferrite crystals than they do other crystal forms. For this reason, these alloys are considered to be soft. Without enough carbon, these low carbon alloys will not become very hard and brittle even if they are quenched. The inability to become hard even alongside a weld makes them easily machinable after welding.

2b. **Pearlite** is a mechanical mixture of ferrite and cementite in much the same way that one might mix two different colors of sand, **Figure 25-18**. Each grain of sand still retains its unique color even though it has been mixed with the other colored grains. Unlike sand, each grain of ferrite and cementite fit together like a puzzle's parts because they grew together.

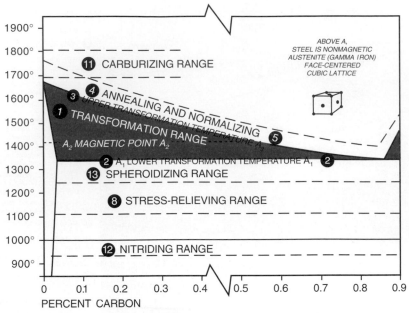

FIGURE 25-17 Transformation range. © Cengage Learning 2012

FIGURE 25-18 Microstructure of pearlitic ductile iron containing cementite (white). Buehler Ltd.

2c. **Cementite,** or iron carbide, is a compound of iron and carbon, Fe_2C.

2d. **Austenite** is the nonmagnetic form of iron and has the power to dissolve carbon and alloying elements. An iron-carbon alloy at a low temperature can dissolve or absorb very little additional alloys. Like water and salt in Experiment 25-3, however, the higher the temperature and the longer the time the more salt that is dissolved. In the same manner, iron will accept more alloys at higher temperatures. These new alloys may stay trapped within the iron-carbon crystal when they are cooled or they may separate. Once dissolved at high temperatures, it is usually the cooling rate or time that affects the final makeup of the iron-carbon crystals.

3. **Upper transformation temperature (A_3)**—Termed Ac_3 on heating, Ar_3 on cooling. Above this temperature the structure consists wholly of austenite, which coarsens with increasing time and temperature. The upper transformation temperature is lowered as carbon increases to 0.85% (eutectoid point).[1]

The upper transformation temperature (A_3) line slopes downward to the right. That is because alloying carbon to iron lowers the temperature for the allotropic crystal transformation from BCC to FCC. This temperature reduction is similar to what is observed as the mixture of lead-tin approached the 80.5% lead and 19.5% tin ratios, Figure 25-14. Figure 25-15, which represents carbon content ranging from 0.0% up to 1.50%, shows over a 200°F (100°C) drop in the melting temperature of iron-carbon alloys. The transition range starts at about 1675°F (912°C) and lowers to about 1335°F (724°C) when 0.85% carbon is present.

The temperature at which the allotropic crystal transition of ferrite to austenite occurs is so important in all heat treatments of steel that metallurgists call it the *critical temperature.*

STRENGTHENING MECHANISMS

Perhaps the most important physical characteristic of a metal is strength. Most pure metals are relatively weak. Structures built of pure metals would be more massive and heavier than those built with metals strengthened by alloying and heat-treating. Welders must understand

---

[1]*Basic Guide to Ferrous Metallurgy*, Copyright 1954 by Tempil, Hamilton Boulevard, South Plainfield, NJ 07080.

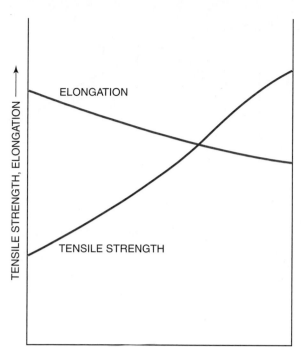

FIGURE 25-19 The general change in properties caused by the addition of other atoms in alloys, producing solid solutions. © Cengage Learning 2012

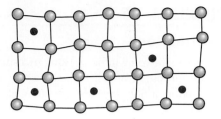

KEY

○ ATOMS OF THE BASE METAL

● ATOMS OF THE HARDENING ALLOY

(A)

(B)

FIGURE 25-20 Solid solution.
(A) Illustration of solid solution matrix. © Cengage Learning 2012
(B) Microstructure of solid solution
George F. VanderVoort, Consultant

numerous methods used to strengthen metals. Some of the strengthening methods used and how welding affects them are described next.

Solid-Solution Hardening

It is possible to replace some of the atoms in the crystal lattice with atoms of another metal in a process called *solid-solution hardening*. In such cases, the lattices of the solid solution and the pure metal are the same except that the lattice of the solid solution is strained as more foreign atoms dissolve. These alloyed prestressed crystals are stronger and less ductile. The amount of change in these properties depends on the number of second atoms introduced, **Figure 25-19.**

Although an important metallurgical tool, solid-solution hardening has its drawbacks. Not all metals have lattice dimensions that allow significant substitution of other atoms. The amount that can be introduced this way is thus limited. Solid-solution hardening does have the advantage of not changing lattice structure as the result of thermal treatments. Thus, solid-solution-strengthened alloys are generally considered very weldable, **Figure 25-20.** Many aluminum alloys are strengthened in this way.

Precipitation Hardening

Alloy systems that show a partial solubility in the solid phase generally have very low solubility at room temperature. Solubility increases with temperature until the

alloy system reaches its solubility limit. For example, the aluminum-copper system has a solid-solution solubility of 0.25% copper at room temperature, which increases to a maximum of 5.6% copper at 1019°F (548°C). For this reason, very few, if any, aluminum alloys have more than 4.5% copper as an alloying element. These alloys reach their saturation just like a sponge saturated with water; when more water is added, it just runs off. When more alloy is added to metals, it will not be combined with the base metal.

Precipitation hardening is a heat treatment involving three steps: (1) heating the alloy enough to dissolve the second phase and form a single solid solution; (2) quenching the alloy rapidly from the solution temperature to keep the second phase in solution, thus producing a supersaturated solution; and (3) reheating the alloy with careful control of time and temperature to precipitate the second phase as very fine crystals that strengthen the lattice in which they had dissolved. This process, called *precipitation hardening* or *age hardening,* is the heat treatment used to strengthen many aluminum alloys.

EXPERIMENT 25-3

Crystal Formation

In this experiment you will be working in a small group to observe the formation of salt crystals. Using a beaker, water, salt, a spoon, a hot plate, gloves, safety glasses, and any other required safety protection, you are going to observe the ability of salt to be dissolved into water as the water is heated.

The dissolving of salt in water shows an increase in solubility as the temperature increases. Add a spoonful of salt to the beaker of cold water and stir vigorously until the salt is dissolved. This is an example of a liquid solution. If more salt is slowly added and stirring is continued, at some point the salt will no longer dissolve into the solution. The water has now reached its solubility limit. No more salt will dissolve no matter how long the mixture is stirred.

If the beaker is heated, the salt crystals that did not dissolve at room temperature soon disappear. Now even more salt could be added, and it would dissolve. The increasing of the solubility limits by heating has resulted in a supersaturated solution. A supersaturated solution is one that, because of heating, a higher percentage of another material can be dissolved in.

When the beaker of water is cooled to room temperature, with time the reverse action occurs. The supersaturated solution can no longer hold the salt in the solution and rejects it from the solution. In time, salt crystals reappear in the beaker.

Complete a copy of the "Student Welding Report" listed in Appendix I or provided by your instructor. ◆

Mechanical Mixtures of Phases

Two phases or constituents may exist in equilibrium, depending on the alloy's temperature and composition. The mixture may consist of two different crystals, which are solid solutions of the two metals of the alloy (see the α & β area in Figure 25-14), or the mixture may consist of a single solid solution and an intermetallic compound,

FIGURE 25-21 Pearlite. George F. VanderVoort, Consultant

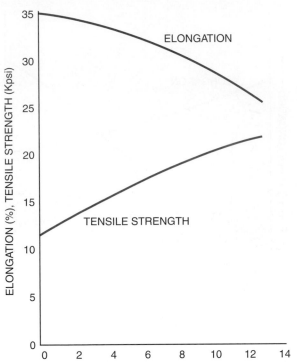

FIGURE 25-22 Change in mechanical properties caused by beta (silicon phase) in mechanical mixture with alpha (aluminum phase). © Cengage Learning 2012

such as the pearlite in **Figure 25-21**. The properties of alloys that are mechanical mixtures of two phases are generally related linearly to the relative amounts of the metal's two constituents, **Figure 25-22**.

At room temperature, the iron-carbon alloy has two forms: alpha iron ferrite and cementite. The ferrite is very ductile but weak; the cementite is very strong but brittle. In a careful combination, the cementite stiffens the soft ferrite crystals, increasing their strength without losing too much ductility.

Quench, Temper, and Anneal

Quenching is the process of rapidly cooling a metal by one of several methods. The quicker the metal cools, the greater the quenching effect. The most common methods of quenching are listed from the slowest to the fastest:

- Molten salt quenching—Very slow cooling.

- Air quenching—Air is blown across the part, cooling it only slightly faster than it would cool in still air.

- Oil quenching—The part is immersed in a bath of oil; because oil is a poor conductor of heat, the part cools more slowly than it would in water.

- Water quenching—The part is immersed in a bath of water and cools rapidly.

- Brine quenching—The part is immersed in a saltwater solution. The salt does not allow a steam pocket to form around the part, as happens with straight water. This results in the fastest quenching time, **Figure 25-23**.

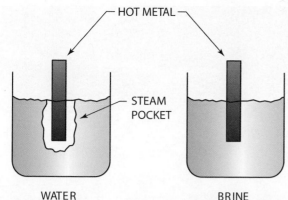

FIGURE 25-23 Effect of brine on quenching rate.
© Cengage Learning 2012

NOTE: Agitating the metal (moving it around rapidly) when it is immersed in a cooling liquid will speed up the cooling rate.

Tempering is the process of reheating a part that has been hardened through heating and quenched. The reheating reduces some of the brittle hardness caused by the quenching, replacing it with toughness and increased tensile strength.

EXPERIMENT 25-4

Effect of Quenching and Tempering on Metal Properties

In this experiment you will be working in a small group to identify the effect of quenching and tempering on metal properties. Using three or more pieces of mild steel approximately 1/4 in. (6 mm) thick, 1 in. (25 mm) wide, and 6 in. (13 mm) long; a vise with a hammer or tensile tester; a hacksaw; a file; water for quenching; two or more firebricks; a safely assembled and properly lit oxyfuel torch; pliers; safety glasses, gloves, and any other required safety equipment; you are going to observe the effect that quenching and tempering has on metal.

Heat the pieces of metal one at a time to a bright red color. Place one of them, while still red hot, between hot firebricks. Immerse the other two into the water while they are still glowing bright red. Moving the metal in the water will ensure a faster quench.

**///// CAUTION **

The steam given off from the quenching of the hot metal can cause severe burns. Use your gloves and be careful not to allow the steam to burn you.

Set one of the pieces aside. File a smooth, clean area on the other part. Slowly heat this piece on the opposite side from the filed area using the oxyfuel torch, **Figure 25-24**. Watch the clear spot, and it will begin to change colors.

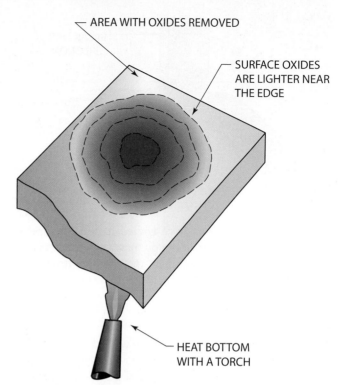

FIGURE 25-24 Surface oxide temper colors.
© Cengage Learning 2012

It will first become a very pale yellow and gradually change to a dark blue. When the surface is dark blue, stop heating it and allow it to cool as slowly as possible between hot firebricks. The surface colors formed are called temper colors, and they indicate the temperature of the metal's surface, **Table 25-4**. The color comes from the layer of oxide formed on the hot metal's surface. The bluing on a gun barrel is the same thing, and it is used both to make the barrel stronger and to protect it from rusting. You could also use an 800°F (425°C) colored temperature indicating crayon such as a Tempil Temperature Indicator.

Degrees Fahrenheit	Color of Steel
430	Very pale yellow
440	Light yellow
450	Pale straw-yellow
460	Straw-yellow
470	Deep straw-yellow
480	Dark yellow
490	Yellow-brown
500	Brown-yellow
510	Spotted red-brown
520	Brown-purple
530	Light purple
540	Full purple
550	Dark purple
560	Full blue
570	Light blue
640	Dark blue

TABLE 25-4 Temperatures Indicated by the Colors of Mild Steel

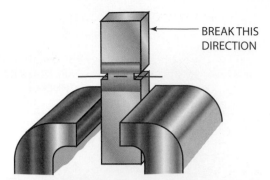

BREAK THIS
DIRECTION

FIGURE 25-25 Nick-break test. © Cengage Learning 2012

After both specimens have cooled to room temperature, they can be tested. If you have a tensile testing machine, test each specimen and record its failure strength. If you do not have a tensile tester, make a 1/4-in. (6-mm) deep saw cut on both edges of each specimen, **Figure 25-25.** Place the specimens in a vise and break them. Note how they break.

Look at the fractured surface and record which has the light-colored surface and which has the darker surface. Also note which one has the smallest grain sizes shown.

This experiment can be repeated using different metals and alloys.

Complete a copy of the "Student Welding Report" listed in Appendix I or provided by your instructor. ◆

Martensitic Reactions

Heating a carbon-iron alloy above the temperature at which austenite forms, see line A₃, **Figure 25-26,** and then quenching it rapidly in water is certainly not an equilibrium condition. The FCC crystals are unable to change to BCC since the rapid cooling was too fast to allow change.

FIGURE 25-27 Microstructure of type 416 martensitic stainless steel. George F. VanderVoort, Consultant

Martensite is the hardest of the transformation products of austenite and is formed only on cooling below a certain temperature known as the M. temperature (about 400°F to 600°F [200°C to 315°C] for carbon steels). Cooling to this temperature must be sufficiently rapid to prevent austenite from transforming to softer constituents at higher temperatures.[1] As formed, martensite is very hard and brittle and useless for most engineering applications.

When martensite is viewed under a microscope, it has a **(needle-like) acicular structure, Figure 25-27.** During welding of medium and high carbon steels, the cooling

_____

[1]*Basic Guide to Ferrous Metallurgy*, Copyright 1954 by Tempil, Hamilton Boulevard, South Plainfield, NJ 07080.

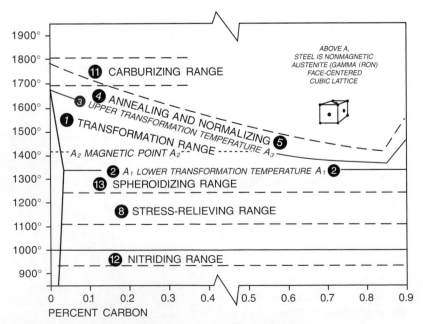

FIGURE 25-26 Upper transformation temperature. © Cengage Learning 2012

WITHOUT PREHEAT—The temperature starts close together and spreads out quicker (a large temperature change between weld and sides of plate).

WITH PREHEAT—The temperature starts further apart and spacing changes slowly (little temperature difference between weld and sides of plate).

FIGURE 25-28 © Cengage Learning 2012

rates can be fast enough to produce the undesirable martensite. Martensite formation can be minimized by preheating the steel to slow the cooling rates. If the surrounding metal is warmed by preheating, the weld does not lose its temperature as quickly, **Figure 25-28**.

Martensite can be tempered to a more useful structure at a temperature below the lower critical temperature, see line A_1 on **Figure 25-29**. The exact temperature, determined by the carbon content and other alloying elements, is generally furnished by the steelmaker.

When martensite is tempered, some of the carbon atoms that were trapped interstitially (small holes in the crystalline lattice) between the iron atoms are allowed to defuse out, **Figure 25-30**. The precipitation of the trapped carbon atoms allows the body-centered tetragonal crystal to reform into the body-centered cube crystal shape, relieving the internal strain. The microstructures of tempered and untempered martensite look very much alike. In this condition, the tempered martensite is very strong and tough.

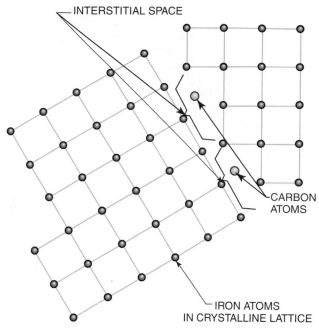

FIGURE 25-30 Interstitial spacing. © Cengage Learning 2012

As the tempering time and/or temperature is increased, the structure changes to a **spheroidized microstructure.** In this state, the steel is in the soft or annealed condition. For example, when a chisel, drill bit, or slag hammer is over ground so that the sharp edge has turned blue, this indicates that a temperature of about 600°F (316°C) has been reached. A temperature this high causes overtempering and softens the edge. Thus, time and temperature are critical variables in the loss of strength and hardness, **Figure 25-31**.

NOTE: For all practical purposes, plain carbon steels with less than 0.30% carbon cannot be hardened through heat-treating and quenching.

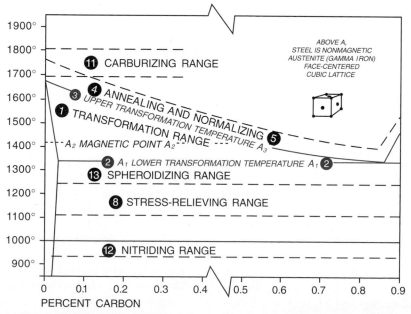

FIGURE 25-29 Transformation range. © Cengage Learning 2012

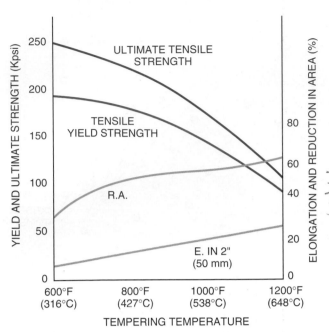

FIGURE 25-31 Changes in properties of SAE 1050 martensite caused by tempering. © Cengage Learning 2012

Cold Work

When metals are deformed at room temperature by cold rolling, drawing, or swaging, the grains are flattened and elongated. Complex movements occur within and between the grains. These movements distort and disrupt the crystalline structure of the metal by markedly increasing its strength and decreasing its ductility. The presence of impurities or alloying elements in metals causes them to work harder more quickly. Sheets, bars, and tubes are intentionally cold-worked to increase their strength, since cold working will strengthen almost all metals and their alloys.

The cold-worked structure can be annealed by heating the metal above the **recrystallization temperature.** Above that temperature new crystals grow. The size of the new grains depends on the severity of the prior cold working and the time above the recrystallization temperature. The final annealed structure is weaker and more ductile than the cold-worked structure. The final properties depend primarily on the alloy and on the grain size. The coarse-grained metals are weaker.

Grain Size Control

One of the few metallurgical effects common to all metals and their alloys is grain growth. When metals are heated, grain growth is expected. The rate of growth increases with temperature and the length of time at that temperature. There are charts that show the effect of time and temperature on the austenite grain growth. These are called time-temperature-transformation (TTT) charts. The process involves larger grains devouring smaller grains. Coarse grains are weaker and tend to be more ductile than fine grains.

The longer the metal is held at a high temperature, above A_3, the larger the grain size. (See the left vertical strip on Figure 25-16.) The austenite crystals will grow so large that when they are cooled and transformed into pearlite their large size can be seen easily in a fracture. Some welders know that this large grain is detrimental to the metal's strength and often refer to it as having been "crystallized."

The production of fine grains is not as simple because large grains cannot be shrunk. Reducing the grain size requires the creation of fresh grains by heating the metal to the austenite range and quenching it to recrystallize it. This technique is common to all metals and alloys. The same results can be obtained with an allotropic transformation.

Allotropic transformation also requires the creation of fresh grains. Since the temperatures of allotropic transformation are high, the new grains begin to grow almost immediately. To obtain a fine-grained structure, the metal must be heated quickly above the critical temperature, line A_3 on Figure 25-26, and cooled quickly in a process called **grain refinement.** Not all metals exhibit allotropic transformation. Fortunately, iron does, and this is one reason that steels are such versatile materials.

Heat Treatments Associated with Welding

Welding specifications frequently call for heat-treating joints before welding or after fabrication. To avoid mistakes in their application, welders should understand the reasons for these **heat treatments.**

Preheat

Preheat is used to reduce the rate at which welds cool. Generally, it provides two beneficial effects—lower residual stresses and reduced cracking. The lowest possible temperature should be selected because preheat increases the size of the heat-affected zone and can damage some grades of quenched and tempered steels. The amount of preheat generally is increased when welding stronger plates or in response to higher levels of hydrogen contamination.

Area 10, "Preheating Range for Welding," **Figure 25-32,** shows the temperature range for preheating various iron-carbon alloys. As the preheat area moves to the higher carbon percentage side, it rises in temperature. This is because as the percentage of carbon increases, the alloy becomes more susceptible to hardening and cracking due to rapid cooling. Preheating charts are available for recommended temperatures for various steels and nonferrous metals, **Table 25-5.**

The most commonly used preheat temperature range is between 250°F and 400°F (121°C and 204°C) for structural steel. The preheat temperatures can be as high as 600°F (316°C) when welding cast irons.

Care must be taken to soak heat into the region of the intended weld. Superficial heating is not enough because

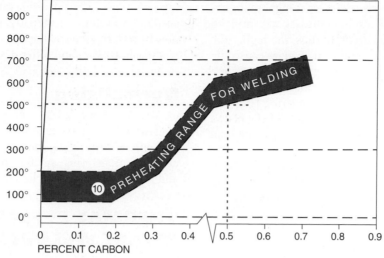

FIGURE 25-32 Preheating range for welding. Tempil, An Illinois Tools Works Co.

	Plate Thickness (inches)			
ASTM/ AISI	1/4 or less	1/2	1	2 or more
A36	70°F	70°F	70°F	150°F to 225°F
A572	70°	70°	150°	225° to 300°
1330	95°	200°	400°	450°
1340	300°	450°	500°	550°
2315	Room temp.	Room temp.	250°	400°
3140	550°	625°	650°	700°
4068	750°	800°	850°	900°
5120	600°	200°	400°	450°

TABLE 23-5 Recommended Minimum Preheat Temperatures for Carbon Steel

the purpose is to affect the rate at which a relatively large weld cools. To prevent problems and ensure uniform heating, the temperature of the preheated section should be measured at least 10 and 20 minutes after heating.

Stress Relief, Process Annealing

Residual stresses are unsuitable in welded structures, and their effects can be significant. The maximum stresses generally equal the yield strength of the weakest material associated with a specific weld. Such stresses can cause distortion, especially if the component is to be machined after welding. They can also reduce the fracture strength of welded structures under certain conditions.

Area 8 of **Figure 25-33**,[1] "Stress-Relieving Range," consists of heating to a point below the lower transformation temperature, A_1, holding for a sufficiently long period to relieve locked-up stresses, then slowly cooling. This process is sometimes called process annealing.

---

[1]*Basic Guide to Ferrous Metallurgy*, Copyright 1954 by Tempil, Hamilton Boulevard, South Plainfield, NJ 07080.

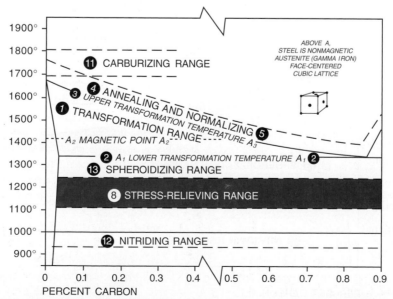

FIGURE 25-33 Stress-relieving range. Tempil, An Illinois Tools Works Co.

The yield strength of steels decreases at higher temperatures. When heated, the residual stresses will drop to conform to the lower yield strength; thus, the higher the temperature, the better. But significant changes caused by overheating, overtempering, grain growth, or even a phase change must be avoided. Therefore, the temperatures selected must be less than the tempering temperature used to heat-treat the plate. Regardless of the plate's metallurgical structure, these temperatures must be kept below the critical temperature.

The most commonly used temperature range for stress relief steel is between 1100°F and 1150°F (593°C and 620°C). This range is high enough to drop the yield residual stresses by 80% and low enough to prevent any harmful metallurgical changes in most steels. While risky, heating to just under the critical temperature does offer a stress reduction of about 90%. Caution should be exercised, however, because some steel will become brittle after thermal stress relief.

Time at temperature is also an important factor since it takes time to bring a weldment to temperature. One hour at temperature is the minimum, while six hours offers an additional, costly 10% drop in stress. Before cooling, the component must be uniformly heated. That requirement alone generally ensures the component will be at temperature long enough to relieve the stresses. The rate of cooling must also be considered. Rapid cooling results in uneven cooling (as in welding) and causes a new set of residual stresses. Ideally, the cooling rate is slow enough to cool the entire mass of the metal uniformly.

Annealing

Annealing, frequently referred to as full annealing, involves heating the structure of a metal to a high enough temperature, slightly above Ac_3, to turn it completely austenitic. See **Figure 25-34**, area 4. After soaking to equalize the temperature throughout the part, it is cooled in the furnace at the slowest possible rate. On cooling slowly, the austenite transforms to ferrite and pearlite. The metal is now its softest, with small grain size, good ductility, excellent machinability, and other desirable properties.

Normalizing

Area 5 of Figure 25-34, "Annealing and Normalizing," consists of heating steels to slightly above Ac_3, holding for austenite to form, then cooling (in still air). On cooling, austenite transforms, giving somewhat higher strength and hardness and slightly less ductility than in annealing.

THERMAL EFFECTS CAUSED BY ARC WELDING

During arc welding, liquid weld metal is deposited on the base metal, which is at or near room temperature. In the process, some of the base metal melts from contact with the liquid weld metal and arc, flame, and so on. Conduction, convection, and radiation pull the heat out of the weld metal. This causes the temperature of the deposited weld metal to cool until it becomes solid. Within a few seconds, the weld and base metal have gone from being a solid at room temperature to a liquid and back to a solid near room temperature.

Metallurgical changes in the heated region are inevitable. The lowest temperature at which any such changes occur defines the outer extremity of the zone of change. That zone of change is called the **heat-affected zone**, abbreviated **HAZ, Figure 25-35**. The exact size and shape of the HAZ are affected by a number of factors:

- *Type of metal or alloy:* Some metals are easily affected even by small temperature changes, while others are more resistant. In work-hardened metals, the heat-affected zone is defined by the recrystallization

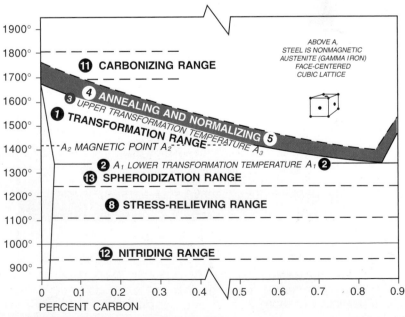

FIGURE 25-34 Annealing and normalizing. © Cengage Learning 2012

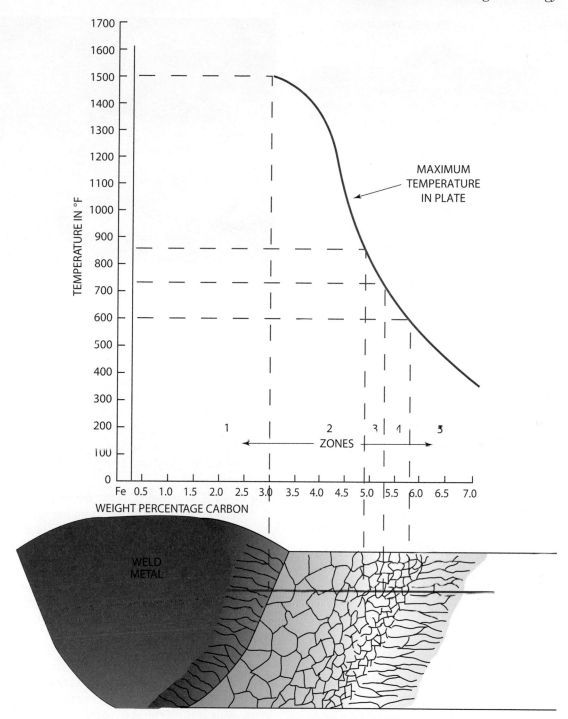

Zone 1—Liquid metal and the beginning of grain growth
Zone 2—Austenitic; grain growth at high temperature,
      fine grain at low temperature
Zone 3—Austenite + ferrite; grain refined and grain growth
Zone 4—Recrystallization
Zone 5—Cold-worked steel 0.2% carbon

FIGURE 25-35 Changes in grain structure caused by heating steel plate into different zones of the iron-carbon phase diagram. © Cengage Learning 2012

temperature, **Figure 25-36.** In age-hardened alloys, this temperature is the lowest temperature at which evidence of overaging is seen, **Figure 25-37.** In quenched and tempered steels, it is the temperature at which overtempering is seen.

- *Method of applying the welding heat:* Some heat sources, such as plasma arc welding (PAW), are very concentrated. This high-intensity welding process can have an HAZ area that is only a few thousandths of an inch wide. Oxyacetylene welding (OFW) is a

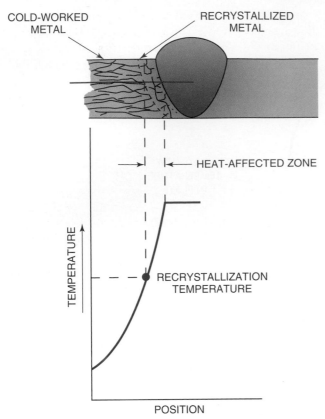

FIGURE 25-36 A heat-affected zone in cold-worked metal. The zone is defined by a change from the cold-worked grains to the recrystallized grains. © Cengage Learning 2012

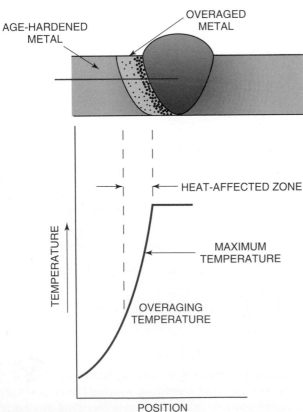

FIGURE 25-37 Heat-affected zone in age-hardened metal. © Cengage Learning 2012

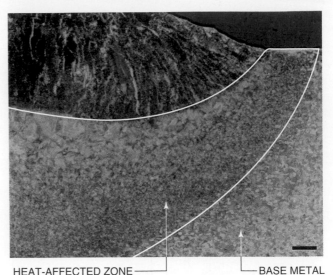

HEAT-AFFECTED ZONE —————— BASE METAL

FIGURE 25-38 Heat-affected zone.

George F. VanderVoort, Consultant

much less intense heating source, and the resulting HAZ will be very large.

- *Mass of the part:* The larger the piece of metal being welded, the greater its ability to absorb heat without a significant change in temperature. Very large weldments may not have a noticeable temperature rise, while small parts may almost completely reach the melting temperature. The more metal that gets hot, the larger the HAZ.

- *Pre- and postheating:* As the temperature of the base metal increases—whether from pre- or postheating or the welding process itself—the larger the HAZ. A cold plate may have an extremely narrow HAZ.

The HAZ need not always be small. A larger HAZ is desirable for welding steels that can produce martensite at the same cooling rates as those produced when welding heavy plate. The welder can slow the cooling rates to tolerable levels by preheating the plate and thus increasing the size of the HAZ. In this case, a large HAZ is safer than contending with a brittle martensitic structure.

An important feature of the HAZ, caused by high temperatures, is the severe grain growth at the fusion line. With steels at this critical temperature, the HAZ also produces fine grains as a result of the **allotropic transformation. Figure 25-38** is an example of a weld that shows the HAZ in cross section. Regardless of the alloy, the welder must control the HAZ, whether that control is to make it large or small.

GASES IN WELDING

Many welding problems and defects result from undesirable gases that can dissolve in the weld metal, **Figure 25-39.** Except for the inert gases argon and helium, discussed in Section 4, "Gas Shielded Welding," the other common gases either react with or dissolve in the molten weld pool. Those that dissolve have a very high solubility in liquid metal but

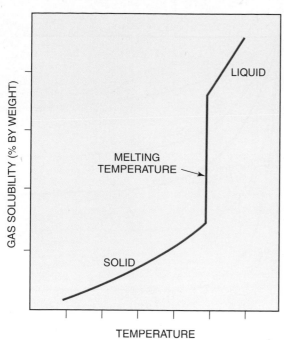

FIGURE 25-39 Typical change of solubility of "active" gases in metals. © Cengage Learning 2012

a very low solubility in solidified weld metal. Thus, during the freezing process, the dissolved gases try to escape. With very slow solidification rates, the gases escape. With very high solidification rates, they become trapped in the metal as supersaturated solutes. At intermediate rates of solidification, they become trapped as gas bubbles, causing porosity. Some, such as hydrogen, produce other problems as well.

Hydrogen

Hydrogen has many sources, including moisture in electrode coatings, fluxes, very humid air, damp weld joints, organic lubricants, rust on wire or on joint surfaces or in weld joints, organic items such as paint, cloth fibers, dirt in weld joints, and others.

Hydrogen is the principal cause of porosity in aluminum welds and with GMAW welds on stainless steels. It can cause random porosity in most metals and their alloys.

Hydrogen can be troublesome in steels. Even in amounts as low as 5 parts per million, it can cause underbead cracking in high-strength steels. This type of cracking, called delayed cracking, cold cracking, or hydrogen-induced cracking, requires three conditions: (1) a high-stress state, (2) a martensitic microstructure, and (3) a critical level of hydrogen. The first two conditions are typical of quenched and tempered steels. With them, the critical amount of hydrogen is inversely proportional to the stress state; that is, the stronger steels can tolerate less hydrogen. The welder must use dry electrodes and dry submerged arc fluxes to avoid the hydrogen, in the form of moisture, when welding steels with strength levels above 80,000 psi (5624 kg/cm^2).

Hydrogen problems are avoidable by keeping organic materials away from weld joints, keeping the welding consumables dry, and preheating the components to be welded. Preheating slows the cooling rate of weldments, allowing more time for hydrogen to escape. Generally, the recommended preheat temperatures range between 250°F and 350°F (121°C and 176°C). In the case of very high-strength materials, the acceptable preheat temperatures will exceed 400°F (204°C). Low-temperature preheating of materials before welding can also remove the moisture condensed on the weld joint.

Nitrogen

Nitrogen comes from air drawn into the arc stream. In GMAW processes, it results from poor shielding or strong drafts that disrupt the shield. In SMAW processes, nitrogen can result from carrying an excessively long arc. The primary problem with nitrogen is porosity. In high-strength steels, it can cause a degree of embrittlement that results in marginal toughness.

Some alloys, such as austenitic stainless steels and metals such as copper, have a high solubility for nitrogen in the solid state. This high solubility does not cause porosity in those materials. Since nitrogen improves the strength of stainless steels, it sometimes is added intentionally to shield gases used when making GTA or GMA welds in them. In Europe, it is used for welding copper to increase the arc energy with GTA welding.

Oxygen

As with nitrogen, the common source of oxygen contamination, air, reaches the weld because of poor shielding or excessively long arcs. But metallurgical changes, not porosity, cause most of the effects of oxygen. Oxygen causes the loss of oxidizable alloys such as manganese and silicon, which reduces strength; produces inclusions in weld metals, which reduce their toughness and ductility; and causes an oxide formation on aluminum welds, which affects appearance and complicates multipass welding.

About 2% of oxygen is added intentionally to stabilize the GMAW process when welding steels with argon shielding. At this concentration, oxygen does not cause the metallurgical problems listed previously. Nevertheless, the amount used must be very carefully controlled because the amount needed varies, depending on the alloys being welded. Mixtures containing only enough oxygen to do an effective stabilizing job should be used. Any more could cause problems, particularly with sensitive alloys.

Carbon Dioxide

Carbon dioxide is an oxygen substitute for stabilizing GMAW processes using argon shields, although about 5% to 8% carbon dioxide is usually added to produce the same effects achieved with 2% oxygen. However, the carbon in carbon dioxide is a potential contaminant that can cause problems with corrosion resistance in the low carbon grades of stainless steels. Even straight carbon dioxide shielding has this problem. In most cases, carbon dioxide

FIGURE 25-40 Underbead cracking. © Cengage Learning 2012

levels below 5% do not seem to increase the carbon content of stainless steels enough to cause difficulty.

METALLURGICAL DEFECTS

Cold Cracking

Cold cracking is the result of hydrogen dissolving in the weld metal and then diffusing into the heat-affected zone. The cracks develop long after the weld metal solidifies. For that reason, it is also known as hydrogen-induced cracking and delayed cracking. This cracking is most commonly found in the course grains of the HAZ, just under the fusion zone. For this reason, it is also called underbead cracking. Generally, these cracks do not surface and can be seen only in radiographs or by sectioning welds. Sometimes they surface in a region of the weld running parallel to the fusion zone. With high-strength steels, hydrogen-induced cracking occurs as transverse cracks in the weld metal that are seen easily with very low magnification, **Figure 25-40**.

Hydrogen-induced cracking requires (1) a high stress, (2) a microstructure sensitive to hydrogen, and, of course, (3) hydrogen. The first two are almost always satisfied with high-strength quenched and tempered steels. The third depends on the welding process, the filler metals, and the preheat, as discussed previously.

Another factor is time. Under very severe conditions, cracking can occur in minutes. With very marginal conditions, cracks might not appear for weeks. For this reason, welds often are not radiographed for weeks to allow time for any potential cracks to develop.

Figure 25-41 illustrates how stress, time, and hydrogen interact to produce such cracks. These cracks never occur with 0% hydrogen, even with very high yield strength steels. To prevent cracking at hydrogen level H_5, the residual stresses must be reduced to the level indicated by the horizontal section of the curve H_5. Low residual stresses are possible with either a very weak steel or a strong steel that has had a stress relief anneal. (The stress relief anneal also will help by dropping the hydrogen level to level H_0.)

Hot Cracking

Hot cracks differ from the cold cracks discussed previously. The hot cracks are caused by tearing the metal along partially fused grain boundaries of welds that have not completely solidified. Unlike those caused by hydrogen, these longitudinal cracks are located in the centers of weld beads. They develop immediately after welding, unlike the delayed nature of hydrogen-induced cracking. In the process of freezing, low-melting materials in the

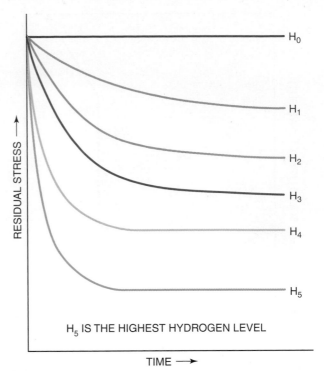

FIGURE 25-41 The time needed to develop hydrogen-induced cracking decreases at higher stresses and at higher hydrogen levels. © Cengage Learning 2012

weld metal are rejected as the columnar grains solidify, leaving a high concentration of the low-melting materials where the grains intersect at the center of the weld. These partially melted and weak grain boundaries are stressed while the weld metal shrinks, causing them to rupture.

A high sulfur content is most often responsible for hot cracking in steel. It forms a low-melting iron sulfide on the grain boundaries. Hot cracking is more likely to occur in steels that contain higher levels of carbon and phosphorus and in steels that are high in sulfur and low in manganese, **Figure 25-42**.

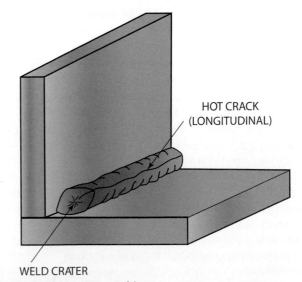

FIGURE 25-42 Hot cracking. © Cengage Learning 2012

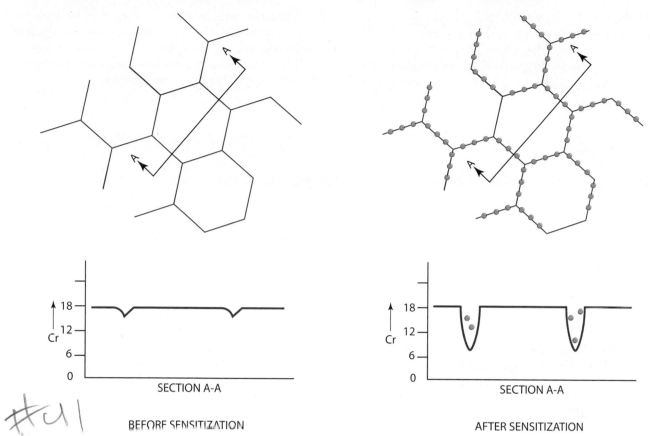

FIGURE 25-43 Depletion of chromium in grain boundaries due to the precipitation of chromium carbide. © Cengage Learning 2012

Severely concave welds may also cause hot cracking because the welds are not as strong. As welds cool, they shrink. If the weld is not thick enough, when it shrinks it cannot pull the metal in, and so it cracks. Even if the weld is larger, it may crack if the weldment is rigid and cannot be pulled inward by the weld. In these situations, even large welds can have hot cracks.

Carbide Precipitation

Stainless steels rely on free chromium for their resistance to corrosion. When carbon is present and the steel is heated to temperatures between about 800°F (427°C) and 1500°F (816°C), carbon combines with the chromium to form chromium carbides in the grain boundaries. The formation of chromium carbides depletes the steel of the free chromium needed for protection. Thus, for steels, every effort is made to use low carbon stainless steel or a special stabilized grade made for welding.

Figure 25-43 illustrates how the chromium in a solution can drop from the nominal 18% in the most commonly used stainless steels to levels well under the 12% needed for even minimal protection. When in contact with corrosive materials, the low chromium grain boundaries can dissolve and the weld can fail, **Figure 25-44**.

Problems are minimized by using very low carbon steel called extra low carbon (ELC) steels. Without carbon, the chromium carbides cannot form. Another way is to tie up the carbon by forming very stable carbides of titanium or columbium. This also prevents chromium carbides from forming. Weld metals are alloyed similarly with low carbon levels or by stabilizing the carbides. Titanium is not generally used for that purpose in electrodes because it is

FIGURE 25-44 Microstructure of solution annealed type 304 stainless steel revealing austenite grains' boundaries.
George F. VanderVoort, Consultant

lost in the transfer. Instead, most electrodes are alloyed with columbium to avoid the corrosion problems associated with carbide precipitation.

Carbon dioxide shield gases can cause a similar problem, especially with the ELC grades. These gases supply the carbon that impairs corrosion resistance by depleting the free chromium. The small amounts used in argon to stabilize the arc are often tolerable. But carbon dioxide additions should be avoided unless used with caution and an awareness that problems can develop.

Summary

Welding engineers' understanding of science, physics, and metallurgy enables them to design better weldments. Welding engineers must know how the chemical elements that make up a metal alloy will react to changes in physical and thermal stresses. Such stresses are applied to metal products during welding fabrication and as part of their normal service. As metals are thermally cycled, their physical and mechanical properties change. Thermal cycling, of course, occurs every time a metal is welded. Often, as part of a post-weld procedure, a welding engineer will have the part heat-treated. Welding engineers use charts and graphs to determine the optimal thermal cycling that will allow for the greatest strength in the weldment design.

You do not have to have as deep an understanding of the thermal effects on metals as a welding engineer does, but you must know the importance of controlling temperature cycles during welding. Weldment failures may be a result of welder-created problems or welding metallurgical problems. A good understanding of metallurgy will aid you in avoiding welding problems.

Shot Peening and Heat Treatment Reduce Stress

Designers of weldments must consider ways to relieve residual stresses created by welding.

Residual stress directly affects a component's fatigue life. Harmful residual tensile stress decreases fatigue life, while beneficial residual compressive stress increases fatigue life. Welding induces residual tensile stress that is harmful to fatigue life. Tensile stresses stretch or pull apart the surface of the material. With enough load at a high enough tensile stress, a metal surface will initiate a crack. Significant improvements in fatigue life can be obtained by modifying the residual stress levels in the material. Two methods of performing this are through heat treatment and shot peening.

Explaining Metal Fatigue and Fracture

Metal fatigue is the phenomenon that leads to fracture when stresses are greater than the tensile strength of the material. The type of stress of most concern is tensile stress. Tensile stress stretches or pulls apart the surface of the material.

Fatigue fractures are progressive, beginning as minute cracks, which grow under the action of fluctuating tensile stress. A fatigue crack usually starts at the surface. Under the repeated action of the tensile stresses, the crack grows, weakening the section.

Residual stress, however, can be beneficial. In particular, residual compressive stress from shot peening directly reduces the tensile stress. Therefore, more compressive stress results in greater improvements in fatigue properties.

Residual Stresses from the Welding Process

Residual stresses are those that exist in a part independent of external force. Neglect of these residual tensile stresses created during welding can lead to stress corrosion, cracking, distortion, fatigue cracking, and premature failures in components. When residual stresses accumulate, the tensile stress experienced by the components is much greater than the actual load applied. This is why welded components fail at the weld.

The residual tensile stress from welding is created because the welding consumable is applied in its molten state. The weld is applied in its hottest, most expanded state. It then joins to the base material, which is much cooler. The weld cools rapidly and attempts to shrink from the cooling. Since it has already joined to the cooler, stronger base material, it is unable to shrink. The net result is a weld that is essentially being "stretched" by the base material. The heat-affected zone is usually most affected by the residual stress and is, hence, where failure will usually occur. Inconsistency in the welding consumable, chemistry, weld, geometry, porosity, and so on, acts as a stress riser for residual and applied tensile stress to initiate fatigue failure.

Residual Stresses from Grinding

Residual tensile stresses and surface brittleness can be caused by the generation of high surface temperatures by friction during grinding operations. It has been found that residual tensile stresses created by grinding can approach the ultimate tensile strength of the material itself. Residual tensile stresses will dramatically affect fatigue resistance of ground parts.

In grinding, tensile stress is created from the generation of excessive, localized heat. The localized area attempts to expand. It cannot because it is surrounded by cooler, stronger metal. This metal yields in compression due to the resistance to its expansion. When it cools, the yielded material attempts to contract. The surrounding material resists this contraction, thus creating a residual tensile stress.

Because heat is the major cause of residual stresses in grinding, it is important to use a coolant to control these stresses. A coolant is easily applied using automatic grinders, but it is rarely used with manual or hand grinders. If an automatic grinder with coolant were used with a gentle grind, this would, in effect, cold-work the material and induce a slight amount of beneficial compressive stress at the surface.

Reducing Residual Stress with Shot Peening

Shot peening is a cold-working process used to increase the fatigue properties of metal components. During the peening process, the surface of the component is bombarded with small, round particles called shot. Each piece of shot striking the material acts as a tiny peening hammer, imparting to the surface a small indentation, or dimple.

Overlapping dimples develop a uniform layer of residual *compressive* stress in the metal. It is well known that cracks will not develop in a compressively stressed area. Since nearly all fatigue failures originate at the surface of a part, compressive stresses formed by shot peening provide considerable increase in part life. When controlled properly, all surface area is formed into a uniform layer of high-magnitude, compressive stress.

Magnification of shot-peened weld surface (30X).
American Welding Society

Relief of Residual Stress through High Temperature

Complex parts that have been welded, cast, or machined often develop internal stresses that could cause distortion of their shape or affect their fatigue strength. To relieve these stresses, parts made from carbon steel are heated to between 1000°F and 1200°F (538°C and 649°C) and are held at that temperature for a period of time determined by their cross-sectional thickness. Parts are then cooled slowly to minimize the development of new residual stresses.

Metal is weaker when it is at higher temperatures. This is why residual stress (compressive or tensile) is relieved at higher temperatures. The strength of the metal is reduced and stresses are able to relieve themselves. Higher temperatures and longer exposure times are able to relieve more residual stress. The stress-relieving process is beneficial for welding stress, as it reduces it, but harmful for shot peening because beneficial compressive stress is removed.

Shot-Peened Welds

This is a 30x magnification of the shot-peened surface. One can see the uniform overlapping dimples from the peening process. In the welding process, it is important to minimize any overlaps, folds, or undercuts that would be geometrically too small or difficult for the peening shot to cover with dimples.

Single-Peened vs. Dual-Peened

A dual shot-peen operation is used to improve fatigue properties beyond that of a single-peened surface. This is accomplished by shot peening the same surface a second time at a reduced energy level (intensity). One can see that the surface finish is more uniform than the single-peened surface finish. This is because the smaller shot is able to obliterate the high points from the first peening operation.

Fatigue testing has shown that dual-intensity peening will produce an even greater improvement in fatigue resistance than single peening. The purpose of dual peening is to increase the compressive stress at the surface fibers with a secondary

Magnification of dual-peened weld surface (30X).
American Welding Society

peening operation. Compressing the top surface fibers makes it even more difficult for a fatigue crack to begin.

Conclusion

When weldments are used in a fatigue environment, the designer needs to consider stresses from both applied loads and manufacturing processes. Welding and grinding processes create residual tensile stresses. These stresses can be reduced and fatigue properties improved through heat treatment and shot peening.

Article courtesy of the American Welding Society.

Review

1. What gives metals their desirable properties?

2. What is heat?

3. What are the basic units of measure for heat?

4. What is sensible heat?

5. What is latent heat?

6. What does the color of light given off from a hot object indicate?

7. What is a crystal lattice?

8. Why does metal not form into one large, single crystal?

9. What is an alloy?

10. Using Figure 25-14, answer the following questions:
    a. At approximately what temperature does a 70% lead and 30% tin mixture become solid?
    b. What crystal structure is formed first when a 20% lead and 80% tin mixture cools down from a 100% liquid phase?

11. What is a eutectic composition?

12. Using Table 25-3, what are the lowest and highest carbon contents of iron-carbon alloys?

13. Approximately how many degrees wide is the transition range at the 0.1% carbon alloy?

14. Referring to Figure 25-16, what color would low carbon steel appear when it is in the transition range?

15. Referring to Figure 25-16, what is the approximate temperature of the *lower transformation temperature?*

16. What three factors affect the size and type of crystals formed in an iron-carbon alloy?

17. Referring to Figure 25-16, above what temperature is a 0.1% iron-carbon alloy nonmagnetic?

18. What is the most significant factor that affects how much of an alloying element remains trapped in the iron-carbon crystals?

19. What is known as the critical temperature of iron-carbon alloys?

20. Can a metal have all the mechanical properties at ideal levels? Why or why not?

21. What other properties can a metal's hardness reveal?

22. Which property, brittleness or ductility, will let a metal deform without breaking? Why?

23. What is toughness?

24. What are the common types of strength measurements?

25. What is a major advantage of solid-solution–strengthened alloys?

26. What are the three steps in precipitation hardening?

27. How do ferrite and cementite work together to form a strong ductile steel?

28. Why is brine quenching faster than water quenching?

29. What can be done to speed up the quenching rate in any liquid?

30. Why is the formation of martensite a problem when welding on high carbon steels?

Review (continued)

31. How can the effects of cold working be removed?

32. How can large-grain crystals be made into smaller sizes?

33. Referring to Figure 25-16, what would the preheat temperature range be for an iron-carbon alloy with 0.6% carbon?

34. Why must the stress-relieving temperature be kept below the critical temperature of the plate?

35. What properties can annealing produce in metals?

36. How long does it take the weld metal to go through its thermal cycle during a weld?

37. What are some sources of hydrogen that can contaminate a weld?

38. How can nitrogen get into an SMA weld?

39. What are some of the problems that oxygen can cause in welds?

40. When do cold cracks develop?

41. What is carbide precipitation?

Chapter 26

Weldability of Metals

OBJECTIVES

After completing this chapter, the student should be able to

- list the methods used to weld most ferrous metals.
- list the methods used to weld four nonferrous metals.
- explain the precautions that must be taken when welding various metals and alloys.
- describe the effects of preheating and postheating on welding.

KEY TERMS

alloy steels

aluminum

cast iron

chromium-molybdenum steel

copper and copper alloys

high manganese steel

low alloy steels

magnesium

malleable cast iron

plain carbon steels

stainless steels

steel classification

titanium

tool steel

weldability

INTRODUCTION

All metals can be welded, although some metals require far more care and skill to produce acceptably strong and ductile joints. The term *weldability* has been coined to describe the ease with which a metal can be welded properly. Good weldability means that almost any process can be used to produce acceptable welds and that little effort is needed to control the procedures. Poor weldability means that the processes used are limited and that the preparation of the joint and the procedure used to fabricate it must be controlled very carefully or the weldment will not function as intended.

Weldability is defined by the American Welding Society (AWS) as "the capacity of a metal to be welded under the fabrication conditions imposed into a specific, suitably designed structure and to perform satisfactorily in the intended service." Books have been written about this subject. Weldability involves the metallurgy of both the metal to be welded and the filler metal, the welding processes, the joint design, the weld preparation, the heat treatments before and after welding, and many other factors, depending on the complexity of the welded system. This chapter does not discuss the selection of methods and procedures for joining difficult-to-weld metals. Those with no previous experience should seek expert help before attempting to weld these metals. Otherwise, they run the risk of failures in service that could physically harm nearby people.

Therefore, an attempt will not be made in this text to cover the weldability of all metals. The welding done in the average school shop is limited to a small number of metals. However, information will be given in this chapter on a number of other metals that will be found in common use in industry.

Most welding processes produce a thermal cycle in which the metals are heated over a range of temperatures. Cooling of the metal to ambient temperatures then follows. The heating and cooling cycles can set up stresses and strains in the weld. Whatever the welding process used, certain metallurgical, physical, and chemical changes also take place in the metal. A wide range of welding conditions can exist for welding methods when joining metals with good weldability. However, if weldability is a problem, adjustments usually will be necessary in one or more of the following factors:

- Filler metal—If the wrong filler metal is selected, the weld can have major defects and not be fit for service. Common defects include porosity, cracks, and filler metal that just will not stick, **Figure 26-1**. The cracks can be in the filler metal or in the base metal alongside the weld. Chapter 27 lists various types of filler metals and their applications. If you are not sure which filler metal to use, a good general rule is that a little stronger or higher grade can be used successfully, but a lower strength or grade seldom works.

- Shielding atmosphere—Metals such as aluminum, copper, magnesium, and titanium are very easily contaminated by the atmosphere when heated. Therefore, it is important that the correct shielding gas be used in sufficient quantity and with the correct application to protect the molten weld metal.

- Fluxing material—Some metals are reactive to the atmosphere when heated, and others have surface oxides that prevent or limit the filler metal from bonding. Most processes such as SMAW and FCAW have fluxes incorporated with the process, but others such as OFW and TB do not. With these

	Plate Thickness (inches)			
ASTM/ AISI	1/4 or less	1/2	1	2 or more
A36	70°F	70°F	70°F	150°F to 225°F
A572	70°	70°	150°	225° to 300°
1330	95°	200°	400°	450°
1340	300°	450°	500°	550°
2315	Room temp.	Room temp.	250°	400°
3140	550°	625°	650°	700°
4068	750°	800°	850°	900°
5120	600°	200°	400°	450°

TABLE 26-1 Recommended Minimum Preheat Temperatures for Carbon Steel

processes, it is as important to select the correct flux as it is to select the correct filler metal.

- Preheat and postheat—Cracking is a common problem when welding on brittle metals such as cast iron or some high-strength alloys. Preheating the part before starting the weld will reduce the stress caused by the weld and will help the filler metal flow. The most commonly used preheat temperature range is between 250°F and 400°F (120°C and 200°C) for most steel, **Table 26-1**. The preheat temperature can be as high as 1200°F (650°C) when welding cast iron. Preheating is required anytime the metal to be welded is below 70°F (20°C) because the cold metal quenches the weld. Postheating slows the cooldown rate following welding, which will prevent postweld cracking of brittle metals. Postheating also reduces weld stresses that can result in cracks forming some time after the part is repaired.

- Welding procedure—The size of the weld bead, the number of welds, and the length of the welds all affect the weld. When large welds are needed, it is better to make more small welds than a few large welds. The small welds serve to postheat the weld. They reduce stresses and result in less distortion. Sometimes a series of short back-stepping welds can be made, **Figure 26-2**. For example, these short welds can be used on very brittle metals like cast iron.

- Welding process—The selection of the welding process to be used for any metal, thickness, and application often requires some research. For example, **Table 26-2** is a summary of acceptable welding methods for joining a variety of ferrous metals.

FIGURE 26-1 Improper welding technique for filler metal resulting in lack of fusion. *Larry Jeffus*

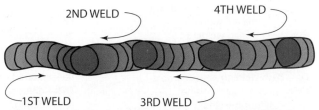

2ND WELD 4TH WELD

1ST WELD 3RD WELD

FIGURE 26-2 Welding sequence to produce the minimum weld stresses. © Cengage Learning 2012

Material	Thick-ness	Joining Process*																						
		Welding															Brazing							
		SMAW	SAW	GMAW	FCAW	GTAW	PAW	ESW	EGW	RW	FW	OFW	DFW	FRW	EBW	LBW	TB	FB	IB	RB	DB	IRB	DFB	S
Carbon steel	S	x	x	x		x			x	x	x			x	x	x	x	x	x	x	x	x	x	x
	I	x	x	x	x	x			x	x	x		x	x	x	x	x	x	x	x	x		x	x
	M	x	x	x	x				x	x	x		x	x	x	x	x	x					x	
	T	x	x	x	x			x	x	x	x		x	x	x		x						x	
Low alloy steel	S	x	x	x		x			x	x	x	x		x	x	x	x	x	x	x	x	x	x	x
	I	x	x	x	x	x			x	x		x	x	x	x	x	x	x					x	x
	M	x	x	x	x				x		x	x	x	x	x	x		x					x	
	T	x	x	x	x			x	x		x	x	x	x	x		x				x			
Stainless steel	S	x	x	x		x	x		x	x	x	x		x	x	x	x	x	x	x	x	x	x	x
	I	x	x	x	x	x	x		x	x		x	x	x	x	x	x	x					x	x
	M	x	x	x	x		x		x		x	x	x	x	x	x		x					x	
	T	x	x	x	x		x		x		x	x	x		x		x						x	
Cast iron	I	x							x					x	x	x						x	x		
	M	x	x	x	x				x					x	x	x						x	x		
	T	x	x	x	x				x					x								x	x		
Nickel and alloys	S	x		x		x	x		x	x	x			x	x	x	x	x	x	x	x	x	x	x	x
	I	x	x	x		x	x		x	x			x	x	x	x	x	x	x				x	x
	M	x	x	x			x		x			x	x	x	x	x		x					x	
	T	x		x			x		x			x	x	x			x						x	
Aluminum and alloys	S	x		x		x	x		x	x	x	x	x	x	x	x	x	x	x	x	x	x	x	x	x
	I	x		x		x			x	x		x	x	x	x	x		x		x			x	x	x
	M	x		x		x			x			x	x	x	x			x		x			x		
	T	x		x			x	x	x			x			x			x					x		
Titanium and alloys	S			x		x	x		x		x		x	x	x		x	x				x	x		
	I			x		x	x		x			x	x	x	x		x						x		
	M			x		x	x		x			x	x	x	x		x						x		
	T			x					x			x		x		x	x						x		
Copper and alloys	S			x		x	x		x				x	x	x	x	x					x	x		
	I			x			x		x			x	x	x	x	x	x						x	x	
	M			x					x			x	x	x	x	x							x		
	T			x					x				x	x	x								x		
Magnesium and alloys	S			x		x			x				x	x	x	x	x				x		x		
	I			x		x			x	x		x	x	x	x	x	x				x		x		
	M			x					x			x	x	x	x		x						x		
	T			x					x				x			x							x		
Refractory alloys	S			x		x	x		x	x				x			x		x	x	x			x	x
	I			x			x		x					x			x		x					x	
	M							x	x					x											
	T																								

This table presented as a general survey only. In selecting processes to be used with specific alloys, the reader should refer to other appropriate sources of information.

*See legend:

Legend

Process Code		Thickness
SMAW—shielded metal arc welding	FRW—friction welding	S—sheet: up to 3 mm (1/8 in.)
SAW—submerged arc welding	EBW—electron beam welding	I—intermediate: 3 to 6 mm (1/8 to 3/4 in.)
GMAW—gas metal arc welding	LBW—laser beam welding	M—medium: 6 to 19 mm (1/4 to 3/4 in.)
FCAW—flux cored arc welding	B—brazing	T—thick: 19 mm (3/4 in.) and up
GTAW—gas tungsten arc welding	TB—torch brazing	X—commercial process
PAW—plasma arc welding	FB—furnace brazing	
ESW—electroslag welding	IB—induction brazing	
EGW—electrogas welding	RB—resistance brazing	
RW—resistance welding	DB—dip brazing	
FW—flash welding	IRB—infrared brazing	
OFW—oxyfuel gas welding	DFB—diffusion brazing	
DFW—diffusion welding	S—soldering	

TABLE 26-2 Overview of Joining Processes *American Welding Society*

STEEL CLASSIFICATION AND IDENTIFICATION

SAE and AISI Classification Systems

Two primary numbering systems have been developed to classify the standard construction grades of steel, including both carbon and alloy steels. These systems classify the types of steel according to their basic chemical composition. One classification system was developed by the Society of Automotive Engineers (SAE). The other system is sponsored by the American Iron and Steel Institute (AISI).

The numbers used in both systems are now just about the same. However, the AISI system uses a letter before the number to indicate the method used in the manufacture of the steel.

Both numbering systems usually have a four-digit series of numbers. In some cases, a five-digit series is used for certain alloy steels. The entire number is a code to the approximate composition of the steel.

In both **steel classification** systems, the first number often, but not always, refers to the basic type of steel, as follows:

1XXX Carbon

2XXX Nickel

3XXX Nickel chrome

4XXX Molybdenum

5XXX Chromium

6XXX Chromium vanadium

7XXX Tungsten

8XXX Nickel chromium vanadium

9XXX Silicomanganese

The first two digits together indicate the series within the basic alloy group. There may be several series within a basic alloy group, depending upon the amount of the principal alloying elements. The last two or three digits refer to the approximate permissible range of carbon content. For example, the metal identified as 1020 would be 1XXX carbon steel with a XX20 0.20% range of carbon content, and 5130 would be 5XXX chromium steel with an XX30 0.30% range of carbon content.

The letters in the AISI system, if used, indicate the manufacturing process as follows:

- C—Basic open-hearth or electric furnace steel and basic oxygen furnace steel
- E—Electric furnace alloy steel

Table 26-3 shows the AISI and SAE numerical designations of alloy steels.

Unified Numbering System (UNS)

A unified numbering system is currently being promoted for all metals. This system eventually will replace the AISI and other systems.

CARBON AND ALLOY STEELS

Steels alloyed with carbon and only a low concentration of silicon and manganese are known as **plain carbon steels.** These steels can be classified as low carbon, medium carbon, and high carbon steels. The division is based upon the percentage of carbon present in the material.

Plain carbon steel is basically an alloy of iron and carbon. Small amounts of silicon and manganese are added to improve their working quality. Sulfur and phosphorus are present as undesirable impurities. All steels contain some carbon, but steels that do not include alloying elements other than low levels of manganese or silicon are classified as plain carbon steels. **Alloy steels** contain specified larger proportions of alloying elements.

The AISI has adopted the following definition of *carbon steel:* "Steel is classified as carbon steel when no minimum content is specified or guaranteed for aluminum, chromium, columbium, molybdenum, nickel, titanium, tungsten, vanadium, or zirconium; and when the minimum content of copper which is specified or guaranteed does not exceed 0.40%; or when the maximum content which is specified or guaranteed for any of the following elements does not exceed the respective percentages hereinafter stated: manganese 1.65%, silicon 0.60%, copper 0.60%." Under this classification will be steels of different composition for various purposes.

Many special alloy steels have been developed and sold under various trade names. These alloy steels usually have special characteristics, such as high tensile strength, resistance to fatigue, corrosion resistance, or the ability to perform at high temperatures. Basically, the ability of carbon steel to be welded is a function of the carbon content, **Table 26-4.** (Other factors to be considered include thickness and the geometry of the joint.) All carbon steels can be welded by at least one method. However, the higher the carbon content of the metal, the more difficult it is to weld the steel. Special precautions must be followed in the welding process.

Low Carbon Also Called Mild Steel

These steels have carbon content of less than 0.30%. These steels can be welded easily by all welding processes. The resulting welds are of extremely high quality. Oxyacetylene welding of these steels can be done by using a neutral flame. Joints welded by this process are of high quality, and the fusion zone is not hard or brittle.

13XX	Manganese 1.75
23XX*	Nickel 3.50
25XX*	Nickel 5.00
31XX	Nickel 1.25; chromium 0.65
E33XX	Nickel 3.50; chromium 1.55; electric furnace
40XX	Molybdenum 0.25
41XX	Chromium 0.50 or 0.95; molybdenum 0.12 or 0.20
43XX	Nickel 1.80; chromium 0.50 or 0.80; molybdenum 0.25
E43XX	Same as above, produced in basic electric furnace
44XX	Manganese 0.80; molybdenum 0.40
45XX	Nickel 1.85; molybdenum 0.25
47XX	Nickel 1.05; chromium 0.45; molybdenum 0.20 or 0.35
50XX	Chromium 0.28 or 0.40
51XX	Chromium 0.80, 0.88, 0.93, 0.95, or 1.00
E5XXXX	High carbon; high chromium; electric furnace bearing steel
E50100	Carbon 1.00; chromium 0.50
E51100	Carbon 1.00; chromium 1.00
E52100	Carbon 1.00; chromium 1.45
61XX	Chromium 0.60, 0.80, or 0.95; vanadium 0.12, or 0.10, or 0.15 minimum
7140	Carbon 0.40; chromium 1.60; molybdenum 0.35; aluminum 1.15
81XX	Nickel 0.30; chromium 0.40; molybdenum 0.12
86XX	Nickel 0.55; chromium 0.50; molybdenum 0.20
87XX	Nickel 0.55; chromium 0.50; molybdenum 0.25
88XX	Nickel 0.55; chromium 0.50; molybdenum 0.35
92XX	Manganese 0.85; silicon 2.00; 9262-chromium 0.25 to 0.40
93XX	Nickel 3.25; chromium 1.20; molybdenum 0.12
98XX	Nickel 1.00; chromium 0.80; molybdenum 0.25
14BXX	Boron
50BXX	Chromium 0.50 or 0.28; boron
51BXX	Chromium 0.80; boron
81BXX	Nickel 0.33; chromium 0.45; molybdenum 0.12; boron
86BXX	Nickel 0.55; chromium 0.50; molybdenum 0.20; boron
94BXX	Nickel 0.45; chromium 0.40; molybdenum 0.12; boron

Note: The elements in this table are expressed in percent.

Consult current AISI and SAE publications for the latest revisions.

*Nonstandard steel.

TABLE 26-3 AISI and SAE Numerical Designation of Alloy Steels

Common Name	Carbon Content	Typical Use	Weldability
Ingot iron	0.03% max.	Enameling, galvanizing, and deep drawing sheet and strip	Excellent
Low carbon (mild) steel	0.00% to 0.30%	Welding electrodes, special plate and shapes, sheet and strip Structural shapes, plate and bar	Excellent to Good
Medium carbon steel	0.30% to 0.50%	Machinery parts	Fair*
High carbon steel	0.50% to 1.00%	Springs, dies, and railroad rails	Poor**

*Preheat and frequently postheat required.

**Difficult to weld without adequate preheat and postheat.

TABLE 26-4 Iron-Carbon Alloys, Uses, and Weldabilities

Low carbon (mild) steels can be welded readily by the shielded metal arc method. The selection of the correct electrode for the particular welding application helps to ensure high strength and ductility in the weld.

The gas metal arc and flux cored arc welding processes are used for welding both low and medium carbon steels due to the ease of welding and because they prevent contamination of the weld. The high productivity and lower cost make them increasingly popular welding processes.

The gas tungsten arc process is slow and will cause severe porosity in the weld if the steel is not fully degassed (metal that has had all the gas normally dissolved in metal during manufacturing removed). GTAW can make superior welds with a very clear X-ray quality.

Medium Carbon Steel

The welding of medium carbon steels, having 0.30% to 0.50% carbon content, is best accomplished by the various fusion processes, depending upon the carbon content of the base metal. The welding technique and materials used are dictated by the metallurgical characteristics of the metal being welded. For steels containing more than 0.40% carbon, preheating and subsequent heat treatment generally are required to produce a satisfactory weld. Shielded arc electrodes of the type used on low carbon steels can be used for welding this type of steel. The use of an electrode with a special low hydrogen coating may be necessary to reduce the tendency toward underbead cracking.

Medium carbon steels can be resistance welded. However, special techniques may be required. Other welding methods that produce sound welds on medium carbon steel include submerged arc welding, thermite welding, pressure gas welding, and spot, flash, and seam welding.

High Carbon Steel

High carbon steels usually have a carbon content of 0.50% to 0.90%. These steels are much more difficult to weld than either the low or medium carbon steels. Because of the high carbon content, the heat-affected zone can transform to very hard and brittle martensite. The welder can avoid this by using preheat and by selecting procedures that produce high-energy inputs to the weld. Refer to **Figure 26-3** for the preheat temperature for the specific carbon content. The martensite that does form is tempered by postweld heat treatments such as stress-relief anneal. Refer to **Figure 26-4** for the temperature for stress-relief annealing between 1125°F and 1250°F (600°C and 675°C).

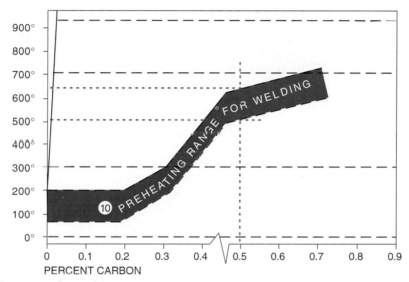

FIGURE 26-3 Preheating range for welding. © Cengage Learning 2012

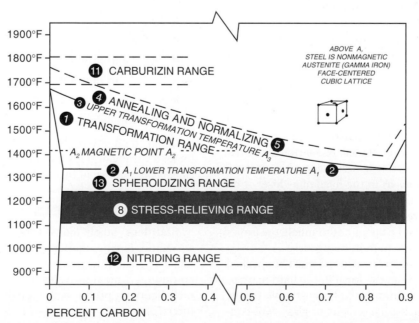

FIGURE 26-4 Stress-relieving range. © Cengage Learning 2012

In arc welding high carbon steel, mild-steel shielded arc electrodes are generally used. However, the weld metal does not retain its normal ductility because it absorbs some of the carbon from the steel. Austenitic steel electrodes can sometimes be used to obtain better ductility in the weld.

Welding on high carbon steels is often done to build up a worn surface to original dimensions or to develop a hard surface. In this type of welding, preheating or heat treatment may not be needed because of the way in which heat builds up in the part during continuous welding.

Tool Steel

Because **tool steel** has a carbon content of from 0.8% to 1.50%, it is very difficult to weld. Gas welding can be employed if the material is in the lower carbon ranges. However, when welding tool steel by the gas welding method, the rods selected for welding repairs should have a carbon content matching that of the base metal. Gas welding requires correct flame adjustment and careful manipulation of the flame and rod. Recommended practice is to preheat the metal, followed by a slow annealing after the welding. A carburizing or neutral flame is desirable in obtaining a strong weld.

In arc welding tool steels, one of the following procedures should be observed:

- Anneal the parts, preheat and weld with a proper electrode, and then heat-treat to restore the original properties to the metal.

- Preheat the parts and deposit one or more layers on the kerf surfaces of the joint with a covered electrode before depositing the weld beads that tie the joint together. These procedures dilute the carbon content, and the surfaces of the joint can then be more readily welded together.

High Manganese Steel

High manganese steel contains 12% or more manganese and a carbon content ranging from 1% to 1.4%. These steels are used for wear resistance in applications involving impact. They are used for such items as power shovels, rock crushers, mine equipment, augers, switch frogs, and so forth.

These alloys are austenitic and, therefore, tough. When inspected, they form a hard martensitic layer on the surface, which provides wear resistance. As martensite develops during work hardness it reaches a stage where it starts to roll over or mushrooms as in the head of a chisel or on the edges of a railroad track. As the components wear in service, they are repaired with electrodes having similar compositions. Care must be taken not to inhale the fumes produced because they are high in manganese.

Martensite is very hard and brittle and must be checked for cracks periodically. If cracks form in the hard martensite layer, they can move into the softer austenitic below, **Figure 26-5**. There cracks, like cracks in glass, will continue moving through the part until it breaks. Removing

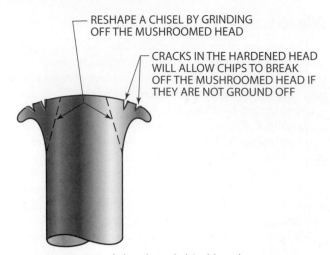

— RESHAPE A CHISEL BY GRINDING OFF THE MUSHROOMED HEAD

— CRACKS IN THE HARDENED HEAD WILL ALLOW CHIPS TO BREAK OFF THE MUSHROOMED HEAD IF THEY ARE NOT GROUND OFF

FIGURE 26-5 Work-hardened chisel head. © Cengage Learning 2012

the cracks by welding them up or cutting them out is the only way to prevent them from causing the part to fail.

Low Alloy, High Tensile Strength Steels

Low alloy steels are used increasingly because of requirements for high strength with less weight. These types of steel are readily weldable by all of the common welding processes. The methods and types of electrodes used depend upon the end results required. Steel manufacturers can usually supply welding information.

The following types of steel can be found in this class:

- Austenitic manganese
- Chromium alloy
- Carbon molybdenum
- Chromium molybdenum
- Chromium vanadium
- Manganese alloy
- Manganese molybdenum
- Nickel chromium
- Nickel chromium molybdenum
- Nickel copper
- Nickel molybdenum

Other high strength, low alloy steels are available under a variety of trade names and are listed in the *Metals Handbook,* published by the American Society for Metals.

Stainless Steels

Stainless steels consist of four groups of alloys: austenitic, ferritic, martensitic, and precipitation hardening. The austenitic group is by far the most common. Its chromium content provides corrosion resistance, while its nickel content produces the tough austenitic microstructure. These steels are relatively easy to weld, and a large variety of electrode types are available.

The most widely used stainless steels are the chromium-nickel austenitic types and are usually referred to by their chromium-nickel content as 18-8, 25-12, 25-20, and so on. For example, 18-8 contains 18% chromium and 8% nickel, with 0.08% to 0.20% carbon. To improve weldability, the carbon content should be as low as possible. Carbon should not be more than 0.03%, with the maximum being less than 0.10%.

Keeping the carbon content low in stainless steel will also help reduce carbide precipitation. Carbide precipitation occurs when alloys containing both chromium and carbon are heated. The chromium and carbon combine to form chromium carbide (Cr_3C_2).

The combining of chromium and carbon lowers the chromium that is available to provide corrosion resistance in the metal. This results in a metal surrounding the weld that will oxidize, or rust. The amount of chromium carbide formed is dependent on the percentage of carbon, the time that the metal is in the critical range, and the presence of stabilizing elements.

If the carbon content of the metal is very low, little chromium carbide can form. Some stainless steel types have a special low carbon variation. These low carbon stainless steels are the same as the base type but with much lower carbon content. To identify the low carbon from the standard AISI number, the L is added as a suffix. See examples 304 and 304L, **Table 26-5.**

Chromium carbides form when the metal is between 800°F and 1500°F (625°C and 815°C). The quicker the metal is heated and cooled through this range, the less time that chromium carbides can form. Since austenitic stainless steels are not hardenable by quenching, the weld can be cooled using a chill plate. The chill plate can be water-cooled for larger welds.

Some filler metals have stabilizing elements added to prevent carbide precipitation. Columbium and titanium are both commonly found as chromium stabilizers. Examples of the filler metals are E310Cb and ER309Cb.

In fusion welding, stainless austenitic steels may be welded by all of the methods used for plain carbon steels.

Since *ferritic stainless steels* contain almost no nickel, they are cheaper than austenitic steels. They are used for ornamental or decorative applications such as architectural trim and at elevated temperatures such as used for heat exchanging. However, ferritic stainless steels also tend to be brittle unless specially deoxidized. Special high-purity, high-toughness ferritic stainless steels have been developed, but careful welding procedures must be used with them to prevent embrittlement. This means very carefully controlling nitrogen, carbon, and hydrogen.

Martensitic stainless steels are also low in nickel but contain more carbon than the ferritic. They are used in applications requiring both wear resistance and corrosion resistance. Items such as surgical knives and razor blades are made of them. Quality welding requires very careful control of both preheat and tempering immediately after welding.

Precipitation hardening stainless steels can be much stronger than the austenitic, without losing toughness. Their strength is the result of a special heat treatment used to develop the precipitate. They can be solution-treated prior to welding and given the precipitation treatment after welding.

The closer the characteristics of the deposited metal match those of the material being welded, the better is the corrosion resistance of the welded joint. The following precautions should be noted:

- Any carburization or increase in carbon must be avoided, unless a harder material with improved wear resistance is actually desired. In this case, there will be a loss in corrosion resistance.

- It is important to prevent all inclusions of foreign matter, such as oxides, slag, or dissimilar metals.

In welding with the metal arc process, direct current is more widely used than alternating current. Generally, reverse polarity is preferred where the electrode is positive and the workpiece is negative.

The diameter of the electrode used to weld steel that is thinner than 3/16 in. (4.8 mm) should be equal to, or slightly less than, the thickness of the metal to be welded.

When setting up for welding, material 0.050 in. (1.27 mm) and less in thickness should be clamped firmly to prevent distortion or warpage. The edges should be butted tightly together. All seams should be accurately aligned at the start. It is advisable to tack weld the joint at short intervals as well as to use clamping devices.

The electrode should always point into the weld in a backhand or drag angle, **Figure 26-6.** Avoid using a figure-8 pattern or excessive side weaving motion such as that used in welding carbon steel. Best results are obtained with a stringer bead with little or very slight weaving motion and steady forward travel, with the electrode holder leading the weld pool at about 60° in the direction of travel.

Nominal Composition of Stainless Steels						
AISI Type	C	Mn Max	Nominal Composition %		Ni	Other
			Si Max	Cr		
304	0.08 max.	2.0	1.0	18–20	8–12	
304L	0.03 max.	2.0	1.0	18–20	8–12	
316	0.08 max.	2.0	1.0	16–18	10–14	2.0–3.0 Mo
316L	0.03 max.	2.0	1.0	16–18	10–14	2.0–3.0 Mo

TABLE 26-5 Comparison of Standard-Grade and Low-Carbon Stainless Steels

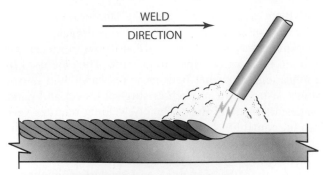

WELD DIRECTION →

FIGURE 26-6 Backhand or drag angle. © Cengage Learning 2012

To weld stainless steels, the arc should be as short as possible. **Table 26-6** can be used as a guide.

Chromium-Molybdenum Steel

Chromium-molybdenum steel is used for high-temperature service and for aircraft parts. It can be welded by the following processes: shielded metal arc, gas tungsten arc, gas metal arc, flux cored, and submerged arc. The welds have a tensile strength of 60,000 psi (49 kg/cm^2) to 80,000 psi (5625 kg/cm^2) as welded. This type of material lends itself to joints in thin sections.

Cast Iron

Cast iron is widely used for engine components such as blocks, heads, and manifolds; for drive components such as transmission cases, gearboxes, transfer cases, and differential cases; and for equipment such as pumps, planters, drills, and pulleys. Cast iron is hard and ridged, which makes it ideal for any size casing or frame that must hold its shape even under heavy loads. For example, if a transmission case were to bend under a load, the gears and shafts inside would bind and stop turning.

All five types of cast iron have high carbon contents, usually ranging from 1.7% to 4%. The most common grades contain about 2.5% to 3.5% total carbon. The carbon in cast iron can be combined with iron or be in a free state. As more of the carbon atoms in the cast iron combine with iron atoms, the cast iron becomes harder and more brittle. The five common types of cast iron are as follows:

- *Gray cast iron* is the most widely used type. It contains so much free carbon that a fracture surface has a uniform dark gray color. Gray cast iron is easily

welded, but because it is somewhat porous it can absorb oils into the surface, which must be baked out before welding.

- *White cast iron* is the hardest and most brittle of the cast irons because almost all of the carbon atoms are combined with the iron atoms. The surface of a fractured piece of this cast iron looks silvery white and may appear shiny. White cast iron is practically unweldable.

- *Malleable cast iron* is white cast iron that has undergone a transformation as the result of a long heat-treating process to reduce the brittleness. The fractured surface of **malleable cast iron** has a light, almost white, thin rim around the dark gray center. If malleable cast iron is heated above its critical temperature, about 1700°F (925°C), the carbon will recombine with the iron, transforming back into white cast iron. Malleable cast iron can easily be welded. To prevent it from reverting back to white cast iron, do not preheat above 1200°F (650°C).

- *Alloy cast iron* has alloying elements such as chromium, copper, manganese, molybdenum, or nickel added to obtain special properties. Various quantities and types of alloys are added to improve alloy cast iron's tensile strength and heat and corrosion resistance. Almost all grades of alloy cast iron can be easily welded if care is taken to slowly preheat and postcool the part to prevent changes in the carbon and iron structure.

- *Nodular cast iron,* sometimes called ductile cast iron, has its carbon formed into nodules or tiny round balls. These nodules are formed by adding an alloy. Nodular cast iron has greater tensile strength than gray cast iron and some of the corrosion resistance of alloy cast iron. Nodular cast iron is weldable, but proper preheating and postweld cooling temperatures and rates must be maintained or the nodular properties will be lost.

Not all cracks or breaks in cast iron present the same degree of difficulty to making welded repairs. Breaks across ears or tabs do not have nearly as much stresses as a crack in a surface, **Figure 26-7**. Cracks may increase in length when you try to weld them unless you drill a small hole, about 1/8 in. (3 mm), at both ends of the crack, **Figure 26-8**.

Preweld and Postweld Heating of Cast Iron The major purpose of preheating and postheating of cast iron is

Metal Thickness		Electrode Diameter		Current	Voltage
in.	(mm)	in.	(mm)	(Amperes)	Open Circuit
0.050	(1.27)	5/64	(1.98)	25–50	30–35
0.050–0.0625	(1.27–1.58)	3/32	(2.38)	30–90	35–40
0.0625–0.1406	(1.58–3.55)	3/32–1/8	(2.38–3.17)	50–100	40–45
0.1406–0.1875	(3.55–4.74)	1/8–5/32	(3.17–3.96)	80–125	45–50
0.250 and up	(6.35 and up)	3/16	(4.76)	100–175	55–60

TABLE 26-6 Shielded Metal Arc Welding Electrode Setup for Stainless Steel

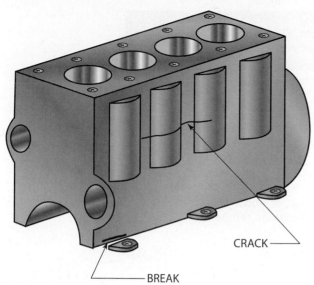

CRACK

BREAK

FIGURE 26-7 Cracks on engines can occur in the water jacket due to freezing, and ears can be broken off as a result of accidents or overtightening of misaligned parts. © Cengage Learning 2012

DRILL A 1/8" (3-MM) HOLE AT BOTH ENDS OF THE CRACK

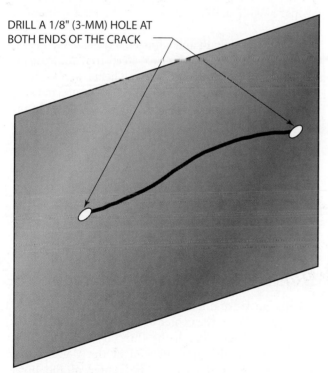

FIGURE 26-8 Stop drill the ends of a crack. © Cengage Learning 2012

to control the rate of temperature change. The level of temperature and the rate of change of temperature affect the hardness, brittleness, ductility, and strength of iron-carbon–based metals such as steel and cast iron, **Table 26-7**.

Preheating the casting before welding reduces the internal stresses caused by the rapid or localized heating resulting from welding, **Table 26-8**. Welding stresses occur because as metal is heated and cooled, it expands and contracts. Unless the heating and cooling cycles are slow and uniform, stresses within brittle materials will cause them to crack. In some aspects, the brittleness of cast iron is much like that of glass. Both cast iron and glass will crack if they are heated or cooled unevenly or too quickly.

Postweld heating changes the rate of cooling. Rapid cooling of a metal from a high temperature is called quenching. The faster an iron-carbon metal is quenched, the harder, more brittle, less ductile, and higher in strength the metal will become. The slow cooling of an iron-carbon metal from a high temperature is called annealing. The slower an iron-carbon metal is cooled from a high temperature, the softer, less brittle, more ductile, and lower in strength the metal will become.

To reduce welding stresses, maintain the casting at the same temperature used for preheat or higher for 30 minutes following welding. The casting should cool slowly over the next 24 hours. Cover the casting to prevent the part from being cooled too rapidly by the surrounding air following welding. A firebrick or heavy metal box can be used to keep cool air away from the casting.

Practice Welding Cast Iron

Because there are a few differences between repairing a break and a crack, the following practices alternate between repairing breaks and cracks in cast iron. For example, other than not clamping the parts together, there would be little difference between using Practice 26-1 for welding a break and for welding a crack. In the field, you can use whichever welding procedure is most appropriate for repairing breaks or cracks.

PRACTICE 26-1

Arc Welding a Cast Iron Break with Preheating and Postheating

Using a properly set-up stick welding station, a gas torch with a heating tip, firebricks, a right angle grinder,

Property	Description
Hardness	The resistance to penetration or shaping by machining or drilling.
Brittleness	The ease with which a metal will crack or break apart without noticeable deformation.
Ductility	The ability of a metal to be permanently twisted, drawn out, bent, or changed in shape without cracking or breaking.
Strength	The ability of a metal to resist deformation or reshaping due to tensile, compression, shear, or torsional forces.

TABLE 26-7 Mechanical Properties of Metal

Preheat Temperatures for Weldable Cast Irons			
Joining Process	Temperature Range	Preferred Temperature	Minimum Temperature
Stick Welding	600°F to 1500°F (315°C to 815°C)	900°F to 1200°F (480°C to 650°C)	400°F (200°C)
Gas Welding	400°F to 1100°F (200°C to 600°C)	500°F (260°C)	400°F (200°C)
Braze Welding	500°F to 900°F (260°C to 480°C)	900°F (480°C)	500°F (260°C)

Note: Maintain the preheat temperature for 30 minutes after it is first reached to allow the core of the casting to reach this temperature.

TABLE 26-8 Welding Should Be Performed at or Above the Preferred Temperature (It can be performed at the minimum temperature, but some hardening and cracking may occur.)

ENi electrodes, a C-clamp, a 900°F (482°C) temperature marking crayon, a chipping hammer, a wire brush, a broken piece of gray cast iron, a welding helmet, gas welding goggles, safety glasses, and all other required safety equipment, you are going to repair a cast iron break.

Grind the brake into a V-groove, leaving a 1/8-in. (3-mm) root face on the broken surface, **Figure 26-9**.

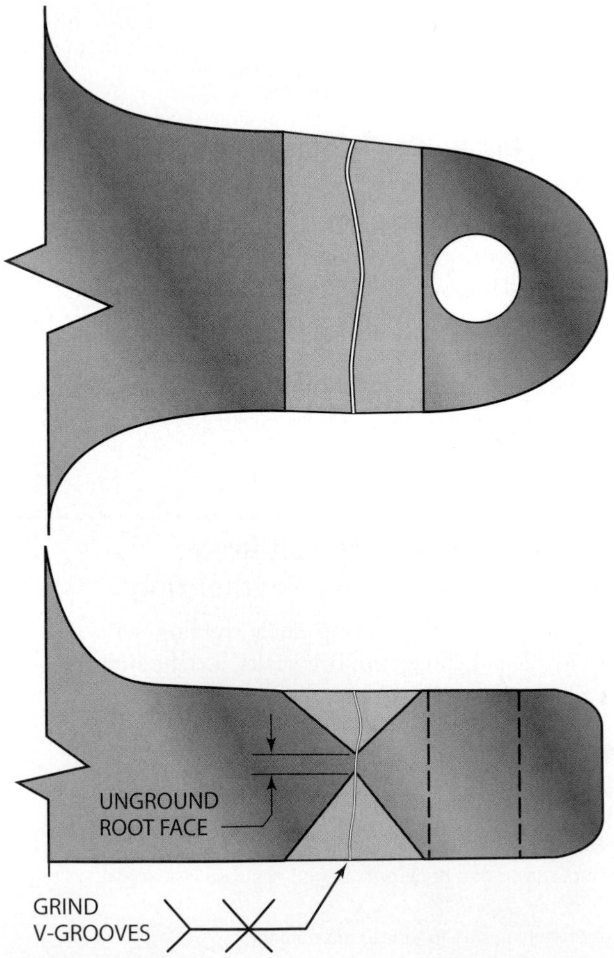

FIGURE 26-9 Grind the V-groove, leaving a small root face so the part can be realigned. © Cengage Learning 2012

> NOTE: Because cast iron is brittle it does not bend before it breaks, so broken parts can usually be fitted back together like the pieces of a puzzle.

Use a C-clamp to hold the broken piece in place. Mark the parts with a 900°F (482°C) temperature marking crayon.

Using a properly lit and adjusted oxyacetylene heating torch, begin preheating the part. Keep the flame moving all around the cast iron part so that it heats evenly. Do not point the flame directly onto the temperature indicator mark or it will turn color before the part is actually preheated. When the temperature mark turns black, the part is properly preheated. With the flame off the part, re-mark the part to check it for proper preheat.

On thick castings, keep the flame on the part for 30 more minutes so the inside of the part is properly preheated.

> NOTE: Starting the welds on the ends of the break or crack and welding to the center concentrates the weld stresses in the center of the weld. This reduces the chances of a crack forming at the end of the weld.

Strike an arc and make the first weld bead, starting at one edge. Weld to the center of the break. Break the arc, and chip and wire brush the flux off the weld. Strike another arc on the opposite end of the V-groove and make another weld back to the center of the break, **Figure 26-10**. Turn the part over and make the same two welds on the back side of the break.

> NOTE: A number of small welds are better than one or two large welds because the small welds do not have as much stress and are less likely to cause postweld cracking.

Repeat the process of making welds by alternating sides between weld passes until the V-groove is filled to no more than 1/8 in. (3 mm) above the surface.

Build a firebrick box around the part and place the torch so the flame will fill the box and the part can be

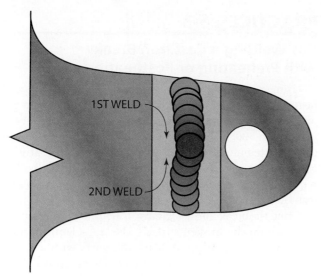

FIGURE 26-10 Start all welding beads on the edge of the break and end them in the middle. © Cengage Learning 2012

postweld heated. Keep heating the part for 30 minutes after completing the weld. Close any gaps in the firebrick box and allow the part to cool slowly over the next 24 hours. Turn off the welder and torch set, and clean up your work area when you are finished welding.

Complete a copy of the "Student Welding Report" listed in Appendix I or provided by your instructor. ◆

Welding without Preheating or Postheating
Cracks in large castings and castings that are to be repaired in place cannot be preheated and postheated to the desired temperatures but can still be repaired. One method of repairing these cracks is to drill and tap a series of overlapping holes along the crack, **Figure 26-11**. This is an excellent way to repair nonload-bearing cracks like those in the water jacket of an engine. Two-part epoxy patch material can also be used on nonload-bearing cracks. Read and follow the manufacturer instructions and safety rules when using epoxy repair kits.

Cracks in parts that cannot be preheated to the desired level can still be welded, but the welds will be very hard and are more likely to recrack. However, welding cracks in engine blocks, pump housings, and other large expensive castings, even if they might crack again, is more desirable than simply buying a new part. The new cracks that might form may even be small enough to be patched with epoxy.

PRACTICE 26-2

Arc Welding a Cast Iron Crack without Preheating or Postheating

Using a properly set-up stick welding station, a gas torch with a heating tip, firebricks, a portable drill with a 1/8-in. (3-mm) drill bit, a right angle grinder, ENi electrodes, a chipping hammer, a wire brush, a cracked piece of gray cast iron, a welding helmet, gas welding goggles, safety glasses, and all other required safety equipment, you are going to repair a cast iron crack.

Locate the ends of the crack and drill stop the crack by drilling 1/8-in. (3-mm) holes at both ends of the crack. Use the edge of the grinding disk to cut a V- or U-groove into the crack.

NOTE: Even though the part cannot be preheated to the desired level, it cannot be welded cold. It must be heated to at least 75°F (24°C) or higher before starting to weld. Engine blocks can be preheated by letting them run for a short time until they are hot.

Strike the arc on the casting just before the end of the crack and make a 1-in. (25-mm) long weld, **Figure 26-12**.

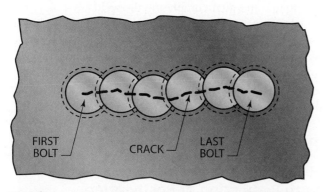

FIGURE 26-11 A crack in cast iron can be plugged by drilling and tapping overlapping holes. Each bolt overlaps the previous bolt. Only the last one must have a locking compound to prevent it from loosening.
© Cengage Learning 2012

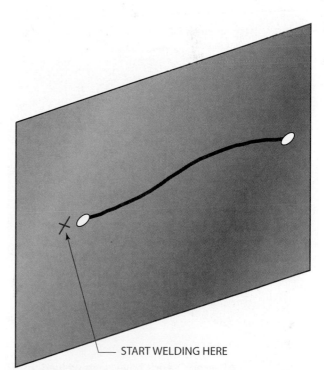

— START WELDING HERE

FIGURE 26-12 Start the weld on the casting's surface outside of the crack. © Cengage Learning 2012

Stop the weld and repeat the process starting at the other end of the crack.

> NOTE: A series of short welds will create less stress than one large weld. The small welds are less likely to crack.

Chip and wire brush the welds.

> NOTE: The next series of welds that are made to close the crack are done in a back-stepping sequence. Back-stepping welds are short welds that start ahead of the ending point of the first weld and go back to the end of the first weld. Back-stepping welds are used because they reduce weld stresses and are less likely to crack.

Start the third weld about 1 in. (25 mm) in front of the end of the first weld and move back to the end of the first weld, **Figure 26-13**. Vigorously chip the slag off the weld immediately after the arc stops. This both cleans the weld and mechanically works the weld surface, called peening, to reduce weld stresses. Repeat the back-step sequence of welding and peening until the crack is completely covered with welds. Turn off the welder and torch set, and clean up your work area when you are finished welding.

Complete a copy of the "Student Welding Report" listed in Appendix I or provided by your instructor. ◆

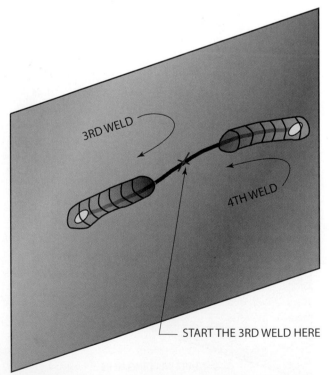

START THE 3RD WELD HERE

FIGURE 26-13 After both ends have been welded, use a back-step welding process to complete the weld.

© Cengage Learning 2012

PRACTICE 26-3

Gas Welding a Cast Iron Break with Preheating or Postheating

Using a properly set-up gas torch with welding and heating tips, firebricks, a right angle grinder, cast iron gas welding rods, high temperature cast iron welding flux, a C-clamp, a 500°F (260°C) temperature marking crayon, a chipping hammer, a wire brush, a broken piece of gray cast iron, gas welding goggles, safety glasses, and all other required safety equipment, you are going to repair a cast iron break.

Grind the break into a V-groove, leaving a 1/8-in. (3-mm) root face on the broken surface. Use a C-clamp to hold the broken piece in place. Mark the parts with a 500°F (260°C) temperature marking crayon. Using a properly lit and adjusted oxyacetylene heating torch, begin preheating the part. Keep the flame moving all around the cast iron part so that it heats evenly. Do not point the flame directly onto the temperature indicator mark or it will turn color before the part is actually preheated. When the temperature mark turns black, the part is properly preheated. With the flame off the part, re-mark the part to check it for proper preheat.

On thick castings, keep the flame on the part for 30 more minutes so that the inside of the part is properly preheated. Flux the end of the cast iron filler rod by heating it with the torch and dipping it into the flux. As the weld progresses, occasionally redip the end of the filler rod back into the flux so new flux can be added to the weld. Start welding at one end of the crack. The molten weld pool formed by cast iron is not bright and shiny like that formed on mild steel. The cast iron molten weld pool looks dull and a little lumpy. As the weld progresses, move the tip of the filler rod around in the molten weld pool to keep it stirred up and to make it flatter.

Fill the crater at the end of the weld with a little extra filler metal. Turn the part over and make the same weld on the back side of the break. Build a firebrick box around the part and place the torch so the flame will fill the box and the part can be postweld heated. Keep heating the part for 30 minutes after completing the weld.

Close any gaps in the firebrick box and allow the part to cool slowly over the next 24 hours. Turn off the torch set and clean up your work area when you are finished welding.

Complete a copy of the "Student Welding Report" listed in Appendix I or provided by your instructor. ◆

PRACTICE 26-4

Braze Welding a Cast Iron Crack with Preheating or Postheating

Using a properly set-up gas torch with welding and heating tips, firebricks, a right angle grinder, BRCuZn rods, brazing flux, a C-clamp, a 900°F (480°C) temperature marking crayon, a chipping hammer, a wire brush, a cracked piece of gray cast iron, gas welding goggles, safety glasses, and all other required safety equipment, you are going to repair a cast iron crack.

Locate the ends of the crack and drill stop the crack by drilling 1/8-in. (3-mm) holes at both ends of the crack. Use the edge of the grinding disk to cut a V- or U-groove into the crack. Mark the parts with a 900°F (480°C) temperature marking crayon. Using a properly lit and adjusted oxyacetylene heating torch, begin preheating the part. Keep the flame moving all around the cast iron part so that it heats evenly. Do not point the flame directly onto the temperature indicator mark or it will turn color before the part is actually preheated. When the temperature mark turns black, the part is properly preheated. With the flame off the part, re-mark the part to check it for proper preheat. On thick castings, keep the flame on the part for 30 more minutes so that the inside of the part is properly preheated.

Flux the end of the brazing rod by heating it with the torch and dipping it into the flux. As the weld progresses, occasionally redip the end of the filler rod back into the flux so new flux can be added to the weld. Start welding at one end of the crack. Heat the groove until it is dull red, to about 1800°F (980°C). Touch the tip of the brazing rod into the groove occasionally to test it for the proper brazing temperature. When the braze metal begins to flow, move the tip of the rod around in the molten braze pool to help it wet the surface of the groove. When the braze reaches the end of the groove, add a little extra fill to the crater at the end of the braze weld. Turn the part over and make the same weld on the back side of the break.

Build a firebrick box around the part and place the torch so that the flame will fill the box and the part can be postweld heated. Keep heating the part for 30 minutes after completing the weld. Close any gaps in the firebrick box and allow the part to cool slowly over the next 24 hours. Turn off the torch set and clean up your work area when you are finished welding.

Complete a copy of the "Student Welding Report" listed in Appendix I or provided by your instructor. ◆

NONFERROUS METALS

Metals that are not primarily composed of iron are known as nonferrous metals. Each nonferrous metal has its unique physical and metallurgical property that must be considered when selecting a welding procedure. The most common nonferrous metals that welders are asked to weld are: copper and copper alloys, aluminum, titanium, and magnesium.

Copper and Copper Alloys

There are many different types of copper alloys. Copper is often alloyed with other metals such as tin, zinc, nickel, silicon, aluminum, and iron. **Copper and copper alloys** can be joined by most of the commonly used methods such as gas welding, arc welding, resistance welding, brazing, and soldering.

For many years, the successful welding of copper was considered impractical. Too much distortion resulted when using a gas torch. That heat source was barely able to melt the metal due to its high thermal conductivity. The more highly intense electric arc overcomes that difficulty.

When you are welding copper, the welding current should be considerably higher than when welding steel. On copper 1/8 in. (3 mm) or more in thickness, the current should not be more than 140 A. The preheat should be 500°F (260°C). By using this method of welding copper, satisfactory results can be obtained for the following factors: economy, speed, ductility, strength, and freedom from distortion. However, the copper must not be electrolytic because excessive amounts of gas in the metal increase porosity.

Aluminum Weldability

One of the characteristics of **aluminum** and its alloys is that it has a great affinity for oxygen. Aluminum atoms combine with oxygen in the air to form a high melting point oxide that covers the surface of the metal. This feature, however, is the key to the high resistance of aluminum to corrosion. It is because of this resistance that aluminum can be used in applications where steel is rapidly corroded.

Pure aluminum melts at 1200°F (650°C). The oxide that protects the metal melts at 3700°F (2037°C). This means that the oxide must be cleaned from the metal before welding can begin.

When the GMA welding process is used, the stream of inert gas covers the weld pool, excluding all air from the weld area. This prevents reoxidation of the molten aluminum. GMA welding does not require a flux.

Aluminum can be arc welded using aluminum welding rods. These rods must be kept in a dry place because the flux picks up moisture easily. Because aluminum melts so easily, use a piece of clean steel plate as a backing to weld on thin sections. The steel backing plate can support the root of the weld without the aluminum weld sticking to the steel plate. Thick aluminum casting must be preheated to about 400°F (200°C) before welding. The preheating helps the weld flow and reduces weld spatter.

Aluminum has high thermal conductivity. Aluminum and its alloys can rapidly conduct heat away from the weld area. For this reason, it is necessary to apply the heat much faster to the weld area to bring the aluminum to the welding temperature. Therefore, the intense heat of the electric arc makes this method best suited for welding aluminum.

When aluminum welds solidify from the molten state, they will shrink about 6% in volume. The stress that results from this shrinkage may create excessive joint distortion unless allowances are made before joining the metal. Cracking can occur because the thermal contraction is about two times that of steel. The heated parent metal expands when welding occurs. This expansion of the metal next to the weld area can reduce the root opening on butt joints during the process. The contraction that results upon cooling, plus the shrinkage of the weld metal, creates a tension and increases cracking.

The shape of the weld groove and the number of beads can affect the amount of distortion. Less distortion occurs with two-pass square butt welds. Other factors that have an influence on the weld are the speed of welding, the use of properly designed jigs and fixtures to support the aluminum while it is being welded, and tack welding to hold parts in alignment.

Titanium

Titanium is a silvery-gray metal weighing about half as much as steel or about one and a half times as much as aluminum. Two of the most important properties of titanium are its extremely high strength-to-weight ratio (in alloy form) and its generally excellent corrosion resistance. Titanium alloys, unlike most other light metals, retain their strength at temperatures up to about 800°F (426°C).

Pure titanium is comparatively soft and weak. It is very difficult to refine. Commercially pure titanium contains trace impurities that increase the strength considerably, while at the same time causing a loss in ductility. Alloy combinations currently in use contain assortments of tin, chromium, iron, aluminum, and vanadium.

The success of welding titanium depends upon complete shielding from the atmospheric gases oxygen, nitrogen, and hydrogen and from sources of carbon. The inert gas welding process seems to be the best method of welding this metal. Shielding gases used are argon, helium, or a combination of the two. It is preferable to perform the welding in a closed chamber.

When welding titanium, the joint must be clean with no traces of contamination. Clean filler metal and perfect shielding are required to eliminate the porosity and embrittlement that can be produced during welding.

Magnesium

Magnesium is an extremely light metal having a silvery-white color. The weight of magnesium is one-fourth that of steel and approximately two-thirds that of aluminum. Its melting point is 1202°F (650°C). Magnesium has considerable resistance to corrosion and compares favorably with some aluminum alloys in this respect.

Magnesium must be alloyed with other elements to provide the necessary strength for most applications. Common alloying elements include zinc, aluminum, manganese, zirconium, and the rare earths. Magnesium alloys may be classified as wrought or casting types. Sheet, plate, and extrusions are part of the first group.

The alloy designations for magnesium consist of one or two letters, which represent the alloying elements. The alloying percentages are next listed in the designation. A letter is also added after the alloying percentages. One example is the ASTM designation AZ91C. This indicates an alloy of 9% aluminum and 1% zinc. The letter following the designation is defined as follows:

- A—aluminum
- C—copper
- E—rare earths
- H—thorium
- K—zirconium
- M—manganese
- Z—zinc

The temper designation is the same as that used for aluminum. Wrought alloys are used to a great extent because of their properties of high strength, ductility, formability, and toughness.

Magnesium can be welded in somewhat the same manner as aluminum. The most widely used processes are GTA and GMA. Spot, seam, and mechanized resistance welding processes can also be used to weld magnesium on a production basis. Spot and seam welding applications consist of welding sheets and extrusions to thicknesses up to 3/16 in. (approximately 4.8 mm).

REPAIR WELDING

Repair, or maintenance, welding is one of the most difficult types of welding. Some of the major problems include preparing the part for welding, identifying the material, and selecting the best repair method.

The part is often dirty, oily, and painted, and it must be cleaned before welding. There are many hazardous compounds that might be part of the material on the part. These compounds may or may not be hazardous on the part, but when they are heated or burned during welding, they can become life threatening.

▰▰▰ **CAUTION** ◤◤◤

It is never safe to weld on any part that has not been cleaned before welding. All surface contamination must be removed before welding to prevent the possibility of injuring your health from exposure to materials released during welding. Some chemicals can be completely safe until they are exposed to the welding. The smoke or fumes they produce can be an irritant to the skin or eyes; and can be absorbed through the skin or lungs. If you are exposed to an unknown contaminant, get professional help immediately.

Contamination can be removed by sandblasting, grinding, or using solvents. If a solvent is used, be sure it does not leave a dangerous residue. Clean the entire part if possible or a large enough area so that any remaining material is not affected by the welding.

Before the joint can be prepared for welding, you must try to identify the type of metal. There are several ways to determine metal type before welding. One method is to use a metal identification kit. These kits use a series of chemical analyses to identify the metal. Some kits can not only identify a type of metal but also tell the specific alloy.

Another way to identify metal is to look at its color, test for magnetism, and do a spark test. The spark test should be done using a fine grinding stone. With experience, it is often possible to determine specific types of alloys with great accuracy. The sparks given off by each metal and its alloy are so consistent that the U.S. Bureau of Mines uses a camera connected to a computer to identify metals. Microprocessor-controlled testing units are used to identify metals for recycling. For the beginner it is best if you use samples of a known alloy and compare the sparks to your unknown. The test specimen and the unknown should be tested using the same grinding wheel and the same force against the wheel.

EXPERIMENT 26-1

Identifying Metal Using a Spark Test

In this experiment you will be working in a small group to identify various metals using a spark test, **Figure 26-14.** You will need to use proper eye safety equipment, a grinder, several different known and unknown samples of metal, and a pencil and paper to identify the unknown metal samples. Starting with the known samples, make several tests and draw the spark patterns as described in the following paragraphs. Next test the unknown samples and compare the drawings with the drawings from the known samples. See how many of the unknowns you can identify.

There are several areas of the spark test pattern that you must observe carefully, **Figure 26-15(1).** The first area is the grinding stone: Are there sparks that are being carried around the wheel, or are all of the sparks leaving the wheel, **Figure 26-15(2)?**

The next area is immediately adjacent to the wheel where the spark stream leaves the wheel. Note the color of this area; it may vary from white to dull red. Also note the stream to see if the sparks are small, medium, or large and whether the column of sparks is tightly packed or spread out. Draw a sketch of what the spark stream looks like here, **Figure 26-15(3).**

As the spark stream moves away from the wheel a few inches it will begin to change. This change may be in color, speed of the sparks, or size of the sparks; the sparks may divide into smaller, separate streams, explode in a burst of

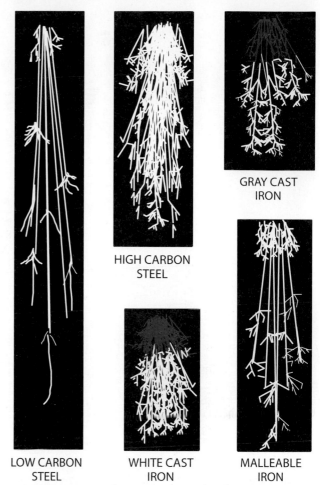

FIGURE 26-14 Spark test patterns for five common metals. © Cengage Learning 2012

GRAY CAST IRON

HIGH CARBON STEEL

LOW CARBON STEEL

WHITE CAST IRON

MALLEABLE IRON

tiny fragments, or just stop glowing, **Figure 26-15(4).** Sketch these changes you see in the sparks.

When the sparks end they may simply stop glowing, change color, change shape, explode, or divide into smaller parts, **Figure 26-15(5).** Sketch these changes you see in the sparks.

Repeat this experiment with other types of metal as they become available.

Once the type of metal to be repaired is determined, to the best of your ability decide on the type of weld groove that is needed. Some breaks may need to be ground into a V- or U-groove and others may not need to be grooved at all, **Table 26-9.**

Often thin sections of most metals can be repaired without the need for the break to be grooved. Thicker sections of most metals will be easier to repair if the crack is ground into a groove. To help in the realignment of the part, it is a good idea to leave a small part of the crack unground so that it can be used to align the parts, **Figure 26-16.**

After the part is ready to be welded, a test tab can be welded onto the part. This test tab should be welded in

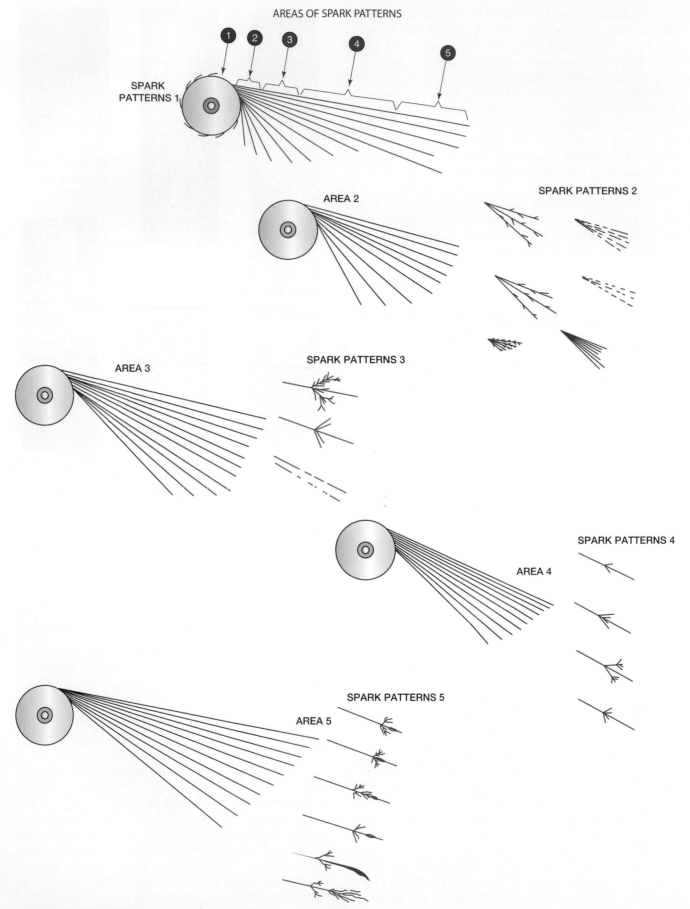

FIGURE 26-15 Spark test. © Cengage Learning 2012

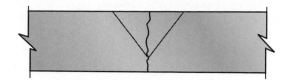

	Thickness in inches			
Metal	**1/8–1/4**	**1/4–1/2**	**1/2–1**	**Over 1**
Mild steel	Square	V	V	V
Aluminum	Square	U	U	U
Magnesium	U	U	U	U
Stainless steel	Square	V	V	V
Cast iron	V	V	U	U

TABLE 26-9 Groove Shapes

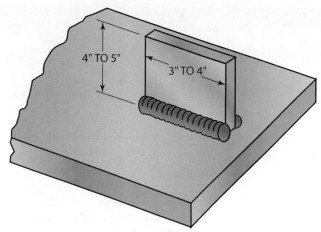

FIGURE 26-17 Tab test. © Cengage Learning 2012

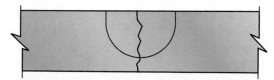

FIGURE 26-16 A small section of the bottom of the break is left so the parts can be realigned. © Cengage Learning 2012

a spot that will not damage the part. Once the tab is welded on, it can be broken off to see if the weld will hold, **Figure 26-17**. If the test tab shows good strength, the repair should continue. If the test tab fails, a new welding procedure should be tried.

Complete a copy of the "Student Welding Report" listed in Appendix I or provided by your instructor. ◆

Summary

All metals are weldable. The only limitation in the fabrication and repair of parts is the cost to our customers. It takes a skilled maintenance and repair welder with an understanding of all of the various characteristics of the metals and types of welding to fix worn or damaged parts. Being able to recognize the differences among the various classifications of metal will allow you to select the most appropriate welding repair procedure. Because of the complexity of this process, you may often be required to research through the original manufacturer of the equipment the types of processes and materials used in the weldment's construction.

Things change, so sometimes once a part is placed in service there is a need to change the welding procedures used in the part's original construction for the repair welding. In addition, the original manufacturer may no longer have the welding procedure. To make a successful weld in such cases, the welder must be able to establish a new welding procedure. As part of your welding procedure you may perform tests to enhance the longevity of your repairs. Your welding experience will help you to be more efficient in producing the new welding procedure.

The diversity and challenges offered by repair and maintenance welding make it one of the most interesting welding jobs available. It can be one of the best paying jobs for a skilled welder.

Welding Offers Answers about New Chrome-Moly Steel

As power plants strive for increased productivity, designers are specifying higher-alloyed steels for tubing networks.

Both welding metallurgy and circumferential welding are being put to the test in hundreds of power plants throughout the world. During shutdowns, pipe and tubing are being replaced with higher-alloyed steels that enable plants to operate at higher temperatures and pressures, therefore increasing operating efficiency. Typically, the old standby steel, P22 (known as 2 1/4 Cr-1Mo steel), is being replaced by a modified (9Cr-1Mo-V) steel, P91.

For new plants the real savings provided by P91 are lighter hangers and support structures, according to David Cable, manager, Welding Operations, Wachs Technical Services Ltd., Charlotte, North Carolina.

Over the years it has been in service (more than 300,000 hours, in some cases), P22 steel has been a very forgiving material. The area to be welded does not have to be purged, and it is not necessary to postweld heat treat on smaller diameters and thicknesses.

There are, however, questions about P91. Unfortunately, in the few years it has been in service in the United States failures have occurred because the steel was not properly postweld heat treated.

"P91 is not just another chrome-moly steel," said R. A. Swain, president of Euroweld, Ltd., Mooresville, North Carolina. He is in constant contact with engineers directly involved in the design and fabrication of P91 steel in numerous North American power plants. "On 2 1/4 chrome-1 moly steel, you might be able to get away without postweld heat treatment, but not with P91. With P91, you do it right or you're in trouble."

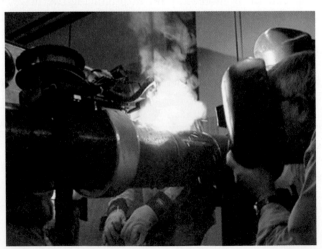

Orbital machine flux cored arc welding of Grade 91 piping spool. Euroweld, Ltd. and Liburdi Dimetrics, Inc.

Induction vs. Resistance Heating

J. F. Henry, director of the Materials Technology Center, Alstom Power, Inc., Chattanooga, Tennessee, discussed some of the issues involving postweld heat treatment (PWHT) of P91 steel. "The power industry has been working with chrome-moly steels like P22 for 35 years or more," Henry said. "It has been found that those steels have high tolerances for variations. If the heat treatment is 150° out of tolerance in the heat treating of P22 steel, it might not affect final properties . . . But P91 is an entirely different story and should be treated as such. No matter which method is used to perform PWHT on this higher-alloyed chrome-moly steel, there should be no deviation whatsoever in the established procedures." At Alstom Power, welded P91 piping is postweld heat treated inside carefully calibrated gas or electric furnaces.

Typically in the field, however, the choice is either resistance or induction heating. Induction heating is the faster of the two processes, but Mannings USA, Dover, New Jersey, a major heat treating subcontractor of P91 welds, uses resistance heating equipment exclusively for this type of work.

It does so because the company uses its own six-way heat-treating console in conjunction with appropriate ceramic pad heaters, thermocouples, plugs, sockets, and other components. Mannings uses its model P256 programmable generator with six channels and either manual or automatic controllers for temperature control when performing PWHT of P91 assemblies.

Miller Electric Mfg. Co., Appleton, Wisconsin, however, uses portable induction heating equipment for this type of application. It has already been used to heat treat orbital welds in cross-country pipelines in the field and is a candidate process for P91 work.

The Development of P91

Today's P91 steel had its beginning in the late 1960s, developed by the Chestnut Run Laboratories of Combustion Engineering in Chattanooga, Tennessee. Metallurgists there modified a version of 9Cr-1Mo steel developed years before by The Timken Co. of Canton, Ohio. The modified grade of steel looked promising, but funding was needed to improve the material. According to C. T. Ward, then a key metallurgist at Combustion Engineering and now a senior consultant at Alstom Power, the search for funding resulted in a proposal submitted to the U.S. Energy Research and Development Administration (later known as the Department of Energy).

The modified grade was a microalloyed steel that contained trace amounts (less than 1%) of carbon, manganese, phosphorus, sulfur, and silicon. The steel also contained

Submerged arc welding of Grade 91 piping using a TipMate® nozzle to increase electrode extension and deposition to improve bead shape and interbead tempering. B. F. Shaw Co.

trace amounts of niobium, vanadium, nickel, nitrogen, and aluminum. The presence of niobium and vanadium improved the strength of the steel at elevated temperatures.

Lightweight and Strong

Weight reduction is key when evaluating P91. Henry of Alstom Power notes that 16-in. (406-mm) diameter steam line piping for headers has a wall thickness in conventional installations of 4 in. (114 mm). By converting to P91, he said, wall thickness can be reduced to 2 in. (63 mm).

The Role of Postweld Heat Treatment (PWHT)

Regardless of thickness or diameter, postweld heat treatment of P91 should be conducted for at least two hours at a temperature to provide appropriate tempering. PWHT temperature ranges between 1375°F and 1400°F (745°C and 760°C). (The better temperature is more likely to be 1400°F.) Postbakes are recommended in cases when the weld cools to room temperature before PWHT is applied.

"With P91," said Euroweld's Swain, "the ASME code will actually allow a minimum postweld heat treat of 1300°F. As you go up in temperature, you go down in yield and the steel becomes softer. The code goes on to state that you can postweld heat treat P91 steel to 1400°F (760°C) or greater, as long as the lower critical temperature is not exceeded."

Welding Processes to Use with P91

Welding processes that can be used to weld P91 steel include SMAW, SAW, FCAW, and GTAW. The SMAW process was the first method used and is still the process of choice when skilled welders are available. However, the shortage of skilled welders and the need to increase weld speed and deposition rates have focused greater attention on other processes, especially FCAW. GTA welding is used to some degree because its weld deposits are high quality, but it is also slow. A faster, hot-wire GTA welding process is being used as well as a cold-wire version of GTAW that produces narrow groove welds.

Flux Cored and Shielded Metal Arc Welding

Automatic flux cored arc welding is being used in the field by J. A. Jones, Inc., Charlotte, North Carolina, according to M. D. Sealey, corporate quality assurance manager. At one of four sites, his company is using FCAW for fill passes. Root passes are made using manual GTA welding, followed by two hot passes, both of which are also made with manual GTAW. Where bends and turns are encountered, manual welding is still employed, although automatic flux cored arc welding is still most cost-effective on long runs of pipe installations. At the other three sites, SMAW is used. Most of the postweld heat treating at installations where J. A. Jones works is subcontracted to companies specializing in electric resistance heating.

Utilities are mostly interested in weld quality and welding speed. Shielded metal arc welding of P91 is not a fast process, but acceptable weld mechanical properties are easy to obtain as long as welders are skilled in working with this process.

Gas tungsten arc welding, another slow process, produces welds of far greater toughness than slag and flux systems. This is due to the comparatively low oxygen content of welds made using GTAW. Gas tungsten arc weld metal using solid wires contains less than 100 ppm (parts per million) of oxygen, whereas the oxygen content in fluxed processes can range from 400 to 800 ppm. GTAW provides weld deposits in P91 of the highest integrity.

There is no doubt that utilities prefer flux cored arc welding, but there is a concern about shielding gas. The high content of CO_2 in shielding gas is beneficial, but if it is too high, the excessive oxygen in the weld metal can reduce impact toughness.

The Future of Alloy Steels

Metallurgists at The National Research Institute for Metals (NRIM), Tsukuba, Japan, predict future ultra-supercritical (USC) boilers will require more highly alloyed steels than even P91. Steels for this next generation of boilers will require heat-resistant alloys with improved strength at elevated temperatures of more than 600°C due to increased operating temperature and steam pressure. One prime candidate is P92, a steel made by Sumitomo. Another is P122, a material also made in Japan. According to NRIM, the USC boilers are more efficient than conventional boilers and release less Co_2 into the atmosphere. P92 may also provide a possible 30% reduction in wall thickness over P91.

Other new steels include E911, made in Europe, which is similar to P91 but has some tungsten, and T23, another steel from Japan and Europe.

"You think P91 is difficult to weld," said Cable, "wait until you see Paralloy, a new casting alloy from Europe. It contains high nickel, high chromium, and resembles Inconel®. It is being used in high-temperature ethylene furnaces and exhibits tremendous shrinkage characteristics that are extremely difficult to weld."

Article courtesy of the American Welding Society.

Review

1. What is meant when a metal is said to have good weldability?

2. What does the term *weldability* involve?

3. What properties of a metal can be affected by the choice of welding process?

4. Referring to Table 26-1, what commercial joining process can be used to weld 1-in. (25-mm) thick cast iron?

5. Referring to Table 26-1, what commercial joining process can be used to braze sheet stainless steel?

6. Referring to Table 26-1, what types of metals can electrogas welding be used to join?

7. What two organizations have developed systems for classifying standard construction grades of steel?

8. Explain the steel classification number 1030.

9. Referring to Table 26-2, what is the composition of the metal identified as 44XX?

10. What is the maximum allowable percentage of manganese in carbon steel as defined by the AISI classification for carbon steel?

11. According to Table 26-3, at what level of carbon content does weldability become poor?

12. What factors other than carbon affect a steel's ability to be welded?

13. Why would some low carbon steels have severe porosity when welded with the GTA welding process?

14. What must be done to steels before welding and after welding if they contain more than 0.40% carbon?

15. Why are high carbon steels preheated before welding?

16. Explain how to weld tool steel with the oxyfuel process.

17. Why must cracks in the held martensite layer of high manganese steels be removed?

18. What properties does chromium and nickel produce in stainless steels?

19. What problems can occur to stainless steel as it is allowed to form carbide precipitation during welding?

20. Why should stainless steel not be held at a temperature between 800°F and 1500°F (425°C and 815°C)?

21. What are the different uses for the three types of stainless steels?

22. Referring to Table 26-5, what diameter SMA welding electrode should be used for 1/4-in. (6-mm) thick stainless steel?

23. Why must cast iron weld metal be ductile?

24. What composition of SMA electrode should be used to make machinable-type weld deposits on cast iron?

25. Why should copper be welded with high currents and preheated?

26. Why can't aluminum oxide be melted off aluminum?

27. Why are aluminum welds likely to cause distortion and cracking?

28. What alloying elements are added to titanium to give it its strength?

29. What do the letters and numbers in a magnesium identification stand for?

30. How can a part be cleaned before it is welded?

31. How can a spark test be used to identify metals?

Chapter 27

Filler Metal Selection

OBJECTIVES

After completing this chapter, the student should be able to

- explain how and when to use each type of filler metal.
- select the best filler metal to fit a specific welding job.
- list the forms filler metals come in.
- explain the significance of the filler metal prefixes.
- explain how to interpret the standard filler metal numbering systems.
- describe the effects alloys have on ferrous metals.

KEY TERMS

alloying elements

arc blow

atomic hydrogen (H)

brazing (B)

carbon equivalent (CE)

Charpy V-notch

core wire

electrode (E)

composite electrode (EC)

elongation

electrode or rod (ER)

tungsten electrode (EW)

flux (F)

fast freezing

filler metals

flux cover

heat-affected zone (HAZ)

hydrogen embrittlement

insert (IN)

minimum tensile strength

molecular hydrogen (H₂)

electrode and/or rod (RB)

oxyfuel (RG)

yield point

INTRODUCTION

Manufacturers of **filler metals** may use any one of a variety of identification systems. There is not a mandatory identification system for filler metals. Manufacturers may use their own numbering systems, trade names, color codes, or a combination of methods to identify filler metal. They may voluntarily choose to use any one of several standardized systems.

The most widely used numbering and lettering system is the one developed by the American Welding Society (AWS). Other numbering and lettering systems have been developed by the American Society for Testing and Materials (ASTM) and the American Iron and Steel Institute (AISI). A system of using colored dots has also been developed by the National

Electrical Manufacturers Association (NEMA). Some manufacturers have produced systems that are similar to the AWS system. Most major manufacturers include both the AWS identification and their own identification on the box, on the package, or directly on the filler metal.

Information that pertains directly to specific filler metals is readily available from most electrode manufacturers. The information given in charts, pamphlets, and pocket electrode guides is specific to their products and may or may not include standard AWS tests, terms, or classifications within their identification systems.

The AWS publishes a variety of books, pamphlets, and charts showing the minimum specifications for filler metal groups. It also publishes comparison charts that include all of the information manufacturers provide the AWS regarding their filler metals. Literature on filler metal specifications and filler metal comparisons may be obtained directly from the AWS or located on the electrode manufacturer's website.

The AWS classification system is for minimum requirements within a grouping. Filler metals manufactured within a grouping may vary but still be classified under that grouping's classification.

A manufacturer may add elements to the metal or flux, such as more arc stabilizers. When one characteristic is improved, another characteristic may also change. The added arc stabilizer may make a smoother weld with less penetration. Other changes may affect the strength and ductility or other welding characteristics.

Because of the variables within a classification, some manufacturers make more than one type of filler metal that is included in a single classification. This and other information may be included in the data supplied by manufacturers.

MANUFACTURERS' ELECTRODE INFORMATION

The type of information given by different manufacturers ranges from general information to technical, chemical, and physical information. A mixture of different types of information may be given.

General information given by manufacturers may include some or all of the following: welding electrode manipulation techniques, joint design, prewelding preparation, postwelding procedures, types of equipment that can be welded, welding currents, and welding positions.

Understanding the Electrode Data

Technical procedures, physical properties, and chemical analysis information given by manufacturers include the following:

- Number of welding electrodes per pound
- Number of inches of weld per welding electrode

- Welding amperage range setting for each size of welding electrode
- Welding codes for which the electrode can be used
- Types of metal that can be welded
- Ability to weld on rust, oil, or paint
- Weld joint penetration characteristics
- Preheating and postheating temperatures
- Weld deposit physical strengths: ultimate tensile strength, yield point, yield strength, elongation, and impact strength
- Percentages of such alloys as carbon, sulfur, phosphorus, manganese, silicon, chromium, nickel, molybdenum, and other alloys

The information supplied by the manufacturer can be used for a variety of purposes, including the following:

- Estimates of the pounds of electrodes needed for a job
- Welding conditions under which the electrode can be used—for example, on clean or dirty metal
- Welding procedure qualification information regarding amperage, joint preparation, penetration, and welding codes
- Physical and chemical characteristics affecting the weld's strengths and metallurgical properties

Data Resulting from Mechanical Tests

Most of the technical information supplied is self-explanatory and easily understood. The mechanical properties of the weld are given as the results of standard tests. The following are some of the standard tests and the meaning of each test:

- **Minimum tensile strength,** psi (N/mm^2)—the load in pounds (megapascal [MPa]) that would be required to break a section of sound weld that has a cross-sectional area of 1 sq in.
- **Yield point,** psi (N/mm^2)—the point in low and medium carbon steels at which the metal begins to stretch when force (stress) is applied after which it will not return to its original length
- **Elongation,** percent in 2 in.(50 mm)—the percentage that a 2-in. (50-mm) piece of weld will stretch before it breaks
- **Charpy V-notch,** ft-lb (m-kgs)—the impact load required to break a test piece of weld metal; this test may be performed on metal below room temperature at which point it may be more brittle

Data Resulting from Chemical Analysis

Chemical analysis of the weld deposit may also be included in the information given by manufacturers. It is not so important to know what the different percentages of the alloys do, but it is important to know how changes in the percentages of the alloys affect the weld. Chemical composition can easily be compared from one electrode to another. The following are the major elements and the effects of their changes on the iron in carbon steel:

- Carbon (C)—As the percentage of carbon increases, the tensile strength increases, the hardness increases, and ductility is reduced. Carbon also causes austenite to form.

- Sulfur (S)—It is usually a contaminant, and the percentage should be kept as low as possible below 0.04%. As the percentage of carbon increases, sulfur can cause hot shortness and porosity.

- Phosphorus (P)—It is usually a contaminant, and the percentage should be kept as low as possible. As the percentage of phosphorus increases, it can cause weld brittleness, reduced shock resistance, and increased cracking.

- Manganese (Mn)—As the percentage of manganese increases, the tensile strength, hardness, resistance to abrasion, and porosity all increase; hot shortness is reduced. It is also a strong austenite former.

- Silicon (Si)—As the percentage of silicon increases, tensile strength increases, and cracking may increase. It is used as a deoxidizer and ferrite former.

- Chromium (Cr)—As the percentage of chromium increases, tensile strength, hardness, and corrosion resistance increase with some decrease in ductility. It is also a good ferrite and carbide former.

- Nickel (Ni)—As the percentage of nickel increases, tensile strength, toughness, and corrosion resistance increase. It is also an austenite former.

- Molybdenum (Mo)—As the percentage of molybdenum increases, tensile strength increases at elevated temperatures; creep resistance and corrosion resistance all increase too. It is also a ferrite and carbide former.

- Copper (Cu)—As the percentage of copper increases, the corrosion resistance and cracking tendency increase.

- Columbium (Cb)—As the percentage of columbium (niobium) increases, the tendency to form chrome-carbides is reduced in stainless steels. It is also a strong ferrite former.

- Aluminum (Al)—As the percentage of aluminum increases, the high temperature scaling resistance improves. It is also a good oxidizer and ferrite former.

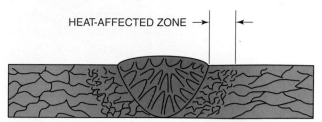

HEAT-AFFECTED ZONE →   ← |

FIGURE 27-1 Heat-affected zone. © Cengage Learning 2012

Carbon Equivalent (CE)

The weldability of an iron alloy is affected by the combination of the alloys used with carbon. To determine the weldability of a specific alloy, formulas have been developed to calculate the carbon equivalence. There are several different formulas, some more complex than others. All of the formulas produce the same product, a **carbon equivalent (CE)** value that can be used to determine the weldability of the alloy. The CE value and the percentage of carbon are used similarly. Carbon equivalence or carbon content should be used to determine whether any special procedures are needed to make an acceptable weld. They are used for selecting such conditions as pre- or postheat temperatures and stress releasing. Carbon and its effects on welding are covered in more depth in Chapter 26.

The CE of an alloy is an indication of how the weld will affect the surrounding metal. This area is called the **heat-affected zone (HAZ)**, **Figure 27-1**. The higher the CE, the more effect the weld can have on the surrounding base metal.

By knowing the CE of the metal, the welder can adjust the welding procedure to control problems with the HAZ. Some of the most common adjustments to the welding procedure are pre- and/or postheating, electrode selection, electrode size, electrode type, and current settings.

Using the following CE formula a welder can make the proper adjustments in the welding procedure for plain carbon steels:

$$CE = \%C + \frac{\%M}{6} + \frac{\%Mo}{4} + \frac{\%C}{5} + \frac{\%Ni}{15} + \frac{\%Cu}{15} + \frac{\%P}{3}$$

- CE = 0.40% or less—no special welding requirements
- CE = 0.40% to 0.60%—low hydrogen welding electrode and the related procedure required
- CE = 0.60% or more—low hydrogen welding electrode, higher welding heat inputs, preheating, and controlled cooling rates

SMAW OPERATING INFORMATION

Shielded metal arc welding electrodes, sometimes referred to as welding rods, stick electrodes, or simply electrodes, have two parts. These two parts are the inner core wire and a flux covering, **Figure 27-2**.

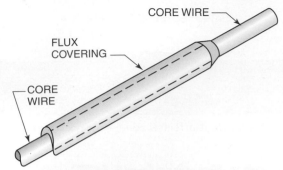

FIGURE 27-2 The two parts of a welding electrode.
© Cengage Learning 2012

The functions of the **core wire** include the following:

- To carry the welding current
- To serve as most of the filler metal in the finished weld

The functions of the **flux covering** include the following:

- To provide some of the alloying elements
- To provide an arc stabilizer (optional)
- To serve as an insulator
- To provide a slag cover to protect the weld bead and slow cooling rate
- To provide a protective gaseous shield during welding

Core Wire

A core wire is the primary metal source for a weld. For fabricating structural and low alloy steels, the core wires of the electrode use inexpensive rimmed or low carbon steel. For more highly alloyed materials, such as stainless steel, high nickel alloys, or nonferrous alloys, the core wires are of the approximate composition of the material to be welded. The core wire also supports the coating that carries the fluxing and alloying materials to the arc and weld pool.

Functions of the Flux Covering

Provides Shielding Gases Heat generated by the arc causes some constituents in the flux covering to decompose and others to vaporize, forming shielding gases. These gases prevent the atmosphere from contaminating the weld metal as it transfers across the arc gap. They also protect the molten weld pool as it cools to form solid metal. In addition, shielding gases and vapors greatly affect both the drops that form at the electrode tip and their transfer across the arc gap, **Figure 27-3**. They also cause the spatter from the arc and greatly determine arc stiffness and penetration. For example, the E6010 electrode contains cellulose. Cellulose decomposes into the hydrogen responsible for the deep electrode penetration so desirable in pipeline welding.

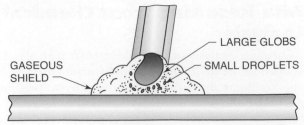

FIGURE 27-3 Methods of metal transfer during an arc.
© Cengage Learning 2012

Alloying Elements Elements in the flux are mixed with the filler metal. Some of these elements stay in the weld metal as alloys. Other elements pick up contaminants in the molten weld pool and float them to the surface. At the surface, these contaminants form part of the slag, **Figure 27-4**.

Effect on Weld Welding fluxes can affect the penetration and contour of the weld bead. Penetration may be pushed deeper if the core wire is made to melt off faster than the flux melts. This forms a small chamber or crucible at the end of the electrode that acts as the combustion chamber of a rocket. As a result, the molten metal and hot gases are forced out very rapidly. The effect of this can be seen on the surface of the molten weld pool as it is blown back away from the end of the electrode. Some electrodes do not use this jetting action, and the resulting molten weld pool is much calmer (less turbulent) and may be rounded in appearance. In addition, the resulting bead may have less penetration, **Figure 27-5**.

Weld bead contour can also be affected by the slag formed by the flux. Some high-temperature slags, called *refractory,* solidify before the weld metal solidifies, forming a mold that holds the molten metal in place. These electrodes are sometimes referred to as **fast freezing** and are excellent for vertical, horizontal, and overhead welding positions.

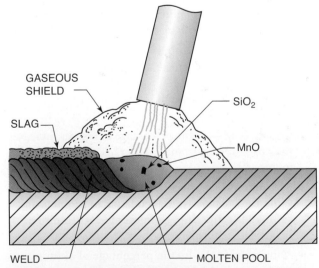

FIGURE 27-4 Silicon and manganese act as scavengers that combine with contaminants and float to the slag on top of the weld. © Cengage Learning 2012

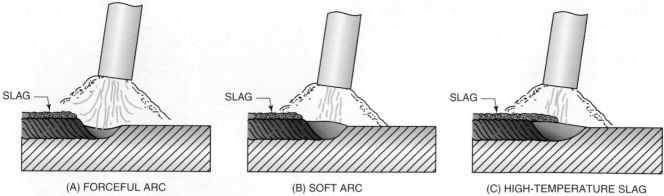

SLAG

SLAG

SLAG

(A) FORCEFUL ARC   (B) SOFT ARC   (C) HIGH-TEMPERATURE SLAG

FIGURE 27-5 Arc force. (A) Forceful arc. (B) Soft arc. (C) High-temperature slag. © Cengage Learning 2012

FILLER METAL SELECTION

Selecting the best filler metal for a job is seldom delegated to the welder in large shops. The selection of the correct process and filler metal is a complex process. If the choice is given to the welder, it is one of the most important decisions the welder will make.

Covering all of the variables for selecting a filler metal would be well beyond the scope of this text. A sample of the types of factors that must be considered for the selection of an SAW electrode follow. To further complicate things, welding electrodes have more than one application, and many welding electrodes may be used for the same type of work.

The following conditions that the welder should consider when choosing a welding electrode are not in order of importance. They are also not all of the factors that must be considered.

Shielded Metal Arc Welding Electrode Selection

- Type of electrode—What electrode has been specified in the blueprints or in the contract for this job?

- Type of current—Can the welding power supply provide AC only, DC only, or both AC and DC?

- Power range—What is the amperage range on the welder and its duty cycle? Different types of electrodes require different amperage settings even for the same size welding electrode. For example, the amperage range for a 1/8-in. (3-mm) diameter E6010 electrode is 75 A to 125 A, and the amperage range for a 1/8-in. (3-mm) diameter E7018 electrode is 90 A to 165 A.

- Type of base metal—Some welding electrodes may be used to join more than one similar type of metal. Other electrodes may be used to join together two different types of metal. For example, an E309-15 electrode can be used to join 305 stainless steel to 1020 mild steel.

- Thickness of metal—The penetration characteristics of each welding electrode may differ. Selecting one electrode that will weld on a specific thickness of material is important. For example, E6013 has very little penetration and is therefore good for welding on sheet metal.

- Weld position—Some welding electrodes can be used to make welds in all positions. Other electrodes may be restricted to making flat, horizontal, and/or vertical position welds, a few electrodes may be used to make flat position welds only.

- Joint design—The type of joint and whether it is grooved or not may affect the performance of the welding electrode. For example, the E7018 electrode does not produce a large, gaseous cloud to protect the molten metal. For this reason, the electrode movement is restricted so that the molten weld pool is not left unprotected by the gaseous cloud.

- Surface condition—It is not always possible to work on new, clean metal. Some welding electrodes will weld on metal that is rusty, oily, painted, dirty, or galvanized.

- Number of passes—The amount of reinforcement needed may require more than one welding pass. Some welding electrodes will build up faster, and others will penetrate deeper. The slag may be removed more easily from some welds than from others. For example, E6013 will build up a weld faster than E6010, and the slag is also more easily removed between weld passes.

- Distortion—Welding electrodes that will operate on low-amperage settings will have less heat input and cause less distortion. Welding electrodes that have a high rate of deposition (fill the joint rapidly) and can travel faster will also cause less distortion. For example, the flux on an E7024 has 50% iron powder, which gives it a faster fill rate and allows it to travel faster, resulting in less distortion of the metal being welded.

■ Preheat or postheat—On low carbon steel plate 1 in. (25 mm) thick or more, preheating is required with most welding electrodes. Postheating may be required to keep a weld zone from hardening or cracking when using some welding electrodes. However, no postheating may be required when welding low alloy steel using E310-15.

■ Temperature service—Weld metals react differently to temperature extremes. Some welds become very brittle and crack easily in cryogenic (low-temperature) service. A few weld metals resist creep and oxidation at high temperatures. For example, E310Mo-15 can weld on most stainless and mild steels without any high-temperature problems.

■ Mechanical properties—Mechanical properties such as tensile strength, yield strength, hardness, toughness, ductility, and impact strength can be modified by the selection of specific welding electrodes.

■ Postwelding cleanup—The hardness or softness of the weld greatly affects any grinding, drilling, or machining. The ease with which the slag can be removed and the quantity of spatter will affect the time and amount of cleanup required.

■ Shop or field weld—The general working conditions such as wind, dirt, cleanliness, dryness, and accessibility of the weld will affect the choice of welding electrode. For example, the E7018 electrode must be kept very dry, but the E6010 electrode is not greatly affected by moisture, **Figure 27-6**.

■ Quantity of welds—If a few welds are needed, a more expensive welding electrode requiring less skill may be selected. For a large production job requiring a higher skill level, a less expensive welding electrode may be best.

After deciding the specific conditions that may affect the welding, the welder has most likely identified more than one condition that needs to be satisfied. Some of the conditions will not interfere with others. For example, the type of current and whether a welder makes one or more weld passes have little or no effect on each other. However, if a welder needs to machine the finished weld, hardness is a consideration. When two or more conditions conflict, the welder is seldom the person who will make the decision. It may be necessary to choose more than one welding electrode. When welding pipe, E6010 and E6011 are often used for the root pass because of their penetration characteristics, and E7018 is used for the cover pass because of its greater strength and resistance to cracking.

Each AWS electrode classification has its own welding characteristics. Some manufacturers have more than one welding electrode in some classifications. In these cases, the minimum specifications for the classification have

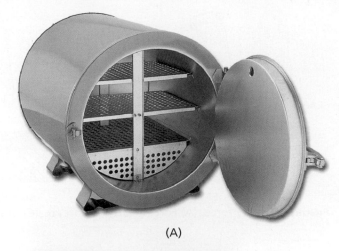

(A)

(B)

FIGURE 27-6 Most codes and standards require all electrodes to be stored in a heated electrode storage box to keep them dry (A). Portable thermos-type containers can be carried to the work area by the welder to keep the electrodes dry and ready to use (B).
Phoenix International

been exceeded. An example of more than one welding electrode in a single classification is Lincoln's Fleetweld 35, Fleetweld 35LS, and Fleetweld 180R. These electrodes are all in AWS classification E6011. For the manufacturer's complete description of these electrodes, consult **Table 27-1**.

The characteristics of each manufacturer's filler metals can be compared to one another by using data sheets supplied by the manufacturer. General comparisons can be made easily using an electrode comparison chart.

When making an electrode selection, many variables must be kept in mind, and the performance characteristics must be compared before making a final choice.

Electrode Identification and Operating Data

Coating Color	AWS Number on Coating	(L)Lincoln	Electrode	Electrode Polarity	Sizes and Current Ranges (Amps) (Electrodes Are Manufactured in These Sizes for Which Current Ranges Are Given)					
					3/32" Size	1/8" Size	5/32" Size	3/16" Size	7/32" Size	1/4" Size
Brick red	6010		Fleetweld 5P	DC+	40–75	75–130	90–175	140–225	200–275	220–325[1]
Gray	6011		Fleetweld 35	AC	50–85	75–120	90–160	120–200	150–260	190–300
				DC+	40–75	70–110	80–145	110–180	135–235	170–270
Red brown	6011	Green	Fleetweld 35LS	AC		80–130	120–160			
				DC±		70–120	110–150			
Brown	6011		Fleetweld 180	AC	40–90	60–120	115–150			
				DC±	40–80	56–110	105–135			
Pink	7010-A1		Shield-arc 85	DC+	50–90	75–130	90–175	140–225		
Pink	7010-A1	Green	Shield-arc 85P	DC+				140–225		
Tan	7010-G		Shield-arc Hyp	DC+		75–130	90–185	140–225	160–250	
Gray	8010-G		Shield-arc 70+	DC+		75–130	90–185	140–225		

[1] Range for 5/16" size is 240–400 amps.   DC+ is Electrode Positive. DC– is Electrode Negative.

All tests were performed in conformance with specifications AWS A5.5 and ASME SFA.5.5 in the aged condition for the E7010-G and E8010-G electrodes, and in the stress-relieved condition for Shield-Arc 85 & 85P. Tests for the other products were performed in conformance with specifications AWS A5.1 and ASME SF A.5.1 for the as-welded condition

Typical Mechanical Properties

Low figures in the stress-relieved tensile and yield strength ranges below for Shield-Arc 85 and 85P are AWS minimum requirements.

Low figures in the as-welded tensile and yield strength ranges below for the other products are AWS minimum requirements.

	Fleetweld 5P	Fleetweld 35	Fleetweld 35LS	Fleetweld 180	Shield-Arc 85	Shield-Arc 85P	Shield-Arc Hyp	Shield-Arc 70+
As welded tensile strength psi	62–69,000	62–68,000	62–67,000	62–71,000	70–78,000	70–78,000	70–84,000	80–92,000
Yield point—psi	52–62,000	50–62,000	50–60,000	50–64,000	60–71,000	57–63,000	60–77,000	67,000–83,000
ductility—% elong. in 2"	22–32	22–30	22–31	22–31	22–26	22–27	22–23	19–24
Charpy V-notch toughness —ft.-lb	20–60 @–20°F	20–90 @–20°F	20–57 @–20°F	20–54 @–20°F	68 @ 70°F	68 @ 70°F	30 @ 20°F	40 @ 50°F
Hardness, Rockwell B (avg)[5]	76–82	76–85	73–88	75–85			83–92	88–93
Stress-relieved @ 1150°F tensile strength—psi	60–69,000	60–66,000	60–65,000		70–83,000	70–74,000	80–82,000	80–84,000
Yield point—psi	46–56,000	46–56,000	46–51,000		57–69,000	57–65,000	72–76,000	71–76,000
ductility—% elong. in 2"	28–36	28–36	28–33		22–28	22–27	24–27	22–26
Charpy V-notch toughness —ft.-lb	71 @ 70°F		120 @ 70°F		64 @ 70°F	68 @ 70°F	30 @ –20°F	30 @ –50°F
Hardness, Rockwell B (avg)[5]					80–89	80–87		

Conformances and Approvals

See *Lincoln Price Book* for certificate numbers, size, and position limitations, and other data.

Conforms to test requirements of AWS—A5.1 and ASME—SFA5.1 AWS—A5.5 and ASME—SFA5.5	FW-5P E6010	FW-35 E6011	FW-35LS E6011	FW-180 E6011	SA-85 E7010-A1[4]	SA-85P E7010-A1	SA-HYP E7010-G	SA-70+ E8010-G
ASME boiler code Group	F3	F3	F3	F3	F3	F3	F3	F3
Analysis	A1		A1	A1	A2	A2	A2	
American Bureau of Shipping and U.S. Coast Guard	Approved	Approved	Approved		Approved			
Conformance certificate available[4]	Yes	Yes	Yes	Yes	Yes	Yes	Yes	Yes
Lloyds	Approved	Approved						
Military specifications	MIL-QQE-450	MIL-QQE-450			MIL-E-22200/7			

[3] Also meets the requirements for E7010-G and E6010 in 3/32" size.

[4] "Certificate of Conformance" to AWS classification test requirements is available. These are needed for Federal Highway Administration projects.

[5] Hardness values obtained from welds made in accordance with AWS A5.1.

TABLE 27-1 Fleetweld 35®, Fleetweld 35LS®, and Fleetweld 180® Lincoln Electrodes Lincoln Electric Company

AWS FILLER METAL CLASSIFICATIONS

The AWS classification system uses a series of letters and numbers in a code that gives the important information about the filler metal. The prefix letter is used to indicate the filler's form, a type of process the filler is to be used with, or both. The prefix letters and their meanings are as follows:

- **E**—Indicates an arc welding **electrode (E).** The filler carries the welding current in the process. We most often think of the *E* standing for an SMA "stick" welding electrode. It also is used to indicate wire electrodes used in GMAW, FCAW, SAW, ESW, EGW, etc.

- **R**—Indicates a **rod (R)** that is heated by some other source other than electric current flowing directly through it. Welding rods are sometimes referred to as being "cut length" or "welding wire." It is often used with OFW and GTAW.

- **ER**—Indicates a filler metal that is supplied for use in either an **electrode or rod (ER)** form. The same alloys are used to produce the electrodes and the rods. This filler metal may be supplied as a wire on a spool for GMAW or as a rod for OFW or GTAW.

- **EC**—Indicates a **composite electrode (EC).** These electrodes are used for SAW. Do not confuse an ECu, copper arc welding electrode, for an ECNi2, which is a composite nickel submerged arc welding wire.

- **B**—Indicates a **brazing (B)** filler metal. This filler metal is usually supplied as a rod, but it can come in a number of other forms. Some of the forms it comes in are powder, sheets, washers, and rings.

- **RB**—Indicates a filler metal that is used as a current carrying **electrode,** as a **rod (RB),** or both. The form the filler is supplied in for each of the applications may be different. The composition of the alloy in the filler metal will be the same for all of the forms supplied. This filler can be used for processes like arc braze welding or oxyfuel brazing.

- **RG**—Indicates a welding rod used primarily with **oxygen (RG)** welding. This filler can be used with all of the oxyfuels, and some of the fillers are used with the GTAW process.

- **IN**—Indicates a consumable **insert (IN).** These are most often used for welding on pipe. They are preplaced in the root of the groove to supply both filler metal and support for the root pass. The inserts may provide for some joint alignment and spacing.

The next two classifications are not filler metal. They are classified under the same system because they are welding consumables. The GTA welding tungsten is not a filler, but it is consumed, very slowly, during the welding process.

- **EW**—Indicates a nonconsumable **tungsten electrode (EW).** The GTAW electrode is obviously not a filler metal, but it falls under the same classification system.

- **F**—Indicates a **flux (F)** used for SAW. The composition of the weld metal is influenced by the flux. There are alloys and agents in the flux used for SAW that are dissolved into the weld metal. For this reason, the filler metal and flux are specified together, with the filler metal identification first and the flux second.

In addition to the prefix, there are some suffix identifiers. The suffix may be used to indicate a change in the alloy in a covered electrode or the type of welding current to be used with stainless steel–covered electrodes.

CARBON STEEL

Carbon and Low Alloy Steel–Covered Electrodes

The AWS specification for carbon steel–covered arc electrodes is A5.1, and for low alloy steel–covered arc electrodes it is A5.5. Filler metals classified within these specifications are identified by a system that uses the letter *E* followed by a series of numbers to indicate the minimum tensile strength of a good weld, the position(s) in which the electrode can be used, the type of flux coating, and the type(s) of welding current, **Figure 27-7.**

The tensile strength is given in pounds per square inch (psi). The actual strength is obtained by adding three zeroes to the right of the number given. For example, E60XX is 60,000 psi, E110XX is 110,000 psi, and so on.

The next number located to the right of the tensile strength—1, 2, or 4—designates the welding position capable—for example,

- 1—in an E601X means all positions flat, horizontal, vertical, and overhead.

- 2—in an E602X means horizontal fillets and flat.

- 3—is an old term no longer used; it meant flat only.

- 4—in an E704X means flat, horizontal, overhead, and vertical down.

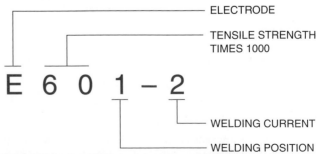

FIGURE 27-7 AWS numbering system for A5.1 and A5.5 carbon and low alloy steel–covered electrodes.
© Cengage Learning 2012

AWS Classification	Type of Covering	Capable of Producing Satisfactory Welds in Positions Shown[a]	Type of Current[b]
E60 Series Electrodes			
E6010	High cellulose sodium	F, V, OH, H	DC, reverse polarity
E6011	High cellulose potassium	F, V, OH, H	AC or DC, reverse polarity
E6012	High titania sodium	F, V, OH, H	AC or DC, straight polarity
E6013	High titania potassium	F, V, OH, H	AC or DC, either polarity
E6020		H-fillets	AC or DC, straight polarity
E6022[c]	High iron oxide	F	AC or DC, either polarity
E6027	High iron oxide, iron powder	H-fillets, F	AC or DC, straight polarity
E70 Series Electrodes			
E7014	Iron powder, titania	F, V, OH, H	AC or DC, either polarity
E7015	Low hydrogen sodium	F, V, OH, H	DC, reverse polarity
E7016	Low hydrogen potassium	F, V, OH, H	AC or DC, reverse polarity
E7018	Low hydrogen potassium, iron powder	F, V, OH, H	AC or DC, reverse polarity
E7024	Iron powder, titania	H-fillets, F	AC or DC, either polarity
E7027	High iron oxide, iron powder	H-fillets, F	AC or DC, straight polarity
E7028	Low hydrogen potassium, iron powder	H-fillets, F	AC or DC, reverse polarity
E7048	Low hydrogen potassium, iron powder	F, OH, H, V-down	AC or DC, reverse polarity

[a]The abbreviations, F, V, V-down, OH, H, and H-fillets indicate the welding positions as follows:

F = Flat
H = Horizontal
H-fillets = Horizontal fillets
V-down = Vertical down
V = Vertical
OH = Overhead

[b]Reverse polarity means the electrode is positive; straight polarity means the electrode is negative.

[c]Electrodes of the E6022 classification are for single-pass welds.

For electrodes 3/16 in. (4.8 mm) and under, except 5/32 in. (4.0 mm) and under for classifications E7014, E7015, E7016, and E7018.

TABLE 27-2 Electrode Classification American Welding Society

The last two numbers together indicate the major type of covering and the type of welding current. For example, EXX10 has an organic covering and uses DCEP polarity. The AWS classification system for A5.1 and A5.5 covered arc welding electrodes is shown in **Table 27-2.** The type of welding current for any electrode may be expanded to include currents not listed if a manufacturer adds additional arc stabilizers to the electrode covering, **Table 27-3.**

On some covered arc electrodes, a suffix may be added to indicate the approximate alloy in the deposit as welded. For example, the letter *A* indicates a 1/2% molybdenum addition to the weld metal deposited. **Table 27-4** is a complete list of the major **alloying elements** in electrodes.

Some of the more popular arc welding electrodes and their uses in these specifications are as follows:

EXXX0—DCRP only
EXXX1—AC and DCRP
EXXX2—AC and DCSP
EXXX3—AC and DC
EXXX4—AC and DC
EXXX5—DCRP only
EXXX6—AC and DCRP
EXXX8—AC and DCRP

TABLE 27-3 Welding Currents

E6010 The E6010 electrodes are designed to be used with DCEP polarity and have an organic-based flux (cellulose, $C_6H_{10}O_5$). They have a forceful arc that results in deep penetration and good metal transfer in the vertical and overhead positions, **Figure 27-8.** The electrode is usually used with a whipping or stepping motion. This motion helps remove unwanted surface materials such as paint, oil, dirt, and galvanizing. Both the burning of the organic compound in the flux to form CO_2, which protects the molten metal, and the rapid expansion of the hot gases force the atmosphere away from the weld. A small amount of slag remains on the finished weld, but it is difficult to remove, especially along the weld edges. E6010 electrodes are commonly used for welding on fired and unfired pressure vessels, on pipe, and in construction jobs, shipyards, and repair work.

E6011 The E6011 electrodes are designed to be used with AC or DCEP reverse polarity and have an organic-based flux. These electrodes have many of the welding characteristics of E6010 electrodes, **Figure 27-9.** In most applications, the E6011 is preferred. The E6011 has added arc stabilizers, which allow it to be used with AC. Using this welding electrode on AC only slightly reduces its penetration but will help control any arc blow problem. **Arc blow,** or arc wander, occurs when using DC current and is the

Suffix Symbol	Molybdenum (Mo) %	Nickel (Ni) %	Chromium (Cr) %	Manganese (Mn) %	Vanadium (Va) %
A 1	0.5				
B 1	0.5		0.50		
B 2	0.5		1.25		
B 3	1.0		2.25		
B 4	0.5		2.00		
C 1		2.5			
C 2		3.5			
C 3		1.0			
D 1	0.3			1.5	
D 2	0.3			1.75	
G	0.2	0.5	0.30	1.00	0.1*
M					

*Only one of these alloys may be used.

TABLE 27-4 Major Alloying Elements in Electrodes

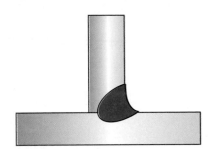

FIGURE 27-8 E6010. © Cengage Learning 2012

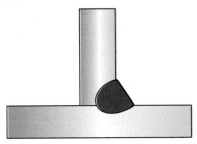

FIGURE 27-9 E6011. © Cengage Learning 2012

magnetic deflection of the arc from its normal path. Arc blow does not occur when alternating current is used. When welding with either E6010 or E6011, the weld pool may be slightly concave from the forceful action of the rapidly expanding gas. This forceful action also results in more spatter and sparks during welding.

E6012 The E6012 electrodes are designed to be used with AC or DCEN polarity and have rutile-based flux (titanium dioxide TiO_2). This electrode has a very stable arc that is not very forceful, resulting in a shallow penetration characteristic, **Figure 27-10.** This limited penetration characteristic helps with poor-fitting joints or thin materials. Thick sections can be welded, but the joint must be grooved. Less smoke is generated with this welding electrode than with E6010 or E6011, but a thicker slag layer is deposited on the weld. If the weld is properly made, the slag can be removed easily and may even free itself after cooling. Spatter can be held to a minimum when using both AC and DC. E6012 electrodes are

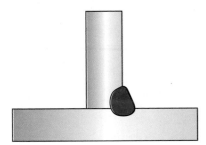

FIGURE 27-10 E6012. © Cengage Learning 2012

commonly used for all new work, including storage tanks, machinery fabrication, ornamental iron, and general repair work.

E6013 The E6013 electrodes are designed to be used with AC or DC, either polarity. They have a rutile-based flux. The E6013 electrode has many of the same characteristics of the E6012 electrode, **Figure 27-11.** The slag layer is usually thicker on the E6013 and is easily removed. The arc of the E6013 is as stable, but there is less penetration, which makes it easier to weld very thin sections. The weld bead will also be built up slightly higher than the E6012. E6013 electrodes are commonly used for sheet metal fabrication, metal buildings, surface buildup, truck and automotive bodywork, and farm equipment.

E7014 The E7014 electrodes are designed to be used with AC or DC, either polarity. They have a rutile-based flux with iron powder added. The E7014 electrode has many arc and weld characteristics that are similar to those

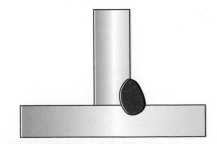

FIGURE 27-11 E6013. © Cengage Learning 2012

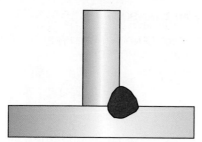

FIGURE 27-12 E7014. © Cengage Learning 2012

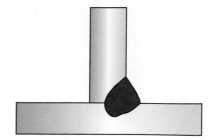

FIGURE 27-14 E7016. © Cengage Learning 2012

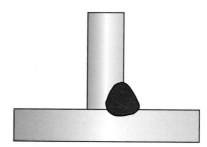

FIGURE 27-15 E7018. © Cengage Learning 2012

of the E6013 electrode, **Figure 27-12.** Approximately 30% iron powder is added to the flux to allow it to build up a weld faster or have a higher travel speed. The penetration characteristic is light. This welding electrode can be used on metal with a light coating of rust, dirt, oil, or paint. The slag layer is thick and hard but can be completely removed with chipping. E7014 electrodes are commonly used for welding on heavy sheet metal, ornamental iron, machinery, frames, and general repair work.

E7024 The E7024 electrodes are designed to be used with AC or DC, either polarity. They have a rutile-based flux with iron powder added. This welding electrode has a light penetration and fast fill characteristic, **Figure 27-13.** The flux contains about 50% iron powder, which gives the flux its high rate of deposition. The heavy flux coating helps control the arc and can support the electrode so that a drag technique can be used. The drag technique allows this electrode to be used by welders with less skill. The slag layer is heavy and hard but can easily be removed. If the weld is performed correctly, the slag may remove itself. Because of the large, fluid molten weld pool, this electrode is equally used in the flat and horizontal position only, although it can be used on work that is slightly vertical. E7024 electrodes are commonly used for welding earth-moving, mining, and railroad equipment.

E7016 The E7016 electrodes are designed to be used with AC or DCEP polarity. They have a low-hydrogen–based (mineral) flux. This electrode has moderate penetration and little buildup, **Figure 27-14.** There is no iron powder in the flux, which helps when welding in the vertical or overhead positions. Welds on high sulfur and cold-rolled metals can be made with little porosity. Low alloy and mild steel heavy plates can be welded with minimum preheating. E7016 electrodes are commonly used for construction, earth-moving and mining equipment, and shipbuilding.

E7018 The E7018 electrodes are designed to be used with AC or DCEP polarity. They have a low-hydrogen–based flux with iron powder added. The E7018 electrodes have moderate penetration and buildup, **Figure 27-15.** The slag layer is heavy and hard but can be removed easily by chipping. The weld metal is protected from the atmosphere primarily by the molten slag layer and not by rapidly expanding gases. For this reason, these electrodes should not be used for open root welds. The atmosphere may attack the root, causing a porosity problem. The E7018 welding electrodes are very susceptible to moisture, which may lead to weld porosity. These electrodes are commonly used for pipe, heavy sections of plate, shipbuilding, boiler work, and low-temperature equipment. E7018 electrodes are sometimes referred to as Lo-Hi rods because they allow very little hydrogen into the weld pool.

WIRE-TYPE STEEL FILLER METALS

Solid Wire

The AWS specification for carbon steel filler metals for gas shielded welding wire is A5.18. Filler metal classified within these specifications can be used for GMAW, GTAW, and PAW processes. Because in GTAW and PAW the wire does not carry the welding current, the letters *ER* are used as a prefix. The *ER* is followed by two numbers to indicate the minimum tensile strength of a good weld. The actual strength is obtained by adding three zeroes to the right of the number given. For example, ER70S-x is 70,000 psi.

The *S* located to the right of the tensile strength indicates that this is a solid wire. The last number—2, 3, 4, 5, 6, or 7—or the letter *G* is used to indicate the filler metal composition and the weld's mechanical properties, **Figure 27-16.**

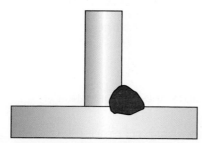

FIGURE 27-13 E7024. © Cengage Learning 2012

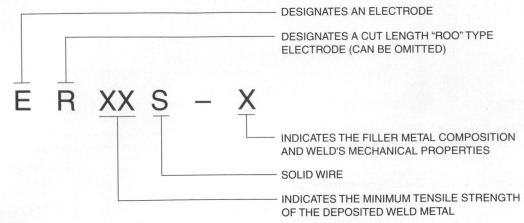

FIGURE 27-16 AWS numbering system for carbon steel filler metal for GMAW. American Welding Society

ER70S-2 This is a deoxidized mild steel filler wire. The deoxidizers allow this wire to be used on metal that has light coverings of rust or oxides. There may be a slight reduction in the weld's physical properties if the weld is made on rust or oxides, but this reduction is only slight, and the weld will usually still pass the classification test standards. This is a general-purpose filler that can be used on killed, semikilled, and rimmed steels. Argon-oxygen, argon-CO_2, and CO_2 can be used as shielding gases. Welds can be made in all positions.

ER70S-3 This is a popular filler wire. It can be used in single or multiple pass welds in all positions. ER70S-3 does not have the deoxidizers required to weld over rust, over oxides, or on rimmed steels. It produces high-quality welds on killed and semikilled steels. Argon-oxygen, argon-CO_2, and CO_2 can be used as shielding gases.

ER70S-6 This is a good general-purpose filler wire. It has the highest levels of manganese and silicon. The wire can be used to make smooth welds on sheet metal or thicker sections. Welds over rust, oxides, and other surface impurities will lower the mechanical properties, but not normally below the specifications of this classification.

Argon-oxygen, argon-CO_2, and CO_2 can be used as shielding gases. Welds can be made in all positions.

Tubular Wire

The AWS specification for carbon steel filler metals for flux cored arc welding wire is A5.20. Filler metal classified within this specification can be used for the FCAW process. The letter *E*, for *electrode*, is followed by a single number to indicate the minimum tensile strength of a good weld. The actual strength is obtained by adding four zeroes to the right of the number given. For example, E6xT-x is 60,000 psi, and E7xT-x is 70,000 psi.

The next number, 0 or 1, indicates the welding position. Ex0T is to be used in a horizontal or flat position only. Ex1T is an all-position filler metal.

The *T* located to the right of the tensile strength and weld position numbers indicates that this is a tubular, flux cored wire. The last number—2, 3, 4, 5, 6, 7, 8, 10, or 11—or the letter *G* or *GS* is used to indicate if the filler metal can be used for single or multiple pass welds. The electrodes with the numbers ExxT-2, ExxT-3, ExxT-10, and ExxT-GS are intended for single pass welds only, **Figure 27-17.**

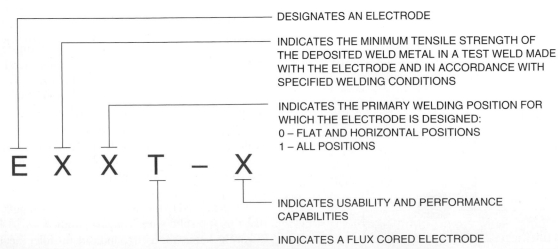

FIGURE 27-17 Identification system for mild steel FCAW electrodes. American Welding Society

large weld pool permits high deposition rates. Weld deposits are ductile and have a high resistance to cracking. E70T-4 can be used to weld joints that have larger-than-usual root openings. Applications include large weldments and earth-moving equipment.

E71T-7 E71T-7 is a self-shielding, all-position, flux cored filler metal. The fluxing system allows the control of the molten weld pool required for out-of-position welds. The high level of deoxidizers reduces the tendency for cracking in the weld. It can be used for single or multiple pass welds.

Stainless Steel Electrodes

The AWS specification for stainless steel–covered arc electrodes is A5.4 and for stainless steel bare, cored, and stranded electrodes and welding rods is A5.9. Filler metal classified within the A5.4 uses the letter *E* as its prefix, and the filler metal within the A5.9 uses the letters *ER* as its prefix, **Table 27-5**.

Following the prefix, the American Iron and Steel Institute's (AISI) three-digit stainless steel number is used. This number indicates the type of stainless steel in the filler metal.

To the right of the AISI number, the AWS adds a dash followed by a suffix number. The number 15 is used to indicate that there is a lime base coating, and the DCEP polarity welding current should be used. The number 16 is used to indicate there is a titania-type coating, and AC or DCEP polarity welding currents can be used. Examples of this classification system are E308-15 and E308-16 electrodes.

The letter *L* may be added to the right of the AISI number before the dash and suffix number to indicate a low carbon stainless welding electrode. E308L-15 and E308L-16 arc welding electrodes and ER308L and ER316L are examples of the use of the letter *L*, **Table 27-6**.

Stainless steel may be stabilized by adding columbium (Cb) as a carbide former. The designation Cb is added after the AISI number for these electrodes, such as E309Cb-16. Stainless steel filler metals are stabilized to prevent chromium-carbide precipitation.

E308-15, E308-16, E308L-15, ER308, and ER308L All are filler metals for 308 stainless steels. 308 stainless steels are used for food or chemical equipment, tanks, pumps, hoods, and evaporators. All E308 and ER308 filler metals can be used to weld on all 18-8-type stainless steels such as 301, 302, 302B, 303, 304, 305, 308, 201, and 202.

E309-15, E309-16, E309Cb-15, E309Cb-16, ER309, and ER309L All are filler metals for 309 stainless steels. 309 stainless steels are used for high-temperatures service, such as furnace parts and mufflers. All E309 filler metals can be used to weld on 309 stainless or to join mild steel to any 18-8-type stainless steel.

E310-15, E310-16, E309Cb-15, E309Cb-16, E310Mo-15, E310Mo-16, and ER310 All are filler metals for 310 stainless steels. 310 stainless steels are used

FIGURE 27-18 Heated storage lockers are available for all types of wire electrodes. *Phoenix International*

Heated storage lockers are available to keep all wire electrodes dry to prevent surface oxidation and keep moisture out of the flux core of FCAW electrodes, **Figure 27-18**.

E70T-1, E71T-1 E70T-1 and E71T-1 have a high-level deoxidizer in the flux core. It has high levels of silicon and manganese, which allow it to weld over some surface contaminations such as oxides or rust. This filler metal can be used for single or multiple pass welds. Argon 75% with 25% CO_2 or 100% CO_2 can be used as the shielding gas. It can be used on ASME A36, A106, A242, A252, A285, A441, and A572 or similar metals. Applications include railcars, heavy equipment, earth-moving equipment, shipbuilding, and general fabrication. The weld metal deposited has a chemical and physical composition similar to that of E7018 low hydrogen electrodes.

E70T-2, E71T-2 E70T-2 and E71T-2 are highly deoxidized flux cored filler metal that can be used for single pass welds only. The high levels of deoxidizers allow this electrode to be used over mill scale and light layers of rust and still produce sound welds. Because of the high level of manganese, if the filler is used for multiple pass welds, there might be manganese-caused centerline cracking of the weld; 100% CO_2 can be used as the shielding gas. E70T-2 can be used on ASME A36, A106, A242, A252, A285, A441, and A572 or similar metals. Applications include repair and maintenance work and general fabrication.

E70T-4, E71T-4 E70T-4 and E71T-4 are self-shielding, flux cored filler metal. The fluxing agents produce a slag, which allows a larger-than-usual molten weld pool. The

AISI TYPE NUMBER	442 446	430F 430FSE	430 431	501 502	416 416SE	403 405 410 420 414	321 348 347	317	316L	316	314	310 310S	309 309S	304 L	303 303SE	201 202 301 302 302B 304 305 308	Mild Steel
201–202–301 302–3028–304 305–308	310 312 309	310 312 309	310 312 309	310 312 309	309 310 312	309 310 312	308	308	308	308	308	308	308	308	308	308	312 310 309
303 **303SE**	310 309 312	310 309 312	310 309 312	310 309 312	309 310 312	309 310 312	308	308	308	308	308	308	308	308	308-15	308	312 310 309
304L	310 309 312	310 309 312	310 309 312	310 309 312	309 310 312	309 310 312	308	308	308-L	308	308	308	308	308 -L	308	308	312 310 309
309 **309S**	310 309 312	310 309 312	310 309 312	310 309 312	309 310 312	309 310 312	308	317 316 309	316	316	309	309	309	308	308	308	309 310 312
310 **310S**	310 309 312	310 309 312	310 309 312	310 309 312	310 309 312	310 309 312	308	317 316 309	316	316	310	310	309 310	308	308	308	310 309 312
314	310 312 309	310 312 309	310 312 309	310 312 309	310 312 309	310 309 312	309 310 308	309 310	309 310	309 310	310-15	310	309 310	309 310	309 310	309 310	310 309 312
316	310 309 312	310 309 312	310 309 312	310 309 312	309 310 312	309 310 312	308	316	316	316	309 310 316	310 309 316	309 310 316	308 316	308 316	308 316	309 310 312
316L	310 309 312	310 309 312	310 309 312	310 309 312	309 310 312	309 310 312	308	316 317 308	316-L	316	309 310 316	310 309 316	316 309	308 316	308 316	308 316	309 310 312
317	310 309 312	310 309 312	310 309 312	310 309 312	309 310 312	309 310 312	308	317	316	316	309 310 317	317 316 309	317 316 309	308 316 317	308 316 317	308 316 317	309 310 312
321 **348** **347**	310 309 312	310 309 312	310 309 312	310 309 312	309 310 312	309 310 312	347	308 347	347 308	347 308	309 310 347	347 308	347 308	347 308 -L	347 308	347 308	309 310 312
403–405 **410–420** **414**	310 309 312	310 309 312	310 309 312	310 309 312	309 310	410[†] 309[††]	309 310	309 310	309 310	309 310	310 309	310 309	309 310	309 310	309 310	309 310	309 310 312
416 **416SE**	310 309	310 309	310 309	310	410-15[†]	410-15[†] 309[††] 310[††]	309 310	309 310 312	309 310 312	309 310 312	309 310 312	310 309 312	309 310 312	309 310 312	309 310 312	309 310 312	309 310 312
501 **502**	310	310	310	502[†] 310[††]	310	310	310 309	310 309	310 309	310 309	310 309	310 309	310 309	310 309	310 309	310 309	310 312 309
430 **431**	310 309	310 309	430-15[†] 310[††] 309[††]	310	310	310 309	310 309	310 309	310 309	310 309	310 309	310 309	310 309	310 309	310 309	310 309	310 309 312
430F **430FSE**	310 309	410-15[†]	310 309	310 309	310 309 312	310 309 312	309 310 312	309 310 312	310 309 312	310 309 312	310 309 312	310 309 312	310 309 312	310 309 312	310 309 312	310 309 312	310 309 312
442 **443**	309 310	309 310 312	310 309 312	310 309 312	310 309 312	310 309 312	310 309 312	310 309 312	310 309 312	310 309 312	310 309 312	310 309 312	310 309 312	310 309 312	310 309 312	310 309 312	310 309 312

[†]Preheat.

[††]No preheat necessary.

Bold numbers indicate first choice; light numbers indicate second and third choices. This choice can vary with specific applications and individual job requirements.

TABLE 27-5 Filler Metal Selector Guide for Joining Different Types of Stainless Steel to the Same Type or Another Type of Stainless Steel Thermacote Welco

UTP Designation	AWS/SFA5.4 Covered	AWS/SFA5.9 TIG and MIG	Description and Applications
6820	E308-16	ER308	For welding conventional 308 type SS
68 Kb	E308-15		Low hydrogen coating
6820 Lc	E308 L-16	ER308L	Low carbon grade, prevents carbide precipitation adjacent to weld
308L Fe Hp	E308 L-16		Fast depositing for maintenance and production coating
68 LcHL	E308 L-16		High-performance electrode with rutile-acid coating, core wire alloyed, for stainless and acid-resisting CrNi steels
68 LcKb	E308 L-15		Low carbon electrode for stainless, acid-resisting CrNi-steels
6824	E309-16	ER309	For welding 309 type SS and carbon steel to SS
6824 Kb	E309-15		Special lime-coated electrode for corrosion and heat-resistant 22/12 CrNi-steels
6824 Lc	E309 L-16	ER309L	Same as 309, but with low carbon content
6824 Nb	E309 Cb-16		Corrosion and heat-resistant 22/12 CrNi-steels
6824 MoNb	E309MoCb-16		Corrosion and heat-resistant 22/12 CrNi-steels
309L Fe Hp	E309 L-16		High deposition rate, easy to use
6824 Mo Lc	E309 L-16		For welding similar and dissimilar SS
68H	E310-16	ER310	For high-temperature service and cladding steel
6820 Mo	E316-16	ER316	For welding acid-resistant stainless steels
6820 Mo Lc	E316 L-16	ER316L	Low carbon grade, prevents intergranular corrosion
68 Tl Mo	E316 L-16		Most efficient type, for maintenance and production, high performance
68 MoLcHL	E316 L-16		High-performance electrode with rutile-acid coating, core wire alloyed, for stainless and acid-resisting CrNi-Mo-steels
68 MoLcKb	E316 L-15		Low carbon electrode for stainless and acid-resisting CrNiMo-steels
317 Lc Titan	E317 L-16	ER317L	Deposit resist sulfuric acid corrosion
317LFe Hp	E 317 L16		Fast melt-off rate, excellent for over-lays, easy to use, high performance
68 Mo	E318-16		Versatile stainless all-position electrode
320 Cb	E320-15		For welding similar acid-resistant SS
3320 Lc	E320		A rutile-coated electrode for welding in all positions except vertical down
6820 N3	E347-16	ER347	Stabilized grade, prevents carbide precipitation
347 FeHp	E347-16		High-performance stainless steel electrode of class E 347-16 for welding stabilized Cr Ni alloys
66	E410-15		Low hydrogen electrode for corrosion and heat-resistant 14% Cr-steels
1915 HST			Low hydrogen, fully austenitic electrode with 0% ferrite content
1925			Extremely corrosion-resistant to phosphoric and sulfuric acids
68 Hcb	E310 Cb	ER310 Cb	For high heat applications and joining steels to stainless steel
2535 NbSn			Electrode is a lime-type special electrode and is used for surfacing and joining heat-resistant base metals, especially cast steel
E 330-16	E330-16		Excellent for welding furnace parts
6805			For welding of base material 17-4 Ph
6808 Mc			Rutile lime-type austenitic-ferritic electrode with low carbon content suited for joining and surfacing on corrosion-resistant steels and cast steel type with austenitic-ferritic structure (Duplex-steels)
6809 Mo			Rutile-basic austenitic-ferritic electrode with low carbon content suited for joining and surfacing on corrosion-resistant steels and cast steel types with an austenitic-ferritic structure (Duplex-steels)

TABLE 27-6 Stainless Steel Electrodes, Filler Metals, and Wires UTP Welding Materials, Inc.

for high-temperature service where low creep is desired, such as for jet engine parts, valves, and furnace parts. All E310 filler metals can be used to weld 309 stainless steel or to join mild steel to stainless steel or to weld most hard-to-weld carbon and alloy steels. E310Mo-15 and 16 electrodes have molybdenum added to improve their strength at high temperatures and to resist corrosive pitting.

E316-15, E316-16, E3116L-15, E316L-16, ER316, ER316L, and ER316L-Si All are filler metals for 316 stainless steels. 316 stainless steels are used for high-temperature service where high strength with low creep is desired. Molybdenum is added to improve these properties and to resist corrosive pitting. E316 filler metals are used for welding tubing, chemical pumps, filters, tanks, and furnace parts. All E316 filler metals can be used on 316 stainless steels or when weld resistance to pitting is required.

NONFERROUS ELECTRODES

The AWS identification system for covered nonferrous electrodes is based on the atomic symbol or symbols of the major alloy(s) or the metal's identification number. The alloy having the largest percentage appears first in the identification. The atomic symbol is prefixed by the letter E. For example, ECu is a covered copper arc welding electrode, and ECuNiAl is a copper-nickel-aluminum alloy–covered arc welding electrode. A letter, number, or letter-number combination may be added to the right of the atomic symbol to indicate some special alloys. For example, ECuAl-A2 is a copper-aluminum welding electrode that has 1.59% iron added.

Aluminum and Aluminum Alloys

The AWS specifications for aluminum and aluminum alloy filler metals are A5.3 for covered arc welding electrodes and A5.10 for bare welding rods and electrodes. Filler metal classified within the A5.3 uses the atomic symbol Al, and in the A5.10 the prefix ER is used with the Aluminum Association number for the alloy, **Table 27-7.**

Aluminum-Covered Arc Welding Electrodes

Al-2 and Al-43 The aluminum electrodes do not use the letter E before the electrode number. Aluminum-covered arc welding electrodes are designed to weld with DCEP polarity. These electrodes can be used on thin or thick sections, but thick sections must be preheated to between 300°F (150°C) and 600°F (315°C). The preheating of these thick sections allows the weld to penetrate immediately when the weld starts. Aluminum arc welding electrodes can be used on 2024, 3003, 5052, 5154, 5454, 6061, and 6063 aluminum. When welding on aluminum, a thin layer of surface oxide may not prevent welding. Thicker oxide layers must be removed mechanically or chemically. The back of the weld can be supported by carbon plates or carbon paste. To prevent

excessive penetration, most aluminum arc welding electrodes can also be used for oxyfuel gas welding of aluminum.

Aluminum Bare Welding Rods and Electrodes

ER1100 1100 aluminum has the lowest percentage of alloy agents of all of the aluminum alloys, and it melts at 1215°F (657°C). The filler wire is also relatively pure. ER1100 produces welds that have good corrosion resistance and high ductility, with tensile strengths ranging from 11,000 to 17,000 psi. The weld deposit has a high resistance to cracking during welding. This wire can be used with OFW, GTAW, and GMAW. Preheating to 300°F (148°C) to 350°F (176°C) is required for GTA welding on plate or pipe 3/8 in. (10 mm) and thicker to ensure good fusion. Flux is required for OFW. 1100 aluminum is commonly used for items such as food containers, food-processing equipment, storage tanks, and heat exchangers. ER1100 can be used to weld 1100 and 3003 grade aluminum.

ER4043 ER4043 is a general-purpose welding filler metal. It has 4.5% to 6.0% silicon added, which lowers its melting temperature to 1155°F. The lower melting temperature helps promote a free-flowing molten weld pool. The welds have high ductility and a high resistance to cracking during welding. This wire can be used with OFW, GTAW, and GMAW. Preheating to 300°F (148°C) to 350°F (176°C) is required for GTA welding on plate or pipe 3/8 in. (10 mm) and thicker to ensure good fusion. Flux is required for OFW. ER4043 can be used to weld on 2014, 3003, 3004, 4043, 5052, 6061, 6062, and 6063 and cast alloys 43, 355, 356, and 214.

ER5356 ER5356 has 4.5% to 5.5% magnesium added to improve the tensile strength. The weld has high ductility but only an average resistance to cracking during welding. This wire can be used for GTAW and GMAW. Preheating to 300°F (148°C) to 350°F (176°C) is required for GTA welding on plate or pipe 3/8 in. (10 mm) and thicker to ensure good fusion. ER5356 can be used to weld on 5050, 5052, 5056, 5083, 5086, 5154, 5356, 5454, and 5456.

ER5556 ER5556 has 4.7% to 5.5% magnesium and 0.5% to 1.0% manganese added to produce a weld with high strength. The weld has high ductility and only average resistance to cracking during welding. This wire can be used for GTAW and GMAW. Preheating to 300°F (148°C) to 350°F (176°C) is required for GTA welding on plate or pipe 3/8 in. (10 mm) and thicker to ensure good fusion. ER5556 can be used to weld on 5052, 5083, 5356, 5454, and 5456.

SPECIAL-PURPOSE FILLER METALS

ENi The nickel arc welding electrodes are designed to be used with AC or DCEP polarity. These arc welding electrodes are used for cast iron repair. The carbon in cast iron will not migrate into the nickel weld metal, thus preventing cracking and embrittlement. The cast iron may or may

Base Metal	319 355	43 356	214	6061 6063 6151	5456	5454	5154 5254	5086	5083	5052 5652	5005 5050	3004	1100 3003	1060
1060	4145 4043 4047	4043 4047 4145	4043 5183 4047	4043 4047	5356 4043	4043 5183 4047	4043 5183 4047	5356 4043	5356 4043	4043 4047	1100 4043	4043	1100 4043	1260 4043 1100
1100 3003	4145 4043 4047	4043 4047 4145	4043 5183 4047	4043 4047	5356 4043	4043 5183 4047	4043 5183 4047	5356 4043	5356 4043	4043 5183 4047	4043 5183 5356	4043 5183 5356	1100 4043	
3004	4043 4047	4043 4047	5654 5183 5356	4043 5183 5356	5356 5183 5556	5654 5183 5356	5654 5183 5356	5356 5183 5556	5356 5183 5556	4043 5183 4047	4043 5183 5356	4043 5183 5356		
5005 5050	4043 4047	4043 4047	5654 5183 5356	4043 5183 5356	5356 5183 5556	5654 5183 5356	5654 5183 5356	5356 5183 5556	5356 5183 5556	4043 5183 4047	4043 5183 5356			
5052 5652	4043 4047	4043 5183 4047	5654 5183 4047	5356 5183 4043	5356 5183 5556	5654 5183 5356	5654 5183 5356	5356 5183 5556	5356 5183 5556	5654 5183 4043				
5083	NR	5356 4043 5183	5356 5183 5556	5356 5183 5556	5183 5356 5556	5356 5183 5556	5356 5183 5556	5356 5183 5556	5183 5356 5556					
5086	NR	5356 4043 5183	5356 5183 5556	5356 5183 5556	5356 5183 5556	5356 5183 5554	5356 5183 5554	5356 5183 5556						
5154 5254	NR	4043 5183 4047	5654 5183 5356	5356 5183 4043	5356 5183 5554	5654 5183 5356	5654 5183 5356							
5454	4043 4047	4043 5183 4047	5654 5183 5356	5356 5183 4043	5356 5183 5554	5554 1013 5183								
5456	NR	5356 4043 5183	5356 5183 5556	5356 5183 5556	5556 5183 5356									
6061 6063 6151	4145 4043 4047	4043 5183 4047	5356 5183 4043	4043 5183 4047										
214	NR	4043 5183 4047	5654 5183 5356											
43 356	4043 4047	4145 4043 4047												
319 355	4145 4043 4047													

Note: First filler alloy listed in each group is the all-purpose choice. NR means that these combinations of base metals are not recommended for welding.

TABLE 27-7 Recommended Filler Metals for Joining Different Types of Aluminum to the Same Type or a Different Type of Aluminum Thermacote Welco

not be preheated. A very short arc length and a fast travel rate should be used with these electrodes.

ECuAl The aluminum bronze welding electrodes are designed to be used with DCEP polarity. This welding electrode has copper as its major alloy. The aluminum content is at a much lower percentage. Iron is usually added but at a percentage that is very low. These electrodes are sometimes referred to as arc brazing electrodes, although this is not an accurate description. Stringer beads and a short arc length should be used with these electrodes. Aluminum bronze welding electrodes are used for overlaying bearing surfaces; welding on castings of manganese, bronze, brass, or aluminum bronze; or assembling dissimilar metals.

Surface and Buildup Electrode Classification

Hardfacing or wear-resistant electrodes are the most popular special-purpose electrodes; however, there are also cutting and brazing electrodes. Specialty electrodes may be identified by manufacturers' trade names. Most manufacturers classify or group hardfacing or wear-resistant electrodes according to their resistance to impact, abrasion, or corrosion. Occasionally, electrode resistance to wear at an elevated temperature is listed. One electrode may have more than one characteristic or type of service listed.

EFeMn-A The EFeMn-A electrodes are designed to be used with AC or DCEP polarity. This electrode is an impact-resistant welding electrode. It can be used on hammers, shovels, and spindles and in other similar applications.

ECoCr-C The ECoCr-C electrodes are designed to be used with AC or DCEP polarity. This electrode is a corrosion- and abrasion-resistant welding electrode. It also maintains its resistance at elevated temperatures. ECoCr-C is commonly used for engine cams, seats and valves, chain-saw bars, bearings, and dies.

Magnesium Alloys

The joining of magnesium alloys by torch welding or brazing is possible without a fire hazard because the melting point of magnesium is 1202°F (651°C) to 858°F (459°C) below its boiling point, where magnesium may start to burn.

ER AZ61A The ER AZ61A filler metal can be used to join most magnesium wrought alloys. This filler has the best weldability and weld strength for magnesium alloys AZ31B, HK31A, and HM21A.

ER AZ92A The ER AZ92A filler metal can be used on cast alloys Mg-Al-Zn and AM 100A. This filler metal has a somewhat higher resistance to cracking.

HYDROGEN EMBRITTLEMENT

Hydrogen embrittlement starts with hydrogen atoms being dissolved in molten weld metal. Then after the weld cools, it causes cracking. It is a potential problem for all steels, but it becomes a greater problem for higher alloyed steels. So, the higher the alloy content in a steel, the greater the potential problem of hydrogen embrittlement. A bonded pair of hydrogen atoms is called **molecular hydrogen (H_2)**, **Figure 27-19**. In nature, hydrogen atoms exist only if they are bonded together or to other types of atoms, such

as oxygen, with which it forms water (H_2O). Hydrogen atoms are also present in many other substances such as grease and oil. The heat of a welding arc can break the hydrogen bond, forming free, unbonded hydrogen atoms called **atomic hydrogen (H)**. Atomic hydrogen is freely dissolved into the molten weld pool. As the weld metal cools, however, most of the dissolved hydrogen comes out of solution to form bubbles of porosity. Because the hydrogen atom is so small, some of the dissolved atomic hydrogen atoms will move into the base metal through the tiny spaces between the grains of solid metal. Sometimes much later this atomic hydrogen can cause underbead cracking, **Figure 27-20**. This hydrogen-caused cracking can appear hours, days, or weeks following welding. The following steps can help to control postweld hydrogen cracking-related problems:

- Use welding filler metal and fluxes classified as low hydrogen.
- Preheat the metal to remove any moisture that may be trapped in the joint or around tack welds.
- Avoid welding when it is precipitating (raining, snowing, or sleeting).
- Follow proper storage and handling procedures for filler metal and fluxes to prevent them from becoming contaminated with sources of hydrogen such as moisture, oil, grease, or other hydrocarbons.
- Avoid welding in strong winds, which may blow the shielding gas cloud away from the molten weld pool, resulting in atmospheric moisture contaminating the weld, **Figure 27-21**.

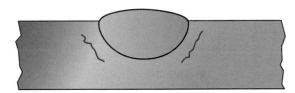

FIGURE 27-20 Underbead cracking. © Cengage Learning 2012

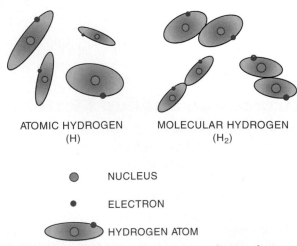

ATOMIC HYDROGEN (H)   MOLECULAR HYDROGEN (H_2)

- NUCLEUS
- ELECTRON
- HYDROGEN ATOM

FIGURE 27-19 Hydrogen atoms are made up of one proton, in the nucleus, and one electron, in the outer shell. © Cengage Learning 2012

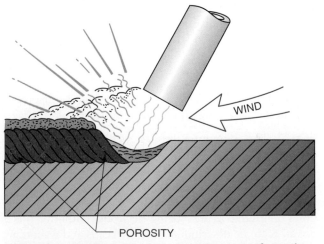

WIND

POROSITY

FIGURE 27-21 Wind can blow shielding away from the molten weld pool, allowing porosity to form in the weld. © Cengage Learning 2012

Summary

Proper filler metal selection is one of the most important factors affecting the successful welding of a joint. Many factors affect the selection of the most appropriate filler metal for a job. In some cases, cost is the greatest factor and in others it is structural strength. For example, if you were building an ultralight aircraft, you would be more concerned with strength than cost. However, if you were building an iron fence, you might be more concerned with cost. Every application is different, so it may be a help for you to list the items you feel are most important for selecting a filler metal. This will help you select the most appropriate filler metal for your needs.

Manufacturers' literature on filler metals can be divided into two general sections. One section of the literature is technical and the other is advertisement. In the technical section, you are provided with specific information on each filler metal's operation, performance, and uses. In the advertisement section, you are provided with marketing information and claims regarding performance. Knowing the types of information in both sections will help you evaluate new material as you select filler metal.

If you are considering a large purchase of filler metals, it is advantageous for you to request samples of the various filler metals from manufacturers so that you can test their performance in your applications. Pretesting of the products in your applications will give you an opportunity to determine which filler metal is going to give you the best value for your money. It may also be necessary to qualify the filler metal for your welding certification program before you make the purchase and begin using the product.

Filler Metal Made Easy

Finding standard, commodity-type welding consumables, such as E7018, E308L-15, or ER316, is usually straightforward. However, when weld metal toughness, properties after postweld heat treatment, specified ferrite (a pure form of iron) criteria, hardness, or other out-of-the-ordinary requirements are needed, communication between the user and supplier becomes critical. In cases in which the lowest bid rules, sole sourcing is difficult. Simply stating the AWS filler metal specification and classification is not enough to obtain exactly what is needed. Frustration can result, deliveries can be delayed, or inadequate products may be purchased. There is a remedy to this situation—the AWS A5.01, Filler Metal Procurement Guidelines.

The Need for Guidelines

The AWS A5.01-93 is the current edition of the original standard, introduced in 1978. Before the introduction of the Filler Metal Procurement Guidelines, the individual AWS filler metal specifications always indicated that "at the option and expense of the purchaser . . . the foregoing tests may be used as a basis of acceptance of electrodes." However, there were no established schedules for testing and no definitions of lot classes.

Fabricated mild steel work center welded with 0.035-in. ER70S-6 wire. American Welding Society

It became apparent in the 1970s, that there was a definite need for such requirements to be published. In addition, a clear definition of the "Manufacturer's Quality Assurance System" needed to be established. Out of this need, the first edition of Filler Metal Procurement Guidelines was developed. This AWS standard was soon adopted by ASME Boiler and Pressure Vessel Code as ASME SFA 5.01 and is included in Section II, Part C: Welding Rods, Electrodes, and Filler Metals. It is now also an ANSI-approved standard.

Half of AWS members and an even higher percentage of fabricators likely do not know this standard exists or when it is needed. When it is used, obtaining filler metal can be streamlined. Although it has been revised, the current scope (A5.01-93) reflects the original purpose of the guideline when it was first issued.

Scope of the Document

This document, together with an AWS filler metal specification, is intended to describe a set method for providing specific details needed for obtaining filler metal. Details consist of the following:

1. The filler metal classification (selected from the relevant AWS filler metal specification)

2. The lot classification (selected from Section 5 of the document)

3. The level of testing schedule (selected from Table 1, Section 6, Level of Testing)

For example, when welding on a highly corrosive austenitic stainless steel piping system exposed to a highly corrosive medium, the engineer needs to specify an E308L-15 class electrode. Another contract requirement, 5-9 FN (ferrite number), is listed for corrosion control of the deposited weld metal. In many cases, the material is ordered simply by its AWS classification, E308L-15. This would more than likely result in delivery of a type-E308 product with 0.04 maximum percentage carbon in the weld deposit. A typical test certificate or certificate of conformance may accompany the product, but it would not necessarily represent testing done on the exact lot delivered. The certificate would indicate only that the product supplied meets the requirements for classification as E308L-15 in accordance with AWS A5.4-92, meaning it was produced by the manufacturer within the annual time frame for a product made to this classification and in accordance with AWS A5.4. Whether the product delivered was actually tested or not remains uncertain. Whether the product will deposit a weld metal that is

5-9 FN is also uncertain. Therefore, what was required for the application may or may not be what is delivered. How can this uncertainty be avoided? By using the AWS A5.01 standard, the purchaser could have specified a higher level of testing to ensure there would be little room for uncertainty, since the welding procedure specification (WPS) would be the one used in production and the actual measurement of delta ferrite (iron) in the weld deposit would be a certifiable result and not just a prediction.

Easy to Understand

The AWS A5.01 specification is organized in a logical order and is user friendly. Whether or not all or part of the criteria is listed in the document for actual lot testing depends on the extent to which special criteria are needed to adequately describe the products(s) needed and to reduce the risk of receiving a product that may not meet the specific need. As a minimum, the manufacturer is required to have an established quality assurance system and is required to trace the product to a known lot that is unique to that manufacturer. This requirement also applies to those who repackage, relabel, and resell another manufacturer's product that is identified as meeting AWS specifications.

The document standard also provides detailed definitions and criteria for items including dry batch, dry blend, wet mix, heat, heat number, and controlled chemical compositions and lot classification for covered electrodes, bare welding wires, and cored electrodes and fluxes, in addition to specifying the level of testing schedule.

Using the suggested forms and referring to precise definitions leave little to chance. In many cases, using A5.01 relieves a company's purchasing personnel of having to make technical decisions. Another benefit is the ability to obtain items that do not have an AWS classification while maintaining complete control of the procurement process. A5.01 provides a means to specify exactly what is required. To improve your ability to get the exact welding filler metal for a job, get a copy of the A5.01 specification from the American Welding Society.

Article courtesy of the American Welding Society.

Review

1. What groups have developed electrode identification systems?

2. What types of general information about electrodes may be given by different electrode manufacturers?

3. Define *tensile strength*.

4. What chemicals alloys are

    a. considered to be contaminants to the weld metal?

    b. used to increase tensile strength in the weld metal?

    c. used to increase corrosion resistance?

    d. used to reduce creep?

5. What should CE be used for?

6. What welding parameters should be used for a metal that has a CE of more than 0.60%?

7. What functions can the flux covering of an SMA electrode provide to the weld?

8. How does an SMA welding electrode's flux covering produce the shielding gas to protect the weld?

9. What fluxing agents act as scavengers in the molten weld pool?

10. How can an SMA welding electrode's flux help with deeper penetration?

11. What are the advantages of refractory-type stages?

12. List the things that must be considered before selecting an electrode for a specific job.

13. Why can there be more than one electrode for each classification manufactured by the same company?

14. What do the following filler metal designations stand for?

    a. E

    b. ER

    c. RG

    d. IN

15. Explain the parts of the AWS classified system for SMAW electrodes such as E7018.

16. Which SMA welding electrode(s) can be used to weld on metal that has a light covering of paint?

17. Which SMA welding electrodes are commonly used to weld on sheet metal fabrications?

18. Which SMA welding electrodes can be used with a dig technique?

19. Referring to Figure 27-8 through Figure 27-15, which electrode has the deepest penetration?

20. How is the E7018 molten weld pool protected?

21. What is the purpose of the deoxidizers in ER70S-2?

22. What alloying element used in FCA welding electrodes causes the electrode to produce centerline cracks in the weld if it is used for multipass welds?

23. What does the 15 and 16 stand for in SMA stainless steel welding electrodes?

24. What stainless steel(s) would

    a. have low creep at high temperatures?

    b. be used for food service equipment?

25. Referring to Table 27-6, what stainless steel filler metal would be selected for welding

    a. 304L stainless steel to 314 stainless steel?

    b. 321 stainless steel to 348 stainless steel?

    c. 316 stainless to mild steel?

26. Referring to Table 27-7, what would be an excellent filler metal for a weld on furnace parts?

27. What forms the basis for the AWS identification system for nonferrous covered electrodes?

28. Why must thick sections of aluminum be preheated before an Al-2 electrode is used for welding?

29. For what types of items would the purest aluminum alloy be used?

30. What are aluminum *arc brazing electrodes* used for?

31. How do most manufacturers classify or group hardfacing or wear-resistant electrodes?

Chapter 28

Welding Automation and Robotics

OBJECTIVES

After completing this chapter, the student should be able to

- explain the difference among manual (MA), semiautomatic (SA), machine (ME), automatic (AU), and automated welding.
- list the major factors to be considered in establishing a robotic welding station.
- list robotic safety considerations.
- explain the need for interaction among various components of robotic workstations.

KEY TERMS

automated joining

automatic joining

computer-aided design (CAD)

computer-aided manufacturing (CAM)

cycle time

industrial robot

machine joining

manipulator

manual joining

multifunctional

pick and place

reprogrammable

semiautomatic joining

sensors

work cell

x-axis

y-axis

z-axis

INTRODUCTION

The use of technology to reduce production time, increase output, and reduce costs began as early as the 1920s. The automotive industry first used the automatic welding process to produce high-quality welded swing-arm supports and other parts. The process used a continuous-feed, unshielded, bare wire. Since that time, advances in automatic welding technology have continued to take place.

In the 1960s, American technology produced the first industrial robots. The first industrial robots were used by the nuclear industry to handle radioactive materials. These early units were mainly **pick and place** robots used to move material, with little repetitive accuracy required. During the 1970s, computers were applied in increasing numbers by large industry to serve as intelligent controllers for automation. By the 1980s, the decreasing cost of computers and advancements in robotics had made this technology possible, even for small businesses.

More and more modern businesses are using **computer-aided design (CAD)** to improve products. CAD technology can help in selecting materials, specifying thicknesses, and locating supports to ensure good engineering design of products.

Computer-aided manufacturing (CAM) can then be used to improve production. CAM technology can aid in selecting assembly methods, planning the product flow through the manufacturing steps, and scheduling the various operations to reduce the actual production costs.

Computers and microprocessors assist modern industry in producing high-quality products with a minimum waste of materials and time. The use of high technology to monitor part flow through production alerts management to potential problem areas before they cause slowdowns, lost manufacturing time, and increased costs, as well as possible hazardous situations.

Automation is revolutionizing industry worldwide. Some manufacturing plants using robots and other automatic equipment require only supervisory personnel to manage large portions of the production. These facilities are only the beginning.

Many workers whose jobs are modernized by automation find that there are new and more technical opportunities for operating, maintaining, designing, and installing automated equipment and robots. Skilled technicians are needed to set up and operate such automatic manufacturing equipment. The technician needs to understand the manufacturing processes, including welding, to ensure that the operating guidelines established will work consistently. Most robot manufacturers recommend that a highly skilled welder be selected to operate the robot. It is much easier to train a welder to operate the robot than it is to train a computer technician to make good welds.

In welding applications, the increased use of robots, however, will never replace the tremendous need for skilled welders. Welders will always be needed to produce high-quality welds in production and maintenance in locations that are too restricting for robotics.

Students who have mastered hands-on welding techniques and those who understand the various automated welding procedures will have the advantage when seeking either type of employment opportunity.

This chapter will give the welding student a general overview of automatic welding processes and robotics with guidelines on implementing automation in welding applications.

MANUAL JOINING PROCESS

A **manual joining** process is one that is completely performed by hand. The welder controls all of the manipulation, rate of travel, joint tracking, and, in some cases, the rate at which filler metal is added to the weld. The manipulation of the electrode or torch in a straight line or oscillating pattern affects the size and shape of the weld, **Figure 28-1**. The manipulation pattern may also be used to control the size of the weld pool during out-of-position welding. The rate of travel or speed at which the weld progresses along the joint affects the width, reinforcement, and penetration of the weld, **Figure 28-2**. The placement or location of the weld bead within the weld joint affects the strength, appearance, and possible acceptance of the

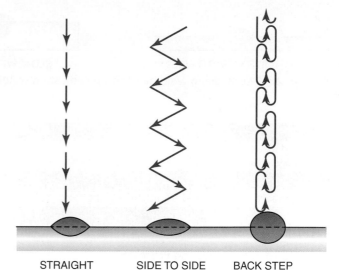

STRAIGHT   SIDE TO SIDE   BACK STEP

FIGURE 28-1 The electrode manipulation affects the size and shape of the weld bead. © Cengage Learning 2012

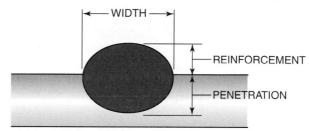

FIGURE 28-2 The travel rate of the weld affects width, reinforcement, and penetration of the weld bead. © Cengage Learning 2012

joint. The rate at which filler metal is added to the weld affects the reinforcement, width, and appearance of the weld, **Figure 28-3**.

The most commonly used manual arc (MA) welding process is shielded metal arc welding (SMAW). This process was described in Section 2 of this text. The flexibility the welder has in performing the weld makes this process one of the most versatile. By changing the manipulation, rate of travel, or joint tracking, the welder can make an acceptable weld on a variety of material thicknesses, **Figure 28-4**.

The most commonly used manual arc welding, gas welding, and brazing processes are listed in **Table 28-1**.

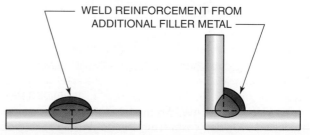

WELD REINFORCEMENT FROM ADDITIONAL FILLER METAL

FIGURE 28-3 Addition of filler metal. © Cengage Learning 2012

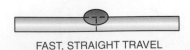

FAST, STRAIGHT TRAVEL

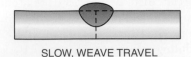

SLOW, WEAVE TRAVEL

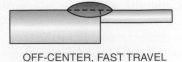

OFF-CENTER, FAST TRAVEL

FIGURE 28-4 Controlling weld bead size by adjusting welding parameters. © Cengage Learning 2012

Arc
Shielded metal arc welding (SMAW)
Gas tungsten arc welding (GTAW)
Gas
Oxyacetylene welding (OAW)
Brazing
Torch brazing (TB)

TABLE 28-1 Manual Joining Processes

SEMIAUTOMATIC JOINING PROCESSES

A **semiautomatic joining** process is one in which the filler metal is fed into the weld automatically. Most other functions are controlled manually by the welder. The addition of filler metal to the weld by an automatic wire-feeder system enables the welder to increase the uniformity of welds, productivity, and weld quality. The distance of the welding gun or torch from the work remains constant. This gives the welder better manipulative control as compared to, for example, shielded metal arc welding, in which the electrode holder starts at a distance of 14 in. (356 mm) from the work. This distance exaggerates the slightest accidental movement made during the first part of the weld, **Figure 28-5**. In the SMAW process, the electrode holder must be lowered steadily as the weld progresses to feed the electrode and maintain the correct arc length, **Figure 28-6**. This constant changing of

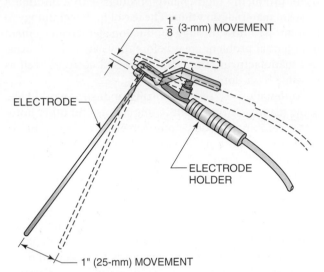

FIGURE 28-5 A 1/8-in. (3-mm) movement of the electrode holder results in a 1-in. (25-mm) movement of the electrode at the surface of the work. © Cengage Learning 2012

the distance above the work causes the welder to shift body position frequently. This change, too, may affect the consistency of the weld.

Because the filler metal is being fed from a large spool, the welder does not have to stop welding to change filler electrodes or filler metal. SMA electrodes cannot be used completely as they have a waste stub of approximately 2 in. (51 mm). This waste stub represents approximately 15% of the filler metal that must be discarded. The frequent

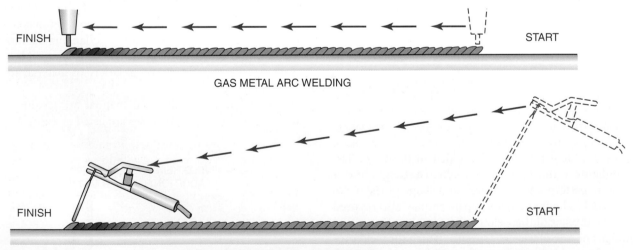

GAS METAL ARC WELDING

SHIELDED METAL ARC WELDING

FIGURE 28-6 In GMAW, the torch height remains constant above the work surface. In SMAW, the height of the electrode holder steadily decreases from the beginning to end of the weld. © Cengage Learning 2012

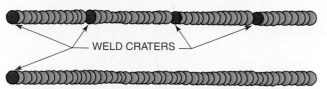

GAS TUNGSTEN ARC WELDING – MANUAL
14" (356 mm) LONG – 4 STOPS – 3 MIN 15 SEC

WELD CRATERS

GAS TUNGSTEN ARC WELDING – COLD WIRE
14" (356 mm) LONG – 1 STOP – 1 MIN 30 SEC

FIGURE 28-7 Fewer weld craters occur in continuous welding. © Cengage Learning 2012

Arc
Gas metal arc welding (GMAW)
Flux cored arc welding (FCAW)
Submerged arc welding (SAW)
Gas tungsten arc welding (GTAW)
Cold-hot wire feed

TABLE 28-2 Semiautomatic Joining Processes

stopping for rod and electrode changes, followed by restarting, wastes time and increases the number of weld craters. These craters are often a source of cracks and other discontinuities, **Figure 28-7**. In some welding procedures, each weld crater must be chipped and ground before the weld can be restarted. These procedures can take up to 10 minutes—time that can be used for welding in a semiautomatic welding process.

The most commonly used semiautomatic arc (SA) welding process is gas metal arc welding (GMAW). This process was fully described in Section 4. **Table 28-2** lists several other semiautomatic processes.

MACHINE JOINING PROCESSES

A **machine joining** process is one in which the joining is performed by equipment requiring the welding operator to observe the progress of the weld and make adjustments as required. The parts being joined may or may not be loaded and unloaded automatically. The operator may monitor the joining progress by watching it directly, observing instruments only, or using a combination of both methods. Adjustments in travel speed, joint tracking, work-to-gun or work-to-torch distance, and current settings may be needed to ensure that the joint is made according to specifications.

The work may move past a stationary welding or joining station, **Figure 28-8**, or it may be held stationary and the welding machine moves on a beam or track along the joint, **Figure 28-9**. On some large machine welds, the operator may ride with the welding head along the path of the weld. During the assembly of the external fuel tanks used for the space shuttle, two operators

were required for a few of the machine welds. One operator watched the root side of the weld while the other observed the face side of the weld. They were able to communicate with each other so that any needed changes could be made.

To minimize adjustments during machine welds, a test weld is often performed just before the actual weld is produced. This practice weld helps increase the already high reliability of machine welds.

AUTOMATIC JOINING PROCESSES

An **automatic joining** process is a dedicated process (designed to do only one type of welding on a specific part) that does not require adjustments to be made by the operator during the actual welding cycle. All operating guidelines are preset, and parts may or may not be loaded or unloaded by the operator. Automatic equipment is often dedicated to one type of product or part. A large investment is usually required in jigs and fixtures used to hold the parts to be joined in the proper alignment. The operational cycle can be controlled mechanically or numerically (computer). The cycle may be as simple as starting and stopping points, or it may be more complex. A more complex cycle may include such steps as prepurge time, hot start, initial current, pulse power, downslope, final current, and postpurge time, **Figure 28-10**.

Automatic welding or brazing is best suited to large-volume production runs because of the expense involved in special jigs and fixtures.

AUTOMATED JOINING

Automated joining processes are similar to automatic joining except that they are flexible and more easily adjusted or changed. Unlike automatic joining, there is no dedicated machine for each product. The equipment can be easily adapted or changed to produce a wide variety of high-quality welds.

The industrial robot is rapidly becoming the main component in automated welding or joining stations. The welding or joining cycles are often controlled by computers or microprocessors. The flexibility provided by automated workstations makes it possible for even small companies with limited production runs to invest in automated equipment. The equipment is controlled by programs, or a series of machine commands expressed in numerical codes, that direct the welding, cutting, assembling, or any other activities. The programs can be stored and quickly changed. Some systems can store and retrieve many different programs internally. Other systems are controlled by a host computer. Both types of systems can speed up production when frequent changes are required.

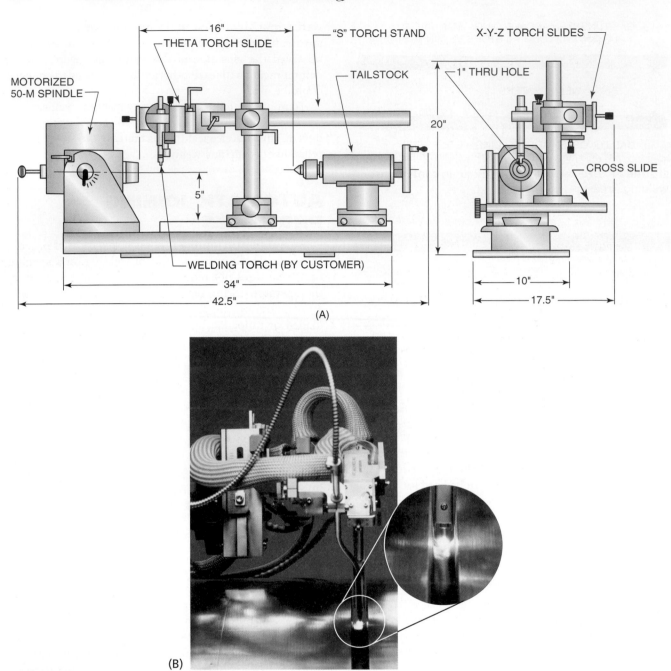

FIGURE 28-8 Two precision bench-welding systems. (A) © Cengage Learning 2012, (B) ARC Machines, Inc.

INDUSTRIAL ROBOTS

An **industrial robot** is a "**reprogrammable, multifunctional manipulator** designed to move material, parts, tools, or specialized devices through variable programmed motions for the performance of a variety of tasks." Industrial robots are primarily powered by electric stepping motors, hydraulics, or pneumatics and are controlled by a program.

Robots can be used to perform a variety of industrial functions, including grinding, painting, assembling, machining, inspecting, flame cutting, product handling, and welding.

Robots range in size and complexity from small desktop units capable of lifting only a few ounces (grams) to large floor models capable of lifting tons. Most robots can perform movements in three basic directions: longitudinal **(x-axis)**, transverse **(y-axis)**, and vertical **(z-axis)**, **Figure 28-11**. The tool end of the robot arm may also be jointed so that it can tilt and rotate, **Figure 28-12**.

The robot may be used with other components to increase production and the flexibility of the system. A computer or microprocessor can synchronize the robot's operation to positioners, conveyors, automatic fixtures, and other production machines. Parallel or multiple workstations increase the duty cycle (the fraction of time during which welding or work is being done) and reduce **cycle time** (the period of time from starting one operation to starting another), **Figure 28-13**. Parts can be loaded or

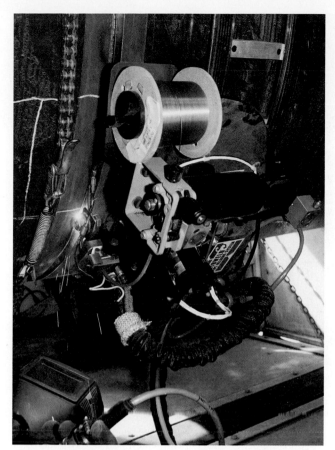

FIGURE 28-9 Automatic GTA machine welding along the seam of a stationary pipe. Lincoln Electric Company

unloaded by the operator at one station while the robot welds at another station.

Robot Programming

The programming that directs the robot through the desired welding or cutting path can be accomplished in any number of ways. Early robot programming took a great deal of time and involved a lot of trial and error. Trial and error is the process of first programming and then making a weld to see if it works. The results are examined, and then any needed changes are made in the programming. This process is still used by some today, but it is a lot faster because of improvements in the programming process.

With the integration of computer-aided design (CAD) to most of today's manufacturing, the robot can simply follow the weld joint design directly from the CAD program. This process is called computer-aided manufacturing (CAM). With this process, there may be only a little adjusting required by the welding technician to make the weld. Another method of programming the robot is to teach it to move its arm along the desired path. The arm can be moved through the complete cycle by physically directing its movements by hand or by using a remote control panel, **Figure 28-14.** The remote control panel allows the operator to direct the arm to the desired points along the weld path. In both cases, the robot will follow a straight line between the points established during the training session, **Figure 28-15.** A circular path or curve can be obtained by using teach points, **Figure 28-16.** Some units can be given the radius to be followed. A weave pattern can be added to the basic straight line weld at this time.

After the robot has been programmed by the teaching session, a test run should be made. Using a properly located part, with the welding current off, watch the robot arm to see if it follows the correct path. If the test run appears to be satisfactory, the welding parameters are set. A part should then be welded and the welds inspected for acceptability.

Another method of programming a robot is to use a computer keyboard to produce the program. These programs can be longer and more detailed than the teaching program. Information that can be given to the robot, such as the total length of weld and the number of welding parameter changes, is restricted by the memory capacity of the controller. Programmed robots use less internal memory (memory required for the teaching machines) because they understand basic programming language.

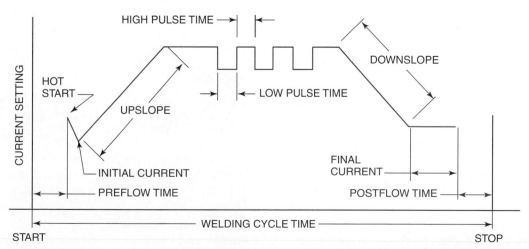

FIGURE 28-10 Typical GTAW automatic welding program. © Cengage Learning 2012

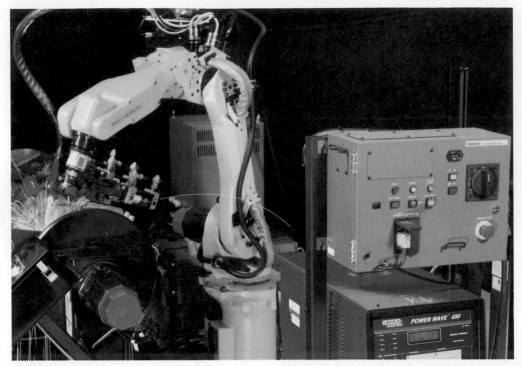

FIGURE 28-11 Machine axes. Reproduced with permission from FANUC Robotics America Corporation © FANUC Robotics America Corporation. All rights reserved.

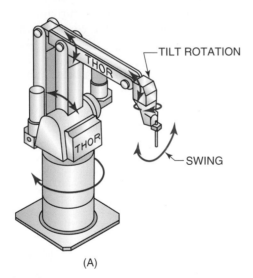

(A)

After the program has been tested and determined to be acceptable, it can be stored for future use. The CD, USB flash drive, and computer disc storage are the three most common storage methods. CDs are easily scratched; however, unlike computer discs, they cannot be erased accidentally, damaged, or changed by other electrical equipment. If the CDs are properly handled and stored, they can be a permanent program record. Programs stored on computer discs can be modified easily for production design changes. Care must be taken to prevent accidental erasure or damage due to the strong magnetism found around welding machines unless CDs are used.

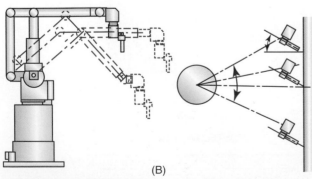

(B)

FIGURE 28-12 The tool end may be jointed so that it can tilt and rotate. © Cengage Learning 2012

FIGURE 28-13 Rotating worktable increases the work zone. Eutectic Corporation

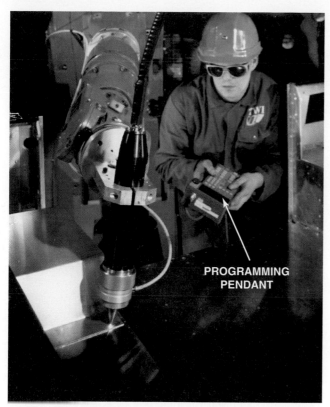

FIGURE 28-14 Remote teaching pendant for programming robot. ESAB Welding & Cutting Products

System Planning

Evaluating the need for the use of a robot and selecting a robot to meet that need is a complex process. Some of the factors that must be considered are the following:

- Present and future needs
- Parts design
- Equipment selection
- Safety

Present and Future Needs A system using a robot can be 80% more productive than a system using manual welders. This higher rate of productivity means that there must be a market for the higher volume of product produced. Premature investment in automation can be more costly than it can be productive. A comprehensive market analysis of present and future needs must be the first step to ensure that your business is not overextended.

Parts Design The design of the weldment must be compatible with a robotic system. The components should be selected and assembled so that the accessibility of the robot's arm and torch is not restricted, **Figure 28-17.** Although the torch can be moved along several axes, it cannot reach restricted areas, as a welder can when using manual

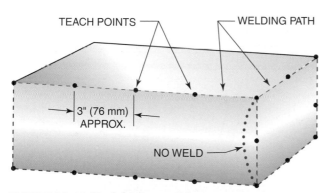

FIGURE 28-15 Straight-line welds. © Cengage Learning 2012

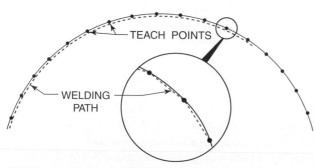

FIGURE 28-16 Curved welding path. © Cengage Learning 2012

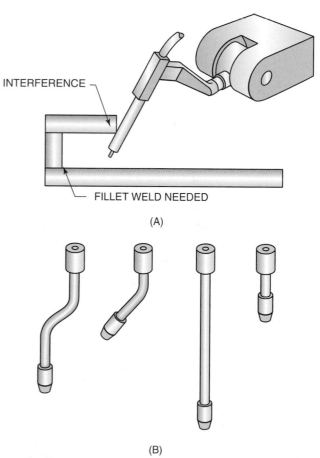

FIGURE 28-17 (A) Parts should be designed to permit access to the joint; (B) automatic gun exchanger, which permits different gun designs to be used when needed.
© Cengage Learning 2012

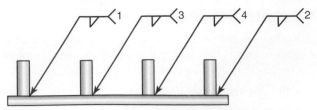

FIGURE 28-18 Staggering weld positions to prevent distortion of the workpiece. © Cengage Learning 2012

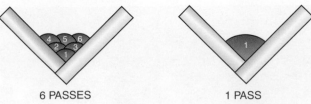

FIGURE 28-20 One large weld can be made faster and is less critical to produce. © Cengage Learning 2012

processes. Proper design and assembly sequences can reduce the amount of manual pickup welding required to complete the weldment. Mockups and CAD/CAM can be used before the robot is installed to reduce possible problems.

The parts design must also take into consideration the higher heat input from the almost constant welding. This heat could result in distortion of the weldment if some compensation were not provided. One method of controlling this distortion is to stagger the weld locations, **Figure 28-18.** However, this practice means that more arm articulation (movement) is required, resulting in a slowdown in production. The best method of eliminating distortion is to use a combination of jigs, tack welds, and preset angles, **Figure 28-19.** This technique can be very

successful, but it requires more initial design and setup time than other methods.

When possible, all welds should be performed in the flat position. Large weld sizes are less critical than small weld sizes in their exact location, **Figure 28-20.** Small weld sizes require better tolerances, since even the slightest mislocation may be unacceptable.

Equipment Selection The robot is the prime component in an automated **work cell** (a manufacturing unit consisting of two or more workstations), **Figure 28-21.** The selection of the system must be considered in connection with the other pieces of equipment being used. To obtain the most effective system, the robot, welding

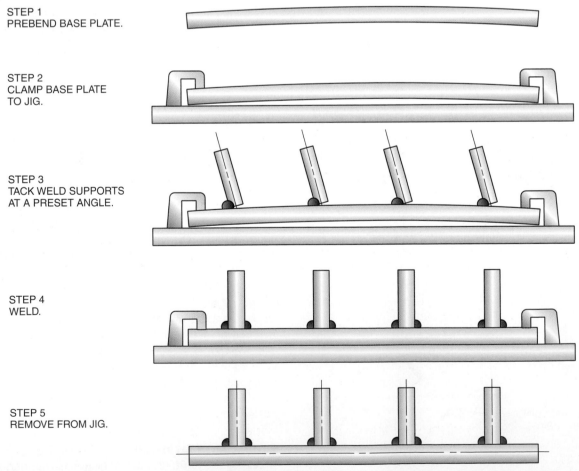

STEP 1
PREBEND BASE PLATE.

STEP 2
CLAMP BASE PLATE
TO JIG.

STEP 3
TACK WELD SUPPORTS
AT A PRESET ANGLE.

STEP 4
WELD.

STEP 5
REMOVE FROM JIG.

FIGURE 28-19 Suggested steps to eliminate heat distortion in completed part. © Cengage Learning 2012

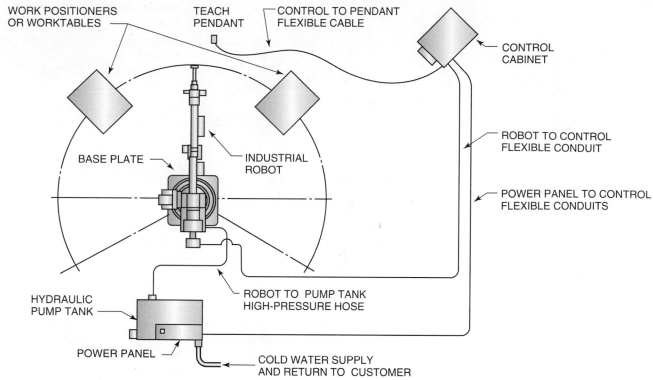

WORK POSITIONERS
OR WORKTABLES

TEACH
PENDANT

CONTROL TO PENDANT
FLEXIBLE CABLE

CONTROL
CABINET

BASE PLATE

INDUSTRIAL
ROBOT

ROBOT TO CONTROL
FLEXIBLE CONDUIT

POWER PANEL TO CONTROL
FLEXIBLE CONDUITS

HYDRAULIC
PUMP TANK

ROBOT TO PUMP TANK
HIGH-PRESSURE HOSE

POWER PANEL

COLD WATER SUPPLY
AND RETURN TO CUSTOMER

FIGURE 28-21 Typical work cell layout. © Cengage Learning 2012

power supply, positioners, conveyors, and other equipment should all be computer-controlled.

The robot itself must be evaluated for its ability to work in a welding environment. Other factors of equipment selection to be considered include speeds, work zone configuration, accuracy, programmability, weave capability, and interaction capability with other equipment. The welding environment, unlike other manufacturing areas, is hot. The parts and machines in the welding environment are frequently exposed to spatter and subject to electronic noise (RF) from high-frequency starts. Weld spatter can stick to unprotected machine surfaces, causing the arm to jam or resulting in excessive wear. Electronic noise can damage the program, interrupt input or output signals, and cause erratic operations.

The speeds at which the robot arm moves must be compatible with the welding process. The rate of movement must be constant when making a specific size weld, but travel speed will have to be varied to permit different weld sizes and positions. A higher speed during arm articulation to a new position can increase productivity.

A welding tip service station can be included to prevent spatter buildup from clogging the welding torch. The service station can include a reamer to remove spatter and an automatic spray of anti-spatter.

The size and configuration of the work zone depends upon the number of axes and reach of the robot, **Figure 28-22A** through **D.** The articulation of the torch may limit the overall reach of the robot, **Figure 28-23A** and **B.**

The use of the robot together with a rotating table or positioner can increase its effective work zone (refer to Figure 28-13).

The accuracy with which a work cycle can be repeated and the joint tolerances that can be accepted must be considered when selecting equipment. To make acceptable welds, it is important that the arm follow the same path within a few thousandths of an inch (millimeters) each time. Some units have **sensors** that track the joint even if it varies from the programmed path. This feature permits the welding of parts that otherwise might have been rejected.

A variety of programming options is available from robot manufacturers. The greater the programming flexibility available, the more versatile is the unit. A typical welding program cycle can include the components shown in **Figure 28-24.**

A signal from the positioner indicates that the parts are in place. The arm moves to the weld starting point and waits as preflow gas starts to flow. Both the current and wire feed are started, and a molten weld pool is established before the arm begins to move along the weld joint. At the end of the joint, the arm stops, and the current and wire feed continue to fill the weld crater. The wire feed stops, followed in one second by the current to burn back the wire. Gas flow continues for postflow. During the welding cycle, the program may perform checks to ensure that the gas flow, voltage, amperage, and other external factors are operating correctly.

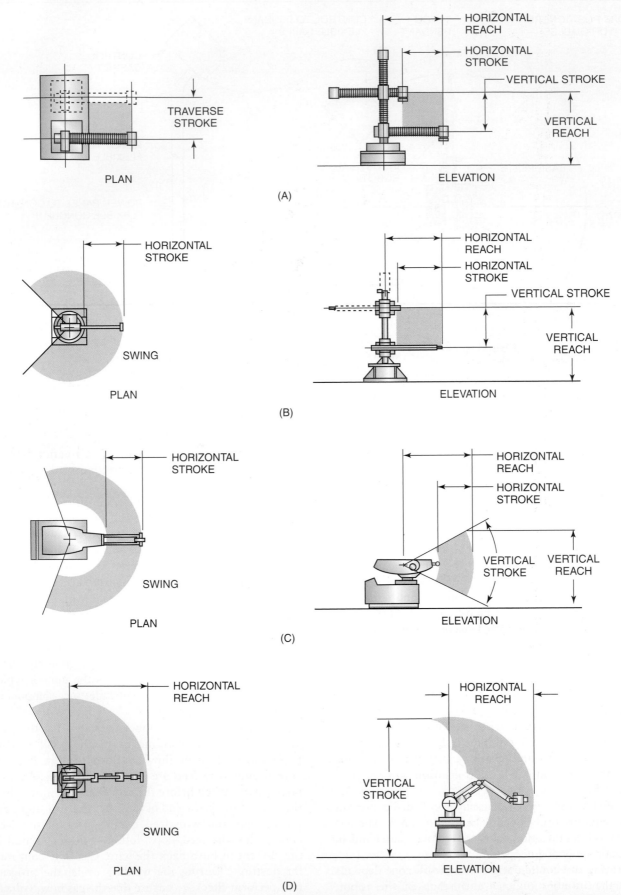

FIGURE 28-22 The work envelope for various types of robots is different: (A) rectangular coordinate robot, (B) cylindrical coordinate robot, (C) spherical coordinate robot, and (D) jointed arm robot. © Cengage Learning 2012

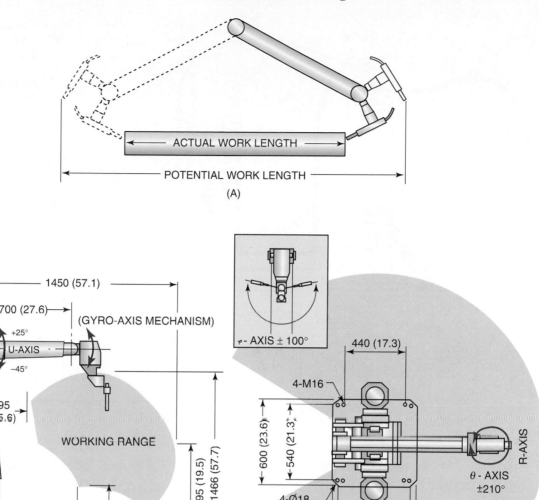

FIGURE 28-23 The actual work length and potential work length (A) often are different but must fit into the working range (B). © Cengage Learning 2012

The program may also allow the torch to be manipulated in several different weave patterns, **Figure 28-25.** This feature means that large single pass welds or out-of-position welds can be performed successfully.

To ensure that the system achieves maximum productivity, it is important that all components be able to be remotely controlled. Preferably, it should be adjustable to various power settings. The wire feeder should have various speed settings in addition to remote starting and stopping. Sensors should provide the robot with an "arc start" signal in addition to continuous monitoring of voltage and amperage readings. Product moving and positioning equipment should have sensors to indicate that the part is "in position" and ready for welding. Safety sensors should immediately stop all

movement of equipment if unauthorized persons are in the work zone.

Safety The following precautions are recommended for the use of automatic welding equipment and robots:

- All personnel should be instructed in the safe operation of the robot.
- All personnel should be instructed in the location of an emergency power shutoff.
- The work area should be restricted to authorized persons only.
- The work area should have fences, gates, or other restrictions to prevent access by unauthorized personnel.

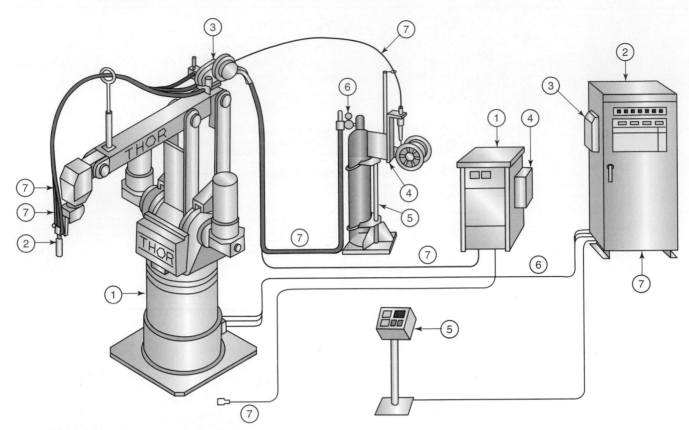

◆ COMPUTERIZED ARC WELDING ROBOT "THOR-K"

SYMBOL	WELDING ROBOT
①	ARC WELDING ROBOT
②	ROBOT CONTROL UNIT
③	TEACHING BOX
④	WELDING SIGNAL UNIT
⑤	CONTROL BOX
⑥	CONTROL CABLES
⑦	EXTERNAL MEMORY UNIT (LOCATED INSIDE OF THE ROBOT CONTROL UNIT)

SYMBOL	WELDING EQUIPMENT
①	WELDING POWER SUPPLY
②	WELDING TORCH
③	WIRE FEEDER
④	WIRE REEL UNIT
⑤	STAND FOR GAS CYLINDER
⑥	CO_2 GAS FLOW REGULATOR
⑦	CABLE AND HOSE ASSEMBLY

COMBINATION WITH OTC CO_2 GAS SHIELDED ARC WELDING MACHINE

FIGURE 28-24 All components must interact. © Cengage Learning 2012

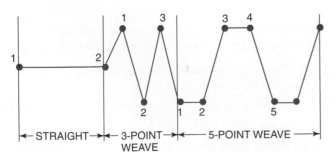

FIGURE 28-25 Weave patterns. © Cengage Learning 2012

■ Sensors should be mounted around the floor and work area to stop all movement when unauthorized personnel are detected in the work area during the operation.

■ The arc welding light should be screened from other work areas.

■ A breakaway toolholder should be used in case of accidental collision with the part, **Figure 28-26.**

■ A signal should sound or flash before the robot starts moving.

FUTURE AUTOMATION

The industrial robot today is in its infancy. Advancements we expect in the future are limited only by our imagination. Robots are getting smarter. They can "see" using fiber optics and are beginning to process what they see to make program adjustments. They can also use optical or magnetic sensors to locate parts and touch sensors to pick them up without damaging them. Advancements in welding processes will allow greater weld control. Improved and faster computers will allow more complex programming, and improved sensors will give better feedback. The outcome of these advancements will be increased production at lower costs with higher quality.

FIGURE 28-26 When using automatic welding equipment and robots, a breakaway collision sensor should be used.
Applied Robotics, Inc.

Summary

The automation of welding does not necessarily have to be in large shops. You may find that in some production applications it would be beneficial to use turntables or positioners. These can either fully or partially automate the welding process so that welder fatigue will be reduced. As you reduce welder fatigue with simple automation equipment, you can increase productivity significantly. It is important, therefore, not to look at automation as a major undertaking in every case. Many welding applications can be improved to increase their speed and reduce weld defects by automating the process.

As the number of companies using automated and robotic equipment has increased, the cost of equipment has significantly decreased and the versatility of the equipment has improved. Some equipment companies provide you with test runs of your product to help you determine the appropriateness of their equipment for your application. Such sample welds will help you determine which equipment is the most cost effective and appropriate for your application.

Improving Productivity with Robotic Welding

Choosing the right robotic welding work cell for an application enhances productivity and boosts the return on investment. These days, manufacturers often rely on outside supplier partners to play a variety of roles when it is time to upgrade production systems.

If you are considering an automation project, selecting the right systems integrator and involving that company early in the planning stages can be crucial to success. Successful automation can improve part quality, increase throughput, and reduce costs. Choosing a part or process that is a good automation candidate also increases the likelihood of success, but not all parts or processes can—or should—be automated. A capable integrator recognizes what to automate and what not to automate and makes sure that the simple parts meet necessary tolerances for automation.

As you begin planning, you should answer the following questions:

- What parts are to be run, now and in the future?
- What is the simple part repeatability?
- What type of positioning is required?
- What is the flow through of the manufacturing facility?
- What is the part mix being run across the machine?
- Are there any advanced process requirements?

Production efficiency and a speedy return on investment are some of the benefits you can receive from planning early and planning well.

Operator Nirmal Prasad unloads a Herman Miller frame from a robotic work cell at the company's facility. The parts Prasad loads and unloads are on the work cell that runs the highest volume of parts and has the highest productivity in the plant. A successful automation project—the result of careful planning—can improve part quality, increase throughput and reduce costs. American Welding Society

The Integrator's Resume

Using a robotics integrator fits with the recent trend toward outsourcing; you get the best an integrator has to offer without adding to your payroll.

An experienced robotics integrator brings unique experience and knowledge to a project. In evaluating which systems integrator is the best fit for specific manufacturing challenges, there are many factors critical to making the final choice. Consider the expertise of the integrator organization and its process development capabilities, safety equipment, and training and service support.

You need to feel confident that the robotics integrator can provide the best resources to handle the job. The most successful robotic welding system integrators offer their customers the following personnel resources:

- Sales engineers to serve as the first point of contact. They are familiar with the latest and most innovative solutions available.
- Project managers who have backgrounds in manufacturing and how the robotic system plays a part in those processes. These engineers are responsible for providing the concept for the robotic solution to the manufacturer. They offer direction in choosing the equipment mix that best matches production goals. Once the concept is approved, they manage the project from the beginning, through implementation, and beyond.
- Application engineers who understand the welding process and how the robotic work cell's components interact with one another. They are responsible for tying the hardware components together so they work as one. This entails robot programming and process development. Once the work cell

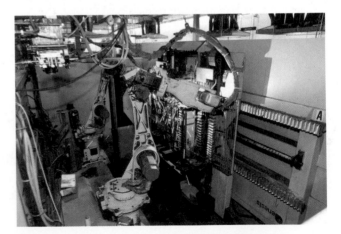

Flexible, modifiable robotic work cells can lead to increased productivity. This Versa 4 work cell from Genesis features a Ferris wheel style positioner, fast indexing time and servo tooling that helped customer Herman Miller weld 6 of its 50 standard frames. These 6 frame styles, however, comprise 50% of the company's product. American Welding Society

is shipped to the manufacturer's site, the application engineer can provide installation and startup assistance.

- Once the system is up and running in the plant, competent technical service engineers help customers deal with any hardware and software problems that arise. The integrator should be able to respond quickly with qualified service technicians to any customer requests.

Planning Ahead

Not every welding process is a good candidate for automation. When the robotic systems integrator has been part of the planning from the earliest stages, the likelihood of success increases.

As a customer, you need to provide the primary goals you want to accomplish with automation (reducing costs, increasing throughput, improving quality and consistency, and so on). The integrator's goal is to gain as much information as possible about quantities needed, number of shifts worked, inches of weld per part, expected return on investment, and the quality and repeatability of parts entering the weld cell. This fact gathering ensures that the integrator fully understands the customer's goals, manufacturing process, and priorities. Gathering the customer's input is a lengthy process that begins to pay off when the integrator produces a work cell concept.

Equipment

Another asset a systems integrator can offer is access to multiple equipment options and configurations that ultimately lead to better productivity. Some integrators offer pre-engineered or standard work cells. These systems provide affordability but are not always the right solution when customization is required. Integrators with changeable work cells can provide a cost-effective solution for each application.

Following are some equipment suggestions for productive robotic welding:

- **Positioners.** The positioner determines how the part is presented to the robot for welding. Typical positioner types include stationary tables, turntables, headstocks, tailstocks, and Ferris wheel types.

- **Fixtures.** These hold the part to proper tolerances for welding. Unique to each automation project, fixtures are only as good as the quality and innovation the systems integrator puts into them. When selecting a robot, the integrator considers accuracy, speed, reach, operator interface, and the various options available with each robot.

- **Base platform.** A common base platform, which the systems integrator designs and constructs, connects all components. It is also a base for controls, flash screens, barriers, and safety components.

- **Operator interface.** This is the final component of the work cell. Simple push buttons or a touch-screen panel are often recommended interfaces.

Safety

Safety should lead the priority list of any systems integrators you interview. Robotic welding systems integrators most concerned about reducing the risk of injuries have adopted the Robotic Industries Association's (RIA) Code for Robotic Safety, a guideline widely accepted as the industry standard.

You should expect systems integrators to conduct a risk analysis and then design and adjust systems for the safest operation that meets or exceeds all plant, federal, state, and local codes. These safety measures include fence barriers, gate interlocks, floor safety switch mats, photo cell and light curtains, fixed guards, and emergency stop hardware and software, as well as frames and screens to protect workers from welding arc flash.

Training

One of the best methods for ensuring worker safety is adequate training conducted by the systems integrator. A hands-on mentoring approach works well with employees in the welding industry who are comfortable working with their hands. While some formal classroom training is needed, the majority of the teaching should be done on the equipment when it is set up at the systems integrator's facility.

Service and Support

Service and support can make or break the relationship between you and your systems integrator. The high capital costs of robotic welding systems mean you have the right to demand efficient, fast service when something goes wrong.

Sound Investment

A competent robotic systems integrator is one of the best investments you can make when it is time to improve the production process and productivity. Checking references and visiting other plants where the systems integrator has handled an automation changeover can help you choose the best systems integrator for fulfilling your requirements.

Article courtesy of the American Welding Society.

Review

1. What were the first industrial robots in America used for?

2. How can CAM technology aid in manufacturing?

3. Why do most robot manufacturers recommend that a skilled welder operate the robot?

4. What must a welder control for a process to be considered manual?

5. List the commonly used manual processes.

6. What must a welder control for a process to be considered semiautomatic?

7. List the commonly used semiautomatic processes.

8. What makes a process a machine process?

9. What makes a process an automatic process?

10. What makes a process an automated process?

11. What types of power provide the industrial robot with movement?

12. In what axes, or directions, can a robot move?

13. What is meant by "teaching" a robot?

14. What is the advantage of storing a robot's program on a disc or CD?

15. Why is it important to reduce restricted areas in a design that will be welded using a robot?

16. Why is it important to make small welds more accurately than large ones?

17. Why must a robot be evaluated for its fitness to work in a welding area?

18. Why is joint tracking important for a robot?

19. Why is it important for a robot to weave the torch?

20. List the safety considerations that must be followed when operating a robotic workstation.

Chapter 29

Other Welding Processes

OBJECTIVES

After completing this chapter, the student should be able to

■ explain the operating principles for the different special welding processes.

■ list the reasons that a particular process should be selected to make a special weld.

■ list the operational limitations of each special welding process explained in this chapter.

■ explain the thermite welding process for rails.

■ describe the characteristics of austenitic manganese steel.

■ list the steps required to repair cracks in rails and rail components.

■ explain the reason for keeping thermite welding materials dry.

KEY TERMS

electrical resistance

electron beam welding (EBW)

evacuated (vacuum) chamber

flash welding (FW)

hardfacing

inertia welding

laser beam welding (LBW)

optical viewing system

percussion welding (PEW)

plasma arc welding (PAW)

resistance welding (RW)

stud welding (SW)

thermal spraying (THSP)

thermite welding (TW)

ultrasonic welding (USW)

upset welding (UW)

INTRODUCTION

More than 80 different welding and allied processes are listed by the American Welding Society (AWS). This text covers 9 of the most commonly used processes that require the welder to have a special skill. This chapter covers 17 additional processes that call for special equipment and techniques. Some of these processes require less skill or knowledge to set up and operate, such as resistance spot welding (RSW). Others demand a great deal of technical information and training, such as electron beam welding (EBW).

The actual operating procedures vary greatly from one manufacturer's machine to another. The specific settings also change from one material to another. Because of these factors, only the general theory, procedures, and applications are discussed in this chapter. More information can be obtained from the AWS or directly from the manufacturer of the equipment being operated. The skill needed to operate this equipment can be learned on the job or in classes taught by the specific equipment manufacturer.

RESISTANCE WELDING (RW)

Resistance welding is a process where fusion between the parts' surfaces occurs as a result of the heat produced by the electric resistance between the parts' contacting surfaces. The welding current required to make a resistance weld must be at a very low voltage but high amperage. Pressure is always applied to ensure a continuous electrical circuit and to forge the heated parts together. Heat is developed in the assembly to be welded, and pressure is applied by the welding machine through the electrodes. During the welding cycle, the mating surfaces of the parts are heated to a plastic state just before melting, and are forced together. The surfaces do not have to melt for a weld to occur. The parts are usually joined as a result of heat and pressure and not their being melted together. Fluxes or filler metals are not needed for this welding process. The heat produced in the weld may be expressed in the following formula:

$$H = I^2R \quad \text{where} \quad \begin{aligned} H &= \text{Heat} \\ I^2 &= \text{Welding current squared} \\ R &= \text{Resistance} \end{aligned}$$

The current for resistance welding is usually supplied by either a transformer or a transformer/capacitor arrangement. The transformer, in both power supplies, is used to convert the high line voltage (low-amperage) power to the welding high-amperage current at a low voltage. A capacitor, when used, stores the welding current until it is used. This storage capacity allows such machines to use a smaller-size transformer. The required pressure, or electrode force, is applied to the workpiece by pneumatic, hydraulic, or mechanical means. The pressure applied may be as little as a few ounces (grams) for very small welders to tens of thousands of pounds (kilos) for large spot welders.

Most resistance welding machines consist of the following three components:

- The mechanical system to hold the workpiece and to apply the electrode force
- The electrical circuit made up of a transformer and, if needed, a capacitor, a current regulator, and a secondary circuit to conduct the welding current to the workpiece
- The control system to regulate the time of the welding cycle

There are several basic resistance welding processes. These processes include spot (RSW), seam (RSEW), high-frequency seam (RSEW-HF), projection (PW), flash (FW), upset (UW), and percussion (PEW) welding.

Resistance welding is one of the most useful and practical methods of joining metal. This process is ideally suited to high production methods.

Resistance Spot Welding (RSW)

Spot welding is the most common of the various resistance welding processes. In this process, the weld is produced by the heat obtained at the interface between the workpieces. This heat is due to the resistance to the flow of electric current through the workpieces, which are held together by pressure from the electrode, **Figure 29-1**. The size and shape of the formed welds are controlled somewhat by the size and contour of the electrodes.

The welding time is controlled by a timer built into the machine or by a computer program. The timer controls four different steps, **Figure 29-2**. The steps are as follows:

- Squeeze time, or the time between the first application of electrode force and the first application of welding current
- Weld time, or the actual time the current flows
- Hold time, or the period during which the electrode force is applied and the welding current is shut off
- Off period, or the time during which the electrodes are not contacting the workpieces

Tables supplied by the machine manufacturer provide information for the exact time for each stage for different types and thicknesses of metal.

Material from 0.001 in. (0.025 mm) to 1 in. (25 mm) thick may be joined by spot welding.

Spot Welding Machines The three types of spot welding machines commonly used are rocker arm, press type, and portable type.

In rocker-arm spot welders, the lower electrode is stationary while movement is transferred to the upper electrode by means of an arm, which moves about a pivot

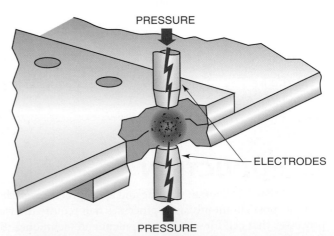

FIGURE 29-1 Heat resulting from resistance of the current through the metal held under pressure by the electrodes creates fusion of the two workpieces during spot welding. © Cengage Learning 2012

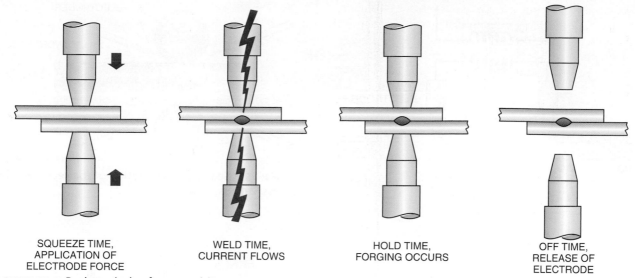

SQUEEZE TIME,
APPLICATION OF
ELECTRODE FORCE

WELD TIME,
CURRENT FLOWS

HOLD TIME,
FORGING OCCURS

OFF TIME,
RELEASE OF
ELECTRODE

FIGURE 29-2 Basic periods of spot welding. © Cengage Learning 2012

point. The rocker arm can be moved by one of three methods: foot pedal, air or hydraulic cylinder, or electric motor. Rocker-arm welders, **Figure 29-3**, are available with throat depths from 12 in. to 48 in. (30 cm to 122 cm) and transformer capacities from 10 kVA to 20 kVA. kVA stands for kilovolt amps and is roughly equivalent to watts. Both kVA and watts are power measurements.

In press-type spot welders the movement of the upper electrode is controlled by a pneumatic or hydraulic cylinder. This type of welder is used for welding heavy sections. Throats having a depth of 60 in. (152 cm) are available with capacities up to 500 kVA.

Portable spot welders are used where work is too large to be moved. These machines are complete with a pressure cylinder, transformers, electrode holders, electrodes, and the necessary controls.

Portable spot welders are used in the mass production of automobiles, aircraft, railroad cars, home appliances, and similar products where relatively thin sheets are to be welded, **Figure 29-4.**

Multiple-Spot Welders Multiple-spot welders are used when a high production rate is a requirement. The multiple-spot welder shown in **Figure 29-5** is especially

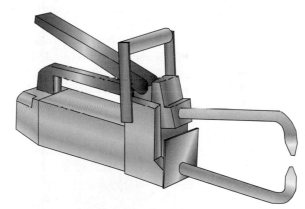

FIGURE 29-4 Small welder for fine, detailed work.
© Cengage Learning 2012

WELDING TRANSFORMER INSIDE

MOVABLE ARM

ELECTRODES

POWER
SWITCH

WELDING
CURRENT
ADJUSTMENT

WELDING TIME
ADJUSTMENT

FIXED ARM

WELDER BASE

FOOT LEVER

FIGURE 29-3 Rocker-arm spot welder operates by depressing the foot switch, which brings the tongs onto the work, and the weld cycle steps through a preset time adjustment. © Cengage Learning 2012

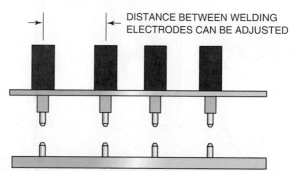

DISTANCE BETWEEN WELDING
ELECTRODES CAN BE ADJUSTED

FIGURE 29-5 Multiple-spot welder, which is a special welder designed as an auto instrument panel welder.
© Cengage Learning 2012

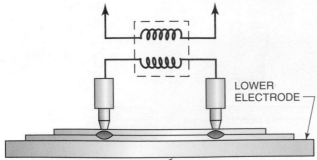

FIGURE 29-6 The arrangement of electrodes in multiple-spot welding. © Cengage Learning 2012

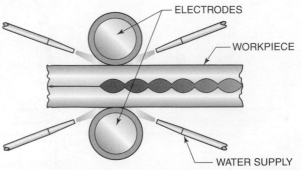

FIGURE 29-8 Schematic illustration of the seam welding process. © Cengage Learning 2012

designed for auto body panels. Some welds can make as many as 20 or more welds at one time. Most multiple-spot welders have a series of air-operated or hydraulically operated guns mounted on a header and use a common bar for the lower electrode. The welding guns are connected by flexible wires to individual transformers or to a common bus bar attached to the transformer, **Figure 29-6.**

All guns make contact with the workpiece at the same time. They can be fired either in a certain sequence or all at the same time. These types of welders are widely used in the automobile industry.

Seam Welding (RSEW)

Seam welding is similar in some ways to spot welding except that the spots are spaced so closely together that they actually overlap one another to make a continuous seam weld.

Seam welding is accomplished by using roller-type electrodes in the form of wheels that are 6 in. (152 mm) to 9 in. (229 mm) or more in diameter, **Figure 29-7.** These roller-type electrodes are usually copper alloy discs 3/8 in. (10 mm) to 5/8 in. (16 mm) thick. Cooling is achieved by

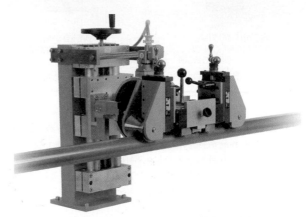

FIGURE 29-9 Tweezer-type seam welder with automatic cycling for circular flat pack. Foerster Instruments, Inc.

a constant stream of water directed to the electrode near the weld, **Figure 29-8.**

Welding is done either with the roller electrodes in motion or while they are stopped for an instant. If continuous motion is used, the rate of welding usually varies between 1 ft and 5 ft (30 cm to 152 cm) per minute. The greatest welding speed is obtained on the thinnest materials. An indexing mechanism can be used when the wheels are to be stopped for each weld.

Electric timing equipment is useful when it is necessary to provide the precise control required for highest-quality welds. Spot welds can be positioned at almost any interval desired by simply adjusting the timing and rate of electrode motion.

It is possible to change the welding current and electrode pressure to control the surface condition and width of the weld. The seam width should be about two times the thickness of the sheet plus 1/16 in. (2 mm). **Figure 29-9** shows a tweezer-type seam welder with automatic cycling.

Types of Resistance Seam Welds **Figure 29-10** illustrates the type of seams used in most seam welding processes. The lap seam is the most common of these seams. The tops and bottoms of containers are usually fastened together with a flanged seam. For a metal thickness of 1/16 in. (2 mm), the mash seam is often used. The electrode face should be wide enough to cover the overlap by approximately one and a half times the thickness of the sheet.

FIGURE 29-7 Seam welder. Foerster Instruments, Inc.

FIGURE 29-10 Types of seam welds. © Cengage Learning 2012

High-Frequency Resistance Seam Welding (RSEW-HF)

This process is similar to RSEW in some ways. The major differences are that the welding current is supplied as high-frequency, 200 to 500 kHz as opposed to DC or 60-cycle AC. The high-frequency power provides very localized heating. This heating is a result of the resistance to the current induced in the metal and not to the resistance between parts. As a consequence of this localized heating, pipe and tubing can be welded with little loss of power flowing around the back side of the joint.

RSEW-HF is used in the production of welded pipe, tubing, and structural shapes. It works very well for the fabrication of I-beams, H-beams, channels, and so on. These welded structural shapes are lighter and stronger than their roll-formed counterparts.

Resistance Projection Welding (RPW)

Projection welding is somewhat similar to spot welding in that it involves joining parts by a resistance welding process.

To make a projection weld, projections are formed on at least one of the workpieces at the points where welds are desired. The projections—small, raised areas—can be any shape, such as round, oval, circular, oblong, or diamond. They can be formed by embossing, casting, stamping, or machining.

The workpieces that have the projections and the other workpiece are placed between plain, large-area electrodes in the welding machine, **Figure 29-11**. The current is turned on, and pressure is applied. Since nearly all of the resistance is in the projections, most of the heating occurs at the points where welds are desired. The heat causes the projections and the opposing metal to soften. The pressure from the electrodes causes the softened projections to flatten, resulting in fusion of the workpieces.

With this type of welding, a complicated electrode shape is not required. In cases where the weld area does not have a regular shape, the electrode can be shaped to fit the surface.

There are many variables in this type of welding. Some of these variables are thickness of metal, number of projections, and kind of material. These variables make it hard to predetermine the current and electrode pressure required.

Steel plate, galvanized sheet steel, and stainless steels can be joined using projection welding.

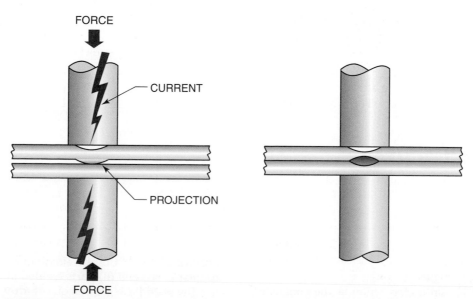

FIGURE 29-11 Projection welding. © Cengage Learning 2012

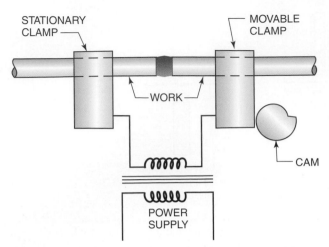

FIGURE 29-12 Schematic diagram of the flash welding process. © Cengage Learning 2012

Flash Welding (FW)

Flash welding may be considered a resistance welding process. Fusion is produced over the ends of stock by heat produced from the resistance to the flow of electric current between the two surfaces. Pressure is applied after heating is completed. As a result, the material is forced together, and fusion takes place. The basic steps in flash welding are as follows:

1. Clamp the parts together in dies (these dies conduct the electric current to the workpieces).

2. Move one part toward the other part until an arc is established.

3. When the ends of the parts reach a plastic temperature, pressure is applied, causing the softened end surfaces to be fused together and deformed. This deformation is referred to as upset.

4. Cut off the welding current when fusion and upset are complete, **Figure 29-12.**

Flash welding may be used to join dissimilar aluminum alloys and to join aluminum to other metals. Flash welding is a high-volume production process. Unless a large number of pieces are to be welded, it is very costly to prepare the work-holding dies. The time and materials involved in setting up the job are also cost factors. The production rate is high since the welding time is short.

Upset Welding (UW)

Welding small areas is usually done by the **upset welding (UW)** method. In this form of welding, two pieces of metal having the same cross section are gripped and pressed together. Heat is generated in the contact surfaces by **electrical resistance.** The upset at the interface extrudes the contaminated contact area to the outside, where it is trimmed, **Figure 29-13.**

Flash welding has an arcing action at the contacting surfaces. No flashing occurs in the upset process. The main

(A)

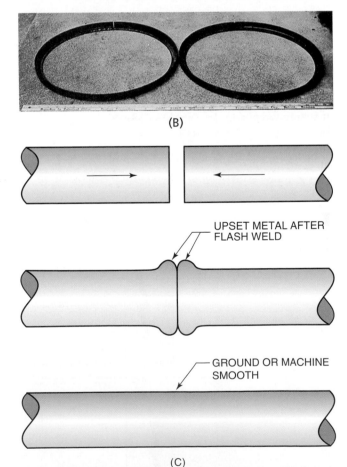

(B)

UPSET METAL AFTER FLASH WELD

GROUND OR MACHINE SMOOTH

(C)

FIGURE 29-13 (A) Upset welding machine for welding circular parts, and (B) parts welded in this machine. (C) The weld flash is trimmed off after an upset weld. **(A) and (B)** National Machine Company

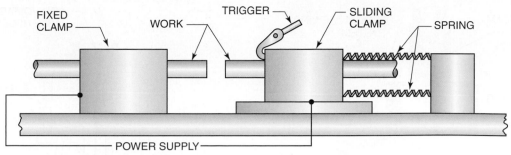

FIGURE 29-14 Principle of percussion welding. © Cengage Learning 2012

difference between the upset and flash processes is that less current is needed in the upset process and welding time is extended.

Percussion Welding (PEW)

Percussion welding is a resistance welding process and is actually a variation of flash welding. Fusion is obtained simultaneously over the entire area of the contacting surfaces. The heat that causes fusion is obtained from an electric arc produced by the rapid discharge of stored energy. Immediately after or during the electrical discharge, impact pressure is applied by a hammer blow or by the snap releasing of a spring, **Figure 29-14**. A percussion weld is similar to a flash weld. A short application of high-intensity energy instantly heats the surfaces to be joined. This rapid heating is almost immediately followed by a quick blow to make the weld.

The action of this process is very rapid. There is little heating effect upon the metal adjacent to the weld. Thus, it is possible for heat-treated parts to be welded without becoming annealed. Examples of this type of welding include welding copper to aluminum or stainless steel, and adding Stellite tips to tools, silver contact tips to copper, cast iron to steel, and zinc to steel.

ELECTRON BEAM WELDING (EBW)

Electron beam welding is used in the auto, electronics, pipeline, shipbuilding, aerospace, and high-speed welded tubing industries. Temperatures up to 180,000°F (100,000°C) are generated by electron beam welding machines. At such high temperatures, the machine can accurately vaporize metals and ceramics. Deep, strong welds with no heat deformation are produced between close-fitting parts. The welds have high strength and purity. This process can be used to join dissimilar metals.

Electron beam welding utilizes the energy from a fast-moving beam of electrons focused on the base material. The electrons strike the metal surface, giving up their kinetic energy almost completely in the form of heat. Welds are made in a vacuum (10^{-3} mm Hg to 10^{-5} mm Hg), which practically eliminates contamination of the weld material by the gases left in the vacuum chamber. The high vacuum is necessary to produce and focus a stable,

uniform electron beam. Welds produced by this process are coalesced from vacuum-melted material, which eliminates the usual fusion weld contaminants caused by water vapor, oxygen, nitrogen, hydrogen, and slag.

Early electron beam welders needed hard vacuums for both the beam and welding chambers. New units need only a soft vacuum in the welding chamber (attained in only a few seconds), or, in the case of some new units, welds can be made 1/4 in. (6 mm) outside the **evacuated (vacuum) chamber.** Nonvacuum welding is thought of as a supplement to, not as a replacement for, conventional EB welding.

Electron Beam Welding Gun

The electron gun consists of a filament, a cathode, an anode, and a focusing coil. These parts are mounted above the work chamber, as shown in **Figure 29-15**. The electrons from the heated filament carry a negative charge and are emitted by the cathode and attracted by the anode. The anode has an opening through which the electrons pass.

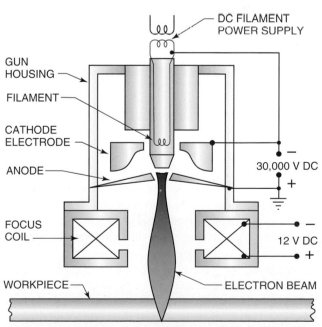

FIGURE 29-15 Electron beam gun produces an accurately controlled heat source adjustable for length and point of focus. © Cengage Learning 2012

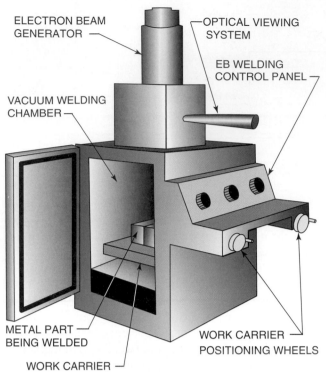

FIGURE 29-16 7.5-kW electron beam welder.
© Cengage Learning 2012

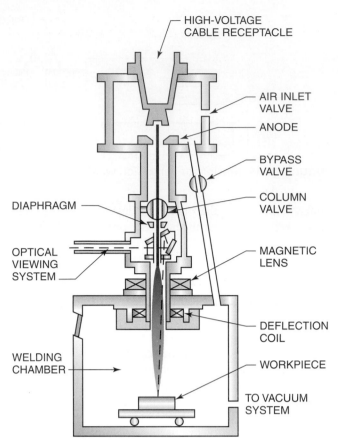

FIGURE 29-17 Schematic illustration of electron beam gun column. © Cengage Learning 2012

The electrons then move through a magnetic field, which is produced by an electromagnetic focusing coil. The machine is equipped with an **optical viewing system**, **Figure 29-16**. This system provides a line of sight down the path of the electron beam centerline to the weld area when the beam is off, **Figure 29-17**.

It is possible to vary the current to the focusing coil so that the operator can focus the beam from a sharp focus to a beam 1/4 in. (6 mm) in diameter. Focal points above, on, or below the work surface are provided to modify the weld pattern.

Electron Beam Seam Tracking

Electron beam equipment is designed and manufactured so that the worktable can be rotated beneath the gun for circular welds or driven along a path that corresponds to, or is parallel to, the centerline of the chamber for linear welds. Sometimes the gun itself is moved.

Automatic seam tracking is available to track any seam. The amount of misalignment of the placement of the joint may be a few thousandths of an inch or a few inches. The seam tracking device continuously checks the actual position of the seam during the welding operation and precisely corrects for the degree of misalignment.

Assuming the piece is driven along the x-axis, **Figure 29-18**, electromagnetic deflection will move the electron beam transversely to the welding motion. As the misaligned seam moves in a linear direction along the axis, the seam tracker interprets the amount of deviation from the theoretical centerline and corrects the deflection of the beam to match the exact position of the seam at any instant.

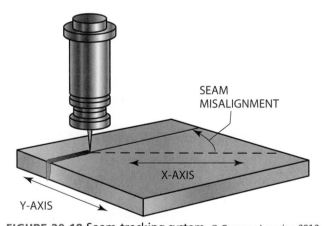

FIGURE 29-18 Seam-tracking system. © Cengage Learning 2012

Basically, the work is moving on the welding axis, and the beam is scanning on the y-axis. However, all motions are interchangeable as selected so that the welding motion may be on the gun, and correction may still be accomplished by moving the beam.

ULTRASONIC WELDING (USW)

Ultrasonic welding is a process for joining similar and dissimilar metals by introducing high-frequency vibrations into the overlapping metals in the area to be joined.

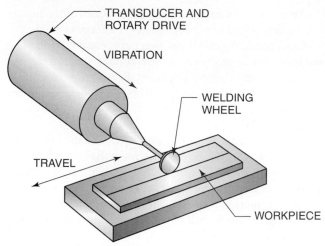

FIGURE 29-19 Ultrasonic seam welding.
© Cengage Learning 2012

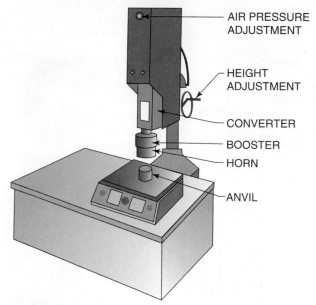

FIGURE 29-20 Ultrasonic spot welder. Larry Jeffus

Fluxes and filler metals are not required, electrical current does not pass through the weld metal, and only localized heating is generated. The temperature produced is below the melting point of the materials being joined. Thus, no melting occurs during the welding cycle.

The most common welds made with this process include spot, overlapping spot, continuous seam, and ring welds. The pieces to be welded are clamped between the welding tip and a hard surface, which serves as an anvil. High-frequency energy is fed to a transducer, which converts it to vibrations. The welding tip is attached to the transducer. The coupling system and welding tip form a unit that is referred to as a sonotrode, **Figure 29-19.** The welding tip oscillates in a plane essentially parallel to the joint interface. Transverse (shear) waves in the material produce the weld. Vertical (compression) waves are not effective in ultrasonic welding of metals.

The ultrasonic vibrations combined with the static clamping force cause dynamic shear stress in workpieces. When these shear stresses are great enough, local plastic deformation of the metal occurs at the interface. Surface films of oxide coatings are shattered and dispersed, and pure base metal contact is achieved. A true metallurgical bond is then formed in the solid state without melting the parent metal.

Continuous seam welding is done with a rotating, disk-shaped sonotrode tip. Either a counter-rotating roller anvil or table-type anvil is also used. The entire tip assembly rotates so that the peripheral speed of the sonotrode tip matches the traversing speed of the workpiece.

The equipment used in this type of welding consists of the following: (1) a power supply or frequency converter that converts 60-hertz (60 cycles per second) line power into high-frequency electrical power and (2) a transducer that changes high-frequency electrical power to rapid vibrations. Welding speeds can range from a few feet per minute to 400 ft per minute (0.3 m per minute to 131 m per minute) for continuous seam welding. Times range from 0.005 sec to 1.0 sec for spot welding.

Ultrasonic Welding Applications

Ultrasonic welding has many applications in the assembly of electrical products. Typical applications include the following:

- Attaching oxide-resistant contact surface buttons to switches
- Attaching leads to coils of foil, sheet, or wire made of aluminum
- Attaching very fine wire leads and elements to other components

Figure 29-20 shows an ultrasonic spot welder used to perform the types of welds just listed. In the plastics field (packaging), ultrasonic welding is used for both spot and continuous seam fabrication and for closures on various types of foil or plastic envelopes and pouches.

Various positive drive tip configurations can be designed for specific joint requirements. Applications of the process to aluminum welding include attaching thin ribs to cylinders, welding two pairs of leads at once, and joining two gears and one spacer in a single weld. Attachment of aluminum parts or brackets to stainless steel walls is another job readily performed by ultrasonic welding.

Ultrasonic welding is used for materials having thicknesses up to about 1/8 in. (3 mm).

INERTIA WELDING PROCESS

Inertia welding is a form of friction welding. In **inertia welding,** one workpiece is fixed in a stationary holding device, **Figure 29-21.** The other is clamped in a spindle chuck. When the spindle motor is energized, it spins rapidly. At a predetermined speed, power to the spindle

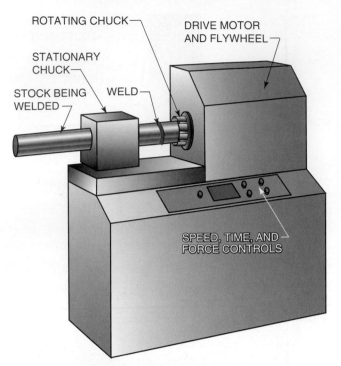

FIGURE 29-21 Inertia welder. © Cengage Learning 2012

motor is cut, as shown in **Figure 29-22A.** One part is then pressed against the other piece. Friction between the parts causes the spindle to decelerate and quickly stop, converting stored energy to frictional heat. Enough heat is formed to soften, but not melt, the faces of the part, **Figure 29-22B.**

An accurate heating rate can be obtained since inertia can be controlled to supply whatever energy is needed.

Until rotation ceases, the two parts cannot bond. The compression force upsets the metal interface, forcing out any impurities or voids. The heat-affected zone is narrow, **Figure 29-22C.** The flash may or may not be cut off. The weld is formed when the spindle stops, as shown in **Figure 29-22D.**

A graph of speed, torque, and upset (change in workpiece length) during the weld period will show what happens, **Figure 29-23.** The curves start when the two pieces come together, after flywheel acceleration.

At first, there is a small torque peak and a corresponding change in the length of the parts. This is when initial temperature buildup occurs.

Torque then drops back and is fairly constant. During this period, a state of near equilibrium exists when energy from the rotating mass is being converted to heat at the same approximate rate as it is being conducted away. Little upset occurs during this time.

Finally, the speed drops to a point where heat penetration does not keep up with heat dissipation and the faces cool slightly. The torque peaks sharply as the weld bonds are formed and broken. Most of the upset occurs just before the spindle stops. A solid state weld is created as the now stationary parts are pressed together.

The inertia welding process produces a superior-quality, complete interface weld. The welding conditions must be consistent so that human judgment is removed in production work. The technique has been applied successfully to super alloys as well as to standard metals.

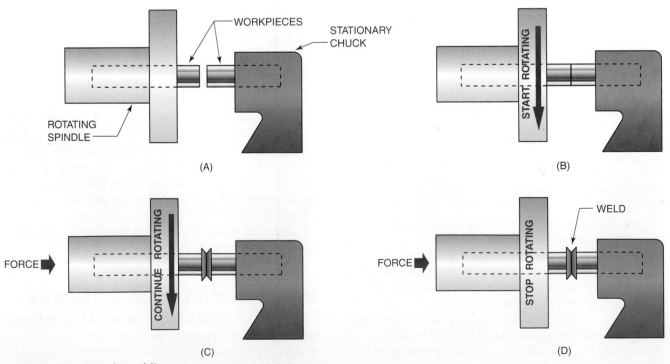

FIGURE 29-22 Inertia welding process. American Welding Society

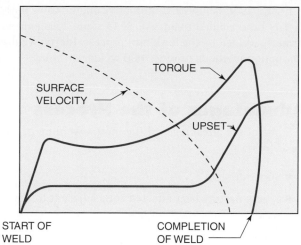

FIGURE 29-23 Chart shows what happens during inertia welding process. © Cengage Learning 2012

Inertia Weld Bond Characteristics

Microscopic examination of the bond resulting from the inertia welding process shows the following three metallurgical characteristics:

- The weld zone is very narrow and has a fine-grained structure with no melt product or grain growth. These conditions indicate a hot-worked structure. When dissimilar metals are welded, often there are streaks of intermixed material near the outside diameter.

- Hardening phases from the rapid chilling are seen throughout the structure. The degree of hardening can be controlled but is often about the same as that achieved with a mild water quench.

- There is a zone of varying grain structure between the heat-affected zone and the parent structure, **Figure 29-24**.

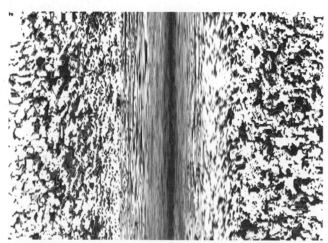

FIGURE 29-24 Mechanical mixing and grain refinement can be seen in a photomicrograph of the weld zone of a inertia weld. Larry Jeffus

Similar Metals	
Carbon steels	Molybdenum
Sintered steels	Waspalloy
Stainless steels	Cobalt alloys
Tool steels	Titanium
Alloy steels	Zircalloy
Aluminum alloys	Inconel®
Copper	Nickel alloys
Brasses, bronzes*	

Dissimilar Metals
High-speed steel to various steels
Sintered steels to wrought steels
316 stainless steel to Inconel®
Stainless steel to medium and low carbon steels
1100 or 6061 aluminum to medium carbon steel
Cobalt-base alloys to steel
347 stainless steel to 17-4 PH
Pure titanium to 302 stainless steel
Copper to 1100 or 6061 aluminum
Copper to medium carbon steel
Copper to various brasses
Aluminum bronze to medium carbon steel
Nickel-base alloys to steel

*Except bearing types.

TABLE 29-1 Partial List of Metals Joined by Inertia Welding. © Cengage Learning 2012

Table 29-1 lists some of the metals that can be joined by inertia welding.

Figure 29-25 shows a hydraulic piston rod assembly. The material is prechromed, cold-drawn tubing with the eye cut from tubing. In the past, this part was manufactured as a one-piece forging. Inertia welding of the parts produced an estimated savings of 50% per rod.

Some parts that have been joined by the inertia welding process are shown in **Figure 29-26**, **Figure 29-27**, and **Figure 29-28**.

A fan bracket assembly is shown in Figure 29-26 with an SAE 1045 shaft welded to an SAE 1020 bracket using the inertia welding process. The steel shaft is pressed into a shrink-fit forging and then welded on the back side. A considerable savings per assembly is realized.

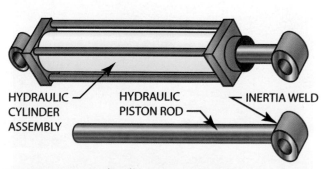

FIGURE 29-25 Hydraulic piston rod inertia welded assembly. © Cengage Learning 2012

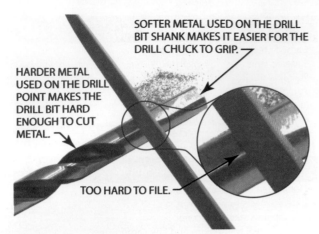

FIGURE 29-26 Fan bracket. Larry Jeffus

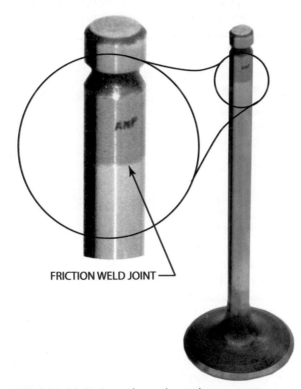

FIGURE 29-27 Automatic engine valves. Larry Jeffus

FIGURE 29-28 Gas filter, laser welded. Preco, Inc.

Figure 29-27 shows an automotive engine valve with a 21-4-N head welded to an SAE 8645 stem. This part was formerly welded by the flash butt process. Inertia welding production methods produce 600 welds per hour.

Advantages of the Process

Some of the advantages of the inertia welding process are as follows:

- Superior weld.
- A very narrow heat-affected zone adjacent to the weld.
- Uniform production welds.
- Fast production welds.
- Clean operation.
- Lowest cost of energy.
- Minimum skill required to operate the welder.
- The amount of upset of parts can be controlled to close tolerances.
- A complete interface weld can be obtained.

Applications of this process include welding dry bearing materials such as oxides and leaded bronzes. Metals that cannot be hot-forged can also be joined by inertia welding.

LASER BEAM WELDING (LBW)

In **laser beam welding**, Figure 29-28, fusion is obtained by directing a highly concentrated beam of coherent light to a very small spot. Laser beams combine low-heat input (0.1 joule to 10 joules) with high-power intensity of more than 10,000 watts per square centimeter (considerably more than the electron beam). Due to the fact that the heat is provided by a beam of light, there is no physical contact between the workpiece and the welding equipment. It is possible to make welds through transparent materials.

The ease with which the beam can be directed to any area of the work makes laser welding very flexible. In the manufacture of complex forms, for example, it is possible to move the focused laser beam under digital control to seam weld any desired shape. The proven high quality of laser welds, along with the flexibility and the comparatively moderate cost of laser welding equipment, has helped to increase their role in microelectronics and other light-gauge metal welding applications.

Laser welding of high-thermal conductivity materials, such as copper, is not difficult to do. The extremely concentrated laser heat will melt the metal locally to make a weld or will vaporize the metal to drill a keyhole. The keyhole allows the laser light to extend deeply into the base metal for deeper penetration.

Laser Welding Advantages and Disadvantages

Laser welding has some distinct advantages and disadvantages when compared to other welding processes. Electron beam welding is the only method that rivals the heat output of a laser. Generally, however, electron beam welding must operate in a vacuum. Since the laser beam is a light beam, it can operate in air or any transparent material, and the source of the beam need not even be close to the work. The material being welded need not be an electrical conductor that limits most other processes or even part of an electrical or a mechanical circuit. However, the light may be diffused by the welding vapors, so techniques to bypass the vapors have been developed.

Laser welds are small, sometimes being less than 0.001 in. (0.0254 mm). Laser welding is used to connect leads to elements in integrated circuitry for electronics. Lead wires insulated with polyurethane can be welded without removing the insulation.

Using the laser welding process, it is possible to weld heat-treated alloys without undoing the heat treatment.

This method of welding can be used to join dissimilar metals. Metals that are difficult to weld, such as tungsten, stainless steels, titanium alloys, Kovar, nickel alloy, aluminum, and tin-plated steels, can also be successfully welded by this process.

An optical system is used to focus the beam on the workpiece. The actual control of the welding energy is done by a switch.

Laser Beam

The laser is based on the principle that atoms in certain crystals and gases can be made to release a coherent, monochromatic (single-wavelength) energy when they are excited. The output is self-amplified in the laser because the excited atoms release their energy much more rapidly when stimulated by light emitted by neighboring atoms, **Figure 29-29**.

The early lasers all used a synthetic ruby rod as the material that produced the laser light. Today, a large number of materials can be used to produce the laser. These materials include such common items as glass and such exotic items as neodymium-doped, yttrium aluminum garnet (Nd:YAG), often referred to as a YAG laser.

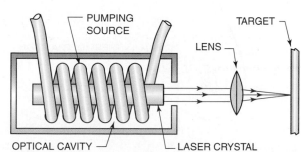

FIGURE 29-29 Schematic diagram of laser welder.
© Cengage Learning 2012

PLASMA ARC WELDING (PAW) PROCESS

The term *plasma* should be defined in its electrical sense. A gas, or plasma, is present in any electrical discharge if sufficient energy is present. The plasma consists of charged particles that transport the charge across the gap.

The two outstanding advantages of plasmas are higher temperature and better heat transfer to other objects. The higher the temperature differential between the heating fluid and the object to be heated, the faster the object can be heated.

In **plasma arc welding (PAW)**, a plasma jet is produced by forcing gas to flow along an arc restricted electromagnetically as it passes through a nozzle, **Figure 29-30**. The stiffness of the arc is increased by its decreased cross-sectional area. As a result, the welder has better control of the weld pool. By forcing the gas into the arc stream, it is heated to its ionization temperature, where it forms free electrons and positively charged ions. The plasma jet produced resembles a brilliant flame. The tip of the electrode is situated above the opening in the torch nozzle that constricts the arc. A plasma welding installation is shown in **Figure 29-31**. The plasma jet passing through the restraining orifice has an accelerated velocity.

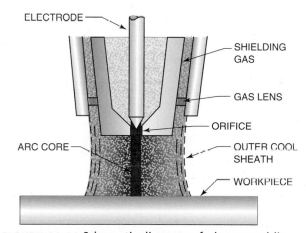

FIGURE 29-30 Schematic diagram of plasma welding process. © Cengage Learning 2012

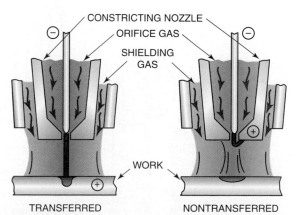

FIGURE 29-31 Transferred and nontransferred plasma arc modes. American Welding Society

Plasma arc welding can have 15 amps or less for extremely fine welds in thin metals or it can have over 100 amps, which allows it to make full penetration welds in thick sections. For example, it is possible to make a welded butt joint in metal having a thickness of up to 1/2 in. (13 mm) in a single pass. Plasma arc welding is similar to GTA welding, but plasma's high temperature makes it much faster than GTAW.

Any known metal can be melted, even vaporized, by the plasma jet process, making it useful for many welding operations. This process can be used to weld carbon steels, stainless steels, Monel, Inconel, aluminum, copper, and brass alloys.

The plasma process is most often used for cutting; see Chapter 8.

STUD WELDING (SW)

Stud welding is a semiautomatic or automatic arc welding process. An arc is drawn between a metal stud and the surface to which it is to be joined. When the end of the stud and the underlying spot on the surface of the work have been properly heated, they are brought together under pressure.

The process uses a pistol-shaped welding gun, which holds the stud or fastener to be welded. When the trigger of the gun is pressed, the stud is lifted to create an arc and is then forced against the molten pool by a backing spring. The operation is controlled by a timer. The arc is shielded by surrounding it with a ceramic ferrule, which confines the metal to the weld area.

In the welding operation, a stud is loaded in the chuck of the gun, and the ferrule is fastened over the stud. The gun is then placed on the workpiece. The action of the gun when the trigger is squeezed causes the stud to pull away from the workpiece, resulting in an arc. The arc melts the end of the stud and an area on the workpiece. At the correct

FIGURE 29-32 Rebuilding a worn crankshaft bearing with a Mogul Tube jet thermal spraying gun. Eutectic Corporation

moment, a timing device shuts off the current and causes the spring to plunge the stud into the molten pool, which freezes instantly. The gun is then released from the stud and the ferrule knocked off.

THERMAL SPRAYING (THSP)

Thermal spraying is the process of spraying molten metal onto a surface to form a coating. Pure or alloyed metal is melted in a flame or an electric arc and atomized by a blast of compressed air. The resulting fine spray builds up on a previously prepared surface to form a solid metal coating. Because the molten metal is accompanied by a blast of air, the object being sprayed does not heat up very much. Therefore, thermal spraying is known as a "cold" method of building up metal, **Figure 29-32**.

Thermal Spraying Equipment

A thermal spraying installation requires, at a minimum, the following equipment: air compressor, air control unit, air flowmeter, oxyfuel gas or arc equipment, and exhaust equipment, **Figure 29-33**.

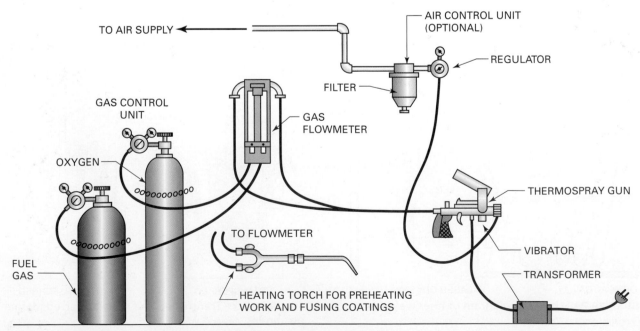

FIGURE 29-33 Complete thermal spray installation. © Cengage Learning 2012

The three types of guns available for use in the thermal spraying process are wire guns, powder guns, and crucible guns.

The wire gun uses metal in the form of wire. Wire sizes range in diameter from 20 gauge to 3/16 in. (4.8 mm). These guns can spray from 4 lb (2 kg) to 12 lb (6 kg) of metal per hour.

The flame gun consists of the following parts: (1) an oxyfuel gas torch with a hole for wire through the center of the tip, (2) a high-speed turbine that drives a pair of knurled rolls equipped with reduction gears to feed wire into the flame at the correct rate, and (3) an "air cap" that encloses the tip of the torch and directs a blast of air to pick up and project the fine molten particles against the workpiece, as shown in **Figure 29-34**. Oxyacetylene gas is most commonly used for the oxyfuel flame. However, other fuel gases such as hydrogen, propane, or natural gas may be used.

FIGURE 29-34 Powder-type thermal spray gun.
Eutectic Corporation

Thermospray (Powder) Process

During the past several years, the development of equipment, materials, and methods has greatly broadened the scope of powder spraying. Now a wide range of alloys and ceramics can be applied at speeds and costs that are economically feasible.

The wire type of flame spray equipment is limited to those materials that can be formed into wires or rods. In contrast, the thermospray process permits the use of metals, alloys, and ceramics in powder form.

Thermospray Gun This type of gun, **Figure 29-35**, usually requires no air. Two lightweight hoses are used to supply oxygen and fuel gas. The powder is fed from a reservoir attached directly to the gun, thus eliminating separate hoppers and hoses. A small reservoir is attached to the gun for hand use, and a larger one is provided for lathe-mounted guns or for large-scale production work. An air cooler may be attached to the gun to reduce overheating

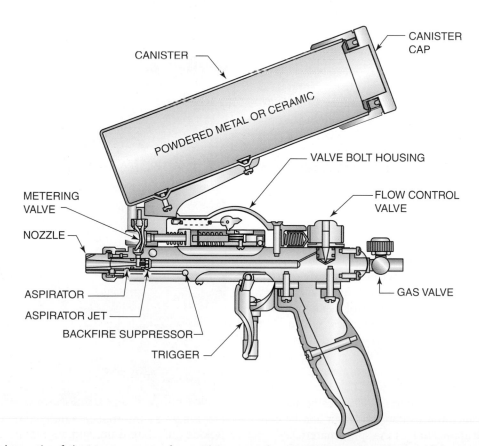

CANISTER

CANISTER CAP

POWDERED METAL OR CERAMIC

VALVE BOLT HOUSING

FLOW CONTROL VALVE

METERING VALVE

NOZZLE

GAS VALVE

ASPIRATOR

ASPIRATOR JET

BACKFIRE SUPPRESSOR

TRIGGER

FIGURE 29-35 Schematic of thermospray gun for applying metals, ceramics, and hardfacing alloys. © Cengage Learning 2012

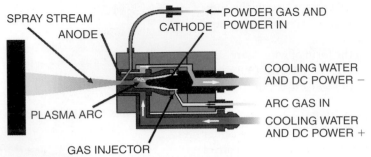

SPRAY STREAM
ANODE
CATHODE
← POWDER GAS AND POWDER IN

COOLING WATER AND DC POWER −

ARC GAS IN

COOLING WATER AND DC POWER +

PLASMA ARC

GAS INJECTOR

FIGURE 29-36 Schematic of plasma arc thermal spraying gun. Praxair, Inc.

of small work or thin sections. Extensions are available for coating inside diameters. A trigger-actuated vibrator, which is attached to the gun, is used with ceramic powders and with some metal powders.

By changing powder orifices, any required powder feed may be obtained. This permits spraying the entire range of metals, alloys, ceramics, and cements that can be applied by the thermospray process. Several different types of nozzles can be used for various purposes.

Acetylene is generally used, but hydrogen is required for some applications. The gun may be used as a torch for preheating light work. For large work and for the fusing of coatings of self-fusing alloys, high-capacity acetylene torches are used.

Torch Spraying A spray metal torch is used for spraying small parts. The torch is made up of a hopper and a spray-control mechanism that can be attached to any conventional acetylene torch. Metal powder is placed in the hopper. The torch is lit and is then moved over the part to create the overlay, **Figure 29-36**. The powder passes into the acetylene flame, which converts it to a fluid state.

Applying Sprayed Metal Metal that is sprayed is usually applied in layers less than 0.010 in. (0.25 mm) thick. Each layer is applied to a surface and bonds to the preceding layer. Greater reliability can be obtained by not trying to lay a heavy coating in one single pass.

Mechanized operation can usually be accomplished by mounting the workpiece in a lathe. The gun is then mounted on the tool post and traverses the workpiece by mechanized operation.

Plasma Spraying Process

Plasma is the term used to describe vapors of materials that are raised to a higher energy level than the ordinary gaseous state. Whereas gases consist of separate molecules, plasma consists of these same gases, which have been broken up and dissociated into ions and electrons.

The plasma spraying process (PSP), **Figure 29-37**, makes use of a spray gun that uses an electric arc contained within a water-cooled jacket. The plasma flame permits the selection of an inert, or chemically inactive, gas for the flame medium so that oxidation can be controlled during the application of the spray material.

The powder is fed into the plasma flame through the side of the nozzle. The high velocity of the flame propels the powder toward the surface to be coated. As this occurs, the ions and electrons of the plasma are recombining into atoms and releasing energy as heat. This heat is sufficient to melt the powder.

Coatings resulting from this process add extra life and superior resistance to heat, wear, and erosion on parts and products of almost any base material.

Thermal spraying techniques are used in applications that involve wear and high-temperature problems. Typical

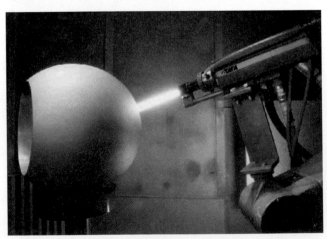

FIGURE 29-37 Internal ceramic coating with Mogul Rokide gun, extension, and angular air cap. Praxair, Inc.

applications include missile nose cones, rocket nozzles, jet turbine cases, electrical contacts, jet engine burner cam clamps, and many aircraft parts.

COLD WELDING (CW)

Cold welding is a solid state process of welding. There is no heating or melting of the metal that forms the bond in this process. The weld takes place at room temperature.

The coalescence of the metal surfaces occurs as a result of the force applied. It is possible to join most soft, ductile metals using this method. Also, dissimilar metals such as aluminum to copper, iron to copper, and so on can be joined.

Early cold welding was primarily done as spot welds, but today it is possible to make lap, edge, and butt joints. Spot welds can be made using portable, hand-operated tools, **Figure 29-38**. Other types of joints require special machines.

A major factor in the success of the cold weld is that the surface oxides and other contaminations be removed before welding. The best method to clean the surface is with a power wire brush. The surface must not be touched after it is cleaned, and the welding should be done as soon as possible after cleaning.

THERMITE WELDING (TW)

Thermite welding is a process that employs an exothermic reaction to develop a high temperature. This process is based on the great attraction of aluminum for oxygen. Aluminum can be used as a reducing agent for many oxides.

Thermite welding of rails is used throughout the world to join lengths of rails into continuous track. None of the other methods of joining and welding rails offer the advantages and service provided by thermite welding. The thermite welding process is widely used because of its high-quality welds, relative simplicity, portability, and economy.

Several companies make thermite welding products. The exact welding procedure must contain preheat temperature, chemical reaction time, charge size, joint spacing, and other essential welding variables and must be obtained from the manufacturer's WPS for each product being used and rail type being welded.

Thermite compound consists of finely divided aluminum and iron oxide mixed in a ratio of 1 to 3 by weight. Alloys are added to the mixture to obtain the desired weld metal properties. The mixture is not explosive and can be ignited only at a temperature of 2800°F (1537°C). A special ignitor is used to start the reaction. After a chemical reaction has taken place, the melt attains a temperature of 4500°F (2482°C).

The products of the reaction are superheated iron alloy and alumina slag. The slag floats to the top and is not used in making the weld.

The rail ends must be cleaned, prepared, and aligned in preparation for welding. Cleaning must include the removal of all dirt, rust, oil, grease, paint, and other sources of possible contamination. Metal flaws such as burrs, chips, and rolled-over edges must be removed, **Figure 29-39**.

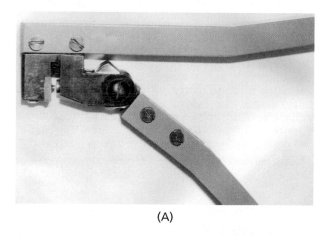

(A)

(B)

FIGURE 29-38 (A) Cold welding tool; (B) cold welds made between aluminum, copper, and mild steel.
Larry Jeffus

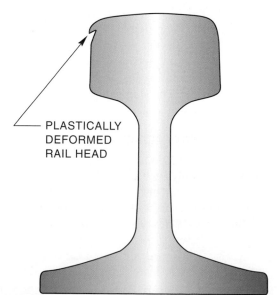

PLASTICALLY DEFORMED RAIL HEAD

FIGURE 29-39 Plastic deformation of rail head occurs over time with use. © Cengage Learning 2012

FIGURE 29-40 Rail clamp used to align and hold rail ends for welding. Larry Jeffus

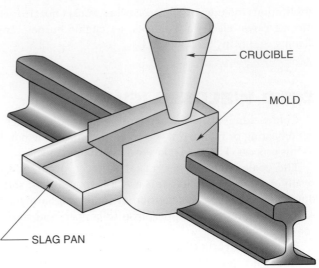

FIGURE 29-41 Thermite welding setup.
© Cengage Learning 2012

Rail end spacing, smoothness, and squareness are all important. The rail ends must be properly prepared for welding. The rail end face must be cut square, smooth, and parallel. It can be cut by sawing, abrasive disk, or oxy-fuel gas cutting.

The rails must be secured in proper alignment to prevent their movement during the welding process. Vertical, horizontal, and joint gap alignments as well as the correct offset of the rails can be provided by special clamping and alignment devices, **Figure 29-40**.

Thermite welds are essentially steel castings. The hardware parts used in making a thermite weld consist of a crucible, mold, and slag pan and are all made from a refractory material. The **crucible** is a container designed to hold the thermite powder while the chemical reaction heats up before welding occurs. The **mold** is placed around the rail ends and sealed to hold the weld metal in place until it cools. The **slag pan** provides a container for the overflow of slag and any excess weld metal. The crucible can be designed as either reusable or for a one-time use. **Figure 29-41** illustrates the arrangement of a thermite welding setup.

Moisture of any kind such as rain, snow, ice, or even the slightest dampness can make the thermite process hazardous to use. If the powder, mold, crucible, slag pan, or track is damp, a steam explosion may occur during the weld.

////// **CAUTION** \\\\\\

A steam explosion can cause serious injury or even death.

Refractory materials are designed to withstand the heat of welding but have relatively low mechanical strength. Care must be taken during their installation so as not to crack or break them by forcing, overtightening, or binding.

////// **CAUTION** \\\\\\

Never use damaged or improperly fitted molds. Damaged molds present a serious hazard to everyone in the area. If the damaged mold fails during the thermite welding process, superheated metal and slag will be released. Because of the temperature and quantity of this material, it can cause steam explosions even from relatively dry ballast or soil. Such explosions can throw large quantities of molten material several feet from the weld.

Following the preheating of the rail ends to WPS recommended temperature, the thermite mixture in the crucible is ignited. When the metal is molten, the plug in the crucible collapses, and the metal flows into the mold. The weld metal temperature is about two times the melting temperature of the base metal. When the weld metal flows into the joint, therefore, fusion takes place. After the deposited metal has cooled, the mold is removed and the riser and sprues are cut off, leaving the desired sound weld, **Figure 29-42**.

Thermite welding is also used to weld together ends of large reinforcement rods in concrete and to make welds in large sections where other methods of welding prove difficult.

The most unique application of thermite welding is used by the military. It uses a process to sabotage moving parts of equipment that were captured or that were going to be abandoned during battles.

(A)

(B)

FIGURE 29-42 (A) Thermite welding. (B) Thermite weld after grinding. Larry Jeffus

HARDFACING

Hardfacing is defined as the process of obtaining desired properties or dimensions by applying, using oxyfuel or arc welding, an integral layer of metal of one composition onto a surface, an edge, or the point of a base metal of another composition. The hardfacing operation makes the surface highly resistant to abrasion.

There are various techniques of hardfacing. Some apply a hard surface coating by fusion welding. In other techniques, no material is added but the surface metal is changed by heat treatment or by contact with other materials.

Several properties are required of surfaces that will be subjected to severe wearing conditions, including hardness, abrasion resistance, impact resistance, and corrosion resistance.

Hardfacing may involve building up surfaces that have become worn. Therefore, it is necessary to know how the part will be used and the kind of wear to expect. In this way, the proper type of wear-resistant material can be selected for the hardfacing operation.

When a part is subjected to rubbing or continuous grinding, it undergoes abrasion wear. When metal is deformed or lost by chipping, crushing, or cracking, impact wear results.

Selection of Hardfacing Metals

Many different types of metals and alloys are available for hardfacing applications. Most of these materials can be deposited by any conventional manual or automatic arc or oxyfuel welding method. Deposited layers may be as thin as 1/32 in. (0.79 mm) or as thick as necessary. The proper selection of hardfacing materials will yield a wide range of characteristics.

Steel or special hardfacing alloys should be used where the surface must resist hard or abrasive wear. Where surfacing is intended to withstand corrosion-type or friction-type wear, bronze or other suitable corrosion-resistant alloys may be used.

Most hardfacing metals have a base of iron, nickel, copper, or cobalt. Other elements that can be added include carbon, chromium, manganese, nitrogen, silicon, titanium, and vanadium. The alloying elements have a tendency to form carbides. Hardfacing metals are provided in the form of rods for oxyacetylene welding, electrodes for shielded metal arc welding, or in hard-wire form for automatic welding. Tubular rods containing a powdered metal mixture, powdered alloys, and fluxing ingredients can be purchased from various manufacturers.

Many hardfacing materials are designated by manufacturers' trade names. Some of the materials have AWS designations. AWS materials are classified into the following designations:

- High-speed steel
- Austenitic manganese steel
- Austenitic high chromium iron
- Cobalt-base metals
- Copper-base alloy metals
- Nickel-chromium boron metal
- Tungsten carbides

The coding system identifies the important elements of the hardfacing metal. The prefix *R* is used to designate a welding rod, and the prefix *E* indicates an electrode. Certain materials are further identified by the addition of digits after a suffix.

Hardfacing Welding Processes

Oxyfuel Welding In hardfacing operations, oxyfuel welding permits the surfacing layer to be deposited by flowing molten filler metal into the underlying surface. This method of surfacing is called sweating or tinning, **Figure 29-43**.

With the oxyacetylene flame, small areas can be hardfaced by applying thin layers of material. In addition, the alloy can be easily flowed to the corners and edges of the workpiece without overheating or building up deposits that are too thick. Placement of the metal can be controlled accurately.

The size of the weld is affected by many factors. These factors include the rate of travel, degree of preheat, type of metal being deposited, and thickness of the work.

Figure 29-44 shows the approximate relationship of the tip, rod, molten pool, and base metal during the hardfacing operation.

Iron, nickel, and cobalt-base alloys require a carburizing flame. Copper alloys and bronze call for a neutral

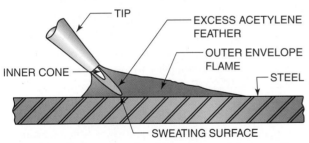

FIGURE 29-43 An example of how to produce sweating.
© Cengage Learning 2012

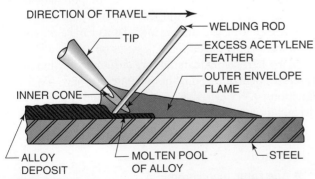

FIGURE 29-44 Approximate relationship of the tip, rod, and molten weld pool for forehand hardfacing.
© Cengage Learning 2012

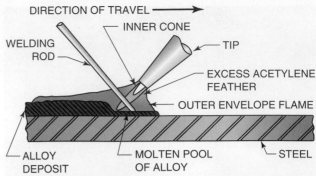

FIGURE 29-45 Backhand method of hardfacing.
© Cengage Learning 2012

or slightly oxidizing flame. Laps, blowholes, and poor adhesion of deposits can be prevented by a flame characteristic that is soft and quiet.

In all types of surfacing operations, the metal should be cleaned of all loose scale, rust, dirt, and other foreign substances before the alloy is applied. The best method of removing these impurities is by grinding or machining the surface. Fluxes may be used to maintain a clean surface. They also help to overcome oxidation that may develop during the operation.

Conventional methods may be used in holding the torch and rod. **Figure 29-45** shows the backhand method of hardfacing.

If the base metal is cast iron, it will not "sweat" like steel. Therefore, slightly less acetylene should be used. Alloys do not flow as readily on cast iron as they do on steel. Usually, it is necessary to break the surface crust on the metal with the end of the rod. A cast iron welding flux is generally necessary. The best method is to apply a thin layer of the alloy and then build on top of it.

The oxyacetylene process is preferred for small parts. Cracking can be minimized by using adequate preheat, postheat, and slow cooling. Shielded metal arc welding is preferred for large parts.

Arc Welding Hardfacing by arc welding may be accomplished by shielded metal arc, gas metal arc, gas tungsten arc, submerged arc, plasma arc, or other processes.

The techniques employed for any one of these processes are similar to those used in welding for joining. The factor of dilution must be carefully considered because the composition of the added metal will differ from the base metal. The least amount of dilution of filler metal with base metal is an important goal, especially where the two metals differ greatly. Little dilution means that the deposited metal maintains its desired characteristics. When using high-melting point alloys, dilution of the weld metal is usually kept well below 15%.

Hardfacing by the arc welding method has many advantages, including high rates of deposition, flexibility of operation, and ease of mechanization.

Hardfacing may be applied to many types of metals, including low and medium carbon steels, stainless steels, manganese steels, high-speed steels, nickel alloys, white cast iron, malleable cast iron, gray and alloy cast iron, brass, bronze, and copper.

Quality of Surfacing Deposit

The type of service to which a part is to be exposed governs the degree of quality required of the surfacing deposit. Some applications require that the deposited metal contain no pinholes or cracks. For other applications, these requirements are of little importance. In most cases, the quality of the deposited metals can be very high. Steel-base alloys do not tend to crack, while other materials, such as high-alloy cast steels, are subject to cracking and porosity.

Hardfacing Electrodes

The proper type of surfacing electrode must be selected, as one type of electrode will not meet all requirements. Most electrodes are sold under manufacturers' trade names.

Electrodes may be classified into the following three general groups:

- Resistance to severe abrasion
- Resistance to both impact and moderate abrasion
- Resistance to severe impact and moderately severe abrasion

Tungsten carbide and chromium electrodes are included in the first group. The material deposited is very hard and abrasive resistant. These electrodes can be one of two types, either coated tubular or regular coated cast alloy. The tubular types contain a mixture of powdered metal, powdered ferro alloys, and fluxing materials. The tubes are the coated type. These electrodes are used with the electric arc.

Electrodes contain small tungsten carbide crystals embedded in the steel alloy. After this material is applied to a surface, the steel wears away with use, leaving the very hard tungsten carbide particles exposed. This wearing away of steel results in a self-sharpening ability of the surfacing material. Cultivator sweeps and scraping tools are among parts that are surfaced with this material, **Figure 29-46.**

Chromium carbide electrodes are tougher than tungsten carbide–type electrodes. However, chromium carbide electrodes are not as hard and are less abrasion resistant. This material is too hard to be machined, but it has good corrosion-resistant qualities.

The electrodes in the second group are the high carbon type. When used for surfacing, these electrodes leave

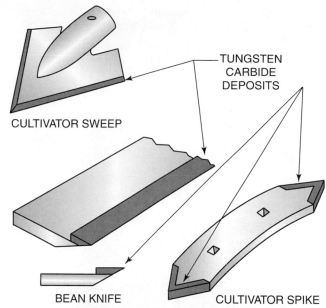
FIGURE 29-46 Farm tools that can be hardfaced with tungsten carbide electrodes to increase the life of the tools. © Cengage Learning 2012

a tough and very hard deposit. Examples of hardfaced products in this group include gears, tractor lugs, and scraper blades.

The third group of electrodes is used for surfacing rock-crusher parts, links, pins, railroad track components, and parts where severe abrasion resistance is a requirement, **Figure 29-47** and **Figure 29-48.** Deposits

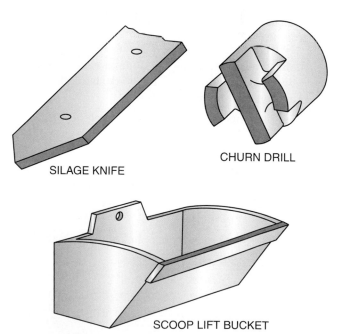
FIGURE 29-47 Products that are hardfaced to produce moderate impact resistance and severe abrasion resistance. © Cengage Learning 2012

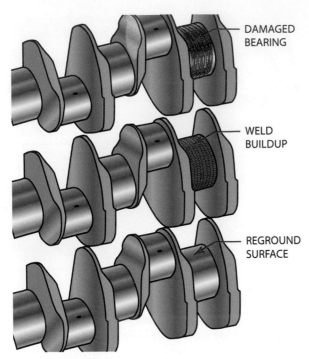

DAMAGED BEARING

WELD BUILDUP

REGROUND SURFACE

FIGURE 29-48 Bearing surfaces are built up on this crankshaft. © Cengage Learning 2012

from these electrodes are very tough but not hard. It is this quality that seems to work-harden the hardfacing material but leaves the material underneath in a softened condition. Therefore, cracking generally is not a problem.

Shielded Metal Arc Method

1. Start the process by cleaning the surface.

2. Since most hardfacing electrodes are too fluid for out-of-position welding, the work should be arranged in the flat position.

3. Set the amperage so that just enough heat is provided to maintain the arc. Too much heat will cause excessive dilution.

4. Hold a medium-long arc, using either a straight or weave pattern. When a thin bead is required, use the weave pattern and keep the weave to a width of 3/4 in. (19 mm).

5. If more than one layer is required, remove all slag before placing other layers.

Hardfacing with Gas Shielded Arc

GTA, GMA, and FCA welding processes may be used in hardfacing operations. These three processes, in many instances, are better methods of hardfacing because of the ease with which the metal can be deposited. In addition, the hardfacing materials may be deposited to form a porosity-free, smooth, and uniform surface.

Where the job calls for cobalt-base alloys, the GTA method does an effective job. Very little preheating of the base is required. The GMA and FCA welding processes are somewhat faster than surfacing by GTA due to the fact that continuous wire is used.

Care must be exercised when using the GMA, FCA, and GTA welding processes for hardfacing to avoid dilution of the weld. Helium or a mixture of helium-argon normally produces a higher arc voltage than pure argon. For this reason, the dilution of the weld metal increases. An argon and oxygen mixture should be used for surfacing with the gas metal arc processes and argon used with gas tungsten arc processes. When using FCAW, shielding may be provided as either shielding gas or self-shielding. The self-shielding hardsurfacing process is used when working outdoors because of its ability to better resist the effect of light winds.

Carbon Arc Method

1. Clean the surface of all rust, scale, dirt, and other foreign particles. Place the workpiece in a flat position.

2. Using a commercial hardfacing paste, spread the paste evenly over the area to be surfaced. Allow the paste to dry somewhat.

3. Regulate the heat carefully so that just enough heat is provided to obtain a free-flowing molten pool. Too high a welding current tends to create an excessive dilution of the base metal, thus lowering the hardness of the finished surface.

4. The carbon electrode should be moved in a circular motion. When sufficient heat is obtained, the paste will melt and fuse with the base metal.

Summary

Of the nearly 70 welding processes in use today, only a few are commonly used. Often there are processes that, if applied to a weldment, could increase productivity and reduce cost. However, lack of knowledge of the various processes often limits their selections. Understanding the opportunities that are afforded by these various processes will make you far more competitive in the labor market and business world. For example, a company may be using a torch brazing process, when a furnace braze may be more effective. Sometimes we become comfortable with our knowledge and abilities with a single process and fail to look at all of the emerging technology's opportunities.

Also, the welding industry is constantly improving existing welding processes and developing new ones. An example of an improvement to an existing welding process is the way laser beam welding and gas metal arc welding have been combined to form a hybrid process.

Laser beam welding has a number of advantages such as deeply penetrating welds with a very narrow heat-affected zone. But it does not add filler metal to the weld. Gas metal arc welding has a number of advantages such as its ability to accurately deliver a wide range of filler metal types to the weld. But it does produce a normal weld width and heat-affected zone.

By combining the two processes welding engineers have been able to develop a new hybrid process that makes welds that are both narrow and deeply penetrating and have weld buildup, **Figure 29-49**.

An example of a new welding process that has recently been developed is friction stir welding. Friction stir welding (FSW) is a solid state welding process. A weld is formed between the joining surfaces as the result of the mechanical stirring of the metal without the metal melting. Although the metal is not melted, friction between the tool and the metal does heat it up to its plastic state. The process uses a wear-resistant tool capable of withstanding the heat of friction generated during the weld, **Figure 29-50**.

To make an FS weld, the base metal must be tightly clamped together with a backing plate held tightly behind the joint. The backing plate keeps the weld

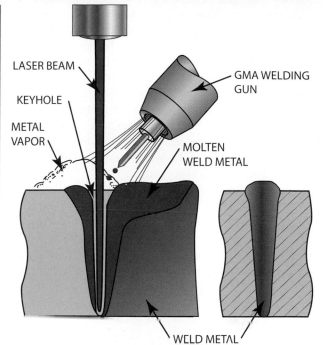

FIGURE 29-49 Hybrid process of laser beam welding and gas metal arc welding. © Cengage Learning 2012

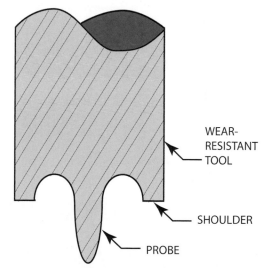

FIGURE 29-50 Cutaway of a friction stir welding tool. © Cengage Learning 2012

Summary (continued)

metal from being pushed out of the back side of the joint. The spinning tool is slowly moved or plunged into the base metal, **Figure 29-51A**. The tool is moved at a constant rate along the joint, **Figure 29-51B** and C. The weld can be completed by moving the tool all the way across the joint or the tool can be raised up at the end of the weld. Either way of ending the weld will leave a probe cavity, **Figure 29-51D**. There are several techniques that can be used to solve the problem of the probe cavity. One technique is to use a runoff tab at the end of the joint. Another technique is to move the tool off of the joint onto an area of scrap before it is stopped or to move the tool off the joint where the probe cavity will not affect the welded joint.

Staying current on new processes and applications is important, and some good sources of current knowledge are welding applications, manufacturers' literature, and the Internet. You should become a member of a professional organization, such as the American Welding Society; this will provide you with up-to-date process information. Learning to weld is a lifelong activity.

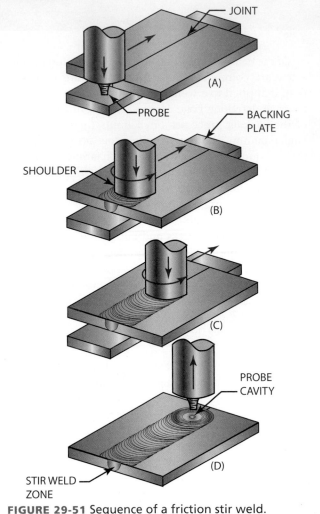

FIGURE 29-51 Sequence of a friction stir weld.
© Cengage Learning 2012

Computer Numeric Controlled Plasma—A Cut above the Rest

In today's economy, businesses must be efficient, accurate, and innovative in order to be successful. In the metalworking industry, getting that edge can be challenging. Although many of the industry's techniques and technologies are tried and true, today they are not always the most effective. For example, the traditional processes of fabricating various flat parts by sawing, drilling, shearing, punching, grinding, and machining can prove very time consuming and costly. But with modern technology, many such parts can now be made much more quickly and accurately with a lot less equipment and setup, Figure 1.

To meet this need equipment manufacturers such as PlasmaCAM have developed computer controlled machines. These machines plug into a personal computer and with the software controlling it, the cutting torch moves across the surface of the metal cutting out the exact shape. The software can be used to draw accurate detailed parts, or drawings can be imported or scanned. Details such as sign

FIGURE 1 Larry Jeffus

FIGURE 3 Larry Jeffus

lettering can easily be added to drawings using the program's various fonts. This company has a large library of clip art that can be easily customized. The PlasmaCAM and machines like it are giving welding, fabricating, and machine shops a significant edge in the market.

The machine uses a plasma cutting torch for optimum speed, accuracy, and simplicity, Figure 2. Although plasma cutting technology has existed for years, advancements in its equipment and torches continue to change the world of metalworking. The high-velocity, high-temperature jet of ionized gas streams out from the torch tip to the metal, melting it and blowing it away. Plasma cuts five times faster than oxyfuel; its smaller kerf width and lower heat input make intricate cutting with precision possible. Plasma can cut almost any type of electrically conductive metal, including steel, stainless steel, aluminum, brass, copper, and titanium. This fully integrated computer numeric control (CNC) cutting system enables you to cut shapes out of sheet metal or plate with precision, achieving detail not possible by hand.

CNC machines like this can cut a variety of metals and thicknesses, and it is proving useful in a wide range of industries. They include welding and repair shops, machine shops, HVAC contractors, manufacturing plants, engineering and rapid prototyping, and metal artists, Figure 3. "You can measure a part that needs to be replaced, draw it in the software, and cut a brand new replacement," said Marty,

owner and operator of AFAB Metalworks. He was able to start his business using his PlasmaCam machine to cut out various parts for snowplows.

Not only have the CMC machines revolutionized normal metal fabrication, their flexibility is providing new opportunities. For example, designers of metal products are learning about CNC plasma cutting and how it is changing manufacturing processes. As a result, new product designs are increasingly utilizing flat parts with complex profiles that are easy to manufacture. These parts may be formed and welded or bolted together with other parts for ultimate aesthetics and functionality at lower manufacturing costs.

With the ability to cut so precisely, the moneymaking possibilities are endless for companies such as ornamental iron, sign, and landscaping businesses. "You can make products like railings, gates, benches, tables, furniture, truck racks, weather vanes, wind chimes, security doors, fireplace screens, company signs, ranch entries, address and welcome signs, key and coat racks, hardware, garden ornaments, wall hangings, and so on," said Ken of KLI Landscaping, Figure 4.

Plasma cut metal is a new medium for creating beautiful sculptures and works of art, Figure 5. Doyle Goebel started his art business using casting as the technique of choice. His artwork has always been amazing, but he was having some trouble with the process and lead times from the foundry where he sent his art. Goebel explains, "I had art shows

FIGURE 2 Larry Jeffus

FIGURE 4 Larry Jeffus

learn how to do CNC manufacturing in a way that is surprisingly easy and fun. Many students are learning a whole new approach to design and metal manufacturing. Only time will tell what innovations are in the future of metal manufacturing, Figures 6 and 7.

FIGURE 5 Larry Jeffus

FIGURE 6 Larry Jeffus

coming up, I needed to have product by a certain time, and with casting it was never there on time, because I wasn't that important to them." That is when he decided to take control of the process himself. Now Goebel saves a lot of time using a CNC plasma cutter. Once one sculpture is done, he has all the needed parts programmed into the computer ready to cut. "The time factor is getting better with each piece I produce. Like a bald eagle I produced with a 6 ft wingspan. The first one took six months, and the fourth one only took six weeks." The sky really is the limit when it comes to creating with this machine.

PlasmaCAM machines provide such a key competitive edge in the metalworking industry that thousands of schools now use them in their metal shop classes. Students

FIGURE 7 Larry Jeffus

Review

1. What generates the heat for fusion in resistance welding?

2. What can be used to produce the force needed to hold the work together for resistance welds?

3. What are the basic resistance welding processes?

4. What steps can be included in RSW?

5. What are the three types of spot welding machines in common use?

6. What are some of the uses for a portable spot welder?

7. How is seam welding similar to spot welding?

8. What is the most common joint for seam welds?

9. What is RSEW-HF used for?

10. What metals are readily welded by the RPW process?

11. Why is FW not usually cost effective for short production runs?

12. What is the main difference between upset welding and flash welding?

13. What unusual metal combination can be welded using percussion welding?

14. What are the characteristics of an EB weld?

15. What is the least amount of vacuum that a part must be subjected to in order to make an EB weld?

16. How is the beam focus changed in the EB welding process?

17. How can a misaligned seam be tracked automatically for EB welds?

18. What types of welds are most commonly made with the US welding process?

19. What equipment is needed to make US welds?

20. What are the typical applications for US welding?

21. List the steps of the inertia welding process.

22. Referring to Table 29-1, what dissimilar metals can be joined to the metals listed below with inertia welding

    a. to 1100 or 6061 aluminum?

    b. to pure titanium?

23. What difficult metals can be welded with a laser?

24. How is the plasma arc stiffened?

25. How does stud welding work?

26. Why is THSP known as a cold buildup process?

27. Which thermal spray process can be used to apply ceramics?

28. Why should thermal spray coats be applied as thin coats?

29. What is the advantage of using an inert gas for plasma spraying?

30. What causes the coalescence during a cold weld?

31. Why is thermite welding used so extensively for welding of rail ends?

32. How can rail ends be secured for thermite welding?

33. Why must thermite welding supplies and equipment be kept dry?

34. What types of wear can hardfacing protect against?

35. What base metals are used for most hardfacing alloys?

36. What are the advantages of oxyacetylene flame hardfacing?

37. Why should there be as little dilution as possible of the base metal when hardfacing?

38. How can wear provide a self-sharpening effect on some hardfaced parts?

39. What hardfacing alloy is best applied with the gas tungsten arc welding process?

SECTION 7

OXYFUEL

Success Story

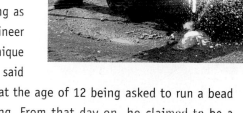

Jay Jones has been associated with welding as long as he can remember. His father is a mechanical engineer and was also a hot-rodder who exposed Jay to many unique mechanical, machining, and welding opportunities. He said his father's eyes were never very good and remembers at the age of 12 being asked to run a bead on the frame of a hot-rod that his dad was fabricating. From that day on, he claimed to be a "rod burner." He had no formal education but a lot of OJT (on-the-job training).

Starting the eighth grade, Jay was faced with a tough decision that at the time he had no idea would impact the rest of his life. The school offered metal shop for eighth graders, and the previous two years he had taken band. He said, "The rub was that the band class and metal shop were both electives, and I was not allowed to take them both, so I chose 'SHOP!'" He continued his metal shop training in high school and joined an industrial arts club.

Jay took welding classes at college where he learned an array of technologies he never imagined related to welding. While attending college, he purchased the equipment to set up his first welding rig. This allowed him to earn money around his class schedule. His welding instructor said, "I recognized Jay's abilities and passions for welding and was pleased he agreed to be my welding lab assistant." Later, Jay became an adjunct instructor and taught in college for the next 18 years.

In 1981, Jay joined the AWS as a student member, and this was another life changing decision for him because of the people he came to know through the AWS. Many of them were pillars in the welding industry who cared enough to take time with a "punk kid" like himself.

Jay earned his Lifetime Texas Teaching Certificate and taught welding in high school before accepting a job with Victor Equipment Co. in 2000. Jay said, "That decision to work for Victor has been one of the best in my life." He has become one of the leading authorities for gas apparatus and has had responsibilities around the world.

He started as the training specialist for Victor equipment company where he conducted product seminars globally on all types of gas apparatus. Jay is a graduate of Eastfield College and Texas A&M University, an author, a past chairman of the North Texas Section of the America Welding Society, AWS Board of Directors. He holds degrees in welding technology and education and is a AWS Certified Welding Educator. He received the AWS District Educator Award in 1999–2003 and the National Instructor Award in 2004–2005.

Jay says, "I'm still active in the AWS, enjoy watching students learn, and love to weld." ■

Chapter 30

Oxyfuel Welding and Cutting Equipment, Setup, and Operation

OBJECTIVES

After completing this chapter, the student should be able to

- describe how to maintain the major components of oxyfuel welding equipment.
- explain the method of testing an oxyfuel system for leaks.
- demonstrate how to set up, light, adjust, extinguish, and disassemble oxyfuel welding equipment safely.

KEY TERMS

absolute pressure	*gauge pressure*	*regulators*
atmospheric pressure	*injector chamber*	*safety disc*
Bourdon tube	*leak-detecting solution*	*safety release valve*
combination welding and cutting torch	*line pressure drop*	*seat*
creep	*manifold system*	*Siamese hose*
cutting torches	*mixing chamber*	*spark lighter*
cylinder pressure	*oxyfuel gas torch*	*two-stage regulators*
diaphragm	*purged*	*valve packing*
flashback arrestor	*regulator gauge*	*working pressure*

INTRODUCTION

Oxyfuel welding, cutting, brazing, hardsurfacing, heating, and other processes use the same basic equipment. When storing, handling, assembling, testing, adjusting, lighting, shutting off, and disassembling this basic equipment, the same safety procedures must be followed for each process. Improper or careless work habits can cause serious safety hazards. Proper attention to all details makes these processes safe.

Certain basic equipment is common to all gas welding. Cylinders, regulators, hoses, hose fittings, safety valves, torches, and tips are some of the basic equipment used. Although numerous manufacturers produce a large variety of gas equipment, it all works on the same principle. When welders are not sure how new equipment is operated, they should seek professional help. A welder should never experiment with any equipment.

All oxyfuel processes use a high-heat, high-temperature flame produced by burning a fuel gas mixed with pure oxygen. The gases are supplied in pressurized cylinders. The gas pressure from the cylinder must be reduced by using a regulator.

The gas then flows through flexible hoses to the torch. The torch controls this flow and mixes the gases in the proper proportion for good combustion at the end of the tip.

Acetylene is the most widely used fuel gas, but about 25 other gases are available. The regulator and tip are usually the only equipment changes required in order to use another fuel gas. The adjustment and skill required are often different, but the storage, handling, assembling, and testing are the same. When changing gases, make sure the tip can be used safely with the new gas.

Because of the similarities in the way equipment is operated and assembled, the following information can be easily applied to all oxyfuel welding systems.

PRESSURE REGULATORS

All pressure **regulators** reduce the high cylinder or system pressure to the proper lower working pressure. It is important that the regulator keep the lower pressure constant over a range of flow rates. Some of the various types of pressure regulators are low-pressure regulators, high-pressure regulators, single-stage regulators, dual-stage regulators, cylinder regulators, manifold regulators, line

regulators, and station regulators. Although they all work the same, they are not interchangeable.

//// **CAUTION** \\\\

Although all regulators work the same way, they cannot be safely used interchangeably on different types of gas or for different pressure ranges without the possibility of a fire or an explosion.

Regulator Operation

A regulator works by holding the forces on both sides of a **diaphragm** in balance, **Figure 30-1**. As the adjusting screw is turned inward, it increases the force of a spring on the flexible diaphragm and bends the diaphragm away. As the diaphragm is moved, the small high-pressure valve is opened, allowing more gas to flow into the regulator. The gas pressure cancels the spring pressure, and the diaphragm returns back to its original position, closing the high-pressure valve, **Figure 30-2**. When the regulator is used, the gas pressure on the back side of

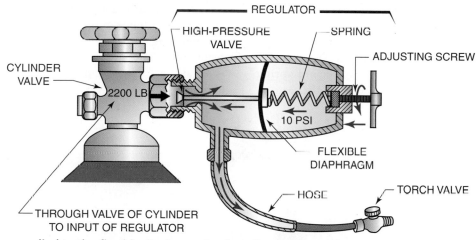

FIGURE 30-1 Force applied to the flexible diaphragm by the adjusting screw through the spring opens the high-pressure valve. © Cengage Learning 2012

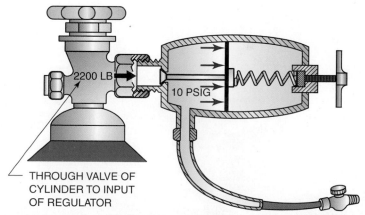

FIGURE 30-2 When the gas pressure against the flexible diaphragm equals the spring pressure, the high-pressure valve closes. © Cengage Learning 2012

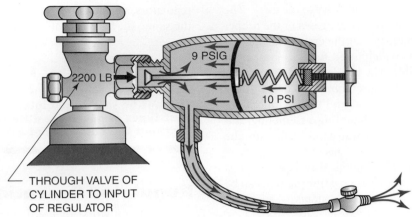

FIGURE 30-3 A drop in the working pressure occurs when the torch valve is opened and gas flows through the regulator at a constant pressure. © Cengage Learning 2012

the diaphragm is reduced, the spring again forces the valve open, and gas flows. The drop in the internal pressure can be seen on the working pressure gauge, **Figure 30-3**.

The size of a regulator determines its ability to hold the working pressure constant over a wider range of flow rates. **Two-stage regulators**, Figure 30-4, are able to keep the pressure constant at very low or high flow rates as the cylinder empties. This type of regulator has two sets of springs, diaphragms, and valves. The first spring is preset at the factory to reduce the cylinder pressure to 225 psig (1550 kPag). The second spring is adjusted like other regulators. Because the second

high-pressure valve has to control a maximum pressure of only around 225 psig (1550 kPag), it can be larger, thus allowing a greater flow.

Regulator Gauges

There may be one or two pressure gauges on a regulator. One pressure gauge shows the working pressure, and the other indicates the cylinder pressure, **Figure 30-5**. The **working pressure** gauge shows the pressure at the regulator and not at the torch. The pressure at the torch is always less than the pressure shown on the working pressure gauge. This pressure difference results from the

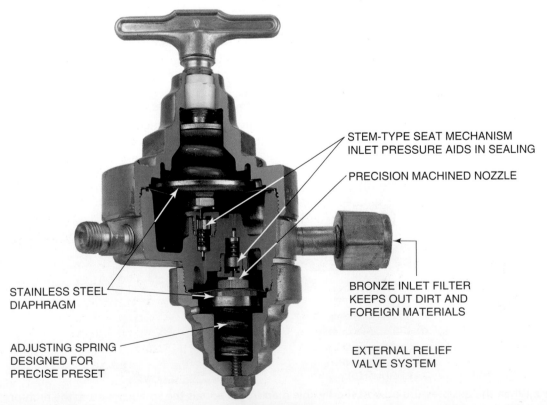

STEM-TYPE SEAT MECHANISM
INLET PRESSURE AIDS IN SEALING

PRECISION MACHINED NOZZLE

STAINLESS STEEL
DIAPHRAGM

BRONZE INLET FILTER
KEEPS OUT DIRT AND
FOREIGN MATERIALS

ADJUSTING SPRING
DESIGNED FOR
PRECISE PRESET

EXTERNAL RELIEF
VALVE SYSTEM

FIGURE 30-4 Two-stage oxygen regulator. Thermadyne Holding Corporation

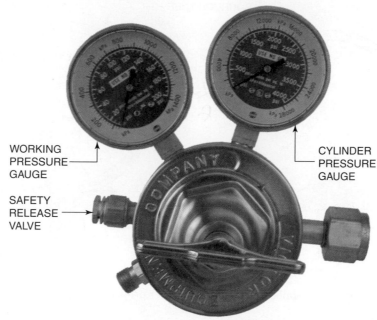

FIGURE 30-5 Safety release valve on an oxygen regulator. Victor Equipment Company

Tip Pressure psig (kg/cm² G)	Regulator Pressure* for Hose Lengths ft (m)				
	10 ft (3 m)	25 ft (7.6 m)	50 ft (15.2 m)	75 ft (22.9 m)	100 ft (30.5 m)
1 (0.1)	1 (0.1)	2.25 (0.15)	3.5 (0.27)	4.75 (0.35)	6 (0.4)
5 (0.35)	5 (0.35)	6.25 (0.4)	7.5 (0.52)	8.75 (0.6)	10 (0.7)
10 (0.7)	10 (0.7)	11.25 (0.75)	12.5 (0.85)	13.75 (0.95)	15 (1.0)

*These values are for hose with a diameter of 1/4 in. (6 mm); larger or smaller hose diameters or high flow rates will change these pressures.

TABLE 30-1 Regulator Pressure for Various Lengths of Hose

resistance to the gas flow, which is referred to as line pressure drop. The smaller in diameter or longer a line is, the greater the pressure drop will be, **Table 30-1**.

EXPERIMENT 30-1

Line Resistance

In this experiment, two pieces of the same diameter hose are required. One piece of hose is to be a short length, less than 10 ft (3 m) long. The other piece of hose is to be more than 25 ft (8 m) long. Blow through the short piece and then blow through the long piece. Observe the difference.

Complete a copy of the "Student Welding Report" listed in Appendix I or provided by your instructor. ◆

The high-pressure gauge on a regulator shows **cylinder pressure** only. This gauge is used to indicate the amount of gas that remains in a cylinder. However, cylinders containing liquefied gases, such as CO_2, propane, and MPS, must be weighed to determine the amount of gas remaining.

Inside a **regulator gauge** there is a **Bourdon tube**. This tube is bent in the shape of the letter *C*, with one end attached

solidly to the gauge body and the other end attached through a gear to a needle, **Figure 30-6**. As the pressure inside the tube increases, the tube tries to straighten out.

The pressure shown on a gauge is read as pounds per square inch gauge (psig) or kilopascals (kPag). The **atmospheric pressure,** 14.7 psi (101.35 kPa), must be added to the **gauge pressure** to find the **absolute pressure,** psia (kPaa). In welding, psig and psi (kPag and kPa) are used interchangeably.

Regulator Safety Pressure Release Device

Regulators may be equipped with either a **safety release valve** or a **safety disc** to prevent excessively high pressures from damaging the regulator. A safety release valve is made up of a small ball held tightly against a **seat** by a spring. The release valve will reseat itself after the excessive pressure has been released.

A safety disc is a thin piece of metal held between two seals, **Figure 30-7**. When a safety disc bursts to release excessive pressure, all of the gas in the cylinder will be

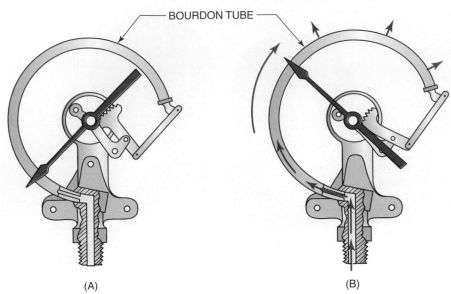

FIGURE 30-6 Gauge before pressure is applied (A) and gauge after pressure is applied (B). © Cengage Learning 2012

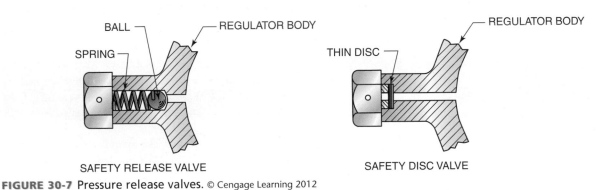

FIGURE 30-7 Pressure release valves. © Cengage Learning 2012

released. A safety disc does not reseal, so it must be replaced before the regulator can be used again.

Cylinder and Regulator Fittings

A variety of inlet or cylinder fittings are available to ensure that the regulator cannot be connected to the wrong gas or pressure, **Figure 30-8A** through D. A few adapters are available that will allow some regulators to be attached to a different type of fitting. The following are the two most common types: (1) adapt a left-hand male acetylene cylinder fitting to a right-hand female regulator fitting, or vice versa and (2) adapt an argon or mixed gas

FIGURE 30-8 (A) Acetylene cylinder valve (left-hand thread). Larry Jeffus

FIGURE 30-8 (B) Oxygen cylinder valve. Larry Jeffus

FIGURE 30-8 (C) Argon cylinder valve. Larry Jeffus

FIGURE 30-8 (D) Carbon dioxide (CO_2) cylinder valve.
Larry Jeffus

male fitting to a female flat washer-type CO_2 fitting, **Figure 30-9.**

The connections to the cylinder and to the hose must be kept free of dirt and oil. Fittings should screw together freely by hand and require only light wrench pressure to be leak tight. If the fitting does not tighten freely on the connection, both parts should be cleaned. If the joint leaks after it has been tightened with a wrench, the seat should be checked. Examine the seat and threads for damage.

FIGURE 30-9 (A) Acetylene cylinder adapter. Larry Jeffus

FIGURE 30-9 (B) Carbon dioxide-to-argon adapter.
Larry Jeffus

If the seat is damaged, it can be repaired by a manufacturer-authorized regulator repair shop. Severely damaged connections must be replaced.

Regulator Safety Precautions

The regulator pressure adjusting screw should be backed off each time the oxyfuel system is being shut down. This is done to release the spring and diaphragm pressures, which, over time, may cause damage. Keeping a spring compressed and the diaphragm stretched can cause the spring to weaken and the diaphragm to be permanently distorted.

In addition, when the cylinder valve is reopened, some high-pressure gas can pass by the open high-pressure valve before the diaphragm can close it. This condition may cause the diaphragm to rupture, the low-pressure gauge to explode, or both.

High-pressure valve seats that leak result in a **creep** or rising pressure on the working side of the regulator. This usually occurs when the gas pressure is set but no gas is flowing. If the leakage at the seat is severe, the maximum safe pressure can be exceeded on the working side, resulting in damage to the diaphragm, gauge, hoses, or other equipment.

////// **CAUTION** \\\\\\

Regulators that creep excessively or beyond the safe working pressure must not be used.

A diaphragm can be tested for leaks by first setting the regulator to 14 psig (95 kPag) for fuel gases or 45 psig (310 kPag) for oxygen and other gases. Once the pressure is set, place a finger over the vent hole and spray it with a **leak-detecting solution, Figure 30-10.** Slowly move the finger from the hole and watch for bubbles, which indicate a leak.

A gauge that gives a faulty reading or that is damaged can result in dangerous pressure settings. Gauges that do not read "0" (zero) pressure when the pressure is released, or those that have damaged glass or case, must be repaired or replaced.

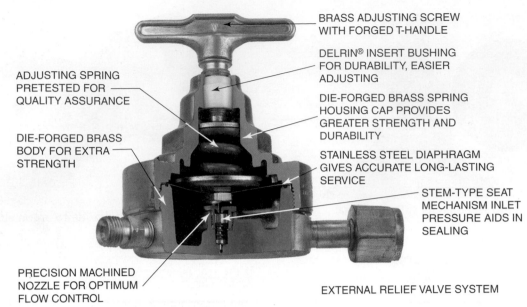

BRASS ADJUSTING SCREW
WITH FORGED T-HANDLE

DELRIN® INSERT BUSHING
FOR DURABILITY, EASIER
ADJUSTING

DIE-FORGED BRASS SPRING
HOUSING CAP PROVIDES
GREATER STRENGTH AND
DURABILITY

STAINLESS STEEL DIAPHRAGM
GIVES ACCURATE LONG-LASTING
SERVICE

STEM-TYPE SEAT
MECHANISM INLET
PRESSURE AIDS IN
SEALING

EXTERNAL RELIEF VALVE SYSTEM

ADJUSTING SPRING
PRETESTED FOR
QUALITY ASSURANCE

DIE-FORGED BRASS
BODY FOR EXTRA
STRENGTH

PRECISION MACHINED
NOZZLE FOR OPTIMUM
FLOW CONTROL

FIGURE 30-10 Single-stage oxygen regulator. Thermadyne Holding Corporation

---

> ///// **CAUTION** \\\\\
>
> **All work on regulators must be done by properly trained repair technicians.**

> ///// **CAUTION** \\\\\
>
> **Regulators should be located far enough from the actual work that flames or sparks cannot reach them.**

The outlet connection on a regulator is either a right-hand fitting for oxygen or a left-hand fitting for fuel gases. A left-hand threaded fitting has a notched nut, **Figure 30-11.**

Regulator Care and Use

There are no internal or external moving parts on a regulator or a gauge that require oiling, **Figure 30-12.**

> ///// **CAUTION** \\\\\
>
> **Oiling a regulator is unsafe and may cause a fire or an explosion.**

If the adjusting screw becomes tight and difficult to turn, it can be removed and cleaned with a dry, oil-free rag. When replacing the adjusting screw, be sure it does not become cross-threaded. Many regulators use a nylon nut in the regulator body, and the nylon is easily cross-threaded.

When welding is finished and the cylinders are turned off, the gas pressure must be released and the adjusting screw backed out. This is required both by federal regulation and to prevent damage to the diaphragm, gauges, and adjusting spring if they are left under a load. A regulator that is left pressurized causes the diaphragm to stretch, the Bourdon tube to straighten, and the adjusting spring to compress. These changes result in a less accurate regulator with a shorter life expectancy.

FIGURE 30-11 Left-hand threaded fittings are identified with a notch. Larry Jeffus

NOTE: NO OIL

FIGURE 30-12 Never oil a regulator. Thermadyne Holding Corporation

WELDING AND CUTTING TORCHES: DESIGN AND SERVICE

The oxyacetylene hand torch is the most common type of **oxyfuel gas torch** used in industry. The hand torch may be either a combination welding and cutting torch or a cutting torch only, **Figure 30-13** and **Figure 30-14**.

The **combination welding and cutting torch** offers more flexibility because a cutting head, welding tip, or heating tip can be attached quickly to the same torch body, **Figure 30-15**. Combination torch sets are often used in schools, automotive repair shops, auto body shops, small welding shops, or any other situation where flexibility is needed. The combination torch sets usually are more practical for portable welding since the one unit can be used for both cutting and welding.

Straight or dedicated **cutting torches** are usually longer than combination torches. The longer length helps keep the operator farther away from heat and sparks. In addition, thicker material can be cut with greater comfort.

FIGURE 30-13 A torch body or handle used for welding or cutting. Victor Equipment Company

FIGURE 30-14 A torch used for cutting only. Victor Equipment Company

FIGURE 30-15 A combination welding and cutting torch kit. Victor Equipment Company

FIGURE 30-16 Medium-duty torch for smaller jobs. Victor Equipment Company

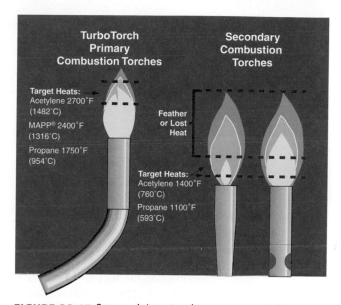

FIGURE 30-17 Some air/gas torches use a special tip that improves the combustion for a hotter, more effective flame. Victor Equipment Company

Most manufacturers make torches in a variety of sizes for different types of work. There are small torches for jewelry work, **Figure 30-16**, and large torches for heavy plates. Specialty torches for heating, brazing, or soldering are also available. Some of these use a fuel-air mixture, **Figure 30-17**. Fuel-air torches are often used by plumbers and air-conditioning technicians for brazing and soldering copper pipe and tubing. There are no industrial standards for tip size identification, tip threads, or seats. Therefore, each style, size, and type of torch can be used only with the tips made by the same manufacturer to fit the specific torch.

Mixing the Gases

Two basic methods are used for mixing the oxygen and fuel gas to form a hot, uniform flame. The two gases must be mixed completely before they leave the tip and create the flame. If the gases are not mixed completely, the torch will have a greater tendency to backfire or flash back. One method uses equal or balanced pressures, and the gases are mixed in a **mixing chamber**. The other method uses a higher oxygen pressure, and the gases are mixed in an **injector chamber**.

The mixing chamber of the equal-pressure torch may be located in the torch body, attached to the tip, or in the

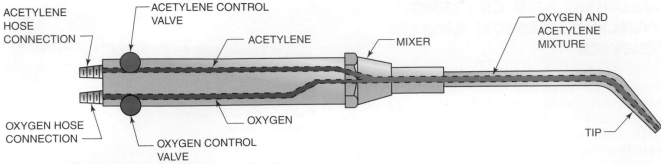

ACETYLENE HOSE CONNECTION

ACETYLENE CONTROL VALVE

ACETYLENE

MIXER

OXYGEN AND ACETYLENE MIXTURE

OXYGEN HOSE CONNECTION

OXYGEN CONTROL VALVE

OXYGEN

TIP

FIGURE 30-18 Schematic drawing of an oxyacetylene welding torch. © Cengage Learning 2012

tip, **Figure 30-18**. Both gases must enter the enlarged mixing chamber through small, separate openings. Because the mixing chamber is larger than the total size of both entrance holes and the exit hole, the gases experience a rapid drop in pressure as they enter the chamber. The rapid drop in pressure causes the gases to become turbulent and mix thoroughly, **Figure 30-19**. Some manufacturers spin the gases before they enter the chamber to ensure complete mixing, **Figure 30-20**.

The injector torches work both with equal gas pressures or low fuel-gas pressures, **Figure 30-21**. The injector allows oxygen at the higher pressure to draw the fuel gas into the chamber, even when the fuel-gas pressure is as low as 6 oz/in.^{2} (26.3 g/cm^2). The injector works by passing the oxygen through a venturi, which creates a vacuum to pull the fuel gas in and then mixes the gases together. An injector-type torch must be used if a low-pressure acetylene generator or low-pressure residential natural gas is used as a fuel-gas supply.

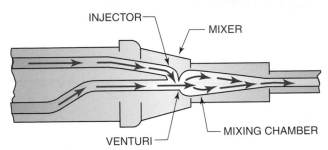

INJECTOR

MIXER

VENTURI

MIXING CHAMBER

FIGURE 30-21 Injector mixing system. © Cengage Learning 2012

Torch Care and Use

The torch body contains threaded connections for the hoses and tips. These connections must be protected from any damage. Most torch connections are external and made of soft brass that is easily damaged. Some connections, however, are more protected because they have either internal threads or stainless steel threads for the tips. The best protection against damage and dirt is to leave the tip and hoses connected when the torch is not in use.

Because the hose connections are close to each other, a wrench should never be used on one nut unless the other connection is protected with a hose fitting nut, **Figure 30-22**.

The hose connections should not leak after they are tightened with a wrench. If leaks are present, the seat should

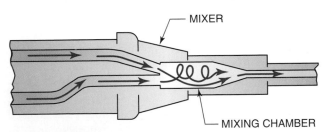

MIXER

MIXING CHAMBER

FIGURE 30-19 Equal-pressure mixing chamber. © Cengage Learning 2012

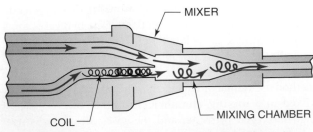

MIXER

COIL

MIXING CHAMBER

FIGURE 30-20 A metal coil in the oxygen tube spins the gas, ensuring a complete mixing of gases. © Cengage Learning 2012

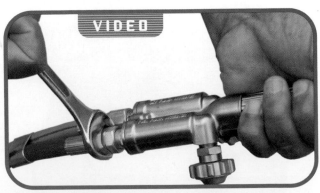

VIDEO

FIGURE 30-22 One hose fitting nut will protect the threads when the other nut is loosened or tightened. Larry Jeffus. ***See DVD Oxyacetylene Welding.***

FIGURE 30-23 Check all connections for possible leaks and tighten if necessary. Larry Jeffus. *See DVD Oxyacetylene Welding.*

be repaired or replaced. Some torches have removable hose connection fittings so that replacement is possible.

The valves should be easily turned on and off and should stop all gas flowing with minimum finger pressure. To find leaking valve seats, set the regulators to a working pressure. With the torch valves off, spray the tip with a leak-detecting solution. The presence of bubbles indicates a leaking valve seat, **Figure 30-23.** The gas should not leak past the valve stem packing when the valve is open or when it is closed. To test leaks around the valve stem, set the regulator to a working pressure. With the valves off, spray the valve stem with a leak-detecting solution and watch for bubbles, indicating a leaking **valve packing.** The valve stem packing can now be tested with the valve open. Place a finger over the hole in the tip and open the valve. Spray the stem and watch for bubbles, which would indicate a leaking valve packing, **Figure 30-24.** If either test

indicates a leak, the valve stem packing nut can be tightened until the leak stops. After the leak stops, turn the valve knob. It should still turn freely. If it does not, or if the leak cannot be stopped, replace the valve packing.

The valve packing and valve seat can be easily repaired on most torches by following the instructions given in the repair kit. On some torches, the entire valve assembly can be replaced, if necessary.

WELDING AND HEATING TORCH TIPS

Because no industrial standard tip size identification system exists, the student must become familiar with the size of the orifice (hole) in the tip and the thickness range for which it can be used. Comparing the overall size of the tip can be done only for tips made by the same manufacturer for the same type and style of torch, **Figure 30-25.** Learning a specific manufacturer's system is not always the answer because on older, worn tips the orifice may have been enlarged by repeated cleaning.

Tip sizes can be compared to the numbered drill bit size used to make the hole, **Table 30-2.** The sizes of tip cleaners are given according to the drill bit size of the hole

FIGURE 30-25 A variety of tip sizes are available for each torch body. Victor Equipment Company

FIGURE 30-24 The torch valves should be checked for leaks and the valve packing nut tightened if necessary. Larry Jeffus. *See DVD Oxyacetylene Welding.*

Tip Cleaner Standard Set			
	Use Cleaner	For Drill	
Smallest	1	77-76	77 = 0.0160" (0.4064 mm)
	2	75-74	
	3	73-72-71	
	4	70-69-68	
	5	67-66-65	
	6	64-63-62	
	7	61-60	
	8	59-58	
	9	57	
	10	56	
	11	55-54	
	12	53-52	49 = 0.0730"
Largest	13	51-50-49	(1.8542 mm)

TABLE 30-2 Tip Cleaner Size Compared to Drill Size Found on Most Standard Tip-Cleaning Sets.

they fit. By knowing the tip cleaner size commonly used to clean a tip, the welder can find the same size tip made by a different manufacturer. The tip size can also be determined by trial and error.

On some torch sets, each tip has its own mixing chamber. On other torch sets, however, one mixing chamber may be used with a variety of tip sizes.

Torch Tip Care and Use

Torch tips may have metal-to-metal seals, or they may have an O-ring or a gasket between the tip and the torch seat. Metal-to-metal seal tips must be tightened with a wrench. Tips with an O-ring or a gasket may be tightened by hand. Using the wrong method of tightening the tip fitting may result in damage to the torch body or the tip.

Dirty tips can be cleaned using a set of tip cleaners. Using the file provided in the tip-cleaning set, **Figure 30-26**, file the end of the tip smooth and square. Next, select the size of tip cleaner that fits easily into the orifice. The tip cleaner is a small, round file and should be moved in and out of the orifice only a few times, **Figure 30-27**. Be sure the tip cleaner is straight and that it is held steady to prevent it from bending or breaking off in the tip. Excessive use of the tip cleaner tends to ream the orifice, making it too large. Therefore, use the tip cleaner only as required.

FIGURE 30-28 Tools used to repair tips. Larry Jeffus

Once the tip is cleaned, turn on the oxygen for a moment to blow out any material loosened during the cleaning.

Damaged tips or tips with cleaners broken in them can be reconditioned, but they require a good deal of work and some specialized tools, **Figure 30-28**.

BACKFIRES

A backfire occurs when a flame goes out with a loud snap or pop. A backfire may be caused by one or more of the following:

- Touching the tip against the workpiece
- Overheating the tip
- Operating the torch when the flame settings are too low
- Loose tip
- Damaged seats
- Dirt in the tip

The problem that caused the backfire must be corrected before relighting the torch. A backfire may cause a flashback.

FIGURE 30-26 Standard set of tip cleaners. Larry Jeffus

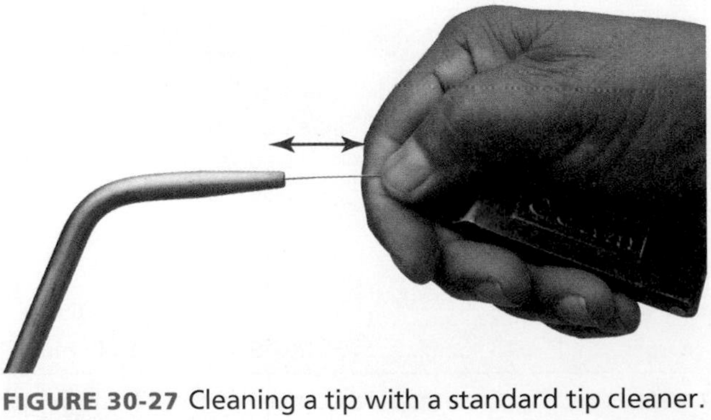

FLASHBACKS

When a flashback occurs, the flame is burning back inside the tip, torch, hose, or regulator. A flashback produces a high-pitched whistle. If the torch does flash back, close the oxygen valve at once and then close the fuel valve. The order in which the valves are closed is not as important as the speed at which they are closed. A flashback that reaches the cylinder may cause a fire or an explosion.

Closing the oxygen valve on the torch stops the flame inside at once. Then the fuel-gas valve should be closed and the torch allowed to cool off before repairing the problem. When a flashback occurs, there is usually a serious

FIGURE 30-27 Cleaning a tip with a standard tip cleaner. Larry Jeffus

problem with the equipment, and a qualified technician should be called. After locating and repairing the problem, blow gas through the tip for a few seconds to clear out any soot that may have accumulated in the passages. A flash-back that burns in the hose leaves a carbon char inside that may explode and burn in a pressurized oxygen system. A fuel gas is not required to kindle a hot, severe fire inside such hose sections. Discard hose sections in which a flash-back has occurred and obtain new hose.

REVERSE FLOW AND FLASHBACK VALVES

The purpose of the reverse flow valve is to prevent gases from accidentally flowing through the torch and into the wrong hose. If the gases being used are allowed to mix in the hose or regulator, they might explode. The reverse flow valve is a spring-loaded check valve that closes when gas tries to flow backward through the torch valves, **Figure 30-29.** Some torches have reverse flow valves built into the torch body, but most torches must have these safety devices added. If the torch does not come with a reverse flow valve, it must be added to either the torch end or regulator end of the hose.

A reverse flow of gas will occur if the torch is not turned off or bled properly. The torch valves must be opened one at a time so that the gas pressure in that hose will be vented into the atmosphere and not through the torch into the other hose, **Figure 30-30.**

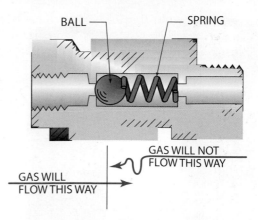

FIGURE 30-29 Reverse flow valve only. © Cengage Learning 2012

▰▰▰ **CAUTION** ▰▰▰

If both valves are opened at the same time, one gas may be pushed back up the hose of the other gas.

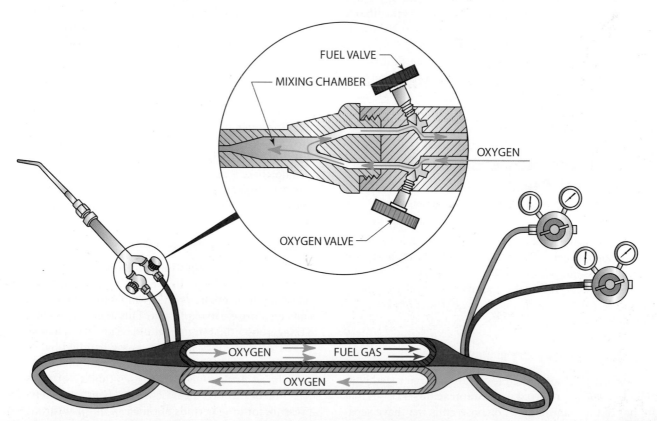

FIGURE 30-30 Gas may flow back up the hose if both valves are opened at the same time when the system is being bled down after use. Installing reverse flow valves on the torch can prevent this from occurring. © Cengage Learning 2012

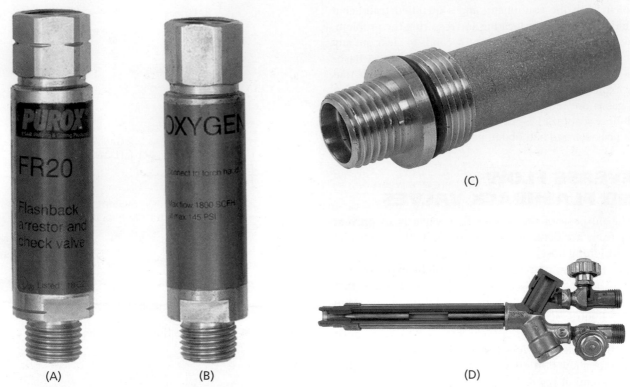

FIGURE 30-31 Combination flashback arrestors and check valves for (A) acetylene and (B) oxygen. (C) Replacement cartridge for flashback arrestor. (D) Torch designed with replaceable flashback arrestors and check valves built into the torch body. (A & B) ESAB Welding & Cutting Products; (C & D) Victor Equipment Company

A reverse flow valve will not stop the flame from a flashback from continuing through the hoses. A **flashback arrestor** will do the job of a reverse flow valve, and it will also stop the flame of a flashback, **Figure 30-31**. The flashback arrestor is designed to quickly stop the flow of gas during a flashback. These valves work on a similar principle as the gas valve at a service station. They are very sensitive to any back pressure in the hose and stop the flow if any back pressure is detected.

Care of the Reverse Flow Valve and Flashback Arrestor

Both devices must be checked on a regular basis to see that they are working correctly. The internal valves may become plugged with dirt, or they may become sticky and not operate correctly. To test the reverse flow valve, you can try to blow air backward through the valve. To test the flashback arrestor, follow the manufacturer's recommended procedure. If the safety device does not function correctly, it must be replaced.

HOSES AND FITTINGS

Most welding hoses used today are molded together as one piece and are referred to as **Siamese hose.** Hoses that are not of the Siamese type, or hose ends that have separated, may be taped together. When taping the hoses, they must not be taped solidly. They should be wrapped for about 2 in. (51 mm) out of every 12 in. (305 mm) of hose length, allowing the colors of the hose to be seen.

Fuel-gas hoses must be red and have left-hand threaded fittings. Oxygen hoses must be green and have right-hand threaded fittings.

Hoses are available in four sizes: 3/16 in. (4.8 mm), 1/4 in. (6 mm), 5/16 in. (8 mm), and 3/8 in. (10 mm). The size given is the inside diameter of the hose. Larger sizes offer less resistance to gas flow and should be used where long hose lengths are required. The smaller sizes are more flexible and easier to handle for detailed work.

The three sizes of hose end fittings available are A (small), B (standard), and C (large). The three sizes are made to fit all hose sizes.

Hose Care and Use

When hoses are not in use, the gas must be turned off and the pressure bled off. Turning off the equipment and releasing the pressure prevents any undetected leaks from causing a fire or an explosion. This action also eliminates a dangerous situation that would be created if a hose were cut by equipment or materials being handled by workers who were unfamiliar with welding equipment. In addition, hoses are permeable to gases (ability of the gas to pass into or through the hose walls). Thus, gases left under pressure for long periods of time can migrate through the hose walls and mix with each other, **Figure 30-32**. If the gases mix and the torch is lit without first purging the lines,

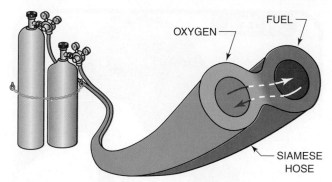

FIGURE 30-32 Gas left under pressure may migrate through the hose walls. © Cengage Learning 2012

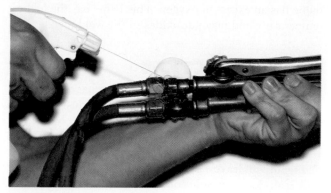

FIGURE 30-34 Screwing the hose nut onto a fitting will help when pushing the nipple into the hose. Larry Jeffus

the hoses can explode. For this reason, if the welder is not certain that the hoses were bled, it is recommended that they be purged before the torch is lit.

Hoses are resistant to burns, but they are not burn-proof. They should be kept out of direct flame, sparks, and hot metal. You must be especially cautious when using a cutting torch. If it becomes damaged, the damaged section should be removed and the hose repaired with a splice. Damaged hoses should never be taped to stop leaks.

Hoses should be checked periodically for leaks. To test a hose for leaks, adjust the regulator to a working pressure with the torch valves closed. Wet the hose with a leak-detecting solution by rubbing it with a wet rag, spraying it, or dipping it in a bucket. Then watch for bubbles, which indicate that the hose leaks.

The hose fittings can be changed if the old ones become damaged. Several kits are available that have new nuts, nipples, ferrules, a ferrule crimping tool, and any other supplies required to replace the hose ends, **Figure 30-33**.

FIGURE 30-35 Crimping hose ferrule. Larry Jeffus

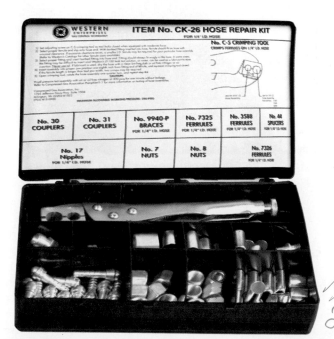

FIGURE 30-33 Hose repair kit. Larry Jeffus

To replace the hose end, the hose is first cut square. The correct-size ferrule is inserted. Then both the hose end and nipple are sprayed with a leak-detecting solution. This will help the nipple slide in more easily. Screw the nipple and nut on a torch body. This will hold the nipple deep inside the nut, and the body will act as a handle for leverage as the nipple is pushed inside the hose, **Figure 30-34**. After the hose is slid up to the nut, crimp the ferrule until it is tight. The crimping tool should be squeezed twice, the second time at right angles to the first, **Figure 30-35**. When the crimping is complete, install the hose on a torch and regulator. Then adjust the regulator to a working pressure and spray the fitting with a leak-detecting solution. Watch for any bubbles, which indicate a leaking fitting.

TYPES OF FLAMES

There are three distinctly different oxyacetylene flame settings. A carburizing flame has an excess of fuel gas. This flame has the lowest temperature and may put extra carbon in the weld metal. A neutral flame has a balance of fuel gas and oxygen. It is the most commonly used flame because it adds nothing to the weld metal. An oxidizing flame has an excess of oxygen. This flame has the highest temperature and may put oxides in the weld metal.

LEAK DETECTION

A leak-detecting solution can be purchased premixed and ready to use or as a concentrate that must be mixed with water. It can also be mixed in the shop by using a small quantity of liquid dishwashing detergent in water. Use only enough detergent to produce bubbles; too much detergent will leave a soapy film.

A leak-detecting solution must be free flowing so that it can seep into small joints, cracks, and other areas that may have a leak. The solution must produce a good quantity of bubbles without leaving a film. The solution can be dipped, sprayed, or brushed on the joints.

///// **CAUTION** \\\\\

Some detergents are not suitable for O₂ because of an oil base. Use only O₂-approved leak-detection solutions on oxygen fittings.

FIGURE 30-37 Unscrew the valve protector caps. Put the caps in a safe place as they must be replaced on empty cylinders before they are returned. Larry Jeffus. *See DVD Oxyacetylene Welding.*

PRACTICE 30-1

Setting Up an Oxyfuel Torch Set

This practice requires a disassembled oxyfuel torch set consisting of two regulators, two reverse flow valves, one set of hoses, a torch body, a welding tip, two cylinders, a portable cart or supporting wall, and a wrench. You will assemble the equipment in a safe manner.

1. Safety chain the cylinders separately to the cart or to a wall, **Figure 30-36**. Then remove the valve protection caps, **Figure 30-37**.
2. Crack the cylinder valve on each cylinder for a second to blow away dirt that may be in the valve, **Figure 30-38**.

FIGURE 30-38 Cracking the oxygen and fuel cylinder valves to blow out any dirt lodged in the valves. Larry Jeffus. *See DVD Oxyacetylene Welding.*

FIGURE 30-36 Safety chain cylinder. Larry Jeffus. *See DVD Oxyacetylene Welding.*

///// **CAUTION** \\\\\

If a fuel-gas cylinder does not have a valve hand wheel permanently attached, you must use a non-adjustable wrench to open the cylinder valve. The wrench must stay with the cylinder as long as the cylinder is on, Figure 30-39.

(A)

(B)

(C)

FIGURE 30-39 Nonadjustable wrenches for acetylene cylinders. ESAB Welding & Cutting Products

(A)

(B)

FIGURE 30-40 Attach the oxygen regulator (A) to the oxygen cylinder valve. Using a wrench (B), tighten the nut. Larry Jeffus. **See DVD Oxyacetylene Welding.**

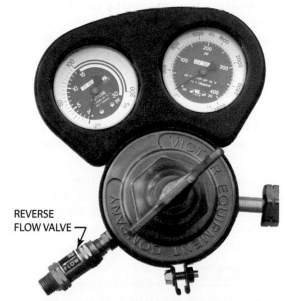

REVERSE
FLOW VALVE

FIGURE 30-41 Install a reverse flow valve if one is not built into the torch body before attaching the hose to the regulator. Larry Jeffus

3. Attach the regulators to the cylinder valves, **Figure 30-40A**. The nuts can be started by hand and then tightened with a wrench, **Figure 30-40B**.

4. Attach a reverse flow valve or flashback arrestor, if the torch does not have them built in, to the hose connection on the regulator or to the hose connection on the torch body, depending on the type of reverse flow valve in the set, **Figure 30-41**.

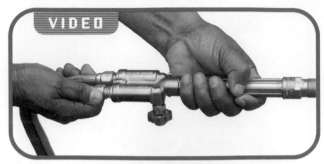

FIGURE 30-42 Connect the free ends of the oxygen (green) and the acetylene (red) hoses to the welding torch. Larry Jeffus. *See DVD Oxyacetylene Welding.*

Occasionally test each reverse flow valve by blowing through it to make sure it works properly.

5. Connect the hoses. The red hose has a left-hand grooved nut and attaches to the fuel-gas regulator. The green hose has a right-hand nut without grooves and attaches to the oxygen regulator.

6. Attach the torch to the hoses, **Figure 30-42.** Connect both hose nuts and tighten by hand before using a wrench to tighten either one.

7. Check the tip seals for nicks or the O-rings, if used, for damage. In most cases, tips that have O-ring–type seals are hand tightened, and tips that have metal-to-metal seals are wrench tightened, but it is best to check the owner's manual, or a supplier, to determine if the torch tip should be tightened, **Figure 30-43.**

///// **CAUTION** \\\\\

Always stand to one side. Point the valve away from anyone in the area and be sure there are no sources of ignition when cracking the valve.

///// **CAUTION** \\\\\

Tightening a tip the incorrect way may be dangerous and might damage the equipment.

Check all connections to be sure they are tight. The oxyfuel equipment is now assembled and ready for use.

Complete a copy of the "Student Welding Report" listed in Appendix I or provided by your instructor. ◆

PRACTICE 30-2

Turning On and Testing a Torch

Using the oxyfuel equipment that was properly assembled in Practice 30-1, a nonadjustable tank wrench, and

FIGURE 30-43 Select the proper tip or nozzle and install it on the torch body. Larry Jeffus. *See DVD Oxyacetylene Welding.*

a leak-detecting solution, you will pressurize the system and test for leaks.

1. Back out the regulator pressure adjusting screws until they are loose, **Figure 30-44.**

2. Standing to one side of the regulator, open the cylinder valve slowly so that the pressure rises on the gauge slowly, **Figure 30-45.**

FIGURE 30-44 Back out both regulator adjusting screws before opening the cylinder valve. Larry Jeffus. *See DVD Oxyacetylene Welding.*

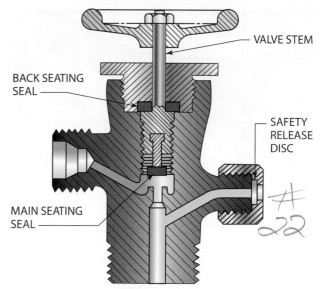

VALVE STEM

BACK SEATING SEAL

SAFETY RELEASE DISC

MAIN SEATING SEAL

FIGURE 30-46 Cutaway of an oxygen cylinder valve showing the two separate seals. The back seating seal prevents leakage around the valve stem when the valve is open. © Cengage Learning 2012

FIGURE 30-45 Stand to one side when opening the cylinder valve. Larry Jeffus. *See DVD Oxyacetylene Welding.*

3. Open the oxygen valve all the way until it is sealed at the top, **Figure 30-46.**

4. Open the acetylene or other fuel gas valve one-quarter turn, or just enough to get gas pressure, **Figure 30-47.** If the cylinder valve does not have a handwheel, use a nonadjustable wrench and leave it in place on the valve stem while the gas is on.

5. Open one torch valve and point the tip away from any source of ignition, including the cylinders, regulators, and hoses. Slowly turn in the pressure adjusting screw until gas can be heard escaping from the torch. The gas should flow long enough to allow the hose to be completely **purged** (emptied) of air and replaced by the gas before the torch valve is closed. Repeat this process with the other gas.

6. After purging is completed, and with both torch valves off, adjust both regulators to read 5 psig (35 kPag), **Figure 30-48.**

7. Spray a leak-detecting solution on each hose and regulator connection and on each valve stem on the torch and cylinders. Watch for bubbles,

FIGURE 30-47 Open the cylinder valve slowly. Larry Jeffus. *See DVD Oxyacetylene Welding.*

FIGURE 30-48 Adjust the regulator to read 5 psig (0.35 kg/cm²g) working pressure. Larry Jeffus. *See DVD Oxyacetylene Welding.*

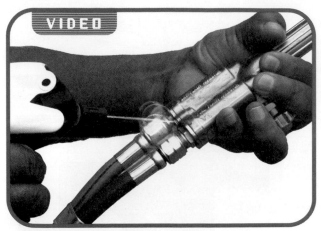

FIGURE 30-49 Spray fittings with a leak-detecting solution. Larry Jeffus. **See DVD Oxyacetylene Welding.**

FIGURE 30-50 Identify any cylinder that has a problem by marking it. Larry Jeffus

which indicate a leak. Turn off the cylinder valve before tightening any leaking connections, Figure 30-49.

///CAUTION\\\

If the valve is opened quickly, the regulator or gauge may be damaged, or the gauge may explode.

///CAUTION\\\

The acetylene valve should never be opened more than one and a half turns so that in an emergency it can be turned off quickly.

///CAUTION\\\

Connections should not be overtightened. If they do not seal properly, repair or replace them.

///CAUTION\\\

Leaking cylinder valve stems should not be repaired. Turn off the valve, disconnect the cylinder, mark the cylinder, move it outdoors or to a very well-ventilated area, and notify the supplier to come and pick up the bad cylinder, Figure 30-50.

The assembled oxyfuel welding equipment is now tested and ready to be ignited and adjusted.

Complete a copy of the "Student Welding Report" listed in Appendix I or provided by your instructor. ◆

PRACTICE 30-3

Lighting and Adjusting an Oxyacetylene Flame

Using the assembled and tested oxyfuel welding equipment from Practice 30-2, a **spark lighter,** gas welding goggles, gloves, and proper protective clothing, you will light and adjust an oxyacetylene torch for welding.

1. Wearing proper clothing, gloves, and gas welding goggles, turn both regulator adjusting screws in until the working pressure gauges read 5 psig (35 kPag). If you mistakenly turn on more than 5 psig (35 kPag), open the torch valve to allow the pressure to drop as the adjusting screw is turned outward.

2. Turn on the torch fuel-gas valve just enough so that some gas escapes.

3. Using a spark lighter, light the torch. Hold the lighter near the end, **Figure 30-51**, of the tip but not covering the end, **Figure 30-52**.

FIGURE 30-51 Correct position to hold a spark lighter. Larry Jeffus. **See DVD Oxyacetylene Welding.**

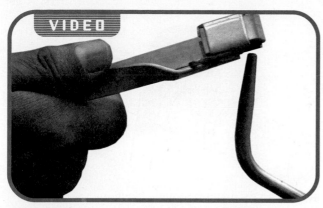

FIGURE 30-52 Spark lighter held too close over the end of the tip. Larry Jeffus. **See DVD Oxyacetylene Welding.**

4. With the torch lit, increase the flow of acetylene until the flame stops smoking.

5. Slowly turn on the oxygen and adjust the torch to a neutral flame.

///// **CAUTION** \\\\\

Be sure the torch is pointed away from any sources of ignition or any object or person that might be damaged or harmed by the flame when it is lit.

///// **CAUTION** \\\\\

A spark lighter is the only safe device to use when lighting any torch.

This flame setting uses the minimum gas flow rate for this specific tip. The fuel flow should never be adjusted to a rate below the point where the smoke stops. This is the minimum flow rate at which the cool gases will pull the flame heat out of the tip, **Figure 30-53**. If excessive heat is allowed to build up in a tip, it can cause a backfire or flashback.

The maximum gas flow rate gives a flame enough flow so that, when adjusted to the neutral setting, it does not settle back on the tip. This will help keep the tip cooler so it is less likely to backfire.

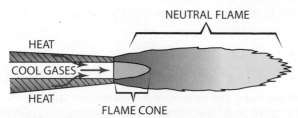

FIGURE 30-53 Enough cool gas flowing through the tip will help prevent popping. © Cengage Learning 2012

Complete a copy of the "Student Welding Report" listed in Appendix I or provided by your instructor. ◆

PRACTICE 30-4

Shutting Off and Disassembling Oxyfuel Welding Equipment

Using the properly lit and adjusted torch from Practice 30-3 and a wrench, you will extinguish the flame and disassemble the torch set.

1. First, quickly turn off the torch fuel-gas valve. This action blows the flame out and away from the tip, ensuring that the fire is out. In addition, it prevents the flame from burning back inside the torch. On large tips or hot tips, turning the fuel off first may cause the tip to pop. The pop is caused by a lean fuel mixture in the tip.

    If you find that the tip pops each time you turn the fuel off first, turn the oxygen off first to prevent the pop. Be sure that the flame is out before putting the torch down.

2. After the flame is out, turn off the oxygen valve.

3. Turn off the cylinder valves.

4. Open one torch valve at a time to bleed off the pressure.

5. When all of the pressure is released from the system, close both torch valves and back both regulator adjusting screws out until they are loose.

6. Loosen both ends of both hoses and unscrew them.

7. Loosen both regulators and unscrew them from the cylinder valves.

8. Replace the valve protection caps.

Complete a copy of the "Student Welding Report" listed in Appendix I or provided by your instructor. ◆

MANIFOLD SYSTEMS

A **manifold system** can be used if there are a number of workstations or if a high volume of gas will be used. The manifold can also be used to keep the cylinders out of the work area and to reduce the number of cylinders needed at one time.

Manifolds must be located 20 ft (6 m) or more from the actual work, or they must be located so that sparks cannot reach them. Oxygen manifolds must be at least 20 ft (6 m) from fuel-gas manifolds and flammable materials or be separated from them by a noncombustible wall 5 ft (1.5 m) high, with a 1/2-hour, fire-resistant rating. Inert gas manifolds can be placed with either oxygen or fuel-gas manifolds.

The rooms that are used for manifolds can also be used for cylinder storage, but full and empty cylinders must be kept separated within the room. Fuel-gas manifold rooms must have good ventilation and explosion-proof lights,

FIGURE 30-54 Explosion-proof light fixtures suitable for a manifold room. Cooper Crouse-Hinds®

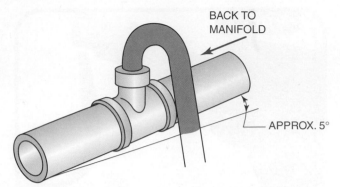

FIGURE 30-55 A slight angle will allow moisture to flow back away from the station regulators. © Cengage Learning 2012

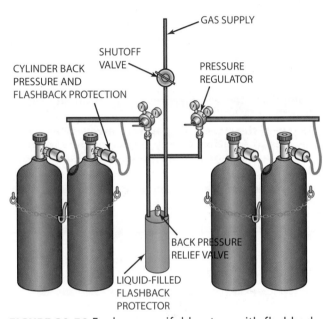

FIGURE 30-56 Fuel gas manifold system with flashback and back pressure safety equipment. © Cengage Learning 2012

Figure 30-54. They must also have a sign on the door that reads "Danger: No Smoking, Matches, or Open Lights," or similar wording.

Piping for the high-pressure side of a manifold must be steel, stainless steel, or alloyed copper. Piping for the low-pressure side, except acetylene, can be stainless steel, copper (type L or K), brass, steel, or wrought iron. All acetylene piping must be steel or wrought iron. Unalloyed copper used with acetylene forms an unstable and explosive compound, copper acetylide.

The pipe joints in copper or brass lines can be welded, brazed, threaded, or flanged. Joints in steel or iron pipe can be welded, threaded, or flanged. Nitrogen should be purged through pipes being welded or brazed. All piping must be clean and oil free and should have a slight angle back toward the manifold so that any moisture will run back toward it, **Figure 30-55.**

Manifold systems should be tested for leaks at one and a half (1 1/2) times the operating pressure. The fuel-gas lines must be protected from oxygen flowing back into the lines by installing a reverse flow valve. In addition, fuel-gas lines must have flashback protection and back pressure release. One device can be installed to satisfy all three requirements, **Figure 30-56.**

Manifold Operation

The pipes should be cleaned with an oil-free, noncombustible fluid before the regulators are attached. Solutions of caustic soda or trisodium phosphate are good for this purpose. Once the system is cleaned, install the regulators and purge the system with nitrogen. Nitrogen will pick up any moisture in the system while providing a noncombustible gas in the system.

After the system has been filled with nitrogen, start filling the pipes with the oxygen or fuel gas. Allow the gas to escape freely from the station farthest away from the manifold. Continue purging until all the nitrogen is removed from the manifold.

CAUTION

Make sure there is good ventilation during the purging and that there are no sources of ignition near the escaping gas.

FIGURE 30-57 Label on gauge notes the maximum pressure to be used. Larry Jeffus.
See DVD Oxyacetylene Welding.

FIGURE 30-58 Manifold station regulators. Larry Jeffus.
See DVD Oxyacetylene Welding.

Set the line pressure as low as possible, still ensuring that the type of work being done at the workstations can be performed satisfactorily, **Figure 30-57**. When work is completed, the system must be turned off and each station, manifold, and cylinder valve closed. If a station regulator is removed, a cap must be put on the line so that air cannot enter the system, **Figure 30-58**. A complete list of operating, maintenance, and emergency procedures should be clearly posted for all manifolds.

///// **CAUTION** \\\\\

In case of fire or severe threatening weather, such as a tornado, all cylinder valves must be turned off.

Summary

After you have developed your skills and found a welding job, you will be exposed to welders with many years of experience. These welders will have developed many good shortcuts through the years that can help you on the job. However, there are few safe shortcuts in equipment setup and maintenance. The safe way of setting up and testing a system should always be followed. Be careful to avoid questionable practices. Always refer to the manufacturer's operating instructions and safety recommendations for the type of equipment you are using.

Worcester Technical High School Welds on Snowflakes

The welding program at Worcester Technical High School in Massachusetts has a long history of student projects. According to a blueprint hanging on the shop wall, lighthouses like the one shown in Figure 1 have been built by many students over the years since 1930. Some of the other projects built by students include a replica of the home of one of Worcester's first welding instructor, Mr. John Kassabulan, Figure 2; replicas of houses at Old Sturbridge Village, Figure 3; flowers, Figure 4; and many others. So it was no surprise to anyone when the students took on a project suggested by the City Manager to build 100 snowflake holiday decorations for Worcester's downtown streetlights.

Mr. Michael V. O'Brien, Worcester's City Manager, said one day he realized that all great cities decorate their downtown streets for the holidays, but Worcester had not had a wreath or candy cane since the 1980s. With tight budgets, Worcester's City Manager knew they could not just shell out a lot of money for holiday decorations.

The City Manager found a way to make his idea come to fruition; it involved using a great local resource—the students, faculty, and administration of Worcester Technical

FIGURE 2 Larry Jeffus

FIGURE 1 Larry Jeffus

FIGURE 3 Larry Jeffus

FIGURE 4 Larry Jeffus

FIGURE 6 Larry Jeffus

FIGURE 7 Larry Jeffus

High School along with some great support from some of Worcester's businesses.

The professional designers at Sunshine Sign Co. of Grafton, Massachusetts, took Mr. O'Brien's sketches and polished them into a spectacular lighted wreath of snowflakes design with an ornate "W" for Worcester, Figure 5.

Once the design was completed and the aluminum stock and other supplies obtained, the students and faculty at Worcester Technical High School took over. Using a plasma cutter, Figure 6, they cut out more than 1000 snowflakes, Figure 7, and 100 W's, Figure 8. Students painstakingly ground and finished each snowflake, Figure 9, laid them out, Figure 10, to be welded to the circular ring, Figure 11.

Because of the tight time frame, GMA welding was used to help speed up the process of attaching all the snowflakes and W's to the 44 four-foot diameter and 56 three-foot diameter decorations, Figure 12. Once all of the welding and grinding was done, the decorations were sent out to be powder-coated a light pewter-gray color. The electrostatic painting process

FIGURE 8 Larry Jeffus

FIGURE 5 Larry Jeffus

was selected because it produces a uniform finish that is very durable. When the wreathes were returned to the shop, white LED lights were attached as the final finishing touch.

Mr. Kevin Casavoy, the Welding Department Head at Worcester Technical High School, believes that this project gave his students some valuable real world experiences. "It's not often that we are able to give students opportunities like this where they are truly working like a production welding

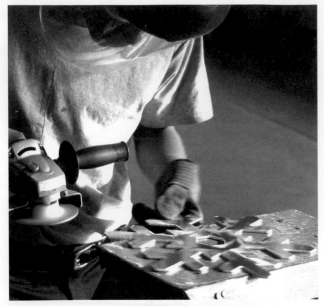

FIGURE 9 Larry Jeffus

FIGURE 11 Larry Jeffus

FIGURE 10 Larry Jeffus

FIGURE 12 Larry Jeffus

shop. They all learned about production standards and tight deadlines from this major fabrication project."

The students took $30,000 worth of materials and turned it into 100 snowflake wreaths valued at around $2500 each. They created approximately $250,000 worth of holiday decorations that will last for years. Mr. Casavoy told his students, "Twenty years from now when you're driving down the street, you can look up at the wreathes and say, 'We built those back in 2010."

On a cold windy December 15th, the Mayor, City Council, City Staff, and citizens of Worcester showed their gratitude to the Worcester Technical High School students and their parents in a ceremony downtown at Worchester City Hall. All in all this was an experience that everyone will remember for years to come.

Review

1. What is the purpose of a pressure regulator?

2. What may result if a pressure regulator is not used on the type of gas or pressure range for which it was designed?

3. Describe how a single-stage regulator operates.

4. Is the torch pressure always the same as the working gauge pressure? Why or why not?

5. Why does the high-pressure gauge on a regulator not always indicate the amount of gas in the cylinder?

6. How does the operation of a safety relief valve differ from the operation of a safety disc valve?

7. Describe the difference between an argon cylinder valve fitting and a carbon dioxide cylinder valve fitting.

8. What is meant by *regulator creep*?

9. Who can repair regulators?

10. Why must the pressure be released from a regulator when work is finished?

11. Why are combination welding and cutting torches considered to be more versatile?

12. What is the advantage of using an injector-type mixing chamber?

13. What should be done to the valve packing if the valve knob does not turn freely after it has been tightened to stop a leak?

14. What may happen to a tip seat if it is incorrectly tightened?

15. What can happen to a tip if it is excessively cleaned with a tip cleaner?

16. What is the difference between a reverse flow valve and a flashback arrestor?

17. What are Siamese hoses?

18. Why must the pressure be bled off hoses when work is complete?

19. What is the difference between a backfire and a flashback?

20. Why is a neutral flame the most commonly used oxyacetylene flame used?

21. What properties should a good leak-detecting solution have?

22. Why must the oxygen cylinder valve be opened all the way to the top?

23. How long should hoses be purged?

24. What should be done with cylinders that have leaking valve stems?

25. How should the spark lighter be held to light a torch?

26. Once the torch is lit, why must the acetylene flow be increased until the flame stops smoking before the oxygen is turned on for adjustment?

27. What type of piping can be used for a manifold system?

Chapter 31

Oxyfuel Gases and Filler Metals

OBJECTIVES

After completing this chapter, the student should be able to

- explain the chemical reaction that takes place in any oxyfuel flame.
- list the major advantages and disadvantages of the different fuel gases.
- demonstrate an ability to choose correct filler metals.
- explain what conditions affect the selection of filler metal.

KEY TERMS

acetone

acetylene (C_2H_2)

atoms

backfire

carbonizing (carburizing)

combustion

combustion rate

ferrous filler metals

filler metals

flashback

heat energy

hydrocarbons

inner cone

liquefied fuel gases

MAPP

methylacetylene-propadiene (MPS)

molecules

neutral flame

optical pyrometer

outer envelope

oxidizing flame

oxyfuel flame

oxyhydrogen

oxyhydrogen flame

piccolo tube

primary combustion

secondary combustion

INTRODUCTION

The general grouping of processes known as oxyfuel consists of a number of separate processes, all of which burn a fuel gas with oxygen. The oxyfuel flame was used for fusion welding as early as the first half of the 1800s when scientists developed the oxyhydrogen torch. Before that time, air fuel torches were used, but because the flame was not hot enough they had limited success. The early use of pure oxygen with hydrogen or acetylene as the fuel gas often resulted in flashbacks and explosions. The use of water traps helped prevent most flashbacks from becoming explosions. But until the early development of the torch mixing chamber, welding was a very dangerous occupation. The mixing chamber gave a more uniform flame that was less likely to flash back.

During the early 1900s, the oxyacetylene flame became more popular as the primary means of welding. Since 1900, when the first shielded metal arc welding (stick) electrodes were introduced by Strohmeyer in Britain, the use of the oxyacetylene flame for welding has declined. During oxyacetylene welding's prime, plates 1 in. (25 mm) thick or more were gas welded to build everything from large, seagoing ships to massive machines used during the Industrial Revolution. Today, because of improvements in other processes, the oxyacetylene flame is seldom used on metal thicker than 1/16 in. (2 mm).

The advances in shielded metal arc welding, gas tungsten arc welding, plasma arc cutting, and gas metal arc welding have overshadowed oxyfuel welding. These processes are faster and cleaner and cause less distortion than oxyfuel welding. Currently, oxyfuel welding is used mainly for farm repairs, maintenance, and in smaller shops.

OXYFUEL FLAME

The usefulness of the **oxyfuel flame** as a primary means of cutting ferrous metals has increased. In 1887, the flame was used to melt through thin metal. Around 1900, the oxygen lance was introduced. The oxygen lance allowed high-pressure oxygen to be directed onto metal heated by a torch, resulting in much improved cuts. The later development of a cutting torch with preheating flames surrounding a central oxygen hole in the tip brought oxyfuel cutting into its own. The cutting torch, used by hand or as part of a machine, is used to rapidly cut out steel parts, **Figure 31-1.**

Today, a large number of manufactured items are touched in some way by the oxyfuel flame. The parts may have been cut out, heated, hardened, or joined with an oxyfuel flame. The expansion of the role of the oxyfuel flame in industry has led to the introduction of new fuel gases and gas mixtures. Each of the new gases has certain advantages and disadvantages. The choice of which gas to use must be based on cost, availability, welder skills and skill changes required, equipment changes, safety, handling, performance, and other concerns. The information supplied for a special gas often points out its strengths and may use comparisons to prove its advantages.

Characteristics of the Fuel-Gas Flame

The data available for fuel-gas flame characteristics are not gathered in a consistent manner. Temperature and heat, for example, are very basic physical facts about the flame of a specific gas. But even these facts can be misleading. The flame condition, such as neutral, oxidizing, or **carbonizing (carburizing),** and/or the purity of the gases being used will affect the temperature of the flame. The method by which the temperature is measured also affects its value. The highest potential temperature values are determined by chemical analysis and by the calculation of the theoretical energy released. However, no combustion is perfect, so this temperature is never attained. An infrared analysis or **optical pyrometer** of the flame gives the highest temperature in the flame. But this temperature is concentrated in a thin layer around the inner cone and is so small that it is not of practical use.

The optical pyrometer and the infrared analysis both give an accurate temperature reading, but where the temperature is measured makes a difference.

Differences in heat values may also be misleading, depending upon how they are obtained. The temperatures, heat, and other flame characteristics noted in this chapter are as close as possible to those that occur during use. They are not necessarily the highest or lowest possible values and, therefore, should not be considered absolute. But they can, however, be compared with each other for the purpose of selecting a fuel gas.

FUEL GASES

Most fuel gases used for welding are **hydrocarbons.** The gases are made up of hydrogen (H) and carbon (C) atoms. The atoms are bound together tightly to form **molecules.** Each molecule of a specific gas has the same type, number, and arrangement of **atoms.** The number of atoms in a molecule of gas varies from one gas to another. A molecule of acetylene is made up of two hydrogen (H) and two carbon (C) atoms (C_2H_2). A molecule of propane is made up of eight hydrogen (H) and three carbon (C) atoms (C_3H_8). The chemical formulas for these two fuel gases are shown in **Figure 31-2.**

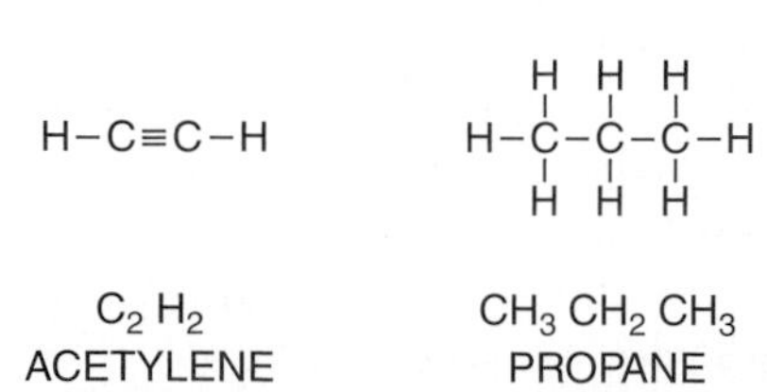

FIGURE 31-1 Modern handheld cutting torch. Larry Jeffus

H—C≡C—H

C_2H_2
ACETYLENE

H—C—C—C—H (with H H H above and H H H below)

CH_3 CH_2 CH_3
PROPANE

FIGURE 31-2 Chemical formulas for two hydrocarbons used as fuel gases. © Cengage Learning 2012

	Cu ft/cu ft			
	Total O₂ Required	Supplied O₂ through Torch*	% Total through Torch	Cu ft O₂ per lb of Fuel
Acetylene— 5589°F/1470 Btu	2.5	1.3	50.0	18.9
MAPP gas— 5301°F/2406 Btu	4.0	2.5	62.5	22.1
Natural gas— 4600°F/1000 Btu	2.0	1.9	95.0	44.9
Propane— 4579°F/2498 Btu	5.0	4.3	85.0	37.2
Propylene— 5193°F/2371 Btu	4.5	3.5	77.0	31.0

*Balance of total oxygen demand is entrained in the fuel-gas flame from the atmosphere.

TABLE 31-1 Oxygen Consumption for Fuels (Neutral Flame) MAPP Products

If a mixture of acetylene (C_2H_2) and oxygen (O_2) is ignited, the acetylene molecule bonds are broken, and new bonds among the hydrogen, carbon, and oxygen atoms are formed. The breaking down of acetylene molecules releases energy in the form of heat and light. The rate at which this reaction occurs results in a change in temperature. The formation of new bonds with the oxygen releases still more energy. This chemical reaction, known as **combustion,** is rapid oxidation.

The number of oxygen atoms required to completely combust the fuel gas varies, depending upon its molecular makeup. **Table 31-1** lists some fuel gases and the amount of oxygen required to complete the reaction in the flame.

Combustion of acetylene is divided into two separate chemical reactions. The first reaction is referred to as **primary combustion,** and the second reaction is referred to as **secondary combustion,** Figure 31-3.

In the primary reaction of acetylene (C_2H_2), the oxygen (O_2) atoms unite with the carbon (C) atoms. This reaction liberates energy and forms carbon monoxide (CO) and free hydrogen, **Figure 31-4.**

In the secondary reaction, oxygen (O_2) from the surrounding air unites with free hydrogen (H) to form water vapor (H_2O) and liberates more heat. Also, the carbon monoxide (CO) unites with additional oxygen (O_2) from the air to form carbon dioxide (CO_2), **Figure 31-5.**

The final products of all clean-burning hydrocarbon flames are the same: water vapor and carbon dioxide.

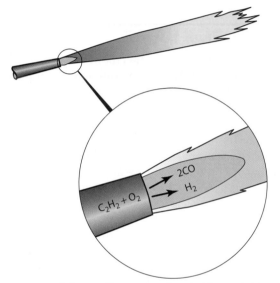

FIGURE 31-4 Primary flame reaction (with acetylene as the fuel gas). © Cengage Learning 2012

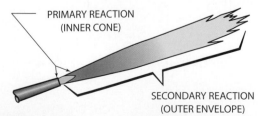

FIGURE 31-3 Parts of an oxyfuel flame.
© Cengage Learning 2012

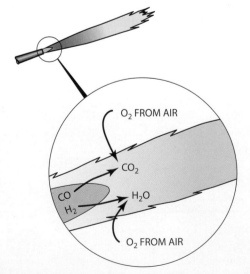

FIGURE 31-5 Secondary flame reaction.
© Cengage Learning 2012

Fuel Gas and Oxygen Mixtures Burn Rates						
Burning Velocity	Acetylene	MAPP	Natural Gas	Propane	Propylene	Hydrogen
ft/sec	22.7	15.4	15.2	12.2	15.1	36
mm/sec	6097	4694	4633	3718	4602	6540

TABLE 31-2 Combustion Rates

The temperature, rate of combustion, and quantity of heat release are the characteristics that make each flame different. These characteristics make some gases better than others for certain operations.

Flame Rate of Burning

The **combustion rate** or rate of propagation of a flame is the rate or speed at which the flame burns. The combustion or burn rate of a fuel gas is given in feet per second (meters per second).

The combustion rate of a gas is determined by the amount of **heat energy** required to break the bonds between the atoms of its molecules and by the amount of heat energy liberated as the bonds are broken. The ratio of fuel to oxygen and the homogeneity of their mixture also affect the propagation rate.

A homogeneous mixture of 50% acetylene and 50% oxygen has a burn rate of 22.7 ft per second (6.9 m per second). As the percentage of oxygen in the mixture increases, the burn rate increases; as the percentage of oxygen decreases, the burn rate also decreases.

The higher the combustion rate of an oxygen fuel mixture, the more prone the mixture is to **backfire** or **flashback**. Table 31-2 lists the combustion rates for most fuel gases.

EXPERIMENT 31-1

Burn Rate

Using properly set-up and adjusted oxyfuel welding equipment, striker, goggles, gloves, and any other required safety equipment, you are going to observe how changing the mixture ratios of oxygen and acetylene affects the combustion rate.

Set both regulators at 5 psig (35 kPag) and purge the lines as required. With only the acetylene gas torch valve on, use the striker to light the torch. Adjust the gas valve until the flame stops smoking, **Figure 31-6**. Turn on the oxygen torch valve and note the immediate reduction in the flame size. Continue opening the valve until the flame is at a neutral setting, **Figure 31-7**. With the flame at a neutral setting, the total flame size is smaller than it was with just the acetylene even though a larger total volume of gas is coming from the tip. Now increase the oxygen flow by opening the valve more and notice that the size of the flame again decreases, **Figure 31-8**.

In this experiment you have observed that, as the percentage of oxygen in the fuel gas increases, the flame propagation or burning rate also increases. This increase is indicated because, even with the increase in total volume of gas, the actual size of the flame decreases.

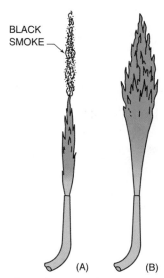

FIGURE 31-6 Adjust the gas valve until the flame stops smoking. © Cengage Learning 2012

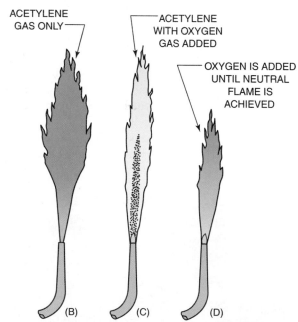

FIGURE 31-7 Achieving the neutral setting of the flame. © Cengage Learning 2012

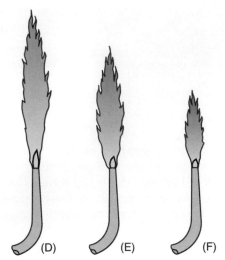

FIGURE 31-8 An increase in oxygen flow again decreases the size of the flame. © Cengage Learning 2012

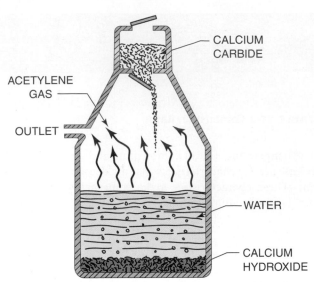

FIGURE 31-10 An acetylene generator is used to mix calcium carbide with water. © Cengage Learning 2012

Complete a copy of the "Student Welding Report" listed in Appendix I or provided by your instructor. ◆

///// **CAUTION** \\\\\

An extremely oxygen-rich fuel-gas mixture may have a combustion rate that is greater than the gas flow rates of large tips, resulting in a backfire or flashback. If this occurs, close the oxygen valve immediately.

Acetylene (C_2H_2)

Acetylene (C_2H_2) is the most frequently used fuel gas. The mixture of oxygen and acetylene produces a high heat and high temperature flame that is widely used for welding, cutting, brazing, heating, metallizing, and hardsurfacing.

Acetylene is produced by mixing calcium carbide (CaC_2) with water (H_2O). Calcium carbide is produced by smelting coke (a coal by-product) and lime in airtight electric arc furnaces. After smelting, the calcium carbide is cooled, crushed, and packed in dry, airtight containers. Either a fixed or portable acetylene generator, **Figure 31-9**,

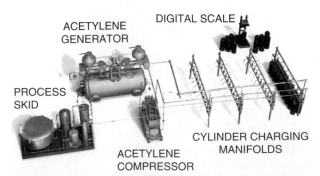

FIGURE 31-9 Acetylene generator. Rexarc International, Inc.

is used to mix the calcium carbide with water. In the generator, crushed calcium carbide is dropped into a large tank of water. The carbon (C) in calcium carbide (CaC_2) unites with the hydrogen (H) in water (H_2O) to form two products. These two products are acetylene (C_2H_2), which bubbles out, and calcium hydroxide [$Ca(OH)_2$], which drops to the bottom, **Figure 31-10**. The acetylene gas is drawn off and may be used immediately or pumped into storage cylinders for later use.

Acetylene is colorless, is lighter than air, and has a strong garlic smell. Acetylene is also unstable at pressures above 30 psig (200 kPag), or at temperatures above 1435°F (780°C). Above the critical pressure or temperature, an explosion may occur if acetylene rapidly decomposes (explodes). This explosion can occur without the presence of oxygen and may occur as a result of electrical shock or extreme physical shock.

///// **CAUTION** \\\\\

Because of the instability of acetylene, it must never be used at pressures above 15 psig (100 kPag) or subjected to any possible electrical shock, excessive heat, or rough handling. Acetylene must not be manifolded or distributed through copper lines because it forms copper acetylide, which is explosive.

Acetone, which is a liquid solvent, is used inside acetylene cylinders to absorb the gas and make it more stable. The cylinder is filled with a porous material, and then acetone is added to the cylinder, where it absorbs about 24 times its own weight in acetylene. Acetone absorbs acetylene in the same manner as water absorbs carbon dioxide in a carbonated drink. As the acetylene is drawn off for use, it bubbles out of the acetone. Excessively high withdrawal

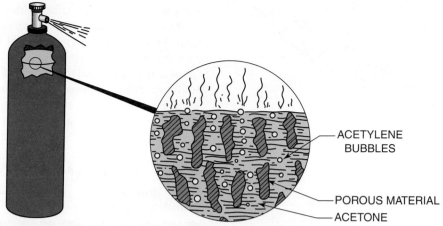

ACETYLENE
BUBBLES

POROUS MATERIAL

ACETONE

FIGURE 31-11 Acetylene coming out of solution. © Cengage Learning 2012

Tip Orifice Size—Drill No.	Thickness of Metal—in. (mm)		Oxygen and Acetylene Pressure—psi (kg/cm²)		Acetylene Flow Rate—cfh (L/min)*		Oxygen Flow Rate—cfh (L/min)	
70	1/64	(0.4)	1	(0.0703)	0.1	(0.0471)	0.1	(0.0471)
65	1/32	(0.8)	1	(0.0703)	0.4	(0.188)	0.4	(0.188)
60	1/16	(1.6)	1	(0.0703)	1	(0.4719)	1.1	(0.519)
59	3/32	(2.4)	2	(0.1406)	2	(0.943)	2.2	(1.038)
53	1/8	(3.2)	3	(0.2109)	8	(3.775)	8.8	(4.152)
49	3/16	(4.8)	4	(0.2812)	17	(8.022)	18	(8.494)
43	1/4	(6.4)	5	(0.3515)	25	(11.797)	27	(12.741)
36	5/16	(8.0)	6	(0.4218)	34	(16.044)	37	(17.460)
30	3/8	(9.5)	7	(0.4921)	43	(20.291)	47	(22.179)

*This flow rate must not exceed 1/7 of the cylinder's capacity per hour. For example, a large acetylene cylinder contains approximately 275 cu ft (7788 L) and its withdrawal rate must not exceed 39 cfh (18.40 L/min).

TABLE 31-3 Flow Rates for Various Tip Sizes

rates will cause the acetylene to boil out of the acetone. This rapid boiling can cause a portion of the acetone to be carried out of the cylinder with the acetylene, **Figure 31-11**. Withdrawing the acetone-acetylene mixture and burning it in the flame will contaminate the weld and may damage the rubber seals and internal parts of the regulator and other equipment. In addition to the danger caused by damaged equipment, the cylinder itself may be damaged, resulting in another hazard.

The withdrawal rate of gas from a cylinder should not exceed one-seventh of the total cylinder capacity per hour. **Table 31-3** lists the flow rates in cubic feet per hour (liters per minute) for various torch tip sizes. It also lists the minimum cylinder size to be used with a single torch. If more than one torch is to be used with a cylinder, the draw rates of the tips should be added together to obtain the minimum cylinder size or the number of cylinders that must be manifolded together. As the gas from the cylinder(s) is emptied, a new maximum withdrawal rate must be calculated based upon current cylinder capacity.

Heat and Temperature The neutral oxyacetylene flame burns at a temperature of approximately 5589°F (3087°C). The maximum temperature of a strongly **oxidizing flame** is approximately 5615°F (3102°C). The flame burns in two

parts, the **inner cone** and the **outer envelope;** refer to **Figure 31-12.** The heat produced by the flame can be divided into portions produced in the inner cone and outer envelope. The inner cone produces 507 Btu per cubic foot of gas (19 kilogram-calories per cubic meter), and the outer envelope produces 963 Btu per cubic foot of gas (36 kg-cal/m³). The total heat produced by the flame is 1470 Btu/ft³ (55 kg-cal/m³).

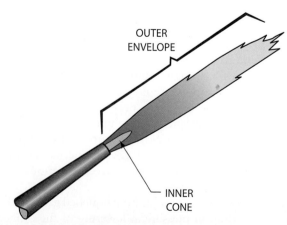

OUTER
ENVELOPE

INNER
CONE

FIGURE 31-12 Parts of a flame. © Cengage Learning 2012

FIGURE 31-13 Thermal gradients of a flame.
© Cengage Learning 2012

The high temperature produced by the oxyacetylene flame is concentrated around the inner cone, **Figure 31-13**. As the flame is moved back away from the work, the localized heating is reduced quickly. For welding, this highly concentrated temperature is the greatest advantage of the oxyacetylene flame over other oxyfuel gases. The molten weld pool is easily controlled by the torch angle and position of the inner cone to the work.

Although more heat is produced in the secondary flame than in the primary flame, the temperature is much lower. The lower temperature and the relatively low concentration of the heat produced by the outer flame are distinct disadvantages of the oxyacetylene flame when it is used for heating. The flame must be held close to the work and moved constantly to obtain uniform heating of large parts.

Liquefied Fuel Gases

Some fuel gases are available in liquefied form in pressurized cylinders. **Liquefied fuel gases** can be obtained in either individual cylinders or bulk tanks. The high-volume use of fuel gas often requires excessive cylinder inventory and handling. Setting up a bulk system will eliminate much of the cylinder inventory and handling.

Pressure The pressure in a cylinder containing a liquefied gas is not an indication of the level of gas in the tank. The pressure is based on the type of gas and the gas temperature. The gas pressure of a liquid increases as the temperature increases and decreases as the temperature decreases. At high temperatures, approximately 200°F (93°C), the pressure may cause the release valve to open to release excessive pressures, over 375 psig (2690 kPag). At extremely low temperatures, the cylinder may not have any working pressure. For example, at temperatures below 31°F (−1°C), butane has no pressure. **Table 31-4** lists the gas pressures for various gases at different temperatures.

High withdrawal rates of gas from liquefied gas cylinders will cause a drop in pressure, a lowering of the cylinder temperature, and the possibility of freezing the regulator.

As a gas is drawn out of the cylinder, the liquid inside absorbs heat from the outside to produce more gas vapor, **Figure 31-14**. If the gas is drawn out faster than heat can be absorbed, the cylinder will begin to cool off. A ring of frost may appear around the bottom of the cylinder. If high withdrawal rates continue, the regulator may also start to ice up. For applications requiring high withdrawal rates, the cylinder should be placed in a warm area.

> ### ⚠ CAUTION
> **Heat should never be applied directly to a cylinder.**

Methylacetylene-Propadiene (MPS)

Many different **methylacetylene-propadiene (MPS)** gases are in use today as fuel gases for oxyfuel cutting, heating, brazing, metallizing, and to a limited extent, welding. MPS gases are mixtures of two or more of the following gases: propane (C_3H_8), butane (C_4H_{10}), butadiene (C_4H_6), methyl acetylene (C_3H_4), and propadiene (C_3H_4). The mixtures vary in composition and characteristics from one manufacturer to another. However, all manufacturers provide MPS gases as liquefied gases in pressurized cylinders.

Production Approximately 26 MPS gases are marketed: **MAPP**, Chem-O-Lean, Apachigas, FG-2, Gulf HP, Flamex, and Hy Temp, among others. The gases may be mixed by a local supplier as the cylinders are filled, or the supplier may receive the gas premixed. In cylinders that are stored for a long time, some mixtures tend to separate. Before using the cylinder, it should be moved enough to remix the gases inside. In some cylinders a **piccolo tube** is used, **Figure 31-15**, to improve the mixing of the gas by causing the liquid to be agitated as it is used. Without a means of stirring the gas, some MPS gases may burn differently when the cylinder is full than when it is nearly empty.

Temperature and Heat MPS gases have a neutral oxyfuel flame temperature of about 5301°F (2927°C). The exact temperature varies with the specific mixture. The flame temperature and heat are high enough to be used for welding, but MPS gases are seldom used for this purpose.

The heat produced by the primary flame, approximately 570 Btu/ft³ (21 kg-cal/m³), is about the same as that produced by acetylene. The secondary flame produces approximately 1889 Btu/ft³ (70 kg-cal/m³) of heat, or twice the heat of acetylene. These values make MPS gases much better than acetylene for heating, brazing, and some types of cutting. The slower burn rate of these gases results in a poorly concentrated flame, which is difficult to use for welding.

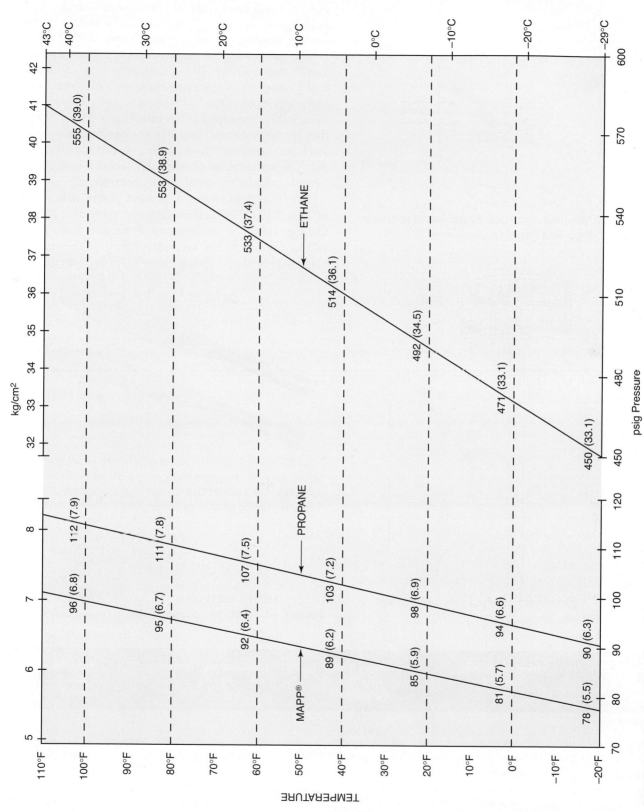

TABLE 31-4 Gas Pressures Vary with Temperature Changes

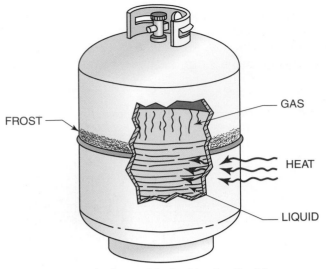

FIGURE 31-14 The heat absorbed by the liquid propane causes it to change to a gas. © Cengage Learning 2012

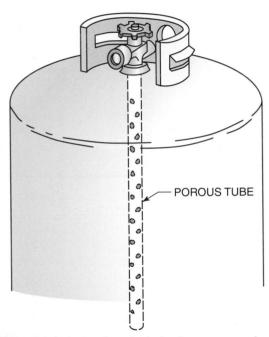

FIGURE 31-15 A piccolo tube helps keep gases mixed during use. © Cengage Learning 2012

MAPP

MAPP gas is the trade name for the stabilized liquefied mixture of methylacetylene (CH_3:C:CH) and propadiene (CH_2:C:CH_2) gases. Oxy MAPP combusts with a high-heat, high-temperature flame that works well for cutting, heating, brazing, and metallizing, **Figure 31-16**.

The gases mixed to produce MAPP have the same atomic composition. This means that three carbon and four hydrogen atoms are present in each molecule of gas. Molecules of each gas have the same mass and size even though they are shaped differently, **Figure 31-17**. Because they are the same mass and size, the gas molecules form a stable mixture that remains uniformly mixed in the cylinder. The stabilization of mixture ensures a uniform and consistent flame for easier quality control.

Oxy MAPP produces a **neutral flame** temperature of 5301°F (2927°C) that yields a heat value of 517 Btu/ft³ (19 kg-cal/m³) in the primary flame and 1889 Btu/ft³ (70 kg-cal/m³) in the secondary flame. The total heat value is 2406 Btu/ft³ (90 kg-cal/m³). **Table 31-5** compares

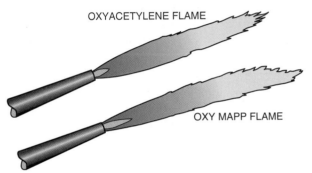

FIGURE 31-16 The inner cone and outer envelope of a MAPP flame are longer than an equal-volume flame of acetylene because of the slower burn rate of MAPP.
© Cengage Learning 2012

METHYLACETYLENE   PROPADIENE

FIGURE 31-17 MAPP gas molecules. © Cengage Learning 2012

Fuel	Neutral Flame Temp °F (°C)	Primary Flame Btu/ft³ (kg-cal/m³)	Secondary Flame Btu/ft³ (kg-cal/m³)	Total Heat Btu/ft³ (kg-cal/m³)
Acetylene	5589 (3087)	507 (4510)	963 (8570)	1470 (13,090)
MAPP gas	5301 (2927)	517 (4600)	1889 (16,820)	2406 (21,420)
Natural gas	4600 (2538)	11 (98)	989 (8810)	1000 (8900)
Propane	4579 (2526)	255 (2270)	2243 (19,970)	2498 (22,240)
Propylene	5193 (2867)	438 (3900)	1962 (17,470)	2371 (21,110)

TABLE 31-5 Heating Value of Major Industrial Fuel Gases (Combusted with Pure Oxygen) Courtesy of BOC Gases

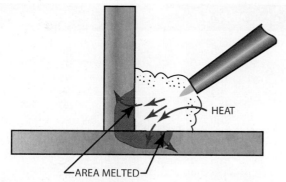

FIGURE 31-18 Secondary heat of an oxyMAPP flame causes the sides of a tee joint to melt before the root of the joint melts. © Cengage Learning 2012

the temperatures and heat produced by five different fuel gases. The difference between the temperature produced by a neutral oxyacetylene flame and a neutral oxy MAPP flame is only 288°F (142°C). Both flames are well above the approximate 2800°F (1536°C) temperature required to melt mild steel. Although MAPP is not normally recommended for use in gas welding, it can be used successfully. Because the secondary flame of MAPP produces almost twice the heat of the acetylene flame, distortion is more of a problem with MAPP than with acetylene. Also, it is more difficult to melt the root of a joint using MAPP.

Because of the higher heat, the sides of the joint melt first, **Figure 31-18**. This is not as much of a problem with other joint designs.

All gases used as alternatives to acetylene are safer to use, store, and handle. MAPP has each one of the safety features of the other fuel gases. These safety features include shock stability, narrow explosive limits in air, no pressure limitation, and slow burning velocities, **Figure 31-19**. Another safety advantage of MAPP is its smell. The odor of MAPP can be detected when there are as few as 100 parts per million (ppm) of the gas or 1/340 of its lower explosive limit in air. By law, propane, natural gas, and propylene must have an odor added to them so that they can be detected at a concentration of 1/15 of their lower explosive limit in air. The ability to detect even small leaks can save gas and avoid the possibility of explosions. The foul odor of MAPP allows leaks to be found more than 22 times faster than acetylene leaks can be found.

Table 31-6 compares acetylene, MAPP gas, and propylene for various procedures. The higher the number given, the better the performance of the gas for that procedure.

Propane and Natural Gas

Propane and natural gas find limited use in the welding industry. Because of their relatively low-temperature, low-heat flames, they are seldom used for purposes other than heating and as preheat fuels for cutting.

The major advantage of propane and natural gas is that they are often used for heating the shop. Therefore, a supply of these gases is readily available. The handling of cylinders is reduced or eliminated because natural gas

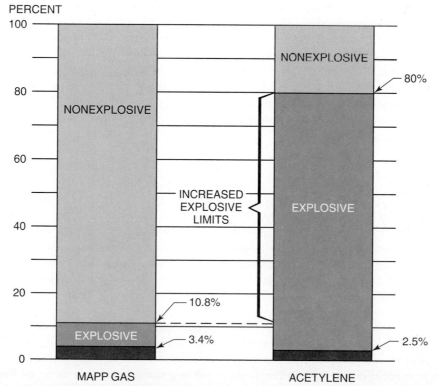

FIGURE 31-19 Explosive limits of MAPP gas in air. MAPP Products

Application	Acetylene	MAPP Gas	Propylene
Cutting			
Under 3/8 in. thick	100	95	90
5/8 in. to 5 in. thick	95	100	95
Over 5 in. thick	80	100	95
Cutting dirty or scaled surfaces	100	95	80
Repetitive cutting	100	100	80
Stack cutting	90	100	95
Cutting low alloy specialty steels	100	90	80
Beveling	100	100	85
Cutting rounds	95	100	85
Piercing	100	100	85
Blind-hole piercing	100	90	80
Rivet washing	100	95	80
Gouging	100	100	85
Wire metallizing	80	100	90
Powder metallizing	100	0	0
Heating, stress relief, bending	70	100	90
Deep flame hardening	90	100	90
Shallow flame hardening	95	100	80
Cobalt-base hardsurfacing	100	0	0
Other alloy hardsurfacing	100	85	70
Welding	100	70	0
Braze welding	100	90	70
Brazing	100	100	90

TABLE 31-6 Average Performance Ratings of Some Oxyfuel Flames Courtesy of BOC Gases

is piped directly to the shop and propane can be delivered in bulk tanks. Both gases are easily piped through the shop.

Propane and natural gas are both obtained from the petroleum industry. Natural gas comes from gas wells, and propane is produced from oil and gas at refineries. An artificial odor, a mercaptan chemical, must be added so that leaks can be detected for safety purposes.

Chemically, propane is C_3H_8; natural gas is mostly methane (CH_4) and ethane (C_2H_6). The major disadvantage is that both gases consume a large amount of oxygen. Propane requires 85% of the flame's oxygen from the cylinder, and natural gas requires 95% of its oxygen from the cylinder, compared with an oxygen consumption of as little as 50% for the oxyacetylene flame.

Hydrogen

Oxyhydrogen produces only a primary combustion flame, unlike hydrocarbon gases, which have both primary and secondary combustion. The hydrogen flame is almost colorless and can be seen only when dirt, dust, and other contaminants from the air glow while burning in the flame. Hydrogen is not widely used in welding because of its expense, its limited availability, and some myths about its safety.

Hydrogen has the fastest burning velocity of any of the fuel gases at 36 ft/sec (10.9 m/sec). Acetylene has a burn rate of less than one-half that of hydrogen. Hydrogen has a very slight tendency to backfire, yet it does not flash back. Unlike acetylene, which can explosively decompose

without oxygen, hydrogen cannot be made to react without the presence of sufficient oxygen.

Hydrogen is much lighter than air. Therefore, when it is released, it diffuses quickly, reducing the possibility of accidental combustion. If a large quantity of hydrogen is allowed to burn uncontrolled, the gas rises into the flame. This means it burns in an upward direction, away from people in an area. Most other gases burn in a downward direction, which can trap people in an area. The chance of large quantities of hydrogen exploding is limited. For example, when the hydrogen-filled airship, the Hindenburg, caught fire and burned in 1931, no explosion occurred, and most of the people on board the airship survived.

The low flame temperature restricts the use of the **oxyhydrogen flame** to cutting, usually underwater, and to gas welding and brazing on low-temperature metals such as aluminum. The flame can be made reducing (needing oxygen) to help protect the aluminum from oxidation, without having excessive carbon to contaminate the weld. The finished flame product is water, H_2O. Only one-quarter of the flame oxygen comes from the cylinder. The rest comes from the air surrounding the flame.

Two major safety problems exist when hydrogen is used as a fuel gas. First, hydrogen has no smell, which makes it difficult to detect leaks. Second, the molecule is extremely small so that it leaks easily. When hydrogen is used, an active leak-checking schedule must be followed to find small problems before they develop into disasters. It is possible for a leak to be on fire and not be noticed because the hydrogen flame is almost invisible.

THINK GREEN
Nonpolluting Fuel-Gas Flame

The oxygen and hydrogen flame is the only 100% nonpolluting fuel-gas flame. The only by-product from this flame is water vapor. In addition, both oxygen and hydrogen gases can be produced with electrolysis of water by using renewable energy such as solar or wind.

EXPERIMENT 31-2

Oxyfuel Flames

Using an identical torch set with each available fuel gas, you are going to observe the flame as each fuel gas is safely lit, adjusted, and extinguished.

Set all fuel and oxygen regulators at approximately 5 psig (35 kPag). Each torch should have the same size tip. The tip should have an orifice equal to a number 53 to 60 drill. Place the torches on a table with the tips pointed up, **Figure 31-20**.

Starting with the oxyacetylene torch, turn on the fuel-gas valve slightly. Using a flint lighter, light the torch and adjust the gas valve so that the flame is not smoking. After securely placing the lit torch back on the table, repeat the process with all of the other torches. Adjust the flame of each torch to the same size as the acetylene flame, **Figure 31-21**.

Pick up the acetylene torch and move it back and forth, **Figure 31-22A**. The flame should be stable, with only the top deflecting as the torch is moved. Replace the torch carefully on the table.

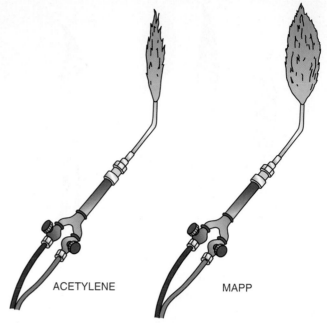

FIGURE 31-21 The second fuel-gas flame may be slightly off the tip. © Cengage Learning 2012

One at a time, pick up each of the other torches and move them just as you did with the acetylene torch. The flames on these torches will deflect more, and some flames may go out, **Figure 31-22B**. The reason for the difference is that less gas is flowing with each of the other gases. The acetylene flame is more stable because it has the highest flow rate and the highest burn rate. Thus, the flame is more compact.

Turn on the oxygen valve slowly until the acetylene flame is adjusted to a neutral setting. Repeat this procedure with each of the other flames. The flames may blow

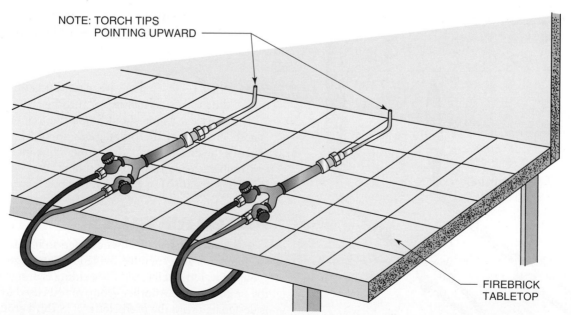

NOTE: TORCH TIPS POINTING UPWARD

FIREBRICK TABLETOP

FIGURE 31-20 Torches set up to compare fuel gases. © Cengage Learning 2012

(A) ACETYLENE FLAME   (B) MAPP® FLAME

FIGURE 31-22 Move the torches back and forth and watch the flame. (A) Acetylene flame. (B) MAPP flame.
© Cengage Learning 2012

out as the oxygen is turned on. If this happens, direct the flame against the firebrick top of the welding table and readjust the oxygen until a neutral flame is reached with each torch. The flames may blow out because of the slow burning velocities of the gases. The flames should all look nearly the same in color, but the inner cone on the acetylene flame will be the shortest, **Figure 31-23.**

Line up the torches and hold the end of a gas welding rod with a 1/8-in. (3-mm) diameter in each flame. The welding rods should all be put in the flames at the same time. They should all be held at the same height above the inner cone, about 1/4 in. (6 mm), **Figure 31-24.** Watch as the welding rods melt, each one at a different rate. Cool the welding rods and repeat the experiment with the ends of the welding rods 1/4 in. (6 mm) higher than before. After the welding rods melt, cool and raise the welding

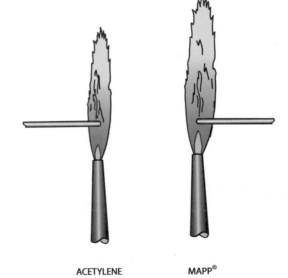

ACETYLENE   MAPP®

FIGURE 31-24 Observe the degree and rate of heating produced by each flame. © Cengage Learning 2012

rods another 1/4 in. (6 mm). Keep repeating this step until the welding rods stop melting or turning red.

The acetylene flame should melt the low welding rod fastest, but one of the other gases should heat the rod faster as the distance above the inner cone increases.

Complete a copy of the "Student Welding Report" listed in Appendix I or provided by your instructor. ◆

FILLER METALS

Filler metals specifically designed to be used with an oxyfuel torch are generally divided into three groups. One group of welding rods used for welding is designated with the prefix letter *R.* Another group of rods used for brazing is designated with the prefix letter *B.* A third group is used for buildup, wear resistance surfacing, or both. Welding rods in this group may also be classified in one of the

NOTE SHORT
INNER CONE
OF
ACETYLENE
FLAME →

FIGURE 31-23 Compare the flames. © Cengage Learning 2012

THINK GREEN
Conserve Filler Metal

The short ends of both welding and brazing rods can be fused together so that the amount of scrap filler metal can be minimized, unlike most other filler metal such as SMAW, GMAW, and FCAW. Being able to use all of the filler metal can be a significant saving of shop funds and reduces waste materials.

other groups, may be patented and use a trade name, or may be tubular with a granular material in the center. The tubular welding rods are designated with an RWC prefix. Some filler metals, for example BRCuZn, are classified both as a braze welding rod (R) and a brazing rod (B) because they can be used either way.

Ferrous Metals

Ferrous filler metals are welding rods that are mainly iron. They may have other elements added to change their strength, corrosion resistance, weldability, or another physical property. There are three major American Welding Society (AWS) specifications for ferrous filler metals. These three specifications are A5.2 low carbon, low alloy steel; A5.15 cast iron; and A5.9 stainless steel. Within each specification, there are classes; for example, in group A5.2 the classes are RG45, RG60, and RG65.

The specifications and classes have minimum and maximum limits for the alloys that are added to provide the required physical properties of the weld they produce. Each manufacturer is free to make changes in the wire composition within the specified limits. The changes generally involve weldability, tensile strength, ductility, cracking, appearance, porosity, impact strength, and hardness.

Physical changes are most often affected by changes in the percentages of alloys of carbon (C), silicon (Si), manganese (Mn), chromium (Cr), vanadium (V), nickel (Ni), and molybdenum (Mo). Contaminants from fuels used in the production of iron, such as phosphorus (P) and sulfur (S), have a negative effect on many of the desired properties, and the amount of contaminants should be kept as low as possible.

Small shops sometimes use other types of wire for weld filler metals. The most popular substitution is often coat-hanger wire. Using such substitutes can cause weld failure. Coat-hanger wires were not manufactured for welding purposes, and their chemistry varies greatly. Porosity inside the weld deposit is common due to higher-than-acceptable levels of phosphorus (P) and sulfur (S). The painted finish of the wire will burn, causing further weld contamination and fumes that can be hazardous to the welder. Another type of substitute wire is often cadmium (Cd), plated to prevent rust. However, when cadmium is burned or vaporized, poisonous fumes are produced. The only safe filler metals to use are the ones specifically designed for welding.

Mild Steel

Ferrous metal filler rods are generally classified by the AWS as mild steel, low alloy steel, or cast iron. Mild steel and low alloy steel are the materials that are most frequently gas welded. They are easily welded without a flux. Cast iron and stainless steels require fluxes and special techniques.

Mild steel and low alloy gas welding rods are classified by the AWS as RG45, RG60, and RG65. The R refers to the welding rod, and the G refers to the ability to use gas for welding. The two digits indicate the minimum tensile strength range of the weld metal deposited.

Class RG45 is a general-purpose gas welding rod that is often used for training welders. It has a smooth, shiny molten weld pool that leaves a nice looking bead. The low carbon content helps make low-strength 45,000 psi to 55,000 psi ($3163 kg/cm^2$ to $3866 kg/cm^2$) welds in the tensile strength range that are ductile. Automotive, auto body, wrought iron, and general welding shops use this welding rod for most or all of their gas welding.

Class RG60 welding rods, compared to RG45 welding rods, have a slightly higher carbon content and produce a higher weld strength in the range of 50,000 psi to 65,000 psi ($3515 kg/cm^2$ to $4569 kg/cm^2$). The addition of silicon, manganese, and other metallic elements may improve the retention of carbon or other easily oxidized material. The molten weld pool is not as clear or shiny as that obtained with RG45 welding rods, and the weld bead also may not look as nice. The RG60 welding rod can be used on low alloy steels requiring good strength and ductility. It is frequently used for mild steel pipe welds, structural shapes, chrome-moly aircraft tubing, American Society of Mechanical Engineers (ASME) code welds, and gas tungsten arc welding.

Class RG65 welding rods are low alloy, high-strength, 65,000 psi to 75,000 psi ($4569 kg/cm^2$ to $5272 kg/cm^2$), low-creep, corrosion-resistant welding rods. The high content of silicon and manganese helps to retain carbon but may cause a thick, crusty layer of manganese silicate flux ($MnSiO_3$) to float on top of the molten weld pool. Although this molten weld pool has the roughest look, it is the purest and strongest. The RG65 welding rod is used in ASME high-pressure piping, tubing, and gas tungsten arc welding.

Cast Iron

Cast iron filler rods for gas welding are small, round, or square iron castings. The prefix R, which refers to the welding rod, is used in front of CI, which stands for cast iron. A high-temperature, borax-based flux must be used to prevent the carbon from burning out.

Class RCI is the lower-strength filler metal, which has a tensile strength of 20,000 psi to 26,000 psi ($1406 kg/cm^2$ to $1757 kg/cm^2$). This class is recommended for most general cast iron repair and for the buildup and fill-in of damaged castings. The weld can be remachined if necessary.

Compared to RCI welding rods, class RCI-A welding rods have a higher tensile strength, in the range of 35,000 psi to 40,000 psi ($2460 kg/cm^2$ to $2612 kg/cm^2$). An RCI-A welding rod is used on alloy cast irons or where higher strengths are required.

Summary

In many applications, such as home and farm work, the oxyacetylene welding process is by far the most desirable because of its flexibility, portability, and cost. These factors will keep this process in the forefront of welding for the foreseeable future.

The wide variety of fuel gases available for oxyfuel welding, cutting, and brazing has offered welders some unique challenges in determining the most appropriate fuel gas for their processes. Although acetylene and oxygen are the most common, acetylene is not always the most appropriate gas for a number of reasons, primarily cost. When treated properly the waste product from acetylene production has little or no environmental impact, but such treatments can be expensive. This significantly increases the cost of acetylene and has made other fuel gases more desirable. Other fuel gases do not have all of the characteristics of acetylene. Each gas has unique advantages and disadvantages. You must look at the advantages and disadvantages of all the gases before selecting the fuel gas that will be most appropriate for your applications.

Filler metal selection for the oxyfuel welding processes used in a home hobby application is not often given much thought. In industrial applications, the proper selection of an oxyfuel filler metal is critical to the success of your product.

Welding with the Right Shielding Gas

Quality and consistency are key to any quality assurance program. You must be confident your processes are consistent and repeatable and that the products being used in production meet specific requirements and specifications.

To maintain a stable, consistent welding process and ensure repeatable results, you should implement a welding specification (WPS). A WPS gives the welder or welding operator the recipe for producing acceptable welds on a given application, time after time. In developing a WPS for gas metal arc or gas tungsten arc welding, many variables must be taken into account, such as voltage, amperage, electrode extension, and shielding gas. These and other variables are called "essential variables" as defined in AWS D1.1, Structural Welding Code—Steel.

The Classification System

With respect to shielding gases, AWS D1.1 states that a procedure qualification record (PQR) requires requalification when there is "a change in shielding gas from a single gas to any other single gas or mixture of gases, or in the specified nominal percentage composition of a gas mixture, or to no gas." Therefore, it is essential that your shielding gas composition be accurate and consistent to ensure WPSs are being followed and the desired weld quality is maintained.

It is the responsibility of your gas supplier to ensure you receive an accurate, consistent shielding gas supply that conforms to AWS A5.32, Specification for Welding Shielding Gases. AWS A5.32 sets the standards for the classification of shielding gases, similar to the way AWS 5.18, Specification for Carbon Steel Electrodes and Rods for Gas Metal Arc Welding, prescribes a classification system for identifying carbon steel electrodes and rods. These

The label on your gas cylinders should state whether your supplier is complying with the AWS specification on shielding gases. American Welding Society

specifications were developed by the American Welding Society and approved by ANSI as a way of identifying products based on chemical composition. When these specifications are utilized and implemented by the manufacturer of each product, the end user is assured of a consistent, quality product.

The classification system outlined in the specification clearly identifies the chemical composition of the shielding gas in question, similar to the way welding wires are identified. For instance, if you order E70S-6 welding wire, you should feel confident you will receive a welding wire that contains certain percentages of silicon, manganese, and so on. Similarly, when you order SG-AC-10 shielding gas you should be confident it contains 10% carbon dioxide and 90% argon and that the product is consistent from cylinder to cylinder.

AWS A5.32 not only establishes an identification system for shielding gases; it also specifies the purity and dew point levels required for individual gases. The specification also covers the requirements for dew point, purity, and mix accuracy for gas mixtures. To adhere to this specification, your shielding gas supplier must test individually filled cylinders or one cylinder from each filling manifold to verify mix accuracy, purity, and dew point.

Look on the cylinder to make sure your gases comply with A5.32. The attached label should state that the gases meet the requirements of the specification. If your shielding gas cylinder contains such a label, you can be assured your gas supplier is taking all the necessary precautions to supply your company with consistent, quality gas mixtures.

Article courtesy of the American Welding Society.

Review

1. What elements make up all hydrocarbons?

2. What are the separate parts that make up an oxyacetylene flame?

3. Use Table 31-1 to determine which fuel gas requires the largest amount of oxygen from the torch.

4. Approximately how long would it take a 50/50 mixture of oxygen and acetylene to flash back through a 21-ft (7.62-m) long hose?

5. Use Table 31-2 to determine which fuel gas has the highest burning velocity when mixed with oxygen.

6. How is acetylene produced?

7. Why is it not safe to use acetylene above 15 psig (100 kPag)?

8. Use Table 31-3 to determine the largest tip orifice size (drill number) that could be used with an acetylene cylinder containing 175 cu ft (4956 L).

9. Where is the highest temperature and where is the greatest heat produced in a neutral oxyacetylene flame?

10. Use Table 31-4 to determine what would be the pressures in cylinders containing (a) MAPP, (b) propane, and (c) ethane on a 100°F (38°C) day.

11. What are methylacetylene-propadiene fuel gases used for?

12. Use Table 31-5 to determine which oxygen fuel-gas mixture produces the highest total heat.

13. Which fuel gas has the strongest odor and is easiest to detect?

14. What is the major advantage of using propane or natural gas?

15. Use Table 31-6 to determine which fuel gas would be best for (a) cutting material thicker than 5 in. (125 mm), (b) powdered metallizing, and (c) stack cutting.

16. What two major safety problems does hydrogen present?

17. Why should coat hangers not be used as gas welding filler metal?

18. Explain the significance of the AWS filler metal classification RG45.

Chapter 32

Oxyacetylene Welding

OBJECTIVES

After completing this chapter, the student should be able to

- ■ explain how to set up and weld mild steel.
- ■ make a variety of welded joints in any position on thin-gauge, mild steel sheet.
- ■ make a satisfactory weld on small diameter pipe and tubing in any position.
- ■ explain the effects of torch angle, flame height, filler metal size, and welding speed on gas welds.

KEY TERMS

1G position	*keyhole*	*shelf*
2G position	*kindling point*	*tee joint*
5G position	*lap joint*	*torch angle*
6G position	*molten weld pool*	*torch manipulation*
burnthrough	*out-of-position welding*	*trailing edge*
flashing	*outside corner joint*	*undercut*
flat butt joint	*overhead weld*	*vertical weld*
heat sink	*overlap*	*weld crater*
horizontal welds	*penetration*	

INTRODUCTION

Oxyacetylene welding is limited to thin metal sections or to times when portability is important. During the early years of welding, oxyacetylene was used to weld thick plate, 1 in. (25 mm) and thicker. Today, it is used almost exclusively on thin metal, 11 gauge or thinner. One of the arc welding processes is most often used today for welding metal thicker than 16 gauge. Some of the arc welding processes, such as GMAW, are replacing the gas welding processes on metals as thin as 28 gauge, **Figure 32-1**. Because of the expanded use of arc welding processes on thinner sections, we will concentrate on the use of gas welding on metal having a thickness of 16 gauge (approximately 1/16 in. [2 mm]) or thinner.

798

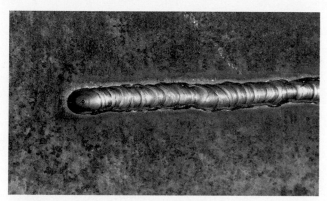

FIGURE 32-1 Gas metal arc welded (GMAW) on 16-gauge mild steel. Larry Jeffus

MILD STEEL WELDS

Mild steel is the easiest metal to gas weld. With this metal, it is possible to make welds with 100% integrity (lack of porosity, oxides, or other defects) and that have excellent strength, good ductility, and other positive characteristics. The secondary flame shields the molten weld pool from the air, which would cause oxidation. The atmospheric oxygen combines with the carbon monoxide (CO) from the outer flame envelope to produce carbon dioxide (CO_2). The carbon dioxide will not react with the molten weld pool. In addition, the carbon dioxide forces the surrounding atmosphere away from the weld.

Factors Affecting the Weld

Torch Tip Size The torch tip size should be used to control the weld bead width, penetration, and speed. **Penetration** is the depth into the base metal that the weld

fusion or melting extends from the surface, excluding any reinforcement. Because each tip size has a limited operating range in which it can be used, tip sizes must be changed to suit the thickness and size of the metal being welded. Never lower the size of the torch flame when the correct tip size is unavailable. If the flame size or volume is lowered below the correct size by lowering the gas flow, the tip will overheat. If the flame is lowered significantly, the tip can overheat even without using it to weld. Overheated tips will backfire. Backfiring can cause dangerous flashback. Other factors that can be changed to control the weld size are the torch angle, the flame-to-metal distance, the welding rod size, and the way the torch is manipulated.

> ### /// CAUTION \\\
>
> **It is never safe to lower the flame size if the tip's flame produces too much heat for your welding job. Get a smaller tip or change your welding technique.**

Torch Angle The **torch angle** and the angle between the inner cone and the metal have a great effect on the speed of melting and size of the molten weld pool. The ideal angle for the welding torch is 45°. As this angle increases toward 90°, the rate of heating increases. As the angle decreases toward 0° to the plate's surface, the rate of heating decreases, as illustrated in **Figure 32-2**. The distance between the inner cone and the metal ideally should be 1/8 in. to 1/4 in. (3 mm to 6 mm). As this distance increases, the rate of heating decreases; as the distance decreases, the heating rate increases, **Figure 32-3**.

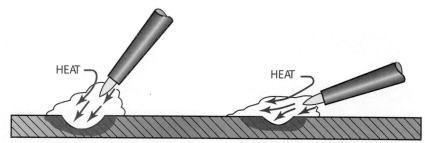

FIGURE 32-2 Changing the torch angle changes the percentage of heat that is transferred into the metal.
© Cengage Learning 2012

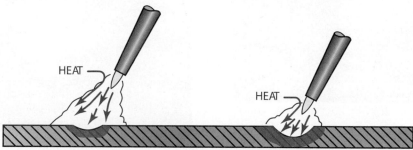

FIGURE 32-3 Changing the distance between the torch tip and the metal changes the percentage of heat input into the metal. © Cengage Learning 2012

Welding Rod Size Welding rod size and **torch manipulation** can be used to control the weld bead characteristics. A larger welding rod can be used to cool the molten weld pool, increase buildup, and reduce penetration, **Figure 32-4A, B,** and **C.** The torch can be manipulated so that the direct heat from the flame is flashed off the molten weld pool for a moment to allow it to cool, **Figure 32-5.**

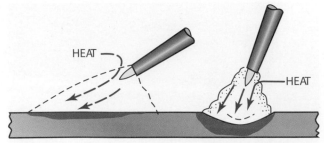

FIGURE 32-5 Flashing the flame off the metal will allow the molten weld pool to cool and reduce in size.
© Cengage Learning 2012

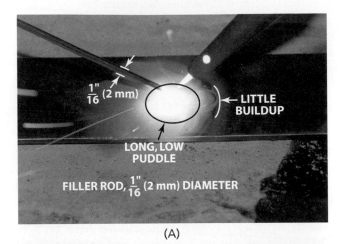

(A)

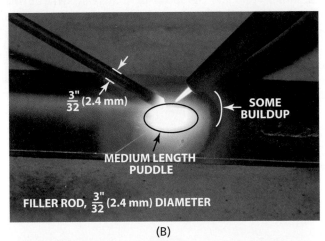

(B)

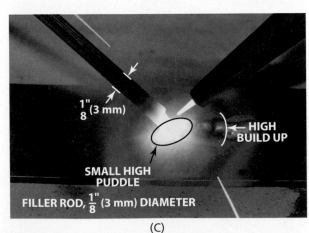

(C)

FIGURE 32-4 If all other conditions remain the same, changing the size of the filler rod will affect the weld as shown in (A), (B), and (C). Larry Jeffus

CHARACTERISTICS OF THE WELD

The **molten weld pool** must be protected by the secondary flame to prevent the atmosphere from contaminating the metal. If the secondary flame is suddenly moved away from a molten weld pool, the pool will throw off a large number of sparks. These sparks are caused by the rapid burning of the metal and its alloys as they come into contact with oxygen in the air. This is particularly a problem when a weld is stopped. The **weld crater** is especially susceptible to cracking. This tendency is greatly increased if the molten weld pool is allowed to burn out, **Figure 32-6.** To prevent burnout, the torch should be raised or tilted, keeping the outer flame envelope over the molten weld pool until it solidifies. As the molten weld pool is being cooled, the molten weld pool should also be filled with welding rod so that it is a uniform height compared to the surrounding weld bead.

The sparks that occur as the weld progresses are due to metal components that are being burned out of the weldment. Silicon oxides make up most of the sparks, and extra silicon can be added by the filler metal so that the weldment retains its desired soundness. A change

FIGURE 32-6 Building up the molten weld pool before it is ended will help prevent crater cracking. Larry Jeffus

in the number of sparks given off by the weld as it progresses can be used as an indication of changes in weld temperature. An increase in sparks on clean metal means an increase in weld temperature. A decrease in sparks indicates a decrease in weld temperature. Often the number of sparks in the air increases just before a **burnthrough** takes place—that is, burning out the molten metal that appears on the back side of the plate. This burnout does not happen to molten metal until it reaches the **kindling point** (the temperature that must be attained before something begins to burn). Small amounts of total penetration usually will not cause a burnout. When the sparks increase quickly, the torch should be pulled back to allow the metal to cool and prevent a burnthrough.

EXPERIMENT 32-1

Flame Effect on Metal

The first experiment examines how the flame affects mild steel. Use a piece of 16-gauge mild steel and the proper-size torch tip. Light and adjust the flame by turning down the oxygen so that the flame has excessive acetylene. Hold the flame on the metal until it melts and observe what happens, **Figure 32-7**. Now adjust the flame by turning up the oxygen so that the flame is neutral. Hold this flame on the metal until it melts and observe what takes place, **Figure 32-8**. Next, adjust the flame by turning up the oxygen so that the flame has excessive oxygen. Hold this flame on the metal until it melts and observe what happens, **Figure 32-9**. Repeat this experiment until you can easily identify each of the three flame settings by the flame, molten weld pool, sound, and sparks.

Before actual welding is started, it is a good idea to find a comfortable position. The more comfortable or relaxed you are, the easier it will be for you to make uniform

FIGURE 32-8 Neutral flame (balanced oxygen and acetylene). Larry Jeffus

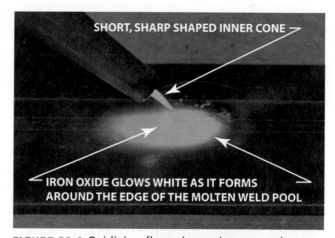

FIGURE 32-9 Oxidizing flame (excessive oxygen). Larry Jeffus

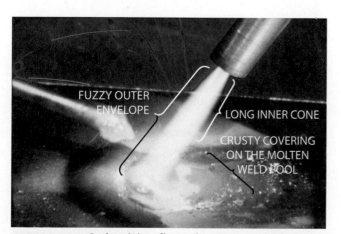

FIGURE 32-7 Carbonizing flame (excessive acetylene). Larry Jeffus

welds. The angle of the plate to you and the direction of travel are important.

Place a plate on the table in front of you and, with the torch off, practice moving the torch in one of the suggested patterns along a straight line, as illustrated in **Figure 32-10**. Turn the plate and repeat this step until you determine the most comfortable direction of travel. Later, when you have mastered several joints, you should change this angle and try to weld in a less comfortable position. Welding in the field or shop must often be done in positions that are less than comfortable, so the welder needs to be somewhat versatile.

It is important to feed the welding wire into the molten weld pool at a uniform rate. **Figure 32-11A and B** show some suggested methods of feeding the wire by hand. It is also suggested that you not cut the welding wire into two pieces for welding. Short lengths are easy to use, but this practice often results in wasted filler metal. The end of the welding wire may be rested on your shoulder so that it is easy to handle.

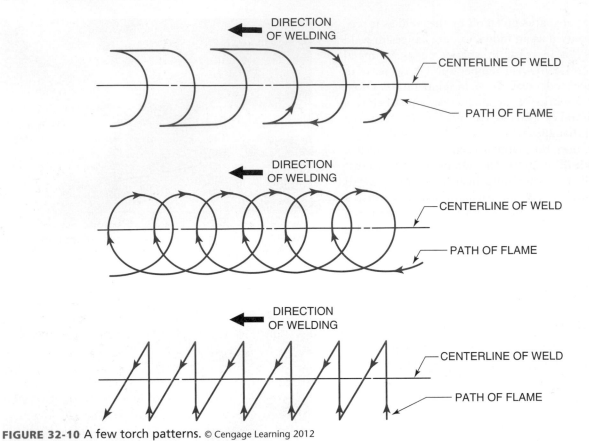

DIRECTION OF WELDING

CENTERLINE OF WELD

PATH OF FLAME

DIRECTION OF WELDING

CENTERLINE OF WELD

PATH OF FLAME

DIRECTION OF WELDING

CENTERLINE OF WELD

PATH OF FLAME

FIGURE 32-10 A few torch patterns. © Cengage Learning 2012

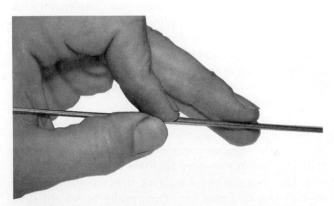

FIGURE 32-11 (A) Feed the filler rod by using your index finger. Larry Jeffus

FIGURE 32-11 (B) Move your finger back each time to be ready to feed the filler wire again. Larry Jeffus

FIGURE 32-12 The end of the filler rod should be bent for safety and easy identification. Larry Jeffus

////CAUTION\\\\

The end of the filler rod should have a hook bent in it so that you can readily tell which end may be hot and so that the sharp end will not be a hazard to a welder who is working next to you, Figure 32-12.

The torch hoses may be stiff and may therefore cause the torch to twist. Before you start welding, move the hoses so that there is no twisting of the torch. This will make the torch easy to manipulate and will be more relaxing for you.

Complete a copy of the "Student Welding Report" listed in Appendix I or provided by your instructor. ◆

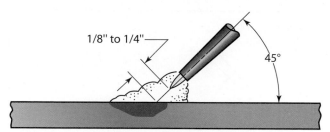

FIGURE 32-13 Hold the torch at a 45° angle in the direction of the weld. © Cengage Learning 2012

PRACTICE 32-1

Pushing a Molten Weld Pool

Using a clean piece of 16-gauge mild steel sheet, approximately 6 in. (152 mm) long, and a torch that is lit and adjusted to a neutral flame, push a molten weld pool in a straight line down the sheet. Start at one end and hold the torch at a 45° angle in the direction of the weld, **Figure 32-13**. When the metal starts to melt, move the torch in a circular pattern down the sheet toward the other end. If the size of the molten weld pool changes, speed up or slow down to keep it the same size all the way down the sheet, **Figure 32-14**. Repeat this practice until you can keep the width of the molten weld pool uniform and the direction of travel in a straight line.

Uniformity in width shows that you have control of the molten weld pool. A straight line indicates that you can see more than the molten weld pool itself. Students just learning to weld usually see only the molten weld pool. As you master the technique of welding, your visual range will increase. A broad visual range is important so that later you will be able to follow the joint, watch for distortion or see if other adjustments are needed, and relax as you weld. Turn off the cylinders, bleed the hoses, back out the regulator adjusting screws, and clean up your work area when you are finished.

Complete a copy of the "Student Welding Report" listed in Appendix I or provided by your instructor. ◆

EXPERIMENT 32-2

Effect of Torch Angle and Torch Height Changes

Use a clean piece of 16-gauge mild steel and a torch adjusted to a neutral flame. You will be experimenting with different torch angles and torch heights to change the size of the molten weld pool, **Figure 32-15**.

To start this experiment, hold the torch at a 45° angle to the metal with the inner cone at about 1/8 in. (3 mm) above the metal surface, **Figure 32-16A, B,** and **C**. As the metal starts to melt, move the torch slowly down the sheet. As you move, increase and decrease the angle of the torch to the sheet and observe the change in the size of the molten weld pool. Repeat the experiment, but this time as you move down the sheet, raise and lower the

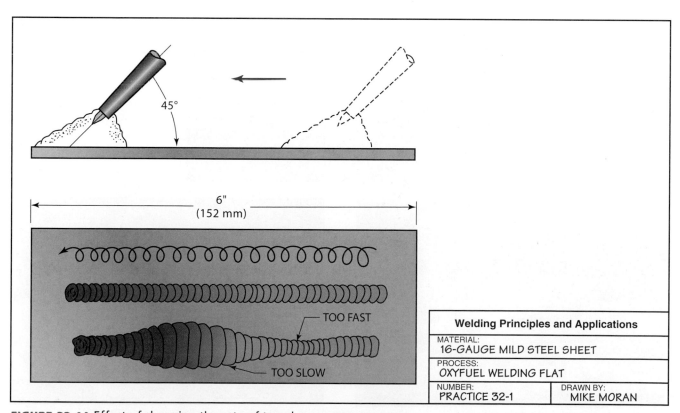

FIGURE 32-14 Effect of changing the rate of travel. © Cengage Learning 2012

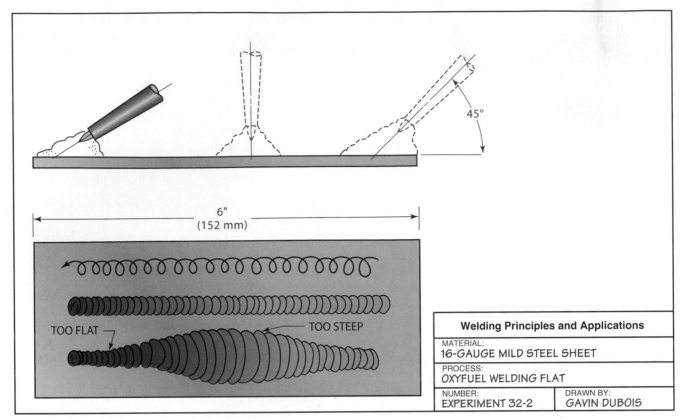

Welding Principles and Applications	
MATERIAL: *16-GAUGE MILD STEEL SHEET*	
PROCESS: *OXYFUEL WELDING FLAT*	
NUMBER: *EXPERIMENT 32-2*	DRAWN BY: *GAVIN DUBOIS*

FIGURE 32-15 Effect of changing torch angle. © Cengage Learning 2012

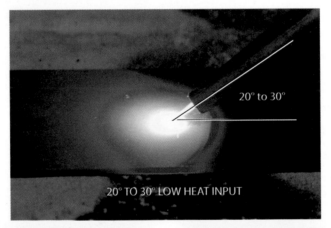

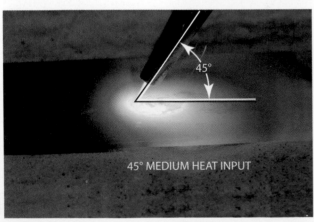

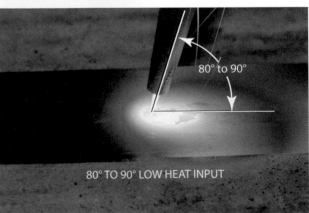

FIGURE 32-16 Changes in the angle between the torch and the work will change the molten weld pool produced.

Larry Jeffus

torch and observe the effect on the size of the molten weld pool.

Complete a copy of the "Student Welding Report" listed in Appendix I or provided by your instructor. ◆

PRACTICE 32-2

Beading

Repeat Experiment 32-2 until you can control the change in the size of the molten weld pool, as shown in **Figure 32-17**. After the control of the molten weld pool has been mastered, you are ready to start adding filler rod. When selecting and adding filler metal, the following are some facts to remember.

Selecting the proper size or correct diameter of filler metal will help control the weld bead width, buildup, and penetration. A large diameter filler rod can be used as a **heat sink** (something to draw off excessive heat) to keep the molten weld pool narrow, with little penetration and high buildup. As the diameter of the filler wire is decreased, the heat-absorbing effect decreases, and the molten weld pool becomes wider, with deeper penetration and reduced buildup. On thin metal, it may be necessary to use a large diameter filler rod to control burnthrough. On thick metal, a small diameter filler rod may be used to increase penetration.

Another thing to remember when adding filler metal to a weld is to keep a smooth and uniform rhythm. Each weld joint has its own indicator that will tell you when to add welding rod, **Figure 32-18**. As a new welder, watch for these indicators and try to anticipate adding the rod so that the indicators do not appear. At the end of a joint, it may be necessary to apply the welding rod faster and cool the molten weld pool to keep the weld bead uniform.

The end of the welding rod should always be kept inside the protective envelope of the flame, **Figure 32-19**. The hot end of the welding rod oxidizes each time it is removed from the protection of the flame. This oxide is deposited in the molten weld pool, causing sparks and a weak or brittle weld.

When the rod is added to the molten weld pool, the flame can be moved back as the end of the rod is dipped

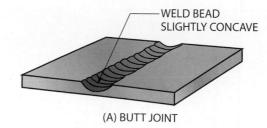

WELD BEAD SLIGHTLY CONCAVE

(A) BUTT JOINT

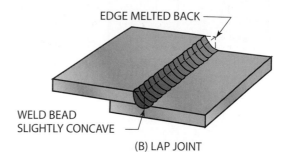

EDGE MELTED BACK

WELD BEAD SLIGHTLY CONCAVE

(B) LAP JOINT

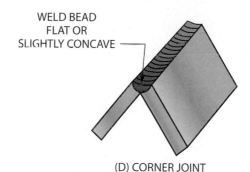

EDGE MELTED BACK

WELD BEAD SLIGHTLY CONCAVE

(C) TEE JOINT

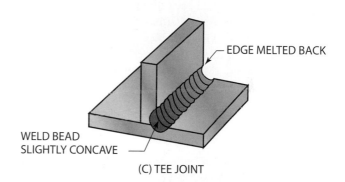

WELD BEAD FLAT OR SLIGHTLY CONCAVE

(D) CORNER JOINT

FIGURE 32-18 Indications that more filler metal should have been added. (A) Butt joint. (B) Lap joint. (C) Tee joint. (D) Corner joint. © Cengage Learning 2012

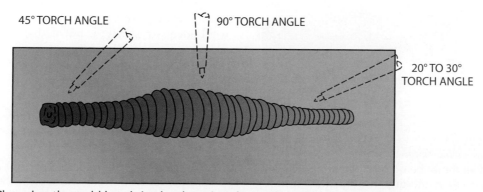

45° TORCH ANGLE   90° TORCH ANGLE   20° TO 30° TORCH ANGLE

FIGURE 32-17 Changing the weld bead size by changing the torch angle. © Cengage Learning 2012

FIGURE 32-19 The hot end of the filler rod is protected from the atmosphere by the outer flame envelope.
Larry Jeffus

into the leading edge of the molten weld pool, **Figure 32-20**. If the torch is not moved back, the rod may melt and drip into the molten weld pool; this is an incorrect way of adding filler metal, **Figure 32-21**. The major problems with adding rod in this manner are (1) the drop of metal tends to overheat, resulting in important alloys being burned out; (2) the metal cannot always be added where it is needed; and (3) the method works only in the flat

FIGURE 32-20 Filler metal being correctly added by dipping the rod into the leading edge of the molten weld pool. Larry Jeffus

FIGURE 32-21 Filler metal being incorrectly added by allowing the rod to melt and drip into the molten weld pool. Larry Jeffus

position. When dipping the rod into the molten weld pool, if the rod touches the hot metal around the molten weld pool, it will stick. When this happens, move the flame directly to the end of the rod to melt and free it. Turn off the cylinders, bleed the hoses, back out the regulator adjusting screws, and clean up your work area when you are finished.

Complete a copy of the "Student Welding Report" listed in Appendix I or provided by your instructor. ◆

EXPERIMENT 32-3

Effect of Rod Size on the Molten Weld Pool

Use a properly lit and adjusted torch; 6 in. (152 mm) of 16-gauge mild steel; and three different diameters of RG45 gas welding rods, 1/8 in. (3 mm), 3/32 in. (2.4 mm), and 1/16 in. (2 mm), by 36 in. (914 mm) long. In this experiment, you will observe the effect on the molten weld pool of changing the size of filler metal. You also will practice adding the filler metal to the molten weld pool.

Starting with the 1/8-in. (3-mm) filler metal, make a weld 6 in. (152 mm) long. Next to this weld, make another one with the 3/32-in. (2.4-mm) rod, then one with the 1/16-in. (2-mm) rod. Try to keep the angle, speed, height, and pattern of the torch the same during each of the welds. Observe the differences in each of the weld sizes.

Complete a copy of the "Student Welding Report" listed in Appendix I or provided by your instructor. ◆

PRACTICE 32-3

Stringer Bead, Flat Position

Repeat Experiment 32-3 until you have mastered a straight and uniform weld with any of the three sizes of filler rod.

Joining two or more clean pieces of metal to form a welded joint is the next step in learning to weld. The joints must be uniform in width and buildup so that they will have maximum strength. For each type of gas welded joint, the amount of penetration required to give maximum strength will vary and may not be 100%, **Figure 32-22**. For example, in thin sheet metal there is usually enough reinforcement on the weld to give the weld adequate strength. But if that reinforcement has

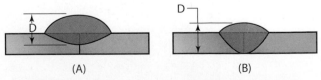

FIGURE 32-22 Welds (A) and (B) both have approximately the same strength—but only if the reinforcement does not have to be removed. © Cengage Learning 2012

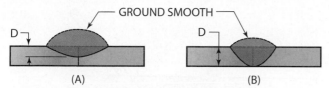

FIGURE 32-23 If the buildup on both welds is removed, weld (B) would be the stronger weld.
© Cengage Learning 2012

to be removed, then 100% penetration is important, **Figure 32-23**. Some joints, such as the lap joint, may never need 100% penetration. However, they do need 100% fusion. Turn off the cylinders, bleed the hoses, back out the regulator adjusting screws, and clean up your work area when you are finished.

Complete a copy of the "Student Welding Report" listed in Appendix I or provided by your instructor. ◆

FLAT POSITION WELDING

Outside Corner Joint

The flat **outside corner joint** can be made with or without the addition of filler metal. This joint is one of the easiest welded joints to make. If the sheets are tacked together properly, the addition of filler metal is not needed. However, if filler metal is added, it

should be added uniformly, as in the stringer beads in Practice 32-3.

PRACTICE 32-4

Outside Corner Joint, Flat Position

Using a properly lit and adjusted torch, two clean pieces of 16-gauge mild steel 6 in. (152 mm) long, and filler metal, you will make a flat outside corner welded joint, **Figure 32-24**.

Place one of the pieces of metal in a jig or on a firebrick and hold or brace the other piece of metal vertically on it, as shown in **Figure 32-25** and **Figure 32-26**. Tack the ends of the two sheets together. Then set it upright and

FIGURE 32-25 Angle iron jig for holding metal so that it can be tack welded. © Cengage Learning 2012

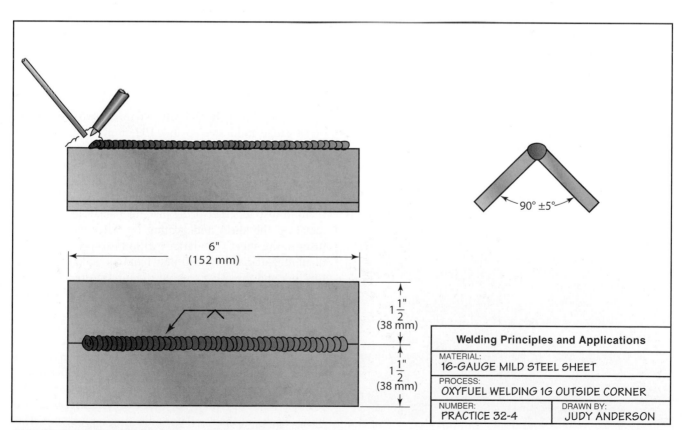

FIGURE 32-24 Outside corner. © Cengage Learning 2012

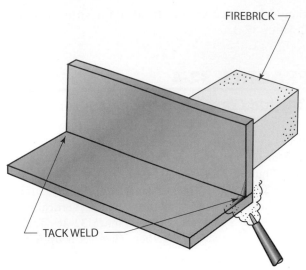

FIGURE 32-26 Tack welding the outside corner joint using a firebrick to support the metal. © Cengage Learning 2012

put two or three more tacks on the joint, **Figure 32-27.** Holding the torch as shown in **Figure 32-28,** make a uniform weld along the joint. Repeat this until the weld can be made without defects. Turn off the cylinders, bleed the hoses, back out the regulator adjusting screws, and clean up your work area when you are finished.

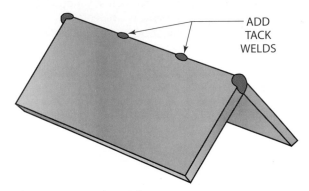

FIGURE 32-27 Tack welding. © Cengage Learning 2012

FIGURE 32-28 Outside corner joint. Larry Jeffus

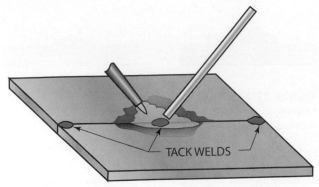

FIGURE 32-29 Making a tack weld. © Cengage Learning 2012

Complete a copy of the "Student Welding Report" listed in Appendix I or provided by your instructor. ◆

Butt Joint

The **flat butt joint** is a welded joint and one of the easiest to make. To make the butt joint, place two clean pieces of metal flat on the table and tack weld both ends together, as illustrated in **Figure 32-29.** Tack welds may also be placed along the joint before welding begins. Point the torch so that the flame is distributed equally on both sheets. The flame is to be in the direction that the weld is to progress. If the sheets to be welded are of different sizes or thicknesses, the torch should be pointed so that both pieces melt at the same time, **Figure 32-30.**

When both sheet edges have melted, add the filler rod in the same manner as in Practice 32-3.

PRACTICE 32-5

Butt Joint, Flat Position

Using a properly lit and adjusted torch, two clean pieces of 16-gauge mild steel 6 in. (152 mm) long, and filler metal, you will make a welded butt joint, **Figure 32-31.**

Place the two pieces of metal in a jig or on a firebrick and tack weld both ends together. The tack on the ends can be made by simply heating the ends and allowing them to fuse together or by placing a small drop of filler metal on the sheet and heating the filler metal until it fuses to the sheet. The latter method is especially convenient if you have to use one hand to hold the sheets together and the other to hold the torch. After both ends are tacked together, place one or two small tacks along the joint to prevent warping during welding.

With the sheets tacked together, start welding from one end to the other using the technique learned in Practice 32-3. Repeat this weld until you can make a welded butt joint that is uniform in width and reinforcement and has no visual defects. The penetration of this practice weld may vary. Turn off the cylinders, bleed the hoses, back out the regulator adjusting screws, and clean up your work area when you are finished.

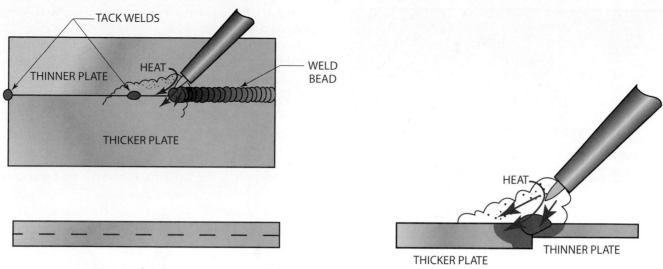

FIGURE 32-30 Direct the flame on the thicker plate. © Cengage Learning 2012

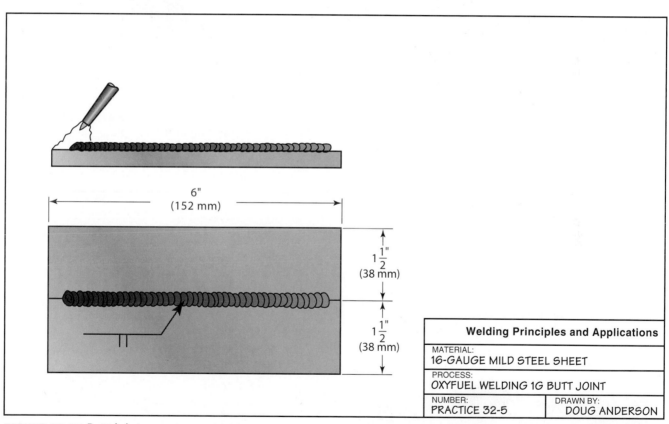

6"
(152 mm)

1 1/2"
(38 mm)

1 1/2"
(38 mm)

Welding Principles and Applications

MATERIAL:
16-GAUGE MILD STEEL SHEET

PROCESS:
OXYFUEL WELDING 1G BUTT JOINT

NUMBER:
PRACTICE 32-5

DRAWN BY:
DOUG ANDERSON

FIGURE 32-31 Butt joint. © Cengage Learning 2012

Complete a copy of the "Student Welding Report" listed in Appendix I or provided by your instructor. ◆

PRACTICE 32-6

Butt Joint with 100% Penetration

Using the same equipment, materials, and setup as described in Practice 32-5, make a welded butt joint with 100% penetration along the entire 6 in. (152 mm) of the welded joint and then visually inspect (VT) the root (back) to see if it has complete penetration. Turn off the cylinders, bleed the hoses, back out the regulator adjusting screws, and clean up your work area when you are finished.

Complete a copy of the "Student Welding Report" listed in Appendix I or provided by your instructor. ◆

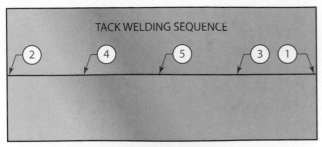

FIGURE 32-32 Tack welding sequence used to minimize distortion. © Cengage Learning 2012

PRACTICE 32-7

Butt Joint with Minimum Distortion

Using a properly lit and adjusted torch, two clean pieces of 16-gauge mild steel 6 in. (152 mm) long, and filler metal, you will make a welded butt joint while controlling distortion and penetration.

Distortion can be controlled by back stepping a weld, proper tacking, and clamping. For this weld, back stepping and proper tacking will be used to control distortion. The tacking sequence to be used is shown in **Figure 32-32**. The back-stepping method to be used is illustrated in **Figure 32-33**. Back stepping will also eliminate the problem of burning away the end of the sheet. Practice this weld until you can pass a visual inspection (VT) for distortion. Turn off the cylinders, bleed the hoses, back out the regulator adjusting screws, and clean up your work area when you are finished.

Complete a copy of the "Student Welding Report" listed in Appendix I or provided by your instructor. ◆

Lap Joint

The flat **lap joint** can be easily welded if some basic manipulations are used. When heating the two clean sheets, caution must be exercised to ensure that both sheets start melting at the same time. Heat is not distributed uniformly in the lap joint, **Figure 32-34**. Because of

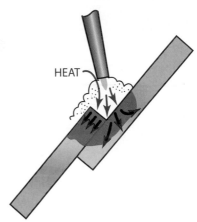

HEAT

FIGURE 32-34 Heat is conducted away faster in the bottom plate, resulting in the top plate melting more quickly. © Cengage Learning 2012

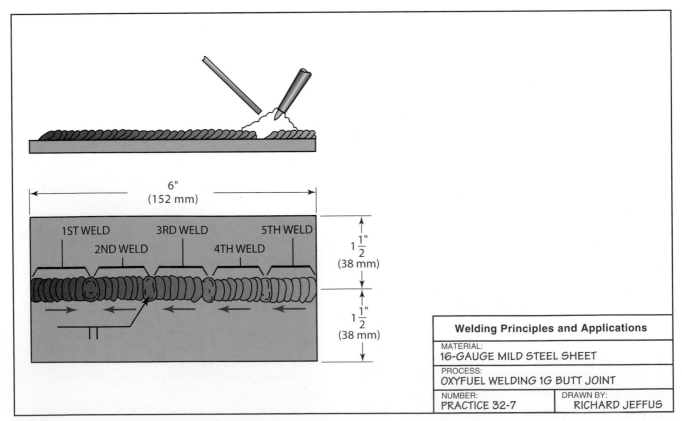

6" (152 mm)

1ST WELD   3RD WELD   5TH WELD
2ND WELD   4TH WELD

$1\frac{1}{2}$" (38 mm)

$1\frac{1}{2}$" (38 mm)

Welding Principles and Applications

| MATERIAL: |
| 16-GAUGE MILD STEEL SHEET |

| PROCESS: |
| OXYFUEL WELDING 1G BUTT JOINT |

| NUMBER: | DRAWN BY: |
| PRACTICE 32-7 | RICHARD JEFFUS |

FIGURE 32-33 Welding with minimum distortion. © Cengage Learning 2012

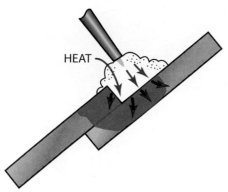

FIGURE 32-35 Flame heat should be directed at the bottom plate to compensate for thermal conductivity.
© Cengage Learning 2012

THE EDGE OF THE TOP PLATE CAN MELT BACK FORMING A NOTCH IF IT BECOMES TOO HOT. ➞

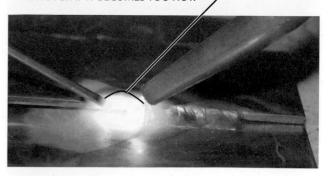

INCREASING THE AMOUNT OF FILLER METAL BEING ADDED TO THE TOP EDGE WILL FILL IN A NOTCH IF ONE FORMS. ➞

FIGURE 32-36 Adding filler metal to a lap joint weld.
Larry Jeffus

this difference in heating rate, the flame must be directed on the bottom sheet and away from the metal top sheet, **Figure 32-35**. The filler rod should be added to the top sheet. Gravity will pull the molten weld pool down to the bottom sheet, so it is therefore not necessary to put metal on the bottom sheet. If the filler metal is not added to the top sheet or if it is not added fast enough, surface tension will pull the molten weld pool back from the joint, **Figure 32-36**. When this happens, the rod should be added directly into this notch, and it will close. The weld appearance and strength will not be affected.

PRACTICE 32-8

Lap Joint, Flat Position

Using a properly lit and adjusted torch, two clean pieces of 16-gauge mild steel 6 in. (152 mm) long, and filler metal, you will make a welded lap joint, **Figure 32-37**. Place the two pieces of metal on a firebrick and tack both

ends, as shown in **Figure 32-38** and **Figure 32-39**. Make two or three more tack welds. Starting at one end, make a uniform weld along the joint. Both sides of the joint can be welded. Repeat this practice until the weld can be made without defects. If you want to test your skill, shear out a

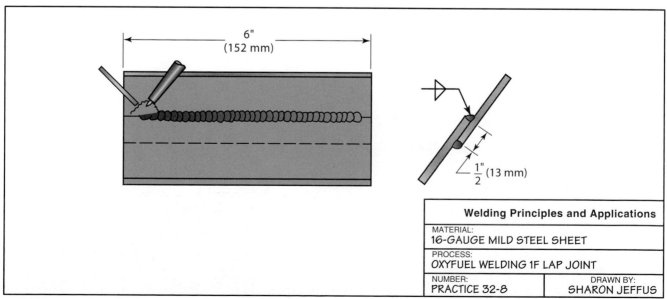

6"
(152 mm)

$\frac{1}{2}$" (13 mm)

Welding Principles and Applications

| MATERIAL: |
| 16-GAUGE MILD STEEL SHEET |

| PROCESS: |
| OXYFUEL WELDING 1F LAP JOINT |

| NUMBER: | DRAWN BY: |
| PRACTICE 32-8 | SHARON JEFFUS |

FIGURE 32-37 Lap joint. © Cengage Learning 2012

FIGURE 32-38 Heating the joint before tacking. Larry Jeffus

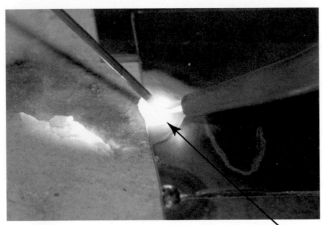

TO HEAT THE JOINT EVENLY THE FLAME IS DIRECTED
TOWARD THE BOTTOM PLATE.

FIGURE 32-39 Filler rod is added after both pieces are heated to a melt. Larry Jeffus

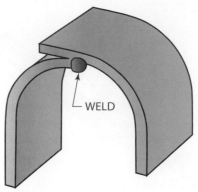

FIGURE 32-41 180° bend to test lap weld quality.
© Cengage Learning 2012

strip 1 in. (25 mm) wide, as shown in **Figure 32-40**, and test it for 100% root penetration, **Figure 32-41**. Turn off the cylinders, bleed the hoses, back out the regulator adjusting screws, and clean up your work area when you are finished.

Complete a copy of the "Student Welding Report" listed in Appendix I or provided by your instructor. ◆

Tee Joint

The flat **tee joint** is more difficult to make than the butt or lap joint. The tee joint has the same problem with uneven heating that the lap joint does. It is important to hold the flame so that both sheets melt at the same time, **Figure 32-42**. Another problem that is unique to the tee joint is that a large percentage of the welding heat is reflected back on the torch. This reflected heat can cause even a properly cleaned and adjusted torch to backfire or pop. To help prevent this from happening, angle the torch more in the direction of the weld travel. Because of the

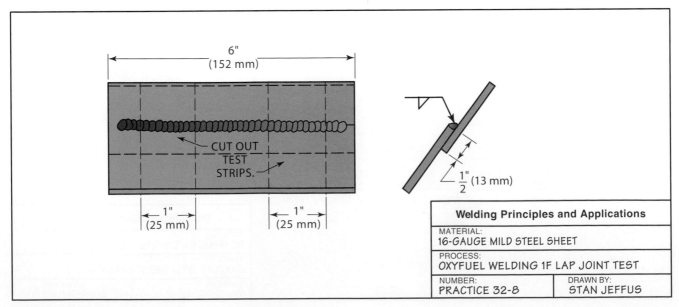

FIGURE 32-40 Cut out test strips. © Cengage Learning 2012

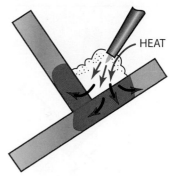

FIGURE 32-42 Direct the heat on the bottom plate to equalize the heating rates. © Cengage Learning 2012

slightly restricted atmosphere of the tee joint, it may be necessary to adjust the flame so that it is somewhat oxidizing. The beginning student should not be overly concerned with this.

PRACTICE 32-9

Tee Joint, Flat Position

Using a properly lit and adjusted torch, two clean pieces of 16-gauge mild steel 6 in. (152 mm) long, and filler metal, you will weld a flat tee joint, **Figure 32-43**.

Place the first piece of metal flat on a firebrick and hold or brace the second piece vertically on the first piece.

The vertical piece should be within 5° of square to the bottom sheet. Tack the two sheets at the ends. Then put two or three more tacks along the joint and brace the tee joint in position. If you want to test your skill, cut out a strip 1 in. (25 mm) wide, as shown in **Figure 32-44**, and test it for 100% root penetration, **Figure 32-45**. Turn off the cylinders, bleed the hoses, back out the regulator adjusting screws, and clean up your work area when you are finished.

Complete a copy of the "Student Welding Report" listed in Appendix I or provided by your instructor. ◆

OUT-OF-POSITION WELDING

A part to be welded cannot always be positioned so that it can be welded in the flat position. Whenever a weld is performed in a position other than flat, it is said to be **out-of-position welding.** Welds made in the vertical, horizontal, or overhead position are out of position and somewhat more difficult than flat welds.

VERTICAL WELDS

A **vertical weld** is the most common out-of-position weld that a welder is required to perform. When making a vertical weld, it is important to control the size of the molten weld pool. If the molten weld pool size increases beyond that which the **shelf** will support,

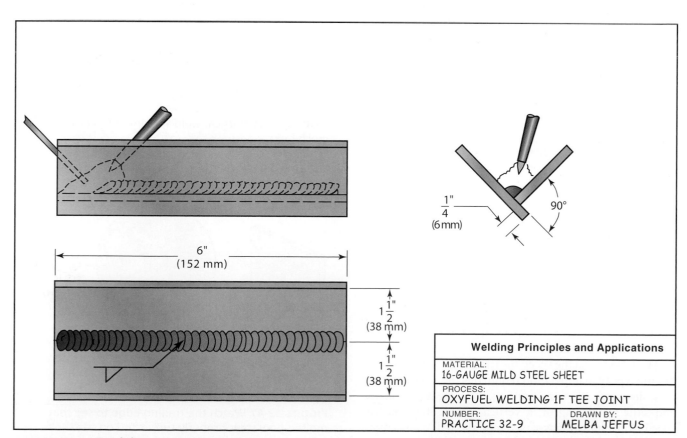

FIGURE 32-43 Tee joint. © Cengage Learning 2012

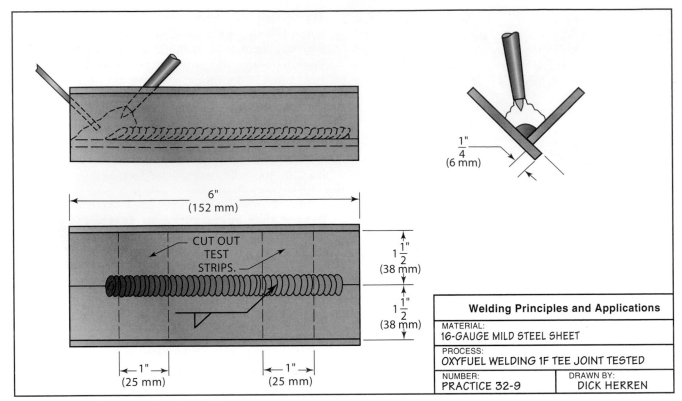

FIGURE 32-44 Cut out strips for testing. © Cengage Learning 2012

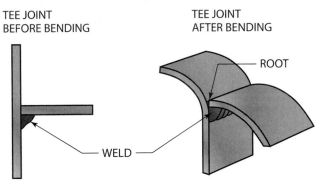

FIGURE 32-45 Bend the test strip to be sure that the weld had good root fusion. © Cengage Learning 2012

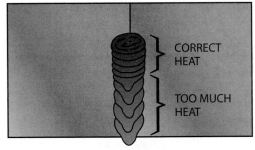

FIGURE 32-46 Vertical weld showing effect of too much heat. © Cengage Learning 2012

Figure 32-46, the molten weld pool will overflow and drip down the weld. These drops, when cooled, look like the drips of wax on a candle. To prevent the molten weld pool from dripping, the trailing edge of the molten weld pool must be watched. The **trailing edge** will constantly be solidifying, forming a new shelf to support the molten weld pool as the weld progresses upward, **Figure 32-47**. Small molten weld pools are less likely than large ones to drip.

The less vertical the sheet, the easier the weld is to make, but the type of manipulation required is the same. Welding on a sheet at a 45° angle requires the same manipulation and skill as welding on a vertical sheet. However, the speed of manipulation is slower, and the skill is less critical than at 90° vertical. This welding technique should be

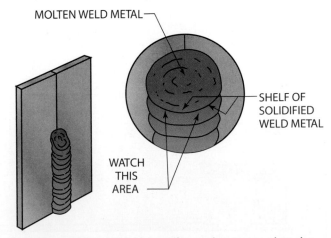

FIGURE 32-47 Watch the trailing edge to see that the molten pool stays properly supported on the shelf.
© Cengage Learning 2012

mastered at a 45° angle. Then the angle of the sheet is increased until it is possible to make totally vertical welds. Each practice weld in this section should be started on an incline, and, as skill is gained, the angle should be increased until the sheets are vertical.

PRACTICE 32-10

Stringer Bead at a 45° Angle

Using a properly lit and adjusted torch, two clean pieces of 16-gauge mild steel 6 in. (152 mm) long, and filler metal, you will make a bead at a 45° angle.

The filler metal should be added as you did in Practice 32-3. It may be necessary to flash the torch off the molten weld pool to allow it to cool, **Figure 32-48A, B,** and **C. Flashing** the torch off allows the molten weld pool to cool by moving the hotter inner cone away from the molten weld pool itself. While still protecting the molten metal with the outer flame envelope, a rhythm of moving the torch and adding the rod should be established. This rhythm helps make the bead uniform. Repeat this practice until the weld can be made without defects. Turn off the cylinders, bleed the hoses, back out

the regulator adjusting screws, and clean up your work area when you are finished.

Complete a copy of the "Student Welding Report" listed in Appendix I or provided by your instructor. ◆

PRACTICE 32-11

Stringer Bead, Vertical Position

Repeat Practice 32-10 until you have mastered a straight and uniform weld bead in a vertical position. Turn off the cylinders, bleed the hoses, back out the regulator adjusting screws, and clean up your work area when you are finished.

Complete a copy of the "Student Welding Report" listed in Appendix I or provided by your instructor. ◆

Butt Joint
PRACTICE 32-12

Butt Joint at a 45° Angle

Using a properly lit and adjusted torch, two clean pieces of 16-gauge mild steel 6 in. (152 mm) long, and filler

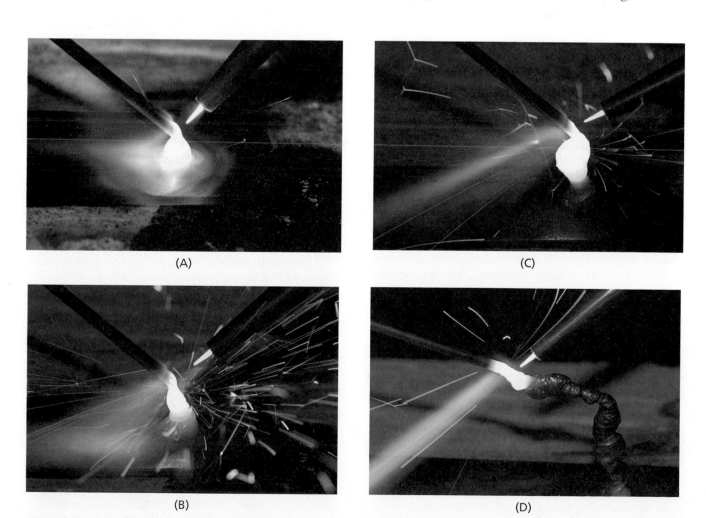

(A)

(B)

(C)

(D)

FIGURE 32-48 By flashing the flame off and controlling the pool size, a weld can be built up (A), and up (B), and over (C), and over (D). Larry Jeffus

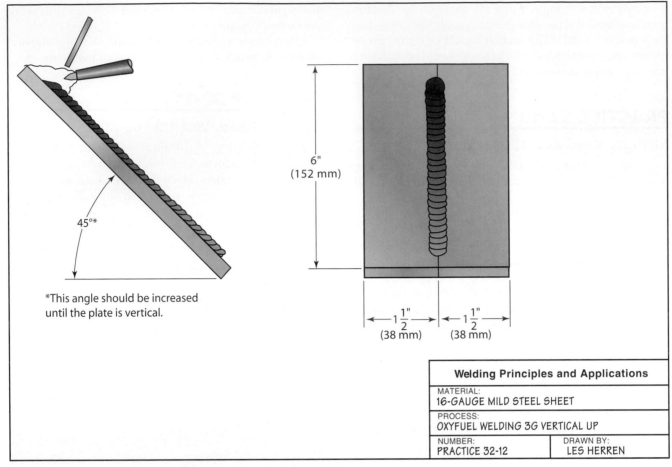

45°*

*This angle should be increased until the plate is vertical.

6"
(152 mm)

1 1/2"
(38 mm)

1 1/2"
(38 mm)

Welding Principles and Applications	
MATERIAL: 16-GAUGE MILD STEEL SHEET	
PROCESS: OXYFUEL WELDING 3G VERTICAL UP	
NUMBER: PRACTICE 32-12	DRAWN BY: LES HERREN

FIGURE 32-49 Butt joint at a 45° angle. © Cengage Learning 2012

metal, you will make a welded butt joint at a 45° angle, **Figure 32-49.**

Tack the sheets together and support them at a 45° angle. Weld using the method of flashing the torch off the molten weld pool to control penetration and weld contour. Make a weld that has uniform width and reinforcement. Repeat this practice until the weld can be made without defects. Turn off the cylinders, bleed the hoses, back out the regulator adjusting screws, and clean up your work area when you are finished.

Complete a copy of the "Student Welding Report" listed in Appendix I or provided by your instructor. ◆

PRACTICE 32-13

Butt Joint, Vertical Position

Using the same equipment, materials, and setup as described in Practice 32-12, make a welded butt joint in the vertical position. Make a weld that is uniform in width and buildup and has no visual defects. The penetration of this practice weld may vary. Turn off the cylinders, bleed the hoses, back out the regulator adjusting screws, and clean up your work area when you are finished.

Complete a copy of the "Student Welding Report" listed in Appendix I or provided by your instructor. ◆

PRACTICE 32-14

Butt Joint, Vertical Position, with 100% Penetration

Using the same equipment, materials, and setup as listed in Practice 32-12, weld a butt joint in the vertical position with 100% penetration along the entire 6 in. (152 mm) of the welded joint. Repeat this practice until this weld can be made without defects. If you want to test your skill, shear out a strip 1 in. (25 mm) wide and test it for 100% root penetration, **Figure 32-50.** Turn off the cylinders, bleed the hoses, back out the regulator

FIGURE 32-50 Bend strips to check a weld.
© Cengage Learning 2012

adjusting screws, and clean up your work area when you are finished.

Complete a copy of the "Student Welding Report" listed in Appendix I or provided by your instructor. ◆

Lap Joint
PRACTICE 32-15

Lap Joint at a 45° Angle

Using a properly lit and adjusted torch, two clean pieces of 16-gauge mild steel 6 in. (152 mm) long, and filler metal, you will weld a lap joint at a 45° angle.

After tacking the sheets together and supporting them at a 45° angle, use the same method of adding rod as you did for the flat lap joint. Again, flash off the torch as needed to control the molten weld pool.

Repeat this weld until you can make a weld that is uniform in width and reinforcement and has no visual defects. Both sides of the joint can be welded. Turn off the cylinders, bleed the hoses, back out the regulator adjusting screws, and clean up your work area when you are finished.

Complete a copy of the "Student Welding Report" listed in Appendix I or provided by your instructor. ◆

PRACTICE 32-16

Lap Joint, Vertical Position

Using the same equipment, materials, and setup as listed in Practice 32-15, weld a lap joint in the vertical position. Make a weld that is uniform in width and reinforcement and has no visual defects. Both sides of the joint can be welded. Repeat this practice until the weld can be made

without defects. If you want to test your skill, shear out a strip 1 in. (25 mm) wide and test it for 100% root penetration. Turn off the cylinders, bleed the hoses, back out the regulator adjusting screws, and clean up your work area when you are finished.

Complete a copy of the "Student Welding Report" listed in Appendix I or provided by your instructor. ◆

Tee Joint

The vertical tee joint has a right and a left side. **Figure 32-51** shows the best way to place the sheets depending upon whether the welder is right-handed or left-handed. It is a good idea to try both the right-hand and left-hand joints because in the field you may not be able to change the joint direction. Use the same method of adjusting the torch and torch angle that was practiced for the flat tee joint. In addition, use the method of flashing the torch off for molten weld pool control. Surface tension in the molten weld pool of a tee joint enables a larger weld to be made than either the butt or lap joint without as severe a problem with dripping.

PRACTICE 32-17

Tee Joint at a 45° Angle

Using a properly lit and adjusted torch, two clean pieces of 16-gauge mild steel 6 in. (152 mm) long, and filler metal, you will weld a tee joint at a 45° angle.

After tacking the sheets together and supporting them at a 45° angle, make a fillet weld that has uniform width and reinforcement and no visual defects. It is often best to weld only one side of the practice tee joint unless the oxides can be easily removed from the back side of the

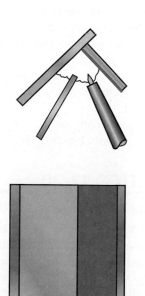

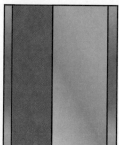

(A) RIGHT-HAND TEE

(B) LEFT-HAND TEE

FIGURE 32-51 Some vertical tee joints are easier for right-handed or left-handed welders. © Cengage Learning 2012

previous weld. Repeat this practice until the weld can be made without defects. Turn off the cylinders, bleed the hoses, back out the regulator adjusting screws, and clean up your work area when you are finished.

Complete a copy of the "Student Welding Report" listed in Appendix I or provided by your instructor. ◆

PRACTICE 32-18

Tee Joint, Vertical Position

Using the same equipment, materials, and setup as listed in Practice 32-17, make a fillet weld. Cut out a strip 1 in. (25 mm) wide (refer to Figure 32-44) and test it for 100% root penetration. Repeat this practice until the weld passes the test. Turn off the cylinders, bleed the hoses, back out the regulator adjusting screws, and clean up your work area when you are finished.

Complete a copy of the "Student Welding Report" listed in Appendix I or provided by your instructor. ◆

HORIZONTAL WELDS

Horizontal welds, like vertical welds, must rely on some part of the weld bead to support the molten weld pool as the weld is made. The shelf that supports a horizontal weld must be built up under the molten weld pool and at the same time must keep the weld bead uniform. The weave pattern required for a horizontal weld is completely different from that of any of the other positions. The pattern, **Figure 32-52,** builds a shelf on the bottom side of the bead to support the molten weld pool, which is elongated across the top. The sheet may be tipped back at a 45° angle for the stringer bead. Doing this allows the student to acquire the needed skills before proceeding to the more difficult, fully horizontal position. As with the vertically inclined sheet, the skills required are the same.

Horizontal Stringer Bead

When starting a horizontal bead, it is important to start with a small bead and build it to the desired size. If too large a molten weld pool is started, the shelf does not have time to form properly. The weld bead will tend to sag downward and not be uniform. As a result, there may be an **undercut** of the top edge and an **overlap** on the bottom edge, **Figure 32-53.**

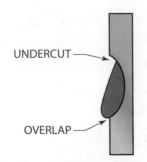

FIGURE 32-53 Too large a molten weld pool. © Cengage Learning 2012

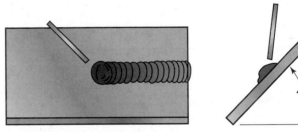

FIGURE 32-54 Horizontal stringer bead at a reclining angle. © Cengage Learning 2012

PRACTICE 32-19

Horizontal Stringer Bead at a 45° Sheet Angle

Using a properly lit and adjusted torch, one clean piece of 16-gauge mild steel 6 in. (152 mm) long, and filler metal, you will make a horizontal bead at a 45° reclining angle, **Figure 32-54.**

Add the filler metal along the top leading edge of the molten weld pool. Surface tension will help hold it on the top. The weld should be uniform in width and reinforcement and have no visual defects. Repeat this practice until the weld can be made without defects. Turn off the cylinders, bleed the hoses, back out the regulator adjusting screws, and clean up your work area when you are finished.

Complete a copy of the "Student Welding Report" listed in Appendix I or provided by your instructor. ◆

PRACTICE 32-20

Stringer Bead, Horizontal Position

Using the same equipment, materials, and setup as listed in Practice 32-19, make a stringer bead in the horizontal position. The stringer bead should be uniform in width and reinforcement and have no visual defects. Turn off the cylinders, bleed the hoses, back out the regulator adjusting screws, and clean up your work area when you are finished.

Complete a copy of the "Student Welding Report" listed in Appendix I or provided by your instructor. ◆

CENTERLINE OF WELD

FIGURE 32-52 A "J" weave pattern for horizontal welds. © Cengage Learning 2012

Butt Joint
PRACTICE 32-21

Butt Joint, Horizontal Position

Using a properly lit and adjusted torch, two clean pieces of 16-gauge mild steel 6 in. (152 mm) long, and filler metal, you will weld a butt joint in the horizontal position.

After tacking the sheets together and supporting them in the horizontal position, make a weld using the same technique as practiced in the horizontal beading, Practice 32-20. The weld must be uniform in width and buildup and have no visual defects. If you want to test your skill, shear out a strip 1 in. (25 mm) wide and test it for 100% root penetration. Repeat this practice until the weld can be made without defects. Turn off the cylinders, bleed the hoses, back out the regulator adjusting screws, and clean up your work area when you are finished.

Complete a copy of the "Student Welding Report" listed in Appendix I or provided by your instructor. ◆

Lap Joint
PRACTICE 32-22

Lap Joint, Horizontal Position

Using a properly lit and adjusted torch, two clean pieces of 16-gauge mild steel 6 in. (152 mm) long, and filler metal, you will weld a lap joint in the horizontal position.

After tacking the sheets together, support the assembly as illustrated in **Figure 32-55**. The weld must be uniform in width and buildup and have no visual defects. The sheet can be turned over, and the other side can be welded. If you want to test your skill, shear out a strip 1 in. (25 mm) wide and test it for 100% root penetration. Repeat this practice until the weld can be made without defects. Turn off the cylinders, bleed the hoses, back out the regulator adjusting screws, and clean up your work area when you are finished.

Complete a copy of the "Student Welding Report" listed in Appendix I or provided by your instructor. ◆

FIGURE 32-55 Horizontal lap joint. © Cengage Learning 2012

Tee Joint

The horizontal and flat tee joints are very similar in relation to the types of skills required to perform metal welds. The horizontal fillet weld tends to flow down toward the horizontal sheet from the vertical sheet. To correct this problem, the filler metal should be added to the top edge of the molten weld pool.

PRACTICE 32-23

Tee Joint, Horizontal Position

Using a properly lit and adjusted torch, two clean pieces of 16-gauge mild steel 6 in. (162 mm) long, and filler metal, you will weld one side of a tee joint in the horizontal position. After tacking the sheets together, make a fillet weld that is uniform in width and reinforcement and has no visual defects. If you want to test your skill, shear out a strip 1 in. (25 mm) wide and test it for 100% root penetration. Repeat this practice until the weld can be made without defects. Turn off the cylinders, bleed the hoses, back out the regulator adjusting screws, and clean up your work area when you are finished.

Complete a copy of the "Student Welding Report" listed in Appendix I or provided by your instructor. ◆

OVERHEAD WELDS

When welding in the overhead position, it is important to wear the proper personal protection, including leather gloves, leather sleeves, a leather apron, and a cap. The possibility of being burned increases greatly when welding in the overhead position. However, with the proper protective clothing you should avoid being burned.

With the **overhead weld,** the molten weld pool is held to the sheet by surface tension in the same manner that a drop of water is held to the bottom of a glass sheet. If the molten weld pool gets too large, big drops of metal may fall. If the welding rod is not dipped into the molten weld pool, but is allowed to melt in the flame, it also may drip. As long as the molten weld pool is controlled and the rod is added properly, overhead welding is safe.

The direction that you choose to weld in the overhead position is one of personal preference. It is a good idea to try several directions before deciding on one. The height of the sheet also affects your skill and progress. Welders often prefer to stand while overhead welding so that sparks do not land in their laps. If you decide to stand, you need to somehow brace yourself to help your stability.

Stringer Bead

Place the metal at a height recommended by your instructor. With the torch off, your goggles down, and a rod in your hand, try to progress across the sheet in a straight line. Use several directions until you find the direction that best suits you. Change the height of the

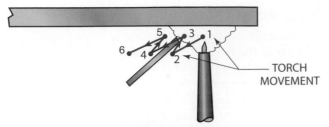

TORCH
MOVEMENT

FIGURE 32-56 Overhead weld. © Cengage Learning 2012

sheet up and down to determine the height at which welding is most comfortable.

PRACTICE 32-24

Stringer Bead, Overhead Position

Using a properly lit and adjusted torch and one clean piece of 16-gauge mild steel 6 in. (152 mm) long, you will make a bead in the overhead position.

Heat the sheet until it melts and forms a molten weld pool. Put the welding rod into the molten weld pool as the torch tip is moved away from the molten weld pool. Return the flame to the molten weld pool as you remove the rod, **Figure 32-56.** Continue repeating this sequence as you move along the sheet. The weld bead should be uniform in width and buildup and have no visual defects. Repeat this practice until the weld can be made without defects. Turn off the cylinders, bleed the hoses, back out the regulator adjusting screws, and clean up your work area when you are finished.

Complete a copy of the "Student Welding Report" listed in Appendix I or provided by your instructor. ◆

Butt Joint

PRACTICE 32-25

Butt Joint, Overhead Position

Using a properly lit and adjusted torch, two clean pieces of 16-gauge mild steel 6 in. (152 mm) long, and filler metal, you will weld a butt joint in the overhead position.

After tacking the sheets together, put them in the overhead position. Following the sequence used in Practice 32-24 for the overhead stringer bead, make a weld along the joint. The weld should be uniform in width and reinforcement and have no visual defects. If you want to test your skill, shear out a strip 1 in. (25 mm) wide and test it for 100% root penetration. Repeat this practice until the weld can be made without defects. Turn off the cylinders, bleed the hoses, back out the regulator adjusting screws, and clean up your work area when you are finished.

Complete a copy of the "Student Welding Report" listed in Appendix I or provided by your instructor. ◆

Lap Joint

PRACTICE 32-26

Lap Joint, Overhead Position

Using a properly lit and adjusted torch, two clean pieces of 16-gauge mild steel 6 in. (152 mm) long, and filler metal, you will weld a lap joint in the overhead position.

After tacking the sheets together, put them in the overhead position. Using the sequence for the overhead stringer bead, make a weld down the joint. The filler metal should be added to the leading edge of the molten weld pool on the top sheet. The weld should be uniform in width and buildup and have no visual defects. If you want to test your skill, shear out a strip 1 in. (25 mm) wide and test it for 100% root penetration. Repeat this practice until the weld can be made without defects. Turn off the cylinders, bleed the hoses, back out the regulator adjusting screws, and clean up your work area when you are finished.

Complete a copy of the "Student Welding Report" listed in Appendix I or provided by your instructor. ◆

Tee Joint

PRACTICE 32-27

Tee Joint, Overhead Position

Using a properly lit and adjusted torch, two clean pieces of 16-gauge mild steel 6 in. (152 mm) long, and filler metal, you will weld one side of a tee joint in the overhead position.

After tacking the sheets together, put them in the overhead position and make a fillet weld. The filler metal should be added to the top sheet. The weld should be uniform in width and buildup and have no visual defects. If you want to test your skill, cut out a strip 1 in. (25 mm) wide and test it for 100% root penetration. Repeat this practice until the weld can be made without defects. Turn off the cylinders, bleed the hoses, back out the regulator adjusting screws, and clean up your work area when you are finished.

Complete a copy of the "Student Welding Report" listed in Appendix I or provided by your instructor. ◆

MILD STEEL PIPE AND TUBING

Mild steel pipe and tubing, both small diameter and thin wall, can be gas welded. The welding process for both pipe and tubing is usually the same. Thin-wall material does not require a grooved preparation. Gas welding is very seldom used to manufacture piping systems. It is used on both pipe and tubing to make structures, such as bicycle and motorcycle frames, gates, works of art, handrails, and light aircraft frames, **Figure 32-57.**

Horizontal Rolled Position 1G

The experiments and practices that follow (through Practice 32-29) will give the student the opportunity to gain skill in making welds in the **1G position, Figure 32-58.**

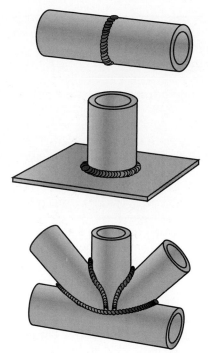

FIGURE 32-57 Examples of tube joints commonly used in industry. © Cengage Learning 2012

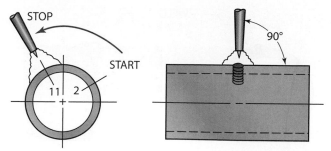

FIGURE 32-59 When the weld is finished, stop and roll the pipe so that the end of the weld is at the 2 o'clock position. © Cengage Learning 2012

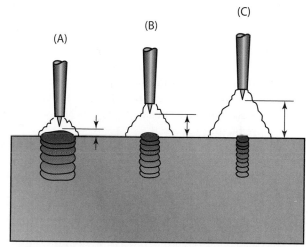

FIGURE 32-60 As the distance between the inner core and pipe changes, the size of the molten weld pool changes. © Cengage Learning 2012

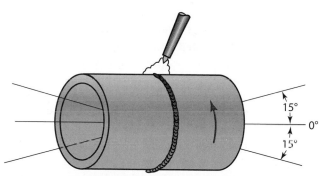

FIGURE 32-58 1G position. The pipe is rolled horizontally. The weld is made in the flat position (approximately 12 o'clock as the pipe is rolled). © Cengage Learning 2012

Keeping the torch at the correct angle to the pipe surface requires some practice. Changes in the relative torch angle will greatly affect the size of the molten weld pool. The distance of the inner core from the pipe surface is also important, **Figure 32-60.** You must be comfortable and able to freely move around. Your free hand should not be used for supporting or steadying the torch or yourself. Later, when you need this hand to add filler metal, you will not have to learn how to be steady without the use of your hand.

Complete a copy of the "Student Welding Report" listed in Appendix I or provided by your instructor. ◆

EXPERIMENT 32-4

Effect of Changing Angle on Molten Weld Pool

This experiment will show how the molten weld pool is affected by changing the surface angle. With a piece of pipe having a diameter of approximately 2 in. (51 mm), you will push a molten weld pool across the top of the pipe. The pipe is in the 1G position, **Figure 32-59.** Starting at the 2 o'clock position, weld upward and across to the 11 o'clock position. Use the same torch angle and distances that you learned in Practice 32-2 and Experiment 32-3. Repeat this experiment until you can make a straight weld bead that is uniform in width.

EXPERIMENT 32-5

Stringer Bead, 1G Position

Using a properly lit and adjusted torch, one clean piece of pipe approximately 2 in. (51 mm) in diameter, and three different sizes of filler metal, you will make a stringer bead on the pipe.

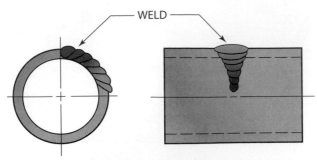

FIGURE 32-61 The weld bead shape is affected by its relative position on the pipe. © Cengage Learning 2012

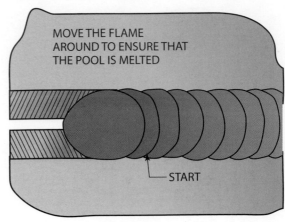

MOVE THE FLAME AROUND TO ENSURE THAT THE POOL IS MELTED

START

FIGURE 32-62 When restarting a weld pool, be sure that the entire pool area is remelted before starting to add weld metal. © Cengage Learning 2012

With the pipe in the 1G position, start a molten weld pool at the 2 o'clock position and weld to the 11 o'clock position. When you are at the 2 o'clock position, gravity will pull the molten metal outward, making the bead high and narrow, **Figure 32-61**. As the weld progresses toward the 12 o'clock position, if you keep using the same technique, the weld metal is pulled down, making the bead flat and wide. Repeat this experiment until you can make a straight weld bead that is uniform in width and reinforcement.

To keep the contour and width of the weld bead uniform, you must adjust your technique as the weld progresses. The following are some methods of keeping the buildup uniform:

- Move the flame farther from the surface of the pipe.
- Decrease the torch angle relative to the pipe surface.
- Travel at an increasing rate of speed.

A combination of these methods can be used to control the spreading weld bead. The spreading weld bead should be treated as if it is becoming too hot and you must cool it down. However, gravity rather than temperature is the problem, but decreasing the temperature can solve the problem.

Complete a copy of the "Student Welding Report" listed in Appendix I or provided by your instructor. ◆

EXPERIMENT 32-6

Stops and Starts

Using a properly lit and adjusted torch, one clean piece of pipe having a diameter of about 2 in. (51 mm), and filler metal, you will learn how to make good starts and stops. When welding pipe, you will frequently need to stop and restart. With practice and the proper technique, it is possible to make uniform stops and starts that are as strong as the surrounding weld bead.

To make a proper stop, you should slightly taper down the molten weld pool by flashing the torch off. This allows the molten weld pool to solidify before the flame is totally removed from the weld pool. If the flame is removed

too soon from the molten weld pool, it will rapidly oxidize, throwing out a burst of sparks. The pocket of oxides will greatly weaken the weld at this point.

Restarting the weld requires that a molten weld pool equal in size to the original one be reestablished. To restart the molten weld pool, point the flame slightly ahead of the crater at the end of the weld bead, **Figure 32-62**. When this metal starts to melt, move the flame back to the weld crater and melt the entire crater. Once the crater melts, start adding filler metal and continue with the weld. If the metal ahead of the crater is not heated up first when the weld metal is added to the crater, it may form a cold lap over the base metal.

Practice stops and starts until you can make them so that they are uniform and unnoticeable on the weld bead.

Complete a copy of the "Student Welding Report" listed in Appendix I or provided by your instructor. ◆

PRACTICE 32-28

Stringer Bead, 1G Position

Using a properly lit and adjusted torch and one clean piece of pipe approximately 2 in. (51 mm) in diameter, you will make a stringer bead around the pipe.

With the pipe in the 1G position, start a welding bead at the 2 o'clock position and weld toward the 12 o'clock position. When you must stop to change positions, roll the pipe so the ending crater is at the 2 o'clock position. Start the weld bead as practiced in Experiment 32-6 and proceed with the weld as before, **Figure 32-63**. Repeat this process until the weld bead extends all the way around the pipe. Repeat this practice until you can produce a straight weld bead that is uniform in width and reinforcement and has no visual defects. Turn off the cylinders, bleed the hoses, back out the regulator adjusting screws, and clean up your work area when you are finished.

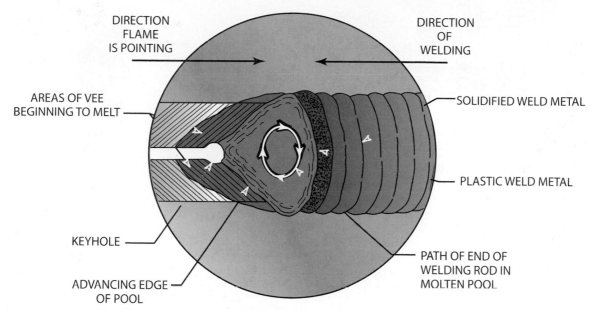

FIGURE 32-63 Pipe welding. © Cengage Learning 2012

Complete a copy of the "Student Welding Report" listed in Appendix I or provided by your instructor. ◆

PRACTICE 32-29

Butt Joint, 1G Position

Using a properly lit and adjusted torch, two clean pieces of schedule 40 pipe approximately 2 in. (51 mm) in diameter, and filler metal, you will weld a butt joint in the 1G position.

The pipe ends should be prepared as shown in **Figure 32-64**. The weld will be made with one pass. Tack weld the pipe together and place it on a firebrick. Using the principles and applications you learned in Practice 32-28, make this weld. Repeat this weld until it can be made without defects. Turn off the cylinders, bleed the hoses, back out the regulator adjusting screws, and clean up your work area when you are finished.

Complete a copy of the "Student Welding Report" listed in Appendix I or provided by your instructor. ◆

Horizontal Fixed Position 5G

The horizontal fixed position 5G weld, **Figure 32-65**, requires little skill development after completing the horizontal rolled position 1G welds. The torch height and angle skills you have developed will help you master the **5G position**.

As the weld changes from overhead (6 o'clock) to vertical (3 o'clock), the weld contour changes. At the overhead position, the weld bead tends to be shaped as shown in **Figure 32-66, Section A-A**. Note that the bead is high in the center and recessed and possibly undercut on the sides. The vertical position has a high and narrow bead shape, **Figure 32-66, Section B-B**.

The overhead bead shape can be controlled by stepping the molten weld pool and moving the flame and rod back and forth at the same time. This process will deposit the metal and allow it to cool before it can sag. It allows the surface tension to hold the metal in place.

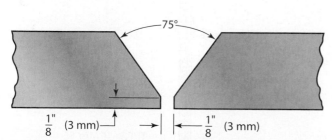

FIGURE 32-64 The bevel on the pipe may be oxyfuel flame cut, ground, or machined. © Cengage Learning 2012

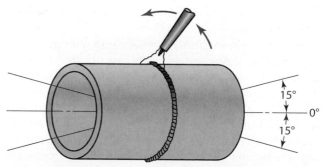

FIGURE 32-65 5G position. The pipe is fixed horizontally. © Cengage Learning 2012

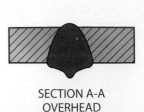
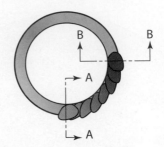

SECTION A-A
OVERHEAD

SECTION B-B
VERTICAL

FIGURE 32-66 Changes in weld bead shape at different locations. © Cengage Learning 2012

As the weld progresses toward the vertical section, the need to step the weld decreases. When the weld reaches the vertical section, the bead shape should be controlled by torch angle and flame distance.

EXPERIMENT 32-7

5G Position

In this experiment, you are going to push a molten weld pool from the bottom of a fixed pipe up to the side of the pipe to see how torch manipulation affects the bead shape.

Using a properly lit and adjusted torch and one clean piece of pipe approximately 2 in. (51 mm) in diameter, start by establishing a molten weld pool at the 6 o'clock position. Then move the molten weld pool forward toward the 3 o'clock position. Observe what effect torch angle, flame distance, and stepping have on the molten weld pool. Turn the pipe and repeat this experiment until you can control the bead width.

Complete a copy of the "Student Welding Report" listed in Appendix I or provided by your instructor. ◆

PRACTICE 32-30

Stringer Bead, 5G Position

Using a properly lit and adjusted torch, one clean piece of pipe having a diameter of about 2 in. (51 mm), and filler metal, you will make a stringer bead upward round one side of the pipe.

With the pipe in the 5G position, start a welding bead at the 6 o'clock position and weld toward the 12 o'clock position. Use all procedures necessary to keep the weld bead uniform. Repeat this practice until you can produce a straight weld bead that is uniform in width and reinforcement and has no visual defects. Turn off the cylinders, bleed the hoses, back out the regulator adjusting screws, and clean up your work area when you are finished.

Complete a copy of the "Student Welding Report" listed in Appendix I or provided by your instructor. ◆

PRACTICE 32-31

Butt Joint, 5G Position

Using a properly lit and adjusted torch, two clean pieces of pipe having a diameter of approximately 2 in. (51 mm), and filler metal, you will weld a butt joint in the 5G position.

The pipe ends should be beveled as shown in Figure 32-64 and tack welded together. Secure the pipe on a stand, such as the one shown in **Figure 32-67**, and start welding at the 6 o'clock position, moving toward the 12 o'clock position. When you reach the top, stop, restart back at the 6 o'clock position, and continue up the other side. Repeat this weld until it can be made without defects. Turn off the cylinders, bleed the hoses, back out the regulator adjusting screws, and clean up your work area when you are finished.

Complete a copy of the "Student Welding Report" listed in Appendix I or provided by your instructor. ◆

Vertical Fixed Position 2G

The vertically fixed pipe requires a horizontal weld, **Figure 32-68**. The welding manipulative skill required for

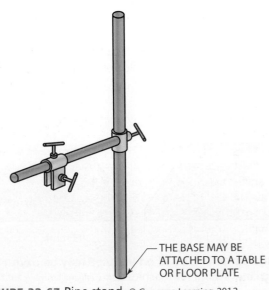

THE BASE MAY BE
ATTACHED TO A TABLE
OR FLOOR PLATE

FIGURE 32-67 Pipe stand. © Cengage Learning 2012

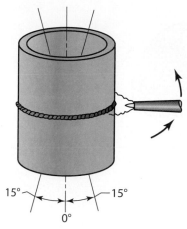

FIGURE 32-68 2G position. The pipe is fixed vertically and welded horizontally. © Cengage Learning 2012

FIGURE 32-69 2G vertical fixed position.
© Cengage Learning 2012

the **2G position** is similar to that for a horizontal butt joint. The pipe may be rotated around its vertical axis but may not be turned end for end after you have started welding, **Figure 32-69**.

PRACTICE 32-32

Stringer Bead, 2G Position

Using a properly lit and adjusted torch and one clean piece of pipe having a diameter of approximately 2 in. (51 mm), you will make a stringer bead around the pipe.

With the pipe in the 2G position, start with a small bead, as illustrated in **Figure 32-70**, and then increase the bead to the desired size. Starting small will allow you to build a shelf to support the molten weld pool. It also will let you tie the end of the weld into the start of the weld so that a stronger weld is obtained. Repeat this weld until it can be made without defects. Turn off the cylinders, bleed the hoses, back out the regulator adjusting screws, and clean up your work area when you are finished.

Complete a copy of the "Student Welding Report" listed in Appendix I or provided by your instructor. ◆

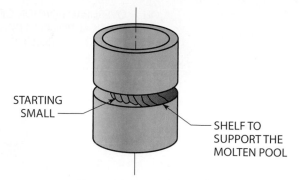

FIGURE 32-70 The proper starting technique will aid with tying in the weld when it is completed around the pipe. © Cengage Learning 2012

PRACTICE 32-33

Butt Joint, 2G Position

Using a properly lit and adjusted torch, two clean pieces of pipe having a diameter of approximately 2 in. (51 mm), and filler metal, you will weld a butt joint in the 2G position. The pipe ends should be beveled and tack welded together. Secure the pipe in the vertical position and start welding. Repeat this weld until it can be made without defects. Turn off the cylinders, bleed the hoses, back out the regulator adjusting screws, and clean up your work area when you are finished.

Complete a copy of the "Student Welding Report" listed in Appendix I or provided by your instructor. ◆

45° Fixed Position 6G

The 45° fixed pipe position, **Figure 32-71**, requires careful manipulation of the molten weld pool to ensure a uniform and satisfactory weld. The weld progresses around the pipe, changing from vertical to horizontal to

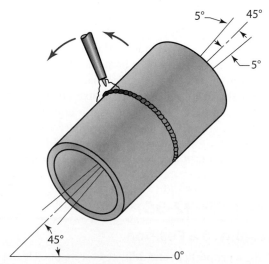

FIGURE 32-71 6G position. The pipe is inclined at a 45° angle. © Cengage Learning 2012

OH = OVERHEAD PORTION OF WELD
V = VERTICAL PORTION OF WELD
H = HORIZONTAL PORTION OF WELD

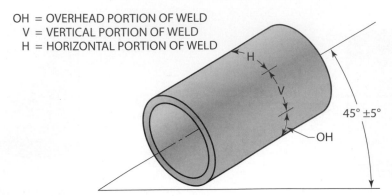

FIGURE 32-72 6G position. © Cengage Learning 2012

overhead to flat and not completely in any one position. It is the combination of compound angles that makes the **6G position** particularly difficult.

PRACTICE 32-34

Stringer Bead, 6G Position

Using a properly lit and adjusted torch, one clean piece of pipe having a diameter of approximately 2 in. (51 mm), and filler metal, you will make a stringer bead around the pipe with the pipe in the 6G position.

Start at the bottom and weld upward toward the top of the pipe, **Figure 32-72.** The weld bead shape will change as you move around the pipe. To prevent the bottom from pulling to one side and slight movement to the high side, the side movement will create a shelf to hold the metal in place.

The side of the bead will pull to one side, but not as severely as at the bottom. To keep the side from being pulled out of shape, simply add the filler metal on the high side of the joint. Some side movement may be required, but not as much as for the bottom.

The top will pull down to one side and tend to be flatter than the other parts of the bead, especially along the side of the pipe. To prevent one-sidedness, the filler metal is added to the top side. To add to the buildup of the bead, change the torch angle or the flame height.

Repeat this weld until it can be made straight and uniform in width and reinforcement. Turn off the cylinders, bleed the hoses, back out the regulator adjusting screws, and clean up your work area when you are finished.

Complete a copy of the "Student Welding Report" listed in Appendix I or provided by your instructor. ◆

PRACTICE 32-35

Butt Joint, 6G Position

Using a properly lit and adjusted torch, one clean piece of pipe approximately 2 in. (51 mm) in diameter, and filler metal, you will weld a butt joint in the 6G position.

The pipe ends should be beveled and tack welded together. Secure the pipe at a 45° angle and start welding. Repeat this weld until it can be made without defects. Turn off the cylinders, bleed the hoses, back out the regulator adjusting screws, and clean up your work area when you are finished.

Complete a copy of the "Student Welding Report" listed in Appendix I or provided by your instructor. ◆

Thin-Wall Tubing

Welding thin-wall tubing requires a technique similar to welding a stringer bead around pipe. If penetration is not a concern, the welding proceeds as if you were making a stringer bead on pipe. However, if penetration is required, then the weld will probably have a **keyhole** to ensure 100% penetration, **Figure 32-73.** This will be similar to welding the root pass.

If you want to practice on tubing, but it is not readily available, then 16-gauge sheet metal can be rolled up and tack welded. Since 16 gauge is a standard wall thickness for tubing, fabricating your own will give you a realistic experience.

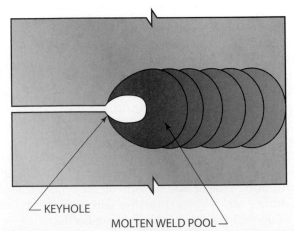

KEYHOLE

MOLTEN WELD POOL

FIGURE 32-73 The keyhole in the root of the joint helps to ensure 100% root penetration. © Cengage Learning 2012

Summary

In an ideal world we would all have the correct tip size for every job we attempted; however, in reality often the oxyfuel welding process is used for repair work when it is not possible to have an infinite selection of tip sizes. Learning to control the heat input to the weld by changing the torch angle, height, or travel speed is important so that you can produce satisfactory welds under these service conditions. It is not safe or recommended that you turn down the flame below the tip's optimal operating capacity. Therefore, you must learn to control the heat by changing the angle, tip height, or travel speed.

Oxyacetylene welding is the process of preference for part-time and amateur welders. It allows them the greatest flexibility and is often the most forgiving. The most common problem OF welding presents is heat and weld distortion on large weldments. Because of the difficulty in controlling distortion and the time required to make large welds, other welding processes are often selected for such welds. That is not to say that large welds are not possible with OF welding; they just take more time and skill.

Confined Space Monitors: Tough Choices for Tight Spots

Reliable operation and proper features in a gas monitor are a must if you are working in a confined space.

Confined space—just the sound of it appears challenging to those who are even slightly claustrophobic. Read the Occupational Safety and Health Administration's (OSHA's) definition of a permit-required confined space—". . . an area with limited or restricted means of entry and exit: not designed for continuous human occupancy; potential to contain a hazardous atmosphere"—and it is apparent that these areas are somewhat less than desirable places to spend the workday. Unfortunately for many, entry and work in confined spaces are a fact of everyday work life.

Confined spaces exist in almost every industry and in every workplace. If you are an electrician, a plumber, a pipe fitter, a welder, a boiler maker, a carpenter, or an emergency service worker, chances are that at some point your daily duties will require you to enter a confined space and be challenged by the hazards that await there.

Know the Hazards

Some of the hazards in confined spaces are easily recognized. It can be relatively easy to see the potential for falls or entrapment from cave-ins or falling equipment. However, it is the unseen atmospheric hazards that present the greatest danger in a confined space. The lack of breathable oxygen, the potential for explosion due to dangerous levels of combustible gases, or the presence of deadly poisonous gas vapors, such as carbon monoxide, cannot be seen. Therefore, you must rely on the readings of a portable gas-monitoring instrument to alert you of potential danger.

Before you drop down a manhole or crawl into a tank car, following are some considerations concerning the gas monitor you choose and the procedures you follow to ensure your safety.

Choosing the Right Sensor

Make sure the sensors in the instrument you are using are appropriate for the confined space. Many people consider a confined space monitor to be the same as a four-gas instrument containing sensors for oxygen, combustible gas, carbon monoxide, and hydrogen sulfide. Although it is true that most confined spaces potentially contain some hazards related to one of these gases, it is not true for all. If you are entering a space that may contain chlorine, an instrument with carbon monoxide and hydrogen sulfide sensors is of little use. Many instruments offer you the ability to change the sensors to match the hazard you will encounter. These instruments provide greater flexibility and value in a wide variety of applications.

A gas-monitoring instrument must be durable and able to withstand harsh conditions. Although a portable gas detector is certainly a sophisticated piece of electronic

Working in confined spaces poses potential hazards that require special precautions. American Welding Society

equipment, it is still a tool. As such, it is subjected to the rigors of the environment it is used in, just as is any other tool. It is dropped, dunked, or caught up in other equipment, and all the while it must still provide a potentially life-saving service. Be certain the instrument you choose is constructed in a manner that will not let its, or your, survival in a tough environment be left to chance.

Confined space preentry tests require remote sampling of the atmosphere from outside the space. Therefore, the instrument you choose must have the capability of using a remote sample pump. The pump must be able to draw an adequate sample flow over the distance required to cover the entire space. The pump also should be able to detect and clearly inform you if the sample line is blocked, thereby preventing gas flow to the instrument and its sensors.

Be certain that the sample tube material you are using with the pump is of high quality and is compatible with the vapors that you expect to encounter. Some highly reactive gases, such as chlorine, nitrogen dioxide, and hydrogen chloride, may be absorbed into the walls of the tubing and scrubbed from the sample stream. Never use sample tubing made of materials containing silicone rubber compounds. The silicone vapors may off-gas from the tubing and poison catalytic bead-type combustible gas sensors, leaving the instrument unable to detect explosive gases. Sample tubing made of urethane or FEP (Teflon) is generally suitable for most applications.

Check the Power

A gas monitor should be able to operate from a variety of power supplies. Rechargeable batteries are generally best for portable monitoring instruments because the combustible gas sensors used in the detectors consume large amounts of battery power. Extended confined space operations often require instrument run times beyond the capacity of the rechargeable batteries. The ability to replace the rechargeable battery pack with disposable alkaline or lithium battery cells will certainly come in handy in these situations. In addition, make sure that the instrument is capable of running the required amount of time when the

sample pump is being used. Some sample pumps use the instrument's battery for power and reduce the run time of the instrument much more than you might expect.

Know the detection limits of the instrument. There are clear differences in the measuring ranges of sensors in various instruments. As a rule of thumb, the monitor should be capable of measuring concentrations approaching the immediately dangerous to life and health (IDLH) level of the target gas. All catalytic combustible gas sensors require a minimum background oxygen concentration be present to respond accurately. Be aware of what that level is and make sure the instrument you choose can be used with a dilution apparatus. By doing so, you will ensure the accuracy of the combustible detector when sampling from an inert or oxygen-deficient atmosphere.

Calibrate Often

Beware of claims that instruments do not need to be tested or calibrated on a regular basis. Know one thing for sure: the only way to ensure a gas monitor will respond to gas is to test and verify its operation with known concentrations of the target gases prior to each use. For example, someone using the instrument before you might have damaged one of the sensors, resulting in no difference in the response of a catalytic or an electrochemical gas sensor used in a clean atmosphere from one that has failed catastrophically. Regular calibration and functional testing will guarantee that the instrument and its sensors are working properly. Whether or not it is done is up to you. If calibration and testing are too difficult and cumbersome for you to do on your own, docking systems and service programs are available to perform these tasks automatically on your instruments and to document the results.

Keep Monitoring

Whether to monitor continuously or not is an often-asked question. Usually, confined space atmospheres are tested before entry to complete the required permits, and

Monitoring should be performed before entering the confined space and during the work operation.
American Welding Society

then the gas monitor is put back on the truck until the next test. Unfortunately, the work done while in the confined space may create an unseen hazard. Chemical reactions with solvents during cleaning processes in tanks or vapors produced during maintenance operations, such as welding, can build up dangerous gases in a confined space while you work. Continuous monitoring of the space's atmosphere during the entire entry will ensure there is no danger.

An instrument capable of activating a remote alarm will also alert your partner watching from outside to hazardous conditions.

There are so many choices and so little room for error. Make sure you take time to know and understand the hazards of confined space operations and ensure you have the right equipment. It could save your life.

Article courtesy of the American Welding Society.

Review

1. What protects the molten weld pool from oxidation during a gas weld?

2. How does tip size affect a gas weld?

3. How does torch angle affect a gas weld?

4. How does welding rod size affect a gas weld?

5. What is a good indication that the molten weld pool is not being protected from oxidation?

6. Why should the end of the gas filler rod be bent?

7. What is the purpose of a heat sink?

8. Why should the end of the welding rod be kept inside the flame envelope?

9. Does a weld have to have 100% penetration to be strong?

10. What technique can be used to minimize weld distortion on a flat butt joint?

11. Why does the edge of the top plate on a lap joint tend to melt more easily?

12. What is meant by the term *out-of-position?*

13. What holds the molten weld pool in place on a vertical weld?

14. What additional personal safety protection should be used for overhead welds?

15. What force holds the molten weld pool in place on an overhead weld?

16. What is the welding sequence for a 1G weld?

17. What is the proper method for stopping a weld pool so it can be easily restarted?

18. What welding positions are required when making a 5G weld on pipe?

19. What welding positions are required when making a 6G weld on pipe?

20. How can 100% penetration be ensured when making welds on thin-wall tubing?

Chapter 33

Soldering, Brazing, and Braze Welding

OBJECTIVES

After completing this chapter, the student should be able to

- define the terms *soldering, brazing,* and *braze welding.*
- explain the advantages and disadvantages of liquid-solid phase bonding.
- demonstrate an ability to properly clean, assemble, and perform required practice joints.
- describe the functions of fluxes in making proper liquid-solid phase bonded joints.

KEY TERMS

braze buildup	*eutectic composition*	*paste range*
braze welding	*fatigue failures*	*phase*
brazing	*fatigue resistance*	*resistance brazing*
brazing alloys	*fluxes*	*shear strength*
capillary action	*furnace brazing*	*silver braze*
corrosion resistance	*induction*	*soldering*
dip brazing	*liquid-solid phase bonding processes*	*soldering alloys*
ductility		*tensile strength*
elastic limit	*low-fuming alloys*	*torch*

INTRODUCTION

Soldering and brazing are both classified by the American Welding Society as **liquid-solid phase bonding processes.** Liquid means that the filler metal is melted; solid means that the base material or materials are not melted. The **phase** is the temperature at which bonding takes place between the solid base material and the liquid filler metal. The bond between the base material and filler metal is a metallurgical bond because no melting or alloying of the base metal occurs. If done correctly, this bond results in a joint having four or five times the tensile strength of that of the filler metal itself.

Soldering and **brazing** differ only in that soldering takes place at a temperature below 840°F (450°C) and brazing occurs at a temperature above 840°F (450°C). Because only the temperature separates the two processes, it is possible to do both soldering and brazing using different mixtures of the same metals, depending upon the alloys used and their melting temperatures.

FIGURE 33-1 A brazed lap joint (A) and a braze welded lap joint (B). © Cengage Learning 2012

Brazing is divided into two major categories, brazing and braze welding. In brazing, the parts being joined must be fitted so that the joint spacing is very small, approximately 0.025 in. (0.6 mm) to 0.002 in. (0.06 mm), **Figure 33-1**. This small spacing allows capillary action to draw the filler metal into the joint when the parts reach the proper phase temperature.

Capillary action is the force that pulls water up into a paper towel or pulls a liquid into a very fine straw, **Figure 33-2**. **Braze welding** does not need capillary action to pull filler metal into the joint. Examples of brazing and braze welding joint designs are shown in **Figure 33-3**.

ADVANTAGES OF SOLDERING AND BRAZING

Some advantages of soldering and brazing as compared to other methods of joining include the following:

- Low temperature—Since the base metal does not have to melt, a low-temperature heat source can be used.

- May be permanently or temporarily joined—Since the base metal is not damaged, parts may be disassembled at a later time by simply reapplying

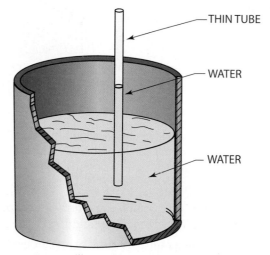

FIGURE 33-2 Capillary action pulls water into a thin tube. © Cengage Learning 2012

heat. The parts then can be reused. However, the joint is solid enough to be permanent, **Figure 33-4**.

- Dissimilar materials can be joined—It is easy to join dissimilar metals, such as copper to steel, aluminum to brass, and cast iron to stainless steel,

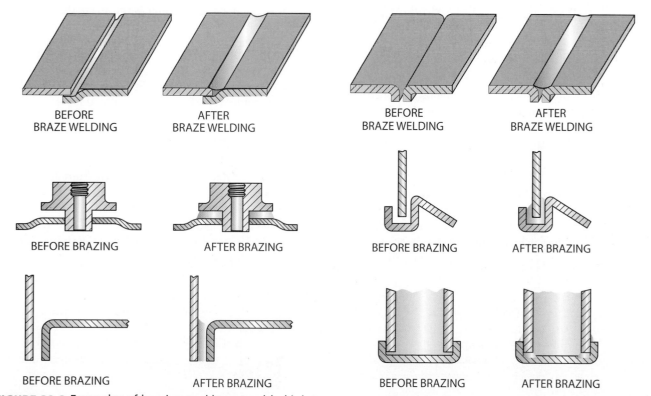

FIGURE 33-3 Examples of brazing and braze welded joints. © Cengage Learning 2012

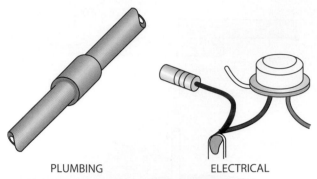

PLUMBING   ELECTRICAL

FIGURE 33-4 Examples of permanent joints that can easily be disassembled and the parts reused. © Cengage Learning 2012

FIGURE 33-6 Furnace brazed part. Larry Jeffus

Figure 33-5. It is also possible to join nonmetals to each other or nonmetals to metals. Ceramics are easily brazed to each other or to metals.

- Speed of joining
  a. Parts can be preassembled and dipped or furnace soldered or brazed in large quantities, **Figure 33-6.**
  b. A lower temperature means less time in heating.

- Less chance of damaging parts—A heat source can be used that has a maximum temperature below the temperature that may cause damage to the parts. With the controlled temperature sufficiently low, even damage from unskilled or semiskilled workers can be eliminated, **Figure 33-7.**

- Slow rate of heating and cooling—Because it is not necessary to heat a small area to its melting temperature and then allow it to cool quickly to a solid, the internal stresses caused by rapid temperature changes can be reduced.

- Parts of varying thicknesses can be joined—Very thin parts or a thin part and a thick part can be joined without burning or overheating them.

- Easy realignment—Parts can easily be realigned by reheating the joint and then repositioning the part.

PHYSICAL PROPERTIES OF THE JOINT

Tensile Strength The **tensile strength** of a joint is its ability to withstand being pulled apart, **Figure 33-8.** A brazed joint can be made that has a tensile strength four to five times higher than the filler metal itself. If a few drops of water are placed between two smooth and flat panes of glass and the panes are pressed together, a tensile load is required to pull the panes of glass apart. The water, which has no tensile strength itself, has added tensile strength to the glass joint.

The glass is being held together by the surface tension of the water. As the space between the pieces of glass decreases, the tensile strength increases. The same action takes place with a soldered or brazed joint. As the joint spacing decreases, the surface tension increases the tensile strength of the joint, **Table 33-1.**

Shear Strength

The **shear strength** of a joint is its ability to withstand a force parallel to the joint, **Figure 33-9.** For a solder or braze joint, the shear strength depends upon the amount of overlapping area of the base parts. The greater the area that is overlapped, the greater is the strength.

Ductility

Ductility of a joint is its ability to bend without failing. Most soldering and brazing alloys are ductile metals, so the joint made with these alloys is also ductile.

Fatigue Resistance

The **fatigue resistance** of a metal is its ability to be bent repeatedly without exceeding its **elastic limit** and without failure. For most soldered or brazed joints, fatigue resistance is usually fairly low. As a joint is bent, the less ductile base materials cause a shear force to be applied to the filler metal, **Figure 33-10,** resulting in joint failure. **Fatigue failures** may also occur as a result of vibration.

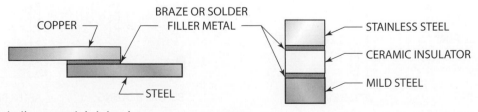

BRAZE OR SOLDER
COPPER — FILLER METAL — STAINLESS STEEL
CERAMIC INSULATOR
STEEL
MILD STEEL

FIGURE 33-5 Dissimilar materials joined. © Cengage Learning 2012

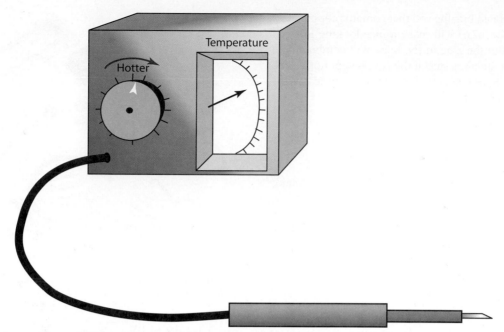

FIGURE 33-7 Control console for resistance soldering. © Cengage Learning 2012

FIGURE 33-8 Joint in tension. © Cengage Learning 2012

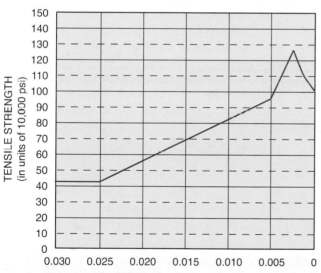

TABLE 33-1 Tensile Strength of Brazed Joint Increases as Joint Space Decreases

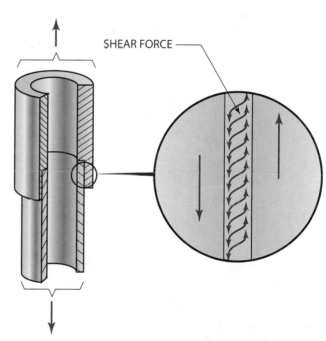

FIGURE 33-9 Effect of shear on a joint. © Cengage Learning 2012

Corrosion Resistance

Corrosion resistance of a joint is its ability to resist chemical attack. The compatibility of the base materials to the filler metal will determine the corrosion resistance. Using the proper filler metal with the base materials that are listed in this chapter will result in corrosion-free joints. However, using filler metals on base materials that are not recommended in this chapter may result in a joint that looks good when completed but will eventually corrode.

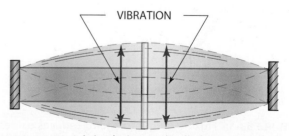

FIGURE 33-10 A joint being cyclically bent. © Cengage Learning 2012

For example, a brass brazing rod that contains copper (Cu) and zinc (Zn) (BCuZn) will make a nice-looking joint on stainless steel. But the zinc in the brass will combine with the nickel in the stainless steel if the part is kept hot for too long. As a result, an embrittled structure is formed in the joint, reducing strength.

FLUXES

Flux

Fluxes used in soldering and brazing have three major functions:

- They must remove any oxides that form as a result of heating the parts.
- They must promote wetting.
- They should aid in capillary action.

The flux, when heated to its reacting temperature, must be thin and flow through the gap provided at the joint. As it flows through the joint, the flux dissolves and absorbs oxides, allowing the molten filler metal to be pulled in behind it, **Figure 33-11**. After the joint is complete, the flux residue should be removed.

Fluxes are available in many forms, such as solids, powders, pastes, liquids, sheets, rings, and washers, **Figure 33-12**. They are also available mixed with the filler

FIGURE 33-12 Braze/solder forms that can be preplaced in a braze/solder joint. Prince & Izant Co.

metal, inside the filler metal, or on the outside of the filler metal, **Figure 33-13**. Sheets, rings, and washers may be placed within the joints of an assembly before heating so that a good bond inside the joint can be ensured. Paste and liquids can be injected into a joint from tubes using a special gun, **Figure 33-14**. Paste, powders, and liquids may be brushed on the joint before or after the material is heated. Paste and powders may also be applied to the end of the rod by heating the rod and dipping it in the flux. Most powders can be made into a paste, or a paste can be thinned by adding distilled water or alcohol; see manufacturers' specifications for details. If water is used, it should be distilled because tap water may contain minerals that will weaken the flux.

Some liquid fluxes may also be added to the gas when using an oxyfuel gas torch for soldering or brazing. The flux is picked up by the fuel gas as it is bubbled through the flux container and is then carried to the torch where it becomes part of the flame.

Flux and filler metal combinations are most convenient and easy to use, **Figure 33-15**. It may be necessary to stock more than one type of flux-filler metal combination for different jobs. These combinations are more expensive than buying the filler and flux separately. In cases where

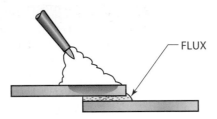
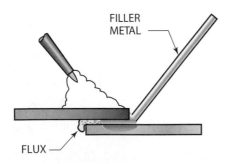

FIGURE 33-11 Flux flowing into a joint reduces oxides to clean the surfaces and gives rise to a capillary action that causes the filler metal to flow behind it. © Cengage Learning 2012

THINK GREEN

Most of the fluxes used for brazing and soldering are not harmful to the environment. However, large quantities of even the most benign materials introduced accidentally or intentionally into the environment can cause damage. Even lemon juice, a common electronic soldering flux, in large enough quantities can cause harm to the environment. Before disposing of any soldering or brazing fluxes, read the material safety data sheet (MSDS) carefully and follow its recommended procedures. Keeping our environment clean and safe is everyone's responsibility.

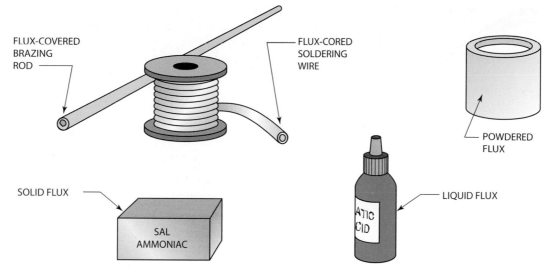

FLUX-COVERED BRAZING ROD

FLUX-CORED SOLDERING WIRE

POWDERED FLUX

SOLID FLUX

SAL AMMONIAC

LIQUID FLUX

FIGURE 33-13 Flux can be purchased with the filler metal or separately. © Cengage Learning 2012

FIGURE 33-14 Gun for injecting flux into joint. Larry Jeffus

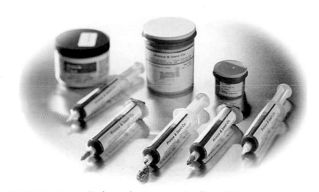

FIGURE 33-15 Tubes that contain flux filler metal mixtures. Prince & Izant Co.

the flux covers the outside of the filler metal, it may be damaged by humidity or chipped off during storage.

Using excessive flux in a joint may result in flux being trapped in the joint, weakening the joint or causing the joint to leak or fail.

Fluxing Action

Soldering and brazing fluxes will remove light surface oxides, promote wetting, and aid in capillary action. The use of fluxes does not eliminate the need for good joint cleaning. Fluxes will not remove oil, dirt, paint, glues, heavy oxide layers, or other surface contaminants.

Soldering fluxes are chemical compounds such as muriatic acid (dilute hydrochloric acid), sal ammoniac (ammonium chloride), or rosin. Brazing fluxes are chemical

compounds such as fluorides, chlorides, boric acids, and alkalies. These compounds react to dissolve, absorb, or mechanically break up thin surface oxides that are formed as the parts are being heated. They must be stable and remain active through the entire temperature range of the solder or braze filler metal. The chemicals in the flux react with the oxides as either acids or bases. Some dip fluxes are salts.

The reactivity of a flux is greatly affected by temperature. As the parts are heated to the soldering or brazing temperature, the flux becomes more active. Some fluxes are completely inactive at room temperature. Most fluxes have a temperature range within which they are most effective. Care should be taken to avoid overheating fluxes. If they become overheated or burned, they will stop working as fluxes, and they become a contamination in the joint. If overheating has occurred, the welder must stop and clean off the damaged flux before continuing.

Fluxes that are active at room temperature must be neutralized (made inactive) or washed off after the job is complete. If these fluxes are left on the joint, premature failure may result due to flux-induced corrosion. Fluxes that are inactive at room temperature do not have to be cleaned off the part. However, if the part is to be painted or auto body plastic is to be applied, fluxes must be removed.

SOLDERING AND BRAZING METHODS

Method Grouping

Soldering and brazing methods are grouped according to the method with which heat is applied: **torch, furnace, induction, dip,** or **resistance.**

Torch Soldering and Brazing

Oxyfuel or air fuel torches can be used either manually or automatically, **Figure 33-16.** Acetylene is often used as the fuel gas, but it is preferable to use one of the

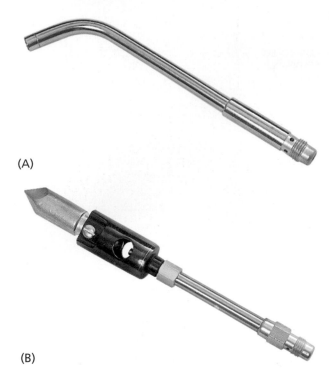

(A)

(B)

FIGURE 33-16 An air propane torch can be used in soldering joints. ESAB Welding & Cutting Products

other fuel gases having a higher heat level in the secondary flame, **Figure 33-17**. The oxyacetylene flame has a very high temperature near the inner cone, but it has little heat in the outer flame. This often results in the

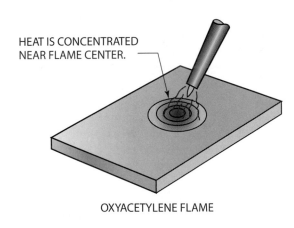

HEAT IS CONCENTRATED NEAR FLAME CENTER.

OXYACETYLENE FLAME

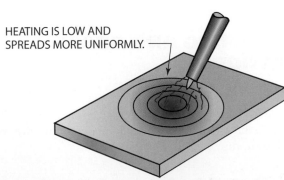

HEATING IS LOW AND SPREADS MORE UNIFORMLY.

OXYPROPANE FLAME

FIGURE 33-17 The high temperature of an oxyacetylene flame may cause localized overheating. © Cengage Learning 2012

parts being overheated in a localized area. Fuel gases such as MAPP, propane, butane, and natural gas have a flame that will heat parts more uniformly. Often torches are used that mix air with the fuel gas in a swirling or turbulent manner to increase the flame's temperature, **Figure 33-18**. The flame may even completely surround a small diameter pipe, heating it from all sides at once, **Figure 33-19**.

Some advantages of using a torch include the following:

■ Versatility—Using a torch is the most versatile method. Both small and large parts in a wide variety of materials can be joined with the same torch.

■ Portability—A torch is very portable. Anyplace a set of cylinders can be taken or anywhere the hoses can be pulled into can be soldered or brazed with a torch.

■ Speed—The flame of the torch is one of the quickest ways of heating the material to be joined, especially on thicker sections.

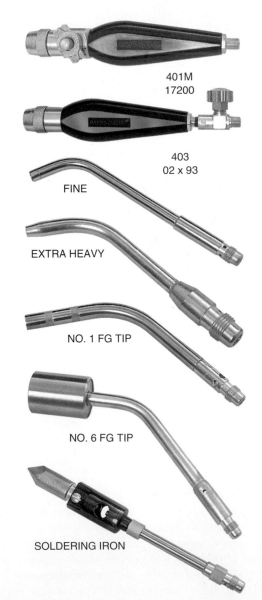

401M
17200

403
02 x 93

FINE

EXTRA HEAVY

NO. 1 FG TIP

NO. 6 FG TIP

SOLDERING IRON

FIGURE 33-18 Examples of torch tips and handles that use air fuel mixtures for brazing. ESAB Welding & Cutting Products

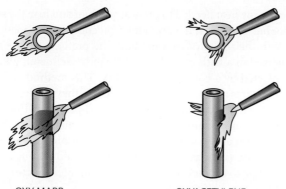

OXY MAPP OXYACETYLENE

FIGURE 33-19 Heating characteristics of oxy MAPP compared with oxyacetylene on round materials.
© Cengage Learning 2012

Some of the disadvantages of using a torch include the following:

- Overheating—When using a torch, it is easy to overheat or burn the parts, flux, or filler metal.
- Skill—A high level of skill with a torch is required to produce consistently good joints.
- Fires—It is easy to start a fire if a torch is used around combustible (flammable) materials.

Furnace Soldering and Brazing

In this method, the parts are heated to their soldering or brazing temperature by passing them through or putting them into a furnace. The furnace may be heated by electricity, oil, natural gas, or any other locally available fuel. The parts may be passed through the furnace on a conveyor belt in trays or placed on the belt itself, **Figure 33-20**. The parts also may be loaded in trays to be placed in a furnace that does not use a conveyor belt, **Figure 33-21**.

Some of the advantages of using a furnace include the following:

- Temperature control—The furnace temperature can be accurately controlled to ensure that the parts will not overheat.
- Controlled atmosphere—The furnace can be filled with an inert gas to prevent oxides from forming on the parts.

FIGURE 33-21 Small furnace brazed part. Larry Jeffus

- Uniform heating—The uniform heating of the parts reduces stresses and distortion.
- Mass production—By using a furnace, it is easy to produce a high quantity of parts.

Some of the disadvantages of using a furnace include the following:

- Size—Unless parts are small, the length of time required to heat them is extremely long.
- Heat damage—The entire part must be able to withstand heating without burning.

Induction Soldering and Brazing

The induction method of heating uses a high-frequency electrical current to establish a corresponding current on the surface of the part, **Figure 33-22**. The current on the part causes rapid and very localized heating of the surface only. There is little, if any, internal heating of the part except by conductivity of heat from the surface.

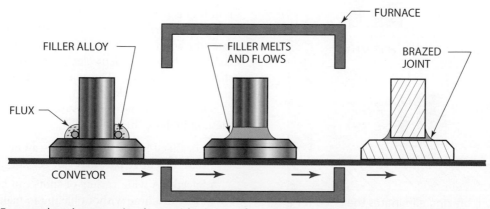

FIGURE 33-20 Furnace brazing permits the rapid joining of parts on a production basis. © Cengage Learning 2012

FIGURE 33-22 Induction brazing and soldering machine.
Prince & Izant Co.

Advantages of the induction method are the following:

- Speed—Very little time is required for the part to reach the desired temperature.

Some of the disadvantages of the induction method include the following:

- Distortion—The very localized heating may result in some distortion.
- Lack of temperature control—The electrical resistance of the part increases as the part heats up. This, in turn, increases the temperature produced.
- Incomplete penetration—Because the inside of the part is not directly heated, it may be too cool to permit the filler metal to flow fully through the joint.

Dip Soldering and Brazing

Two types of dip soldering or brazing are used: molten flux bath and molten metal bath. With the molten flux method, the soldering or brazing filler metal in a suitable form is preplaced in the joint, and the assembly is immersed in a bath of molten flux, as shown in **Figure 33-23**. The bath supplies the heat needed to preheat the joint and fuse the solder or braze metal, and it provides protection from oxidation.

With the molten metal method, the prefluxed parts are immersed in a bath of molten solder or braze metal, which is protected by a cover of molten flux. This method is confined to wires and other small parts. Once they are removed from the bath, the ends of the wires and parts must not be allowed to move until the solder or braze metal has solidified. As with all soldering or brazing operations, any movement of the parts as they cool from a liquid through the paste range to become a solid will result in microfractures in the filler metal. In electronic parts these microfractures cause resistance to the electron flow and may render the part unfit for service. Reheating can be used to re-fuse the joint only if reheating will not damage the part beyond use.

Some of the advantages of dip processing include the following:

- Mass production—It is possible to dip many small parts at one time.
- Corrosion protection—The entire surface of the part can be covered with the filler metal at the same time that it is being joined. If a corrosion-resistant filler metal is used, the thin layer provided will help protect the part from corrosion.
- Distortion minimized—The entire part is heated uniformly, which reduces distortion.

Some of the disadvantages of dip processing include the following:

- Steam explosions—Moisture trapped in the joint may cause a steam explosion that can scatter molten metal.
- Corrosion—If any of the salt is trapped in the joint or is left on the surface, corrosion may cause the part to fail at some time in the future.
- Size—Parts must be small to be effectively joined.
- Quantity—Only a large quantity of parts can justify heating the large amount of molten filler metal or salt required for dipping.

Resistance Soldering and Brazing

The resistance method of heating uses an electric current that is passed through the part. The resistance of the part to the current flow results in the heat needed to produce the bond. The flux is usually preplaced, and the material must have sufficient electrical resistance to produce the desired heating. The machine used in this method resembles a spot welder.

Some of the advantages of the resistance heating method include the following:

- Localized heating—The heat can be localized so that the entire part may not get hot.
- Speed—A wide variety of spots can be made on the same machine without having to make major adjustments on the machine.

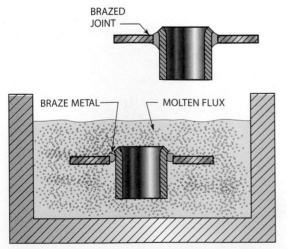

FIGURE 33-23 Dip brazing eliminates the need for a separate fluxing operation. © Cengage Learning 2012

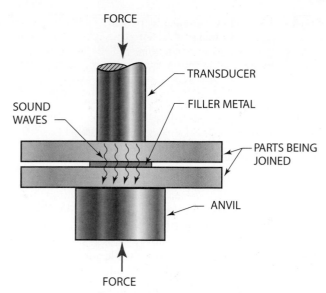

FORCE

TRANSDUCER

SOUND
WAVES

FILLER METAL

PARTS BEING
JOINED

ANVIL

FORCE

FIGURE 33-24 Ultrasonic bonding. © Cengage Learning 2012

■ Multiple spots—Many spots can be joined in a small area without disturbing joints that are already made.

Some of the disadvantages of the resistance heating method include the following:

■ Distortion—Localized heating may result in distortion.

■ Conductors—Parts must be able to conduct electricity.

■ Joint design—Lap joints in plate are the only joint designs that can be made.

Special Methods

A few other methods of producing soldered or brazed parts are used that do not entirely depend upon heat to produce the joint. The ultrasonic method uses high-frequency sound waves to produce the bond or to aid with heat in the bonding, **Figure 33-24.** Another method is diffusion, which uses pressure and may use heat or ultrasound to form a bond. Still another process uses infrared light to heat the part for soldering or brazing.

FILLER METALS

Types of Filler Metals

The type of filler metal used for any specific joint should be selected by considering as many of the criteria listed in **Figure 33-25** as possible. It would be impossible to consider each of these items with the same importance. Welders must decide which things they feel are the most important and then base their selection on that decision.

Soldering and brazing metals are alloys—that is, a mixture of two or more metals. Each alloy is available in a variety of percentage mixtures. Some mixtures are stronger, and some melt at lower temperatures than other mixtures. Each one has specific properties. Almost all of the alloys used for soldering or brazing have a paste range. A **paste range**

■ Material being joined
■ Strength required
■ Joint design
■ Availability and cost
■ Appearance
■ Service (corrosion)
■ Heating process to be used
■ Cost

FIGURE 33-25 Criteria for selecting filler metal. © Cengage Learning 2012

is the temperature range in which a metal is partly solid and partly liquid as it is heated or cooled. As the joined part cools through the paste range, it is important that the part not be moved. If the part is moved, the solder or braze metal may crumble like dry clay, destroying the bond.

EXPERIMENT 33-1

Paste Range

This experiment shows the effect on bonding of moving a part as the filler metal cools through its paste range. The experiment also shows how metal can be "worked" using its paste range. You will need tin-lead solder composed of 20% to 50% tin, with the remaining percentage being lead. You also will need a properly lit and adjusted torch, a short piece of brazing rod, and a piece of sheet metal. Using a hammer, make a dent in the sheet metal about the size of a quarter (25¢), **Figure 33-26.**

In a small group, watch the effects of heating and cooling solder as it passes through the paste range. Using the torch, melt a small amount of the solder into the dent and allow it to harden. Remelt the solder slowly, frequently flashing the torch off and touching the solder with the brazing rod until it is evident the solder has all melted. Once the solder has melted, stick the brazing rod in the solder and remove the torch. As the solder cools, move the brazing rod in the metal and observe what happens, **Figure 33-27.**

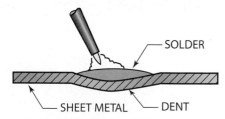

SOLDER

SHEET METAL

DENT

FIGURE 33-26 Partially fill the drilled hole with solder. © Cengage Learning 2012

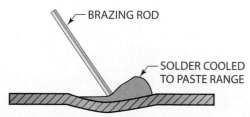

BRAZING ROD

SOLDER COOLED
TO PASTE RANGE

FIGURE 33-27 Solder being shaped as it cools to its paste range. © Cengage Learning 2012

As the solder cools to the uppermost temperature of its paste range, it will have a rough surface appearance as the rod is moved. When the solder cools more, it will start to break up around the rod. Finally, as it becomes a solid, it will be completely broken away from the rod.

Now slowly reheat the solder and work the surface with the rod until it can be shaped like clay. If the surface is slightly rough, a quick touch of the flame will smooth it. This is the same way in which "lead" is applied to some body panel joints on a new car so that the joints are not seen on the car when it is finished. The lead used is actually a tin-lead alloy or solder. A large area can be made as smooth as glass without sanding by simply flashing the area with the flame.

Complete a copy of the "Student Welding Report" listed in Appendix I or provided by your instructor. ◆

Soldering Alloys

Soldering alloys are usually identified by their major alloying elements. **Table 33-2** lists the major types of solder and the materials they will join. In many cases, a base material can be joined by more than one solder alloy. In addition to the considerations for selecting filler metal listed in Figure 33-25, specific factors are listed in the following sections for the major soldering alloys.

Tin-lead	Copper and copper alloys Mild steel Galvanized metal
Tin-antimony	Copper and copper alloys Mild steel
Cadmium-silver	High strength for copper and copper alloys Mild steel Stainless steel
Cadmium-zinc	Aluminum and aluminum alloys

TABLE 33-2 Soldering Alloys

Tin-Lead This is the most popular solder and is the least expensive one. An alloy of 61.9% tin and 38.1% lead melts at 362°F (183°C) and has no paste range. This is the **eutectic composition** (lowest possible melting point of an alloy) of the tin-lead solder. An alloy of 60% tin and 40% lead is commercially available and is close enough to the eutectic alloy to have the same low melting point with only a 12°F (7.8°C) paste range. The widest paste range is 173°F (78°C) for a mixture of 19.5% tin and 80.5% lead. This mixture begins to solidify at 535°F (289°C) and is totally solid at 362°F (193°C). The closest mixture that is commercially available is a 20% tin and 80% lead alloy. **Table 33-3** lists the percentages, temperatures, and paste ranges for tin-lead solders. Tin-lead solders are most

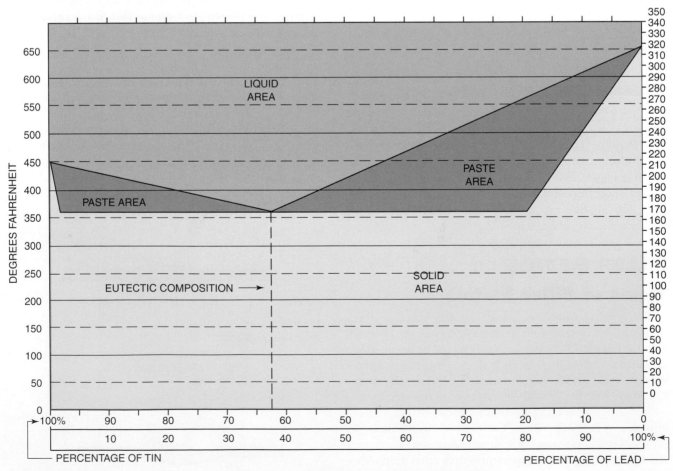

TABLE 33-3 Melting, Solidification, and Paste Range Temperatures for Tin-Lead Solders

Cadmium	Zinc	Completely Liquid	Completely Solid	Paste Range
82.5%	17.5%	509°F (265°C)	509°F (265°C)	No paste range
40.0%	60.0%	635°F (335°C)	509°F (265°C)	126°F (52°C)
10.0%	90.0%	750°F (399°C)	509°F (265°C)	241°F (116°C)

TABLE 33-4 Cadmium-Zinc Alloys

commonly used on electrical connections but must never be used for water piping. Most health and construction codes will not allow tin-lead solders for use on water or food-handling equipment.

Tin-lead solders must not be used where lead could become a health hazard in things such as food and water.

Tin-Antimony This family of solder alloys has a higher tensile strength and lower creep than the tin-lead solders. The most common alloy is 95/5, 95% tin and 5% antimony. This is often referred to as "hard solder." This is the most common solder used in plumbing because it is lead free. The use of "C" flux, which is a mixture of flux and small flakes of solder, makes it much easier to fabricate quality joints. This mixture of flux and solder will draw additional solder into the joint as it is added.

Cadmium-Silver These solder alloys have excellent wetting, flow, and strength characteristics, but they are expensive. The silver in this solder helps improve wetting and strength. Cadmium-silver alloys melt at a temperature of around 740°F (393°C); they are called high-temperature solders because they retain their strength at temperatures above most other solders. These solder alloys can be used to join aluminum to itself or other metals—for example, to piping that is used in air-conditioning equipment.

When silver soldering on food-handling equipment, use a cadmium-free silver solder.

If cadmium is overheated, the fumes can be hazardous unless the area is properly ventilated.

Cadmium-Zinc Cadmium-zinc alloys have good wetting action and corrosion resistance on aluminum and aluminum alloys. The melting temperature is high, and some alloys have a wide paste range, **Table 33-4.**

Brazing Alloys

The American Welding Society's classification system for **brazing alloys** uses the letter *B* to indicate that the alloy is to be used for brazing. The next series of letters in the classification indicates the atomic symbol of metals used to make up the alloy, such as CuZn (copper and zinc). There may be a dash followed by a letter or number to indicate a specific alloyed percentage. The letter *R* may be added to indicate that the braze metal is in rod form. An example of a filler metal designation is BRCuZn-A, which indicates a copper-zinc brazing rod with 59.25% copper, 40% zinc, and 0.75% tin; **Table 33-5** is a list of the base metals and the most common alloys used to join the base metals. Not all of the available brazing alloys have an AWS classification. Some special alloys are known by registered trade names.

Copper-Zinc Copper-zinc alloys are the most popular brazing alloys. They are available as regular and

Base Metal	Brazing Filler Metal
Aluminum	BAlSi, aluminum silicon
Carbon steel	BCuZn, brass (copper-zinc)
	BCu, copper alloy
	BAg, silver alloy
Alloy steel	BAg, silver alloy
	BNi, nickel alloy
Stainless steel	BAg, silver alloy
	BAu, gold base alloy
	BNi, nickel alloy
Cast iron	BCuZn, brass (copper-zinc)
Galvanized iron	BCuZn, brass (copper-zinc)
Nickel	BAu, gold base alloy
	BAg, silver alloy
	BNi, nickel alloy
Nickel-copper alloy	BNi, nickel alloy
	BAg, silver alloy
	BCuZn, brass (copper-zinc)
Copper	BCuZn, brass (copper-zinc)
	BAg, silver alloy
	BCuP, copper-phosphorus
Silicon bronze	BCuZn, brass (copper-zinc)
	BAg, silver alloy
	BCuP, copper-phosphorus
Tungsten	BCuP, copper-phosphorus

TABLE 33-5 Base Metals and Common Brazing Filler Metals Used to Join the Base Metals

low-fuming alloys. The zinc in this braze metal has a tendency to burn out if it is overheated. Overheating is indicated by a red glow on the molten pool, which gives off a white smoke. The white smoke is zinc oxide. If zinc oxide is breathed in, it can cause zinc poisoning. Using a low-fuming alloy will help eliminate this problem. Examples of low-fuming alloys are RCuZn-B and RCuZn-C.

///// **CAUTION** \\\\\

Breathing zinc oxide can cause zinc poisoning. If you think you have zinc poisoning, get medical treatment immediately.

Copper-Zinc and Copper-Phosphorus A5.8

The copper-zinc filler rods are often grouped together and known as brazing rods. The copper-phosphorus rods are referred to as phos-copper. Both terms do not adequately describe the metals in this group. There are vast differences among the five major classifications of the copper-zinc filler metals, as well as among the five major classifications of the copper-phosphorus filler metals. The following material describes the five major classifications of copper-zinc filler rods.

Class BRCuZn is used for the same application as BCu fillers. The addition of 40% zinc (Zn) and 60% copper (Cu) improves the corrosion resistance and aids in this rod's use with silicon-bronze, copper-nickel, and stainless steel.

///// **CAUTION** \\\\\

Care must be exercised to prevent overheating this alloy, as the zinc will vaporize, causing porosity and poisonous zinc fumes.

Class BRCuZn-A is commonly referred to as naval brass and can be used to fuse weld naval brass. The addition of 17% tin (Sn) to the alloy adds strength and corrosion resistance. The same types of metal can be joined with this rod as could be joined with BRCuZn.

Class BRCuZn-B is a manganese-bronze filler metal. It has a relatively low melting point and is free flowing. This rod can be used to braze weld steel, cast iron, brass, and bronze. The deposited metal is higher than BRCuZn or BRCuZn-A in strength, hardness, and corrosion resistance.

Class BRCuZn-C is a low fuming, high silicon (Si) bronze rod. It is especially good for general purpose work due to the low-fuming characteristic of the silicon on the zinc.

Class BRCuZn-D is a nickel-bronze rod with enough silicon to be low fuming. The nickel gives the deposit a silver-white appearance and is referred to as white brass. This rod is used to braze and braze weld steel, malleable iron, and cast iron and for building up wear surfaces on bearings.

Copper-Phosphorus This alloy is sometimes referred to as phos-copper. It is a good alloy to consider for joints where silver braze alloys may have been used in the past. Phos-copper has good fluidity and wettability on copper and copper alloys. The joint spacing should be from 0.001 in. (0.03 mm) to 0.005 in. (0.12 mm) for the strongest joints. Heavy buildup of this alloy may cause brittleness in the joint. Phosphorus forms brittle iron phosphide at brazing temperatures on steel. Copper-phos or copper-phos-silver should not be used on copper-clad fittings with ferrous substrates because the copper can easily be burned off, exposing the underlying metal to phosphorus embrittlement.

The copper-phosphorus (BCuP group) rods are used in air-conditioning applications and in plumbing to join copper piping. The phosphorus makes the rod self-fluxing on copper. This feature is one of the major advantages of copper-phosphorus rods. The addition of a small amount of silver, approximately 2%, helps with wetting and flow into joints.

Class BCuP-1 has a low wetting characteristic and a lower flow rate than the other phos-copper alloys. This type of filler metal should be preplaced in the joint. The major advantage of this type of filler metal is its increased ductility.

Classes BCuP-2 and BCuP-4 both have good flow into the joint. The high phosphorus content of the rods makes them self-fluxing on copper. Both of these classes are used often for plumbing installations.

Classes BCuP-3 and BCuP-5 both have high surface tension and low flow so that they are used when close fit ups are not available.

Copper-Phosphorus-Silver This alloy is sometimes referred to as sil-phos. Its characteristics are similar to those of copper-phosphorus except the silver gives this alloy a little better wetting and flow characteristic. Often it is not necessary to use flux with alloys containing 5% or more of silver when joining copper pipe. This is the most common brazing alloy used in air-conditioning and refrigeration work. When sil-phos is used on air-conditioning compressor fittings that are copper-clad steel, care must be taken to make the braze quickly. If the fitting is heated too much or for too long, the copper cladding can be burned off. With this burn-off, the phosphorus can make the steel fitting very brittle, and embrittlement can cause the fitting to crack and leak sometime later.

Silver-Copper Silver-copper alloys can be used to join almost any metal, ferrous or nonferrous, except aluminum, magnesium, zinc, and a few other low-melting metals. This alloy is often referred to as **silver braze** and is the most versatile. It is among the most expensive alloys, except for the gold alloys.

Nickel Nickel alloys are used for joining materials that need high strength and corrosion resistance at an elevated temperature. Some applications of these alloys include joining turbine blades in jet engines, torch parts, furnace parts, and nuclear reactor tubing. Nickel will wet and flow acceptably on most metals. When used on copper-based alloys, nickel may diffuse into the copper, stopping its capillary flow.

Nickel and Nickel Alloys A5.14 Nickel and nickel alloys are increasingly used as a substitute for silver-based alloys. Nickel is generally more difficult to use than silver because it has lower wetting and flow characteristics. However, nickel has much higher strength than silver.

Class BNi-1 is a high-strength, heat-resistant alloy that is ideal for brazing jet engine parts and for other similar applications.

Class BNi-2 is similar to BNi-1 but has a lower melting point and a better flow characteristic.

Class BNi-3 has a high flow rate that is excellent for large areas and close-fitted joints.

Class BNi-4 has a higher surface tension than the other nickel filler rods, which allows larger fillets and poor-fitted joints to be filled.

Class BNi-5 has a high oxidation resistance and high strength at elevated temperatures and can be used for nuclear applications.

Class BNi-6 is extremely free flowing and has good wetting characteristics. The high corrosion resistance gives this class an advantage when joining low chromium steels in corrosive applications.

Class BNi-7 has a high resistance to erosion and can be used for thin or honeycomb structures.

Aluminum-Silicon (BAlSi) brazing filler metals can be used to join most aluminum sheet and cast alloys. The AWS type number 1 flux must be used when brazing aluminum. It is very easy to overheat the joint. If the flux is burned by overheating, it will obstruct wetting. Use standard torch brazing practices but guard against overheating.

Copper and Copper Alloys A5.7 Although pure copper (Cu) can be gas fusion welded successfully using a neutral oxyfuel flame without a flux, most copper filler metals are used to join other metals in a brazing process.

Class BCu-1 can be used to join ferrous, nickel, and copper-nickel metals with or without a flux. BCu-1 is also available as a powder that is classified as BCu-1a. This material has the same applications as BCu-1. The AWS type number 3B flux must be used with metals that are prone to rapid oxidation or with heavy oxides such as chromium, titanium, manganese, and others.

Class BCu-2 has applications similar to those for BCu-1. However, BCu-2 contains copper oxide suspended in an organic compound. Since copper oxides can cause porosity, tying up the oxides with the organic compounds reduces the porosity.

Silver and Gold Silver and gold are both used in small quantities when joining metals that will be used under corrosive conditions, when high joint ductility is needed, or when low electrical resistance is important. Because of the ever-increasing price and reduced availability of these precious metals, other filler metals should first be considered. In many cases, other alloys can be used with great success. When substituting a different filler metal for one that has been used successfully, the new metal and joint should first be extensively tested.

JOINT DESIGN

Joint Spacing

The spacing between the parts being joined greatly affects the tensile strength of the finished part. **Table 33-6** lists the spacing requirements at the joining temperature for the most common alloys. As the parts are heated, the initial space may increase or decrease, depending upon the joint design and fixturing. The changes due to expansion can be calculated, but trial and error also works.

The strongest joints are obtained when the parts use lap or scarf joints where the joining area is equal to three times the thickness of the thinnest joint member, **Figure 33-28**. The strength of a butt joint can be increased if the area being joined can be increased. Parts that are 1/4 in. (6 mm) thick should not be considered for brazing or soldering if another process will work successfully.

Some joints can be designed so that the flux and filler metal may be preplaced. When this is possible, visual checking for filler metal around the outside of the joint is easy. Evidence of filler metal around the outside is a good indication of an acceptable joint.

Joint preparation is also very important to a successful soldered or brazed part. The surface must be cleaned of all oil, dirt, paint, oxides, or any other contaminants. The surface can be mechanically cleaned by using a wire brush or by sanding, sandblasting, grounding, scraping, or filing, or it can be cleaned chemically with an acid, alkaline, or salt bath. Soldering or brazing should start as soon as possible after the parts are cleaned to prevent any additional contamination of the joint.

Filler Metal	Joint Spacing	
	in.	mm
BAlSi	0.006–0.025	(0.15–0.61)
BAg	0.002–0.005	(0.05–0.12)
BAu	0.002–0.005	(0.05–0.12)
BCuP	0.001–0.005	(0.03–0.12)
BCuZn	0.002–0.005	(0.05–0.12)
BNi	0.002–0.005	(0.05–0.12)

TABLE 33-6 Brazing Alloy Joint Tolerance

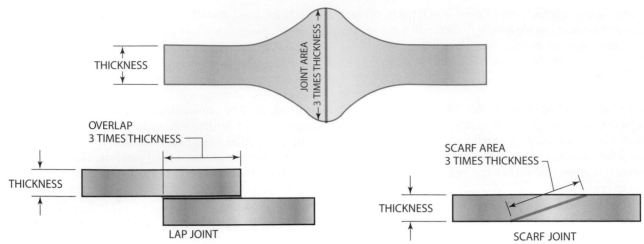

FIGURE 33-28 The joining area should be three times the thickness of the thinnest joint member. © Cengage Learning 2012

EXPERIMENT 33-2

Fluxing Action

In this experiment, as part of a small group, you will observe oxide removal by a flux as the flux reaches its effective temperature. For this experiment, you will need a piece of copper, either tubing or sheet, rosin or C flux, and a properly lit and adjusted torch.

Any paint, oil, or dirt must first be removed from the copper. Do not remove the oxide layer unless it is blue-black in color. Put some flux on the copper and start heating it with the torch. When the flux becomes active, the copper that is covered by the flux will suddenly change to a bright coppery color. The copper that is not covered by the flux will become darker and possibly turn blue-black, **Figure 33-29**. Continue heating the copper until the flux is burned off and the once-clean spot quickly builds an oxide layer.

Repeat this experiment, but this time hold the torch farther from the metal's surface. When the flux begins to clean the copper, flash the torch off the metal (quickly move the flame off and back onto the same spot). Try to control the heat so that the flux does not burn off.

Complete a copy of the "Student Welding Report" listed in Appendix I or provided by your instructor. ◆

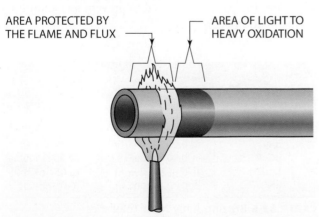

AREA PROTECTED BY
THE FLAME AND FLUX

AREA OF LIGHT TO
HEAVY OXIDATION

FIGURE 33-29 Copper pipe fluxed and exposed to heat.
© Cengage Learning 2012

EXPERIMENT 33-3

Uniform Heating

In this experiment, as part of a small group, you will learn how to control the flame direction so that two pieces of metal of unequal size are heated at the same rate to the same temperature. For this experiment, you will need, two pieces of mild steel, one 16 gauge and the other 1/8 in. (3 mm) thick, and a properly lit and adjusted torch.

Place the two pieces of metal on a firebrick to form a butt joint. Then take the torch and point the flame toward the thicker piece of metal, moving it as needed so that both plates turn red at the same time. Now move the torch so that the red area is equal in size on both plates. Keep the spot red but do not allow it to melt. Repeat this experiment until you can control the area and rate of heating of both plates at the same time.

Complete a copy of the "Student Welding Report" listed in Appendix I or provided by your instructor. ◆

EXPERIMENT 33-4

Tinning or Phase Temperature

In this experiment, as part of a small group, you will observe the wetting of a piece of metal by a filler metal. For this experiment, you will need one piece of 16-gauge mild steel, BRCuZn filler metal rod, powdered flux, and a properly lit and adjusted torch.

Place the sheet flat on a firebrick. Heat the end of the rod and dip it in the flux so that some flux sticks on the rod, **Figure 33-30A, B,** and **C.** BRCuZn brazing rods are available as both bare rods and prefluxed, **Figure 33-30D.** Direct the flame onto the plate. When the sheet gets hot, hold the brazing rod in contact with the sheet, directing the flame so that a large area of the sheet is dull red and the rod starts to melt, **Figure 33-31.** After a molten pool of braze metal is deposited on the sheet, remove the rod and continue heating the sheet and molten pool until the braze metal flows out. Repeat this experiment until you can get the braze metal to flow out in all directions equally at the same time.

(A)

(C)

(B)

(D)

FIGURE 33-30 (A) Heating a brazing rod. (B) Dipping the heated rod into the flux. (C) Flux stuck to rod ready for brazing. (D) Prefluxed brazing rods. Larry Jeffus

FIGURE 33-31 Deposit a spot of braze on the plate and continue heating the plate until the braze flows onto the surface. Larry Jeffus

Complete a copy of the "Student Welding Report" listed in Appendix I or provided by your instructor. ◆

BRAZING PRACTICES

The brazing practices that follow use copper-zinc alloys (BRCuZn) for brazing joints on mild steel. Using prefluxed rods is easier for students than using powdered flux, which has to have the hot tip of the brazing rod dipped to apply.

Because you may be asked to braze with bare brazing rods, the practices explain how to use them. If you are using prefluxed rods, just ignore this step.

PRACTICE 33-1

Brazed Stringer Bead

Using a properly lit and adjusted torch, 6 in. (152 mm) of clean 16-gauge mild steel, brazing flux, and BRCuZn brazing rod, you will make a straight bead the length of the sheet.

Place the sheet flat on a firebrick and hold the flame at one end until the metal reaches the proper temperature. Then touch the flux-covered rod to the sheet and allow a small amount of brazing rod to melt onto the hot sheet, **Figure 33-32.** Once the molten brazing metal wets the sheet, start moving the torch in a circular pattern while dipping the rod into the molten braze pool as you move along the sheet. If the size of the molten pool increases, you can control it by reducing the torch angle, raising the torch, traveling at a faster rate, or flashing the flame off the molten braze pool, **Figure 33-33A, B,** and **C.** Flashing the torch off a braze joint will not cause oxidation problems

FIGURE 33-32 Checking the surface temperature with a spot of braze metal. Larry Jeffus

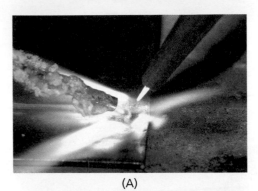

(A)

(B)

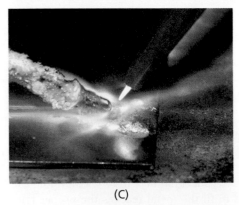

(C)

FIGURE 33-33 (A) Once the plate is up to temperature, start adding more filler. (B) Dip the brazing rod into the leading edge of the molten weld pool. (C) Remove the rod from the flame area when it is not being added to the molten weld pool. Larry Jeffus

as it does when welding because the molten metal is protected by a layer of flux.

As the braze bead progresses across the sheet, dip the end of the rod back in the flux, if a powdered flux is used,

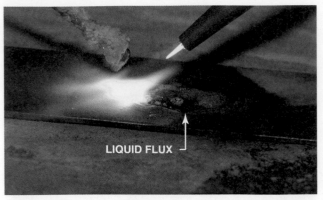

LIQUID FLUX

FIGURE 33-34 Observe the molten flux flowing ahead of the molten weld pool. Larry Jeffus

as often as needed to keep a small molten pool of flux ahead of the bead, **Figure 33-34.**

The object of this practice is to learn how to control the size and direction of the braze bead. Controlling the width, buildup, and shape shows that you have a good understanding and control of the process. Keeping the braze bead in a straight line indicates that you have mastered the technique well enough to watch the bead and the direction at the same time. Turn off the cylinders, bleed the hoses, back out the regulator adjusting screws, and clean up your work area when you are finished.

Complete a copy of the "Student Welding Report" listed in Appendix I or provided by your instructor. ◆

PRACTICE 33-2

Brazed Butt Joint

Using the same equipment and setup as listed in Practice 33-1, make a braze butt joint on two pieces of 16-gauge mild steel sheet, 6 in. (152 mm) long.

Place the metal flat on a firebrick, hold the plates tightly together, and make a tack braze at both ends of the joint. If the plates become distorted, they can be bent back into shape with a hammer before making another tack weld in the center. Align the sheets so that you can comfortably make a braze bead along the joint. Starting as you did in Practice 33-1, make a uniform braze along the joint. Repeat this practice until a uniform braze can be made without defects. Turn off the cylinders, bleed the hoses, back out the regulator adjusting screws, and clean up your work area when you are finished.

Complete a copy of the "Student Welding Report" listed in Appendix I or provided by your instructor. ◆

PRACTICE 33-3

Brazed Butt Joint with 100% Penetration

Using the same equipment, material, and setup as described in Practice 33-2, make a brazed butt joint with 100% penetration, **Figure 33-35.** To ensure that 100% penetration is obtained, a little additional heat is required to flow the braze metal through the joint. Apply the additional

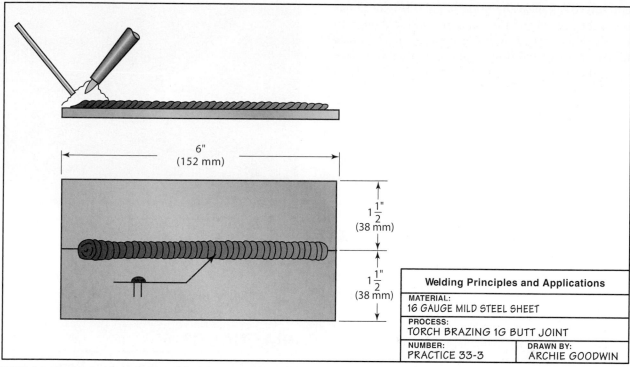

FIGURE 33-35 Brazed butt joint with 100% penetration. © Cengage Learning 2012

heat just ahead of the bead on the base sheets. After the braze is completed, turn the plate over and look for a small amount of braze showing along the entire joint. Repeat this practice until the joint can be made without defects. Turn off the cylinders, bleed the hoses, back out the regulator adjusting screws, and clean up your work area when you are finished.

Complete a copy of the "Student Welding Report" listed in Appendix I or provided by your instructor. ◆

PRACTICE 33-4

Brazed Tee Joint

Using the same equipment, material, and setup as listed in Practice 33-2, you will make a brazed tee joint with 100% root penetration, **Figure 33-36**.

Tack the pieces of metal into a tee joint as you did in Practice 33-2. To obtain 100% penetration, direct the flame on the sheets just ahead of the braze bead, being

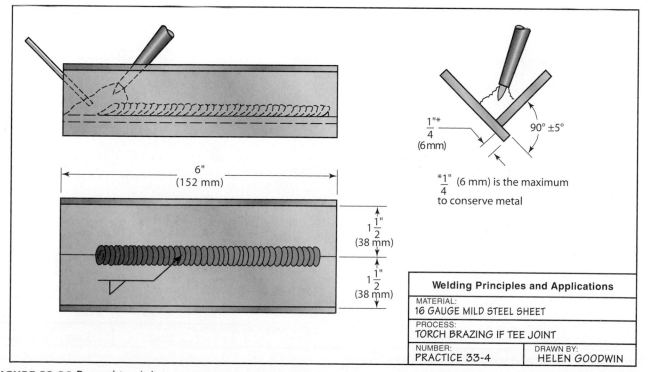

FIGURE 33-36 Brazed tee joint. © Cengage Learning 2012

careful not to overheat the braze metal. If the bead has a notch, the root of the joint is still not hot enough to allow the braze metal to flow properly. If the braze metal appears to have flowed properly after you have completed the joint, look at the back of the joint for a line of braze metal that flowed through. Repeat this practice until the joint can be made without defects. Turn off the cylinders, bleed the hoses, back out the regulator adjusting screws, and clean up your work area when you are finished.

Complete a copy of the "Student Welding Report" listed in Appendix I or provided by your instructor. ◆

FIGURE 33-37 Tack braze lap joint. Larry Jeffus

PRACTICE 33-5

Brazed Lap Joint

Using the same equipment, material, and setup as listed in Practice 33-2, you will make a brazed lap joint in the flat position.

Place the pieces of sheet metal on a firebrick so that they overlap each other by approximately 1/2 in. (13 mm). It is important that the pieces be held flat relative to each other, **Figure 33-37**. Make a small tack braze on both ends and then one or two tack brazes along the joint. Hold the torch so the flame moves along the joint and heats up both pieces at the same time. When the sheets are hot, touch the rod to the sheets and make a bead similar to the butt joint. After completing the brazed joint, it should be uniform in width and appearance. Repeat this practice until the joint can be made without defects. Turn off the cylinders, bleed the hoses, back out the regulator adjusting screws, and clean up your work area when you are finished.

Complete a copy of the "Student Welding Report" listed in Appendix I or provided by your instructor. ◆

PRACTICE 33-6

Brazed Lap Joint with 100% Penetration

Using the same equipment, material, and setup as listed in Practice 33-2, you will make a lap joint having 100% penetration in the flat position, **Figure 33-38**.

Tack the plates together and start the bead the same way as you did in Practice 33-5. Next move the torch back onto the top plate, as you did for Practice 33-5, so that the braze metal will be drawn into the joint. After the joint is completed, check the back side of the joint for braze metal showing along the joint. Repeat this

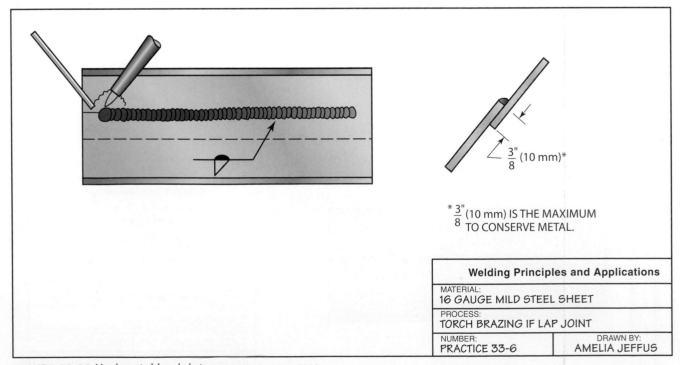

$\frac{3"}{8}$ (10 mm)*

* $\frac{3"}{8}$ (10 mm) IS THE MAXIMUM TO CONSERVE METAL.

Welding Principles and Applications	
MATERIAL: 16 GAUGE MILD STEEL SHEET	
PROCESS: TORCH BRAZING IF LAP JOINT	
NUMBER: PRACTICE 33-6	DRAWN BY: AMELIA JEFFUS

FIGURE 33-38 Horizontal lap joint. © Cengage Learning 2012

practice until the joint can be made without defects. Turn off the cylinders, bleed the hoses, back out the regulator adjusting screws, and clean up your work area when you are finished.

Complete a copy of the "Student Welding Report" listed in Appendix I or provided by your instructor. ◆

PRACTICE 33-7

Brazed Tee Joint, Thin to Thick Metal

Using a properly lit and adjusted torch, two pieces of mild steel 6 in. (152 mm) long, one 16 gauge and the other 1/4 in. (6 mm) thick, brazing flux, and BRCuZn brazing rod, you will make a tee joint in the flat position.

Hold the 16-gauge metal vertically on the 1/4-in. (6-mm) plate and tack braze both ends. The vertical member of a tee joint heats up faster than the flat member because the heat on the vertical member can be conducted in only one direction, **Figure 33-39**. The thin plate will heat up faster than the thick plate because there is less mass (metal).

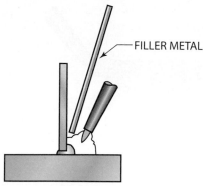

FIGURE 33-40 Torch and rod positions to balance heating between parts of unequal thickness. © Cengage Learning 2012

For the braze bead to be equal, both plates must be heated equally. Direct the flame on the thicker plate, as shown in **Figure 33-40**, and add the brazing rod on the thinner plate. This action will keep the thin plate from overheating. Make a braze along the joint that is uniform in appearance. Repeat this practice until the joint can be made without defects. Turn off the cylinders, bleed the hoses, back out the regulator adjusting screws, and clean up your work area when you are finished.

Complete a copy of the "Student Welding Report" listed in Appendix I or provided by your instructor. ◆

PRACTICE 33-8

Brazed Lap Joint, Thin to Thick Metal

Using the same equipment, materials, and setup as described in Practice 33-7, make a lap joint in the horizontal position, **Figure 33-41**.

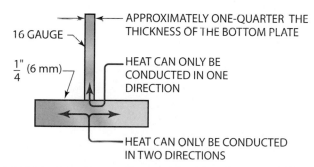

FIGURE 33-39 Unequal rate of heating due to a difference in mass. © Cengage Learning 2012

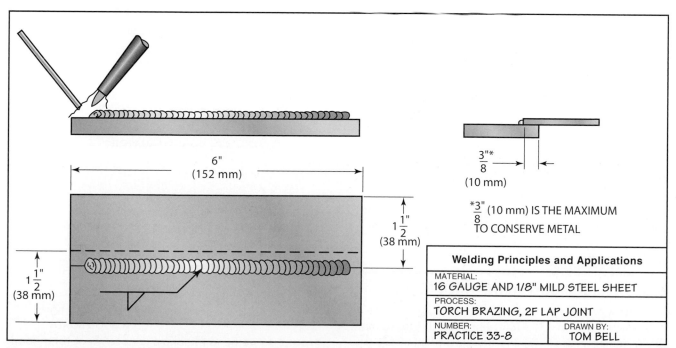

FIGURE 33-41 Brazed lap joint, thin to thick metal. © Cengage Learning 2012

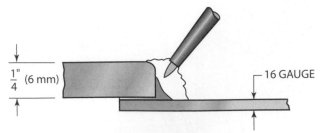

FIGURE 33-42 Direct the heat on the thicker section to ensure proper bonding. © Cengage Learning 2012

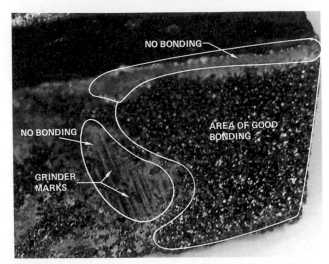

FIGURE 33-43 Only part of this braze joint is bonded properly. Grinder marks can be seen casted into the braze. The base metal was too cool for proper bonding to occur. Larry Jeffus

Tack braze the pieces together, being sure that they are held tightly together. Place the metal on a firebrick with the thin metal up. Apply heat to the exposed thick metal and more slowly to the overlapping thin metal so that conduction from the thin metal will heat the thick metal at the lap, **Figure 33-42**. If the braze is started before the thick metal is sufficiently heated, the filler metal will be chilled, and a bond will not occur. After the joint is completed and cooled, tap the joint with a hammer to see if there is a good bonded joint. **Figure 33-43** shows a broken braze joint that bonded properly only in certain areas. Repeat this practice until the joint can be made without defects. Turn off the cylinders, bleed the hoses, back out the regulator adjusting screws, and clean up your work area when you are finished.

Complete a copy of the "Student Welding Report" listed in Appendix I or provided by your instructor. ◆

PRACTICE 33-9

Braze Welded Butt Joint, Thick Metal

Using a properly lit and adjusted torch, two pieces of mild steel plate 6 in. (152 mm) long, one 1/4 in. (6 mm) thick, one 16-gauge flux, and BRCuZn filler metal, you will make a flat braze welded butt joint.

FIGURE 33-44 Joint preparation. © Cengage Learning 2012

Grind the edges of the 1/4-in. (6-mm) plate so that they are slightly rounded, **Figure 33-44**. The rounded edges are better when used for brazing because they distribute the strain on the braze more uniformly. After the plates have been prepared, tack braze both ends together. Because of the mass of the plates, it may be necessary to preheat the plates. This helps the penetration and eliminates cold lap at the root. The flame should be moved in a triangular motion so that the root is heated, as well as the top of the bead, **Figure 33-45**. When the joint is complete and cool, bend the joint to check for complete root bonding, **Figure 33-46**. Repeat this practice until the joint can be made without defects. Turn off the cylinders, bleed the hoses, back out the regulator adjusting screws, and clean up your work area when you are finished.

Complete a copy of the "Student Welding Report" listed in Appendix I or provided by your instructor. ◆

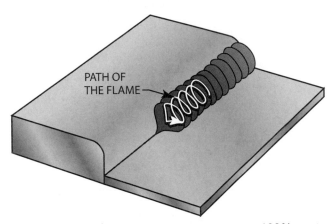

FIGURE 33-45 Flame movement to ensure a 100% root penetration. © Cengage Learning 2012

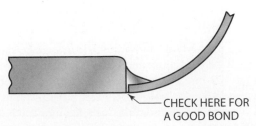

FIGURE 33-46 Bend the braze joint to check for a good bond. © Cengage Learning 2012

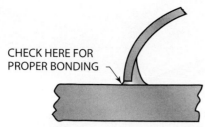

FIGURE 33-47 Bend test to check for root bonding.
© Cengage Learning 2012

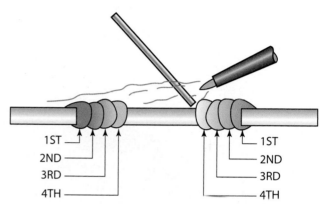

FIGURE 33-48 Filling a hole with braze. First run a bead around the outside of the hole. © Cengage Learning 2012

PRACTICE 33-10

Braze Welded Tee Joint, Thick Metal

Using the same equipment, materials, and setup as listed in Practice 33-9, you will make a braze welded tee joint in the flat position.

As with the butt joint, the edge of the plate is to be slightly rounded, and the metal may have to be preheated to get a good bond at the root. Direct the flame toward the bottom plate and into the root and add the filler metal back into the root. Watch for a notch, which indicates lack of bonding at the root. When the braze is complete and cool, bend it with a hammer to check for a bond at the root, **Figure 33-47.** Repeat this practice until the joint can be made without defects. Turn off the cylinders, bleed the hoses, back out the regulator adjusting screws, and clean up your work area when you are finished.

Complete a copy of the "Student Welding Report" listed in Appendix I or provided by your instructor. ◆

SURFACE BUILDUP AND HOLE FILL PRACTICES

Surfaces on worn parts are often built up again with braze metal. Braze metal is ideal for parts that receive limited abrasive wear because the buildup is easily machinable. Unlike welding or hardsurfacing, **braze buildup** has no hard spots that make remachining difficult. Braze buildups are good both for flat and round stock. The lower temperature used in brazing does not tend to harden or soften the base metal as much as in welding.

Holes in light-gauge metal can be filled using braze metal. The filled hole can be ground flush if it is required for clearance, leaving a strong patch with minimum distortion.

PRACTICE 33-11

Braze Welding to Fill a Hole

Using a properly lit and adjusted torch, one piece of 16-gauge mild steel, flux, and BRCuZn filler rod, you will fill a 1-in. (25-mm) hole. Place the piece of metal on two firebricks so that the hole is between them. Start by running a stringer bead around the hole, **Figure 33-48.** Once the bead is complete, turn the torch at a very sharp angle and point it at the edge of the hole nearest the torch.

FIGURE 33-49 Keep running beads around the hole until it is closed. © Cengage Learning 2012

```
1ST
2ND
3RD
4TH
```
```
1ST
2ND
3RD
4TH
```

Hold the end of the filler rod in the flame so that both the bead around the hole and the rod meet at the same time, **Figure 33-49.** Put the rod in the molten bead and flash the torch off to allow the molten braze pool to cool. When it has cooled, repeat this process. Surface tension will hold a small piece of molten metal in place. If the piece of molten metal becomes too large, it will drop through. Progress around the hole as many times as needed to fill the hole. When the braze weld is complete, it should be fairly flat with the surrounding metal. Repeat this practice until it can be made without defects. Turn off the cylinders, bleed the hoses, back out the regulator adjusting screws, and clean up your work area when you are finished.

Complete a copy of the "Student Welding Report" listed in Appendix I or provided by your instructor. ◆

PRACTICE 33-12

Flat Surface Buildup

Using a properly lit and adjusted torch, a 3-in. (76-mm) square of 1/4-in. (6-mm) mild steel, flux, and BRCuZn filler rod, you will build up a surface.

Place the square plate flat on a firebrick. Start along one side of the plate and make a braze weld down that side. When you get to the end, turn the plate 180° and braze back alongside the first braze, **Figure 33-50,** covering about one-half of the first braze bead. Repeat this procedure until the side is covered with braze metal.

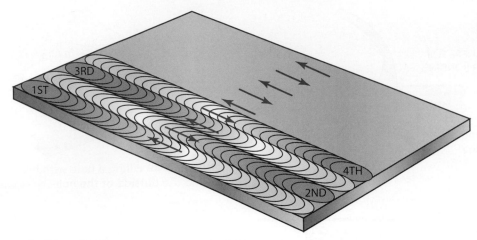

FIGURE 33-50 Braze buildup, first layer. © Cengage Learning 2012

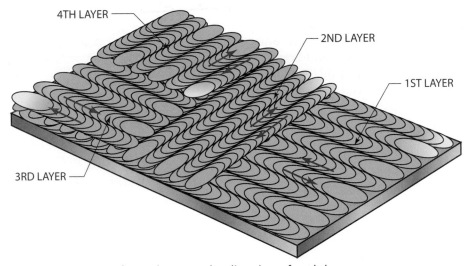

FIGURE 33-51 When building up a surface, alternate the direction of each layer. © Cengage Learning 2012

Turn the plate 90° and repeat the process, **Figure 33-51.** Be sure that you are getting good fusion with the first layer and that there are no slag deposits trapped under the braze. Be sure to build up the edges so that they can be cut back square. This process should be repeated until there is at least 1/4 in. (6 mm) of buildup on the surface. The surfacing can be checked visually or by machining the square to a 3-in. (76-mm) × 1/2-in. (13-mm) block and checking for slag inclusions. The plate will warp as a result of this buildup. If the plate is going to be used, it should be clamped down to prevent distortion. Turn off the cylinders, bleed the hoses, back out the regulator adjusting screws, and clean up your work area when you are finished.

Complete a copy of the "Student Welding Report" listed in Appendix I or provided by your instructor. ◆

PRACTICE 33-13

Round Surface Buildup

Using a properly lit and adjusted torch, one piece of mild steel rod 1/2 in. (13 mm) in diameter × 3 in. (76 mm)

long, flux, and BRCuZn brazing rod, you will build up a round surface.

In the flat position, start at one end and make a braze weld bead, 1 1/2 in. (38 mm) long, along the side of the steel rod. Turn the rod and make another bead next to the first bead, covering about one-half of the first, **Figure 33-52** and **Figure 33-53.** Repeat this procedure until the rod is 1 in. (25 mm) in diameter. It may be necessary to make a braze bead around both ends of the buildup to keep it square. The buildup can be visually inspected, or it can be turned down in a lathe. Repeat this practice until it can

FIGURE 33-52 Round shaft built up with braze. Larry Jeffus

FIGURE 33-53 Shaft turned down to check for slag inclusions or poor bonding. Larry Jeffus

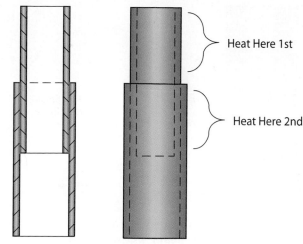

Heat Here 1st

Heat Here 2nd

FIGURE 33-54 2G vertical down position. © Cengage Learning 2012

be made without defects. Turn off the cylinders, bleed the hoses, back out the regulator adjusting screws, and clean up your work area when you are finished.

Complete a copy of the "Student Welding Report" listed in Appendix I or provided by your instructor. ◆

SILVER BRAZING PRACITCES

The silver brazing practices that follow will use BCuP-2 to BCuP-5 alloys to make brazed copper pipe joints. The melting temperature for the alloys is around 1400°F (760°C). At these temperatures the copper pipe will be glowing a dull red. The best types of flame to use for this type of brazing are air acetylene, air MAPP, air propane, or any air fuel-gas mixture. The most popular types are air acetylene and air MAPP. If an oxyacetylene torch is used, it is easy to overheat the alloy. To prevent overheating with an oxyacetylene flame, keep the torch moving and hold the flame so that the inner cone is about 1 in. (25 mm) from the surface.

When using BCuP silver brazing alloys on clean copper, it is not necessary to use a flux. The phosphorus in these alloys makes them self-fluxing. It is the phosphorus that promotes the wetting and enhances the flow of the alloy into the joint space.

PRACTICE 33-14

Silver Brazing Copper Pipe, 2G Vertical Down Position

Using a properly lit and adjusted torch with air acetylene, air MAPP, air propane, or any air fuel-gas mixture; two or more pieces of 1/2 in. to 1 in. (13 mm to 25 mm) copper pipe with matching copper pipe fittings; BCuP-2 to BCuP-5 brazing metal; steel wool, sand cloth, and/or wire brush; safety glasses with side shields; gloves; proper protective clothing; pliers; and any other required personal safety equipment; you will make a silver brazed joint in copper pipe in the 2G vertical down position, **Figure 33-54.**

1. Set the regulator pressures according to the manufacturer's specifications for your fuel type and tip size.
2. Clean the pipe and fitting using the steel wool, sand cloth, or a wire brush. *Note:* Do not touch the cleaned surfaces with your hands. Oils from your skin can prevent the braze metal from wetting. This can result in leaks.
3. Slide the fitting onto the pipe. Be sure that the pipe is completely seated at the bottom of the fitting.
4. Heat the brazing rod and make a bend in the wire about 3/4 in. (19 mm) from the end. This will give you a gauge so that you do not put too much brazing metal in the joint.
5. Heat the pipe first, but not too much. *Note:* As the pipe is heated it will expand, forming a tighter fit in the fitting. This will aid in conducting heat deeper into the fitting socket, and the tighter fit will aid in capillary attraction of the braze metal.
6. Once the pipe is hot but not glowing red, start heating the fitting. Keep the torch moving to uniformly heat the entire joint.
7. The joint is at the correct temperature for brazing when the braze metal starts to wet the surface. Touch the tip of the brazing rod to check the parts so that you know when they reach the correct temperature.
8. Move the flame to the back side of the joint and feed the brazing rod into the joint space. *Note:* A slight pressure directly into the joint with the brazing rod will help it flow deeper into the fitting. This is especially helpful on larger diameter pipes.
9. Next, move the torch and brazing rod around the pipe and fitting so that the entire joint will be filled. There should be a slight fillet showing all the way around the joint, **Figure 33-55.** *Note:* The braze metal fillet adds very little strength to the joint, but it is an easy way to make sure that the joint is leak free.

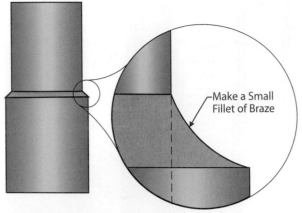

FIGURE 33-55 Making a smooth brazed fillet is a good way to prevent pinhole leaks. © Cengage Learning 2012

FIGURE 33-58 Flatten the sawed quarters. Larry Jeffus

After the pipe has cooled, test the joint by sawing the fitting in to a point just past the end of the inside pipe, **Figure 33-56**. Place the pipe vertically in a vise and saw it into quarters, **Figure 33-57**. Use a hammer and flat surface to flatten out each quarter section, **Figure 33-58**.

With the joint prepared, check for complete penetration of the braze alloy, **Figure 33-59**. Check the sides of the flattened quarters for braze alloy, **Figure 33-60**, or other defects.

Repeat this practice until you can make defect-free joints. Turn off the cylinder, bleed the hoses, back out the regulator adjusting screws, and clean up your work area when you are finished.

Complete a copy of the "Student Welding Report" listed in Appendix I or provided by your instructor. ◆

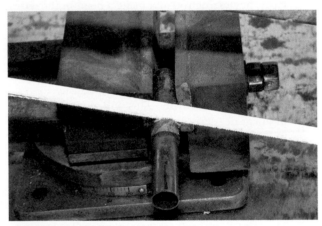

FIGURE 33-56 Saw the joint in two. Larry Jeffus

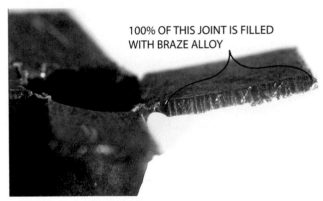

100% OF THIS JOINT IS FILLED WITH BRAZE ALLOY

FIGURE 33-59 Complete joint fill. Larry Jeffus

FIGURE 33-57 Quarter saw the joint. Larry Jeffus

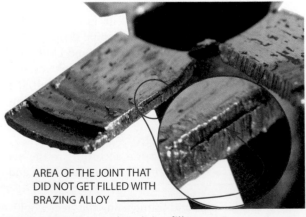

AREA OF THE JOINT THAT DID NOT GET FILLED WITH BRAZING ALLOY

FIGURE 33-60 Incomplete joint fill. Larry Jeffus

PRACTICE 33-15

Silver Brazing Copper Pipe, 5G Horizontal Fixed Position

Using the same equipment, materials, setup, and procedures as described in Practice 33-14, make a silver brazing copper pipe joint in the 5G horizontal fixed position.

1. Mount the pipe in a horizontal position, **Figure 33-61**.

2. Start by heating entirely around the pipe. Move the flame back and forth around the entire diameter of the pipe. *Note:* On pipe that is larger than 1 in. (25 mm), concentrate most of your heating on the top of the pipe because the flame may not have enough heat to heat up the entire pipe to the brazing temperature.

3. Once the pipe is hot but not glowing red, start heating the fitting. Keep the torch moving to heat the entire joint uniformly.

4. The joint is at the correct temperature for brazing when the braze metal starts to wet the surface. Touch the tip of the brazing rod to check the parts so that you know when they reach the correct temperature.

5. Move the flame to the back side of the joint and feed the brazing rod into the joint space with a slight pressure.

6. Gravity and capillary action will pull the braze metal down to the bottom of the joint if the entire diameter of the pipe is at the correct temperature. If necessary, move the torch slowly down the sides of the pipe to bring this part of the joint up to the brazing temperature. Add braze metal as needed, but do not overfill the joint. An indication that the joint has been overfilled is a bump of braze metal hanging from the bottom of the finished joint.

Repeat this practice until you can make defect-free joints. Turn off the cylinder, bleed the hoses, back out the regulator adjusting screws, and clean up your work area when you are finished.

Complete a copy of the "Student Welding Report" listed in Appendix I or provided by your instructor. ◆

PRACTICE 33-16

Silver Brazing Copper Pipe, 2G Vertical Up Position

Using the same equipment, materials, setup, and procedures as described in Practice 33-14, make a silver brazing copper pipe joint in the 2G vertical up position.

1. Mount the pipe in a vertical up position, **Figure 33-62**.

2. Start by heating entirely around the pipe. Move the flame back and forth around the entire diameter of the pipe.

3. Once the pipe is hot but not glowing red, move the flame up onto the fitting to start it heating. Keep the torch moving to uniformly heat the entire joint.

4. The joint is at the correct temperature for brazing when the braze metal starts to wet the surface. Touch the tip of the brazing rod to check the parts so that you know when they reach the correct temperature.

5. Move the flame to the top of the joint and feed the brazing rod into the joint space with a slight pressure.

6. The heat and capillary action will pull the braze metal up into the top of the joint space if the pipe is at the correct temperature. Add braze metal as needed, but do not overfill the joint. An indication that the joint has been overfilled is drips of braze metal running down the vertical pipe below the joint.

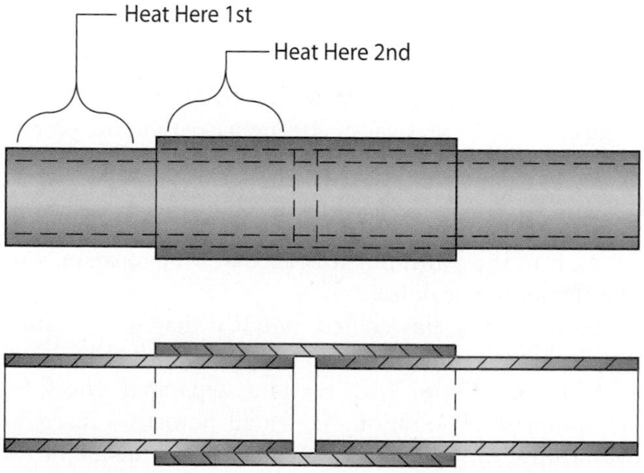

Heat Here 1st

Heat Here 2nd

FIGURE 33-61 5G horizontal fixed position. © Cengage Learning 2012

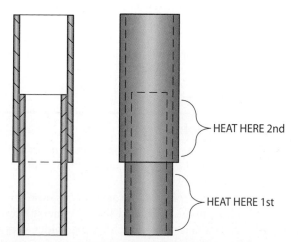

HEAT HERE 2nd

HEAT HERE 1st

FIGURE 33-62 2G vertical up position. © Cengage Learning 2012

Repeat this practice until you can make defect-free joints. Turn off the cylinder, bleed the hoses, back out the regulator adjusting screws, and clean up your work area when you are finished.

Complete a copy of the "Student Welding Report" listed in Appendix I or provided by your instructor. ◆

SOLDERING PRACTICES

The soldering practices that follow use tin-lead or tin-antimony solders. Both solders have a low melting temperature. If an oxyacetylene torch is used, it is very easy to overheat the solder. Caution is necessary because most of the fluxes used with this type of solder are easily overheated. The best type of flame to use for this type of soldering is air acetylene, air MAPP, air propane, or any air fuel-gas mixture. The most popular types are air acetylene and air propane. If galvanized metal is used, additional ventilation should be used to prevent zinc oxide poisoning.

PRACTICE 33-17

Soldered Tee Joint

Using a properly lit and adjusted torch, two pieces of 18-gauge to 24-gauge mild steel sheet 6 in. (152 mm) long, flux, and tin-lead or tin-antimony solder wire, you will solder a flat tee joint.

Hold one piece of metal vertical on the other piece and spot solder both ends. If flux cored wire is not being used, paint the flux on the joint at this time. Hold the torch flame so it moves down the joint in the same direction you will be soldering. Continue flashing the torch off and touching the solder wire to the joint until the solder begins to melt. Keeping the molten pool small enough to work with is a major problem with soldering. The flame must be flashed off frequently to prevent overheating. When the joint is completed, the solder should be uniform. Repeat this practice until it can be made without defects. Turn off the cylinders, bleed the hoses, back out the regulator adjusting screws, and clean up your work area when you are finished.

Complete a copy of the "Student Welding Report" listed in Appendix I or provided by your instructor. ◆

PRACTICE 33-18

Soldered Lap Joint

Using the same equipment, materials, and setup as listed in Practice 33-17, you will solder a lap joint in the horizontal position.

Tack the pieces of metal together as shown in **Figure 33-63**. Apply the flux and heat the metal slowly, checking the temperature by touching the solder wire to the metal often. When the work gets hot enough, flash the

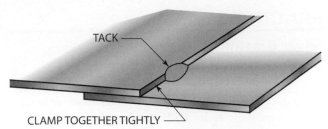

TACK

CLAMP TOGETHER TIGHTLY

FIGURE 33-63 Tacking metal together for soldering.
© Cengage Learning 2012

flame off frequently to prevent overheating and proceed along the joint. When the joint is completed, the solder should be uniform. Repeat this practice until it can be made without defects. Turn off the cylinders, bleed the hoses, back out the regulator adjusting screws, and clean up your work area when you are finished.

Complete a copy of the "Student Welding Report" listed in Appendix I or provided by your instructor. ◆

PRACTICE 33-19

Soldering Copper Pipe, 2G Vertical Down Position

Using a properly lit and adjusted torch, a piece of 1/2-in. to 1-in. (13-mm to 25-mm) copper pipe, a copper pipe fitting, steel wool, flux, and tin-lead or tin-antimony solder wire, you will solder a pipe joint in the vertical down position, **Figure 33-64**.

Clean the pipe and fitting using steel wool and apply the flux to both parts. Slide the fitting onto the pipe and twist the fitting to ensure that the flux is applied completely around the inside of the joint.

Make a bend in the solder wire about 3/4 in. (19 mm) from the end. This will give you a gauge so that you do not put too much solder in the joint. Excessive solder will flow inside the pipe, and it may cause problems to the system later.

Heat the pipe and the fitting with the torch. As the parts become hot, keep checking the parts with the solder wire so that you know when they reach the correct temperature. When the solder starts to wet, remove the flame and wipe the joint with the end of the solder wire, **Figure 33-65**. Next, heat the fitting more than the pipe so that the solder will be drawn into the joint. Rewipe the joint with the solder as needed. There should be a small fillet of solder around the joint. This fillet adds very little strength to the joint, but it is an easy way to make sure that the joint is leak free.

After the pipe has cooled, notch it diagonally with a hacksaw, put a screwdriver in the cut, and twist the joint apart, **Figure 33-66**. With the joint separated, check for (1) complete penetration, (2) small porosity caused by overheating the solder, (3) drops of solder inside of the pipe, or (4) other defects. Repeat this practice until it can

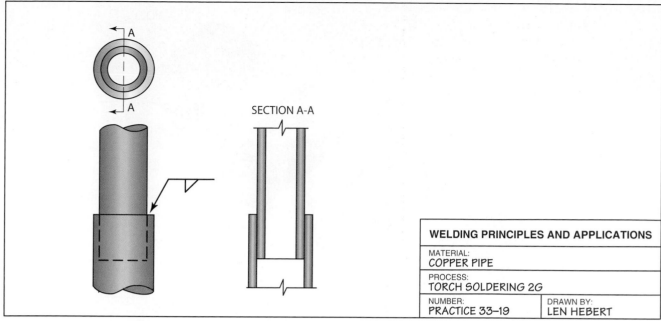

WELDING PRINCIPLES AND APPLICATIONS

MATERIAL: COPPER PIPE

PROCESS: TORCH SOLDERING 2G

NUMBER: PRACTICE 33-19 | DRAWN BY: LEN HEBERT

FIGURE 33-64 2G vertical down position. © Cengage Learning 2012

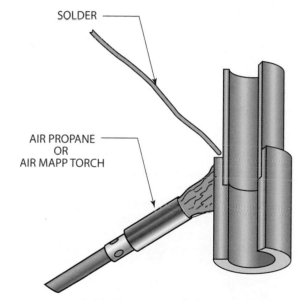

SOLDER

AIR PROPANE OR AIR MAPP TORCH

FIGURE 33-65 Soldering copper fitting to copper pipe. © Cengage Learning 2012

be done without defects. Turn off the cylinders, bleed the hoses, back out the regulator adjusting screws, and clean up your work area when you are finished.

Complete a copy of the "Student Welding Report" listed in Appendix I or provided by your instructor. ◆

PRACTICE 33-20

Soldering Copper Pipe, 1G Position

Using the same equipment and setup as listed in Practice 33-19, make a soldered joint with the pipe held

horizontally, **Figure 33-67**. Test the joint as before and repeat as necessary until it passes the test. Turn off the cylinders, bleed the hoses, back out the regulator adjusting screws, and clean up your work area when you are finished.

Complete a copy of the "Student Welding Report" listed in Appendix I or provided by your instructor. ◆

PRACTICE 33-21

Soldering Copper Pipe, 4G Vertical Up Position

Using the same equipment and setup as listed in Practice 33-19, make a soldered joint with the pipe held in the vertical position with the solder flowing uphill. Test the joint as before and repeat as necessary until it can be made without defects. Turn off the cylinders, bleed the hoses, back out the regulator adjusting screws, and clean up your work area when you are finished.

Complete a copy of the "Student Welding Report" listed in Appendix I or provided by your instructor. ◆

PRACTICE 33-22

Soldering Aluminum to Copper

Using a properly lit and adjusted torch, a piece of aluminum plate, a copper penny, steel wool, flux, and tin-lead or tin-antimony solder wire, you will tin both the aluminum and the copper with solder and then join both together.

The surface of the aluminum must be clean and free of paint, oils, dirt, and coatings such as anodizes. Hold the

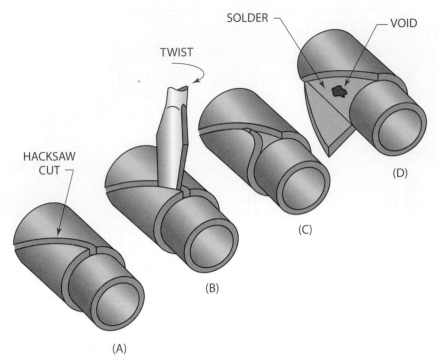

FIGURE 33-66 (A) Use a hacksaw to cut a groove through the outside copper pipe. (B) Carefully push a flat blade screwdriver into the saw cut groove. (C) Twist the blade to open up the groove. (D) Unwrap the outer copper pipe to reveal the solder in the joint. © Cengage Learning 2012

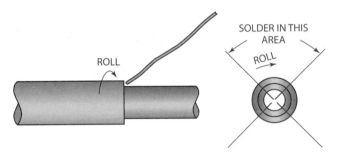

FIGURE 33-67 1G soldering. © Cengage Learning 2012

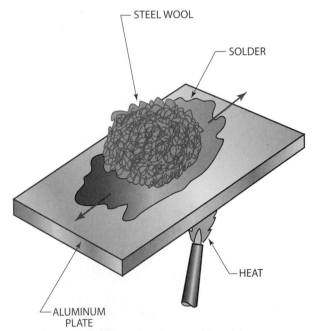

FIGURE 33-68 Tinning aluminum with solder. © Cengage Learning 2012

flame on the aluminum until it warms up slightly. Hold the solder in the flame and allow a small amount to melt and drop on the aluminum plate; do not add flux, **Figure 33-68.** Move the flame off the plate and rub the liquid solder with the steel wool. Be careful not to burn your fingers or allow the flame to touch the steel wool. The solder should be stuck in the steel wool when it is lifted off the plate. Alternately heat the plate and rub it with the steel wool–solder. When the plate becomes hot enough, it will melt the solder and the solder will tin the aluminum surface, **Figure 33-69.**

Use some flux and solder, and tin the penny. Place the penny on the aluminum plate so the areas of solder on both the penny and plate are touching each other. Heat the two until the solder melts and flows out from between the penny and plate. When the parts cool, to check the bond, try to break the joint apart.

This process will work on other types of metals that have a strong oxide layer that prevents the solder from bonding. By breaking the oxide layer free with the mechanical action of the steel wool, the metals can join. This process can allow a copper patch over an aluminum tube such as those used in air-conditioning or refrigeration,

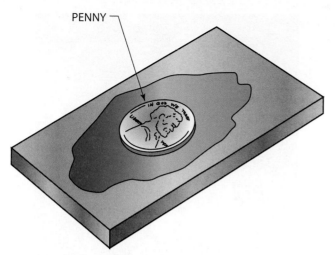

PENNY

FIGURE 33-69 Copper penny soldered to aluminum.
© Cengage Learning 2012.

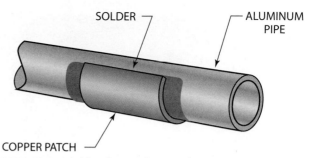

SOLDER — ALUMINUM PIPE

COPPER PATCH

FIGURE 33-70 Aluminum pipe patch. © Cengage Learning 2012

Figure 33-70. Turn off the cylinders, bleed the hoses, back out the regulator adjusting screws, and clean up your work area when you are finished.

Complete a copy of the "Student Welding Report" listed in Appendix I or provided by your instructor. ◆

Summary

Brazing and soldering are processes that have many great advantages but are often overlooked when a joining process is being selected. For example, brazing and soldering are excellent processes for portable applications. In addition, their versatility makes them great choices for many jobs in which good joint design will result in joint strength equal to or higher than that of welding. The ability to join many different materials with a limited variety of fluxes and filler metals reduces the need for a large inventory of materials, which can result in great cost savings for a small business, home shop, or farm.

For example, solder can be used on the threads of a bolt on an off-road vehicle to act as a locknut to keep it from vibrating off. To remove the nut, one need only heat the threads, and it unscrews easily. Soldering can then be either a permanent or a temporary attaching process.

Be creative in the way you apply these processes. They can be very beneficial to you and your employer.

Greater Lowell Tech Welding Students Restore Historic Statue

The Chief of the Penacooks proudly stands tall again thanks to the hard work of the welding students and faculty at Greater Lowell Technical High School in Massachusetts, **Figure 1**. The statue of Passaconaway was dedicated on August 19, 1899, honoring the memory of this 17th century leader of the Penacooks for his friendship and the help he gave the early New England settlers.

Over the years, thieves stole the left hand, headdress, spear, bow, and tomahawk, desecrating this memorial for the few dollars they would receive for the scrap metal, **Figure 2**. The students and faculty at Greater Lowell Technical High School came to the rescue. The students did extensive historical research of Indian tribes in the region. To aid them in making the correct sizes and shapes of the missing parts, they made models such as the various spear points shown in **Figure 3**. They then made drawings and fabricated the parts, **Figure 4**. Parts such as the feathers for the spear were cut out of copper sheets and compared to the design, **Figure 5**.

FIGURE 1 Larry Jeffus

FIGURE 2 Larry Jeffus

Students made mockups of the missing hand out of aluminum foil to ensure that the part they fabricated would be realistic and the correct size for the Chief's larger-than-life statue. With the design size and layout set, the students learned the art of hammering copper sheet to form the three-dimensional shapes needed for the hands, spear, and

tomahawk, **Figure 6.** Once the parts were shaped, they were cut out. Then the various parts needed for each fabrication were assembled, **Figure 7,** and soldered together, **Figure 8.**

After months of design and fabrication work in the high school welding shop, the students were ready to replace the stolen items. Because of the size of the statue and the tall granite pedestal it stands on, scaffolding had to be used to

FIGURE 3 Larry Jeffus

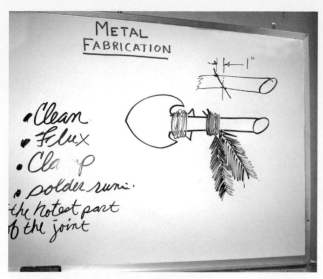

FIGURE 4 Larry Jeffus

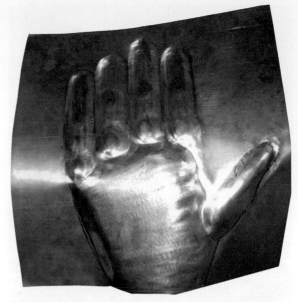

FIGURE 6 Larry Jeffus

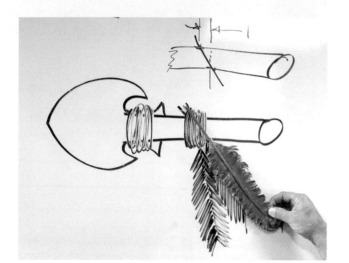

FIGURE 5 Larry Jeffus

FIGURE 8 Larry Jeffus

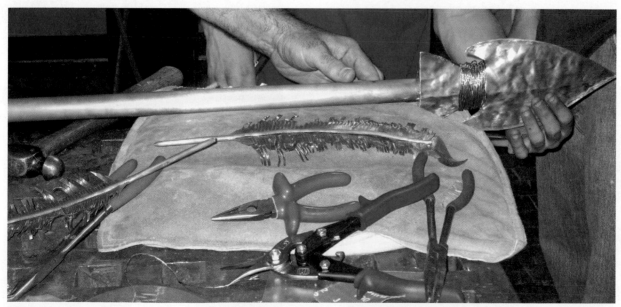

FIGURE 7 Larry Jeffus

FIGURE 9 Larry Jeffus

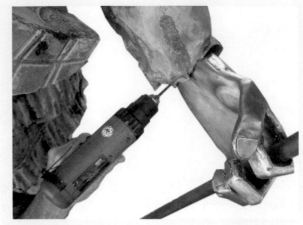

FIGURE 11 Larry Jeffus

FIGURE 12 Larry Jeffus

make the restoration work safe, **Figure 9.** Everyone working on the project had to make untold trips up and down the scaffolding to complete the detailed work, **Figure 10.** The replacement parts were primarily attached by drilling and pop riveting, **Figures 11** and **12.**

After lots of long hours and hard work, the reconstruction work was completed; and after a thorough cleaning, Passaconaway, Chief of the Penacooks looked as good as new, **Figure 13.**

FIGURE 10 Larry Jeffus

FIGURE 13 Larry Jeffus

Review

1. Explain the difference between brazing and soldering.

2. How does capillary action separate brazing from braze welding?

3. Why can brazing be both a permanent and a temporary joining method?

4. Why is it less likely that a semiskilled worker would damage a part with brazing than with welding?

5. What is the effect of joint spacing on joint tensile strength?

6. Why are braze joints subject to fatigue failure?

7. Do all braze joints resist corrosion? Give an example.

8. What are the three primary functions that a flux must perform?

9. In what forms are fluxes available?

10. How can liquid fluxes be delivered to the joint through the torch?

11. How do fluxes react with the base metal?

12. How are soldering and brazing methods grouped?

13. What are the advantages of torch soldering?

14. What are the advantages of furnace brazing?

15. How does the induction brazing method heat the part being brazed?

16. What soldering process can be used to join parts and provide a protective coating to the part at the same time?

17. What soldering or brazing process uses a machine similar to a spot welder to produce the heat required to make a joint?

18. What is a metal alloy?

19. Why must parts not be moved as they cool through the paste range?

20. Why must you not use tin-lead solders on water piping?

21. Using Table 33-3, determine the approximate temperature at which a mixture of 90% tin with 10% lead would become 100% solder.

22. What is a major use of tin-antimony solder?

23. What solder alloy can retain its strength at elevated temperatures?

24. Why do brazing alloys use letters such as CuZn to identify the metal?

25. What is the white smoke that can be given off from copper-zinc brazing alloy?

26. Which copper-zinc brazing alloy could be used to join or repair each of the following?
    a. cast iron
    b. build up wear surfaces
    c. malleable iron
    d. naval brass

27. Why should copper-phosphorus not be used on fittings with ferrous substrates?

28. Silver-copper alloys can be used to join which metals?

29. Which nickel alloy would be best for joining the following?
    a. poor-fitting joints
    b. honeycomb structures
    c. jet engine parts
    d. corrosive applications

30. Why does BCu-2 brazing alloy use an organic compound mixed with copper?

31. Using Table 33-6, determine the ideal joint spacings for the following braze alloys.
    a. BCuZn
    b. BAlSi
    c. BCuP

32. What indicates that you have overheated the solder flux on copper?

33. How can the size of a braze bead be controlled?

34. Why is brazing a better process than welding to rebuild some surfaces?

Appendix

Appendix I

Student Welding Report

I. STUDENT WELDING REPORT

Student Name:_____ Date:_____

Instructor:_____ Class:_____

Experiment or Practice #:_____ Process:_____

Briefly describe task: _____

_____

_____

_____

INSPECTION REPORT			
Inspection	Pass/Fail	Inspector's Name	Date
Safety:			
Equip. Setup:			
Equip. Operation:			
Welding	Pass/Fail	Inspector's Name	Date
Accuracy:			
Appearance:			
Overall Rating:			
Comments:			

Student Grade:_____ Instructor Initials:_____ Date:_____

Appendix II

Conversion of Decimal Inches to Millimeters and Fractional Inches to Decimal-Inches and Millimeters

Inches dec	mm	Inches dec	mm	Inches frac	dec	mm	Inches frac	dec	mm
0.01	0.2540	0.51	12.9540	1/64	0.015625	0.3969	33/64	0.515625	13.0969
0.02	0.5080	0.52	13.2080	1/32	0.031250	0.7938	17/32	0.531250	13.4938
0.03	0.7620	0.53	13.4620						
0.04	1.0160	0.54	13.7160	3/64	0.046875	1.1906	35/64	0.546875	13.8906
0.05	1.2700	0.55	13.9700						
0.06	1.5240	0.56	14.2240	1/16	0.062500	1.5875	9/16	0.562500	14.2875
0.07	1.7780	0.57	14.4780	5/64	0.078125	1.9844	37/64	0.578125	14.6844
0.08	2.0320	0.58	14.7320						
0.09	2.2860	0.59	14.9860	3/32	0.093750	2.3812	19/32	0.593750	15.0812
0.10	2.5400	0.60	15.2400						
0.11	2.7940	0.61	15.4940	7/64	0.109375	2.7781	39/64	0.609375	15.4781
0.12	3.0480	0.62	15.7480	1/8	0.125000	3.1750	5/8	0.625000	15.8750
0.13	3.3020	0.63	16.0020						
0.14	3.5560	0.64	16.2560	9/64	0.140625	3.5719	41/64	0.640625	16.2719
0.15	3.8100	0.65	16.5100	5/32	0.156250	3.9688	21/32	0.656250	16.6688
0.16	4.0640	0.66	16.7640						
0.17	4.3180	0.67	17.0180	11/64	0.171875	4.3656	43/64	0.671875	17.0656
0.18	4.5720	0.68	17.2720	3/16	0.187500	4.7625	11/16	0.687500	17.4625
0.19	4.8260	0.69	17.5260						
0.20	5.0800	0.70	17.7800	13/64	0.203125	5.1594	45/64	0.703125	17.8594
0.21	5.3340	0.71	18.0340						
0.22	5.5880	0.72	18.2880	7/32	0.218750	5.5562	23/32	0.718750	18.2562
0.23	5.8420	0.73	18.5420	15/64	0.234375	5.9531	47/64	0.734375	18.6531
0.24	6.0960	0.74	18.7960						
0.25	6.3500	0.75	19.0500	1/4	0.250000	6.3500	3/4	0.750000	19.0500
0.26	6.6040	0.76	19.3040						
0.27	6.8580	0.77	19.5580	17/64	0.265625	6.7469	49/64	0.765625	19.4469
0.28	7.1120	0.78	19.8120	9/32	0.281250	7.1438	25/32	0.781250	19.8437
0.29	7.3660	0.79	20.0660						
0.30	7.6200	0.80	20.3200	19/64	0.296875	7.5406	51/64	0.796875	20.2406
0.31	7.8740	0.81	20.5740	5/16	0.312500	7.9375	13/16	0.812500	20.6375
0.32	8.1280	0.82	20.8280						
0.33	8.3820	0.83	21.0820	21/64	0.328125	8.3344	53/64	0.828125	21.0344
0.34	8.6360	0.84	21.3360	11/32	0.343750	8.7312	27/32	0.843750	21.4312
0.35	8.8900	0.85	21.5900						
0.36	9.1440	0.86	21.8440	23/64	0.359375	9.1281	55/64	0.859375	21.8281
0.37	9.3980	0.87	22.0980						
0.38	9.6520	0.88	22.3520	3/8	0.375000	9.5250	7/8	0.875000	22.2250
0.39	9.9060	0.89	22.6060	25/64	0.390625	9.9219	57/64	0.890625	22.6219
0.40	10.1600	0.90	22.8600						
0.41	10.4140	0.91	23.1140	13/32	0.406250	10.3188	29/32	0.906250	23.0188
0.42	10.6680	0.92	23.3680	27/64	0.421875	10.7156	59/64	0.921875	23.4156
0.43	10.9220	0.93	23.6220						
0.44	11.1760	0.94	23.8760	7/16	0.437500	11.1125	15/16	0.937500	23.8125
0.45	11.4300	0.95	24.1300	29/64	0.453125	11.5094	61/64	0.953125	24.2094
0.46	11.6840	0.96	24.3840						
0.47	11.9380	0.97	24.6380	15/32	0.468750	11.9062	31/32	0.968750	24.6062
0.48	12.1920	0.98	24.8920	31/64	0.484375	12.3031	62/64	0.984375	25.0031
0.49	12.4460	0.99	25.1460						
0.50	12.7000	1.00	25.4000	1/2	0.500000	12.7000	1	1.000000	25.4000

For converting decimal-inches in "thousandths," move decimal point in both columns to left.

Appendix III

Conversion Factors: U.S. Customary (Standard) Units and Metric Units (SI)

TEMPERATURE
    Units

°F (each 1° change)	= 0.555°C (change)
°C (each 1° change)	= 1.8°F (change)
32°F (ice freezing)	= 0°Celsius
212°F (boiling water)	= 100°Celsius
–460°F (absolute zero)	= 0° Rankine
–273°C (absolute zero)	= 0° Kelvin

    Conversions

°F to °C	_____ °F – 32	=	_____ × 0.555	=	_____ °C
°C to °F	_____ °C × 1.8	=	_____ + 32	=	_____ °F

LINEAR MEASUREMENT
    Units

1 inch	= 25.4 millimeters
1 inch	= 2.54 centimeters
1 millimeter	= 0.0394 inch
1 centimeter	= 0.3937 inch
12 inches	= 1 foot
3 feet	= 1 yard
5280 feet	= 1 mile
10 millimeters	= 1 centimeter
10 centimeters	= 1 decimeter
10 decimeters	= 1 meter
1000 meters	= 1 kilometer

    Conversions

in. to mm	_____ in.	× 25.4	=	_____ mm
in. to cm	_____ in.	× 2.54	=	_____ cm
ft to mm	_____ ft	× 304.8	=	_____ mm
ft to m	_____ ft	× 0.3048	=	_____ m
mm to in.	_____ mm	× 0.0394	=	_____ in.
cm to in.	_____ cm	× 0.3937	=	_____ in.
mm to ft	_____ mm	× 0.00328	=	_____ ft
m to ft	_____ m	× 3.28	=	_____ ft

AREA MEASUREMENT
    Units

1 sq in.	= 0.0069 sq ft
1 sq ft	= 144 sq in.
1 sq ft	= 0.111 sq yd
1 sq yd	= 9 sq ft
1 sq in.	= 645.16 sq mm
1 sq mm	= 0.00155 sq in.
1 sq cm	= 100 sq mm
1 sq m	= 1000 sq cm

    Conversions

sq in. to sq mm	_____ sq in. × 645.16	=	_____ sq mm
sq mm to sq in.	_____ sq mm × 0.00155	=	_____ sq in.

VOLUME MEASUREMENT
    Units

1 cu in.	= 0.000578 cu ft
1 cu ft	= 1728 cu in.
1 cu ft	= 0.03704 cu yd
1 cu ft	= 28.32 L
1 cu ft	= 7.48 gal (U.S.)
1 gal (U.S.)	= 3.737 L
1 cu yd	= 27 cu ft
1 gal	= 0.1336 cu ft
1 cu in.	= 16.39 cu cm
1 L	= 1000 cu cm

Appendix III

Conversion Factors: U.S. Customary (Standard) Units and Metric Units (SI) (Continued)

1 L	= 61.02 cu in.
1 L	= 0.03531 cu ft
1 L	= 0.2642 gal (U.S.)
1 cu yd	= 0.769 cu m
1 cu m	= 1.3 cu yd

Conversions

cu in. to L	_____ cu in.	× 0.01638	=	_____ L
L to cu in.	_____ L	× 61.02	=	_____ cu in.
cu ft to L	_____ cu ft	× 28.32	=	_____ L
L to cu ft	_____ L	× 0.03531	=	_____ cu ft
L to gal	_____ L	× 0.2642	=	_____ gal
gal to L	_____ gal	× 3.737	=	_____ L

WEIGHT (MASS) MEASUREMENT

Units

1 oz	= 0.0625 lb
1 lb	= 16 oz
1 oz	= 28.35 g
1 g	= 0.03527 oz
1 lb	= 0.0005 ton
1 ton	= 2000 lb
1 oz	= 0.283 kg
1 lb	= 0.4535 kg
1 kg	= 35.27 oz
1 kg	= 2.205 lb
1 kg	= 1000 g

Conversions

lb to kg	_____ lb	× 0.4535	=	_____ kg
kg to lb	_____ kg	× 2.205	=	_____ lb
oz to g	_____ oz	× 0.03527	=	_____ g
g to oz	_____ g	× 28.35	=	_____ oz

PRESSURE and FORCE MEASUREMENTS

Units

1 psig	= 6.8948 kPa
1 kPa	= 0.145 psig
1 psig	= 0.000703 kg/sq mm
1 kg/sq mm	= 6894 psig
1 lb (force)	= 4.448 N
1 N (force)	= 0.2248 lb

Conversions

psig to kPa	_____ psig	× 6.8948	=	_____ kPa
kPa to psig	_____ kPa	× 0.145	=	_____ psig
lb to N	_____ lb	× 4.448	=	_____ N
N to lb	_____ N	× 0.2248	=	_____ lb

VELOCITY MEASUREMENTS

Units

1 in./sec	= 0.0833 ft/sec
1 ft/sec	= 12 in./sec
1 ft/min	= 720 in./sec
1 in./sec	= 0.4233 mm/sec
1 mm/sec	= 2.362 in./sec
1 cfm	= 0.4719 L/min
1 L/min	= 2.119 cfm

Conversions

ft/min to in./sec	_____ ft/min	× 720	=	_____ in./sec
in./min to mm/sec	_____ in./min	× 0.4233	=	_____ mm/sec
mm/sec to in./min	_____ mm/sec	× 2.362	=	_____ in./min
cfm to L/min	_____ cfm	× 0.4719	=	_____ L/min
L/min to cfm	_____ L/min	× 2.119	=	_____ cfm

Appendix IV

Abbreviations and Symbols

U.S. Customary (Standard) Units

°F	= degrees Fahrenheit
°R	= degrees Rankine = degrees absolute F
lb	= pound
psi	= pounds per square inch = lb per sq in.
psia	= pounds per square inch absolute = psi + atmospheric pressure
in.	= inches = i = "
ft	= foot or feet = f = '
sq in.	= square inch = in.^{2}
sq ft	= square foot = ft^2
cu in.	= cubic inch = in.^{3}
cu ft	= cubic foot = ft^3
ft-lb	= foot-pound
ton	= ton of refrigeration effect
qt	= quart

Metric Units (SI)

°C	= degrees Celsius
°K	= degrees Kelvin
mm	= millimeter
cm	= centimeter
cm^2	= centimeter squared
cm^3	= centimeter cubed
dm	= decimeter
dm^2	= decimeter squared
dm^3	= decimeter cubed
m	= meter
m^2	= meter squared
m^3	= meter cubed
L	= liter
g	= gram
kg	= kilogram
J	= joule
kJ	= kilojoule
N	= newton
Pa	= pascal
kPa	= kilopascal
W	= watt
kW	= kilowatt
MW	= megawatt

Miscellaneous Abbreviations

P	= pressure
h	= hours
sec	= seconds
D	= diameter
r	= radius of circle
A	= area
π	= 3.1416 (a constant used in determining the area of a circle)
V	= volume
∞	= infinity

Appendix V

Metric Conversion Approximations

Metric Conversion Approximations
1/4 inch = 6 mm
1/2 inch = 13 mm
3/4 inch = 18 mm
1 inch = 25 mm
2 inches = 50 mm
1/2 gal = 2 L
1 gal = 4 L
1 lb = 1/2 K
2 lb = 1 K
1 psig = 7 kPa
1°F = 2°C

Appendix VI

Pressure Conversion

psi	kPa	psi	kPa
1	7 (6.9)	100	690 (689)
2	14 (13.7)	110	760 (758)
3	20 (20.6)	120	820 (827)
4	30 (27.5)	130	900 (896)
5	35 (34.4)	140	970 (965)
6	40 (41.3)	150	1030 (1034)
7	50 (48.2)	160	1100 (1103)
8	55 (55.1)	170	1170 (1172)
9	60 (62.0)	180	1240 (1241)
10	70 (69.9)	190	1310 (1310)
15	100 (103)	200	1380 (1379)
20	140 (137)	225	1550 (1551)
25	170 (172)	250	1720 (1723)
30	200 (206)	275	1900 (1896)
35	240 (241)	300	2070 (2068)
40	280 (275)	325	2240 (2240)
45	310 (310)	350	2410 (2413)
50	340 (344)	375	2590 (2585)
55	380 (379)	400	2760 (2757)
60	410 (413)	450	3100 (3102)
65	450 (448)	500	3450 (3447)
70	480 (482)	550	3790 (3792)
75	520 (517)	600	4140 (4136)
80	550 (551)	650	4480 (4481)
85	590 (586)	700	4830 (4826)
90	620 (620)	750	5170 (5171)
95	660 (655)	800	5520 (5515)
		850	5860 (5860)
		900	6210 (6205)
		950	6550 (6550)
		1000	6890 (6894)

Pounds per square inch (psi) converted to kilopascals (kPa). One psi equals 6.8948 kPa. In most applications the conversion from standard units of pressure to SI units can be rounded to an even number. The number in parentheses is the value before it is rounded off.

Appendix VII

Welding Codes and Specifications

A *welding code* is a detailed listing of the rules or principles that are to be applied to a specific classification or type of product.

A *welding specification* is a detailed statement of the legal requirements for a specific classification or type of product. Products manufactured to code or specification requirements commonly must be inspected and tested to ensure compliance.

A number of agencies and organizations publish welding codes and specifications. The application of the particular code or specification to a weldment can be the result of one or more of the following requirements:

- Local, state, or federal government regulations
- Bonding or insuring company
- End-user (customer) requirements
- Standard industrial practices

The three most popular codes are

#1104, American Petroleum Institute—Used for pipelines
Section IX, American Society of Mechanical Engineers— Used for pressure vessels
D1.1, American Welding Society—Used for bridges and buildings

The following organizations publish welding codes and/or specifications.

AAR
  Association of American Railroads
  425 Third Street, SW, Suite 1000
  Washington, DC 20024

AASHTO
  American Association of State Highway
    and Transportation Officials
  444 N. Capitol Street, NW
  Washington, DC 20001

AISC
  American Institute of Steel Construction
  1 East Wacker Drive, Suite 700
  Chicago, IL 60601

ANSI
  American National Standards Institute
  1899 L Street, NW, 11th Floor
  Washington, DC 20036

API
  American Petroleum Institute
  1220 L Street, NW
  Washington, DC 20005

AREMA
  American Railway Engineering and Maintenance-of-Way
    Association
  10003 Derekwood Lane, Suite 210
  Lanham, MD 20706

ASME
  American Society of Mechanical Engineers
  3 Park Avenue
  New York, NY 10016

AWS
  American Welding Society
  550 NW LeJeune Road
  Miami, FL 33126

AWWA
  American Water Works Association
  6666 West Quincy Avenue
  Denver, CO 80235

MIL
  Department of Defense
  Washington, DC 20301-0001

SAE
  Society of Automotive Engineers
  400 Commonwealth Drive
  Warrendale, PA 15096

Appendix VIII

Welding Associations and Organizations

AA
  Aluminum Association
  1525 Wilson Boulevard, Suite 600
  Arlington, VA 22209

AASHTO
  American Association of State Highway
    and Transportation Officials
  444 N. Capitol Street, NW
  Washington, DC 20001

AGA
  American Gas Association
  400 N. Capitol Street, NW
  Washington, DC 20001

AIME
  American Institute of Mining, Metallurgical
    and Petroleum Engineering
  8307 Shaffer Parkway
  Littleton, CO 80127

AISC
  American Institute of Steel Construction
  1 East Wacker Drive, Suite 700
  Chicago, IL 60601

AISI
  American Iron and Steel Institute
  1140 Connecticut Avenue, NW, Suite 705
  Washington, DC 20036

ANS
  American Nuclear Society
  555 North Kensington Avenue
  La Grange Park, IL 60526

ANSI
  American National Standards Institute
  1899 L Street, NW, 11th Floor
  Washington, DC 20036

API
  American Petroleum Institute
  1220 L Street, NW
  Washington, DC 20005

ASCE
  American Society of Civil Engineers
  1801 Alexander Bell Drive
  Reston, VA 20191

ASME
  American Society of Mechanical Engineers
  3 Park Avenue
  New York, NY 10016-5990

ASNT
  American Society for Nondestructive Testing
  1711 Arlingate Lane
  Columbus, OH 43228

ASTM
  American Society for Testing and Materials
  100 Barr Harbor Drive
  West Conshohocken, PA 19428

AWS
  American Welding Society
  550 NW LeJeune Road
  Miami, FL 33126

CWB
  Canadian Welding Bureau
  8260 Parkhill Drive
  Milton, ON L9T 5V7
  Canada

EWI
  Edison Welding Institute
  1250 Arthur E. Adams Drive
  Columbus, OH 43221

IIW
  International Institute of Welding
  550 NW LeJeune Road
  Miami, FL 33126-5699

NSC
  National Safety Council
  1121 Spring Lake Drive
  Itasca, IL 60143

PFI
  Pipe Fabrication Institute
  511 Avenue of Americas, Suite 601
  New York, NY 10011

SAE
  Society of Automotive Engineers
  400 Commonwealth Drive
  Warrendale, PA 15096

Appendix VIII

Welding Associations and Organizations (Continued)

SME
Society of Manufacturing Engineers
1 SME Drive
Dearborn, MI 48128

TWI
The Welding Institute
Granta Park
Great Abington, Cambridge CB21 6AL
United Kingdom

WRC
Welding Research Council
P.O. Box 1942
New York, NY 10156

Glossary

The terms and definitions in this glossary are extracted from the American Welding Society publication AWS A3.0-80 Welding Terms and Definitions. The terms with an asterisk are from a source other than the American Welding Society. *Note:* The English term and definition are given first, followed by the same term and definition in Spanish.

A

*abrasives. Materials that are usually sharp and are used to clean or grind a surface. They may be used as a powder such as sand to blast the surface or they may be formed into disks or stones to be used by a grinder.
abrasivos. Materiales que son por lo general ásperos y que se utilizan para limpiar o pulir superficies. Pueden venir en polvo, como arena, para bruñir las superficies, o en forma de discos o piedras para ser usados por una esmeriladora.

*absolute pressure. The sum of the gauge pressure and the atmospheric pressure.
presión absoluta. La suma de la presión manómetro y la presión atmosférica.

absorptive lens. A filter lens designed to attenuate the effects of transmitted and reflected light.
lente absorbente. Un lente de filtro diseñado para disminuir los efectos de la luz y la reflexión de la luz extraviada.

acceptable criteria. Agreed upon standards that must be satisfactorily met.
criterios aceptables. Las normas sobre las que se ha llegado a un acuerdo y que deben cumplirse en forma satisfactoria.

acceptable weld. A weld that meets all the requirements and the acceptance criteria.
soldadura aceptable. Una soldadura que satisface los requisitos y el criterio aceptable prescribida por las especificaciónes de la soldadura.

*acetone. A fragrant liquid chemical used in acetylene cylinders. The cylinder is filled with a porous material and acetone is then added to fill. Acetylene is then added and absorbed by the acetone, which can absorb up to 28 times its own volume of the gas.
acetona. Un liquido fragante químico que se usa en los cilindros del acetileno. El cilindro se llena de un material poroso y luego se le agrega la acetona hasta que se llene. El acetileno es absorbido por la acetona, la cual puede absorber 28 veces el propio volumen del gas.

*acetylene. A fuel gas used for welding and cutting. It is produced as a result of the chemical reaction between calcium carbide and water. The chemical formula for acetylene is C_2H_2. It is colorless, is lighter than air, and has a strong garlic-like smell. Acetylene is unstable above pressures of 15 psig (1.05 kg/cm^2 g). When burned in the presence of oxygen, acetylene produces one of the highest flame temperatures available.
acetileno. Un gas combustible que se usa para soldar y cortar. Es producido a consecuencia de una reacción química de agua y calcio y carburo. La fórmula química para el acetileno es C_2H_2. No tiene color, es más ligero que el aire, y tiene un olor fuerte como a ajo. El acetileno es inestable en presiones más altas de 15 psig (1.05 kg/cm^2 g). Cuando se quema en presencia del oxígeno, el acetileno produce una de las llamás con una temperatura más alta que la que se utiliza.

*acicular structure. A fine micrograin structure found in rapidly cooled steel.
*estructura acicular. Una estructura micro granulada fina que se encuentra en el acero que se ha enfriado con rapidez.

actual throat. See throat of a fillet weld.
garganta actual. Vea garganta de soldadura filete.

*adaptable. Capable of making self-directed corrections; in a robot, this is often accomplished with visual, force, or tactile sensors.
adaptable. Capaz de hacer correcciones por instrucción propia de un robot, esto se lleva a cabo con sensores tangibles visuales, o de fuerza.

air acetylene welding (AAW). An oxyfuel gas welding process that uses an air-acetylene flame. The process is used without the application of pressure. This is an obsolete or seldom used process.
soldadura de aire acetileno (AAW). Un proceso de soldar con gas (oxi/combustible) que usa aire-acetileno sin aplicarse presión. Un proceso anticuado que es una rareza.

air carbon arc cutting (CAC-A). A carbon arc cutting process variation removing molten metal with a jet of air.
arco de carbón con aire (CAC-A). Un proceso de cortar con arco de carbón variante que quita el metal derretido con un chorro de aire.

*allotropic metals. Metals that have specific lattice or crystal structures that form when the metal is cool and that change within the solid metal as it is heated and before it melts.
metales alotrópicos. Metales que tienen un determinado enrejado o estructuras cristalinas que se forman cuando el metal está frío y cambian dentro del metal sólido mientras se lo calienta y antes de que éste se derrita.

allotropic transformation. A change in the crystalline lattice pattern of a metal due to a change in temperature.
transformación alotropico. Un cambio en el modelo cristalino enrejado del metal debido a un cambio en la temperatura.

*alloy. A metal with one or more elements added to it, resulting in a significant change in the metal's properties.
aleación. Un metal en que se le agrega uno o más elementos resultando en un cambio significante en las propiedades del metal.

*alloying elements. Elements in the flux that mix with the filler metal and become part of the weld metal. Major alloying elements are molydenum, nickel, chromium, manganese, and vanadium.

elementos de mezcla. Elementos en el flujo que se mezclan con el metal para rellenar y formar parte del metal soldado. Los elementos principales de mezcla son molibdeno, niquel, cromo, manganeso y vanadio.

*alloy steels. Steels that may contain any number of a variety of elements that are used to change the mechanical properties of steel. The amount of the added elements can range from 1% to 50%.

aleación de acero. Acero que contiene cualquier número de una variedad de elementos que se usan para cambiar las propiedades mecánicas del acero. La cantidad de elementos agregados puede variar entre el 1% al 50%.

*all-weld-metal test specimen. A test specimen with the reduced section composed wholly of weld metal.

prueba de metal soldado. Una prueba con una sección reducida compuesta totalmente del metal de la soldadura.

*alphabet of lines. Lines are the language of drawing; they are used to represent various parts of the object being illustrated. The various line types are collectively known as the Alphabet of Lines.

alfabeto de líneas. Las líneas son el idioma del dibujo; se usan para representar las diferentes partes de un objeto que se ilustra. El conjunto de los distintos tipos de líneas se conoce como Alfabeto de Líneas.

*aluminum. A soft silvery-gray highly thermally and electrically conductive metal that naturally resists corrosion.

aluminio. Metal blando de color gris plateado que es buen conductor del calor y de la electricidad y que resiste naturalmente la corrosión.

*American Welding Society. (AWS). Organization that promotes the art and science of welding and that publishes international codes and standards.

Sociedad Americana de Soldadores. (AWS). Organización que promueve el arte y la ciencia de la soldadura y que publica códigos y estándares internacionales.

*amperage. A measurement of the rate of flow of electrons; amperage controls the size of the arc.

amperaje. Una medida de la proporción de la corriente de electrones; el amperaje controla el tamaño del arco.

amperage range. The lower and upper limits of welding power, in amperage, that can be produced by a welding machine or used with an electrode or by a process.

rango de amperaje. Los límites máximos y mínimos de poder de soldadura (en amperaje) que puede tener una máquina para soldar o que pueden usarse con un electrodo o a través de un proceso.

angle of bevel. See preferred term bevel angle.

ángulo del bisel. Es preferible que vea el término ángulo del bisel.

*anode. Material with a lack of electrons; thus, it has a positive charge.

ánodo. Un material que carece electrones; por eso tiene una carga positiva.

arc blow. The deflection of arc from its normal path because of magnetic forces.

soplo del arco. Desviación de un arco eléctrico de su senda normal a causa de fuerzas magnéticas.

arc braze welding (ABW). A braze welding process variation using an electric arc as the heat source.

soldadura fuerte aplicada por arco (ABW). Un proceso de soldadura fuerte donde el calor requerido es obtenido de un arco eléctrico.

arc cutting (AC). A group of thermal cutting processes that severs or removes metal by melting with the heat of an arc between an electrode and the workpiece.

corte con arco (AC). Un grupo de procesos termales para cortar que desúne o quita el metal derretido con el calor del arco en medio del electrodo y la pieza de trabajo.

arc cutting electrodes. An electrode that has a flux covering that allows the core wire to burn back inside so it creates a strong jetting action which blows out the molten metal created by the arc. See arc cutting.

electrodos para corte con arco. Electrodos con fundente que permiten que el núcleo de éstos se queme desde adentro para crear presiones que soplan el metal derretido que se crea por el arco. Ver corte con arco.

arc force. The axial force developed by a plasma.

fuerza del arco. La fuerza axial desarrollada por la plasma.

arc gouging. Thermal gouging that uses an arc cutting process variation to form a bevel or groove.

gubiadura con arco. Gubiadura termal que usa un proceso variante de corte con arco para formar un bisel o ranura.

arc length. The length from the tip of the welding electrode to the adjacent surface of the weld pool.

largura del arco. La distancia de la punta del electrodo a la superficie que colinda con el charco de la soldadura.

arc plasma. A gas heated by an arc to at least a partially ionized condition, enabling it to conduct an electric current. See also plasma.

arco de plasma. Un estado de la materia encontrado en la región de una descarga eléctrica (arco). Vea también **plasma.**

arc spot weld. A spot weld made by an arc welding process.

soldadura de puntos por arco. Una soldadura de punto hecha por un proceso de soldadura de arco.

arc strike. A discontinuity consisting of any localized remelted metal, heat-affected metal, or change in the surface profile of any part of a weld or base metal resulting from an arc.

golpe del arco. Una discontinuidad que consiste de cualquier rederretimiento del metal localizado, metal afectado por el calor, o cambio en el perfil de la superficie de cualquier parte de la soldadura o metal base resultante de un arco.

arc time. The time during which an arc is maintained in making an arc weld.

tiempo del arco. El tiempo durante el que el arco se mantiene al hacer una soldadura de arco.

arc voltage. The voltage across the welding arc.

voltaje del arco. El voltaje a través del arco de soldar.

arc welding (AW). A group of welding processes producing coalescence of workpieces by melting them with an arc. The processes are used with or without the application of pressure and with or without filler metal.

soldadura de arco (AW). Un grupo de procesos de soldadura que producen una unión de piezas de trabajo calentándolas con un arco. Los procesos se usan con o sin la aplicación de presión y con o sin metal para rellenar.

arc welding deposition efficiency. The ratio of the weight of filler metal deposited in the weld metal to the weight of filler metal melted, expressed in percent.

eficiencia de deposición de soldadura de arco. La relación del peso del metal depositado en la soldadura al peso del metal para rellenar, y el metal derretido, expresado en por ciento.

arc welding electrode. A component of the welding circuit through which current is conducted between the electrode holder and the arc. See also **arc welding (AW)**.

electrodo para soldadura de arco. Un componente del circuito de soldadura que conduce la corriente a través del portaelectrodo y el arco. Vea también **soldadura de arco**.

arc welding gun. A device used to transfer current to a continuously fed consumable electrode, guide the electrode, and direct the shielding gas.

pistola de soldadura de arco. Aparato que se usa para transferir corriente eléctrica continuamente a un alimentador de electrodo consumible. También se usa para guiar al electrodo y dirigir el gas de protección.

*arm. An interconnected set of links and powered joints comprising a manipulator, which supports or moves a wrist and hand or end effector.

brazo. Una entreconexión que une un juego de eslabones y coyunturas de potencia conteniendo un manipulador, que apolla o mueve una muñeca o mano o el que efectúa al final.

as-welded. Pertaining to the condition of weldments prior to subsequent thermal, mechanical, or chemical treatments.

como-soldado. Pertenece a metal soldado, juntas soldadas, o soldaduras ya soldadas pero antes de que se les hagan tratamientos termales, mecánicos, o químicos.

*atmospheric pressure. The pressure at sea level resulting from the weight of a column of air on a specified area; expressed for an area of 1 sq in. or sq cm; normally given as 14.7 psi (1.05 kg/cm^2).

presion atmosferica. La presión al nivel del mar que resulta del peso de una columna de aire en una area especificada; expresada para una área de 1 pulgada cuadrada o un centímetro cuadrado. Normalmente dado como 14.7 psi (1.05 kg/cm^2).

*atomic hydrogen. A single free, unbounded hydrogen atom (H) usually formed when molecular hydrogen is exposed to an arc.

hidrógeno atómico. Un solo átomo de hidrógeno (H) libre, que normalmente se forma cuando se expone hidrógeno molecular a un arco.

atomic hydrogen welding (AHW). An arc welding process that uses an arc between two metal electrodes in a shielding atmosphere of hydrogen and without the application of pressure. This is an obsolete or seldom-used process.

soldadura atómica hidrógena (AHW). Un proceso de soldadura de arco que usa un arco entre dos electrodos de metal en una atmósfera de protección de hidrógeno y sin la aplicación de presión. Esto es un proceso anticuado y es rara la vez que se use.

*atoms. Smallest particle of matter that still contains the properties of that matter.

átomos. Las partículas más pequeñas de materia que aún mantienen las propiedades de dicha materia.

*austenite. A nonmagnetic course micro structured crystalline form of some iron and iron-carbon alloys that form as the result of the rate of cooling.

austenita. Forma no magnética de microestructura cristalina de algunos tipos de hierro y aleaciones de hierro y carbono que se forma como resultado del índice de enfriamiento.

*austenitic manganese steel. A steel alloy with a high carbon content containing 10% or more manganese that is very tough and that will harden when cold worked.

acero al manganeso austenítico. Aleación de acero con un alto contenido de carbono, que contiene un 10% o más de manganeso, y que es muy tenaz y endurece cuando se lo trabaja en frío.

autogenous weld. A fusion weld made without filler metal.

soldadura autógena. Una soldadura fundida sin metal de rellenar.

*automated operation. Welding operations performed repeatedly by a robot or another machine that is programmed to perform a variety of processes.

operación automatizada. Operaciones de soldaduras que se ejecutan repetidamente por un robot u otra maquina que está programada para hacer una variedad de procesos.

*automatic arc welding downslope time. The time during which the current is changed continuously from final taper current or welding current to final current.

tiempo del pendiente con descenso en soldadura de arco automático. El tiempo durante en que la corriente cambia continuamente disminuyendo la corriente final o la corriente de la soldadura a la corriente final.

*automatic arc welding upslope time. The time during which the current changes continuously from the initial current to the welding current.

tiempo de pendiente con ascenso en soldadura de arco automático. El tiempo durante en que la corriente cambia continuamente de donde se inició la corriente a la corriente de la soldadura.

*automatic operation. Welding operations performed repeatedly by a machine that has been programmed to do an entire operation without the interaction of the operator.

operación automática. Operaciones de soldadura que se ejecutan repetidamente por una maquina que ha sido programada para hacer una operación entera sin influencia del operador.

*automatic oxygen cutting. Oxygen cutting with equipment that performs the cutting operation without constant observation and the adjustment of the controls by an operator. The equipment may or may not perform loading and unloading of the work.

corte del oxígeno automático. Cortadura del oxígeno con equipo que hace la operación de cortar sin observación y ajuste de los controles por el operador. El equipo puede que ejecute o no el trabajo de cargar o descargar las piezas.

*automatic welding. Welding with equipment that requires only occasional or no observation of the welding and no manual adjustment of the equipment controls. Variations of this term are *automatic brazing, automatic soldering, automatic thermal cutting,* and *automatic thermal spraying.*

soldadura automática. Soldadura con equipos que requieren ocasional o ninguna observación, y ningun ajuste manual de los controles de los equipos. Variaciones de éste término son *soldadura fuerte automática, soldadura automática, corte termal automático, rociadura termal automático.*

***axial spray metal transfer.** Method of metal transfer used by GMAW and FCAW processes. See **spray transfer.**

traslado del metál por rociado axial. Método de transferencia de metal usado en los procesos GMAW y FCAW. Ver traslado rociado.

axis of a weld. A line through the length of a weld, perpendicular to and at the geometric center of its cross section.

eje de la soldadura. Una linea a lo largo de la soldadura perpendicula a y al centro geométrico de su corte transversal.

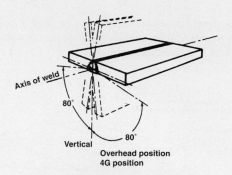

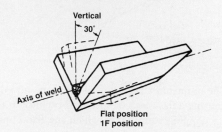

Positions of welding – groove welds

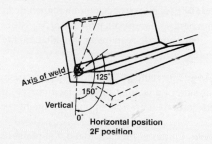

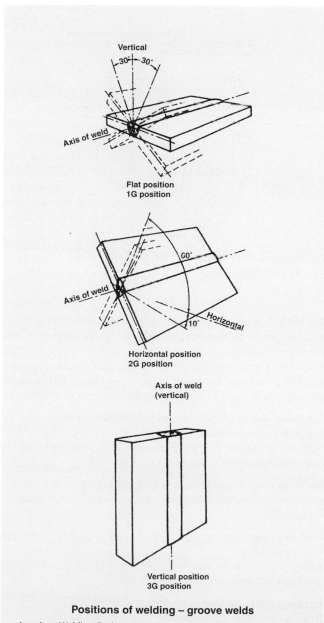

Positions of welding – groove welds

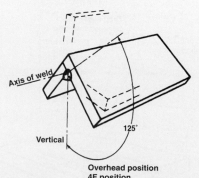

Positions of welding – fillet welds

B

back weld. A weld deposited at the back of a single groove weld.
soldadura de atrás. Una soldadura que se deposita en la parte de atrás de una soldadura de ranura sencilla.

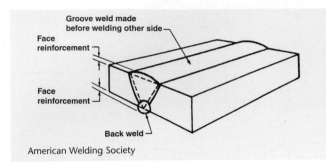

American Welding Society

backfire. The momentary recession of the flame into the torch, potentially causing a flashback or sustained backfire. It is usually signaled by a popping sound after which the flame may either extinguish or reignite at the end of the tip.
llama de retroceso. El retroceso momentaneo de la llama dentro de la punta para soldar, punta para cortar, o pistola para rociar con llama. La llama puede reaparecer inmediatamente o apagarse completamente.

back gouging. The removal of weld metal and base metal from the weld root side of a welded joint to facilitate complete fusion and complete joint penetration upon subsequent welding from that side.
gubia trasera. Quitar el metal soldado y el metal base del lado de la raíz de una junta soldada para facilitar una fusión completa y penetración completa de la junta soldada subsecuente a soldar de ese lado.

backhand welding. A welding technique in which the welding torch or gun is directed opposite to the progress of welding. Sometimes referred to as the "pull gun technique" in GMAW and FCAW. See also **travel angle**, **work angle**, and **drag angle**.
soldadura en revés. Una técnica de soldar la cual el soplete o pistola es guiada en la dirección contraria al adelantamiento de la soldadura. A veces se refiere como una tecnica de "estirar la pistola" en GMAW y FCAW. Vea también **ángulo de avance**, **ángulo de trabajo**, y **ángulo del tiro**.

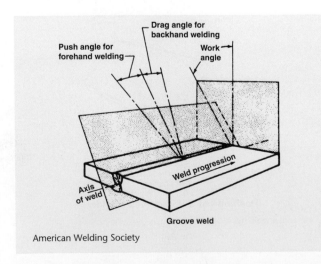

American Welding Society

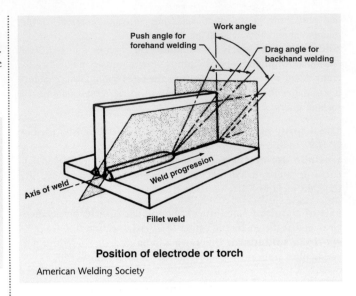

Position of electrode or torch

American Welding Society

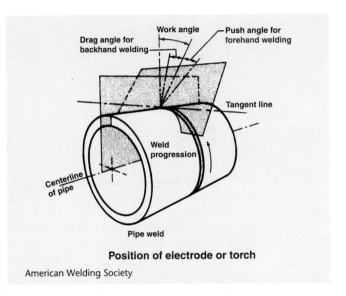

Position of electrode or torch

American Welding Society

backing. A material (base metal, weld metal, carbon, or granular material) placed at the root of a weld joint for the purpose of supporting molten weld metal.
respaldo. Un material (metal base, metal de soldadura, carbón o material granulado) puesto en la raíz de la junta soldada con el proposito de sostener el metal de la soldadura que está derretido.

*backing gas. A gas provided to protect the root of a weld from atmospheric contamination.
*gas de respaldo. Gas proporcionado para proteger la raíz de una soldadura de la contaminación atmosférica.

backing pass. A pass made to deposit a backing weld.
pasada de respaldo. Una pasada hecha para depositar la pasada del respaldo.

backing ring. Backing in the form of a ring, generally used in the welding of piping.
anillo o argolla de respaldo. Respaldo en forma de argolla, generalmente se usa en soldaduras de tubos.

backing strip. Backing in the form of a strip.
tira de respaldo. Un respaldo en la forma de una tira.

backing weld. Backing in the form of a weld.
soldadura de respaldo. Respaldo en la forma de soldadura.

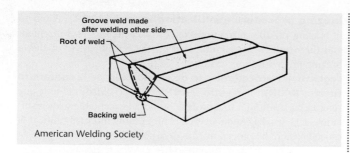

American Welding Society

backstep sequence. A longitudinal sequence in which the weld bead increments are deposited in the direction opposite to the progress of welding the joint.

secuencia a la inversa. Una serie de soldaduras en secuencia longitudinal hechas en la dirección opuesta del progreso de la soldadura.

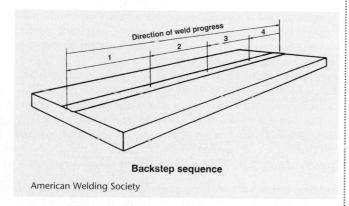

Backstep sequence

American Welding Society

bare electrode. A filler metal electrode that has been produced as a wire, strip, or bar with no coating or covering other than that which is incidental to its manufacture or preservation.

electrodo descubierto. Un electrodo de metal para rellenar que se ha producido como alambre, tira, o barra sin revestimiento o cubierto con solo lo necesario para su fabricación y conservación.

base material. The material that is welded, brazed, soldered, or cut. See also **base metal** and **substrate**.

material base. El material que está soldado, soldado con soldadura fuerte, soldado con soldadura blanda, o cortado. Vea también **metal base** y **substrato**.

base metal. The metal or alloy that is welded, brazed, soldered, or cut. See also **base material** and **substrate**.

metal base. El metal que está soldado con soldadura fuerte, soldado con soldadura blanda, o cortado. Vea también **material base** y **substrato**.

base metal test specimen. A test specimen composed wholly of base metal.

probeta para metal base. Una probeta totalmente compuesta de metal base.

bend test. A test in which a specimen is bent to a specified bend radius. See also **face-bend test**, **root-bend test**, and **side-bend test**.

prueba de dobléz. Una prueba donde la probeta se dobla a una vuelta con un radio especificado. Vea también **prueba de dobléz de cara**, **prueba de dobléz de raíz**, y **prueba de dobléz de lado**.

bevel. An angular type of edge preparation.

bisel. Una preparación de tipo angular con filo.

bevel angle. The angle formed between the prepared edge of a member and a plane perpendicular to the surface of the member. Refer to drawings for **bevel**.

ángulo del bisel. El ángulo formado entre el corte preparado de un miembro y la plana perpendicular a la superficie del miembro. Refiera a los dibujos del **bisel**.

*bill of materials. A document that lists all of the material required to construct a project. It may or may not contain the prices.

*lista de materiales. Documento que lista todos los materiales requeridos para construir un proyecto. Puede o no contener los precios.

*bird nesting. A ball of tangled wire that results when welding wire gets stuck in the gun liner and the feeder keeps feeding on either GMA or FCA welding.

*nido de ave. En las soldaduras GMA o FCA, bola de alambres entrelazados que se produce cuando el alambre de la soldadura se queda atorado en la pistola y el alimentador continúa suministrando más alambre.

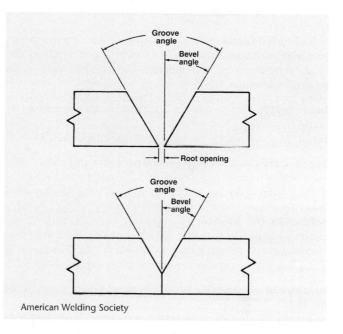

American Welding Society

blind joint. A joint, no portion of which is visible.

junta ciega. Una junta en que no hay porción visible.

block sequence. A combined longitudinal and cross-sectional sequence for a continuous multiple pass weld in which separated segments are completely or partially welded before intervening segments are welded.

secuencia de bloques. Una soldadura continua de pasadas multiples en sucesión combinadas longitudinal y sección transversa donde los incrementos separados son completamente o parcialmente soldados antes que los incrementos sean soldados.

*body-centered cubic (BCC). A crystalline structure called ferrite that iron forms as it changes from a liquid to a solid. See also **crystalline structures**.

*red cúbica centrada en las caras (BCC). Estructura cristalina llamada ferrita que el hierro forma a medida que cambia de líquido a sólido. Ver **también estructura espacial cristalina**.

bonding force. The force that holds two atoms together; it results from a decrease in energy as two atoms are brought closer to one another.

fuerza ligamentosa. La fuerza que detiene dos átomos juntos; es el resultado del decrecimiento en la energía cuando dos átomos son traídos cerca del uno al otro.

braze. A bond produced as a result of heating an assembly to the brazing temperature, 840°F (450°C), using a brazing filler metal distributed and retained between the closely fitted faying surfaces of the joint by capillary action.
soldadura de latón. Una soldadura producida cuando se calienta un montaje a una temperatura conveniente usando un metal de relleno que se liquida arriba de 840°F (450°C) y abajo del estado sólido del metal base. El metal de relleno es distribuido por acción capilar en una junta entre las superficies empalmadas montadas muy cerca.

*braze buildup. Braze metal added to the surface of a part to repair wear.
formación con bronce. Reparación de partes gastadas donde se agrega bronce.

brazeability. The capacity of a material to be brazed under the imposed fabrication conditions into a specific, suitably designed structure capable of performing satisfactorily in the intended service.
soldabilidad fuerte. La capacidad de un metal de refuerzo bajo las condiciones impuestas en la fabricación de una estructura diseñada especificamente para funcionar satisfactoriamente en los servicios intentados.

braze metal. That portion of a braze that has been melted during brazing.
latón. La porción del bronce que se derrite cuando se solda.

braze welding. A welding process variation that uses a filler metal with a liquidus above 840°F (450°C) and below the solidus of the base metal. Unlike brazing, in braze welding the filler metal is not distributed in the joint by capillary action.
soldadura con bronce. Es un proceso de soldar variante que usa un metal de relleno con un liquido arriba de 840°F (450°C) y abajo del estado del metal base. Distinto a la soldadura fuerte, el metal de relleno no es distribuido por acción capilar.

brazement. An assembly whose component parts are joined by brazing.
montaje de soldadura fuerte. Un montaje donde las partes son unidas por soldadura fuerte.

brazing (B). A group of welding processes that produces coalescence of materials by heating them to the brazing temperature in the presence of a filler metal with a liquidus above 840°F (450°C) and below the solidus of the base metal. The filler metal is distributed between the closely fitted faying surfaces of the joint by capillary action.
soldadura fuerte (B). Un grupo de procesos de soldadura que produce coalescencia de materiales calentándolos a una temperatura de soldar en la presencia de un material de relleno el cual se derrite a una temperatura de 840°F (450°C) y bajo del estado sólido del metal base. El metal de relleno se distribuye por acción capilar de una junta entre las superficies empalmadas montadas muy cerca.

*brazing alloys. A nonstandard term for brazing filler metal.
*aleaciones para soldadura fuerte. Término no estandarizado de metal de relleno para soldadura fuerte.

brazing filler metal. The metal or alloy to be added in making a brazed joint. The filler metal has a liquidus above 840°F (450°C) but below the solidus of the base material.
metal de relleno para soldadura fuerte. El metal que rellena el espacio libre en la junta capilar y se derrite a una temperatura de 840°F (450°C) y bajo del estado sólido del metal base.

brazing procedure qualification record (BPQR). A record of brazing variables used to produce an acceptable test brazement and the results of tests conducted on the brazement to qualify a brazing procedure specification.
registro del procedimiento de calificación de la soldadura fuerte (BPQR). Un registro de variables de la soldadura fuerte que se usan para producir una probeta bronceada aceptable y los resultados de la prueba conducidas en el bronceamiento para calificar la especificación del procedimiento de la soldadura fuerte.

brazing procedure specification (BPS). A document specifying the required brazing variables for a specific application.
especificación del procedimiento de la soldadura fuerte (BPS). Un documento especificando los variables requeridos de la soldadura fuerte para una aplicación especificada.

brazing rod. Filler metal used in the brazing process is supplied in the form of rods; the filler metal is usually an alloy of two or more metals; the percentages and types of metals used in the alloy impart different characteristics to the braze being made. Several classification systems are in use by manufacturers of filler metals. See also **brazing filler metal.**
varilla de latón. Metal de relleno usado en procesos de soldadura fuerte es surtida en forma de varillas; el metal para rellenar es usualmente un aleación de dos o más metales; los porcentajes y tipos de metales usados en la aleación imparten diferentes características a la soldadura que se está haciendo. Varios sistemás de clasificación se están usando por los fabricantes de metales de relleno. Vea también **metal de relleno para soldadura fuerte.**

brazing temperature. The temperature to which the base metal is heated to enable the filler metal to wet the base metal and form a brazed joint.
temperatura de soldadura fuerte. La temperatura a la cual se calienta el metal base para permitir que el metal de relleno moje al metal base y forme la soldadura fuerte.

buildup. The material deposited by the welding filler metal to a weld. Also, surfacing variation in which surfacing material is deposited to achieve the required dimensions. See also **buttering, cladding,** and **hardfacing.**
recubrimiento. La materia depositada por el metal de masilla de soldadura a una soldadura. También, una variación en la superficie donde el metal es depositado para que pueda obtener las dimensiones requeridas. Vea también **recubrimiento antes de terminar una soldadura, capa de revestimiento,** y **endurecimiento de caras.**

*buried arc transfer. In gas metal arc welding, a method of transfer in which the wire tip is driven below the surface of the weld pool due to the force of the carbon dioxide shielding gas. The shorter arc reduces the size of the drop, and any spatter is trapped in the cavity produced by the arc.
traslado de arco enterrado. En soldaduras de arco metalico con gas, un método de transferir en la cual la punta del alambre es enterrado debajo de la superficie del metal derretido debido a la fuerza del carbón bióxido del gas protector. Lo corto del arco reduce el tamaño de la gota y la salpicadura es atrapada en el hueco producido por el arco.

burnthrough. A hole or depression in the root bead of a single-groove weld due to excessive penetration. A nonstandard term when used for melt-through.
metal quemado que pasa al otro lado. Metal derretido que se quema en el lado de atrás del plato.

butt joint. A joint between two members aligned approximately in the same plane.

junta a tope. Una junta entre dos miembros alineados aproximadamente en el mismo plano.

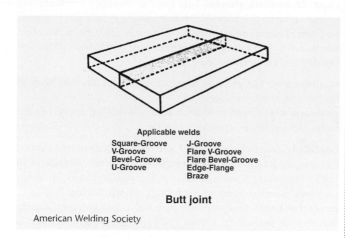

Applicable welds

Square-Groove	J-Groove
V-Groove	Flare V-Groove
Bevel-Groove	Flare Bevel-Groove
U-Groove	Edge-Flange
	Braze

Butt joint

American Welding Society

buttering. A surfacing variation depositing surfacing metal on one or more surfaces to provide metallurgically compatible weld metal for the subsequent completion of the weld.

recubrimiento antes de terminar una soldadura. Una variación de la superficie donde se deposita una o más capas de metal soldado en la ranura de un miembro (por ejemplo, un deposito de soldadura de aleación alta en un metal base de acero la cual será soldada a un metal base diferente). El recubrimiento proporciona una transición conveniente a la soldadura depositada para el acabamiento subsiguiente de una junta tope.

C

capillary action. The force by which liquid, in contact with a solid, is distributed between closely fitted faying surfaces of the joint to be brazed or soldered.

acción capilar. La fuerza por la que el liquido, en contacto con un sólido, es distribuido entre el empalme de las juntas del superficie para soldadura fuerte o blanda.

*carbon. A nonmetallic element that can be found in all organic and many inorganic compounds, the most common allowing element used in iron to change its mechanical properties.

*carbón. Elemento no metálico que está presente en todos los compuestos orgánicos y en muchos compuestos inorgánicos; elemento adicional más comúnmente usado con el hierro para cambiarle sus propiedades mecánicas.

carbon arc cutting (CAC). An arc cutting process that uses a carbon electrode. See also **air carbon arc cutting.**

corte con arco y carbón (CAC). Un proceso de corte con arco que usa un electrodo de carbón. Vea también **arco de carbón con aire.**

carbon arc gouging (CAG). A thermal gouging process using heat from a carbon arc and the force of compressed air or other nonflammable gas.

gubiadura con arco de carbón (CAG). Proceso de gubiadura térmica que utiliza el calor del arco de carbón y la fuerza del aire comprimido u otro gas no inflamable.

carbon arc welding (CAW). An arc welding process that uses an arc between a carbon electrode and the weld pool. The process is used with or without shielding and without the application of pressure.

soldadura con arco de carbón (CAW). Un proceso de soldar de arco en que se usa un arco entre el electrodo de carbón y el metal derretido. El proceso se usa con o sin protección y sin aplicación de presión.

carbon electrode. A nonfiller metal electrode used in arc welding and cutting, consisting of a carbon or graphite rod, which may be coated with copper or other materials.

electrodo de carbón. Un electrodo de metal que no se rellena usado en soldaduras de arco y para cortes consistiendo de varillas de carbón o grafito que pueden ser cubiertas de cobre u otros materiales.

*carbon steel. Steel whose physical properties are primarily the result of the percentage of carbon contained within the alloy. Carbon content ranges from 0.04% to 1.4%, often referred to as *plain carbon steel, low carbon steel,* or *straight carbon steel.*

acero al carbono. Acero cuyas propiedades físicas son primariamente el resultado del porcentaje de carbón que es contenido dentro de la aleación. El contenido del carbón es clasificado entre 0.04% a 1.4%, frecuentemente es referido *como el carbono de acero liso, acero de bajo carbón,* o *carbon de acero recto.*

carbonizing (carburizing). A reducing oxyfuel gas flame in which there is an excess of fuel gas, resulting in a carbon-rich zone extending around and beyond the cone.

llama carburante. Una llama minorada de gas combustible a oxígeno donde hay un exceso de gas combustible, resultando en una zona rica de carboón extendiendose alrededor y al otro lado del cono.

*cast. The natural curve in the electrode wire for gas metal arc welding as it is removed from the spool; cast is measured by the diameter of the circle that the wire makes when it is placed on a flat surface without any restraint.

distancia. La curva natural en el alambre electrodo para soldadura de arco metálico para gas cuando se aparta del carrete; la distancia es medida en el círculo que hace el alambre cuando es puesto en una superficie plana sin restricción.

*cast iron. A combination of iron and carbon. The carbon may range from 2% to 4%. Approximately 0.8% of the carbon is combined with the iron. The remaining free carbon is found as graphite mixed throughout the metal. Gray cast iron is the most common form of cast iron.

acero vaciado. Una combinación de acero y carbono. El carbono puede ser clasificado de 2% a 4%. Aproximadamente 0.8% del carbono es combinado con el hierro. El resto de carbono libre es encontrado como grafito mezclado por todo el metal. El acero fundido gris es la forma más común.

cathode. A natural curve material with an excess of electrons, thus having a negative charge.

cátodo. Un material de curva natural con un exceso de electrones, por eso tiene una carga negativa.

*cell. A manufacturing unit consisting of two or more workstations and the material transport mechanisms and storage buffers that interconnect them.

celda. Una unidad manufacturera la cual consiste de dos o más estaciones de trabajo y mecanismos para trasladar el material y los amortiguadores del almacén que los entreconecta.

cellulose-based electrode fluxes. Fluxes that use an organic-based cellulose ($C_6H_{10}O_5$) (a material commonly used to make paper) held together with a lime binder. When this flux is exposed to the heat of the arc, it burns and forms a rapidly expanding gaseous cloud of CO_2 that protects the molten weld pool from oxidation. Most of the fluxing material is burned, and little slag is deposited on the weld. E6010 is an example of an electrode that uses this type of flux.

fundentes para electrodos celulósicos. Fundentes que usan celulosa de base orgánica ($C_6H_{10}O_5$) (un material normalmente utilizado para fabricar papel), y que se mantienen unidos con un aglomerante de cal. Cuando a este fundente se lo expone al calor del arco, se consume y forma una nube gaseosa de CO_2 que se expande rápidamente y protege de la oxidación al charco de soldadura derretido. La mayor parte del material del fundente se consume, y se deposita poca escoria en la soldadura. El E6010 es un ejemplo de un electrodo que utiliza este tipo de fundente.

*cementite. A crystalline form of iron and carbon that is hard and brittle.
*cementita. Forma cristalina del hierro y del carbón que es dura y quebradiza.

*center. A manufacturing unit consisting of two or more cells and the materials transport and storage buffers that interconnect them.
centro. Una unidad manufacturera la cual consiste de dos o más celdas y el traslado de materiales y los amortiguadores del almacén que los entreconecta.

*centerline. Lines on a drawing that show the center point of circles and arcs and round or symmetrical objects. They also locate the center point for holes, irregular curves, and bolts.
*línea central. Líneas de un dibujo que muestran el punto central de círculos y arcos y de objetos redondos o simétricos. También sirven para ubicar los puntos centrales de hoyos, curvas irregulares, y pernos.

*certification. See certified welders.
*certificación. Ver soldador certificados.

certified welders. Individuals who have demonstrated their welding skills for a process by passing a specific welding test.
soldador certificados. Personas que han demostrado, mediante una prueba específica de soldadura, su habilidad para soldar en un proceso.

chain intermittent welds. An intermittent weld on both sides of a joint in which the weld segments on one side are approximately opposite those on the other side.
soladura intermitente de cadena. Soldadura intermitente en los dos lados de una junta en cual los incrementos de soldadura están aproximadamente opuestos a los del otro lado.

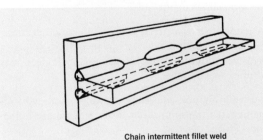

Chain intermittent fillet weld

American Welding Society

*chill plate. A large piece of metal used in welding to correct overheating.
plato desalentador. Una pieza de metal grande que se usa para corregir el sobrecalentamiento.

cladding. A relatively thick layer greater than 1 mm (0.04 in.) of material applied by surfacing for the purpose of improved corrosion resistance or other properties. See also **coating**, **surfacing**, and **hardfacing**.
capa de revestimiento. Una capa de material relativamente grueso superior a 1 mm (0.04 pulgadas) aplicada por la superficie con el objeto de mejorar la resistencia a la corrosión u otras propiedades. Vea también **revestimiento**, **recubrimiento superficial**, y **endurecimiento de caras**.

coalescence. The growing together or growth into one body of the materials being welded.
coalescencia. El crecimiento o desarrollo de un cuerpo de los materiales los cuales se están soldando.

coating. A relatively thin layer 1 mm (0.04 in.) of material applied by surfacing for the purpose of corrosion prevention, resistance to high-temperature scaling, wear resistance, lubrication, or other purposes. See also **cladding**, **surfacing**, and **hardfacing**.
revestimiento. Una capa de material relativamente delgado material de menos de 1 mm (0.04 pulgadas) aplicada por la superficie con el propósito de prevenir corrosión, resistencia a las altas temperaturas, resistencia a la deterioración, lubricación, o para otros propósitos. Vea también **capa de revestimiento**, **recubrimiento superficial**, y **endurecimiento de caras**.

cold crack. A crack occuring in a metal at or near ambient temperatures. Cold cracks can occur in base metal, heat-affected zones, and weld metal zones.
quebradura en frío. Quebradura que ocurre en un metal a temperatura ambiente o a temperaturas cercanas a ésta. Las quebraduras en frío pueden ocurrir en base de metal, zonas afectadas por el calor y zonas de metal de soldadura.

cold soldered joint. A joint with incomplete coalescence caused by insufficient application of heat to the base metal during soldering.
junta soldada fría. Una junta con coalescencia incompleta causada por no haber aplicado suficiente calor al metal base durante la soldadura.

*combustion. A rapid chemical reaction between oxygen and another substance that gives off heat and light.
combustión. Reacción química rápida entre el oxígeno y otra sustancia que emite calor y luz.

*combustion rate. Also known as rate of propagation of a flame, this is the speed at which the fuel gas burns, in ft/sec (m/sec). The ratio of fuel gas to oxygen affects the rate of burning: a higher percentage of oxygen increases the burn rate.
velocidad de combustión. También es conocida como velocidad de propagación de una llama, ésta es la velocidad en la cual se quema el gas combustible, en pies/sec (m/sec). La proporción del gas combustible al oxígeno afecta la proporción de quemadura: un porcentaje más alto del oxígeno aumenta la proporción de quemarse.

complete fusion. Fusion over the entire fusion faces and between all adjoining weld beads.
fusión completa. Fusión sobre todas las caras de fusión y en medio de todos los cordónes de soldadura inmediatos.

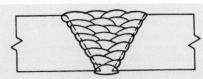

Multiple pass groove weld.

American Welding Society

composite electrode. A generic term for multicomponent filler metal electrodes in various physical forms, such as stranded wires, tubes, and covered wire. See also **covered electrode, flux cored electrode,** and **stranded electrode.**

electrodo compuesto. Un término genérico para componentes múltiples para electrodos de metal de aporte en varias formas físicas, como cable de alambre, tubos, y alambre cubierto. Vea también **electrodo cubierto, electrodo de núcleo de fundente,** y **electrodo cable.**

*computer-aided design (CAD). Computer software programs that typically use vector lines to produce a mechanical type drawing.

*diseño asistido por computadora (CAD). Programa de computadora que por lo general utiliza líneas vectoriales para producir un tipo mecánico de dibujo.

*computer-aided manufacturing (CAM). Computer software programs used to aid in the automated manufacturing of parts.

*fabricación asistida por computadora (CAM). Programa de computadora que se utiliza para asistir en la fabricación automatizada de piezas.

*computer control. Control involving one or more electronic digital computers.

control de computadora. Un control que incluye una o más computadoras electrónicas dactilares.

concave fillet weld. A fillet weld with a concave face.

soldadura de filete cóncava. Soldadura de filete con cara cóncava.

concave root surface. A root surface that is concave.

superficie raíz cóncava. La superficie del cordón raíz con cara cóncava.

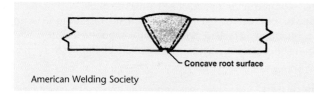

Concave root surface

American Welding Society

concavity. The maximum distance from the face of a concave fillet weld perpendicular to a line joining the toes.

concavidad. La distancia máxima de la cara de una soldadura de filete cóncava perpendicular a una linea que une con los pies.

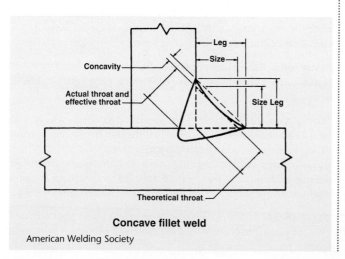

Concave fillet weld

American Welding Society

conduit liner. A flexible steel tube that guides the welding wire from the feed rollers through the welding lead to the gun used for GMAW and FCAW welding. The steel conduit liner may have a nylon or Teflon inner surface for use with soft metals such as aluminum.

revestimiento de conducto. Un tubo flexible de acero que guía el alambre para soldar desde los rodillos de alimentación, a través de los cables para soldar, hasta la pistola, usado en soldaduras de tipo GMAW y FCAW. El revestimiento del conducto de acero puede tener una superficie interior de Teflon o nylon para su uso con metales blandos como el aluminio.

cone. The conical part of an oxyfuel gas flame adjacent to the orifice of the tip.

cono. La parte cónica de la llama del gas de oxígeno combustible que colinda con la abertura de la punta.

constricted arc. A plasma arc column shaped by the constricting orifice in the nozzle of the plasma arc torch or plasma spraying gun.

arco constreñido. Una columna de arco plasma que está formada por el constreñimiento del orificio en la lanza de la antorcha del arco plasma o pistola de rociado plasma.

constricting nozzle. A device at the exit end of a plasma arc torch or plasma spraying gun containing the constricting orifice.

boquilla de constreñimiento. Un aparato a la salida de la antorcha de un arco plasma o la pistola de rociado plasma que contiene la boquilla de constreñimiento.

constricting orifice. The hole in the constricting nozzle of the plasma arc torch or plasma spraying gun through which the arc plasma passes.

orificio de constreñimiento. El agujero en la boquilla del constreñimiento en la antorcha de arco plasma o de la pistola de rociado plasma por donde pasa el arco de plasma.

consumable electrode. An electrode that provides filler metal.

electrodo consumible. Un electrodo que surte el metal de relleno.

consumable insert. Filler metal placed at the joint root before welding and intended to be completely fused into the root of the joint and become part of the weld.

inserción consumible. Metal de relleno antepuesto que se funde completamente en la raíz de la junta y se hace parte de la soldadura.

contact tube. A device that transfers current to a continuous electrode.

tubo de contacto. Un aparato que traslada corriente continua a un electrodo.

*contamination. Any undesirable material that might enter the molten weld metal.

*contaminación. Cualquier material no deseado que pueda ingresar dentro del metal de soldadura derretido.

continuous weld. A weld that extends continuously from one end of a joint to the other. Where the joint is essentially circular, it extends completely around the joint.

soldadura continua. Una soldadura que se extiende continuamente de una punta de la junta a la otra. Donde la junta es esencialmente circular, se extiende completamente alrededor de la junta.

convex fillet weld. A fillet weld with a convex face.

soldadura de filete convexa. Una soldadura de filete con una cara convexa.

convexity. The maximum distance from the face of a convex fillet weld perpendicular to a line joining the toes.

convexidad. La distancia máxima de la cara de la soldadura convexa filete perpendicula a la linea que une los pies.

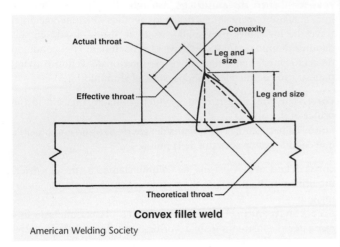

Convex fillet weld

American Welding Society

convex root surface. A root surface that is convex.

raíz superficie convexa. La raíz que es convexa.

copper. A pinkish or peach colored highly thermally and electrically conductive soft metal that resists corrosion.

cobre. Metal blando de color rosado o aduraznado resistente a la corrosión que es buen conductor del calor y de la electricidad.

copper alloys. An alloy primarily containing copper. The two most common copper alloys are brass, a copper tin alloy; and bronze, a copper zinc alloy.

aleación de cobre. Aleación que contiene principalmente cobre. Los dos tipos más comunes de aleaciones de cobre son el latón, un tipo de aleación de cobre y estaño; y el bronce, una aleación de cobre y zinc.

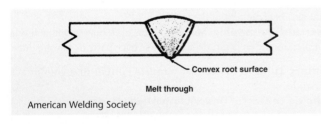

Convex root surface

Melt through

American Welding Society

*core wire. The wire portion of the coated electrode for shielded metal arc welding. The wire carries the welding current and adds most of the filler metal required in the finished weld. The composition of the core wire depends upon the metals to be welded.

alambre del centro. La porción del alambre del electrodo forrado para proteger el metal de la soldadura de arco. El alambre lleva la corriente de la soldadura y añade casi todo el metal para rellenar que es requerido para terminar la soldadura. La composición del alambre del centro depende de los metales que se van a usar para soldar.

cored solder. A solder wire or bar containing flux as a core.

soldadura de núcleo. Un alambre o barra para soldar que contiene fundente en el núcleo.

corner joint. A joint type in which butting or nonbutting ends of one or more workpieces converge approximately perpendicular to one another.

junta de esquina. Una junta dentro de dos miembros localizados aproximadamente a ángulos rectos de unos a otros.

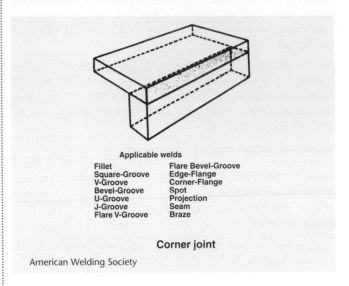

Applicable welds

Fillet	Flare Bevel-Groove
Square-Groove	Edge-Flange
V-Groove	Corner-Flange
Bevel-Groove	Spot
U-Groove	Projection
J-Groove	Seam
Flare V-Groove	Braze

Corner joint

American Welding Society

*corrosion resistance. The ability of the joint to withstand chemical attack; determined by the compatibility of the base materials to the filler metal.

resistencia a la corrosión. La abilidad de una junta de resistir ataques químicos; determinado por la compatibilidad de los materiales bases al metal de relleno.

corrosive flux. A flux with a residue that chemically attacks the base metal. It may be composed of inorganic salts and acids, organic salts and acids, or activated rosins or resins.

fundente corrosivo. Un fundente con un residuo que ataca químicamente al metal base. Puede estar compuesto de sales y ácidos inorgánicos, sales y ácidos orgánicos, o abelinotes o resinas activados.

cosmetic pass. A weld pass made primarily to enhance appearance.

pasada cosmética. Una pasada que se le hace a la soldadura para mejorar la apariencia.

*coupling distance. The distance to be maintained between the inner cones of the cutting flame and the surface of the metal being cut, in the range of 1/8 in. (3 mm) to 3/8 in. (10 mm).

distancia de acoplamiento. La distancia que debe mantenerse entre los conos internos de la llama y la superficie del metal que se está cortando, varía de 1/8 pulgadas (3 mm) a 3/8 pulgadas (10 mm).

cover lens. A round cover plate.

lente para cubrir. Un plato redondo de vidrio para cubrir el lente obscuro.

*cover pass. The last layer of weld beads on a multipass weld. The final bead should be uniform in width and reinforcement, not excessively wide, and free of any visual defects.

pasada para cubrir. La última capa de cordónes soldadura de pasadas múltiples. La pasada final debe ser uniforme en anchura y refuerzo, no excesivamente ancha, y libre de defectos visuales.

cover plate. A removable pane of colorless glass, plastic-coated glass, or plastic that covers the filter plate and protects it from weld spatter, pitting, and scratching.

plato para cubrir. Una hoja removible de vidrio claro, vidrio cubierto con plástico o plástico que cubre el plato filtrado y lo protege de salpicadura, picaduras y de que se rayen.

covered electrode. A composite filler metal electrode consisting of a bare or metal cored electrode with a flux covering to provide a slag layer and/or alloying elements. The covering may contain materials providing such functions as shielding from the atmosphere, deoxidation, and arc stabilization and can serve as a source of metallic additions to the weld.

electrodo cubierto. Un electrodo compuesto de metal para rellenar que consiste de un núcleo de un electrodo liso o electrodo con núcleo de metal el cual se le agrega cubrimiento suficiente para proveer una capa de escoria sobre el metal de la soldadura que se le aplicó. El cubierto puede contener materiales que pueden proveer funciones como protección de la atmósfera, deoxidación, y estabilización del arco, y también puede servir como fuente para añadir metales adicionales a la soldadura.

crack. A fracture-type discontinuity characterized by a sharp tip and high ratio of length and width to opening displacement.

grieta. Una desunión discontinuada de tipo fractura caracterizada por una punta filoza y proporción alta de lo largo y de lo ancho al desplazamiento de la abertura.

crater. A depression in the weld face at the termination of a weld bead.

crater. Una depresión en la superficie de la soldadura a donde se termina el cordón de soldadura.

crater crack. A crack initiated and localized within a crater.

grieta de crater. Una grieta en el crater del cordón de soldar.

*creep. A property of metal that allows it to be deformed under a load that is below the metal's yield point.

fluencia. Propiedad de los metales que permite su deformación bajo una carga menor al punto de deformación.

crevice corrosion. Oxidation that occurs in the small space (crevice) between two pieces of metal as the result of moisture being trapped in the small space.

corrosión de grieta. Oxidación que ocurre en espacios pequeños (grietas) entre dos piezas de metal como resultado de que la humedad quedara atrapada en el espacio pequeño.

critical weld. A weld so important to the soundness of the weldment that its failure could result in the loss or destruction of the weldment and injury or death.

soldadura crítica. Una soldadura tan importante para la calidad del conjunto de partes soldadas, que su fracaso podría ocasionar la pérdida o destrucción de dicho conjunto, así como también lesiones o muerte.

cross-sectional sequence. The order in which the weld passes of a multiple pass weld are made with respect to the cross section of the weld. See also **block sequence.**

secuencia del corte transversal. La orden en la cual se hacen las pasadas de la soldadura en una soldadura de pasadas múltiples hechas al respecto al corte transversal de la soldadura. Vea también **secuencia de bloque.**

crucible. A high-temperature container that holds the thermite welding mixture as it begins its thermal reaction before the molten metal is released into the mold.

crisol. Recipiente de alta temperatura que contiene la mezcla de la soldadura con termita en el momento en que comienza su reacción térmica antes de que el metal fundido se vierta en el molde.

*crystal lattices. See crystalline structures.

celosías cristalinas. Ver estructura espacial cristalina.

*crystalline structures. Orderly arrangements of atoms in a solid in a specific geometric pattern; sometimes called *lattices.*

estructura espacial cristalina. Un arreglo metódico de átomos en un modelo geométrico preciso. A veces se llaman *celosías.*

cup. A nonstandard term for gas nozzle.

tazón. Un término que no es la norma para de boquilla de gas.

cutting attachment. A device for converting an oxyfuel gas welding torch into an oxygen cutting torch.

equipo para cortar. Un aparato para convertir una antorcha para soldar en una antorcha para cortar con oxígeno.

*cutting gas assist. A nonreactive or exothermic gas jet directed on the metal to aid in laser cutting.

corte con asistencia de gas. Chorro de gas exotérmico no reactivo dirigido hacia el metal para asistir en cortes con láser.

cutting head. The part of a cutting machine in which a cutting torch or tip is incorporated.

cabeza de la antorcha para cortar. La parte de una maquina para cortar en donde una antorcha para cortar o una punta es incorporada.

*cutting plane line. Lines on a drawing that represent an imaginary cut through the object. They are used to expose the details of internal parts that would not be shown clearly with hidden lines.

Línea del plano de corte. Líneas de un dibujo que representan un corte imaginario a través del objeto. Se usan para exponer los detalles de las piezas internas que no se verían claramente con líneas escondidas.

cutting tip. The part of an oxygen cutting torch from which the gases issue.

punta para cortar. Esa parte de la antorcha para cortar con oxígeno por donde salen los gases.

cutting tip, high speed. Designed to provide higher oxygen pressure, thus allowing the torch to travel faster.

punta para cortar a alta velocidad. Diseñada para proveer presión más alta de oxígeno, asi puede caminar la antorcha más rápidamente.

cutting torch. A device used for plasma arc cutting to control the position of the electrode, to transfer current to the arc, and to direct the flow of plasma and shielding gas.

antorcha para cortar. Un aparato que se usa para cortes de arco de plasma para el control de la posición del electrodo.

*cycle time. The period of time from starting one machine operation to starting another (in a pattern of continuous repetition).

tiempo del ciclo. El período de tiempo de cuando se empieza la operación de una maquina y cuando se empieza otra (en una norma de repetición continua).

cylinder. A portable container used for transportation and storage of a compressed gas.

cilindro. Un recipiente portátil que se usa para transportar y guardar un gas comprimido.

cylinder manifold. A multiple header for interconnection of gas sources with distribution points.

conexión de cilindros múltiple. Una tuberia con conexiones múltiples que sirve como fuente de gas con puntos de distribución.

*cylinder pressure. The pressure at which a gas is stored in approximately 2200 lb per square inch (psi), and acetylene is stored at approximately 225 psi.

presión del cilindro. La presión del cilindro en el cual un gas se guarda en aproximadamente 2200 libras por pulgada cuadrada (psi), y el acetilino se guarda a aproximadamente 225 psi.

D

defect. A discontinuity or discontinuities that by nature or accumulated effect (for example, total crack length) render a part or product unable to meet minimum applicable acceptance standards or specifications. This term designates rejectability and flaw. See also **discontinuity** and **flaw.**

defecto. Una desunión o desuniónes que por la naturaleza o efectos acumulados (por ejemplo, distancia total de una grieta) hace que una parte o producto no esté de acuerdo con las normas o especificaciones mínimas para aceptarse. Este término designado resectabilidad y falta. Vea también **discontinuidad** y **falta.**

defective weld. A weld containing one or more defects.

soldadura defectuosa. Una soldadura que contiene uno o más defectos.

*deoxidizers. An element such as silicon that is added to a flux for the purpose of removing oxygen from the molten weld pool.

desoxidantes. Elemento como el silicio que se añade a un fundente con el propósito de extraer el oxígeno de la soldadura derretida.

deposited metal. Filler metal that has been added during a soldering, brazing, or welding operation.

metal depositado. Metal de relleno que se ha agregado durante una operación de soldadura.

deposition efficiency (arc welding). The ratio of the weight of deposited metal to the net weight of filler metal consumed, exclusive of stubs.

eficiencia de deposición (soldadura por arco). La relación del peso del metal depositado al peso neto del metal de relleno consumido, excluyendo los tacones.

deposition rate. The weight of material deposited in a unit of time. It is usually expressed as kilograms per hour (kg/hr) (pounds per hour [lb/h]).

relación de deposición. El peso del material depositado en una unidad de tiempo. Es regularmente expresado en kilogramos por hora (kg/hora) (libras por hora [lb/hora]).

depth of fusion. The distance that fusion extends into the base metal or previous bead from the surface melted during welding.

grueso de fusión. La distancia en que la fusión se extiende dentro del metal base o del cordón anterior de la superficie que se derretió durante la soldadura.

*destructive testing. Mechanical testing of weld specimens to measure strength and other properties. The tests are made on specimens that duplicate the material and weld procedures required for the job.

prueba destructiva. Pruebas mecánicas de probetas de soldadura para medir la fuerza y otras propiedades. Las pruebas se hacen en probetas que duplican el material y los procedimientos de la soldadura requeridos para el trabajo.

*dimension line. Lines on a drawing that are drawn so that their ends touch the object being measured, or they may touch the extension line extending from the object being measured.

línea de dimensión. Líneas de un dibujo que se trazan con el propósito de que sus extremos toquen el objeto que se mide. También pueden tocar la línea de extensión que se extiende desde el objeto que se mide.

*dimensioning. The measurements of an object, such as its length, width, and height, or the measurements for locating such things as parts, holes, and surfaces.

acotación. Las medidas de un objeto, tal como su longitud, ancho, y altura, o las medidas para ubicar cosas como piezas, agujeros o superficies.

dip brazing (DB). A brazing process that uses heat from a molten salt bath or a metal. When using the molten salt bath may act as flux. When using the molten metal, the bath provides the filler metal.

soldadura fuerte por inmersión (DB). Es un proceso de soldadura fuerte que usa el calor de una sal fundida o un baño de metal. Cuando se usa la sal fundida el baño puede actuar como flujo. Cuando se usa el metal fundido, el baño proporciona el metal de relleno.

dip soldering (DS). A soldering process using the heat, oil, or salt bath in which it is immersed.

soldadura blanda de bajo punto de fusión por inmersión (DS). Un proceso que usa calor proporcionado por un baño de metal derretido que provee el metal de relleno para soldar.

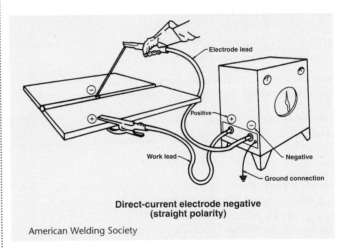

Direct-current electrode negative
(straight polarity)

American Welding Society

direct-current electrode negative (DCEN). The arrangement of direct-current arc welding leads on which the electrode is the negative pole and the workpiece is the positive pole of the welding arc.

corriente directa con electrodo negativo (DCEN). El arreglo de los cables para soldar con la soldadura de arco donde el electrodo es polo negativo y la pieza de trabajo es polo positivo de la soldadura de arco.

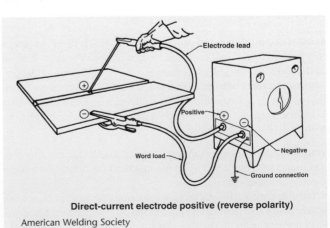

Direct-current electrode positive (reverse polarity)

American Welding Society

direct-current electrode positive (DCEP). The arrangement of direct-current arc welding leads on which the electrode is the positive pole and the workpiece is the negative pole of the welding arc.

corriente directa con el electrodo positivo (DCEP). El arreglo de los cables para soldar con la soldadura de arco con el electrodo es el polo positivo y la pieza de trabajo es el polo negativo de la soldadura de arco.

discontinuity. An interruption of the typical structure of a material, such as a lack of homogeneity in its mechanical, metallurgical, or physical characteristics. A discontinuity is not necessarily a defect. See also **defect** and **flaw.**

discontinuidad. Una interrupción de la estructura típica de un material, el que falta de homogenidad en sus caracteristicas mecánicas, metalúrgicas, o fisica. Vea también **defecto** y **falta.**

*distortion. Movement or warping of parts being welded, from the prewelding position and condition compared to the postwelding condition and position.

deformacion. Movimiento o torcimiento de las partes que se están soldando, comparando la posición antes de soldar a la posicion despues de soldar.

double bevel-groove weld. A type of groove weld. Refer to drawing for **groove weld.**

soldadura de ranura con doble bisel. Es un tipo de soldadura de ranura. Refiérase al dibujo para **soldadura de ranura.**

double-flare bevel-groove weld. A type of groove weld. Refer to drawing for **groove weld.**

soldadura de ranura con doble bisel acampanada. Un tipo de soldadura de ranura. Refiérase al dibujo para **soldadura de ranura.**

double-flare V-groove weld. A weld in grooves formed by two members with curved surfaces. Refer to drawing for **groove weld.**

soldadura de ranura de doble V acampanada. Una soldadura de ranura formada por dos miembros con superficies curvados. Refiérase al dibujo para **soldadura de ranura.**

double J-groove weld. A type of groove weld. Refer to drawing for **groove weld.**

soldadura de ranura de doble-J-. Un tipo de soldadura de ranura. Refiérase al dibujo para **soldadura de ranura.**

double U-groove weld. A type of groove weld. Refer to drawing for **groove weld.**

soldadura de ranura de doble-U-. Un tipo de soldadura de ranura. Refiérase al dibujo para **soldadura de ranura.**

double V-groove weld. A type of groove weld. Refer to drawing for **groove weld.**

soldadura de ranura de doble-V-. Un tipo de soldadura de ranura. Refiérase al dibujo para **soldadura de ranura.**

downslope time. See **automatic arc welding downslope time** and **resistance welding downslope time.**

tiempo de caída del pendiente. Vea **tiempo del pendiente con descenso en soldadura de arco automático** y **tiempo del pendiente descenso en soldadura de resistencia.**

drag (thermal cutting). The offset distance between the actual and straight-line exit points of the gas stream or cutting beam measured on the exit surface of the base metal.

tiro (corte termal). La distancia desalineada entre la actual y la linea recta del punto de salida del chorro de gas o el rayo de cortar medido a la salida de la superficie del metal base.

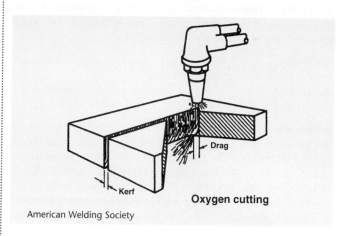

American Welding Society

Drag — Kerf — **Oxygen cutting**

drag angle. The travel angle when the electrode is pointing in a direction opposite to the progression of welding. This angle can also be used to partially define the position of guns, torches, rods, and beams.

ángulo del tiro. El ángulo de avance cuando el electrodo está apuntando en una dirección opuesta del progreso de la soldadura. Este ángulo también se puede usar para parcialmente definir la posición de pistolas, antorchas, varillas, y rayos.

*drag lines. High-pressure oxygen flow during cutting forms lines on the cut faces. A correctly made cut has up and down drag lines (zero drag); any deviation from the pattern indicates a change in one of the variables affecting the cutting process; with experience the welder can interpret the drag lines to determine how to correct the cut by adjusting one or more variables.

lineas del tiro. La salida del oxígeno a presión elevada durante el corte forma lineas en las caras del corte. Un corte hecho correctamente tiene lineas hacia arriba y hacia abajo (zero tiro); cualquier desviación de la norma indica un cambio en uno de los variables que afectan el proceso de cortar; con experiencia el soldador puede interpretar las lineas de tiro y determinar como corregir el corte ajustando uno o más variables.

*drift. The tendency of a system's response to gradually move away from the desired response.

deriva. La tendencia de la respuesta del sistema de retirarse gradualmente de la respuesta deseada.

*drooping output. Volt-ampere characteristic of the shielded metal arc process power supply where the voltage output decreases as increasing current is required of the power supply. This characteristic provides a reasonably high voltage at a constant current.

reducción de potencia de salida. Característica de voltio-amperios de la alimentación de poder de un proceso de soldadura de arco protegido donde la salida del voltaje disminuye mientras un aumento de la corriente es requerida de la alimentación de poder. Está característica proporciona un voltaje razonable alto con corriente constante.

*dross. The material expelled from the plasma arc and oxygen assist laser cutting processes, which contains 40% or more of unoxidized base metal.

escoria. Material expelido en los procesos de corte de arco de plasma y de corte láser asistido con oxígeno que contiene 40% o más de base de metal no oxidada.

*dual shield. FCA welding process that uses both the flux core and an external shielding gas to protect the molten weld pool. See self-shielding.

protección doble. Proceso de soldadura FCA que utiliza un fundente de núcleo y protección externa de gas para proteger la soldadura derretida. Ver **autoprotección**.

*ductility. As applied to a soldered or brazed joint, it is the ability of the joint to bend without failing.

ductilidad. Como aplicada a junta de soldadura fuerte o soldadura blanda, es la abilidad de la junta de doblarse sin fallar.

duty cycle. The percentage of time during a specified test period that a power source or its accessories can be operated at rated output without overheating. The test periods for arc welding and resistance welding are ten (10) minutes and one (1) minute, respectively.

ciclo de trabajo. El porcentaje de tiempo durante un período a prueba arbitraria de una fuente de poder y sus accesorios que pueden operarse a la capacidad de carga de salida sin sobrecalentarse.

E

*eddy current inspection (ET). See nondestructive testing.

inspección de corriente parásita (ET). Ver **pruebas no destructivas**.

edge joint. A joint type in which the nonbutting ends of one or more workpieces lie approximately parallel.

junta de orilla. Una junta en medio de las orillas de dos o más miembros paralelos o casi paralelos.

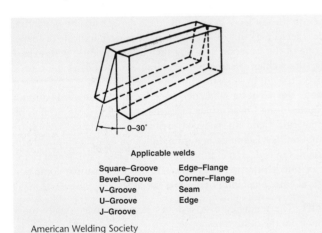

0–30°

Applicable welds

Square–Groove	Edge–Flange
Bevel–Groove	Corner–Flange
V–Groove	Seam
U–Groove	Edge
J–Groove	

American Welding Society

edge preparation. The surface prepared on the edge of a member for welding.

preparación de orilla. La superficie preparada en la orilla de un miembro que se va a soldar.

effective length of weld. The length of weld throughout which the correctly proportioned cross section exists. In a curved weld, it shall be measured along the axis of the weld.

distancia efectiva de soldadura. La distancia de una sección transversa correctamente proporcionada que existe por toda la soldadura. En una soldadura en curva, debe medirse por el axis de la soldadura.

effective throat. The minimum distance from the root of a weld to its face, less any reinforcement. See also **joint penetration**. Refer to drawing for **convexity**.

garganta efectiva. La distancia mínima de la raíz a la cara de una soldadura, menos el refuerzo. Vea también **penetración de junta**. Refiérase al dibujo para **convexidad**.

*elastic limit. The maximum force that can be applied to a material or joint without causing permanent deformation or failure.

limite elástico. La fuerza máxima que se le puede aplicar a un material o junta sin causar deformación o falta permanente.

*elbow. The joint that connects a robot's upper arm and forearm.

codo. La junta que conecta al brazo de arriba con el brazo de enfrente en un robot.

electrode. A component of the secondary circuit terminating at the arc, molten conductive slag, or base metal.

electrodo. Un componente del circuito eléctrico que termina al arco, escoria derretida conductiva, o metal base.

*electrode angle. The angle between the electrode and the surface of the metal; also known as the direction of travel (leading angle or trailing angle); leading angle pushes molten metal and slag ahead of the weld; trailing angle pushes the molten metal away from the leading edge of the molten weld pool toward the back, where it solidifies.

ángulo del electrodo. El ángulo en medio del electrodo y la superficie del metal; también conocido como la dirección de avance (apuntado hacia adelante o apuntado hacia atras); el ángulo apuntado empuja el metal derretido y la escoria enfrente de la soldadura; y el ángulo apuntado hacia atrás empuja el metal derretido lejos de la orilla delantera del charco del metal derretido hacia atrás, donde se solidifica.

*electrode classification. Any of several systems developed to identify shielded metal arc welding electrodes. The most widely used identification system was developed by the American Welding Society (AWS). The information represented by the classification generally includes the minimum tensile strength of a good weld, the position(s) in which the electrode can be used, the type of flux coating, and the type(s) of welding currents with which the electrode can be used.

clasificación de electrodo. Cualquiera de los varios sistemas desarollados para identificar electrodos protegidos para soldadura de arco. El sistema de identificación que se usa mucho más fue desarrollado por la Sociedad de Soldadura Americana (AWS). La información representada por la clasificación generalmente incluye la fuerza tensible mínima de una soldadura, la posición(es) donde se puede usar el electrodo, el tipo de recubrimiento de fundente y los tipo(s) de corrientes para soldar con la cual se puede usar el electrodo.

electrode extension (GMAW, FCAW, SAW). The length of unmelted electrode extending beyond the end of the contact tube during welding.

extensión del electrodo (GMAW, FCAW, SAW). La distancia de extensión del electrodo que no está derretido más allá de la punta del tubo de contacto durante la soldadura.

electrode holder. A device used for mechanically holding and conducting current to an electrode or electrode adapter.

porta electrodo. Un aparato usado para detener mecánicamente y conducir corriente a un electrodo durante la soldadura.

electrode lead. The secondary circuit conductor transmitting energy from the power supply to the electrode holder, gun, or torch. Refer to drawing for **direct-current electrode negative**.

cable de electrodo. Un conductor eléctrico en medio de la fuente para la corriente de soldar con arco y el portaelectrodo. Refiérase al dibujo de **corriente directa con electrodo negativo**.

electrode setback. The distance the electrode is recessed behind the constricting orifice of the plasma arc torch or thermal spraying gun, measured from the outer face of the nozzle.

retroceso del electrodo. La distancia del hueco del electrodo que está detrás del orificio constringente de la antorcha de arco plasma o pistola de rocio termal, se mide de la cara de afuera a la boquilla.

*electrode tip. The end of an electrode where the arc jumps from the electrode to the work.

punta del electrodo. Extremo de un electrodo donde el arco salta desde el electrodo hasta la zona de la pieza de trabajo.

electron beam cutting (EBC). A thermal cutting process severing metals by melting them with the heat from a concentrated beam composed primarily of high-velocity electrons, impacting upon the workpieces.

cortes a rayo de electron (EBC). Un proceso de cortar que usa calor obtenido de un rayo concentrado primeramente de electrones de alta velocidad que choca sobre la pieza de trabajo la cual se va a cortar; puede o no usar un gas surtido externamente.

electron beam welding (EBW). A welding process that produces coalescence with a concentrated beam, composed primarily of high-velocity electrons, impinging on the joint. The process is used without shielding gas and without the application of pressure.

soldadura a rayo de electron (EBW). Un proceso de soldadura la cual produce coalescencia de un rayo concentrado, compuesto primeramente de electrones de alta velocidad al chocar con la junta. Este proceso no usa gas de protección y sin la aplicación de presión.

*electrons. Small particle of an atom whose movement is associated with electrical flow.

electrones. Partículas pequeñas de un átomo cuyo movimiento está asociado con el flujo eléctrico.

electroslag welding electrode. A filler metal component of the welding circuit through which current is conducted from the electrode guiding member to the molten slag.

electrodo para soldadura de electroescoria. Un componente de metal de relleno del circuito para soldar por donde la corriente es conducida del miembro que guía el electrodo a la escoria derretida.

emissive electrode. A filler metal electrode consisting of a core of a bare electrode or a composite electrode to which a very light coating has been applied to produce a stable arc.

electrodo emisivo. Un electrodo de metal de relleno consistiendo de un electrodo liso o un electrodo compuesto de una capa ligera que se le aplica para producir un arco estable.

*end effector. An actuator, a gripper, or a mechanical device attached to the wrist of a manipulator by which objects can be grasped or acted upon.

punta que efectúa. Un movedor, el que agarra, o un aparato mecánico fijo a la muñeca de un manipulador por el cual objetos se pueden agarrar u obrar en impulso.

entry-level welder. A person just entering the welding profession.

soldador principiante. Una persona que acaba de comenzar en la profesión de la soldadura.

*error signal. The difference between desired response and actual response.

señal equivocada. La diferencia entre la respuesta deseada y la respuesta efectiva.

*etching. The process of chemically preparing the surface of a specimen so that the metal's grain can be examined.

decapado. Proceso para preparar químicamente la superficie de un espécimen de modo que el grano del metal pueda ser examinado.

*eutectic composition. The composition of an alloy that has the lowest possible melting point for that mixture of metals.

composición de tipo eutectico. La composición de un aleado que tenga el punto de fusión lo más bajo posible para esa mezcla de metales.

*evacuated (vacuum) chamber. A device that is used to reduce or remove the atmosphere around a part(s) being welded to prevent weld contamination.

cámara de evacuación (vacío). Dispositivo que se usa para reducir o extraer el aire que rodea la(s) pieza(s) que se suelda(n) para prevenir la contaminación.

*exhaust pickup. A component of a forced ventilation system that has sufficient suction to pick up fumes, ozone, and smoke from the welding area and carry the fumes, etc., outside of the area.

recogedor de extracción. Un componente de un sistema de ventilación forzada que tiene suficiente succión para recoger vaho, ozono, y humo de la área de soldadura y lleva al vaho, etc., a fuera de la area.

*exothermic gases. Gases that react with the material being cut and produce additional heat.

gases exotérmicos. Gases que reaccionan con el material que se corta y producen calor adicional.

*extension line. Lines in a drawing that extend from an object which locate the points being dimensioned.

*línea de extensión. Líneas de un dibujo que se extienden desde un objeto y que ubican los puntos que se miden.

F

*fabrication. An assembly whose parts may be joined by a combination of methods, including welds, bolts, screws, adhesives, etc.

conjunto. Un conjunto cuyas partes pueden estar unidas por una combinación de métodos que incluyen soldaduras, pernos, tornillos, adhesivos, etc.

face-bend test. A test in which the weld face is on the convex surface of a specified bend radius.

prueba de dobléz de cara. Una prueba donde la cara de la soldadura está en la superficie convexa al radio de dobléz especificado.

face-centered cubic (FCC). Crystal form of iron above 1675°F (913°C) and called austenite.

red cúbica centrada en las caras (FCC). Forma cristalina del hierro a más de 1675°F (913°C) denominada austenita.

face of weld. The exposed surface of a weld on the side from which welding was done.

cara de la soldadura. La superficie expuesta de una soldadura del lado de donde se hizo la soldadura.

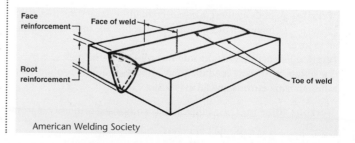

American Welding Society

face reinforcement. Reinforcement of a weld at the side of the joint from which welding was done. See also **root reinforcement**. Refer to drawing for **face of weld**.

refuerzo de cara. Refuerzo de una soldadura en el lado de la junta de donde se hizo la soldadura. Vea también **refuerzo de raíz**. Refiérase al dibujo para **cara de la soldadura**.

face shield. A device positioned in front of the eyes and a portion of, or all of, the face, whose predominant function is protection of the eyes and face. See also **helmet**.

protector de cara sostenido a mano. Un aparato puesto en frente de los ojos y una porción, o en toda la cara, cuya función predominante es de proteger los ojos y la cara. Vea también **casco**.

*fast-freezing electrode. An electrode whose flux forms a high-temperature slag that solidifies before the weld metal solidifies, thus holding the molten metal in place. This is an advantage for vertical, horizontal, and overhead welding positions.

electrodo de congelación rápida. Un electrodo cuyo flujo forma una escoria a temperaturas altas que se puede solidificar antes de que el metal de soldadura se pueda solidificar, asi detiene el metal derretido en su lugar. Está es una ventaja en soldaduras de posiciones vertical, horizontal y sobrecabeza.

*fatigue failures. The failure of a part that is subjected to repeated forces below the normal static breaking point.

fallas por fatiga. Falla de una pieza que está sujeta a fuerzas repetitivas por debajo del punto de ruptura estático normal.

*fatigue resistance. As applied to a soldered or brazed joint, it is the ability of the joint to be bent repeatedly without exceeding its elastic limit and without failure. Generally, fatigue resistance is low for most soldered and brazed joints.

resistencia a la fatiga. Como aplicada a una junta de soldadura con metales de bajo punto de fusión y soldadura fuerte, es la abilidad de una junta de ser doblada repetidamente sin exceder los limites elásticos y sin fracaso. Generalmente, la resistencia a la fatiga es muy baja para la mayoría de las juntas de soldadura con bajo punto de fusión y soldadura fuerte.

faying surface. The mating surface of a workpiece in contact with or in close proximity to another workpiece to which it is to be joined.

superficie de unión. La superficie de apareamiento de un miembro del que está en contacto con otro miembro o está en proximidad cercana a otro miembro que está para ser unido.

feed rollers. A set of two or four individual rollers that, when pressed tightly against the filler wire and powered up, feed the wire through the conduit liner to the gun for GMAW and FCAW welding.

rodillos de alimentación. Un conjunto de dos o cuatro rodillos individuales que al ser presionados fuertemente contra el alambre de relleno y ser accionados alimentan al alambre a través del revestimiento de canal hasta la pistola, en soldaduras tipo GMAW y FCAW.

*ferrite. A body-centered cubic crystalline structure that iron forms as it changes from a liquid to a solid. See also **crystalline structures**.

ferrita. Estructura cristalina de red cúbica centrada en el cuerpo que el hierro forma a medida que cambia de líquido a sólido. Ver también **estructura espacial cristalina**.

*ferrous filler metals. Filler metals comprised primarily of iron (steel).

metal de aporte ferroso. Metal de aporte compuesto principalmente de hierro (acero).

filler metals. The metals or alloys to be added in making a brazed, soldered, or welded joint.

metales de aporte. Los metales o aleados que se agregan cuando se hace una soldadura blanda o soldadura fuerte.

*filler pass. One or more weld beads used to fill the groove with weld metal. The bead must be cleaned after each pass to prevent slag inclusions.

pasada para rellenar. Uno o más cordones de soldadura usados para llenar la ranura con el metal de soldadura. El cordón debe ser limpiado después de cada pasada para prevenir inclusiones de escoria.

fillet weld. A weld of approximately triangular cross section joining two surfaces approximately at right angles to each other in a lap joint, tee joint, or corner joint. Refer to drawing for **convexity**.

soldadura de filete. Una soldadura de filete de sección transversa aproximadamente triangular que une dos superficies aproximadamente en ángulos rectos de uno al otro en junta de traslape, junta en-T-o junta de esquina. Refiérase al dibujo para **convexidad**.

fillet weld break test. A test in which the specimen is loaded so that the weld root is in tension.

prueba de rotura en soldadura de filete. Una prueba en donde la probeta es cargada de manera en que la tensión esté sobre la soldadura.

fillet weld leg. The distance from the joint root to the toe of the fillet weld.

pierna de soldadura filete. La distancia de la raíz de la junta al pie de la soldadura filete.

fillet weld size. For equal leg fillet welds, the leg lengths of the largest isosceles right triangle that can be inscribed within the fillet weld cross section. For unequal leg fillet welds, the leg lengths of the largest right triangle that can be inscribed within the fillet weld cross section.

tamaño de soldadura filete. Para soldaduras de filete que tienen piernas iguales, lo largo de las piernas del isósceles más grande del triángulo recto que puede ser inscribido dentro de la sección. Para soldaduras de filete con piernas desiguales, lo largo de las piernas del triángulo recto más grande puede inscribirse dentro de la sección transversal.

filter plate. An optical material that protects the eyes against excessive ultraviolet, infrared, and visible radiation.

lente filtrante. Un material óptico que protege los ojos contra ultravioleta excesiva, infrarrojo, y radiación visible.

final current. The current after downslope but prior to current shutoff.

corriente final. La corriente después del pendiente en descenso pero antes de que la corriente sea cerrada.

fisheye. A discontinuity, attributed to the presence of hydrogen in the weld, observed on the fracture surface of a weld in steel that consists of a small pore or inclusion surrounded by an approximately round, bright area.

ojo de pescado. Una discontinuidad que se encuentra en una fractura de superficie en una soldadura de acero que consiste de poros pequeños o inclusiones rodeadas de áreas aproximadamente redondas y brillantes.

*fissure. A small, crack-like discontinuity with only slight separation (opening displacement) of the fracture surfaces. The prefixes *macro* and *micro* indicate relative size.

hendemiento. Una pequeña, discontinuidad como una grieta con solamente una separación (abertura desalojada) de las superficies fracturadas. El prefijo marco y micro indica el tamaño relativo.

*fixed inclined (6G) position. For pipe welding, the pipe is fixed at a 45° angle to the work surface. The effective welding angle changes as the weld progresses around the pipe.

posición (6G) inclinado fijo. Para soldadura de tubo, el tubo se fija a un ángulo de 45° de la superficie del trabajo. El ángulo efectivo de la soldadura cambia cuando la soldadura progresa alrededor del tubo.

fixture. A device designed to maintain the fit of a workpiece(s) in proper relationship.

fijación. Una devisa diseñada para detener partes que se van a unir en relación propia de una a la otra.

*flame propagation rate. The speed at which flame travels through a mixture of gases.

cantidad de propagación de la llama. La rapidez en que la llama camina a través de una mezcla de gas.

flame spraying (FLSP). A thermal spraying process in which an oxyfuel gas flame is the source of heat for melting the surfacing material. Compressed gas may or may not be used for atomizing the propellant and surfacing material to the substrate.

rociado a llama (FLSP). Un proceso de rociado termal en donde la llama del gas oxicombustible es la fuente de calor para derretir el material de revestimiento. El gas comprimido se puede o no se puede usar para automizar el propulsor y el material de revestimiento al substrato.

flange weld. A weld made on the edges of two or more joint members, at least one of which is flanged.

soldadura de reborde. Una soldadura que se hace en las orillas de dos o más miembros de junta, donde por lo menos uno tiene reborde.

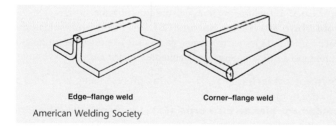

Edge–flange weld   Corner–flange weld

American Welding Society

flash. The material that is expelled or squeezed out of a weld joint and that forms around the weld.

ráfaga. El material que es despedido o exprimido fuera de una junta de soldadura y se forma alrededor de la soldadura.

*flash burn. An arc-caused burn typically on the eye that results from an extremely brief exposure to the direct ultraviolet light produced by an arc welding process.

quemadura de relámpago. Quemadura causada por arco, por lo general en el ojo, y que resulta de la exposición extremadamente breve a la luz ultravioleta directa producida por un proceso de soldadora con arco.

*flash glasses. Lightly tinted safety glasses, usually a number two shade, worn by welders to protect themselves from flash burns as well as flying debris.

anteojos de relámpago. Anteojos de seguridad sombreados, por lo general con sombra n°2, que los soldadores usan para protegerse de las quemaduras de relámpago y fragmentos de soldadura.

flash welding (FW). A resistance welding process producing a weld at the faying surfaces of a butting member by the rapid upsetting of the workpiece after a controlled period of flashing action.

soldadura de relámpago (FW). Un proceso de soldadura de resistencia que produce una soldadura en el empalme de la superficie de una junta tope por una acción de relampagueo y por la aplicación de presión después que el calentamiento este substancialmente acabado. La acción del relampagueo, causado por densidades de corrientes altas a unos puntos de contacto pequeños en medio de las piezas de trabajo, despiden fuertemente el material de la junta cuando las piezas de trabajo se mueven despacio. La soldadura es terminada por un acortamiento rápido de las piezas de trabajo.

flashback. A recession of the flame through the torch and hose, regulator, and/or cylinder, potentially causing an explosion.

llamarada de retroceso. Una recesión de la llama adentro o atrás de la cámara mezcladora de una antorcha de gas oxicombustible o pistola de rociar a llama.

flashback arrester. A device to limit damage from a flashback by preventing propagation of the flame front beyond the point at which the arrester is installed.

válvula de retención. Un aparato para limitar el daño de una llamarada de retroceso para prevenir la propagación del frente de la llama más allá del punto donde se instala la válvula de retención.

flat position. The position used to weld from the upper side of the joint resulting in the face of the weld being oriented approximately horizontal.

posición plana. La posición de soldadura que se usa para soldar del lado de arriba de una junta; la cara de la soldadura está aproximadamente horizontal.

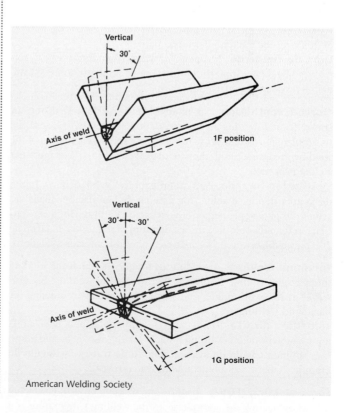

American Welding Society

flaw. An undesirable discontinuity.
falta. Una discontinuidad indeseable.

flow rate. The rate at which a given volume of shielding gas is delivered to the weld zone. The units used for welding are cubic feet, inches, meters, and centimeters.
caudal. Velocidad a la cual llega un determinado volumen de gas protector a la zona de soldadura. Las unidades usadas para la soldadura son pies cúbicos, pulgadas, metros, y centímetros.

flux. A material applied to the workpiece(s) before or during joining or surfacing to cause interactions that remove oxides and other contaminants, improve wetting, and affect the final surface profile. Welding flux may also affect the weld metal chemical composition.
flujo. Un material que se usa para impedir o prevenir la formación de óxidos y otras substancias indeseables en el metal derretido y en las superficies del metal sólido, y para desolver o de otra manera facilitar el removimiento de dichas substancias.

flux cored arc welding (FCAW). An arc welding process using an arc between a continuous filler metal electrode and the weld pool. The process is used with shielding gas from a flux contained within the tubular electrode, with or without additional shielding from an externally supplied gas, and without the application of pressure.
soldadura de arco con núcleo de fundente (FCAW). Un proceso de soldadura de arco que usa un arco entre medio de un electrodo de metal rellenado continuo y el charco de la soldadura. El proceso es usado con gas de protección del flujo contenido dentro del electrodo tubular, y sin usarse protección adicional de abastecimiento de gas externo, y sin aplicarse presión.

flux cored electrode. A composite tubular filler metal electrode consisting of a metal sheath and a core of various powdered materials, producing an extensive slag cover on the face of a weld bead.
electrodo de núcleo de fundente. Un electrodo tubular de metal para rellenar con una compostura que consiste de una envoltura de metal y un núcleo con varios materiales de polvo, que producen un forro extensivo de escoria en la superficie del cordón de soldadura. Protección externa puede ser requerida.

flux cover. In metal bath dip brazing and dip soldering, a cover of flux over the molten filler metal bath.
tapa de fundente. En metal de baño soldadura fuerte y soldadura blanda por inmersión, una tapa de fundente sobre el baño del metal de relleno derretido.

*forced ventilation. To remove excessive fumes, ozone, or smoke from a welding area, a ventilation system may be required to supplement natural ventilation. Where forced ventilation of the welding area is required, the rate of 200 cu ft (56 m^3) or more per welder is needed.
ventilación forzada. Para quitar excesivo vaho, ozono y humo de la área donde se solda, un sistema de ventilación puede ser requerido para suplementar la ventilación natural. Donde la ventilación forzada de la área de la soldadura es requerida, la rázon de 200 pies cúbicos (56 m^3) o más es requerido por cada soldador.

forehand welding. A welding technique in which the welding torch or gun is directed toward the progress of welding. See also **travel angle**, **work angle**, and **push angle**. Refer to drawing for **backhand welding**.
soldadura directa. Una técnica de soldar en cual la pistola o la antorcha para soldar es dirigida hacia al progreso de la soldadura. Vea también **ángulo de avance**, **ángulo de trabajo**, y **ángulo de empuje**. Refiérase al dibujo **soldadura en revés**.

forge welding (FOW). A solid state welding process producing a weld by heating the workpieces to the welding temperature and applying sufficient blows to cause permanent deformation at the faying surfaces.
soldadura por forjado. Un proceso de soldadura de estado sólido que produce una soldadura calentando las piezas de trabajo a una temperatura de soldadura y aplicando golpes suficientes para causar una deformación permanente en las superficies del empalme.

*frequency. As it relates to alternating current, this refers to the rate that the current reverses direction.
frecuencia. En relación a corriente alterna. Velocidad a la que la corriente revierte su polaridad.

*frogs. The large rail structures, usually castings, that form the center of a crossing or the rail crossing point of a switch.
corazónes de cruzamiento. Las grandes estructuras de rieles, normalmente piezas moldeadas, que forman el centro de un cruzamiento o el punto de cruzamiento de rieles en el sitio de convergencia.

fuel gases. A gas, when mixed with air or oxygen and ignited, producing heat for cutting, joining, or thermal spraying.
gases combustibles. Gases como acetileno, gas natural, hidrógeno, propano, metilacetileno propodieno estabilizado, y otros combustibles normalmente usados con oxígeno en uno de los procesos de oxicombustible y para calentar.

*full face shield. Personal protective device to protect the eyes and face from flying debris; it may be clear or tinted.
protector facial completo. Dispositivo de protección personal que protege los ojos y el rostro de fragmentos volantes; puede ser transparente o sombreado.

full fillet weld. A fillet weld whose size is equal to the thickness of the thinner member joined.
soldadura de filete llena. Una soldadura de filete cuyo tamaño es igual de grueso como el miembro más delgado de la junta.

full penetration. A nonstandard term for complete joint penetration.
penetración llena. Un término fuera de la norma en vez de la penetración de junta.

furnace brazing (FB). A brazing process in which assemblies are heated to the brazing temperature in a furnace.
soldadura fuerte en horno (FB). Un proceso de soldadura fuerte en donde las partes que se van a unir se ponen en un horno calentado a una temperatura adecuada.

furnace soldering (FS). A soldering process using heat from a furnace or oven.
soldadura blanda en horno (FS). Un proceso de soldadura blanda en donde las partes que se van a unir se ponen en un horno calentado a una temperatura adecuada.

fusion. The melting together of filler metal and base metal, or of base metal only, to produce a weld. See also **depth of fusion**.
fusión. El derretir el metal de relleno y el metal base juntos o el metal base solamente, para producir una soldadura. Vea **también grueso de fusión**.

fusion welding. Any welding process or method that uses fusion to complete the weld.
soldadura de fusión. Cualquier proceso de soldadura o método que usa fusión para completar la soldadura.

fusion zone. The area of base metal melted as determined on the cross section of a weld. Refer to drawing for **depth of fusion**.
zona de fusión. La área del metal base que se derritió como determinada en la sección transversa de la soldadura. Refiérase al dibujo **grueso de fusion**.

G

gap. A nonstandard term when used for arc length, joint clearance, and root opening.
abertura. Un término fuera de norma cuando se usa en lugar del arco, despejo de junta, y abertura de raíz.

gas cup. A nonstandard term for gas nozzle.
tazón de gas. Un término fuera de norma en vez de boquilla de gas.

gas cylinder. A portable container used for transportation and storage of compressed gas.
cilindro de gas. Un recipiente portátil que se usa para transportación y deposito de gas comprimido.

*****gas laser.** A laser in which the lasing medium is a gas.
láser de gas. Tipo de láser cuyo medio de transmisión es un gas.

gas lens. One or more fine mesh screens located in the torch nozzle to produce a stable stream of shielding gas. Primarily used for gas tungsten arc welding.
lente para gas. Uno o más cedazos de malla fina localizados en la lanza de la antorcha para producir un chorro estable de gas de protección primeramente usada para soldaduras de arco tungsteno y gas.

gas metal arc cutting (GMAC). An arc cutting process employing a continuous consumable electrode and a shielding gas.
cortes de arco metálico con gas (GMAC). Un proceso de corte con arco que usa un alambre consumible continuo y un gas de protección.

gas metal arc welding (GMAW). An arc welding process using an arc between a continuous filler metal electrode and the weld pool. The process is used with shielding from an externally supplied gas and without the application of pressure.
soldadura de arco metálico con gas (GMAW). Un proceso de soldar con arco que usa un arco en medio de un electrodo de metal para rellenar continuo y el charco de soldadura. El proceso usa protección de un abastecedor externo de gas y sin la aplicación de presión.

gas metal arc welding-pulsed arc (GMAW-P). A gas metal arc welding process variation in which the current is pulsed.
soldadura con arco metálico con gas arco pulsado (GMAW-P). Un proceso de soldadura de arco metálico con gas con variación en cual la corriente es de pulsación.

gas metal arc welding-short-circuit arc (GMAW-S). A gas metal arc welding process variation in which the consumable electrode is deposited during repeated short circuits. Sometimes this process is referred to as MIG or CO_2 welding (nonpreferred terms).
soldadura con arco metálico con gas-arco de corto circuito (GMAW-S). Un proceso de soldadura de arco metálico con gas variación en cual el electrodo consumible es depositado durante los cortos circuitos repetidos. A veces el proceso es referido como soldadura MIG o CO_2 (términos que no son preferidos).

gas nozzle. A device at the exit end of the torch or gun that directs shielding gas.
boquilla de gas. Un aparato a la salida de la punta de la antorcha o pistola que dirige el gas protector.

gas regulator. A device for controlling the delivery of gas at some substantially constant pressure.
regulador de gas. Un aparato para controlar la salida de gas a una presión substancialmente constante.

gas tungsten arc cutting (GTAC). An arc cutting process employing a single tungsten electrode with gas shielding.

corte de arco con tungsteno y gas (GTAC). Un proceso de corte de arco que usa un electrodo de tungsteno sencillo con gas de protección.

gas tungsten arc welding (GTAW). An arc welding process using an arc between a tungsten electrode (nonconsumable) and the weld pool. The process is used with shielding gas and without the application of pressure.
soldadura de arco de tungsteno con gas (GTAW). Un proceso de soldadura de arco que usa un arco en medio del electrodo tungsteno (no consumible) y el charco de la soldadura. El proceso es usado con gas de protección y sin aplicación de presión.

gas tungsten arc welding-pulsed arc (GTAW-P). A gas tungsten arc welding process variation in which the current is pulsed.
soldadura de arco de tungsteno con gas-arco pulsado (GTAW-P). Un proceso de soldadura de arco tungsteno con variación en cual la corriente es de pulsación.

*****gauge** (regulator). A device mounted on a regulator to indicate the pressure of the gas passing into the gauge. A regulator is provided with two gauges—one (high-pressure gauge) indicates the pressure of the gas in the cylinder; the second gauge (low-pressure gauge) shows the pressure of the gas at the torch.
manómetro (regulador). Un aparato montado en un regulador para indicar la presión del gas que está pasando por el manómetro. El regulador tiene dos manómetros—uno (manómetro de alta presión) indica la presión del gas en el cilindro; el segundo manómetro (manómetro de presión baja) enseña la presión del gas en la antorcha.

*****gauge pressure.** The actual pressure shown on the gauge; does not take into account atmospheric pressure.
manómetro para presión. La presión actual que se enseña en el manómetro; no toma en cuenta la presión atmosférica.

*****GFCI.** Ground fault circuit interrupters are fast-acting circuit breakers that shut off the power to an electrical circuit when they detect a small imbalance in the circuit's electrical flow.
ICFCT. Interruptor del circuito de fallos de conexión a tierra. Tipo de interruptor de circuito de acción rápida que corta el suministro eléctrico a un circuito cuando detecta un pequeño desequilibrio en el flujo eléctrico del circuito.

*****globular transfer.** The transfer of molten metal in large drops from a consumable electrode across the arc.
traslado globular. El traslado del metal derretido en gotas grandes de un electrodo consumible a través del arco.

*****goggles.** Personal protective device to protect the eyes from flying debris; may be clear or tinted.
gafas. Equipo de protección personal para proteger los ojos de fragmentos volantes; pueden ser transparentes o sombreadas.

gouging. The forming of a bevel or groove by material removal. See also **back gouging, arc gouging,** and **oxygen gouging.**
escopleando con gubia. Formando un bisel o ranura removiendo el material. Vea también **gubia trasera, gubia dura con arco,** y **escopleando con la gubia con oxígeno.**

*****grain refinement.** Is the process that occurs when larger metallic grain structures break down in size. Grain refinement can be associated with a change in temperature, mechanical working, or time.
refinamiento por grano. Proceso que ocurre cuando estructuras metálicas de granos mayores se rompen. El refinamiento por grano puede estar asociado con un cambio en temperatura, procesos mecánicos, o tiempo.

*****graphite.** A form of carbon.
grafito. Una de las formas del carbón.

groove. An opening or a channel in the surface of a part or between two components, that provides space to contain a weld.
ranura. Una abertura o un canal en la superficie de una parte o en medio de dos componentes, la cual provee espacio para contener una soldadura.

groove angle. The total included angle of the groove between parts to be joined by a groove weld.
ángulo de ranura. El ángulo total incluido de la ranura entre partes para unirse por una soldadura de ranura.

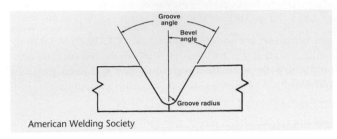

American Welding Society

groove face. The surface of a joint member included in the groove.
cara de ranura. La superficie de un miembro de una junta incluido en la ranura.

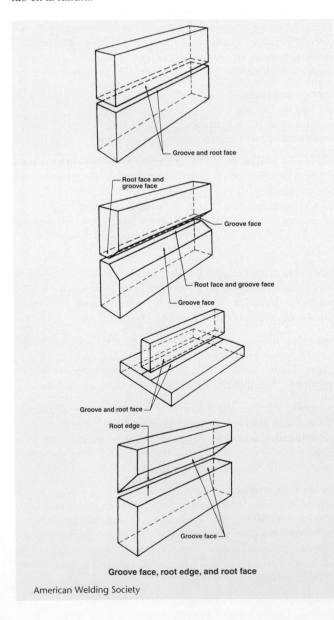

Groove face, root edge, and root face

American Welding Society

groove radius. The radius used to form the shape of a J- or U-groove weld joint. Refer to drawings for **bevel**.
radio de ranura. La radio que se usa para formar una junta de una soldadura con una ranura de forma U o J. Refiérase a los dibujos para **bisel**.

groove weld. A weld made in the groove between two members to be joined. The standard types of groove welds are shown in the drawings.
soldadura de ranura. Una soldadura hecha en la ranura dentro de dos miembros que se unen. Los tipos normales de soldadura de ranura se ven en los dibujos.

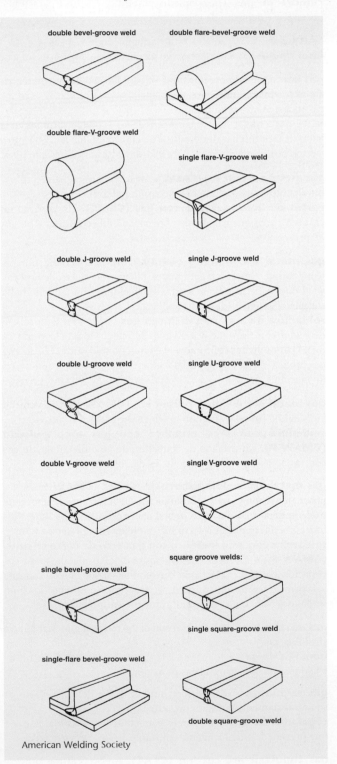

American Welding Society

ground connection. An electrical connection of the welding machine frame to the earth for safety.

conexión a tierra. Una conexión eléctrica del marco de la máquina de soldar a la tierra para seguridad.

ground lead. A nonstandard and incorrect term for workpiece lead.

cable de tierra. Un término fuera de norma e incorrecto que se usa en vez de cable de pieza de trabajo.

*group technology. A system for coding parts based on similarities in geometrical shape or other characteristics of parts. The grouping of parts into families based on similarities in their production so that parts of a particular family could be processed together.

codificador. Un sistema para codificar partes basadas en similaridades en forma geométrica y otras características de las partes. La agrupación de partes en familias basadas en similaridades en su producción para que partes de una familia en particular puedan ser procesadas juntas.

*guided-bend specimen. Any bend specimen that will be bend-tested in a fixture that controls the bend radii, such as the AWS bend-test fixture.

probeta de dobléz guiada. Cualquier probeta de dobléz en la cual se va a hacer un dobléz guiado en una máquina que controla el radio del dobléz, como la máquina de dobléz guiado del AWS.

H

hardfacing. A surfacing variation in which surfacing material is deposited to reduce wear.

endurecimiento de caras. Una variación superficial donde el material superficial es depositado para reducir el desgastamiento.

heat-affected zone. The portion of the base metal whose mechanical properties or microstructure has been altered by the heat of welding, brazing, soldering, or thermal cutting.

zona afectada por el calor. La porción del metal base cuya propiedad mecánica o microestructura ha sido alterada por el calor de soldadura, soldadura fuerte, soldadura blanda, o corte termal.

*heat treatments. Postweld heat treatment to reduce weld stresses is the most common type of heat treatment used on weldments.

tratamiento con calor. Tratamiento térmico posterior a la soldadura para reducir la tensión de la soldadura. Es el tipo más común de tratamiento térmico que se utiliza en soldaduras.

heating torch. A device for directing the heating flame produced by the controlled combustion of fuel gases.

antorcha de calentamiento. Un aparato para dirigir la llama de calentamiento que es producida por una combustión controlada de gases de combustión.

helmet. A device designed to be worn on the head to protect eyes, face, and neck from arc radiation, radiated heat, spatter, or other harmful matter expelled during arc welding, arc cutting, and thermal spraying.

casco. Un aparato diseñado para usarse sobre la cabeza para proteger ojos, cara y cuello de radiación del arco, calor radiado, salpicadura, u otra materia dañosa despedida durante la soldadura de arco, corte por arco, y rociado termal.

*hidden line. Lines on a drawing that show the same features as object lines except that the corners, edges, and curved surfaces cannot be seen because they are hidden behind the surface of the object.

línea oculta. Líneas de un dibujo que muestran las mismas características como líneas de objetos, excepto que las esquinas, los bordes y las superficies curvas no se pueden visualizar porque están ocultas detrás de la superficie del objeto.

*high-frequency alternating current. Electric current that changes polarity at a rate higher than 3 million cycles a second (3 MHz).

corriente alterna de alta frecuencia. Corriente eléctrica que cambia la polaridad a una velocidad de 3 millones de ciclos por segundo (3 MHz).

horizontal fixed position (pipe welding). The position of a pipe joint in which the axis of the pipe is approximately horizontal and the pipe is not rotated during welding.

posición fija horizontal (soldadura de tubos). La posición de una junta de tubo la cual el axis del tubo es aproximadamente horizontal, y el tubo no da vueltas durante la soldadura.

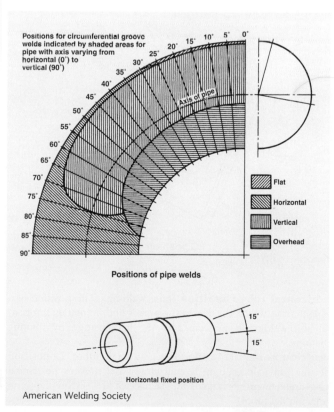

Positions of pipe welds

American Welding Society

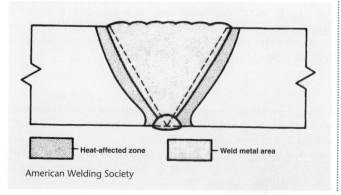

■ Heat-affected zone   ■ Weld metal area

American Welding Society

*horizontal fixed (5G) position weld. For pipe welding, the pipe is fixed horizontally (cannot be rolled). The weld progresses from overhead, to vertical, to flat position around the pipe.

soldadura de posición fija horizontal (5G). Para soldadura de tubos, el tubo está fijo horizontalmente (no se pueder rodar). La soldadura progresa de sobre cabeza, a vertical, a la posición plana alrededor del tubo.

horizontal position (fillet weld). The position in which welding is performed on the upper side of an approximately horizontal surface and against an approximately vertical surface.

posición horizontal (soldadura de filete). La posición de la soldadura la cual es hecha en el lado de arriba de una superficie horizontal aproximadamente y junto a una superficie vertical aproximadamente.

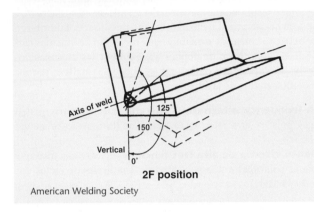

2F position

American Welding Society

horizontal position (groove weld). The position of welding in which the weld axis lies in an approximately horizontal plane and the weld face lies in an approximately vertical plane.

posición horizontal (de ranura). La posición para soldar en la cual el axis de la soldadura está en una plana horizontal aproximadamente, y la cara de la soldadura está en una plana vertical aproximadamente.

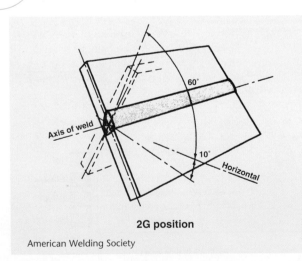

2G position

American Welding Society

horizontal rolled position (pipe welding). The position of a pipe joint in which the axis of the pipe is approximately horizontal and welding is performed in the flat position by rotating the pipe.

posición horizontal rodada (soldadura de tubo). La posición de una junta de tubo en la cual el axis del tubo es horizontal aproximadamente, y la soldadura se hace en la posición plana con rotación del tubo.

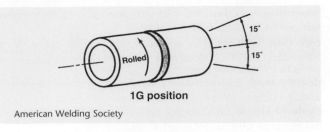

1G position

American Welding Society

*horizontal rolled (1G) position. For pipe welding, this position yields high-quality and high-quantity welds. Pipe to be welded is placed horizontally on the welding table in a fixture to hold it steady and permit each rolling. The weld proceeds in steps, with the pipe being rolled between each step, until the weld is complete. For plate, see **axis of a weld**.

posición (1G) horizontal rodada. Para soldadura de tubo, está posición produce soldaduras de alta calidad y alta cantidad. El tubo que se va a soldar se pone horizontalmente sobre la mesa de soldadura en una instalación fija que lo detiene seguro y permite cada rodadura. La soldadura procede en pasos, con el tubo siendo rodado entre cada paso, hasta que la soldadura esté completa. Para plato, vea **eje de la soldadura**.

hot crack. A crack occurring in a metal during solidification or at elevated temperatures. Hot cracks can occur in both heat-affected (HAZ) and weld metal (WMZ) zones.

grieta caliente. Una grieta formada a temperaturas cerca de la terminación de la solidificación.

*hot pass. The welding electrode is passed over the root pass at a higher-than-normal amperage setting and travel rate to reshape an irregular bead and turn out trapped slag. A small amount of metal is deposited during the hot pass so the weld bead is convex, promoting easier cleaning.

pasada caliente. El electrodo de soldadura se pasa sobre la pasada de raíz poniendo el amperaje más alto que lo normal y proporción de avance para reformar un cordón irregular y sacar la escoria atrapada. Una cantidad pequeña de metal es depositada durante la pasada caliente para que el cordón soldado sea convexo, promoviendo más fácil la limpieza.

hot start current. A very brief current pulse at arc initiation to stabilize the arc quickly. Refer to drawing for **upslope time**.

corriente caliente para empezar. Un pulso muy breve de corriente a iniciación de arco para estabilizar el arco aprisa. Refiérase al dibujo **tiempo del pendiente en ascenso**.

hydrogen embrittlement. The delayed cracking in steel that may occur hours, days, or weeks following welding. It is a result of hydrogen atoms that dissolved in the molten weld pool during welding.

fragilidad causada por el hidrógeno. El fisuramiento retardado en el acero que puede ocurrir horas, días o semanas después de la soldadura. Es el resultado de la disolución de átomos de hidrógeno en el charco de soldadura derretido durante la soldadura.

I

inclined position. The position of a pipe joint in which the axis of the pipe is at an angle of approximately 45° to the horizontal, and the pipe is not rotated during welding.

posición inclinada. Una posición de junta de tubo en la cual el axis del tubo está a un ángulo de aproximadamente 45° a la horizontal, y no se le da vueltas al tubo durante la soldadura.

inclined position, with restriction ring. The position of a pipe joint in which the axis of the pipe is at an angle of approximately 45° to the horizontal, and a restriction ring is located near the joint. The pipe is not rotated during welding.

posición inclinada con argolla de restricción. La posición de una junta de tubo en la cual el axis del tubo está a un ángulo de aproximadamente 45° a la horizontal, y una argolla de restricción está localizada cerca de la junta. No se le da vuelta al tubo durante la soldadura.

included angle. A nonstandard term for groove angle.

ángulo incluido. Un término fuera de norma para ángulo de ranura.

inclusion. Entrapped foreign solid material, such as slag, flux, tungsten, or oxide.

inclusión. Material extraño atrapado sólido, como escoria, flujo, tungsteno, u óxido.

incomplete fusion. A weld discontinuity in which fusion did not occur between weld metal and fusion faces or adjoining weld beads.

fusión incompleta. Una discontinuidad en la soldadura en la cual no ocurrió fusión entre el metal soldado y caras de fusión o cordones soldados inmediatos.

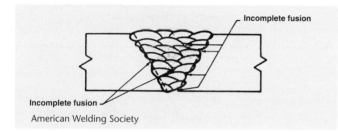

Incomplete fusion

Incomplete fusion

American Welding Society

incomplete joint penetration. Joint penetration that is unintentionally less than the thickness of the weld joint.

penetración de junta incompleta. Penetracion de la junta que no es intencionalmente menos de lo grueso de la junta de soldar.

*induction. The transfer of heat obtained from the resistance of the work pieces to the flow of induced high frequency welding current.

inducción. Transferencia del calor obtenido de la resistencia de las piezas de trabajo al flujo de corriente inducida de soldadura de alta frecuencia.

induction brazing (IB). A brazing process using heat from the resistance of the assembly to induced electric current.

soldadura fuerte por inducción (IB). Un proceso de soldadura blanda que usa calor de la resistencia de las piezas de trabajo para inducir la corriente eléctrica.

induction soldering (IS). A soldering process in which the heat required is obtained from the resistance of the workpieces to induced electric current.

soldadura blanda por inducción (IS). Un proceso de soldadura blanda en el cual el calor requerido es obtenido de la resistencia de las piezas de trabajo a la corriente eléctrica inducida.

inert gas. A gas that does not react chemically with materials.

gas inerte. Un gas que normalmente no se combina químicamente con materiales.

*inertia welding. A welding process in which one workpiece revolves rapidly and one is stationary. At a predetermined speed the power is cut, the rotating part is thrust against the stationary part, and frictional heating occurs. The weld bond is formed when rotation stops.

soldadura inercial. Un proceso en el cual una pieza de trabajo voltea rápidamente y la otra está fija. A una velocidad predeterminada se le corta la potencia, la parte que está volteando es empujada en contra de la parte fija, y el calor de fricción ocurre. La unión soldada es formada cuando se para la rotación.

*infrared light. Light that has a wavelength longer than visible light with a wavelength ranging approximately from 1 THz to 430 THz.

luz infrarroja. Luz que tiene una longitud de onda mayor que la luz visible. Las longitudes de onda varían entre 1 THz a 430 THz.

infrared radiation. Electromagnetic energy with wavelengths from 770 to 12,000 nanómeters.

radiación infrarrojo. Energia electromagnética con longitud de ondas de 770 a 12,000 nanómetros.

*injector chamber. One method of completely mixing the fuel gas and oxygen to form a flame. High-pressure oxygen is passed through a narrowed opening (venturi) to the mixing chamber. This action creates a vacuum, which pulls the fuel gas into the chamber and ensures thorough mixing. Used for equal gas pressures and is particularly useful for low-pressure fuel gases.

cámara de inyector. Un método de mezclar completamente el gas de combustión y el oxígeno para formar una llama. Oxígeno a alta presión es pasado por una abertura angosta (venturi) a la cámara de mezcla. Está acción hace un vacuo, la cual estira el gas combustible para dentro de la cámara y asegura una mezcla completa. Usada para presiones de gas que son iguales y es particularmente útil para gases combustibles de presión baja.

*inner cone. The portion of the oxyacetylene flame closest to the welding tip. The primary combustion reaction occurs in the inner cone. The size and color of the cone serve as indicators of the type of flame (carburizing, oxidizing, neutral).

cono interno. La porción de la llama de oxiacetileno más cerca de la punta para soldar. La reacción de combustión principal ocurre en el cono interno. El tamaño y el calor del cono sirve como indicadores del tipo de la llama (carburante, oxidante, neutral).

*intelligent robot. A robot that can be programmed to make performance choices contingent on sensory inputs.

robot inteligente. Un robot que puede ser programado para hacer preferencias contigentes en sensorios de entrada.

*interface. A shared boundary. An interface might be a mechanical or an electrical connection between two devices; it might be a portion of computer storage accessed by two or more programs; or it might be a device for communication to or from a human operator.

interface. Un limite repartido. Un interface puede ser una conexión mecánica o eléctrica entre dos aparatos; puede ser una porción de deposito con accesión a dos o más programas; o puede ser un aparato para comunicarse a o con un operador humano.

interpass temperature. In a multipass weld, the temperature of the weld area between weld passes.

temperatura de pasada interna. En una soldadura de pasadas multiples, la temperatura en la área de la soldadura entre pasadas de soldaduras.

*ionized gas. A gas that is heated to a point where it becomes conductive. See **plasma.**
gas ionizado. Gas que pasa a ser conductor una vez que se lo calienta a cierto punto. Ver **plasma.**

*iron. An element. Very seldom used in its pure form. The most common element alloyed with iron is carbon.
hierro. Un elemento. Es muy raro que se use en forma pura. El elemento más común del aleado con hierro es carbón.

J

J-groove weld. A type of groove weld.
soldadura con ranura-J. Es un tipo de soldadura de ranura.

joint. The junction of the workpiece(s) that are to be joined or have been joined.
junta. El punto en que se unen dos miembros o las orillas de los miembros que están para unirse o han sido unidos.

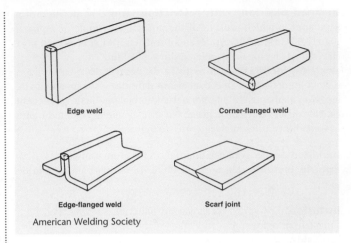

Edge weld Corner-flanged weld

Edge-flanged weld Scarf joint

American Welding Society

joint buildup sequence. The order in which the weld beads of a multiple pass weld are deposited with respect to the cross section of the joint.
secuencia de formación de una junta. La orden en la cual los cordones de soldadura en una soldadura de pasadas múltiples son depositadas con respecto a la sección transversa de la junta.

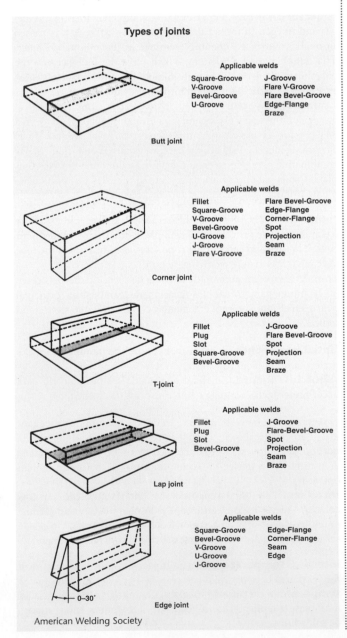

Types of joints

Applicable welds

Square-Groove	J-Groove
V-Groove	Flare V-Groove
Bevel-Groove	Flare Bevel-Groove
U-Groove	Edge-Flange
	Braze

Butt joint

Applicable welds

Fillet	Flare Bevel-Groove
Square-Groove	Edge-Flange
V-Groove	Corner-Flange
Bevel-Groove	Spot
U-Groove	Projection
J-Groove	Seam
Flare V-Groove	Braze

Corner joint

Applicable welds

Fillet	J-Groove
Plug	Flare Bevel-Groove
Slot	Spot
Square-Groove	Projection
Bevel-Groove	Seam
	Braze

T-joint

Applicable welds

Fillet	J-Groove
Plug	Flare-Bevel-Groove
Slot	Spot
Bevel-Groove	Projection
	Seam
	Braze

Lap joint

Applicable welds

Square-Groove	Edge-Flange
Bevel-Groove	Corner-Flange
V-Groove	Seam
U-Groove	Edge
J-Groove	

0–30°

Edge joint

American Welding Society

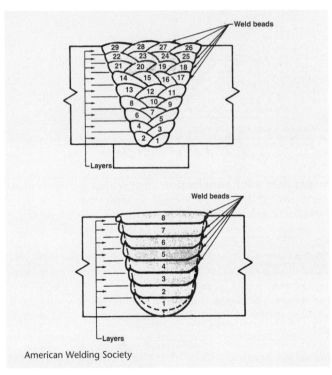

Weld beads

Layers

Weld beads

Layers

American Welding Society

joint clearance. The distance between the faying surfaces of a joint.
despejo de junta. La distancia entre las superficies del empalme de una junta.

joint design. The joint geometry together with the required dimensions of the welded joint.
diseño de junta. La geometría de la junta junto con las dimensiones requeridas de la junta de la soldadura.

joint efficiency. The ratio of the strength of a joint to the strength of the base metal.
eficiencia de junta. La razón de la fuerza de una junta a la fuerza del metal base, expresada en por ciento.

joint geometry. The shape, dimensions, and configuration of a joint prior to welding.

geometría de junta. La figura y dimensión de una junta en sección transversa antes de soldarse.

joint penetration. The distance the weld metal extends from its face into a joint, exclusive of weld reinforcement.

penetración de junta. La distancia del metal soldado que se extiende de su cara hacia adentro de la junta, exclusiva de la soldadura de refuerzo.

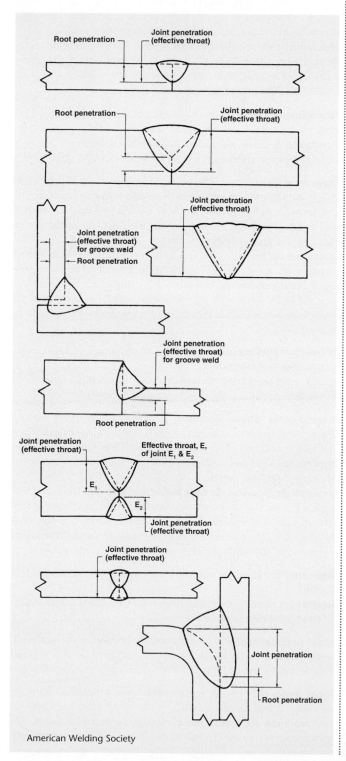

American Welding Society

joint root. The portion of a joint to be welded where the members approach closest to each other. In cross section, the joint root may be either a point, a line, or an area.

raíz de junta. Esa porción de una junta que está para soldarse donde los miembros están más cercanos uno del otro. En la sección transversa, la raíz de la junta puede ser una punta, una línea, o una área.

joint type. A weld joint classification based on the five basic arrangements of the component parts such as a butt joint, corner joint, edge joint, lap joint, and tee joint.

tipo de junta. Una clasificación de una junta de soldadura basada en los cinco arreglos del componente de partes como junta a tope, junta en esquina, junta de orilla, junta de solape, y junta en T.

joint welding sequence. See preferred term **joint buildup sequence.**

secuencia para soldar una junta. Vea el término preferido **secuencia de formación de una junta.**

*joules. SI unit of heat.

joules. Unidad de calor del SI.

K

kerf. The width of the cut produced during a cutting process. Refer to drawing for **drag.**

cortadura. La anchura del corte producido durante un proceso de cortar. Refiérase al dibujo de **tiro.**

keyhole welding. A technique in which a concentrated heat source penetrates completely through a workpiece, forming a hole at the leading edge of the weld pool. As the heat source progresses, the molten metal fills in behind the hole to form the weld bead.

soldadura con pocillo. Una técnica en la cual una fuente de calor concentrado se penetra completamente a través de la pieza de trabajo, formando un agujero en la orilla del frente del charco de la soldadura. Asi como progresa la potencia de calor, el metal derretido rellena detrás del agujero para formar un cordón de soldadura.

*kindling point. The lowest temperature at which a material will burn.

punto de ignición. La temperatura más baja la cual un material se puede quemar.

L

lack of fusion. A nonstandard term for incomplete fusion.

falta de fusión. Un término fuera de norma para fusión incompleta.

lack of penetration. A nonstandard term for incomplete joint penetration.

falta de penetración. Un término fuera de norma para penetración de junta incompleta.

lamellar tear. A subsurface terrace and steplike crack in the base metal with a basic orientation parallel to the wrought surface caused by tensile stresses in the through-thickness direction of the base metals weakened by the presence of small, dispersed, planar-shaped, nonmetallic inclusions parallel to the metal surface.

rasgadura laminar. Una terraza subsuperficie y una grieta como un escalón en el metal base con una orientación paralela a la superficie forjada. Es causada por tensión en la dirección de lo grueso-continuo de los metales de base debilitados por la presencia de pequeños, dispersados, formados como plano, inclusiones no metálicas paralelas a la superficie del metal.

land. See preferred term root face.
hombro. Vea el término preferido **cara de raíz.**

lap joint. A joint in which the nonbutting ends of one or more workpieces overlap approximately parallel to one another.
junta de solape. Una junta entre dos miembros traslapadas.

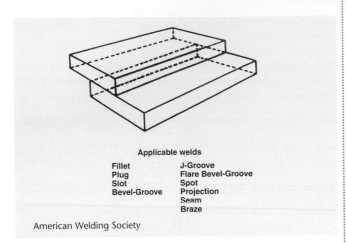

Applicable welds

Fillet	J-Groove
Plug	Flare Bevel-Groove
Slot	Spot
Bevel-Groove	Projection
	Seam
	Braze

American Welding Society

laser beam cutting (LBC). A thermal cutting process severing metal by locally melting or vaporizing it with the heat from a laser beam. The process is used with or without assist gas to aid the removal of molten and vaporized material.
cortes con rayo laser (LBC). Un proceso de cortes termal que separa al metal vaporizado o derretido localmente con el calor de un rayo laser. El proceso es usado sin gas que asiste a remover el material vaporizado o derretido.

*laser beam drilling. A thermal cutting process used to produce holes in metal as accurately as if they had been drilled.
perforación por haz de láser. Proceso de corte térmico usado para producir hoyos de alta precisión en metales como si hubieran sido perforados con taladro.

laser beam welding (LBW). A welding process producing coalescence with the heat from a laser beam impinging on the joint.
soldadura con rayo laser (LBW). Un proceso de soldar que produce coalescencia con calor de un rayo laser al golpear contra la junta.

lattice. An orderly geometric pattern of atoms within a solid metal. The lattice structure is responsible for many of the mechanical properties of the metal.
enrejado. Es una forma geométrica bien arreglada de átomos dentro de un metal sólido. La estructura de enrejado es responsable por muchas propiedades mecánicas del metal.

layer. A stratum of weld metal or surfacing material. The layer may consist of one or more weld beads laid side by side. Refer to drawing for **joint buildup sequence.**
capa. Un estrato de metal de soldadura o material de superficie. La capa puede consistir de uno o más cordones de soldadura depositados o puestos de lado. Refiérase al dibujo **secuencia de formación de una junta.**

*leaders and arrows. Leaders are the straight part and arrows are the pointed end that points to a part to identify it, show the location, and/or are the basis of a welding symbol.
guías y flechas. Las guías son la parte recta y las flechas son el extremo con el que se apunta.

*leak-detecting solution. A solution, usually soapy water, that is brushed or sprayed on the hose fittings at the regulator and torch to detect gas leaks. If a small leak exists, soap bubbles form.
solución para descubrir escape. Una solución, por lo regular de agua enjabonada, que se acepilla o se rocía sobre las conexiones de las mangueras y los reguladores y antorcha para detectar escape de gas. Si existe un escape pequeño, se forman burbujas de jabón.

leg of a fillet weld. See fillet weld leg.
pierna de soldadura filete. Vea pierna de soldadura filete.

lightly coated electrode, *shielded metal arc welding.* A filler metal electrode consisting of a metal wire with a light coating applied subsequent to the drawing operation, primarily for stabilizing the arc. This is an obsolete or seldom used term.
electrodo con recubrimiento ligero. Un electrodo de metal de aporte consistiendo de un alambre de metal con un recubrimiento ligero aplicado subsecuente a la operación del dibujo, principalmente para estabilizar el arco.

*lime-based flux. These alkaline fluxes are commonly used on both SMA and FCA welding electrodes.
fundente a base de cal. Estos fundentes alcalinos se usan comúnmente en electrodos SMA y FCA.

*line drop. The difference between the pressure at the low-pressure gauge and the pressure at the torch; results from the resistance to gas flow offered by the hose and how it is affected by the diameter and length of the hose. The smaller the hose diameter, or the longer the hose, the greater is the line drop.
descenso de línea. La diferencia de la presión en el manómetro de baja presión y la presión en la antorcha; resultados de la resistencia de la corriente del gas causada por la manguera, y como es afectada por el diámetro, y lo largo de la manguera. Si el diámetro de la manguera es más chica o es más larga la manguera, más grande es el descenso.

*liquefied fuel gases. A gas that is stored under adequate pressure so that it is a liquid.
gases combustibles licuados. Gas almacenado que toma forma líquida debido a la presión de almacenamiento.

*liquid-solid phase bonding process. Soldering or brazing where the filler metal is melted (liquid) and the base material does not melt (solid); the phase is the state at which bonding takes place between the solid base material and liquid filler metal. There is no alloying of the base metal.
proceso de ligación de fase líquido-sólido. Soldando con soldadura blanda o soldadura fuerte donde el metal de relleno se derrite (líquido) y el material base no se derrite (sólido); la fase es el estado la cual el ligamento se lleva a cabo entre el material base sólido y el metal de relleno (líquido). No se mezcla con el metal base.

liquidus. The lowest temperature at which a metal is completely liquid.
liquidus. La temperatura más baja en la cual un metal o un aleado es completamente líquido.

local preheating. Preheating a specific portion of a structure.
precalentamiento local. El precalentamiento de una porción especificada de un estructura.

local stress relief heat treatment. Stress relief heat treatment of a specific portion of a structure.
tratamiento de calor para relevar la tensión local. Un tratamiento de calor el cual releva la tensión de una porción especificada de una estructura.

longitudinal sequence. The order in which the weld passes of a continuous weld are made in respect to its length. See also **backstep sequence.**

secuencia longitudinal. La orden en que las pasadas de un soldadura continua son hechas en respecto a su longitud. Vea también **secuencia a la inversa.**

*low-fuming alloys. As it relates to brazing filler rods it indicates that there is enough deoxidizers in the alloy to reduce the problem of zinc forming oxides during torch brazing.

aleación poco humeante. En relación a barras de relleno de soldadura fuerte. Indica que hay suficientes desoxidantes en la aleación para reducir el problema de que el zinc forme óxidos durante la soldadura fuerte con antorcha.

M

*machine operation. Welding operations are performed automatically under the observation and correction of the operator.

operación de máquina. Operaciones de soldadura son ejecutadas automáticamente bajo la observación y corrección del operador.

machine welding. Welding with equipment that performs the welding operation under the constant observation and control of a welding operator. The equipment may or may not perform the loading and unloading of the work. See also **automatic welding.**

máquina para soldadura. Soldadura con equipo que ejecutan la operación de soldadura bajo la observación constante de un operador de soldadura. El equipo pueda o no ejecutar el cargar o descargar del trabajo. Vea también **soldadura automática.**

*macro structure. A structure large enough to be seen with the naked eye or low magnification, usually under 30 power.

estructura macro. Una estructura suficientemente grande que puede verse con el puro ojo o con un amplificador de aumento bajo, regularmente abajo de poder 30.

macroetch test. A test in which a specimen is prepared with a fine finish, etched, and examined under low magnification.

prueba con grabado al agua fuerte y examinado por magnificación. Una prueba en una probeta preparada con acabado fino, grabada al agua fuerte, y examinado debajo de un amplificador de aumento bajo.

*magnetic flux lines. Parallel lines of force that always go from the north pole to the south pole in a magnet, and surround a DC current–carrying wire.

líneas magnéticas de flujo. Líneas paralelas de fuerza que siempre van del polo norte al polo sur en un magneto, y rodea un alambre que lleva corriente DC.

*malleable cast iron. See **cast iron.**

acero vaciado maleable. Ver **acero vaciado.**

manifold. A multiple header for interconnection of gas or fluid sources with distribution points.

conexión múltiple. Una tuberia con conexiones múltiples que sirve como fuente de gas o fluído con puntos de distribución.

*manifold system. Used when there are a number of workstations or a high volume of gas is required. A piping system that allows several oxygen and fuel-gas cylinders to be connected to several welding stations. Normally regulators are provided at the manifold and at the stations to provide control of the oxygen and fuel-gas pressures. Safety features such as reverse flow valves, flashback arrestors, and back pressure release must be provided at the manifold.

sistema de conexiones múltiples. Usado cuando hay un número de estaciones de trabajo o cuando se requiere un alto volumen de gas. Un sistema de tubos que permite que se conecten varios cilindros de oxígeno y gas combustible a varias estaciones de soldadura. Normalmente se usan reguladores en el tubo de conexiones múltiples y en las estaciones para mantener el control de la presión del oxígeno y el gas combustible. Normas de seguridad como válvulas de retención, protector de agua contra retroceso de llama, y escape de presión deben usarse en el tubo de conexiones múltiples.

*manipulator. A mechanism, usually consisting of a series of segments, joined or sliding relative to one another for the purpose of grasping and moving objects, usually in several degrees of freedom. It may be remotely controlled by a computer or by a human.

manipulador. Un mecanismo, que consiste regularmente de una serie de segmentos unidos o corredizos con relación del uno al otro con el propósito de que agarre y mueva objetos, por lo regular en varios grados de libertad. Puede ser controlado remotamente por una computadora o un humano.

*manual operation. The entire welding process is manipulated by the welding operator.

operación manual. Todo el proceso de soldadura es manipulado por un operador de soldadura.

manual welding. Welding with the torch, gun, or electrode holder held and manipulated by hand. Variations of this term are *manual brazing, manual soldering, manual thermal cutting,* and *manual thermal spraying.*

soldadura manual. Soldando con la antorcha, pistola, porta electrodo detenido y manipulado por la mano. Variaciones de este término son *soldadura fuerte manual, soldadura blanda manual, cortes termal manual,* y *rociado termal manual.*

*MAPP. One manufacturer's trade name for a specific stabilized, liquefied MPS mixture. MAPP® has a distinctive odor, which makes it easy to detect; used for welding and cutting. See also **methylacetylene-propadiene (MPS).**

MAPP. Un nombre comercial de un fabricante para una específica estabilizada, licuada, mezcla MPS. MAPP tiene un olor distintivo, el cual es muy fácil de descubrir; es usado para cortes y soldaduras. Vea también **metilacetileno y propadieno.**

*martensite. A very hard and brittle solid-solution phase that is found in medium and high carbon steels.

martensita. Una solución sólida con un aspecto muy duro y quebradizo que se encuentra en aceros medianos y de alto carbón.

*mechanical testing (DT). See **destructive testing.**

prueba mecánica (DT). Ver **prueba destructiva.**

meltback time. The time interval at the end of crater fill time to arc outage during which electrode feed is stopped. Arc voltage and arc length increase and current decreases to zero to prevent the electrode from freezing in the weld deposit.

tiempo de refundición. El tiempo de intervalo al fin del tiempo en que se llena el crater hasta que se apaga el arco durante el cual el alimento del electrodo se detiene. El voltaje del arco y lo largo del arco aumenta y la corriente empieza a desminuir hasta llegar a cero para prevenir la congelación del electrodo en el deposito de la soldadura.

melting range. The temperature range between solidus and liquidus.

variación de derretimiento. La variación de temperatura entre solidus y liquidus.

melting rate. The weight or length of electrode, wire, rod, or powder, melted in a unit of time.
cantidad de derretimiento. El peso o lo largo de un electrodo, alambre, varilla, o polvo derretido en una unidad de tiempo.

melt through. Complete joint penetration for a joint welded from one side. Visible root reinforcement is produced.
derretir de un lado a otro. Una junta con penetración completa para una junta que está soldada de un lado. Refuerzo de raíz visible es producido.

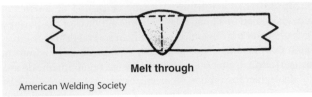

Melt through

American Welding Society

*metal. An opaque, lustrous, elemental, chemical substance that is a good conductor of heat and electricity, usually malleable, ductile, and more dense than other elemental substances.
metal. Una opaca, brillante, elemental, substancia química que es una buena conductora de calor y electricidad, por lo regular es maleable, ductil, y es más densa que otras substancias elementales.

metal arc cutting (MAC). Any of a group of arc cutting processes that sever metals by melting them with the heat of an arc between a metal electrode and the base metal. See also **shielded metal arc cutting** and **gas metal arc cutting**.
cortes de metal con arco (MAC). Cualquiera de un grupo de procesos de cortes con arco que corta metales derritiéndolos con el calor de un arco entre un electrodo de metal y el metal base. Vea también **cortes de arco metálico protegido** y **cortes de arco metálico con gas**.

metal cored electrode. A composite tubular filler metal electrode consisting of a metal sheath and a core of various powdered materials, producing no more than slag islands on the face of a weld bead. External shielding may be required.
electrodo de metal de núcleo. Un electrodo de metal para rellenar tubular compuesto consistiendo de una envoltura de metal y núcleo de varios materiales en polvo, que producen nada más que islas de escoria en la cara del cordón de soldadura. Protección externa puede ser requerida.

metal electrode. A filler or nonfiller metal electrode used in arc welding or cutting that consists of a metal wire or rod manufactured by any method and either bare or covered.
electrodo de metal. Un electrodo de metal que se usa para rellenar o para no rellenar la soldadura de arco o para cortar, que consiste de un alambre de metal o varilla que ha sido fabricada por cualquier método ya sea liso o cubierto con un cubierto o revestimiento propio.

*metallurgy. The scientific study of metals.
metalurgia. Estudio científico de los metales.

*methylacetylene-propadiene (MPS). A family of fuel gases that are mixtures of two or more gases (propane, butane, butadiene, methylacetylene, and propadiene). The neutral flame temperature is approximately 5031°F (2927°C), depending upon the actual gas mixture. MPS is used for oxyfuel cutting, heating, brazing, and metallizing; rarely used for welding.
metilacetileno y propadieno (MPS). Una familia de gases de combustión que son mezclas de dos o más gases (propano, butano,

butadiano, metilacetileno, propadieno). La temperatura de la llama natural es aproximadamente 5031°F (2927°C), dependiendo de la mezcla actual del gas. MPS es usado como gas de combustión para cortar, calentar, soldadura fuerte, y metalizar; es muy raro que se use para soldar.

*micro structure. A structure that is visible only with high magnification or with the aid of a microscope.
estructura micronesia. Una estructura que es visible solamente con un amplificador de poder muy alto o con la ayuda de un microscopio.

*microcomputer. A computer that uses a microprocessor as its basic element.
computadora micronesia. Una computadora que usa un procesor micronesio como su elemento básico.

microetch test. A test in which the specimen is prepared with a polished finish, etched, and examined under high magnification.
prueba con grabado al agua fuerte y examinada por un amplificador de alto poder. Una prueba en una probeta preparada con acabado fino, grabada al agua fuerte y examinado bajo un amplificador de alto poder.

*microprocessor. The principal processing element of a microcomputer, made as a single, integrated circuit.
procesor micronesio. El elemento principal de un procedimiento de una computadora micronesia, hecha con un solo circuito integrado.

mineral-based electrode fluxes. Fluxes that use inorganic compounds such as the rutile-based flux (titanium dioxide, TiO_2). These mineral compounds do not contain hydrogen, and electrodes that use these fluxes are often referred to as low hydrogen electrodes. Less smoke is generated with this welding electrode than with cellulose-based fluxes, but a thicker slag layer is deposited on the weld. E7018 is an example of an electrode that uses this type of flux.
fundentes para electrodos de base mineral. Fundentes que usan compuestos inorgánicos, como por ejemplo, el fundente a base de rutilo (bióxido de titanio, TiO_2). Estos compuestos minerales no contienen hidrógeno, y a los electrodos que usan estos fundentes se los llama con frecuencia electrodos de bajo hidrógeno. En la soldadura con electrodos se producen menos humos que en la que se realiza con fundentes celulósicos, pero se deposita una capa de escoria más gruesa en la soldadura. El E7018 es un ejemplo de un electrodo que usa este tipo de fundente.

mixing chamber. That part of a welding or cutting torch in which a fuel gas and oxygen are mixed.
cámara mezcladora. Esa parte de una antorcha para soldar y cortar por la cual el gas combustible y el oxígeno son mezclados.

mold. A high-temperature container into which liquid metal from the thermite welding process is poured and held until it cools and hardens into the container's interior shape.
molde. un contenedor de alta temperatura en el cual se vierte y se mantiene metal líquido del proceso de soldadura con termita hasta que éste se enfríe y se solidifique tomando la forma interior del contenedor.

molecular hydrogen. A bonded pair of hydrogen atoms (H_2). This is the configuration that all hydrogen atoms try to form.
hidrógeno molecular. Un par de átomos de hidrógeno (H_2) unidos. Ésta es la configuración que tratan de formar todos los átomos de hidrógeno.

molten weld pool. The liquid state of a weld prior to solidification as weld material.

charco de soldadura derretido. El estado líquido de una soldadura antes de solidificarse como material de soldadura.

*multipass weld. A weld requiring more than one pass to ensure complete and satisfactory joining of the metal pieces.

soldadura de pasadas múltiples. Una soldadura que requiere más de una pasada para asegurar una completa y satisfactoria unión de las piezas de metal.

N

*natural ventilation. Ventilation usually resulting from the heat-generated convection currents that cause welding fumes to rise.

ventilación natural. Ventilación que por lo general resulta de corrientes de convección generadas por el calor que causa la elevación del humo de la soldadura.

*needle-like (acicular structure). See **acicular structure.**

con forma de aguja (estructura acicular). Ver **estructura acicular.**

neutral flame. An oxyfuel gas flame that is neither oxidizing nor reducing.

llama neutral. Una llama de gas oxicombustible que no tiene características de oxidación ni de reducción. Refiérase al dibujo para cono.

*noble inert gas. See preferred term **inert gas.**

gas inerte noble. Ver el término preferido **gas inerte.**

nonconsumable electrode. An electrode that does not provide filler metal.

electrodo no consumible. Un electrodo que no provee metal de relleno.

noncorrosive flux, *brazing and soldering.* A soldering flux that in either its original or its residual form does not chemically attack the base metal.

flujo no corrosivo, *soldadura fuerte y soldadura.* Un flujo para soldadura blanda que ni en su forma original ni en su forma restante químicamente ataca el metal base. Regularmente es compuesto de materiales de colofonia o resino de base.

*nondestructive testing. Methods that do not alter or damage the weld being examined; used to locate both surface and internal defects. Methods include visual inspection, penetrant inspection, magnetic particle inspection, radiographic inspection, and ultrasonic inspection.

pruebas no destructivas. Métodos que no alteran ni dañan la soldadura que se está examinando. Se usa para encontrar ambos defectos internos y de superficie. Incluye métodos como inspección visual, inspección penetrante, inspección de partículas magnéticas, inspección de radiografía, inspección ultrasónica.

nontransferred arc. An arc established between the electrode and the constricting nozzle of the plasma arc torch or thermal spraying. The workpiece is not in the electrical circuit. See also **transferred arc.**

arco no transferible. Un arco establecido entre el electrodo y la boquilla constrictiva de la antorcha del arco de plasma o pistola termal para rociar. La pieza de trabajo no está en el circuito eléctrico. Vea también **arco transferido.**

nozzle. A device that directs shielding media.

boquilla. Un aparato que dirige el medio de protección.

nugget. The weld metal joining the workpieces in spot, seam, and projection welds.

botón. El metal de soldadura que une a las piezas de trabajo en soldadura de puntos, costura, y proyección de soldaduras.

nugget size. A nonstandard term when used for projection weld size, resistance weld size, or seam weld size.

tamaño del botón. El diámetro o lo ancho del botón medido en el plano del interfaze entre las piezas unidas.

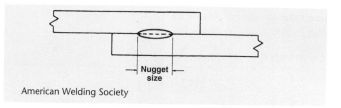

American Welding Society

O

*object line. Lines on a drawing that show the edge of an object, the intersection of surfaces that form corners or edges, and the extent of a curved surface, such as the sides of a cylinder.

línea de objeto. Líneas de un dibujo que muestran el borde de un objeto, la intersección de las superficies que forman las esquinas o los bordes y la extensión de una superficie curva, como los lados de un cilindro.

open circuit voltage. The voltage between the output terminals of the power source when the rated primary voltage is applied and no current is flowing in the secondary circuit.

voltaje de circuito abierto. El voltaje entre los terminales de salida de una fuente de poder cuando la corriente no está corriendo a la antorcha o pistola.

open-root joint. An unwelded joint without backing or consumable insert.

junta de raíz abierta. Una junta que no está para soldarse sin respaldo o inserto consumible.

*operating voltage. It is the actual voltage across the arc or closed-circuit voltage.

tensión operativa. Tensión real en el arco o tensión de circuito cerrado.

orifice. See **constricting orifice.**

orifice. Vea **orifice de constreñimiento.**

orifice gas. The gas that is directed into the plasma arc torch or thermal spraying gun to surround the electrode. It becomes ionized in the arc to form the arc plasma and issues from the constricting orifice of the nozzle as a plasma jet.

gas para orifice. El gas que es dirigido dentro de la antorcha de plasma o la pistola de rociado termal para rodear el electrodo. Se vuelve ionizado dentro del arco para formar el arco de plasma y sale de la orifice de constreñimiento a la boquilla como chorro de plasma.

orifice throat length. The length of the constricting orifice in the plasma arc torch or thermal spraying gun.

largo de garganta del orifice. Lo largo de la orifice constreñida en la antorcha de plasma o en la pistola de plasma para rociar.

*out-of-position welding. Any welding position other than the flat position; includes vertical, horizontal, and overhead positions.

soldadura fuera de posición. Cualquier posición de soldadura menos la de la posición plana; incluye vertical, horizontal, y posiciones de sobrecabeza.

*outer envelope. The outer boundary of the oxyacetylene flame. The secondary combustion reaction occurs in the outer envelope.

envoltura externa. El límite de afuera de la llama de oxiacetileno. La reacción de la combustión secundaria ocurre en la envoltura externa.

*outside corner joint. See joint.

junta de esquina externa. Ver junta.

overhead position. The position in which welding is performed from the underside of the joint.

posición de sobrecabeza. La posición en la cual se hace la soldadura por el lado de abajo de la junta.

overlap. The protrusion of weld metal beyond the toe, face, or root of the weld; in resistance seam welding, the area in the preceding weld remelted by the succeeding weld.

traslapo. El metal de la soldadura que sobresale más allá del pie, cara, o de la raíz de una soldadura; en soldaduras de costuras por resistencia, la área de la soldadura anterior se rederrite por la soldadura subsiguiente.

*oxide layer. A layer of oxidized metal on the surface, on steel it can be called rust.

capa de óxido. Capa de metal oxidado que está sobre la superficie. En el acero se puede denominar herrumbre.

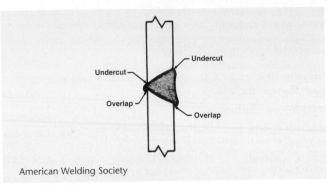

American Welding Society

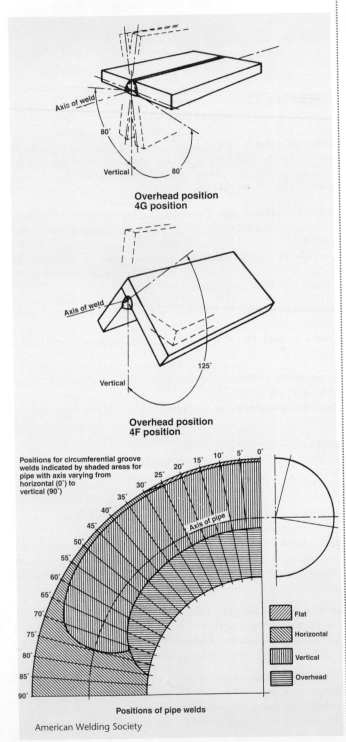

Overhead position
4G position

Overhead position
4F position

Positions for circumferential groove welds indicated by shaded areas for pipe with axis varying from horizontal (0°) to vertical (90°)

Axis of pipe

Flat

Horizontal

Vertical

Overhead

Positions of pipe welds

American Welding Society

oxidizing flame. An oxyfuel gas flame in which there is an excess of oxygen, resulting in an oxygen-rich zone extending around and beyond the cone.

llama oxidante. Una llama de gas oxicombustible en la cual hay un exceso de oxígeno, resultando en una zona rica de oxígeno extendiéndose alrededor y más allá del cono.

oxyacetylene cutting (OFC-A). An oxyfuel cutting process variation employing acetylene as the gas.

cortes con oxiacetileno (OFC-A). Un proceso de cortes con gas con variación que usa acetileno como gas de combustión.

*oxyacetylene hand torch. Most commonly used oxyfuel gas cutting torch; may be a cutting torch only or a combination welding and cutting torch set. On the combination set different tips can be attached to the same torch body. The torch mixes the oxygen and fuel gas and directs the mixture to the tip. The torch can be an equal pressure type (equal pressures of oxygen and fuel gas) or an injector type (equal pressures of high-pressure oxygen and low-pressure fuel gas).

antorcha de mano oxiacetileno. La antorcha de gas oxicombustible es la que se usa más frecuentemente; puede ser una antorcha para hacer cortes solamente o una combinación de un juego de antorcha para soldar y hacer cortes. En el juego de combinación diferentes puntas pueden ser conectadas al mismo mango de la antorcha. La antorcha mezcla el oxígeno y gas combustible y dirige la mezcla a la punta. La antorcha puede ser de tipo de presión igual (presiones iguales de oxígeno y gas combustible) o de tipo inyector (presiones iguales de oxígeno de alta presión y baja presión de gas combustible).

oxyacetylene welding (OAW). An oxyfuel gas welding process employing acetylene as the fuel gas. The process is used without the application of pressure.

soldadura con oxiacetileno (OAW). Un proceso de soldadura de gas oxicombustible que usa acetileno con gas de combustión. El proceso se usa sin aplicación de presión.

*oxyfuel flame. A flame resulting from the combustion of oxygen mixed with a fuel gas. This intense flame is applied to two pieces of metal to cause them to melt to form weld pools. When the edges of the weld pools run together and fuse, the two pieces of metal are joined.

llama oxicombustible. Una llama que resulta de una combustión de oxígeno mezclado con gas combustible. Está llama intensa es aplicada a dos piezas de metal para hacer que se derritan para formar charcos de soldadura. Cuando las orillas de los charcos de soldadura se juntan y se derriten, las dos piezas de metal se unen.

*oxyfuel gas. A general term covering all the different fuel gases such as acetylene, MAPP, natural gas, propane, hydrogen, etc., that can be used with oxygen.

gas oxicombustible. Término general que cubre todos los diferentes gases combustibles como el acetileno, MAPP, gas natural, propano, hidrógeno, etc., que se pueden utilizar con el oxígeno.

oxyfuel gas cutting (OFC). A group of oxygen cutting processes that uses heat from an oxyfuel gas flame. See also oxygen cutting, oxyacetylene cutting, oxyhydrogen cutting, and oxypropane cutting.

gas para cortar oxicombustible (OFC). Un grupo de procesos para cortar con oxígeno que usa calor de una llama de gas oxicombustible. Vea también cortes con oxigeno, cortes con oxiacetileno, cortes con oxihidrógeno, y cortes con oxipropano.

oxyfuel gas cutting torch. A device used for directing the preheating flame produced by the controlled combustion of fuel gases and to direct and control the cutting oxygen.

antorcha para cortes de gas oxicombustible. Un aparato que se usa para dirigir la llama precalentada producida por la combustión controlada de los gases de combustión y para dirigir y controlar el oxígeno para cortar.

oxyfuel gas welding (OFW). A group of welding processes producing coalescence of workpieces by heating them with an oxyfuel gas flame. The processes are used with or without the application of pressure and with or without filler metal.

soldadura con gas oxicombustible (OFW). Un grupo de procesos de soldadura que produce coalescencia de las piezas de trabajo calentándolas con una llama de gas oxicombustible. Los procesos son usados sin la aplicación de presión y con o sin el metal para rellenar.

oxyfuel gas welding torch. A device used in oxyfuel gas welding, torch brazing, and torch soldering for directing the heating flame produced by the controlled combustion of fuel gases.

antorcha para soldar con gas oxicombustible. Un aparato que se usa para soldar con gas oxicombustible, soldadura fuerte con antorcha, soldadura blanda con antorcha y para dirigir la llama calentada producida por combustión controlada de gases de combustión.

oxygen arc cutting (OAC). An oxygen cutting process using an arc between the workpiece and a consumable tubular electrode, through which oxygen is directed to the workpiece.

cortes de oxígeno con arco (OAC). Es un proceso de cortar con oxígeno que usa un arco entre la pieza de trabajo y un electrodo tubular consumible, por el cual el oxígeno es dirigido a la pieza de trabajo.

oxygen cutting (OC). A group of thermal cutting processes severing or removing metal by means of the chemical reaction between oxygen and the base metal at elevated temperature. The necessary temperature is maintained by the heat from an arc, an oxyfuel gas maintained by the heat from an arc, an oxyfuel gas flame, or other sources. See also oxyfuel gas cutting.

cortes con oxígeno (OC). Un grupo de procesos termales que corta y quita el metal por medio de una reacción química entre el oxígeno y el metal base a una temperatura elevada. La temperatura necesaria es mantenida por el calor del arco, un gas oxicombustible mantenido por el calor del arco, una llama de gas oxicombustible, o de otras fuentes. Vea también gas para cortar oxicombustible.

oxygen gouging. Thermal gouging using an oxygen cutting process variation to form a bevel or groove.

escopleando con la gubia con oxígeno. Gubia termal que usa un proceso de variación de corte con oxígeno para formar un bisel o ranura.

oxygen lance. A length of pipe used to convey oxygen to the point of cutting in oxygen lance cutting.

lanza de oxígeno. Un tramo de tubo usado para conducir oxígeno al punto de cortar en cortes con lanza de oxígeno.

oxygen lance cutting (OLC). An oxygen cutting process employing oxygen supplied through a consumable lance. The preheat to start the cutting is obtained by other means.

cortes con lanza de oxígeno (OLC). Un proceso de cortar con oxígeno que usa oxígeno surtido por una lanza consumible. El precalentamiento para empezar a cortar es obtenido por otros medios.

oxyhydrogen cutting (OFC-H). An oxyfuel gas cutting process variation that uses hydrogen as the fuel gas.

cortes con oxihidrógeno (OFC-H). Un proceso de cortar de gas oxicombustible con variación que usa hidrógeno como gas combustible.

*oxyhydrogen flame. A specific flame resulting from the combustion of oxygen and hydrogen; consists of primary combustion region only; used for welding and cutting.

llama oxihidrógeno. Una llama específica que resulta de la combustión del oxígeno e hidrógeno; consiste solamente de la región de combustión primaria; usada para cortar y soldar.

oxyhydrogen welding (OHW). An oxyfuel gas welding process that uses hydrogen as the fuel gas. The process is used without the application of pressure.

soldadura oxihidrógeno (OHW). Un proceso de soldar con gas oxicombustible que usa hidrógeno como gas de combustible. El proceso se usa sin la aplicación de presión.

oxypropane cutting (OFC-P). An oxyfuel gas cutting process variation employing propane as the fuel gas.

cortes con oxipropano (OFC-P). Un proceso de cortar con gas combustible con variación que usa propano como gas combustible.

P

parent metal. See preferred term base metal.

metal de origen. Vea término preferido metal base.

*paste range. The temperature range of soldering and brazing filler metal alloys in which the metal is partly solid and partly liquid as it is heated or cooled.

grados de la pasta. Los grados de la temperatura del metal para rellenar aleados para soldadura blanda o soldadura fuerte cuando se calienta o se enfria.

*pearlite. A two-phased lamellar iron carbon crystalline structure that forms during slow cooling.

perlita. Estructura cristalina de hierro y carbono laminar de dos fases que se forma durante el enfriamiento lento.

peel test. A destructive method of inspection that mechanically separates a lap joint by peeling.
prueba por pelar. Un método de inspección destructivo de pelar que separa mecánicamente una junta de solape.

peening. The mechanical working of metals using impact, often with a ball peen hammer.
martillazos (con martillo de bola). Metales que se trabajan mecánicamente con golpes de impacto.

*penetration. The depth into the base metal (from the surface) that the weld metal extends, excluding any reinforcement.
penetración. La profundidad de adentro del metal base (de la superficie) que el metal de soldadura se extiende, excluyendo cualquier refuerzo.

percussion welding (PEW). A welding process that produces coalescence with an arc resulting from a rapid discharge of electrical energy. Pressure is applied percussively during or immediately following the electrical discharge.
soldadura a percusión (PEW). Un proceso de soldadura que produce coalescencia con un arco resultando de una descarga rápida de energía. Presión es aplicada a percusión durante o inmediatamente después de la descarga eléctrica.

*phantom lines. Lines on a drawing that show an alternate position of a moving part or the extent of motion, such as the on/off position of a light switch. They can also be used as a place holder for a part that will be added later.
líneas fantasmas. Líneas de un dibujo que muestran una posición alternativa de una pieza móvil o el rango del movimiento de ésta, como las posiciones de encendido/apagado de un interruptor de luz. También se pueden usar como marcador de posición para una pieza que se añadirá posteriormente.

*phase diagrams. Provide information on the crystalline constituents of metal alloys at different temperatures in three different phases: pure metal, solid solutions of two or more metals, and intermetallic compounds.
diagramas de equilibrio. Proporciona información en los constituyentes cristalinos de metales aleados a diferentes temperaturas en tres aspectos: metal puro, soluciones sólidas de dos o más metales, y mezclas intermetálicas.

*pick-and-place robot. A simple robot, often with only two or three degrees of freedom, which transfers items from place to place by means of point-to-point moves. Little or no trajectory control is available. Often referred to as a "bank-bank" robot.
robot de escoger y atar. Un robot simple, frecuentemente con solo dos o tres grados de libertad, el cual traslada artículos de un lugar a otro por medio de movidas de punto a punto. Un poco o nada de control trayectoria es utilizado. A veces es referido como un "banco-banco" robot.

*pictorial drawings. A type of mechanical drawing that represents an object as a picture.
dibujo ilustrativo. Tipo de dibujo mecánico que representa un objeto como una ilustración.

pilot arc. A low-current arc between the electrode and the constricting nozzle of the plasma arc torch to ionize the gas and facilitate the start of the welding arc.
piloto del arco. Un arco de corriente baja en medio del electrodo y la boquilla constreñida de la antorcha de arco de plasma para ionizar el gas y facilitar el arranque del arco para soldar.

*pitch. The angular rotation of a moving body about an axis perpendicular to its direction of motion and in the same plane as its top side.
grado de inclinación. La rotación angular de un cuerpo en movimiento alrededor de un eje perpendicular a su dirección y en el mismo plano como el del lado de arriba.

*plain carbon steels. See carbon steel.
acero de carbón puro. Ver acero al carbono

plasma. A gas that has been heated to an at least partially ionized condition, enabling it to conduct an electric current.
plasma. Un gas que ha sido calentado a lo menos parcialmente a una condición ionizada permitiendo que conduzca una corriente eléctrica.

*plasma arc. See plasma.
arco de plasma. Ver plasma.

plasma arc cutting (PAC). An arc cutting process employing a constricted arc and removes the molten metal with a high-velocity jet of ionized gas issuing from the constricting orifice.
cortes con arco de plasma (PAC). Un proceso de cortar con el arco que usa un arco constreñido y quita el metal derretido con un chorro de alta velocidad de gas ionizado que sale de la orifice constringente.

*plasma arc gouging. See plasma arc cutting.
gubiadura con arco de plasma. Ver cortes con arco de plasma.

plasma arc welding (PAW). An arc welding process employing a constricted arc between a nonconsumable electrode and the weld pool (transferred arc) or between the electrode and the constricting nozzle (nontransferred arc). Shielding is obtained from the ionized gas issuing from the torch, which may be supplemented by an auxiliary source of shielding gas. The process is used without the application of pressure.
soldadura con arco de plasma (PAW). Un proceso de soldadura de arco que usa un arco constreñido entre un electrodo que no se consume y el charco de la soldadura (arco transferido) o entre el electrodo y la lanza constreñida (arco no transferido). La protección es obtenida del gas ionizado que sale de la antorcha, el cual puede ser suplementado por una fuente auxiliar de gas para protección. El proceso es usado sin la aplicación de presión.

plug weld. A weld made in a circular hole in one member of a joint fusing that member to another member. A fillet-welded hole should not be construed as conforming to this definition.
soldadura de tapón. Una soldadura que se hace en un agujero circular en un miembro de una junta uniendo ese miembro con otro miembro. Un agujero de soldadura de filete no debe ser interpretado como confirmación de está definición.

*point-to-point control. A control scheme whereby the inputs or commands specify only a limited number of points along a desired path of motion. The control system determines the intervening path segments.
control de punto a punto. Una esquema de control con que las entradas o las ordenes especifican solamente un número limitado de puntos a lo largo de la senda de moción deseada. El sistema de control determina el intervenio de los segmentos de la senda.

porosity. Cavity-type discontinuities formed by gas entrapment during solidification or in a thermal spray deposit.
porosidad. Un tipo de cavidad de desuniones formadas por gas atrapado durante la solidificación o en un deposito rociado termal.

postflow time. The time interval from current shutoff to shielding gas and/or cooling water shutoff.

tiempo de poscorriente. El intervalo de tiempo de cuando se cierra la corriente a cuando se cierra el gas de protección y o cuando se cierra el agua para enfriar.

postheat current (resistance welding). The current through the welding circuit during postheat time in resistance welding.

corriente de poscalentamiento (soldadura de resistencia). La corriente que va de un lado a otro del circuito durante el tiempo de poscalentamiento en la soldadura de resistencia.

postheating. The application of heat to an assembly after welding, brazing, soldering, thermal spraying, or thermal cutting. See also **postweld heat treatment.**

poscalentamiento. La aplicación de calor a una asamblea después de la soldadura, soldadura fuerte, soldadura blanda, rociado termal o corte termal. Vea también **tratamiento de calor postsoldadura.**

postheat time (resistance welding). The time from the end of weld heat time to the end of weld time. Refer to drawing for **downslope time.**

tiempo de poscalentamiento (soldadura de resistencia). El tiempo del fin del calor de la soldadura al tiempo al fin del tiempo de la soldadura. Refiérase al dibujo de **tiempo de cadía del pendiente.**

*postpurge. Once welding current has stopped in gas tungsten arc welding, this is the time during which the gas continues to flow to protect the molten pool and the tungsten electrode as cooling takes place to a temperature at which they will not oxidize rapidly.

pospurgante. Cuando la corriente de soldar se ha dentenido en la soldadura de arco gas tungsteno, este es el tiempo durante en que el gas continua a salir para proteger el charco de soldadura derretido y el electrodo de tungsteno se enfrian a una temperatura donde no se oxidan rápidamente.

postweld heat treatment. Any heat treatment subsequent to welding.

tratamiento de calor postsoldadura. Cualquier tratamiento de calor subsiguiente a la soldadura.

powder flame spraying. A thermal spraying process variation in which the material to be sprayed is in powder form. See also **flame spraying (FLSP).**

rociado de polvo con llama. Un proceso termal para rociar con variación el cual el material que está para rociar se está en forma de polvo. Vea también **rociado a llama.**

power source. An apparatus for supplying current and voltage suitable for welding, thermal cutting, or thermal spraying.

fuente de poder. Un aparato para surtir corriente y voltaje conveniente para soldar, para hacer cortes termales, o rociado termal.

power supply. Nonstandard term for power source.

fuente de alimentación. Término no estandarizado de fuente de poder.

*preflow. Shielding gas that flows before the GTA welding current starts to force the air away from the arc zone, which protects both the hot tungsten and weld from atmospheric contamination when welding starts.

preflujo. Gas protector que fluye antes de que la corriente de soldadura GTA corra para forzar el aire a salir de la zona del arco. El gas protege tanto el tungsteno caliente como la soldadura de la contaminación atmosférica una vez iniciada la soldadura.

preheat. The heat applied to the base metal or substrate to attain and maintain preheat temperature.

precalentamiento. El calor aplicado al metal base o substrato para obtener y mantener temperatura de precalentamiento.

*preheat flame. Brings the temperature of the metal to be cut above its kindling point, after which the high-pressure oxygen stream causes rapid oxidation of the metal to perform the cutting.

llama para precalentamiento. Sube la temperatura del metal que está para cortarse a una temperatura de encendimiento, después que la corriente del oxígeno de alta presión cause una oxidación rápida del metal para hacer el corte.

*preheat holes. The cutting tip has a central hole through which the oxygen flows. Surrounding this central hole are a number of other holes called preheat holes. The differences in the type or number of preheat holes determine the type of fuel gas to be used in the tip.

agujeros para precalentamiento. La boquilla para cortar tiene un agujero central por donde corre el oxígeno. Rodeando este agujero central hay un numero de otros agujeros que se llaman agujeros para precalentar. Las diferencias en el tipo o número de agujeros percalentados determina el tipo de gas combustible que se usará en la boquilla.

preheat temperature. The temperature of the base metal or substrate in the welding, brazing, soldering, thermal spraying, or thermal cutting area immediately before these operations are performed. In a multipass operation, it is also the temperature in the area immediately before the second and subsequent passes are started.

temperatura de precalentamiento. La temperatura del metal base o substrato en la soldadura, soldadura fuerte, soldadura blanda, rociado termal, o en la área de los cortes termal inmediatamente antes de que estas operaciones sean ejecutadas. En una operación multipasada, es también la temperatura en la área inmediatamente antes de empezar la segunda pasada y pasadas subsiguientes.

preheating. The application of heat to the base metal immediately before welding, brazing, soldering, thermal spraying, or cutting.

precalentamiento. La aplicación de calor al metal base inmediatamente antes de la soldadura, soldadura fuerte, soldadura blanda, rociado termal o cortes.

*prepurge. In gas tungsten arc welding, the time during which gas flows through the torch to clear out any air in the cup or surrounding the weld zone. Prepurge time is set by the operator and is completed before the welding current is started.

prepurgar. En soldadura de arco de tungsteno con gas, el tiempo durante el cual el gas corre por la antorcha para quitar el aire en la boquilla o la zona de soldadura. El tiempo de prepurgar es determinado por el operador y es acabado antes de que la corriente de soldadura es empezada.

*primary combustion. The first reaction in the chemical reaction resulting when a mixture of acetylene and oxygen is ignited. This reaction frees energy and forms carbon monoxide (CO) and free hydrogen.

combustión primaria. La primera reacción en una reacción química resulta cuando una mezcla de oxígeno y acetileno es encendida. Está reacción libra la energia y forma carbón monóxido (CO) e hidrógeno libre.

procedure qualification. The demonstration that welds made by a specific procedure can meet prescribed standards.
calificación de procedimiento. La demostración en que las soldaduras hechas por un procedimiento específico conformen con las normas prescribidas.

projection weld. A weld made by projection welding.
soldadura de proyección. Una soldadura hecha con soldadura de proyección.

protective atmosphere. A gas envelope surrounding the part to be brazed, welded, or thermal sprayed, with the gas composition controlled with respect to chemical composition, dew point, pressure, flow rate, etc. Examples are inert gases, combusted fuel gases, hydrogen, and vacuum.
atmósfera protectora. Una envoltura de gas que está alrededor de la parte que está para soldarse con soldadura fuerte, soldadura o rociada termal, con la composición del gas controlado con respecto a la química compuesta, punto de rocío, presión, cantidad de corriente, etc. Ejemplos son gas inerto, gases de combustión que ya están encendidos, hidrógeno, y vacuo.

*proximity sensor. A device that senses that an object is only a short distance (e.g., a few inches or feet) away and/or measures how far away it is. Proximity sensors work on the principles of triangulation of reflected light, lapsed time for reflected sound, or intensity-induced eddy currents, magnetic fields, back pressure from air jets, and others.
sensor de proximidad. Un aparato que siente que un objeto está solamente a una corta distancia (e.g., unas pulgadas o pies) afuera, y/o mide que tan lejos está. Sensores de proximidad trabajan en los fundamentos de triangulación de luz reflejada, tiempo lapso del sonido reflejado, o en las corrientes de Fancault inducidas con intensidad, campos magnéticos, contrapresión trasera del chorro de aire, y otras.

puddle. See preferred term **weld pool**.
charco. Vea término preferido **charco de soldadura**.

pulse start delay time. The time interval from current initiation to the beginning of current pulsation, if pulsation is used. Refer to drawing for **upslope time**.
tiempo de dilación en empezar la pulsación. El tiempo del intervalo de donde se inicia la corriente al principio de la pulsación de la corriente, si es que se use pulsación. Refiérase al dibujo de **tiempo del pendiente en ascenso**.

*pulsed-arc metal transfer. In gas metal arc welding, pulsing the current from a level below the transition current to a level above the transition current to achieve a controlled spray transfer at lower average currents; spray transfer occurs at the higher current level.
transferir el metal por arco pulsado. En la soldadura de arco metálico con gas, se pulsa la corriente de un nivel más alto de la corriente de transición para lograr un traslado de rocío controlado a una corriente media baja; el traslado del rocío ocurre al nivel más alto de la corriente.

*purged. The process of opening first one cylinder valve and then the other to replace all air in the hoses with the appropriate gas prior to welding.
limpidor. El proceso de abrir primero una válvula de un cilindro y luego el otro para reemplazar todo el aire en las mangueras con un gas apropiado antes de empezar a soldar.

push angle. The travel angle when the electrode is pointing in the direction of weld progression. This angle can also be used to partially define the position of guns, torches, rods, and beams.
ángulo de empuje. El ángulo de avance cuando el electrodo apunta en la dirección en que la soldadura progresa. Este ángulo también puede ser usado para parcialmente definir la posición de pistolas, antorchas, varillas, y rayos.

push welding. A resistance welding process variation in which spot or projection welds are produced by manually applying force to one electrode.
soldadura de empuje (soldadura de resistencia). Una soldadura de botón o proyección hecha por soldadura de empuje.

Q

qualification. See preferred terms **welder performance qualification** and **procedure qualification**.
calificación. Vea términos preferidos **calificación de ejecución del soldador** y **calificación de procedimiento**.

*quality control. A procedure of tests set up by shops to inspect weldments as they are produced to ensure that they meet the standards set up by the manufacturer.
control de calidad. Procedimiento de pruebas diseñadas por los talleres para inspeccionar las soldaduras a medidas que son producidas, con el propósito de asegurarse de que cumplan con los estándares fijados por el fabricante.

R

*radiographic inspection (RT). See **nondestructive testing**.
inspección radiográfica (RT). Ver **prueba no destructiva**.

reactor (arc welding). A device used in arc welding circuits for the purpose of minimizing irregularities in the flow of welding current.
reactor (soldadura de arco). Un aparato usado en los circuitos de la soldadura de arco con el propósito de reducir a lo mínimo las irregularidades en la manera que corre la corriente de soldadura de arco.

reduced section tension test. A test in which a transverse section of the weld is located in the center of the reduced section of the specimen.
prueba de tensión de sección reducida. Una prueba en la cual la sección transversa de la soldadura está ubicada en el centro de la sección reducida de la probeta.

reducing atmosphere. A chemically active protective atmosphere, which at elevated temperature will reduce metal oxides to their metallic state. (Reducing atmosphere is a relative term, and such an atmosphere may be reducing to one oxide but not to another oxide.)
atmósfera de reducción. Una atmósfera protectiva activa, la cual a una temperatura elevada reduce los óxidos del metal a sus estados metalicos. (Atmósfera de reducción es un término relativo, y cierta atmósfera puede reducir a un óxido pero no al otro óxido.)

reducing flame. An oxyfuel gas flame with an excess of fuel gas.
llama de reducción. Una llama de gas oxicombustible con un exceso de gas combustible.

regulator. A device for controlling the delivery of gas at some substantially constant pressure.
regulador. Un aparato para controlar la expedición de gas a una presión substancialmente constante.

residual stress. Stress present in a joint member or material that is free of external forces or thermal gradients.

fuerza residual. Fuerza presente en un miembro de una junta o material que está libre de fuerzas externas o ambulantes termales.

resistance brazing (RB). A brazing process using heat from the resistance to electric current flow in a circuit that includes the assembly.

soldadura fuerte por resistencia (RB). Un proceso de soldadura fuerte que usa calor de la resistencia al correr de la corriente eléctrica en un circuito en las cuales las piezas de trabajo forman parte.

resistance soldering (RS). A soldering process that uses heat from the resistance to electric current flow in a circuit of which the workpieces are a part.

soldadura blanda por resistencia (RS). Un proceso de soldadura blanda que usa calor de la resistencia al correr de la corriente eléctrica en un circuito en las cuales las piezas de trabajo forman parte.

resistance spot welding (RSW). A resistance welding process that produces a weld at the faying surfaces of a joint by the heat obtained from resistance to the flow of welding current through the workpieces from electrodes that concentrate the welding current and pressure at the weld area.

soldadura de puntos por resistencia (RSW). Un proceso de soldar por resistencia que produce una soldadura en los empalmes de la superficie de una junta por el calor obtenido de la resistencia al correr la corriente a través de las piezas de trabajo de los electrodos que sirven para concentrar la corriente para soldar y la presión en la área de la soldadura.

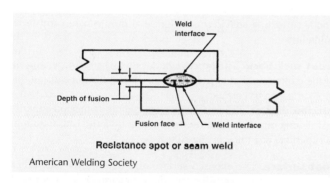

Resistance spot or seam weld

American Welding Society

resistance welding (RW). A group of welding processes that produces coalescence of the faying surfaces with the heat obtained from resistance of the workpieces to the flow of the welding current of which the workpieces are a part and by the application of pressure.

soldadura por resistencia (RW). Un grupo de procesos para soldar que producen coalescencia de las superficies empalmadas con el calor obtenido de la resistencia de las piezas de trabajo al correr la corriente de soldadura en un circuito en las cuales las piezas de trabajo forman parte, y por la aplicación de presión.

resistance welding downslope time. The time during which the welding current is continuously decreasing.

tiempo del pendiente de descenso en soldadura de resistencia. El tiempo durante el cual la corriente está continuamente disminuyendo.

resistance welding electrode. The part of a resistance welding machine through which the welding current and, in most cases, force are applied directly to the workpiece. The electrode may be in the form of a rotating wheel, rotating roll, bar, cylinder, plate, clamp, chuck, or modification thereof.

electrodo para soldadura por resistencia. La parte de una máquina para soldar por resistencia por cual la corriente de soldar y, en muchos casos, la fuerza es aplicada directamente a la pieza de trabajo. El electrodo puede ser en la forma de una rueda que da vueltas, rollo rotativo, barra, cilindro, plato, empalme, calzo, o modificación de ello.

reverse polarity. The arrangement of direct-current arc welding leads with the work as the negative pole and the electrode as the positive pole of the welding arc. A synonym for direct-current electrode. Refer to drawing for **direct-current electrode positive.**

polaridad invertida. El arreglo de los cables para soldar con el arco con corriente directa con el cable de tierra como el polo negativo y el electrodo como polo positivo del arco para soldar. Un sinónimo para corriente directa electrodo. Refiérase al dibujo para **corriente directa con el electrodo positivo.**

*robot. A reprogrammable, multifunctional manipulator designed to move material, parts, tools, or specialized devices through variable programmed motions for the performance of a variety of tasks.

robot. Un manipulador reprogramable, multifuncional diseñado para mover material, partes, herramienta, o aparatos especializados por medio de mociones programadas variables para la ejecución de una variedad de tareas.

*robot programming language. A computer language especially designed for writing programs for controlling robots.

lenguaje para programación del robot. Un lenguaje para computadoras con un diseño especial para escribir programas para el control de los robots.

root. See preferred terms of **root of joint** and **root of weld.**

raíz. Vea las términos preferidos de **raíz de junta** y **raíz de soldadura.**

root bead. A weld bead extending into, or including part or all of, the joint root.

cordón de raíz. Un cordón de soldadura que se extiende adentro, o incluye parte o toda la junta de raíz.

root-bend test. A test in which the weld root is on the convex surface of a specified bend radius.

prueba de dobléz de raíz. Una prueba en la cual la raíz de la soldadura está en una superficie convexa de un radio especificado para el dobléz.

root crack. A crack in the weld or heat-affected zone occurring at the root of a weld.

grieta de raíz. Una grieta en la soldadura o en la zona afectada por el calor que ocurre en la raíz de la soldadura.

root edge. A root face of zero width. See also **root face.** Refer to drawing for **groove face.**

orilla de raíz. Una cara de raíz con una anchura de cero. Vea también **cara de raíz.** Refiérase al dibujo para **cara de ranura.**

root face. The portion of the groove face adjacent to the root of the joint. Refer to drawing for **groove face.**

cara de raíz. La porción de la cara de la ranura adyacente a la raíz de la junta. Refiérase al dibujo para **cara de ranura.**

root gap. See preferred term **root opening.**

rendija de raíz. Vea el término preferido **abertura de raíz.**

root of joint. The portion of a joint to be welded where the members approach closest to each other. In a cross section, the root of the joint may be a point, a line, or an area.
raíz de junta. Saporción de una junta que está para soldarse donde los miembros se acercan muy cerca del uno al otro. En sección transversal, la raíz de una junta puede ser una punta, una línea, o una área.

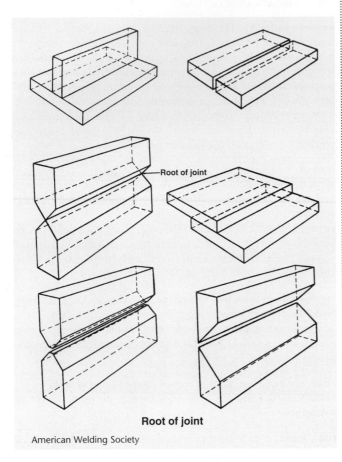

Root of joint

American Welding Society

root of weld. The points, as shown in a cross section, at which the back of the weld intersects the base metal surfaces.
raíz de soldadura. Las puntas, como ensena la sección transversa, donde la parte de atrás cruza con la superficie del metal base.

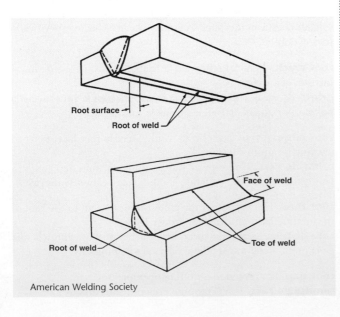

American Welding Society

root opening. The separation between the members to be joined at the root of the joint. Refer to drawings for **bevel**.
abertura de raíz. La separación entre los miembros que están para unirse a la raíz de la junta. Refiérase al dibujo para **bisel**.

*root pass. The first weld of a multipass weld. The root pass fuses the two pieces together and establishes the depth of weld metal penetration.
pasada de raíz. La primera soldadura de una soldadura de pasadas múltiples. La pasada de raíz funde las dos piezas juntas y establece la profundidad de la penetración del metal soldado.

root penetration. The distance the weld metal extends into the joint root.
penetración de raíz. La distancia que se extiende el metal de soldadura adentro de la junta de raíz.

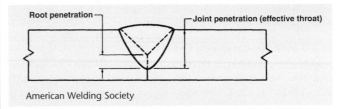

American Welding Society

root radius. See preferred term **groove radius**.
radio de raíz. Vea el término preferido **radio de ranura**.

root reinforcement. Reinforcement of weld at the side other than that from which welding was done. Refer to drawing for **face of weld**.
refuerzo de raíz. Refuerzo de soldadura en el lado opuesto de donde se hizo la soldadura. Refiérase al dibujo para **cara de la soldadura**.

*root suck back. The process that occurs when surface tension of the molten weld pool root surface is drawn inward causing the root surface to be concave.
succión de raíz. Proceso que ocurre cuando la tensión superficial de una raíz de soldadura derretida se invierte, causando que la superficie raíz tome forma cóncava.

root surface. The exposed surface of a weld on the side other than that from which welding was done. Refer to drawings for **root of weld**.
superficie de raíz. La superficie expuesta de una soldadura en el lado opuesto de donde se hizo la soldadura. Refiérase a los dibujos para **raíz de soldadura**.

runoff weld tab. Additional material that extends beyond the end of the joint, on which the weld is terminated.
solera de carrera final de soldadura. Material adicional que se extiende más allá de donde se acaba la junta, en la cual la soldadura es terminada.

*rutile-based fluxes. Acidic fluxes that produce smooth stable arcs and fast-freeze slags.
flujos por destello. Flujos acídicos que producen arcos parejos y estables y escorias de solidificación rápida.

S

scarf joint. A form of a butt joint.
junta de echarpe. Una forma de junta a tope.

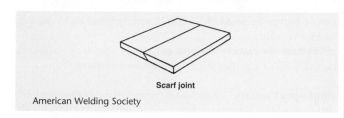

Scarf joint

American Welding Society

*scavenger. Elements in the flux that pick up contaminants in the molten weld pool and float them to the surface where they become part of the slag.

limpiadores o (expulsadores). Elementos en el flujo que levantan los contaminantes en el charco de soldadura derretida y los flotan a la superficie donde se forman parte de la escoria.

seam weld. A continuous weld produced between overlapping members with coalescence initiating and occuring at faying surfaces proceeding from the outer surface of one member. The weld can consist of either a weld bead, multiple overlapping nuggets, or a single nugget formed by the simultaneous application of resistance heating and forging force along the weld joint.

soldadura de costura. Una soldadura continua hecha en medio o encima de los miembros traslapados, en la cual la coalescencia puede empezar y ocurrir en la superficie del empalme, o puede haber procedido de la superficie de un miembro. La soldadura continua puede consistir de un solo cordón de soldadura o una serie de puntos traslapados en las soldaduras.

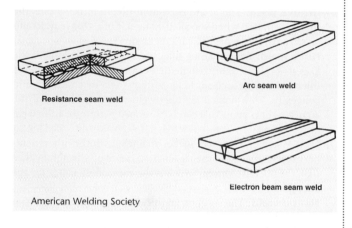

Resistance seam weld

Arc seam weld

Electron beam seam weld

American Welding Society

*secondary combustion. In the combustion of acetylene and oxygen, the secondary reaction unites oxygen and the free hydrogen to form water vapor (H_2O) and liberate more heat. The carbon monoxide unites with more oxygen to form carbon dioxide (CO_2).

combustión secundaria. En la combustión de acetileno y oxígeno, la reacción secundaria une el oxígeno y el hidrógeno libre para formar vapor de agua (H_2O) y liberar más calor. El carbón monóxido se une con más oxígeno para formar carbón bióxido (CO_2).

*section line. Lines on a drawing that show the surface that has been imaginarily cut away with a cutting plane line to show internal details.

línea de sección. Líneas de un dibujo que muestran la superficie que ha sido cortada imaginariamente con una línea de plano de corte para mostrar los detalles internos.

*section view. A special view of an object that shows the internal components as if the object were cut apart.

vista de sección. Vista especial de un objeto que muestra los componentes internos como si el objeto hubiera sido cortado y separado.

*self-shielding. As it relates to FCA welding, it refers to electrodes that contain enough fluxing agents inside the electrode to provide complete weld protection without the need to provide a shielding gas. See **dual shield**.

autoprotección. En relación a la soldadura FCA. Se refiere a los electrodos que contienen suficiente agentes fundentes en su interior como para proporcionar protección completa de la soldadura sin la necesidad de usar gas de protección. Ver **protección doble**.

semiautomatic arc welding. Arc welding with equipment that controls only the filler metal feed. The advance of the welding is manually controlled.

soldadura de arco semiautomático. La soldadura de arco con equipo que controla solamente la alimentación del metal de relleno. El avance de la soldadura es controlado manualmente.

*semiautomatic operation. During the welding process, the filler metal is added automatically, and all other manipulation is performed manually by the operator.

operación semiautomática. Durante el proceso de la soldadura, el metal de relleno es añadido automáticamente, y todas las otras manipulaciones son ejecutadas manualmente por el operador.

*sensor. A transducer whose input is a physical phenomenon and whose output is a quantitative measure of the physical phenomenon.

sensor. Un transducor cuya entrada es un fenómeno físico y cuya medida es una medida cuantitativa del fenómeno físico.

sequence time (automatic arc welding). See preferred term **welding cycle**.

tiempo de secuencia (soldadura de arco automático). Vea el término preferido **ciclo de soldadura**.

*shear strength. As applied to a soldered or brazed joint, it is the ability of the joint to withstand a force applied parallel to the joint.

fuerza cizallada. Asi como es aplicada a una junta de soldadura fuerte o soldadura blanda, es la habilidad de la junta de resistir una fuerza aplicada al paralelo de la junta.

shielded metal arc cutting (SMAC). An arc cutting process employing a covered electrode.

cortes de arco metálico protegido (SMAC). Un proceso de cortar con arco que usa un electrodo cubierto.

shielded metal arc welding (SMAW). An arc welding process with an arc between a covered electrode and the weld pool. The process is used with shielding from the decomposition of the electrode covering, without the application of pressure, and with filler metal from the electrode.

soldadura de arco metálico protegido (SMAW). Un proceso de soldadura de arco con un arco en medio de un electrodo cubierto y el charco de soldadura. El proceso se usa con protección de descomposición del cubrimiento del electrodo sin la aplicación de presión, y con el metal de relleno del electrodo.

shielding gas. A gas used to produce a protective atmosphere.
gas protector. El gas protector se usa para prevenir o reducir la contaminación atmosférica.

short arc. A nonstandard term for short-circuiting transfer arc welding.
arco corto. Un término fuera de la norma para transferir por corto circuito (soldadura de arco).

short-circuiting arc welding. A nonstandard term for short-circuiting transfer (arc welding).
soldadura de arco con corto circuito. Un término fuera de la norma para transferir por corto circuito (soldadura de arco).

short-circuiting transfer (arc welding). Metal transfer in which molten metal from a consumable electrode is deposited during repeated short circuits.
transferir por corto circuito (soldadura de arco). Transferir metal el cual el metal derretido del electrodo consumible es depositado durante repetidos cortos circuitos.

shoulder. See preferred term **root face.**
hombro. Vea término preferido **cara de raíz.**

shrinkage void. A cavity-type discontinuity formed as a metal contracts during solidification.
vacío de encogimiento. Una discontinuidad tipo cavidad normalmente formada por encogimiento durante solidificación.

side-bend test. A test in which the side of a transverse section of the weld is on the convex surface of a specified bend radius.
prueba de dobléz de lado. Una prueba en la cual el lado de una sección transversa de la soldadura está en la superficie convexa de un radio de dobléz especificado.

*silver braze. A brazing process using an alloyed brazing rod which contains some percentage of silver. Silver is used as an alloy to promote wetting and strength.
soldadura fuerte con plata. Proceso de soldadura fuerte que utiliza una vara de aleación que contiene un porcentaje de plata. La plata se usa como aleación para promover la exudación y la fortaleza.

silver soldering, silver alloy brazing. Nonpreferred terms used to denote brazing with a silver-base filler metal. See preferred term **furnace brazing.**
soldadura blanda con plata, soldadura fuerte con aleación de plata. Términos no preferidos que se usan para denotar soldadura fuerte con metal para rellenar con base de plata. Vea el término preferido **soldadura fuerte en horno.**

single bevel-groove weld. A type of groove weld. Refer to drawing for **groove weld.**
soldadura de ranura de un solo bisel. Tipo de soldadura de ranura. Refiérase al dibujo para **soldadura de ranura.**

single-flare bevel-groove weld. A type of groove weld. Refer to drawing for **groove weld.**
soldadura de ranura de un solo bisel acampanado. Un tipo de soldadura de ranura. Refiérase al dibujo para **soldadura de ranura.**

single-flare V-groove weld. A type of groove weld. Refer to drawing for **groove weld.**
soldadura de ranura de una sola V acampanada. Un tipo de soldadura de ranura. Refiérase al dibujo para **soldadura de ranura.**

single J-groove weld. A type of groove weld. Refer to drawing for **groove weld.**
soldadura de ranura de una sola J. Un tipo de soldadura de ranura. Refiérase al dibujo para **soldadura de ranura.**

single-port nozzle. A constricting nozzle of the plasma arc torch that contains one orifice, located below and concentric with the electrode.
boquilla de una sola abertura. Es una boquilla constreñida de la antorcha de arco de plasma que contiene un orificio, situado debajo y concéntrico al electrodo.

single square-groove weld. A type of groove weld. Refer to drawing for **groove weld.**
soldadura de ranura de una sola escuadra. Un tipo de soldadura de ranura. Refiérase al dibujo de **soldadura de ranura.**

single U-groove weld. A type of groove weld. Refer to drawing for **groove weld.**
soldadura de ranura de una sola U. Un tipo de soldadura de ranura. Refiérase al dibujo para **soldadura de ranura.**

single V-groove weld. A type of groove weld. Refer to drawing for **groove weld.**
soldadura de ranura de una sola V. Un tipo de soldadura de ranura. Refiérase al dibujo de **soldadura de ranura.**

size of weld.

groove weld. The joint penetration (depth of bevel plus the root penetration when specified). The size of a groove weld and its effective throat are one and the same.

fillet weld. For equal leg fillet welds, the leg lengths of the largest isosceles right triangle that can be inscribed within the fillet weld cross section. Refer to drawings for concavity and convexity. For unequal leg fillet welds, the leg lengths of the largest right triangle that can be inscribed within the fillet weld cross section.
Note: When one member makes an angle with the other member greater than 105°, the leg length (size) is of less significance than the effective throat, which is the controlling factor for the strength of a weld.

flange weld. The weld metal thickness measured at the root of the weld.

tamaño de la soldadura.

soldadura de ranura. La penetración de la junta (profundidad del bisel más la penetración de la raíz cuando está especificada). El tamaño de la soldadura de ranura y la garganta efectiva son una y la misma.

soldadura filete. Para soldaduras con piernas iguales de filete, lo largo de las piernas del triángulo recto con el isosceles más grande que puede ser inscrito dentro de la sección transversal de la soldadura de filete. Refiérase al dibujo para concavidad y convexidad. Para piernas de soldadura de filete desiguales, lo largo de las piernas del triángulo recto más grande que puede ser inscrito dentro de la sección transversa de la soldadura de filete.
Nota: Cuando un miembro hace un ángulo con otro miembro más grande de 105 grados, lo largo de la pierna (tamaño) es de menor significado que la garganta efectiva, la cual es el factor de control para la fuerza de una soldadura.

soldadura de brida. Lo grueso del metal de soldadura se mide a la raíz de la soldadura.

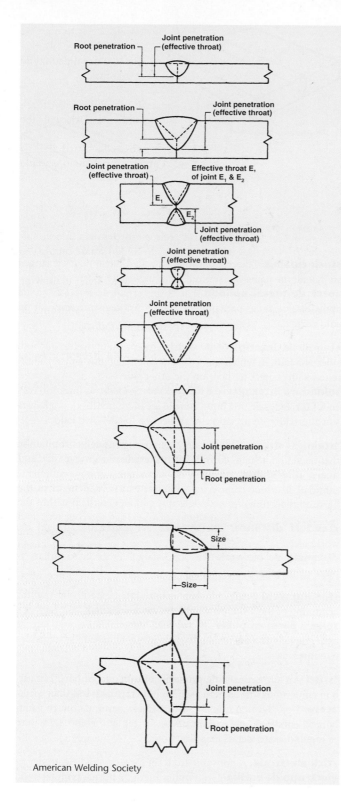

American Welding Society

slag. A nonmetallic product resulting from the mutual dissolution of flux and nonmetallic impurities in some welding and brazing processes.
escoria. Un producto que no es metálico resultando de una disolución mutua del flujo y las impuridades no metálicas en unos procesos de soldadura y soldadura fuerte.

slag inclusion. Nonmetallic solid material entrapped in weld metal or between weld metal and base metal.

inclusion de escoria. Material sólido no metálico atrapado en el metal de soldadura o entre el metal de soldadura y el metal base.

slag pan. A high-temperature container that holds the slag from a thermite weld until it cools.
bandeja para escoria. Un recipiente de alta temperatura que contiene la escoria de una soldadura con termita hasta que se enfría.

*slope. For gas metal arc welding, the volt-ampere curve of the power supply indicates that there is a slight decrease in voltage as the amperage increases; the rate of voltage decrease in the slope.
pendiente. Para soldadura de arco de metal con gas, la curva voltio-amperio de la fuente de poder indica que si hay un ligero decremento en voltaje cuando los amerios aumentan; la proporción del voltaje decrementa en el pendiente.

slot weld. A weld made in an elongated hole in one member of a lap or tee joint joining that member to that portion of the surface of the other member that is exposed through the hole. The hole may be open at one end and may be partially or completely filled with weld metal. (A fillet-welded slot should not be construed as conforming to this definition.)
soldadura de ranura alargada. Una soldadura hecha en un agujero alargado en un miembro de una junta en solape o T uniendo ese miembro a esa porción de la superficie del otro miembro que está expuesto a través del agujero. El agujero puede ser abierto en una punta y puede ser parcialmente o completamente rellenado con metal de soldadura. (Una ranura alargada con soldadura de filete no debe de interpretarse como conforme a está definición.)

slugging. The act of adding a separate piece or pieces of material in a joint before or during welding that results in a welded joint not complying with design, drawing, or specification requirements.
usar trozos de metal. El acto de agregar una pieza o piezas separadas de material en una junta antes o durante la soldadura que resulta en una junta soldada que no cumple con diseño, dibujo, o las especificaciones requeridas.

solder. A filler metal used in soldering that has a liquidus not exceeding 840°F (450°C).
soldadura (material para soldar). Un metal de relleno usado para soldadura blanda que tiene un liquidus que no excede de 840°F (450°C).

soldering (S). A group joining process in which the workpiece(s) and solder are heated to the soldering temperature to form a solder joint.
soldadura blanda (S). Un grupo de procesos de soldadura que produce coalescencia de materiales calentándolos a una temperature de soldar y usando un metal para rellenar con un liquidus que no exceda de 840°F (450°C) y más abajo del solidus de los metales base. El metal para rellenar es distribuido en medio de las superficies empalmadas acopladas muy cerca de la junta por acción capilar.

soldering gun. An electrical heated soldering iron with a pistol grip.
pistola de soldar. Un fierro eléctrico para soldar con mango de pistola, rápido para calentarse, y tiene una punta relativamente pequeña.

solid state welding (SSW). A group of welding processes producing coalescence by the application of pressure without melting any of the joint components.

soldadura de estado sólido (SSW). Un grupo de procesos para soldar que produce coalescencia cuando se le aplica presión a una temperatura de soldadura más baja que las temperatures que se usan para derretir el metal base y el metal de relleno.

solidus. The highest temperature at which a metal is completely solid.
solidus. La temperatura más alta cuando un metal o una aleación está completamente sólido.

spatter. The metal particles expelled during welding and that do not form a part of the weld.
salpicadura. Las partículas de metal que se despidan cuando se está soldando y que no forman parte de la soldadura.

spatter loss. Metal lost due to spatter.
pérdida causa salpicadura. El metal perdido debido a la salpicadura.

spool. A filter metal package configuration in which the wire is wound on a cylinder (called a barrel), which is flanged at both ends. The flange contains a spindle hole centered inside the barrel.
carrete. Un paquete de metal tipo filtro consistiendo de una extensión continua de un electrodo enrollado en un cilindro (llamado el barril), el cual tiene una brida en los dos extremos. La brida se extiende debajo del diámetro de adentro del barril y contiene un agujero huso.

spot weld. A weld produced between or upon overlapping members with coalescence initating and occurring at faying surfaces or proceeding from the outer surface of one member. The weld typically has a round cross section in the plane of the faying surfaces. See also **arc spot weld** and **resistance spot welding**.
soldadura de puntos. Una soldadura hecha en medio o sobre miembros traslapados en la cual la coalescencia puede empezar y ocurrir en las superficies empalmadas o puede continuar en la superficie de un miembro. La sección transversa (plan de vista) es aproximadamente circular. Vea también **soldadura de puntos por arco** y **soldadura de puntos por resistencia**.

spray arc. A nonstandard term for spray transfer.
arco para rociar. Un término fuera de norma para traslado rociado.

spray transfer (arc welding). Metal transfer in which molten metal from a consumable electrode is propelled axially across the arc in small droplets.
traslado rociado (soldadura de arco). Transferir el metal el cual el metal derretido de un electrodo consumible es propelado axialmente a traves del arco en gotitas pequeñas.

*square butt joint. A joint made when two flat pieces of metal face each other with no edge preparation. See also **square groove weld**.
junta escuadra de tope. Una junta hecha cuando dos piezas planas de metal se enfrentan una a la otra sin preparación de orilla. Vea también **soldadura de ranura escuadra**.

square butt weld. See butt joint.
junta escuadra de tope. Vea **junta a tope**.

square-groove weld. A type of groove weld.
soldadura de ranura escuadra. Un tipo de soldadura de ranura.

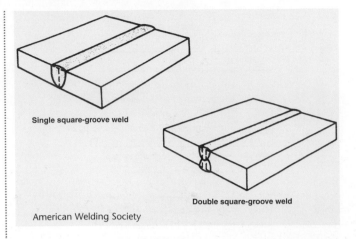

Single square-groove weld
Double square-groove weld
American Welding Society

stack cutting. Thermal cutting of stacked metal plates arranged so that all the plates are severed by a single cut.
corte de metal apilado. Un corte termal de hojas de metal apilados arregladas para que todas las hojas sean cortadas por un solo corte.

staggered intermittent welds. Intermittent welds on both sides of a joint in which the weld increments on one side are alternated with respect to those on the other side.
soldadura intermitente de cadena. Soldaduras intermitentes en los dos lados de una junta en cual los incrementos de soldadura son alternados de un lado con respecto a los del otro lado.

*stainless steels. Alloys of steel containing enough chromium so that they do not stain, corrode, or rust easily.
acero inoxidable. Aleación de acero que contiene suficiente cantidad de cromo como para evitar que ésta se manche, corroa, o herrumbre fácilmente.

standoff distance. The distance between a nozzle and the workpiece.
distancia de alejamiento. La distancia entre la boquilla y la pieza de trabajo.

starting weld tab. Additional material that extends beyond the beginning of the joint, on which the weld is started.
solera para empezar a soldar. Material adicional que se extiende más allá del principio de la junta, en donde la soldadura es empezada.

*steel. An alloy consisting primarily of iron and carbon. The carbon content may be as high as 2.2% but is usually less than 1.5%.
acero. Una aleación que consiste primeramente de hierro y carbón. El contenido del carbón puede ser tan alto como 2.2% pero es regularmente menos de 1.5%.

stick electrode. A nonstandard term for a covered electrode.
electrodo de varilla. Un término fuera de norma por electrodo cubierto.

stickout. See preferred term **electrode extension**.
sobresalga. Vea término preferido **extensión del electrodo**.

straight polarity. The arrangement of direct-current arc welding leads in which the work is the positive pole and the electrode is the negative pole of the welding arc. A synonym for direct-current electrode negative. Refer to drawing for **direct-current electrode negative**.
polaridad directa. El arreglo de los cables de soldadura de arco con corriente directa donde el cable de la tierra es el polo positivo

y el porta electrodo es el polo negativo del arco de soldadura. Un sinónimo para corriente directa con electrodo negativo. Refiérase al dibujo para **corriente directa con electrodo negativo**.

stranded electrode. A composite filler metal electrode consisting of stranded wires that may mechanically enclose materials to improve properties, stabilize the arc, or provide shielding.

electrodo cable. Electrodo de metal para rellenar compuesto que consiste de cable de alambres que pueden encerrar materiales mecánicamente para mejorar propiedades, estabilizar el arco, o proveer protección.

*stress point. Any point in a weld where incomplete fusion of the weld on one or both sides of the root gives rise to stress, which can result in premature cracking or failure of the weld at a load well under the expected strength of the weld.

punto de tensión. Cualquier punto en una soldadura donde la fusión incompleta en la soldadura en uno o en los dos lados de la raíz le aumenta la tensión, la cual puede resultar en una grieta o falta prematura en la soldadura con una carga mucho menos que la fuerza de la soldadura que se esperaba.

stress relief heat treatment. Uniform heating of a structure or a portion thereof to a sufficient temperature to relieve the major portion of the residual stresses, followed by uniform cooling.

tratamiento de calor para relevar la tensión. Calentamiento uniforme de una estructura o una porción a una temperatura suficiente para relevar la mayor porción de las tensiones restantes, seguido por enfriamiento uniforme.

stringer bead. A type of weld bead made without appreciable weaving motion. See also **weave bead**.

cordón encordador. Un tipo de cordón de soldadura sin movimiento del tejido apreciable. Vea también **cordón tejido**.

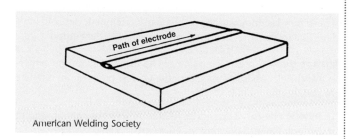

Path of electrode

American Welding Society

stud arc welding (SW). An arc welding process that uses an arc between a metal stud, or similar part, and the other workpiece. The process is used with or without shielding gas or flux, with or without partial shielding from a ceramic ferrule surrounding the stud, with the application of pressure after the faying surfaces are sufficiently heated, and without filler metal.

esparrago (tachón) para soldadura de arco (SW). Un proceso de soldadura de arco que usa un arco entre un esparrago (tachón) de metal, o parte similar, y la otra pieza de trabajo. El proceso es usado con o sin flujo o protección de gas, con o sin protección parcial del casquillo cerámico que rodea el esparrago (tachón), con la aplicación de presión después que las superficies empalmadas tengan suficiente calor, y sin metal de relleno.

stud welding. A general term for the joining of a metal stud or similar part to a workpiece. Welding may be accomplished by arc, resistance, friction, or other suitable process with or without external gas shielding.

soldadura de esparrago (tachón). Un término general para la unión de esparragos (tachones) o parte similar a una pieza de trabajo. La soldadura se puede efectuar por arco, resistencia, fricción, u otro proceso conveniente con o sin gas externo para protección.

submerged arc welding (SAW). An arc welding process using an arc or arcs between a bare metal electrode or electrodes and the weld pool. The arc and molten metal are shielded by a blanket of granular flux on the workpieces. The process is used without pressure and with filler metal from the electrode and sometimes from a supplemental source (welding rod, flux, or metal granules).

soldadura por arco sumergido (SAW). Un proceso de soldar con arco que usa un arco o arcos entre un electrodo de metal liso o electrodos y el charco de la soldadura. El arco y el metal derretido son protegidos por una capa de flujo granular sobre la pieza de trabajo. El proceso es usado sin presión y con metal para rellenar del electrodo y a veces de una fuente suplementaria (varilla para soldar, flujo, o gránulos de metal).

substrate. Any base material to which a thermal sprayed coating or surfacing weld is applied.

substrato. Cualquier material base al cual se le aplica una capa termal o una soldadura de superficie.

suck back. See preferred term **concave root surface**.

succión del cordón de raíz. Vea el término preferido **superficie raíz concavo**.

surface preparation. The operations necessary to produce a desired or specified surface condition.

preparación de la superficie. Las operaciones necesarias para producir una deseada o una especificada condición de la superficie.

surfacing. The application by welding, brazing, or thermal spraying of a layer of material to a surface to obtain desired properties or dimensions, as opposed to making a joint. See also **buttering**, **cladding**, **coating**, and **hardfacing**.

recubrimiento superficial. La aplicación a la soldadura, a la soldadura fuerte o rociado termal de una capa de material a la superficie para obtener las deseadas propiedades o dimensiones, contrario a la hechura de una junta. Vea también **recubrimiento antes de terminar una soldadura**, **capa de revestimiento**, **revestimiento**, y **endurecimiento de caras**.

*synergic system. Pulsed-arc metal transfer system in which the power supply and wire-feed settings are made by adjusting a single knob.

sistema sinérgico. Sistema de transferir metal por arco pulsado en cual la fuente de poder y los ajustes del alimentador de alambre son hechos por el ajuste de un botón solamente.

T

tab. See runoff weld tab, starting weld tab, and weld tab.

solera. Vea **solera de carrera final de soldadura**, **solera para empezar a soldar**, y **solera para soldar**.

tack weld. A weld made to hold parts of a weldment in proper alignment until the final welds are made.

soldadura de puntos aislados. Una soldadura hecha para detener las partes en su propio alineamiento hasta que se hagan las soldaduras finales.

*tactile sensor. A transducer that is sensitive to touch.

sensor táctil. Un transducer que es sensitivo al tocar.

taps. Connections to a transformer winding that are used to vary the transformer turns ratio, thereby controlling welding voltage and current.
grifo. Conexiones al arrollamiento de un transformador que se usan para variar la proporción de vueltas del transformador, asi se puede controlar la corriente y el voltaje para soldar.

*teach. To program a manipulator arm by guiding it through a series of points or in a motion pattern that is recorded for subsequent automatic action by the manipulator.
enseñar. Para programar un manipulador de brazo guiándolo por una serie de puntos o en una muestra de movimiento que está registrada para acción automática subsiguiente por el manipulador.

*teaching interface. The mechanisms or devices by which a human operator teaches a machine.
enseñanza de interfaze. Los mecanismos o aparatos por los cuales un operador humano enseña a la máquina.

tee joint. A joint between two members located approximately at right angles to each other in the form of a T.
junta en T. Una junta en medio de dos miembros que están localizados aproximadamente a ángulos rectos de uno al otro en la forma de T.

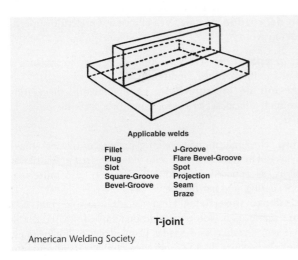

Applicable welds

Fillet	J-Groove
Plug	Flare Bevel-Groove
Slot	Spot
Square-Groove	Projection
Bevel-Groove	Seam
	Braze

T-joint

American Welding Society

*tempering. Reheating hardened metal before it cools to room temperature to make it tough, not brittle.
templar. Recalentando un metal endurecido antes de que se enfríe a la temperatura del ambiente para hacerlo duro, no frágil.

*tensile strength. As applied to a brazed or soldered joint, the ability of the joint to withstand being pulled apart.
resistencia a la tensión. Como es aplicada a una junta de soldadura fuerte o soldadura blanda, la capacidad de una junta que resista ser estirada hasta que se rompa en dos pedazos.

tension test. A test in which a specimen is loaded in tension until failure occurs. See also **reduced section tension test.**
prueba de tensión. Una prueba en la cual la probeta está cargada de tensión hasta que ocurra el fracaso. Vea también **prueba de tensión de sección reducida.**

thermal cutting (TC). A group of cutting processes severing or removing metal by localized melting, burning, or vaporizing of the workpiece. See also **arc cutting, electron beam cutting, laser beam cutting,** and **oxygen cutting.**
corte termal (TC). Un grupo de procesos para cortar que desúne o quita el metal para derretir o quemar o vaporizar localmente las

piezas de trabajo. Vea también **corte con arco, cortes rayo de electron, corte de rayo laser,** y **cortes de oxígeno.**

thermal spraying (THSP). A group of processes in which finely divided metallic or nonmetallic surfacing materials are deposited in a molten or semimolten condition on a substrate to form a thermal spray deposit. The surfacing material may be in the form of powder, rod, cord, or wire.
rociado termal (THSP). Un grupo de procesos el cual los materiales metálicos o no metálicos de superficie que son depositados en una condición derretida o semiderretida sobre el substrato para formar un depósito de rociado termal. El material de superficie puede ser en forma de polvo, varilla, cordón, o alambre.

thermal stresses. Stresses in a material or assembly resulting from nonuniform temperature distribution or differential thermal expansion.
tensión termal. Tensiones en el metal resultando cuando la distribución de la temperatura no está uniforme.

thermite welding (TW). A welding process producing coalescence of metals by heating them with superheated liquid metal from a chemical reaction between a metal oxide and aluminum, with or without the application of pressure.
soldadura termita (TW). Un proceso de soldadura que produce coalescencia del metal calentándolos con un metal supercalentado líquido da una reacción química entre un metal óxido y el aluminio, con o sin la aplicación de presión. El metal de relleno, cuando es usado, es obtenido del metal líquido.

throat area, *resistance welding.* The region bounded by the physical components of the secondary circuit in a welding machine.
área de garganta, *soldadura por resistencia.* La área limitada por las partes físicas del circuito secundario en un punto de resistencia, costura, o una máquina de soldar de proyección. Se usa para determinar las dimensiones de una parte que puede ser soldada y determinar, en parte, la impedancia secundaria del equipo.

throat depth, *resistance welding.* The distance from the centerline of the electrodes or platens to the nearest point of interference for flat sheets.

actual throat. The shortest distance from the root of weld to its face. Refer to drawing for **convexity.**
effective throat. The minimum distance minus any reinforcement from the root of weld to its face. Refer to drawing for **convexity.**
theoretical throat. The distance from the beginning of the root of the joint perpendicular to the hypotenuse of the largest right triangle that can be inscribed within the fillet weld cross section. This dimension is based on the assumption that the root opening is equal to zero. Refer to drawing for **convexity.**
profundidad de garganta, *soldadura por resistencia.* En una punta de resistencia, costura, o máquina de soldar de proyección, la distancia de la línea del centro del electrodo o platinas al punto más cercano de interferencia para las hojas planas.

garganta actual. La distancia más corta de la raíz de una soldadura a su cara. Refiérase al dibujo para **convexidad.**
garganta efectiva. La distancia mínima menos cualquier refuerzo de la raíz de la soldadura a su cara. Refiérase al dibujo para **convexidad.**
garganta teórica. La distancia de donde empieza la raíz de la junta perpendicular a la hipotenusa del triángulo recto más grande que puede ser inscrito adentro de la sección transversa

de una soldadura de filete. Esta dimensión está basada en la proposición que la abertura de la raíz es igual a cero. Refiérase al dibujo para **convexidad**.

TIG welding. A nonstandard term when used for gas tungsten arc welding.
soldadura TIG. Un término fuera de norma cuando es usado por soldadura de arco de tungsteno con gas.

*titanium. This silver colored metal has a high strength to weight ratio which is why it is used extensively in the aerospace industry.
titanio. Metal de color plateado con proporción dureza-peso alta y por lo cual se usa extensamente en la industria aeroespacial.

toe crack. A crack in the base metal occurring at the toe of a weld.
grieta de pie. Una grieta en el metal base que ocurre al pie de la soldadura.

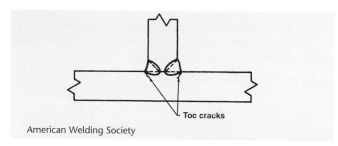

Toe cracks

American Welding Society

toe of weld. The junction between the face of a weld and the base metal. Refer to drawing for **face of weld**.
pie de la soldadura. La unión entre la cara de la soldadura y el metal base. Refiérase al dibujo para **cara de la soldadura**.

tolerances. The allowable deviation in accuracy or precision between the measurement specified and the part as laid out or produced.
tolerancias. Desviación permitida en la precisión entre la medida especificada y la pieza instalada o producida.

torch. See preferred terms **cutting torch** and **welding torch**.
antorcha. Vea el término preferido **antorcha para cortar** y **antorcha para soldar**.

*torch angle. The angle between the centerline of the torch and the work surface; the ideal torch angle is 45°. The torch angle affects the percentage of heat input into the metal, thus affecting the speed of melting and the size of the molten weld pool.
ángulo de antorcha. El ángulo en medio de la línea del centro de la antorcha y la superficie del trabajo; el ángulo ideal de la antorcha es 45°. El ángulo de la antorcha afecta el por ciento de calor que entra dentro del metal, asi afectando la rapidez de derretimiento y el tamaño del charco del metal derretido de la soldadura.

torch brazing (TB). A brazing process that uses heat from a fuel-gas flame.
soldadura fuerte con antorcha (TB). Un proceso de soldadura fuerte que usa calor de una llama de gas combustible.

*torch manipulation. The movement of the torch by the operator to control the weld bead characteristics.
manipulacion de la antorcha. El movimiento de la antorcha por el operador para el control de las características del cordón de la soldadura.

torch soldering (TS). A soldering process that uses heat from a fuel-gas flame.
soldadura blanda con antorcha (TS). Un proceso de soldadura blanda que usa calor de una llama de gas combustible.

*trailing edge. See electrode angle.
borde del flujo. Ver ángulo del electrodo.

transducer. A device that transforms one form of energy into another.
transducor. Un aparato que convierte una forma de energía a otra.

transferred arc. A plasma arc established between the electrode of the plasma arc torch and the workpiece.
arco transferido. Un arco de plasma establecido entre el electrodo de la antorcha de arco de plasma y la pieza de trabajo.

*transition current. In gas metal arc welding, current above a critical level to permit spray transfer; the rate at which drops are transferred changes in relationship to the current. Transition current depends upon the alloy bearing welded and is proportional to the wire diameter.
corriente de transición. En soldadura de arco y metal con gas, corriente arriba de un nivel crítico para permitir el traslado del rociado; la proporción en la cual las gotas son transferidas cambia en relación a la corriente. La corriente de transición depende del aleado que se está soldando y es proporcional al diámetro del alambre.

transverse face bend. See face bend.
doblez de cara transversal. Vea **doblez de cara**.

transverse root bend. See root bend.
cordón de raíz transversal. Vea **cordón de raíz**.

travel angle. The angle less than 90° between the electrode axis and a line perpendicular to the weld axis, in a plane determined by the electrode axis and the weld axis. The angle can also be used to partially define the position of guns, torches, rods, and beams. See also **drag angle** and **push angle**. Refer to drawing for **backhand welding**.
ángulo de avance. El ángulo menos de 90° entre el eje del electrodo y una línea perpendicular al eje de la soldadura, en un plano determinado por el eje del electrodo y el eje de la soldadura. El ángulo también puede ser usado para parcialmente definir la posición de las pistolas, antorchas, varillas, y rayos. Vea también **ángulo del tiro** y **ángulo de empuje**. Refiérase al dibujo para **soldadura en revés**.

travel angle (pipe). The angle less than 90° between the electrode axis and a line perpendicular to the weld axis at its point of intersection with the extension of the electrode axis, in a plane determined by the electrode axis and a line tangent to the pipe surface of the same point. This angle can also be used to partially define the position of guns, torches, rods, and beams. Refer to drawing for **backhand welding**.
ángulo de avance (tubo). El ángulo menos de 90° entre el eje del electrodo y la línea perpendicular al eje a su punto de intersección con la extensión del eje del electrodo, en un plano determinado por el eje del electrodo y una línea tangente a la superficie del tubo del mismo punto. Este ángulo también puede ser usado para parcialmente definir la posición de las pistolas, varillas, y rayos. Refiérase al dibujo para **soldadura en revés**.

*tungsten. This steel-gray metal has a melting temperature of 6170°F (3410°C) and it freely emits electrons which makes it ideal for the nonconsumable electrodes for PAC, PAW, and GTAW processes.
tungsteno. Metal de color gris que posee una temperatura de fusión de 6170°F (3410°C) y que emite libremente electrodos,

lo cual lo hace ideal para los electrodos no consumibles de los procesos PAC, PAW, y GTAW.

tungsten electrode. A nonfiller metal electrode used in arc welding, arc cutting, and plasma spraying, made principally of tungsten.
electrodo de tungsteno. Un electrodo de metal que no se rellena que se usa para soldadura de arco, cortes por arco, rociado por plasma, y hecho principalmente de tungsteno.

*type A fire extinguisher. An extinguisher used for combustible solids, such as paper, wood, and cloth. Identifying symbol is a green triangle enclosing the letter *A*.
extinguidor para incendios tipo A. Un extinguidor que se usa para combustibles sólidos como papel, madera, y tela. El símbolo de identificación es un triángulo verde con la letra *A* adentro.

*type B fire extinguisher. An extinguisher used for combustible liquids, such as oil and gas. Identifying symbol is a red square enclosing the letter *B*.
extinguidor para incendios tipo B. Un extinguidor que se usa para liquidos combustibles, como aceite y gas. El símbolo de identificación es un cuadro rojo con la letra *B* adentro.

*type C fire extinguisher. An extinguisher used for electrical fires. Identifying symbol is a blue circle enclosing the letter *C*.
extinguidor para incendios tipo C. Un extinguidor que se usa para incendios eléctricos. El símbolo de identificación es un círculo azul con la letra *C* adentro.

*type D fire extinguisher. An extinguisher used on fires involving combustible metals, such as zinc, magnesium, and titanium. Identifying symbol is a yellow star enclosing the letter *D*.
extinguidor para incendios tipo D. Un extinguidor que se usa para incendios de metales combustibles, como zinc, magnesio, y titanio. El símbolo de identificación es una estrella amarilla con una letra *D* adentro.

U

U-groove weld. A type of groove weld.
soldadura de ranura en U. Un tipo de soldadura de ranura.

ultrasonic coupler (ultrasonic soldering and ultrasonic welding). Elements through which ultrasonic vibration is transmitted from the transducer to the tip.
acoplador ultrasónico (soldadura blanda ultrasónica y soldadura ultrasónica). Los elementos por los cuales la vibración ultrasónica es transmitida del transducor a la punta.

*ultrasonic inspection (UT). See nondestructive testing.
inspección ultrasónica (UT). Ver prueba no destructiva.

ultrasonic welding (USW). A solid state welding process that produces a weld by the local application of high-frequency vibratory energy as the workpieces are held together under pressure.
soldadura ultrasónica. Un proceso de soldadura de estado sólido que produce una soldadura por la aplicación local de energía vibratoria de alta frecuencia asi cuando las piezas de trabajo están agarradas juntas bajo presión.

*ultraviolet light. A very short wavelength light that is above the visible light range.
luz ultravioleta. Luz de longitud de onda muy corta que se encuentra por encima del espectro de la luz visible.

underbead crack. A crack in the heat-affected zone, generally not extending to the surface of the base metal.

grieta entre o bajo cordones. Una grieta en la zona afectada por el calor, generalmente no se extiende a la superficie del metal base.

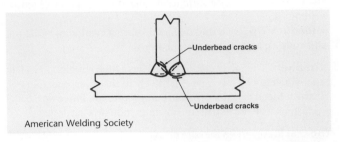

American Welding Society

undercut. A groove melted into the base metal adjacent to the toe or root of a weld and left unfilled by weld metal. Refer to drawing for **overlap**.
socavación. Una ranura dentro del metal base adyacente al pie o raíz de la soldadura y se deja sin rellenar con el metal de soldadura. Refiérase al dibujo para **traslapo**.

underfill. A depression on the face of the weld or root surface extending below the surface of the adjacent base metal.
faltante de material. Una depresión en la cara de la soldadura o la superficie de la raíz extendiéndose más abajo de la superficie del adyacente metal base.

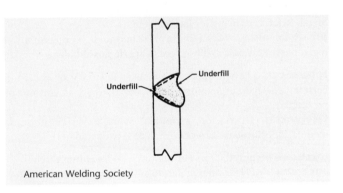

American Welding Society

uphill. Welding with an upward progression.
soldando hacia arriba. Solando con progresión hacia arriba.

*upper arm (robot). The portion of a jointed arm that is connected to the shoulder.
brazo de arriba (robot). La porción que úne al brazo que conecta con el hombro.

upset. Bulk deformation of a workpiece(s) resulting from the application of pressure with or without added heat, expressed in terms of increase in transverse section area, reduction in length, reduction in thickness, or reduction of the cross wire weld stack height.
recalada. Deformación en bulto resultado de la aplicación de presión en la soldadura. El recalado puede ser medido como un aumento en porciento en una área interfacial, una reducción en lo largo, o un porciento de reducción en lo grueso para juntas de solape.

upset welding (UW). A resistance welding process producing a weld over the entire area of faying surfaces or progressively along a butt joint.
soldadura recalada (UW). Un proceso de soldadura por resistencia que produce coalescencia sobre toda la área de superficie empalmada o progresivamente sobre una junta a tope con el calor obtenido de la resistencia del flujo de la corriente de la soldadura

por la área donde esas superficies están en contacto. Presión es usada para completar la soldadura.

upslope time (automatic arc welding). The time during which the current changes continuously from initial current valve to the welding value.

tiempo del pendiente en ascenso (soldadura automática de arco). El tiempo durante el cual la corriente cambia continuamente el valor de la corriente inicial al valor de la soldadura.

V

vector lines. Lines the computer draws between two points. These lines always stay crisp and sharp even when the drawing is zoomed in (magnified) hundreds of times.

líneas vectoriales. Líneas que la computadora traza entre dos puntos. Estas líneas siempre se mantienen visibles y nítidas incluso cuando se aumenta (magnifica) el tamaño del dibujo cientos de veces.

vertical position. The position of welding in which the axis of the weld is approximately vertical.

posición vertical. La posición de la soldadura en la cual el eje para soldarse es aproximadamente vertical.

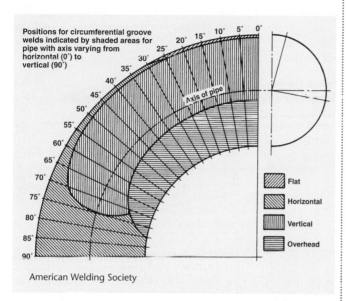

American Welding Society

vertical position (pipe welding). The position of a pipe joint in which welding is performed in the horizontal position and the pipe may or may not be rotated.

posición vertical (soldadura de tubo). La posición de una junta de tubo en la cual la soldadura se hace en la posición horizontal y el tubo puede o no dar vueltas.

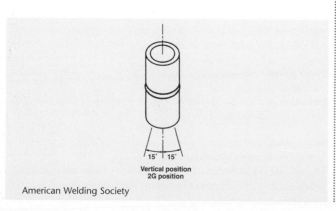

American Welding Society

V-groove weld. A type of groove weld.

soldadura de ranura V. Un tipo de soldadura de ranura.

visible light. The light range that we see.

luz visible. El espectro de luz que podemos ver.

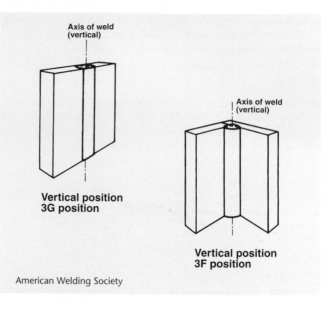

American Welding Society

voltage range. The lower and upper limits of welding power, in volts, that can be produced by a welding machine or used with an electrode or by a process.

rango de voltaje. Los límites máximos y mínimos de poder de soldadura (en voltios) que puede tener una máquina para soldar o que pueden usarse con un electrodo o a través de un proceso.

voltage regulator. An automatic electrical control device for maintaining a constant voltage supply to the primary of a welding transformer.

regulador de voltaje. Un aparato de control eléctrico automático para mantener y proporcionar un voltaje constante a la primaria de un transformador de una soldadura.

W

wagon tracks. A pattern of trapped slag inclusions in the weld that show up as discontinuities in X-rays of the weld.

huellas de carreta. Una muestra de inclusiones de escoria atrapadas en la soldadura que enseña que hay discontinuidades en los rayos-x de la soldadura.

water-arc plasma cutting (PAC). In this process, nitrogen is used as the plasma gas. The plasma is created at a high temperature in an arc between the electrode and the orifice. A tap water spray applied to the plasma causes it to constrict and accelerate. This column causes the cutting action and melts a very narrow kerf in the material.

cortes de arco-agua plasma (PAC). En este proceso, nitrogeno es usado como el gas de plasma. La plasma es creada a una temperatura alta en un arco en medio del electrodo y la orifice. La aplicación de agua del grifo a la plasma la hace que se constriñe y acelera. Está columna causa la acción de cortar y derritir un corte que es muy estrecho en el material.

water jet cutting. A process using extremely high pressure water to cut through material.

corte con chorro de agua. Proceso que usa agua a presión extremadamente alta para cortar un material.

*wattage. A measurement of the amount of power in the arc; the wattage of the arc controls the width and depth of the weld bead.
número de vatios. Una medida de la cantidad de poder en el arco; el número de vatios del arco controla lo ancho y hondo del cordón de la soldadura.

wax pattern (thermit welding). Wax molded around the workpieces to the form desired for the completed weld.
soldadura termita (molde de cera). Un molde de cera alrededor de las piezas de trabajo a la forma deseada para la soldadura terminada.

weave bead. A type of weld bead made with transverse oscillation.
cordón tejido. Un tipo de cordón de soldadura hecha con oscilación transversa.

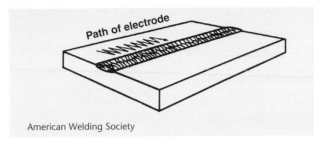

Path of electrode

American Welding Society

*weave pattern. The movement of the welding electrode as the weld progresses; common weave patterns include circular, square, zigzag, stepped, C, J, T, and figure 8.
muestra de tejido. El movimiento del electrodo para soldar a como progresa la soldadura; las muestras de tejidos comunes incluyen circular, de cuadro, zigzag, de pasos, C, J, T, y la figura 8.

weld. A localized coalescence of metals or nonmetals produced either by heating the materials to suitable temperatures, with or without the application of pressure, or by the application of pressure alone and with or without the use of the filler material.
soldar. Una coalescencia localizada de metales o metaloides producida al calentar los materiales a una temperatura adecuada, con o sin la aplicación de presión, o por la aplicación de presión solamente y con o sin el uso del material de relleno.

weld axis. A line through the length of the weld, perpendicular to and at the geometric center of its cross section.
eje de la soldadura. Una línea a través de lo largo de la soldadura, perpendicular a y al centro geométrico de su sección transversa.

weld bead. A weld resulting from a single welding pass. See also **stringer bead** and **weave bead.**
cordón de soldadura. Una soldadura resultante de una soldadura de una sola pasada. Vea también **cordón encordador** y **cordón tejido.**

weld brazing. Brazing using heat from a welding process such that the preplaced brazing filler metal is melted to form a braze augmenting the weld by increasing joint strength, or creating a seal between spot or intermittent welds.
soldadura y soldadura fuerte. Un método de unir que combina soldadura de resistencia con soldadura fuerte.

weld crack. A crack located in the weld metal or heat-affected zone.
grieta en la soldadura. Una grieta en el metal de soldadura.

weld face. The exposed surface of a weld on the side from which welding was done.
cara de la soldadura. La superficie expuesta de una soldadura en el lado de donde se hizo la soldadura.

weld gauge. A device designed for measuring the shape and size of welds.
instrumento para medir la soldadura. Un aparato diseñado para comprobar la forma y tamaño de las soldaduras.

weld groove. A channel in the surface of a workpiece or an opening between two joint members that provides space to contain a weld.
soldadura de ranura. Un canal en la superficie de una pieza de trabajo o una abertura entre dos miembros de junta que provee espacio para contener una soldadura.

weld interface. The interface between weld metal and base metal in a fusion weld, between base metals in a solid state weld without filler metal, or between filler metal and base metal in a solid state weld with filler metal. Refer to drawing for **depth of fusion.**
interfaze de la soldadura. La interfase entre el metal de soldadura y el metal base en una soldadura de fusión, entre metales de base en una soldadura de estado sólido sin metal para rellenar, o entre metal de rellenar y metal base en una soldadura de estado sólido con metal para rellenar. Refiérase al dibujo **grueso de fusión.**

weld length. See effective length of weld.
largura de la soldadura. Vea **distancia efectiva de soldadura.**

weld metal. The portion of a fusion weld that has been completely melted during welding.
metal de soldadura. La porción de una soldadura de fusión que se ha derretido completamente durante la soldadura.

weld metal area. The area of the weld metal as measured on the cross section of a weld. Refer to drawing for **heat-affected zone.**
área de metal de soldadura. La área del metal de la soldadura la cual fue medida en la sección transversa de la soldadura. Refiérase al dibujo para **zona afectada por el calor.**

weld pass. A single progression of welding along a joint. The result of a pass is a weld bead or layer.
pasada de soldadura. Una progresión singular de la soldadura a lo largo de una junta. El resultado de una pasada es un cordón o una capa.

weld pass sequence. The order in which the weld passes are made. See also **longitudinal sequence** and **cross-sectional sequence.**
secuencia de pasadas de soldadura. La orden en que las pasadas de soldadura se hacen. Vea también **secuencia longitudinal** y **secuencia del corte transversal.**

weld penetration. A nonstandard term for joint penetration and root penetration.
penetracion de soldadura. Un término fuera de norma para penetración de junta y penetración de raíz.

weld pool. The localized volume of molten metal in a weld prior to its solidification as weld metal.
charco de soldadura. El volumen localizado del metal derretido en una soldadura antes de su solidificación como metal de soldadura.

weld puddle. A nonstandard term for weld pool.

charco de soldadura. Un término fuera de norma para charco de soldadura.

weld reinforcement. Weld metal in excess of the quantity required to fill a joint. See also **face reinforcement** and **root reinforcement**.

refuerzo de soldadura. Metal de soldar en exceso de la cantidad requerida para llenar una junta. Vea también **refuerzo de cara** y **refuerzo de raíz**.

weld root. The points, shown in a cross section, at which the root surface intersects the base metal surfaces.

raíz de soldadura. Los puntos, enseñados en una sección transversa, la cual la superficie de la raíz se interseca con las superficies del metal base.

weld size. See preferred term **size of weld**.

tamaño de soldadura. Vea el término preferido tamaño de la soldadura.

*weld specimen. A sample removed from a welded plate according to AWS specifications, which detail the preparation of the plate, the cutting of the plate, and the size of the specimen to be tested.

probeta de soldadura. Una prueba apartada del plato soldado de acuerdo con las especificaciones del AWS, las cuales detallan la preparación del plato, el corte del plato, y el tamaño de la probeta que se va a probar.

weld symbol. A graphical character connected to the welding symbol indicating the type of weld.

símbolo de soldadura. Un signo gráfico conectado al símbolo de soldadura indicando el tipo de soldadura.

weld tab. Additional material that extends beyond either end of the joint, on which the weld is started or terminated.

solera para soldar. Material adicional que se extiende más allá de cualquier punto de la junta en la cual la soldadura es empezada o terminada.

weld test. A welding performance test to a specific code or standard.

prueba de soldadura. Una prueba de ejecución de soldadura según una norma o código específico.

weld time (automatic arc welding). The time interval from the end of start time or end of upslope to beginning of crater fill time or beginning of downslope. Refer to the drawings for **upslope time**.

tiempo de soldadura (soldadura de arco automática). El intervalo de tiempo del fin del tiempo de arranque o el fin del pendiente en ascenso al principio del tiempo de llenar el crater o el principio del tiempo del pendiente en descenso. Refiérase al dibujo para **tiempo del pendiente en ascenso**.

weld timer. A device that controls only the weld time in resistance welding.

contador de tiempo para soldadura. Un aparato que controla solamente el tiempo de soldar en soldaduras de resistencia.

weld toe. The junction of the weld face and the base metal.

pie de la soldadura. La unión de la cara de la soldadura y el metal base.

weldability. The capacity of a material to be welded under the imposed fabrication conditions into a specific, suitably designed structure performing satisfactorily in the intended service.

soldabilidad. La capacidad de un material para soldarse bajo las condiciones de fabricación impuestas en un específico, en un diseño de estructura adecuada y para ejecutar satisfactoriamente los servicios intentados.

welder. One who performs manual or semiautomatic welding.

soldador. Uno que ejecuta soldadura manual o semiautomática.

welder certification. Written verification that a welder has produced welds meeting a prescribed standard of welder performance.

certificación del soldador. Verificación escrita de que un soldador ha producido soldaduras que cumplen con la norma prescrita de la ejecución del soldador.

welder performance qualification. The demonstration of a welder's ability to produce welds meeting prescribed standards.

calificación de ejecución del soldador. La demostración de la habilidad del soldador de producir soldaduras que cumplen con las normas prescritas.

welder registration. The act of registering a welder certification or a photostatic copy thereof.

registración del soldador. El acto de registrar una certificación del soldador o una copia fotostata de ello.

welding. A joining process producing coalescence of materials by heating them to the welding temperature, with or without the application of pressure or by the application of pressure alone, and with or without the use of filler metal.

soldadura. Un proceso de unión que produce coalescencia de materiales calentándolos a la temperatura de soldadura, con o sin la aplicación de presión o por la aplicación de presión solamente, y con o sin el uso del metal de relleno.

welding arc. A controlled electrical discharge between the electrode and the workpiece that is formed and sustained by the establishment of a gaseous conductive medium, called an arc plasma.

arco de soldadura. Una descarga eléctrica controlada entre el electrodo y la pieza de trabajo que es formada y sostenida por el establecimiento de un medio conductivo gaseoso, llamado un arco de plasma.

welding cables. The work cable and electrode cable of an arc welding circuit. Refer to drawing for **direct-current electrode positive**.

cables para soldar. Los cables de pieza de trabajo y el portelectrodo de un circuito de soldadura de arco. Refiérase al dibujo **orriente directa con el electrodo positivo**.

welding current. The current in the welding circuit during the making of a weld.

corriente para soldadura. La corriente en el circuito de soldar durante la hechura de una soldadura.

welding current (automatic arc welding). The current in the welding circuit during the making of a weld, but excluding upslope, downslope, start, and crater fill current. Refer to drawing for **upslope time**.

corriente de soldadura (soldadura de arco automático). La corriente en el circuito de soldar durante la hechura de una soldadura, pero excluyendo el pendiente en ascenso, pendiente en descenso, empiezo, y corriente par llenar el crater. Refiérase al dibujo para **tiempo del pendiente en ascenso**.

welding cycle. The complete series of events involved in the making of a weld. Refer to drawings for **downslope time** and **upslope time**.

ciclo de soldadura. Una serie completa de eventos envueltos en hacer una soldadura. Refiérase al dibujo para **tiempo de cáida del pendiente** y **tiempo del pendiente en descenso.**

welding electrode. A component of the welding circuit through which current is conducted and that terminates at the arc, molten conductive slag, or base metal. See also **arc welding electrode, bare electrode, carbon electrode, composite electrode, covered electrode, electroslag welding electrode, emissive electrode, flux cored electrode, lightly coated electrode, metal cored electrode, metal electrode, resistance welding electrode, stranded electrode,** and **tungsten electrode.**

soldadura con electrodo. Un componente del circuito de soldar por donde la corriente es conducida y que termina en el arco, en la escoria derretida conductiva, o en el metal base. Vea también **electro para soldar de arco, electrodo descubierto, electrodo de carbón, electrodo compuesto, electrodo cubierto, electrodo para soldadura de electroescoria, electrodo emisivo, electrodo de núcleo de fundente, electrodo con recubrimiento ligero, electrodo de metal de núcleo, electrodo de metal, electrodo para soldadura por resistencia, electrodo cable,** y **electrodo de tungsteno.**

welding filler metal. The metal or alloy to be added in making a weld joint that alloys with the base metal to form weld metal in a fusion welded joint.

metal de soldadura para rellenar. El metal o aleación que se va a agregar en la hechura de una junta de soldadura que se mezcla con el metal base para formar metal de soldadura en una junta de fusión de soldadura.

welding generator. A generator used for supplying current for welding.

generador para soldar. Un generador que se usa para proporcionar la corriente para la soldadura.

welding ground. A nonstandard and incorrect term for workpiece connection.

tierra de soldadura. Un término fuera de norma e incorrecto para conexión de pieza de trabajo.

welding head. The part of a welding machine in which a welding gun or torch is incorporated.

cabeza de soldar. La parte de una máquina para soldar la cual una pistola de soldadura o una antorcha se puede incorporar.

*welding helmet. See **helmet**

casco para soldar. Ver **casco.**

welding leads. The work lead and electrode lead of an arc welding circuit. Refer to drawing for **direct-current electrode positive.**

cables para soldar. Los cables de pieza de trabajo y el portelectrodo de un circuito de soldadura de arco. Refiérase al dibujo **corriente directa con el electrodo positivo.**

welding machine. Equipment used to perform the welding operation. For example, spot welding machine, arc welding machine, seam welding machine, etc.

máquina para soldar. El equipo que se usa para ejecutar la operación de soldadura. Por ejemplo, máquina de soldadura por puntos, máquina de soldadura de arco, máquina de soldadura de costura, etc.

welding operator. One who operates adaptive control, automatic, mechanized, or robotic welding equipment.

operador de soldadura. Uno que opera control adaptivo, automático, mecanizado, o equipo robótico para soldar.

welding position. See flat position, horizontal position, horizontal fixed position, horizontal rolled position, overhead position, and vertical position.

posición de soldadura. Vea posición plana, posición horizontal, posición fija horizontal, posición horizontal rodada, posición de sobrecabeza, y posición vertical.

```
POSITION OF WELDING

Flat. See flat position
Horizontal. See horizontal position, horizontal fixed position,
    and horizontal rolled position.
Vertical. See vertical position.
Overhead. See overhead position.

POSITION FOR QUALIFICATION

Plate welds

  Groove welds

    1G.    See flat position.
    2G.    See horizontal position.
    3G.    See vertical position.
    4G.    See overhead position.

  Fillet welds

    1F.    See flat position.
    2F.    See horizontal position.
    3F.    See vertical position.
    4F.    See overhead position.

Pipe welds

  Groove welds

    1G.    See horizontal rolled position.
    2G.    See vertical position.
    3G.    See horizontal fixed position.
    5G.    Inclined position.
    5GR.   Inclined position.
```

welding power source. An apparatus for supplying current and voltage suitable for welding. See also **welding generator, welding rectifier,** and **welding transformer.**

fuente de poder para soldar. Un aparato para surtir corriente y voltaje adecuado para soldar. Vea también **generador para soldar, rectificador para soldar,** y **transformador para soldar.**

welding procedure qualification record (WPQR). A record of welding variables used to produce an acceptable test weldment and the results of tests conducted on the weldment to qualify a welding procedure specification.

registro de calificación de procedimiento de la soldadura (WPQR). Un registro de los variables usados para producir una probeta aceptable y los resultados de la prueba conducida en la probeta para calificar el procedimiento de especificación.

Welding Procedure Specification (WPS). A document providing in detail the required variables for specific application to ensure repeatability by properly trained welders and welding operators.

calificación de procedimiento de soldadura (WPA). Un documento que provee en detalle los variables requeridos para la aplicación específica para asegurar la habilidad de repetir el procedimiento por soldadores y operadores que estén propiamente preparados.

welding process. A materials joining process that produces coalescence of materials by heating them to suitable temperatures, with or without the application of pressure or by the application of pressure alone, and with or without the use of filler metal.

proceso para soldar. Un proceso para unir materiales que produce coalescencia calentándolos a una temperatura adecuada con o sin la aplicación de presión solamente y con o sin usarse material para rellenar.

welding rectifier. A device in a welding machine for converting alternating current to direct current.

rectificador para soldar. Un aparato en una máquina para soldar para convertir la corriente alterna a corriente directa.

welding rod. A form of welding filler metal, normally packaged in straight lengths, that does not conduct the welding current.

varilla para soldar. Una forma de metal de soldadura para rellenar, normalmente empaquetada en piezas derechas, que no conduce la corriente para soldar.

welding sequence. The order of making the welds in a weldment.

orden de sucesión. La orden de hacer las soldaduras de una estructura soldada.

welding symbol. A graphical representation of a weld.

símbolo de soldadura. Una representación gráfica de una soldadura.

welding technique. The details of a welding procedure that are controlled by the welder or welding operator.

ejecución de soldadura. Los detalles del procedimiento que son controlados por el soldador u operador de soldadura.

welding tip. A welding torch tip designed for welding.

boquilla (punta) para soldar. Una boquilla en la antorcha de soldadura que está diseñada para soldar.

welding torch (arc). A device used in the gas tungsten and plasma arc welding processes to control the position of the electrode, to transfer current to the arc, and to direct the flow of shielding and plasma gas.

antorcha para soldar (arco). Un aparato usado en los procesos de soldadura del gas tungsteno y arco plasma para controlar la posición del electrodo, para transferir corriente al arco, y para dirigir la corriente del gas protector y gas de la plasma.

welding torch (oxyfuel gas). A device used in oxyfuel gas welding, torch brazing, and torch soldering for directing the heating flame produced by the controlled combustion of fuel gases.

antorcha para soldar (gas oxicombustible). Un aparato usado en soldadura de gas oxicombustible, soldadura blanda con antorcha y soldadura fuerte con antorcha y para dirigir la llama para calentar producida por la combustión controlada de gases de combustión.

welding transformer. A transformer used to supplying current for welding. See also **reactor (arc welding)**.

transformador para soldar. Un transformador que se usa para dar corriente para la soldadura. Vea también **reactor (soldadura de arco)**.

welding voltage. See arc voltage.

voltaje para soldar. Vea voltaje del arco.

welding wire. A form of welding filler metal, normally packaged as coils or spools, that may or may not conduct electrical current, depending upon the filler metal and base metal in a solid state weld with filler metal.

alambre para soldar. Una forma de metal para rellenar con soldadura, normalmente empaquetado en rollos o en carretes que pueda o pueda que no conducir corriente eléctrica, dependiendo en el metal de relleno y el metal base en una soldadura que está en estado sólido con metal de relleno.

weldment. An assembly whose component parts are joined by welding.

conjunto de partes soldadas. Una asamblea cuyas partes componentes están unidas por la soldadura.

weldor. See preferred term **welder**.

soldador. Vea el término preferido **soldador**.

wetting. The phenomenon whereby a liquid filler metal or flux spreads and adheres in a thin, continuous layer on a solid base metal.

exudación. El fenómeno de que un metal para rellenar líquido o un flujo se puede desparramar y adherirse en una capa delgada, capa continua en un sólido metal base.

wire-feed speed. The rate at which wire is consumed in arc cutting, thermal spraying, or welding.

velocidad de alimentador de alambre. La velocidad que el alambre es consumido en cortes de arco, rociado termal, o soldadura.

work angle. The angle less than 90° between a line perpendicular to the major workpiece surface and a plane determined by the electrode axis and the weld axis. In a tee joint or a corner joint, the line is perpendicular to the nonbutting member. This angle can also be used to partially define the position of guns, torches, rods, and beams.

ángulo de trabajo. El ángulo menos de 90° entre una línea perpendicular a la superficie de pieza de trabajo mayor y una plana determinada por el eje del electrodo y el eje de la soldadura. En una junta-T o en una junta de esquina, la línea es perpendicular a un miembro que no topa. Este ángulo puede ser usado también para parcialmente definir la posición de pistolas, antorchas, varillas, y rayos.

work angle (pipe). The angle less than 90° between a line perpendicular to the cylindrical pipe surface at the point of intersection of the weld axis and the extension of the electrode axis, and a plane determined by the electrode axis and a line tangent to the pipe at the same point. In a T-joint, the line is perpendicular to the nonbutting member. This angle can also be used to partially define the position of guns, torches, rods, and beams.

ángulo de trabajo (tubo). El ángulo menos de 90° entre una línea, la cual es perpendicular a la superficie de un tubo cilíndrico al punto de intersección del eje de la soldadura y la extensión del eje del electrodo, y un plano determinado por el eje del electrodo y una línea tangente al tubo al mismo punto. En una junta-T, la línea es perpendicular a un miembro que no topa. Este ángulo puede también usarse para definir parcialmente la posición de pistolas, antorchas, varillas, y rayos.

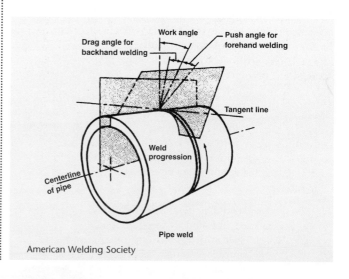

American Welding Society

work lead. The electric conductor between the source of arc welding current and the work. Refer to drawing for **direct-current electrode negative**.
cable de tierra. Un conductor eléctrico entre la fuente de la corriente del arco y la pieza de trabajo. Refiérase al dibujo **corriente directa con electrodo negativo**.

work connection. The connection of the work lead to the work. Refer to drawing for **direct-current electrode negative**.
pinza de tierra. La conexión del cable de trabajo (tierra) al trabajo. Refiérase al dibujo para **corriente directa con electrodo negativo**.

working envelope. The set of points representing the maximum extent or reach of the robot hand or working tool in all directions.
alcance de operación. Un juego de puntos que representan la máxima extensión o alcance de la mano del robot o la herramienta del trabajo en todas las direcciones.

*working pressure. The pressure at the low-pressure gauge, ranging from 0 psi to 45 psi (depending on the type of gas), used for welding and cutting.
presión de trabajo. La presión en el manómetro de baja presión, con escala de 0 psi a 45 psi (dependiendo en el tipo de gas), usado para cortar y soldar.

working range. All positions within the working envelope. The range of any variable within which the system normally operates.
extensión de trabajo. Todas las posiciones dentro del alcance del trabajo. El alcance de cualquier variable dentro del sistema que opera normalmente.

workpiece. An assembly, component, member, or part in the process of being manufactured.
pieza de trabajo. La parte que está soldada, con soldadura fuerte, soldadura blanda, corte termal, o rociado termal.

workpiece connection. A device (clamp) used to provide an electrical connection between the workpiece and the workpiece lead.
conexion de pieza de trabajo (abrazadera). La conexión del cable de la pieza de trabajo a la pieza de trabajo.

workpiece lead. A secondary circuit conductor transmitting energy from the power source to the workpiece connection.
cable de pieza de trabajo. El conductor eléctrico entre la fuente de corriente de soldadura de arco y la conexión de la pieza de trabajo.

workstation. A manufacturing unit consisting of one or more numerically controlled machine tools serviced by a robot.
estación de trabajo. Una unidad manufacturera de una o más herramienta numerada que es controlada por una máquina y abatecida por un robot.

wrist. A set of rotary joints between the arm and hand that allows the hand to be oriented to the workpiece.
muñeca. Un juego de coyunturas rotatorias entre el brazo y la mano que permite a la mano ser orientada a la pieza de trabajo.

X

*X-axis. Machine, automated, or robotic movement in a longitudinal direction.
eje X. Movimiento mecánico, automatizado o robótico en dirección longitudinal.

Y

yaw. The angular displacement of a moving body about an axis that is perpendicular to the line of motion and to the top side of the body.
guiñada. El desalojamiento de un cuerpo en movimiento alrededor de un eje que está perpendicular a la línea de movimiento y al lado más alto del cuerpo.

*Y-axis. Machine, automated, or robotic movement in a transverse direction.
eje Y. Movimiento mecánico, automatizado o robótico en dirección transversal.

Z

*Z-axis. Machine, automated, or robotic movement in a vertical direction.
eje Z. Movimiento mecánico, automatizado o robótico en dirección vertical.

Index

A

AAR. See Association of American Railroads
AASHTO. *See* Association of American of State Highway and Transportation Officials
Abbreviations, 19t
Abrasives, 227
Absolute pressure, 759
AC. *See* Alternating current
Acetone, 35, 786
Acetylene (C_2H_2), 786–788, 786f–788f, 787t
    cylinders, 35, 36f
    for flame cutting, 161t
Acicular structure, needle-like, 654, 654f
Adjustable wrench, 38, 38f
Advanced shielded metal arc welding, 128–156.
        See also Shielded metal arc welding
    cover pass in, 138–139, 138f–139f
    filler pass in, 136–138, 137f–138f
    hot pass in, 134–136, 135f–136f
    introduction to, 129
    plate preparation in, 139, 140f, 141, 141f
    poor fit-up in, 152–154, 153f
    postheating in, 144, 152
    preheating, 144, 144t
    preparing specimens for testing in, 141–142, 141f–142f
    restarting a weld bead in, 142–144, 143f–144f, 600–601, 600f–601f
    root pass in, 129–134, 129f–134f, 134t
Aerospace Industries Association (AIA), 551
AIA. *See* Aerospace Industries Association
Air carbon arc cutting (CAC-A), 11, 219, 219t, 220t
    applications of, 221–222, 222f–223f
    electrodes in, 220–221, 221f
    equipment and, 223
    gouging in, 222, 222f
    safety in, 222–223
    washing in, 222, 223f
Air Conditioning National Contractors Association, 547
Air-cooled welding guns, 299–300, 300f
Air Force, 478–479
Airports, 358–359
Air supply, 221
AISI. *See* American Iron and Steel Institute
Al-2, 704
Al-43, 704
Allen, Jim, 103–104
Allotropic metal, 647
Allotropic transformation, 660
Alloying elements, 304, 698t
Alloys. *See also specific alloys*
    not specified EWG electrode, 377
    steels, 670t, 671–674
        low, 674

All-position welding, 301
Alphabet of lines, 482, 484t, 485f
Alternate bend test, 580
Alternating current (AC), 57, 58f
Alternators, 63–64, 64f
Aluminum, 220t, 405–406, 405f, 670t, 691
    filler metal for, 405t
    filler metal selection with, 704, 705t
    weldability of, 681–682
Aluminum bare welding rods and electrodes, 704
Aluminum-covered arc welding electrodes, 704
Aluminum-silicon, 843
American Bureau of Shipping, 103, 141, 568t
American Iron and Steel Institute (AISI), 370
    classification system of, 671, 672t
American Museum of Natural History, 394–396
American National Standards Institute (ANSI), 47, 551
American Petroleum Institute (API), 551, 568t
American Railway Engineering and Maintenance-of-Way Association (AREMA), 551
American Society for Testing and Materials, 568t
American Society of Mechanical Engineers (ASME), 141, 568t
    ASME Section IX, 551, 555
American Water Works Association (AWWA), 551
American Welding Society (AWS), 5, 129, 141, 259, 568t
    AWS A5.01-93 document of, 707–708
    AWS A5.32, 796–797
    AWS D1.1, 551, 555
    filler metal selection classifications of, 696, 697t
    welder certification levels of, 15
Ammonium persulphate, 582
Amperage, 244, 336–337
    of current, 56
    in plasma arc cutting, 196
    range, 75, 75f, 77, 337
Annealing, 652, 658, 658f
ANSI. *See* American National Standards Institute
Antarctica, 20–21
API. *See* American Petroleum Institute
API Standard 1104, 551
Apollo-5, 478, 519
Ar. *See* Argon
Arc blow, 59, 59f–60f
Arc cutting, 192
Arc cutting electrodes, 227, 228, 228f
Arc length, 77–78, 77f–78f
Arc plasma, 192
Arc stabilizers, 304
Arc starting methods, 365, 365f
Arc strikes, 572, 572f
Arc-voltage and amperage characteristics, GMAW, 270–271, 271t
Arc welding, 746–747
    thermal effects of, 658–660, 659f–660f

F